SOME PHYSICAL PROPERTIES

Air (dry, at 20°C and 1 atm)	
Density	1.21 kg/m^3
Specific heat capacity at constant pressure	1010 J/kg·K
Ratio of specific heat capacities	1.40
Speed of sound	343 m/s
Electrical breakdown strength	3×10^6 V/m
Effective molecular mass	0.0289 kg/mol
Water	
Density	1000 kg/m^3
Speed of sound	1460 m/s
Specific heat capacity at constant pressure	4190 J/kg·K
Heat of fusion (0°C)	333 kJ/kg
Heat of vaporization (100°C)	2260 kJ/kg
Index of refraction ($\lambda = 589$ nm)	1.33
Molecular mass	0.0180 kg/mol
Earth	
Mass	5.98×10^{24} kg
Mean radius	6.37×10^6 m
Free fall acceleration at the earth's surface	9.81 m/s^2
Standard atmosphere	1.01×10^5 Pa
Period of satellite at 100 km altitude	86.3 min
Radius of the geosynchronous orbit	42,200 km
Escape speed	11.2 km/s
Magnetic dipole moment	8.0×10^{22} $A \cdot m^2$
Mean electric field at surface	150 V/m, down
Distance to:	
Moon	3.82×10^8 m
Sun	1.50×10^{11} m
Nearest star	4.04×10^{16} m
Galactic center	2.2×10^{20} m
Andromeda galaxy	2.1×10^{22} m
Edge of the observable universe	$\sim 10^{26}$ m

THE GREEK ALPHABET

Alpha	A	α	Iota	I	ι	Rho	P	ρ
Beta	B	β	Kappa	K	κ	Sigma	Σ	σ
Gamma	Γ	γ	Lambda	Λ	λ	Tau	T	τ
Delta	Δ	δ	Mu	M	μ	Upsilon	Y	υ
Epsilon	E	ϵ	Nu	N	ν	Phi	Φ	ϕ, φ
Zeta	Z	ζ	Xi	Ξ	ξ	Chi	X	χ
Eta	H	η	Omicron	O	o	Psi	Ψ	ψ
Theta	Θ	θ	Pi	Π	π	Omega	Ω	ω

SUPPLEMENTS

FUNDAMENTALS OF PHYSICS, THIRD EDITION EXTENDED is accompanied by a complete supplementary package.

STUDY GUIDE

Stanley Williams, Iowa State University
Kenneth Brownstein, University of Maine
Thomas Marcella, University of Lowell

Chapter introductions outline specific types of problems and tie together work done in previous chapters. Quick reviews of definitions, laws, and concepts are combined with detailed explanations of how to apply them to problems.

SELECTED SOLUTIONS

Edward Derringh, Wentworth Institute of Technology

Provides solutions to selected exercises and problems.

WONDERING ABOUT PHYSICS . . . Using Spreadsheets to Find Out

Dewey I. Dykstra, Jr., Boise State University
Robert G. Fuller, U.S. Air Force Academy and University of Nebraska-Lincoln

Composed of 53 investigations, this specially developed supplement leads students to explore the real world of physical phenomena by using spreadsheet software on a personal computer.

LABORATORY PHYSICS, SECOND EDITION

Harry F. Meiners, Rensselaer Polytechnic Institute
Walter Eppenstein, Rensselaer Polytechnic Institute
Kenneth Moore, Rensselaer Polytechnic Institute
Ralph A. Oliva, Texas Instruments, Inc.

This laboratory manual offers a clear introduction to laboratory procedures and instrumentation, including errors, graphing, apparatus handling, calculators, and computers, in addition to over 70 different experiments grouped by topic.

FOR THE INSTRUCTOR

A complete supplementary package of teaching and learning materials is available for instructors. Contact your local Wiley representative for further information.

FUNDAMENTALS OF PHYSICS

SUPPLEMENTS

FUNDAMENTALS OF PHYSICS, THIRD EDITION EXTENDED is accompanied by a complete supplementary package.

STUDY GUIDE

Stanley Williams, Iowa State University
Kenneth Brownstein, University of Maine
Thomas Marcella, University of Lowell

Chapter introductions outline specific types of problems and tie together work done in previous chapters. Quick reviews of definitions, laws, and concepts are combined with detailed explanations of how to apply them to problems.

SELECTED SOLUTIONS

Edward Derringh, Wentworth Institute of Technology

Provides solutions to selected exercises and problems.

WONDERING ABOUT PHYSICS . . . Using Spreadsheets to Find Out

Dewey I. Dykstra, Jr., Boise State University
Robert G. Fuller, U.S. Air Force Academy and University of Nebraska–Lincoln

Composed of 53 investigations, this specially developed supplement leads students to explore the real world of physical phenomena by using spreadsheet software on a personal computer.

LABORATORY PHYSICS, SECOND EDITION

Harry F. Meiners, Rensselaer Polytechnic Institute
Walter Eppenstein, Rensselaer Polytechnic Institute
Kenneth Moore, Rensselaer Polytechnic Institute
Ralph A. Oliva, Texas Instruments, Inc.

This laboratory manual offers a clear introduction to laboratory procedures and instrumentation, including errors, graphing, apparatus handling, calculators, and computers, in addition to over 70 different experiments grouped by topic.

FOR THE INSTRUCTOR

A complete supplementary package of teaching and learning materials is available for instructors. Contact your local Wiley representative for further information.

FUNDAMENTALS OF PHYSICS

Third Edition Extended

Volume Two

David Halliday
University of Pittsburgh

Robert Resnick
Rensselaer Polytechnic Institute

with the assistance of

John Merrill
Brigham Young University

JOHN WILEY & SONS New York · Chichester · Brisbane · Toronto · Singapore

Text and cover design: Karin Gerdes Kincheloe
Production supervised by Lucille Buonocore
Illustration supervised by John Balbalis
Copy editing supervised by Deborah Herbert
Photo Research: Anita Duncan

Library of Congress Cataloging-in-Publication Data:

Halliday, David
Fundamentals of physics.

Includes index.
1. Physics. I. Resnick, Robert
II. Title.
QC21.2.H35 1988c 530 88-17380
ISBN 0-471-61917-5

Printed in the United States of America

10 9 8 7 6 5 4 3 2

PREFACE

This third (1988) edition of *Fundamentals of Physics* is a major revision of both the second (1981) edition of that text and of its revised printing (1986). Although we have retained the basic framework of these earlier versions, we have virtually rewritten the entire book. Users of the earlier editions can appreciate the changes better if we list them in some detail.

(a) In the words of one reviewer, "you have succeeded in maintaining the overall level throughout but have substantially lowered the learning threshold." Many new techniques have been used to achieve this. For example, many hints on problem solving are sprinkled throughout the early chapters of Volume One, each one focusing on a chronic student hang-up. A larger set of worked examples — now called Sample Problems to reflect their consistent focus — is included to provide problem-solving models for all aspects of each chapter; several are put in extended question and answer format to reveal directly the pathways followed by experienced problem solvers. There are more but shorter sections per chapter for easier digestion of the material, and more use of subheads is made within sections for greater clarity and emphasis. In the body of the text, relationships are typically displayed and discussed before they are formally derived, a more inductive procedure that we think will prove effective. Often these formal derivations appear in separate sections or subsections. And, we have greatly expanded the set of confidence-building exercises for homework while increasing the number of problems as well.

(b) Greater clarity has been achieved in many ways. A more student-oriented style is employed than before and a two-column format has been adopted for easier reading. Chapter-head photographs are now included along with captions that make a valid attention-grabbing point about the contents of each chapter. Opening sections discuss the relevance of the topics to be treated in each chapter in order to motivate students from the start. Throughout the chapter, photographs and diagrams are featured in greater numbers, with self-contained captions, to reinforce the text material. In each chapter there are examples that deal with practical and applied situations. Chapters conclude with a detailed Review and Summary section for student reference and study.

(c) The sets of chapter-ending questions, exercises, and problems are by far the largest and most varied of any introductory physics text. We have edited the highly praised sets of the earlier edition to achieve even greater clarity and interest and have added a substantial number of new applied and conceptual ones. A more generous use of figures and photographs serves better to illustrate the questions, exercises, and problems than before.

The thought questions have always been a special feature of our books. They are used as sources of classroom discussion and for clarification of homework concepts. There are nearly 30 per chapter. Their total number, now over 1400 in the entire book, is greater than before and they relate even more to everyday phenomena, serve to arouse curiosity and interest, and stress conceptual aspects of physics.

Exercises typically involve one step or formula or represent a single application and are used for building student confidence. They now constitute about 45 percent of the exercise-problem sets. In preparing the new set of problems, we have been careful not to discard the many tried and true problems that have survived the test of the classroom for many years. Long-time users of our text will not find their favorites missing. Of the substantial number of new problems, many fit the "real world" category of student interest and these and the others serve different pedagogic objectives as well. Amongst the problems are a small number of advanced ones, as well, identified by stars* next to their number. A typical chapter has about 31 exercises and 37 problems, the total number of exercises and problems in the entire book being about 3400.

By labeling exercises "**E**" and problems "**P**" and organizing them in order of difficulty for each section of the chapter, we have simplified the selection process for teachers from the voluminous material now made available. The variation of level and the breadth of scope have been enlarged. Hence, teachers can vary the content emphasis and the level of difficulty to suit their tastes and the preparation of the student body while still having a very adequate supply for many years of instruction. Indeed, the book is now somewhat longer principally because of all the self-study and learning features that are now included.

(d) Our treatment of modern physics has been enhanced. There are two entirely new modern physics chapters, one on Relativity and the other on Quarks, Leptons and the Big Bang. And, in rewriting the earlier chapters, we have sought to pave the way more effectively than in previous editions for the systematic study of modern physics presented in the later chapters. We have done this in three ways. (i) In appropriate places we have called attention—by specific example—to the impact of relativistic and quantum ideas on our daily lives. (ii) We have stressed those concepts (conservation principles, symmetry arguments, reference frames, role of aesthetics, similarity of methods, use of models, field concepts, wave concepts, etc.) that are common to both classical and modern physics. (iii) Finally, we have included a number of short optional sections in which selected relativistic and quantum ideas are presented in ways that lay the foundation for the detailed and systematic treatments of relativity, atomic, nuclear, solid state, and particle physics given in later chapters.

(e) To emphasize the relevance of what physicists do, and further motivate the student, we include, within the chapters, numerous applications of physics in engineering, technology, medicine, and familiar everyday phenomena. In addition, we feature 21 separate, self-contained essays, written by distinguished scientists and distributed at appropriate locations in the text, on the application of physics to special topics of student interest such as sports, toys, amusement parks, medicine, lasers, holography, space, superconductivity, concert-hall acoustics, and many more. (See the Table of Contents.)

(f) In the interests of simplification and of greater clarity for students, certain rearrangements of material have been made. For example, motion in one dimension is now treated before vectors. A better balance of the material on rotational motion in mechanics is achieved over two chapters by presenting the simpler concepts in kinematics and dynamics first, and then the more difficult concepts, enabling the instructors to more easily choose the depth desired. Similarly, formerly-scattered material—such as on the Doppler effect or on special relativity—has been drawn together in one place for greater conceptual unity. Material on elasticity, now somewhat longer, fits more naturally into the chapter on equilibrium. There are, of course, many other smaller rearrangements too numerous to mention here.

Like the second edition, this edition is available in a single volume of 42 chapters, ending with relativity, and in an Extended Version of 49 chapters that contains in addition a development of quantum physics and its applications to atoms, solids, nuclei, and particles. The Extended Version is also available as a two-volume set: Volume One covers Mechanics and Thermodynamics (Chapters 1–22); Volume Two covers Electricity and Magnetism, Optics, and Modern Physics (Chapters 23–49). The former is meant for introductory courses that treat modern quantum physics in a subsequent separate course or semester. There are also numerous optional sections throughout the text that are of an advanced, historical, general, or specialized nature.

Indeed, just as a textbook alone is not a course, so a course does not include the entire textbook. We have consciously made available much more material than any one course or instructor is expected to "cover." More can be "uncovered" by doing less. The process of physics and its essential unity can be revealed by judicious selective coverage of many fewer chapters than are contained here and by coverage of only portions of many included chapters. Rather than give numerous examples of such coherent selections, we urge the instructor to be guided by his or her own interests and circumstances and to plan ahead so that some topics in modern physics are always included.

A textbook contains far more contributions to the elucidation of a subject than those made by the authors alone. As before, John Merrill (Brigham Young University) has been of special service for all aspects of this work, as has Edward Derringh (Wentworth Institute of Technology). Albert Bartlett (University of Colorado) has been of particular help with the essays and Benjamin Chi (SUNY Albany) with the figures and photographs. At John Wiley, publishers, we have been fortunate to receive strong coordination and support from Robert McConnin and Catherine Faduska, physics editors, with notable contributions from John Balbalis, Lucille Buonocore, Deborah Herbert, Karin Kincheloe, Safra Nimrod, and other members of the production team. We are grateful to all these persons.

Our external reviewers have been outstanding and we acknowledge here our debt to each member of that team, namely, Joseph Buschi (Manhattan College), Philip A. Casabella (Rensselaer Polytechnic Institute), Randall Caton (Christopher Newport College), Roger Clapp (University of South Florida), William P. Crumment (Montana College of Science and Technology), Robert Endorf (University of Cincinnati), E. Paul Esposito (University of Cincinnati), Andrew L. Gardner (Brigham Young University), John Gieniec (Central Missouri State University), Leonard Kleinman (University of Texas at Austin), Kenneth Krane (Oregon State University), Howard C. McAllister (University of Hawaii at Manoa), Manuel Schwartz (University of Louisville), John Spangler (St. Norbert College), Ross Spencer (Brigham Young University), Harold Stokes, (Brigham Young University), David Toot (Alfred University), Donald Wieber (Contra Costa College) and George U. Williams (University of Utah).

We thank all the essayists for their valuable contributions and cooperative spirit. Kathaleen Guyette has been superb in providing the wide range of secretarial services required.

We hope that the final product proves worthy of the effort and that this Third Edition of *Fundamentals of Physics* will contribute to the enhancement of physics education.

DAVID HALLIDAY
5110 Kenilworth Place, NE
Seattle, WA. 98105

ROBERT RESNICK
Rensselaer Polytechnic Institute
Troy, NY 12180-3590

January, 1988

THE ESSAYISTS

ALBERT A. BARTLETT

Albert A. Bartlett (Essays 4 and 11) is a professor of physics at the University of Colorado, Boulder, where he has been a member of the faculty since 1950. He received his B. A. from Colgate University and his Ph. D. from Harvard in nuclear physics. His interests are centered on teaching physics; he was President of the American Association of Physics Teachers in 1978. He is a founding member of PLAN-Boulder—an environmental organization that is largely responsible for Boulder's Greenbelt and open space land acquisition program.

CHARLES P. BEAN

Charles P. Bean (Essay 14) is Institute Professor of Science at Rensselaer Polytechnic Institute. He received his Ph. D. in physics from the University of Illinois in 1952. For more than 33 years he was a research scientist in the General Electric Research Laboratory and its successor, the General Electric Research and Development Center. While there he made research contributions to the fields of ionic crystals, magnetism, superconductivity, and membrane biophysics. He is a member of the National Academy of Sciences and the American Academy of Arts and Sciences. As an avocation he studies the physics of phenomena in nature.

PETER J. BRANCAZIO

Peter J. Brancazio (Essay 6) is a professor of physics at Brooklyn College, City University of New York. He received his Ph. D. in astrophysics from New York University in 1966. He is the author of two books: *The Nature of Physics* (Macmillan, 1975) and *Sport Science* (Simon & Schuster, 1984). His articles on the physics of baseball, football, and basketball have appeared in *Discover, Physics Today, New Scientist, The Physics Teacher,* and the *American Journal of Physics.* A lifelong athlete and sports fan, he is equally at home on the basketball court and in the classroom.

PATRICIA ELIZABETH CLADIS

Patricia Elizabeth Cladis (Essay 20) was born in Shanghai, China and grew up in Vancouver, British Columbia. She received her Ph. D. in physics from the University of Rochester with a thesis on the dc superconducting transformer. Before joining AT&T Bell Laboratories, she did postdoctoral research at the University of Paris, Orsay, France, where she first learned about liquid crystals and discovered "escape into the third dimension" and point defects in nematics. At Bell Labs she discovered the "reentrant nematic" phase. Currently she uses liquid crystals to study general problems in nonlinear physics. She has published nearly 100 scientific papers and is on the editorial board of the journal *Liquid Crystals.*

ELSA GARMIRE

Elsa Garmire (Essay 19) is professor of electrical engineering and physics, and Director of the Center for Laser Studies at the University of Southern California. Garmire received the A. B. in physics from Harvard University in 1961 and the Ph. D. in physics from M. I. T. in 1965 for research in nonlinear optics under Nobel Prizewinner C. H. Townes. The author of over 120 papers with seven patents, she has been a researcher in quantum electronics and in linear and nonlinear optical devices for 25 years. She is a fellow of the Optical Society of America and of IEEE and has been on the board of both societies. She is associate editor of the journals *Optics Letters* and *Fiber and Integrated Optics,* and was U.S. delegate to the International Commission for Optics.

RUSSELL K. HOBBIE

Russell K. Hobbie (Essays 8 and 21) is a professor of physics at the University of Minnesota. He received his B. S. from M. I. T. and his Ph. D. from Harvard. His research interests include diagnostic radiology, magnetic resonance imaging, impedance cardiography, and computerized medical diagnosis. He is the author of *Intermediate Physics for Medicine and Biology,* published by Wiley, 1988.

TUNG H. JEONG

Tung H. Jeong (Essay 18) received his B. S. from Yale University in 1957, and Ph. D. in nuclear physics from the University of Minnesota in 1963. Presently, he is chairman of the physics department and holder of the Albert Blake Dick endowed chair in Lake Forest College. Besides directing annual summer holography workshops since 1972, he consults and lectures on holography in hundreds of institutions around the world. He is a Fellow of the Optical Society of America and a recipient of the Robert A. Millikan Medal from the American Association of Physics Teachers. His hobbies include skiing, tennis, and playing the violin in the Lake Forest Symphony Orchestra.

KENNETH LAWS

Kenneth Laws (Essay 3) is professor of physics at Dickinson College in Carlisle, Pa., where he has been teaching since 1962. He earned his B. S., M. S., and Ph. D. degrees from Caltech, the University of Pennsylvania, and Bryn Mawr College, respectively. For the last dozen years he has been studying classical ballet at the Central Pennsylvania Youth Ballet, and he has recently been applying the principles of physics to dance movement. This work has lead to numerous lectures and classes around the country, and has culminated in a book, *The Physics of Dance,* published in 1984 (paperback 1986) by Schirmer Books.

PETER LINDENFELD

Peter Lindenfeld (Essay 12) has degrees in electrical engineering and engineering physics from the University of British Columbia and a Ph. D. in physics from Columbia University. Since his graduation he has been at Rutgers University where he is professor of physics. His research and publications are on low temperature physics and superconductivity as well as on activities related to physics teaching. He is a fellow of the American Physical Society and an honorary life member of the New Jersey Section of the American Association of Physics Teachers. He has received awards for his booklet "Radioactive Radiations and their Biological Effects," for his work on a solar calorimeter, and for some of his photographs.

RICHARD L. MORIN

Richard L. Morin (Essays 8 and 21) is an associate professor in the Department of Radiology and Director of the Physics Section in Radiology at the University of Minnesota. He received his undergraduate training in chemistry at Emory University and his Ph. D. in radiological sciences from the University of Oklahoma. His research interests are in the area of computer applications in radiology and nuclear medicine.

SUZANNE R. NAGEL

Suzanne R. Nagel (Essay 17) is head of the Glass Research and Engineering Department at AT&T-Bell Laboratories, Murray Hill, N.J. She received her Ph. D. in ceramic engineering from the University of Illinois in 1972 after her undergraduate studies at Rutgers University. She has authored 30 technical papers in the area of glass science and lightguide technology, and her research has involved the processing and property optimization of optical fibers for communications, as well as close interaction with their manufacture. She is actively involved in promoting careers in science and engineering for women and minorities, and was recently featured in the Chicago Museum of Science and Technology Exhibit "My Daughter the Scientist," which is touring the country.

GERARD K. O'NEILL

Gerard K. O'Neill (Essay 13) holds a bachelor's degree from Swarthmore College, a Ph. D. from Cornell University (1954), and an honorary D. Sc. from Swarthmore. He was a member of the faculty at Princeton University from 1954 to 1985, and was made full professor of physics in 1965. In 1985, he retired early from Princeton, becoming Professor Emeritus of Physics. He is the author of several books and many articles in his field. In March 1985, he was appointed by President Reagan to the National Commission on Space. In 1983 he founded the Geostar Corporation, a communications and navigation satellite company based on patents issued to him. His latest commercial start-up company is O'Neill Communications, Inc.

SALLY K. RIDE

Sally K. Ride (Essay 5) is a NASA Space Shuttle astronaut. She earned a B. S. in physics and a B. A. in English from Stanford University in 1973, and a Ph. D. in physics from Stanford in 1978. After graduate school, she was selected for the Astronaut Corps. She has flown in space twice: on the seventh Space Shuttle mission (STS-7, the second flight of the *Challenger,* launched in June, 1983), and the thirteenth Shuttle mission (STS-41G, launched in October, 1984). In 1986, she was appointed to the Presidential Commission investigating the Space Shuttle *Challenger* accident. Since the completion of the investigation, she has acted as Special Assistant to the Administrator of NASA, helping to develop NASA's long-range plans for human exploration of space. She is currently affiliated with the Center for International Security and Arms Control at Stanford University.

JOHN S. RIGDEN

John S. Rigden (Essay 7) received his Ph. D. from Johns Hopkins University in 1960. After postdoctoral work at Harvard University, he started the physics department at Eastern Nazarene College. After one year at Middlebury College, he moved to St. Louis where he is now professor of physics at the University of Missouri–St. Louis. He is the author of *Physics and the Sound of Music* (Wiley, 1977; Second Edition, 1985). More recently, he has written the definitive biography of the great American physicist, I. I. Rabi: *Rabi: Scientist and Citizen* (Basic Books, 1987). Since 1978 he has been the editor of the *American Journal of Physics.*

JOHN L. ROEDER

John L. Roeder (Essay 2) began investigating the physics of the amusement park with an article in the September 1975 issue of *The Physics Teacher* and now takes his physics classes at The Calhoun School in New York City on field trips to Six Flags Great Adventure, the site of his "original research." In addition to serving as a double Resource Agent — for both the American Association of Physics Teachers and the New York Energy Education Project — John is a cofounder of and the newsletter editor for the Teachers Clearinghouse for Science and Society Education, Inc. He received his A. B. from Washington University and his M. A. and Ph. D. from Princeton University.

WILLIAM A. SHURCLIFF

William A. Shurcliff, Physics Department, Emeritus, Harvard University (Essay 9) received his Ph.D. from Harvard in 1934. He has held positions such as senior scientist and research fellow in numerous government, industrial, and university laboratories, including American Cyanamid Company, Polaroid Corporation, Office of Scientific Research and Development, and the Cambridge Electron Accelerator. He is the author of books on polarized light, solar heated houses, and superinsulated houses.

RAYMOND C. TURNER

Raymond C. Turner (Essay 15) is well known for his work with the physics of toys. He received his B. S. in physics from Carnegie Institute of Technology and his Ph. D. in solid state physics from the University of Pittsburgh in 1966. He is now a professor of physics at Clemson University in South Carolina where he conducts research on electron-spin-resonance studies of polymers. He has presented numerous workshops and lectures at national teachers' meetings on the use of toys in physics education, and he has served on local and national committees of the American Association of Physics Teachers. He has published articles on physics and toys in the *American Journal of Physics* and *The Physics Teacher.*

JEARL WALKER

Jearl Walker (Essays 1, 2, 10 and 16) is professor of physics at Cleveland State University. He received a B. S. in physics from M. I. T. and a Ph. D. in physics from the University of Maryland. Since 1977 he has conducted "The Amateur Scientist" department of *Scientific American,* where he is read in 10 languages in world-wide publication. His book *The Flying Circus of Physics with Answers* is also published in 10 languages.

CONTENTS

CHAPTER 26
ELECTRIC POTENTIAL
592

CHAPTER 27
CAPACITANCE
618

CHAPTER 28
CURRENT AND RESISTANCE
640

CHAPTER 29
ELECTROMOTIVE FORCE AND CIRCUITS
662

CHAPTER 30
THE MAGNETIC FIELD
686

CHAPTER 35
ELECTROMAGNETIC OSCILLATIONS
800

CHAPTER 36
ALTERNATING CURRENTS
814

CHAPTER 37
MAXWELL'S EQUATIONS
832

CHAPTER 38
ELECTROMAGNETIC WAVES
843

CHAPTER 39
GEOMETRICAL OPTICS
866

CHAPTER 40
INTERFERENCE
900

CHAPTER 41
DIFFRACTION
922

CHAPTER 42
RELATIVITY
952

CHAPTER 43
QUANTUM PHYSICS—I
978

CHAPTER 47

NUCLEAR PHYSICS

1076

CHAPTER 48

ENERGY FROM THE NUCLEUS

1102

CHAPTER 49

QUARKS, LEPTONS, AND THE BIG BANG

1124

APPENDICES

A1

CHAPTER 23

ELECTRIC CHARGE

Electric charges over Seattle. Perhaps 10 coulombs of charge are exchanged in this lightning flash. By contrast, a quarter contains about 250,000 coulombs of positive charge, neatly balanced by the same amount of negative charge. There is an enormous amount of electric charge locked up in ordinary matter.

23-1 Electromagnetism

The early Greek philosophers knew that if you rubbed a piece of amber it could pick up bits of straw. There is a direct line of development from this ancient observation to the electronic age in which we live. The strength of the connection is indicated by our word "electron," which is derived from the Greek word for amber.

The Greeks also knew that some naturally occurring "stones," which we know today as the mineral magnetite, would attract iron. Such were the modest origins of the sciences of electricity and magnetism. These two sciences developed quite separately for centuries, until 1820 in fact, when Hans Christian Oersted found a connection between them: An electric current in a wire can deflect a magnetic compass needle. Interestingly enough, Oersted made this discovery while preparing a demonstration lecture for his physics students.

The new science of electromagnetism was developed further by workers in many countries. One of the very best was Michael Faraday,* a truly gifted experimenter with a talent for physical intuition and visualization. His collected laboratory notebooks, for example, do not contain a single equation. James Clerk Maxwell†

* See "Michael Faraday," by Herbert Kondo, *Scientific American,* October 1953. For the definitive biography see L. Pearce Williams, *Michael Faraday,* Basic Books, New York, 1964.

† See "James Clerk Maxwell," by James R. Newman, *Scientific American,* June 1955.

put Faraday's ideas into mathematical form, introduced many new ideas of his own, and put electromagnetism on a sound theoretical basis.

Table 2 of Chapter 37 shows the basic laws of electromagnetism, called Maxwell's equations. We plan to work our way through them in the chapters that follow but you might want to glance at them now, just to see where we are headed. These equations play the same role in electromagnetism that Newton's laws of motion do in classical mechanics or that the laws of thermodynamics do in the study of heat.

Maxwell's great discovery in electromagnetism was that light is an electromagnetic wave and that you can measure its speed by making purely electrical and magnetic measurements. With this discovery, Maxwell linked the ancient science of optics to those of electricity and magnetism. Heinrich Hertz* took a gient step forward when he produced electromagnetic "Maxwellian waves" in his laboratory. We now call them short radio waves. It remained for Marconi and others to push forward with the practical applications. Today, Maxwell's equations are used the world over in the solution of a wide variety of practical engineering problems.

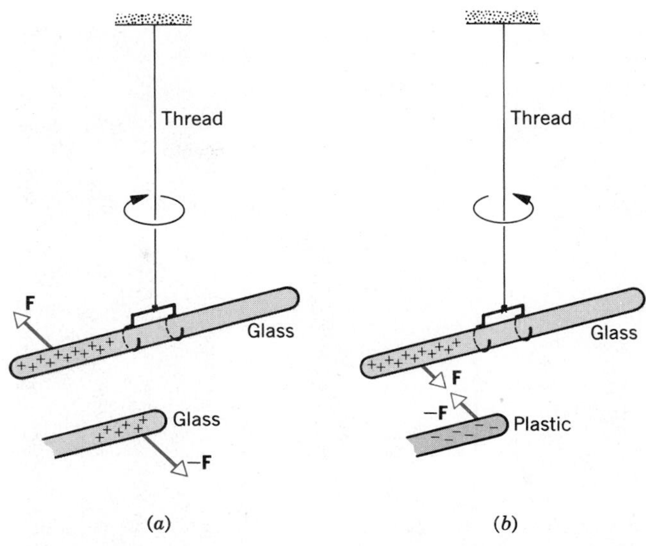

Figure 1 (*a*) Two similarly charged rods repel each other. (*b*) Two oppositely charged rods attract each other.

23-2 Electric Charge

If you walk across a carpet in dry weather, you can draw a spark by touching a metal door knob. Television advertising has alerted us to problem of "static cling." On a grander scale, lightning is familiar to everyone. All these phenomena represent the merest glimpse of the vast amount of *electric charge* that is stored up in the familiar objects that surround us and—indeed—in our own bodies.

The electrical neutrality of most objects in our visible and tangible world conceals the fact that such objects contain enormous amounts of positive and negative electric charge that largely cancel each other in their external effects. Only when this nice electrical balance is slightly disturbed does nature reveal to us the effects of uncompensated positive or negative charge. When we say that a body is "charged" we mean that it has a slight charge imbalance.

Charged bodies exert forces on each other. To show this, let us charge a glass rod by rubbing it with silk. The process of rubbing transfers a tiny amount of charge from one body to the other, thus slightly upsetting the electrical neutrality of each. If you suspend this charged rod from a thread, as in Fig. 1*a*, and if you bring a second charged glass rod nearby, the two rods will repel each other. However, if you rub a plastic rod with fur it will attract the charged end of the hanging glass rod; see Fig. 1*b*.

We explain all this by saying there are two kinds of charge, one of which (the one on the glass rubbed with silk) we have come to call *positive* and the other (the one on the plastic rubbed with fur) we have come to call *negative*. These simple experiments can be summed up by saying:

Like charges repel and unlike charges attract.

In Section 23-4, we put this rule into quantitative form, as Coulomb's law of force. We shall consider only charges that are either at rest with respect to each other or moving very slowly, a restriction that defines the subject of *electrostatics*.

The positive and negative labels for electric charge are due to Benjamin Franklin who, among many other accomplishments, was a scientist of international reputation. It has even been said that Franklin's triumphs in diplomacy in France during the American War of Inde-

* See "Heinrich Hertz," by Philip and Emily Morrison, *Scientific American*, December 1957.

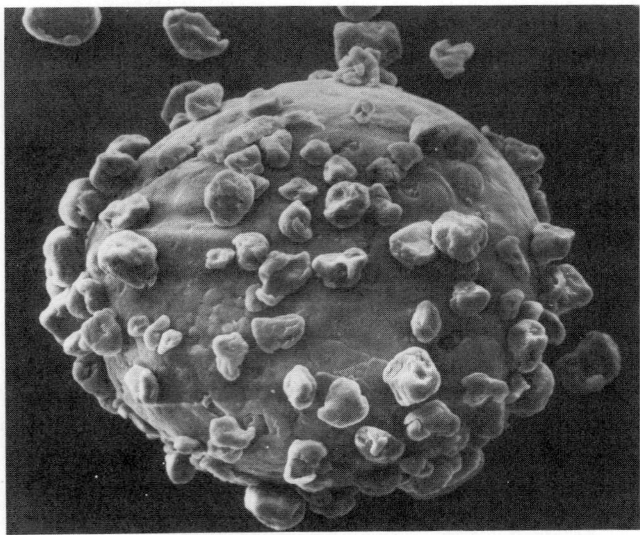

Figure 2 A carrier bead from a Xerox copying machine, covered with toner particles that cling to it by electrostatic attraction.

pendence were facilitated, and perhaps even made possible, because he was so highly regarded as a scientist.

Franklin introduced the words "charge" and "battery" into the language of electricity. When the *battery* pack in your pocket calculator loses its *charge* and, in charging it, you see the *plus* sign marking the *positive* battery terminal, think of Franklin.

Electrical forces between charged bodies have many industrial applications, among them being electrostatic paint spraying and powder coating, fly-ash precipitation, nonïmpact ink-jet printing, and photocopying. Figure 2, for example, shows a tiny carrier bead in a Xerox copying machine, covered with particles of black powder, called *toner,* that stick to it by electrostatic forces. These negatively charged toner particles are eventually attracted from their carrier beads to a positively charged latent image of the document to be copied, formed on a rotating drum. A charged sheet of paper then attracts the toner particles from the drum to itself, after which they are heat-fused in place and you have your copy.

23-3 Conductors and Insulators

You cannot seem to charge up a copper rod, no matter how hard you rub it or with what you rub it. However, if you fit the rod with a plastic handle, you will be able to build up a charge. The explanation is that charges placed on some materials—we call them *insulators*—are not free to move around; they stay where you put them. In other materials—we call them *conductors*—charges can move around more or less freely. If you touch a charged isolated copper rod with your finger, the charges will quickly move from the rod through your body to the ground.

Glass, chemically pure water, and plastics are common examples of insulators. Although there are no perfect insulators, fused quartz is pretty good, its insulating ability being about 10^{25} times greater than that of copper.

Copper, metals in general, tap water, and the human body, as Fig. 3*a* shows, are common examples of conductors, In metals, a fairly subtle experiment called the *Hall effect* shows that it is the negative charges that are free to move; we discuss this effect in Section 30-4. When copper atoms come together to form solid copper, their outer electrons do not remain attached to the individual atoms but become free to wander about within the rigid lattice structure formed by the positively charged ion cores. We call these mobile electrons the *conduction electrons.* The positive charges in a copper rod are just as immobile as they are in a glass rod.

(*a*)

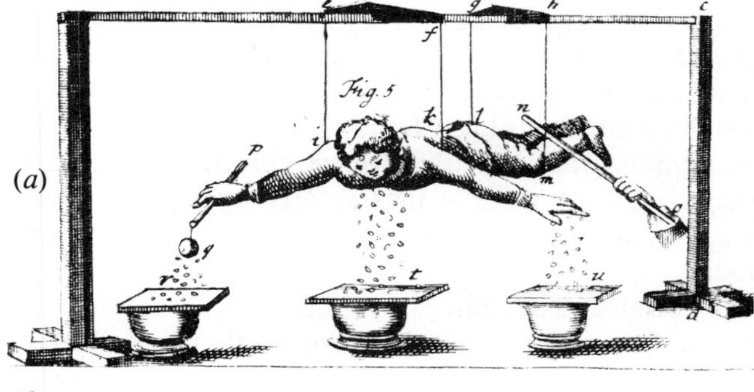

(*b*)

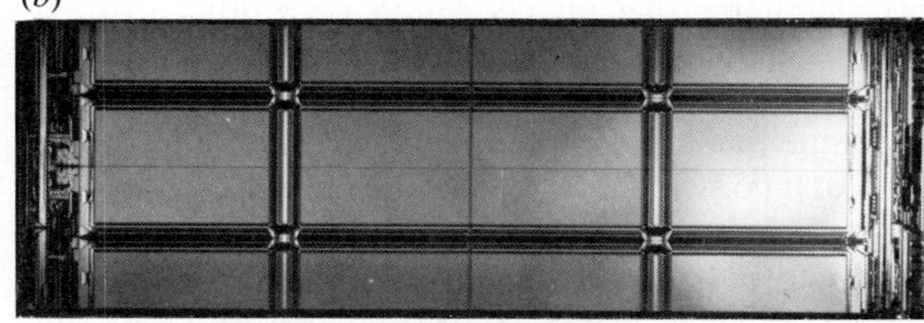

Figure 3 Progress in electricity. (*a*) Not a parlor stunt but a serious experiment carried out in 1774 to prove that the human body is a conductor of electricity. (*b*) A megabyte chip.

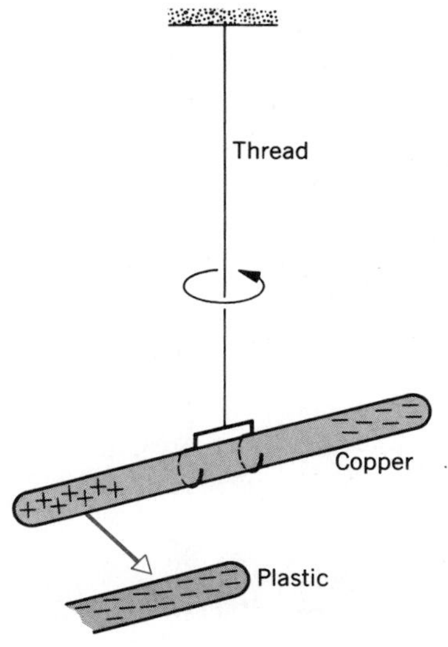

Figure 4 Either end of an isolated uncharged copper rod will be attracted by a charged rod of either sign. In this case, conduction electrons in the copper rod are repelled to the far end of that rod, leaving the near end positive.

The experiment of Fig. 4 demonstrates the mobility of charge in a conductor. A negatively charged plastic rod will attract either end of a suspended but uncharged copper rod. The (mobile) conduction electrons in the copper rod are repelled by the negative charge on the plastic rod to the far end of the copper rod, leaving the near end of the copper rod with a positive charge. An uncharged copper rod will also be attracted by a positively charged glass rod. In this case, the conduction electrons in the copper are attracted by the positively charged glass rod to the near end of the copper rod; the far end of the copper rod is then left with a positive charge.

There are also *semiconductors,* such as silicon and germanium. The microelectronic revolution that has transformed our lives in so many ways rests solidly on semiconducting devices; see Fig. 3*b*. We describe the operation of semiconductors, whose electrical characteristics can often be controlled to suit the requirements of the problem at hand, in Chapter 46 of the extended version of this text.

Finally, there are *superconductors.* The resistance of these materials to the flow of electricity is not just small; it is absolutely zero. If you set up a current in a superconducting ring, it persists without change for as long as you care to watch it, no battery or other source of energy being needed in the circuit.

Superconductivity was discovered in 1911 by the Dutch physicist Kammerlingh Onnes, who observed that solid mercury lost its electrical resistance completely at temperatures below 4.2 K. Until 1986, superconductivity was limited in its usefulness because superconducting materials needed to be cooled to temperatures below about 20 K. For example, the magnet coils on the large particle accelerator at Fermilab are made of a superconducting alloy kept cool with liquid helium. The central feature of the particle accelerator called the Superconducting Supercollider, or SSC, will be a ring of superconducting magnets some 52 miles in circumference.

In recent years, however, alloys have been developed that become superconducting at much higher temperatures so that a new era of useful applications seems to be upon us. Superconductivity at room temperature has not been ruled out as a possibility. See Essay 12.

It is not possible to understand semiconductors or superconductors without some background in quantum physics; we discuss both subjects in greater detail from this point of view in Chapter 46 of the extended version of this book.

23-4 Coulomb's Law

The law that gives the electrostatic force acting between the charges of two particles has exactly the same form as the law that gives the gravitational force acting between the masses of two particles. The two laws are

$$F_{\text{grav}} = G\frac{m_1 m_2}{r^2} \quad \text{(Newton's law—gravitation)} \quad (1)$$

and

$$F_{\text{elec}} = C\frac{q_1 q_2}{r^2} \quad \text{(Coulomb's law—electrostatics),} \quad (2)$$

in which G and C are constants.

Coulomb's law has survived every experimental test, no exceptions to it having ever been found. It holds deep within the atom, correctly describing the force between the positively charged nucleus and each extranuclear electron. Although classical Newtonian mechanics fails in that realm—where it is replaced by quantum physics—Coulomb's simple law continues to give correct answers. This law also correctly accounts for the forces that bind atoms together to form molecules and the forces that bind atoms or molecules together to form solids or liquids. We ourselves are assemblies of nuclei and electrons held together by forces arising from Eq. 2. There is much more to Coulomb's law than accounting for the forces between two charged rods.

In Eq. 2, F is the magnitude of the force acting on either particle owing to the charge on the other; q_1 and q_2 are the absolute values of the charges of the two particles and r is the distance between them. The constant C, by analogy with the gravitational constant G, may be called the *electrostatic constant.* Both laws are inverse square laws and both involve a property of the interacting particles—the mass in one case and the charge in the other.

The laws differ in that gravitational forces are always attractive but electrostatic forces may be either attractive or repulsive, depending on the signs of the two charges. These differences arise from the fact that, although there are two kinds of charge, there is apparently only one kind of mass.

For practical reasons having to do with the accuracy of measurements, the SI unit of charge is derived from the SI unit of electric current. If you connect the ends of a long wire to the terminals of a battery, a current is set up in the wire. We visualize this current as a flow of charge. The SI unit of current is the *ampere* (abbr. A). It is an SI base unit and in Section 31–4 we shall describe how it is defined in terms of experimental operations.

The SI unit of charge is the *coulomb* (abbr. C). A coulomb is defined as the amount of charge that passes through any cross section of a wire in 1 second if there is a current of 1 ampere in the wire. In general,

$$dq = i\,dt, \tag{3}$$

where dq (in coulombs) is the charge transferred by a current i (in amperes) during the interval dt (in seconds).

The constant G in Eq. 1 determines the absolute magnitude of the gravitational force. In the same way, the electrostatic constant C in Eq. 2 determines the absolute magnitude of the electrostatic force. For historical reasons, this constant is written, not simply as C, but in a more complex form, so that Coulomb's law appears as

$$\boxed{F = \frac{1}{4\pi\epsilon_0}\frac{q_1 q_2}{r^2}} \quad \text{(Coulomb's law).} \tag{4}$$

Certain equations that are derived from Eq. 4, but are

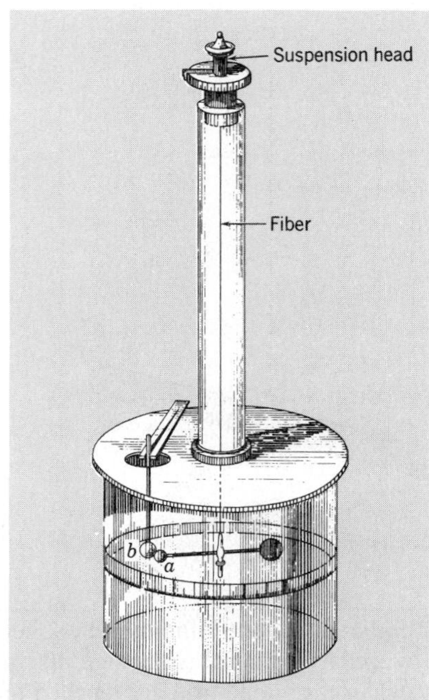

Figure 5 Coulomb's torsion balance, from his 1785 memoir to the Paris Academy of Sciences.

used much more often than it is, are simpler in form if we do it this way. The electrostatic constant in Eq. 4 has the value

$$\frac{1}{4\pi\epsilon_0} = 8.99 \times 10^9\ \mathrm{N\cdot m^2/C^2}. \tag{5}$$

The quantity ϵ_0, called the *permittivity constant,* sometimes appears separately in equations; its value is

$$\epsilon_0 = 8.85 \times 10^{-12}\ \mathrm{C^2/N\cdot m^2}. \tag{6}$$

Coulomb's law is named for Charles Augustus Coulomb who, in 1785, measured the electrical forces between small charged spheres, using the apparatus shown in Fig. 5. It is a torsion balance, operating on much the same principle as the torsion balance used by Henry Cavendish in 1798 to measure gravitational forces; see Section 15–3.

Although torsion balance methods are direct, they are not very accurate. We could not be convinced, on the basis of Coulomb's data, that the exponent 2 in Eq. 4 was not really, say, 2.0003. Fortunately, there are indirect methods that convince us that this exponent, if it is not 2 exactly, differs from it by a number that is less than

3×10^{-16}. We shall describe these powerful methods in Section 25-8.

Another parallel between the gravitational and the electrostatic force is that they both obey the principle of superposition. If we have n point charges, they interact independently in pairs and the force on any one of them, let us say q_1, is given by the vector sum

$$\mathbf{F}_1 = \mathbf{F}_{12} + \mathbf{F}_{13} + \mathbf{F}_{14} + \mathbf{F}_{15} + \cdots + \mathbf{F}_{1n}, \qquad (7)$$

in which, for example, $\mathbf{F}_{14}$ is the force acting on particle 1 owing to the presence of particle 4. An identical formula holds for the gravitational force.

Finally, the two shell theorems that we found so useful in our study of gravitation hold equally well in electrostatics:

Theorem 1 A uniform spherical shell of charge behaves, for external points, as if all its charge were concentrated at its center.

Theorem 2 A uniform spherical shell of charge exerts no force on a charged particle placed inside the shell.

The proof of these theorms follows exactly the proof for the gravitational case in Section 15-5. All you have to do is replace the mass m in that proof by the charge q, wherever it appears, and replace the gravitational constant G by the corresponding electrostatic constant $1/4\pi\epsilon_0$. The formal identity of the proofs follows from the fact that the two fundamental laws, Eqs. 1 and 2, have exactly the same mathematical form.

Sample Problem 1 Figure 6 shows three charged particles, held in place by forces not shown. What electrostatic force, owing to the other two charges, acts on q_1? Take $q_1 = -1.2\ \mu C$, $q_2 = +3.7\ \mu C$, $q_3 = -2.3\ \mu C$, $r_{12} = 15$ cm, $r_{13} = 10$ cm, and $\theta = 32°$.

This problem calls for the superposition principle. We start by computing the magnitudes of the forces that q_2 and q_3 exert on q_1. We substitute only the absolute values of the charges into Eq. 4, disregarding—for the time being—their signs. We then have

$$F_{12} = \frac{1}{4\pi\epsilon_0}\frac{q_1 q_2}{r_{12}^2}$$

$$= \frac{(8.99 \times 10^9\ \mathrm{N \cdot m^2/C^2})(1.2 \times 10^{-6}\ \mathrm{C})(3.7 \times 10^{-6}\ \mathrm{C})}{(0.15\ \mathrm{m})^2}$$

$$= 1.77\ \mathrm{N}.$$

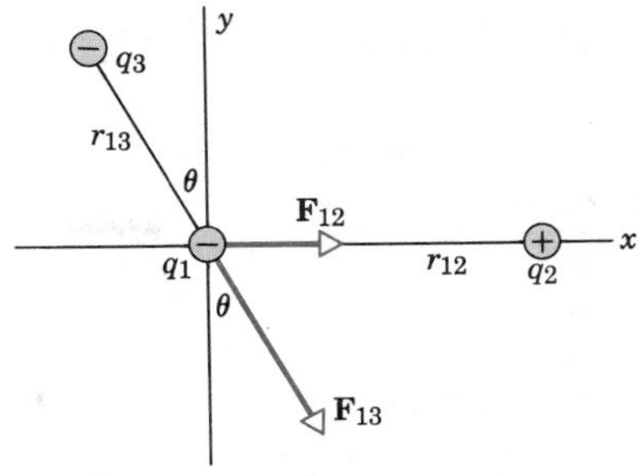

Figure 6 Sample Problem 1. The three charges exert three action-reaction pairs of forces on each other. Only the two forces acting on charge q_1 are shown.

The charges q_1 and q_2 have opposite signs so that the force between them is attractive. Hence, $\mathbf{F}_{12}$ points to the right in Fig. 6.

We also have

$$F_{13} = \frac{(8.99 \times 10^9\ \mathrm{N \cdot m^2/C^2})(1.2 \times 10^{-6}\ \mathrm{C})(2.3 \times 10^{-6}\ \mathrm{C})}{(0.10\ \mathrm{m})^2}$$

$$= 2.48\ \mathrm{N}.$$

These two charges have the same (negative) sign so that the force between them is repulsive. Thus, $\mathbf{F}_{13}$ points as shown in Fig. 6.

The components of the resultant force $\mathbf{F}_1$ acting on q_1 are

$$F_{1x} = F_{12x} + F_{13x} = F_{12} + F_{13} \sin\theta$$
$$= 1.77\ \mathrm{N} + (2.48\ \mathrm{N})(\sin 32°) = 3.08\ \mathrm{N}$$

and

$$F_{1y} = F_{12y} + F_{13y} = 0 - F_{13}\cos\theta$$
$$= -(2.48\ \mathrm{N})(\cos 32°) = -2.10\ \mathrm{N}.$$

From these components, you can show that the magnitude of $\mathbf{F}_1$ is 3.73 N and that this vector makes an angle of $-34°$ with the x axis.

23-5 Charge is Quantized

In Franklin's day, electric charge was thought to be a continuous fluid, an idea that was useful for many purposes. However, we now know that fluids themselves, such as air or water, are not continuous but are made up of atoms and molecules; matter is discrete. Experiment shows that the "electrical fluid" is not continuous either but that it is made up of multiples of a certain elementary charge. That is, any charge q that can be observed and

Table 1 Some Properties of Three Particles

Particle	*Symbol*	*Charge*[a] e	*Mass*[b] m_e	*Angular Momentum*[c] $h/2\pi$
Electron	e	-1	1	$\frac{1}{2}$
Proton	p	$+1$	1836.15	$\frac{1}{2}$
Neutron	n	0	1838.68	$\frac{1}{2}$

[a] In units of the elementary charge e.
[b] In units of the electron mass m_e.
[c] The intrinsic spin angular momentum, in units of $h/2\pi$. We introduced this concept in Section 12–12 and give a fuller treatment in Chapter 45 of the extended version of this book.

measured* directly can be written

$$q = ne \qquad n = 0, \pm 1, \pm 2, \pm 3, \cdots, \tag{8}$$

in which e, the *elementary charge*, has the value

$$e = 1.60 \times 10^{-19}\ \text{C}. \tag{9}$$

The elementary charge is one of the important constants of nature.

When a physical quantity such as charge exists only in discrete "packets" rather than in continuously variable amounts, we say that quantity is *quantized*. We have already seen that matter, energy, and angular momentum are quantized; charge adds one more important physical quantity to the list. Equation 8 tells us that it is possible, for example, to find a particle that carries a charge of zero or of $+10e$ or of $-6e$ but not a particle with a charge of, say, $3.57e$. Table 1 shows the charges and some other properties of the three particles that can be said to make up the material world around us.

The quantum of charge is small. In an ordinary 100-W, 120-V light bulb, for example, about 10^{19} elementary charges enter and leave the bulb every second. The graininess of electricity does not show up in large-scale phenomena, just as you cannot feel the individual molecules of water when you move your hand through it.

Sample Problem 2 A penny, being electrically neutral, contains equal amounts of positive and negative charge. What is the magnitude of these equal charges?

The charge q is given by NZe, in which N is the number of atoms in a penny and Ze is the magnitude of the positive and the negative charges carried by each atom.

The number N of atoms in a penny, assumed for simplicity to be made of copper, is $N_A m/M$, in which N_A is the Avogadro constant. The mass m of the coin is 3.11 g and the atomic weight M of copper is 63.5 g/mol. We find

$$N = \frac{N_A m}{M} = \frac{(6.02 \times 10^{23}\ \text{atoms/mol})(3.11\ \text{g})}{63.5\ \text{g/mol}}$$
$$= 2.95 \times 10^{22}\ \text{atoms}.$$

Every neutral atom has a negative charge of magnitude Ze associated with its electrons and a positive charge of the same magnitude associated with its nucleus. Here e is the magnitude of the charge on the electron, which is 1.60×10^{-19} C, and Z is the atomic number of the element in question. For copper, Z is 29. The magnitude of the total negative, or positive, charge in a penny is then

$$q = NZe$$
$$= (2.95 \times 10^{22})(29)(1.60 \times 10^{-19}\ \text{C})$$
$$= 137{,}000\ \text{C}. \qquad \text{(Answer)}$$

This is an enormous charge. By comparison, the charge that you might get by rubbing a plastic rod is perhaps 10^{-9} C, smaller by a factor of about 10^{14}. For another comparison, it would take about 38 h for a charge of 137,000 C to flow through the filament of a 100-W, 120-V light bulb. There is a lot of electric charge in ordinary matter.

Sample Problem 3 In Sample Problem 2 we saw that a copper penny contains both positive and negative charges, each of magnitude 1.37×10^5 C. Suppose that these charges could be concentrated into two separate bundles, held 100 m apart. What attractive force would act on each bundle?

From Eq. 4 we have

$$F = \frac{1}{4\pi\epsilon_0}\frac{q^2}{r^2} = \frac{(8.99 \times 10^9\ \text{N}\cdot\text{m}^2/\text{C}^2)(1.37 \times 10^5\ \text{C})^2}{(100\ \text{m})^2}$$
$$= 1.69 \times 10^{16}\ \text{N}. \qquad \text{(Answer)}$$

This is about 2×10^{12} tons of force! Even if the charges were separated by one earth diameter, the attractive force would still be about 120 tons. In all of this, we have sidestepped the problem of forming each of the separated charges into a "bundle" whose dimensions are small compared to their separation. Such bundles, if they could ever be formed, would be blasted apart by mutual Coulomb repulsion forces.

The lesson of this sample problem is that you cannot disturb the electrical neutrality of ordinary matter very much. If you try to pull out any sizable fraction of the charge contained in a body, a large Coulomb force appears automatically, tending to pull it back.

* Quarks carry charges of $-\frac{1}{3}e$ or $+\frac{2}{3}e$ but these particles, thought to be the constituent particles of protons and neutrons, have never been detected as free particles; see Section 2–8.

Sample Problem 4 The average distance r between the electron and the central proton in the hydrogen atom is 5.3×10^{-11} m. (a) What is the magnitude of the average electrostatic force that acts between these two particles?

From Eq. 4 we have, for the electrostatic force,

$$F_e = \frac{1}{4\pi\epsilon_0}\frac{q_1 q_2}{r^2} = \frac{(8.99 \times 10^9\ \mathrm{N \cdot m^2/C^2})(1.60 \times 10^{-19}\ \mathrm{C})^2}{(5.3 \times 10^{-11}\ \mathrm{m})^2}$$

$$= 8.2 \times 10^{-8}\ \mathrm{N}. \qquad \text{(Answer)}$$

(b) What is the magnitude of the average gravitational force that acts between these particles?

From Eq. 1 we have, for the gravitational force,

$$F_g = G\frac{m_e m_p}{r^2}$$

$$= \frac{(6.67 \times 10^{-11}\ \mathrm{m^3/kg \cdot s^2})(9.11 \times 10^{-31}\ \mathrm{kg})(1.67 \times 10^{-27}\ \mathrm{kg})}{(5.3 \times 10^{-11}\ \mathrm{m})^2}$$

$$= 3.6 \times 10^{-47}\ \mathrm{N}. \qquad \text{(Answer)}$$

We see that the gravitational force is weaker than the electrostatic force by the enormous factor of about 10^{39}. Although the gravitational force is weak, it is always attractive. Thus, it can act to build up very large masses, as in the formation of stars and planets, so that large gravitational forces can develop. The electrostatic force, on the other hand, is repulsive for like charges so that it is not possible to accumulate large concentrations of either positive or negative charge. We must always have the two together, so that they largely compensate for each other. The charges that we are accustomed to in our daily experiences are slight disturbances of this overriding balance.

Sample Problem 5 The nucleus of an iron atom, which has a radius of about 4×10^{-15} m, contains 26 protons. What repulsive electrostatic force acts between two protons in such a nucleus if they are that far apart?

From Eq. 4 we have

$$F = \frac{1}{4\pi\epsilon_0}\frac{q_p q_p}{r^2}$$

$$= \frac{(8.99 \times 10^9\ \mathrm{N \cdot m^2/C^2})(1.60 \times 10^{-19}\ \mathrm{C})^2}{(4 \times 10^{-15}\ \mathrm{m})^2}$$

$$= 14\ \mathrm{N}. \qquad \text{(Answer)}$$

This enormous force, more than 3 lb and acting on a single proton, must be more than balanced by the attractive strong nuclear force that binds the nucleus together. The strong force, whose range is so short that its effects cannot be felt very far outside the nucleus, is very well named; see Section 6-5.

23-6 Charge Is Conserved

If you rub a glass rod with silk, a positive charge appears on the rod. Measurement shows that a negative charge of equal magnitude appears on the silk. This suggests that rubbing does not create charge but only transfers it from one body to another, upsetting the electrical neutrality of each during the process. This hypothesis of *conservation of charge,* first put forward by Benjamin Franklin, has stood up under close examination, both for macroscopic charges and at the level of interactions between elementary particles. No exceptions have ever been found. Thus, we add electric charge to our growing list of quantities—including energy and both linear and angular momentum—that obey a conservation law.

Radioactive decay gives us ready examples of charge conservation at the nuclear level. The common uranium isotope uranium-238, or ^{238}U, is typical:

$$^{238}\mathrm{U} \rightarrow {}^{234}\mathrm{Th} + {}^{4}\mathrm{He} \quad \text{(radioactive decay)}. \qquad (10)$$

The atomic number of the radioactive *parent* nucleus, ^{238}U, is 92, which tells us that this nucleus contains 92 protons. It decays spontaneously by emitting an alpha particle, ^{4}He, for which $Z = 2$, leaving behind the *daughter* nucleus ^{234}Th, with $Z = 90$. Thus, the amount of charge present before the decay, $92e$, is equal to the amount present after the decay, $90e + 2e$ or $92e$. Charge is conserved.

Another example of charge conservation occurs when an electron, whose charge is $-e$, happens to find itself close to a positive electron, called a *positron,* charge $+e$, both particles being essentially at rest. The two particles may simply disappear, in an *annihilation process,* converting all the energy associated with their masses, which is $2mc^2$, into the radiant energy of two oppositely directed gamma rays. Thus,

$$e^- + e^+ \rightarrow \gamma + \gamma \quad \text{(annihilation)}. \qquad (11)$$

In applying the conservation-of-charge principle, we must add the charges algebraically, with due regard for their signs. In the annihilation process of Eq. 11, the net charge of the system is zero, both before and after the event. Charge is conserved.

The converse of annihilation also occurs. Here an energetic gamma ray, passing near a heavy nucleus that serves as a catalyst, simply disappears, converting its radiant energy into the creation of an electron and a positron. Thus,

$$\gamma \rightarrow e^- + e^+ \quad \text{(pair production)}. \qquad (12)$$

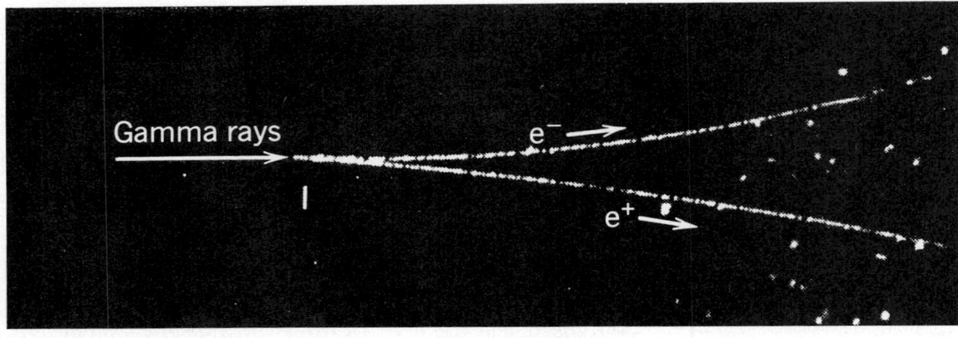

Figure 7 An energetic gamma ray coming in from the left generates an electron-positron pair. The two electrons leave tracks of bubbles in the chamber in which they were created. The tracks curve, in opposite directions, because of the action of a strong external magnetic field.

Again, charge is conserved. The two particles produced in the pair-production process carry off any energy that the incoming gamma ray may have left after providing the energy $2mc^2$ needed to create the pair of electrons.

Figure 7 shows such a pair-production event occurring in a bubble chamber at the Lawrence Berkeley Laboratory. Gamma rays entering the chamber from the left, being uncharged, leave no bubble tracks in the chamber. The bubble tracks left in the wakes of the electron and the positron are curved because of the action of a magnetic field in the chamber. The fact that the particle tracks curve in opposite directions is evidence that the electron and the positron carry charges of opposite sign.

An interesting consequence of the conservation of charge is that it ensures that the electron is an absolutely stable particle. From the point of view of the conservation of energy, there is nothing to stop a resting electron from decaying into two* oppositely-directed gamma rays, in this way

$$e^- \rightarrow \gamma + \gamma \quad \text{(doesn't happen!)}. \qquad (13)$$

If such a process occurred, there would be a charge of $-e$ before the event but a charge of zero after the event, which would violate the law of charge conservation. Because charge must be conserved, we conclude that the process does not happen. There is no lighter charged particle into which the electron can decay; we conclude that a single isolated electron must be absolutely stable.

23-7 The Constants of Physics: An Aside

In this chapter we have introduced yet another of the fundamental constants of physics, the elementary charge e. Perhaps it is time to step aside and to review the role that these constants play in the structure of physics. Table 2 lists four of them that are particularly central.

We note at once how precisely these constants are known. Although we have been using only three significant figures in our illustrative Sample Problems, we see that the constants are typically known to at least seven or eight significant figures. An exception is the gravitational constant, the least well known of all the important physical constants. Experiments seeking improved values of the various constants are going on, in laboratories all over the world, on a continuing basis. A particular constant may be involved, either alone or with other constants, in a wide variety of experiments. Unraveling all these data is no simple task. Every decade or so it seems appropriate to survey the accumulated measurements and, with the help of an elaborate computer program, extract from this vast array of data a set of "best values" of the physical constants. The last such survey was in 1985.*

The improvement in our knowledge of the constants over time is impressive. Table 3, for example, shows how the precision of measurement of the speed of light has improved over the years. Note the variety of methods and the geographical spread of the effort. The measurements finally reached a point at which the precision was limited by the practical reproducibility of the standard of length that was in use at that time. As a result, it was decided to assign a value to the speed of light *by definition* and to redefine the length standard in terms of the speed of light; see Section 1-4.

Each of the constants in Table 2 plays an important role in the structure of physics. We discuss each in turn.

* We need *two* gamma rays so that linear momentum can be conserved.

* See Appendix B. For a very readable account of the physical constants, see also B. W. Petley, *The Fundamental Physical Constants and the Frontiers of Measurement,* A. Hilger, Boston, 1985.

Table 2 Four Fundamental Constants of Physics

Constant	*Symbol*	*Value (1985)*	*Uncertainty*[a]	*Section*
Gravitational constant	G	6.67260×10^{-11} m^3/kg · s^2	100	15–3
Speed of light	c	2.99792458×10^{8} m/s	Exact	17–7
Planck constant	h	$6.6260754 \times 10^{-34}$ J · s	0.6	8–10
Elementary charge	e	$1.60217733 \times 10^{-19}$ C	0.3	23–5

[a] In parts per million.

The Gravitational Constant G. This constant, which appears in Newton's law of gravity, is the central constant in both Newton's theory of gravity and Einstein's general theory of relativity. Any theory of the large-scale structure and development of the universe must involve this constant at a deep level.

The Speed of Light c. This constant, which appears in all relativistic equations, is the foundation stone of Einstein's special theory of relativity. The speed of light is large by ordinary standards but it is not infinitely great. If it were, the world would be governed by the laws of classical physics and would be an unimaginably different place.

The Planck Constant h. This constant is the central constant of quantum physics. The Planck constant is small but it is not zero. If it were, the world would be governed by the laws of classical physics and would, again, be unimaginably different. We introduced this constant briefly in Section 8–10. In Chapters 43 and 44 of the extended version of this book—in which we develop the concepts of quantum physics from their origins—the Planck constant will play a central role.

The Elementary Charge e. The basic importance of this constant lies in the fact that it can be combined with two other constants to form a dimensionless number, called the *fine structure constant,** α. Thus,

$$\alpha = \frac{e^2}{2\epsilon_0 hc} \approx \frac{1}{137}. \qquad (14)$$

This dimensionless constant is central to the theory of quantum electrodynamics, or QED as it is called.† This theory, which combines quantum physics with the theory of relativity, is perhaps the most successful theory in physics in terms of predicting results that agree with experiment. The number 137 has fascinated physicists for decades as they sought—and seek—to explore the significance of the fine structure constant. It is an unusual physicist who, coming upon page 137, does not have a fleeting thought of this constant.

* It received its name for historical reasons, having to do with the detailed structure of the spectrum lines. The quantity ϵ_0 that appears in Eq. 14 has a value that is exact by definition and does not play a fundamental role.

† See Richard P. Feynman, *QED—The Strange Theory of Light and Matter,* Princeton University Press, Princeton, NJ, 1985.

Table 3 The Speed of Light: Some Selected Measurements

Date	*Experimenter*	*Country*	*Method*	*Speed* (10^8 m/s)	*Uncertainty* (m/s)
1600	Galileo	Italy	Lanterns and shutters	"Fast"	?
1676	Roemer	France	Moons of Jupiter	2.14	?
1729	Bradley	England	Aberration of light	3.08	?
1849	Fizeau	France	Toothed wheel	3.14	?
1879	Michelson	United States	Rotating mirror	2.99910	75,000
	Michelson	United States	Rotating mirror	2.99798	22,000
1950	Essen	England	Microwave cavity	2.997925	1,000
1958	Froome	England	Interferometer	2.997925	100
1972	Evenson et al.	United States	Laser method	2.997924574	1.1
1974	Blaney et al.	England	Laser method	2.997924590	0.6
1976	Woods et al.	England	Laser method	2.997924588	0.2
1983	Internationally adopted value:			2.99792458	Exact

Sample Problem 6 It is possible to combine the three constants G, h, and c in such a way as to yield a quantity that has the dimensions of a time. This *Planck time* is given by

$$T_P = \sqrt{\frac{hG}{2\pi c^5}}. \tag{15}$$

Show that this quantity does indeed have the dimensions of time and find its value.

From Table 2 we write the three constants as

$$h = (6.63 \times 10^{-34}\ \mathrm{J \cdot s})\left(\frac{1\ \mathrm{kg \cdot m^2/s^2}}{1\ \mathrm{J}}\right)$$
$$= 6.63 \times 10^{-34}\ \mathrm{kg \cdot m^2/s},$$

$$G = 6.67 \times 10^{-11}\ \mathrm{m^3/kg \cdot s^2},$$

and

$$c = 3.00 \times 10^8\ \mathrm{m/s}.$$

To find the Planck time we substitute these values into Eq. 15, obtaining

$$T_P = \sqrt{\frac{(6.63 \times 10^{-34}\ \mathrm{kg \cdot m^2/s})(6.67 \times 10^{-11}\ \mathrm{m^3/kg \cdot s^2})}{(2\pi)(3.00 \times 10^8\ \mathrm{m/s})^5}}$$

$$= 5.38 \times 10^{-44}\ \mathrm{s}. \qquad \text{(Answer)}$$

Check this result carefully, to convince yourself that the units do indeed reduce to those of time.

It is perhaps not surprising that the Planck time, which is built up from the fundamental constants of three great theories, should have a fundamental significance. It turns out to be the time following the Big Bang before which we cannot have confidence that our present theories of physics are valid.

The constants h, G, and c can also be arranged to form quantities that have the dimensions of length and of mass. In Exercise 41 and Problem 42 we ask you to evaluate these quantities, which are called the *Planck length* and the *Planck mass.* Like the Planck time, each has physical significance in studies bearing on the origin and evolution of the universe.

REVIEW AND SUMMARY

Electromagnetism

Electromagnetism is a description of the interactions involving electric charge. Classical electromagnetism, summarized by Maxwell's equations, includes the phenomena of electricity, magnetism, electromagnetic induction (electric generators), and electromagnetic radiation (including all of classical optics).

Electric Charge

The strength of a particle's electromagnetic interaction is partly determined by its electric charge, which can be either positive or negative. Like charges repel and unlike charges attract each other. Objects with equal amounts of the two kinds of charge are electrically neutral, whereas those with an imbalance are electrically charged.

Conducting Materials

Conductors are materials in which a significant number of charged particles (electrons in metals) are free to move. The charged particles in insulators are not free. Semiconductors are intermediate between conductors and insulators in this respect. Superconductors offer no resistance to the flow of electrons.

The Coulomb and Ampere

The SI unit of charge is the coulomb (C). It is defined in terms of the unit of current, the ampere (A), as the charge passing a particular point in one second when a current of one ampere is flowing; see Eq. 3. Sample Problem 2 illustrates the large amount of electric charge in ordinary samples of matter.

Coulomb's Law

Coulomb's law describes *electrostatic* forces; the forces between small (point) electric charges at rest. In SI form,

$$F = \frac{1}{4\pi\epsilon_0}\frac{q_1 q_2}{r^2} \quad \text{(Coulomb's law).} \qquad [4]$$

Here $\epsilon_0 = 8.85 \times 10^{-12}\ \mathrm{C^2/N \cdot m^2}$ is the *permittivity constant;* $1/4\pi\epsilon_0 = 8.99 \times 10^9\ \mathrm{N \cdot m^2/C^2}$. Sample Problem 1 demonstrates a representative calculation.

The force of attraction or repulsion between point charges at rest acts along the line joining the two charges. If more than two charges are present, Eq. 4 holds for the forces between each pair. The resultant force on each charge is then found, using the superposition principle, as the vector sum of the forces exerted on it by each of the others. Sample Problem 1 shows how such forces can be calculated.

The Shell Theorems

Electrostatic forces obey the same shell theorems proved earlier for gravitational forces:

Theorem 1. A uniform spherical shell of charge behaves, for external points, as if all its charge were concentrated at its center.

Theorem 2. A uniform spherical shell of charge exerts no force on a charged particle placed inside the shell.

The Elementary Charge e

Electric charge is *quantized.* This means that any charge found in nature can be written as ne, where n is a positive or negative integer and e is a constant of nature called the *elementary charge;* its value is approximately 1.60×10^{-19} C.

Electric Charges in Matter

Matter as we ordinarily encounter it can be regarded as composed of protons, neutrons, and electrons whose properties are summarized in Table 1. Atoms contain a small, dense, positively charged nucleus (composed of neutrons and protons) surrounded by a cloud of electrons attracted to the nucleus by the electrical force. Most forces we ordinarily deal with, such as friction, fluid pressure, contact forces between solid surfaces, and elastic forces, are all manifestations of the electrical interactions between the charged particles of which atoms are made. Sample Problems 4 and 5 show that the electrical force is much stronger than gravity, when both interactions occur simultaneously, and that the nuclear force is still stronger.

Conservation of Charge

Electric charge is conserved. This means that the (algebraic) net charge of an isolated system of charges does not change, no matter what interactions occur within the system. Section 23–6 describes three illustrations of this important conservation law.

QUESTIONS

1. You are given two metal spheres mounted on portable insulating supports. Find a way to give them equal and opposite charges. You may use a glass rod rubbed with silk but may not touch it to the spheres. Do the spheres have to be of equal size for your method to work?

2. In Question 1, find a way to give the spheres equal charges of the same sign. Again, do the spheres need to be of equal size for your method to work?

3. A charged rod attracts bits of dry cork dust which, after touching the rod, often jump violently away from it. Explain.

4. The experiments described in Section 23–2 could be explained by postulating four kinds of charge, that is, on glass, silk, plastic, and fur. What is the argument against this?

5. A positive charge is brought very near to an uncharged insulated conductor. The conductor is grounded while the charge is kept near. Is the conductor charged positively or negatively or not at all if (*a*) the charge is taken away and then the ground connection is removed and (*b*) the ground connection is removed and then the charge is taken away?

6. A charged insulator can be discharged by passing it just above a flame. Explain why.

7. If you rub a coin briskly between your fingers, it will not seem to become charged by friction. Why?

8. If you walk briskly across a carpet, you often experience a spark upon touching a door knob. (*a*) What causes this? (*b*) How might it be prevented?

9. Why do electrostatic experiments not work well on humid days?

10. An insulated rod is said to carry an electric charge. How could you verify this and determine the sign of the charge?

11. If a charged glass rod is held near one end of an insulated uncharged metal rod as in Fig. 8, electrons are drawn to one end, as shown. Why does the flow of electrons cease? After all, there is an almost inexhaustible supply of them in the metal rod.

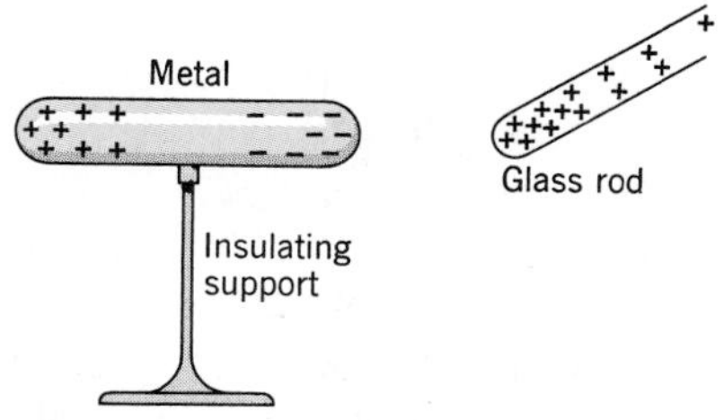

Figure 8 Questions 11 and 12.

12. In Fig. 8, does any electric force act on the metal rod? Explain.

13. A person standing on an insulated stool touches a charged, insulated conductor. Is the conductor discharged completely?

14. (*a*) A positively-charged glass rod attracts a suspended object. Can we conclude that the object is negatively charged? (*b*) A positively charged glass rod repels a suspended object. Can we conclude that the object is positively charged?

15. Is the electric force that one charge exerts on another changed if other charges are brought nearby?

16. A solution of copper sulfate is a conductor. What particles serve as the charge carriers in this case?

17. If the electrons in a metal such as copper are free to move about, they must often find themselves headed toward the metal surface. Why don't they keep on going and leave the metal?

18. Would it have made any important difference if Benjamin Franklin had chosen, in effect, to call electrons positive and protons negative?

19. Coulomb's law predicts that the force exerted by one point charge on another is proportional to the product of the two charges. How might you go about testing this aspect of the law in the laboratory?

20. An electron (charge $= -e$) circulates around a helium nucleus (charge $= +2e$) in a helium atom. Which particle exerts the larger force on the other?

21. The charge of a particle is a true characteristic of the particle, independent of its state of motion. Explain how you can test this statement by making a rigorous experimental check of whether the hydrogen atom is truly electrically neutral.

22. Earnshaw's theorem says that no particle can be in stable equilibrium under the action of electrostatic forces alone. Consider, however, point P at the center of a square of four equal positive charges, as in Fig. 9. If you put a positive test charge there it might seem to be in equilibrium. Every one of the four external charges pushes it toward P. Yet Earnshaw's theorem holds. Can you explain how?

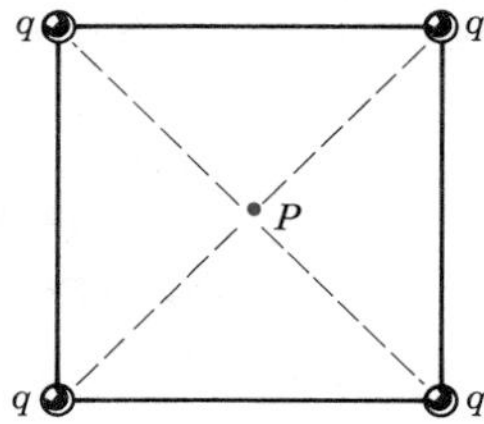

Figure 9 Question 22.

23. The quantum of charge is 1.60×10^{-19} C. Is there a corresponding single quantum of mass?

24. What does it mean to say that a physical quantity is (*a*) quantized or (*b*) conserved? Give some examples.

25. In Sample Problem 4 we show that the electrical force is about 10^{39} times stronger than the gravitational force. Can you conclude from this that a galaxy, or a star, or a planet must be essentially neutral electrically?

26. How do we know that electrostatic forces are not the cause of gravitational attraction, between the earth and moon, for example?

EXERCISES AND PROBLEMS

Section 23-4 Coulomb's Law

1E. What would be the force of attraction between two 1.0-C charges separated by a distance of (*a*) 1.0 m and (*b*) 1.0 km?

2E. A point charge of $+3.0 \times 10^{-6}$ C is 12 cm distant from a second point charge of -1.5×10^{-6} C. Calculate the magnitude of the force on each charge.

3E. What must be the distance between point charge $q_1 = 26\ \mu$C and point charge $q_2 = -47\ \mu$C in order that the attractive electrical force between them has a magnitude of 5.7 N?

4E. In the return stroke of a typical lightning bolt (see photo at the beginning of this chapter), a current of 2.5×10^4 A flows for 20 μs. How much charge is transferred in this event?

5E. Two equally-charged particles, held 3.2×10^{-3} m apart, are released from rest. The initial acceleration of the first particle is observed to be 7.0 m/s^2 and that of the second to be 9.0 m/s^2. If the mass of the first particle is 6.3×10^{-7} kg, what are (*a*) the mass of the second particle and (*b*) the magnitude of the common charge?

6E. Figure 10*a* shows two charges, q_1 and q_2, held a fixed distance d apart. (*a*) What is the strength of the electric force that acts on q_1? Assume that $q_1 = q_2 = 20\ \mu$C and $d = 1.5$ m. (*b*) A third charge $q_3 = 20\ \mu$C is brought in and placed as shown in Fig. 10*b*. What now is the strength of the electric force on q_1?

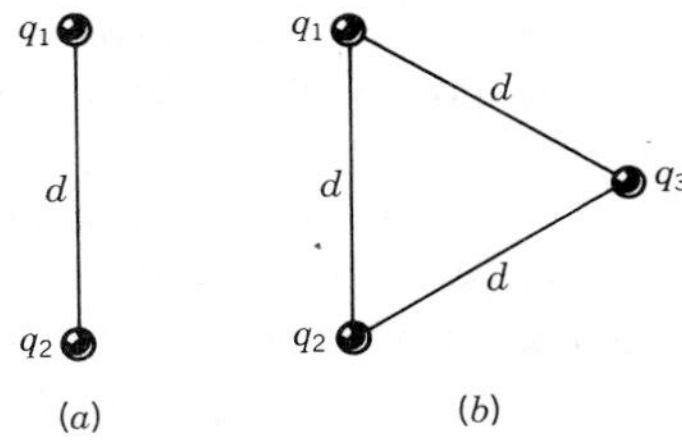

Figure 10 Exercise 6.

7E. Two identical conducting spheres, 1 and 2, carry equal amounts of charge and are separated by a distance large compared with their diameters. They repel each other with an electrical force **F**. Suppose now that a third identical sphere 3, having an insulating handle and initially uncharged, is touched first to sphere 1, then to sphere 2, and finally removed. What now is the force **F**′ between spheres 1 and 2, in terms of **F**? See Fig. 11.

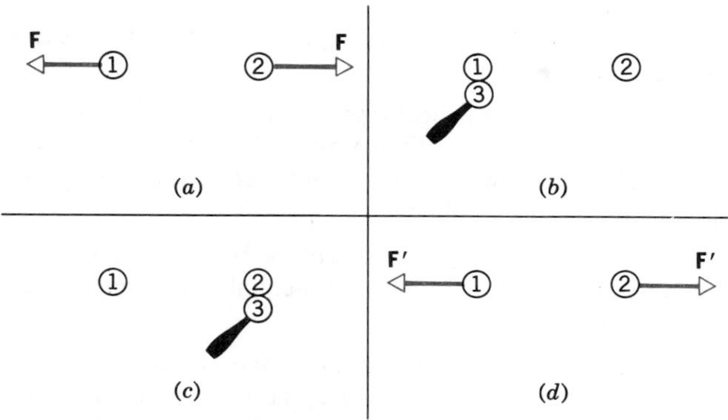

Figure 11 Exercise 7.

8P. Three charged particles lie on a straight line and are separated by a distance d as shown in Fig. 12. Charges q_1 and q_2 are held fixed. Charge q_3, which is free to move, is found to be in equilibrium under the action of the electric forces. Find q_1 in terms of q_2.

Figure 12 Problem 8.

9P. Charges q_1 and q_2 lie on the x axis at points $x = -a$ and $x = +a$, respectively. (*a*) How must q_1 and q_2 be related for the net force on charge $+Q$, placed at $x = +a/2$, to be zero? (*b*) Answer the same question if the $+Q$ charge is placed at $x = +3a/2$.

10P. In Fig. 13, what are the horizontal and vertical components of the resultant electric force on the charge in the lower left corner of the square? Assume that $q = 1.0 \times 10^{-7}$ C and $a = 5.0$ cm. The charges are at rest.

11P. Each of two small spheres is charged positively, the combined charge being 5.0×10^{-5} C. If each sphere is repelled from the other by a force of 1.0 N when the spheres are 2.0 m apart, calculate the charge on each sphere.

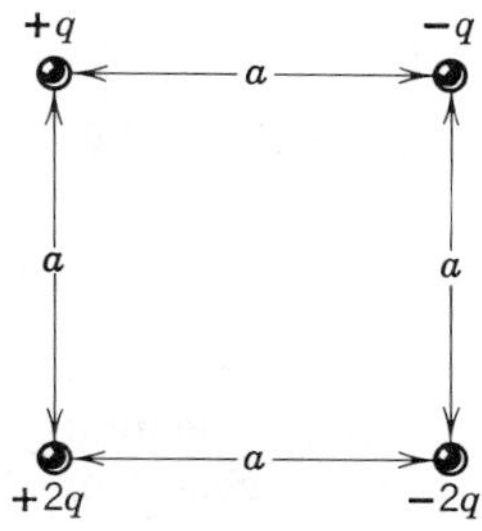

Figure 13 Problem 10.

12P. Two identical conducting spheres, having charges of opposite sign, attract each other with a force of 0.108 N when separated by 50.0 cm. The spheres are connected by a thin conducting wire, which is then removed, and thereafter the spheres repel each other with a force of 0.036 N. What were the initial charges on the spheres?

13P. Two fixed charges, $+1.0\ \mu$C and $-3.0\ \mu$C, are 10 cm apart. Where may a third charge be located so that no force acts on it?

14P. The charges and coordinates of two charged particles held fixed in the xy plane are: $q_1 = +3.0\ \mu$C, $x = 3.5$ cm, $y = 0.50$ cm, and $q_2 = -4.0\ \mu$C, $x = -2.0$ cm, $y = 1.5$ cm. (*a*) Find the magnitude and direction of the electrical force on q_2. (*b*) Where could you locate a third charge $q_3 = +4.0\ \mu$C such that the total electrical force on q_2 is zero?

15P. Two *free* point charges $+q$ and $+4q$ are a distance L apart. A third charge is so placed that the entire system is in equilibrium. (*a*) Find the location, magnitude, and sign of the third charge. (*b*) Show that the equilibrium is unstable.

16P. (*a*) What equal positive charges would have to be placed on the earth and on the moon to neutralize their gravitational attraction? Do you need to know the lunar distance to solve this problem? Why or why not? (*b*) How many thousand kilograms of hydrogen would be needed to provide the positive charge calculated in part (*a*)?

17P. A charge Q is fixed at each of two opposite corners of a square. A charge q is placed at each of the other two corners. (*a*) If the resultant electrical force on Q is zero, how are Q and q related? (*b*) Could q be chosen to make the resultant electrical force on *every* charge zero? Explain your answer.

18P. A certain charge Q is to be divided into two parts $(Q - q)$ and q. What is the relation of Q to q if the two parts, placed a given distance apart, are to have a maximum Coulomb repulsion?

19P. Two similar tiny conducting balls of mass m are hung from silk threads of length L and carry similar charges q as in

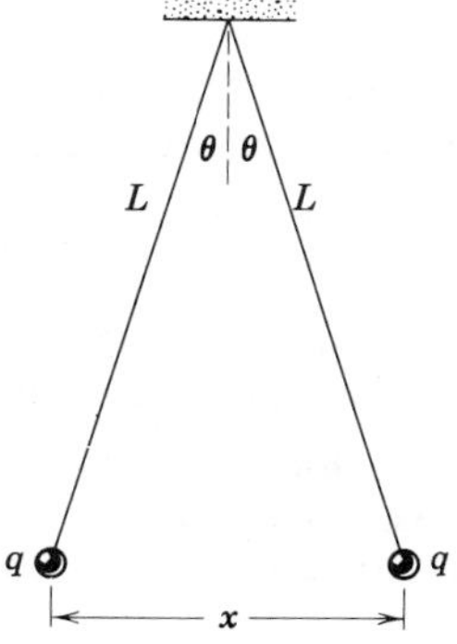

Figure 14 Problems 19 and 20.

Fig. 14. Assume that θ is so small that $\tan\theta$ can be replaced by its approximate equal, $\sin\theta$. (*a*) To this approximation show that, for equilibrium,

$$x = \left(\frac{q^2 L}{2\pi\epsilon_0 mg}\right)^{1/3},$$

where x is the separation between the balls. (*b*) If $L = 120$ cm, $m = 10$ g, and $x = 5.0$ cm, what is q?

20P. If the balls of Fig. 14 are conducting, (*a*) what happens to them after one is discharged? Explain your answer. (*b*) Find the new equilibrium separation.

21P. Figure 15 shows a long, insulating, massless rod of length L, pivoted at its center and balanced with a weight W at a distance x from the left end. At the left and right ends of the rod are attached positive charges q and $2q$, respectively. A distance h directly beneath each of these charges is a fixed positive charge Q. (*a*) Find the distance x for the position of the weight when the rod is balanced. (*b*) What value should h have so that the rod exerts no vertical force on the bearing when balanced? Neglect the interaction between charges at the opposite ends of the rod.

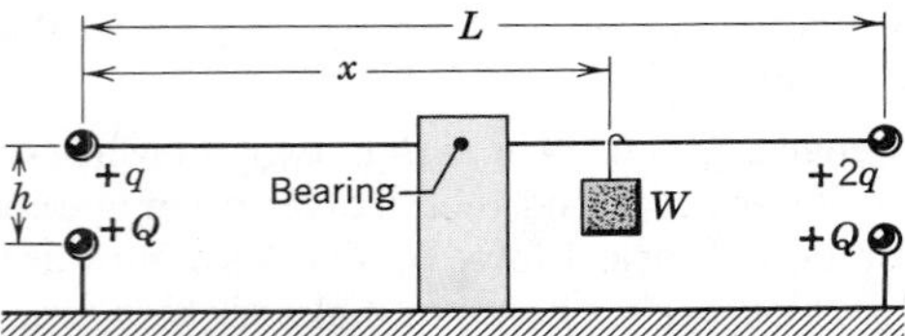

Figure 15 Problem 21.

Section 23-5 Charge Is Quantized

22E. What is the force of attraction between a singly charged sodium ion and an adjacent singly charged chlorine ion in a salt crystal if their separation is 2.82×10^{-10} m?

23E. A neutron is thought to be composed of one "up" quark of charge $+\frac{2}{3}e$ and two "down" quarks each having charge $-\frac{1}{3}e$. If the down quarks are 2.6×10^{-15} m apart inside the neutron, what is the repulsive electrical force between them?

24E. What is the total charge in coulombs of 75 kg of electrons?

25E. How many coulombs of charge, positive and negative, are there in 1 mol of molecular hydrogen gas?

26E. The electrostatic force between two identical ions that are separated by a distance of 5.0×10^{-10} m is 3.7×10^{-9} N. (*a*) What is the charge on each ion? (*b*) How many electrons are missing from each ion?

27E. Two small water droplets in air are 1.0 cm apart. Each has acquired a charge of 1.0×10^{-16} C. (*a*) Find the magnitude of the electric force on each droplet. (*b*) How many excess electrons are on each droplet?

28E. (*a*) How many electrons would have to be removed from a penny to leave it with a charge of $+1.0 \times 10^{-7}$ C? (*b*) To what fraction of the electrons in the penny does this correspond? See Sample Problem 2.

29E. How far apart must two protons be if the electrical repulsive force acting on either one is equal to its weight at the earth's surface?

30E. An electron is in a vacuum near the surface of the earth. Where should a second electron be placed so that the net force on the first electron, owing to the other electron and to gravity, is zero?

31P. A 100-W lamp operated on a 120-V circuit has a current (assumed steady) of 0.83 A in its filament. How long does it take for one mole of electrons to pass through the lamp?

32P. Protons in cosmic rays strike the earth's atmosphere at a rate, averaged over the earth's surface, of 1500 protons/$m^2 \cdot s$. What total current does the earth receive from beyond its atmosphere in the form of incident cosmic ray protons?

33P. Calculate the number of coulombs of positive charge in a glass of water. Assume the volume of the water to be 250 cm^3.

34P. In the compound CsCl (cesium chloride), the Cs atoms are situated at the corners of a cube with a Cl atom at the cube's center. The edge length of the cube is 0.40 nm; see Fig. 16. The Cs atoms are each deficient in one electron and the Cl atom carries one excess electron. (*a*) What is the strength of the net electric force on the Cl atom resulting from the eight Cs atoms shown? (*b*) Suppose that the Cs atom marked with an arrow is missing (crystal defect). What now is the net electric force on the Cl atom resulting from the seven remaining Cs atoms?

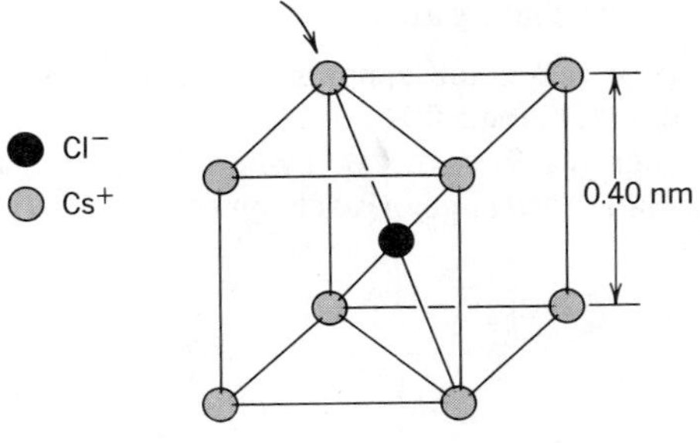

Figure 16 Problem 34.

35P. We know that, within the limits of measurement, the magnitudes of the negative charge on the electron and the positive charge on the proton are equal. Suppose, however, that these magnitudes differed from each other by as little as 0.00010%. With what force would two copper pennies, placed

1.0 m apart, then repel each other? What do you conclude? (*Hint:* See Sample Problems 2 and 3.)

36P. Two engineering students (John at 200 lb and Mary at 100 lb) are 100 ft apart. Let each have a 0.01% imbalance in their amount of positive and negative charge, one student being positive and the other negative. Estimate *roughly* the electrostatic force of attraction between them. (*Hint:* Replace the students by equivalent spheres of water.)

Section 23-6 Charge Is Conserved

37E. In beta decay a heavy fundamental particle changes to another heavy particle and an electron or a positron is emitted. (*a*) If a proton becomes a neutron, which particle is emitted? (*b*) If a neutron becomes a proton, which particle is emitted?

38E. Identify X in the following nuclear reactions:

(*a*) $^{1}H + {}^{9}Be \rightarrow X + n$;

(*b*) $^{12}C + {}^{1}H \rightarrow X$;

(*c*) $^{15}N + {}^{1}H \rightarrow {}^{4}He + X$.

(*Hint:* See Appendix D.)

39E. In the radioactive decay of ^{238}U (see Eq. 10), the center of the emerging ^{4}He particle is, at a certain instant, 9.0×10^{-15} m from the center of the residual nucleus ^{234}Th. At this instant, (*a*) what is the force on the ^{4}He particle and (*b*) what is its acceleration?

Section 23-7 The Constants of Physics: An Aside

40E. Verify that the fine structure constant is dimensionless and that its numerical value can be expressed as shown in Eq. 14.

41E. (*a*) Arrange the quantities h, G, and c to form a quantity with dimensions of length. (*Hint:* Combine the Planck time with the speed of light; see Sample Problem 6.) (*b*) Evaluate this "Planck length" numerically.

42P. (*a*) Arrange the quantities h, G, and c to form a quantity with dimensions of mass. Do not include any dimensionless factors. (*Hint:* Consider the units of h, G, and c as displayed in Sample Problem 6.) (*b*) Evaluate this "Planck mass" numerically.

CHAPTER 24
THE ELECTRIC FIELD

In a fly ash precipitator, an electric field exerts a force on a charged particle, deflecting it sideways. On the right, the electric field is ON. On the left, it has been turned OFF momentarily. The electric field moves a lot of fly ash.

24-1 Charges and Forces: A Closer Look

During the 1986 flyby of Uranus by the spacecraft *Voyager 2,* the round trip radio communication time from earth was about 5.5 h. What has this to do with charges, forces, and Coulomb's law? As it turns out, a great deal.

Coulomb's law tells us that two point charges exert forces on each other. The law says nothing, however, about how one charge "senses" the distant presence of the other. If the charges are stationary, we do not need to raise this question because we can solve all problems that arise if we know the magnitudes and the positions of the charges. Suppose, however, that one charge suddenly moves a little closer to the second charge. According to Coulomb's law, the force on the second charge must increase. Speaking loosely, how does the second charge "know" that the first charge has moved?

In our *Voyager 2* example, the first charge might be in the transmitting antenna of the space probe and the second in the antenna of the receiver back on earth. We know that the second charge responds to the motion of the first only after a time L/c, where L is the distance between the charges and c is the speed of light.

The key to understanding this kind of communication between charges is the concept of the *electromagnetic field.* We say that the second charge "learns" that the first charge has moved by means of an electromagnetic field disturbance that travels through the intervening space at the speed of light. This concept leads to the view that light is an electromagnetic wave and that the once separate sciences of electricity, magnetism, and optics can be joined together into a single comprehensive body of knowledge. Among the many practical consequences of the electromagnetic field idea were the invention of radio, the development of microwave radar and television, and a full understanding of a host of electromagnetic devices such as motors, generators, and transformers.

Our plan in these chapters is to establish separately the concepts of the electric field (for stationary charges) and of the magnetic field (for constant currents) and then to use what we learn to develop the concept of light as a traveling electromagnetic wave.

24-2 The Electric Field

The temperature has a definite value at every point in space in the room in which you may be sitting. You can measure it by putting a thermometer at that point. We call such a distribution of temperatures a *temperature field.* In much the same way, you can think of a *pressure field* extending throughout the atmosphere.

These two examples are *scalar fields,* temperature and pressure being scalar quantities. The gravitational field near the earth is a ready example of a *vector field.* To every point of space we can assign a vector **g** that represents the acceleration of a test body—released at that point—resulting from the earth's gravitational attraction. If m is the mass of the test body and **F** the gravitational force that acts on it, **g** is given by

$$\mathbf{g} = \mathbf{F}/m \quad \text{(gravitational field).} \tag{1}$$

If we place a test body carrying a positive electric charge q near a charged rod, an electrostatic force **F** will act on it. We speak of an *electric field* in this region, and we represent it by a vector **E**, defined by

$$\boxed{\mathbf{E} = \mathbf{F}/q} \quad \text{(electric field).} \tag{2}$$

The direction of the vector **E** is that of the vector **F**. That is, it is the direction in which a resting *positive* test charge, placed at the point, would be accelerated.

The SI unit for the electric field is the newton/coulomb (N/C). Although we usually write the unit for the gravitational field as the meter/second^2, we could just as easily write it as the newton/kilogram. Thus, both E and g are expressed as a force divided by a property—charge or mass—of the test body. Table 1 shows the electric fields that occur in a few situations.

Table 1 Some Electric Fields[a]

Field	*Value* (N/C)
At the surface of a uranium nucleus	3×10^{21}
Within a hydrogen atom, at the electron orbit	5×10^{11}
Electric breakdown occurs in air	3×10^{6}
At the charged drum of a photocopier	10^{5}
The electron beam accelerator in a TV set	10^{5}
Near a charged plastic comb	10^{3}
In the lower atmosphere	10^{2}
Inside the copper wire of household circuits	10^{-2}

[a] Approximate values.

If the primary charges that set up the electric field that we are examining were fixed in position, we could use a test charge of any size. If the primary charges are not fixed, however, the test charge should be as small as possible. Otherwise, it might cause the primary charges to change their positions. Put another way, we do not want the act of measurement to change the thing being measured. Because charge is quantized, we cannot, of course, use a test charge smaller than the elementary charge e.

The force acting between charged particles was originally thought of as a direct and instantaneous interaction between the charges. We can represent this action-at-a-distance view as

$$\text{charge} \rightleftharpoons \text{charge.} \tag{3}$$

Today, we think of the electric field as an intermediary between the charges. Thus, (1) charge q_1 in Fig. 1a sets up an electric field in the surrounding space, suggested by the dots in the figure and (2) this field acts on charge q_2, the action showing up as the force $\mathbf{F}_2$ that q_2 experiences.

As Fig. 1b suggests, we could just as well say that q_1 is immersed in an electric field set up by q_2 and for that reason experiences a force $\mathbf{F}_1$. Note in Fig. 1 that the fields set up by q_1 and q_2 are different but that the forces acting on the two charges have the same magnitude and form an action–reaction pair. That is, $\mathbf{F}_2 = -\mathbf{F}_1$.

In thinking about how charges exert forces on each other, we see our task divided into two parts: (1) calculating the field set up by a given distribution of charge and (2) calculating the force that a given field will exert on a charge placed in it. We think in terms of

$$\text{charge} \rightleftharpoons \text{field} \rightleftharpoons \text{charge} \tag{4}$$

rather than in terms of direct action between the charges, as Eq. 3 suggests.

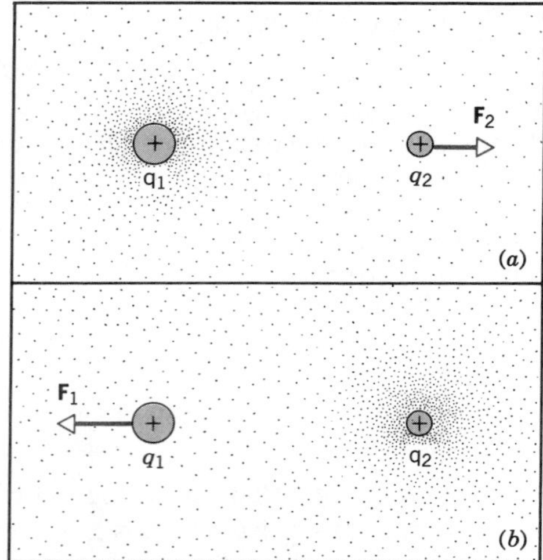

Figure 1 (*a*) Charge q_1 sets up an electric field that exerts a force $\mathbf{F}_1$ on charge q_2. (*b*) Charge q_2 sets up a field that exerts a force $\mathbf{F}_2$ on charge q_1. If the charges are different, the fields they set up will be different. The forces, however, are always equal in magnitude and form an action–reaction pair. That is, $\mathbf{F}_1 = -\mathbf{F}_2$.

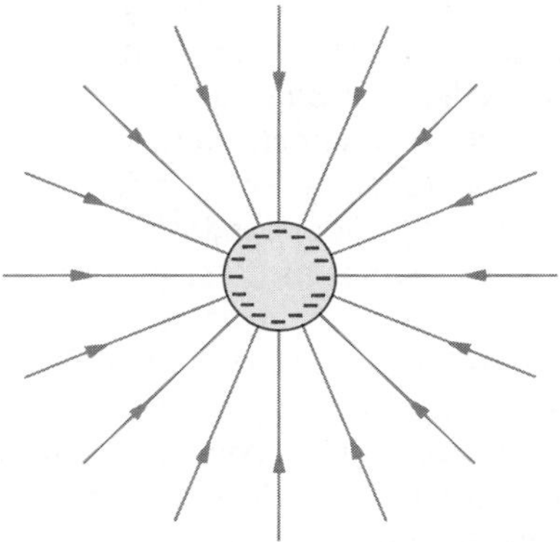

Figure 2 Field lines for a uniform sphere of negative charge. The lines originate on positive charges on distant bodies that are not shown, perhaps the walls of the room. The direction of the electric field at all points near the sphere is radially inward.

Sample Problem 1 A proton is placed in a uniform electric field **E**. What must be the magnitude and direction of this field if the electrostatic force acting on the proton is to just balance its weight?

From Eq. 2, replacing q by e and F by mg we have

$$E = \frac{F}{q} = \frac{mg}{e} = \frac{(1.67 \times 10^{-27}\ \text{kg})(9.8\ \text{m/s}^2)}{1.60 \times 10^{-19}\ \text{C}}$$

$$= 1.0 \times 10^{-7}\ \text{N/C, directed up.} \qquad \text{(Answer)}$$

This is a very weak field indeed. **E** must point vertically upward to float the (positively charged) proton.

24–3 Lines of Force

Michael Faraday (1791–1867), who introduced the field concept, thought of the space around a charged body as filled with *lines of force.* Although we no longer attach the same kind of reality to these lines that Faraday did, they are still a nice way to visualize field patterns and that is how we shall use them.

Figure 2 shows the field lines, as we shall call them, for a uniform sphere of negative charge. The lines point radially inward because a positive test charge would be accelerated in that direction. You can tell that the electric field becomes weaker as you move away from the central charge because the field lines become progressively farther apart.

In interpreting field line patterns, bear in mind that the field lines in Fig. 2, and in similar figures in this chapter, are representations on the plane of the page of what are really three-dimensional patterns. The relation between the field lines and the electric field vector is this: (1) The tangent to a field line at any point gives the *direction* of **E** at that point, and (2) the field lines are drawn so that the number of lines per unit area, in a plane at right angles to the lines, is proportional to the *magnitude* of **E**. This means that where the field lines are close together E is large and that where they are far apart E is small. Field lines always originate on positive charges and terminate on negative charges. When field lines are shown "hanging in the air," as in Fig. 2 and in other figures of this chapter, we must imagine the lines extended to originate or terminate on charges located on objects not shown, such as the walls of the room.

Figure 3 shows the field lines for a section of an infinitely large, uniform sheet of positive charge. No real sheet can be infinitely large but if we consider only points that are close to a real sheet and not near its edges the pattern of Fig. 3 will hold.

A positive test charge, released in front of such a sheet, would move away from it along a perpendicular line, all other directions being ruled out by considerations of symmetry. Thus, the electric field vector at any point near the sheet must point away from the sheet at right angles. The lines are uniformly spaced, which

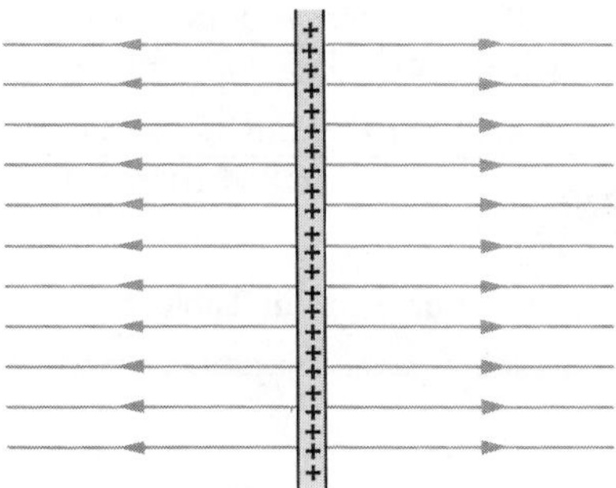

Figure 3 Field lines close to a section of a very large sheet of uniformly distributed positive charge. The lines terminate on negative charges on distant bodies that are not shown. The electric field points away from the sheet and at right angles to it.

means that **E** has the same magnitude for all points near the sheet.

Figure 4 shows the field lines for two equal positive charges. Figure 5 shows the pattern for two charges that are equal in magnitude but of opposite sign, a configuration that we call an *electric dipole.* Although we do not often use field lines quantitatively, they are very useful in visualizing what is going on. Can you not almost "see" the charges being pushed apart in Fig. 4 and pulled together in Fig. 5?

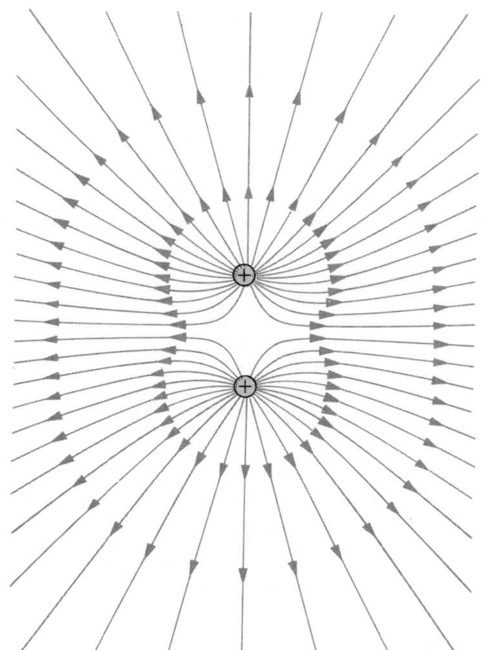

Figure 4 Field lines for two equal positive point charges. Two such charges repel each other. the pattern has rotational symmetry about a vertical axis. The lines terminate on negative charges on distant bodies that are not shown.

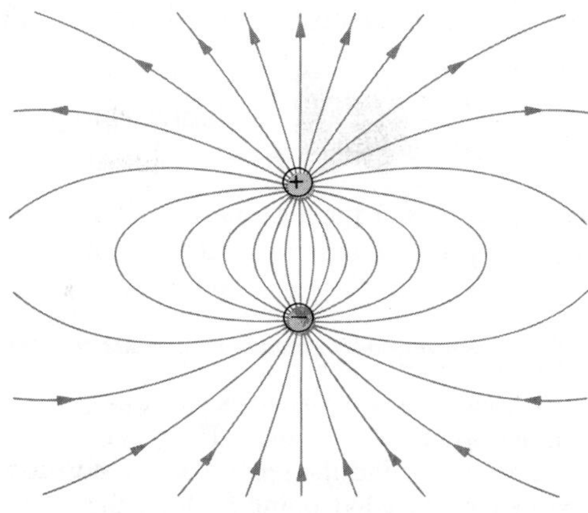

Figure 5 Field lines for a positive and a negative point charge of equal magnitude. Two such charges attract each other. The pattern has rotational symmetry about a vertical axis.

Sample Problem 2 In Fig. 2, how does the magnitude of the electric field vary with the distance from the center of the uniformly charged sphere?

Suppose that N field lines terminate on the sphere of Fig. 2. Draw an imaginary concentric sphere of radius r. The number of lines per unit area at any point on this sphere is $N/4\pi r^2$. Because E is proportional to this quantity, we can write $E \propto 1/r^2$. Thus, the electric field set up by a uniform sphere of charge varies as the inverse square of the distance from the center of the sphere. In much the same way, you can show that the electric field set up by an infinitely long uniformly charged rod will vary as $1/r$, where r is the perpendicular distance from the axis of the rod.

24-4 Calculating the Field: A Point Charge

Given a charged object, what electric field does it set up at nearby points? We look first at the case of a single point charge q. Put a test charge q_0 a distance r from this charge. The magnitude of the force acting on q_0, from Coulomb's law (see Eq. 4 of Chapter 23), is

$$F = \frac{1}{4\pi\epsilon_0}\frac{qq_0}{r^2}. \tag{5}$$

The magnitude of the electric field follows from Eq. 2, or

$$E = \frac{F}{q_0} = \frac{1}{4\pi\epsilon_0}\frac{q}{r^2} \quad \text{(point charge).} \qquad (6)$$

The direction of **E** is along a radial line from q, pointing outward if q is positive and inward if q is negative.

Sample Problem 3 Figure 6 shows a charge q_1 of $+1.5\ \mu C$ and a charge q_2 of $+2.3\ \mu C$. The first charge is at the origin of an x axis and the second is at a position $x = L$, where $L = 13$ cm. At what point P along the x axis is the electric field zero?

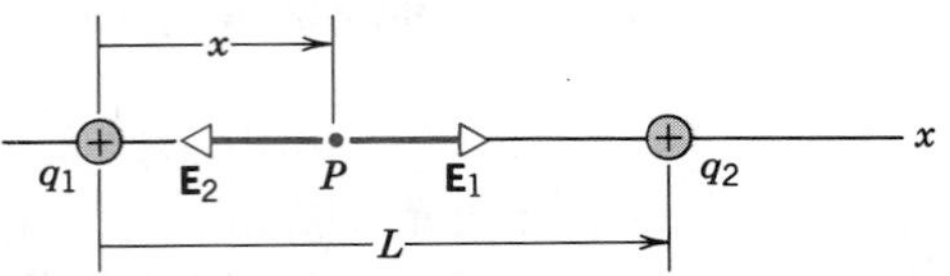

Figure 6 Sample Problem 3. Two positive point charges are separated by a distance L. For what value of x does the electric field vanish?

The point must lie between the charges because only in this region do the forces exerted by q_1 and by q_2 on a test charge oppose each other. If $\mathbf{E}_1$ is the electric field due to q_1 and $\mathbf{E}_2$ is that due to q_2, the magnitudes of these vectors must be equal, or

$$E_1 = E_2.$$

From Eq. 6 we then have

$$\frac{1}{4\pi\epsilon_0}\frac{q_1}{x^2} = \frac{1}{4\pi\epsilon_0}\frac{q_2}{(L-x)^2},$$

where x is the position of point P. Taking the square root of each side and solving for x, we obtain

$$x = \frac{L}{1+\sqrt{q_2/q_1}} = \frac{13\ \text{cm}}{1+\sqrt{2.3\ \mu C/1.5\ \mu C}} = 5.8\ \text{cm}. \quad \text{(Answer)}$$

This result is positive and is less than L, confirming that the zero-field point lies between the two charges, as we know it must.

Sample Problem 4 The nucleus of a uranium atom has a radius R of 6.8 fm. What is the magnitude of the electric field at its surface?

The nucleus has a positive charge of Ze where the atomic number $Z(=92)$ is the number of protons in the nucleus and e $(= 1.60 \times 10^{-19}$ C) is the charge of the proton. This charge, which we assume to be distributed with spherical symmetry, behaves electrically for external points as if it were concentrated at the nuclear center. From Eq. 6 then,

$$E = \frac{1}{4\pi\epsilon_0}\frac{Ze}{R^2} = \frac{(8.99 \times 10^9\ \text{N}\cdot\text{m}^2/\text{C}^2)(92)(1.60 \times 10^{-19}\ \text{C})}{(6.8 \times 10^{-15}\ \text{m})^2}$$

$$= 2.9 \times 10^{21}\ \text{N/C}. \qquad \text{(Answer)}$$

This value appears as an entry in Table 1.

24-5 Calculating the Field: An Electric Dipole

Figure 7 shows two charges of magnitude q but of opposite sign, separated by a distance d. As we indicated in connection with Fig. 5, we call this configuration an *electric dipole.* What is the electric field due to the dipole of Fig. 7 at a point P, a distance z from the midpoint of the dipole on its central axis?

From symmetry, the electric field **E** at point P— and also the fields $\mathbf{E}_+$ and $\mathbf{E}_-$ due to the separate charges that make up the dipole—must lie along the dipole axis, which we take to be a z axis. The superposition principle

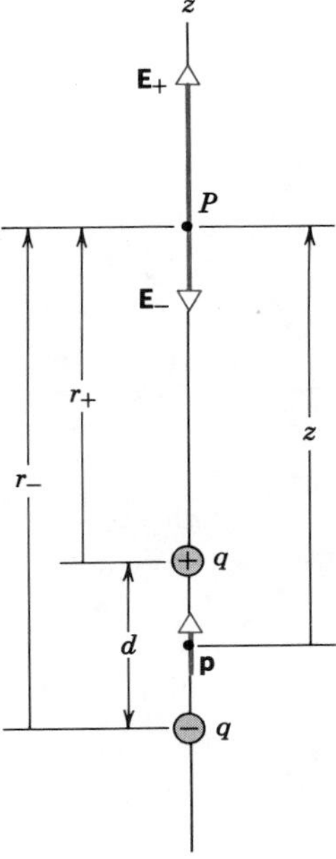

Figure 7 An electric dipole. the electric fields at point P on the dipole axis resulting from the separate charges are shown. How does the resultant field vary with z, the distance from the center of the dipole?

applies to electric fields so that we can write for the magnitude of the field at P,

$$\begin{aligned} E &= E_+ - E_- \\ &= \frac{1}{4\pi\epsilon_0}\frac{q}{r_+^2} - \frac{1}{4\pi\epsilon_0}\frac{q}{r_-^2} \\ &= \frac{q}{4\pi\epsilon_0(z-\frac{1}{2}d)^2} - \frac{q}{4\pi\epsilon_0(z+\frac{1}{2}d)^2}. \end{aligned} \quad (7)$$

After a little algebra, we can rewrite this equation as

$$E = \frac{q}{4\pi\epsilon_0 z^2}\left(1-\frac{d}{2z}\right)^{-2} - \frac{q}{4\pi\epsilon_0 z^2}\left(1+\frac{d}{2z}\right)^{-2}. \quad (8)$$

Physically, we are usually interested in the electrical effect of a dipole only at distances that are large compared with the dimensions of the dipole, that is, at distances such that $z \gg d$. At such large distances, we have $d/2z \ll 1$ in Eq. 8. We can then expand the two quantities in the parentheses in that equation by the binomial theorem, obtaining

$$E = \frac{q}{4\pi\epsilon_0 z^2}\left[\left(1+\frac{d}{z}-\cdots\right)-\left(1-\frac{d}{z}+\cdots\right)\right]. \quad (9)$$

As an approximate result that holds at large distances, we then have

$$E \approx \frac{q}{4\pi\epsilon_0 z^2}\frac{2d}{z} = \frac{1}{2\pi\epsilon_0}\frac{qd}{z^3}. \quad (10)$$

The product qd, which involves the two intrinsic properties of the dipole, is called the *electric dipole moment* p of the dipole. Thus, we can write Eq. 10 as

$$\boxed{E = \frac{1}{2\pi\epsilon_0}\frac{p}{z^3}} \quad \text{(electric dipole).} \quad (11)$$

By defining the electric dipole moment as a vector $\mathbf{p}$, we can use it to specify the direction of the dipole axis. The magnitude of $\mathbf{p}$, as we have seen, is qd and its direction is taken to be from the negative to the positive end of the dipole. The dipole moment vector is indicated in Fig. 7.

Equation 11 shows that, if we measure the electric field of a dipole only at distant points, we can never find q and d separately, only their product. The field at distant points would be unchanged if, for example, q were doubled and d simultaneously cut in half.

Although Eq. 11 holds only for distant points along the dipole axis, it turns out that E for a dipole varies as $1/r^3$ for *all* distant points, whether or not they lie on the z axis; here r is the radial distance of the point in question from the dipole center. In Problem 31, for example, we ask you to show this by deriving an expression for $E(r)$ at distant points in the equatorial plane of the dipole.

Inspection of Fig. 7 and of the field lines in Fig. 5 shows that the direction of $\mathbf{E}$ for distant points on the dipole axis is always in the direction of the dipole moment vector $\mathbf{p}$. This is true no matter whether point P in Fig. 7 is on the upper or the lower part of the dipole axis.

Inspection of Eq. 11 shows that if you were to double your distance from a dipole, the electric field would drop by a factor of 8. If you were to double your distance from a single point charge, however (see Eq. 6), the electric field would drop only by a factor of 4. It is not hard to see why the electric field of a dipole should decrease more rapidly with distance than does the field of a single charge. For distant points a dipole looks like two equal but opposite charges that are almost—but not quite—on top of each other. We are not surprised then to learn that their electric fields at distant points almost—but not quite—cancel each other.

Sample Problem 5 A molecule of water vapor (H_2O) contains 10 protons and 10 electrons and is thus electrically neutral. However, the centers of positive and of negative charge do not quite coincide so that, for external points, the molecule behaves as if it were an electric dipole like that of Fig. 7, its dipole moment p being $6.2 \times 10^{-30}\ \text{C}\cdot\text{m}$. What is the electric field at a distance z of 1.1 nm from the molecule on the dipole axis? This distance is about 10 diameters of a hydrogen atom.

From Eq. 11

$$E = \frac{1}{2\pi\epsilon_0}\frac{p}{z^3} = \frac{6.2\times10^{-30}\ \text{C}\cdot\text{m}}{(2\pi)(8.85\times10^{-12}\ \text{C}^2/\text{N}\cdot\text{m}^2)(1.1\times10^{-9}\ \text{m})^3}$$

$$= 8.4 \times 10^7\ \text{N/C}. \quad \text{(Answer)}$$

24-6 Calculating the Field: A Ring of Charge

Sometimes we find it desirable—or even necessary—to express the charge on a body in terms other than its total magnitude q. In situations involving charge spread out along a line, for example, we may report the *linear charge density* λ, whose SI unit is coulombs/meter. Table 2, to which we shall refer in future problems, summarizes the possibilities.

Table 2 Some Measures of Electric Charge

Name	*Symbol*	*SI Unit*
Charge	q	C
Linear charge density	λ	C/m
Surface charge density	σ	C/m^2
Volume charge density	ρ	C/m^3

Figure 8 shows a thin ring of radius R, charged to a constant linear charge density λ around its circumference. We may imagine the ring to be made of plastic or some other insulator, so that the charges can be regarded as fixed in place. What is the electric field at a point P, a distance z from the plane of the ring along its central axis?

We cannot apply Coulomb's law directly because the ring is not a point charge. We must, instead, break up the ring into charge elements that are small enough so that we can apply Coulomb's law to *them.* We will then find the electric field due to the ring by adding up the field contributions of all these charge elements.

Consider a differential element of the ring of length ds located at an arbitrary position on the ring in Fig. 8. It contains an element of charge given by

$$dq = \lambda\, ds. \tag{12}$$

This element sets up a differential field $d\mathbf{E}$ at point P. From Eq. 6 we have

$$dE = \frac{1}{4\pi\epsilon_0}\frac{\lambda\, ds}{r^2} = \frac{\lambda\, ds}{4\pi\epsilon_0(z^2 + R^2)}. \tag{13}$$

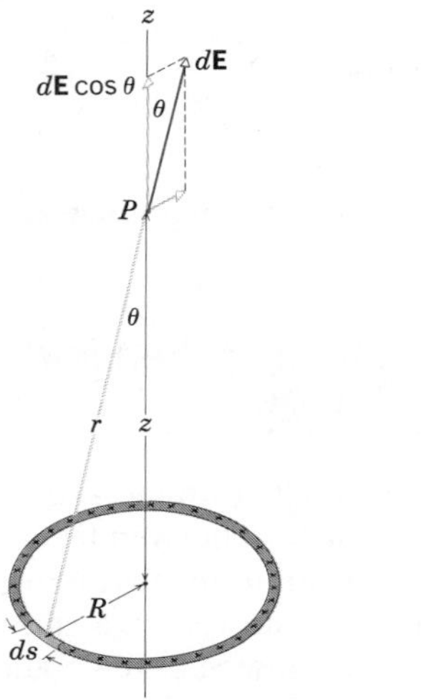

Figure 8 A uniform ring of charge. What is the electric field at point P, located on the axis of the ring?

Note that all charge elements that make up the ring are the same distance r from point P.

To find the resultant field at P we must add up, vectorially, all the field contributions $d\mathbf{E}$ made by the differential elements of the ring. This may seem like a hard job but, as so often happens, symmetry comes to our rescue. From symmetry, we know that the resultant field $\mathbf{E}$ must lie along the axis of the ring. Thus, only the components of $d\mathbf{E}$ parallel to this axis need be counted. Components of $d\mathbf{E}$ at right angles to the axis will cancel in pairs, the contribution from a charge element at any ring location being canceled by the contribution from the diametrically opposite charge element. Thus, our vector addition becomes a scalar addition of parallel axial components.

The axial component of $d\mathbf{E}$ is $dE \cos\theta$. From Fig. 8 we see that

$$\cos\theta = \frac{z}{r} = \frac{z}{(z^2 + R^2)^{1/2}}. \tag{14}$$

If we combine Eq. 14 and Eq. 13, we find

$$dE \cos\theta = \frac{z\lambda\, ds}{4\pi\epsilon_0(z^2 + R^2)^{3/2}}. \tag{15}$$

To add the various contributions, we need add only the ds elements, because all other quantities in Eq. 15 have the same value for all charge elements. Thus,

$$E = \int dE \cos\theta = \frac{z\lambda}{4\pi\epsilon_0(z^2 + R^2)^{3/2}} \int ds = \frac{z\lambda(2\pi R)}{4\pi\epsilon_0(z^2 + R^2)^{3/2}}, \tag{16}$$

in which the integral is simply $2\pi R$, the circumference of the ring. But $\lambda(2\pi R)$ is q, the total charge on the ring, so that we can write Eq. 16 as

$$\boxed{E = \frac{qz}{4\pi\epsilon_0(z^2 + R^2)^{3/2}}} \quad \text{(charged ring)}. \tag{17}$$

For points far enough away from the ring so that $z \gg R$, we can put $R = 0$ in Eq. 17. Doing so yields

$$E \approx \frac{1}{4\pi\epsilon_0}\frac{q}{z^2}, \tag{18}$$

which (with z replaced by r) is Eq. 6. We are not surprised because, at large enough distances, the ring behaves electrically like a point charge. We note also from Eq. 17 that $E = 0$ for $z = 0$. This is also not surprising because a test charge at the center of the ring would be pushed or pulled equally in all directions in the plane of the ring and would experience no net force. In general, if symmetry considerations do not permit a (single) preferred direction for a force to act on a test charge, the field at that site must be zero. If there *is* a preferred direction, however, that is the direction of the field.

24-7 Calculating the Field: A Charged Disk

Figure 9 shows a circular plastic disk of radius R, carrying a uniform surface charge of density σ on its upper surface; see Table 2. What is the electric field at point P, a distance z from the disk along its axis?

Our plan is to divide the disk up into concentric rings and then to calculate the electric field by adding up, that is, by integrating, the contributions of the various rings. Figure 9 shows a flat ring with radius s and of width ds, its total charge being

$$dq = \sigma\, dA = \sigma(2\pi s)\, ds, \qquad (19)$$

where dA is the differential area of the ring.

We have already solved the problem of the electric field due to a ring of charge. Substituting dq from Eq. 19 for q in Eq. 17, and replacing R in Eq. 17 by s, we obtain

$$dE = \frac{z\sigma 2\pi s\, ds}{4\pi\epsilon_0(z^2+s^2)^{3/2}} = (\sigma z/4\epsilon_0)(z^2+s^2)^{-3/2}(2s)ds.$$

We can now find E by integrating over the surface of the disk, that is, by integrating with respect to the variable s. Note that z remains constant during this process. Thus,

$$E = \int dE = \frac{\sigma z}{4\epsilon_0}\int_0^R (z^2+s^2)^{-3/2}\,(2s)ds, \qquad (20)$$

the limits on the variable being $s = 0$ and $s = R$. We see that this integral is of the form $\int X^m dX$, in which $X = (z^2+s^2)$, $m = -3/2$, and $dX = (2s)ds$; see Appendix G. Integrating and substituting the limits, we obtain

$$E(r) = \frac{\sigma}{2\epsilon_0}\left(1 - \frac{z}{\sqrt{z^2+R^2}}\right) \quad \text{(charged disk)} \qquad (21)$$

as the final result.

For $R \gg z$, the second term in the parentheses in Eq. 21 approaches zero and this equation reduces to

$$E = \frac{\sigma}{2\epsilon_0} \quad \text{(infinite sheet).} \qquad (22)$$

This is the electric field set up by a uniform sheet of charge of infinite extent; see Fig. 3. Note that Eq. 22 also follows as $z \to 0$ in Eq. 21; for such nearby points the charged disk does indeed behave as if it were infinite in extent. In Exercise 40 we ask you to show that Eq. 21 reduces to the field of a point charge for $z \gg R$.

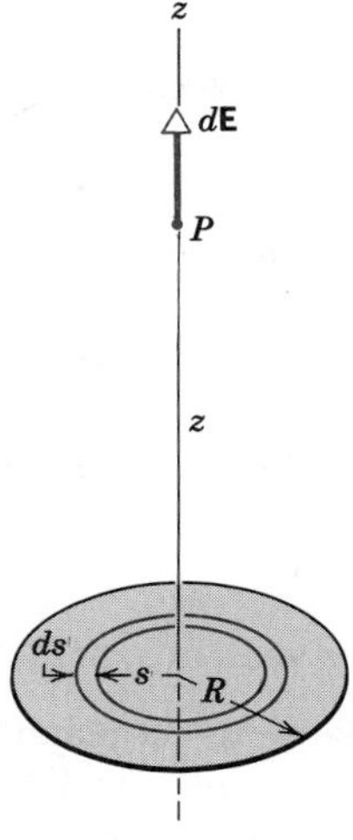

Figure 9 A uniformly charged disk of radius R. What is the electric field at point P, on the axis of the disk?

Sample Problem 6 The disk of Fig. 9 has a radius R of 2.5 cm and a surface charge density σ of $= +5.3\ \mu C/m^2$ on its upper face. (This, incidentally, is a possible value for the surface charge density on the photosensitive cylinder of a photocopying machine.) (a) What is the electric field at an axial point a distance $z = 12$ cm from the disk?

From Eq. 21 we have

$$E = \frac{\sigma}{2\epsilon_0}\left(1 - \frac{z}{\sqrt{z^2+R^2}}\right)$$

$$= \frac{5.3\times 10^{-6}\ C/m^2}{(2)(8.85\times 10^{-12}\ C^2/N\cdot m^2)}\left(1 - \frac{12\ cm}{\sqrt{(12\ cm)^2 + (2.5\ cm)^2}}\right)$$

$$= 6.3\times 10^3\ N/C. \qquad \text{(Answer)}$$

Note that the values of R and r in this problem can be left in centimeters because the units cancel.

(b) What is the electric field at the surface of the disk?

From Eq. 22 we have

$$E = \frac{\sigma}{2\epsilon_0} = \frac{5.3\times 10^{-6}\ C/m^2}{(2)(8.85\times 10^{-12}\ C^2/N\cdot m^2)}$$

$$= 3.0\times 10^5\ N/C. \qquad \text{(Answer)}$$

This value holds for all points close to the surface of the disk and not near its edge. Comparison with Table 1 shows that this value is less than the value at which air breaks down electrically so that the charged disk in this problem will not spark.

24-8 A Point Charge in an Electric Field

In the preceding four sections we looked into the first half of the charge-field-charge interaction: Given a charge distribution, what electric field does it set up in the surrounding space? Here we look into the second half of the interaction: What happens to a charged particle placed in a known external electric field?

The answer is that an electric force given by $\mathbf{F} = \mathbf{E}q$ acts on the particle. Unless this force is counterbalanced by other forces, the particle will accelerate. We look at two situations of special interest, an important example chosen from the history of physics and another example chosen from modern technology.

Measuring the Elementary Charge.* Figure 10 shows the apparatus used by the American physicist Robert A. Millikan in 1910–1913 to measure the elementary charge e. Oil droplets are introduced into chamber A by an atomizer, some of them becoming charged, either positively or negatively, in the process. Consider a drop that finds its way through a small hole in plate P_1 and drifts into chamber C. Let us assume that this drop carries a charge q, which we take to be negative.

If there is no electric field, two forces act on the drop, its weight mg and an upwardly directed viscous drag force, whose magnitude F is proportional to the speed of the falling drop. The drop quickly comes to a constant terminal speed v at which these two forces are just balanced, as Fig. 11a shows.

An electric field $\mathbf{E}$ is now set up in the chamber, by connecting battery B between plates P_1 and P_2. A third force, $q\mathbf{E}$, now acts on the drop. If q is negative, this force will point upward, and—we assume—the drop will now drift upward, at a new terminal speed v'. Figure 11b shows how the upward electric force qE is just balanced by the weight mg and the new drag force F'. Note that, in each case, the drag force points in the direction opposite to that in which the drop is moving and has a magnitude proportional to the speed of the drop. The charge q on the drop can be found from measurements of v and v'.

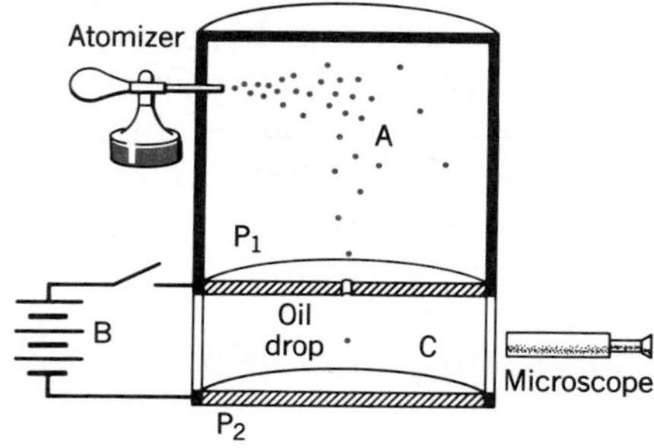

Figure 10 The Millikan oil-drop apparatus for measuring the elementary charge e. In chamber C the drop is acted on by the gravitational force, a force due to the electric field, and, if the drop is moving, a viscous drag force.

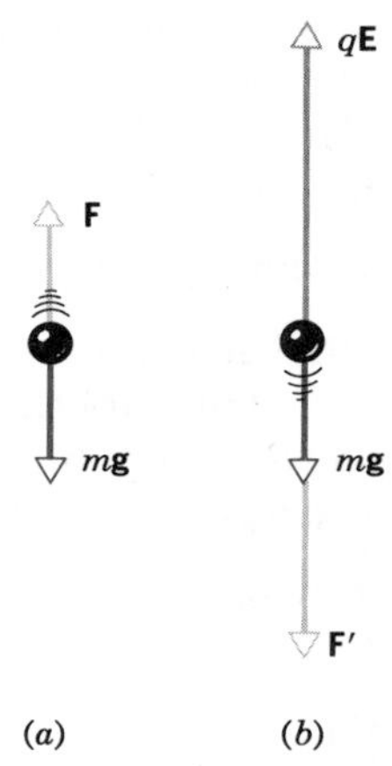

Figure 11 (a) An oil drop falls at terminal speed v in a field-free region. Its weight is balanced by an upward drag force. (b) An electric field force acts upward on the drop, which now rises with a terminal speed v'. The drag force, which always opposes the velocity, now acts downward.

Millikan found that the values of q were all consistent with the relation

$$q = ne \qquad n = 0, \pm 1, \pm 2, \pm 3, \cdots, \tag{23}$$

in which e is a fundamental constant called the *elementary charge* and whose value is 1.60×10^{-19} C. Millikan's experiment is convincing proof that charge is quantized. He earned the 1923 Nobel Prize in physics in

* For details of Millikan's experiments, see Henry A. Boorse and Lloyd Motz (eds.), *The World of the Atom,* Basic Books, New York, 1966, Chapter 40. To see Millikan from the point of view of two physicists who knew him as graduate students, see "Robert A. Millikan, Physics Teacher," by Alfred Romer, *The Physics Teacher,* February 1978, and "My Work with Millikan on the Oil-Drop Experiment," by Harvey Fletcher, *Physics Today,* June 1982.

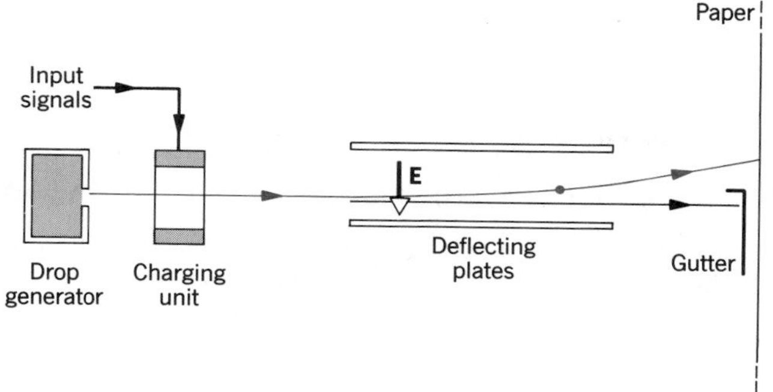

Figure 12 The essential features of an ink-jet printer. An input signal from the computer controls the charge given to the drop and thus the position on the page at which the drop lands. It takes about 100 drops to form a single character.

ABCDEFGHIJKLMNOPQRSTUVWXYZ
abcdefghijklmnopqrstuvwxyz
±@#$%¢&*()_+°":?.,1234567890

MNO

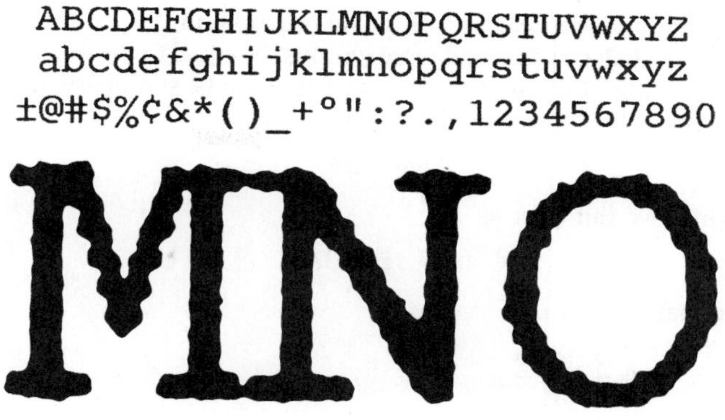

Figure 13 A sample of ink-jet printing, with three letters shown enlarged.

Figure 14 Ink drops issuing from a multijet drop generator in an ink-jet printer. The vertical height of the array of jets is that of a single character. The drops are moving from left to right.

part for this work. Modern measurements of the elementary charge rely on a variety of interlocking experiments, all more precise than the pioneering experiment of Millikan.

Ink-Jet Printing.* The need for high-quality, high-speed printing has caused a search for an alternative to impact printing, such as occurs in a standard typewriter. Building up letters by squirting tiny droplets of ink at the paper has proved to be one way to do it.

Figure 12 shows a charged drop moving between two conducting plates, between which an electric field E has been set up. The drop will be deflected upward and, after passing the edge of the plates, will proceed to the paper and strike it at a position that is determined by the value of E and of the charge q on the drop.

In practice, E is held constant and the position of the drop is determined by the charge q delivered to the drop in the charging unit, through which the drop must pass before entering the deflecting system. The charging unit, in turn, is activated by electronic signals that encode the material to be printed. The printing head moves sideways, from left to right, and each individual character is built up as the motion proceeds.

The drops, which are about 30 μm in diameter, are produced at the rate of about 100,000 per second and move toward the paper at about 18 m/s. Figure 13 shows a sample of ink-jet printing. It takes about 100 drops to form a single character. Using multiple jets, printing speeds exceeding 45,000 lines/min have been achieved. Figure 14 shows the drops issuing from such a multijet drop generator; the height of the nozzle array is only that of a single printed character.

Sample Problem 7 In the Millikan oil-drop apparatus of Fig. 10, a drop of radius $R = 2.76$ μm carries a charge q of three electrons. What electric field (magnitude and direction) is re-

* See "Ink-Jet Printing," by Larry Kuhn and Robert A. Myers, *Scientific American,* April 1979.

quired to balance the drop so that it remains stationary in the apparatus? The density ρ of the oil is 920 kg/m³.

At balance, the field force qE must be equal to the weight of the drop. Because the drop is assumed to be stationary, no viscous drag force acts on it. Thus, we have

$$mg = qE,$$

which we can write as

$$\tfrac{4}{3}\pi R^3 \rho g = (3e)E,$$

so that

$$E = \frac{4\pi R^3 \rho g}{9e}$$
$$= \frac{(4\pi)(2.76 \times 10^{-6}\text{ m})^3(920\text{ kg/m}^3)(9.80\text{ m/s}^2)}{(9)(1.60 \times 10^{-19}\text{ C})}$$
$$= 1.65 \times 10^6\text{ N/C}. \qquad \text{(Answer)}$$

Because the drop is negatively charged, the electric field must point down. This will give an upward force to balance the weight of the drop.

Sample Problem 8 Figure 15 shows the deflecting electrode system of an ink-jet printer. An ink drop whose mass m is 1.3×10^{-10} kg carries a charge q of 1.5×10^{-13} C and enters the deflecting plate system with a speed $v = 18$ m/s. The length L of these plates is 1.6 cm and the electric field E between the plates is 1.4×10^6 N/C. What will be the vertical deflection of the drop at the far edge of the plates? Ignore the varying electric field at the edges of the plates.

Let t be the time of passage of the drop through the deflecting system. The vertical and the horizontal displacements are given by

$$y = \tfrac{1}{2}at^2 \quad \text{and} \quad L = vt,$$

respectively, in which a is the acceleration of the drop.

The electric force acting on the drop, qE, is much greater than the gravitational force mg so that the acceleration of the drop can be taken to be qE/m. Eliminating t between the two equations above and substituting this value for a leads to

$$y = \frac{qEL^2}{2mv^2}$$
$$= \frac{(1.5 \times 10^{-13}\text{ C})(1.4 \times 10^6\text{ N/C})(1.6 \times 10^{-2}\text{ m})^2}{(2)(1.3 \times 10^{-10}\text{ kg})(18\text{ m/s})^2}$$
$$= 6.4 \times 10^{-4}\text{ m} = 0.64\text{ mm}. \qquad \text{(Answer)}$$

The deflection at the paper will be larger than this. To aim the ink drops so that they form the characters well, it is necessary to control the charge q on the drops—to which the deflection is proportional—to within a few percent. In our treatment, we have neglected the viscous drag forces that act on the drop; they are substantial at these high drop speeds. The analysis is the same as for the deflection of the electron beam in an electrostatic cathode ray tube.

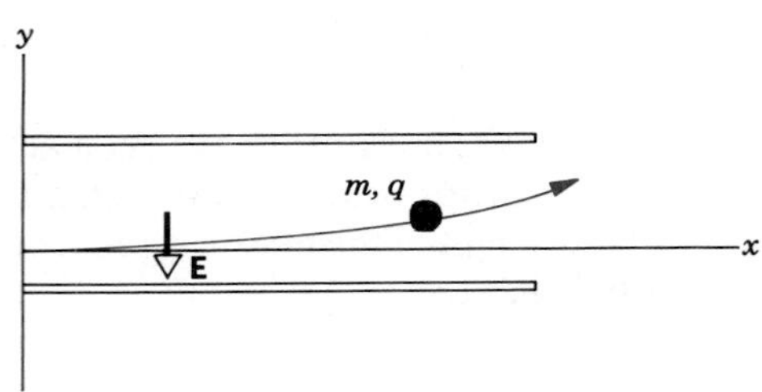

Figure 15 Sample Problem 8. An ink drop of mass m and charge q is deflected in the electric field of an ink-jet printer.

24-9 A Dipole in an Electric Field

We have defined the electric dipole moment **p** of an electric dipole to be a vector whose direction is along the dipole axis, pointing from the negative to the positive charge. We shall see that the behavior of a dipole in a uniform external electric field can be described completely in terms of the two vectors **E** and **p**, with no need to give any details about the structure of the dipole.

As we have seen in Sample Problem 5, a molecule of water vapor (H_2O) has an electric dipole moment of magnitude $p = 6.2 \times 10^{-30}$ C·m. Figure 16 is a representation of this molecule, showing the three nuclei and the surrounding electron cloud. The electric dipole moment **p** is represented by the vector on the axis of symmetry. This moment arises because the effective center

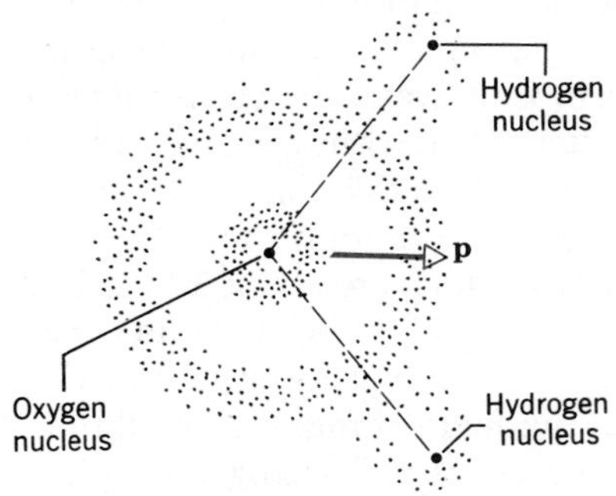

Figure 16 A molecule of H_2O, showing the three nuclei, the electron cloud, and the electric dipole moment vector **p**.

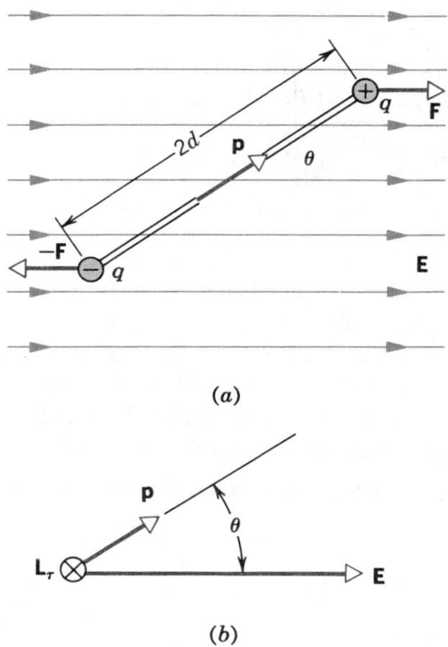

Figure 17 (*a*) An electric dipole in a uniform electric field. (*b*) Illustrating the relation $\tau = \mathbf{p} \times \mathbf{E}$.

of positive charge does not coincide with the effective center of negative charge.

Figure 17*a* shows a dipole in a uniform electric field **E**, its dipole moment **p** making an angle θ with the field direction. Two equal and opposite forces **F** and $-\mathbf{F}$ act as shown, where $F = qE$. The net force on the dipole is zero but there is a net torque about the center of mass of the molecule, given by

$$\tau = -Fd \sin \theta.$$

The minus sign indicates that the torque, which is exerted *by* the electric field *on* the dipole, is clockwise in Fig. 17, acting to reduce the value of θ. Substituting qE for F in the above and recalling that $p = qd$, we have

$$\tau = -qdE \sin \theta = -pE \sin \theta. \quad (24)$$

We can generalize Eq. 24 to vector form as

$$\boxed{\tau = \mathbf{p} \times \mathbf{E}} \quad \text{(torque on a dipole)}, \quad (25)$$

the appropriate vectors being shown in Fig. 17*b*. This torque tends to orient the dipole in the field direction. In the same way, the torque exerted by a *magnetic* field tends to orient a compass needle (a *magnetic* dipole) in the direction of that field.

An electric dipole in an electric field has its smallest potential energy when it is lined up with the field, just as a pendulum has its smallest gravitational potential energy when it is at the bottom of its swing. To increase its potential energy we must do work on the dipole, the increase being the work that we do.

In any potential energy situation, we are free to define the zero potential energy configuration in a perfectly arbitrary way because only differences in potential energy have physical meaning. It turns out that our final result will be simplest if we choose the potential energy of a dipole in an external field to be zero when the angle θ in Fig. 17*b* is 90°. The potential energy at any other orientation is then given by

$$U(\theta) = W = \int_{90^\circ}^{\theta} (-\tau)d\theta = \int_{90^\circ}^{\theta} pE \sin \theta \, d\theta, \quad (26)$$

in which W is the work done *by an external agent* to turn the dipole from its reference orientation to angle θ.* Evaluating the integral between the indicated limits leads to

$$U(\theta) = -pE \cos \theta. \quad (27)$$

We can generalize this to vector form as

$$\boxed{U(\theta) = -\mathbf{p} \cdot \mathbf{E}} \quad \text{(energy of a dipole)}. \quad (28)$$

This equation confirms that U is a minimum (most negative) when the **p** points in the field direction ($\theta = 0$) and is a maximum (most positive) when **p** points opposite to that direction ($\theta = 180°$).

Sample Problem 9 A water molecule (H_2O) in its vapor state has an electric dipole moment of 6.2×10^{-30} C·m. (a) How far apart are the effective centers of positive and negative charge in this molecule?

There are 10 electrons and, correspondingly, 10 positive charges in this molecule; see Fig. 16. We can write, for the magnitude of the dipole moment,

$$p = qd = (10e)(d),$$

in which d is the separation we are seeking and e is the elemen-

* The torque applied *by the external agent* must be opposite in sign to the torque applied *by the electric field* (see Eq. 24)—hence the minus sign in Eq. 26.

tary charge. Thus,

$$d = \frac{p}{10e} = \frac{6.2 \times 10^{-30}\ \mathrm{C \cdot m}}{(10)(1.60 \times 10^{-19}\ \mathrm{C})}$$
$$= 3.9 \times 10^{-12}\ \mathrm{m} = 3.9\ \mathrm{pm}. \qquad \text{(Answer)}$$

This is about 4% of the OH bond distance in this molecule.

(b) If the molecule is placed in an electric field of 1.5×10^4 N/C, what maximum torque can the field exert on it? Such a field can easily be set up in the laboratory.

As Eq. 25 shows, the torque is a maximum when $\theta = 90°$. Substituting this value in that equation yields

$$\tau = pE \sin\theta$$
$$= (6.2 \times 10^{-30}\ \mathrm{C \cdot m})(1.5 \times 10^4\ \mathrm{N/C})(\sin 90°)$$
$$= 9.3 \times 10^{-26}\ \mathrm{N \cdot m}. \qquad \text{(Answer)}$$

(c) How much work must an external agent do to turn this molecule end for end in this field, starting from its fully aligned position, for which $\theta = 0$?

The work is the difference in potential energy between the positions $\theta = 180°$ and $\theta = 0$. Thus, from Eq. 28,

$$W = U(180°) - U(0)$$
$$= (-pE \cos 180°) - (-pE \cos 0)$$
$$= 2pE = (2)(6.2 \times 10^{-30}\ \mathrm{C \cdot m})(1.5 \times 10^4\ \mathrm{N/C})$$
$$= 1.9 \times 10^{-25}\ \mathrm{J}. \qquad \text{(Answer)}$$

By comparison, the average translational energy of a molecule at room temperature ($= \frac{3}{2}kT$) is 6.2×10^{-21} J, which is 33,000 times larger. For the conditions of this problem, thermal agitation would swamp the tendency of the dipoles to align themselves with the field. If we wish to align the dipoles, we must use much stronger fields and/or much lower temperatures.

REVIEW AND SUMMARY

The Field Interpretation of Electric Force

One way to explain the electric force between charges is to presume that each charge sets up an electric field in the space surrounding itself. Each charge then interacts with the resulting field of the other charges at its own location.

Definition of Electric Field

We define the electric field at a point in space in terms of the electric force exerted on a small test charge placed at that point. The defining equation is

$$\mathbf{E} = \mathbf{F}/q \quad \text{(electric field).} \qquad [2]$$

See Sample Problem 1. If the value obtained using Eq. 2 is different for different test charges, the field is defined as the limit of this ratio as the test charge q approaches zero.

Lines of Force

Lines of force are a convenient representation of an electric field. They are drawn so that the tangent to a given line is in the direction of the field and so that strength of the field is indicated by the closeness of the lines, closer lines indicating stronger fields. Section 24–3 discusses field lines for a large sheet of charge, a charged sphere (see also Sample Problem 2), two equal charges, and two equal but opposite charges.

The Electric Field Due to a Point Charge

Any electric field can be calculated from Coulomb's law by knowing the locations and charges of the charged particles that cause it. The field due to a point charge has magnitude

$$E = \frac{1}{4\pi\epsilon_0}\frac{q}{r^2} \quad \text{(point charge).} \qquad [6]$$

The Electric Field Due to a Collection of Charges

The field points directly away from a positive charge and directly toward a negative one. The field due to a collection of point charges is obtained by superposition—adding vectorially the contributions from the individual charges. See Sample Problems 3 and 4. We also use superposition to calculate fields due to continuous distributions of charge; the sum of fields due to point charges becomes an integral. We often reduce the complexity of a vector sum by taking advantage of symmetry. We illustrate the procedure in Sections 24–6 (a ring of charge) and 24–7 (a charged disk).

Electric Dipoles

An assembly of two equal but opposite charges $\pm q$ separated by a distance d is called an *electric dipole.* It may be described in terms of a vector dipole moment $\mathbf{p}$ of magnitude qd and direction parallel to a line from the negative to the positive charge. The field due to an electric dipole is described in Section 24–5; see also Problem 31. Many naturally occurring molecules, such as the water molecule discussed in Sample Problem 5, have dipole electrical properties.

Force and Acceleration in an Electric Field

A charged particle in an electric field **E** experiences an electric force $\mathbf{F} = \mathbf{E}q$ and an acceleration in accord with Newton's second law. Sample Problems 7 and 8 illustrate the application of these principles to charged particles in uniform electric fields.

Torque on an Electric Dipole

A dipole in a uniform electric field experiences no net force but does experience a torque given by

$$\boldsymbol{\tau} = \mathbf{p} \times \mathbf{E} \quad \text{(torque on a dipole).} \qquad [25]$$

The torque tends to align the dipole in a direction parallel to the field. The potential energy of a dipole in an electric field may conveniently be taken to be

Potential Energy of an Electric Dipole

$$U(\theta) = -\mathbf{p}\cdot\mathbf{E} \quad \text{(energy of a dipole),} \qquad [28]$$

in which the zero of potential energy occurs when the dipole is perpendicular to the electric field. See Sample Problem 9 for an example.

QUESTIONS

1. Name as many scalar fields and vector fields as you can.

2. (*a*) In the gravitational attraction between the earth and a stone, can we say that the earth lies in the gravitational field of the stone? (*b*) How is the gravitational field due to the stone related to that due to the earth?

3. A positively charged ball hangs from a long silk thread. We wish to measure **E** at a point in the same horizontal plane as that of the hanging charge. To do so, we put a positive test charge q_0 at the point and measure $\mathbf{F}/q_0$. Will F/q_0 be less than, equal to, or greater than E at the point in question?

4. In exploring electric fields with a test charge, we have often assumed, for convenience, that the test charge was positive. Does this really make any difference in determining the field? Illustrate in a simple case of your own devising.

5. Electric lines of force never cross. Why?

6. In Fig. 4, why do the lines of force around the edge of the figure appear, when extended backward, to radiate uniformly from the center of the figure?

7. A point charge q of mass m is released from rest in a nonuniform field. Will it necessarily follow the line of force that passes through its release point?

8. A point charge is moving in an electric field at right angles to the lines of force. Does any force act on it?

9. What is the origin of "static cling," a phenomenon that sometimes affects clothes as they are removed from a dryer?

10. Two point charges of unknown magnitude and sign are a distance d apart. The electric field is zero at one point between them, on the line joining them. What can you conclude about the charges?

11. In Sample Problem 3, a charge placed at point P in Fig. 6 is in equilibrium because no force acts on it. Is the equilibrium stable (*a*) for displacements along the line joining the charges and (*b*) for displacements at right angles to this axis?

12. Two point charges of unknown sign and magnitude are fixed a distance L apart. Can we have $\mathbf{E} = 0$ for off-axis points (excluding ∞)? Explain.

13. In Fig. 5, the force on the lower charge points up and is finite. The crowding of the lines of force, however, suggests that E is infinitely great at the site of this (point) charge. A charge immersed in an infinitely great field should have an infinitely great force acting on it. What is the solution to this dilemma?

14. Three small spheres x, y, and z carry charges of equal magnitude and with signs as shown in Fig. 18. They are placed at the vertices of an isosceles triangle with the distance between x and y equal to the distance between x and z. Spheres y and z are held in place but sphere x is free to move on a frictionless surface. Which path will sphere x take when released?

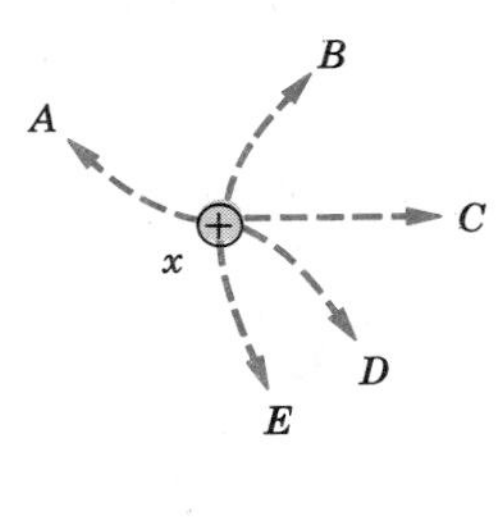

Figure 18 Question 14.

15. A positive and a negative charge of the same magnitude lie on a long straight line. What is the direction of **E** for points on this line that lie (*a*) between the charges, (*b*) outside the charges in the direction of the positive charge, (*c*) outside the charges in the direction of the negative charge, and (*d*) off the line but in the median plane of the charges?

16. In the median plane of an electric dipole, is the electric field parallel or antiparallel to the electric dipole moment **p**?

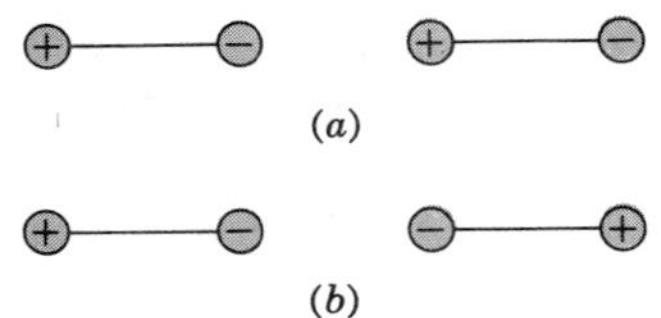

Figure 19 Question 17.

17. (*a*) Two identical electric dipoles are placed in a straight line, as shown in Fig. 19*a*. What is the direction of the electric force on each dipole owing to the presence of the other? (*b*) Suppose that the dipoles are rearranged as in Fig. 19*b*. What now is the direction of the force?

18. Compare the way E varies with r for (*a*) a point charge (Eq. 6), (*b*) a dipole (Eq. 11), and (*c*) a quadrupole (Problem 32).

19. What mathematical difficulties would you encounter if you were to calculate the electric field of a charged ring (or disk) at points *not* on the axis?

20. Figure 3 shows that **E** has the same value for all points in front of an infinite uniformly charged sheet. Is this reasonable? One might think that the field should be stronger near the sheet because the charges are so much closer.

21. Describe, in your own words, the purpose of the Millikan oil-drop experiment.

22. How does the sign of the charge on the oil drop affect the operation of the Millikan experiment?

23. You turn an electric dipole end for end in a uniform electric field. How does the work you do depend on the initial orientation of the dipole with respect to the field?

24. For what orientations of an electric dipole in a uniform electric field is the potential energy of the dipole (*a*) the greatest and (*b*) the least?

25. An electric dipole is placed in a nonuniform electric field. Is there a net force on it?

26. An electric dipole is placed at rest in a uniform external electric field, as in Fig. 17*a*, and released. Discuss its motion.

EXERCISES AND PROBLEMS

Section 24–2 The Electric Field

1E. An electron is released from rest in a uniform electric field of magnitude 2.0×10^4 N/C. Calculate the acceleration of the electron. (Ignore gravity.)

2E. An electron is accelerated eastward at 1.8×10^9 m/s^2 by an electric field. Determine the magnitude and direction of the electric field.

3E. Humid air breaks down (its molecules become ionized) in an electric field of 3.0×10^6 N/C. What is the magnitude of the electric force on (*a*) an electron and (*b*) an ion (with a single electron missing) in this field?

4E. An α particle, the nucleus of a helium atom, has a mass of 6.7×10^{-27} kg and a charge of $+2e$. What are the magnitude and direction of the electric field that will balance its weight?

5E. In a uniform electric field near the surface of the earth, a particle having a charge of -2.0×10^{-9} C is acted on by a downward electric force of 3.0×10^{-6} N. (*a*) What is the magnitude of the electric field? (*b*) What is the magnitude and direction of the electric force exerted on a proton placed in this field? (*c*) What is the gravitational force on the proton? (*d*) What is the ratio of the electric force to the gravitational force in this case?

6E. An electric field **E** with an average magnitude of about 150 N/C points downward in the earth's atmosphere. We wish to "float" a sulfur sphere weighing 1.0 lb (= 4.4 N) in this field by charging it. (*a*) What charge (sign and magnitude) must be used? (*b*) Why is the experiment not practical?

Section 24–3 Lines of Force

7E. Figure 20 shows field lines of an electric field. (*a*) If the magnitude of the field at A is 40 N/C, what force does an electron at that point experience? (*b*) What is the magnitude of the field at B?

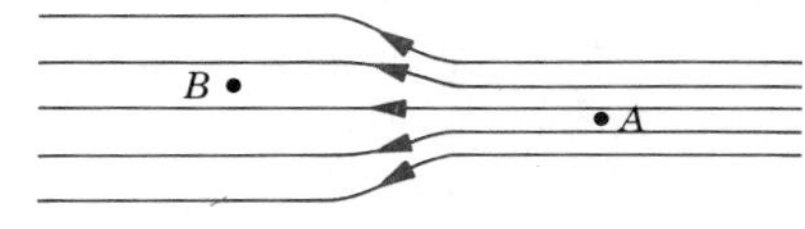

Figure 20 Exercise 7.

8E. Sketch qualitatively the lines of force associated with two separated point charges $+q$ and $-2q$.

9E. Three charges are arranged in an equilateral triangle as in Fig. 21. Consider the lines of force due to $+Q$ and $-Q$, and

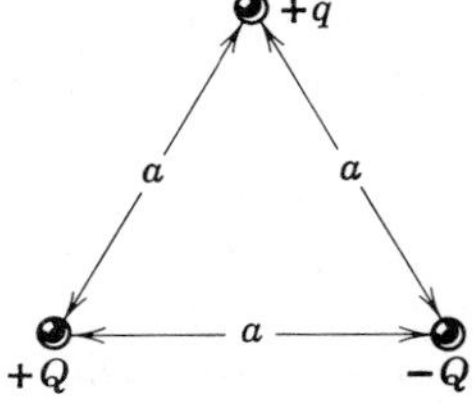

Figure 21 Exercise 9.

from them identify the direction of the force that acts on $+q$ because of the presence of the other two charges. (*Hint:* See Fig. 5.)

10E. Sketch qualitatively the lines of force both between and outside of two concentric conducting spherical shells, a positive charge q_1 being placed on the inner shell and a negative charge $-q_2$ on the outer.

11E. Sketch qualitatively the lines of force associated with a thin, circular, uniformly-charged disk of radius R. (*Hint:* Consider as limiting cases points very close to the disk, where the electric field is perpendicular to the surface, and points very far from it, where the electric field is like that of a point charge.)

12P. Sketch qualitatively the lines of force associated with three long parallel lines of charge, in a perpendicular plane. Assume that the intersections of the lines of charge with such a plane form an equilateral triangle (Fig. 22) and that each line of charge has the same linear charge density λ.

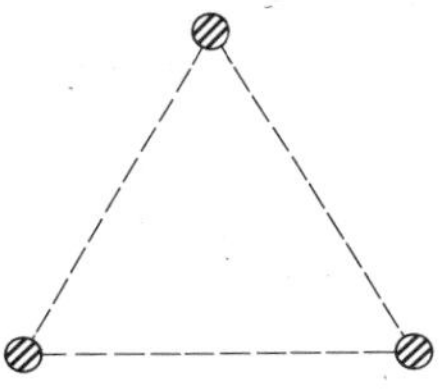

Figure 22 Problem 12.

13P. Assume that the exponent in Coulomb's law is not 'two' but n. Show that for $n \neq$ 'two' it is impossible to construct lines that will have the properties listed for lines of force in Section 24-3. For simplicity, treat an isolated point charge.

Section 24-4 Calculating the Field: A Point Charge

14E. What is the magnitude of a point charge that would create an electric field of exactly 1 N/C at points exactly 1 m away?

15E. In Fig. 23, charges are placed at the vertices of an equilateral triangle. For what value of Q (both sign and magnitude) does the total electric field vanish at C, the center of the triangle?

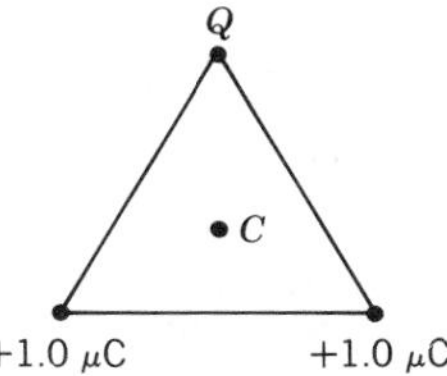

Figure 23 Exercise 15.

16E. What is the magnitude of a point charge chosen so that the electric field 50 cm away has the magnitude 2.0 N/C?

17E. Two point charges of magnitude 2.0×10^{-7} C and 8.5×10^{-8} C are 12 cm apart. (*a*) What electric field does each produce at the site of the other? (*b*) What force acts on each?

18E. Two equal and opposite charges of magnitude 2.0×10^{-7} C are held 15 cm apart. (*a*) What are the magnitude and direction of **E** at a point midway between the charges? (*b*) What force (magnitude and direction) would act on an electron placed there?

19P. The nucleus of an atom of uranium-238 has a radius of 6.8 fm ($= 6.8 \times 10^{-15}$ m) and carries a (positive) charge of Ze in which $Z (= 92)$ is the atomic number of uranium and e is the elementary charge. What is the electric field at the surface of such a nucleus? In what direction does it point? Does its numerical value surprise you?

20P. Two point charges are fixed at a distance d apart (Fig. 24). Plot $E(x)$, assuming $x = 0$ at the left-hand charge. Consider both positive and negative values of x. Plot E as positive if **E** points to the right and negative if **E** points to the left. Assume $q_1 = +1.0 \times 10^{-6}$ C, $q_2 = +3.0 \times 10^{-6}$ C, and $d = 10$ cm.

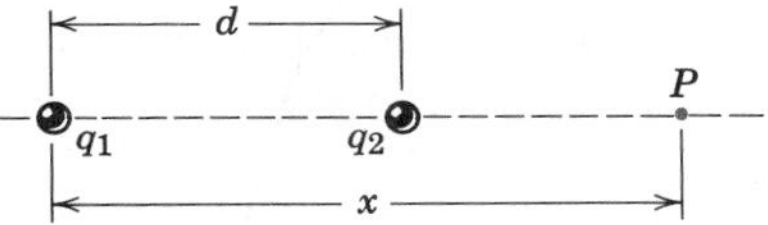

Figure 24 Problem 20.

21P. (*a*) In Fig. 25, locate the point (or points) at which the electric field is zero. (*b*) Sketch qualitatively the lines of force.

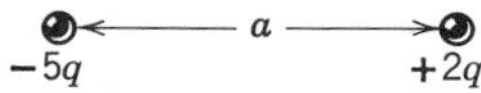

Figure 25 Problem 21.

22P. Charges $+q$ and $-2q$ are fixed a distance d apart as in Fig. 26. (*a*) Find **E** at points A, B, and C. (*b*) Sketch roughly the lines of force.

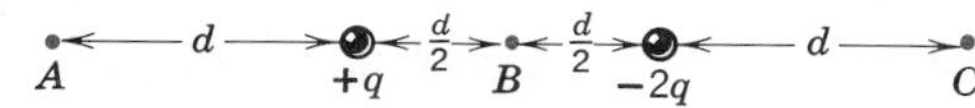

Figure 26 Problem 22.

23P. Two charges $q_1 = 2.1 \times 10^{-8}$ C and $q_2 = -4q_1$ are placed 50 cm apart. Find the point along the straight line passing through the two charges at which the electric field is zero.

24P. A clock face has negative point charges $-q$, $-2q$, $-3q$, . . . , $-12q$ fixed at the positions of the corresponding numerals. The clock hands do not perturb the field. At what time does the hour hand point in the same direction as the electric field at the center of the dial? (*Hint:* Consider diametrically opposite charges.)

25P. An electron is placed at each corner of an equilateral triangle having sides 20 cm long. (*a*) What is the electric field at the midpoint of one of the sides? (*b*) What force would another electron placed there experience?

26P. Calculate **E** (direction and magnitude) at point P in Fig. 27.

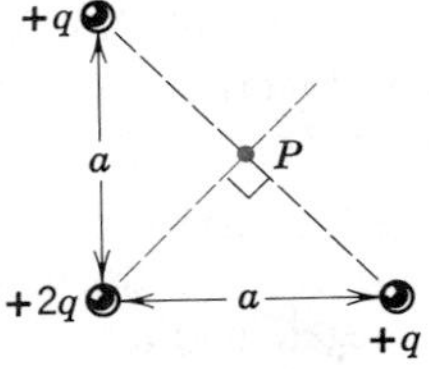

Figure 27 Problem 26.

27P. What is **E** in magnitude and direction at the center of the square of Fig. 28? Assume that $q = 1.0 \times 10^{-8}$ C and $d =$ 5.0 cm.

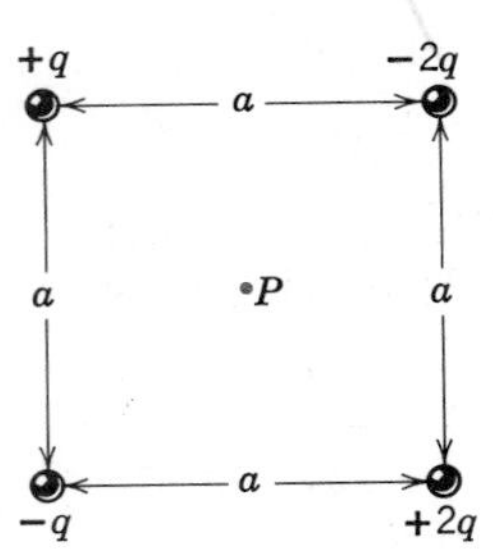

Figure 28 Problem 27.

Section 24-5 Calculating the Field: An Electric Dipole

28E. Calculate the dipole moment of an electron and a proton 4.3 nm apart.

29E. Calculate the magnitude of the force, due to a small electric dipole of dipole moment 3.6×10^{-29} C·m, on an electron 25 nm away along the dipole axis.

30E. In Fig. 7, assume that both charges are positive. Show that E at point P in that figure, assuming $z \gg d$, is given by

$$E = \frac{1}{4\pi\epsilon_0}\frac{2q}{z^2}.$$

31P. Calculate the electric field, magnitude and direction, due to an electric dipole, at a point P located at a distance $r \gg d$ along the perpendicular bisector of the line joining the charges; see Fig. 29. Express your answer in terms of the electric dipole moment **p**.

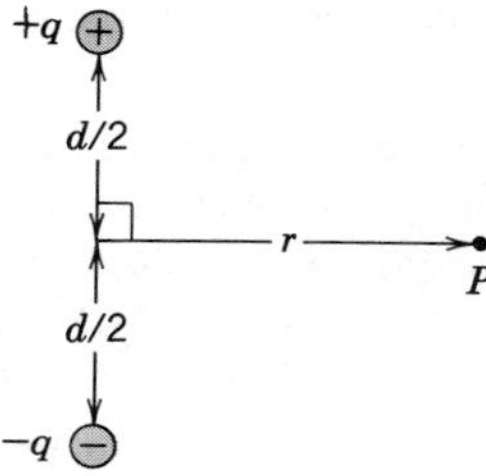

Figure 29 Problem 31.

32P*. *Electric quadrupole.* Figure 30 shows a typical electric quadrupole. It consists of two dipoles whose effects at external points do not quite cancel. Show that the value of E on the axis of the quadrupole for points a distance z from its center (assume $z \gg d$) is given by

$$E = \frac{3Q}{4\pi\epsilon_0 z^4},$$

where Q ($= 2qd^2$) is called the *quadrupole moment* of the charge distribution.

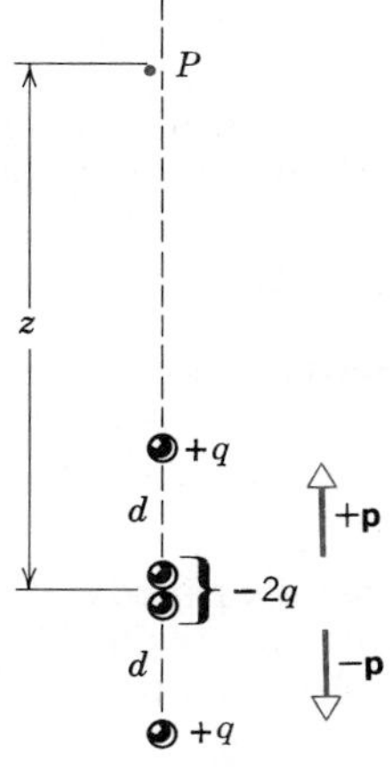

Figure 30 Problem 32.

Section 24-6 Calculating the Field: A Ring of Charge

33E. Make a quantitative plot of the electric field on the axis of a charged ring having a diameter of 6.0 cm and a uniformly distributed charge of 1.0×10^{-8} C.

34P. At what distance along the axis of a charged ring of radius R is the axial electric field strength a maximum?

35P. An electron is constrained to move along the axis of the ring of charge discussed in Section 24-6. Show that the electron can perform small oscillations, through the center of the ring, with a frequency given by

$$\omega = \sqrt{\frac{eq}{4\pi\epsilon_0 mR^3}}.$$

36P. A thin glass rod is bent into a semicircle of radius r. A charge $+Q$ is uniformly distributed along the upper half and a charge $-Q$ is uniformly distributed along the lower half, as shown in Fig. 31. Find the electric field **E** at P, the center of the semicircle.

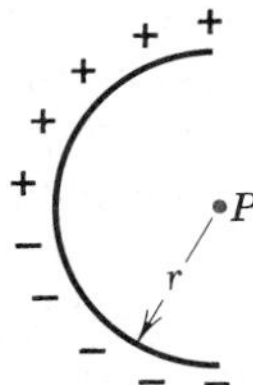

Figure 31 Problem 36.

37P. A thin nonconducting rod of finite length L carries a total charge q, spread uniformly along it. Show that E at point P on the perpendicular bisector in Fig. 32 is given by

$$E = \frac{q}{2\pi\epsilon_0 y} \frac{1}{(L^2 + 4y^2)^{1/2}}.$$

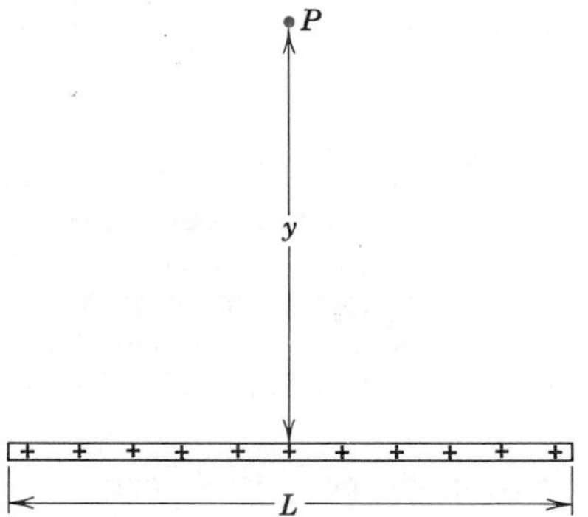

Figure 32 Problem 37.

38P. An insulating rod of length L has charge $-q$ uniformly distributed along its length, as shown in Fig. 33. (*a*) What is the linear charge density of the rod? (*b*) What is the electric field at point P a distance a from the end of the rod? (*c*) If P were very far from the rod compared to L, the rod would look like a point charge. Show that your answer to (*b*) reduces to the electric field of a point charge for $a \gg L$.

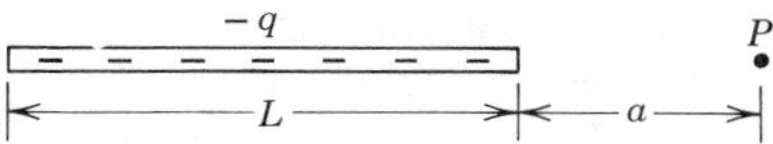

Figure 33 Problem 38.

39.* A "semi-infinite" insulating rod (Fig. 34) carries a constant charge per unit length of λ. Show that the electric field at the point P makes an angle of 45° with the rod and that this result is independent of the distance R.

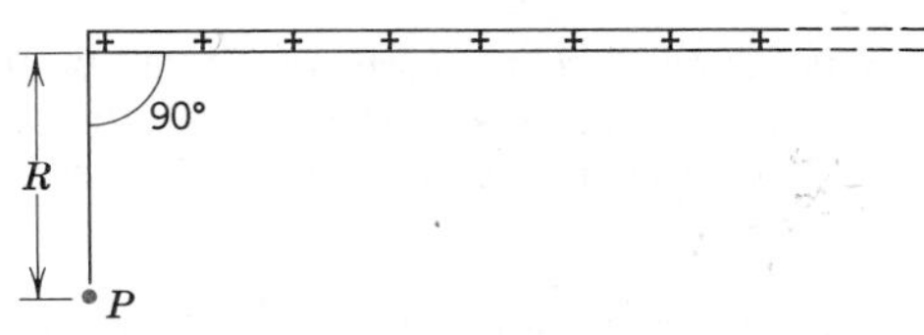

Figure 34 Problem 39.

Section 24-7 Calculating the Field: A Charged Disk

40E. Show that Eq. 21, for the electric field of a charged disk at points on its axis, reduces to the field of a point charge for $z \gg R$.

41P. (*a*) What total charge q must the disk in Sample Problem 6 carry in order that the electric field on the surface of the disk at its center equals the value at which air breaks down electrically, producing sparks? See Table 1. (*b*) Suppose that each atom at the surface has an effective cross-sectional area of 0.015 nm². How many atoms are at the disk's surface? (*c*) The charge in (*a*) results from some of the surface atoms carrying one excess electron. What fraction of the surface atoms must be so charged?

42P. At what distance along the axis of a charged disk of radius R is the electric field strength equal to one-half the value of the field at the surface of the disk at the center?

Section 24-8 A Point Charge in an Electric Field

43E. (*a*) What is the acceleration of an electron in a uniform electric field of 1.4×10^6 N/C? (*b*) How long would it take for the electron, starting from rest, to attain one-tenth the speed of light? (*c*) How far would it travel? Assume that Newtonian mechanics holds.

44E. One defensive weapon being considered for the Strategic Defense Initiative (Star Wars) uses particle beams. For example, a proton beam striking an enemy missile could render it harmless. Such beams can be produced in "guns" using electric fields to accelerate the charged particles. (*a*) What acceleration would a proton experience if the electric field is 2.0×10^4 N/C? (*b*) What speed would the proton attain if the field acts over a distance of one centimeter?

45E. An electron moving with a speed of 5.0×10^8 cm/s is shot parallel to an electric field of strength 1.0×10^3 N/C arranged so as to retard its motion. (*a*) How far will the electron travel in the field before coming (momentarily) to rest and (*b*) how much time will elapse? (*c*) If the electric field ends abruptly after 0.8 cm, what fraction of its initial kinetic energy will the electron lose in traversing it?

46E. A spherical water droplet 1.2 μm in diameter is suspended in calm air owing to a downward-directed atmospheric electric field $E = 462$ N/C. (*a*) What is the weight of the drop? (*b*) How many excess electrons does it carry?

47E. In Millikan's experiment, a drop of radius 1.64 μm and density 0.851 g/cm³ is balanced when an electric field of 1.92×10^5 N/C is applied. Find the charge on the drop, in terms of *e*.

48P. In a particular early run (1911), Millikan observed that the following measured charges, among others, appeared at different times on a single drop:

6.563×10^{-19} C	13.13×10^{-19} C	19.71×10^{-19} C
8.204×10^{-19} C	16.48×10^{-19} C	22.89×10^{-19} C
11.50×10^{-19} C	18.08×10^{-19} C	26.13×10^{-19} C

What value for the elementary charge *e* can be deduced from these data?

49P. An object having a mass of 10 g and a charge of $+8.0 \times 10^{-5}$ C is placed in an electric field defined by $E_x = 3.0 \times 10^3$ N/C, $E_y = -600$ N/C, and $E_z = 0$. (*a*) What are the magnitude and direction of the force on the object? (*b*) If the object starts from rest at the origin, what will be its coordinates after 3.0 s?

50P. A uniform electric field exists in a region between two oppositely charged plates. An electron is released from rest at the surface of the negatively charged plate and strikes the surface of the opposite plate, 2.0 cm away, in a time 1.5×10^{-8} s. (*a*) What is the speed of the electron as it strikes the second plate? (*b*) What is the magnitude of the electric field **E**?

51P. At some instant the velocity components of an electron moving between two charged parallel plates are $v_x = 1.5 \times 10^5$ m/s and $v_y = 3.0 \times 10^3$ m/s. If the electric field between the plates is given by $\mathbf{E} = (120 \text{ N/C})\mathbf{j}$, (*a*) what is the acceleration of the electron? (*b*) What will be the velocity of the electron after its *x* coordinate has changed by 2.0 cm?

52P. Two large parallel copper plates are 5.0 cm apart and have a uniform electric field between them as depicted in Fig. 35. An electron is released from the negative plate at the same time that a proton is released from the positive plate. Neglect the force of the particles on each other and find their distance from the positive plate when they pass each other. Does it surprise you that you need not know the electric field to solve this problem?

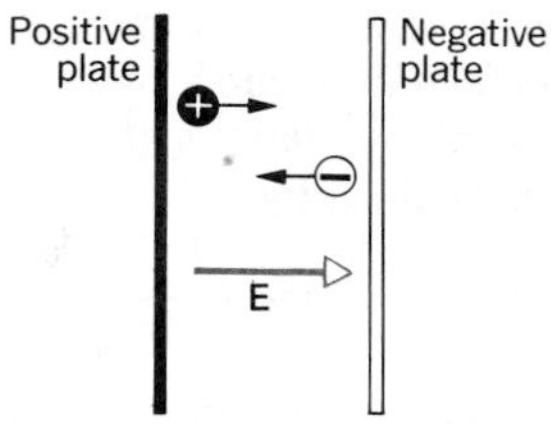

Figure 35 Problem 52.

53P. A uniform vertical field **E** is established in the space between two large parallel plates. In this field one suspends a small conducting sphere of mass *m* from a string of length *l*. Find the period of this pendulum when the sphere is given a charge $+q$ if the lower plate (*a*) is charged positively and (*b*) is charged negatively.

54P. An electron is projected as in Fig. 36 at a speed of 6.0×10^6 m/s and at an angle θ of 45°; $\mathbf{E} = 2.0 \times 10^3$ N/C (directed upward), $d = 2.0$ cm, and $L = 10$ cm. (*a*) Will the electron strike either of the plates? (*b*) If it strikes a plate, where does it do so?

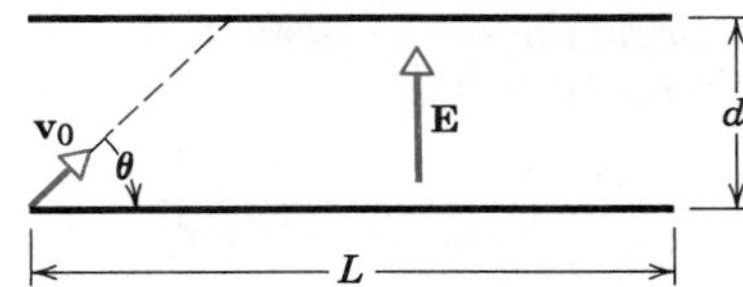

Figure 36 Problem 54.

Section 24-9 A Dipole in an Electric Field

55E. An electric dipole, consisting of charges of magnitude 1.5 nC separated by 6.2 μm, is in an electric field of strength 1100 N/C. (*a*) What is the magnitude of the electric dipole moment? (*b*) What is the difference in potential energy corresponding to a dipole orientation parallel and antiparallel to the field?

56E. An electric dipole consists of charges $+2e$ and $-2e$ separated by 0.78 nm. It is in an electric field of strength 3.4×10^6 N/C. Calculate the magnitude of the torque on the dipole when the dipole moment is (*a*) parallel, (*b*) at a right angle, and (*c*) opposite to the electric field.

57P. Find the work required to turn an electric dipole end for end in a uniform electric field **E**, in terms of the dipole moment **p** and the initial angle θ_0 between **p** and **E**.

58P. Find the frequency of oscillation of an electric dipole, of moment *p* and rotational inertia *I*, for small amplitudes of oscillation about its equilibrium position in a uniform electric field *E*.

CHAPTER 25

GAUSS' LAW

This bronze sculpture by Henry Moore graces the campus of Princeton University. The sculptor's art lies in the shaping of the surfaces of three-dimensional objects. Physicists and mathematicians are also interested in surfaces. In this chapter we introduce Gaussian surfaces, *which have a beauty of their own. It lies in the elegance and simplicity that the use of such constructions brings to the solution of many problems in electrostatics.*

25-1 A New Look at Coulomb's Law

If you want to find the center of mass of a potato, you can do so by experiment or by laborious calculation, involving the numerical evaluation of a triple integral. However, if the potato happens to be a perfect ellipsoid, there is no problem. You know exactly where the center of mass is without calculation. Such are the advantages of symmetry. Symmetrical situations arise in all fields of physics and, when possible, it makes sense to cast the laws of physics in forms that take full advantage of this fact.

Coulomb's law is the governing law in electrostatics, but it is not cast in a form that particularly simplifies the work in cases of high symmetry. In this chapter we introduce a new formulation of Coulomb's law, called *Gauss' law,** that *can* take easy advantage of such special cases. Gauss' law — for electrostatic problems — is the equivalent of Coulomb's law; which of them we choose depends on the problem at hand. Although both laws are valid for all electrostatic problems, it is one thing for a law to be valid and quite another thing for it to be useful.

* This law was originally derived for another inverse square force, gravity, by the German mathematician and physicist Carl Friedrich Gauss (1777 – 1855). For information about the life of this great scientist, see "Gauss," by Ian Stewart, *Scientific American,* July 1977.

We use Coulomb's law, the workhorse of electrostatics, for all problems in which the degree of symmetry is low. Even the most complicated of these can be solved, given enough computer capacity. We use Gauss' law when the symmetry is appropriately high. In such cases, this law not only tremendously simplifies the work but also—because of its simplicity—often provides new insights.

As Table 2 of Chapter 37 shows, Gauss' law is one of Maxwell's four equations. It is the entire purpose of Chapters 23–38 of this book to work toward a full understanding of these equations, which govern all classical electromagnetism and optics.

25-2 What Gauss' Law Is All About

Gauss' law is a relation between electric charge and the electric field that it sets up. Coulomb's law, which we write in the form of Eq. 6 of Chapter 24, namely,

$$E = \frac{1}{4\pi\epsilon_0} \frac{q}{r^2} \quad \text{(Coulomb's law)}, \tag{1}$$

also connects these two quantities. Gauss' law is simply a second, equivalent, way of doing so. Before presenting it formally, we want to give some notion of its nature.

Central to Gauss' law is a hypothetical closed surface—called a *Gaussian surface*—that you can set up in space. The Gaussian surface can be of any shape that you wish, but you will find it most useful to draw it in such a way as to conform to the symmetry of the problem you are facing. Thus, the Gaussian surface will often turn out to be a sphere, a cylinder, or some other symmetrical form. It must always be a *closed* surface, so that there can be a clear distinction between points that are inside the surface, on the surface, and outside the surface.

Imagine now that you wander over the Gaussian surface with an electric field meter in hand; you may—or may not—encounter electric fields at various points. You can note how strong they are and in what direction they point. Imagine also that you wander throughout the volume enclosed by this surface with a charge meter in hand; you may—or may not—encounter electric charges. You can note their sign and magnitude.

Gauss' law tells how the fields at the Gaussian surface are related to the charges contained within that surface.

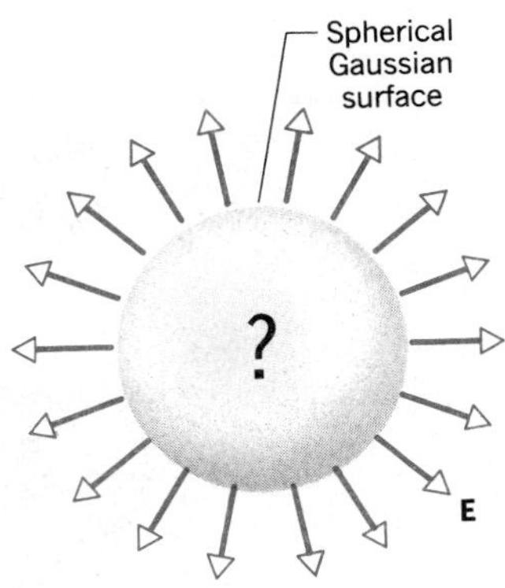

Figure 1 A spherical Gaussian surface. If the electric field vectors are of uniform magnitude and point radially outward at all surface points, you can conclude that a net positive distribution of charge must lie within the surface and that it must have spherical symmetry.

Figure 1 shows a simple example, in which the Gaussian surface is a sphere. Suppose that you find, as you explore this surface, that there is an electric field at every point on the surface, of the same magnitude and pointing radially outward. Without knowing anything about Gauss' law, what are you likely to conclude? That there must be a distribution of net positive charge inside the spherical Gaussian surface, symmetrically distributed about its center. It could be a point charge located at the center. It could be a uniform sphere or shell of charge, centered within the Gaussian sphere. Without further information, you cannot select from these possibilities.

Before we can make these ideas quantitative, we must pause to develop a new concept, that of the *flux* of a vector field.

25-3 Flux

Suppose that you dip a square wire loop into a uniformly flowing stream, the plane of the loop being at right angles to the flow as in Fig. 2*a*. The rate Φ at which water flows through the loop, measured perhaps in m^3/s, is given by

$$\Phi = Av, \tag{2}$$

in which A is the area of the loop and v is the speed of the water. We call Φ the *flux*. If you turn the loop to the position shown in Fig. 2*b*, the flux becomes

$$\Phi = Av \cos\theta, \tag{3}$$

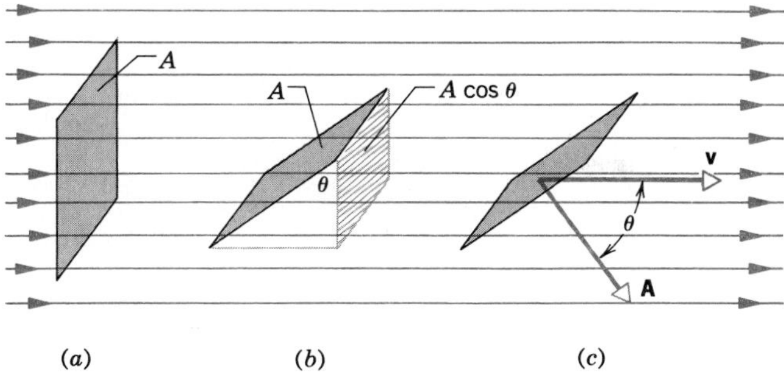

Figure 2 A wire loop of area A is immersed in a flowing stream. (*a*) The loop is at right angles to the flow. (*b*) The loop is turned at an angle θ to the flow, its projected area being $A \cos \theta$. (*c*) The area of the loop is represented by a vector **A** and the stream velocity by **v**, the angle between them being θ.

because only the projected area of the loop counts.

Figure 2*c* shows how we can express Eq. 3 in vector language. We represent the area of the loop by a vector **A**, at right angles to the plane of the loop. The angle θ is the angle between the vector **A** and the velocity vector **v**. We may write

$$\Phi = Av \cos \theta = \mathbf{A} \cdot \mathbf{v}. \quad (4)$$

Equation 4 shows that flux, being the scalar product of two vectors, is itself a scalar quantity. The SI unit of flux in this case is m^3/s.

The word "flux" comes from the Latin word meaning "to flow," and its use is appropriate in our flowing stream model. However, it is most useful for our purpose not to think about the water—which is what is doing the flowing—but about the array of velocity vectors, one for every point in the stream, that describes the flowing stream. From this point of view we regard the stream velocity as a *vector field,* and we say that Eq. 3 gives the flux of the *velocity field* through the surface of the wire loop.

The flux concept can be applied to *any* vector field, including the electric field, which is our special concern in this chapter. In the case of the electric field, nothing is actually flowing but we retain the useful word flux just the same. One more point: The surfaces shown in Fig. 2 are open surfaces, because they do not define an enclosed volume. Gauss' law deals only with closed surfaces; from now on, unless we specifically say otherwise, we shall be dealing only with the flux of an *electric field* through a *closed* (Gaussian) surface.

25-4 Flux of the Electric Field

To define the flux of the electric field, consider Fig. 3*a*, which shows an arbitrary Gaussian surface immersed in a nonuniform electric field. Let us divide the surface into small squares, of area ΔA, each square being small enough so that we can consider it to be a plane. We represent each such element of area by a vector $\Delta\mathbf{A}$, whose magnitude is the area ΔA. The direction of $\Delta\mathbf{A}$ is defined to be that of the *outward-drawn normal* to the surface.

At every square in Fig. 3*a* we can consider the electric field vector **E**. Because the squares have been taken to be arbitrarily small, **E** may be taken as constant for all points on a given square.

The vectors $\Delta\mathbf{A}$ and **E** that characterize each square make an angle θ with each other. Figure 3*b* shows an enlarged view of the three squares on the Gaussian surface marked x, y, and z. Table 1 summarizes the information about each square needed to calculate the flux.

A provisional definition for the flux of the electric field for the Gaussian surface of Fig. 3 is

$$\Phi = \sum \mathbf{E} \cdot \Delta\mathbf{A}. \quad (5)$$

Equation 5 instructs us to visit each square on the Gaussian surface, to evaluate the scalar product $\mathbf{E} \cdot \Delta\mathbf{A}$ for the two vectors **E** and $\Delta\mathbf{A}$ that we find there, and to add up the results for all the squares that make up the surface. As Table 1 shows, squares like x in which **E** points inward make a negative contribution to the sum of Eq. 5. Squares like y, in which **E** lies in the surface, make zero contribution and squares like z, in which **E** points outward, make a positive contribution.

The exact definition of the flux of the electric field through a closed surface is found by allowing the area of the squares shown in Fig. 3*a* to become smaller and smaller, approaching a differential limit. The sum of Eq. 5 then becomes an integral and we have, for the definition of flux,

$$\boxed{\Phi = \oint \mathbf{E} \cdot d\mathbf{A}} \quad \text{(flux through a Gaussian surface).} \quad (6)$$

Table 1 Three Squares on a Gaussian Surface

Square	θ	*Direction of* **E**	*Sign of* $\mathbf{E} \cdot \Delta\mathbf{A}$
x	$>90°$	Into the surface	Negative
y	$=90°$	Parallel to the surface	Zero
z	$<90°$	Out of the surface	Positive

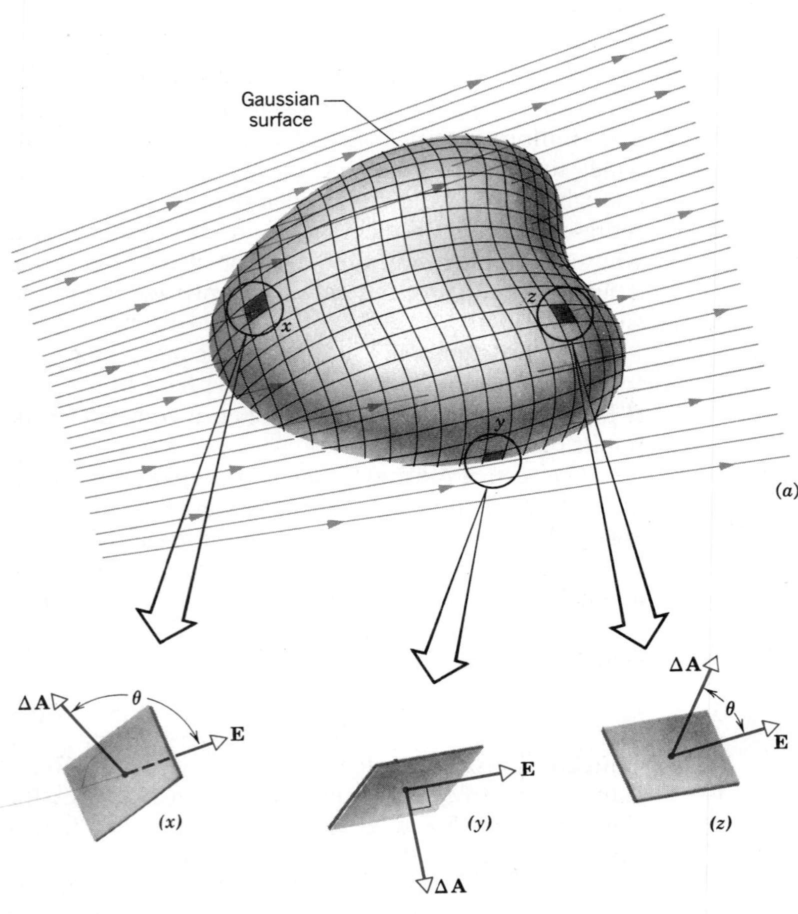

Figure 3 (*a*) A Gaussian surface of arbitrary shape immersed in an electric field. Its surface is divided into small squares of area ΔA. (*b*) The electric field vectors **E** and the area vectors $\Delta\mathbf{A}$ for three representative squares, marked x, y, and z.

The circle on the integral sign reminds us that the integration is to be taken over the entire (closed) surface. The flux of the electric field is a scalar, its directly derivable SI unit being $\text{N}\cdot\text{m}^2/\text{C}$.

Sample Problem 1 Figure 4 shows a Gaussian surface in the form of a cylinder of radius R immersed in a uniform electric field **E**, the cylinder axis being parallel to the field. What is the flux Φ of the electric field through this closed surface?

We can write the flux as the sum of three terms, an integral over (a) the left cylinder cap, (b) the cylindrical surface, and (c) the right cap. Thus, from Eq. 6,

$$\Phi = \oint \mathbf{E}\cdot d\mathbf{A}$$
$$= \int_a \mathbf{E}\cdot d\mathbf{A} + \int_b \mathbf{E}\cdot d\mathbf{A} + \int_c \mathbf{E}\cdot d\mathbf{A}. \tag{7}$$

For the left cap, the angle θ for all points is 180°, **E** is constant, and all the vectors $d\mathbf{A}$ are parallel. Thus,

$$\int \mathbf{E}\cdot d\mathbf{A} = \int E(\cos 180°)\, dA = -E\int dA = -EA,$$

where A, which is πR^2, is the cap area. Similarly, for the right cap,

$$\int \mathbf{E}\cdot d\mathbf{A} = +EA,$$

the angle θ for all points being zero there. Finally, for the cylinder wall,

$$\int \mathbf{E}\cdot d\mathbf{A} = 0,$$

the angle θ being 90° for all points on the cylindrical surface. Substituting these terms into Eq. 7 leads us to

$$\Phi = -EA + 0 + EA = 0. \qquad \text{(Answer)}$$

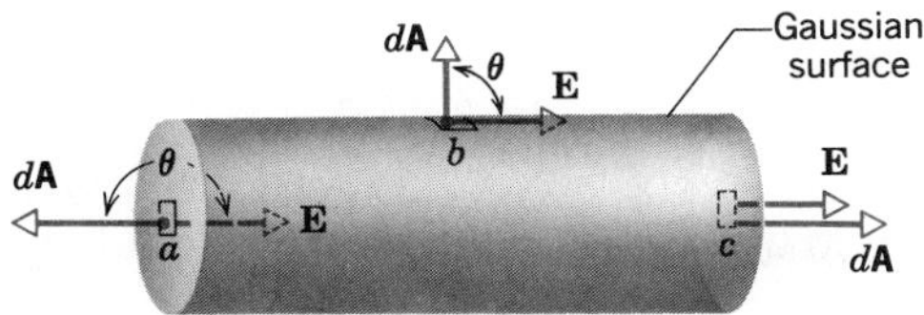

Figure 4 Sample Problem 1. A cylindrical Gaussian surface, closed by end caps, is immersed in a uniform electric field. The cylinder axis is parallel to the field direction.

This result is perhaps not surprising because the lines of force that represent the field pass right through the Gaussian surface, entering through the left end cap and leaving through the right end cap. If the field lines were taken to be stream lines of flowing water, we would say that water flows in at the left cap, out at the right cap, and doesn't flow at all through the cylindrical surface. The total flux, or net flow into the closed surface, is zero.

25-5 Gauss' Law

Now that we have defined the flux of the electric field, we are ready to state Gauss' law, namely,

$$\boxed{\epsilon_0 \Phi = q} \quad \text{(Gauss' law).} \tag{8}$$

Equation 8 refers to a Gaussian surface. The quantity Φ is the flux of the electric field over that surface and q is the *net* charge enclosed by that surface. The quantity ϵ_0 is the *permittivity constant,* its value being $8.85 \times 10^{-12}\ \mathrm{C^2/N \cdot m^2}$.

By introducing Eq. 6, the definition of flux, we can also write Gauss' law as*

$$\boxed{\epsilon_0 \oint \mathbf{E} \cdot d\mathbf{A} = q} \quad \text{(Gauss' law).} \tag{9}$$

Note that q in Eqs. 8 and 9 is the *net* charge within the Gaussian surface, taking its algebraic sign into account. Charge outside the surface, no matter how large or how nearby it may be, is not included in the term q in Gauss' law. The exact form or location of the charges inside the Gaussian surface is also of no concern; the only thing that matters, on the right side of Eq. 9, is the magnitude and sign of the net enclosed charge. The **E** on the left side of Eq. 9, however, is the electric field resulting from *all* charges, both those inside and those outside the Gaussian surface.

Let us apply these ideas to Fig. 5, which shows two charges, equal in magnitude but opposite in sign, and also the lines of force describing the electric fields that they set up in the surrounding space. Four irregularly shaped Gaussian surfaces are also shown, in cross section. Let us consider each in turn.

Surface S_1. The electric field is outward for all points on this surface. Thus, the flux of the electric field is positive. So is the net charge within the surface, as Gauss' law requires.

Surface S_2. The electric field is inward for all points. Thus, the flux of the electric field is negative and so is the enclosed charge, as Gauss' law requires.

Surface S_3. This surface contains no charge. Gauss' law then requires that the flux of the electric field be zero for this surface. This is reasonable because, as we see, the lines of force pass right through the surface, entering it at the top and leaving at the bottom.

Surface S_4. This surface encloses no *net* charge, the positive and the negative charges being of equal magnitude. Gauss' law then requires that the flux for this surface be zero. Again, as for surface S_3, that is reasonable. The lines of force that leave surface S_4 at the top all curve around and reenter it at the bottom.

What would happen if we were to bring an enormous charge Q up close to the surface S_4 in Fig. 5? The pattern of the lines of force would certainly change but the net flux for the four Gaussian surfaces in Fig. 5 would not change. We can understand this because the lines of force associated with the added Q alone would pass right through each of the four Gaussian surfaces they encounter, making no contribution to the total flux through any of them. The value of Q would not enter Gauss' law in any way because Q lies outside all four of the Gaussian surfaces that we are considering.

Sample Problem 2 Figure 6 shows three lumps of plastic, each carrying an electric charge, and a dime, carrying no charge. The cross sections of two Gaussian surfaces are indicated. What is the flux of the electric field for each of these

* Gauss' law as presented here is for the important special case in which the charges are in a vacuum or, for most practical purposes, in air. In Section 27-8, we extend Gauss' law to include cases in which other materials, such as mica, oil, or glass, are present.

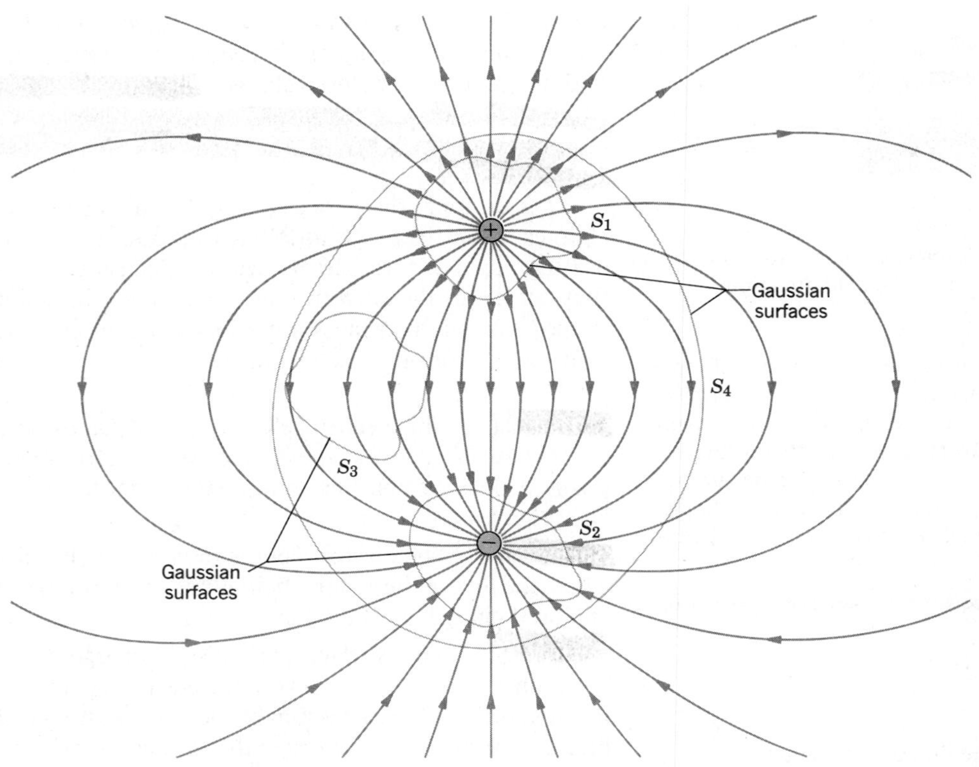

Figure 5 Two point charges, equal in magnitude but opposite in sign, and the lines of force that represent their electric field pattern. Four Gaussian surfaces are shown, in cross section.

surfaces? Assume $q_1 = +3.1$ nC, $q_2 = -5.9$ nC, and $q_3 = -3.1$ nC.

For surface S_1, the net enclosed charge q is q_1. The uncharged coin makes no contribution even though the positive and negative charges it contains may be separated by the action of the field in which the coin is immersed. Charges q_2 and q_3 are outside the surface S_1 and are therefore not included in q. From Eq. 8, we then have

$$\Phi = \frac{q}{\epsilon_0} = \frac{q_1}{\epsilon_0} = \frac{+3.1 \times 10^{-9}\ \text{C}}{8.85 \times 10^{-12}\ \text{C}^2/\text{N}\cdot\text{m}^2}$$
$$= +350\ \text{N}\cdot\text{m}^2/\text{C}. \qquad \text{(Answer)}$$

The plus sign indicates that the net charge within the surface is positive and also that the net flux through the surface is outward.

For surface S_2, the net charge q is $q_1 + q_2 + q_3$ so that

$$\Phi = \frac{q}{\epsilon_0} = \frac{q_1 + q_2 + q_3}{\epsilon_0}$$
$$= \frac{+3.1 \times 10^{-9}\ \text{C} - 5.9 \times 10^{-9}\ \text{C} - 3.1 \times 10^{-9}\ \text{C}}{8.85 \times 10^{-12}\ \text{C}^2/\text{N}\cdot\text{m}^2}$$
$$= -670\ \text{N}\cdot\text{m}^2/\text{C}. \qquad \text{(Answer)}$$

The minus sign shows that the net charge within the surface is negative and that the net flux through the surface is inward.

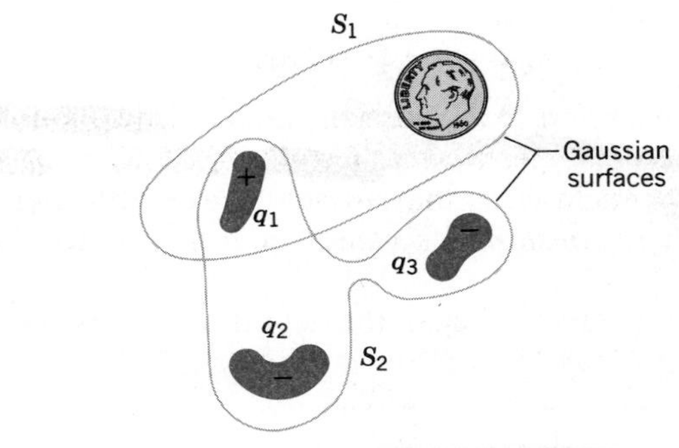

Figure 6 Sample Problem 2. Three plastic objects, each carrying an electric charge, and a coin, which carries no charge. The outlines of two possible Gaussian surfaces are shown.

25-6 Gauss' Law and Coulomb's Law

If Gauss' law and Coulomb's law are equivalent, we should be able to derive each from the other. Here we

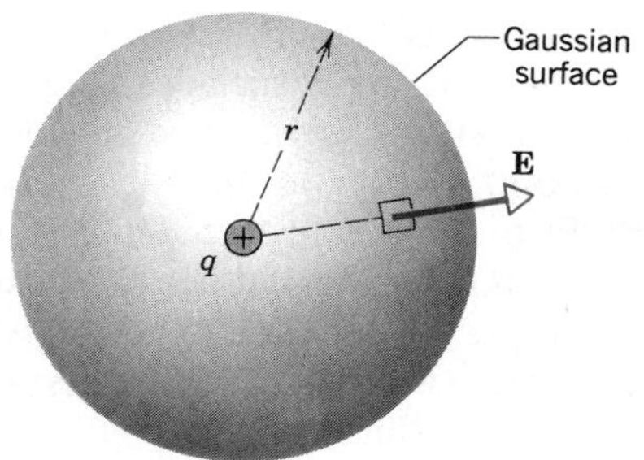

Figure 7 A spherical Gaussian surface centered on a point charge q.

derive Coulomb's law from Gauss' law and from some symmetry considerations.*

Figure 7 shows a point charge q, around which we have drawn a concentric spherical Gaussian surface of radius r. If we divide this surface into differential squares, we know from symmetry that both $\mathbf{E}$ and $d\mathbf{A}$ will be at right angles to the surface, the angle θ between them being zero. Thus, the quantity $\mathbf{E} \cdot d\mathbf{A}$ becomes simply $E\,dA$ and Gauss' law (see Eq. 9) becomes

$$\epsilon_0 \oint \mathbf{E} \cdot d\mathbf{A} = \epsilon_0 \oint E\,dA = q.$$

Because $\mathbf{E}$ has the same magnitude for all points on the sphere, we can factor it out of the integral, leaving

$$\epsilon_0 E \oint dA = q. \tag{10}$$

However, the integral in Eq. 10 is just the area of the spherical surface, or $4\pi r^2$, so that the equation becomes

$$\epsilon_0 E(4\pi r^2) = q$$

or

$$E = \frac{1}{4\pi\epsilon_0}\frac{q}{r^2}, \tag{11}$$

which is Coulomb's law, in the form in which we have written it in Eq. 1.

* These two laws are totally equivalent when—as in these Chapters—we apply them to problems involving charges that are either stationary or slowly moving. Gauss' law is more general in that it also covers the case of a rapidly moving charge. For such charges the electric lines of force become compressed in a plane at right angles to the direction of motion, thus losing their spherical symmetry.

25-7 A Charged Isolated Conductor

Gauss' law permits us to prove an important theorem about isolated conductors:

> *If you put an excess charge on an isolated conductor, that charge will move entirely to the surface of the conductor. None of the excess charge will be found within the body of the conductor.*

This might not seem unreasonable considering that like charges repel each other. You might imagine that, by moving to the surface, the added charges are getting as far away from each other as they can. We turn to Gauss' law for a quantitative proof of this qualitative speculation.

Figure 8a shows, in cross section, an isolated lump of copper hanging from a thread and carrying an added charge q. The light gray line shows a Gaussian surface that lies just below the actual surface of the conductor.

The key to our proof is the realization that, under equilibrium conditions, the electric field inside the conductor must be zero. If this were not so, the field would exert a force on the conduction electrons that are ever present in the metal conductor, and internal currents

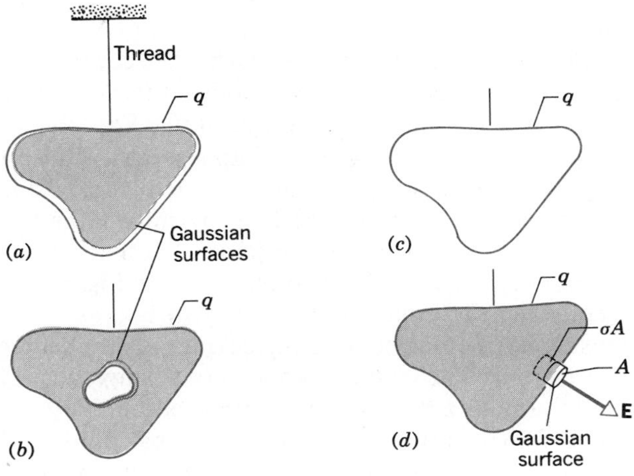

Figure 8 (a) A lump of copper, carrying a charge q, hangs from a thread. A Gaussian surface is drawn within the metal, just below the actual surface. (b) The lump of copper now has an internal cavity. A Gaussian surface lies within the metal, close to the cavity wall. (c) Nothing changes if the cavity is enlarged to include practically the entire lump of metal. (d) A squat cylindrical Gaussian surface pierces the surface of the conductor. It contains a charge σA.

would be set up. However, we know that there are no such enduring currents in an isolated lump of copper. Electric fields will appear inside a conductor during the process of charging it but these fields will not last long. Internal currents will act quickly to redistribute the added charge in such a way that the electric fields inside the conductor vanish, the currents stop, and equilibrium (electrostatic) conditions prevail.

If **E** is zero everywhere inside the conductor, it must be zero for all points on the Gaussian surface because that surface, though close to the surface of the conductor, is definitely inside it. This means that the flux through the Gaussian surface must be zero. Gauss' law then tells us that the charge inside the Gaussian surface must also be zero. If the added charge is not inside the Gaussian surface it can only be outside that surface, which means that it must lie on the actual surface of the conductor.

An Isolated Conductor with a Cavity. Figure 8*b* shows the same hanging conductor in which a cavity has been scooped out. It is perhaps reasonable to suppose that scooping out the electrically neutral material to form the cavity should not change the distribution of charge or the pattern of the electric field that exists in Fig. 8*a*. Again, we must turn to Gauss' law for a quantitative proof.

Draw a Gaussian surface surrounding the cavity, close to its walls but inside the conducting body. Because $\mathbf{E} = 0$ inside the conductor, there can be no flux through this new Gaussian surface. Therefore, from Gauss' law, that surface can enclose no net charge. We conclude that there is no charge on the cavity walls; it remains on the outer surface of the conductor, as in Fig. 8*a*.

The Conductor Removed. Suppose that, by some magic, the excess charges could be "frozen" into position on the conductor surface, perhaps by embedding them in a thin plastic coating, and suppose that then the conductor could be removed completely, as in Fig. 8*c*. This is equivalent to enlarging the cavity of Fig. 8*b* until it consumes the entire conductor, leaving only the charges. The electric field pattern would not change at all; it would remain zero inside the thin shell of charge and would remain unchanged for all external points. The electric field is set up by the charges and not by the conductor. The conductor simply provides a pathway so that the charges can change their positions.

The External Electric Field. Although the excess charge on an isolated conductor moves entirely to its surface, that charge—except for a spherical conductor —does not distribute itself uniformly over that surface. Put another way, the surface charge density σ, whose SI unit is C/m^2, will vary from point to point over the surface.

We can use Gauss' law to find a relation—at any surface point—between the surface charge density σ at that point and the electric field **E** just outside the surface at that same point. Figure 8*d* shows a squat cylindrical Gaussian surface, the (small) area of its two end caps being A. The end caps are parallel to the surface, one lying entirely inside the conductor and the other entirely outside. The short cylindrical walls are perpendicular to the surface of the conductor.

The electric field just outside a charged isolated conductor in electrostatic equilibrium must be at right angles to the surface of the conductor. If this were not so, there would be a component of **E** lying in the surface and this component would set up surface currents that would redistribute the surface charges—hence violating our assumption of electrostatic equilibrium. Thus, **E** is perpendicular to the surface and the flux through the exterior end cap of the Gaussian surface of Fig. 8*d* is EA. The flux through the interior end cap is zero because $\mathbf{E} = 0$ for all interior points of the conductor. The flux through the cylindrical walls is also zero because the lines of **E** are parallel to the surface, so they cannot pierce it. The charge enclosed by the Gaussian surface is σA.

To sum up: The flux through the entire Gaussian surface of Fig. 8*d* is EA and the charge enclosed by that surface is σA. Gauss' law then yields

$$\epsilon_0 \Phi = q$$

or

$$\epsilon_0 EA = \sigma A.$$

Thus, we find

$$\boxed{E = \frac{\sigma}{\epsilon_0}} \quad \text{(conducting surface).} \tag{12}$$

Earlier (see Eq. 22 of Chapter 24), we found that the electric field caused by a charged sheet has a magnitude $\sigma/2\epsilon_0$. Equation 12 shows a value twice as large just outside a charged conductor. Where does the factor of 2 come from? Let us explain.

Consider a point P just outside the surface of the charged conductor and a corresponding point P' just inside that surface. It is convenient to divide the surface charge on the conductor into two parts: (1) the *local charges,* which are those charges on the surface very close to these two points, and (2) the *distant charges,* which are the charges on the rest of the surface of the conductor. The electric fields, both at P and at P', are the superposition of the fields set up by these two charge systems.

The field set up by the local charges is just like that set up by the plane sheet of charge in Fig. 3 of Chapter 24; it has a magnitude of $\sigma/2\epsilon_0$ (see Eq. 22 of Chapter 24) and points *away from* the surface, both at P and at P'.

Because we must have $\mathbf{E} = 0$ inside the conductor, the distant charges must arrange themselves so that they produce a field of magnitude $\sigma/2\epsilon_0$ at P', just canceling the field of the local charges at that point. At P, however, the fields set up by the local charges and the distant charges add, producing a resultant field of magnitude $\sigma/2\epsilon_0 + \sigma/2\epsilon_0$ or σ/ϵ_0, in agreement with Eq. 12.

Sample Problem 3 The electric field just above the surface of the charged drum of a photocopying machine has a magnitude E of 2.3×10^5 N/C. What is the surface charge density on the drum if it is a conductor?

From Eq. 12 we have

$$\sigma = \epsilon_0 E = (8.85 \times 10^{-12}\ \mathrm{C^2/N \cdot m^2})(2.3 \times 10^5\ \mathrm{N/C})$$
$$= 2.0 \times 10^{-6}\ \mathrm{C/m^2} = 2.0\ \mu\mathrm{C/m^2}. \qquad \text{(Answer)}$$

Sample Problem 4 The magnitude of the average electric field normally present in the earth's atmosphere just above the surface of the earth is about 150 N/C, directed downward. What is the total net surface charge carried by the earth? Assume the earth to be a conductor.

Lines of force end on negative charges so that, if the earth's electric field points downward, the earth's average surface charge density σ must be negative. From Eq. 12 we find

$$\sigma = \epsilon_0 E = (8.85 \times 10^{-12}\ \mathrm{C^2/N \cdot m^2})(-150\ \mathrm{N/C})$$
$$= -1.33 \times 10^{-9}\ \mathrm{C/m^2}.$$

The earth's total charge q is the surface charge density multiplied by $4\pi R^2$, the surface area of the (presumed spherical) earth. Thus,

$$q = \sigma 4\pi R^2$$
$$= (-1.33 \times 10^{-9}\ \mathrm{C/m^2})(4\pi)(6.37 \times 10^6\ \mathrm{m})^2$$
$$= -6.8 \times 10^5\ \mathrm{C} = -680\ \mathrm{kC}. \qquad \text{(Answer)}$$

25-8 A Sensitive Test of Coulomb's Law

If an excess charge on an isolated conductor does *not* move entirely to its surface—as we proved it did in the preceding section—then Gauss' law cannot be true because our proof was based on that law. If Gauss' law is not true, then Coulomb's law cannot be true. In particular, the exponent 2 in the inverse square law might not be exactly 2. Thus, this law might be

$$E = \frac{1}{4\pi\epsilon_0}\frac{q}{r^{2\pm\delta}}, \qquad (13)$$

in which δ—if not zero—is a small number.

Coulomb's law is vitally important in physics and if δ in Eq. 13 is not zero, there are serious consequences for our understanding of electromagnetism and quantum physics. The best way to measure δ is to find out *by experiment* whether an excess charge, placed on an isolated conductor, does or does not move *entirely* to its outside surface.

Benjamin Franklin seems to have been the first to carry out experiments along these lines. Figure 9 shows his simple arrangements. Charge a metal ball and lower it by a thread deep inside a metal can. Touch the ball to the inside of the can. When the ball touches the can, the *ball + can* form a single "isolated conductor." If the charge does indeed flow *entirely* to the outside of the can, the ball should be found to be entirely uncharged when removed from the can. Within the accuracy of his experiments, Franklin found it so.

Franklin recommended this "singular fact" to Jo-

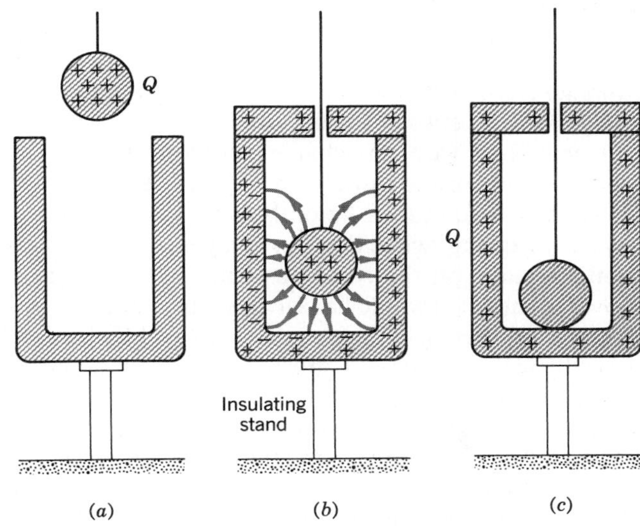

Figure 9 An arrangement conceived by Benjamin Franklin to show that charge placed on a conductor moves to its surface. (*a*) A charged metal ball is lowered into an uncharged metal can. (*b*) The ball is inside the can and an (almost) closed cover is added. (*c*) The ball is touched to the can, thus forming a single conducting object as in Fig. 8*a*. When the ball is removed from the can it is found to be completely uncharged, thus showing that the charge must have been transferred entirely to the can.

Table 2 Tests of Coulomb's Inverse Square Law

Experimenters	*Date*	δ*(Eq. 13)*
Franklin	1755	
Priestley	1767	. . . according to the squares . . .
Robison	1769	<0.06
Cavendish	1773	<0.02
Coulomb	1785	a few percent at most
Maxwell	1873	$<5 \times 10^{-5}$
Plimpton and Lawton	1936	$<2 \times 10^{-9}$
Bartlett, Goldhagen, and Phillips	1970	$<1.3 \times 10^{-13}$
Williams, Faller, and Hill	1971	$<3.0 \times 10^{-16}$

seph Priestly, who checked Franklin's experiments and realized that Coulomb's inverse square law followed from them. Many others, including Cavendish* and Maxwell, repeated the experiments, with ever increasing precision. Modern experiments, carried out with remarkable precision, have shown that if δ in Eq. 13 is not zero it is certainly very very small. Table 2 summarizes the most important of these experiments.

Figure 10 is a sketch of the apparatus used by Plimpton and Lawton to measure δ in Eq. 13 and thus check up on Coulomb's law. The apparatus consists of two concentric metal shells, A and B, the former being 5 ft in diameter. The inner shell contains a sensitive electrometer E connected so that it will indicate whether any charge moves between shells A and B. If the shells are connected by a wire, any charge placed on the shell assembly should reside entirely on the outside of shell A if Gauss' law—and thus Coulomb's law—holds.

By throwing switch S to the left, you can put a substantial charge on the sphere assembly. If any charge moves to shell B, it would have to pass through electrometer E and cause a deflection, which can be observed optically using telescope T, mirror M, and windows W.

However, when the switch S was thrown alternately from left to right, thus connecting the shell assembly alternately to the battery and to the ground, no effect was observed. Knowing the sensitivity of their electrometer, the experimenters concluded that if δ in Eq. 13 is not zero it is no more than ~0.000 000 002, a very small number indeed. The inverse square law seems to be on safe ground.

* The same Cavendish who "weighed the earth" with a torsion balance; see Section 15-3.

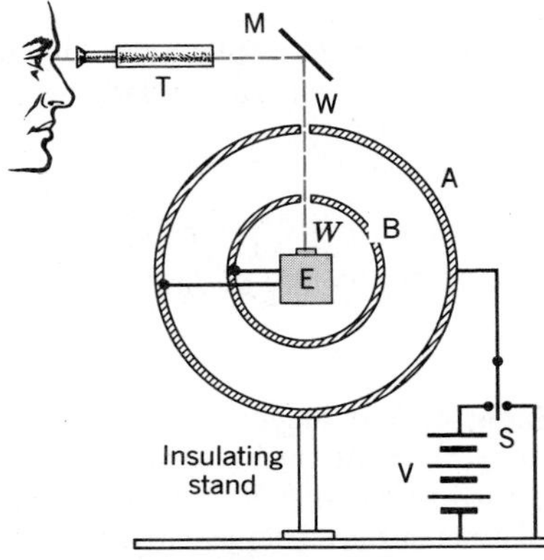

Figure 10 A modern and more precise version of the apparatus of Fig. 9, also designed to verify that charge resides on the outside surface of a metal object. If a charge is placed on sphere A, by throwing switch S to the left, any charge that moved to the inner sphere B would be detected by the sensitive electrometer E. No such charge transfer was found; the charge remained on the outside surface of A.

25-9 Gauss' Law: Linear Symmetry

Figure 11 shows a section of an infinitely long charged plastic rod, its linear charge density, or charge per unit length, being λ. Let us find an expression for the magnitude of the electric field **E** at a distance r from the axis of the rod.

Our Gaussian surface should match the symmetry of the problem, which is cylindrical. We choose a circular cylinder of radius r and length h, coaxial with the rod. The Gaussian surface must be closed so we include two end caps as part of the surface.

Imagine now that, while you are not watching, someone rotates the plastic rod around its cylindrical axis and/or turns it end for end. When you look again at the rod, you will not be able to detect any change. We conclude that, from symmetry, the only uniquely specified direction in this problem is a radial line. Thus, **E** must have a constant magnitude E and (for a positively charged rod) must be directed radially outward at every point on the cylindrical Gaussian surface.

The flux of **E** through this surface is $(E)(2\pi r)(h)$, where $2\pi r$ is the circumference of the cylinder and h is its height, so that $2\pi rh$ is the area of the cylindrical surface. There is no flux through the end caps because **E**, being radially directed, lies parallel to the surface at every point.

The charge enclosed by the surface is λh so that Gauss' law (Eq. 9),

$$\epsilon_0 \oint \mathbf{E} \cdot d\mathbf{A} = q,$$

reduces to

$$\epsilon_0 E(2\pi rh) = \lambda h,$$

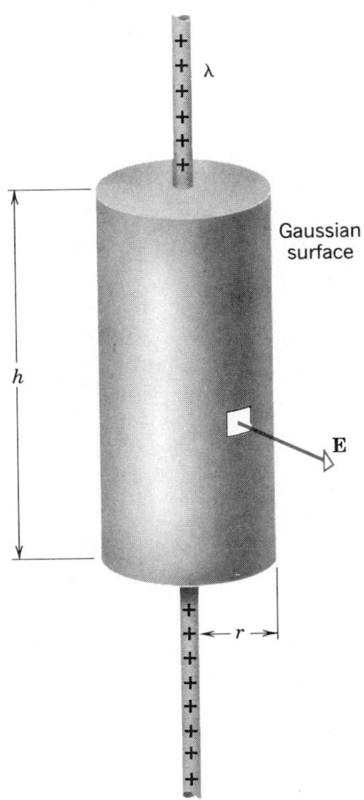

Figure 11 A Gaussian surface in the form of a closed cylinder surrounds a section of a very long, uniformly charged, plastic rod.

yielding

$$\boxed{E = \frac{\lambda}{2\pi\epsilon_0 r}} \quad \text{(line of charge).} \tag{14}$$

The direction of **E** is radially outward if the line of charge is positive and radially inward if it is negative.

Sample Problem 5 A plastic rod, whose length L is 220 cm and whose radius R is 3.6 mm, carries a negative charge q of magnitude 3.8×10^{-7} C, spread uniformly over its surface. What is the electric field near the midpoint of the rod, at a point on its surface?

Although the rod is not infinitely long, from a point on its surface and near its midpoint it is effectively very long so that we are justified in using Eq. 14. The linear charge density for the rod is

$$\lambda = \frac{q}{L} = \frac{-3.8 \times 10^{-7}\,\text{C}}{2.2\ \text{m}} = -1.73 \times 10^{-7}\ \text{C/m}.$$

From Eq. 14 we then have

$$E = \frac{\lambda}{2\pi\epsilon_0 r}$$

$$= \frac{-1.73 \times 10^{-7}\ \text{C/m}}{(2\pi)(8.85 \times 10^{-12}\ \text{C}^2/\text{N}\cdot\text{m}^2)(0.0036\ \text{m})}$$

$$= -8.6 \times 10^5\ \text{N/C}. \qquad \text{(Answer)}$$

The minus sign tells us that, because the rod is negatively charged, the direction of the electric field is radially inward, toward the axis of the rod. Sparking occurs in dry air at atmospheric pressure at an electric field strength of about 3×10^6 N/C. The field strength we calculated above is lower than this by a factor of about 3.4 so that sparking should not occur.

25-10 Gauss' Law: Planar Symmetry

Figure 12 shows a portion of a thin infinite sheet of charge, its surface charge density, or charge per unit area, being σ. A sheet of thin plastic wrap, uniformly charged on one side by friction, can serve as a prototype. Let us find the electric field **E** a distance r in front of the sheet.

A useful Gaussian surface is a closed cylinder of cross-sectional area A and height $2r$, arranged to pierce the sheet at right angles as shown. From symmetry, the only preferred direction for **E** is at right angles to the end cap. Furthermore, **E** must point *away* from the posi-

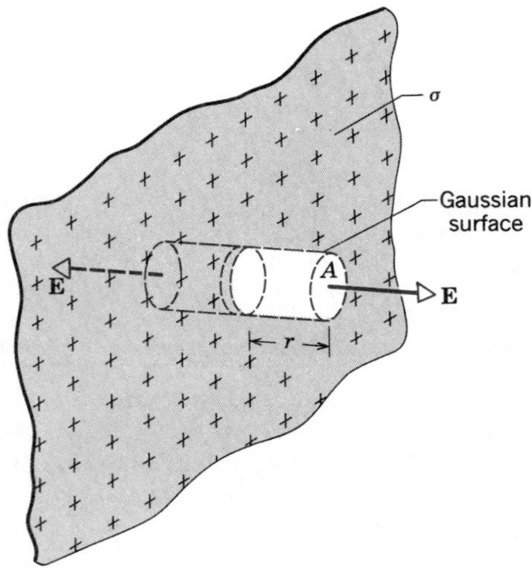

Figure 12 A portion of a very large thin plastic sheet, uniformly charged on one side to surface charge density σ. A Gaussian surface, in the form of a short cylinder closed at each end, pierces the sheet at right angles to it.

tively charged sheet of Fig. 12, thus piercing the Gaussian surface in an outward direction. Because **E** does not pierce the cylinder walls, there is no flux through this portion of the Gaussian surface. Thus, Gauss' law

$$\epsilon_0 \oint \mathbf{E} \cdot d\mathbf{A}$$

becomes

$$\epsilon_0(EA + EA) = \sigma A,$$

where σA is the enclosed charge. This gives

$$\boxed{E = \frac{\sigma}{2\epsilon_0}} \quad \text{(sheet of charge).} \qquad (15)$$

This result (Eq. 15) agrees with our earlier result (Eq. 22 of Chapter 24), which we obtained by direct integration.

You may wonder why we introduce such seemingly unrealistic problems as the field set up by an infinite line of charge or by an infinite sheet of charge. It is not enough to say that we do so because it is simple to solve such problems, although that is indeed true. The proper answer is that the solutions for "infinite" problems apply to real world problems to a very good approximation. Thus, Eq. 15 holds quite well for a finite sheet as long as you are close to the sheet and not too near its edges. After all, an ant in the center of a large parking lot probably thinks that it is walking on an infinite sheet, and it might as well be.

Sample Problem 6 Figure 13*a* shows portions of two large sheets of charge with uniform surface charge densities of $\sigma_+ = +6.8\ \mu\text{C/m}^2$ and $\sigma_- = -4.3\ \mu\text{C/m}^2$. Find the electric field **E** (a) to the left of the sheets, (b) between the sheets, and (c) to the right of the sheets.

Our strategy is to deal with each sheet separately and then to add the resulting electric fields algebraically, using the superposition principle.

For the positive sheet we have, from Eq. 15,

$$E_+ = \frac{\sigma_+}{2\epsilon_0} = \frac{6.8 \times 10^{-6}\ \text{C/m}^2}{(2)(8.85 \times 10^{-12}\ \text{C}^2/\text{N}\cdot\text{m}^2)} = 3.84 \times 10^5\ \text{N/C}.$$

Similarly, for the negative sheet

$$E_- = \frac{|\sigma_-|}{2\epsilon_0} = \frac{4.3 \times 10^{-6}\ \text{C/m}^2}{(2)(8.85 \times 10^{-12}\ \text{C}^2/\text{N}\cdot\text{m}^2)} = 2.43 \times 10^5\ \text{N/C}.$$

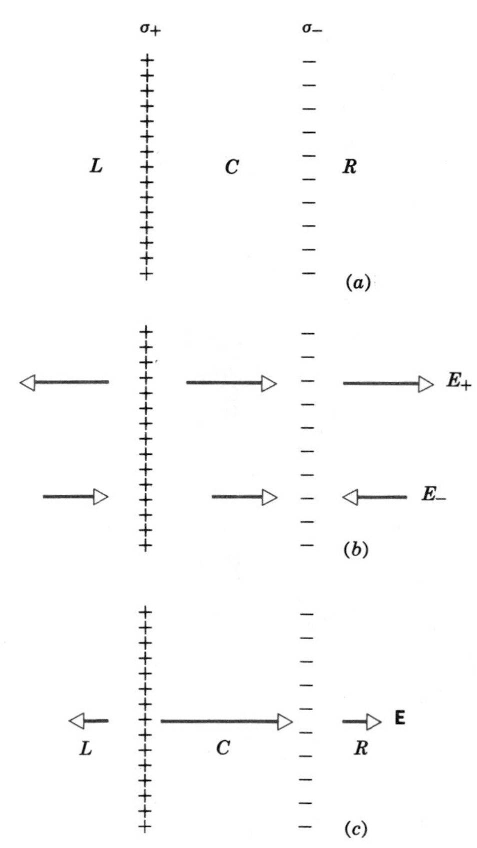

Figure 13 Sample Problem 6. (*a*) Two large thin sheets of charge face each other. (*b*) Showing, separately, the electric fields resulting from each charge sheet. (*c*) The resultant field due to both charge sheets, found from the principle of superposition.

Figure 13b shows the two sets of fields calculated above, to the left of the sheets, between them, and to the right of the sheets.

The resultant fields in these three regions follow from the superposition principle. To the left of the sheets, we have

$$E_L = E_+ - E_- = 3.84 \times 10^5 \text{ N/C} - 2.43 \times 10^5 \text{ N/C} = 1.4 \times 10^5 \text{ N/C}. \quad \text{(Answer)}$$

The resultant electric field in this region points to the left, as Fig. 13c shows. To the right of the sheets, the electric field has this same magnitude but points to the right in Fig. 13c.

Between the sheets, the two fields add and we have

$$E_C = E_+ + E_- = 3.84 \times 10^5 \text{ N/C} + 2.43 \times 10^5 \text{ N/C} = 6.3 \times 10^5 \text{ N/C}. \quad \text{(Answer)}$$

Outside the sheets, the electric field behaves like that from a single sheet whose surface charge density is $\sigma_+ + \sigma_-$ or $+2.5 \times 10^{-6}$ C/m^2. The field pattern of Fig. 13c bears this out. In Exercise 30 and Problem 35 we ask you to investigate the case in which the two surface charge densities are equal in magnitude but opposite in sign and also the case in which they are equal in both magnitude and sign.

25–11 Gauss' Law: Spherical Symmetry

Here we use Gauss' law to prove the two theorems presented without proof in Section 23–4, namely:

A uniform spherical shell of charge behaves, for external points, as if all its charge were concentrated at its center,

and

A uniform spherical shell of charge exerts no electrical force on a charged particle placed inside the shell.

These two shell theorems are the electrostatic analogs of the two gravitational shell theorems presented in Chapter 15. We shall see how much simpler is our Gauss' law proof than the detailed proof of Section 15–5, in which full advantage of the spherical symmetry was not taken.

Figure 14 shows a spherical shell of charge q and radius R and two concentric Gaussian surfaces, S_1 and S_2. Applying Gauss' law to surface S_2, for which $r > R$, leads directly to

$$\boxed{E = \frac{1}{4\pi\epsilon_0}\frac{q}{r^2}} \quad \text{(spherical shell, } r > R), \quad (16)$$

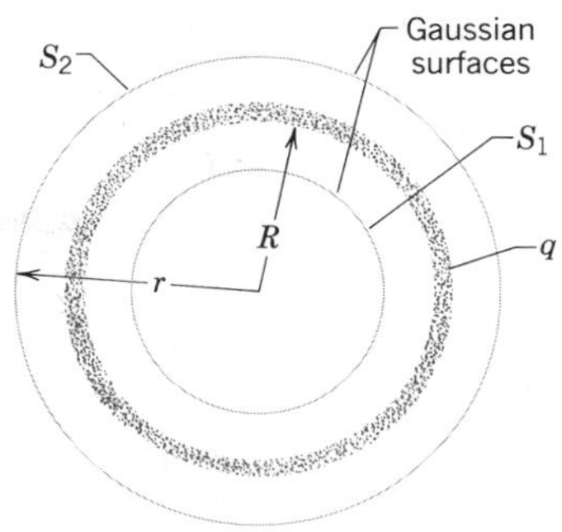

Figure 14 A thin uniformly charged shell. Two Gaussian surfaces are shown in cross section, one enclosing the shell and one enclosed by it.

just as it did in connection with Fig. 7.

Applying Gauss' law to surface S_1, for which $r < R$, leads directly to

$$\boxed{E = 0} \quad \text{(spherical shell, } r < R), \quad (17)$$

because this Gaussian surface encloses no charge.

Any spherically symmetric charge distribution, such as that of Fig. 15, can be made up of a nest of such shells. The volume charge density ρ, whose SI units are C/m^3, must have a constant value for each shell but need not be the same from shell to shell. That is, for the charge distribution as a whole, the volume charge density ρ need not be constant but can vary only with r, the radial distance from the center.

In Fig. 15a we show a Gaussian surface with $r > R$ so that the entire charge lies within the surface. In this case the charge behaves as if it were a point charge located at the center and Eq. 16 holds.

Figure 15b shows a Gaussian surface of radius r, where $r < R$. Here, by virtue of Eq. 16, that part of the charge distribution that lies outside the Gaussian surface makes no contribution to the field at radius r, which is determined only by the charge q' that lies within that surface. Thus, from Eq. 16

$$E = \frac{1}{4\pi\epsilon_0}\frac{q'}{r^2}, \quad (18)$$

where q' is that portion of the charge within the Gaussian surface at radius r.

Table 3 summarizes the relations between charge and field for the symmetrical charge distributions that we have considered in this chapter.

Table 3 A Summary of Formulas

Situation	*Formula for E*	*Remarks*	*Equation*
Charged conductor	σ/ϵ_0	On the conductor surface	12
	0	Inside the conductor	
Point charge	$q/4\pi\epsilon_0 r^2$		11
Spherical shell	$q/4\pi\epsilon_0 r^2$	Outside the shell	16
	0	Inside the shell	17
Infinite line of charge	$\lambda/2\pi\epsilon_0 r$		14
Infinite sheet of charge	$\sigma/2\epsilon_0$		15

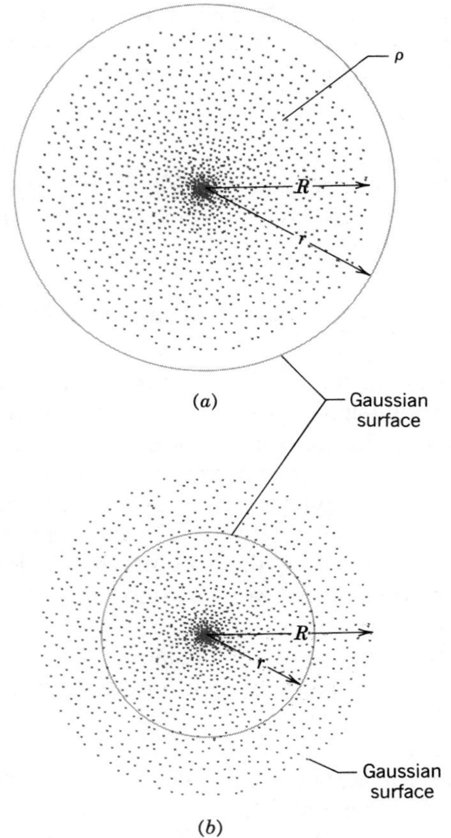

Figure 15 The dots represent a spherically symmetric distribution of charge of radius R, the volume charge density ρ being a function of distance from the center only. The object is not a conductor and the charges are assumed to be held fixed in position. (a) A concentric Gaussian surface with $r > R$. (b) A Gaussian surface with $r < R$.

Sample Problem 7 The nucleus of an atom of gold has a radius $R = 6.2 \times 10^{-15}$ m and carries a positive charge $q = Ze$, where Z (= 79) is the atomic number of gold and e is the elementary charge. Plot the magnitude of the electric field from the center of the gold nucleus outward to a distance of about twice its radius. Assume that the nucleus is spherical and that the charge is distributed uniformly throughout its volume.

The total charge q on the nucleus is

$$q = Ze = (79)(1.60 \times 10^{-19}\ \text{C}) = 1.264 \times 10^{-17}\ \text{C}.$$

Outside the nucleus, the situation is represented by Fig. 15a and by Eq. 16. From this equation we have, for a point on the surface of the nucleus,

$$E = \frac{1}{4\pi\epsilon_0}\frac{q}{r^2}$$

$$= \frac{1.264 \times 10^{-17}\ \text{C}}{(4\pi)(8.85 \times 10^{-12}\ \text{C}^2/\text{N}\cdot\text{m}^2)(6.2 \times 10^{-15}\ \text{m})^2}$$

$$= 3.0 \times 10^{21}\ \text{N/C}. \qquad \text{(Answer)}$$

Inside the nucleus, Fig. 15b and Eq. 18 apply. The charge q' contained within a sphere of radius r, where $r < R$, is proportional to the respective volumes. The volume of a sphere of radius r is $\frac{4}{3}\pi r^3$ so that we have

$$q' = q\frac{\frac{4}{3}\pi r^3}{\frac{4}{3}\pi R^3} = q\frac{r^3}{R^3}.$$

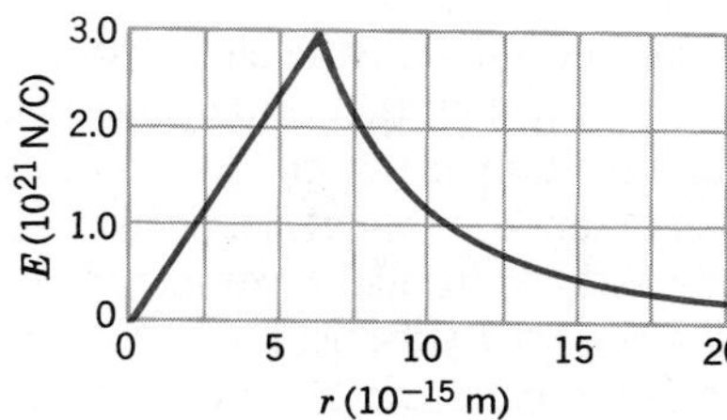

Figure 16 The variation of electric field with distance from the center for the nucleus of a gold atom. The positive charge is assumed to be distributed uniformly throughout the volume of the nucleus.

If we substitute this result into Eq. 18, we find

$$E = \frac{1}{4\pi\epsilon_0}\frac{q'}{r^2} = \left(\frac{q}{4\pi\epsilon_0 R^3}\right) r. \qquad (19)$$

The quantity in parentheses is a constant so that, within the nucleus, E is directly proportional to r, being zero at the nuclear center. Comparison of Eqs. 19 and 16 shows that they give the same result—that calculated above—for $r = R$. This simply tells us that the "inside" equation and the "outside" equation must yield the same value for the electric field at the nuclear surface, where the two equations both apply. Figure 16 shows these results graphically.

REVIEW AND SUMMARY

Gauss' law and Coulomb's law, although expressed in different forms, are equivalent ways of describing the relation between charge and the electric field in static situations. Gauss' law is

Gauss' Law

$$\epsilon_0 \Phi = q \quad \text{(Gauss' law)}, \qquad [8]$$

in which q is the net charge inside an imaginary closed surface (a *Gaussian surface)* and Φ is the outward flux of the electric field through the surface:

The Flux of the Electric Field

$$\Phi = \oint \mathbf{E} \cdot d\mathbf{A} \quad \text{(flux through a Gaussian surface)}. \qquad [6]$$

See Sample Problems 1 and 2 and the discussions related to Figs. 3 and 4.

Coulomb's Law and Gauss' Law

Coulomb's law can readily be derived from Gauss' law, the details being given in Section 25-6. Experimental verification of Gauss' law—and thus of Coulomb's law—shows that the exponent of r in Coulomb's law is now known to be exactly 2 within an experimental uncertainty of less than 1×10^{-16}. See Section 25-8 and Table 2.

Using Gauss' law and, in some cases, symmetry arguments, we can derive several important results. Among these are:

Charge on a Conductor

a. An excess charge on an *isolated conductor* is, in equilibrium, entirely on its outer surface (Section 25-7).

Field Near a Charged Conductor

b. The electric field near the *surface of a charged conductor* in equilibrium is perpendicular to the surface and has magnitude

$$E = \frac{\sigma}{\epsilon_0} \quad \text{(conducting surface)} \qquad [12]$$

(Section 25-7 and Sample Problems 3 and 4).

c. The electric field due to an infinite *line of charge* with uniform charge per unit length, λ, is in a direction perpendicular to the line of charge and has magnitude

Field Due to an Infinite Line of Charge

$$E = \frac{\lambda}{2\pi\epsilon_0 r} \quad \text{(line of charge)} \qquad [14]$$

(Section 25-9 and Sample Problem 5).

Field Due to an Infinite Sheet of Charge

d. The electric field due to an infinite sheet of charge is perpendicular to the plane of the sheet and has magnitude

$$E = \frac{\sigma}{2\epsilon_0} \quad \text{(sheet of charge)} \qquad [15]$$

(Section 25-10 and Sample Problem 6).

e. The electric field outside a *spherically symmetrical shell* with radius R and total charge q is directed radially and has magnitude

Field of a Spherical Shell

$$E = \frac{1}{4\pi\epsilon_0}\frac{q}{r^2} \quad \text{(spherical shell, } r > R\text{)}. \qquad [16]$$

The charge behaves, for external points, as if it were all at the center of the sphere. The field *inside* a uniform spherical shell is exactly zero:

$$E = 0 \quad \text{(spherical shell, } r < R\text{)}. \qquad [17]$$

The electric field *inside a uniform sphere of charge* is directed radially and has magnitude

Field Inside a Uniform Spherical Charge

$$E = \left(\frac{q}{4\pi\epsilon_0 R^3}\right) r \qquad [19]$$

(Section 25–11 and Sample Problem 7).

QUESTIONS

1. What is the basis for the statement that lines of electric force begin and end only on electric charges?

2. Positive charges are sometimes called "sources" and negative charges "sinks" of electric field. How would you justify this terminology? Are there sources and/or sinks of gravitational field?

3. Does Gauss' law hold for the flow of water? Consider various Gaussian surfaces intersecting a fountain or a waterfall in different ways. What would correspond to positive and negative charges in this case?

4. By analogy with Φ, how would you define the flux Φ_g of a gravitational field? What is the flux of the earth's gravitational field through the boundaries of a room, assumed to contain no matter? Through a sphere closely surrounding the earth? Through a sphere the size of the moon's orbit? See Problem 14.

5. Consider a Gaussian surface that surrounds part of the charge distribution shown in Fig. 17. (*a*) Which of the charges contribute to the electric field at point *P*? (*b*) Would the value obtained for the flux through the surface, calculated using only the electric field due to q_1 and q_2, be greater than, equal to, or less than that obtained using the total field?

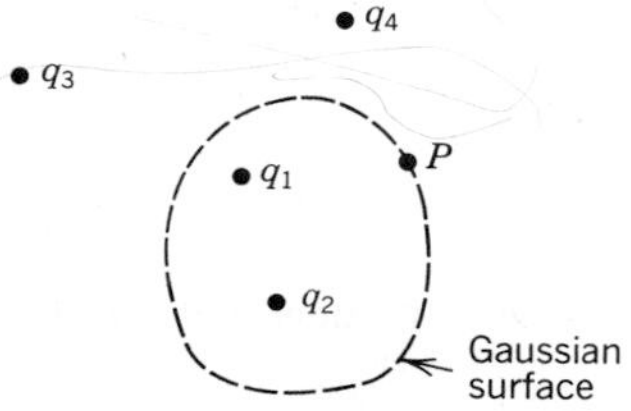

Figure 17 Question 5.

6. Suppose that an electric field in some region is found to have a constant direction but to be decreasing in strength in that direction. What do you conclude about the charge in the region? Sketch the lines of force.

7. A point charge is placed at the center of a spherical Gaussian surface. Is Φ changed (*a*) if the surface is replaced by a cube of the same volume, (*b*) if the sphere is replaced by a cube of one-tenth the volume, (*c*) if the charge is moved off-center in the original sphere, still remaining inside, (*d*) if the charge is moved just outside the original sphere, (*e*) if a second charge is placed near, and outside, the original sphere, and (*f*) if a second charge is placed inside the Gaussian surface?

8. In Gauss' law,

$$\epsilon_0 \oint \mathbf{E} \cdot d\mathbf{A} = q,$$

is **E** necessarily the electric field attributable to the charge *q*?

9. A surface encloses an electric dipole. What can you say about Φ_E for this surface?

10. Suppose that a Gaussian surface encloses no net charge. Does Gauss' law require that **E** equal zero for all points on the surface? Is the converse of this statement true; that is, if **E** equals zero everywhere on the surface, does Gauss' law require that there be no net charge inside?

11. Is Gauss' law useful in calculating the field due to three equal charges located at the corners of an equilateral triangle? Explain why or why not.

12. A total charge *Q* is distributed uniformly throughout a cube of edge length *a*. Is the resulting electric field at an external point *P*, a distance *r* from the center of the cube, given by $E = Q/4\pi\epsilon_0 r^2$? See Fig. 18. If not, can *E* be found by constructing a "concentric" cubical Gaussian surface? If not, explain why not. Can you say anything about *E* if $r \gg a$?

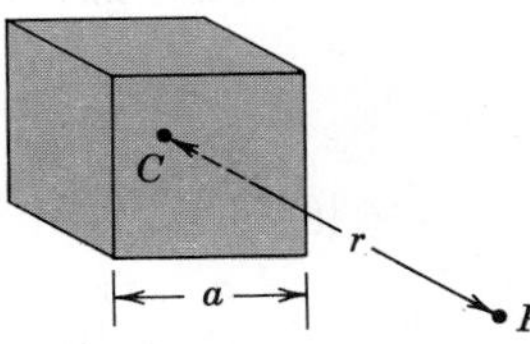

Figure 18 Question 12.

13. Is **E** necessarily zero inside a charged rubber balloon if the balloon is (*a*) spherical or (*b*) sausage shaped? For each shape, assume the charge to be distributed uniformly over the surface. How would the situation change, if at all, if the balloon has a thin layer of conducting paint on its outside surface?

14. A spherical rubber balloon carries a charge that is uniformly distributed over its surface. As the balloon is blown up, how does E vary for points (*a*) inside the balloon, (*b*) at the surface of the balloon, and (*c*) outside the balloon?

15. In Section 25–6 we have seen that Coulomb's law can be derived from Gauss' law. Does this necessarily mean that Gauss' law can be derived from Coulomb's law?

16. A large, insulated, hollow conductor carries a positive charge. A small metal ball carrying a negative charge of the same magnitude is lowered by a thread through a small opening in the top of the conductor, allowed to touch the inner surface, and then withdrawn. What is then the charge on (*a*) the conductor and (*b*) the ball?

17. Can we deduce from the argument of Section 25–7 that the electrons in the wires of a house wiring system move along the surfaces of those wires? If not, why not?

18. Does Gauss' law, as applied in Section 25–7, require that all the conduction electrons in an insulated conductor reside on the surface?

19. Suppose that you have a Gaussian surface in the shape of a donut and that there is a single point charge inside. Does Gauss' law hold? If not, why not? If so, is there enough symmetry in the situation to apply it usefully?

20. A positive point charge q is located at the center of a hollow metal sphere. What charges appear on (*a*) the inner surface and (*b*) the outer surface of the sphere? (*c*) If you bring an (uncharged) metal object near the sphere, will it change your answers in (*a*) or (*b*) above? Will it change the way charge is distributed over the sphere?

21. Explain why the symmetry of Fig. 11 restricts us to a consideration of **E** that has only a radial component at any point. Remember, in this case, that the field must not only look the same at any point along the line but must also look the same if the figure is turned end for end.

22. In Section 25–9, the *total* charge on the infinite rod is infinite. Why is not E also infinite? After all, according to Coulomb's law, if q is infinite, so is E.

23. Explain why the symmetry of Fig. 12 restricts us to a consideration of **E** that has only a component directed away from the sheet. Why, for example, could **E** not have components parallel to the sheet? Remember, in this case, that the field must not only look the same at any point along the sheet in any direction but must also look the same if the sheet is rotated about any line perpendicular to the sheet.

24. The field due to an infinite sheet of charge is uniform, having the same strength at all points no matter how far from the surface charge. Explain how this can be, given the inverse square nature of Coulomb's law.

25. Explain why the spherical symmetry of Fig. 7 restricts us to a consideration of **E** that has only a radial component at any point. (*Hint:* Imagine other components, perhaps along the equivalent of longitude or latitude lines on the earth's surface. Spherical symmetry requires that these look the same from any perspective. Can you invent such field lines that satisfy this criterion?)

26. As you penetrate a uniform sphere of charge, E should decrease because less charge lies inside a sphere drawn through the observation point. On the other hand, E should increase because you are closer to the center of this charge. Which effect dominates and why?

27. Given a spherically symmetric charge distribution (not of uniform density of charge radially), is E necessarily a maximum at the surface? Comment on various possibilities.

EXERCISES AND PROBLEMS

Section 25–3 Flux

1E. Water in an irrigation ditch of width $w = 3.22$ m and depth $d = 1.04$ m flows with a speed of 0.207 m/s. Find the mass flux through the following surfaces: (*a*) a surface of area wd, entirely in the water, perpendicular to the flow; (*b*) a surface with area $3wd/2$, of which wd is in the water, perpendicular to the flow; (*c*) a surface of area $wd/2$, entirely in the water, perpendicular to the flow; (*d*) a surface of area wd, half in the water and half out, perpendicular to the flow; (*e*) a surface of area wd, entirely in the water, with its normal 34° from the direction of flow.

Section 25–4 Flux of the Electric Field

2E. The square surface shown in Fig. 19 measures 3.2 mm on each side. It is immersed in a uniform electric field with $E = 1800$ N/C. The field lines make an angle of 35° with the "outward pointing" normal, as shown. Calculate the flux through the surface.

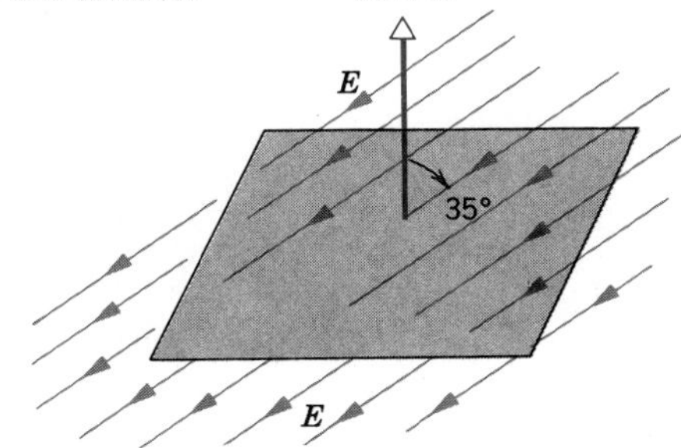

Figure 19 Exercise 2.

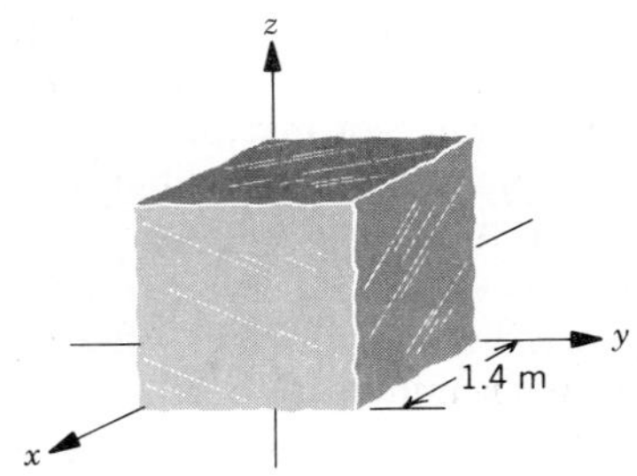

Figure 20 Exercise 3 and Problem 12.

3E. A cube with 1.4-m edges is oriented as shown in Fig. 20 in a region of uniform electric field. Find the electric flux through the right face if the electric field, in newtons per coulomb, is given by (*a*) $6\mathbf{i}$, (*b*) $-2\mathbf{j}$, and (*c*) $-3\mathbf{i} + 4\mathbf{k}$. (*d*) What is the total flux through the cube for each of these fields?

4P. Calculate Φ through (*a*) the flat base and (*b*) the curved surface of a hemisphere of radius R. The field $\mathbf{E}$ is uniform, is parallel to the axis of the hemisphere, and the lines of $\mathbf{E}$ enter through the flat base.

Section 25-5 Gauss' Law

5E. Four charges, $2q$, q, $-q$, and $-2q$ are arranged at the corners of a square as shown in Fig. 21. If possible, describe a closed surface that encloses the charge $2q$ and through which the net electric flux is (*a*) 0, (*b*) $+3q/\epsilon_0$, and (*c*) $-2q/\epsilon_0$.

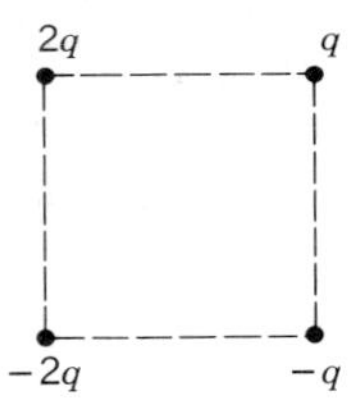

Figure 21 Exercise 5.

6E. Charge on an originally uncharged insulated conductor is separated by holding a positively charged rod very closely nearby, as in Fig. 22. Calculate the flux for the five Gaussian surfaces shown. Assume that the induced negative charge on the conductor is equal to the positive charge q on the rod.

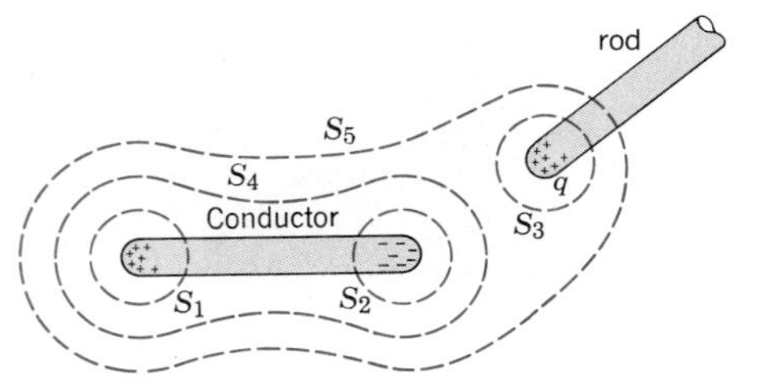

Figure 22 Exercise 6.

7E. A point charge of 1.8 μC is at the center of a cubical Gaussian surface 55 cm on edge. What is Φ_E through the surface?

8E. The net electric flux through each face of a die (singular of dice) has magnitude in units of $10^3\ \mathrm{N\cdot m^2/C}$ equal to the number of spots on the face (1 through 6). The flux is inward for N odd and outward for N even. What is the net charge inside the die?

9E. A point charge $+q$ is a distance $d/2$ from a square surface of side d and is directly above the center of the square as shown in Fig. 23. What is the electric flux through the square? (*Hint:* Think of the square as one face of a cube with edge d.)

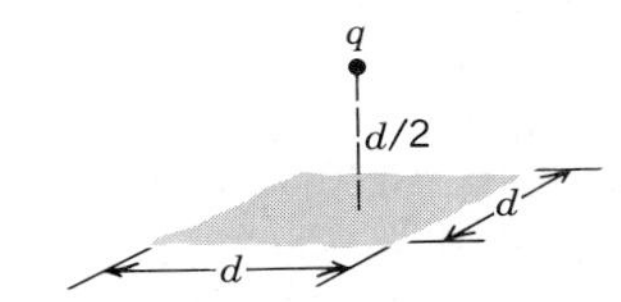

Figure 23 Exercise 9.

10E. A butterfly net is in a uniform electric field as shown in Fig. 24. The rim, a circle of radius a, is aligned perpendicular to the field. Find the electric flux through the netting.

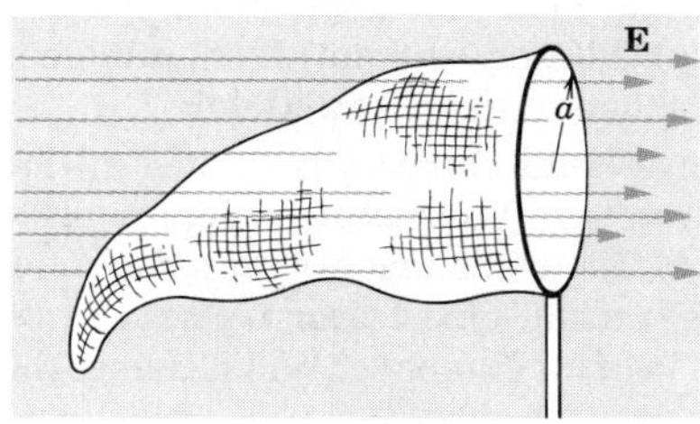

Figure 24 Exercise 10.

11P. It is found experimentally that the electric field in a certain region of the earth's atmosphere is directed vertically down. At an altitude of 300 m the field is 60 N/C and at an altitude of 200 m it is 100 N/C. Find the net amount of charge contained in a cube 100 m on edge located at an altitude between 200 and 300 m. Neglect the curvature of the earth.

12P. Find the net flux through the cube of Exercise 3 and Fig. 20 if the electric field is given by (*a*) $\mathbf{E} = 3y\mathbf{j}$ and (*b*) $\mathbf{E} = -4\mathbf{i} + (6 + 3y)\mathbf{j}$. E is in newtons per coulomb if y is in meters. (*c*) In each case, how much charge is inside the cube?

13P. A point charge q is placed at one corner of a cube of edge a. What is the flux through each of the cube faces? (*Hint:* Use Gauss' law and symmetry arguments.)

14P. "Gauss' law for gravitation" is

$$\frac{1}{4\pi G}\Phi_g = \frac{1}{4\pi G}\oint \mathbf{g}\cdot d\mathbf{A} = -m,$$

where m is the enclosed mass and G is the universal gravitation constant (Section 15–6). Derive Newton's law of gravitation from this. What is the significance of the minus sign?

Section 25–7 A Charged Isolated Conductor

15E. A uniformly charged conducting sphere of 1.2-m diameter has a surface charge density of 8.1 $\mu C/m^2$. (*a*) Find the charge on the sphere. (*b*) What is the total electric flux leaving the surface of the sphere?

16E. Space vehicles traveling through the earth's radiation belts collide with trapped electrons. Since in space there is no ground, the resulting charge buildup can become significant and can damage electronic components, leading to control-circuit upsets and operational anomalies. A spherical metallic satellite 1.3 m in diameter accumulates 2.4 μC of charge in one orbital revolution. (*a*) Find the surface charge density. (*b*) Calculate the resulting electric field just outside the surface of the satellite.

17E. A conducting sphere carrying charge Q is surrounded by a spherical conducting shell. (*a*) What is the net charge on the inner surface of the shell? (*b*) Another charge q is placed outside the shell. Now what is the net charge on the inner surface of the shell? (*c*) If q is moved to a position between the shell and the sphere, what is the net charge on the inner surface of the shell? (*d*) Are your answers valid if the sphere and shell are not concentric?

18P. An insulated conductor of arbitrary shape carries a net charge of $+10 \times 10^{-6}$ C. Inside the conductor is a hollow cavity within which is a point charge $q = +3.0 \times 10^{-6}$ C. What is the charge (*a*) on the cavity wall and (*b*) on the outer surface of the conductor?

19P. An irregularly-shaped conductor has an irregularly-shaped cavity inside. A charge q is placed on the conductor but there is no charge inside the cavity. Show that there is no net charge on the cavity wall.

Section 25–9 Gauss' Law: Linear Symmetry

20E. An infinite line of charge produces a field of 4.5×10^4 N/C at a distance of 2.0 m. Calculate the linear charge density.

21E. (*a*) The drum of the photocopying machine in Sample Problem 3 has a length of 42 cm and a diameter of 12 cm. What is the total charge on the drum? (*b*) The manufacturer wishes to produce a desktop version of the machine. This requires reducing the size of the drum to a length of 28 cm and a diameter of 8.0 cm. The electric field at the drum surface must remain unchanged. What must be the charge on this new drum?

22E. A very long straight wire carries -3.6 nC/m of fixed negative charge. The wire is to be surrounded by a uniform cylinder of positive charge, radius 1.5 cm, coaxial with the wire. The charge density ρ of the cylinder is to be selected so that the net electric field outside the cylinder is zero. Calculate the required positive charge density ρ.

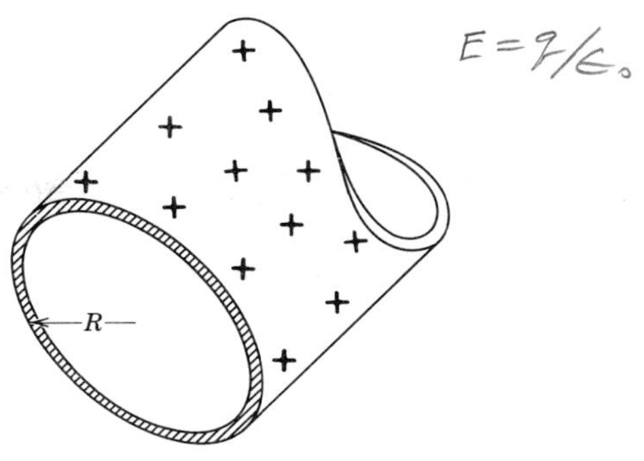

Figure 25 Problem 23.

23P. Figure 25 shows a section through a long, thin-walled metal tube of radius R, carrying a charge per unit length λ on its surface. Derive expressions for E for various distances r from the tube axis, considering both (*a*) $r > R$ and (*b*) $r < R$. Plot your results for the range $r = 0$ to $r = 5.0$ cm, assuming that $\lambda = 2.0 \times 10^{-8}$ C/m and $R = 3.0$ cm. (*Hint:* Use cylindrical Gaussian surfaces, coaxial with the metal tube.)

24P. Figure 26 shows a section through two long thin concentric cylinders of radii a and b. The cylinders carry equal and opposite charges per unit length λ. Using Gauss' law, prove (*a*) that $E = 0$ for $r < a$ and (*b*) that between the cylinders E is given by

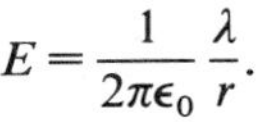

$$E = \frac{1}{2\pi\epsilon_0}\frac{\lambda}{r}.$$

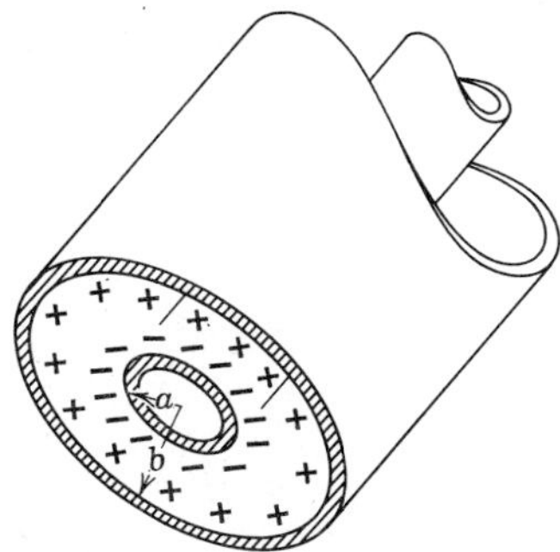

Figure 26 Problem 24.

25P. Figure 27 shows a Geiger counter, used to detect ionizing radiation. The counter consists of a thin central wire, carrying positive charge, surrounded by a concentric circular conducting cylinder, carrying an equal negative charge. Thus, a strong radial electric field is set up inside the cylinder. The cylinder contains a low-pressure inert gas. When a particle of radiation enters the tube through the cylinder walls, it ionizes a few of the gas atoms. The resulting free electrons are drawn to the positive wire. However, the electric field is so intense that,

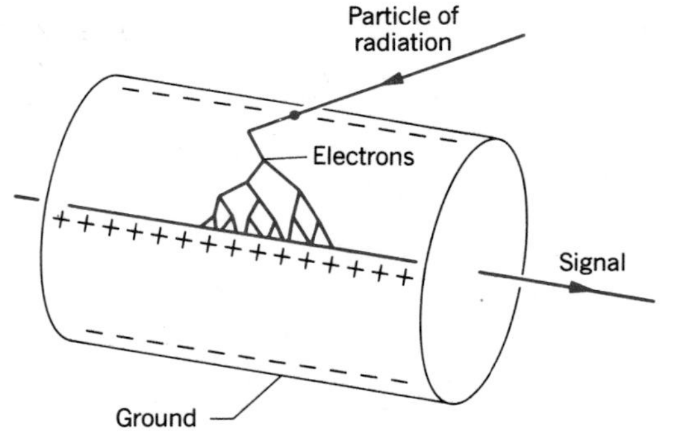

Figure 27 Problem 25.

between collisions with the gas atoms, they have gained energy sufficient to ionize these atoms also. More free electrons are thereby created, and the process is repeated until the electrons reach the wire. The "avalanche" of electrons is collected by the wire, generating a signal recording the passage of the incident particle of radiation. Suppose that the radius of the central wire is 25 μm, the radius of the cylinder 1.4 cm, and the length of the tube 16 cm. The electric field at the cylinder wall is 2.9×10^4 N/C. Calculate the amount of positive charge on the central wire. (*Hint:* See Problem 24.)

26P. A very long conducting cylinder (length L) carrying a total charge $+q$ is surrounded by a conducting cylindrical shell (also of length L) with total charge $-2q$, as shown in cross section in Fig. 28. Use Gauss' law to find (*a*) the electric field at points outside the conducting shell, (*b*) the distribution of the charge on the conducting shell, and (*c*) the electric field in the region between the cylinders.

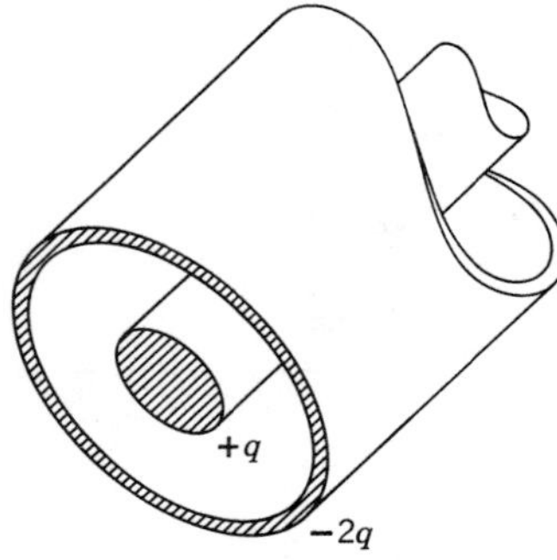

Figure 28 Problem 26.

27P. Two long charged concentric cylinders have radii of 3.0 cm and 6.0 cm. The charge per unit length on the inner cylinder is 5.0×10^{-6} C/m and that on the outer cylinder is -7.0×10^{-6} C/m. Find the electric field at (*a*) $r = 4.0$ cm and (*b*) $r = 8.0$ cm.

28P. In Problem 24 a positron revolves in a circular path of radius r, between and concentric with the cylinders. What must be its kinetic energy K in electron-volts? Assume that $a = 2.0$ cm, $b = 3.0$ cm, and $\lambda = 30$ nC/m.

29P. Charge is distributed uniformly throughout an infinitely long cylinder of radius R. (*a*) Show that E at a distance r from the cylinder axis ($r < R$) is given by

$$E = \frac{\rho r}{2\epsilon_0},$$

where ρ is the density of charge. (*b*) What result do you expect for $r > R$?

Section 25-10 Gauss' Law: Planar Symmetry

30E. Two large sheets of positive charge face each other as in Fig. 29. What is **E** at points (*a*) to the left of the sheets, (*b*) between them, and (*c*) to the right of the sheets? Assume the same surface charge density σ for each sheet. Consider only points not near the edges whose distance from the sheets is small compared to the dimensions of the sheet. (*Hint:* See Sample Problem 6.)

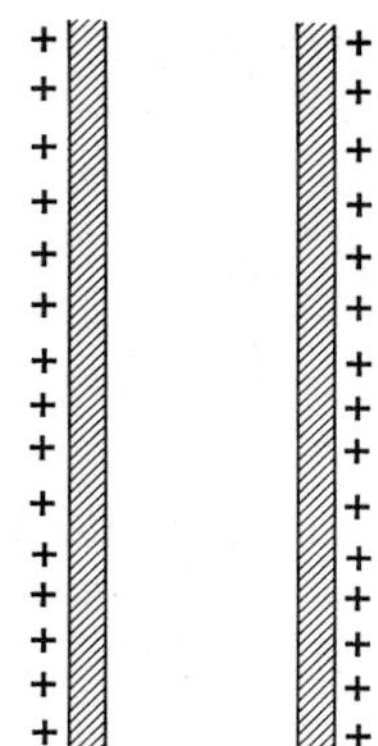

Figure 29 Exercise 30.

31E. A metal plate 8.0 cm on a side carries a total charge of 6.0×10^{-6} C. (*a*) Using the infinite plate approximation, calculate the electric field 0.50 mm above the surface of the plate near the plate's center. (*b*) Estimate the field at a distance of 30 m.

32E. A large flat nonconducting surface carries a uniform charge density σ. A small circular hole of radius R has been cut in the middle of the sheet, as shown in Fig. 30. Ignore fringing of the field lines around all edges and calculate the electric field at point P, a distance z from the center of the hole along its axis. (*Hint:* See Eq. 21 of Chapter 24 and use the principle of superposition.)

33P. A small sphere whose mass m is 1.0 mg carries a charge $q = 2.0 \times 10^{-8}$ C. It hangs in the earth's gravitational field

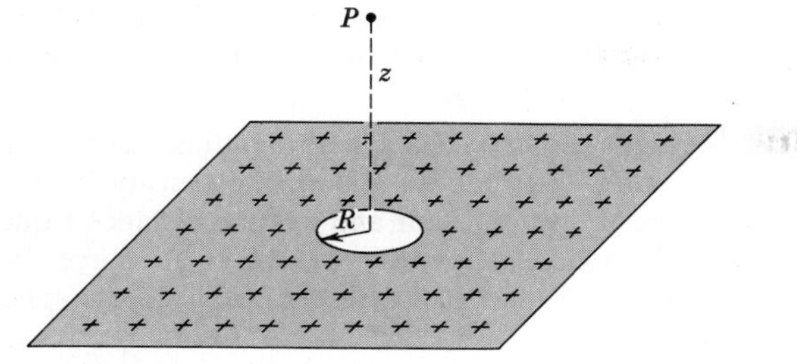

Figure 30 Exercise 32.

from a silk thread that makes an angle $\theta = 30°$ with a large, uniformly-charged nonconducting sheet as in Fig. 31. Calculate the uniform charge density σ for the sheet.

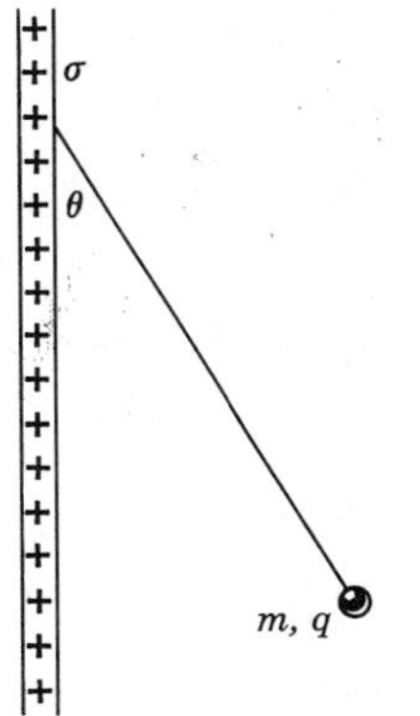

Figure 31 Problem 33.

34P. A 100-eV electron is fired directly toward a large metal plate that has a surface charge density of -2.0×10^{-6} C/m². From what distance must the electron be fired if it is to just fail to strike the plate?

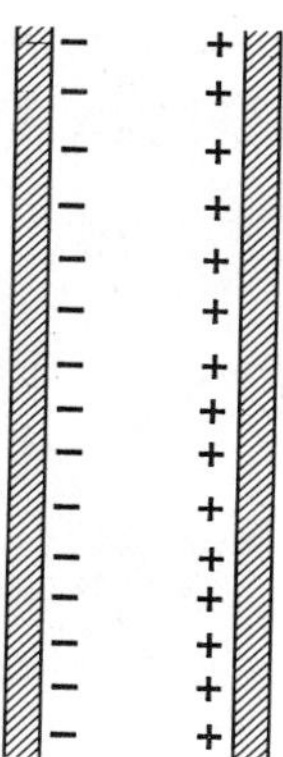

Figure 32 Problem 35.

35P. Two large metal plates face each other as in Fig. 32 and carry charges with surface charge density $+\sigma$ and $-\sigma$, respectively, on their inner surfaces. Find **E** at points (*a*) to the left of the sheets, (*b*) between them, and (*c*) to the right of the sheets. Consider only points not near the edges whose distances from the sheets are small compared to the dimensions of the sheet. (*Hint:* See Sample Problem 6.)

36P. Two large metal plates of area 1.0 m² face each other. They are 5.0 cm apart and carry equal and opposite charges on their inner surfaces. If E between the plates is 55 N/C, what is the charge on the plates? Neglect edge effects.

37P. An electron remains stationary in an electric field directed downward in the earth's gravitational field. If the electric field is due to charge on two large parallel conducting plates, oppositely charged and separated by 2.3 cm, what is the surface charge density, assumed to be uniform, on the plates?

38P.* A plane slab of thickness d has a uniform volume charge density ρ. Find the magnitude of the electric field at all points in space both (*a*) inside and (*b*) outside the slab, in terms of x, the distance measured from the median plane of the slab.

Section 25-11 Gauss' Law: Spherical Symmetry

39E. A conducting sphere of radius 10 cm carries an unknown net charge. If the electric field 15 cm from its center is 3.0×10^3 N/C and points radially inward, what is the net charge on the sphere?

40E. A point charge at the origin causes an electric flux of -750 N·m²/C to pass through a spherical Gaussian surface of 10-cm radius centered at the origin. (*a*) If the radius of the Gaussian surface is doubled, how much flux would then pass through the surface? (*b*) What is the value of the point charge?

41E. A thin-walled metal sphere has a radius of 25 cm and carries a charge of 2.0×10^{-7} C. Find E for a point (*a*) inside the sphere, (*b*) just outside the sphere, and (*c*) 3.0 m from the center of the sphere.

42E. Two charged concentric spheres have radii of 10 cm and 15 cm. The charge on the inner sphere is 4.0×10^{-8} C and that on the outer sphere is 2.0×10^{-8} C. Find the electric field (*a*) at $r = 12$ cm and (*b*) at $r = 20$ cm.

43E. A thin, metallic, spherical shell of radius a carries a charge q_a. Concentric with it is another thin, metallic, spherical shell of radius b ($b > a$) carrying a charge q_b. Find the electric field at radial points r where (*a*) $r < a$, (*b*) $a < r < b$, and (*c*) $r > b$. (*d*) Discuss the criterion one would use to determine how the charges are distributed on the inner and outer surfaces of each shell.

44E. In a 1911 paper, Ernest Rutherford said, "In order to form some idea of the forces required to deflect an α particle through a large angle, consider an atom containing a point positive charge Ze at its centre and surrounded by a distribution of negative electricity, $-Ze$ uniformly distributed within a sphere of radius R. The electric field E . . . at a distance r

from the center for a point *inside* the atom [is]

$$E = \frac{Ze}{4\pi\epsilon_0}\left(\frac{1}{r^2} - \frac{r}{R^3}\right)."$$

Verify this equation.

45P. An uncharged, spherical, thin, metallic shell has a point charge q at its center. Derive expressions for the electric field (*a*) inside the shell and (*b*) outside the shell, using Gauss' law. (*c*) Has the shell any effect on the field due to q? (*d*) Has the presence of q any effect on the shell? (*e*) If a second point charge is held outside the shell, does this outside charge experience a force? (*f*) Does the inside charge experience a force? (*g*) Is there a contradiction with Newton's third law here? Why or why not?

46P. Equation 12 ($E = \sigma/\epsilon_0$) gives the electric field at points near a charged conducting surface. Apply this equation to a conducting sphere of radius r, carrying a charge q on its surface, and show that the electric field outside the sphere is the same as the field of a point charge at the position of the sphere center.

47P. Figure 33 shows a charge $+q$ arranged as a uniform conducting sphere of radius a and placed at the center of a spherical conducting shell of inner radius b and outer radius c. The outer shell carries a charge of $-q$. Find $E(r)$ (*a*) within the sphere ($r < a$), (*b*) between the sphere and the shell ($a < r < b$), (*c*) inside the shell ($b < r < c$), and (*d*) outside the shell ($r > c$). (*e*) What charges appear on the inner and outer surfaces of the shell?

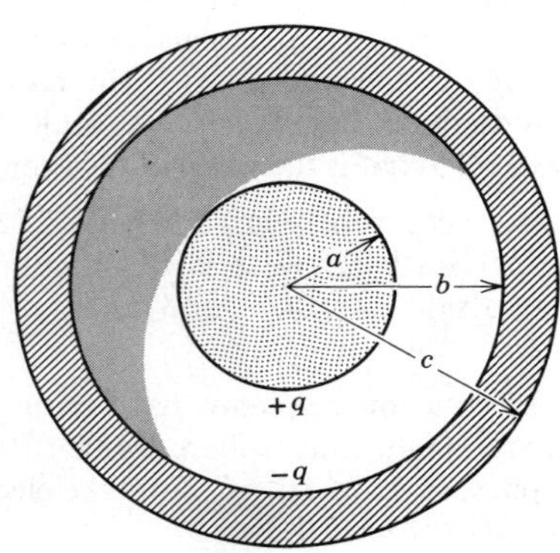

Figure 33 Problem 47.

48P. Figure 34 shows a spherical shell of charge of uniform density ρ. Plot E for distances r from the center of the shell ranging from zero to 30 cm. Assume that $\rho = 1.0 \times 10^{-6}$ C/m^3, $a = 10$ cm, and $b = 20$ cm.

49P. Figure 35 shows a point charge of 1.0×10^{-7} C at the center of a spherical cavity of radius 3.0 cm in a piece of metal. Use Gauss' law to find the electric field (*a*) at point P_1, halfway from the center to the surface and (*b*) at point P_2.

50P. A proton orbits with a speed $v = 3.0 \times 10^5$ m/s just outside a charged sphere of radius $r = 10^{-2}$ m. What is the charge on the sphere?

51P. A solid nonconducting sphere of radius R carries a non-uniform charge distribution, the charge density being $\rho = \rho_s r/R$, where ρ_s is a constant and r is the distance from the center of the sphere. Show that (*a*) the total charge on the sphere is $Q = \pi\rho_s R^3$ and (*b*) the electric field inside the sphere is given by

$$E = \frac{1}{4\pi\epsilon_0}\frac{Q}{R^4}r^2.$$

52P. The spherical region $a < r < b$ carries a charge per unit volume of $\rho = A/r$, where A is constant. At the center ($r = 0$) of the enclosed cavity is a point charge q. What should the value of A be so that the electric field in the region $a < r < b$ has constant magnitude?

53.* A spherical region carries a uniform charge per unit volume ρ. Let **r** be the vector from the center of the sphere to a general point P within the sphere. (*a*) Show that the electric field at P is given by $\mathbf{E} = \rho\mathbf{r}/3\epsilon_0$. (*b*) A spherical cavity is

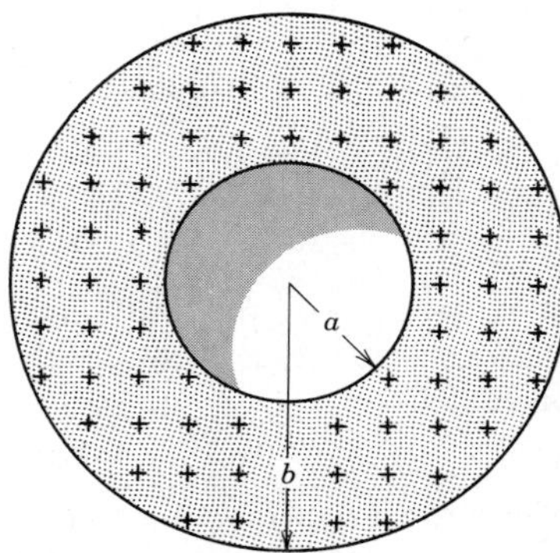

Figure 34 Problem 48.

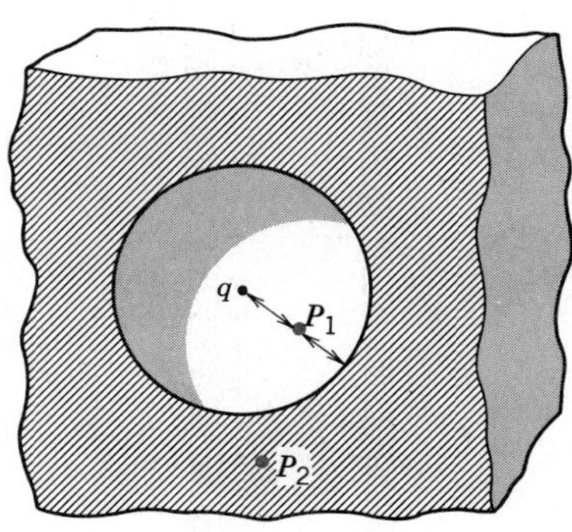

Figure 35 Problem 49.

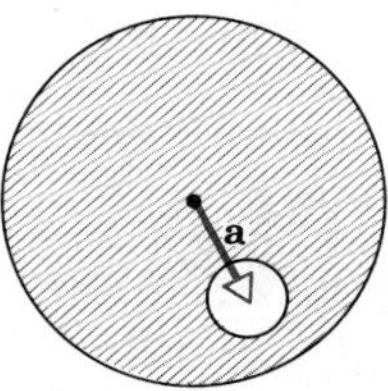

Figure 36 Problem 53.

created in the above sphere, as shown in Fig. 36. Using superposition concepts, show that the electric field at all points within the cavity is $\mathbf{E} = \rho\mathbf{a}/3\epsilon_0$ (uniform field), where **a** is the vector connecting the center of the sphere with the center of the cavity. Note that both these results are independent of the radii of the sphere and the cavity.

54P.* Show that stable equilibrium under the action of electrostatic forces alone is impossible. (*Hint:* Assume that at a certain point P in an electric field **E** a charge $+q$ would be in stable equilibrium if it were placed there. Draw a spherical Gaussian surface about P, imagine how **E** must point on this surface, and apply Gauss' law to show that the assumption leads to a contradiction.) This result is known as Earnshaw's theorem.

CHAPTER 26

ELECTRIC POTENTIAL

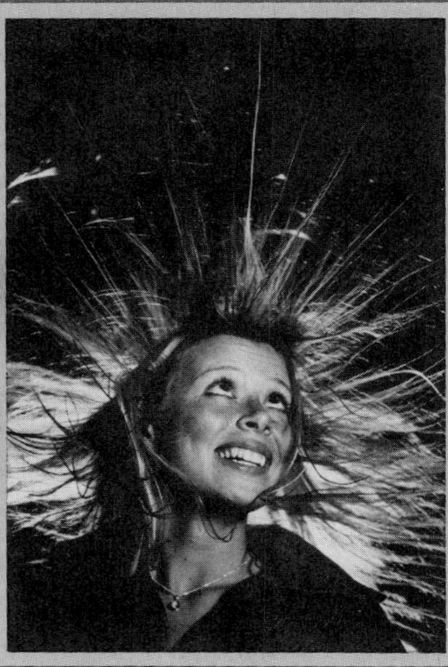

This young lady has been raised to an electric potential of possibly 50,000 volts above the potential of her surroundings. The strands of her hair suggest the beginnings of lines of force that reach out to the walls of the room. Although 120 volts can be lethal, she remains unharmed. Can you explain why?

26-1 Gravitation, Electrostatics, and Potential Energy

There is little point in solving for a second time a problem that you have already solved. That is why physicists are always on the lookout for areas of physics that—different as they may be in terms of the physical quantities involved—are expressed in the same mathematical framework. As we pointed out in Section 23-4, Newton's law of gravitation and Coulomb's law of electrostatics—both inverse square laws—are mathematically identical.* Thus, whatever you can deduce about gravitation by analyzing its basic law you can carry over with full confidence to electrostatics, and conversely. All that you need to do is change the symbols.

Figure 1*a* shows a test particle of mass m_0—perhaps a baseball—in free fall in the earth's gravitational field. We can analyze its motion from the point of view of energy and say that, as the baseball falls, its gravitational potential energy is transformed into kinetic energy.

Figure 1*b* shows the electrostatic parallel, a test charge q_0—perhaps a proton—in "free fall" in an electric field. Drawing on the mathematical identity of the basic laws, we can say with confidence that we must also be able to analyze the motion of the test charge in terms of energy transfers. In particular, we must be able to assign to the test charge an *electric potential energy U*,

* True, the gravitational force is always attractive and the electrostatic force may be either attractive or repulsive. However, we can easily keep this difference straight in our calculations.

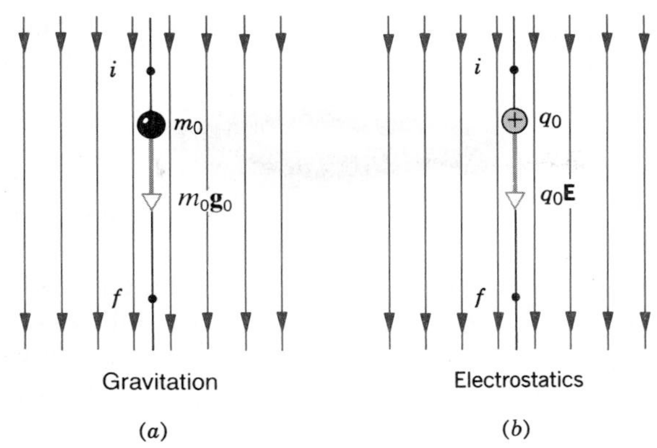

Figure 1 (*a*) A gravitational force $m_0\mathbf{g}_0$ acts on a freely falling test particle of mass m_0. (b) An electrostatic force $q_0\mathbf{E}$ acts on a "freely falling" test particle of charge q_0.

whose value for a given test charge depends only on the position of that charge in the electric field.

Let us review the gravitational case. If the test body in Fig. 1*a* moves from an initial point *i* to a final point *f*, we define the difference in its gravitational potential energy (see Section 8-4) to be

$$\Delta U = U_f - U_i = -W_{if}. \tag{1}$$

Here W_{if} is the work done by the gravitational force on the test body as it moves between these points. Although Fig. 1*a* shows a uniform gravitational field, the definition of Eq. 1 holds for any field configuration whatever, uniform or not. Other forces may act on the moving test body and may—or may not—do work on it. W_{if} in Eq. 1, however, includes only the work done by the gravitational force.

We proved in Sections 8-5 and 15-7 that the difference in the gravitational potential energy of a test particle as it moves between any two points is independent of the path taken between those points. We summarized this conclusion by labeling the gravitational force as a *conservative* force. If the gravitational force were not conservative, the whole notion of potential energy would fall apart.

We can take over Eq. 1 directly as a definition of the difference in the *electric potential energy* of a test charge q_0 as it moves from an initial point *i* to a final point *f* in an electric field:

> *Let a test charge move from one point to another in an electric field. The difference in the electric potential energy of the test charge between those points is the negative of the work done* by *the electric field* on *that charge during its motion.*

Again, the difference in the electric potential energy of a test charge between two points is independent of the path taken between those points. That is, the electrostatic force, like the gravitational field force, is a *conservative* force. We do not need to prove this because we have already done so in the gravitational case.

We now move from the definition of the difference in the electric potential energy of a test charge between two points to the definition of the electric potential energy of a test charge at a single point. To do so we make two decisions that are both arbitrary and independent of each other: (1) We take our initial point *i* to be a standard reference point whose location we specify, and (2) we assign an arbitrary value to the potential energy of the test charge at that point. Specifically, we choose to locate point *i* at a very large—strictly, an infinite—distance from all relevant charges and we assign the value zero to the potential energy of any test charge at such points. If we put $U_i = 0$ and $U_f = U$ in Eq. 1, we can write that equation as

$$U = -W_{\infty}. \tag{2}$$

In words, Eq. 2 can be stated as follows:

> *The potential energy U of a test charge q_0 at any point is equal to the negative of the work W_∞ done on the test charge by the electric field as that charge moves in from infinity to the point in question.*

Keep in mind that *potential energy differences* are fundamental and that *potential energy,* defined by Eq. 2, depends on defining a reference point and assigning a potential energy to it. Instead of choosing $U_i = 0$, we could have chosen $U_i = -137$ J, and it would not have changed the value of the potential energy difference of a given test charge between any pair of field points.

Recall that, in studying the motions of baseballs in the earth's gravity, we chose the earth's surface as the zero level of potential energy. In studying the motions of satellites, however, we found it more convenient to choose an infinite separation as the zero-potential energy configuration.

Sample Problem 1 A child's helium-filled balloon, carrying a charge $q = -5.5 \times 10^{-8}$ C, rises vertically into the air. The electric field that normally exists in the atmosphere near the surface of the earth has a magnitude $E = 150$ N/C and is directed downward. What is the difference in electric potential energy of the balloon between its release position (point i) and its position at an altitude $h = 520$ m (point f)?

The work done on the balloon by the electric field is

$$\begin{aligned} W_{if} &= \mathbf{F} \cdot \mathbf{h} = q\mathbf{E} \cdot \mathbf{h} = qE(\cos 180°)h = -qEh \\ &= -(-5.5 \times 10^{-8}\ \mathrm{C})(150\ \mathrm{N/C})(520\ \mathrm{m}) \\ &= +4.3 \times 10^{-3}\ \mathrm{J} = +4.3\ \mathrm{mJ}. \end{aligned}$$

Let us review the signs carefully. Although the electric field **E** points down, the electric force $q\mathbf{E}$ points up because the charge q on the balloon is negative. Thus, the electric force on the balloon acts in the direction the balloon is moving so that the work done by this force on the balloon is indeed positive, in agreement with our solution.

The difference in the electric potential energy of the balloon follows from Eq. 1, or

$$U_f - U_i = -W_{if} = -4.3\ \mathrm{mJ}. \qquad \text{(Answer)}$$

Can you show that this same answer holds even if the balloon drifts horizontally as it rises to altitude h?

Other forces also do work on the balloon, including the gravitational force, the atmospheric buoyant force, the atmospheric drag force, and possibly the wind force. All these forces are usually much stronger than the electric force but our concern in this problem is only with the electric force. The gravitational force, like the electric force, is conservative so that the balloon has potential energy of two kinds. The minus sign in our calculated result shows that the *electric* potential energy of the balloon decreases as it rises. On the other hand, the *gravitational* potential energy of the balloon increases as it rises.

26-2 The Electric Potential

As we have seen, the potential energy of a point charge in an electric field depends not only on the nature of the field but also on the magnitude of the charge. On the other hand, the potential energy *per unit charge* would have a unique value at any point in the field, independent of the magnitude of the test charge. We call this useful quantity the *electric potential* V (or simply the *potential*) at the point in question.*

We define the difference in potential between any two points as $\Delta U/q_0$ so that, guided by Eq. 1, we can write

$$\Delta V = V_f - V_i = -\frac{W_{if}}{q_0} \qquad \text{(potential difference defined).} \quad (3)$$

The work W_{if} done by the electric field on the positive test charge as it moves from point i to point f may be positive, negative, or zero. Correspondingly, because of the minus sign in Eq. 3, the potential at f will then be less than, greater than, or the same as the potential at i.

We can also regard Eq. 3 from another point of view. If we happen to know the potential difference ΔV between any two points, the work that *we* must do to transport a charge q_0 from one point to the other is given by $\Delta V\, q_0$.

Guided by Eq. 2, we define the potential at a point from

$$V = -\frac{W_\infty}{q_0} \qquad \text{(potential defined)} \qquad (4)$$

in which W_∞ is the work done by the electric field on the test charge as that charge moves in from infinity to the point in question.

Equation 4 tells us that the potential V at a point near an isolated positive charge is positive. To see this, imagine that you push a small positive test charge in from infinity to a point near an isolated positive charge. The electric force acting on the test charge will point away from the central charge, acting to repel the test charge. Thus, the work done *on* the test charge *by* the field force will be negative.* The minus sign in Eq. 4 assures that the potential at the point will be positive. Similarly, the potential for any point near an isolated negative charge will be negative.

The SI unit for potential that follows from Eq. 4 is the joule/coulomb. This combination occurs so often that a special unit, the *volt* (abbr. V) is used to represent it. That is,

$$1\ \text{volt} = 1\ \text{joule/coulomb}. \qquad (5)$$

The word "volt" is familiar, being associated with light

* In earlier chapters we define *gravitational potential energy*. We could also have defined *gravitational potential* but we did not do so.

* The work that the agent pushing the charge does is positive but the W in Eqs. 2–4 is the work done *by the electric field*, not the work done by the agent.

bulbs, electric appliances, electric outlets, and the batteries in your car or your portable stereo. If you touch the probes of a voltmeter to two points in an electric circuit, you are measuring the potential difference between those points. Potential, which is measured in volts, and potential energy, which is measured in joules, are quite different quantities and must not be confused.

We point out that this new unit allows us to adopt a more conventional unit for the electric field **E**, which we have measured up to now in newtons/coulomb. Thus,

$$1\ \mathrm{N/C} = \left(1\ \frac{\mathrm{N}}{\mathrm{C}}\right)\left(\frac{1\ \mathrm{V\cdot C}}{1\ \mathrm{J}}\right)\left(\frac{1\ \mathrm{J}}{1\ \mathrm{N\cdot m}}\right) = 1\ \mathrm{V/m}. \quad (6)$$

The conversion factor in the second set of parentheses is derived from Eq. 4; that in the third set of parentheses is derived from the definition of the joule. From now on, we shall report values of the electric field in volts/meter rather than in newtons/coulomb.

Finally, we are now in a position to define the electron-volt, the energy unit that we introduced in Section 7-2 as a convenient one for energy measurements in the atomic and subatomic domain.

> *One* electron-volt *(abbr. eV) is an energy equal to the work required to move a single elementary charge e, such as that carried by the electron or the proton, through a potential difference of 1 volt.*

Guided by Eq. 3, we can write (since 1 V = 1 J/C)

$$\begin{aligned} 1\ \mathrm{eV} &= e\,\Delta V \\ &= (1.60 \times 10^{-19}\ \mathrm{C})(1\ \mathrm{J/C}) = 1.60 \times 10^{-19}\ \mathrm{J}. \end{aligned}$$

As we pointed out earlier, multiples of this unit, such as the keV, MeV, and GeV, are in common use.

26-3 Equipotential Surfaces

The locus of points, all of which have the same potential, is called an *equipotential surface.* A family of equipotential surfaces, each surface corresponding to a different value of the potential, can be used to represent the electric field throughout a certain region. We have seen earlier that electric lines of force can also be used for this purpose. In later sections, we shall look into the intimate connection between these two equivalent ways of describing the electric field.

No net work is done by the electric field as a charge moves between any two points on the same equipotential surface. This follows from Eq. 3 ($V_f - V_i = -W_{if}/q_0$) because W_{if} must be zero if $V_f = V_i$. Because of the path independence, $W_{if} = 0$ for *any* path connecting points i and f, whether or not that path lies entirely on the equipotential surface.

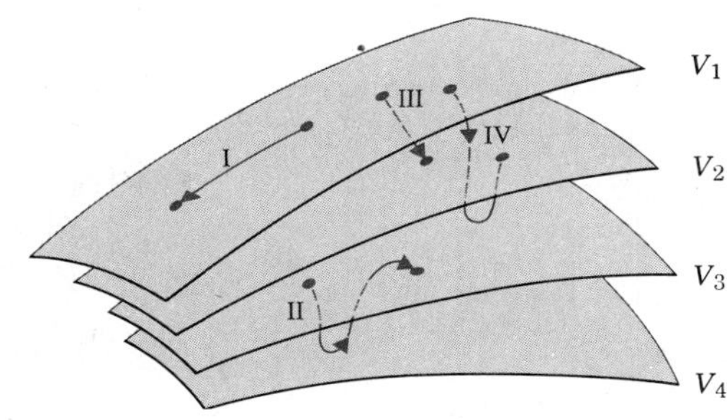

Figure 2 Portions of four equipotential surfaces. Four typical paths along which a test charge may move are also shown.

Figure 2 shows a family of equipotential surfaces, associated with a distribution of charges not shown. The work done by the electric field on a test charge as it moves from one end to the other of paths I and II is zero because each of these paths begins and ends on the same equipotential surface. The work done as a test charge moves from one end to the other of paths III and IV is not zero but has the same value for both of these paths because the initial and also the final potentials are identical for the two paths. Put another way, paths III and IV connect the same pair of equipotential surfaces.

From symmetry, the equipotential surfaces for a point charge or a spherically symmetric charge distribution are a family of concentric spheres. For a uniform field, they are a family of planes at right angles to the field. In all cases, including these two examples, the equipotential surfaces are at right angles to the lines of force and thus to **E**, which is tangent to these lines. If **E** were *not* at right angles to the equipotential surface, it would have a component lying in that surface. This component would then do work on a test charge as it moves about on the surface. But work cannot be done if the surface is to be truly an equipotential; the only conclusion is that **E** must be everywhere perpendicular to the surface. Figure 3 shows the lines of force and the equipotential surfaces for a uniform electric field and for the fields associated with a point charge and with an electric dipole.

Finally, we consider the case of the electrified young woman whose photo is shown on the first page of this chapter. Her safety is assured because all parts of her

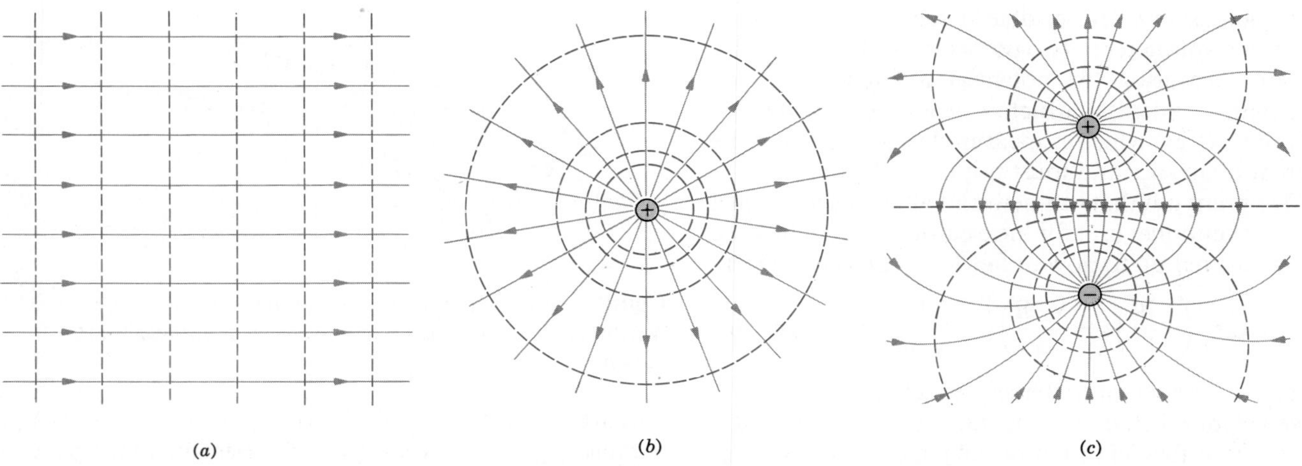

Figure 3 Equipotential surfaces (dashed lines) and lines of force (solid lines) for (*a*) a uniform field, (*b*) the field resulting from a point charge, and (*c*) the field resulting from an electric dipole.

body are raised to the same uniform potential. Damage is only done when a *potential difference* is maintained between two points on the body; in such cases, modest potential differences, such as 120 V or less, can easily be lethal.

26-4 Calculating the Potential from the Field

Here we show how to calculate the potential difference between any two points if you know the electric field **E** at all positions along some path that connects them.

In Fig. 4 a test charge q_0 moves from an initial point i to a final point f in an electric field, along the path shown. As the charge moves a distance $d\mathbf{s}$ along this path, the electric field does an element of work on it given by $(q_0\mathbf{E}) \cdot d\mathbf{s}$, in which $q_0\mathbf{E}$ is the force exerted by the field on the test charge. To find the total work W_{if} done by the field, we add up—that is, we integrate—the contributions to the work from all the differential segments into which the path is divided. Thus,

$$W_{if} = q_0 \int_i^f \mathbf{E} \cdot d\mathbf{s}. \tag{7}$$

Such an integral is called a *line integral*. If we substitute W_{if} from Eq. 7 into Eq. 3 we find

$$V_f - V_i = -\int_i^f \mathbf{E} \cdot d\mathbf{s}. \tag{8}$$

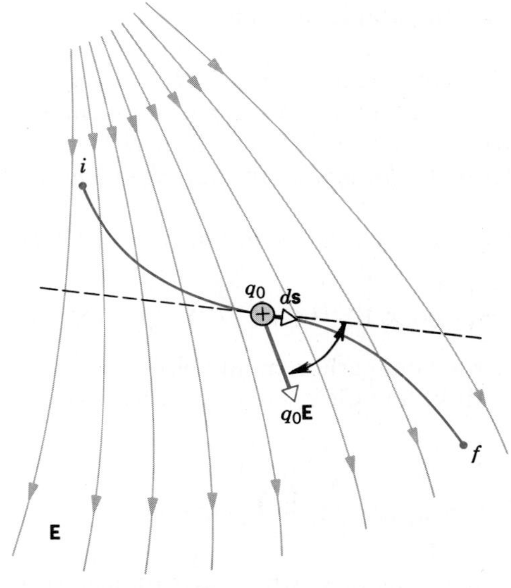

Figure 4 A test charge q_0 moves from point i to point f along the path shown in a nonuniform electric field.

If the initial point i is taken to be at infinity and the potential V_i at the point is set equal to zero, Eq. 8 becomes

$$V = -\int_i^f \mathbf{E} \cdot d\mathbf{s}. \tag{9}$$

If the electric field is known throughout a certain region, Eq. 8 allows us to calculate the difference in potential between any two points in the field. Because the electric force is conservative, all paths will yield the same result. Some paths, of course, are easier to use than others.

Sample Problem 2 Use Eq. 8 to calculate $V_f - V_i$ for the special case of Fig. 5, in which the electric field is uniform and in which the path connecting the initial and final points is a straight line parallel to the field direction.

As the test charge moves from i to f in Fig. 5, its path element $d\mathbf{s}$, which is always in the direction of motion, points down. The electric field $\mathbf{E}$ also points down so that the angle θ between these two vectors is zero. Equation 8 then becomes

$$V_f - V_i = -\int_i^f \mathbf{E} \cdot d\mathbf{s} = -\int_i^f E(\cos 0°)\, ds = -\int_i^f E\, ds.$$

E is constant over the path in this problem and can be removed from the integral, leaving

$$V_f - V_i = -E\int_i^f ds = -Ed, \qquad \text{(Answer)}$$

in which the integral is simply the length of the path. The minus sign shows that the potential at point f in Fig. 5 is lower than the potential at point i.

Sample Problem 3 In Fig. 6, let a test charge q_0 move from i to f over path icf. Calculate the potential difference $V_f - V_i$ for this path.

At all points along the line ic, $\mathbf{E}$ and $d\mathbf{s}$ are at right angles to each other. Thus, $\mathbf{E} \cdot d\mathbf{s} = 0$ everywhere along this part of the path. Equation 8 then tells us that points i and c are at the same potential. In other words, i and c lie on the same equipotential surface.

For path cf we have $\theta = 45°$ and, from Eq. 8,

$$V_f - V_i = -\int_c^f \mathbf{E} \cdot d\mathbf{s} = -\int_c^f E\,(\cos 45°)\, ds = -\frac{E}{\sqrt{2}}\int_c^f ds.$$

The integral in this equation is the length of the line cf, which is $\sqrt{2}d$. Thus,

$$V_f - V_i = -\frac{E}{\sqrt{2}}\sqrt{2}d = -Ed. \qquad \text{(Answer)}$$

This is the same result that we found in Sample Problem 2, as it must be because the potential difference between two points,

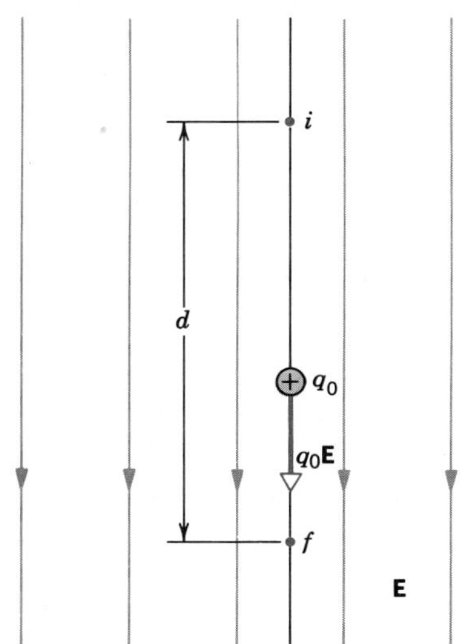

Figure 5 Sample Problem 2. A test charge q_0 moves in a straight line from point i to point f in a uniform electric field.

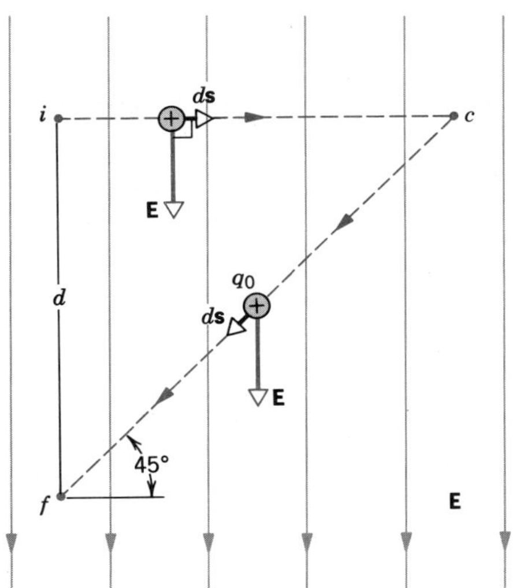

Figure 6 Sample Problem 3. A test charge q_0 moves from point i to point f along path icf in a uniform electric field.

such as i and f in Figs. 5 and 6, does not depend on the path connecting them.

26-5 Calculating the Potential: A Point Charge

Figure 7 shows an isolated positive point charge q. We wish to find the potential at point P, a radial distance r from that charge. Guided by Eq. 9, let us imagine that a test charge q_0 moves from infinity to point P. Because the path followed by the test charge does not matter, we make the simplest choice, a radial line from infinity to q, passing through P.

Let the test charge be at a position r', as Fig. 7 shows. The electric field $\mathbf{E}$ at the site of the test charge points in the direction of increasing r'. The displacement $d\mathbf{s}$ of the test charge is in the direction of decreasing r' (so that $ds = -dr'$) and we have

$$\mathbf{E} \cdot d\mathbf{s} = (E)(\cos 180°)(-dr') = E\, dr'. \quad (10)$$

Substituting this result into Eq. 9 gives us

$$V = -\int_i^f \mathbf{E} \cdot d\mathbf{s} = -\int_\infty^r E\, dr'. \quad (11)$$

The magnitude of the electric field at the site of the test charge is given by Eq. 6 of Chapter 24 as

$$E = \frac{1}{4\pi\epsilon_0} \frac{q}{r'^2}.$$

Substituting this result into Eq. 11 leads to

$$V = -\frac{q}{4\pi\epsilon_0} \int_\infty^r \frac{1}{r'^2}\, dr' = -\frac{q}{4\pi\epsilon_0} \left| -\frac{1}{r'} \right|_\infty^r$$

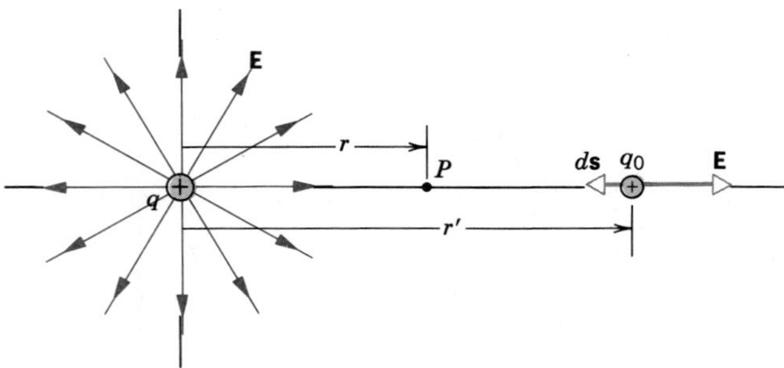

Figure 7 A test charge q_0 moves in from infinity along a radial line to point P. The field is that resulting from a positive point charge q.

or

$$\boxed{V = \frac{1}{4\pi\epsilon_0} \frac{q}{r}.} \quad (12)$$

We see that the sign of V is the same as the sign of q, confirming our earlier conclusion. Figure 8 shows computer-generated plots of Eq. 12 for a positive and a negative point charge.

To find the potential difference between any two points near an isolated point charge, all that is needed is to apply Eq. 12 to each point and subtract one potential from the other.

Sample Problem 4 What must be the magnitude of an isolated positive charge for which the potential V at a distance of 15 cm is +120 V?

Solving Eq. 12 for q yields

$$q = V4\pi\epsilon_0 r = (120\text{ V})(4\pi)(8.85 \times 10^{-12}\text{ C}^2/\text{N} \cdot \text{m}^2)(0.15\text{ m})$$
$$= 2.0 \times 10^{-9}\text{ C}. \quad \text{(Answer)}$$

A charge of this size can easily be produced on a balloon by rubbing it. In sorting out the units in this problem, it helps to recall from Eq. 5 that 1 V = 1 J/C = 1 N · m/C.

Sample Problem 5 What is the potential on the surface of a gold nucleus? The radius R of the nucleus is 6.2 fm and the atomic number Z of gold is 79.

The nucleus, assumed to be spherical, behaves for all outside points as if it were a point charge at the center. The charge q on the nucleus is Ze, where e is the elementary charge. We then have, from Eq. 12,

$$V = \frac{1}{4\pi\epsilon_0} \frac{q}{r} = \frac{1}{4\pi\epsilon_0} \frac{Ze}{R}$$
$$= \frac{(8.99 \times 10^9\text{ N} \cdot \text{m}^2/\text{C}^2)(79)(1.60 \times 10^{-19}\text{ C})}{6.2 \times 10^{-15}\text{ m}}$$
$$= 1.8 \times 10^7\text{ V} = 18\text{ MV}. \quad \text{(Answer)}$$

This large positive potential cannot be detected outside a gold object such as a coin because it is compensated by an equally large negative potential owing to the electrons in the coin.

26-6 Calculating the Potential: An Electric Dipole

We can find the potential for a group of point charges with the help of the superposition principle. We calcu-

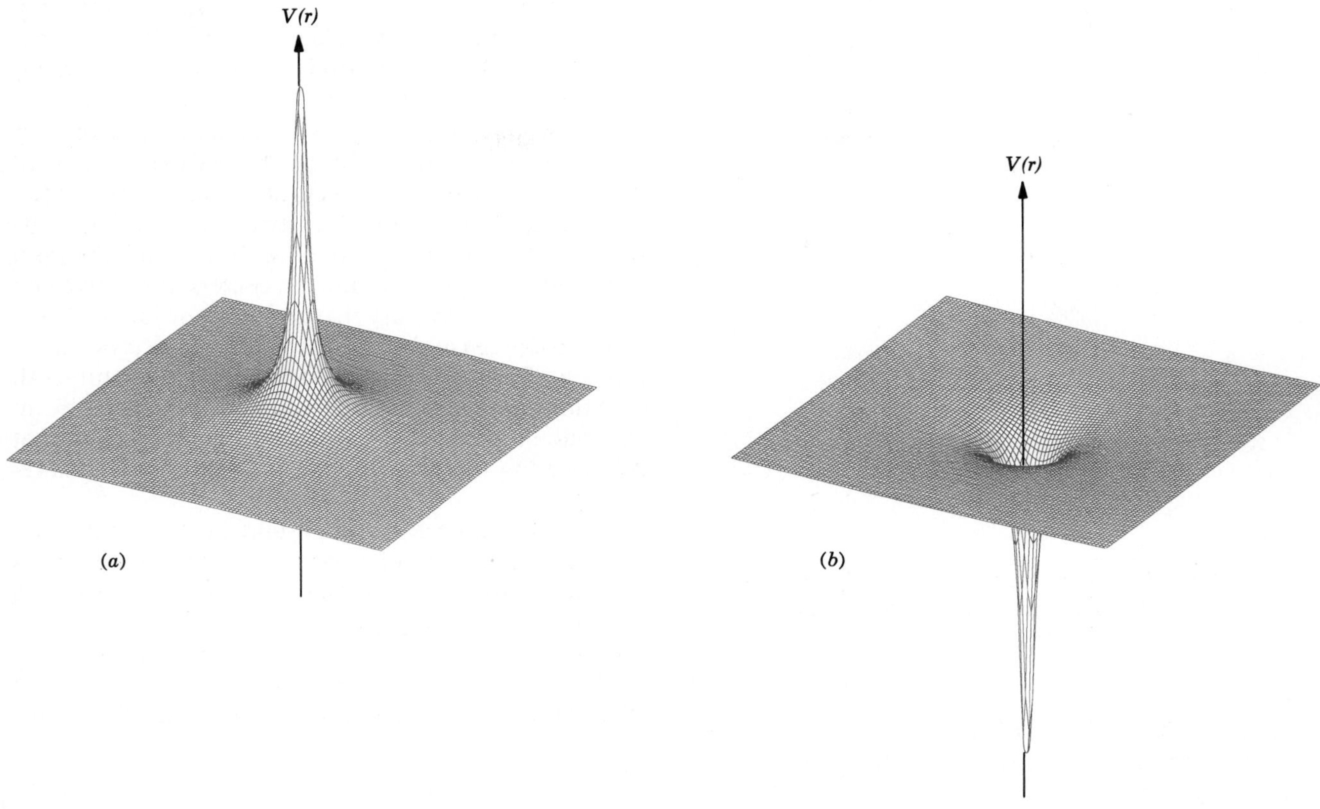

Figure 8 (*a*) A computer-generated plot of the potential $V(r)$ near a positive point charge. (*b*) The same for a negative point charge. The region close to such a charge is often described as a "potential well."

late the potential resulting from each charge at any given point separately, using Eq. 12. Then we add the potentials so calculated algebraically. In equation form, for n charges,

$$V = \sum_n V_n = \frac{1}{4\pi\epsilon_0} \sum \frac{q_n}{r_n}. \quad (13)$$

Here q_n is the value of the nth charge and r_n is the radial distance of the point in question from the nth charge. The sum in Eq. 13 is an algebraic sum and not a vector sum like the sum that would be used to calculate the electric field resulting from a group of point charges. Herein lies an important computational advantage of potential over electric field.

Now let us apply Eq. 13 to two equal charges of opposite sign that constitute an *electric dipole.* We first encountered the electric dipole in Section 24-5, where we calculated the value of the electric field for points along its axis.

A point P in the field of a dipole can be specified by giving the quantities r and θ in Fig. 9. From symmetry, the potential V_P will not change as the point P rotates about the z axis, r and θ being fixed. Thus, we need only find $V(r, \theta)$ for any plane containing this axis; the plane of Fig. 9 is such a plane. Equations 13 and 12 yield

$$V_P = \sum V_n = V_+ + V_- = \frac{1}{4\pi\epsilon_0}\left(\frac{q}{r_+} + \frac{-q}{r_-}\right)$$
$$= \frac{q}{4\pi\epsilon_0} \frac{r_- - r_+}{r_- r_+},$$

which is an exact relation.

Because naturally occurring dipoles—such as those possessed by many molecules—are small, we are usually interested only in points far from the dipole, such that

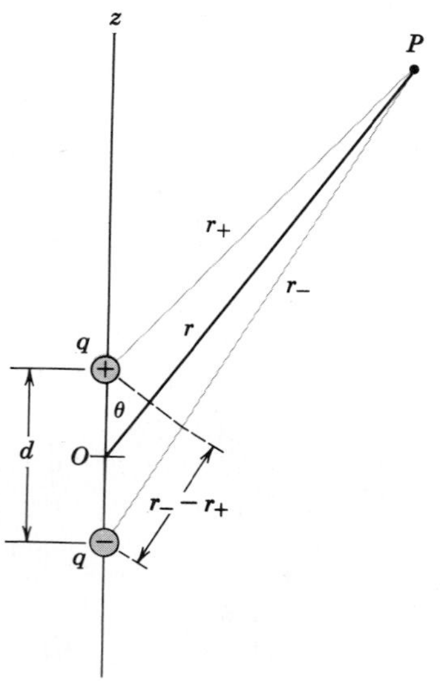

Figure 9 Point P is a distance r from the midpoint O of a dipole, the line OP making an angle θ with the dipole axis.

$r \gg d$, where d is the distance between the charges. Under these conditions, the approximations that follow from a study of Fig. 9 are

$$r_- - r_+ \approx d \cos\theta \quad \text{and} \quad r_- r_+ \approx r^2.$$

If we substitute these quantities into the above expression for V_p, we find

$$V_P \approx \frac{q}{4\pi\epsilon_0} \frac{d\cos\theta}{r^2},$$

or

$$\boxed{V_P = \frac{1}{4\pi\epsilon_0} \frac{p\cos\theta}{r^2},} \tag{14}$$

in which $p(= qd)$ was defined in Section 24-5 as the electric dipole moment. The quantity p is actually the magnitude of a vector $\mathbf{p}$, the direction of which is along the dipole axis, pointing from the negative to the positive charge.

Equation 14 shows that $V = 0$ everywhere in the equatorial plane of the dipole, defined by $\theta = 90°$. This reflects the fact that a test charge lying in this plane is always equidistant from the positive and the negative charges that make up the dipole so that the (scalar) potentials set up by each charge cancel each other. For a given distance, V_P has its greatest positive value for $\theta = 0$ and its greatest negative value for $\theta = 180°$. Note that the potential does not depend separately on q and d but only on their product.

As we saw in Section 24–5, many molecules have electric dipole moments. In the absence of a permanent dipole moment, one can always be induced by placing an atom or a molecule in an external electric field. The action of the field, as Fig. 10 shows, is to stretch the atom or molecule, thus separating the centers of positive and negative charge. We say that the atom or molecule becomes *polarized* and acquires an *induced* electric dipole moment. For solids whose properties are the same in all directions, an electric dipole moment $\mathbf{p}$ induced by an external electric field $\mathbf{E}$ points in the direction of that field (see Fig. 10) and disappears when that field is removed.

Electric dipoles are important in situations other than atomic or molecular ones. Radio and TV antennas are often in the form of a metal wire in which electrons surge back and forth periodically. At a certain time, one end of the wire will be negative and the other end positive. Half a cycle later, the polarity of the ends will be reversed. Such an antenna is an oscillating electric di-

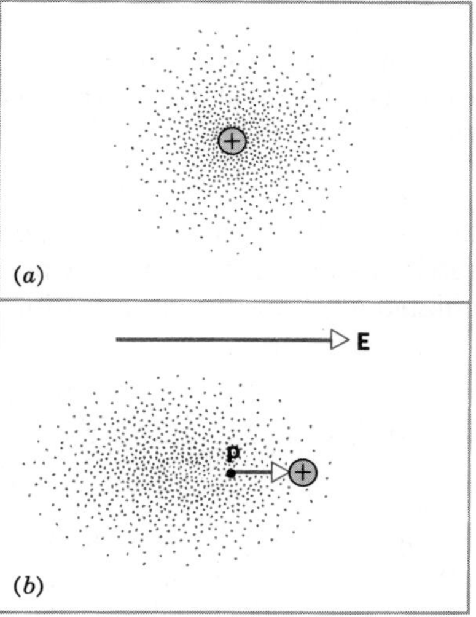

Figure 10 (*a*) An atom, showing the positively charged nucleus and the electron cloud. The centers of positive and of negative charge coincide. (*b*) If the atom is placed in an external electric field, the electron cloud is distorted so that the centers of positive and negative charge no longer coincide. An induced dipole moment appears. The distortion is greatly exaggerated here.

pole, so named because its electric dipole moment changes in a periodic way with time.

Sample Problem 6 What is the potential at point P, located at the center of the square of point charges shown in Fig. 11a? Assume that $d = 1.3$ m and that the charges are

$$q_1 = +12 \text{ nC}, \quad q_3 = +31 \text{ nC},$$
$$q_2 = -24 \text{ nC}, \quad q_4 = +17 \text{ nC}.$$

From Eq. 13 we have

$$V_P = \sum_n V_n = \frac{1}{4\pi\epsilon_0} \frac{q_1 + q_2 + q_3 + q_4}{R}.$$

The distance R of each charge from the center of the square is $d/\sqrt{2}$ or 0.919 m, so that

$$V_P = \frac{(8.99 \times 10^9 \text{ N} \cdot \text{m}^2/\text{C}^2)(12 - 24 + 31 + 17) \times 10^{-9} \text{ C}}{0.919 \text{ m}}$$

$$\approx 350 \text{ V}. \quad \text{(Answer)}$$

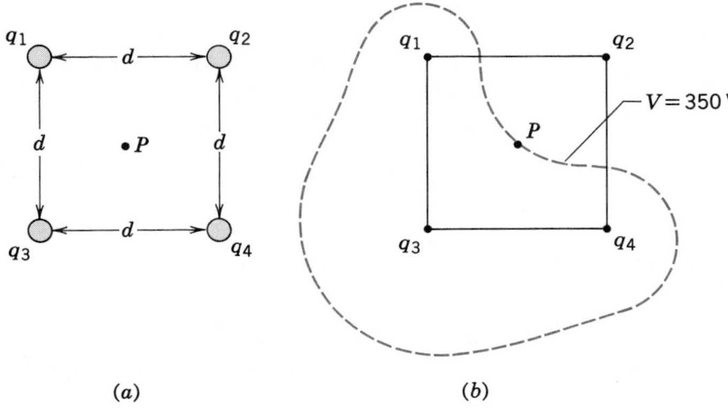

Figure 11 Sample Problem 6. (a) Four charges are held fixed at the corners of a square. What is the potential at P, the center of the square? (b) The curve is the intersection with the plane of the figure of the equipotential surface that contains point P.

Close to the three positive charges in Fig. 11a, the potential can assume very large positive values. Close to the single negative charge in that figure, the potential can assume arbitrarily large negative values. There must then be other points within the boundaries of the square that have the same potential as that at point P. Figure 11b shows the trace of a portion of the equipotential surface that contains this point.

26-7 Calculating the Potential: A Charged Disk

If the charge distribution is continuous, rather than an assembly of point charges, the sum in Eq. 13 must be replaced by an integral and we have

$$V = \int dV = \frac{1}{4\pi\epsilon_0} \int \frac{dq}{r}, \tag{15}$$

where dq is a differential element of the charge distribution and r is its distance from the point at which V is to be calculated. The integral is to be taken over the entire charge distribution.

In Section 24-7, we calculated the magnitude of the electric field for points on the axis of a plastic disk of radius R that is uniformly charged on one face with a surface charge density σ. Here we wish to derive an expression for $V(z)$, the potential for axial points. It will be instructive to compare these two derivations, one involving a vector integration and the other a scalar integration.

In Fig. 12, consider a charge element dq consisting of a flat circular strip of radius s and width ds. We have

$$dq = \sigma(2\pi s)(ds),$$

in which $(2\pi s)(ds)$ is the area of the strip. All parts of this charge element are the same distance r from the axial

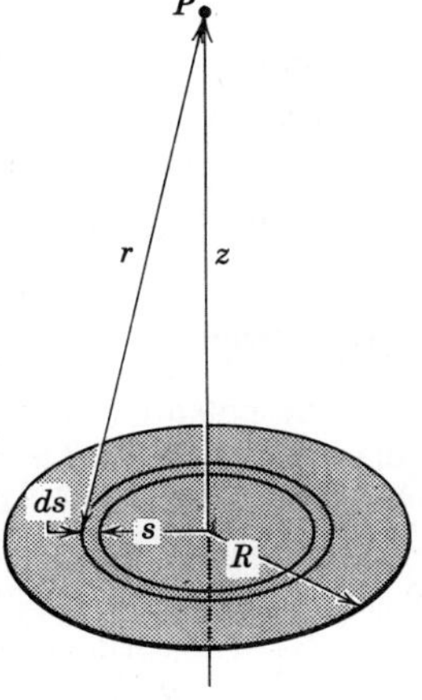

Figure 12 A plastic disk of radius R is charged on its front surface to a surface charge density σ. What is the potential at various points along its axis?

Table 1 Fields and Potentials for Some Charge Configurations

Configuration	*Field E*	*Equation*	*Potential V*	*Equation*
Point charge	$\frac{1}{4\pi\epsilon_0}\frac{q}{r^2}$	(24–6)	$\frac{1}{4\pi\epsilon_0}\frac{q}{r}$	(26–12)
Dipole[a]	$\frac{1}{2\pi\epsilon_0}\frac{p}{z^3}$	(24–11)	$\frac{1}{4\pi\epsilon_0}\frac{p\cos\theta}{r^2}$	(26–14)
Charged disk[b]	$\frac{\sigma}{2\epsilon_0}\left(1-\frac{z}{\sqrt{z^2+R^2}}\right)$	(24–21)	$\frac{\sigma}{2\epsilon_0}(\sqrt{z^2+R^2}-z)$	(26–17)
Infinite sheet	$\frac{\sigma}{2\epsilon_0}$	(25–15)	$V_0-\left(\frac{\sigma}{2\epsilon_0}\right)z$	(Exercise 6)
Insulated conductor	$E=0$, inside $E=\sigma/\epsilon_0$, at surface		$V=$ a constant, inside and on the surface	

[a] The field equation is for distant *axial* points. The potential equation is for *all* distant points.
[b] For axial points only. Note also that we assume $z \geq 0$ in both equations.

point P so that their contribution to the potential at P is given by Eq. 12, or

$$dV = \frac{1}{4\pi\epsilon_0}\frac{dq}{r} = \frac{1}{4\pi\epsilon_0}\frac{\sigma(2\pi s)(ds)}{\sqrt{z^2+s^2}}. \tag{16}$$

We find the potential at P by adding up the contributions of all the strips into which the disk can be divided, or

$$V_P = \int dV = \frac{\sigma}{2\epsilon_0}\int_0^R (z^2+s^2)^{-1/2}\, s\, ds$$

$$= \frac{\sigma}{2\epsilon_0}(\sqrt{z^2+R^2}-z). \tag{17}$$

Note that the variable in this integral is s and not z, which remains constant while the integration over the surface of the disk is carried out. (Note also that, in evaluating the integral, we have assumed that $z \geq 0$.)

Table 1 summarizes the electric field and the electric potential expressions that we have derived in this and previous chapters for various charge configurations.

Sample Problem 7 The potential at the center of a uniformly charged circular disk of radius R is 550 V. (a) What is the total charge on the disk? Take $R = 3.5$ cm.

At the center of the disk, z in Eq. 17 is zero, so that equation reduces to

$$V_0 = \frac{\sigma R}{2\epsilon_0}. \tag{18}$$

The total charge q is $\sigma(\pi R^2)$, where πR^2 is the area of the disk. If we solve Eq. 18 for σ and insert it in the expression for q, we find

$$q = \sigma(\pi R^2) = 2\pi\epsilon_0 R V_0$$
$$= (2\pi)(8.85\times10^{-12}\ \mathrm{C^2/N\cdot m^2})(0.035\ \mathrm{m})(550\ \mathrm{V})$$
$$= 1.1\times10^{-9}\ \mathrm{C} = 1.1\ \mathrm{nC}. \quad \text{(Answer)}$$

Again, it helps to note from Eq. 5 that 1 V = 1 J/C = 1 N · m/C.

(b) What is the potential at a point on the axis of the disk a distance z of 5.0 radii from the center of the disk?

If we put $z = 5R$ in Eq. 17, we find

$$V = \frac{\sigma}{2\epsilon_0}[\sqrt{(5R)^2+R^2} - 5R].$$

But, from Eq. 18, $\sigma = 2\epsilon_0 V_0/R$. Substituting this expression in the above yields

$$V = V_0(\sqrt{5^2+1^2}-5)$$
$$= (550\ \mathrm{V})(0.09902) = 54\ \mathrm{V}. \quad \text{(Answer)}$$

26–8 Calculating the Field from the Potential

In Section 26–4, we showed how to find the potential if you know the electric field. In this section, we propose to go the other way, that is, to show how to find the electric field if you know the potential. As Fig. 3 shows, we have already solved this problem graphically. If you know the potential V for all points near an assembly of charges, you can draw in a family of equipotential surfaces. The lines of force, sketched in at right angles to those surfaces, describe the variation of **E**. What we seek here is the

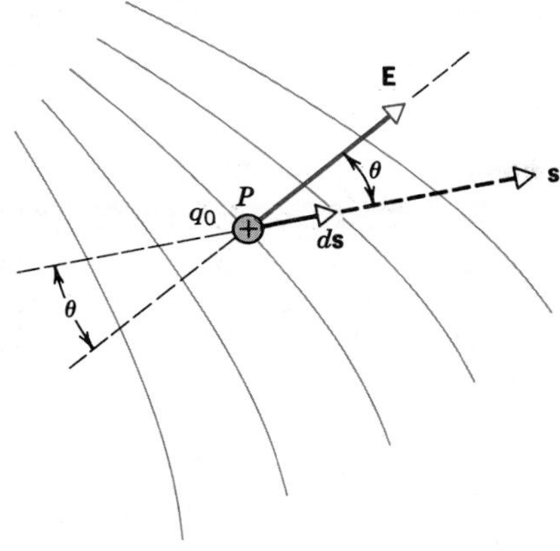

Figure 13 A test charge q_0 moves a distance ds from one equipotential surface to another. The displacement ds makes an angle θ with the direction of the electric field **E**.

mathematical equivalent of this graphical procedure.

Figure 13 shows the intersection with the plane of the page of a family of closely spaced equipotential surfaces, the potential difference between each pair of adjacent surfaces being dV. The figure shows that **E** at a typical point P is at right angles to the equipotential surface through P, as it must be.

Suppose that a test charge q_0 moves through a displacement $d\mathbf{s}$ from one equipotential surface to the adjacent surface. From Eq. 3, we see that the work that the electric field does on the test charge as it moves is $-q_0\,dV$. From another point of view, we can say that the work done by the electric field is $(q_0\mathbf{E}) \cdot d\mathbf{s}$ or $q_0E(\cos\theta)\,ds$. Equating these two expressions for the work yields

$$-q_0\,dV = q_0E(\cos\theta)ds$$

or

$$E\cos\theta = -\frac{dV}{ds}. \qquad (19)$$

But $E\cos\theta$ is the component of **E** in the **s** direction. Equation 19 then becomes*

$$\boxed{E_s = -\frac{\partial V}{\partial s}.} \qquad (20)$$

In words, Eq. 20 (which is essentially the inverse of Eq. 8) states:

The rate of change of the potential with distance in any direction is, when changed in sign, the component of **E** *in that direction.*

If we take the **s** direction in Eq. 20 to be, in turn, the directions of the x, y, and z axes, we can find the three components of **E** at any point from

$$\boxed{E_x = -\frac{\partial V}{\partial x};\quad E_y = -\frac{\partial V}{\partial y};\quad E_z = -\frac{\partial V}{\partial z}.} \qquad (21)$$

Thus, if we know V for all points in the region around a charge distribution, that is, if we know the function $V(x, y, z)$, we can find the components of **E**—and thus **E** itself—at any point by taking partial derivatives.

Sample Problem 8 The potential for points on the axis of a charged disk is given by Eq. 17, which we write as

$$V_z = \frac{\sigma}{2\epsilon_0}[(z^2 + R^2)^{1/2} - z].$$

Starting with this expression, derive an expression for the electric field at axial points.

From symmetry, **E** must lie along the axis of the disk. If we choose the **s** direction in Eq. 20 to be the z direction, we then have

$$E_z = -\frac{\partial V}{\partial z} = -\frac{\sigma}{2\epsilon_0}\frac{d}{dz}[(z^2 + R^2)^{1/2} - z]$$

$$= \frac{\sigma}{2\epsilon_0}\left(1 - \frac{z}{\sqrt{z^2 + R^2}}\right). \qquad \text{(Answer)}$$

This is the same expression that we derived in Section 24-7 by direct integration, using Coulomb's law; compare Eq. 21 of that chapter. Verify that other expressions for E listed in Table 1 can be derived from the corresponding expressions for V that are also listed in that Table.

26-9 Electric Potential Energy

In Section 26-1, we discussed the electric potential energy of a test charge as a function of its position in an electric field. In that section, we assumed that the charges that gave rise to the field were fixed in place, so that the field itself could not be influenced by the presence of the

* The partial derivative symbol indicates that we consider only the variation of V in one specified direction, the **s** direction.

test charge. In this section, we take a broader view of electric potential energy, applying it to the energy that a given charge configuration has because of the magnitude and position of its component charges.

For a simple example, if you pull apart two bodies that carry charges of opposite signs, the work that you must do will be stored as electric potential energy in the two-charge system. If you release the charges, you can recover this stored energy, in whole or in part, as kinetic energy of the charged bodies as they rush toward each other. We define the electric potential energy of a system of point charges, held in fixed positions by forces not specified, as follows:

The electric potential energy of a system of fixed point charges is equal to the work that must be done by an external agent to assemble the system, bringing each charge in from an infinite distance.

We assume that the charges are at rest in their initial infinitely distant positions and also in their final configuration.

Figure 14 shows two point charges, separated by a distance r_{12}. Let us imagine q_2 removed to infinity and at rest. The potential V at the original site of q_2, caused by q_1, is given by Eq. 12 as

$$V = \frac{1}{4\pi\epsilon_0}\frac{q_1}{r_{12}}.$$

As q_2 is moved in from infinity to its original position, the work that an external agent must do is Vq_2. Thus, from our definition, the electric potential energy of the two-charge system of Fig. 14 is

$$\boxed{U = W = \frac{1}{4\pi\epsilon_0}\frac{q_1 q_2}{r_{12}}.} \qquad (22)$$

If the charges have the same sign, an external agent would have to do positive work to push them together against their mutual repulsion. Hence, as Eq. 22 shows, their mutual potential energy will be positive. If the charges have opposite signs, the external agent will have to do negative work to restrain them against their mutual attraction or to bring them to rest at their final positions and the mutual potential energy of the two charges will be negative.

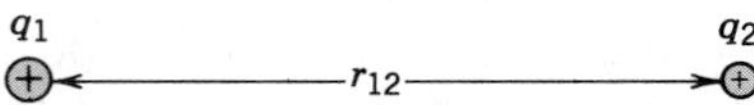

Figure 14 Two charges are held a fixed distance r_{12} apart. What is the electric potential energy of the configuration?

Sample Problem 9 Two protons in the nucleus of a uranium-238 atom are 6.0 fm apart. What is the potential energy associated with the repulsive electric force that acts between these two particles?

From Eq. 22, with $q_1 = q_2 = q$,

$$U = \frac{1}{4\pi\epsilon_0}\frac{q^2}{r} = \frac{(8.99 \times 10^9\ \text{N}\cdot\text{m}^2/\text{C}^2)(1.60 \times 10^{-19}\ \text{C})^2}{6.0 \times 10^{-15}\ \text{m}}$$
$$= 3.8 \times 10^{-14}\ \text{J} = 2.4 \times 10^5\ \text{eV} = 240\ \text{keV}. \qquad \text{(Answer)}$$

The two protons do not fly apart because they are held together by the attractive *strong force* that serves to bind the nucleus together. There is a potential energy associated with this force but it is not an electric potential energy.

Sample Problem 10 Figure 15 shows three charges held in fixed positions by forces that are not shown. What is the electric potential energy of this system of charges? Assume that $d = 12$ cm and that

$$q_1 = +q, \quad q_2 = -4q, \quad \text{and} \quad q_3 = +2q,$$

where $q = 150$ nC.

Imagine that q_1 alone is in place and that you bring up q_2 from infinity. From Eq. 22, the work W_{12} that you must do is U_{12}, where

$$W_{12} = U_{12} = \frac{1}{4\pi\epsilon_0}\frac{q_1 q_2}{d}.$$

If you then bring up charge q_3 and put it in place, the additional work that you must do is $U_{13} + U_{23}$.

We see that the total work that you must do to assemble the three charges is just the sum of the potential energies of the

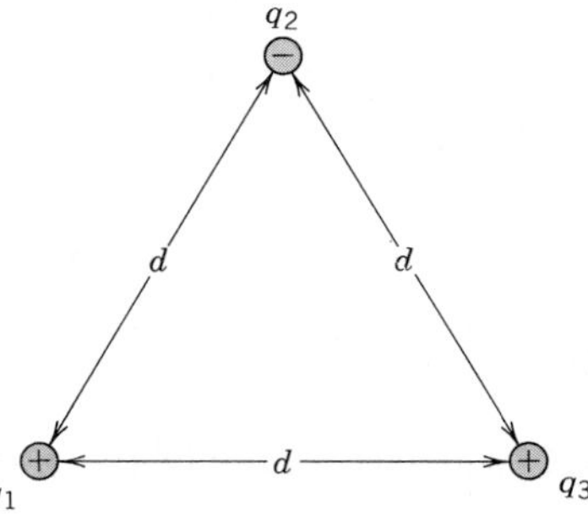

Figure 15 Sample Problem 10. Three charges are held fixed at the vertices of a triangle. What is the electric potential energy of the configuration?

three pairs of charges and is independent of the order in which the charges are brought together. Thus,

$$\begin{aligned} U &= U_{12} + U_{13} + U_{23} \\ &= \frac{1}{4\pi\epsilon_0}\left(\frac{(+q)(-4q)}{d} + \frac{(+q)(+2q)}{d} + \frac{(-4q)(+2q)}{d}\right) \\ &= -\frac{10q^2}{4\pi\epsilon_0 d} \\ &= -\frac{(8.99 \times 10^9\ \mathrm{N \cdot m^2/C^2})(10)(150 \times 10^{-9}\ \mathrm{C})^2}{0.12\ \mathrm{m}} \\ &= -1.7 \times 10^{-2}\ \mathrm{J} = -17\ \mathrm{mJ}. \qquad \text{(Answer)} \end{aligned}$$

The fact that the potential energy in this case is negative means that negative work would have to be done to assemble this structure, starting with the three charges infinitely separated and at rest. Put another way, an external agent would have to do 17 mJ of work to dismantle the structure completely.

26-10 An Insulated Conductor

In Section 25-7, we concluded that $\mathbf{E} = 0$ for all points inside an insulated conductor and we then used Gauss' law to prove the following:

*Once equilibrium has been established, an excess charge placed on an insulated conductor will be found to lie entirely on its surface. This remains true even if the conductor has an empty internal cavity.**

Here we use the fact that $\mathbf{E} = 0$ for all points inside an insulated conductor to prove another fact about such conductors, namely:

An excess charge placed on an insulated conductor will distribute itself on the surface of that conductor so that all points of the conductor—whether on the surface or inside—come to the same potential. This remains true whether or not the conductor has an internal cavity.

Our proof follows directly from Eq. 8, which is

$$V_f - V_i = -\int_i^f \mathbf{E} \cdot d\mathbf{s}.$$

* If the cavity encloses an insulated charge, then some of the conductor's charge will be found on its inner surface as well as on its outer surface.

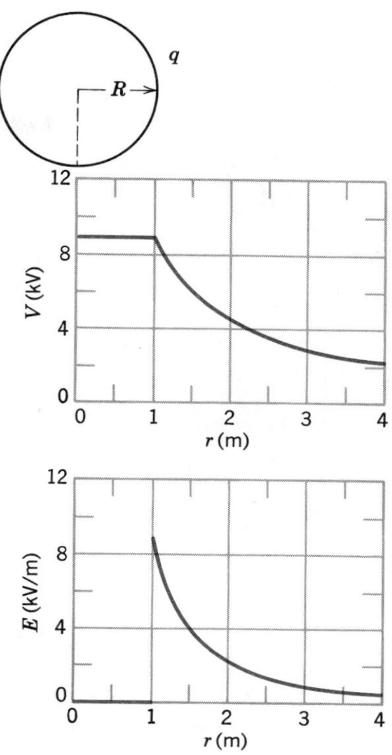

Figure 16 (*a*) A plot of $V(r)$ for a charged spherical shell. (*b*) A plot of $E(r)$ for the same shell.

Since $\mathbf{E} = 0$ for all points within a conductor, it follows directly that $V_f = V_i$ for all possible pairs of points. We might also note that, because the surface of the conductor is an equipotential surface, $\mathbf{E}$ for points on the surface must be at right angles to the surface, as we saw in Chapter 25.

Figure 16*a* is a plot of potential against radial distance for an insulated spherical conducting shell of 1.0-m radius, carrying a charge of 1.0 μC. For points outside the shell, we can calculate $V(r)$ from Eq. 12 because the charge q behaves for such external points as if it were concentrated at the center of the shell. This equation holds right up to the surface of the shell. Now let us push a small test charge right through the shell—assuming a small hole exists—to its center. No extra work is needed because no electric forces act on the test charge once it is inside the shell. Thus, the potential for all points inside the shell has the same value as that on the surface, as Fig. 16*a* shows.

Figure 16*b* shows the variation of electric field with radial distance for the same shell. Note that $E = 0$ every-

Figure 17 A lightning strike has burned the grass on the green of a golf course.

Figure 18 The car is a safe haven in a thunderstorm. Its safety owes more to Gauss' law than to the insulating ability of its rubber tires.

where inside. The lower of these two curves can be derived from the upper by differentiating with respect to r, using Eq. 20; the derivative of a constant, for example, is zero. The upper curve can be derived from the lower by integrating with respect to r, using Eq. 9. The negative of the integral of $1/r^2$, for example, is $1/r$.

Except for spherical conductors, the surface charge does not distribute itself uniformly over the surface of a conductor.* At sharp points or edges, the surface charge density—and thus the external electric field, which is proportional to it—may reach very high values. The air around such sharp points may become ionized, producing the corona discharge that golfers and mountaineers may experience when thunderstorms threaten. Such corona discharges are often the precursors of lightning strikes; see Fig. 17. In such circumstances, it is wise to enclose oneself in a cavity inside a conducting shell, where the electric field is guaranteed to be zero; Fig. 18 shows a suitable mobile arrangement.

If an insulated conductor is placed in an *external electric field,* as in Fig. 19, all points of the conductor still come to a single potential and they do so whether or not the conductor carries an excess charge. The conduction electrons distribute themselves on the surface in such a way that the electric field they produce at interior points cancel the external electric field that would otherwise be present at those points. If the surface charges in Fig. 19 could be somehow frozen in place and the conductor removed, the pattern of the electric field would remain absolutely unchanged, for both exterior and interior points.

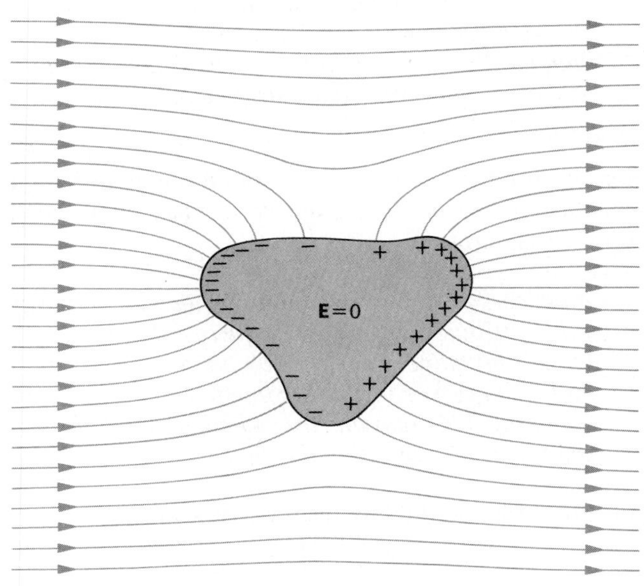

Figure 19 An uncharged conductor is suspended in an external electric field. The conduction electrons distribute themselves on the surface as shown, reducing the electric field inside the conductor to zero.

* See "The Lightning-rod Fallacy," by Richard H. Price and Ronald J. Crowley, *American Journal of Physics,* September 1985, p. 843 for a careful discussion.

26-11 The Van de Graaff Accelerator

The heart of a Van de Graaff accelerator* is an arrangement for generating potential differences of the order of several million volts. By allowing charged particles such as electrons or protons to "fall" through this potential difference, a beam of energetic particles can be produced. In medicine, such beams are widely used in the management of malignancies. In physics, accelerated particle beams can be used in a variety of "atom-smashing" experiments. Figure 20 shows such an accelerator in a nuclear physics laboratory at Purdue University.

Figure 20 A view of a nuclear physics laboratory at Purdue University. The particle beam emerges from the Van de Graaff accelerator, which is housed in the large tank at the rear. The beam, bent through 90° by the magnet in the foreground, enters the experimental area beyond the wall to the left.

Figure 21 suggests how the high potential is generated in a Van de Graaff accelerator. A small conducting shell of radius r is located inside a larger shell of radius R. The two shells carry charges q and Q, respectively. If the inner shell is connected to the outer shell by a conducting path, the two shells then form a single insulated conductor. The charge q then moves *entirely* to the outer surface of the large shell, no matter how much charge there may already be on that shell. Every such charge transfer increases the potential of the outer shell.

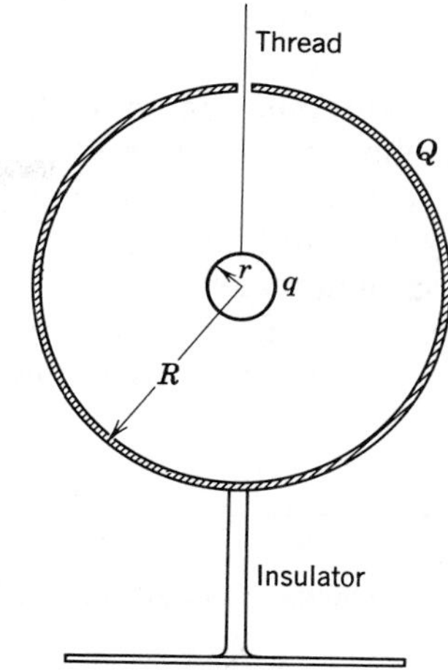

Figure 21 Illustrating the operating principle of a Van de Graaff accelerator.

In practice, charge is carried into the inner shell by a rapidly moving charged belt. Charge is "sprayed" onto the belt outside the machine by a comb of "corona points" and removed from the belt inside the machine in the same way. The motor driving the belt provides the energy needed to move the charge to the higher potential of the shells. The maximum potential that may be achieved with a given accelerator occurs when the rate at which charge is being carried into the inner shell is equal to the rate at which charge leaves the outer shell by leakage along the supports and by corona discharge.

A Clever Idea. Figure 22 shows an arrangement by which a particle can be accelerated *twice* by the same accelerating potential, thus emerging from the machine with double the kinetic energy that it otherwise would have. Let the high-potential terminal, located in the center of the device, be at a positive potential V. If the

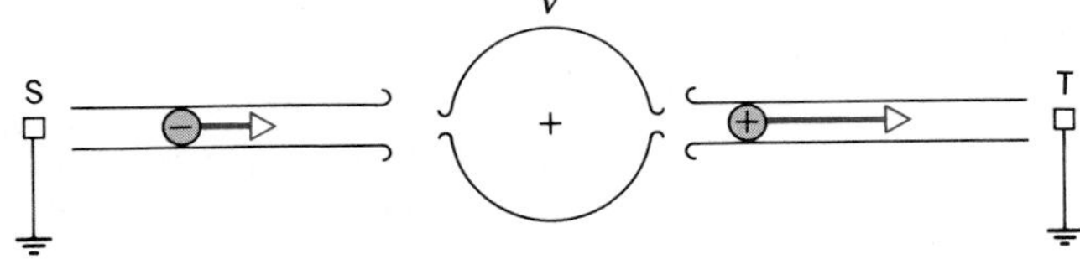

Figure 22 A schematic diagram of an energy-doubling arrangement. Particles leave the source S at low speed, are accelerated twice by the potential difference V maintained in the center of the device, and strike the target T at high speed. The sign of the net charge of the particles changes as they pass through a thin foil in the central terminal.

* So called after Robert J. Van de Graaff, who first put a suggestion by Lord Kelvin into useful practice. See "The Biggest Van de Graaff Machine," by Joe Watson, *The New Scientist,* March 1974. The original Van de Graaff machine, used as a "lightning generator," is on display at the Museum of Science in Boston.

particle approaching from the left carries a charge of $-e$, it will be *attracted* to the terminal, gaining a kinetic energy Ve. If the particle could somehow change its charge to $+e$ as it passes through the central terminal, it would then be *repelled* by this terminal as it emerged from it on the right, gaining an additional Ve of kinetic energy at the time of its arrival at target T.

This "charge switching" can be done! The particle approaching from the left is a hydrogen atom carrying an excess electron, so that its net charge is indeed $-e$. Inside the terminal, the beam is made to pass through a thin "stripping foil" in which its two outer electrons are stripped off by grazing collisions in the foil. The beam then emerges from the terminal on the right as a single proton, with a charge of $+e$.*

* In the early days of its development, an accelerator based on this principle was affectionately called a "Swindletron."

REVIEW AND SUMMARY

Electric Potential Energy

The change ΔU in the electric potential energy U of a charged object as it moves from an initial point i to a final point f is

$$\Delta U = U_f - U_i = -W_{if}. \quad [1]$$

Work W_{if} is that done by the electric field. For zero potential energy at infinity, the *electric potential energy* U of the body *at a point* is

$$U = -W_\infty. \quad [2]$$

W_∞ is the work done by the electric field if the object were to move from infinity to the point. See Sample Problem 1.

Electric Potential Difference

We define the *potential difference* ΔV between two points in an electric field as

$$\Delta V = V_f - V_i = -\frac{W_{if}}{q_0} \quad \text{(potential difference defined)}, \quad [3]$$

q_0 being a test charge on which work is done by the field. The *potential* at a point is

Electric Potential V

$$V = -\frac{W_\infty}{q_0} \quad \text{(potential defined)}. \quad [4]$$

The SI unit of potential is the *volt:* 1 volt = 1 joule/coulomb.

Equipotential Surfaces

The points on an *equipotential surface* all have the same potential. The work done on a test charge moving from one such surface to another is independent of the locations of the initial and terminal points on these surfaces and of the path that joins them; see Fig. 2. The electric field **E** is always at right angles to equipotential surfaces; see Fig. 3.

The potential difference between any two points is

Finding V from **E**

$$V_f - V_i = -\int_i^f \mathbf{E} \cdot d\mathbf{s}. \quad [8]$$

The line integral is taken over any path connecting the points. For i at infinity we have, for the potential at a particular point,

$$V = -\int_\infty^f \mathbf{E} \cdot d\mathbf{s}. \quad [9]$$

See Sample Problems 2 and 3.

The potential due to a single point charge is

Potential Due to Point Charges

$$V = \frac{1}{4\pi\epsilon_0}\frac{q}{r}. \quad [12]$$

The potential due to a collection of point charges is

$$V = \sum_n V_n = \frac{1}{4\pi\epsilon_0} \sum \frac{q_n}{r_n}. \qquad [13]$$

(See Sample Problems 4–6.)

Electric Dipole Potential

The potential due to an electric dipole with dipole moment $p = qd$ is

$$V(r, \theta) = \frac{1}{4\pi\epsilon_0} \frac{p \cos\theta}{r^2} \qquad [14]$$

for $r \gg d$; r and θ are defined in Fig. 9.

Continuous Charge

For a continuous distribution of charge, Eq. 13 becomes

$$V = \frac{1}{4\pi\epsilon_0} \int \frac{dq}{r}. \qquad [15]$$

See Section 26–7, in which we show the potential at a distance z from the center of a charged disk along its axis to be

$$V = \frac{\sigma}{2\epsilon_0} (\sqrt{z^2 + R^2} - z), \qquad [17]$$

and Sample Problem 7.

Calculating **E** from V

Any component of **E** may be found from V by differentiation. The rate of change of potential with distance in any direction is, when changed in sign, the component of **E** in that direction:

$$E_s = -\frac{\partial V}{\partial s}. \qquad [20]$$

The Cartesian components of **E** may be found from

$$E_x = -\frac{\partial V}{\partial x}; \quad E_y = -\frac{\partial V}{\partial y}; \quad E_z = -\frac{\partial V}{\partial z}. \qquad [21]$$

See Sample Problem 8 and Table 1 for examples.

Electric Potential Energy

The electric potential energy of a system of point charges is the work needed to assemble the system with the charges initially at rest and infinitely distant from each other. For two charges,

$$U = W = \frac{1}{4\pi\epsilon_0} \frac{q_1 q_2}{r_{12}}. \qquad [22]$$

Sample Problems 9 and 10 show applications.

A Charged Conductor

An excess charge placed on a conductor will, in equilibrium, be on its outer surface. The charge brings the entire conductor, including both surface and interior points, to a uniform potential. Figure 16 shows $V(r)$ and $E(r)$ for a charged conducting sphere or spherical shell. An electrostatic generator is an important application. Figure 21 illustrates its working principle. If a connection is made between the two conductors in that figure the charge on the inner sphere will move *entirely* to the outer shell, no matter what charge is already on that shell.

QUESTIONS

1. Are we free to call the potential of the earth +100 V instead of zero? What effect would such an assumption have on measured values of (*a*) potentials and (*b*) potential differences?

2. What would happen to you if you were on an insulated stand and your potential was increased by 10 kV with respect to the earth?

3. Why is the electron volt often a more convenient unit of energy than the joule?

4. How would a proton-volt compare with an electron-volt? The mass of a proton is 1840 times that of an electron.

5. Do electrons tend to go to regions of high potential or of low potential?

6. Why is it possible to shield a room against electrical forces but not against gravitational forces?

7. Suppose that the earth has a net charge that is not zero. Why is it still possible to adopt the earth as a standard reference point of potential and to assign the potential $V = 0$ to it?

8. Does the potential of a positively-charged insulated conductor have to be positive? Give an example to prove your point.

9. Can two different equipotential surfaces intersect?

10. An electrical worker was accidentally electrocuted and a newspaper account reported: "He accidentally touched a high-voltage cable and 20,000 V of electricity surged through his body." Criticize this statement.

11. Advice to mountaineers caught in lightning and thunderstorms is (*a*) get rapidly off peaks and ridges and (*b*) put both feet together and crouch in the open, only the feet touching the ground. What is the basis for this good advice?

12. If **E** equals zero at a given point, must V equal zero for that point? Give some examples to prove your answer.

13. If you know **E** only at a given point, can you calculate V at that point? If not, what further information do you need?

14. In Fig. 2, is the electric field E greater at the left or at the right side of the figure?

15. Is the uniformly-charged, nonconducting disk of Section 26–7 a surface of constant potential? Explain.

16. We have seen that, inside a hollow conductor, you are shielded from the fields of outside charges. If you are *outside* a hollow conductor that contains charges, are you shielded from the fields of these charges? Explain why or why not.

17. Distinguish between potential difference and difference of potential energy. Give examples of statements in which each term is used properly.

18. If the surface of a charged conductor is an equipotential, does that mean that charge is distributed uniformly over that surface? If the electric field is constant in magnitude over the surface of a charged conductor, does *that* mean that the charge is distributed uniformly?

19. In Section 26–10 we learned that charge delivered to the *inside* of an isolated conductor is transferred *entirely* to the outer surface of the conductor, no matter how much charge is already there. Can you keep this up forever? If not, what stops you?

20. Ions and electrons act like condensation centers; water droplets form around them in air. Explain why.

21. If V equals a constant throughout a given region of space, what can you say about **E** in that region?

22. In Chapter 15 we saw that the gravitational field strength is zero inside a spherical shell of matter. The electrical field strength is zero not only inside an isolated charged spherical conductor but inside an isolated conductor of any shape. Is the gravitational field strength inside, say, a cubical shell of matter zero? If not, in what respect is the analogy not complete?

23. How can you ensure that the electric potential in a given region of space will have a constant value?

24. Devise an arrangement of three point charges, separated by finite distances, that has zero electric potential energy.

25. We have seen (Section 26–10) that the potential inside a conductor is the same as that on its surface. (*a*) What if the conductor is irregularly shaped and has an irregularly shaped cavity inside? (*b*) What if the cavity has a small "worm hole" connecting it to the outside? (*c*) What if the cavity is closed but has a point charge suspended within it? Discuss the potential within the conducting material and at different points within the cavities.

26. An isolated conducting spherical shell carries a negative charge. What will happen if a positively charged metal object is placed in contact with the shell interior? Discuss the three cases in which the positive charge is (*a*) less than, (*b*) equal to, and (*c*) greater than the negative charge in magnitude.

EXERCISES AND PROBLEMS

Section 26–2 The Electric Potential

1E. The electric potential difference between discharge points during a particular thunderstorm is 1.2×10^9 V. What is the magnitude of the change in the electrical potential energy of an electron that moves between these points?

2E. A particular 12-V car battery is rated to deliver a charge of 84 A·h (ampere · hours). (*a*) How many coulombs of charge does this represent? (*b*) If this entire charge is delivered at 12 V, how much energy is available?

3P. In a typical lightning flash the potential difference between discharge points is about 10^9 V and the quantity of charge transferred is about 30 C. (*a*) How much energy is released? (*b*) If all the energy released could be used to accelerate a 1000-kg automobile from rest, what would be its final speed? (*c*) If it could be used to melt ice, how much ice would it melt at 0°C? The heat of fusion of ice is 3.3×10^5 J/kg.

Section 26–3 Equipotential Surfaces

4E. Two line charges are parallel to the z axis. One, of charge per unit length $+\lambda$, is a distance a to the right of this axis. The other, of charge per unit length $-\lambda$, is a distance a to the left of this axis (the lines and the z axis being in the same plane). Sketch some of the equipotential surfaces.

5E. In moving from A to B along an electric field line, the electric field does 3.94×10^{-19} J of work on an electron in the field illustrated in Fig. 23. What are the differences in the electric potential (*a*) $V_B - V_A$, (*b*) $V_C - V_A$, and (*c*) $V_C - V_B$?

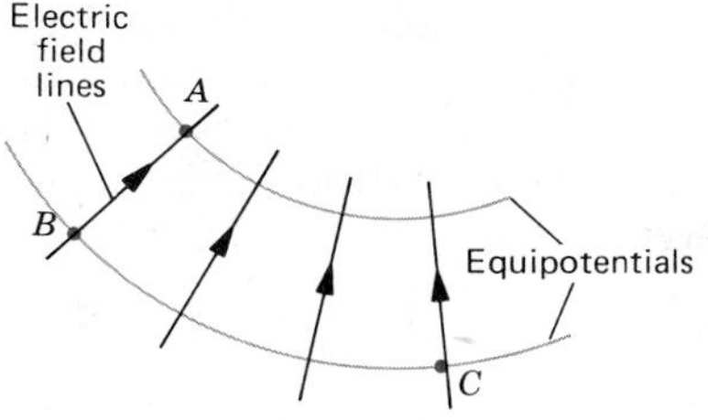

Figure 23 Exercise 5.

6E. Figure 24 shows, edge-on, an "infinite" sheet of positive charge density σ. (*a*) How much work is done by the electric field of the sheet as a small positive test charge q_0 is moved from an initial position on the sheet to a final position located a perpendicular distance z from the sheet? (*b*) Use the result from (*a*) and Eq. 8 to show that, as displayed in Table 1, the electric potential of an infinite sheet of charge can be written

$$V = V_0 - (\sigma/2\epsilon_0)z,$$

where V_0 is the potential at the surface of the sheet.

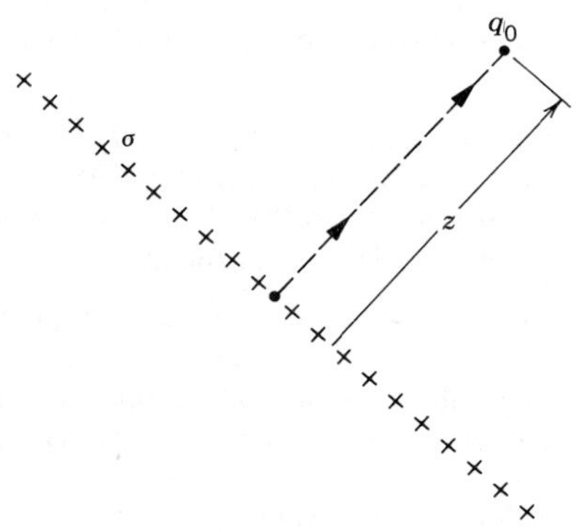

Figure 24 Exercise 6.

7E. In the Millikan oil-drop experiment (see Section 24–8 in Chapter 24), an electric field of 1.92×10^5 N/C is maintained at balance across two plates separated by 1.50 cm. Find the potential difference between the plates.

8E. Two large parallel conducting plates are 12 cm apart and carry equal but opposite charges on their facing surfaces. An electron placed midway between the two plates experiences a force of 3.9×10^{-15} N. (*a*) Find the electric field at the position of the electron. (*b*) What is the potential difference between the plates?

9E. An infinite sheet of charge has a charge density $\sigma = 0.10$ μC/m^2. How far apart are the equipotential surfaces whose potentials differ by 50 V?

10P. Three long parallel lines of charge have the relative linear charge densities shown in Fig. 25. Sketch some lines of force and the intersections of some equipotential surfaces with the plane of this figure.

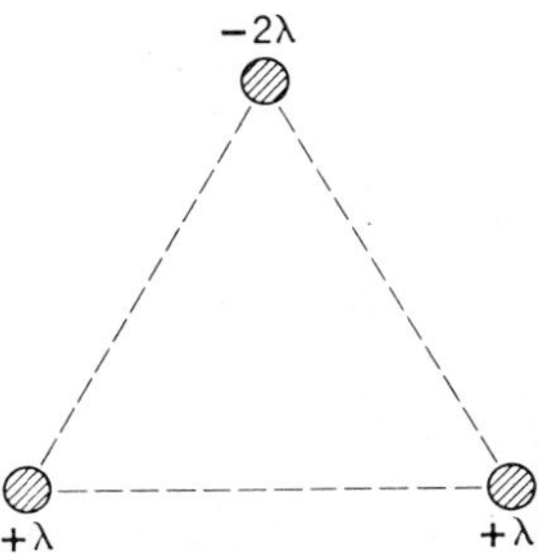

Figure 25 Problem 10.

11P. The electric field inside a nonconducting sphere of radius R, containing uniform charge density, is radially directed and has magnitude

$$E(r) = \frac{qr}{4\pi\epsilon_0 R^3},$$

where q is the total charge in the sphere and r is the distance from the sphere center. (*a*) Find the potential $V(r)$ inside the sphere, taking $V = 0$ at the center of the sphere. (*b*) What is the difference in electric potential between a point on the surface and the sphere center? If q is positive, which point is at the higher potential?

12P. A Geiger counter has a metal cylinder 2.0 cm in diameter along whose axis is stretched a wire 1.3×10^{-4} cm in diameter. If 850 V is applied between them, what is the electric field at the surface of (*a*) the wire and (*b*) the cylinder? (*Hint:* Use the result of Problem 24, Chapter 25.)

13P.* A charge q is distributed uniformly throughout a spherical volume of radius R. (*a*) Show that the potential a distance r from the center, where $r < R$, is given by

$$V = \frac{q(3R^2 - r^2)}{8\pi\epsilon_0 R^3},$$

where the zero of potential is taken at $r = \infty$. (*b*) Why does this result differ from Problem 11?

Section 26–5 Calculating the Potential: A Point Charge

14E. A point charge has $q = +1.0$ μC. Consider point A, which is 2.0 m distant, and point B, which is 1.0 m distant in a direction diametrically opposite, as in Fig. 26*a*. (*a*) What is the potential difference $V_A - V_B$? (*b*) Repeat if points A and B are located as in Fig. 26*b*.

15E. Consider a point charge with $q = 1.5 \times 10^{-8}$ C. (*a*) What is the radius of an equipotential surface having a potential of 30 V? (*b*) Are surfaces whose potentials differ by a constant amount (1.0 V, say) evenly spaced?

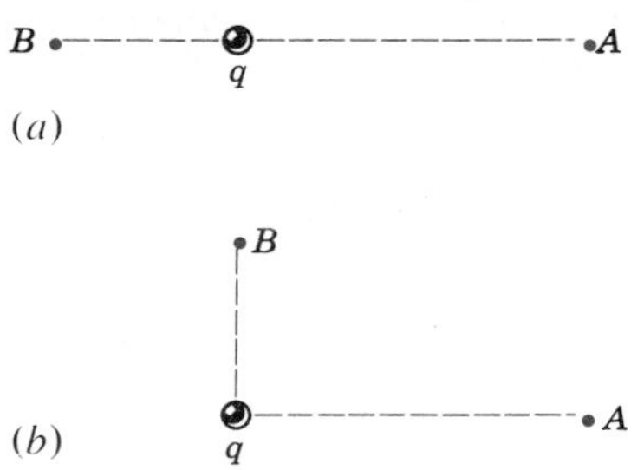

Figure 26 Exercise 14.

16E. A charge of 1.5×10^{-8} C can be produced by simple rubbing. To what potential would such a charge raise an insulated conducting sphere of 16-cm radius? (See Sample Problem 5.)

17E. As a Space Shuttle moves through the dilute ionized gas of the earth's ionosphere, its potential is typically changed by -1.0 V before it completes one revolution. By assuming that the Shuttle is a sphere of radius 10 m, estimate the amount of charge it collects.

18E. Much of the material comprising Saturn's rings (see Fig. 27) is in the form of tiny dust particles having radii on the order of 10^{-6} m. These grains are in a region containing a dilute ionized gas, and they pick up excess electrons. If the electric potential at the surface of a grain is -400 V, how many excess electrons has it picked up?

Figure 27 Exercise 18.

19E. In Fig. 28 sketch qualitatively (*a*) the lines of force and (*b*) the intersections of the equipotential surfaces with the plane of the figure. (*Hint:* Consider the behavior close to each point charge and at considerable distances from the pair of charges.)

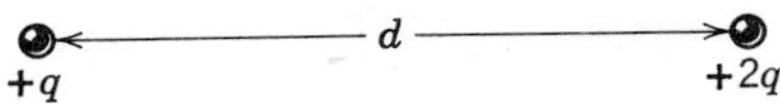

Figure 28 Exercise 19.

20E. Repeat the procedure explained in Exercise 19 for Fig. 29.

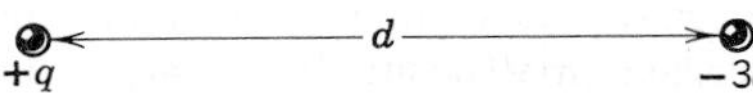

Figure 29 Exercise 20.

21P. Can a conducting sphere 10 cm in radius hold a charge of 4 μC in air without breakdown? The dielectric strength (minimum field required to produce breakdown) of air at 1 atm is 3 MV/m.

22P. What are (*a*) the charge and (*b*) the charge density on the surface of a conducting sphere of radius 0.15 m whose potential is 200 V?

23P. An electric field of approximately 100 V/m is often observed near the surface of the earth. If this field were the same over the entire surface, what would be the electric potential of a point on the surface? See Sample Problem 5.

24P. Suppose that the negative charge in a copper one-cent coin were removed to a very large distance from the earth—perhaps to a distant galaxy—and that the positive charge were distributed uniformly over the earth's surface. By how much would the electric potential at the surface of the earth change? (See Sample Problem 2 in Chapter 23.)

25P. A spherical drop of water carrying a charge of 30 pC has a potential of 500 V at its surface. (*a*) What is the radius of the drop? (*b*) If two such drops of the same charge and radius combine to form a single spherical drop, what is the potential at the surface of the new drop so formed?

26P. In Fig. 30, locate the points, if any, (*a*) where $V = 0$ and (*b*) where $\mathbf{E} = 0$. Consider only points on the axis and choose $d = 1.0$ m.

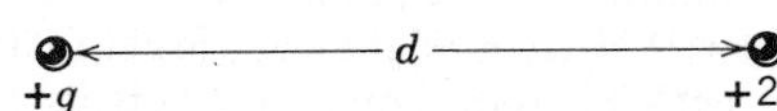

Figure 30 Problem 26.

27P. A copper sphere whose radius is 1.0 cm has a very thin surface coating of nickel. Some of the nickel atoms are radioactive, each atom emitting an electron as it decays. Half of these electrons enter the copper sphere, each depositing 100 keV of energy there. The other half of the electrons escape, each carrying away a charge of $-e$. The nickel coating has an activity of 10 mCi (= 10 millicuries = 3.70×10^8 radioactive decays per second). The sphere is hung from a long, nonconducting string and insulated from its surroundings. (*a*) How long will it take for the potential of the sphere to increase by 1000 V? (*b*) How long will it take for the temperature of the sphere to increase by 5.0°C? The heat capacity of the sphere is 14.3 J/°C.

28P. A point charge $q_1 = +6e$ is fixed at the origin of a rectangular coordinate system, and a second point charge $q_2 = -10e$ is fixed at $x = 8.6$ nm, $y = 0$. The locus of all points in the xy plane with $V = 0$ is a circle centered on the x axis, as shown in Fig. 31. Find (*a*) the location x_c of the center of the circle and (*b*) the radius R of the circle. (*c*) Is the $V = 5$ V equipotential also a circle?

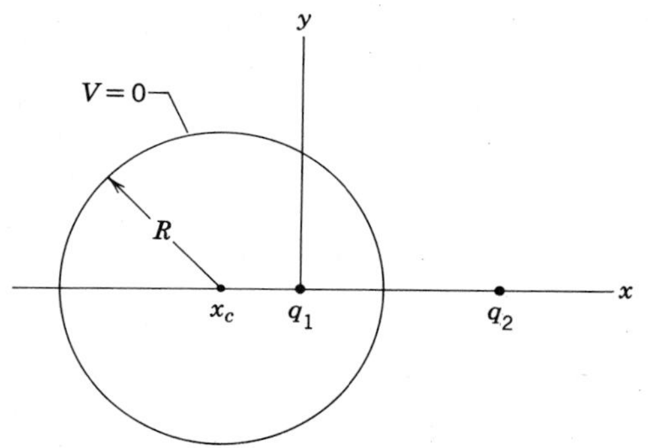

Figure 31 Problem 28.

29P.* A thick spherical shell of charge of uniform charge density is bounded by radii r_1 and r_2, where $r_2 > r_1$. Find the electric potential V as a function of the distance r from the center of the distribution, considering the regions (*a*) $r > r_2$, (*b*) $r_2 > r > r_1$, and (*c*) $r < r_1$. (*d*) Do these solutions agree at $r = r_2$ and at $r = r_1$?

Section 26-6 Calculating the Potential: An Electric Dipole

30E. The ammonia molecule NH_3 has a permanent electric dipole moment equal to 1.47 D, where D = debye unit = 3.34×10^{-30} C·m. Calculate the electric potential due to an ammonia molecule at a point 52 nm away along the axis of the dipole.

31P. For the charge configuration of Fig. 32, show that $V(r)$ for points on the vertical axis, assuming $r \gg d$, is given by

$$V = \frac{1}{4\pi\epsilon_0}\frac{q}{r}\left(1 + \frac{2d}{r}\right).$$

(*Hint:* The charge configuration can be viewed as the sum of an isolated charge and a dipole.)

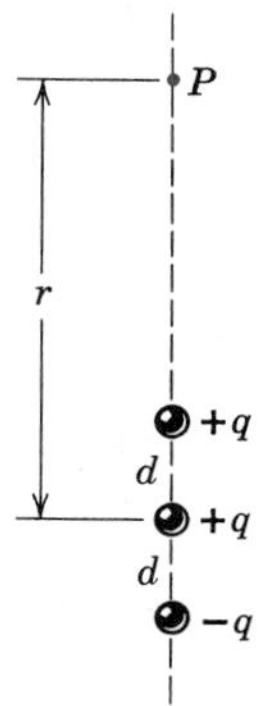

Figure 32 Problem 31.

Section 26-7 Calculating the Potential: A Charged Disk

32P. (*a*) Show that the electric potential at a point on the axis of a ring of charge of radius R, computed directly from Eq. 15, is given by

$$V = \frac{1}{4\pi\epsilon_0}\frac{q}{\sqrt{z^2 + R^2}}.$$

(*b*) From this result derive an expression for E at axial points; compare with the direct calculation of E in Section 24-6 of Chapter 24.

Section 26-8 Calculating the Field from the Potential

33E. Two large parallel metal plates are 1.5 cm apart and carry equal but opposite charges on their facing surfaces. The negative plate is grounded and its potential is taken to be zero. If the potential halfway between the plates is +5.0 V, what is the electric field in this region?

34E. The electric potential varies along the x axis as shown in the graph of Fig. 33. For each of the intervals shown (ignore the behavior at the end points of the intervals), determine the x component of the electric field and plot E_x versus x.

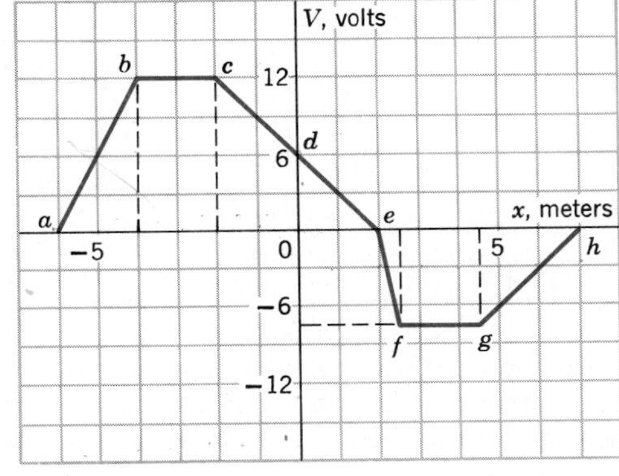

Figure 33 Exercise 34.

35E. Starting from Eq. 14, find E_r, the radial component of the electric field due to a dipole.

36E. In Section 26–7 the potential at an axial point for a charged disk was shown to be

$$V = \frac{\sigma}{2\epsilon_0}(\sqrt{z^2 + R^2} - z).$$

Use Eq. 21 to show that E for axial points is given by

$$E = \frac{\sigma}{2\epsilon_0}\left(1 - \frac{z}{\sqrt{R^2 + z^2}}\right).$$

37E. The electric potential V in the space between the plates of a particular, and now obsolete, vacuum tube is given by $V = 1500x^2$, where V is in volts if x, the distance from one of the plates, is in meters. Calculate the magnitude and direction of the electric field at $x = 1.3$ cm.

38E. Exercise 44 in Chapter 25 deals with Rutherford's calculation of the electric field a distance r from the center of an atom. He also gave the electric potential as

$$V = \frac{Ze}{4\pi\epsilon_0}\left(\frac{1}{r} - \frac{3}{2R} + \frac{r^2}{2R^3}\right).$$

(*a*) Show how the expression for the electric field given in Exercise 44 of Chapter 25 follows from the above expression for V. (*b*) Why does this expression for V not go to zero as $r \rightarrow \infty$?

39P. A charge per unit length λ is distributed uniformly along a straight-line segment of length L. (*a*) Determine the potential (chosen to be zero at infinity) at a point P a distance y from one end of the charged segment and in line with it (see Fig. 34). (*b*) Use the result of (*a*) to compute the component of the electric field at P in the y direction (along the line). (*c*) Determine the component of the electric field at P in a direction perpendicular to the straight line.

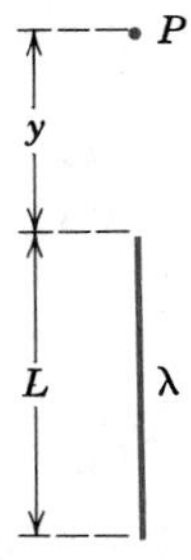

Figure 34 Problem 39.

40P. On a thin rod of length L lying along the x axis with one end at the origin ($x = 0$), as in Fig. 35, there is distributed a charge per unit length given by $\lambda = kx$, where k is a constant. (*a*) Taking the electrostatic potential at infinity to be zero, find V at the point P on the y axis. (*b*) Determine the vertical component, E_y, of the electric field intensity at P from the result of part (*a*) and also by direct calculation. (*c*) Why cannot E_x, the horizontal component of the electric field at P, be found using the result of part (*a*)?

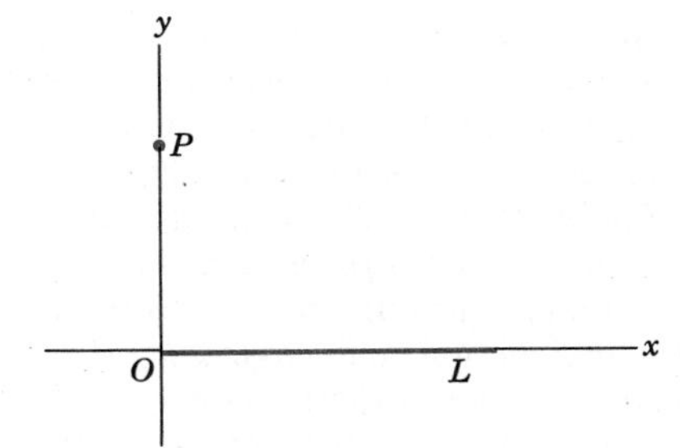

Figure 35 Problem 40.

Section 26–9 Electric Potential Energy

41E. (*a*) For Fig. 36, derive an expression for $V_A - V_B$. (*b*) Does your result reduce to the expected answer when $d = 0$? When $a = 0$? When $q = 0$?

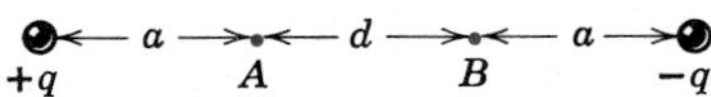

Figure 36 Exercise 41.

42E. Two charges $q = +2.0\ \mu$C are fixed in space a distance $d = 2.0$ cm apart, as shown in Fig. 37. (*a*) What is the electric potential at point C? (*b*) You bring a third charge $q = +2.0\ \mu$C slowly from infinity to C. How much work must you do? (*c*) What is the potential energy U of the configuration when the third charge is in place?

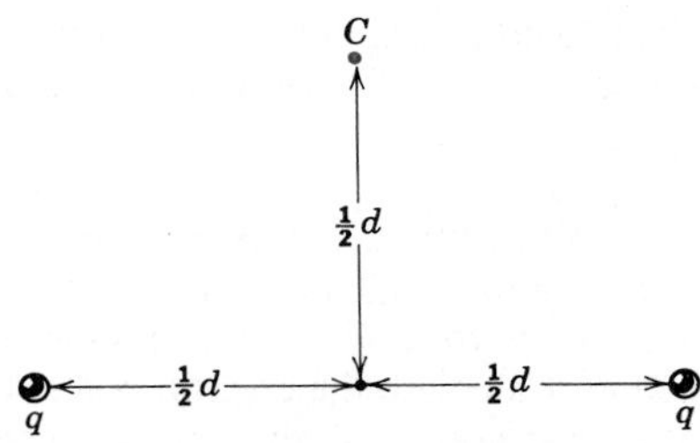

Figure 37 Exercise 42.

43E. The charges and coordinates of two point charges located in the xy plane are: $q_1 = +3.0 \times 10^{-6}$ C, $x = 3.5$ cm, $y = +0.50$ cm; and $q_2 = -4.0 \times 10^{-6}$ C, $x = -2.0$ cm, $y = +1.5$ cm. (*a*) Find the electric potential at the origin. (*b*) How much total work must be done to locate these charges at their given positions, starting from infinite separation?

44E. A decade before Einstein published his theory of relativity, J. J. Thomson proposed that the electron might be made up of small parts and that its mass is due to the electrical interaction of the parts. Furthermore, he suggested that the energy equals mc^2. Make a rough estimate of the electron mass in the following way: Assume that the electron is composed of three identical parts that are brought in from infinity and placed at the vertices of an equilateral triangle having sides equal to the classical radius of the electron, 2.82×10^{-15} m. (*a*) Find the total electrical potential energy of this arrangement. (*b*) Divide by c^2 and compare your result to the accepted electron mass (9.11×10^{-31} kg). The result improves if more parts are assumed.

45E. In the quark model of fundamental particles, a proton is composed of three quarks: two "up" quarks, each having charge $+\frac{2}{3}e$, and one "down" quark, having charge $-\frac{1}{3}e$. Suppose that the three quarks are equidistant from each other. Take the distance to be 1.32×10^{-15} m and calculate (*a*) the potential energy of the interaction between the two "up" quarks and (*b*) the total electrical potential energy of the system.

46E. Derive an expression for the work required to put the four charges together as indicated in Fig. 38.

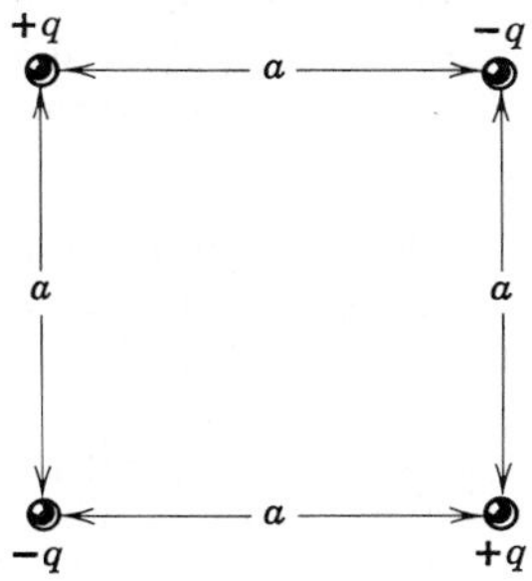

Figure 38 Exercise 46.

47E. What is the electric potential energy of the charge configuration of Fig. 11*a*? Use the numerical values of Sample Problem 6.

48P. Three charges of +0.12 C each are placed on the corners of an equilateral triangle, 1.7 m on a side. If energy is supplied at the rate of 0.83 kW, how many days would be required to move one of the charges onto the midpoint of the line joining the other two?

49P. In the rectangle shown in Fig. 39, the sides have lengths 5.0 cm and 15 cm, $q_1 = -5.0\ \mu$C and $q_2 = +2.0\ \mu$C. (*a*) What are the electric potentials at corner *B* and at corner *A*? (*b*) How much work is required to move a third charge $q_3 = +3.0\ \mu$C from *B* to *A* along a diagonal of the rectangle? (*c*) In this process, is the external work converted into electrostatic potential energy or vice versa? Explain.

Figure 39 Problem 49.

50P. A particle of (positive) charge *Q* is assumed to have a fixed position at *P*. A second particle of mass *m* and (negative) charge $-q$ moves at constant speed in a circle of radius r_1, centered at *P*. Derive an expression for the work *W* that must be done by an external agent on the second particle in order to increase the radius of the circle of motion, centered at *P*, to r_2.

51P. Calculate (*a*) the electric potential established by the nucleus of a hydrogen atom at the average distance of the circulating electron ($r = 5.3 \times 10^{-11}$ m), (*b*) the electric potential energy of the atom when the electron is at this radius, and (*c*) the kinetic energy of the electron, assuming it to be moving in a circular orbit of this radius centered on the nucleus. (*d*) How much energy is required to ionize the hydrogen atom? Express all energies in electron-volts.

52P. An electric charge of −9.0 nC is uniformly distributed around a ring of radius 1.5 m that lies in the *yz* plane with its center at the origin. A point charge of −6.0 pC is located on the *x* axis at $x = 3.0$ m. Calculate the work done in moving the point charge to the origin.

53P. A particle of charge *q* is kept in a fixed position at a point *P* and a second particle of mass *m*, having the same charge *q*, is initially held at rest a distance r_1 from *P*. The second particle is then released and is repelled from the first one. Determine its speed at the instant it is a distance r_2 from *P*. Let $q = 3.1\ \mu$C, $m = 20$ mg, $r_1 = 0.90$ mm, and $r_2 = 2.5$ mm.

54P. Two small metal spheres of mass $m_1 = 5.0$ g and mass $m_2 = 10$ g carry equal positive charges $q = 5.0\ \mu$C. The spheres are connected by a massless string of length $d = 1.0$ m, which is much greater than the sphere radii. (*a*) What is the electrostatic potential energy of the system? (*b*) You cut the string. At that instant what is the acceleration of each of the spheres? (*c*) A long time after you cut the string, what is the speed of each sphere?

55P. Between two parallel, flat, conducting surfaces of spacing $d = 1.0$ cm and potential difference $V = 10$ kV, an electron is projected from one plate directly toward the second. What is the initial velocity of the electron if it comes to rest just at the surface of the second plate?

56P. A gold nucleus contains a positive charge equal to that of 79 protons and has a radius of 6.2 fm; see Sample Problem 5. An α particle (which consists of two protons and two neutrons) has a kinetic energy K at points far from the nucleus and is traveling directly toward it. The α particle just touches the surface of the nucleus where its velocity is reversed in direction. (*a*) Calculate K. (*b*) The actual α particle energy used in the experiment of Rutherford and his collaborators that led to the discovery of the concept of the atomic nucleus was 5.0 MeV. What do you conclude?

57P. A particle of mass m, charge $q > 0$, and initial kinetic energy K is projected (from "infinity") toward a heavy nucleus of charge Q, assumed to have negligible size and a fixed position in our reference frame. If the aim is "perfect," how close to the center of the nucleus is the particle when it comes momentarily to rest?

58P. A thin, spherical, conducting shell of radius R is mounted on an insulated support and charged to a potential $-V$. An electron is fired from point P a distance r from the center of the shell ($r \gg R$) with an initial speed v_0, directed radially inward. What value of v_0 is needed for the electron to just reach the shell?

59P. Two electrons are fixed 2.0 cm apart. Another electron is shot from infinity and comes to rest midway between the two. What was its initial speed?

60P. Compute the escape speed for an electron from the surface of a uniformly charged sphere of radius 1.0 cm and total charge 1.6×10^{-15} C. Neglect gravitational forces.

61P. An electron is projected with an initial speed of 3.2×10^5 m/s directly toward a proton that is essentially at rest. If the electron is initially a great distance from the proton, at what distance from the proton is its speed instantaneously equal to twice its initial value?

Section 26-10 An Insulated Conductor

62E. A hollow metal sphere is charged to a potential of +400 V with respect to ground and carries a charge of 5.0×10^{-9} C. Find the electric potential at the center of the sphere.

63E. A thin conducting spherical shell of outer radius 20 cm carries a charge of +3.0 μC. Sketch (*a*) the magnitude of the electric field **E** and (*b*) the potential V versus the distance r from the center of the shell.

64E. Two identical conducting spheres of radius $r = 0.15$ m are separated by a distance $a = 10$ m. What is the charge on each sphere if the potential of one is +1500 V and if the other is −1500 V? What assumptions have you made?

65E. Consider two widely separated conducting spheres, 1, and 2, the second having twice the diameter of the first. The smaller sphere initially has a positive charge q and the larger one is initially uncharged. You now connect the spheres with a long thin wire. (*a*) How are the final potentials V_1 and V_2 of the spheres related? (*b*) Find the final charges q_1 and q_2 on the spheres in terms of q.

66P. The metal object in Fig. 40 is a figure of revolution about the horizontal axis. If it is charged negatively, sketch roughly a few equipotentials and lines of force. Use physical reasoning rather than mathematical analysis.

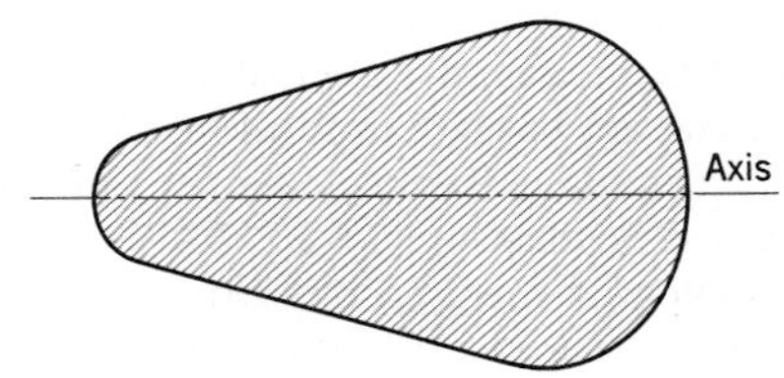

Figure 40 Problem 66.

67P. If the earth had a net charge equivalent to 1 electron/m^2 of surface area (a very artificial assumption), (*a*) what would be the earth's potential? (*b*) What would be the electric field due to the earth just outside its surface?

68P. Two metal spheres are 3.0 cm in radius and carry charges of $+1.0 \times 10^{-8}$ C and -3.0×10^{-8} C, respectively, assumed to be uniformly distributed. If their centers are 2.0 m apart, calculate (*a*) the potential of the point halfway between their centers and (*b*) the potential of each sphere.

69P. A charged metal sphere of radius 15 cm has a net charge of 3.0×10^{-8} C. (*a*) What is the electric field at the sphere's surface? (*b*) What is the electric potential at the sphere's surface? (*c*) At what distance from the sphere's surface has the electric potential decreased by 500 V?

70P. Two thin, insulated, concentric conducting spheres of radii R_1 and R_2 carry charges q_1 and q_2. Derive expressions for $E(r)$ and $V(r)$, where r is the distance from the center of the spheres. Plot $E(r)$ and $V(r)$ from $r = 0$ to $r = 4.0$ m for $R_1 = 0.50$ m, $R_2 = 1.0$ m, $q_1 = +2.0$ μC, and $q_2 = +1.0$ μC. Compare with Fig. 16.

Section 26-11 The Van de Graaff Accelerator

71E. (*a*) How much charge is required to raise an isolated metallic sphere of 1.0-m radius to a potential of 1.0 MV? Repeat for a sphere of 1.0-cm radius. (*b*) Why use a large sphere in an electrostatic accelerator when the same potential can be achieved using a smaller charge with a small sphere?

72E. Let the potential difference between the high-potential inner shell of a Van de Graaff accelerator and the point at which charges are sprayed onto the moving belt be 3.4 MV. If the belt transfers charge to the shell at the rate of 2.8 mC/s, what minimum power must be provided to drive the belt?

73E. An α particle (which consists of two protons and two neutrons) is accelerated through a potential difference of 1 million volts in a Van de Graaff accelerator. (*a*) What kinetic energy does it acquire? (*b*) What kinetic energy would a proton acquire under these same circumstances? (*c*) Which particle would acquire the greater speed, starting from rest?

74P. (*a*) Show, for the electrostatic Van de Graaff accelerator of Fig. 21, that the potential difference between the small sphere and the large sphere is

$$V_r - V_R = \frac{q}{4\pi\epsilon_0}\left(\frac{1}{r} - \frac{1}{R}\right).$$

Note that the potential difference is independent of the charge Q on the outer sphere. (*b*) Assume q is positive. Show that if the spheres are connected by a fine wire, the charge q will flow entirely to the outer sphere, regardless of the charge Q that may already be present.

75P. The high-voltage electrode of an electrostatic accelerator is a charged spherical metal shell having a potential $V = +9.0$ MV. (*a*) Electrical breakdown occurs in the gas in this machine at a field $E = 100$ MV/m. To prevent such breakdown, what restriction must be made on the radius r of the shell? (*b*) A long moving rubber belt transfers charge to the shell at 300 μC/s, the potential of the shell remaining constant because of leakage. What minimum power is required to transfer the charge? (*c*) The belt is of width $w = 0.50$ m and travels at speed $v = 30$ m/s. What is the surface charge density on the belt?

CHAPTER 27
CAPACITANCE

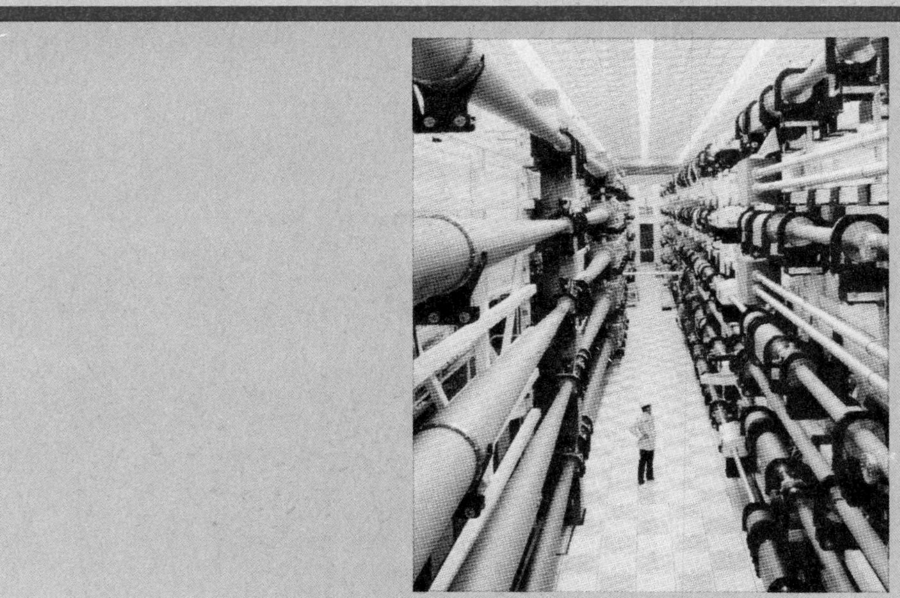

The NOVA laser at the Lawrence Livermore National Laboratory. This powerful laser, used for research in thermonuclear fusion, generates nanosecond pulses of light at a peak power level of $\sim 10^{14}$ W, about 200 times the present power-generating capacity of the United States. This energy for the pulse is accumulated and stored in the electric fields of a large capacitor bank.

27-1 The Uses of Capacitors

You can store potential energy by pulling a bow, stretching a spring, compressing a gas, or lifting a book. In the latter case, we think of the energy as stored in the *gravitational field* of the book + earth system.

You can also store potential energy in an *electrostatic field.* From this point of view, a *capacitor* is a device designed to "package" an electric field for that purpose. The capacitor in your portable battery-operated photoflash unit does just that, accumulating energy relatively slowly during the charging process and releasing it rapidly during the short duration of the flash. On a grander scale, we have the enormous capacitor bank powering the NOVA laser shown above.

Capacitors have many uses in our electronic and microelectronic age beyond serving as storehouses for potential energy. For one example, they are vital components of the electromagnetic oscillators that are central components of radio and TV transmitters and receivers. For another example, microscopic capacitors form the memory banks of computers. The electric fields in these tiny devices are significant—not so much for the stored energy they represent as for the ON-OFF information that their presence or absence provides. Figure 1 shows some capacitors of a kind typically used in electronic circuits.

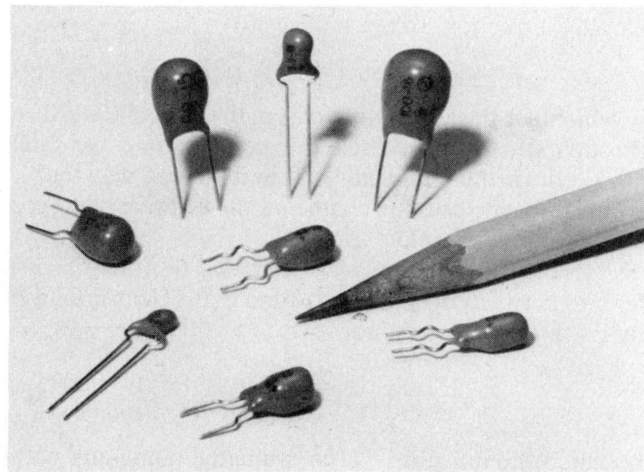

Figure 1 An assortment of capacitors that may be found in electronic circuits. Check the motherboard of your computer.

27-2 Capacitance

Figure 2 shows the elements of a *capacitor*—two isolated conductors of arbitrary shape. No matter what their geometry, we call these conductors *plates.*

Figure 3*a* shows a less general but more conventional arrangement, a *parallel-plate capacitor* formed of two parallel conducting plates of area A separated by a distance d. The symbol that we use to represent a capacitor (⊣⊢) is based on the structure of a parallel-plate capacitor but is used for capacitors of all geometries. We assume for the time being that no material medium such as glass or plastic is present in the region between the plates. In Section 27-6, we shall remove this restriction.

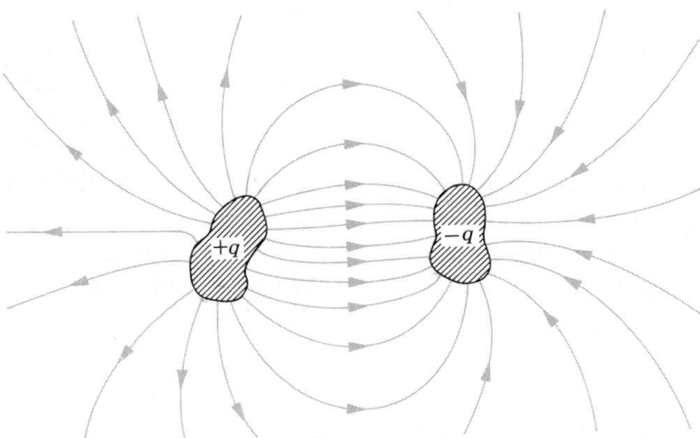

Figure 2 Two conductors, isolated from each other and from their surroundings, form a *capacitor.* When the capacitor is charged, the conductors, or plates as they are called, carry equal but opposite charges of magnitude q.

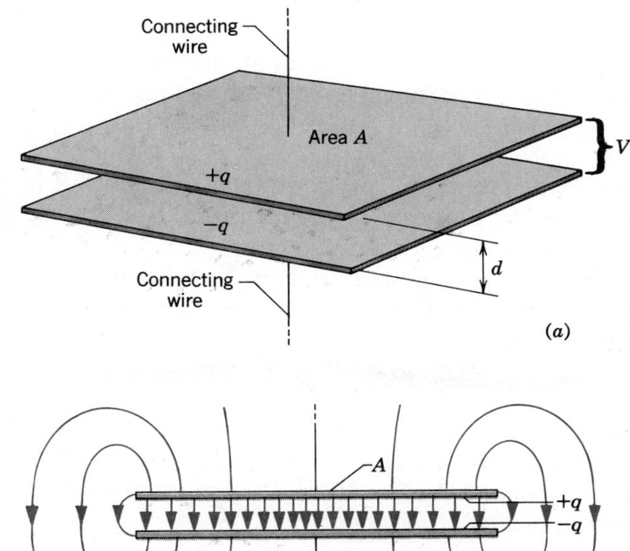

Figure 3 (*a*) A parallel-plate capacitor, made up of two plates of area A separated by a distance d. (*b*) As the field lines show, the electric field is uniform in the central region between the plates. The field lines "fringe" at the edges of the plates, showing that the field is not uniform there.

We describe a capacitor as *charged* if its plates carry equal but opposite charges, of absolute value q.* Note that q is *not* the net charge on the capacitor, which is zero. One way to charge a capacitor is to connect its plates momentarily to the terminals of a battery; equal but opposite charges will then be transferred by the battery to the two plates.

Because the plates are conductors, they are equipotentials. A definite potential difference of absolute value V will exist between the two plates; V is *not* the potential of either conductor but the *potential difference* between

* Up to now we have treated q as a scalar, possessing both absolute value and sign. When dealing with capacitors, we find it convenient to treat q as the absolute value only of the charge, its sign to be specified separately.

them. Both q and V, being absolute values, are positive quantities.

The charge q and the potential difference V for a capacitor are proportional to each other. That is,

$$q = CV, \tag{1}$$

in which the proportionality constant C, whose value is determined by the geometry of the plates, is called the *capacitance* of the capacitor.*

The SI unit of capacitance that follows from Eq. 1 is the coulomb/volt. This unit occurs so often that it is given a special name, the *farad* (abbr. F): That is,

$$1 \text{ farad} = 1 \text{ F} = 1 \text{ coulomb/volt} = 1 \text{ C/V}. \tag{2}$$

As we shall see, the farad is a very large unit. Submultiples of the farad, such as the microfarad ($1\ \mu\text{F} = 10^{-6}$ F) and the picofarad ($1 \text{ pF} = 10^{-12}$ F) are more convenient units in practice.

27-3 Calculating the Capacitance

Our task here is to calculate the capacitance of a capacitor once we know its geometry. Because we are going to consider a number of different geometries, it seems wise to develop a general plan to simplify the work. In brief our plan is as follows: (1) Assume a charge q on the plates. (2) Calculate the electric field $\mathbf{E}$ between the plates in terms of this charge, using Gauss' law. (3) Knowing $\mathbf{E}$, calculate the potential difference V between the plates from Eq. 8 of Chapter 26. (4) Calculate C from $C = q/V$ (Eq. 1).

Before we start, we can simplify the calculation of both the electric field and the potential difference by making certain simplifying assumptions. We discuss each in turn.

Calculating the Electric Field. The electric field is related to the charge on the plates by Gauss' law, or

$$\epsilon_0 \oint \mathbf{E} \cdot d\mathbf{A} = q. \tag{3}$$

Here q is the charge contained within the Gaussian surface and the integral is carried out over that surface. In all cases that we shall consider, the Gaussian surface will be such that whenever flux passes through it $\mathbf{E}$ will have a constant magnitude E and the vectors $\mathbf{E}$ and $d\mathbf{A}$ will be parallel. Equation 3 then reduces to

$$q = \epsilon_0 EA \quad \text{(special case of Eq. 3)}, \tag{4}$$

in which A is the area of that part of the Gaussian surface through which flux passes. For convenience, we shall always draw the Gaussian surface in such a way that it completely encloses the charge on the positive plate; see Fig. 4 for an example.

Calculating the Potential Difference. The potential difference between the plates is related to the electric field $\mathbf{E}$ by Eq. 8 of Chapter 26, or

$$V_f - V_i = -\int_i^f \mathbf{E} \cdot d\mathbf{s}, \tag{5}$$

in which the integral is to be evaluated along any path that starts on one plate and ends on the other. We will always choose a path that follows an electric field line from the positive plate to the negative plate. For this path, the vectors $\mathbf{E}$ and $d\mathbf{s}$ will always point in the same direction, so that the quantity $V_f - V_i$ will be negative. Since we are looking for V, the *absolute value* of the potential difference between the plates, we can set $V_f - V_i = -V$. Thus, we can recast Eq. 5 as

$$V = \int_+^- E\,ds \quad \text{(special case of Eq. 5)}, \tag{6}$$

in which the + and the − signs remind us that our path of integration starts on the positive plate and ends on the negative plate.

We conclude by forming the ratio $C = q/V$, which will always be independent of the value chosen for q. We are now ready to apply Eqs. 4 and 6 to some particular cases.

A Parallel-Plate Capacitor. We assume, as Fig. 4 suggests, that the plates of this capacitor are so large and so close together that we can neglect the "fringing" of the electric field at the edges of the plates, taking $\mathbf{E}$ to be constant throughout the volume between the plates.

Let us draw in a Gaussian surface that includes the charge q on the positive plate, as Fig. 4 shows. From Eq. 4 we can then write

$$q = \epsilon_0 EA, \tag{7}$$

where A is the area of the plate.

Equation 6 yields

$$V = \int_+^- E\,ds = E\int_0^d ds = Ed. \tag{8}$$

* Equation 1 is a theorem whose proof, which is based on Coulomb's law, is beyond the scope of this text.

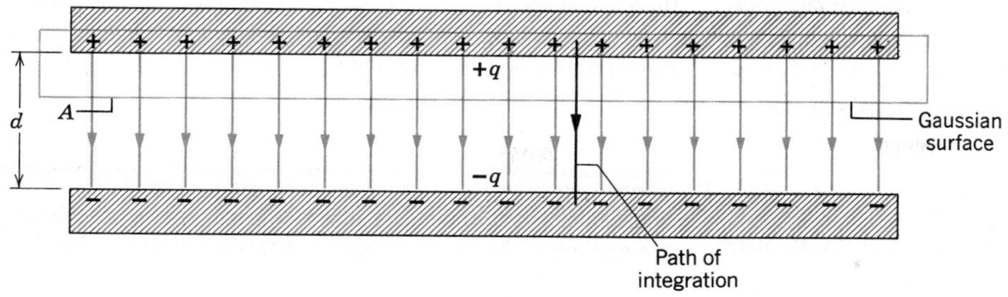

Figure 4 A charged parallel-plate capacitor. The light gray lines represent a Gaussian surface enclosing the charge on the positive plate. The heavy vertical line shows the path of integration along which we apply Eq. 6.

In Eq. 8, E is constant and can be removed from the integral; the second integral above is simply the plate separation d.

If we substitute q from Eq. 7 and V from Eq. 8 into the relation $q = CV$ (Eq. 1), we find

$$C = \epsilon_0 \frac{A}{d} \quad \text{(parallel-plate capacitor).} \tag{9}$$

The capacitance does indeed depend only on geometrical factors, namely, the plate area A and the plate separation d.

As an aside we point out that Eq. 9 suggests one reason why we wrote the electrostatic constant in Coulomb's law in the form $1/4\pi\epsilon_0$. If we had not done so, Eq. 9—which is used more often in engineering practice than is Coulomb's law—would have been less simple in form. We note further that Eq. 9 permits us to express the permittivity constant ϵ_0 in units more appropriate for use in problems involving capacitors, namely,

$$\epsilon_0 = 8.85 \times 10^{-12}\ \text{F/m} = 8.85\ \text{pF/m}. \tag{10}$$

We have previously expressed this constant as

$$\epsilon_0 = 8.85 \times 10^{-12}\ \text{C}^2/\text{N}\cdot\text{m}^2, \tag{11}$$

units that prove useful when dealing with problems that involve Coulomb's law; see Section 23-4. The two sets of units are equivalent.

A Cylindrical Capacitor. Figure 5 shows, in cross section, a cylindrical capacitor of length L formed by two coaxial cylinders of radii a and b. We assume that $L \gg b$ so that we can neglect the "fringing" of the electric field that occurs at the ends of the cylinders.

As a Gaussian surface, we choose a cylinder of length L and radius r, closed by end caps. Equation 4 yields

$$q = \epsilon_0 EA = \epsilon_0 E(2\pi rL)$$

in which $2\pi rL$ is the area of the curved part of the Gaussian surface. Solving for E yields

$$E = \frac{q}{2\pi\epsilon_0 Lr}. \tag{12}$$

Substitution of this result into Eq. 6 yields

$$V = \int_+^- E\,ds = \frac{q}{2\pi\epsilon_0 L}\int_a^b \frac{dr}{r} = \frac{q}{2\pi\epsilon_0 L}\ln\left(\frac{b}{a}\right). \tag{13}$$

From the relation $C = q/V$, we then have

$$C = 2\pi\epsilon_0 \frac{L}{\ln(b/a)} \quad \text{(cylindrical capacitor).} \tag{14}$$

We see that the capacitance of the cylindrical capacitor, like that of a parallel-plate capacitor, depends only on geometrical factors, in this case L, b, and a.

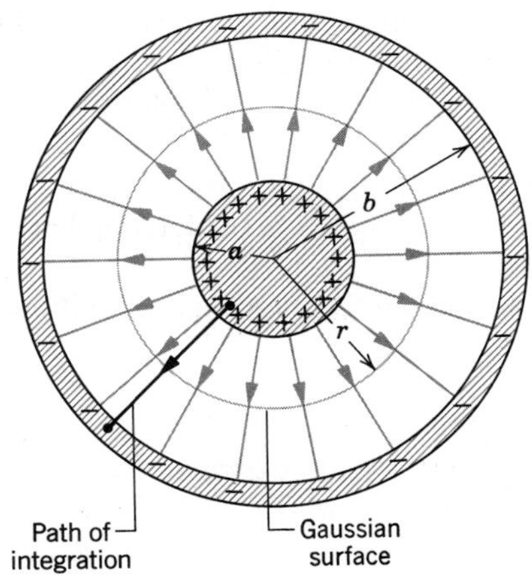

Figure 5 A long cylindrical capacitor shown in cross section. The figure shows a cylindrical Gaussian surface and also the radial path of integration along which Eq. 6 is to be applied. The figure also serves to illustrate a spherical capacitor in central cross section.

A Spherical Capacitor. Figure 4 can also serve as a central cross section of a capacitor that consists of two concentric spherical shells, of radii a and b. As a Gaussian surface we draw a sphere of radius r. Applying Eq. 4 to this surface yields

$$q = \epsilon_0 EA = \epsilon_0 E(4\pi r^2)$$

in which $4\pi r^2$ is the area of the spherical Gaussian surface. We solve this equation for E, obtaining

$$E = \frac{1}{4\pi\epsilon_0}\frac{q}{r^2}, \qquad (15)$$

which we recognize as the expression for the electric field due to a uniform spherical charge distribution.

If we substitute this expression into Eq. 6, we find

$$V = \int_+^- E\,ds = \frac{q}{4\pi\epsilon_0}\int_a^b \frac{dr}{r^2} = \frac{q}{4\pi\epsilon_0}\left(\frac{1}{a} - \frac{1}{b}\right)$$
$$= \frac{q}{4\pi\epsilon_0}\,\frac{b-a}{ab}. \qquad (16)$$

If we substitute Eq. 16 into Eq. 1 and solve for C, we find

$$C = 4\pi\epsilon_0 \frac{ab}{b-a} \quad \text{(spherical capacitor).} \qquad (17)$$

An Isolated Sphere. We can assign a capacitance to a single isolated conductor by assuming that the "missing plate" is a conducting sphere of infinite radius. After all, the field lines that leave the surface of a charged isolated conductor must end somewhere; the walls of the room in which the conductor is housed can serve effectively as our sphere of infinite radius.

If we let $b \to \infty$ in Eq. 17 and substitute R for a, we find

$$C = 4\pi\epsilon_0 R \quad \text{(isolated sphere).} \qquad (18)$$

Table 1 summarizes the various capacitances that we have derived in this section. Note that every formula involves the constant ϵ_0 multiplied by a quantity that has the dimensions of a length.

Table 1 Some Capacitances

Capacitor	*Capacitance*	*Equation*
Parallel plate	$\epsilon_0 \dfrac{A}{d}$	9
Cylindrical	$2\pi\epsilon_0 \dfrac{L}{\ln(b/a)}$	14
Spherical	$4\pi\epsilon_0 \dfrac{ab}{b-a}$	17
Isolated sphere	$4\pi\epsilon_0 R$	18

Sample Problem 1 The plates of a parallel-plate capacitor are separated by a distance $d = 1.0$ mm. What must be the plate area if the capacitance is to be 1.0 F?

From Eq. 9 we have

$$A = \frac{Cd}{\epsilon_0} = \frac{(1.0\text{ F})(1.0 \times 10^{-3}\text{ m})}{8.85 \times 10^{-12}\text{ F/m}}$$
$$= 1.1 \times 10^8\text{ m}^2. \qquad \text{(Answer)}$$

This is the area of a square more than 10 km on edge. The farad is indeed a large unit. Modern technology, however, has permitted the construction of 1-F capacitors of very modest size. These "Supercaps" are used as backup voltage sources for computers; they can maintain the computer memory for up to 30 days in case of power failure.

Sample Problem 2 The space between the conductors of a long coaxial cable, used to transmit TV signals, has an inner diameter $a = 0.15$ mm and an outer diameter $b = 2.1$ mm. What is the capacitance per unit length of this cable?

From Eq. 14 we have (see Eq. 10)

$$\frac{C}{L} = \frac{2\pi\epsilon_0}{\ln(b/a)} = \frac{(2\pi)(8.85\text{ pF/m})}{\ln(2.1\text{ mm}/0.15\text{ mm})}$$
$$= 21\text{ pF/m}. \qquad \text{(Answer)}$$

Sample Problem 3 A storage capacitor on a random access memory (RAM) chip has a capacitance of 55 fF. If it is charged to 5.3 V, how many excess electrons are there on its negative plate?

We can write, using Eq. 1,

$$n = \frac{q}{e} = \frac{CV}{e} = \frac{(55 \times 10^{-15}\text{ F})(5.3\text{ V})}{1.60 \times 10^{-19}\text{ C}}$$
$$= 1.8 \times 10^6\text{ electrons}. \qquad \text{(Answer)}$$

For electrons, this is a very small number. A speck of houshold dust, so tiny that it essentially never settles, contains about 10^{17} electrons (and the same number of protons).

Sample Problem 4 What is the capacitance of the earth, viewed as an isolated conducting sphere of radius 6370 km?

From Eq. 18 we have

$$C = 4\pi\epsilon_0 R = (4\pi)(8.85 \times 10^{-12}\text{ F/m})(6.37 \times 10^6\text{ m})$$
$$= 7.1 \times 10^{-4}\text{ F} = 710\ \mu\text{F}. \qquad \text{(Answer)}$$

A tiny Supercap has a capacitance that is about 1400 times larger than that of the earth.

27-4 Capacitors in Series and in Parallel

We often want to know the equivalent capacitance of two or more capacitors that are connected in a particular way. By "equivalent capacitance" we mean the capacitance of a single capacitor that can be substituted for the actual combination with no change in the operation of the external circuit. We discuss two limiting cases.

Capacitors in Parallel. Figure 6 shows three capacitors connected *in parallel,* a battery B being connected across the output terminals of the combination.

Capacitors are said to be connected in parallel when the same potential difference is applied to each.

We seek the single capacitance C_{eq} that is "equivalent" to this parallel combination.

For each capacitor we can write, from Eq. 1,

$$q_1 = C_1V, \quad q_2 = C_2V, \quad \text{and} \quad q_3 = C_3V.$$

The total charge on the parallel combination is then

$$q = q_1 + q_2 + q_3 = (C_1 + C_2 + C_3)V.$$

The equivalent capacitance is then

$$C_{eq} = \frac{q}{V} = C_1 + C_2 + C_3,$$

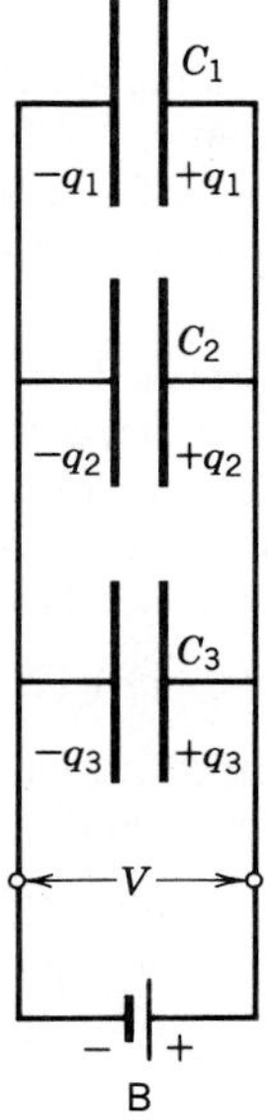

Figure 6 Three capacitors connected in parallel. The criterion for a parallel connection is that each capacitor has the same potential difference V across its plates.

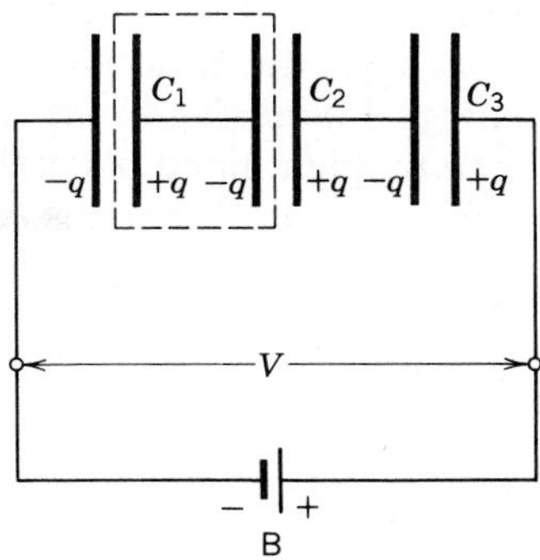

Figure 7 Three capacitors connected in series. The criterion for a series connection is that the sum of the potential differences across each capacitor must equal the potential difference that is applied to the combination.

a result that we can easily extend to any number n of capacitors, as

$$\boxed{C_{eq} = \sum_n C_n} \quad \text{(capacitors in parallel).} \quad (19)$$

Thus, to find the equivalent capacitance of a parallel combination you simply add the individual capacitances.

Capacitors in Series. Figure 7 shows the three capacitors connected *in series,* a battery B being connected across the output terminals of the combination.

*Capacitors are said to be connected in series if the sum of the potential differences across each is equal to the potential difference applied to the combination.**

We seek the single capacitance C_{eq} that is equivalent to this series combination.

We assert that, when the battery is connected, each capacitor in Fig. 7 carries the same charge q. This is true even though the three capacitors may be of different types and may have different capacitances. To understand this, note that the element of the circuit enclosed by the dashed lines in Fig. 7 is "floating;" that is, it is electrically isolated from the rest of the circuit. This element initially carries no net charge and—barring electrical breakdown of the capacitors—there is no way that

* We assume further that the combination of capacitors has no side branches.

charge can be moved onto it. Connecting the battery simply has the effect of producing a charge separation in this element, a charge $+q$ moving to the left-hand plate and a charge $-q$ to the right-hand plate; the net charge within the dashed line in Fig. 7 remains zero.

Application of Eq. 1 to each capacitor yields

$$V_1 = \frac{q}{C_1}, \quad V_2 = \frac{q}{C_2}, \quad \text{and} \quad V_3 = \frac{q}{C_3}.$$

The potential difference for the series combination is then

$$V = V_1 + V_2 + V_3 = q\left(\frac{1}{C_1} + \frac{1}{C_2} + \frac{1}{C_3}\right).$$

The equivalent capacitance is then

$$C_{eq} = \frac{q}{V} = \frac{1}{1/C_1 + 1/C_2 + 1/C_3},$$

which we can easily extend to any number n of capacitors as

$$\frac{1}{C_{eq}} = \sum_n \frac{1}{C_n} \quad \text{(capacitors in series).} \quad (20)$$

From Eq. 20 you can deduce that the equivalent series capacitance is always less than the smallest capacitance in the series of capacitors.

We note that capacitors can be connected in ways that cannot be broken down into series/parallel combinations.

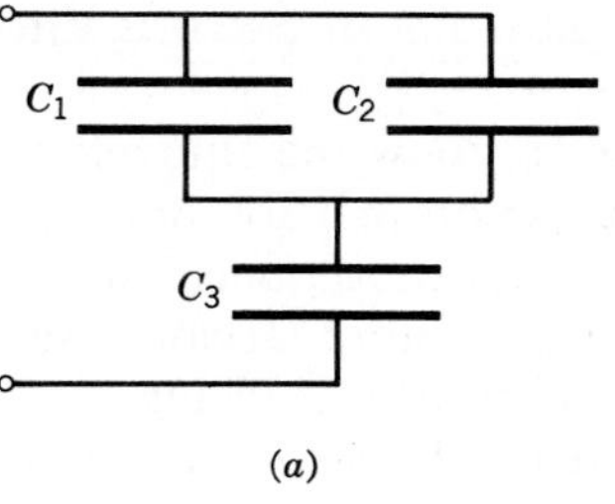

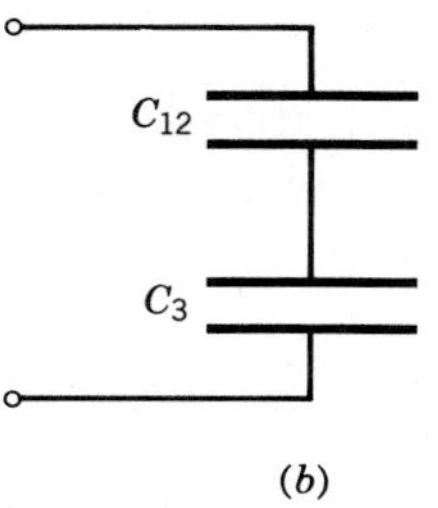

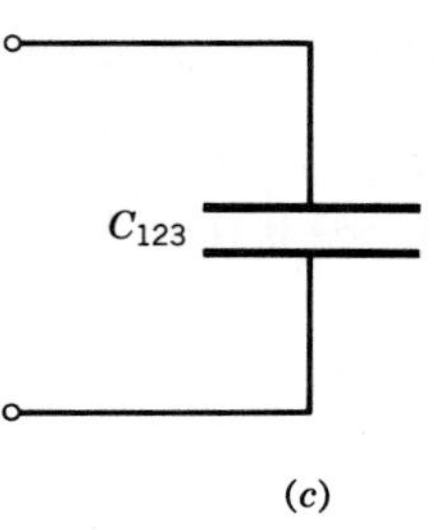

Figure 8 (*a*) Three capacitors. What is the equivalent capacitance of the combination? (*b*) C_1 and C_2, a parallel combination, are replaced by C_{12}. (*c*) C_{12} and C_3, a series combination, are replaced by the equivalent capacitance C_{123}.

Sample Problem 5 (a) Find the equivalent capacitance of the combination shown in Fig. 8*a*. Assume

$$C_1 = 12.0\ \mu F, \quad C_2 = 5.3\ \mu F, \quad \text{and} \quad C_3 = 4.5\ \mu F.$$

Capacitors C_1 and C_2 are in parallel. From Eq. 19, their equivalent capacitance is

$$C_{12} = C_1 + C_2 = 12.0\ \mu F + 5.3\ \mu F = 17.3\ \mu F.$$

As Fig. 8*b* shows, the combination C_{12} and C_3 is in series. From Eq. 20, the final equivalent combination (see Fig. 8*c*) is found from

$$\frac{1}{C_{123}} = \frac{1}{C_{12}} + \frac{1}{C_3} = \frac{1}{17.3\ \mu F} + \frac{1}{4.5\ \mu F} = 0.280\ \mu F^{-1},$$

or

$$C_{123} = \frac{1}{0.280\ \mu F^{-1}} = 3.57\ \mu F. \qquad \text{(Answer)}$$

(b) A potential difference $V = 12.5$ V is applied to the input terminals in Fig. 8*a*. What is the charge on C_1?

We treat the equivalent capacitors C_{12} and C_{123} exactly as we would real capacitors of the same capacitance. For the charge on C_{123} in Fig. 8*c* we then have

$$q_{123} = C_{123}V = (3.57\ \mu F)(12.5\ V) = 44.6\ \mu C.$$

This same charge exists on each capacitor in the series combination of Fig. 8*b*. The potential difference across C_{12} in that figure is then

$$V_{12} = \frac{q_{12}}{C_{12}} = \frac{44.6\ \mu C}{17.3\ \mu F} = 2.58\ V.$$

This same potential difference appears across C_1 in Fig. 8*a*, so that

$$q_1 = C_1V_1 = (12\ \mu F)(2.68\ V) = 31\ \mu C. \qquad \text{(Answer)}$$

27-5 Storing Energy in an Electric Field

Work must be done by an external agent to charge a capacitor. Starting with an uncharged capacitor, for example, imagine that—using "magic tweezers"—you remove electrons from one plate and transfer them one at a time to the other plate. The electric field that builds up in the space between the plates will be in a direction that tends to oppose further transfer. Thus, as charge accumulates on the capacitor plates, you will have to do increasingly larger amounts of work to transfer additional electrons. In practice, this work is not done by "magic tweezers" but by a battery, at the expense of its store of chemical energy.

We visualize the work required to charge a capacitor as stored in the form of *electrical potential energy* U in the electric field between the plates. You can recover this energy at will, by discharging the capacitor, just as you can recover the potential energy stored in a bow by releasing the bow string.

Suppose that, at a given instant, a charge q' has been transferred from one plate to the other. The potential difference V' between the plates at that moment will be q'/C. If an extra increment of charge dq' is transferred, the increment of work required will be

$$dW = V'dq' = \frac{q'}{C}\,dq'.$$

The work required to bring the total capacitor charge up to a final value q is

$$W = \int dW = \frac{1}{C}\int_0^q q'\,dq' = \frac{q^2}{2C}.$$

This work is stored as potential energy in the capacitor, so that

$$\boxed{U = \frac{q^2}{2C}} \quad \text{(potential energy).} \qquad (21)$$

From Eq. 1, we can also write this as

$$\boxed{U = \tfrac{1}{2}CV^2} \quad \text{(potential energy).} \qquad (22)$$

Equations 21 and 22 remain true no matter what the geometry of the capacitor.

To gain some physical insight into energy storage, consider two parallel-plate capacitors C_1 and C_2 that are identical except that C_1 has twice the plate separation of C_2. C_1 will then have twice the volume between its plates and also, from Eq. 9, half the capacitance of C_2. Equation 4 tells us that, if each capacitor carries the same charge q, the electric fields between their plates will be identical. We then see from Eq. 21 that C_1, which has twice the volume between its plates, also has twice the stored potential energy. It is reasonable to reach the following conclusion from arguments like this:

> *The potential energy of a charged capacitor may be viewed as stored in the electric field between its plates.*

Energy Density. In a parallel-plate capacitor, neglecting fringing, the electric field has the same value for all points between the plates. Thus, the *energy density u*, that is, the potential energy per unit volume, should also be uniform. We find u by dividing the energy by the volume Ad of the space between the plates. Thus, using Eq. 22, we obtain

$$u = \frac{U}{Ad} = \frac{CV^2}{2Ad}.$$

From Eq. 9, $C = \epsilon_0 A/d$, this result becomes

$$u = \tfrac{1}{2}\epsilon_0 \left(\frac{V}{d}\right)^2.$$

But V/d is the electric field E so that

$$\boxed{u = \tfrac{1}{2}\epsilon_0 E^2} \quad \text{(energy density).} \qquad (23)$$

Although we derived this result for the special case of a parallel-plate capacitor, it holds generally, whatever may be the source of the electric field. If an electric field **E** exists at any point in space, we can think of that point as the site of potential energy in amount, per unit volume, given by Eq. 23.

Sample Problem 6 A 3.55-μF capacitor C_1 is charged to a potential difference $V_0 = 6.30$ V, using a battery. The charging battery is then removed and the capacitor is connected as in Fig. 9 to an uncharged 8.95-μF capacitor C_2. Charge then starts to flow from C_1 to C_2 until an equilibrium is established, with both capacitors at the same potential difference V. (a) What is this common potential difference?

The original charge q_0 is now shared by two capacitors, or

$$q_0 = q_1 + q_2.$$

Applying the relation $q = CV$ to each term yields

$$C_1V_0 = C_1V + C_2V,$$

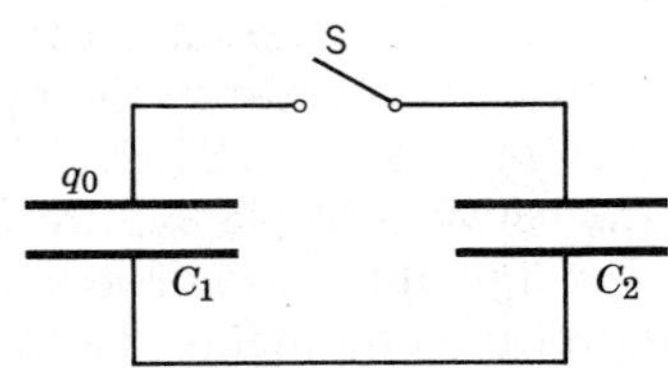

Figure 9 A potential difference V_0 is applied to C_1 and the charging battery is removed. Switch S is then closed so that the charge on C_1 is shared with C_2. What potential difference then appears across the combination?

or

$$V = V_0 \frac{C_1}{C_1 + C_2} = \frac{(6.30 \text{ V})(3.55 \text{ }\mu\text{F})}{3.55 \text{ }\mu\text{F} + 8.95 \text{ }\mu\text{F}}$$
$$= 1.79 \text{ V}. \qquad \text{(Answer)}$$

This suggests a way to measure unknown capacitances.

(b) What is the potential energy before and after the switch S in Fig. 9 is thrown?

The initial potential energy is

$$U_i = \tfrac{1}{2}C_1V_0^2 = (\tfrac{1}{2})(3.55 \times 10^{-6} \text{ F})(6.30 \text{ V})^2$$
$$= 7.05 \times 10^{-5} \text{ J} = 70.5 \text{ }\mu\text{J}. \qquad \text{(Answer)}$$

The final potential energy is

$$U_f = \tfrac{1}{2}C_1V^2 + \tfrac{1}{2}C_2V^2 = \tfrac{1}{2}(C_1 + C_2)V^2$$
$$= (\tfrac{1}{2})(3.55 \times 10^{-6} \text{ F} + 8.95 \times 10^{-6} \text{ F})(1.79 \text{ V})^2$$
$$= 2.00 \times 10^{-5} \text{ J} = 20.0 \text{ }\mu\text{J}. \qquad \text{(Answer)}$$

Thus, $U_f < U_i$, by about 72%.

This is not a violation of energy conservation. The "missing" energy appears as thermal energy in the connecting wires.*

Sample Problem 7 An isolated conducting sphere whose radius R is 6.85 cm carries a charge $q = 1.25$ nC. (a) How much potential energy is stored in the electric field of this charged conductor?

From Eqs. 21 and 18 we have

$$U = \frac{q^2}{2C} = \frac{q^2}{8\pi\epsilon_0 R} = \frac{(1.25 \times 10^{-9} \text{ C})^2}{(8\pi)(8.85 \times 10^{-12} \text{ F/m})(0.0685 \text{ m})}$$
$$= 1.03 \times 10^{-7} \text{ J} = 103 \text{ nJ}. \qquad \text{(Answer)}$$

(b) What is the energy density at the surface of the sphere?

* Some slight amount of energy is also radiated away. For a critical discussion, see "Two-Capacitor Problem: A More Realistic View," by R. A. Powell, *American Journal of Physics,* May 1979.

From Eq. 23

$$u = \tfrac{1}{2}\epsilon_0 E^2,$$

so that we must first find E at the surface of the sphere. This is given by

$$E = \frac{1}{4\pi\epsilon_0}\frac{q}{R^2}.$$

The energy density is then

$$u = \tfrac{1}{2}\epsilon_0 E^2 = \frac{q^2}{32\pi^2\epsilon_0 R^4}$$
$$= \frac{(1.25 \times 10^{-9} \text{ C})^2}{(32\pi^2)(8.85 \times 10^{-12} \text{ C}^2/\text{N}\cdot\text{m}^2)(0.0685 \text{ m})^4}$$
$$= 2.54 \times 10^{-5} \text{ J/m}^3 = 25.4 \text{ }\mu\text{J/m}^3. \qquad \text{(Answer)}$$

(c) What is the radius R_0 of a spherical surface such that one-half of the stored potential energy lies within it?

The problem requires that

$$\int_R^{R_0} dU = \tfrac{1}{2}\int_R^{\infty} dU. \qquad (24)$$

The energy that lies in a spherical shell between radii r and $r + dr$ is

$$dU = (u)(4\pi r^2)(dr),$$

where $(4\pi r^2)(dr)$ is the volume of the spherical shell. The energy density is

$$u = \tfrac{1}{2}\epsilon_0 E^2.$$

At any radius r, the electric field is

$$E = \frac{1}{4\pi\epsilon_0}\frac{q}{r^2}.$$

Therefore,

$$dU = \frac{q^2}{8\pi\epsilon_0}\frac{dr}{r^2}.$$

From Eq. 24 above,

$$\int_R^{R_0}\frac{dr}{r^2} = \tfrac{1}{2}\int_R^{\infty}\frac{dr}{r^2},$$

which becomes

$$\frac{1}{R} - \frac{1}{R_0} = \frac{1}{2R}.$$

Solving for R_0 yields

$$R_0 = 2R = (2)(6.85 \text{ cm}) = 13.7 \text{ cm}. \qquad \text{(Answer)}$$

Thus, half the stored energy is contained within a sphere whose radius is twice the radius of the conducting sphere.

27-6 Capacitor with a Dielectric

If you fill the space between the plates of a capacitor with an insulating material such as mineral oil or plastic, what happens to the capacitance? In 1837, Michael Faraday—to whom the whole concept of capacitance is largely due and for whom the SI unit of capacitance is named—first looked into the matter. Using simple apparatus much like that shown in Fig. 10, he found that the capacitance *increased* by a numerical factor κ, which he called the *dielectric constant* of the introduced material. Table 2 shows some dielectric materials and their dielectric constants. The dielectric constant of a vacuum is unity by definition. Because air is mostly empty space, the measured dielectric constant is only slightly greater than unity; the difference is usually insignificant.

Another effect of the dielectric is to limit the potential difference that can be applied between the plates to a certain value V_{max}. If this value is substantially exceeded, the dielectric material will break down and form a conducting path between the plates. Every dielectric material has a characteristic *dielectric strength,* which is the maximum value of the electric field that it can tolerate without breakdown. A few such values are listed in Table 2.

As Table 1 suggests, the capacitance of any capacitor can be written in the form

$$C = \epsilon_0 L \quad (25)$$

Table 2 Some Properties of Dielectrics[a]

Material	Dielectric Constant κ	Dielectric Strength (kV/mm)
Air (1 atm)	1.00054	3
Polystyrene	2.6	24
Paper	3.5	16
Transformer oil	4.5	
Pyrex	4.7	14
Ruby mica	5.4	
Porcelain	6.5	
Silicon	12	
Germanium	16	
Ethanol	25	
Water (20°C)	80.4	
Water (25°C)	78.5	
Titania ceramic	130	
Strontium titanate	310	8
For a vacuum, κ = unity		

[a] Measured at room temperature.

Figure 10 The simple electrostatic apparatus used by Faraday.

in which L has the dimensions of a length. Table 1 shows, for example, that $L = A/d$ for a parallel-plate capacitor and that $L = 4\pi ab/(b-a)$ for a spherical capacitor. Faraday's discovery was that, with a dielectric *completely* filling the space between the plates, Eq. 25 becomes

$$C = \kappa\epsilon_0 L = \kappa C_{air}, \quad (26)$$

where C_{air} is the value of the capacitance with air between the plates.

Figure 11 provides some insight into Faraday's experiments. In Fig. 11*a* the battery ensures that the potential difference V between the plates will remain constant. As a dielectric slab is inserted, we see that the charge q increases by a factor of κ, the additional charge being delivered to the capacitor plates by the battery. In Fig. 11*b* there is no battery and therefore the charge q must remain constant as the slab is inserted; we see that the potential difference V between the plates decreases by a factor of κ. Both of these observations are consistent—through the relation $q = CV$—with the increase in capacitance caused by the dielectric.

Comparison of Eqs. 25 and 26 suggests that the effect of a dielectric can be summed up in more general terms:

> *In a region completely filled by a dielectric, all electrostatic equations containing the permittivity constant ϵ_0 are to be modified by replacing that constant by $\kappa\epsilon_0$.*

Thus, Coulomb's law (see Eq. 1 of Chapter 25) becomes,

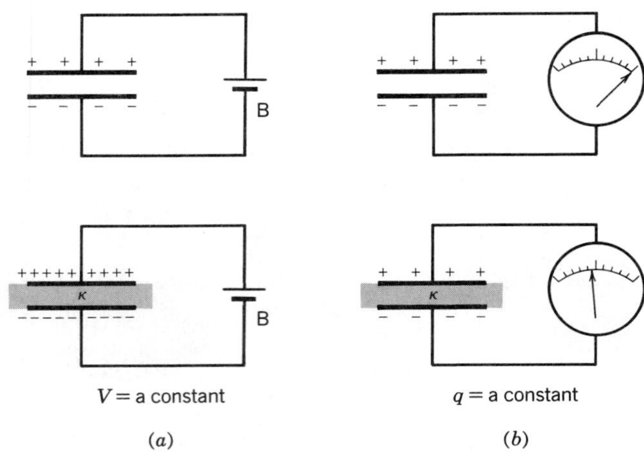

Figure 11 (*a*) If the potential difference between the plates of a capacitor is maintained, as by battery B, the effect of a dielectric is to increase the charge on the plates. (*b*) If the charge on the capacitor plates is maintained, as in this case, the effect of a dielectric is to reduce the potential difference between the plates.

for a point charge inside a dielectric,

$$E = \frac{1}{4\pi\kappa\epsilon_0}\frac{q}{r^2}, \tag{27}$$

and the expression for the electric field just outside an isolated conductor immersed in a dielectric (see Eq. 12 of Chapter 25) becomes

$$E = \frac{\sigma}{\kappa\epsilon_0}. \tag{28}$$

Both of these expressions show that, for a fixed distribution of charges, the effect of a dielectric is to weaken the electric field that would otherwise be present.

Sample Problem 8 A parallel-plate capacitor whose capacitance C is 13.5 pF has a potential difference $V = 12.5$ V between its plates. The charging battery is now disconnected and a porcelain slab ($\kappa = 6.5$) is slipped between the plates as in Fig. 11*b*. What is the potential energy of the unit, both before and after the slab is introduced?

The initial potential energy is given by Eq. 22 as

$$U_i = \tfrac{1}{2}CV^2 = (\tfrac{1}{2})(13.5 \times 10^{-12}\ \text{F})(12.5\ \text{V})^2$$
$$= 1.055 \times 10^{-9}\ \text{J} = 1055\ \text{pJ}. \qquad \text{(Answer)}$$

We can also write the initial potential energy, from Eq. 21, in the form

$$U_i = \frac{q^2}{2C}.$$

We choose to do so because, from the conditions of the problem statement, q (but not V) remains constant as the slab is introduced. After the slab is in place, C increases to κC so that

$$U_f = \frac{q^2}{2\kappa C} = \left(\frac{1}{\kappa}\right) U_i = \frac{1055\ \text{pJ}}{6.5}$$
$$= 162\ \text{pJ}. \qquad \text{(Answer)}$$

The energy after the slab is introduced is smaller by a factor of $1/\kappa$.

The "missing" energy, in principle, would be apparent to the person who introduced the slab. The capacitor would exert a tiny tug on the slab and would do work on it, in amount

$$W = U_i - U_f = (1055 - 162)\ \text{pJ} = 893\ \text{pJ}.$$

If the slab were introduced with no restraint and if there were no friction, the slab would oscillate back and forth between the plates with a (constant) mechanical energy of 893 pJ, the system energy shuttling back and forth between kinetic energy of the moving slab and potential energy stored in the electric field.

27-7 Dielectrics: An Atomic View

What happens, in atomic and molecular terms, when we put a dielectric in an external electric field? There are two possibilities.

Polar Dielectrics. The molecules of some dielectrics, like water, have permanent electric dipole moments. In such materials (called *polar dielectrics*), the electric dipoles tend to line up with an external electric field as in Fig. 12. Because the molecules are in constant thermal agitation, the alignment will not be complete but will increase as the strength of the applied field is increased or as the temperature is decreased.

Nonpolar Dielectrics. Whether or not molecules have permanent electric dipole moments, they acquire dipole moments by induction when placed in an external electric field. In Section 26-6 (see Fig. 10 of that chapter), we saw that this external field tends to "stretch" the molecule, separating slightly the centers of negative and positive charge. Figure 13*a* shows a dielectric slab with no external electric field applied. In Fig. 13*b*, an external field $\mathbf{E}_0$ is applied, its effect being to separate slightly the centers of the positive and negative distributions. The

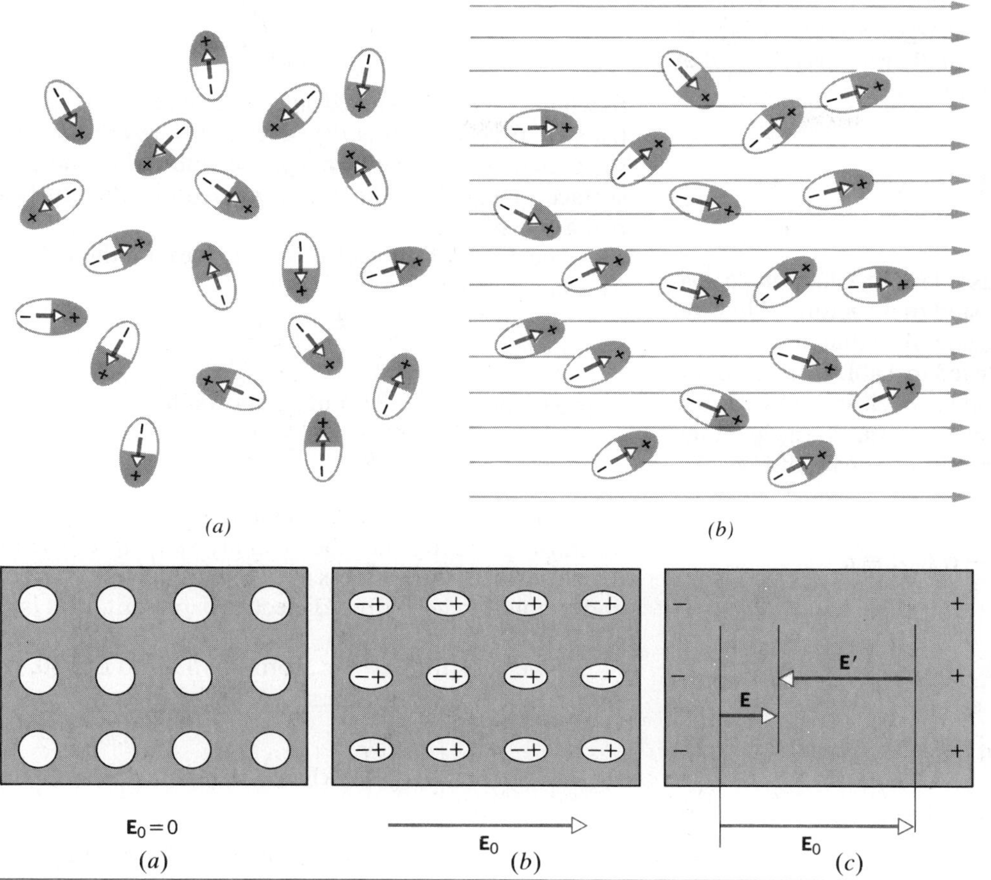

Figure 12 (*a*) Molecules with a permanent electric dipole moment, showing their random orientation in the absence of an external electric field. (*b*) An electric field is applied, producing a partial alignment of the dipoles. Thermal agitation prevents complete alignment.

Figure 13 (*a*) A dielectric slab, suggesting the neutral atoms within the slab. (*b*) An external electric field is applied, stretching out the atoms and separating the centers of positive and negative charge. (*c*) The overall effect is the production of surface charges, as shown. These charges set up a field $\mathbf{E}'$, which opposes the applied external field $\mathbf{E}_0$.

overall effect is a buildup of positive charge on the right face of the slab and of negative charge on the left face. The slab as a whole remains electrically neutral and—within the slab—there is no excess charge in any volume element.

Figure 13*c* shows that the induced surface charges appear in such a way that the electric field $\mathbf{E}'$ set up by them opposes the external electric field $\mathbf{E}_0$. The resultant field $\mathbf{E}$ inside the dielectric, which is the vector sum of $\mathbf{E}_0$ and $\mathbf{E}'$, points in the same direction as $\mathbf{E}_0$ but is smaller in magnitude. Thus, the effect of the dielectric is to weaken the applied field inside the dielectric.

This induced surface charge is the explanation of the most elementary fact of electrostatics, namely, that a charged rod will attract bits of uncharged nonconducting material such as paper. In Fig. 14 we see how surface charges are induced on a bit of paper placed near a charged rod. The attraction of the induced negative charges to the rod exceeds the repulsion of the more distant induced positive charge so that the net effect is an

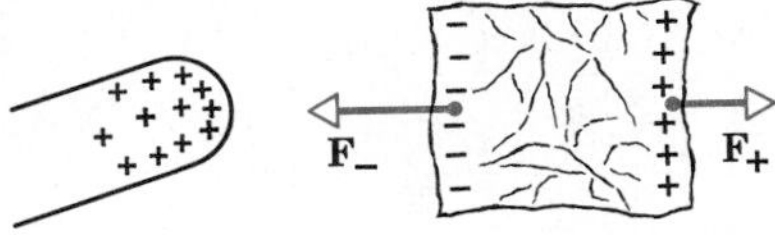

Figure 14 A charged rod attracts an uncharged bit of paper because unbalanced forces act on the induced surface charges.

attraction. If the bit of paper were placed in a *uniform* electric field, induced surface charges would appear but the forces on them would be equal and opposite so that there would be no net attraction.

27-8 Dielectrics and Gauss' Law (Optional)

In our presentation of Gauss' law in Chapter 25, we assumed that the charges existed in a vacuum. Here we shall see how to modify and generalize that law if dielectric materials, such as those listed in Table 2, are present. Figure 15 shows a parallel-plate capacitor both with and without a dielectric. We assume that the charge q on the plates is the same in each case. For the case of Fig. 15*a*, Gauss' law yields

$$\epsilon_0 \oint \mathbf{E} \cdot d\mathbf{A} = \epsilon_0 E_0 A = q$$

or

$$E_0 = \frac{q}{\epsilon_0 A}. \tag{29}$$

If a dielectric is present, as in Fig. 15*b*, Gauss' law yields

$$\epsilon_0 \oint \mathbf{E} \cdot d\mathbf{A} = \epsilon_0 E A = q - q' \tag{30}$$

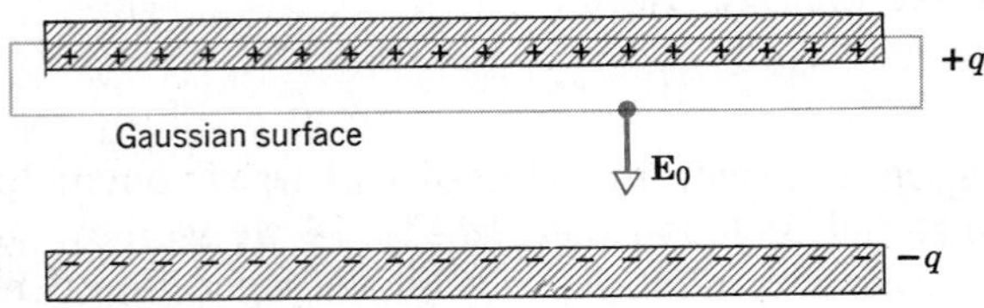

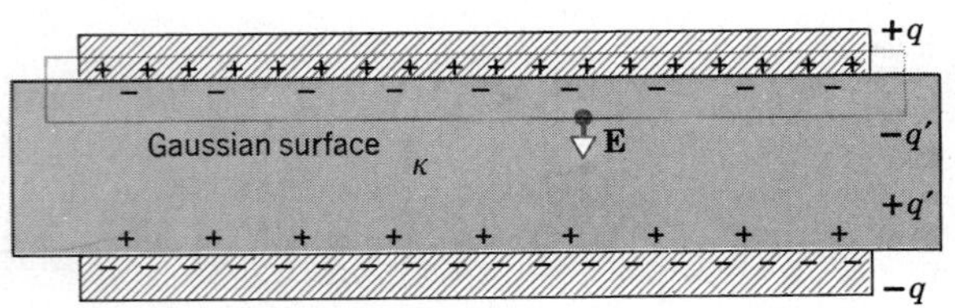

Figure 15 (*a*) A parallel-plate capacitor. (*b*) The same, with a dielectric slab inserted. The charge q on the plates is assumed to be the same in each case.

or

$$E = \frac{q - q'}{\epsilon_0 A}, \tag{31}$$

in which $-q'$, the induced surface charge on the dielectric slab, must be distinguished from q, the *free charge* on the plates. Both of these charges lie within the Gaussian surface of Fig. 15*b*, the *net* charge within that surface being $q - q'$.

The effect of the dielectric is to weaken the field E_0 by a factor κ, or

$$E = \frac{E_0}{\kappa} = \frac{q}{\kappa \epsilon_0 A}. \tag{32}$$

Comparison of Eqs. 31 and 32 shows that

$$q - q' = \frac{q}{\kappa}, \tag{33}$$

is the *net* charge inside the Gaussian surface. Equation 33 shows correctly that the absolute value q' of the induced surface charge is less than that of the free charge q and is equal to zero if no dielectric is present, that is, if $\kappa = 1$ in Eq. 33.

If we substitute $q - q'$ from Eq. 33 into Eq. 30, we see that we can write Gauss' law in the form

$$\boxed{\epsilon_0 \oint \kappa \mathbf{E} \cdot d\mathbf{A} = q}$$

(Gauss' law with dielectric). (34)

This important equation, although derived for a parallel-plate capacitor, is true generally and is the most general form in which Gauss' law can be written. Note the following:

1. The flux integral now deals with $\kappa\mathbf{E}$, not with $\mathbf{E}$.*
2. The charge q enclosed by the Gaussian surface is taken to be the *free charge only.* Induced surface charge is deliberately ignored on the right side of this equation, having been taken fully into account by introducing the dielectric constant κ on the left side.
3. Equation 34 differs from Eq. 9 of Chapter 25, our original statement of Gauss' law, only in that ϵ_0 in the latter equation has been replaced by $\kappa\epsilon_0$ and κ has been taken inside the integral to allow for cases in which κ is not constant over the entire Gaussian surface. This is in full accord with our statement on

* The vector $\epsilon_0 \kappa \mathbf{E}$ is called the *electric displacement* $\mathbf{D}$, which allows us to write Eq. 34 in the simplified form $\oint \mathbf{D} \cdot d\mathbf{A} = q$.

page 627 of the consequences for electrostatics of filling a region of space with a dielectric.

Sample Problem 9 Figure 16 shows a parallel-plate capacitor of plate area A and plate separation d. A potential difference V is applied between the plates. The battery is then disconnected and a dielectric slab of thickness b and dielectric constant κ is placed between the plates as shown. Assume

$$A = 115\text{ cm}^2,\quad d = 1.24\text{ cm},$$
$$b = 0.78\text{ cm},\quad \kappa = 2.61,$$
$$V_0 = 85.5\text{ V}.$$

(a) What is the capacitance C_0 before the slab is inserted? From Eq. 9 we have

$$C_0 = \frac{\epsilon_0 A}{d} = \frac{(8.85 \times 10^{-12}\text{ F/m})(115 \times 10^{-4}\text{ m}^2)}{1.24 \times 10^{-2}\text{ m}}$$
$$= 8.21 \times 10^{-12}\text{ F} = 8.21\text{ pF}. \qquad \text{(Answer)}$$

(b) What free charge appears on the plates? From Eq. 1,

$$q = C_0 V_0 = (8.21 \times 10^{-12}\text{ F})(85.5\text{ V})$$
$$= 7.02 \times 10^{-10}\text{ C} = 702\text{ pC}. \qquad \text{(Answer)}$$

Because the charging battery was disconnected before the slab was introduced, the free charge remains unchanged as the slab is put into place.

(c) What is the electric field E_0 in the gaps between the plates and the dielectric slab?

Let us apply Gauss' law in the form given in Eq. 34 to the Gaussian surface in Fig. 16 that includes the free charge on the upper capacitor plate. We have

$$\epsilon_0 \oint \kappa \mathbf{E} \cdot d\mathbf{A} = (1)\epsilon_0 E_0 A = q$$

or

$$E_0 = \frac{q}{\epsilon_0 A} = \frac{7.02 \times 10^{-10}\text{ C}}{(8.85 \times 10^{-12}\text{ F/m})(115 \times 10^{-4}\text{ m}^2)}$$
$$= 6900\text{ V/m} = 6.90\text{ kV/m}. \qquad \text{(Answer)}$$

Note that we put $\kappa = 1$ in this equation because the Gaussian surface over which Gauss' law was integrated does not pass through any dielectric. Note too that the value of E_0 remains unchanged as the slab is introduced. It depends only on the free charge on the plates.

Table 3 Sample Problem 9: A Summary of Results

Quantity		No Slab	Partial Slab	Full Slab
C	pF	8.21	13.4	21.4
q	pC	702	702	702
q'	pC	—	433	433
V	V	85.5	52.3	32.8
E_0	kV/m	6.90	6.90	6.90[a]
E	kV/m	—	2.64	2.64

[a] Assumes that a very narrow gap is present.

(d) Calculate the electric field E in the dielectric slab.

Again we apply Eq. 34, this time to the lower Gaussian surface in Fig. 16. We find

$$\epsilon_0 \oint \kappa \mathbf{E} \cdot d\mathbf{A} = \kappa \epsilon_0 E A = q$$

or

$$E = \frac{q}{\kappa \epsilon_0 A} = \frac{E_0}{\kappa} = \frac{6.90\text{ kV/m}}{2.61}$$
$$= 2.64\text{ kV/m}. \qquad \text{(Answer)}$$

(e) What is the potential difference between the plates after the slab has been introduced?

Equation 6 yields

$$V = \int_{+}^{-} E\, ds = E_0(d - b) + Eb$$
$$= (6900\text{ V/m})(0.0124\text{ m} - 0.0078\text{ m})$$
$$+ (2640\text{ V/m})(0.0078\text{ m})$$
$$= 52.3\text{ V}. \qquad \text{(Answer)}$$

This contrasts with the original applied potential difference of 85.5 V.

(f) What is the capacitance with the slab in place? From Eq. 1,

$$C = \frac{q}{V} = \frac{7.02 \times 10^{-10}\text{ C}}{52.3\text{ V}}$$
$$= 1.34 \times 10^{-11}\text{ F} = 13.4\text{ pF}. \qquad \text{(Answer)}$$

Table 3 summarizes the results of this sample problem and also includes the results that would have followed if the dielectric slab had completely filled the space between the plates.

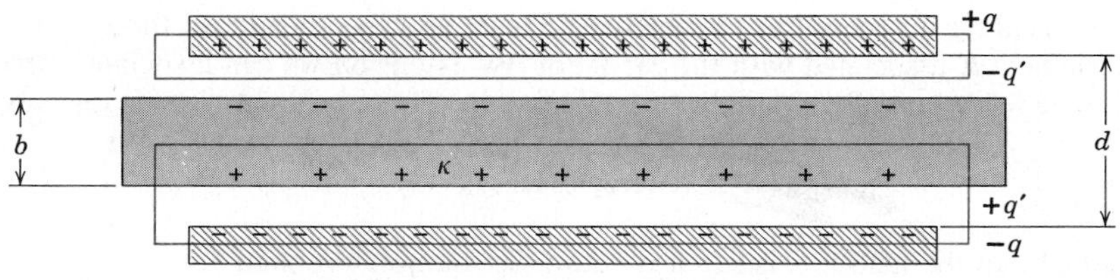

Figure 16 A parallel-plate capacitor containing a dielectric slab that only partially fills the space between the plates.

REVIEW AND SUMMARY

A *capacitor* consists of two isolated conductors (plates) carrying equal and opposite charges $+q$ and $-q$. The capacitance C is defined from

A Capacitor; Capacitance

$$q = CV \quad [1]$$

where V is the potential difference between the plates. The SI unit of capacitance is the farad (1 farad = 1 F = 1 coulomb/volt). See Sample Problem 3.

Evaluating Capacitance

We generally evaluate the capacitance of a capacitor by (1) assuming charge q to have been placed on the plates, (2) finding the electric field **E** due to this charge, (3) evaluating the potential difference V, and (4) calculating C from Eq. 1. Section 27–3 shows several important examples in detail. The results are summarized here and in Table 1.

A *parallel-plate capacitor* with plane parallel plates of area A and spacing d has capacitance

Parallel-Plate Capacitor

$$C = \frac{\epsilon_0 A}{d}; \quad \text{(parallel-plate capacitor)}. \quad [9]$$

See Sample Problem 1. A *cylindrical capacitor* consists of two long coaxial cylinders of length L. The inner and outer radii are a and b, and the capacitance is

Cylindrical Capacitor

$$C = 2\pi\epsilon_0 \frac{L}{\ln(b/a)} \quad \text{(cylindrical capacitor)}. \quad [14]$$

See Sample Problem 2. A *spherical capacitor* with concentric spherical plates of inner and outer radii a and b has capacitance

Spherical Capacitor

$$C = 4\pi\epsilon_0 \frac{ab}{b-a} \quad \text{(spherical capacitor)}. \quad [17]$$

If $b \to \infty$ and $a = R$, as for the earth in Sample Problem 4, we have the capacitance of an isolated sphere:

Isolated Sphere

$$C = 4\pi\epsilon_0 R \quad \text{(isolated sphere)}. \quad [18]$$

The *equivalent capacitances* C_{eq} of combinations of individual capacitors arranged in *series* and in *parallel* are

Capacitors in Series and in Parallel

$$C_{eq} = \sum_n C_n \quad \text{(capacitors in parallel)} \quad [19]$$

and

$$\frac{1}{C_{eq}} = \sum_n \frac{1}{C_n} \quad \text{(capacitors in series)}. \quad [20]$$

Sample Problem 5 illustrates how these can be combined to calculate the capacitance of more complicated series-parallel combinations.

The *potential energy* U of a charged capacitor, given by

Potential Energy

$$U = \frac{q^2}{2C} = \frac{1}{2} CV^2 \quad \text{(potential energy)}, \quad [21,22]$$

is the work required to charge it. (See Sample Problem 6.) This energy is conveniently thought of as stored in the electric field **E** associated with the capacitor. By extension we can associate stored energy with an electric field generally, no matter what its origin. The *energy density* u is given by

Energy Density

$$u = \frac{1}{2} \epsilon_0 E^2 \quad \text{(energy density)} \quad [23]$$

in which it is assumed that the field **E** exists in a vacuum. See Sample Problem 7.

Capacitance with a Dielectric

If the space between the plates of a capacitor is completely filled with a dielectric material, the capacitance C is increased by a factor κ, called the *dielectric constant,* which is characteristic of the material (see Table 2). In a region completely filled by a dielectric, all electrostatic equations containing ϵ_0 are to be modified by replacing ϵ_0 by $\kappa\epsilon_0$. Sample Problem 8 shows some examples.

The effects of adding a dielectric can be understood physically in terms of the action of an electric field on the permanent or induced electric dipoles in the dielectric slab. As Fig. 14 shows, the result is the formation of induced surface charges, which results in a weakening of the field within the body of the dielectric.

When a dielectric is present, Gauss' law may be generalized to

Gauss's Law with a Dielectric

$$\epsilon_0 \oint \kappa \mathbf{E} \cdot d\mathbf{A} = q \quad \text{(Gauss' law with dielectric).} \qquad [18]$$

Here q includes only the free charge, the induced surface charge being accounted for by including κ inside the integral. Sample Problem 9 applies this relation in an important special case.

QUESTIONS

1. A capacitor is connected across a battery. (*a*) Why does each plate receive a charge of exactly the same magnitude? (*b*) Is this true even if the plates are of different sizes?

2. Can there be a potential difference between two adjacent conductors that carry the same amount of positive charge?

3. A sheet of aluminum foil of negligible thickness is placed between the plates of a capacitor as in Fig. 17. What effect has it on the capacitance if (*a*) the foil is electrically insulated and (*b*) the foil is connected to the upper plate?

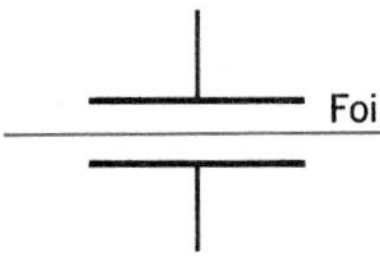

Figure 17 Question 3.

4. You are given two capacitors, C_1 and C_2, in which $C_1 > C_2$. How could things be arranged so that C_2 could hold more charge than C_1?

5. In Fig. 2 suppose that a and b are nonconductors, the charge being distributed arbitrarily over their surfaces. (*a*) Would Eq. 1 ($q = CV$) hold, with C independent of the charge arrangements? (*b*) How would you define V in this case?

6. You are given a parallel-plate capacitor with square plates of area A and separation d, in a vacuum. What is the qualitative effect of each of the following on its capacitance? (*a*) Reduce d. (*b*) Put a slab of copper between the plates, touching neither plate. (*c*) Double the area of both plates. (*d*) Double the area of one plate only. (*e*) Slide the plates parallel to each other so that the area of overlap is, say, 50%. (*f*) Double the potential difference between the plates. (*g*) Tilt one plate so that the separation remains d at one end but is $\frac{1}{2}d$ at the other.

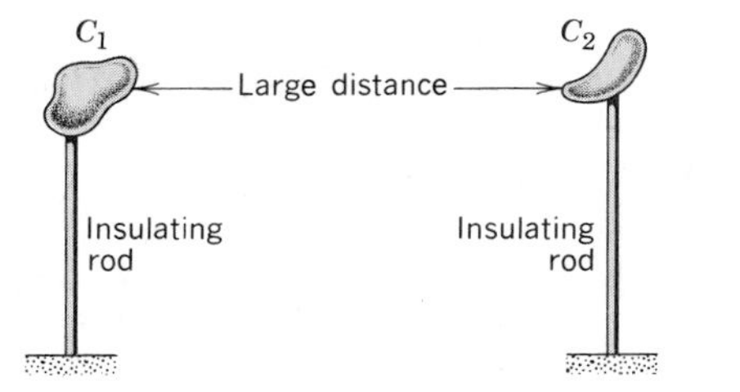

Figure 18 Question 7.

7. You have two isolated conductors, each of which has a certain capacitance; see Fig. 18. If you join these conductors by a fine wire, how do you calculate the capacitance of the combination? In joining them with the wire, have you connected them in parallel or in series?

8. Capacitors often are stored with a wire connected across their terminals. Why is this done?

9. If you were not to neglect the fringing of the electric field lines in a parallel-plate capacitor, would you calculate a higher or a lower capacitance?

10. Two circular copper disks are facing each other a certain distance apart. In what ways could you reduce the capacitance of this combination?

11. Would you expect the dielectric constant of a material to vary with temperature? If so, how? Does whether or not the molecules have permanent dipole moments matter here?

12. Discuss similarities and differences when (*a*) a dielectric slab and (*b*) a conducting slab are inserted between the plates of a parallel-plate capacitor. Assume the slab thicknesses to be one-half the plate separation.

13. An oil-filled, parallel-plate capacitor has been designed to have a capacitance C and to operate safely at or below a certain maximum potential difference V_m without arcing over. However, the designer did not do a good job and the capacitor occasionally arcs over. What can be done to redesign the capacitor, keeping C and V_m unchanged and using the same dielectric?

14. A dielectric object in a nonuniform electric field experiences a net force. Why is there no net force if the field is uniform?

15. A stream of tap water can be deflected if a charged rod is brought close to the stream. Explain carefully how this happens.

16. Water has a high dielectric constant. Why isn't it used ordinarily as a dielectric material in capacitors?

17. A dielectric slab is inserted in one end of a charged parallel-plate capacitor (the plates being horizontal and the charging battery having been disconnected) and then released. Describe what happens. Neglect friction.

18. A parallel-plate capacitor is charged by using a battery, which is then disconnected. A dielectric slab is then slipped between the plates. Describe qualitatively what happens to the charge, the capacitance, the potential difference, the electric field, and the stored energy.

19. While a parallel-plate capacitor remains connected to a battery, a dielectric slab is slipped between the plates. Describe qualitatively what happens to the charge, the capacitance, the potential difference, the electric field, and the stored energy. Is work required to insert the slab?

20. Two identical capacitors are connected as shown in Fig. 19. A dielectric slab is slipped between the plates of one capacitor, the battery remaining connected. Describe qualitatively what happens to the charge, the capacitance, the potential difference, the electric field, and the stored energy for each capacitor.

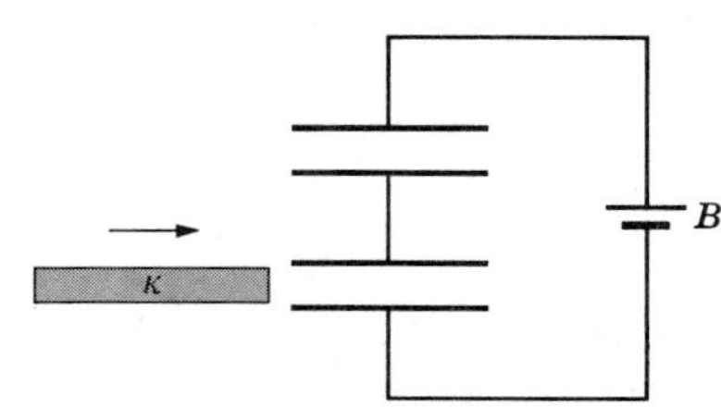

Figure 19 Question 20.

EXERCISES AND PROBLEMS

Section 27-2 Capacitance

1E. An electrometer is a device used to measure static charge. Unknown charge is placed on the plates of a capacitor and the potential difference is measured. What minimum charge can be measured by an electrometer with a capacitance of 50 pF and a voltage sensitivity of 0.15 V?

2E. The two metal objects in Fig. 20 have net charges of +70 pC and −70 pC, and this results in a 20-V potential difference between them. (*a*) What is the capacitance of the system? (*b*) If the charges are changed to +200 pC and −200 pC, what does the capacitance become? (*c*) What does the potential difference become?

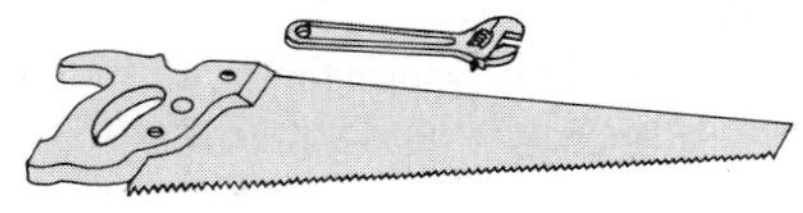

Figure 20 Exercise 2.

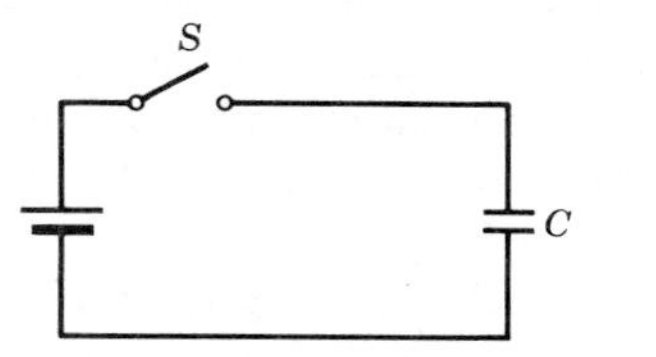

Figure 21 Exercise 3.

3E. The capacitor in Fig. 21 has a capacitance of 25 μF and is initially uncharged. The battery supplies 120 V. After switch S has been closed for a long time, how much charge will have passed through the battery?

Section 27-3 Calculating the Capacitance

4E. If we solve Eq. 9 for ϵ_0, we see that its SI units are farad/meter. Show that these units are equivalent to those obtained earlier for ϵ_0, namely coulomb²/newton · meter².

5E. A parallel-plate capacitor has circular plates of 8.2-cm radius and 1.3-mm separation. (*a*) Calculate the capacitance. (*b*) What charge will appear on the plates if a potential difference of 120 V is applied?

6E. You have two flat metal plates, each of area 1.0 m², with which to construct a parallel-plate capacitor. If its capacitance is to be 1.0 F, what must be the separation between the plates? Could this capacitor actually be constructed?

7E. The plate and cathode of a vacuum tube diode are in the form of two concentric cylinders with the cathode as the central cylinder. The cathode diameter is 1.6 mm and the plate diameter 18 mm with both elements having a length of 2.4 cm. Calculate the capacitance of the diode.

8E. The plates of a spherical capacitor have radii 38 mm and 40 mm. (*a*) Calculate the capacitance. (*b*) What must be the plate area of a parallel-plate capacitor with the same plate separation and capacitance?

9E. After you walk over a carpet on a dry day, your hand comes close to a metal doorknob and a 5-mm spark results. Such a spark means that there must have been a potential difference of possibly 15 kV between you and the doorknob. Assuming this potential difference, how much charge did you accumulate in walking over the carpet? For this extremely rough calculation, assume that your body can be represented by a uniformly charged conducting sphere 25 cm in radius and isolated from its surroundings.

10E. Two sheets of aluminum foil have a separation of 1.0 mm, a capacitance of 10 pF, and are charged to 12 V. (*a*) Calculate the plate area. The separation is now decreased by 0.10 mm with the charge held constant. (*b*) What is the new capacitance? (*c*) By how much does the potential difference change? Explain how a microphone might be constructed using this principle.

11P. Show that the plates of a parallel-plate capacitor attract each other with a force given by

$$F = \frac{q^2}{2\epsilon_0 A}.$$

Prove this by calculating the work necessary to increase the plate separation from x to $x + dx$, the charge q remaining constant.

12P. Using the result of Problem 11 show that the force per unit area (the *electrostatic stress*) acting on either capacitor plate is given by $\frac{1}{2}\epsilon_0 E^2$. Actually, this result is true in general, for a conductor of *any* shape with an electric field **E** at its surface.

13P. In Section 3 the capacitance of a cylindrical capacitor was calculated. Using the approximation (see Appendix G) that $\ln(1 + x) \approx x$ when $x \ll 1$, show that the capacitance approaches that of a parallel plate capacitor when the spacing between the two cylinders is small.

14P. Suppose that the two spherical shells of a spherical capacitor have their radii approximately equal. Under these conditions the device approximates a parallel-plate capacitor with $b - a = d$. Show that Eq. 17 does indeed reduce to Eq. 9 in this case.

15P. A spherical drop of mercury of radius R has a capacitance given by $C = 4\pi\epsilon_0 R$ (see Section 3). If two such drops combine to form a single larger drop, what is its capacitance?

16P. A capacitor is to be designed to operate, with constant capacitance, in an environment of fluctuating temperature. As shown in Fig. 22, the capacitor is a parallel-plate type with plastic "spacers" to keep the plates aligned. (*a*) Show that the rate of change of capacitance C with temperature T is given by

$$\frac{dC}{dT} = C\left(\frac{1}{A}\frac{dA}{dT} - \frac{1}{x}\frac{dx}{dT}\right),$$

where A is the plate area and x the plate separation. (*b*) If the plates are aluminum, what should be the coefficient of thermal

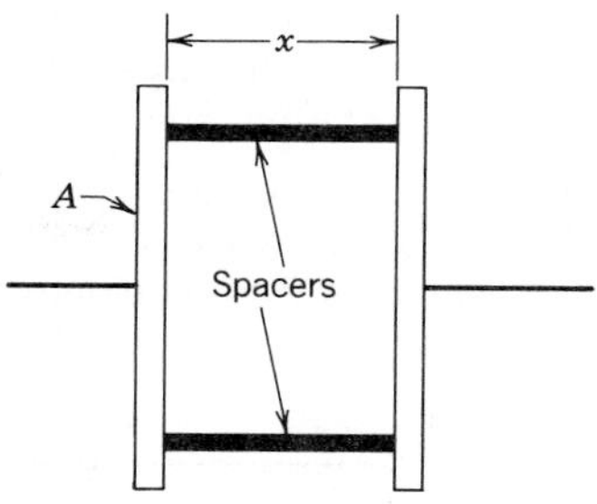

Figure 22 Problem 16.

expansion of the spacers in order that the capacitance not vary with temperature? (Ignore the effect of the spacers on the capacitance.)

17P*. A soap bubble of radius R_0 is slowly given a charge q. Because of mutual repulsion of the surface charges, the radius increases slightly to R. The air pressure inside the bubble drops, because of the expansion, to $p(V_0/V)$ where p is the atmospheric pressure, V_0 is the initial volume, and V the final volume. Show that

$$q^2 = 32\pi^2\epsilon_0 pR(R^3 - R_0^3).$$

(*Hint:* Consider the forces acting on a small area of the charged bubble. These are due to (*i*) gas pressure, (*ii*) atmospheric pressure, (*iii*) electrostatic stress; see Problem 12.)

Section 27-4 Capacitors in Series and Parallel

18E. How many 1.0-μF capacitors must be connected in parallel to store a charge of 1.0 C with a potential of 110 V across the capacitors?

19E. In Fig. 23 find the equivalent capacitance of the combination. Assume that $C_1 = 10\ \mu$F, $C_2 = 5.0\ \mu$F, and $C_3 = 4.0\ \mu$F.

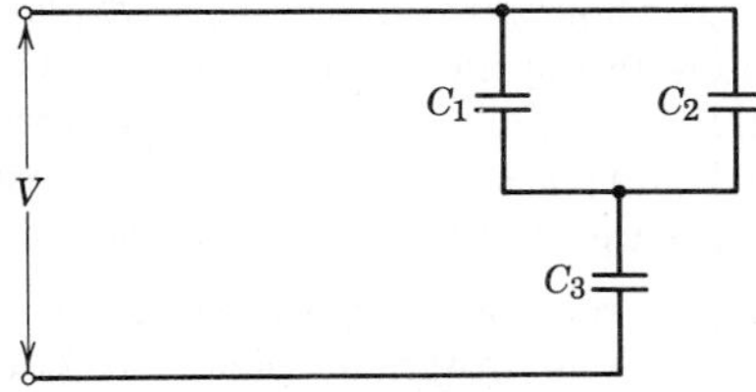

Figure 23 Exercise 19; Problems 27 and 48.

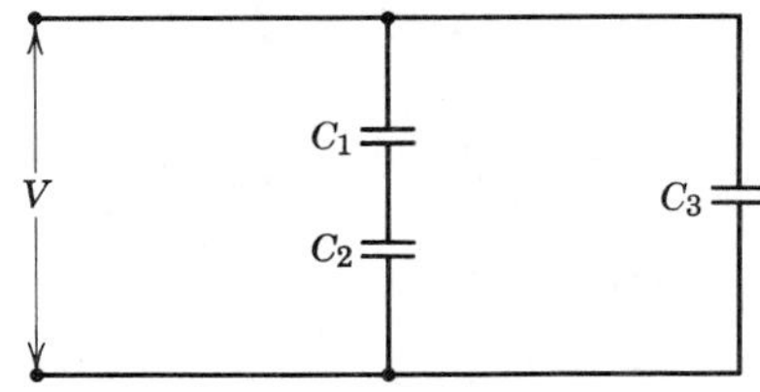

Figure 24 Exercise 20 and Problem 28.

20E. In Fig. 24 find the equivalent capacitance of the combination. Assume that $C_1 = 10\ \mu F$, $C_2 = 5.0\ \mu F$, and $C_3 = 4.0\ \mu F$.

21E. Each of the uncharged capacitors in Fig. 25 has a capacitance of 25 μF. A potential difference of 4200 V is established when the switch is closed. How many coulombs of charge then pass through the meter A?

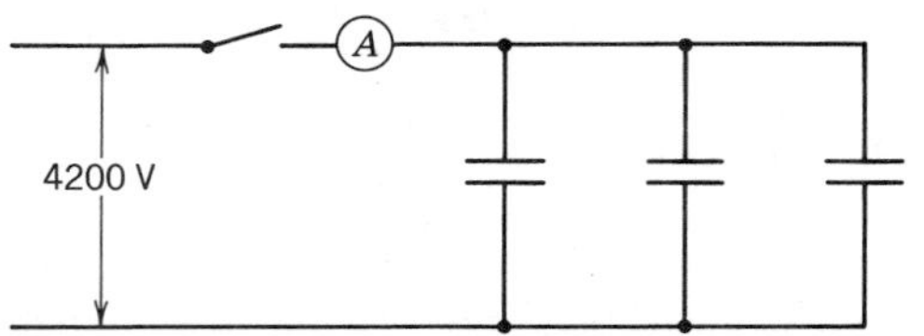

Figure 25 Exercise 21.

22E. A 6.0-μF capacitor is connected in series with a 4.0-μF capacitor; a potential difference of 200 V is applied across the pair. (*a*) Calculate the equivalent capacitance. (*b*) What is the charge on each capacitor? (*c*) What is the potential difference across each capacitor?

23E. Work the previous problem for the same two capacitors connected in parallel.

24P. (*a*) Three capacitors are connected in parallel. Each has plate area A and plate spacing d. What must be the spacing of a single capacitor of plate area A if its capacitance equals that of the parallel combination? (*b*) What must be the spacing if the three capacitors are connected in series?

25P. A potential difference of 300 V is applied to a 2.0-μF capacitor and an 8.0-μF capacitor connected in series. (*a*) What are the charge and the potential difference for each capacitor? (*b*) The charged capacitors are disconnected from each other and from the battery. They are then reconnected with their positive plates together and their negative plates together, no external voltage being applied. What are the charge and the potential difference for each? (*c*) The charged capacitors in (*a*) are reconnected with plates of *opposite* sign together. What are the steady-state charge and potential difference for each?

26P. In Fig. 26 a variable air capacitor of the type used in tuning radios is shown. Alternate plates are connected together, one group being fixed in position, the other group being capable of rotation. Consider a pile of n plates of alternate polarity, each having an area A and separated from adjacent plates by a distance d. Show that this capacitor has a maximum capacitance of

$$C = \frac{(n-1)\epsilon_0 A}{d}.$$

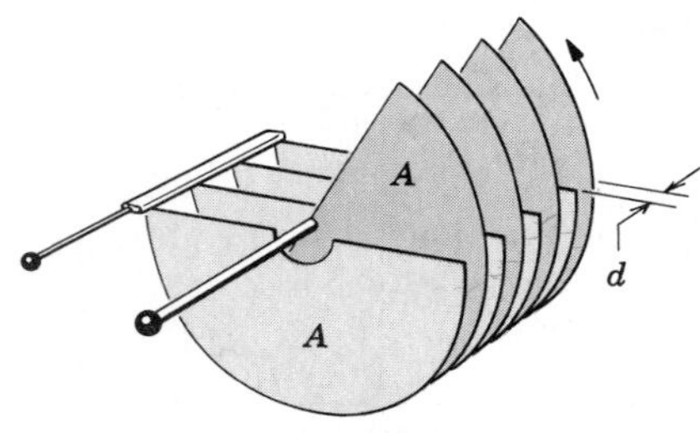

Figure 26 Problem 26.

27P. In Fig. 23 suppose that capacitor C_3 breaks down electrically, becoming equivalent to a conducting path. What *changes* in (*a*) the charge and (*b*) the potential difference occur for capacitor C_1? Assume that $V = 100$ V.

28P. In Fig. 24 find (*a*) the charge, (*b*) the potential difference, and (*c*) the stored energy for each capacitor. Assume the numerical values of Problem 20, with $V = 100$ V.

29P. You have several 2.0-μF capacitors, each capable of withstanding 200 V without breakdown. How would you assemble a combination having an equivalent capacitance of (*a*) 0.40 μF or of (*b*) 1.2 μF, each capable of withstanding 1000 V?

30P. Figure 27 shows two capacitors in series, the rigid center section of length b being movable vertically. Show that the equivalent capacitance of the series combination is independent of the position of the center section and is given by

$$C = \frac{\epsilon_0 A}{a - b}.$$

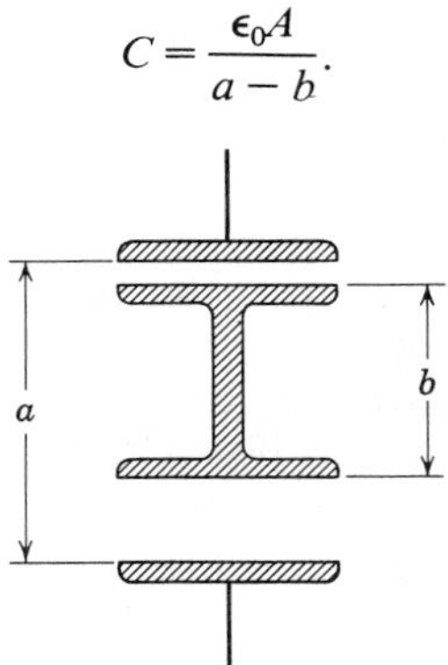

Figure 27 Problem 30.

31P. A 100-pF capacitor is charged to a potential difference of 50 V, the charging battery then being disconnected. The capacitor is then connected in parallel with a second (initially uncharged) capacitor. If the measured potential difference drops to 35 V, what is the capacitance of this second capacitor?

32P. In Figure 28, capacitors $C_1 = 1.0\ \mu F$ and $C_2 = 3.0\ \mu F$ are each charged to a potential $V = 100$ V but with opposite polarity, so that points a and c are on the side of the respective positive plates of C_1 and C_2, and points b and d are on the side of the respective negative plates. Switches S_1 and S_2 are now closed. (*a*) What is the potential difference between points e and f? (*b*) What is the charge on C_1? (*c*) What is the charge on C_2?

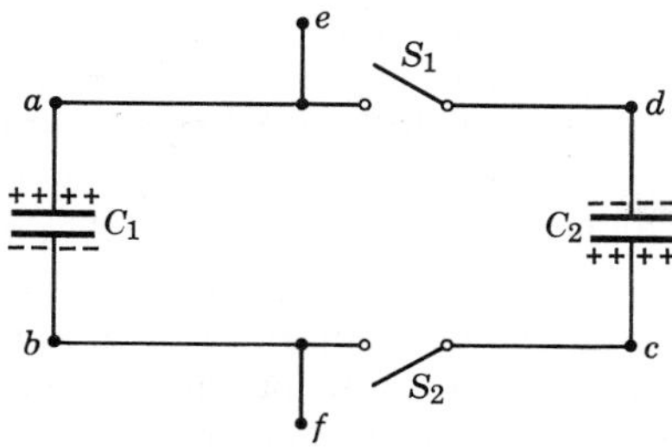

Figure 28 Problem 32.

33P. When switch S is thrown to the left in Fig. 29, the plates of the capacitor C_1 acquire a potential difference V_0. C_2 and C_3 are initially uncharged. The switch is now thrown to the right. What are the final charges q_1, q_2, q_3 on the corresponding capacitors?

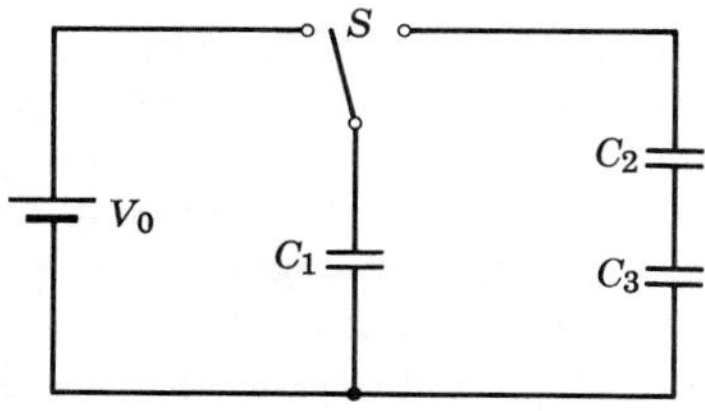

Figure 29 Problem 33.

34P. In Fig. 30 the battery B supplies 12 V. (*a*) Find the charge on each capacitor when switch S_1 is closed and (*b*) when (later) switch S_2 is also closed. Take $C_1 = 1.0\ \mu F$, $C_2 = 2.0\ \mu F$, $C_3 = 3.0\ \mu F$, and $C_4 = 4.0\ \mu F$.

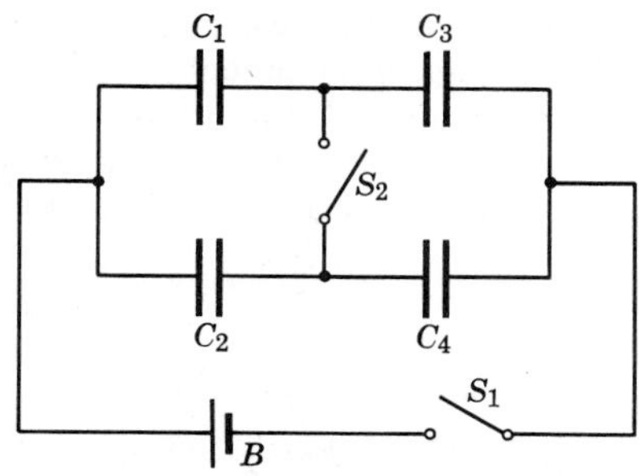

Figure 30 Problem 34.

35P. Figure 31 shows two identical capacitors C in a circuit with two (ideal) diodes D. A 100-V battery is connected to the input terminals, first with terminal a positive and later with terminal b positive. In each case, what is the potential difference across the output terminals? (An ideal diode has the property that positive charge flows through it only in the direction of the arrow and negative charge flows through it only in the opposite direction.)

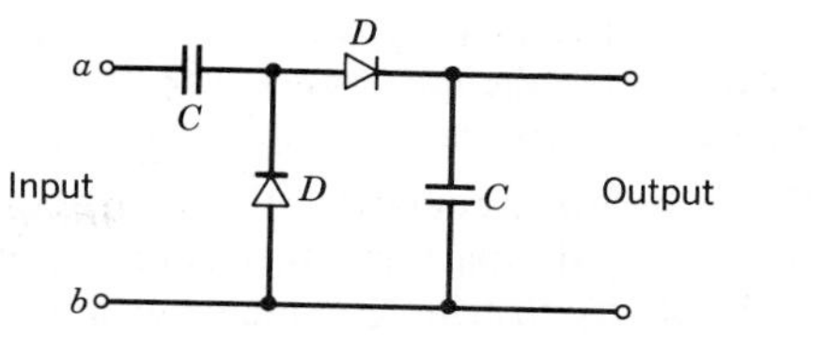

Figure 31 Problem 35.

Section 27-5 Storing Energy in an Electric Field

36E. How much energy is stored in one cubic meter of air due to the "fair weather" electric field of strength 150 V/m?

37E. Attempts to build a controlled thermonuclear fusion reactor which, if successful, could provide the world with a vast supply of energy from heavy hydrogen in sea water, usually involve huge electric currents for short periods of time in magnetic field windings. For example, ZT-40 at Los Alamos Scientific Laboratory has rooms full of capacitors. One of the capacitor banks provides 61 mF at 10 kV. Calculate the stored energy (*a*) in joules, and (*b*) in kW·h.

38E. What capacitance is required to store an energy of 10 kW·h at a potential difference of 1000 V?

39E. A parallel-plate air capacitor has a capacitance of 130 pF. (*a*) What is the stored energy if the applied potential difference is 56 V? (*b*) Can you calculate the energy density for points between the plates?

40E. A parallel-plate air capacitor having area 40 cm^2 and spacing of 1.0 mm is charged to a potential difference of 600 V. Find (*a*) the capacitance, (*b*) the magnitude of the charge on each plate, (*c*) the stored energy, (*d*) the electric field between the plates, and (*e*) the energy density between the plates.

41E. Two capacitors, 2.0 μF and 4.0 μF, are connected in parallel across a 300-V potential difference. Calculate the total energy stored in the capacitors.

42E. Calculate the energy density of the electric field at distance r from the center of an electron at rest. If the electron is a point particle, what does this calculation yield for the energy density at points very close to the electron?

43E. A certain capacitor is charged to a potential V. If you wish to increase its stored energy by 10%, by what percentage should you increase V?

44P. An isolated metal sphere whose diameter is 10 cm has a potential of 8000 V. Calculate the energy density in the electric field near the surface of the sphere.

45P. A parallel-connected bank of 2000 5.0-μF capacitors is used to store electric energy. What does it cost to charge this bank to 50,000 V, assuming a rate of 3.0¢/kW·h?

46P. For the capacitors of Problem 25, compute the energy stored for the three different connections of parts (*a*), (*b*), and (*c*). Compare your answers and explain any differences.

47P. One capacitor is charged until its stored energy is 4.0 J. A second uncharged capacitor is then connected to it in parallel. (*a*) If the charge distributes equally, what is now the total energy stored in the electric fields? (*b*) Where did the excess energy go?

48P. In Fig. 23 find (*a*) the charge, (*b*) the potential difference, and (*c*) the stored energy for each capacitor. Assume the numerical values of Exercise 19, with $V = 100$ V.

49P. A parallel-plate capacitor has plates of area A and separation d, and is charged to a potential difference V. The charging battery is then disconnected and the plates are pulled apart until their separation is $2d$. Derive expressions in terms of A, d, and V for (*a*) the new potential difference, (*b*) the initial and the final stored energy, and (*c*) the work required to separate the plates.

50P. A cylindrical capacitor has radii a and b as in Fig. 5. Show that half the stored electric potential energy lies within a cylinder whose radius is

$$r = \sqrt{ab}.$$

51P. Assume that the electron is not a point but a sphere of radius R over whose surface the electronic charge is uniformly distributed. (*a*) Determine the energy associated with the external electric field in vacuum of the electron as a function of R. (*b*) If you now associate this energy with the mass of the electron, you can, using $E = mc^2$, estimate the value of R. Evaluate this radius numerically; it is often called the *classical radius* of the electron.

Section 27-6 Capacitors with a Dielectric

52E. An air-filled parallel-plate capacitor has a capacitance of 1.3 pF. The separation of the plates is doubled and wax inserted between them. The new capacitance is 2.6 pF. Find the dielectric constant of the wax.

53E. Given a 7.4-pF air capacitor, you are asked to design a capacitor to store up to 7.4 μJ with a maximum potential difference of 652 V. What dielectric in Table 2 will you use to fill the gap in the air capacitor if you do not allow for a margin of error?

54E. For making a parallel-plate capacitor you have available two plates of copper, a sheet of mica (thickness = 0.10 mm, $\kappa = 5.4$), a sheet of glass (thickness = 2.0 mm, $\kappa = 7.0$), and a slab of paraffin (thickness = 1.0 cm, $\kappa = 2.0$). To obtain the largest capacitance, which sheet should you place between the copper plates?

55E. A parallel-plate air capacitor has a capacitance of 50 pF. (*a*) If its plates each have an area of 0.35 m^2, what is their separation? (*b*) If the region between the plates is now filled with material having a dielectric constant of 5.6, what is the capacitance?

56E. A coaxial cable used in a transmission line responds as a "distributed" capacitance to the circuit feeding it. Calculate the capacitance per meter for a cable having an inner radius of 0.10 mm and an outer radius of 0.60 mm. Assume that the space between the conductors is filled with polystyrene.

57P. A certain substance has a dielectric constant of 2.8 and a dielectric strength of 18 MV/m. If it is used as the dielectric material in a parallel-plate capacitor, what minimum area may the plates of the capacitor have in order that the capacitance be 7.0×10^{-2} μF and that the capacitor be able to withstand a potential difference of 4.0 kV?

58P. You are asked to construct a capacitor having a capacitance near 1 nF and a breakdown potential in excess of 10,000 V. You think of using the sides of a tall drinking glass (Pyrex), lining the inside and outside with aluminum foil (neglect the ends). What are (*a*) the capacitance and (*b*) breakdown potential? You use a glass 15 cm tall with an inner radius of 3.6 cm and an outer radius of 3.8 cm.

59P. You have been assigned to design a transportable capacitor that can store 250 kJ of energy. You select a parallel-plate type with dielectric. (*a*) What is the minimum capacitor volume achievable using a dielectric selected from those listed in Table 2 with values of dielectric strength? (*b*) Modern high-performance capacitors that can store 250 kJ have volumes of 0.087 m^3. Assuming that the dielectric used has the same dielectric strength as in (*a*), what must be its dielectric constant?

60P. Two parallel-plate capacitors have the same plate area A and separation d, but the dielectric constants of the materials between their plates are $\kappa + \Delta\kappa$ and $\kappa - \Delta\kappa$, respectively. (*a*) Find the equivalent capacitance when they are connected in parallel. (*b*) If the total charge on the parallel combination is Q, what is the charge on the capacitor with the larger capacitance?

61P. A slab of copper of thickness b is thrust into a parallel-plate capacitor as shown in Fig. 32; it is exactly halfway between the plates. (*a*) What is the capacitance after the slab is introduced? (*b*) If a charge q is maintained on the plates, find the ratio of the stored energy before to that after the slab is inserted. (*c*) How much work is done on the slab as it is inserted? Is the slab sucked in or do you have to push it in?

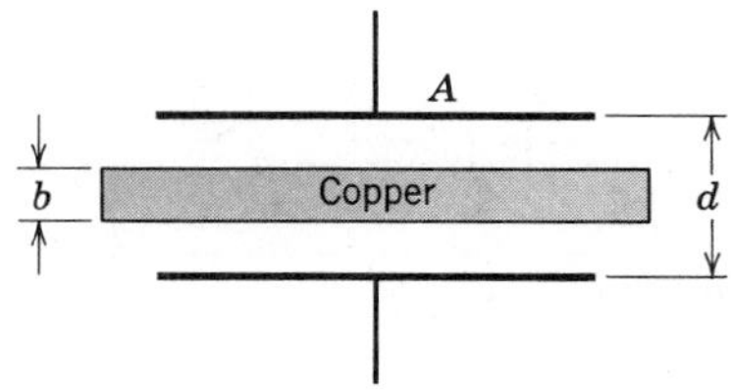

Figure 32 Problems 61 and 62.

62P. Reconsider Problem 61 assuming that the potential difference rather than the charge is held constant.

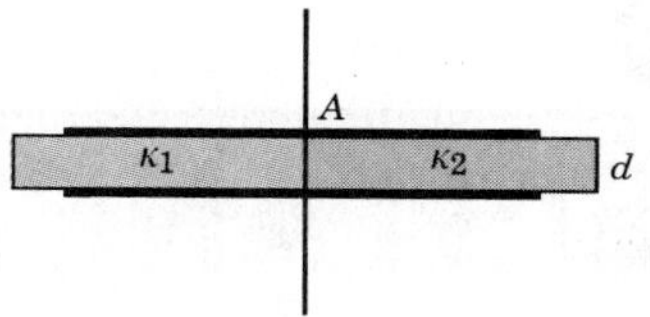

Figure 33 Problem 63.

63P. A parallel-plate capacitor is filled with two dielectrics as in Fig. 33. Show that the capacitance is given by

$$C = \frac{\epsilon_0 A}{d}\left(\frac{\kappa_1 + \kappa_2}{2}\right).$$

Check this formula for all the limiting cases that you can think of. (*Hint:* Can you justify regarding this arrangement as two capacitors in parallel?)

64P. A parallel-plate capacitor is filled with two dielectrics as in Fig. 34. Show that the capacitance is given by

$$C = \frac{2\epsilon_0 A}{d}\left(\frac{\kappa_1 \kappa_2}{\kappa_1 + \kappa_2}\right).$$

Check this formula for all the limiting cases that you can think of. (*Hint:* Can you justify regarding this arrangement as two capacitors in series?)

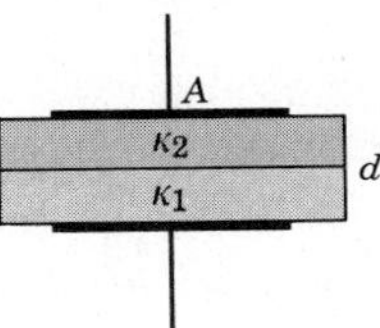

Figure 34 Problem 64.

65P. What is the capacitance of the capacitor in Fig. 35?

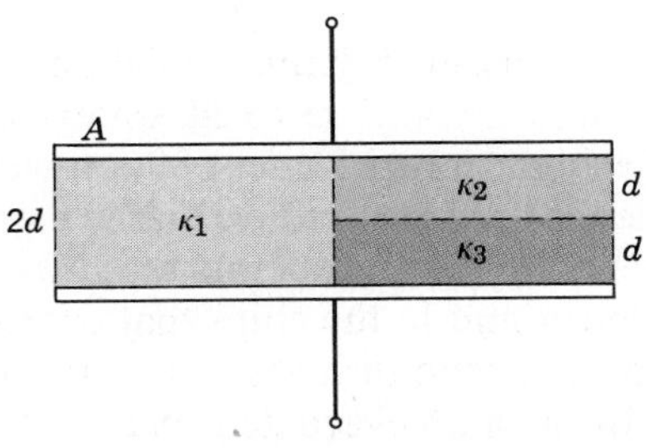

Figure 35 Problem 65.

Section 27-8 Dielectrics and Gauss' Law

66E. A parallel-plate capacitor has a capacitance of 100 pF, a plate area of 100 cm^2, and a mica dielectric ($\kappa = 5.4$). At 50-V potential difference, calculate (*a*) E in the mica, (*b*) the magnitude of the free charge on the plates, and (*c*) the magnitude of the induced surface charge.

67E. In Sample Problem 9, suppose that the battery remains connected during the time that the dielectric slab is being introduced. Calculate (*a*) the capacitance, (*b*) the charge on the capacitor plates, (*c*) the electric field in the gap, (*d*) the electric field in the slab, after the slab is introduced.

68P. Two parallel plates of area 100 cm^2 are each given equal but opposite charges of 8.9×10^{-7} C. The electric field within the dielectric material filling the space between the plates is 1.4×10^6 V/m. (*a*) Calculate the dielectric constant of the material. (*b*) Determine the magnitude of the charge induced on each dielectric surface.

69P. A parallel-plate capacitor has plates of area 0.12 m^2 and a separation of 1.2 cm. A battery charges the plates to a potential difference of 120 V and is then disconnected. A dielectric slab of thickness 40 mm and dielectric constant 4.8 is then placed symmetrically between the plates. (*a*) Find the capacitance before the slab is inserted. (*b*) What is the capacitance with the slab in place? (*c*) What is the free charge q before and after the slab is inserted? (*d*) Determine the electric field in the space between the plates and dielectric. (*e*) What is the electric field in the dielectric? (*f*) With the slab in place what is the potential difference across the plates? (*g*) How much external work is involved in the process of inserting the slab?

70P. In the capacitor of Sample Problem 9 (Fig. 16), (*a*) what fraction of the energy is stored in the air gaps? (*b*) What fraction is stored in the slab?

71P. A dielectric slab of thickness b is inserted between the plates of a parallel-plate capacitor of plate separation d. Show that the capacitance is given by

$$C = \frac{\kappa \epsilon_0 A}{\kappa d - b(\kappa - 1)}.$$

(*Hint:* Derive the formula following the pattern of Sample Problem 9.) Does this formula predict the correct numerical result of Sample Problem 9? Verify that the formula gives reasonable results for the special cases of $b = 0$, $\kappa = 1$, and $b = d$.

CHAPTER 28

CURRENT AND RESISTANCE

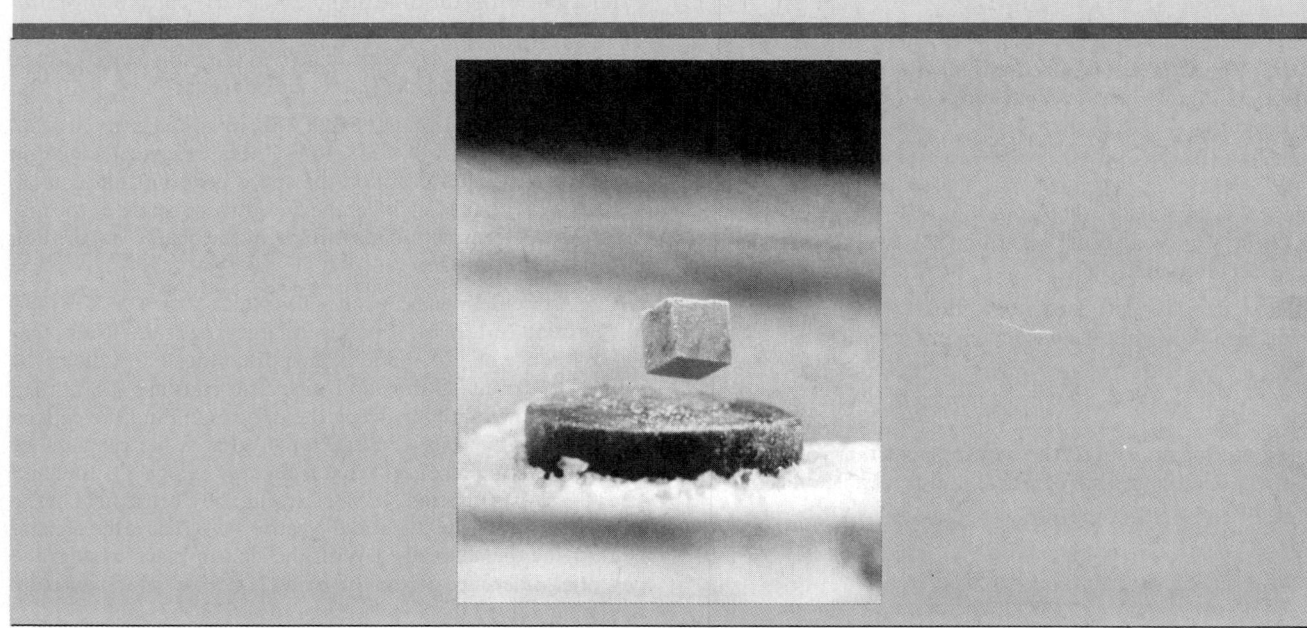

The 1870s saw the dawn of the electrical age as copper wires *began to snake into our homes. The age of the* semiconductor *began in the 1950s. The 1980s may well herald the age of the* superconductor. *These materials—which have no electrical resistance at all—have many tantalizing and potentially useful properties. We see here a small magnet floating in air above a disk of superconducting material that is immersed in a bath of liquid nitrogen. . . . Can widespread use of high-speed levitated trains be far behind?**

28-1 Moving Charges and Electric Currents

The previous five chapters dealt with *electrostatics,* that is, with charges at rest. With this chapter we begin our study of *electric currents,* that is, of charges in motion.

Examples of electric currents abound, ranging from the large currents that constitute lightning strokes to the tiny nerve currents that regulate our muscular activity. The currents in household wiring, in light bulbs, and in electric appliances are familiar to all. Charge carriers of *both* signs flow in the ionized gases of fluorescent lamps, in the batteries of transistor radios, and in car batteries. Electric currents in semiconductors are to be found in pocket calculators and in the chips that control microwave ovens and electric dishwashers. A beam of electrons moves through an evacuated space in the picture tube of a TV set.

On a global scale, charged particles trapped in the Van Allen radiation belts surge back and forth above the atmosphere between the north and the south magnetic poles. We see the effects of occasional "spillover" from

* See Essay 12, *Superconductivity,* by Peter Lindenfeld and Essay 13, *Magnetic Flight,* by Gerard K. O'Neill, for a fuller account of the magnetic properties and applications of superconductors.

these belts in the auroral displays of the northern and southern polar regions. On the scale of the solar system, enormous currents of protons, electrons, and ions fly radially outward from the sun as the *solar wind.* On the galactic scale, cosmic rays, which are largely energetic protons, stream through the galaxy.

Although an electric current is a stream of moving charges, not all moving charges constitute an electric current. If we are to say that an electric current passes through a given surface, there must be a net flow of charge through that surface. Two examples clarify our meaning.

First Example. The conduction electrons in an isolated length of copper wire are in random motion at speeds of the order of 10^6 m/s. If you pass a hypothetical plane through such a wire, conduction electrons pass through it *in both directions* at the rate of many billions per second. However, there is no *net* transport of charge and thus no current. However, if you connect the ends of the wire to a battery, you bias the flow—ever so slightly—in one direction so that there now is a net transport of charge and thus an electric current.

Second Example. The flow of water through a garden hose represents the directed flow of positive charge at a rate of perhaps several million coulombs per second. There is no net transport of charge, however, because there is a parallel flow of negative charge of exactly the same amount moving in exactly the same direction.

In this chapter we restrict ourselves largely to the study—within the framework of classical physics—of *steady* currents of *conduction electrons* moving through *metallic conductors* such as copper wires.

28-2 Electric Current

As Fig. 1*a* reminds us, an isolated conductor in electrostatic equilibrium—whether charged or not—is all at the same potential. No electric field can exist within it or parallel to its surface. Although conduction electrons are available, no net electric force acts on them and thus there is no current.

If, as in Fig. 1*b*, we insert a battery in the loop, the conducting loop is no longer an equipotential. Electric fields act in its interior, exerting forces on the conduction electrons and establishing a current. After a very short time, the electron flow reaches a steady-state condition. The situation is then completely analogous to the steady-state fluid flow that we studied in Chapter 16.

Figure 2 shows a section of a conductor, part of a

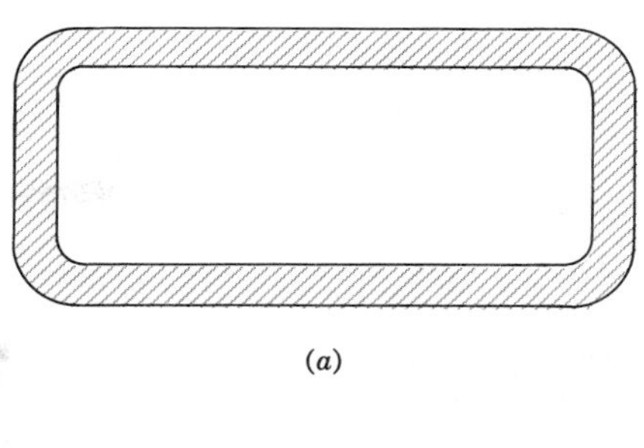

(*a*)

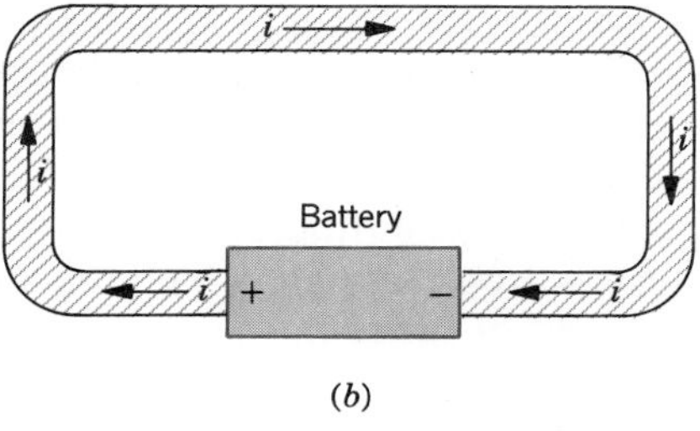

(*b*)

Figure 1 (*a*) A loop of copper in electrostatic equilibrium. The entire loop is at a single potential and the electric field for all internal points is zero. (*b*) Adding a battery imposes a potential difference. An electric field appears inside the conductor and causes charges to move around the loop, constituting a current.

conducting loop in which current has been established. The amount of charge dq that passes through a hypothetical plane, such as xx, is proportional to the length of time dt required for all the charge dq to pass through that plane. The proportionality constant is the current i; therefore,

$$\boxed{dq = i\,dt} \quad \text{(definition of current).} \qquad (1)$$

Using Eq. 1, we can find the charge that passes through plane xx in a time interval extending from 0 to t

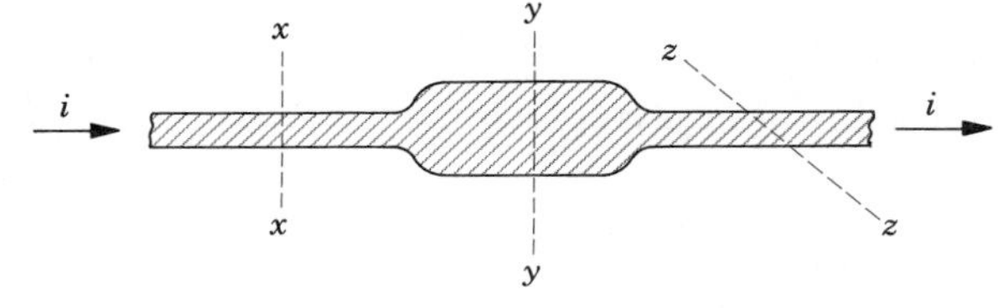

Figure 2 The current i through the conductor has the same value at planes xx, yy, and zz. The current density **J** at plane yy is smaller than at plane xx because the cross-sectional area of the conductor is greater at yy.

from

$$q = \int dq = \int_0^t i\,dt, \tag{2}$$

in which the current i may—or may not—be a function of time.

Under steady-state conditions, the current is the same for planes yy and zz and indeed for all planes that pass completely through the conductor, no matter what their location or orientation. This follows from the fact that charge is *conserved.* Under the steady-state conditions that we have assumed, an electron must enter the conductor at one end for every electron that leaves at the other. In the same way, if we have a steady flow of water through a garden hose, a drop of water must leave the nozzle for every drop that enters the hose at the other end. The amount of water in the hose is a conserved quantity.

The SI unit for current is the coulomb per second or the ampere (abbr. A); that is,

$$1 \text{ ampere} = 1 \text{ A} = 1 \text{ coulomb/second} = 1 \text{ C/s}.$$

The ampere is an SI base unit; the coulomb is defined in terms of the ampere, as we discussed earlier in Chapter 23. We shall present the formal definition of the ampere in Chapter 31.

Current, defined by Eq. 1, is a scalar because both q and t in that equation are scalars. This may cause some difficulty because we often represent a current in a wire by an arrow to indicate the direction in which the charges are moving. Such arrows are not vectors, however, because they do not obey the laws of vector addition. Figure 3a shows a conductor splitting at a junction into two branches. Because charge is conserved, the magnitudes of the currents in the branches must add to yield the magnitude of the current in the original conductor, or

$$i_0 = i_1 + i_2. \tag{3}$$

As Fig. 3b suggests, bending or reorienting the wires in space does not change the validity of Eq. 3. Current arrows are not vectors; they only show a direction (or sense) of flow along a conductor, not a direction in space.

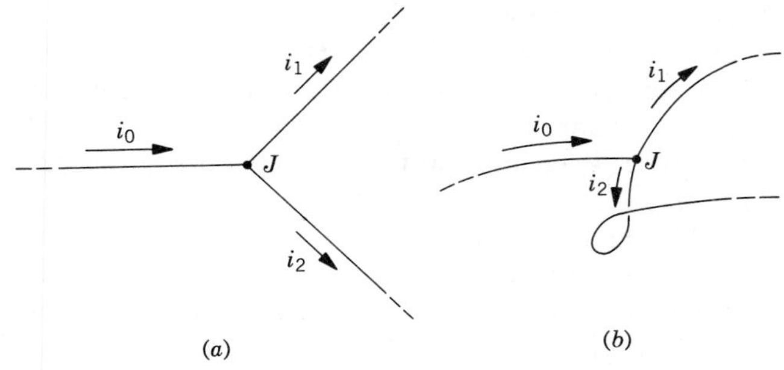

Figure 3 The relation $i_0 = i_1 + i_2$ is true at junction J no matter what the orientation in space of the three wires. Currents are scalars, not vectors.

The Directions of Currents.

In Fig. 1b we drew the current arrows in the direction that a positive charge carrier—repelled by the positive battery terminal and attracted by the negative terminal—would circulate around the loop. Actually, the charge carriers in the copper loop of Fig. 1b are electrons, which carry a negative charge. They circulate in a direction opposite to that of the current arrows. Recall also that in a fluorescent lamp, charge carriers of *both* signs are present. Since positive and negative charge carriers move in opposite directions, we must choose which direction a current arrow represents.

In drawing the current arrows in Fig. 1b in a clockwise direction, we were following this convention:

> *The current arrow is drawn in the direction that positive carriers would move, even if the actual carriers are not positive.*

It is only when we are interested in the detailed mechanism of charge transport that we need to pay attention to what the signs of the charge carriers actually are.

This convention is possible only because a positive charge carrier moving from left to right has the same external effect as a negative carrier moving from right to left.* In Fig. 4, for example, the two previously un-

* This is not entirely true. In Section 30–4 we discuss the *Hall effect,* which can be used in simple cases to establish the actual sign of the charge carriers.

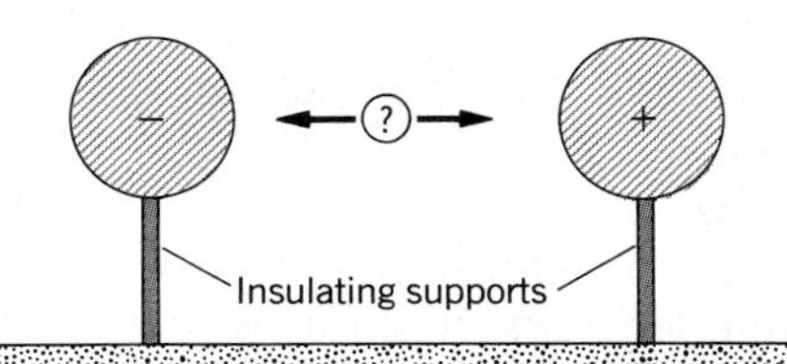

Figure 4 The charges on the two previously uncharged spheres arose by transferring charge from one sphere to the other. You cannot tell, after the event, whether positive charge was transferred from left to right or negative charge from right to left; the end result is the same.

charged spheres may have been charged by transporting positive charge from left to right or negative charge from right to left; both processes lead to the same end result.

Sample Problem 1 Water flows through a garden hose at a rate R of 450 cm³/s. To what current of negative charge does this correspond?

The current of negative charge is the rate at which molecules pass through any plane that cuts across the hose times the amount of negative charge carried by each molecule. If ρ is the density of water and M is its molecular mass, the rate (in moles per second) at which water is flowing through the plane is $R\rho/M$. If N_A is the Avogadro constant, the rate dN/dt at which molecules pass through the plane is

$$\frac{dN}{dt} = \frac{R\rho N_A}{M}$$
$$= \frac{\left(450 \times 10^{-6}\,\frac{\text{m}^3}{\text{s}}\right)\left(1000\,\frac{\text{kg}}{\text{m}^3}\right)\left(6.02 \times 10^{23}\,\frac{\text{molecules}}{\text{mol}}\right)}{0.018\ \text{kg/mol}}$$
$$= 1.51 \times 10^{25}\ \text{molecules/s}.$$

Each water molecule contains 10 electrons, 8 for the oxygen atom and 1 each for the two hydrogens. Each electron carries a charge of $-e$ so that the magnitude of the current corresponding to this movement of negative charge is

$$i = \frac{dq}{dt} = (1.51 \times 10^{25}\ \text{molecules/s})(10\ \text{electrons/molecule})$$
$$\times (1.60 \times 10^{-19}\ \text{C/electron})$$
$$= 2.4 \times 10^7\ \text{C/s} = 2.4 \times 10^7\ \text{A} = 24\ \text{MA}. \quad \text{(Answer)}$$

This current of negative charge is exactly compensated by a current of positive charge associated with the (positively charged) nuclei of the three atoms that make up the water molecule. Thus, there is no net flow of charge through the hose.

28-3 Current Density

Sometimes we are interested in the current i in a particular conductor. At other times we take a localized view and are interested in the flow of charge at a particular point within a conductor. A (positive) charge carrier at a given point will flow in the direction of the electric field **E** at that point. To describe this flow, we introduce the *current density* **J**, a vector quantity that points in the direction of this field.

Figure 5*a* shows a simple case in which a current i is uniformly distributed over the cross section of a uniform conductor. The current density J in this case is constant for all points within the conductor and is related to the current i by

$$J = i/A, \tag{4}$$

in which A is the cross-sectional area of the conductor. The SI unit for current density is the ampere per square meter (abbr. A/m²). As comparison of Figs. 5*a* and 5*b* shows, the direction of the current density **J** is that of the electric field **E** regardless of the sign of the charge carriers.

For any surface—whether plane or not—through which there is a current i, the current density **J** for points on that surface is related to i by*

$$i = \int \mathbf{J} \cdot d\mathbf{A}. \tag{5}$$

For the cross section xx in Fig. 2, **J** is constant in magnitude and direction and Eq. 5 readily reduces to Eq. 4.

In Section 16-9 we showed that, in fluid flow, the vector field represented by the array of velocity vectors of the fluid particles could be represented by a system of streamlines. The vector field represented by the array of current density vectors within a conductor can be represented in the same way. Figure 6, which is actually Fig. 19 of Chapter 16, can represent either the flow of a fluid

* Equation 5 shows that current is the flux of the current density through a surface. You may wish to review Section 25-4, in which we discuss a different flux, namely, the flux of the electric field through a surface.

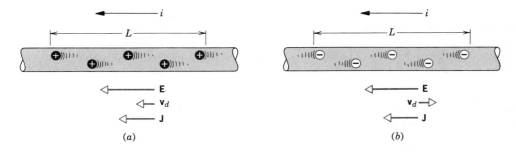

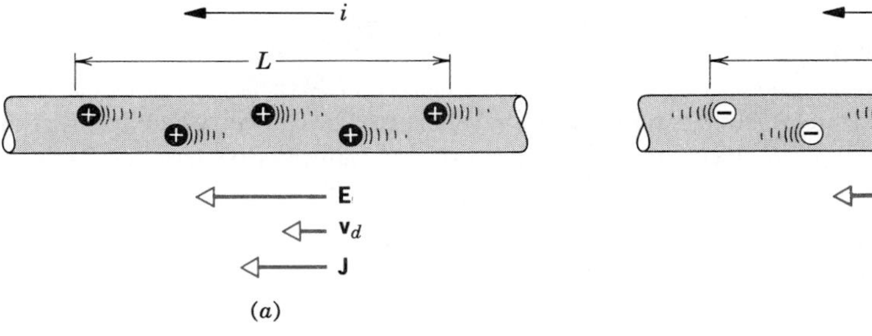

Figure 5 (*a*) Positive carriers drift in the direction of the applied electric field **E**. (*b*) Negative carriers drift in the opposite direction. By convention, the direction of the current density **J** and the sense of the current arrow are drawn *as if* the charge carriers were positive.

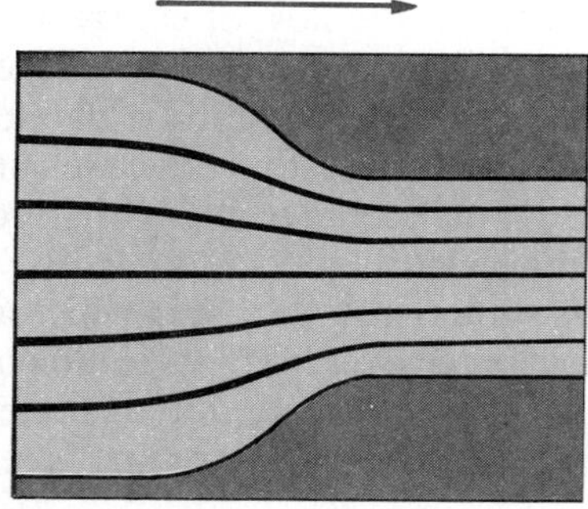

Figure 6 Streamlines representing the current density vectors in the flow of charge through a constricted conductor. Compare with Fig. 19 of Chapter 16, which shows the streamlines representing the velocity vectors in fluid flow.

through a constricted pipe or the flow of charge through a constricted conductor.

Calculating the Drift Speed. The conduction electrons in a copper conductor have *randomly directed* velocities of the order of 10^6 m/s. The *directed* flow of the conduction electrons for a typical current, that in household wiring, for example, is characterized by an average *drift speed* that is very much smaller, possibly of the order of 10^{-3} m/s.

For an apt analogy, consider a large crowd of people running around in random directions and jostling each other constantly. If the crowd is standing on a nearly level surface that slopes ever so slightly in a particular direction, the crowd will work its way slowly down the slope in that direction. A small but directed "drift speed" will be superimposed on the random jostling motion. For the conduction electrons, it is this drift speed that determines the current.

Now let us estimate the drift speed. Figure 5*a* shows the charge carriers in a wire moving to the left with an assumed constant drift speed v_d.* The number of carriers in a length L of the wire is nAL, where n is the number of carriers per unit volume and AL is the volume of a specified length of wire. A charge of magnitude

$$\Delta q = (nAL)e$$

passes out of this volume, through the left end of the conductor, in a time interval given by

$$\Delta t = L/v_d.$$

* Following convention, we assume that the charge carriers are positive, even though in fact they are (negative) conduction electrons.

The current in the wire is

$$i = \frac{\Delta q}{\Delta t} = \frac{nALe}{L/v_d} = nAev_d \tag{6}$$

Solving for v_d and recalling Eq. 4 ($J = i/A$), we obtain

$$v_d = \frac{i}{nAe} = \frac{J}{ne}$$

or, extended to vector form,

$$\boxed{\mathbf{J} = (ne)\mathbf{v}_d.} \tag{7}$$

Here the product ne, whose SI units are C/m^3, is the carrier charge density. For positive carriers, which we always assume, ne is positive and Eq. 7 predicts that $\mathbf{J}$ and $\mathbf{v}_d$ point in the same direction.

Sample Problem 2 One end of an aluminum wire whose diameter is 2.5 mm is welded to one end of a copper wire whose diameter is 1.8 mm. The composite wire carries a steady current i of 1.3 A. What is the current density in each wire?

We may take the current density as constant within each wire except for points near the junction. The current density is given by Eq. 4,

$$J = \frac{i}{A}.$$

The cross-sectional area A of the aluminum wire is

$$A_{\text{Al}} = \tfrac{1}{4}\pi d^2 = (\pi/4)(2.5 \times 10^{-3}\ \text{m})^2 = 4.91 \times 10^{-6}\ \text{m}^2$$

so that

$$J_{\text{Al}} = \frac{1.3\ \text{A}}{4.91 \times 10^{-6}\ \text{m}^2}$$
$$= 2.6 \times 10^5\ \text{A/m}^2 = 26\ \text{A/cm}^2. \qquad \text{(Answer)}$$

As you can verify, the cross-sectional area of the copper wire proves to be 2.54×10^{-6} m^2 so that

$$J_{\text{Cu}} = \frac{1.3\ \text{A}}{2.54 \times 10^{-6}\ \text{m}^2}$$
$$= 5.1 \times 10^5\ \text{A/m}^2 = 51\ \text{A/cm}^2. \qquad \text{(Answer)}$$

The fact that the wires are of different materials does not enter here.

Sample Problem 3 What is the drift speed of the conduction electrons in the copper wire of Sample Problem 2?

The drift speed is given by Eq. 7,

$$v_d = \frac{J}{ne}.$$

In copper, there is very nearly one conduction electron per atom on the average. The number n of electrons per unit volume is therefore the same as the number of atoms per unit volume and is given by

$$\frac{n}{N_A} = \frac{\rho_m}{M} \quad \text{or} \quad \left(\frac{\text{atoms/m}^3}{\text{atoms/mol}} = \frac{\text{mass/m}^3}{\text{mass/mol}}\right).$$

Here ρ_m is the (mass) density of copper, N_A is the Avogadro constant, and M is the atomic mass of copper.* Thus,

$$n = \frac{N_A \rho_m}{M}$$
$$= \frac{(6.02 \times 10^{23}\ \text{mol}^{-1})(9.0 \times 10^3\ \text{kg/m}^3)}{64 \times 10^{-3}\ \text{kg/mol}}$$
$$= 8.47 \times 10^{28}\ \text{electrons/m}^3.$$

We then have

$$v_d = \frac{5.1 \times 10^5\ \text{A/m}^2}{(8.47 \times 10^{28}\ \text{electrons/m}^3)(1.60 \times 10^{-19}\ \text{C/electron})}$$
$$= 3.8 \times 10^{-5}\ \text{m/s} = 14\ \text{cm/h}. \qquad \text{(Answer)}$$

You may well ask: "If the electrons drift so slowly, why do the room lights turn on so quickly after I throw the switch?" Confusion on this point results from not distinguishing between the drift speed of the electrons and the speed at which *changes* in the electric field configuration travel along wires. This latter speed approaches that of light; electrons everywhere in the wire begin drifting almost at once. Similarly, when you turn the valve on your garden hose, with the hose full of water, a pressure wave travels along the hose at the speed of sound in water. The speed at which the water moves through the hose—measured perhaps with a dye marker—is much lower.

Sample Problem 4 A strip of silicon has a width w of 3.2 mm, a thickness t of 250 μm, and carries a current i of 5.2 mA. The silicon is an *n-type semiconductor,* having being "doped" with a controlled phosphorus impurity. The doping has the effect of greatly increasing n, the number of charge carriers per unit volume, as compared with the value for pure silicon. In this case, $n = 1.5 \times 10^{23}\ \text{m}^{-3}$. (a) What is the current density in the strip?

From Eq. 4,

$$J = \frac{i}{wt} = \frac{5.2 \times 10^{-3}\ \text{A}}{(3.2 \times 10^{-3}\ \text{m})(250 \times 10^{-6}\ \text{m})}$$
$$= 6500\ \text{A/m}^2. \qquad \text{(Answer)}$$

(b) What is the drift speed?

* We use the subscript m to make it clear that the density referred to here is a mass density (kg/m³), not a charge density (C/m³).

From Eq. 7,

$$v_d = \frac{J}{ne} = \frac{6500\ \text{A/m}^2}{(1.5 \times 10^{23}\ \text{m}^{-3})(1.60 \times 10^{-19}\ \text{C})}$$
$$= 0.27\ \text{m/s} = 27\ \text{cm/s}. \qquad \text{(Answer)}$$

The drift speed (0.27 m/s) of the electrons in this silicon semiconductor is much greater than the drift speed (3.8×10^{-5} m/s) of the conduction electrons in the metallic copper conductor of Sample Problem 3. This is because the density of charge carriers ($1.5 \times 10^{23}\ \text{m}^{-3}$) in the silicon semiconductor is much smaller than the density of charge carriers ($8.47 \times 10^{28}\ \text{m}^{-3}$) in the copper conductor; the fewer charge carriers that are available, the faster they must drift to transport the same amount of charge per unit time.

28-4 Resistance and Resistivity

If we apply the same potential difference between the ends of geometrically similar rods of copper and of glass, very different currents result. The characteristic of the conductor that enters here is its *resistance.* We determine the resistance of a conductor between any two points by applying a potential difference V between those points and measuring the current i that results. The resistance R is then

$$\boxed{R = V/i} \qquad \text{(definition of } R\text{)}. \qquad (8)$$

The SI unit for resistance that follows from Eq. 8 is the volt/ampere. This combination occurs so often that we give it a special name, the *ohm* (symbol Ω). That is,

$$1\ \text{ohm} = 1\ \Omega = 1\ \text{volt/ampere} = 1\ \text{V/A}. \qquad (9)$$

A conductor whose function in a circuit is to provide a specified resistance is called a *resistor;* see Fig. 7. We represent a resistor in a circuit diagram by the symbol ⋀⋀⋀.

The resistance of a conductor depends on the manner in which the potential difference is applied to it. Figure 8, for example, shows a given potential difference applied in two different ways to the same conductor. As the current density streamlines suggest, the currents in the two cases—and hence the measured resistances—will be quite different.

The flow of charge through a resistor is often compared with the flow of water through a pipe when there is a pressure difference between the ends of the pipe. The pump that establishes the pressure difference between

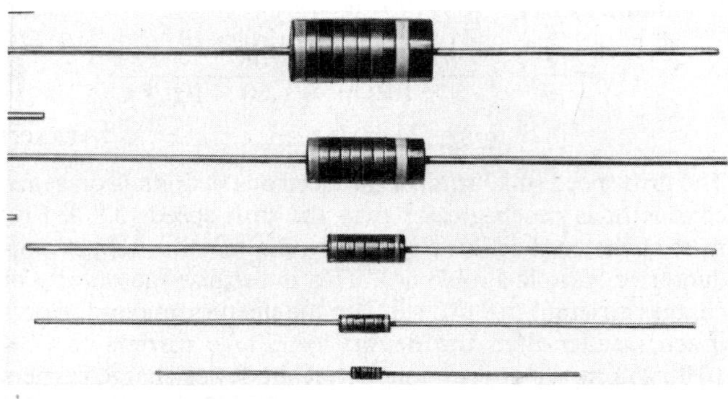

Figure 7 An assortment of resistors. The circular bands are color coding marks that identify the value of the resistance and the maximum allowable power dissipation.

the ends of the pipe can be compared with the battery that establishes the potential difference between the ends of the resistor. The flow of water (m^3/s) can be compared with the flow of charge (C/s or A). The rate of flow of water for a given pressure difference depends on the characteristics of the pipe. Is it long or short? Is it narrow or wide? Is it empty or filled with sand or gravel? These characteristics of the pipe are analogous to those that determine the resistance of the resistor.

As we have done several times in other connections, we often wish to take a local view and deal not with a particular object but with a substance. We do this by focusing not on the potential difference V across a particular resistor but on the electric field $\mathbf{E}$ at a point in a resistive material. Instead of dealing with the current i through the resistor, we deal with the current density $\mathbf{J}$ at the point in question. Instead of the resistance R we define a new quantity, the *resistivity* ρ, from

$$\boxed{\rho = E/J} \quad \text{(definition of } \rho\text{)}. \tag{10}$$

The SI units of E are V/m and those of J are A/m^2. The SI units of ρ can then be seen to be $\Omega\cdot$m, pronounced "ohm · meter."*

$$\left(\frac{\text{V/m}}{\text{A/m}^2}\right)\left(\frac{1\ \Omega}{1\ \text{V/A}}\right) = \Omega\cdot\text{m}.$$

The conversion factor in the second parentheses is based on Eq. 8, the definition of the ohm.

We can write Eq. 10 in vector form as

$$\boxed{\mathbf{E} = \rho\mathbf{J}.} \tag{11}$$

Equations 10 and 11 hold only for isotropic materials—materials in which the electrical properties are the same in all directions.

We often speak of the *conductivity* σ of a material. This is simply the reciprocal of its resistivity, or†

$$\boxed{\sigma = 1/\rho} \quad \text{(definition of } \sigma\text{)}. \tag{12}$$

The SI units of σ are $(\Omega\cdot\text{m})^{-1}$, pronounced "reciprocal ohm · meters."‡ Table 1 lists the resistivities of some materials.

Calculating the Resistance. If we know the resistivity of a substance such as copper, we should be able to calculate the resistance of a length of wire of given diameter made of that substance. Let A (see Fig. 9) be the cross-sectional area of the wire, let L be its length, and let a

* An instrument used to measure resistance is called an *ohmmeter;* it is quite different from the *ohm · meter* ($\Omega\cdot$m), a unit of resistivity. In the same spirit, the *micrometer* (mi · crom′ · e · ter) is an instrument used by machinists; it is quite different from the *micrometer* (mi′ · cro · me · ter), a unit of length.

† We have used ρ earlier to represent both mass density and charge density. Also, we have used σ earlier to represent surface charge density. We use these symbols here with entirely different meanings; there are just not enough good symbols to go around.

‡ The usage mhos/m for conductivity is also not uncommon.

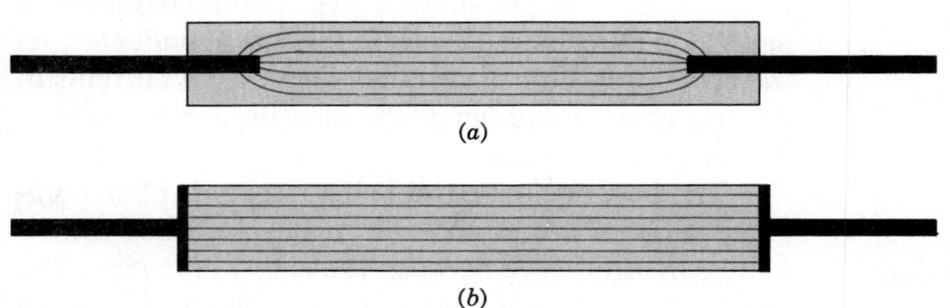

Figure 8 Two ways of applying a potential difference to a conducting rod. The heavy black conductors are assumed to have negligible resistance. Which rod will have the smaller measured resistance?

Table 1 Resistivities of Some Materials at Room Temperature (20°C)

Material	Resistivity ρ $\Omega\cdot$m	Temperature coefficient of resistivity α K^{-1}
	Typical Metals	
Silver	1.62×10^{-8}	4.1×10^{-3}
Copper	1.69×10^{-8}	4.3×10^{-3}
Aluminum	2.75×10^{-8}	4.4×10^{-3}
Tungsten	5.25×10^{-8}	4.5×10^{-3}
Iron	9.68×10^{-8}	6.5×10^{-3}
Platinum	10.6×10^{-8}	3.9×10^{-3}
Manganin[a]	48.2×10^{-8}	0.002×10^{-3}
	Typical Semiconductors	
Silicon pure	2.5×10^{3}	-70×10^{-3}
Silicon *n*-type[b]	8.7×10^{-4}	
Silicon *p*-type[c]	2.8×10^{-3}	
	Typical Insulators	
Glass	$10^{10}-10^{14}$	
Fused quartz	$\sim 10^{16}$	

[a] An alloy specifically designed to have a small value of α.
[b] Pure silicon "doped" with phosphorus impurities to a charge carrier density of 10^{23} m^{-3}
[c] Pure silicon "doped" with aluminum impurities to a charge carrier density of 10^{23} m^{-3}.

potential difference V exist between its ends. If the streamlines representing the current density are uniform throughout the wire, the electric field and the current density will be constant for all points within the wire and will have the values

$$E = V/L \quad \text{and} \quad J = i/A. \tag{13}$$

We can then combine Eqs. 10 and 13 to write

$$\rho = \frac{E}{J} = \frac{V/L}{i/A}. \tag{14}$$

But V/i is the resistance R, which allows us to recast Eq. 14 as

$$\boxed{R = \rho\frac{L}{A}.} \tag{15}$$

We stress that Eq. 15 can only be applied to a homogeneous isotropic conductor of uniform cross section and provided the potential difference is applied as in Fig. 8*b*.

The macroscopic quantities V, i, and R are of greatest interest when we are making electrical measurements on specific conductors. They are the quantities that we read directly on meters. We turn to the microscopic quantities E, J, and ρ when we are interested in the fundamental electrical behavior of matter, as we are in the research area of solid-state physics.

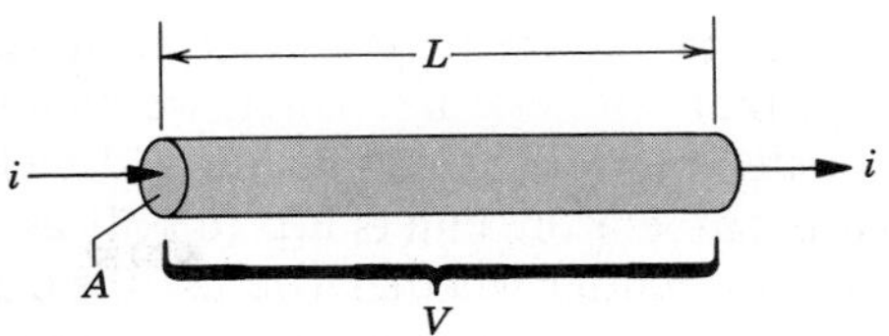

Figure 9 A potential difference V is applied between the ends of a wire of length L and cross section A, establishing a current i.

Variation with Temperature. The values of most physical properties vary with temperature and resistivity is no exception. Figure 10, for example, shows the variation of this property for copper over a wide range. The relation for copper—and for metals in general—is fairly linear over a rather broad range of temperatures. We can write, as an empirical approximation that is good enough for essentially all engineering purposes,

$$\boxed{\rho - \rho_0 = \rho_0\alpha(T - T_0).} \tag{16}$$

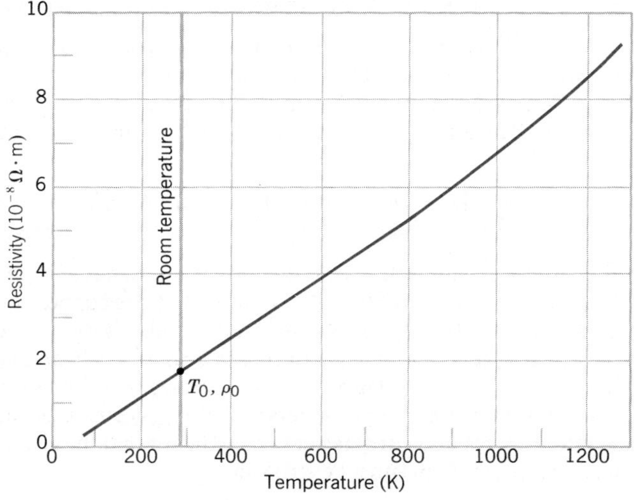

Figure 10 The resistivity of copper as a function of temperature. The dot on the curve marks a convenient room-temperature reference point ($T_0 = 293$ K, $\rho_0 = 1.69 \times 10^{-8}\ \Omega\cdot$m).

Here T_0 is a selected reference temperature and ρ_0 is the resistivity at that temperature. Often we choose $T_0 =$ 293 K (room temperature), for which $\rho_0 = 1.69\ \mu\Omega\cdot\text{cm}$.

Because temperature enters Eq. 16 only as a difference, it does not matter whether you use the Celsius or the Kelvin scales in that equation because the sizes of the degree on these scales are identical. The quantity α in Eq. 16, called the *temperature coefficient of resistivity,* is chosen so that the equation gives the best agreement with experiment for temperatures in the chosen range. Some values of α for different metals are listed in Table 1.

The variation of resistivity with temperature is so reproducible that a *platinum resistance thermometer* has been adopted as a secondary thermometric standard for measuring temperatures in the range 14–900 K on the International Practical Temperature Scale; see Section 19–5. In using this device to measure temperatures, terms proportional to $(T - T_0)^2$ and $(T - T_0)^3$ are added to the right side of Eq. 16, yielding an equation of improved precision.

Sample Problem 5 (a) What is the strength of the electric field present in the copper conductor of Sample Problem 2?

In Sample Problem 2 we found the current density J to be $5.1 \times 10^5\ \text{A/m}^2$; from Table 1 we see that the resistivity ρ for copper is $1.69 \times 10^{-8}\ \Omega\cdot\text{m}$. Thus, from Eq. 11

$$\begin{aligned} E &= \rho J = (1.69 \times 10^{-8}\ \Omega\cdot\text{m})(5.1 \times 10^5\ \text{A/m}^2) \\ &= 8.6 \times 10^{-3}\ \text{V/m} \quad \text{(copper).} \qquad \text{(Answer)} \end{aligned}$$

(b) What is the magnitude of the electric field in the *n*-type silicon semiconductor of Sample Problem 4?

From that Sample Problem we find that $J = 6500\ \text{A/m}^2$ and from Table 1 we see that $\rho = 8.7 \times 10^{-4}\ \Omega\cdot\text{m}$. Thus, from Eq. 11

$$\begin{aligned} E &= \rho J = (8.7 \times 10^{-4}\ \Omega\cdot\text{m})(6500\ \text{A/m}^2) \\ &= 5.7\ \text{V/m} \quad (n\text{-type silicon}). \qquad \text{(Answer)} \end{aligned}$$

We see that the electric field in the silicon semiconductor (5.7 V/m) is considerably higher than that in the copper conductor (8.7×10^{-3} V/m). We can understand this in terms of the much lower concentration of charge carriers in silicon than in copper. From the relation $J = nev_d$, we see that for a given current density, the charge carriers in silicon (because there are so few of them) must drift faster, which means that the electric field acting on them must be stronger.

Sample Problem 6 A rectangular block of iron has dimensions $1.2 \times 1.2 \times 15$ cm. (a) What is the resistance of the block measured between the two square ends? The resistivity of iron at room temperature is $9.68 \times 10^{-8}\ \Omega\cdot\text{m}$.

The area of a square end is $(1.2 \times 10^{-2}\ \text{m})^2$ or $1.44 \times 10^{-4}\ \text{m}^2$. From Eq. 15,

$$\begin{aligned} R &= \frac{\rho L}{A} = \frac{(9.68 \times 10^{-8}\ \Omega\cdot\text{m})(0.15\ \text{m})}{1.44 \times 10^{-4}\ \text{m}^2} \\ &= 1.0 \times 10^{-4}\ \Omega = 100\ \mu\Omega. \qquad \text{(Answer)} \end{aligned}$$

(b) What is the resistance between two opposing rectangular faces?

The area of a rectangular face is $(1.2 \times 10^{-2}\ \text{m})(0.15\ \text{m})$ or $1.80 \times 10^{-3}\ \text{m}^2$. From Eq. 15,

$$\begin{aligned} R &= \frac{\rho L}{A} = \frac{(9.68 \times 10^{-8}\ \Omega\cdot\text{m})(1.2 \times 10^{-2}\ \text{m})}{1.80 \times 10^{-3}\ \text{m}^2} \\ &= 6.5 \times 10^{-7}\ \Omega = 0.65\ \mu\Omega. \qquad \text{(Answer)} \end{aligned}$$

We assume in each case that the potential difference is applied to the block in such a way that the surfaces between which the resistance is desired are equipotentials. Otherwise, Eq. 15 would not be valid.

28-5 Ohm's Law

Figure 11*a* shows a "black box" with a potential difference V applied between its terminals. Figure 11*b* shows the current i that results for various values and polarities of V. The plot is a straight line passing through the origin, which means that the resistance V/i of the device in the box is a constant, independent of the potential difference used to measure it.

Figure 11*c* is a plot for a repetition of the experiment with a different object in the box. The V–i curve in the forward direction in this case is not a straight line and there is a total lack of symmetry when the polarity of the applied potential difference is reversed.

We say that the object in the box of Fig. 11*b*—which turns out to be a 1000-Ω resistor—obeys *Ohm's law.* The object in the box of Fig. 11*c*—which turns out to be a so-called *pn* junction diode—does not.

A conducting device obeys Ohm's law if its resistance between any two points is independent of the magnitude and polarity of the potential difference applied between those points.

Modern microelectronics—and therefore much of the character of our present technological civilization—depends almost totally on devices that do *not* obey Ohm's law. Your pocket calculator, for example, is full of them.

It is a common error to say that Eq. 8 ($V = Ri$) is a statement of Ohm's law. Not true! This equation is sim-

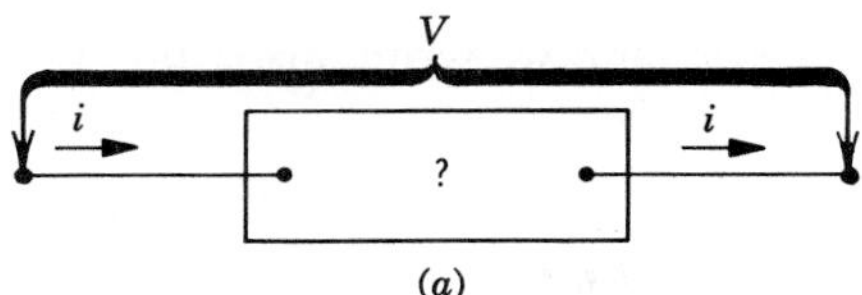

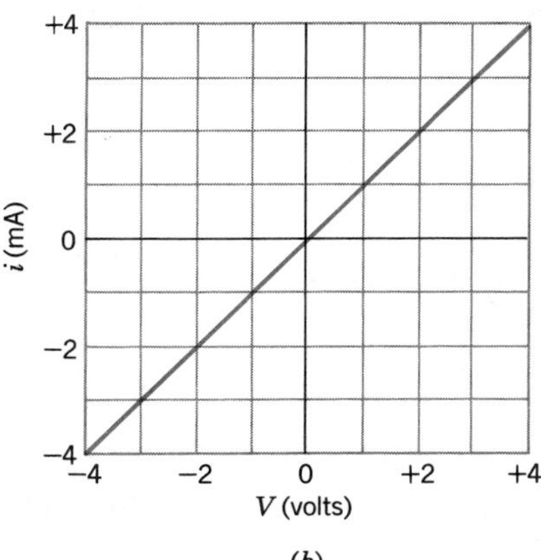

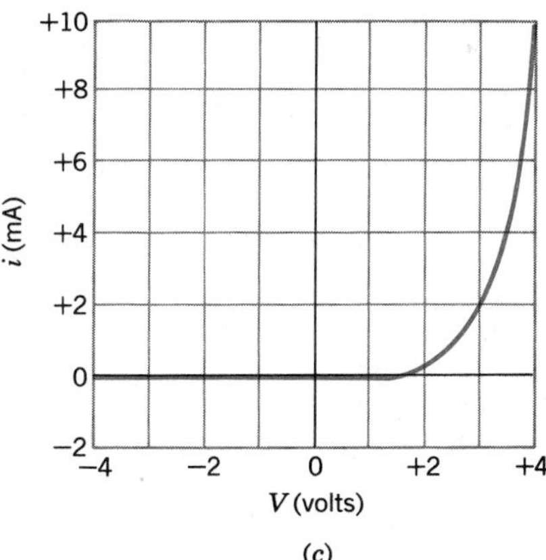

Figure 11 (*a*) A "black box" to whose terminals a potential difference V is applied, establishing a current i. (*b*) A V–i plot when the object in the box is a 1000-Ω resistor. (*c*) A V–i plot for a semiconducting *pn* junction diode.

ply the defining equation for resistance and applies to all conducting devices whether or not they obey Ohm's law. The essence of Ohm's law is that the $V-i$ curve is linear; that is, the value of R is independent of the value of V.

We can also express Ohm's law from the local point of view, in which we focus our interest on conducting materials rather than on conducting devices. The relevant relation is Eq. 11 ($\mathbf{E} = \rho\mathbf{J}$), which is the local analog of Eq. 8 ($V = Ri$).

A conducting material obeys Ohm's law if its resistivity is independent of the magnitude and direction of the applied electric field.

All homogeneous materials, be they conductors like copper or semiconductors like silicon (doped or pure), obey Ohm's law for some range of values of the electric field. If the field is too strong, however, there are departures from Ohm's law in all cases.

28-6 Ohm's Law: A Microscopic View

To find out *why* a given material such as copper or silicon obeys Ohm's law, we must look into the details of the conduction process at the atomic level. Here we consider only conduction in metals, such as copper.* We base our analysis on the *free-electron model,* in which we assume that the conduction electrons in the metal are free to move throughout the volume of the sample, like the molecules of a gas in a closed container.†

According to classical physics, the electrons should have a Maxwellian speed distribution somewhat like that of the molecules in a gas. In such a distribution, as we have seen, the average electron speed would be proportional to the square root of the absolute temperature; see Section 21-7. The motions of the electrons, however, are not governed by the laws of classical physics but by those of quantum physics. As it turns out, an assumption that is much closer to the quantum reality is that the electrons move with a single effective speed v_{eff}. For copper, $v_{\text{eff}} \approx 1.6 \times 10^6$ m/s, essentially independent of the temperature.

When we apply an electric field to the metal specimen, the electrons modify their random motions slightly and drift very slowly—in a direction opposite to that of the field—with an average drift speed v_d. As we saw in

* In Section 28-8 we consider the conduction mechanism in semiconductors.

† We assume that the electrons do not interact with each other but only with the atoms at the lattice sites. We assume further that after such a collision the electron emerges with a random velocity, with no memory—so to speak—of its initial velocity.

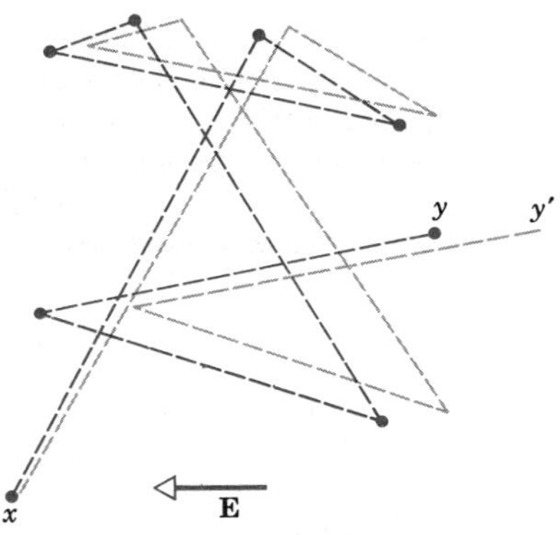

Figure 12 The colored lines show an electron moving from x to y, making six collisions en route. The gray lines show what its path *might* have been in the presence of an applied electric field **E**. Note the steady drift in the direction of $-\mathbf{E}$. (Actually, the gray lines should be slightly curved, to represent the parabolic paths followed by the electrons between collisions under the influence of the electric field.)

Sample Problem 3, the drift speed ($\sim 4 \times 10^{-5}$ m/s) in a typical metallic conductor is less than the effective speed (1.6×10^6 m/s) by many orders of magnitude. Figure 12 suggests the relation between these two speeds. The colored lines show a possible random path for an electron in the absence of an applied field; the electron proceeds from x to y, making six collisions along the way. The light gray lines show how the same events *might* have occurred if an electric field **E** had been applied. We see that the electron drifts steadily to the right, ending at y' rather than at y. In preparing Fig. 12, we assumed that $v_d \approx 0.02\ v_{\text{eff}}$; actually it is more like $v_d \approx 10^{-9}\ v_{\text{eff}}$, so that the drift displayed in the figure is greatly exaggerated.

If an electron of mass m is placed in an electric field E, it will experience an acceleration given by Newton's second law:

$$a = \frac{F}{m} = \frac{eE}{m}. \tag{17}$$

The nature of the collisions experienced by the electrons is such that, after a typical collision the electron will—so to speak—completely lose its memory of its accumulated drift velocity. The electron will then start off afresh after every encounter, moving off in a random direction. In the average time interval τ until the next collision, an average electron will change its velocity by an amount $a\tau$. As the Special Note that appears just before Sample Problem 7 makes clear, we identify this with the drift speed v_d and we write, using Eq. 17,

$$v_d = a\tau = \frac{eE\tau}{m}. \tag{18}$$

Combining this result with Eq. 7 yields

$$v_d = \frac{J}{ne} = \frac{eE\tau}{m},$$

which we can write as

$$E = \left(\frac{m}{e^2 n\tau}\right) J.$$

Comparing this with Eq. 11 ($E = \rho J$) leads to

$$\rho = \frac{m}{e^2 n\tau}. \tag{19}$$

Equation 19 may be taken as a statement that metals obey Ohm's law if we can show that ρ is a constant, independent of the strength of the applied electric field **E**. Because n, m, and e are constant, this reduces to convincing ourselves that τ, the mean free time between collisions, is a constant, independent of the strength of the applied electric field.

Figure 12 reminds us that the speed distribution of the conduction electrons is only minimally affected by even a rather strong electric field. Speaking loosely, a given electron in its random motion is scarcely aware whether the electric field is ON or OFF: its "collision experience" remains essentially unchanged. We may be confident that, whatever the value of τ for a metal such as copper in the absence of a field, its value remains essentially unchanged when a field is applied.

A Special Note.* There is a natural inclination to write Eq. 18 as $v_d = \frac{1}{2}a\tau$, reasoning that $a\tau$ is the electron's *final velocity* and that its *average velocity* is just half that value. Indeed, this assumption was made by the distinguished physicist Paul Drude (1863–1906) who first proposed the free-electron model of conduction. Equation 18, however, is correct as it stands, without the factor of $\frac{1}{2}$.

This factor would be appropriate if we followed a typical electron, averaged its velocity over its mean time τ between collisions, and called that the drift speed.

* We are grateful to Professor Philip A. Casabella, who suggested the substance of this argument to us. For a fuller account, see *Electricity and Magnetism,* 2nd ed., by Edward Purcell, McGraw-Hill, New York, 1985 (Section 4.4). Factors such as 2, $\sqrt{2}$, and 2π that require special understanding are not uncommon in physics.

However, the drift speed ($=J/ne$) is proportional to the current density and must be the velocity averaged over *all* the electrons *at one instant of time.*

When we take an average over all electrons, the random, thermal velocities clearly average to zero and make no contribution to the drift velocity. Thus, we need only concern ourselves with the effect of the electric field on the velocity of the electrons.

At any given instant, each electron has a velocity component at, produced by the electric field, where t is the time since the *last* collision experienced by that electron. The average velocity for all the electrons at our given instant—which is the drift speed—is the average value of at. Since the acceleration $a\,(=eE/m)$ is the same for all electrons, the average value of at is $a\tau$, where τ is the average time since the last collision. It may take some thought to convince yourself that—since you are free to start your stopwatch for each electron at the time it makes a collision—the average time since the last collision, the average time to the next collision, and the average time between collisions, all have the same value τ.

Sample Problem 7 (a) What is the mean free time τ between collisions for the conduction electrons in copper?

From Eq. 19 we have

$$\tau = \frac{m}{ne^2\rho}$$
$$= \frac{9.1 \times 10^{-31}\ \text{kg}}{(8.47 \times 10^{28}\ \text{m}^{-3})(1.60 \times 10^{-19}\ \text{C})^2(1.69 \times 10^{-8}\ \Omega\cdot\text{m})}$$
$$= 2.5 \times 10^{-14}\ \text{s}. \qquad \text{(Answer)}$$

We picked up the value of n, the number of conduction electrons per unit volume in copper, from Sample Problem 3; the value of ρ comes from Table 1.

(b) What is the mean free path λ for these collisions? Assume an effective speed v_{eff} of 1.6×10^6 m/s.

As in Section 21-6, we define the mean free path from

$$\lambda = \tau v_{\text{eff}} = (2.5 \times 10^{-14}\ \text{s})(1.6 \times 10^6\ \text{m/s})$$
$$= 4.0 \times 10^{-8}\ \text{m} = 40\ \text{nm}. \qquad \text{(Answer)}$$

This is about 150 times the distance between nearest-neighbor ions in a copper lattice. A full quantum physics treatment reveals that we cannot view a "collision" as a direct interaction between an electron and an ion. Rather, it is an interaction between an electron and the thermal vibrations of the lattice, lattice imperfections, or lattice impurity atoms. An electron can pass very freely through an "ideal" lattice, that is, a geometrically "perfect" lattice close to the absolute zero of temperature. Mean free paths as large as 10 cm have been observed under such conditions. Charge carriers in superconductors experience no collisions at all.

28-7 Energy and Power in Electric Circuits

Figure 13 shows a circuit consisting of a battery B connected to a "black box." A steady current i exists in the connecting wires and a steady potential difference V_{ab} exists between the terminals a and b. The box might contain a resistor, a storage battery, or a motor, among many other possibilities.

If a charge element dq moves through the box from terminal a to terminal b, its potential energy will be reduced by $dq\ V_{ab}$. The conservation of energy principle tells us that this energy must appear elsewhere in some form or other. What that form will be depends on what is in the box. In a time interval dt, the energy transferred within the box is then

$$dU = dq\ V_{ab} = i\ dt\ V_{ab}.$$

The *rate P* of energy transfer is dU/dt, or

$$\boxed{P = iV_{ab}} \quad \text{(rate of electrical energy transfer).} \qquad (20)$$

If the device in the box is an ideal motor connected to a mechanical load, the energy appears as mechanical work. If the device is a storage battery that is being charged, the energy appears as stored chemical energy in this second battery. If the device is a resistor, the energy

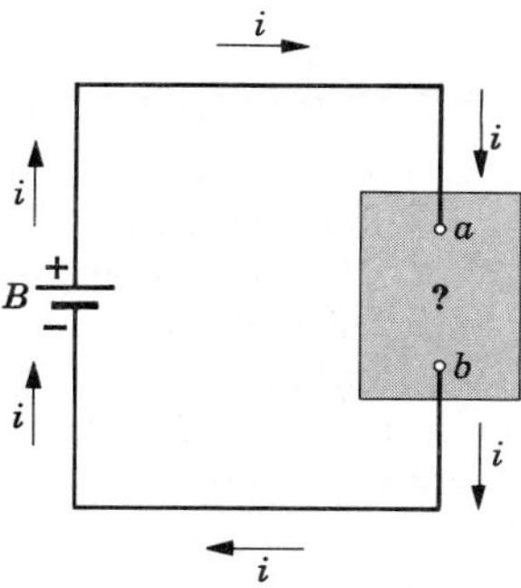

Figure 13 A battery B sets up a current i in a circuit containing a "black box," that is, a box whose contents are unknown to us.

appears as internal thermal energy, revealing itself as a temperature rise of the resistor.

The unit of power that follows from Eq. 20 is the volt · ampere. We can write it as

$$1\ \text{V}\cdot\text{A} = 1\ \text{V}\cdot\text{A}\left(\frac{1\ \text{J}}{1\ \text{V}\cdot\text{C}}\right)\left(\frac{1\ \text{C}}{1\ \text{A}\cdot\text{s}}\right) = 1\ \text{J/s} = 1\ \text{W}.$$

The conversion factor in the first set of parentheses comes from the definition of the volt; that in the second set of parentheses comes from the definition of the coulomb. Recall that we introduced the watt as a unit of mechanical work in Section 7–6.

The course of an electron moving through a resistor at constant drift speed is much like that of a stone falling through water at constant terminal speed. The average kinetic energy of the electron as it moves remains constant so that its lost electric potential energy must appear as thermal energy in the resistor. On a microscopic scale we can understand this since collisions between the electrons and the lattice increase the amplitude of the thermal lattice vibrations; this reveals itself as an increase in the temperature of the lattice.

For a resistor we can combine Eqs. 8 ($R = V/i$) and 20 to obtain, for the rate of electric energy dissipation in a resistor, either

$$\boxed{P = i^2R} \quad \text{(resistive dissipation)} \qquad (21)$$

or

$$\boxed{P = \frac{V^2}{R}} \quad \text{(resistive dissipation).} \qquad (22)$$

Although Eq. 20 applies to electric energy transfers of all kinds, Eqs. 21 and 22 apply only to the transfer of electric potential energy to thermal energy in a resistor.

Sample Problem 8 You are given a length of heating wire made of a nickel–chromium–iron alloy called Nichrome; it has a resistance R of 72 Ω. Can you obtain more heat by winding the wire into a single coil or by cutting the wire in two and winding two separate coils? In each case the coils are to be connected individually across a 120-V line.

The power P for a single coil is, from Eq. 22,

$$P = \frac{V^2}{R} = \frac{(120\ \text{V})^2}{72\ \Omega} = 200\ \text{W}. \qquad \text{(Answer)}$$

The power for a coil of half length (and thus half resistance) is

$$P' = \frac{V^2}{\frac{1}{2}R} = \frac{(120\ \text{V})^2}{36\ \Omega} = 400\ \text{W}. \qquad \text{(Answer)}$$

There are two half-coils so that the power obtained from both of them is 800 W, or four times that for a single coil. This would seem to suggest that you could buy a heating coil, cut it in half, and reconnect it to obtain four times the heat output. Why is this not such a good idea?

Sample Problem 9 A wire whose length L is 2.35 m and whose diameter d is 1.63 mm carries a current i of 1.24 A. The wire dissipates thermal energy at the rate P of 48.5 mW. Of what is the wire made?

We can identify the material by its resistivity. From Eqs. 15 and 21 we have

$$P = i^2R = \frac{i^2\rho L}{A} = \frac{4i^2\rho L}{\pi d^2},$$

in which A ($=\frac{1}{4}\pi d^2$) is the cross-sectional area of the wire. Solving for ρ, the resistivity of the material of which the wire is made, yields

$$\rho = \frac{\pi P d^2}{4i^2L} = \frac{(\pi)(48.5 \times 10^{-3}\ \text{W})(1.63 \times 10^{-3}\ \text{m})^2}{(4)(1.24\ \text{A})^2(2.35\ \text{m})}$$
$$= 2.80 \times 10^{-8}\ \Omega\cdot\text{m}. \qquad \text{(Answer)}$$

Inspection of Table 1 reveals the material to be aluminum.

28–8 Semiconductors (Optional)

Semiconducting devices are at the heart of the microelectronic revolution that has so influenced our lives. Table 2 compares the properties of silicon—a typical semiconductor—with those of copper—a typical metallic conductor. We see that, compared with copper, silicon (1) has many fewer charge carriers, (2) has a much higher resistivity, and (3) has a temperature coefficient of resistivity that is both large and negative. That is, although the resistivity of copper increases with temperature, that of pure silicon decreases.

The resistivity of pure silicon is so high that it is virtually an insulator and is thus of not much direct use in microelectronic circuits. The property that makes it useful is that—as Table 1 shows—its resistivity can be reduced in a controlled way by adding minute amounts of specific foreign "impurity" atoms, a process called *doping*.

We may fairly conclude that, because their electrical properties are so different, the fundamental conduction process in silicon must be quite different from that for copper. We explore these differences in some detail in Chapter 46 of the extended version of this book, restricting ourselves here to a broad outline.

Table 2 Some Electric Properties of Two Elements[a]

Property	*Unit*	*Copper*	*Silicon*
Type of material	—	Metal	Semiconductor
Density of charge carriers	m^{-3}	9×10^{28}	1×10^{16}
Resistivity	$\Omega \cdot m$	2×10^{-8}	3×10^{3}
Temperature coefficient of resistivity	K^{-1}	$+4 \times 10^{-3}$	-70×10^{-3}

[a] Data rounded off to one significant figure for easy comparison.

We saw in Section 8–10 (see Fig. 17 of that chapter) that electrons in isolated atoms occupy quantized energy levels, each level containing a single electron. Electrons in solids also occupy quantized levels, as Fig. 14 shows. These levels—whose number is very great—are tightly compressed into allowed *bands* of closely spaced levels. The bands are separated by *gaps,* which represent ranges of energy that electrons may not possess.

In a metallic conductor such as copper (see Fig. 14*a*), the highest band that contains any electrons—called the *valence band*—is only partially filled. If an applied electric field is to establish a current, it must be possible for the conduction electrons to increase their energies. In a metal such as copper, this poses no problem because many vacant energy levels are readily at hand within the valence band.

In an insulator (Fig. 14*b*), the valence band is completely filled. The next higher available vacant levels lie in an empty band (called the *conduction band*) separated from the valence band by a considerable energy gap. If an electric field is applied, no current can occur because there is no mechanism by which an electron can increase its energy; the energy jump to the nearest vacant energy level is simply too great.

A semiconductor (Fig. 14*c*) is like an insulator except that the energy gap between the conduction band and the valence band is small enough so that the probability that electrons might "jump the gap" by thermal agitation is not vanishingly small. More important is the fact that controlled impurities—deliberately added—can contribute charge carriers to the conduction band.* Most semiconducting devices, such as transistors and junction diodes, are fabricated by the selective doping of different regions of the silicon matrix with different kinds of impurity atoms.

Let us now look again at Eq. 19, the expression for the resistivity of a conductor, with the band-gap picture in mind:

$$\rho = \frac{m}{ne^2 \tau}. \tag{23}$$

Consider how the variables n and τ change as the temperature is increased, n being the number of charge carriers per unit volume and τ the mean time between collisions of the charge carriers.

In a conductor, n is large but very closely constant; that is, its value does not change appreciably with temperature. The increase of resistivity with temperature for metals is caused by an increase in the collision rate of the charge carriers, which shows up in Eq. 23 as a decrease in τ, the mean time between collisions.

In a semiconductor, n is small but increases very rapidly with temperature as the increased thermal agitation makes more charge carriers available. This causes the decrease of resistivity with temperature displayed in Table 2. The same increase in collision rate that we noted

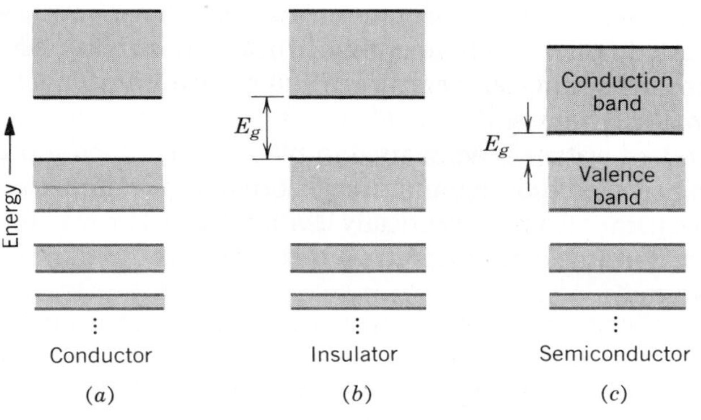

Figure 14 The allowed energy levels for the electrons in a solid form a pattern of allowed bands and forbidden gaps. (*a*) In a metallic conductor, the valence band is only partially filled. (*b*) In an insulator, the valence band is completely filled and the gap between the valence band and the conduction band is relatively large. (*c*) A semiconductor resembles an insulator except that the gap between bands is relatively small.

* Vacancies (called *holes*) in the valence band can also serve as charge carriers. Details are given in Chapter 46.

Figure 15 A model of the first transistor. Today, many thousands of such devices can be placed on a thin wafer a few millimeters on edge.

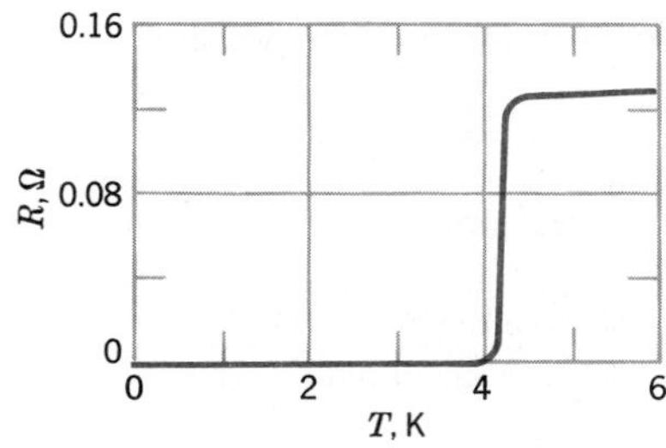

Figure 16 The resistivity of mercury drops to zero at a temperature of about 4 K. Mercury is solid at this low temperature.

for metals also occurs for semiconductors but its effect is swamped by the rapid increase in the number of charge carriers.

We begin to see how the band-gap picture—which is based solidly on quantum physics—can account for the properties of semiconductors. It is no accident that the transistor (see Fig. 15) was discovered by three physicists (William Shockley, John Bardeen, and Walter Brattain) as a specific application of quantum physics to solid materials. These physicists earned the 1956 Nobel Prize for their work.

28-9 Superconductors (Optional)

In 1911, the Dutch physicist Kammerlingh Onnes discovered that the resistivity of mercury absolutely disappears at temperatures below about 4 K; see Fig. 16. This phenomenon of *superconductivity* is of vast potential importance in technology because it means that charges can flow through a conductor without thermal losses. Currents induced in a superconducting ring, for example, have persisted for several years without diminution, no battery of any kind being present in the circuit. A large superconducting ring is now being used in Tacoma, Washington to store electrical energy. It takes in up to five megawatts during peaks of supply and releases the energy during peaks in demand.

The problem with the technological development of superconductivity has always been the low temperatures that were necessary to maintain it. The magnets in the large Fermilab accelerator, for example, are energized by currents in superconducting coils, which must be maintained at ~ 4 K, the temperature of liquid helium.

In 1986, however, new ceramic materials were discovered that become superconducting at considerably higher temperatures. As we write this, temperatures as high as 125 K have been reported and confirmed. As you read this, higher temperatures may well have been reported, with room temperature a distinct possibility. The normal boiling temperature of liquid nitrogen is 77 K so that this inexpensive coolant—which is cheaper than bottled water—can be used in place of the much more expensive liquid helium. Conjectured applications run the gamut from magnetically levitated trains to desk-top main-frame computers to powerful motors the size of a walnut.

Superconductivity must not be thought of as simply a dramatic improvement in the normal conductivity process that we described in Section 28-6. The two processes are completely different. In fact, the best normal conductors, such as silver or copper, do not become superconducting; on the other hand, the recently discovered "supersuperconductors" are ceramic materials, which—as far as normal conduction is concerned—are insulators.

The mechanism of superconductivity remained unexplained for some 60 years after the discovery of the phenomenon. Then John Bardeen,* Leon Cooper, and Robert Schrieffer advanced a theoretical explanation, for which they were jointly awarded the 1972 Nobel Prize. The heart of the BCS theory—as it is called, after the initials of its developers—is the assumption that the charge carriers are not single electrons but pairs of electrons. These *Cooper pairs* behave like single particles, with properties dramatically different from those of single electrons.

* Yes, this is the same John Bardeen who shared the 1956 Nobel Prize for discovering the transistor; see Section 28-8. Professor Bardeen is the only person who has earned two Nobel Prizes in the same field.

Electrons normally repel each other so that some special mechanism is needed to induce them to form a pair. A semiclassical picture that helps in understanding this quantum BCS phenomenon is as follows: An electron plows through the lattice, distorting it slightly and thus leaving in its wake a very short-lived concentration of enhanced positive charge. If a second electron is nearby at the right moment, it may well be attracted to this region by the positive charge, thus forming a pair with the first electron. It is known that the newly discovered superconductors operate by means of Cooper pairs but, as of 1988, there is no universal agreement as to the mechanism by which these pairs are formed.

REVIEW AND SUMMARY

Current i

An *electric current* i in a conductor is defined by

$$dq = i\,dt. \qquad [1]$$

Here dq is the amount of (positive) charge that passes in time dt through a hypothetical surface that cuts across the conductor; see Sample Problem 1. The direction of electric current is the direction in which postive charge carriers would move. The SI unit of electric current is the *ampere* (= 1 C/s; abbr. A).

Current (a scalar) is related to *current density* **J** (a vector) by

Current Density **J**

$$i = \int \mathbf{J} \cdot d\mathbf{A} \qquad [5]$$

where $d\mathbf{A}$ is an element of area and the integral is taken over any surface cutting across the conductor. The direction of **J** at any point is that in which a positive charge carrier would move if placed at that point; see Fig. 5 and Sample Problem 2.

When an electric field **E** is established in a conductor the charge carriers (assumed positive) acquire a *mean drift speed* $\mathbf{v}_d$ in the direction of **E**, related to the current density by

The Mean Drift Speed of the Charge Carriers

$$\mathbf{J} = (ne)\mathbf{v}_d \qquad [7]$$

where (ne) is the charge density; see Sample Problems 3 and 4.

The *resistance R* between any two equipotential surfaces of a conductor is defined from

The Resistance of a Conductor

$$R = V/i \quad \text{(definition of } R) \qquad [8]$$

where V is the potential difference between those surfaces and i is the current. The SI unit of R is the *ohm* (= 1 V/A; abbr. Ω). Similar equations define the *resistivity* ρ and *conductivity* σ of a material:

Resistivity and Conductivity

$$\rho = \frac{1}{\sigma} = \frac{E}{j} \quad \text{(definitions of } \rho \text{ and } \sigma) \qquad [10,12]$$

where E is the applied electric field. The SI unit of resistivity is the ohm-meter (abbr. Ω · m); see Table 1 and Sample Problem 5. Eq. 10 corresponds to the vector equation

$$\mathbf{E} = \rho\mathbf{J}. \qquad [11]$$

The resistance R for a cylindrical conductor of any cross-sectional shape is

$$R=\rho\frac{L}{A}; \qquad [15]$$

see Sample Problem 6.

The resistivity ρ for most materials changes with temperaure. For many materials, including metals, the empirical linear relationship is

The Change of ρ with Temperature

$$\rho-\rho_0=\rho_0\alpha(T-T_0). \qquad [16]$$

Here T_0 is a reference temperature, ρ_0 is the resistivity at T_0, and α is a mean temperature coefficient of resistivity; see Table 1.

Ohm's Law

A given *conductor* obeys *Ohm's law* if its resistance R, defined by Eq. 8, is independent of the applied potential difference V; compare Figs. 11*b* and 11*c*. A given *material* obeys Ohm's law if its resistivity, defined by Eq. 10, is independent of the magnitude and direction of the applied electric field **E**.

By treating the conduction electrons in a metal like the molecules of a gas it is possible to derive for the resistivity of a metal

Resistivity of a Metal

$$\rho=\frac{m}{e^2 n\tau}. \qquad [19]$$

Here n is the number of electrons per unit volume and τ is the mean time between the collisions of an electron with the ion cores of the lattice. The discussion based on Fig. 12 shows that τ is independent of E and thus accounts for the fact that metals obey Ohm's law; see Sample Problem 7.

The power P or rate of energy transfer in an electric device across which a potential difference V_{ab} is maintained is

Power

$$P=iV_{ab} \quad \text{(rate of electrical energy transfer).} \qquad [20]$$

If the device is a resistor we can write this as

Resistive Dissipation

$$P=i^2R=\frac{V^2}{R} \quad \text{(resistive dissipation);} \qquad [21,22]$$

see Sample Problems 8 and 9. In a resistor electrical potential energy is transferred to the lattice by the drifting charge carriers, appearing as internal thermal energy.

Semiconductors

Semiconductors are materials with few conduction electrons but with available conduction-level states close, in energy, to their valence bands. These then become conductors either by thermal agitation of electrons or, more importantly, by *doping* the material with other atoms which contribute electrons or holes to the conduction band.

Superconductors

Superconductors lose all electrical resistance at temperatures below some critical value. Recent research has discovered materials with increasingly higher critical temperatures, leading to the possibility of room temperature (or, at worst, liquid nitrogen temperature) superconducting devices.

QUESTIONS

1. What conclusions can you draw by applying Eq. 5 to a closed surface through which a number of wires pass in random directions, carrying steady currents of different magnitudes?

2. In our convention for the direction of current arrows (*a*) would it have been more convenient, or even possible, to have assumed all charge carriers to be negative? (*b*) Would it have been more convenient, or even possible, to have labeled the electron as positive, the proton as negative, etc?

3. List in tabular form similarities and differences between the flow of charge along a conductor, the flow of water through a horizontal pipe, and the conduction of heat through a slab. Consider such ideas as what causes the flow, what opposes it, what particles (if any) participate, and the units in which the flow may be measured.

4. Explain in your own words why we can have $\mathbf{E} \neq 0$ inside a conductor in this chapter whereas we took $\mathbf{E} = 0$ for granted in Section 7 of Chapter 25.

5. Let a battery be connected to a copper cube at two corners defining a body diagonal. Pass a hypothetical plane completely through the cube, tilted at an arbitrary angle. (*a*) Is the current i through the plane independent of the position and orientation of the plane? (*b*) Is there any position and orientation of the plane for which $\mathbf{J}$ is a constant in magnitude, direction, or both? (*c*) Does Eq. 5 hold for all orientations of the plane? (*d*) Does Eq. 5 hold for a closed surface of arbitrary shape, which may or may not lie entirely within the cube?

6. A potential difference V is applied to a copper wire of diameter d and length L. What is the effect on the electron drift speed of (*a*) doubling V, (*b*) doubling L, and (*c*) doubling d?

7. Why is it not possible to measure the drift speed for electrons by timing their travel along a conductor?

8. A potential difference V is applied to a circular cylinder of carbon by clamping it between circular copper electrodes, as in Fig. 17. Discuss the difficulty of calculating the resistance of the carbon cylinder using the relation $R = \rho L/A$.

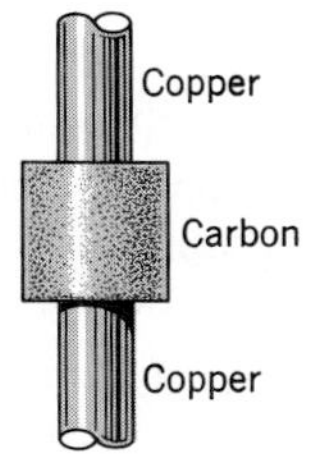

Figure 17 Question 8.

9. How would you measure the resistance of a pretzel-shaped metal block? Give specific details to clarify the concept.

10. Sliding across the seat of an automobile can generate potentials of several thousand volts. Why isn't the slider electrocuted?

11. Discuss the difficulties of testing whether the filament of a light bulb obeys Ohm's law.

12. How does the relation $V = iR$ apply to resistors that do *not* obey Ohm's law?

13. A cow and a man are standing in a meadow when lightning strikes the ground nearby. Why is the cow more likely to be killed than the man? The responsible phenomenon is called "step voltage."

14. The gray lines in Fig. 12 should be curved slightly. Why?

15. A fuse in an electrical circuit is a wire that is designed to melt, and thereby open the circuit, if the current exceeds a predetermined value. What are some characteristics of an ideal fuse wire?

16. Why does an incandescent light bulb grow dimmer with use?

17. The character and quality of our daily lives is influenced greatly by devices that do not obey Ohm's law. What can you say in support of this claim?

18. From a student's paper: "The relationship $R = V/i$ tells us that the resistance of a conductor is directly proportional to the potential difference applied to it." What do you think of this proposition?

19. Carbon has a negative temperature coefficient of resistivity. This means that its resistivity drops as its temperature increases. Would its resistivity disappear entirely at some high enough temperature?

20. What special characteristics must heating wire have?

21. Equation 21 ($P = i^2R$) seems to suggest that the rate of increase of thermal energy in a resistor is reduced if the resistance is made less; Eq. 22 ($P = V^2/R$) seems to suggest just the opposite. How do you reconcile this apparent paradox?

22. Why do electric power companies reduce voltage during times of heavy demand? What is being saved?

23. Is the filament resistance lower or higher in a 500-W light bulb than in a 100-W bulb? Both bulbs are designed to operate on 120 V.

24. Five wires of the same length and diameter are connected in turn between two points maintained at constant potential difference. Will thermal energy be developed at the faster rate in the wire of (*a*) the smallest or (*b*) the largest resistance?

25. Why is it better to send 10,000 kW of electric power long distances at 10,000 volts rather than at 220 volts?

EXERCISES AND PROBLEMS

Section 28-2 Electric Current

1E. A current of 5.0 A exists in a 10-Ω resistor for 4.0 min. (*a*) How many coulombs and (*b*) how many electrons pass through any cross section of the resistor in this time?

2E. The current in the electron beam of a typical video display terminal is 200 μA. How many electrons strike the screen each second?

3P. You are given an isolated conducting sphere of 10-cm radius. One wire carries a current of 1.0000020 A into it. Another wire carries a current of 1.0000000 A out of it. How long would it take for the sphere to increase in potential by 1000 V?

4P. The belt of an electrostatic generator is 50 cm wide and travels at 30 m/s. The belt carries charge into the sphere at a rate corresponding to 100 μA. Compute the surface charge density on the belt. See Section 11 of Chapter 26.

Section 28-3 Current Density

5E. We have 2.0×10^8 doubly charged positive ions per cubic centimeter, all moving north with a speed of 1.0×10^5 m/s. (*a*) What is the current density **J**, in magnitude and direction? (*b*) Can you calculate the total current i in this ion beam? If not, what additional information is needed?

6E. A small but measurable current of 1.2×10^{-10} A exists in a copper wire whose diameter is 2.5 mm. Calculate (*a*) the current density and (*b*) the electron drift speed. See Sample Problem 3.

7E. A fuse in an electrical circuit is a wire that is designed to melt, and thereby open the circuit, if the current exceeds a predetermined value. Suppose that the material composing the fuse melts once the current density rises to 440 A/cm². What diameter of cylindrical wire should be used to limit the current to 0.50 A?

8E. The (United States) National Electric Code, which sets maximum safe currents for rubber-insulated copper wires of various diameters, is given (in part) below. Plot the safe current density as a function of diameter. Which wire gauge has the maximum safe current density?

Gauge[a]	4	6	8	10	12	14	16	18
Diameter (mils)[b]	204	162	129	102	81	64	51	40
Safe current (A)	70	50	35	25	20	15	6	3

[a] A way of identifying the wire diameter.
[b] 1 mil = 10^{-3} in.

9E. A current is established in a gas discharge tube when a sufficiently high potential difference is applied across the two electrodes in the tube. The gas ionizes; electrons move toward the positive terminal and singly charged positive ions toward the negative terminal. What are the magnitude and direction of the current in a hydrogen discharge tube in which 3.1×10^{18} electrons and 1.1×10^{18} protons move past a cross-sectional area of the tube each second?

10E. A *p-n* junction is formed from two different semiconducting materials in the form of identical cylinders with radius 0.165 mm, as depicted in Fig. 18. In one application 3.50×10^{15} electrons per second flow across the junction from the n to the p side while 2.25×10^{15} holes per second flow from the p to the n side. (A hole acts like a particle with charge $+1.6 \times 10^{-19}$ C.) What are (*a*) the total current and (*b*) the current density?

11P. Near the earth, the density of protons in the solar wind (see Section 1) is 8.7 cm^{-3} and their speed is 470 km/s. (*a*) Find the current density of these protons. (*b*) If the earth's magnetic field did not deflect them, the protons would strike the earth. What total current would the earth receive?

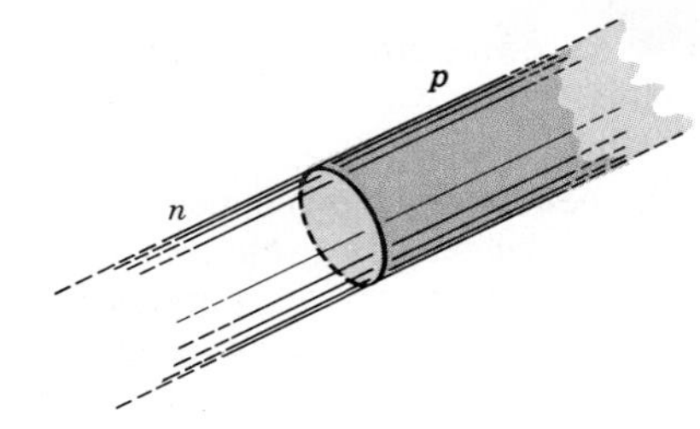

Figure 18 Exercise 10.

12P. In a hypothetical fusion research lab, high temperature helium gas is completely ionized, each helium atom being separated into two free electrons and the remaining positively charged nucleus (alpha particle). An applied electric field causes the alpha particles to drift to the east at 25 m/s while the electrons drift to the west at 88 m/s. The alpha particle density is 2.8×10^{15} cm^{-3}. Calculate the net current density; specify the current direction.

13P. How long does it take electrons to get from a car battery to the starting motor? Assume the current is 300 A and the electrons travel through a copper wire with cross-sectional area 0.21 cm² and length 0.85 m. See Sample Problem 3.

14P. A steady beam of alpha particles ($q = 2e$) traveling with constant kinetic energy 20 MeV carries a current 0.25 μA. (*a*) If the beam is directed perpendicular to a plane surface, how many alpha particles strike the surface in 3.0 s? (*b*) At any instant, how many alpha particles are there in a given 20-cm length of the beam? (*c*) Through what potential difference was it necessary to accelerate each alpha particle from rest to bring it to an energy of 20 MeV?

15P. (*a*) The current density across a cylindrical conductor of radius R varies according to the equation

$$J = J_0(1 - r/R),$$

where r = distance from the axis. Thus, the current density is a maximum J_0 at the axis $r = 0$ and decreases linearly to zero at the surface $r = R$. Calculate the current in terms of J_0 and the conductor's cross-sectional area $A = \pi R^2$. (*b*) Suppose that, instead, the current density is a maximum J_0 at the surface and decreases linearly to zero at the axis, so that

$$J = J_0 r/R.$$

Calculate the current. Why is the result different from (*a*)?

Section 28-4 Resistance and Resistivity

16E. A steel trolley-car rail has a cross-sectional area of 56 cm². What is the resistance of 10 km of rail? The resistivity of the steel is 3.0×10^{-7} $\Omega \cdot$m.

17E. A conducting wire has a 1.0-mm diameter, a 2.0-m

length, and a 50-mΩ resistance. What is the resistivity of the material?

18E. A human being can be electrocuted if a current as small as 50 mA passes near the heart. An electrician working with sweaty hands makes good contact with the two conductors he is holding. If his resistance is 2000 Ω, what might the fatal voltage be?

19E. A coil is formed by winding 250 turns of insulated gauge 16 copper wire (diameter = 1.3 mm) in a single layer on a cylindrical form whose radius is 12 cm. What is the resistance of the coil? Neglect the thickness of the insulation. See Table 1.

20E. A wire 4.0 m long and 6.0 mm in diameter has a resistance of 15 mΩ. If a potential difference of 23 V is applied between the ends, (*a*) what is the current in the wire? (*b*) What is the current density? (*c*) Calculate the resistivity of the wire material. Can you identify the material? See Table 1.

21E. A wire of Nichrome (a nickel-chromium–iron alloy commonly used in heating elements) is 1.0 m long and 1.0 mm^2 in cross-sectional area. It carries a current of 4.0 A when a 2.0-V potential difference is applied between its ends. Calculate the conductivity σ of Nichrome.

22E. (*a*) At what temperature would the resistance of a copper conductor be double its resistance at 20°C? (Use 20°C as the reference point in Eq. 16; compare your answer with Fig. 10.) (*b*) Does this same temperature hold for all copper conductors, regardless of shape or size?

23E. The copper windings of a motor have a resistance of 50 Ω at 20°C when the motor is idle. After running for several hours the resistance rises to 58 Ω. What is the temperature of the windings? Ignore changes in the dimensions of the windings. See Table 1.

24E. Using data from Fig. 11*c*, plot the resistance of the *pn* junction as a function of applied potential difference.

25E. A 4.0-cm-long caterpillar crawls in the direction of electron drift along a 5.2-mm-diameter bare copper wire that carries a current of 12 A. (*a*) What is the potential difference between the two ends of the caterpillar? (*b*) Is its tail positive or negative compared to its head? (*c*) How much time could it take the caterpillar to crawl 1.0 cm and still keep up with the drifting electrons in the wire?

26E. A cylindrical copper rod of length L and cross-sectional area A is reformed to twice its original length with no change in volume. (*a*) Find the new cross-sectional area. (*b*) If the resistance between its ends was R before the change, what is it after the change?

27E. A wire with a resistance of 6.0 Ω is drawn out through a die so that its new length is three times its original length. Find the resistance of the longer wire, assuming that the resistivity and density of the material are not changed during the drawing process.

28E. A certain wire has a resistance R. What is the resistance of a second wire, made of the same material, that is half as long and has half the diameter?

29P. What must be the diameter of an iron wire if it is to have the same resistance as a copper wire 1.2-mm in diameter, both wires being the same length?

30P. Two conductors are made of the same material and have the same length. Conductor A is a solid wire of diameter 1.0 mm. Conductor B is a hollow tube of outside diameter 2.0 mm and inside diameter 1.0 mm. What is the resistance ratio, R_A/R_B, measured between their ends?

31P. A copper wire and an iron wire of the same length have the same potential difference applied to them. (*a*) What must be the ratio of their radii if the current is to be the same? (*b*) Can the current density be made the same by suitable choices of the radii?

32P. A square aluminum rod is 1.3 m long and 5.2 mm on edge. (*a*) What is the resistance between its ends? (*b*) What must be the diameter of a circular 1.3-m copper rod if its resistance is to be the same?

33P. A potential difference V is applied to a wire of cross section A, length L, and resistivity ρ. You want to change the applied potential difference and draw out the wire so the power dissipated is increased by a factor of 30 and the current is increased by a factor of 4. What should be the new values of L and A?

34P. A rod of a certain metal is 1.6 m long and 5.5 mm in diameter. The resistance between its ends (at 20°C) is 1.09×10^{-3} Ω. A round disk is formed of this same material, 2.00 cm in diameter and 1.00 mm thick. (*a*) What is the material? (*b*) What is the resistance between the opposing round faces, assuming equipotential surfaces?

35P. An electrical cable consists of 125 strands of fine wire, each having 2.65-μΩ resistance. The same potential difference is applied between the ends of each strand and results in a total current of 0.75 A. (*a*) What is the current in each strand? (*b*) What is the applied potential difference? (*c*) What is the resistance of the cable?

36P. A common flashlight bulb is rated at 0.30 A and 2.9 V, the values of the current and voltage under operating conditions. If the resistance of the bulb filament when cold is 1.1 Ω, calculate the temperature of the filament when the bulb is on. The filament is made of tungsten.

37P. When 115 V is applied across a 0.30-mm radius, 10-m-long wire, the current density is 1.4×10^4 A/m^2. Find the resistivity of the wire.

38P. A block in the shape of a rectangular solid has a cross-sectional area of 3.50 cm^2, a length of 15.8 cm, and a resistance of 935 Ω. The material of which the block is made has 5.33×10^{22} conduction electrons/m^3. A potential difference of 35.8 V is maintained between its ends. (*a*) What is the current in the block? (*b*) If the current density is uniform, what is its value? (*c*) What is the drift velocity of the conduction electrons? (*d*) What is the electric field in the block?

39P. Copper and aluminum are being considered for a high-voltage transmission line that must carry a current of 60 A. The resistance per unit length is to be 0.15 Ω/km. Compute for each choice of cable material (*a*) the current density and (*b*) the mass per meter of the cable. The densities of copper and aluminum are 8960 and 2700 kg/m³, respectively.

40P. In the lower atmosphere of the earth there are negative and positive ions, created by radioactive elements in the soil and cosmic rays from space. In a certain region, the atmospheric electric field strength is 120 V/m, directed vertically down. Due to this field, singly charged positive ions, 620 per cm³, drift downward and singly charged negative ions, 550 per cm³, drift upward; see Fig. 19. The measured conductivity is $2.7 \times 10^{-14}/\Omega\cdot m$. (*a*) Calculate the ion drift speed, assumed the same for positive and negative ions, and (*b*) the current density.

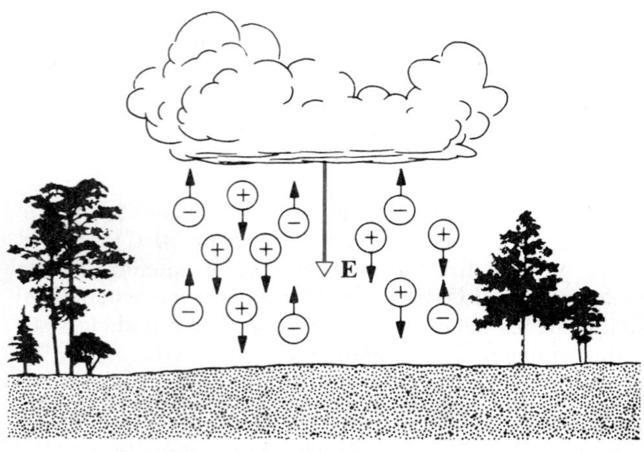

Figure 19 Problem 40.

41P. If the gauge number of a wire is increased by 6, the diameter is halved; if a gauge number is increased by 1, the diameter decreases by the factor $2^{1/6}$ (see the table in Exercise 8). Knowing this, and also knowing that 1000 ft of #10 copper wire has a resistance of approximately 1.00 Ω, estimate the resistance of 25 ft of #22 copper wire.

42P. When a metal rod is heated, not only its resistance but also its length and its cross-sectional area change. The relation $R = \rho L/A$ suggests that all three factors should be taken into account in measuring ρ at various temperatures. (*a*) If the temperature changes by 1.0°C, what percentage changes in R, L, and A occur for a copper conductor? (*b*) What conclusion do you draw? The coefficient of linear expansion is $1.7 \times 10^{-5}/°C$.

43P. A resistor is in the shape of a truncated right circular cone (Fig. 20). The end radii are a and b, the altitude is L. If the taper is small, we may assume that the current density is uniform across any cross section. (*a*) Calculate the resistance of this object. (*b*) Show that your answer reduces to $\rho(L/A)$ for the special case of zero taper ($a = b$).

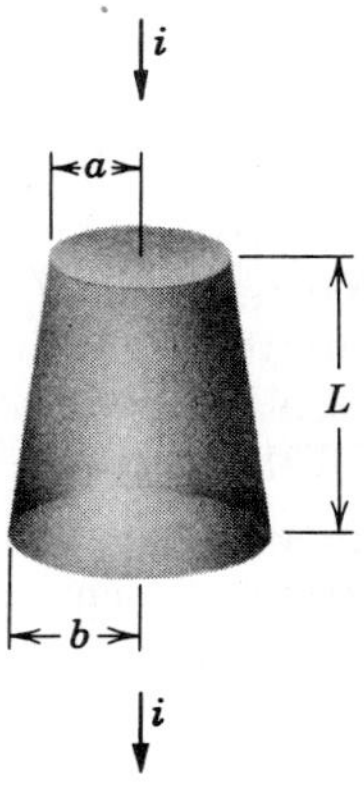

Figure 20 Problem 43.

Section 28-6 Ohm's Law—A Microscopic View

44P. Show that, according to the free-electron model of electrical conduction in metals and classical physics, the resistivity of metals should be proportional to $\sqrt{T}$, where T is absolute temperature. (See Eq. 21 of Chapter 21.)

Section 28-7 Energy and Power in Electric Circuits

45E. A student kept his 9.0-V, 7.0-W portable radio turned on from 9:00 p.m. until 2:00 a.m. How much charge went through it?

46E. An x-ray tube takes a current of 7.0 mA and operates at a potential difference of 80 kV. What power in watts is dissipated?

47E. Thermal energy is developed in a resistor at a rate of 100 W when the current is 3.0 A. What is the resistance?

48E. The headlights of a moving car draw about 10 A from the 12-V alternator, which is driven by the engine. Assume the alternator is 80 percent efficient and calculate the horsepower the engine must supply to run the lights.

49E. A space heater, operating from a 120-V line, has a hot resistance of 14 Ω. (*a*) At what rate is electrical energy transferred into heat? (*b*) At 5.0¢/kW·h, what does it cost to operate the device for 5.0 h?

50E. An unknown resistor is connected between the terminals of a 3.0-V battery. The power dissipated in the resistor is

0.54 W. The same resistor is then connected between the terminals of a 1.5-V battery. What power is dissipated in this case?

51E. A 500-W space heater operates from a 120-V line. (*a*) What is its (hot) resistance? (*b*) At what rate do electrons flow through any cross section of the filament?

52E. The National Board of Fire Underwriters has fixed safe current-carrying capacities for various sizes and types of wire. For #10 rubber-coated copper wire (diameter = 0.10 in.) the maximum safe current is 25 A. At this current, find (*a*) the current density, (*b*) the electric field, (*c*) the potential difference for 1000 ft of wire, and (*d*) the rate at which thermal energy is developed for 1000 ft of wire.

53E. A potential difference of 1.2 V is applied to a 33-m length of #18 copper wire (diameter = 0.040 in.). Calculate (*a*) the current, (*b*) the current density, (*c*) the electric field, and (*d*) the rate at which thermal energy is developed in the wire.

54P. A cylindrical resistor of radius 5.0 mm and length 2.0 cm is made of material that has a resistivity of 3.5×10^{-5} $\Omega \cdot$m. What are (*a*) the current density and (*b*) the potential difference when the power dissipation is 1.0 W?

55P. A heating element is made by maintaining a potential difference of 75 V along the length of a Nichrome wire with a 2.6×10^{-6} m^2 cross-section and a resistivity of 5.0×10^{-7} $\Omega \cdot$m. (*a*) If the element dissipates 5000 W, what is its length? (*b*) If a potential difference of 100 V is used to obtain the same power output, what should the length be?

56P. A 1250-W radiant heater is constructed to operate at 115 V. (*a*) What will be the current in the heater? (*b*) What is the resistance of the heating coil? (*c*) How much thermal energy is generated in one hour by the heater?

57P. A 100-W light bulb is plugged into a standard 120-V outlet. (*a*) How much does it cost per month to leave the light turned on? Assume electric energy costs 6¢/kW·h. (*b*) What is the resistance of the bulb? (*c*) What is the current in the bulb? (*d*) Is the resistance different when the bulb is turned off?

58P. A Nichrome heater dissipates 500 W when the applied potential difference is 110 V and the wire temperature is 800°C. How much power would it dissipate if the wire temperature were held at 200°C by immersion in a bath of cooling oil? The applied potential difference remains the same; α for Nichrome at 800°C is 4.0×10^{-4}/°C.

59P. A beam of 16-MeV deuterons from a cyclotron falls on a copper block. The beam is equivalent to a current of 15 μA. (*a*) At what rate do deuterons strike the block? (*b*) At what rate is thermal energy produced in the block?

60P. An electron linear accelerator produces a pulsed beam of electrons. The pulse current is 0.50 A and the pulse duration 0.10 μs. (*a*) How many electrons are accelerated per pulse? (*b*) What is the average current for a machine operating at 500 pulses/s? (*c*) If the electrons are accelerated to an energy of 50 MeV, what are the average and peak power outputs of the accelerator?

61P. A coil of current-carrying Nichrome wire is immersed in a liquid contained in a calorimeter. When the potential difference across the coil is 12 V and the current through the coil is 5.2 A, the liquid boils at a steady rate, evaporating at the rate of 21 mg/s. Calculate the heat of vaporization of the liquid, in cal/g.

62P. A resistance coil, wired to an external battery, is placed inside an adiabatic cylinder fitted with a frictionless piston and containing an ideal gas. A current i = 240 mA flows through the coil, which has a resistance R = 550 Ω. At what speed v must the piston, mass m = 12 kg, move upward in order that the temperature of the gas remains unchanged? See Fig. 21.

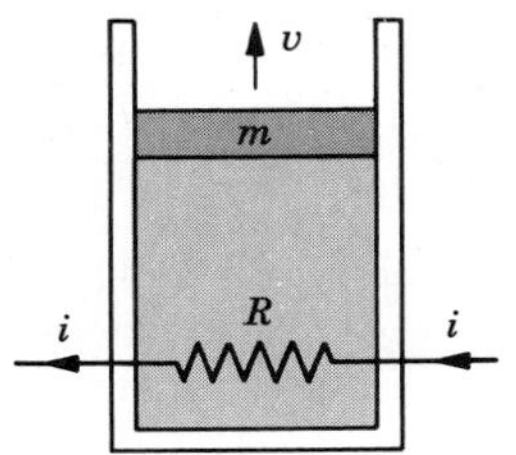

Figure 21 Problem 62.

63P. A 500-W heating unit is designed to operate from a 115-V line. (*a*) By what percentage will its heat output drop if the line voltage drops to 110 V? Assume no change in resistance. (*b*) Taking the variation of resistance with temperature into account, would the actual heat output drop be larger or smaller than that calculated in (*a*)?

64P. An electric immersion heater normally takes 100 min to bring cold water in a well-insulated container to a certain temperature, after which a thermostat switches the heater off. One day the line voltage is reduced by 6.0 percent because of a laboratory overload. How long will it now take to heat the water? Assume that the resistance of the heating element is the same for each of these two modes of operation.

65P. A 400-W immersion heater is placed in a pot containing 2.0 liters of water at 20°C. (*a*) How long will it take to bring the water to boiling temperature, assuming that 80% of the available energy is absorbed by the water? (*b*) How much longer will it take to boil half the water away?

66P. A 30-μF capacitor is connected across a programmed power supply. During the interval from $t = 0$ to $t = 3$ s the output voltage of the supply is given by $V(t) = 6 + 4t - 2t^2$ volts. At $t = 0.5$ s find (*a*) the charge on the capacitor, (*b*) the current into the capacitor, and (*c*) the power output from the power supply.

CHAPTER 29

ELECTROMOTIVE FORCE AND CIRCUITS

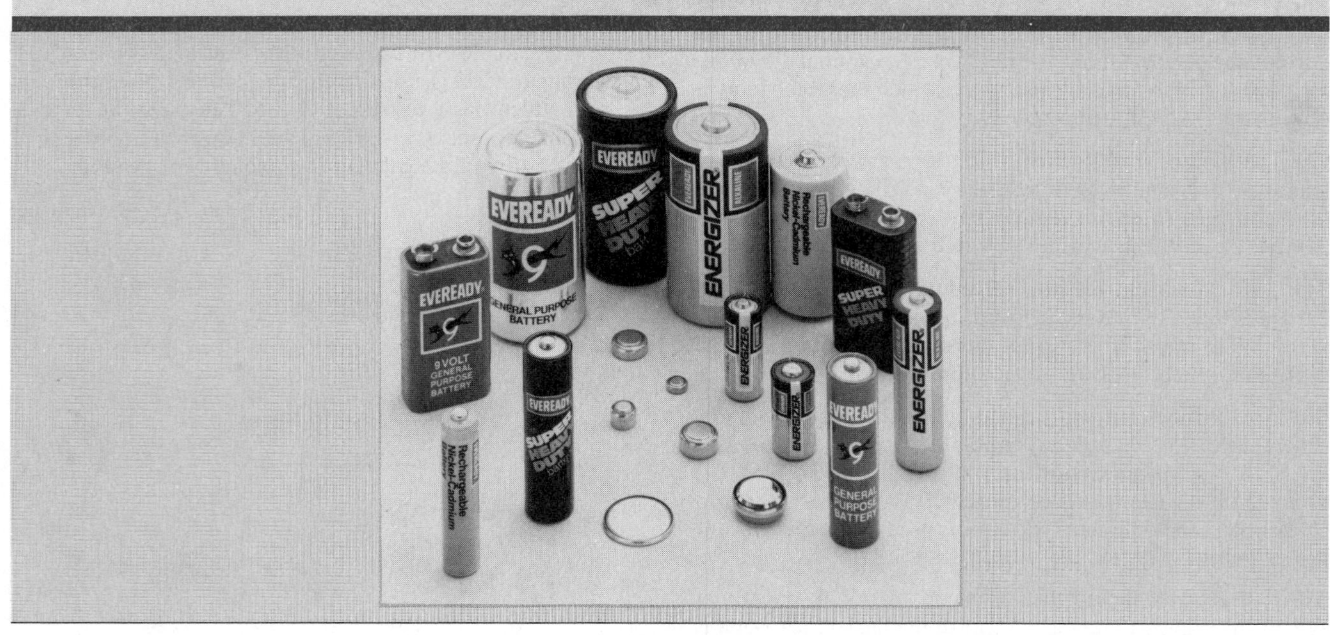

Batteries, whose bewildering variety is indicated by this small sample, free us from the tyranny of the extension cord and make possible activities that we could not otherwise imagine. They bring light and sound to remote places and make it possible to do calculations on a mountain top. Did you know that every Polaroid film pack contains a battery?

29-1 "Pumping" Charges

If you want to make charge carriers flow through a resistor, you must establish a potential difference between its ends. One way to do this is to connect each end of the resistor to a conducting sphere, one sphere being charged negatively and the other positively, as in Fig. 1. The trouble with this scheme is that the flow of charge acts to discharge the spheres, bringing them quickly to the same potential. When that happens, the flow of charge stops.

To maintain a steady flow of coolant in the cooling system of your car, you need a *water pump,* a device that—by doing work on the fluid—maintains a pressure difference between its input and its output ends. In the electrical case, we need a *charge pump,* a device that—by doing work on the charge carriers—maintains a potential difference between its terminals. We call such a device a seat of *electromotive force* (symbol $\mathcal{E}$; abbr. emf).

A common seat of emf is the *battery,* used to power devices from wristwatches to submarines. The seat of

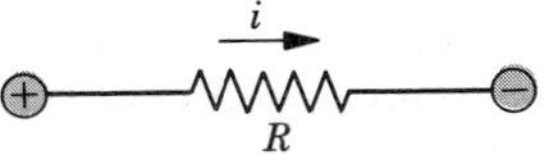

Figure 1 A steady current cannot exist because there is no mechanism to maintain a steady potential difference across the resistor. When the energy stored in the electric fields of the charged spheres has all been transferred to thermal energy in the resistor, the current must stop.

emf that most influences our daily lives, however, is the *electric generator,* whose output potential difference is led into our homes and workplaces from (usually) a remote generating plant. *Solar cells,* long familiar as the winglike panels on spacecraft, also dot the countryside for domestic applications. Less familiar seats of emf are the *fuel cells* that power the Space Shuttles and the *thermopiles* that provide onboard electric power for some spacecraft and for remote stations in Antarctica and elsewhere. Another example is the *electrostatic generator,* in which the potential difference is maintained by the mechanical movement of charge on an insulating belt. Living systems, ranging from electric eels and human beings to plants, are also seats of emf.

Although the devices that we have listed differ widely in their modes of operation, they all perform the basic function of a seat of emf: They are able to do work on charge carriers and thus maintain a potential difference between their output terminals.

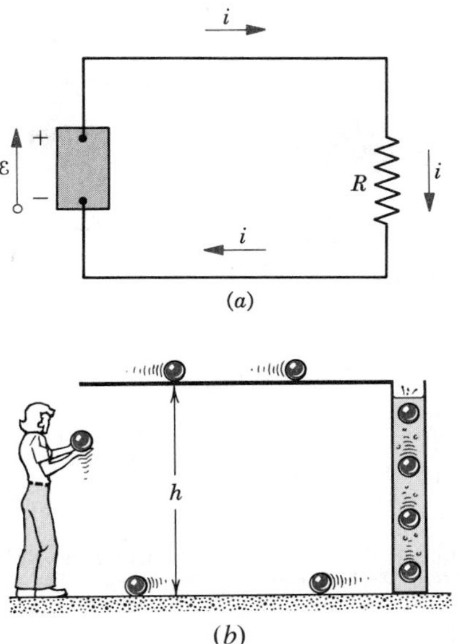

Figure 2 (*a*) A simple electric circuit, in which the emf $\mathscr{E}$ does work on the charge carriers and maintains a steady current through the resistor. (*b*) Its gravitational analog. Work done by the person maintains a steady flow of bowling balls through the viscous medium.

29-2 Work, Energy, and Electromotive Force

Figure 2*a* shows a seat of emf (which, for concreteness, you may think of as a battery) as part of a simple circuit. The seat maintains its upper terminal positive and its lower terminal negative, as shown by the + and − signs. We represent its emf by an arrow placed next to the seat and pointing in the direction in which the seat, acting alone, would cause a positive charge carrier to move in the external circuit. In Fig. 2*a*, this direction is clockwise. We draw a small circle on the tail of an emf arrow so that we do not confuse it with a current arrow.

Within the seat of emf, the circulating positive charge carriers must move from a region of low potential (the negative terminal) to a region of high potential (the positive terminal). This is just opposite to the direction in which the electric field between the terminals would compel them to move.

We conclude that there must be some source of energy within the seat, enabling it to do work on the charges and thus forcing them to move as they do. The energy source may be chemical, as in a battery or a fuel cell. It may involve mechanical forces, as in a conventional generator or an electrostatic generator. Temperature differences may supply the motive power, as in a thermopile; or solar energy may supply it, as in a solar cell.

Let us analyze the circuit of Fig. 2*a* from the point of view of work and energy transfers. In any time interval dt, a charge dq passes through any cross section of this circuit. In particular, this amount of charge must enter the seat of emf at its low-potential end and must leave at its high-potential end. The seat must do an amount of work dW on the charge element dq to force it to move in this way. We define the emf of the seat from

$$\mathscr{E} = \frac{dW}{dq} \quad \text{(definition of } \mathscr{E}\text{).} \tag{1}$$

The SI unit for emf that follows from Eq. 1 is the joule per coulomb, which we have met before (Eq. 5 of Chapter 26) and have called the *volt* (abbr. V). An ideal battery with an emf of 2.1 V would maintain a 2.1-V potential difference between its terminals.* The electromotive force, incidentally, is not a force; we measure its magnitude in volts and not in newtons. The name—like

* A *real* battery—in contrast to an ideal battery—only does so on open circuits, that is, if no current is passing through the battery. We clarify this point in the following section.

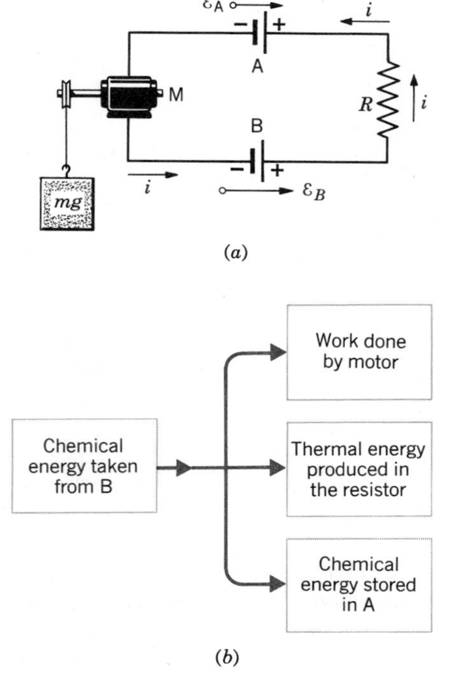

Figure 3 (*a*) $\mathcal{E}_B > \mathcal{E}_A$ so that battery B determines the direction of the current in this single-loop circuit. (*b*) Energy transfers in this circuit.

many other names in physics—is involved with the early history of the subject.

Figure 2*b* shows a gravitational analog to the circuit of Fig. 2*a*. In Fig. 2*a*, the seat of emf—which we take to be a battery—does work on the charge carriers, depleting its store of chemical energy. This energy appears as thermal energy in the resistor. In Fig. 2*b*, the person, in lifting the bowling balls from the floor to the shelf, does work on these "mass carriers." The balls roll slowly along the shelf, dropping from the right end into a cylinder of viscous oil. They sink to the bottom at an essentially constant terminal speed, are removed by a trapdoor mechanism not shown, and roll back along the floor to their starting position. The work done by the person, at the expense of her store of internal biochemical energy, appears as thermal energy in the viscous fluid, whose temperature rises slightly.

The circulation of charges in Fig. 2*a* will eventually stop if the battery does not replenish its store of chemical energy by being recharged. The circulation of bowling balls in the circuit of Fig. 2*b* will also eventually stop if the person does not replenish her store of biochemical energy by eating.

Figure 3*a* shows a circuit containing two storage batteries, A and B, a resistor R, and an (ideal) electric motor used to lift a weight. The batteries are connected so that they tend to send charges around the circuit in opposite directions. The actual direction of current is determined by battery B, which has the larger emf. Figure 3*b* shows the energy transfers in this circuit. The chemical energy in B is steadily depleted, the energy appearing in the three forms shown on the right. Battery A is being charged while battery B is being discharged. That is, the energy of charges traversing battery A decreases as they pass through that battery; therefore, the charges do work on battery A, which is stored internally as chemical energy.

29-3 Calculating the Current

We present here two equivalent ways to calculate the current in the simple circuit of Fig. 4, one method based on considerations of energy conservation and the other on the concept of potential.

Energy Method. Joule's law ($P = i^2R$; see Eq. 21 of Chapter 28) tells us that in a time interval dt an amount of energy given by $i^2R\,dt$ will appear in the resistor of Fig. 4 as thermal energy. During this same interval, a charge dq $(= i\,dt)$ will have moved through the seat of emf and the seat will have done work on this charge, according to Eq. 1:

$$dW = \mathcal{E}\,dq = \mathcal{E}i\,dt.$$

From the principle of conservation of energy, the work done by the seat of emf must equal the thermal energy that appears in the resistor, or

$$\mathcal{E}i\,dt = i^2R\,dt.$$

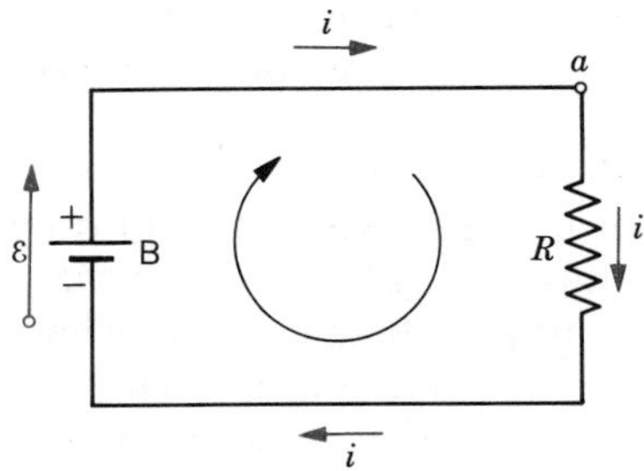

Figure 4 A single-loop circuit, in which $i = \mathcal{E}/R$.

Solving for i, we obtain

$$i = \frac{\mathcal{E}}{R}. \tag{2}$$

Potential Method. If electric potential is to have any meaning, a given point in a circuit can have only a single value of the potential at a given time. Let us start at any point in the circuit of Fig. 4 and, in imagination, go around the circuit in either direction, adding algebraically the potential differences that we encounter. When we arrive at our starting point we must have returned to our starting potential. We formalize this in a statement that holds not only for single-loop circuits such as that of Fig. 4 but for any complete loop in a multiloop circuit:

Loop Rule.* The algebraic sum of the changes in potential encountered in a complete traversal of any closed circuit must be zero.

In a gravitational analog, this is no more than saying that any point on the side of a mountain must have a unique value of the gravitational potential, that is, a unique elevation above sea level. If you start from any point and return to it after walking around on the mountain, the algebraic sum of the changes in elevation that you encounter must be zero.

In Fig. 4, let us start at point a, whose potential is V_a, and traverse the circuit clockwise. In going through the resistor, there is a change in potential of $-iR$. The minus sign shows that the top of the resistor is at a higher potential than the bottom. As we traverse the battery from bottom to top, there is an increase in potential of $+\mathcal{E}$. Adding the potential changes algebraically to the initial potential must yield the initial potential. Thus,

$$V_a - iR + \mathcal{E} = V_a.$$

Solving for i leads directly to Eq. 2.

These two methods for finding the current are completely equivalent because, as we learned in Section 26–2, potential differences are defined in terms of work and energy.

To prepare for the study of circuits more complex than that of Fig. 4, let us set down two "rules" for finding potential differences:

Resistor Rule. If you traverse a resistor in the direction of the current, the change in potential is $-iR$: in the opposite direction it is $+iR$. In a gravitational analog: if you walk downstream in a brook, your elevation decreases; if you walk upstream, it increases.

Emf Rule. If you traverse a seat of emf in the direction of the emf, the change in potential is $+\mathcal{E}$; in the opposite direction it is $-\mathcal{E}$.

These rules, which follow from our previous discussion, are not meant to be memorized. You should understand them so thoroughly that it becomes trivial to rederive them every time you use them.

29-4 Other Single-Loop Circuits

In this section we extend the simple circuit of Fig. 4 in two ways.

Internal Resistance. Figure 5a shows a circuit that emphasizes that all real seats of emf have an intrinsic internal resistance r. This resistance cannot be removed —although we usually wish it could be—because it is an inherent part of the device. Although we represent the emf and the internal resistance by separate symbols in Fig. 5a, they physically occupy the same region of space.

If we apply the loop rule, starting at b and going around clockwise, we obtain

$$V_b + \mathcal{E} - ir - iR = V_b$$

or

$$+\mathcal{E} - ir - iR = 0. \tag{3}$$

Verify that we have properly applied the rules for signs given at the end of the preceding section. Compare Eq. 3 with Fig. 5b, which shows the changes in potential graphically. It is helpful to imagine Fig. 5b folded into a cylinder, with labeled points b connected, to suggest the continuity of the closed loop.

If we solve Eq. 3 for the current, we find

$$i = \frac{\mathcal{E}}{R + r}. \tag{4}$$

As it must, this equation reduces to Eq. 2 for the case of $r = 0$.

* The loop rule is sometimes referred to as *Kirchhoff's loop rule,* after Gustav Robert Kirchhoff (pronounced Keerk-hoff; 1824–1887).

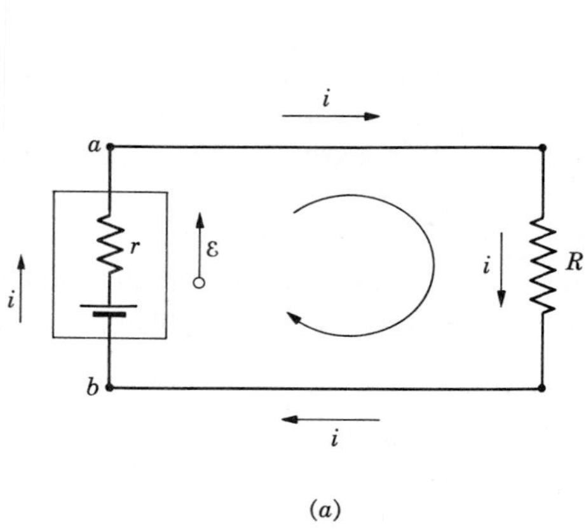

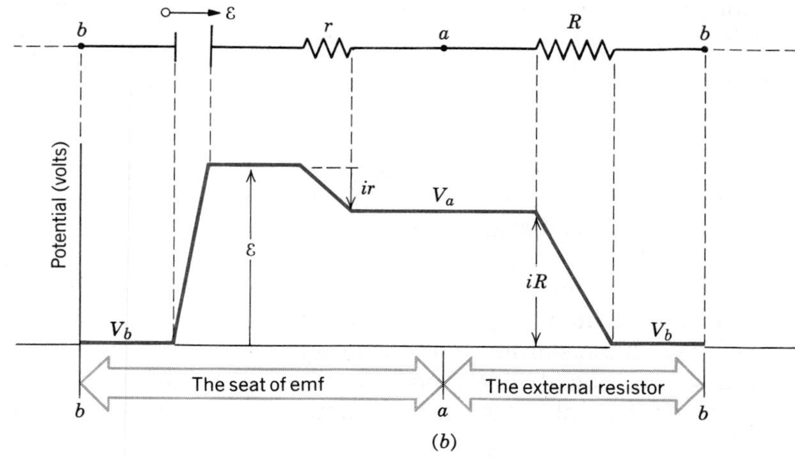

Figure 5 (*a*) A single-loop circuit, containing a seat of emf having an internal resistance *r*. (*b*) The circuit is shown spread out at the top. The potentials encountered in traversing the circuit clockwise from *b* are shown at the bottom.

Resistances in Series. Figure 6 shows three resistors connected *in series,* a battery being connected across the output terminals of the combination.

> *Resistors are said to be connected in series if the sum of the potential differences across each is equal to the potential difference applied to the combination.*

We seek the single resistance R_{eq} that is equivalent to this series combination.

Recall (see Section 27–4) that the definition of a series connection given above also applies to capacitors.* The definition requires that capacitors in series each have the same charge q; it requires that resistors in series each have the same current i. The equivalent resistance that we seek is the single resistance R_{eq} that, substituted for the series combination between the terminals a and b, will leave the current i unchanged.

Let us apply the loop rule, starting from terminal a and going clockwise around the circuit of Fig. 6. We find

$$-iR_1 - iR_2 - iR_3 + \mathscr{E} = 0,$$

* For both resistors and capacitors in series, there is the implicit assumption that there are no side branches between the terminal points of the series combination.

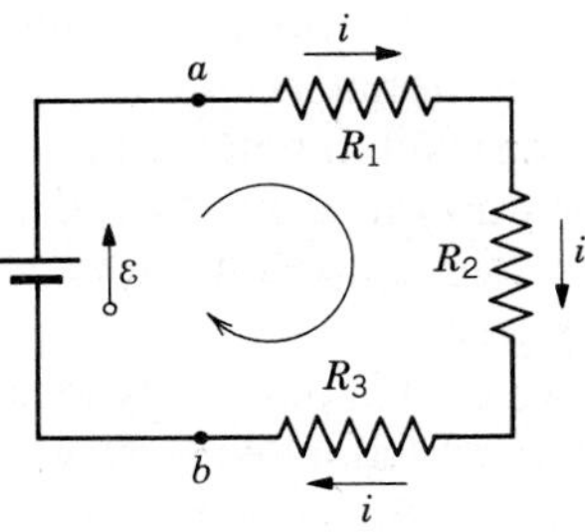

Figure 6 Three resistors are connected in series between points *a* and *b*. What is their equivalent resistance?

or

$$i = \frac{\mathscr{E}}{R_1 + R_2 + R_3}. \tag{5}$$

For the equivalent resistance we have

$$i = \frac{\mathscr{E}}{R_{eq}}. \tag{6}$$

Comparison of Eqs. 5 and 6 shows that

$$R_{eq} = R_1 + R_2 + R_3.$$

The extension to n resistors is straightforward and is

$$\boxed{R_{eq} = \sum_n R_n} \quad \text{(resistances in series).} \tag{7}$$

Comparison with Eq. 19 of Chapter 27 shows that resistors in series follow the same rule as capacitors in parallel; to find the equivalent value of either capacitance or resistance for these arrangements you simply add the individual values.

29-5 Potential Differences

We often want to find the potential difference between two points in a circuit. In Fig. 5*a*, for example, what will a voltmeter read if we touch the positive probe to point *a* and the negative probe to point *b*? To find out, let us start from point *a* and traverse the circuit clockwise to point *b*, passing through resistor *R*. If V_a and V_b are the potentials at *a* and *b*, respectively, we have

$$V_a - iR = V_b$$

because (according to our resistor rule) we experience a decrease in potential in going through a resistor in the direction of the current. We rewrite this as

$$V_a - V_b = +iR, \tag{8}$$

which tells us that point *a* is more positive in potential than point *b*. Combining Eq. 8 with Eq. 4, we have

$$V_a - V_b = \mathcal{E}\frac{R}{R+r}. \tag{9}$$

To find the potential difference between any two points in a circuit, start at one point and traverse the circuit to the other, following any path, and add algebraically the changes in potential that you encounter.

Let us again calculate $V_a - V_b$, starting again from point *a* but this time proceeding counterclockwise to *b* through the seat of emf. We have

$$V_a + ir - \mathcal{E} = V_b$$

or

$$V_a - V_b = \mathcal{E} - ir.$$

Again, combining this relation with Eq. 4 leads to Eq. 9.

Note that $V_a - V_b$ in Fig. 5 is the potential difference of the battery across the battery terminals. We see from Eq. 9 that $V_a - V_b$ is only equal to the emf $\mathcal{E}$ if the battery has no internal resistance ($r = 0$) or if the battery is an open circuit ($R \to \infty$).

Sample Problem 1 What is the current in the circuit of Fig. 7*a*? The emfs and the resistors have the following values:

$$\mathcal{E}_1 = 2.1 \text{ V}, \quad \mathcal{E}_2 = 4.4 \text{ V},$$
$$r_1 = 1.8\ \Omega, \quad r_2 = 2.3\ \Omega, \quad R = 5.5\ \Omega.$$

The two emfs are connected so that they oppose each other but $\mathcal{E}_2$, because it is larger than $\mathcal{E}_1$, controls the direction of the current in the circuit, which is counterclockwise. The loop rule, applied clockwise from point *a*, yields

$$-\mathcal{E}_2 + ir_2 + iR + ir_1 + \mathcal{E}_1 = 0. \tag{10}$$

Check that this same equation results by going around counterclockwise or by starting at some point other than *a*. Also, compare this equation term by term with Fig. 7*b*, which shows the potential changes graphically.

Solving Eq. 10 for the current *i*, we obtain

$$i = \frac{\mathcal{E}_2 - \mathcal{E}_1}{R + r_1 + r_2}$$
$$= \frac{4.4 \text{ V} - 2.1 \text{ V}}{5.5\ \Omega + 1.8\ \Omega + 2.3\ \Omega}$$
$$= 0.2396 \text{ A} \approx 240 \text{ mA}. \qquad \text{(Answer)}$$

It is not necessary to know the direction of the current in advance. To show this, let us assume that the current in Fig. 7*a* is clockwise, that is, reverse the direction of the current arrow in Fig. 7*a*. The loop rule would then yield (going clockwise from *a*)

$$-\mathcal{E}_2 - ir_2 - iR - ir_1 + \mathcal{E}_1 = 0$$

or

$$i = -\frac{\mathcal{E}_2 - \mathcal{E}_1}{R + r_1 + r_2}.$$

Substituting numerical values (see above) yields $i = -240$ mA for the current. The minus sign is a signal that the current is in the opposite direction from that which we have assumed.

In more complex circuits involving many loops and branches, it is often impossible to know in advance the actual directions for the currents in all parts of the circuit. The procedure is to choose initially the current directions for each branch arbitrarily. If you get an answer with a positive sign for a particular current, you have chosen its direction correctly; if you get a negative sign, the current is opposite in direction to that chosen. In all cases, the numerical value will be correct.

Sample Problem 2 (a) What is the potential difference between points *a* and *b* in Fig. 7*a*?

This potential difference is the terminal potential difference of the emf $\mathcal{E}_2$. Let us start at point *b* and traverse the circuit counterclockwise to point *a*, passing directly through

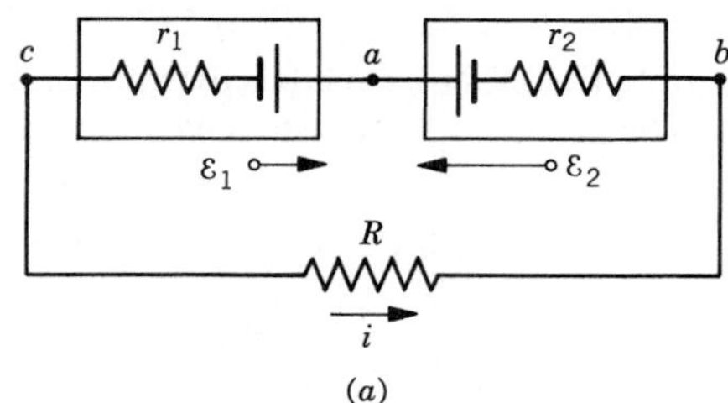

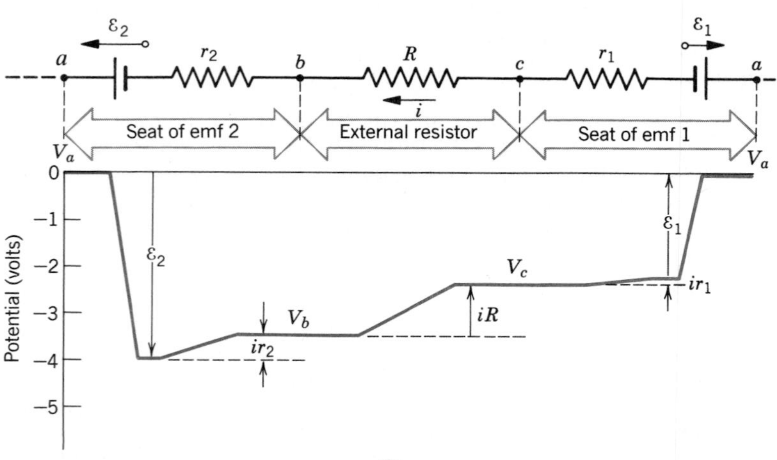

Figure 7 Sample Problems 1 and 2. *(a)* A single-loop circuit containing two seats of emf. *(b)* The potentials encountered in traversing this circuit clockwise from point *a*.

the seat of emf. We find

$$V_b - ir_2 + \mathcal{E}_2 = V_a$$

or

$$\begin{aligned} V_a - V_b &= -ir_2 + \mathcal{E}_2 \\ &= -(0.2396 \text{ A})(2.3\ \Omega) + 4.4 \text{ V} \\ &= +3.85 \text{ V}. \qquad \text{(Answer)} \end{aligned}$$

Thus, *a* is more positive than *b* and the potential difference between them (3.85 V) is *smaller* than the emf (4.4 V); see Fig. 7*b*.

We can verify this result by starting at point *b* in Fig. 7*a* and traversing the circuit clockwise to point *a*. For this different path we find

$$V_b + iR + ir_1 + \mathcal{E}_1 = V_a$$

or

$$\begin{aligned} V_a - V_b &= iR + ir_1 + \mathcal{E}_1 \\ &= (0.2396 \text{ A})(5.5\ \Omega + 1.8\ \Omega) + 2.1 \text{ V} \\ &= +3.85 \text{ V}, \qquad \text{(Answer)} \end{aligned}$$

exactly as before. The potential difference between two points has the same value for all paths connecting those points.

(b) What is the potential difference between points *a* and *c* in Fig. 7*a*?

Note that this potential difference is the terminal potential difference of the emf $\mathcal{E}_1$. Let us start at *c* and traverse the circuit clockwise to point *a*. We find

$$V_c + ir_1 + \mathcal{E}_1 = V_a$$

or

$$\begin{aligned} V_a - V_c &= ir_1 + \mathcal{E}_1 \\ &= (0.2396 \text{ A})(1.8\ \Omega) + 2.1 \text{ V} \\ &= +2.5 \text{ V}. \qquad \text{(Answer)} \end{aligned}$$

This tells us that *a* is at a higher potential than *c*. The terminal potential difference (2.5 V) is in this case *larger* than the emf (2.1 V); see Fig. 7*b*. Charge is being forced through $\mathcal{E}_1$ in a direction opposite to the one in which it would send charge if it were acting by itself; if $\mathcal{E}_1$ were a storage battery it would be charging at the expense of $\mathcal{E}_2$.

29-6 Multiloop Circuits

Figure 8 shows a circuit containing two loops. For simplicity, we have assumed ideal batteries, that is, batteries with no internal resistance. There are two *junctions* in this circuit, at *b* and *d*, and there are three *branches*

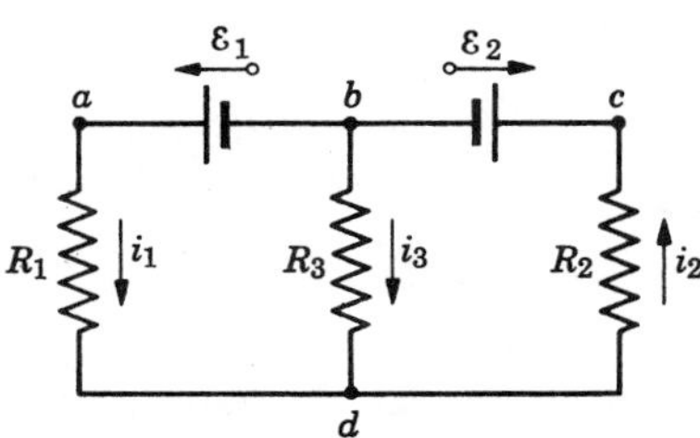

Figure 8 A simple multiloop circuit. What are the currents in the three branches?

connecting these junctions. The branches are the left branch *(bad),* the right branch *(bcd),* and the central branch *(bd).* What are the currents in the three branches?

We label the currents i_1, i_2, and i_3, as shown. Current i_1 has the same value for any cross section of branch *bad.* Similarly, i_2 has the same value everywhere in the right branch and i_3 in the central branch. You can choose the directions of the currents arbitrarily. By studying the signs of the emfs in Fig. 8, you should be able to convince yourself that the current in the central branch must point up, not down as we have drawn it. We have deliberately chosen it pointing in the wrong direction so that we may see how the algebra will automatically correct such wrong guesses.

Consider junction *d.* The three currents carry charge either toward this junction or away from it. In a steady-state condition, charge does not pile up at any junction, nor does it drain away from it. Thus, the rate at which charge is delivered to the junction must equal the rate at which it is taken away, or

$$i_1 + i_3 = i_2. \tag{11}$$

You can easily check that applying this theorem to junction *a* leads to exactly this same equation.* Equation 11 suggests a general principle:

Junction rule.† The sum of the currents approaching any junction must be equal to the sum of the currents leaving that junction.

This rule is simply a statement of the conservation of charge. Thus, our basic tools for solving complex circuits are the *loop rule,* which is based on the conservation of energy, and the *junction rule,* which is based on the conservation of charge.

Equation 11 will give us any one of the currents if we know the other two. To solve the problem completely, we need more information; we can find it by applying the loop rule. If we traverse the left loop of Fig. 8 in a counterclockwise direction, this rule gives

$$\mathcal{E}_1 - i_1R_1 + i_3R_3 = 0. \tag{12}$$

The right loop yields

$$-i_3R_3 - i_2R_2 - \mathcal{E}_2 = 0. \tag{13}$$

Equations 11, 12, and 13 are three simultaneous equations involving the three currents as variables. Solving for the three unknowns we find, after a little algebra,

$$i_1 = \frac{\mathcal{E}_1(R_2 + R_3) - \mathcal{E}_2R_3}{R_1R_2 + R_2R_3 + R_1R_3} \quad \text{(left branch)}, \tag{14}$$

$$i_2 = \frac{\mathcal{E}_1R_3 - \mathcal{E}_2(R_1 + R_3)}{R_1R_2 + R_2R_3 + R_1R_3} \quad \text{(central branch)}, \tag{15}$$

and

$$i_3 = -\frac{\mathcal{E}_1R_2 + \mathcal{E}_2R_1}{R_1R_2 + R_2R_3 + R_1R_3} \quad \text{(right branch)}. \tag{16}$$

Be sure to supply the missing steps.

Equation 16 shows that no matter what the numerical values of the resistances and the emfs, the current i_3 will have a negative sign. Thus—as we knew all along—the current is opposite in direction to that shown in Fig. 8. The currents i_1 and i_2 may be in either direction, depending on the numerical values of the resistances and the emfs.

When we have a complex array of equations such as Eqs. 14, 15, and 16, where there is always a chance that we have made a mistake in algebra, it is a good idea to check the equations to make sure that they give expected results in simple special cases. One such case arises if we put $R_3 = \infty$, which corresponds to clipping the central resistor out of the circuit with a pair of cutters. The circuit then becomes a single-loop circuit and Eqs. 14, 15, and 16 predict that

$$i_1 = i_2 = \frac{\mathcal{E}_1 - \mathcal{E}_2}{R_1 + R_2} \quad \text{and} \quad i_3 = 0.$$

These results are what we expect. How many other simple special cases can you identify?

It might occur to you that you can apply the loop

* It can be shown that, for a circuit with n junctions, there are only $n - 1$ independent junction equations. In this case, $n = 2$ so that there is just one independent equation.

† The junction rule is sometimes referred to as *Kirchhoff's junction rule.*

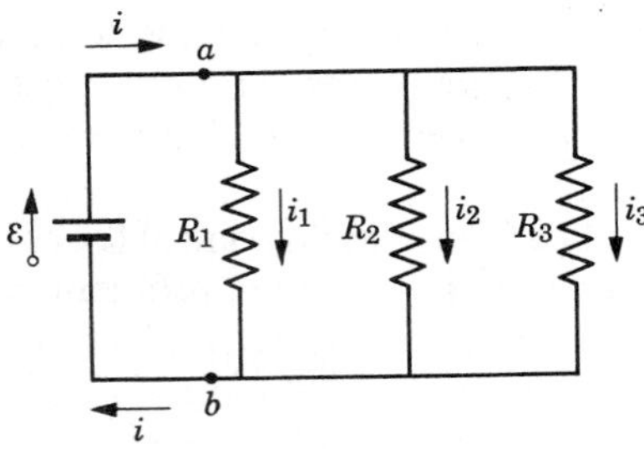

Figure 9 Three resistors are connected in parallel across points a and b. What is their equivalent resistance?

rule to a large loop, consisting of the entire circuit *abcda* in Fig. 8. The rule yields for this loop

$$\mathcal{E}_1 - \mathcal{E}_2 - i_2R_2 - i_1R_1 = 0,$$

which is nothing more than the sum of Eqs. 12 and 13. Thus, the large loop does not yield another independent equation. In solving multiloop circuits, you will never find more independent equations than there are variables, no matter how many times you apply the loop rule and the junction rule. When you have written down as many independent equations as you need, stop.

Resistors in Parallel. Figure 9 shows three resistors connected across the same seat of emf. Resistors across which the same potential difference is applied are said to be *in parallel.* What is the equivalent resistance R_{eq} of this parallel combination? By *equivalent resistance,* we mean the single resistance that, substituted for the parallel combination between terminals a and b, would leave the current i unchanged.

The currents in the three branches of Fig. 9 are

$$i_1 = \frac{V}{R_1}, \quad i_2 = \frac{V}{R_2}, \quad \text{and} \quad i_3 = \frac{V}{R_3},$$

where V is the potential difference between a and b. If we apply the junction rule at point a, we find

$$i = i_1 + i_2 + i_3 = V\left(\frac{1}{R_1} + \frac{1}{R_2} + \frac{1}{R_3}\right). \tag{17}$$

If we replace the parallel combination by the equivalent resistance, we have

$$i = \frac{V}{R_{eq}}. \tag{18}$$

Comparing Eqs. 17 and 18 leads to

$$\frac{1}{R_{eq}} = \frac{1}{R_1} + \frac{1}{R_2} + \frac{1}{R_3}. \tag{19}$$

Table 1 Resistors and Capacitors in Series and in Parallel[a]

	Series	*Parallel*
Resistors	$R_{eq} = \sum_n R_n$ (29–7) Same current	$\frac{1}{R_{eq}} = \sum_n \frac{1}{R_n}$ (29–20) Same potential difference
Capacitors	$\frac{1}{C_{eq}} = \sum_n \frac{1}{C_n}$ (27–20) Same charge	$C_{eq} = \sum_n C_n$ (27–19) Same potential difference

[a] The chapter and equation numbers are displayed below the formulas.

Extending this result to the case of n resistors, we find

$$\boxed{\frac{1}{R_{eq}} = \sum_n \frac{1}{R_n}} \quad \text{(resistors in parallel).} \tag{20}$$

For the case of two resistors, the equivalent resistance is their product divided by their sum.* That is,

$$R_{eq} = \frac{R_1R_2}{R_1 + R_2} = \left(\frac{R_2}{R_1 + R_2}\right) R_1 = \left(\frac{R_1}{R_1 + R_2}\right) R_2.$$

Note that, because the fractions in the parentheses above are each less than unity, the equivalent resistance is smaller than either of the two combining resistors. A little thought will convince you that this remains true for any number of resistors connected in parallel. Also note that the formula for resistors in parallel is identical in form to the formula for capacitors in series. Table 1 summarizes the relations for resistors and capacitors in series and in parallel.

Sample Problem 3 Figure 10 shows a circuit whose elements have the following values:

$$\mathcal{E}_1 = 2.1\ \text{V}, \quad \mathcal{E}_2 = 6.3\ \text{V},$$
$$R_1 = 1.7\ \Omega, \quad R_2 = 3.5\ \Omega.$$

(a) Find the currents in the three branches of the circuit.

Let us draw and label the currents as shown in the figure, choosing the current directions arbitrarily. Applying the junc-

* If you accidentally took the equivalent resistance to be the sum divided by the product, you would notice at once that this result would be dimensionally incorrect.

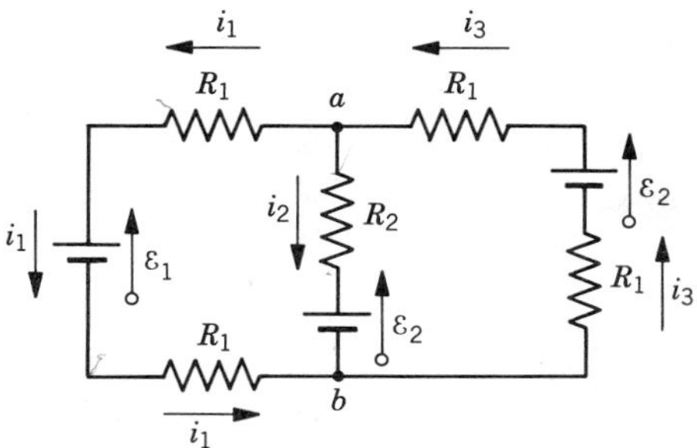

Figure 10 Sample Problem 3. A multiloop circuit. What are the currents in the three branches? What is the potential difference between points a and b?

tion rule at a, we find

$$i_3 = i_1 + i_2. \tag{21}$$

Now let us start at point a and traverse the left-hand loop in a counterclockwise direction. We find

$$-i_1R_1 - \mathcal{E}_1 - i_1R_1 + \mathcal{E}_2 + i_2R_2 = 0$$

or

$$2i_1R_1 - i_2R_2 = \mathcal{E}_2 - \mathcal{E}_1. \tag{22}$$

If we traverse the right-hand loop in a clockwise direction from point a, we find

$$+i_3R_1 - \mathcal{E}_2 + i_3R_1 + \mathcal{E}_2 + i_2R_2 = 0$$

or

$$i_2R_2 + 2i_3R_1 = 0. \tag{23}$$

Equations 21, 22, and 23 are three independent simultaneous equations involving the three variables i_1, i_2, and i_3. We can solve these equations for these variables, obtaining, after a little algebra,

$$\begin{aligned} i_1 &= \frac{(\mathcal{E}_2 - \mathcal{E}_1)(2R_1 + R_2)}{4R_1(R_1 + R_2)}, \\ &= \frac{(6.3\text{ V} - 2.1\text{ V})(2 \times 1.7\ \Omega + 3.5\ \Omega)}{(4)(1.7\ \Omega)(1.7\ \Omega + 3.5\ \Omega)} \\ &= 0.82\text{ A}, \qquad \text{(Answer)} \end{aligned}$$

$$\begin{aligned} i_2 &= -\frac{\mathcal{E}_2 - \mathcal{E}_1}{2(R_1 + R_2)}, \\ &= -\frac{6.3\text{ V} - 2.1\text{ V}}{(2)(1.7\ \Omega + 3.5\ \Omega)} = -0.40\text{ A}, \qquad \text{(Answer)} \end{aligned}$$

and

$$\begin{aligned} i_3 &= \frac{(\mathcal{E}_2 - \mathcal{E}_1)(R_2)}{4R_1(R_1 + R_2)} \\ &= \frac{(6.3\text{ V} - 2.1\text{ V})(3.5\ \Omega)}{(4)(1.7\ \Omega)(1.7\ \Omega + 3.5\ \Omega)} = 0.42\text{ A}. \qquad \text{(Answer)} \end{aligned}$$

The signs of the currents tell us that we have guessed correctly about the directions of i_1 and i_3 but that we are wrong about the direction of i_2; it should point up—and not down—in the central branch of the circuit of Fig. 10.

(b) What is the potential difference between points a and b in the circuit of Fig. 10?

We have, assuming the current directions shown in the figure,

$$V_a - i_2R_2 - \mathcal{E}_2 = V_b,$$

or

$$\begin{aligned} V_a - V_b &= \mathcal{E}_2 + i_2R_2 \\ &= 6.3\text{ V} + (-0.40\text{ A})(3.5\ \Omega) = +4.9\text{ V}. \qquad \text{(Answer)} \end{aligned}$$

The positive sign of this result tells us that a is more positive in potential than b. From study of the circuit, this is what we would expect because all three batteries have their positive terminals on the top side of the figure.

Sample Problem 4 Figure 11a shows a cube made of 12 resistors, each of resistance R. Find R_{12}, the equivalent resistance of a cube edge.

Although this problem can be attacked by "brute force" methods, using the loop and junction rules, the symmetry of the connections suggests that there must be a neater method. The key is the realization that, from considerations of symmetry alone, points 3 and 6 must be at the same potential. So must points 4 and 5.

If two points in a circuit have the same potential, the currents in the circuit do not change if you connect these points by a wire. There will be no current in the wire because there is no potential difference between its ends. Electrically, then, points 3 and 6 are a single point; so are points 4 and 5.

This allows us to redraw the cube as in Fig. 11b. From this point, it is simply a matter of reducing the circuit between the input terminals to a single resistor, using the rules for resistors in series and in parallel. In Fig. 11c, we make a start by replacing five parallel resistor combinations by their equivalents, each of resistance $\frac{1}{2}R$.

In Fig. 11d, we have added the three resistors that are in series in the right-hand loop, obtaining a single equivalent resistance of $2R$. In Fig. 11e, we have replaced the two resistors that now form the right-hand loop by a single equivalent resistor $\frac{2}{3}R$. In so doing, it is useful to recall that the equivalent resistance of two resistors in parallel is equal to their product divided by their sum.

In Fig. 11f, we have added the three series resistors of Fig. 11e, obtaining $\frac{7}{5}R$ and in Fig. 11g we have reduced this parallel combination to the single equivalent resistance that we seek, namely,

$$R_{12} = \tfrac{7}{12}R. \qquad \text{(Answer)}$$

You can also use these methods to find R_{13}, the equivalent resistance of a cube across a face diagonal and R_{17}, the equivalent resistance across a body diagonal.

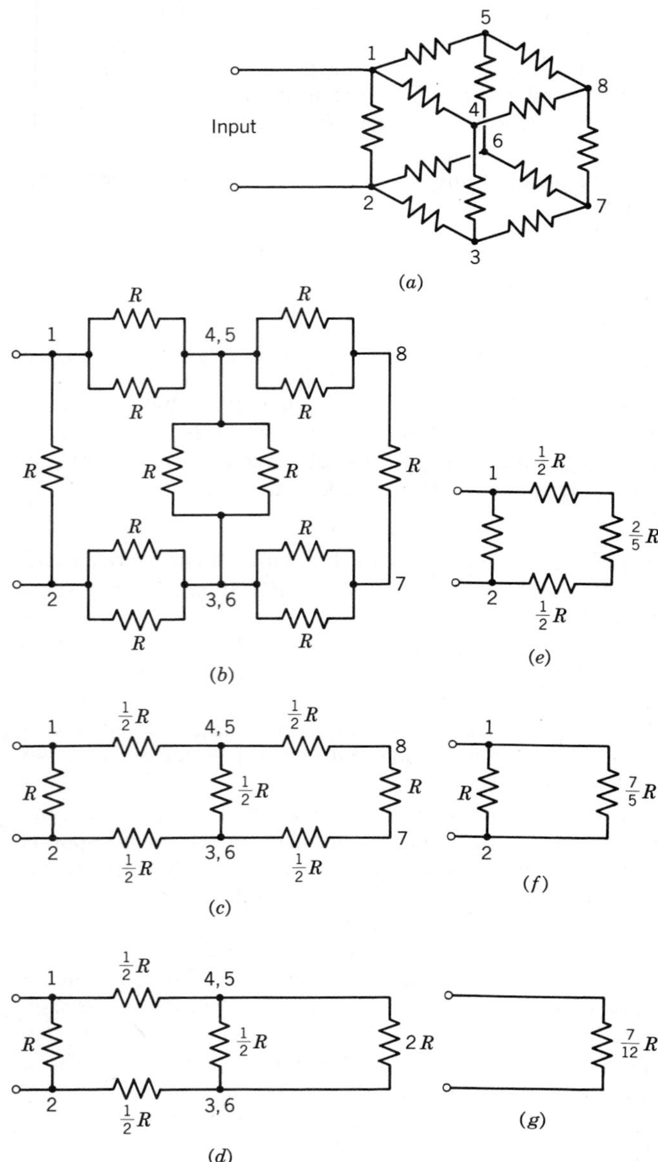

Figure 11 Sample Problem 4. (*a*) A cube formed of identical resistors. The remaining figures show how to reduce these 12 resistors to a single equivalent resistance.

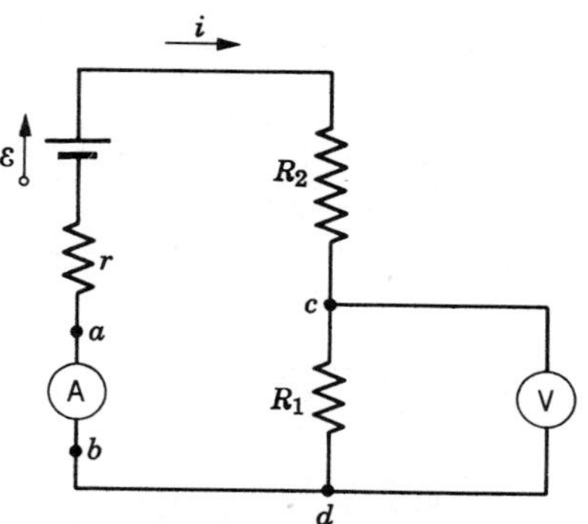

Figure 12 A single-loop circuit, showing how to connect an ammeter (A) and a voltmeter (V).

29-7 Measuring Instruments

Several electric measuring instruments involve circuits that can be analyzed by the methods of this chapter. We discuss three of them.

1. The Ammeter. An instrument used to measure currents is called an *ammeter.* To measure the current in a wire, you usually have to break or cut the wire and insert the ammeter so that the current to be measured passes through the meter; see Fig. 12.

It is essential that the resistance R_A of the ammeter be very small compared to other resistances in the circuit. Otherwise, the very presence of the meter will change the current to be measured. In the circuit of Fig. 12, the required condition, assuming that the voltmeter is not connected, is

$$R_A \ll (r + R_1 + R_2).$$

2. The Voltmeter. A meter to measure potential differences is called a *voltmeter.* To find the potential difference between any two points in the circuit, the voltmeter terminals are connected between those points, without breaking the circuit; see Fig. 12.

It is essential that the resistance R_V of a voltmeter be very large compared to any circuit element across which the voltmeter is connected. Otherwise, the meter itself becomes an important circuit element and will alter the potential difference that is to be measured. In Fig. 12, the required condition is

$$R_V \gg R_1.$$

Often a single unit is packaged so that, by external switching, it can serve as either an ammeter or a voltmeter—and usually also as an *ohmmeter,* designed to measure the resistance of any element connected between its terminals. Such a versatile unit is called a *multimeter;* see Fig. 13. Its output readings may take the form of a pointer moving over a scale or of a digital display.

3. The Potentiometer. A *potentiometer* is a device

Figure 13 A typical multimeter, used to measure currents, potential differences, and resistances.

for measuring an unknown emf $\mathcal{E}_x$ by comparing it with a known standard emf $\mathcal{E}_s$.

Figure 14 shows its rudiments. The resistor that extends from a to e is a carefully made precision resistor with a sliding contact shown positioned at d. The resistance R in the figure is the resistance between points a and d.

When using the instrument, $\mathcal{E}_s$ is first placed in the position $\mathcal{E}$ and the sliding contact is adjusted until the current i is zero as noted on the sensitive ammeter A. The potentiometer is then said to be *balanced,* the value of R at balance being R_s. In this balance condition we have, considering the loop *abcda,*

$$\mathcal{E}_s = i_0 R_s. \tag{24}$$

Because $i = 0$ in branch *abcd,* the internal resistance r of the standard source of emf does not enter.

The process is now repeated with $\mathcal{E}_x$ substituted for $\mathcal{E}_s$, the potentiometer being balanced once more. The current i_0 remains unchanged and the new balance condition is

$$\mathcal{E}_x = i_0 R_x. \tag{25}$$

From Eqs. 24 and 25 we then have

$$\mathcal{E}_x = \mathcal{E}_s \frac{R_x}{R_s}. \tag{26}$$

Thus, the unknown emf can be found in terms of the known emf by making two adjustments of the precision resistor. In practice, potentiometers are conveniently

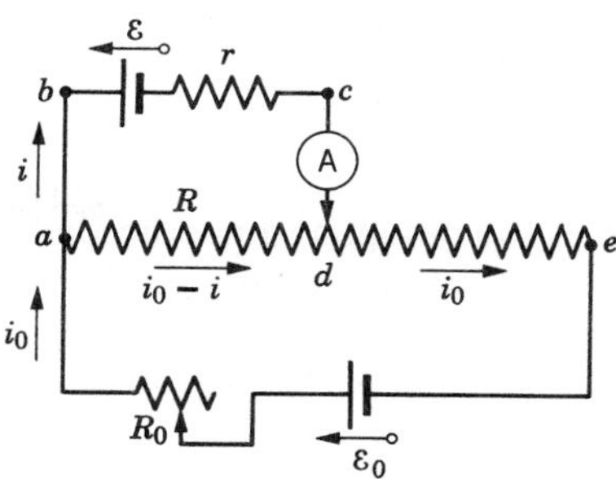

Figure 14 The rudiments of a potentiometer, used to compare emfs.

packaged units, containing a built-in *standard cell* that, after calibration at the National Bureau of Standards or elsewhere, serves as a convenient reference standard $\mathcal{E}_s$. Switching arrangements for interchanging the standard and the unknown emfs are also incorporated. The sliding contact usually moves over a scale on which the value of the unknown emf can be read directly, without having to perform the calculation required by Eq. 26.

29-8 *RC* Circuits

The preceding sections dealt with circuits in which the circuit elements were resistors and in which the currents did not vary with time. Here we introduce the capacitor as a circuit element, which will lead us to the study of time-varying currents.

Charging a Capacitor. In the circuit of Fig. 15, let us throw switch S from the indicated position to position a, thus introducing an emf $\mathcal{E}$ into the circuit and *charging* the capacitor C through the resistor R. How will the charging current i vary with time?

Let us apply the loop rule to the circuit of Fig. 15,

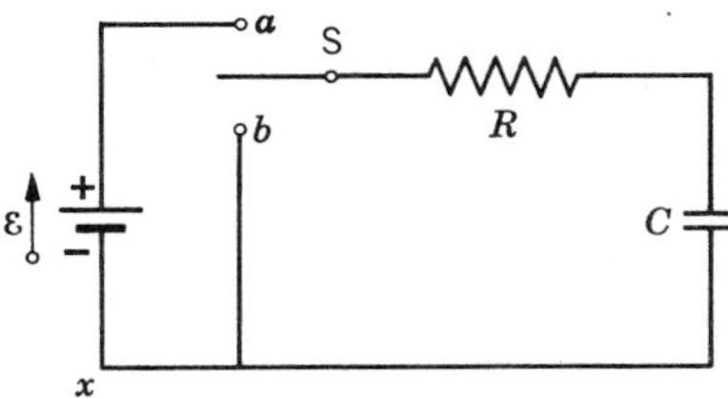

Figure 15 When switch S is closed on a, the capacitor C is *charged* through the resistor R. When the switch is afterward closed on b, the capacitor *discharges* through R.

going around in a clockwise direction. We have

$$\mathcal{E} - iR - \frac{q}{C} = 0,$$

in which q/C is the potential difference between the capacitor plates. We rearrange this equation as

$$iR + \frac{q}{C} = \mathcal{E}. \qquad (27)$$

We cannot immediately solve Eq. 27 because it contains two variables, the current i and the charge q. However, these variables are not independent but are related by

$$i = \frac{dq}{dt}.$$

Substituting for i in Eq. 27, we find

$$\boxed{R\frac{dq}{dt} + \frac{q}{C} = \mathcal{E}} \quad \text{(charging equation).} \qquad (28)$$

This is the differential equation that describes the variation with time of the charge q in the circuit of Fig. 15. Our task is to find the function $q(t)$ that satisfies this equation and also satisfies the requirement that the capacitor be initially uncharged. This requirement, namely that $q = 0$ at $t = 0$, is typical of the *initial conditions* that we impose—in general—on differential equations.

Although Eq. 28 is not hard to solve, we choose to avoid mathematical complexity by simply presenting the solution, which is (we claim)

$$\boxed{q = C\mathcal{E}(1 - e^{-t/RC})} \quad \text{(charge).} \qquad (29)$$

Note that Eq. 29 does indeed satisfy our required initial condition, in that $q = 0$ at $t = 0$. The derivative of $q(t)$ is

$$\boxed{i = \frac{dq}{dt} = \left(\frac{\mathcal{E}}{R}\right) e^{-t/RC}} \quad \text{(current).} \qquad (30)$$

If we substitute q from Eq. 29 and i from Eq. 30 into Eq. 28, the differential equation does indeed reduce to an identity, as you should be certain to verify. Thus, Eq. 29 is indeed a solution of Eq. 28.

We can measure $q(t)$ experimentally by measuring a quantity proportional to it, namely, V_C, the potential difference across the capacitor. From Eq. 29 we have

$$V_C = \frac{q}{C} = \mathcal{E}(1 - e^{-t/RC}) \qquad (31)$$

Similarly, we can measure $i(t)$ by measuring V_R, the potential difference across the resistor. From Eq. 30 we have

$$V_R = iR = \mathcal{E}e^{-t/RC}. \qquad (32)$$

Figure 16 shows plots of V_C and V_R. We note that V_C and V_R add up at each instant to $\mathcal{E}$, as Eq. 27 requires.

The Time Constant. The product RC that appears in Eqs. 29 and 30 has the dimensions of time (because the exponent in those equations must be dimensionless). RC is called the *capacitive time constant* of the circuit and is represented by the symbol τ. It is the time at which the charge on the capacitor has increased to a factor of $(1 - e^{-1})$ or about 63% of its equilibrium value. To show this, let us put $t = RC$ in Eq. 29. We find

$$q = C\mathcal{E}(1 - e^{-1}) = 0.63\ C\mathcal{E}.$$

Because $C\mathcal{E}$ is the equilibrium charge on the capacitor, corresponding to $t \to \infty$ in Eq. 29, we have proved our point.

The Charging Process. It is easy enough to derive equations such as Eqs. 29 and 30 without any real physical understanding of what is going on. Let us therefore look at the charging process in an RC circuit qualitatively and physically.

When switch S in Fig. 15 is first closed on a, there is no charge on the capacitor and therefore no potential difference between its plates. Thus, the full potential difference $\mathcal{E}$ appears across the resistor, setting up an initial current $\mathcal{E}/R$. Once the current starts, charges

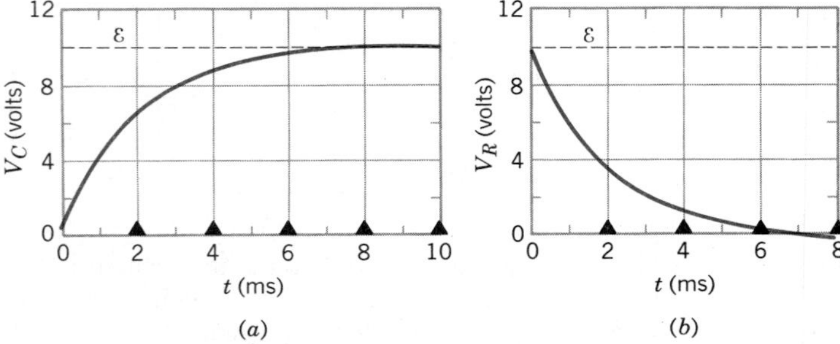

Figure 16 (*a*) A plot of Eq. 31, which traces the buildup of charge on the capacitor of Fig. 15. (*b*) A plot of Eq. 32, which traces the decline of the charging current in the circuit of Fig. 15. The curves are plotted for $R = 2000\ \Omega$, $C = 1\ \mu\text{F}$, and $\mathcal{E} = 10$ V; the filled triangles represent successive time constants.

begin to appear on the capacitor plates and a potential difference q/C builds up between those plates. This in turn means that the potential difference across the resistor must decrease by this same amount. This decrease in the potential difference across R means that the charging current is reduced. Thus, the charge on the capacitor builds up and the charging current through the resistor falls off until the capacitor is fully charged. In this condition, the full emf $\mathcal{E}$ is being applied to the capacitor, there being no potential drop across the resistor. This is precisely the reverse of the initial state, in which the full emf $\mathcal{E}$ appeared across the resistor. Review the derivations of Eqs. 29 and 30 and study Fig. 16 with the arguments of this paragraph in mind.

Discharging a Capacitor. Assume now that the capacitor of Fig. 15 is fully charged to the voltage of the emf and that the switch S is then thrown from a to b so that capacitor C can *discharge* through resistor R. How does the discharge current in this single-loop circuit vary with time?

Equation 28 continues to hold except that now there is no longer an emf in the circuit. Putting $\mathcal{E} = 0$ in this equation, we find

$$R\frac{dq}{dt} + \frac{q}{C} = 0 \quad \text{(discharging equation).} \quad (33)$$

The solution to this differential equation is

$$q = q_0 e^{-t/RC} \quad \text{(charge),} \quad (34)$$

in which q_0 ($= C\mathcal{E}$ in our case) is the initial charge on the capacitor. You can verify by substitution that Eq. 34 is indeed a solution of Eq. 33.

The capacitive time constant RC governs the discharging process as well as the charging process. We see that at $t = RC$ the capacitor charge is reduced to $C\mathcal{E}e^{-1}$, which is about 37% of its initial charge.

The current during discharge follows from differentiating Eq. 34 and is

$$i = \frac{dq}{dt} = -\frac{q_0}{RC}e^{-t/RC} = -i_0 e^{-t/RC} \quad \text{(current).} \quad (35)$$

Here $\mathcal{E}/R$ is the magnitude of the initial current, i_0, corresponding to $t = 0$. This is reasonable because the potential difference across the fully charged capacitor is $\mathcal{E}$. The negative sign shows that the discharge current is in the opposite direction to the charging current, as we expect. Table 2 summarizes the equations for the charging and discharging of a capacitor.

Table 2 ***RC* Charging and Discharging**[a]

	Charging	*Discharging*
Differential equation	$R\frac{dq}{dt} + \frac{q}{C} = \mathcal{E}$ (29–28)	$R\frac{dq}{dt} + \frac{q}{C} = 0$ (29–33)
Charge	$q = C\mathcal{E}(1 - e^{-t/RC})$ (29–29)	$q = q_0 e^{-t/RC}$ (29–34)
Current	$i = (\mathcal{E}/R)e^{-t/RC}$ (29–30)	$i = -(q_0/RC)e^{-t/RC}$ (29–35)

[a] The chapter and equation numbers are displayed below the formulas.

Sample Problem 5 A capacitor C is discharging through a resistor R. (a) After how many time constants will its charge fall to one-half of its initial value?

The charge on the capacitor varies according to Eq. 34,

$$q = q_0 e^{-t/RC},$$

in which q_0 is the initial charge. We are asked to find the time t at which $q = \frac{1}{2}q_0$, or

$$\tfrac{1}{2}q_0 = q_0 e^{-t/RC}.$$

Canceling q_0 and then taking logarithms of each side, we find

$$0 - \ln 2 = -\frac{t}{RC}$$

or

$$t = (\ln 2)\,RC = 0.69\,RC. \quad \text{(Answer)}$$

Thus, the charge will drop to half its initial value after 0.69 time constants.

(b) After how many time constants will the stored energy drop to half of its initial value?

The energy of the capacitor is

$$U = \frac{q^2}{2C} = \frac{q_0^2}{2C}e^{-2t/RC} = U_0 e^{-2t/RC},$$

in which U_0 is the initial stored energy. We are asked to find the time at which $U = \frac{1}{2}U_0$, or

$$\tfrac{1}{2}U_0 = U_0 e^{-2t/RC}.$$

Canceling U_0 and then taking logarithms of each side, we obtain

$$-\ln 2 = -2t/RC$$

or

$$t = RC\frac{\ln 2}{2} = 0.35\,RC. \quad \text{(Answer)}$$

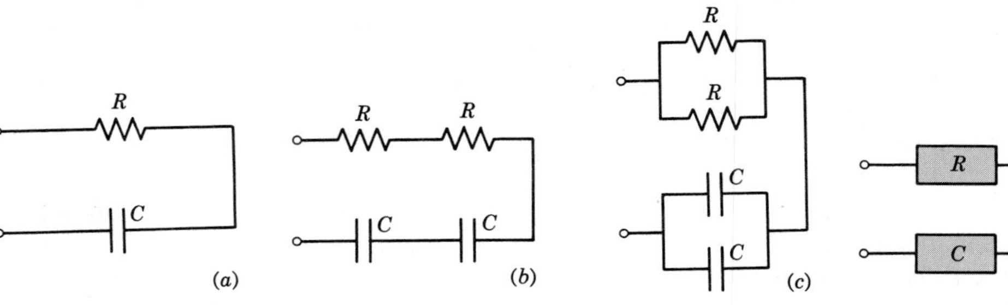

Figure 17 Sample Problem 6. The capacitive time constant of all these circuits is RC.

Thus, the stored energy will drop to half its initial value after 0.35 time constants have elapsed. This remains true no matter what the initial stored energy may be. It takes longer (0.69 RC) for the *charge* to fall to half its initial value than for the *energy* (0.35 RC) to fall to half its initial value. Does this result surprise you?

Sample Problem 6 The time constant τ of the circuit of Fig. 17*a* is clearly RC. (a) What is the time constant of the circuit of Fig. 17*b*?

The equivalent resistance is $2R$ and the equivalent capacitance is $\frac{1}{2}C$ so that

$$\tau = (2R)(\tfrac{1}{2}C) = RC, \qquad \text{(Answer)}$$

just as for the circuit of Fig. 17*a*.

(b) What is the time constant of the circuit of Fig. 17*c*?

The equivalent resistance is $\frac{1}{2}R$ and the equivalent capacitance is $2C$ so that, once again,

$$\tau = (\tfrac{1}{2}R)(2C) = RC. \qquad \text{(Answer)}$$

These results are evidence of the reciprocal nature of the formulas for combining resistors and capacitors in series and in parallel; see Table 1. The time constant of the generalized circuit of Fig. 17*d* is also RC, provided that the box marked R is any configuration whatever of identical resistors and the box marked C is the same configuration of identical capacitors.*

* This generalization was pointed out to us by Professor Andrew L. Gardner.

REVIEW AND SUMMARY

Electromotive Force

A *seat of electromotive force* is a device (a battery or a generator, say) that supplies energy (chemical or mechanical, say) to maintain a potential difference between its output terminals. If dW is the energy the seat provides to force positive charge dq through the seat from the negative to the positive terminal then

$$\mathscr{E} = \frac{dW}{dq} \quad \text{(definition of } \mathscr{E}\text{)}. \qquad [1]$$

The volt is the SI unit of emf as well as of potential difference. Study Fig. 2.

Several *rules* or *methods* are stated here in sufficient generality that they can be used, not only for the circuits of this chapter, but for more complex situations to be encountered later:

Energy Method

Energy Method *The net total energy delivered by seats of emf must be balanced by energy dissipated by or stored in the circuit;* see Fig. 3.

Potential Method

Potential Method *The potential difference between any two points is the algebraic sum of the changes in potential encountered in traversing the circuit from one point to the other along any path.*

The change in potential in traversing a *resistor* in the direction of the current is $-iR$; in the opposite direction it is $+iR$. The change in potential in traversing a *seat of emf* in the direction of the emf is $+\mathscr{E}$; in the opposite direction it is $-\mathscr{E}$. See Sample Problem 2.

Loop Rule

Loop Rule *The algebraic sum of the changes in potential encountered in a complete traversal of any closed circuit must be zero;* see Figs. 5 and 6 and Sample Problem 1.

Junction Rule

Junction Rule *The sum of the currents entering any junction must be equal to the sum of the currents leaving that junction.* See Fig. 8 and the analysis of Section 29–6.

The current in a single-loop circuit containing single resistor R and an emf $\mathcal{E}$ with internal resistance r (see Fig. 5) is

$$i = \frac{\mathcal{E}}{R + r}, \qquad [4]$$

which reduces to Eq. 2 when $r = 0$.

Series Elements

Resistors, and other circuit elements, are in *series* if the sum of the individual potential differences is the potential difference of the combination; see Fig. 6. The equivalent resistance of resistors in series is

$$R_{eq} = \sum_n R_n \quad \text{(resistors in series).} \qquad [7]$$

Parallel Elements

Resistors, and other circuit elements, are in *parallel* if they share a common potential difference; see Fig. 7. The equivalent resistance of resistors in parallel is

$$\frac{1}{R_{eq}} = \sum_n \frac{1}{R_n} \quad \text{(resistors in parallel).} \qquad [20]$$

Sample Problem 3 illustrates the use of the circuit rules and Sample Problem 4 shows an interesting application of the analysis of a circuit using equivalent resistances of series and parallel combinations. Section 29–7 uses the circuit rules to analyze the operation of conventional ammeters, voltmeters, and potentiometers.

RC Circuits

When an emf $\mathcal{E}$ is applied to a resistor R and capacitor C in series, as in Fig. 15, the charge on the capacitor increases according to

$$q = C\mathcal{E}(1 - e^{-t/RC}) \quad \text{(charging capacitor),} \qquad [29]$$

in which $C\mathcal{E} = q_0$ is the equilibrium charge and $RC = \tau$ is the *capacitive time constant* of the circuit. When the emf is replaced by a short circuit, the charge on the capacitor decays according to

$$q = q_0 e^{-t/RC} \quad \text{(discharging capacitor).} \qquad [34]$$

See Sample Problem 5.

QUESTIONS

1. Does the direction of the emf provided by a battery depend on the direction of current flow through the battery?

2. In Fig. 3 discuss what changes would occur if we increased the mass m by such an amount that the "motor" reversed direction and became a "generator," that is, a seat of emf.

3. Discuss in detail the statement that the energy method and the loop theorem method for solving circuits are perfectly equivalent.

4. Devise a method for measuring the emf and the internal resistance of a battery.

5. How could you calculate V_{ab} in Fig. 5a by following a path from a to b that does not lie in the conducting circuit?

6. A 25-W, 120-V bulb glows at normal brightness when connected across a bank of batteries. A 500-W, 120-V bulb glows only dimly when connected across the same bank. How could this happen?

7. Under what circumstances can the terminal potential difference of a battery exceed its emf?

8. Automobiles generally use a 12-V electrical system. Years ago a 6-volt system was used. Why the change? Why not 24 V?

9. The loop theorem is based on the conservation of energy principle and the junction theorem on the conservation of charge principle. Explain just how these theorems are based on these principles.

10. Under what circumstances would you want to connect batteries in parallel? In series?

11. Under what circumstances would you want to connect resistors in parallel? In series?

12. What is the difference between an emf and a potential difference?

13. Referring to Fig. 8 use a qualitative physical argument to convince yourself that i_3 is drawn in the wrong direction.

14. Explain in your own words why the resistance of an ammeter should be very small whereas that of a voltmeter should be very large.

15. Do the junction and loop theorems apply in a circuit containing a capacitor?

16. Show that the product RC in Eqs. 29 and 30 has the dimensions of time, that is, that 1 second = 1 ohm × 1 farad.

17. A capacitor, resistor, and battery are connected in series. The charge that the capacitor stores is unaffected by the resistance of the resistor. What purpose then is served by the resistor?

18. Explain why, in Sample Problem 5, the energy falls to half its initial value more rapidly than does the charge.

19. The light flash in a camera is produced by the discharge of a capacitor across the lamp. Why don't we just connect the photoflash lamp directly across the power supply used to charge the capacitor?

20. Does the time required for the charge on a capacitor in an RC circuit to build up to a given fraction of its equilibrium value depend on the value of the applied emf? Does the time required for the charge to change by a given amount depend on the applied emf?

21. A capacitor is connected across the terminals of a battery. Does the charge that eventually appears on the capacitor plates depend on the value of the internal resistance of the battery?

22. Devise a method whereby an RC circuit can be used to measure very high resistances.

23. In Fig. 15 suppose that switch S is closed on a. Explain why, in view of the fact that the negative terminal of the battery is not connected to resistance R, the initial current in R should be $\mathcal{E}/R$, as Eq. 30 predicts.

EXERCISES AND PROBLEMS

Section 29-2 Work, Energy, and Electromotive Force

1E. (*a*) How much work does a 12-V seat of emf do on an electron as it passes through from the positive to the negative terminal? (*b*) If 3.4×10^{18} electrons pass through each second, what is the power output of the seat?

2E. A 5.0-A current is set up in an external circuit by a 6.0-V storage battery for 6.0 min. By how much is the chemical energy of the battery reduced?

3E. A standard flashlight battery can deliver about 2.0 W·h of energy before it runs down. (*a*) If a battery costs 80 cents, what is the cost of operating a 100-W lamp for 8.0 h using batteries? (*b*) What is the cost if power provided by an electric utility company, at 12 cents per kW·h, is used?

4P. A certain 12-V car battery carries an initial charge of 120 A·h. Assuming that the potential across the terminals stays constant until the battery is completely discharged, for how many hours can it deliver energy at the rate of 100 W?

Section 29-5 Potential Differences

5E. In Fig. 18 $\mathcal{E}_1 = 12$ V and $\mathcal{E}_2 = 8$ V. What is the direction of the current in the resistor? Which emf is doing positive work? Which point, A or B, is at the higher potential?

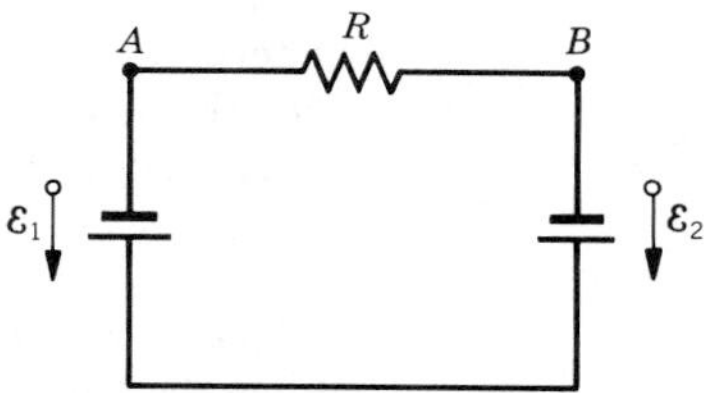

Figure 18 Exercise 5.

6E. A wire of resistance 5.0 Ω is connected to a battery whose emf $\mathcal{E}$ is 2.0 V and whose internal resistance is 1.0 Ω. In 2.0 min (*a*) how much energy is transferred from chemical to electrical form? (*b*) How much energy appears in the wire as thermal energy? (*c*) Account for the difference between (*a*) and (*b*).

7E. In Fig. 5*a* put $\mathcal{E} = 2.0$ V and $r = 100$ Ω. Plot (*a*) the current and (*b*) the potential difference across R, as functions of R over the range 0 to 500 Ω. Make both plots on the same graph. (*c*) Make a third plot by multiplying together, for each value of R, the two curves plotted. What is the physical significance of this plot?

8E. Assume that the batteries in Fig. 19 have negligible internal resistance. Find: (*a*) the current in the circuit, (*b*) the power dissipated in each resistor, and (*c*) the power supplied or absorbed by each emf.

9E. A 12-V emf car battery with an internal resistance of 0.04 Ω is being charged with a current of 50 A. (*a*) What is the potential difference across its terminals? (*b*) At what rate is heat being dissipated in the battery? (*c*) At what rate is electric energy being converted to chemical energy? (*d*) What are the answers to (*a*) and (*b*) when the battery is used to supply 50 A to the starter motor?

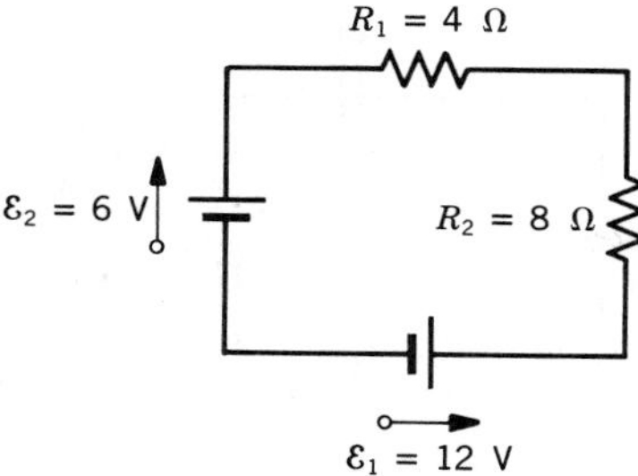

Figure 19 Exercise 8.

10E. In Fig. 20 the potential at point P is 100 V. What is the potential at point Q?

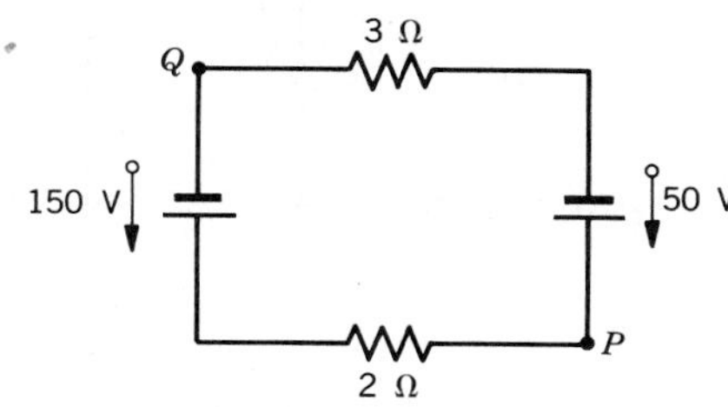

Figure 20 Exercise 10.

11E. The section of circuit AB (see Fig. 21) absorbs 50 W of power when a current $i = 1.0$ A passes through it in the indicated direction. (*a*) What is the potential difference between A and B? (*b*) If the element C does not have internal resistance, what is its emf? (*c*) What is its polarity?

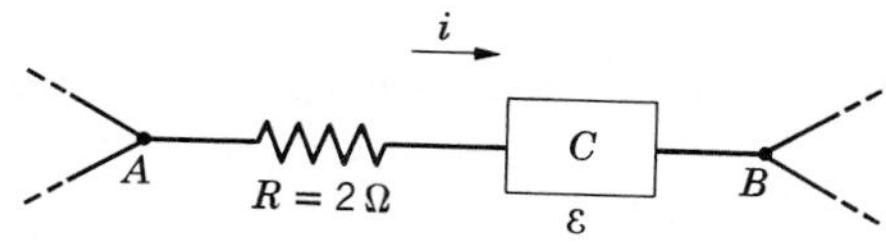

Figure 21 Exercise 11.

12E. In Fig. 6 calculate the potential difference across R_2, using two different paths. Assume $\mathcal{E} = 12$ V, $R_1 = 3.0\ \Omega$, $R_2 = 4.0\ \Omega$, and $R_3 = 5.0\ \Omega$.

13E. In Fig. 7*a* calculate the potential difference between a and c by considering a path that contains R and $\mathcal{E}_2$. (See Sample Problem 2.)

14E. A gasoline gauge for an automobile is shown schematically in Fig. 22. The indicator (on the dashboard) has a resistance of 10 Ω. The tank unit is simply a float connected to a resistor that has a resistance of 140 Ω when the tank is empty, 20 Ω when it is full, and varies linearly with the volume of gasoline. Find the current in the circuit when the tank is (*a*) empty; (*b*) half full; (*c*) full.

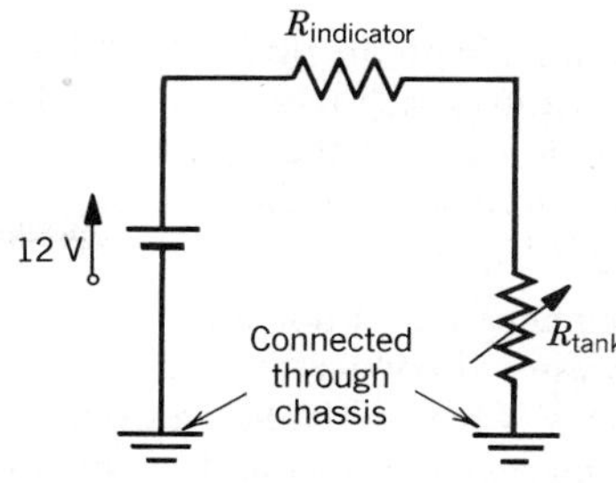

Figure 22 Exercise 14.

15P. (*a*) In Fig. 23 what value must R have if the current in the circuit is to be 1.0 mA? Take $\mathcal{E}_1 = 2.0$ V, $\mathcal{E}_2 = 3.0$ V, and $r_1 = r_2 = 3.0\ \Omega$. (*b*) What is the rate at which thermal energy appears in R?

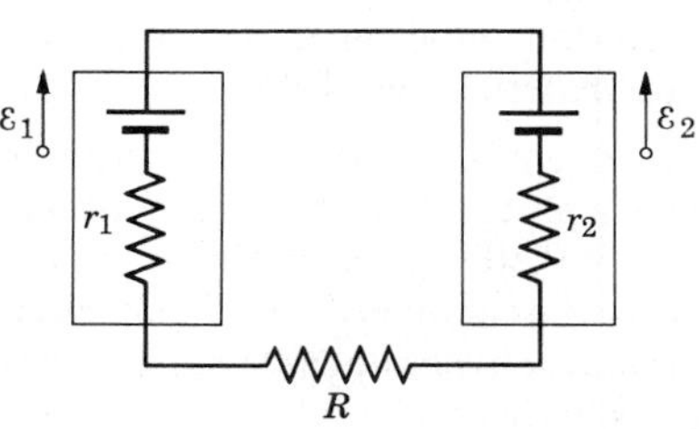

Figure 23 Problem 15.

16P. Thermal energy is to be generated in a 0.10-Ω resistor at the rate of 10 W by connecting it to a battery whose emf is 1.5 V. (*a*) What is the internal resistance of the battery? (*b*) What potential difference exists across the resistor?

17P. The current in a single-loop circuit is 5.0 A. When an additional resistance of 2.0 Ω is inserted in series, the current drops to 4.0 A. What was the resistance in the original circuit?

18P. Power is supplied by an emf $\mathcal{E}$ to a transmission line with resistance R. Find the ratio of the power dissipated in the line for $\mathcal{E} = 110{,}000$ V to that dissipated for $\mathcal{E} = 110$ V, assuming the power supplied is the same for the two cases.

19P. The starting motor of an automobile is turning slowly and the mechanic has to decide whether to replace the motor, the cable, or the battery. The manufacturer's manual says that the 12-V battery can have no more than 0.020 Ω internal resistance, the motor no more than 0.200 Ω resistance, and the cable no more than 0.040 Ω resistance. The mechanic turns on the motor and measures 11.4 V across the battery, 3.0 V across the cable, and a current of 50 A. Which part is defective?

20P. Two batteries having the same emf $\mathcal{E}$ but different internal resistances r_1 and r_2 ($r_1 > r_2$) are connected in series to an external resistance R. (*a*) Find the value of R that makes the potential difference zero between the terminals of one battery. (*b*) Which battery is it?

21P. A solar cell generates a potential difference of 0.10 V when a 500-Ω resistor is connected across it and a potential difference of 0.15 V when a 1000-Ω resistor is substituted. What are (*a*) the internal resistance and (*b*) the emf of the solar cell? (*c*) The area of the cell is 5.0 cm² and the intensity of light striking it is 2.0 mW/cm². What is the efficiency of the cell for converting light energy to heat in the 1000-Ω external resistor?

22P. (*a*) In the circuit of Fig. 5*a* show that the power delivered to R as thermal energy is a maximum when R is equal to the internal resistance r of the battery. (*b*) Show that this maximum power is $P = \mathcal{E}^2/4r$.

23P. Conductors A and B, having equal lengths of 40 m and a common diameter of 2.6 mm, are connected in series. A potential difference of 60 V is applied between the ends of the composite wire. The resistances of the wires are 0.127 and 0.729 Ω, respectively. Determine: (*a*) the current density in each wire; (*b*) the potential difference across each wire. (*c*) Identify the wire materials. See Table 1 in Chapter 28.

24P. A battery of emf $\mathcal{E} = 2.0$ V and internal resistance $r = 0.50$ Ω is driving a motor. The motor is lifting a 2.0-N mass at constant speed $v = 0.50$ m/s. Assuming no power losses, find (*a*) the current i in the circuit and (*b*) the potential difference V across the terminals of the motor. (*c*) Discuss the fact that there are two solutions to this problem.

25P. A temperature-stable resistor is made by connecting a resistor made of silicon in series with one made of iron. If the required total resistance is 1000 Ω at 20°C, what should be the resistances of the two resistors? See Table 1 in Chapter 28.

Section 29-6 Multiloop Circuits

26E. Four 18-Ω resistors are connected in parallel across a 25-V battery. What is the current through the battery?

27E. A total resistance of 3.0 Ω is to be produced by connecting an unknown resistance to a 12 Ω resistance. What must be the value of the unknown resistance and how should it be connected?

28E. By using only two resistors—singly, in series, or in parallel—you are able to obtain resistances of 3, 4, 12, and 16 Ω. What are the separate resistances of the resistors?

29E. In Fig. 24 find the current in each resistor and the potential difference between a and b. Put $\mathcal{E}_1 = 6.0$ V, $\mathcal{E}_2 = 5.0$ V, $\mathcal{E}_3 = 4.0$ V, $R_1 = 100$ Ω, and $R_2 = 50$ Ω.

30E. Figure 25 shows a circuit containing an ammeter and three switches, labeled S_1, S_2, and S_3. Find the readings of the ammeter for all possible combinations of switch settings. Put $\mathcal{E} = 120$ V, $R_1 = 20$ Ω, and $R_2 = 10$ Ω. Assume that the ammeter and battery have no resistance.

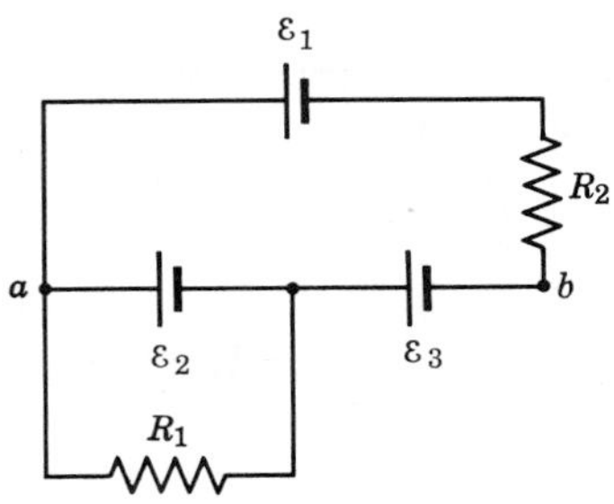

Figure 24 Exercise 29.

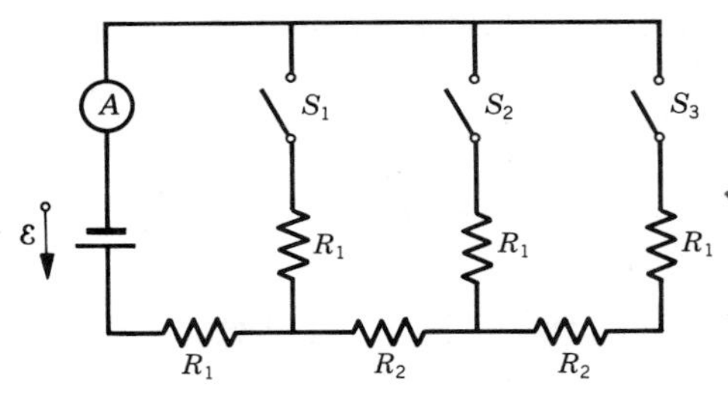

Figure 25 Exercise 30.

31E. In Fig. 26, find the equivalent resistance between points (*a*) A and B, (*b*) A and C, and (*c*) B and C.

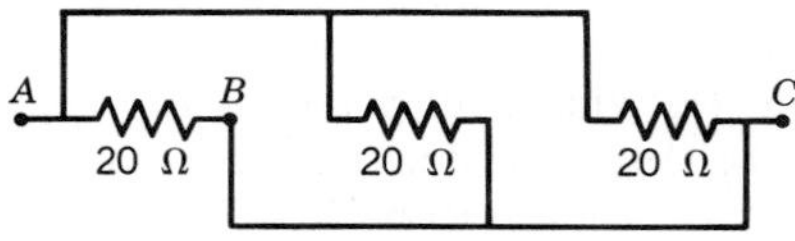

Figure 26 Exercise 31.

32E. In Fig. 27, find the equivalent resistance between points D and E.

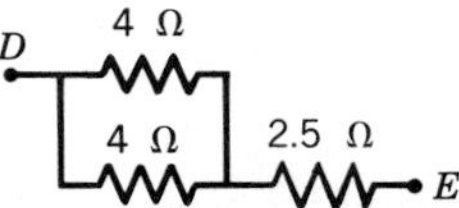

Figure 27 Exercise 32.

33E. Two light bulbs, one of resistance R_1 and the other of resistance R_2 ($<R_1$) are connected (*a*) in parallel and (*b*) in series. Which bulb is brighter in each case?

34E. In Fig. 8 calculate the potential difference between points c and d by as many paths as possible. Assume that $\mathcal{E}_1 = 4.0$ V, $\mathcal{E}_2 = 1.0$ V, $R_1 = R_2 = 10$ Ω, and $R_3 = 5.0$ Ω.

35E. Nine copper wires of length l and diameter d are connected in parallel to form a single composite conductor of resistance R. What must be the diameter D of a single copper wire of length l if it is to have the same resistance?

36E. A 120-V power line is protected by a 15-A fuse. What is the maximum number of 500-W lamps that can be simultaneously operated in parallel on this line?

37E. A circuit containing five resistors connected to a 12-V battery is shown in Fig. 28. What is the voltage drop across the 5.0-Ω resistor?

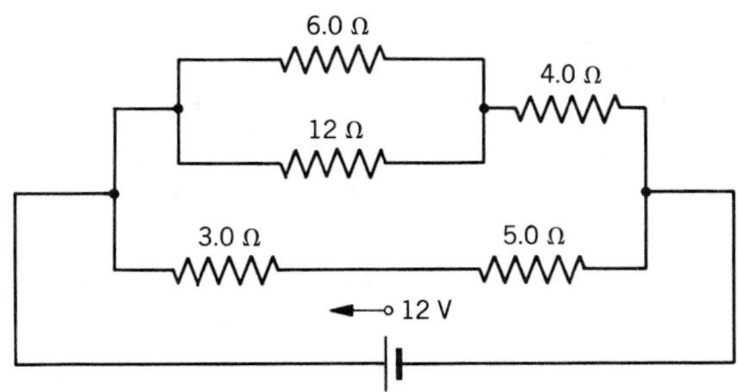

Figure 28 Exercise 37.

38E. For manual control of the current in a circuit, you can use a parallel combination of variable resistors of the sliding contact type, as in Fig. 29, with $R_1 = 20R_2$. (*a*) What procedure is used to adjust the current to the desired value? (*b*) Why is the parallel combination better than a single variable resistor?

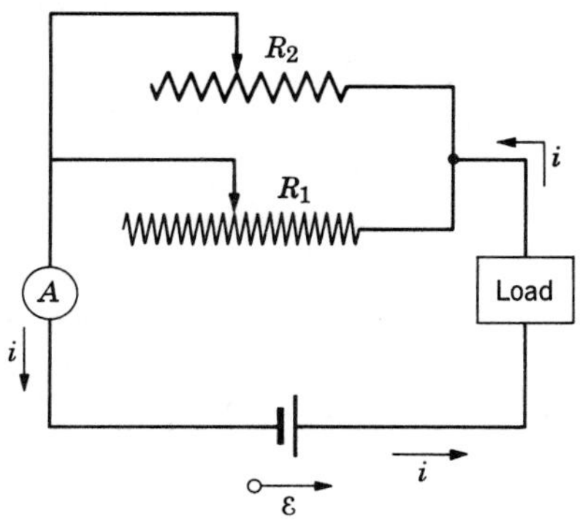

Figure 29 Exercise 38.

39P. Two resistors R_1 and R_2 may be connected either in series or parallel across a (resistanceless) battery with emf $\mathcal{E}$. We desire the thermal energy transfer rate for the parallel combination to be five times that for the series combination. If $R_1 = 100\ \Omega$, what is R_2?

40P. You are given a number of 10-Ω resistors, each capable of dissipating only 1.0 W. What is the minimum number of such resistors that you need to combine in series or parallel combinations to make a 10-Ω resistor capable of dissipating at least 5.0 W?

41P. Two batteries of emf $\mathcal{E}$ and internal resistance r are connected in parallel across a resistor R, as in Fig. 30*a*. (*a*) For what value of R is the thermal energy delivered to the resistor a maximum? (*b*) What is the maximum energy dissipation rate?

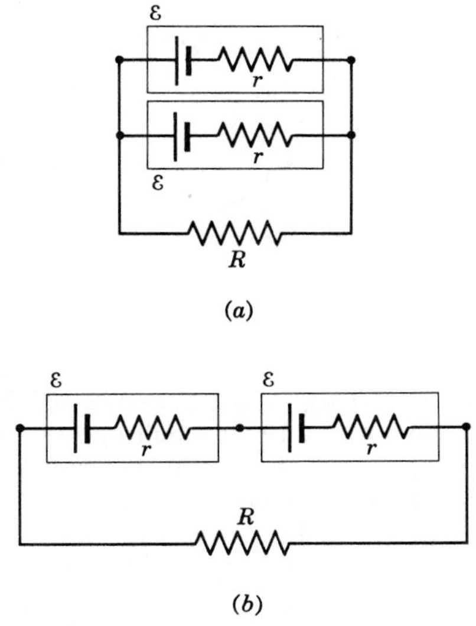

Figure 30 Problems 41 and 43.

42P. In Fig. 31 imagine an ammeter inserted in the branch containing R_3. (*a*) What will it read, assuming $\mathcal{E} = 5.0$ V, $R_1 = 2.0\ \Omega$, $R_2 = 4.0\ \Omega$, and $R_3 = 6.0\ \Omega$? (*b*) The ammeter and the source of emf are now physically interchanged. Show that the ammeter reading remains unchanged.

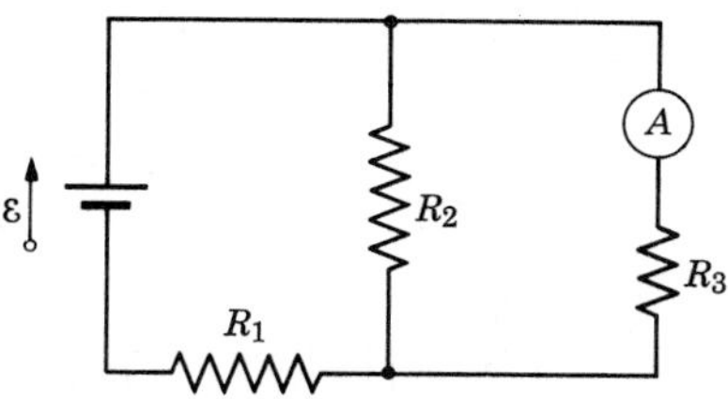

Figure 31 Problem 42.

43P. You are given two batteries of emf $\mathcal{E}$ and internal resistance r. They may be connected either in series or in parallel and are used to establish a current in a resistor R, as in Fig. 30. (*a*) Derive expressions for the current in R for both methods of connection. (*b*) Which connection yields the larger current if $R > r$ and if $R < r$?

44P. N identical batteries of emf $\mathcal{E}$ and internal resistance r may be connected all in series or all in parallel. Show that each arrangement will give the same current in an external resistor R, if $R = r$.

45P. (*a*) Calculate the current through each source of emf in Fig. 32. (*b*) Calculate V_{ab}. Assume that $R_1 = 1.0\ \Omega$, $R_2 = 2.0\ \Omega$, $\mathcal{E}_1 = 2.0$ V, and $\mathcal{E}_2 = \mathcal{E}_3 = 4.0$ V.

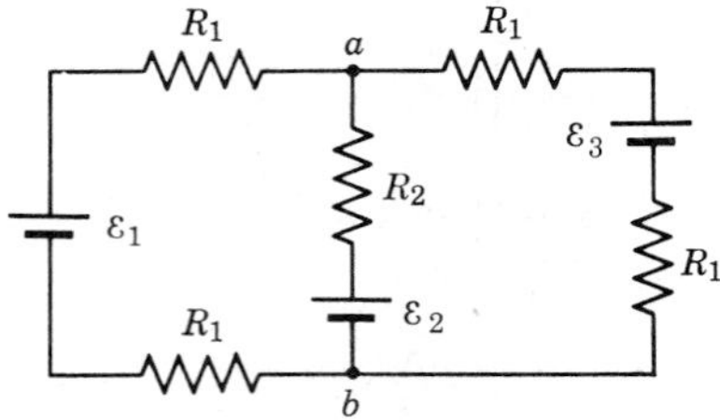

Figure 32 Problem 45.

46P. A three-way 120-V lamp bulb, rated for 100-200-300 W burns out a filament. Afterwards, the bulb operates at the same intensity on its lowest and its highest switch positions but does not operate at all on the middle position. (*a*) How are the filaments wired up inside the bulb and how is the switch constructed? (*b*) Calculate the resistances of the filaments.

47P. What current, in terms of $\mathcal{E}$ and R, does the ammeter A in Fig. 33 read? Assume that A has zero resistance.

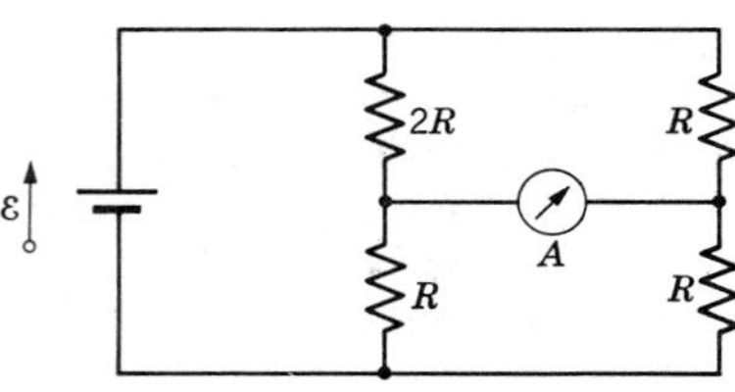

Figure 33 Problem 47.

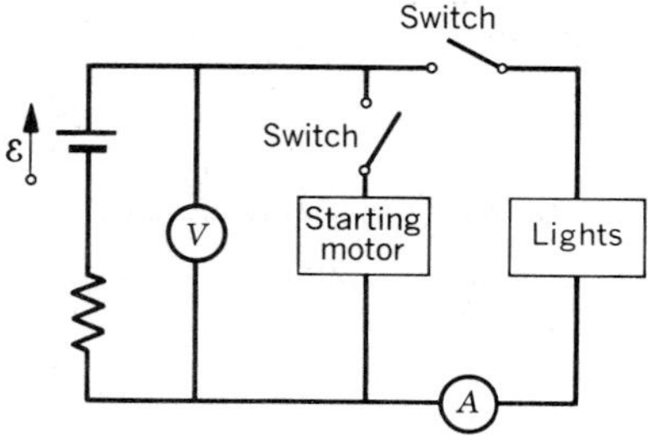

Figure 34 Problem 48.

48P. When the lights of an automobile are switched on, an ammeter in series with them reads 10 A and a voltmeter connected across them reads 12 V. See Fig. 34. When the electric starting motor is turned on, the ammeter reading drops to 8.0 A and the lights dim somewhat. If the internal resistance of the battery is 0.050 Ω and that of the ammeter is negligible, what are (*a*) the emf of the battery and (*b*) the current through the starting motor when the lights are on?

49P. (*a*) In Fig. 35 what is the equivalent resistance of the network shown? (*b*) What are the currents in each resistor? Put $R_1 = 100\ \Omega$, $R_2 = R_3 = 50\ \Omega$, $R_4 = 75\ \Omega$, and $\mathcal{E} = 6.0$ V.

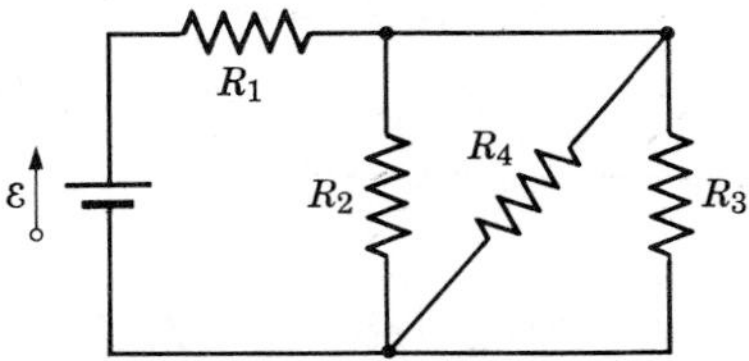

Figure 35 Problem 49.

50P. (*a*) In Fig. 36 what power appears as thermal energy in R_1? In R_2? In R_3? (*b*) What power is supplied by $\mathcal{E}_1$? By $\mathcal{E}_2$? (*c*) Discuss the energy balance in this circuit. Assume that $\mathcal{E}_1 = 3.0$ V, $\mathcal{E}_2 = 1.0$ V, $R_1 = 5.0\ \Omega$, $R_2 = 2.0\ \Omega$, and $R_3 = 4.0\ \Omega$.

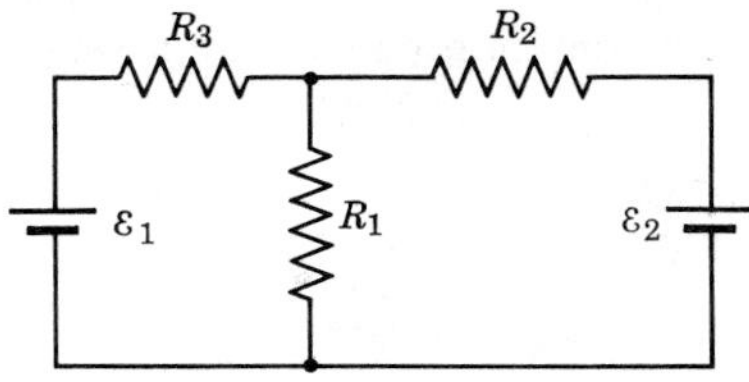

Figure 36 Problem 50.

51P. (*a*) In the circuit of Fig. 37, for what value of R will the battery deliver energy to the circuit at a rate of 60 W? (*b*) What is the maximum power the battery can deliver to the circuit? (*c*) What is the minimum power the battery can deliver?

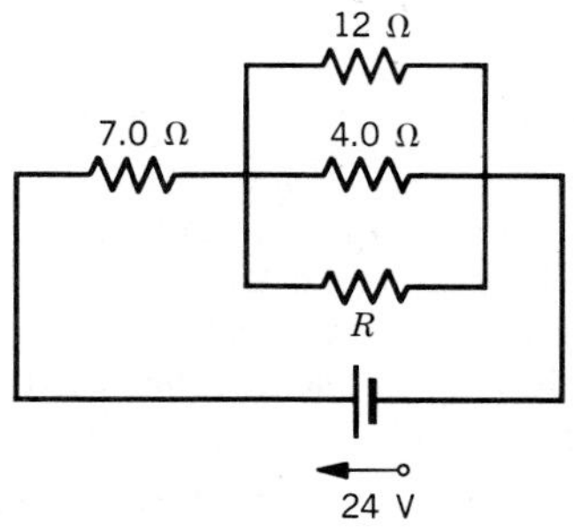

Figure 37 Problem 51.

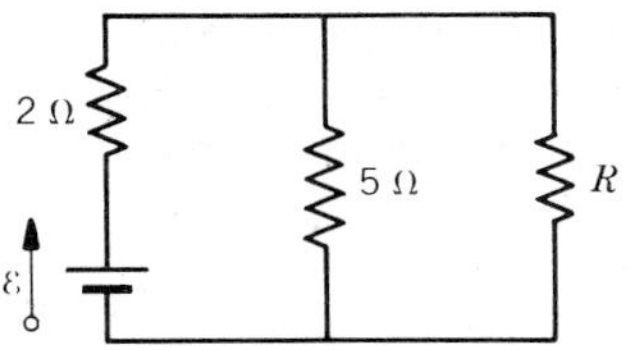

Figure 38 Problem 52.

52P. In the circuit of Fig. 38, $\mathcal{E}$ has a constant value but R can be varied. Find the value of R that results in the maximum heating in that resistor.

53P. In Fig. 39, find the equivalent resistance between points (*a*) F and H and (*b*) F and G.

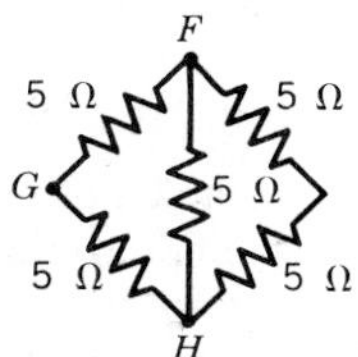

Figure 39 Problem 53.

54P. A copper wire of radius $a = 0.25$ mm has an aluminum jacket of outside radius $b = 0.38$ mm. (*a*) There is a current $i = 2.0$ A in the composite wire. Using Table 1 in Chapter 28, calculate the current in each material. (*b*) What is the wire length if a potential difference $V = 12$ V maintains the current?

55P. Figure 40 shows a battery connected across a uniform resistor R_0. A sliding contact can move across the resistor from $x = 0$ at the left to $x = 10$ cm at the right. Find an expression for the power dissipated in the resistor R as a function of x. Plot the function for $\mathcal{E} = 50$ V, $R = 2000\ \Omega$, and $R_0 = 100\ \Omega$.

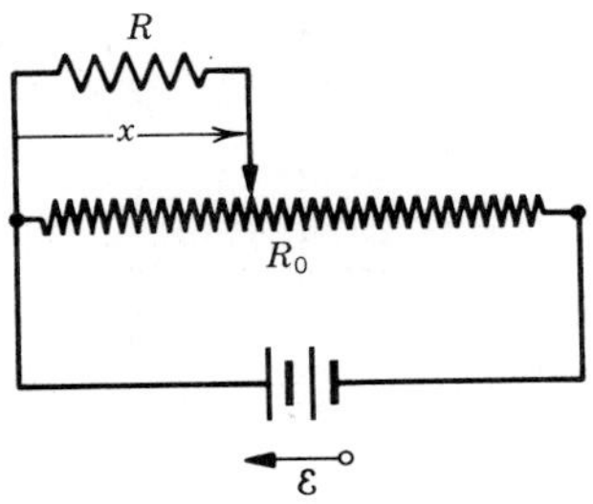

Figure 40 Problem 55.

56P. Twelve resistors, each of resistance R ohms, form a cube (see Fig. 11*a*). (*a*) Find R_{13}, the equivalent resistance of a face diagonal. (*b*) Find R_{17}, the equivalent resistance of a body diagonal. See Sample Problem 4.

Section 29-7 Measuring Instruments

57E. A simple ohmmeter is made by connecting a 1.5-V flashlight battery in series with a resistor R and a 1.0-mA ammeter, as shown in Fig. 41. R is adjusted so that when the circuit terminals are shorted together the meter deflects to its full-scale value of 1.0 mA. What external resistance across the terminals results in a deflection of (*a*) 10%, (*b*) 50%, and (*c*) 90% of full scale? (*d*) If the ammeter has a resistance of 20 Ω and the internal resistance of the battery is negligible, what is the value of R?

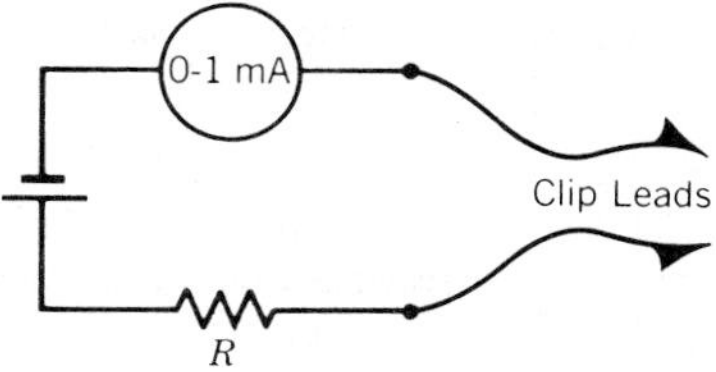

Figure 41 Exercise 57.

58P. In Fig. 12 assume that $\mathcal{E} = 3.0$ V, $r = 100\ \Omega$, $R_1 = 250\ \Omega$, and $R_2 = 300\ \Omega$. If $R_V = 5.0$ kΩ, what percent error is made in reading the potential difference across R_1? Ignore the presence of the ammeter.

59P. In Fig. 12 assume that $\mathcal{E} = 5.0$ V, $r = 2.0\ \Omega$, $R_1 = 5.0\ \Omega$, and $R_2 = 4.0\ \Omega$. If $R_A = 0.10\ \Omega$, what percent error is made in reading the current? Assume that the voltmeter is not present.

60P. *Resistance measurement.* A voltmeter (resistance R_V) and an ammeter (resistance R_A) are connected to measure a resistance R, as in Fig. 42*a*. The resistance is given by $R = V/i$, where V is the voltmeter reading and i is the current in the resistor R. Some of the current registered by the ammeter (i') goes through the voltmeter so that the ratio of the meter readings ($= V/i'$) gives only an *apparent* resistance reading R'.

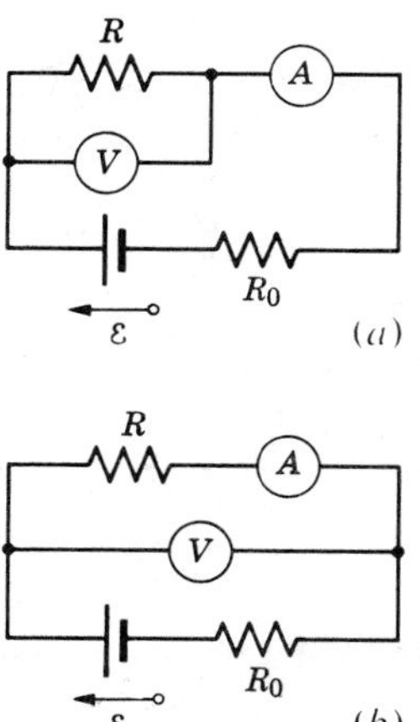

Figure 42 Problems 60, 61, and 62.

Show that R and R' are related by

$$\frac{1}{R} = \frac{1}{R'} - \frac{1}{R_V}.$$

Note that as $R_V \to \infty$, $R' \to R$.

61P. *Resistance measurement.* If meters are used to measure resistance, they may also be connected as they are in Fig. 42*b*. Again the ratio of the meter readings gives only an apparent resistance R'. Show that R' is related to R by

$$R = R' - R_A,$$

in which R_A is the ammeter resistance. Note that as $R_A \to 0$, $R' \to R$.

62P. In Fig. 42 the ammeter and voltmeter resistances are 3.00 Ω and 300 Ω, respectively. If $R = 85$ Ω, (*a*) What will the meters read for the two different connections? (*b*) What apparent resistance R' will be computed in each case? Take $\mathcal{E} =$ 12 V and $R_0 = 100$ Ω.

63P. *The Wheatstone bridge.* In Fig. 43 R_s is to be adjusted in value until points *a* and *b* are brought to exactly the same potential. (One tests for this condition by momentarily connecting a sensitive ammeter between *a* and *b*; if these points are at the same potential, the ammeter will not deflect.) Show that when this adjustment is made, the following relation holds:

$$R_x = R_s(R_2/R_1).$$

An unknown resistor (R_x) can be measured in terms of a standard (R_s) using this device, which is called a Wheatstone bridge.

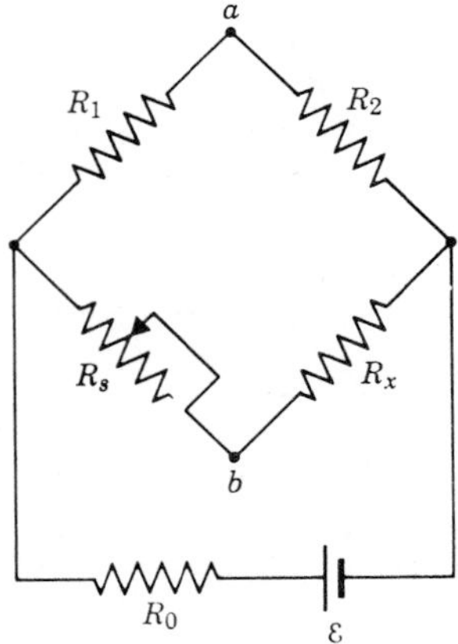

Figure 43 Problems 63 and 64.

64P. If points *a* and *b* in Fig. 43 are connected by a wire of resistance *r*, show that the current in the wire is

$$i = \frac{\mathcal{E}(R_s - R_x)}{(R + 2r)(R_s + R_x) + 2R_sR_x},$$

where $\mathcal{E}$ is the emf of the battery. Assume that R_1 and R_2 are equal ($R_1 = R_2 = R$) and that R_0 equals zero. Is this formula consistent with the result of Problem 63?

Section 29-8 *RC* Circuits

65E. In an *RC* series circuit $\mathcal{E} = 12$ V, $R = 1.4$ MΩ, and $C = 1.8$ μF. (*a*) Calculate the time constant. (*b*) Find the maximum charge that will appear on the capacitor during charging. (*c*) How long does it take for the charge to build up to 16 μC?

66E. How many time constants must elapse before a capacitor in an *RC* circuit is charged to within 1.0 percent of its equilibrium charge?

67E. From Fig. 16, deduce the value of the time constant of the *RC* circuit. Can the values of *R* and *C* be found?

68E. A 15-kΩ resistor and a capacitor are connected in series and a 12-V potential is suddenly applied. The potential across the capacitor rises to 5.0 V in 1.3 μs. (*a*) Calculate the time constant. (*b*) Find the capacitance of the capacitor.

69P. An *RC* circuit is discharged by closing a switch at time $t = 0$. The initial potential difference across the capacitor is 100 V. If the potential difference has decreased to 1.0 V after 10 s, (*a*) calculate the time constant of the circuit. (*b*) What will be the potential difference at $t = 17$ s?

70P. Figure 44 shows the circuit of a flashing lamp, like those attached to barrels at highway construction sites. The fluorescent lamp *L* is connected in parallel across the capacitor *C* of an *RC* circuit. Current passes through the lamp only when the potential across it reaches the breakdown voltage V_L; in this event, the capacitor discharges through the lamp and it flashes for a very short time. Suppose that two flashes per second are needed. Using a lamp with breakdown voltage $V_L = 72$ V, a 95-V battery, and a 0.15-μF capacitor, what should be the resistance *R* of the resistor?

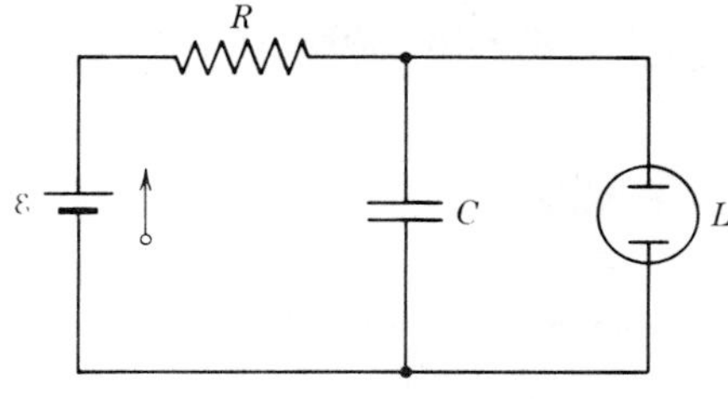

Figure 44 Problem 70.

71P. A 1.0-μF capacitor with an initial stored energy of 0.50 J is discharged through a 1.0-MΩ resistor. (*a*) What is the initial charge on the capacitor? (*b*) What is the current through the resistor when the discharge starts? (*c*) Determine V_C, the voltage across the capacitor, and V_R, the voltage across the resistor, as functions of time. (*d*) Express the rate of generation of thermal energy in the resistor as a function of time.

72P. A 3.0-MΩ resistor and a 1.0-μF capacitor are connected in a single-loop circuit with a seat of emf with $\mathscr{E} = 4.0$ V. At 1.0 s after the connection is made, what are the rates at which (*a*) the charge of the capacitor is increasing, (*b*) energy is being stored in the capacitor, (*c*) thermal energy is appearing in the resistor, and (*d*) energy is being delivered by the seat of emf?

73P. The potential difference between the plates of a leaky 2.0-μF capacitor drops to one-fourth its initial value in 2.0 s. What is the equivalent resistance between the capacitor plates?

74P. Prove that when switch *S* in Fig. 15 is thrown from *a* to *b* all the energy stored in the capacitor is transformed into thermal energy in the resistor. Assume that the capacitor is fully charged before the switch is thrown.

75P. An initially uncharged capacitor *C* is fully charged by a constant emf $\mathscr{E}$ in series with a resistor *R*. (*a*) Show that the final energy stored in the capacitor is half the energy supplied by the emf. (*b*) By direct integration of i^2R over the charging time, show that the thermal energy dissipated by the resistor is also half the energy supplied by the emf.

76P. A controller on an electronic arcade game consists of a variable resistor connected across the plates of a 0.22-μF capacitor. The capacitor is charged to 5.0 V, then discharged through the resistor. The time for the potential difference across the plates to decrease to 0.80 V is measured by an internal clock. If the range of discharge times that can be handled effectively is from 10 μs to 6.0 ms, what should be the range of the resistor?

77P. The circuit of Fig. 45 shows a capacitor *C*, two batteries, two resistors, and a switch *S*. Initially *S* has been open for a long time. It is then closed for a long time. By how much does the charge on the capacitor change over this period? Assume $C = 10$ μF, $\mathscr{E}_1 = 1.0$ V, $\mathscr{E}_2 = 3.0$ V, $R_1 = 0.20$ Ω, and $R_2 = 0.40$ Ω.

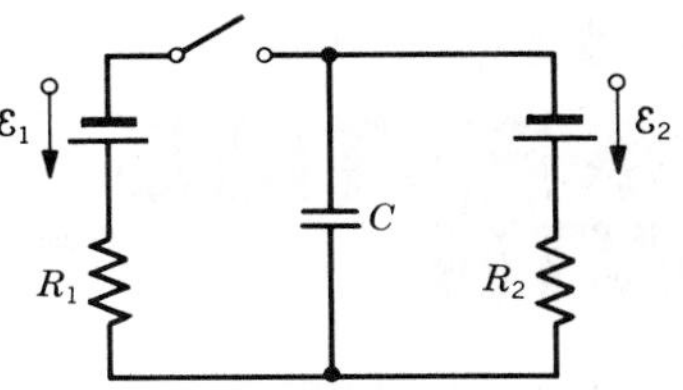

Figure 45 Problem 77.

78P. In the circuit of Fig. 46, $\mathscr{E} = 1.2$ kV, $C = 6.5$ μF, $R_1 = R_2 = R_3 = 0.73$ MΩ. With *C* completely uncharged, the switch *S* is suddenly closed ($t = 0$). (*a*) Determine the currents through each resistor for $t = 0$ and $t = \infty$. (*b*) Draw qualitatively a graph of the potential drop V_2 across R_2 from $t = 0$ to $t = \infty$. (*c*) What are the numerical values of V_2 at $t = 0$ and $t = \infty$? (*d*) Give the physical meaning of "$t = \infty$" in this case.

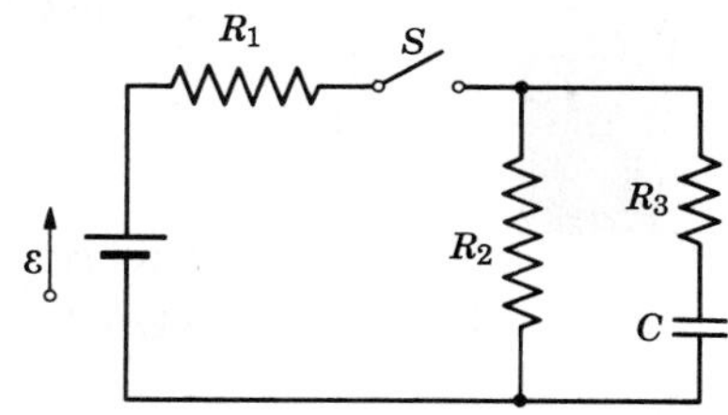

Figure 46 Problem 78.

ESSAY 11 EXPONENTIAL GROWTH

ALBERT A. BARTLETT
UNIVERSITY OF COLORADO, BOULDER

The greatest shortcoming of the human race is our inability to understand the exponential function.

The importance of the exponential function arises, not so much from its appearance in physics (as in the *RC* circuit), but because it describes steady growth. To see this, let us assume a quantity N is growing steadily a fixed fraction (or percent) per unit time. We would describe this by writing

$$\frac{dN}{Ndt} = k \tag{1}$$

where (dN/N) is the fractional change in N in a time dt. A quantity growing 4% per year would have $k = 0.04\ \text{year}^{-1}$.

Let us rearrange Eq. 1:

$$\frac{dN}{dt} = kN. \tag{2}$$

This equation reflects the fact that the rate of change of N is proportional to N. Equation 2 may be integrated to give

$$N = N_0 e^{kt} \tag{3}$$

where N_0 is the size of N at the time $t = 0$. The quantity e is the base of natural logarithms (2.718 . . .). For example, suppose the quantity N is growing 5% per year. Then $(dN/N) = 0.05$ when $dt = 1$ year, and $k = 0.05\ \text{year}^{-1}$. Equation 2 becomes

$$\frac{dN}{dt} = 0.05\ N \tag{2^1}$$

and Eq. 3 becomes

$$N = N_0 e^{0.05t} \tag{3^1}$$

Table 1 shows the increase in size of a quantity that is growing 5% per unit time. We can interpret the table as follows. If a quantity is growing steadily at 5% per year, then in 10 years it will have increased in size by 64.9%. If a quantity is growing 5% per day, then in 100 days it will be 148.4 times as large as it was initially. It is easy to calculate that in 1095 days the quantity growing 5% per day would increase in size by a factor equal to Avogadro's number! We note from Table 1 that as time goes on the size of N becomes very large, ultimately approaching infinity. This leads to a profound conclusion. It is impossible for steady growth of any real thing to continue forever —or even for a long period of time.

Our society worships growth, yet growth for long periods of time is impossible. This is the conflict that is recognized in the opening statement.

Table 1 Steady Growth of 5% per Unit Time. The Time Units Could Be Seconds, Days, Years, or Any Unit of Time.

Time Units	*Value of* N, *as a Multiple of* N_0.
0	1.000
1	1.051
2	1.105
5	1.284
10	1.649
20	2.718
50	12.18
100	148.4
200	2.20×10^4
500	7.20×10^{10}
1000	5.18×10^{21}

Table 2 The Size of the Growing Quantity in Units of the Doubling Time T_2

Time in Units of T_2	*Size in Units of* N_0
0	$2^0 = 1$
1	$2^1 = 2$
2	$2^2 = 4$
3	$2^3 = 8$
4	$2^4 = 16$
5	$2^5 = 32$
10	$2^{10} = 1024$
n	2^n

The Doubling Time, T_2

Let us now calculate the period of time T_2 for the growing quantity to double in size ($N = 2N_0$):

$$2N_0 = N_0 e^{kT_2}$$
$$\ln(2) = kT_2$$
$$T_2 = \frac{\ln 2}{k} = \frac{0.693}{k} = \frac{69.3}{100\,k} \approx \frac{70}{P}. \tag{4}$$

Here P is the percent growth per unit time, which is $100k$. Thus one finds the doubling time by dividing 70 by the percentage growth per unit time. For example, if something is growing at a rate of 5% per year, the doubling time is $(70/5) = 14$ years. If a quantity is growing steadily at a rate of 5% per day, then the quantity will double in size in $(70/5) = 14$ days! This is the origin of the "Rule of 70" (sometimes referred to as the "Rule of 72") that is used by bankers and investment counselors.

It should be clear that the time T_3 for the growing quantity to triple in size will be given by

$$T_3 = \frac{\ln 3}{k} \approx \frac{110}{P}. \tag{5}$$

We can express Eq. 3 in terms of powers of 2 instead of powers of e:

$$N = N_0(2)^{t/T_2}. \tag{6}$$

Table 2 shows the way the growth develops when Eq. 6 is used. In estimating the effects of steady growth it is convenient to remember that

$$2^{10} \approx 10^3 \tag{7}$$

Example

When compound interest is calculated annually it provides an example of growth that is approximately exponential. A sum of money ($N_0 = \$1$) is placed in a savings account that adds 7% interest at the end of each year. The interest in the first year is $0.07 \times \$1$, so that at the end of the year the account contains \$1.07.

If the interest is calculated monthly at a rate of ($7/12 = 0.5833\%$ per month), the account at the end of the year will contain \$1.072, which is slightly larger than the amount when the interest is calculated only once at the end of the year. If one calculates the interest continuously (as many banks do) the account grows in accord with Eq. 3. Thus in 1 year of continuous compounding of the interest the value of the account is

$$N = 1 \times e^{0.07 \times 1} = \$1.0725 \ldots.$$

In 10 years, the value is

$$N = 1 \times e^{0.07 \times 10} = 2.014.$$

In 100 years we have

$$N = 1 \times e^{0.07 \times 100} = \$1096.63$$

and in 1000 years we have

$$N = 1 \times e^{0.07 \times 1000} = \$2.52 \times 10^{30}.$$

The growth of the account is ultimately so rapid that the banking laws prohibit one from leaving sums of money at interest untouched for long periods of time.

This problem can also be used to illustrate the ease with which exponential arithmetic can be worked out in one's head. For $k = 0.07$ ($P = 7\%$) the doubling time is $70/7 = 10$ years. Thus in 200 years we are dealing with 20 doublings in which time \$1 becomes $\$2^{20}$. Equation 7 allows us to estimate the value of 2^{20}:

$$2^{20} = 2^{10} \times 2^{10} \approx 10^3 \times 10^3 = 10^6.$$

Thus \$1 at an annual interest rate of 7% compounded continuously would become approximately one million dollars in 200 years. The actual value is $\$1.20 \times 10^6$.

Problem

In 1985 the assessed valuation of New York City was $\$39.4 \times 10^9$. In the year 1626 Manhattan Island was purchased for \$24. If the \$24 had been placed in a savings account, what interest rate (compounded continuously) would be required so that the account in 1985 would contain a sum of money equal to the assessed valuation of New York City in 1985? In this example, $t = 1985 - 1626 = 359$ years and

$$39.4 \times 10^9 = 24\, e^{k \times 359}.$$

Solving, we find that k has the value 0.0591, which corresponds to an annual interest rate of 5.91 percent compounded continuously.

Example

On July 7, 1986, reports appeared in the press saying that the world population had reached the number 5×10^9 people. The growth rate is reported to be 1.7% per year, ($T_2 = 41$ years). If this growth rate continued unchanged (a) How long a time is required for the world population to increase by one million people? Since $dN \ll N$

we can use Eq. 2:

$$dN = kN\,dt$$
$$10^6 = 0.017 \times 5.0 \times 10^9\,dt$$
$$dt = 1.2 \times 10^{-2}\text{ years} = 4.3\text{ days!}$$

(b) How long a time would be required for the mass of people to equal the mass of the earth? (Assume the mass of a person is 65 kg.)

$$5.98 \times 10^{24} = (5.0 \times 10^9 \times 65)e^{0.017t}$$

Solving for t gives 1.8×10^3 years! The impossibility of steady growth of populations for long periods of time is clear from this example.

Illustration, The Power of Powers of Two

Legend has it that the game of chess was invented by a mathematician who worked for a king. The king was so pleased that he asked the mathematician to name the reward he wanted. The mathematician said, "Take my new chess board and on the first square place one grain of wheat; on the next square double the one and put 2 grains. On the third square double the 2 and put 4 grains . . . Just keep doubling this way until you have doubled for every square. That will be an adequate payment." Table 3 indicates the details of the filling the board. This total number of grains is approximately 500 times the 1976 worldwide harvest of wheat! How did we get such a large number? We started with *one* grain but we let the number grow steadily until it had doubled 63 times. Table 3 illustrates a further important aspect of steady growth: the number of grains required on any square is one grain more than the total of all the grains on *all* of the preceding squares!

Example

In the mid-1970s the American electric power industries placed ads in magazines calling attention to the fact that for a century their industry had been growing steadily at a rate of about 7% per year and they expected this growth to continue steadily in the future. As we saw with the chessboard and Table 3, this would require that the electric generating capacity constructed in the United States in each future decade be equal to the total generating capacity in place at the start of that decade! It would also require that the consumption of electrical energy in the United States in any coming decade be equal to the total of all the electrical energy that had been consumed in the entire previous history of the electrical generating industry in the United States prior to the start of that decade! (See Table 3.) The financial trauma of the electrical

Table 3

Square Number	*Grains on Square*	*Total Grains on Board*
1	1	1
2	2	3
3	4	7
4	8	15
5	16	31
64	2^{63}	$2^{64} - 1 = 1.8 \times 10^{19}$

generating industry in the United States in the late 1970s and early 1980s was due in part to the failure to recognize the simple facts of the arithmetic of steady growth. The industry was not prepared for the inevitable reductions in their rate of growth.

Steady Growth in a Finite Environment

Fossil fuels are finite resources. What happens if we plan to try to maintain steady growth in the rate of consumption of a fossil fuel? We can illustrate this with the story of bacteria in a bottle.

Bacteria grow by doubling; one bacterium divides to become two; the two divide to become four; the four divide to become eight, etc. Suppose we had bacteria that doubled in number this way every minute. Suppose we put one bacterium in an empty bottle at 11:00 A.M. and later we observed that the bottle was full at 12:00 noon.

Here we have our example of ordinary steady growth with a doubling time of one minute in the finite environment of one bottle. Let's examine three questions.

1. At what time was the bottle half full?
   Answer: at 11:59—because the bacteria double in number every minute!
2. If you were an average bacterium in that bottle, at what time would you first realize that you were running out of space?

Table 4 shows the last minutes in the bottle and may help you to decide on your answer to this question.

Table 4

Time	*Fraction Full (used)*	*Fraction Empty (available)*
11:55	$\frac{1}{32}$	$\frac{31}{32}$
11:56	$\frac{1}{16}$	$\frac{15}{16}$
11:57	$\frac{1}{8}$	$\frac{7}{8}$
11:58	$\frac{1}{4}$	$\frac{3}{4}$
11:59	$\frac{1}{2}$	$\frac{1}{2}$
12:00	1	Zero

3. Suppose that at 11:58 some of the bacteria realize that they are running out of space so they launch a great search for new bottles. Their search yields three new bottles, which is three times as much resource as they had known previously. Surely this will make them self-sufficient in space! How long can the growth continue because of this tremendous discovery? The answer is given in Table 5.

If we plan to have continued steady growth in our rates of consumption of finite resources of fossil fuels, then this example shows that enormous discoveries of new reserves will permit the growth to continue for only very short periods of time.

Table 5

Time	*Bottles Filled*	*Bottles Unfilled*
12:00	1	3
12:01	2	2
12:02	4	Zero

Petroleum is perhaps the most widely used fossil fuel. It has no serious competition as *the* fuel for transportation. Petroleum consumption worldwide grew at a steady rate of about 7% per year for a century before leveling off (zero growth) in the late 1970s. In the United States we equate growth of the rate of consumption with prosperity. The less-developed nations aspire to have growth in their rates of consumption of petroleum in order to improve the quality of life for their people (whose numbers are growing steadily). The world and U.S. situations in regard to petroleum, were summarized in *Science.*[2] "In a recent U.S. Geological Survey report, government researchers specializing in the estimation of oil resources issued a strong warning—the amount of oil left to slake the world's thirst for energy is smaller than some optimists would like to think and may be even smaller than conservative estimates. . . . This new oil [yet to be found] plus reserves known to exist in the ground would last the world about 60 years at present rates of consumption [zero growth]. The estimated U.S. supply from undiscovered resources and demonstrated reserves is 36 years at present rates of production [zero growth] or 19 years in the absence of imports."

In 1977 James Schlesinger, the U.S. Secretary of Energy, noted that in the energy crisis, "we have a classic case of exponential growth against a finite source."[3] The economic trauma of the period of the energy crisis in the late 1970s was followed by a period of declining prices and apparent abundance of fossil fuels. During this time the arithmetic seems to have been forgotten.

Summary

We can summarize a few of the facts of steady growth as they are revealed by this review of the exponential function.

1. Steady growth occurs whenever the rate of change of a quantity is proportional to that quantity.
2. Steady growth is characterized by a constant doubling time.
3. In a modest number of doubling times, steady growth can produce very large numbers.
4. In steady growth, the increase in one doubling time is equal to the sum of all preceding growth.
5. When one has steady growth in a finite environment, the end comes frighteningly fast.
6. Growth is always a short-term transient phenomenon.

The greatest shortcoming of the human race is our inability to understand the exponential function.

BIBLIOGRAPHY (1) A. A. Bartlett, *Civil Engineering,* Dec. 1969, p. 71. (2) *Science,* Jan. 27, 1984, p. 382. (3) *Time,* April 25, 1977, p. 27. Extended discussions of this material are available in the magazine, *The Physics Teacher, 14,* 1976, p. 393; *14,* 1976, p. 485; *15,* 1977, p. 37; *15,* 1977, p. 98; *15,* 1977, p. 225; *16,* 1978, p. 23; *16,* 1978, p. 92; *16,* 1978, p. 158; *17,* 1979, p. 23; and in the *American Journal of Physics, 46,* 1978, p. 876; *54,* 1986, p. 398.

CHAPTER 30

THE MAGNETIC FIELD

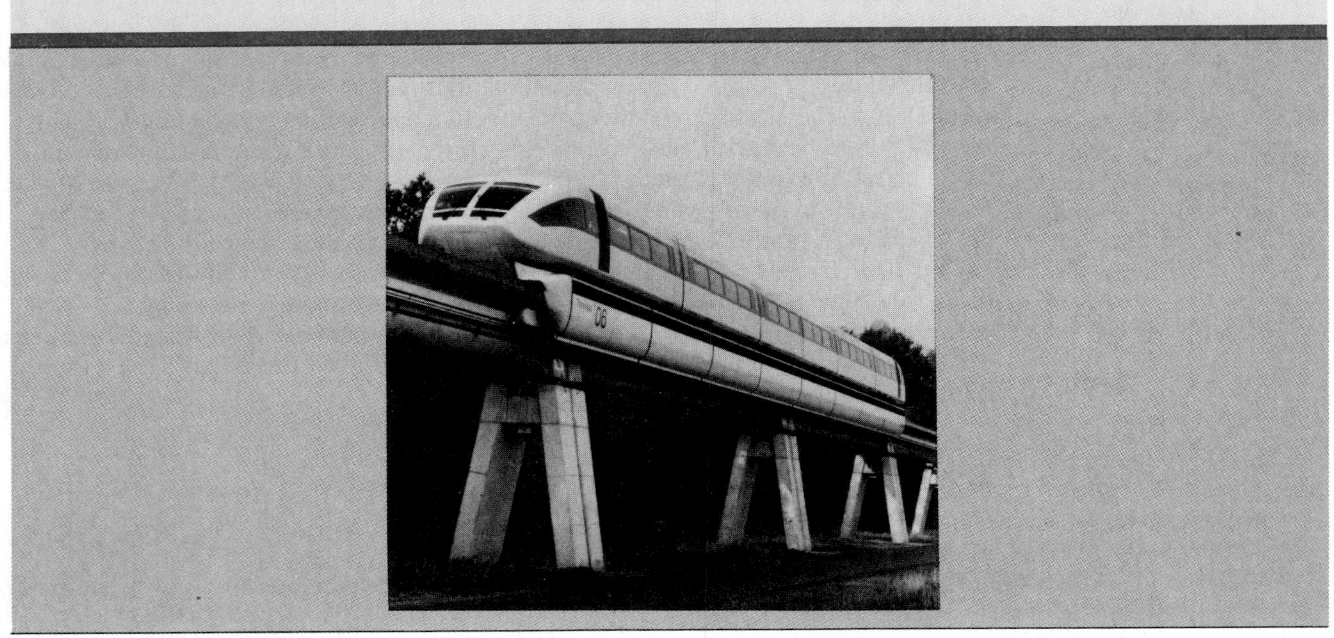

The TRANSRAPID 06, a 200-passenger intercity transit vehicle undergoing tests in West Germany. It has no wheels. Supported and propelled by electromagnetic forces, it has achieved a speed of 300 km/h. New developments in superconducting materials make the prospects for such trains considerably brighter.

30-1 The Magnetic Field

As you fasten a note to the refrigerator door with a small magnet, you become aware with your fingertips that the space around the magnet has special properties. The space near a rubbed plastic rod also has special properties. In that case, we learned to understand them by postulating that there is an *electric field* **E** at all points near the rod. By analogy, it seems logical to postulate that there is a *magnetic field,* which we represent by the symbol **B**, at all points near a magnet.

Figure 1 suggests, by means of iron filings, the magnetic field in the space surrounding a small *permanent magnet.* In another familiar type of magnet, a coil is wound around an iron core, the strength of the external magnetic field being determined by the current in the coil. In industry, such *electromagnets* are used for sorting scrap iron (Fig. 2). Figure 3 shows another type of electromagnet, found in research laboratories the world over.

In electrostatics we represented the relation between electric field **E** and electric charge by

$$\text{electric charge} \rightleftharpoons \mathbf{E} \rightleftharpoons \text{electric charge}. \quad (1)$$

That is, electric charges set up an electric field and this field in turn exerts an (electric) force on another charge that may be placed in it.

Symmetry—a powerful tool that we have used many times before—suggests that we set up a similar relation for magnetism, namely,

$$\text{“magnetic charge”} \rightleftharpoons \mathbf{B} \rightleftharpoons \text{“magnetic charge,”} \quad (2)$$

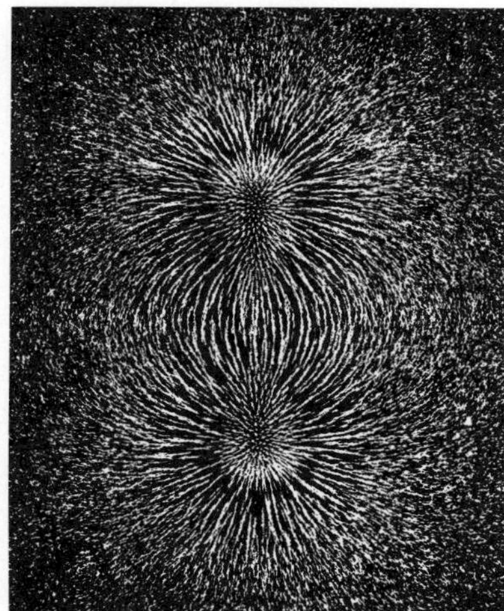

Figure 1 Iron filings sprinkled on a sheet of paper tell us that there is a bar magnet underneath.

Figure 2 An industrial electromagnet, used to load scrap iron.

in which $\mathbf{B}$ is the magnetic field. The only problem with this idea is that there seem to be no magnetic charges. That is, there are no isolated point objects from which magnetic field lines emerge. Certain current theories predict that such *magnetic monopoles* should exist and many physicists have looked for them, but so far their existence has not been confirmed.

From where then does the magnetic field come? Experiment shows that it comes from *moving* electric charges. A charge sets up an *electric* field whether the charge is at rest or is moving. A charge sets up a *magnetic* field, however, *only* if it is moving. Where are these moving electric charges? In the permanent magnet of Fig. 1, they are the spinning and circulating electrons in the iron atoms that make up the magnet. In the electromagnets of Figs. 2 and 3, they are the electrons drifting through the coils of wire that surround these magnets.

In magnetism then we think in terms of

$$\text{moving electric charge} \rightleftharpoons \mathbf{B} \rightleftharpoons \text{moving electric charge.} \quad (3)$$

Because an electric current in a wire is a stream of moving charges, we can also write Eq. 3 as

$$\text{electric current} \rightleftharpoons \mathbf{B} \rightleftharpoons \text{electric current.} \quad (4)$$

Equations 3 and 4 tell us (1) a moving charge or a current sets up a magnetic field and also (2) if we place a moving charge, or a wire carrying a current, in a magnetic field, a magnetic force will act on it. It was the Danish physicist Hans Christian Oersted who (in 1820) first linked the then separate sciences of current electricity and magnetism by showing that an (electric) current in a wire could deflect a (magnetic) compass needle.

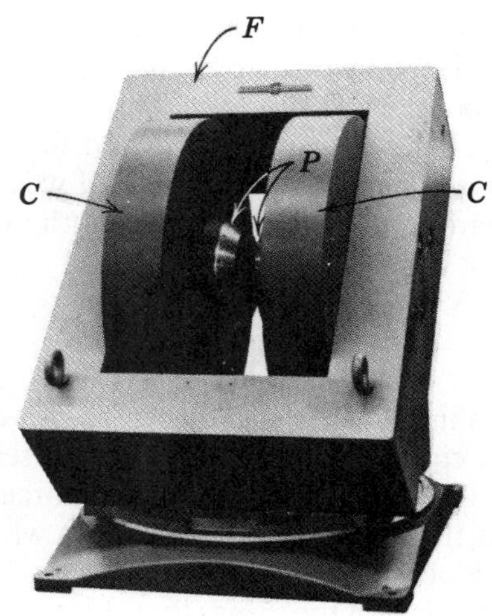

Figure 3 A research-type electromagnet, showing iron frame F, pole faces P, and coils C. The pole faces are 30 cm in diameter.

An electric motor illustrates Eq. 4 very well. In most motors (1) a current in a wire (the field coils) sets up a magnetic field and (2) the magnetic field in turn exerts a force on a second current-carrying wire (the armature windings), causing the shaft to rotate.

In this chapter, we deal with only half of the story summarized by Eqs. 3 and 4. That is, we assume that a magnetic field exists — perhaps in the space between the pole faces of an electromagnet such as that of Fig. 3 — and we ask what force this field exerts on charges that move through it. In the next chapter, we deal with the other half of the story, namely, where does the magnetic field come from in the first place?

30-2 The Definition of B

We defined the electric field **E** at a point by putting a test charge q at rest at that point and measuring the (electric) force $\mathbf{F}_E$ that acted on this charge. We then defined **E** from

$$\mathbf{F}_E = q\mathbf{E}. \tag{5}$$

If we had a magnetic monopole available, we could define **B** in a similar way. Because such particles have not been found in nature, we must define **B** in another way, in terms of the magnetic force exerted on a moving electric charge, as Eq. 3 suggests.

How do we find the magnitude and direction of **B** at any given point?* In principle, we fire a test charge q through the point, in various directions and with various speeds. For every such trial, we record the force $\mathbf{F}_B$ (if any) that acts on the moving charge. If we analyze the results of many such trials, we find that the vector $\mathbf{F}_B$ is related to the velocity vector **v** by

$$\boxed{\mathbf{F}_B = q\,\mathbf{v} \times \mathbf{B}} \tag{6}$$

in which **B** is the magnetic field vector. This equation is our defining equation for **B**.† As we shall see later, the direction of **B** found from analyzing the force on a moving charge is the same as the direction in which a compass needle would point.

Here are some of the things that you can verify from Eq. 6, and from Fig. 4 which represents it.

1. The magnetic force $\mathbf{F}_B$ always acts sideways, that is, at right angles to the velocity vector. This means that a (steady) magnetic field can neither speed up nor slow down a moving charged particle. It can deflect the particle sideways but it cannot change the particle's kinetic energy. Figure 5 shows a beam of electrons in a cathode ray tube being deflected sideways by a magnetic field that has a direction at right angles to the plane of the figure.

2. A magnetic field exerts no force on a charge that moves parallel (or antiparallel) to it. From Eq. 6 we see that the magnitude of the magnetic deflecting force is given by

$$\boxed{F_B = qvB \sin\phi.} \tag{7}$$

This tells us that when **v** is parallel or antiparallel to **B** — which corresponds to $\phi = 0°$ or $180°$ — the magnetic deflecting force is indeed zero.

3. The *maximum* value of the deflecting force

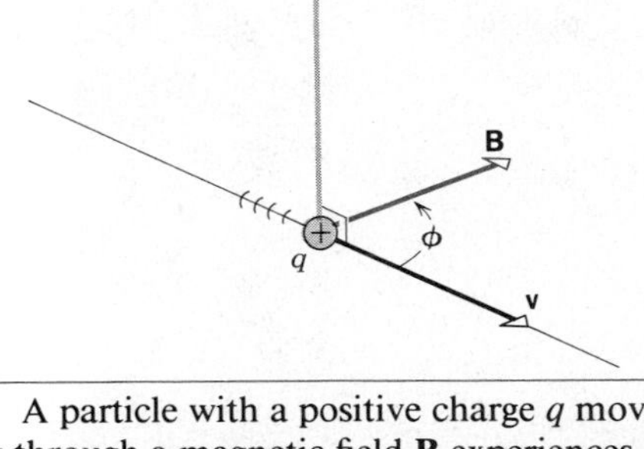

Figure 4 A particle with a positive charge q moving with velocity **v** through a magnetic field **B** experiences a magnetic deflecting force $\mathbf{F}_B$. See Eq. 6.

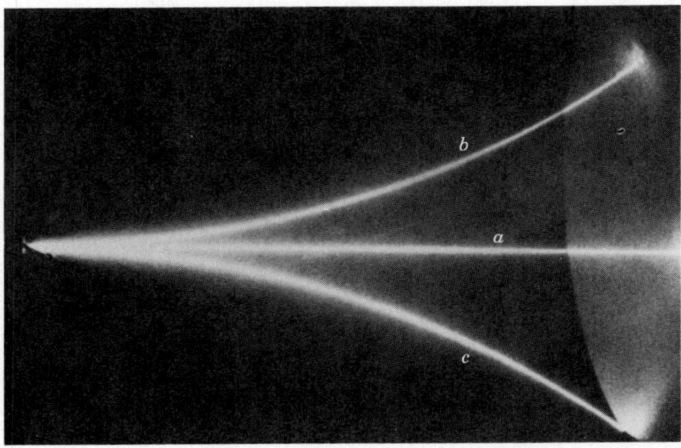

Figure 5 A triple exposure of an electron beam in a cathode ray tube. In (*a*) there is no external magnetic field. In (*b*) a magnetic field points out of the plane of the figure. In (*c*) a magnetic field points into the plane of the figure.

* As a preliminary, we had better make sure that there is no complicating *electric* field acting at that point. We can do this by putting a *stationary* charge there and verifying that no force acts on it.

† You may wish to review Section 3-7, which deals with vector products.

(=qvB) occurs when the test charge is moving at right angles to the magnetic field ($\phi = 90°$).

4. The magnitude of the deflecting force is directly proportional to q and to v. The greater the charge of the test body and the faster it is moving, the greater the magnetic deflecting force.

5. The direction of the magnetic deflecting force depends on the sign of q, positive and negative test charges with equal velocities being deflected in opposite directions.

It is impressive that all this information can be compressed into an equation as modest in structure as Eq. 6. This economy of expression is one of the reasons that we find vectors useful in physics.

To develop a feeling for Eq. 6, consider Fig. 6, which shows some tracks left by fast charged particles moving through a *bubble chamber* at the Lawrence Berkeley Laboratory. The chamber, which is filled with liquid hydrogen, is immersed in a strong uniform magnetic field that points out of the plane of the figure.

At point P in Fig. 6 an incoming gamma ray—which leaves no track because it is uncharged—triggers an event in the chamber that imparts energy to two electrons (tracks marked e^-) and to one positive electron (track marked e^+). Check from Eq. 6 that the three tracks bend in the directions that you expect.

The SI unit for **B** that follows from Eqs. 6 and 7 is the newton/(coulomb · meter/second). For convenience, this is called the *tesla* (abbr. T). Recalling that a coulomb/second is an ampere, we have

$$1 \text{ tesla} = 1 \text{ T} = 1 \text{ N/A} \cdot \text{m}. \tag{8}$$

Table 1 Some Magnetic Fields[a]

At the surface of a neutron star (calculated)	10^8 T
The magnet of Fig. 3	1.5 T
Near a small bar magnet	10^{-2} T
At the surface of the earth	10^{-4} T
In interstellar space	10^{-10} T
Smallest value in a magnetically shielded room	10^{-14} T

[a] Approximate values.

An earlier (non-SI) unit for **B**, still in common use, is the *gauss;* the relation is

$$1 \text{ tesla} = 10^4 \text{ gauss}. \tag{9}$$

Table 1 shows the magnetic fields that occur in a few situations. Note that the earth's magnetic field near the earth's surface is about 10^{-4} T (=100 μT or 1 gauss).

We can represent magnetic fields by field lines, just as we did for the electric field. The same rules apply. That is, (1) the direction of the tangent to a magnet field line at any point gives the direction of **B** at that point, and (2) the spacing of the lines is a measure of the magnitude of **B**. Therefore, the magnetic field is strong where the lines are close together and conversely.

Figure 7 (compare Fig. 1) shows how the magnetic field near a bar magnet can be represented by magnetic field lines. Note that the lines pass right through the magnet, forming closed loops. The external magnetic effects of a bar magnet are strongest near its ends. The end from which the field lines emerge is called a *north pole,* the other end being the *south pole.*

Opposite magnetic poles attract each other. From this fact and from the fact that the north pole of a com-

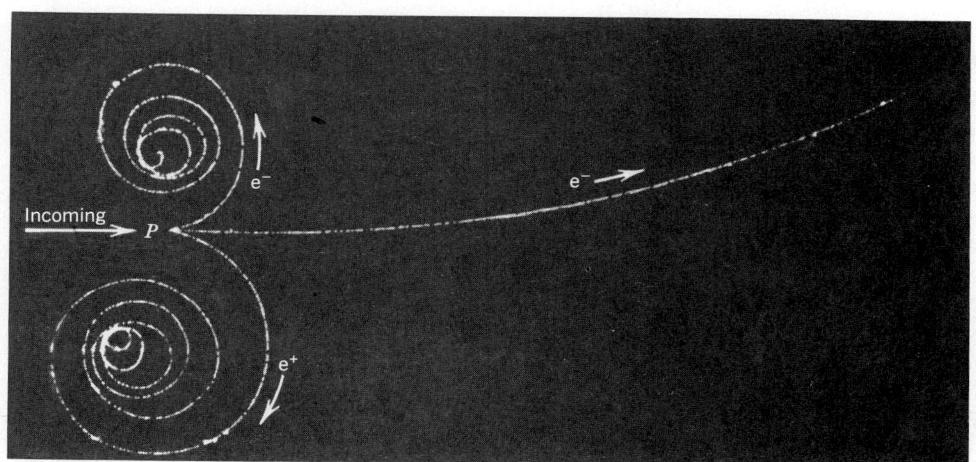

Figure 6 The tracks of two electrons (e^-) and a positron (e^+) in a bubble chamber. A magnetic field fills the chamber, pointing out of the plane of the figure. What can you tell about the particles by examining their tracks?

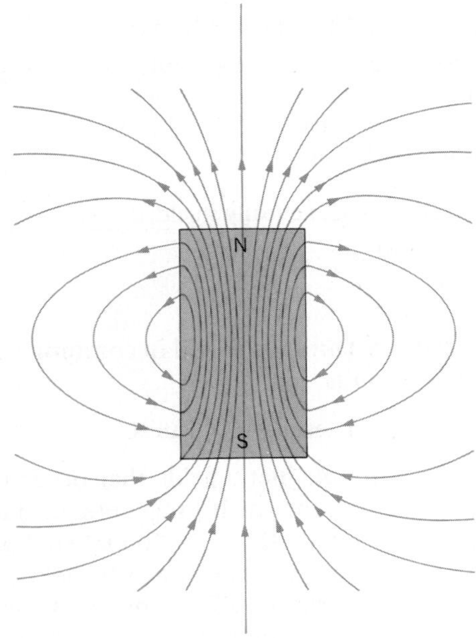

Figure 7 The magnetic field lines for a bar magnet. The lines form closed loops, leaving the magnet at its north pole and entering it at its south pole.

pass needle points north, we conclude that the earth's magnetic pole in the northern hemisphere is a south magnetic pole. That is, in the Arctic regions, the lines of the earth's magnetic field point generally down, into the earth's surface. The earth's magnetic pole in Antarctica is a north magnetic pole. That is, lines of the earth's magnetic field in this region point generally up, out of the earth's surface.

Sample Problem 1 A uniform magnetic field **B**, with magnitude 1.2 mT, points vertically upward throughout the volume of the room in which you are sitting. A 5.3-MeV proton moves horizontally from south to north through a certain point in the room. What magnetic deflecting force acts on the proton as it passes through this point? The proton mass is 1.67×10^{-27} kg.

The magnetic deflecting force depends on the speed of the proton, which we can find from $K = \frac{1}{2}mv^2$. Solving for v, we find

$$v = \sqrt{\frac{2K}{m}} = \sqrt{\frac{(2)(5.3\ \text{MeV})(1.60 \times 10^{-13}\ \text{J/MeV})}{1.67 \times 10^{-27}\ \text{kg}}}$$
$$= 3.2 \times 10^7\ \text{m/s}.$$

Equation 7 then yields

$$\begin{aligned} F_B &= qvB \sin \phi \\ &= (1.60 \times 10^{-19}\ \text{C})(3.2 \times 10^7\ \text{m/s}) \\ &\quad \times (1.2 \times 10^{-3}\ \text{T})(\sin 90°) \\ &= 6.1 \times 10^{-15}\ \text{N}. \end{aligned} \qquad \text{(Answer)}$$

This may seem like a small force but it acts on a particle of small mass, producing a large acceleration, namely,

$$a = \frac{F_B}{m} = \frac{6.1 \times 10^{-15}\ \text{N}}{1.67 \times 10^{-27}\ \text{kg}} = 3.7 \times 10^{12}\ \text{m/s}^2.$$

It remains to find the direction of $\mathbf{F}_B$. In Eq. 6, **v** points horizontally from south to north and **B** points vertically up. The rule for the direction of vector products (see Section 3–7) shows us that the deflecting force $\mathbf{F}_B$ must point horizontally from west to east, as Fig. 8 shows.

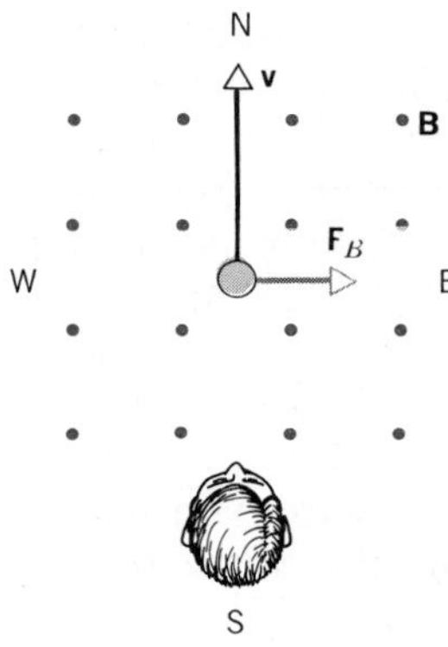

Figure 8 Sample Problem 1. A view from above of a student sitting in a room filled with a magnetic field and watching a moving proton deflect toward the east. The magnetic field points vertically upward in the room.

If the charge of the particle had been negative, the magnetic deflecting force would have pointed in the opposite direction, that is, horizontally from east to west. This is predicted automatically by Eq. 6, if we substitute $-e$ for q.

In this calculation, we used the (approximate) classical expression ($K = \frac{1}{2}mv^2$) for the kinetic energy of the proton rather than the (exact) relativistic expression (see Eq. 27 of Chapter 7). The criterion for when the classical expression may safely be used is that $K \ll mc^2$, where mc^2 is the rest energy of the particle. In this case, $K = 5.3$ MeV and the rest energy of a proton is 938 MeV. This proton passes the test and we were justified in treating it as "slow," that is, in using the classical $K = \frac{1}{2}mv^2$ formula for the kinetic energy. In dealing with energetic particles, we must always be alert to this point.

30-3 Discovering the Electron

A beam of electrons can be deflected by a magnetic field. You are perhaps more familiar with this fact than you realize. Such deflections paint the images on our television screens and trace out the letters on the screens of our word processors. It was not always so.

At the end of the last century, the cathode ray tube, far from being found in every home, was the last word in advanced laboratory research instrumentation. In 1897, J. J. Thomson at Cambridge University showed that the "rays" that cause the glass walls of such tubes to glow were streams of negatively charged particles, which he called *corpuscles.* We now call them *electrons.*

What Thomson did was measure the ratio of the mass m to the charge q of the cathode ray particles. In Fig. 9, which is a modern version of his apparatus, electrons are emitted from hot filament F and accelerated by an applied potential difference V. Passing through a slit in screen C, they then enter a region in which they move at right angles to an electric field $\mathbf{E}$ and a magnetic field $\mathbf{B}$, these two fields being at right angles to each other. The beam, striking fluorescent screen S, is visible as a spot of light.

Study of Fig. 9 shows that, no matter what the sign of the charge carried by the particle, the electric field and the magnetic field will deflect it in opposite directions. In particular, if the particle is negatively charged, the electric field will deflect the particle up and the magnetic field will deflect it down.

Suppose the two fields in Fig. 9 are adjusted so that the two deflecting forces just cancel. From Eqs. 5 and 6 we then have

$$qE = qvB$$

or

$$v = \frac{E}{B}. \tag{10}$$

The arrangement of crossed electric and magnetic fields thus acts as a *velocity filter* and allows us to measure the speed of the particles that pass through it.

Thomson's procedure was equivalent to the following: (1) Set $E = 0$ and $B = 0$ and note the position of the undeflected beam spot. (2) Apply the electric field $\mathbf{E}$, measuring on the fluorescent screen the beam deflection that it causes. (3) Leaving $\mathbf{E}$ in place, apply a magnetic field $\mathbf{B}$ and adjust its value until the beam deflection is restored to zero.

The deflection of an electron in a purely electric field (step 2 above), measured at the far edge of the deflecting plates, is analyzed in Sample Problem 8 of Chapter 24. The deflection is shown there to be

$$y = \frac{qEL^2}{2mv^2}, \tag{11}$$

where v is the electron speed and L is the length of the plates. The deflection y cannot be measured directly but it can be calculated from the measured displacement of the spot on the screen. The *direction* of the deflection allows one to determine the sign of the charge carried by the particle.

Eliminating v between Eqs. 10 and 11 leads to

$$\frac{m}{q} = \frac{B^2L^2}{2yE}, \tag{12}$$

in which all quantities on the right can be measured.

Thomson put forward the daring and important claim—which turned out to be correct—that his corpuscles are a constituent of all matter. He further con-

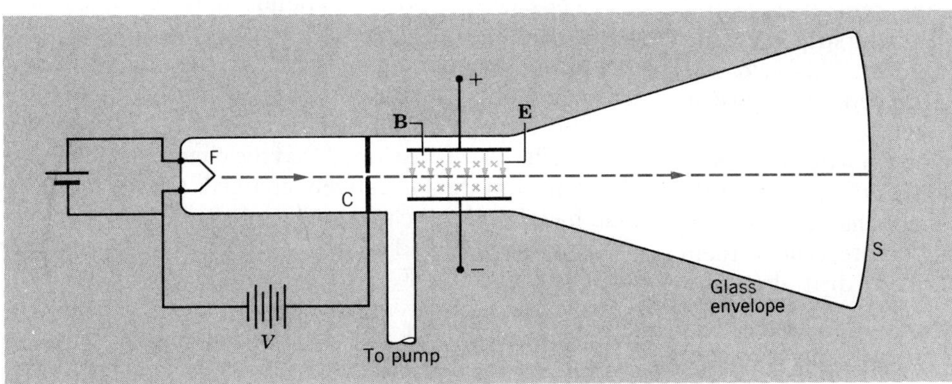

Figure 9 A modern version of J. J. Thomson's apparatus for measuring the ratio of mass to charge for the electron. The electric field **E** is established by connecting a battery across the plate terminals. The magnetic field **B** is set up by means of a current in a system of coils (not shown).

cluded that his corpuscles were lighter than the lightest known atom (hydrogen) by a factor of more than 1000.* It was his m/q measurement, coupled with the boldness and accuracy of these two claims, that constitutes the "discovery of the electron," with which he is generally credited. Direct measurement of the electronic charge soon followed and, within a few years, acceptance of the electron as a particle of nature was firmly established. When you hear our times referred to as the *electronic* age, think of J. J. Thomson and his corpuscles.

30-4 The Hall Effect

A beam of electrons in a vacuum can be deflected sideways by a magnetic field. Do you suppose that it is possible to deflect the drifting conduction electrons in a copper wire sideways by means of a magnetic field? In 1879, Edwin H. Hall, then a 24-year old graduate student of Henry A. Rowland at the Johns Hopkins University, showed that it is. This *Hall effect* allows us to find out whether the charge carriers in a conductor carry a positive or a negative charge. Beyond that, we can measure the number of such carriers per unit volume of the conductor.

Figure 10*a* shows a copper strip of width d, carrying a current i whose conventional direction is from top to bottom. The charge carriers are electrons and, as we know, they drift (with drift speed v_d) in the opposite direction. At the instant shown in Fig. 10*a*, an external magnetic field **B**, pointing into the plane of the figure, has just been turned on. From Eq. 6 we see that a magnetic deflecting force will act on each drifting electron, pushing it toward the right edge of the strip.

As time goes on, electrons will move to the right, mostly piling up on the right edge of the strip, leaving uncompensated positive charges in their fixed positions near the left edge. Thus, a constant transverse electric field **E** will build up at all points within the strip, pointing from left to right as shown in Fig. 10*b*. This field will exert an electric force on each electron, tending to push it to the left.

An equilibrium quickly develops in which the electric force on each electron builds up until it just cancels the magnetic force. When this happens, as Fig. 10*b* shows, the drifting electrons are in balance as far as their sideways motions are concerned. They drift along the strip toward the top of the figure with no net wandering either to the right or to the left.

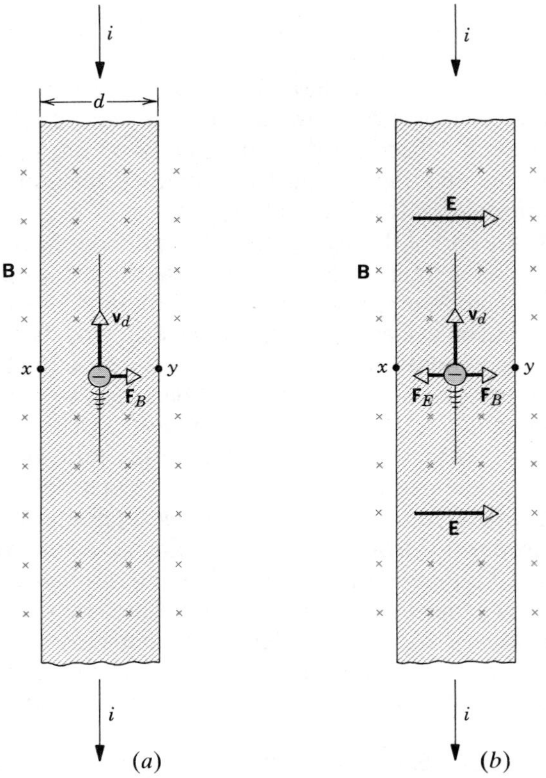

Figure 10 A strip of copper immersed in a magnetic field **B** carries a current i: (*a*) the situation immediately after the magnetic field is turned on and (*b*) the situation at equilibrium, which quickly follows. Note that negative charges pile up on the right side of the strip, leaving uncompensated positive charges on the left. Point x is at a higher potential than point y.

The electric field **E** that builds up is associated with a *Hall potential difference* V, where $E = V/d$. We can measure V by connecting the terminals of a voltmeter between points x and y in Fig. 10*b*. From its polarity, we can find the sign of the charge carriers. Let us see how.

In Fig. 10*b* we assumed that the charge carriers were electrons and thus negatively charged. Check carefully that, had the charge carriers been positive, the directions of the vectors $\mathbf{v}_d$ and **E** in Fig. 10*b* would reverse but those of the vectors $\mathbf{F}_E$ and $\mathbf{F}_B$ would remain unchanged. Thus, drifting positive charges would also be pushed to the right in Fig. 10*b*, leaving uncompensated negative charges on the left. The polarity of the Hall potential

* The exact ratio proved later to be 1836.15.

difference V would be opposite to that for negative charge carriers.

In Chapter 28, which deals with direct currents, we often assumed the charge carriers to be positive when, in fact, we knew that they were negative. We justified this on the basis that it made no difference to our measurements of current and potential difference. The Hall effect, however, is one case in which it *does* make a difference.

Now for the quantitative part. At equilibrium (Fig. 10*b*) the electric and magnetic forces are in balance, or, from Eqs. 5 and 7,

$$(-e)E = (-e)v_dB. \tag{13}$$

The drift speed v_d is given from Eq. 7 of Chapter 28 as

$$v_d = \frac{J}{ne} = \frac{i}{neA}, \tag{14}$$

in which $J(= i/A)$ is the current density in the strip, A being the cross-sectional area of the strip.

Substituting V/d for E in Eq. 13 and eliminating v_d by means of Eq. 14, we obtain

$$\boxed{n = \frac{Bi}{Vte}} \tag{15}$$

in which $t(= A/d)$ is the thickness of the strip. Thus, we can find n, the density of charge carriers, in terms of quantities that we can measure.

It is also possible to use the Hall effect to measure directly the drift speed v_d of the charge carriers, which you may recall is of the order of centimeters per hour.* In this clever experiment, the metal strip is moved mechanically through the magnetic field in a direction opposite to that of the drift velocity of the charge carriers. The speed of the moving specimen is then adjusted until the Hall potential difference vanishes. At this condition, the velocity of the moving specimen is equal in magnitude but opposite in direction to the velocity of the charge carriers; the velocity of the charge carriers *with respect to the magnetic field* is thus zero and there is no Hall effect.

The Hall effect has been, and continues to be, tremendously useful in helping us to understand electrical conduction in metals and semiconductors. To interpret it fully, however, we must replace the classical derivation that we have just given by one based on quantum physics. The 1985 Nobel Prize for physics was awarded for a fundamental discovery about the quantized nature of resistance, based on Hall effect measurements.

* "In Memoriam J. Jaumann: A Direct Demonstration of the Drift Velocity in Metals," by W. Klein, *American Journal of Physics*, January 1987.

Sample Problem 2 A strip of copper 150 μm thick is placed in magnetic field $\mathbf{B} = 0.65$ T and a current $i = 23$ A is set up in the strip. What Hall potential difference V will appear across the width of the strip?

In Sample Problem 3 of Chapter 28, we calculated the number of charge carriers per unit volume for copper, finding

$$n = 8.47 \times 10^{28} \text{ electrons/m}^3.$$

From Eq. 15 then,

$$V = \frac{Bi}{net} = \frac{(0.65 \text{ T})(23 \text{ A})}{(8.47 \times 10^{28} \text{ m}^{-3})(1.60 \times 10^{-19} \text{ C})(150 \times 10^{-6} \text{ m})}$$
$$= 7.4 \times 10^{-6} \text{ V} = 7.4\ \mu\text{V}. \quad \text{(Answer)}$$

This potential difference, though small, is readily measurable.

30-5 A Circulating Charge

If a particle moves in a circle at constant speed, we can be sure that the net force acting on the particle is constant in magnitude and points toward the center of the circle, at right angles to the particle's velocity. Think of a stone tied to a string and whirled in a circle on a smooth horizontal surface, or of a satellite moving in a circular orbit around the earth. In the first case, the tension in the string provides the necessary sideways force. In the second case, it is the earth's gravitational attraction that does so.

Figure 11 shows another example. Here a beam of electrons is projected in the plane of the figure with speed v from an *electron gun* G. The electrons find themselves moving in a region filled with a uniform magnetic field $\mathbf{B}$, whose direction is out of the plane of the figure. A sideways magnetic deflecting force $\mathbf{F}$ $(= q\mathbf{v} \times \mathbf{B})$ acts on them so that they follow a circular path. The path is visible because atoms of the residual gas emit light when some of the circulating electrons collide with them.

If we apply Newton's second law to a particle of charge q moving at right angles to a uniform magnetic

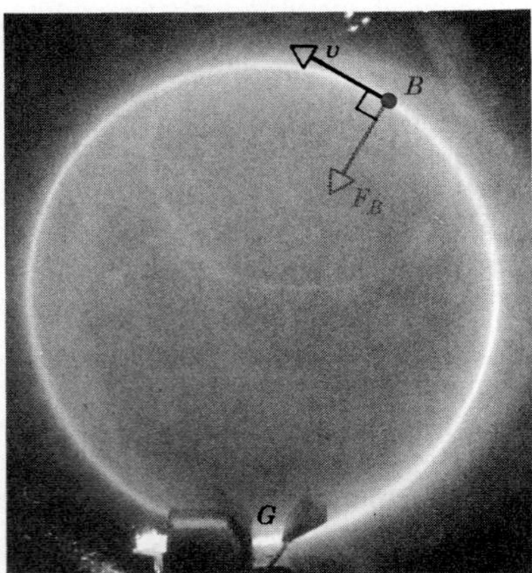

Figure 11 Electrons circulating in a chamber containing gas at low pressure. A uniform magnetic field **B**, pointing out of the plane of the figure at right angles to it, fills the chamber. Note the radially-directed magnetic deflecting force $\mathbf{F}_B$.

field **B**, we find

$$qvB = \frac{mv^2}{r}$$

or

$$r = \frac{mv}{qB} \quad \text{(radius)}, \tag{16}$$

which gives the radius of the path.

The angular frequency ω, the frequency ν, and the period T are given by

$$\omega = \frac{v}{r} = \frac{qB}{m} \quad \text{(angular frequency)}, \tag{17}$$

$$\nu = \frac{\omega}{2\pi} = \frac{qB}{2\pi m} \quad \text{(frequency)}, \tag{18}$$

and

$$T = \frac{1}{\nu} = \frac{2\pi m}{qB} \quad \text{(period)}. \tag{19}$$

Note that none of these three related quantities depends on the speed of the particle.* Fast particles move in large

* This is true only to the extent that the speed of the particle is much less than the speed of light.

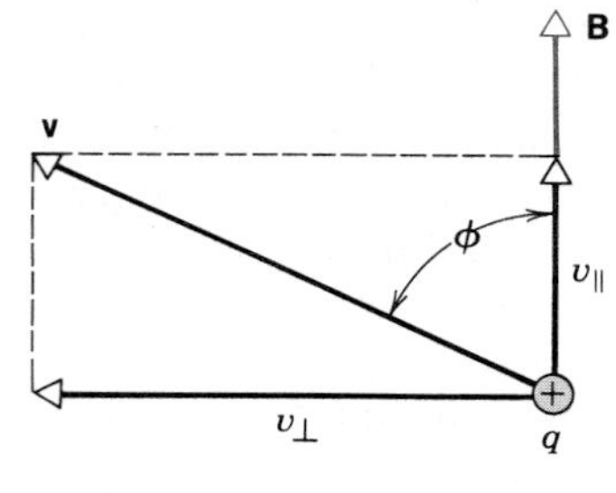

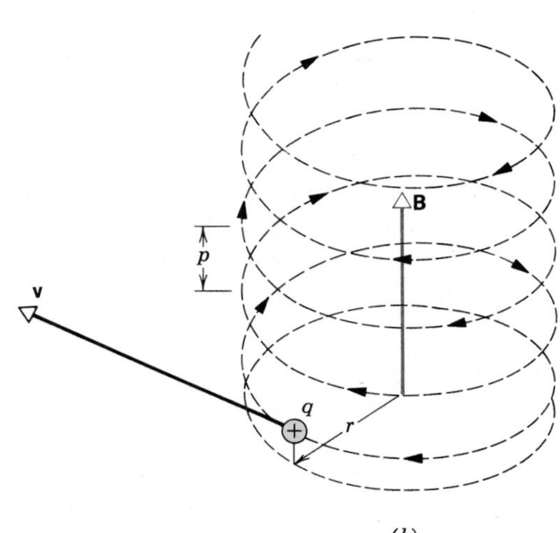

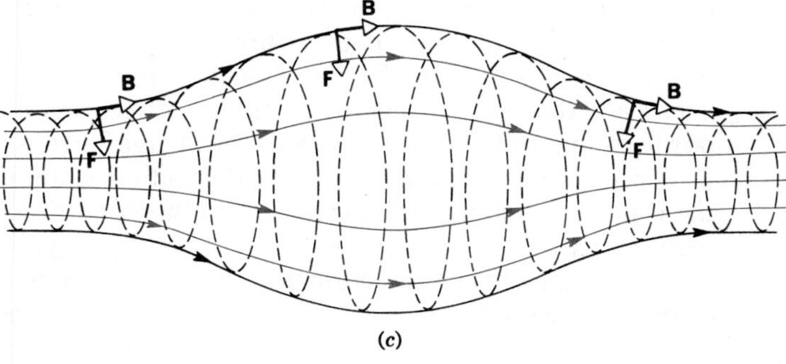

Figure 12 (*a*) A particle moves in a magnetic field, its velocity making an angle ϕ with the field direction. (*b*) The particle follows a helical path, of radius r and pitch p. (*c*) A charged particle spiraling in an inhomogeneous magnetic field. The particle can become trapped, spiraling back and forth between the strong field regions at either end. Note that the magnetic force vectors at each end of the *magnetic bottle* have a component pointing toward the center of the bottle.

circles and slow ones in small circles but all particles with the same value of the ratio q/m take the same time T (the period) to complete one round trip.

If the velocity of a charged particle has a component parallel to the (uniform) magnetic field, the particle will spiral about the direction of the field in a helical path. Figure 12*a*, for example, shows the velocity vector **v** resolved into two components, one parallel to **B** and one at right angles to it. Thus,

$$v_{\parallel} = v \cos \phi \quad \text{and} \quad v_{\perp} = v \sin \phi. \tag{20}$$

The particle, shown as positively charged, moves in an upward spiraling helix, as Fig. 12*b* shows.

Figure 12*c* shows a charged particle spiraling in a highly *nonuniform* magnetic field. The spiraling particle, reflected from the high-field regions at either end of this *magnetic bottle,* can become trapped for substantial periods of time. Figure 13 shows electrons and protons trapped in this way by the earth's magnetic field, forming the Van Allen radiation belts. These belts—naturally occurring magnetic bottles—were discovered in 1958 when a particle counter aboard an early satellite *(Explorer 1)* stopped functioning above a certain altitude. Scientists learned later that it had been jammed by the unexpectedly high counting rate. Electrons and protons spiral back and forth in these belts, bouncing between the strong-field regions near the earth's magnetic poles in a few seconds.

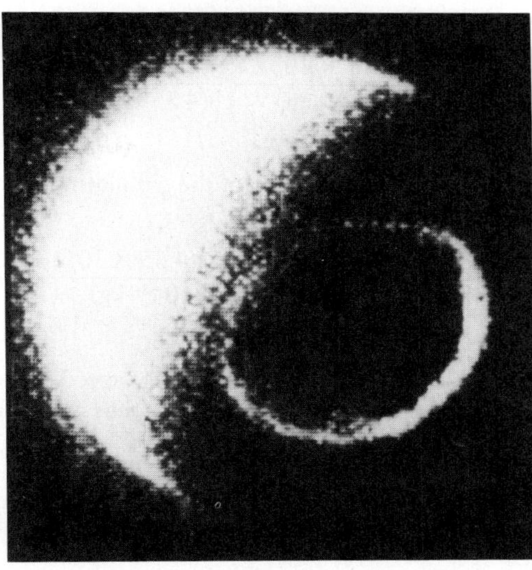

Figure 14 This photo, transmitted from the satellite *Dynamics Explorer I,* clearly shows the auroral oval and also the dark and light hemispheres of the earth.

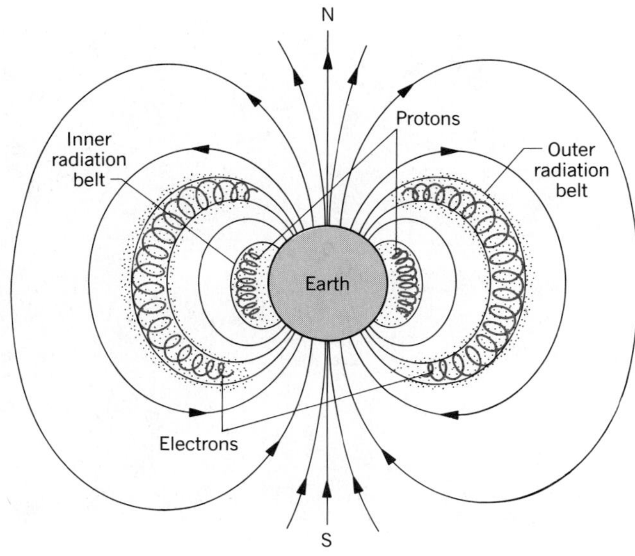

Figure 13 The earth's magnetic field, showing protons and electrons trapped in the Van Allen radiation belts. The magnetic north and south poles of the earth are shown. The figure is a figure of revolution about this N–S axis.

Electrons and protons, streaming out from the sun as a *solar wind,* greatly distort the earth's magnetic field, producing electric currents that stretch it out into a *magnetotail* some 1000 earth radii long. Magnetic substorms in the magnetotail occasionally inject charged particles into the earth's upper atmosphere at high latitudes, forming the familiar aurora. Figure 14, transmitted from the *Dynamics Explorer I* satellite, shows that the aurora is concentrated in an oval around the earth's magnetic pole.

Sample Problem 3 The electrons shown circulating in Fig. 11 have a kinetic energy of 22.5 eV. A uniform magnetic field emerges from the plane of the figure and has a magnitude of 4.55×10^{-4} T. (a) What is the orbit radius?

Calculating the speed from the kinetic energy (just as we did in Sample Problem 1) yields $v = 2.81 \times 10^6$ m/s.* From

* The kinetic energy of the electron (= 22.5 eV) is such a tiny fraction of its rest energy (= 0.511 MeV or 5.11×10^5 eV) that we are easily justified in using the classical (that is, the nonrelativistic) formula for the kinetic energy.

Eq. 16 then,

$$r = \frac{mv}{qB} = \frac{(9.11 \times 10^{-31}\ \text{kg})(2.81 \times 10^{6}\ \text{m/s})}{(1.60 \times 10^{-19}\ \text{C})(4.55 \times 10^{-4}\ \text{T})}$$
$$= 3.52\ \text{cm}. \qquad \text{(Answer)}$$

(b) What is the frequency ν of the circulating electrons? From Eq. 18,

$$\nu = \frac{qB}{2\pi m} = \frac{(1.60 \times 10^{-19}\ \text{C})(4.55 \times 10^{-4}\ \text{T})}{2\pi(9.11 \times 10^{-31}\ \text{kg})}$$
$$= 1.27 \times 10^{7}\ \text{Hz} = 12.7\ \text{MHz}. \qquad \text{(Answer)}$$

Note that this result does not depend on the speed of the particle but only on the nature of the particle (that is, on the ratio q/m) and on the strength B of the magnetic field.

(c) What is the period of revolution T?

$$T = \frac{1}{\nu} = \frac{1}{1.27 \times 10^{7}\ \text{Hz}} = 7.86 \times 10^{-8}\ \text{s}$$
$$= 78.6\ \text{ns}. \qquad \text{(Answer)}$$

Verify, using the right-hand rule for vector products, that Eq. 6 gives the correct direction of rotation for the circulating electrons in Fig. 11, that is, counterclockwise.

Sample Problem 4 Suppose that the velocity vector of the electron in Sample Problem 3 makes an angle ϕ of 65.5° with the direction of the magnetic field, as Fig. 12*a* shows. (a) What is the radius of its helical path?

From Eq. 16, the radius of the helix is

$$r = \frac{mv_\perp}{qB} = \frac{m(v \sin \phi)}{qB}$$
$$= \frac{(9.11 \times 10^{-31}\ \text{kg})(2.81 \times 10^{6}\ \text{m/s})(\sin 65.5°)}{(1.60 \times 10^{-19}\ \text{C})(4.55 \times 10^{-4}\ \text{T})}$$
$$= 3.20\ \text{cm}. \qquad \text{(Answer)}$$

Note that this is less than the value (= 3.52 cm) calculated in the previous problem because we used only a component of $\mathbf{v}$ rather than v in our calculation.

(b) What is the pitch p of the helix (that is, the distance between corresponding points on adjacent loops)?

We note first that the period T of rotation of the particle, being independent of its speed, is the same as in the previous Sample Problem. From Fig. 12*b* then,

$$p = v_\parallel T = (v \cos \phi)T$$
$$= (2.81 \times 10^{6}\ \text{m/s})(\cos 65.5°)(7.86 \times 10^{-8}\ \text{s})$$
$$= 9.16 \times 10^{-2}\ \text{m} = 9.16\ \text{cm}. \qquad \text{(Answer)}$$

30-6 Cyclotrons and Synchrotrons

What is the ultimate structure of matter? This question has always been at the cutting edge of physics. One way of getting at this problem is to allow an energetic charged particle (a proton, for example) to slam into a solid target. Better yet, allow two such energetic protons to collide head-on. Analyzing the debris from such collisions is the most useful way to learn about the nature of the subatomic particles of matter. The Nobel Prizes in Physics for 1976 and 1984 were awarded for just such studies.

How then can a proton acquire kinetic energy? The direct approach is to allow the proton to "fall" through a potential difference V, thereby increasing its kinetic energy by eV. As we want higher and higher energies, however, it becomes more and more difficult to establish the necessary potential difference.

A better way is to arrange for the proton to circulate in a magnetic field and to give it a modest electrical "kick" once per revolution. For example, if a proton circulates 100 times in a magnetic field and if it receives an energy boost of 100 keV every time it completes an orbit, it will end up with a kinetic energy of (100)(100 keV) or 10 MeV. We describe two devices based on this principle.

The Cyclotron. Figure 15 shows a beam of accelerated particles emerging from an early cyclotron through a thin metal foil that separates the vacuum of the cyclotron chamber from the air of the room.

Figure 16 is a top view of the region of a cyclotron in which the particles (protons, say) circulate. The two hol-

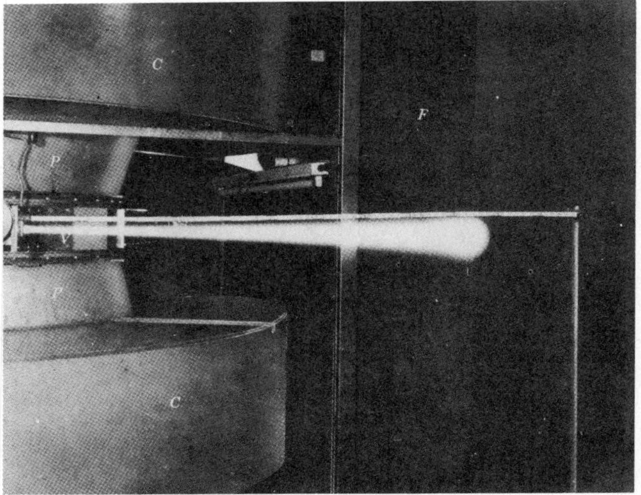

Figure 15 An early cyclotron at the University of Pittsburgh, showing an accelerated beam of deuterons emerging into the air of the laboratory. Note vacuum chamber V, magnet frame F, magnet pole faces P, and magnet coils C. The rule is 6 ft long.

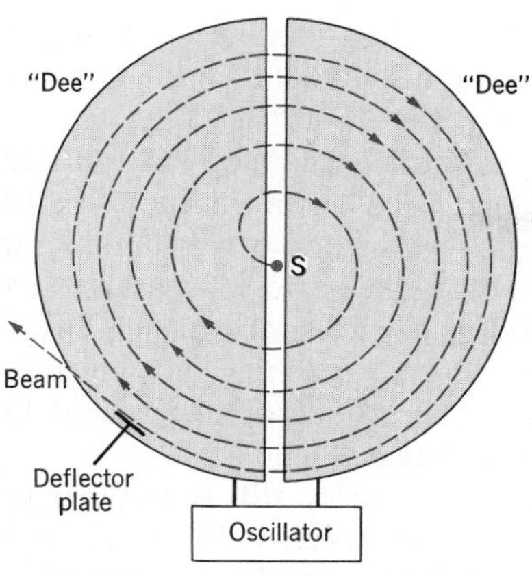

Figure 16 The elements of a cyclotron, showing the ion source S and the dees. A uniform magnetic field emerges from the plane of the figure. The circulating protons spiral outward within the hollow dees, gaining energy every time they cross the gap between the dees.

low D-shaped objects (open on their straight edges) are made of copper sheet. These *dees,* as they are called, form part of an electrical oscillator, which establishes an alternating potential difference across the gap between them. The dees are immersed in a magnetic field ($B =$ 1.5 T) whose direction is out of the plane of the figure. This field is set up by a large electromagnet, not shown in Fig. 16 but apparent in Fig. 15.

Suppose that a proton, injected at the center of the cyclotron in Fig. 16, finds the dee that it faces to be negative. It will accelerate toward this dee and will enter it. Once inside, it is screened from electric fields by the copper walls of the dee. The magnetic field, however, is not screened by the (nonmagnetic) copper dees so that the proton bends in a circular path whose radius, which depends on its speed, is given by Eq. 16, or $r = mv/qB$.

Let us assume that at the instant the proton emerges from the dee the accelerating potential difference has changed sign. Thus, the proton *again* faces a negatively charged dee and is *again* accelerated. This process continues, the circulating proton always being in step with the oscillations of the dee potential, until the proton spirals out to the edge of the dee system.

The key to the operation of the cyclotron is that the frequency ν at which the proton circulates in the field must be equal to the fixed frequency ν_{osc} of the electrical oscillator, or

$$\boxed{\nu = \nu_{osc}} \quad \text{(resonance condition).} \tag{21}$$

This *resonance condition* says that, if the energy of the circulating proton is to increase, energy must be fed to it at a frequency ν_{osc} that is equal to the natural frequency ν at which the proton circulates in the magnetic field.

Combining Eqs. 18 and 21 allows us to write the resonance condition as

$$qB = 2\pi m \nu_{osc}. \tag{22}$$

For the proton, q and m are fixed. The oscillator (we assume) is designed to work at a single fixed frequency ν_{osc}. We then "tune" the cyclotron by varying B until Eq. 22 is satisfied and a beam of energetic protons appears.

The Proton Synchrotron. At proton energies above 50 MeV, the conventional cyclotron begins to fail because one of its assumptions—that the frequency of revolution of a charged particle circulating in a magnetic field is independent of its speed—is true only for speeds that are much less than the speed of light. As the proton speed increases, we must treat the problem by relativistic rules.

We know from relativity theory that as the speed of the circulating proton approaches that of light it takes a longer and longer time to make the trip around its orbit. This means that the frequency of revolution of the circulating proton decreases steadily. Thus, the protons get out of step with the oscillator—whose frequency remains fixed at ν_{osc}—and eventually the energy of the circulating proton stops increasing.

There is another problem. For a 500-GeV proton in a magnetic field of 1.5 T, the radius of curvature is 1.1 km. A magnet of the conventional cyclotron type of this size would be impossibly expensive, the area of its pole faces being about 1000 acres.

The *proton synchrotron* is designed to meet these two difficulties. The magnetic field B and the oscillator frequency ν_{osc}, instead of having fixed values as in the conventional cyclotron, are made to vary with time during the accelerating cycle. If this is done properly, (1) the frequency of the circulating protons remains in step with the oscillator at all times and (2) the protons follow a circular—not a spiral—path. Thus, the magnet need only be ring-shaped, rather than solid. The ring, however, must be large if high energies are to be achieved. Indeed, the area enclosed by the ring of the proton synchrotron at the Fermi National Accelerator Laboratory (Fermilab) is large enough to be used as a test area for the rejuvenation of the primeval midwestern prairie.

Figure 17 is a view inside the accelerator tunnel at Fermilab. Energetic protons, traveling inside a highly evacuated pipe about 2 inches in diameter, curve gently

Figure 17 A view along the tunnel of the proton synchrotron at Fermilab. The tunnel circumference is 6.3 km.

around the 4-mile circumference of the magnet ring. The protons make about 400,000 round trips to reach their full energy of 1 TeV (= 10^{12} eV).

The accelerator, called the *Tevatron,* is operated in a mode such that 1-TeV protons and 1-TeV antiprotons collide head-on with each other, making an energy of 2 TeV available in the center-of-mass reference frame of the two colliding particles. As of 1988, this was the highest collision energy ever achieved by an accelerator. Figure 18 shows an aerial view of the ring and the associated laboratory buildings.

The cry goes up for still more energetic protons. Figure 19 shows (smallest circle) the Fermilab ring and (next largest circle) the accelerator ring at the European Center for Particle Physics (CERN). The largest circle represents the Superconducting Super Collider (SSC), which — now in advanced design stage — is planned to generate proton–antiproton collision energies of 20 TeV. The ring — some 52 miles in circumference and planned to house 10,000 superconducting magnets — is displayed against a satellite photo of Washington, DC to show the scale. The SSC ring will be about the size of the beltway system of highways that surrounds this city.

Figure 18 An aerial view of the Fermi National Accelerator Laboratory at Batavia, Illinois.

Figure 19 The largest circle shows the planned Superconducting Super Collider (SSC), superimposed (for scale purposes only) on a satellite photo of the Washington, DC area. The intermediate circle is the European accelerator at CERN in Switzerland and the smallest circle is the Fermilab accelerator. All circles are drawn to correspond to the same assumed magnetic field.

Sample Problem 5 The cyclotron shown in Fig. 15 operated at an oscillator frequency of 12 MHz and had a dee radius $R = 53$ cm. (a) What value of the magnetic field B was needed to allow deuterons to be accelerated?

A deuteron has the same charge as a proton but approximately twice the mass ($m = 3.34 \times 10^{-27}$ kg). From Eq. 22,

$$B = \frac{2\pi m \nu_{\text{osc}}}{q} = \frac{(2\pi)(3.34 \times 10^{-27}\text{ kg})(12 \times 10^{6}\text{ s}^{-1})}{1.60 \times 10^{-19}\text{ C}}$$

$$= 1.6\text{ T}. \qquad \text{(Answer)}$$

Note that, to allow protons to be accelerated, the magnitude of the magnetic field would have to be reduced by a factor of 2, assuming that the oscillator frequency remained fixed at 12 MHz.

(b) What deuteron energy results?

From Eq. 16, the speed of a deuteron circulating with a radius equal to the dee radius R is given by

$$v = \frac{RqB}{m} = \frac{(0.53\text{ m})(1.60 \times 10^{-19}\text{ C})(1.6\text{ T})}{3.34 \times 10^{-27}\text{ kg}}$$

$$= 4.06 \times 10^{7}\text{ m/s}.$$

This speed corresponds to a kinetic energy of

$$K = \tfrac{1}{2}mv^2$$
$$= \tfrac{1}{2}(3.34 \times 10^{-27}\text{ kg})(4.06 \times 10^{7}\text{ m/s})^2$$
$$\times (1\text{ MeV}/1.60 \times 10^{-13}\text{ J})$$
$$= 17\text{ MeV}. \qquad \text{(Answer)}$$

As Fig. 15 shows, deuterons of this energy, as they emerge from the near vacuum of the cyclotron chamber, have a range in air of about 1.5 m.

30-7 The Magnetic Force on a Current

We have already seen (in connection with the Hall effect) that a magnetic field exerts a sideways force on the conduction electrons in a wire. This force must be transmitted bodily to the wire itself, because the conduction electrons cannot escape sideways out of the wire.

In Fig. 20*a*, a wire, carrying no current, hangs vertically between the pole faces of a permanent magnet. The magnetic field points outward from the page. In Fig. 20*b*, a current is set up in the wire, pointing upward; the wire deflects to the right. In Fig. 20*c*, we reverse the direction of the current and the wire deflects to the left. Note that the wire deflects sideways to the direction of the current, just as we expect.

Let us look at the magnetic deflection of a wire more closely, tying it to the magnetic forces that act on the individual charge carriers. Figure 21 shows an enlarged view of a wire carrying a current i, the axis of the wire being at right angles to a magnetic field **B**. We see one of

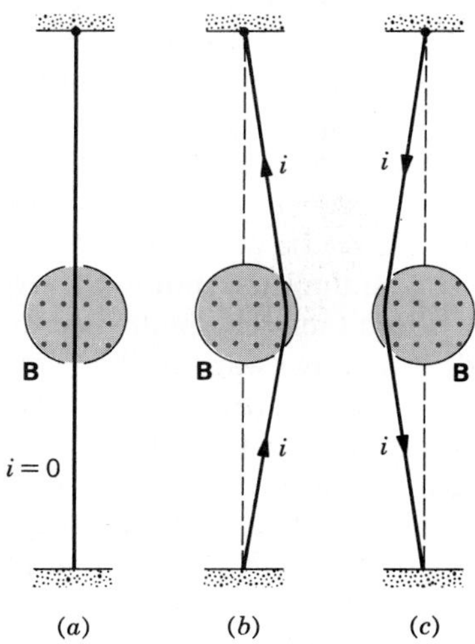

Figure 20 A flexible wire passes between the pole faces of an electromagnet. (*a*) There is no current in the wire. (*b*) A current is established. (*c*) The same as (*b*) except that the direction of the current is reversed. The connections for getting the current into the wire at one end and out of it at the other end are not shown.

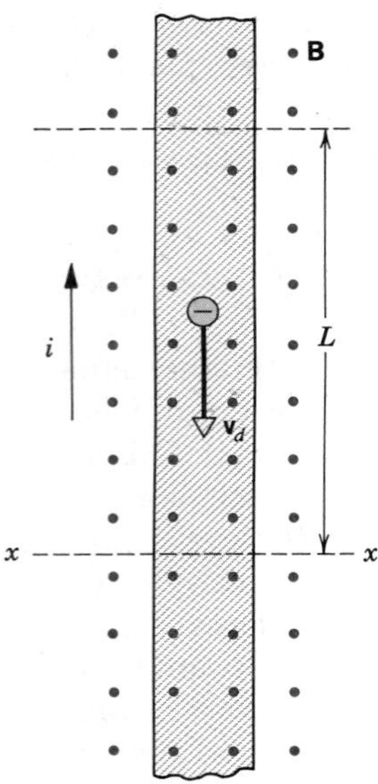

Figure 21 A close-up view of a length L of the wire of Fig. 20*b*. The current direction is upward, which means that electrons drift downward. A magnetic field emerges from the plane of the figure so that the wire is deflected to the right.

the conduction electrons, drifting downward with an assumed drift speed v_d. Equation 7, in which we must put $\phi = 90°$, tells us that a force given by $(-e)v_dB$ must act on each such electron. From Eq. 6 we see that this force must point to the right. We expect then that the wire as a whole will experience a force to the right, in agreement with Fig. 20*b*.

If, in Fig. 21, we were to reverse *either* the direction of the magnetic field *or* the direction of the current, the force on the wire would reverse, pointing now to the left. Note too that it doesn't matter whether we consider negative charges drifting downward in the wire (the actual case) or positive charges drifting upward. The direction of the deflecting force on the wire is the same. We are safe then in dealing with the conventional direction of current, which assumes positive charge carriers.

Consider a length L of the wire of Fig. 21. The electrons in this section of wire will drift past plane xx in a time L/v_d, carrying a charge given by

$$q = i\left(\frac{L}{v_d}\right)$$

through that plane. Substituting this into Eq. 7 yields

$$\begin{aligned} F_B &= qv_dB \sin \phi \\ &= (iL/v_d)(v_d)B \sin 90° \end{aligned}$$

or

$$F_B = iLB. \tag{23}$$

This equation gives the force that acts on a segment of a straight wire of length L, immersed in a magnetic field that is at right angles to the wire.

If the magnetic field is *not* at right angles to the wire, as in Fig. 22, the magnetic force is given by a generalization of Eq. 23, or

$$\boxed{\mathbf{F}_B = i\mathbf{L} \times \mathbf{B}} \quad \text{(force on a current).} \tag{24}$$

Here **L** is a vector that points along the wire segment in the direction of the (conventional) current.

Equation 24 is equivalent to Eq. 6 in that either can be taken as the defining equation for **B**. In practice, we define **B** from Eq. 24. It is much easier to measure the magnetic force acting on a wire than on a single moving charge.

If a wire is not straight, we can imagine it broken up into small straight segments and apply Eq. 24 to each such segment. The force on the wire as a whole is then the vector sum of all the forces on the segments that make it up. In the differential limit, we can write

$$d\mathbf{F}_B = i\, d\mathbf{L} \times \mathbf{B}, \tag{25}$$

and we can find the resultant force on any given structure of currents by integrating Eq. 25 over that structure.

In using Eq. 25, bear in mind that there is no such thing as an isolated, current-carrying wire segment of length dL. There must always be a way to introduce the current into the segment at one end and to take it out at the other.

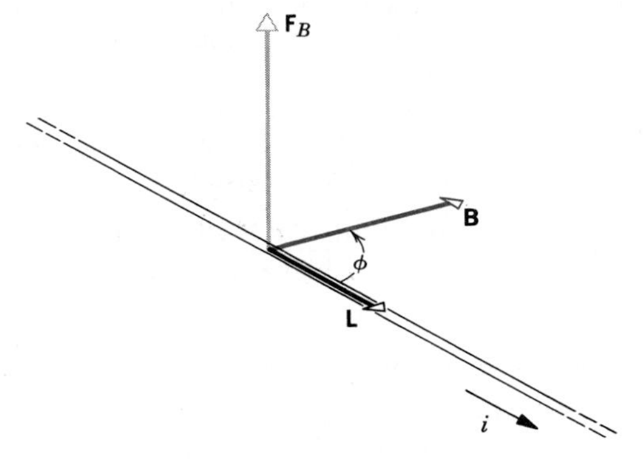

Figure 22 A wire segment of length L makes an angle ϕ with a magnetic field. Compare carefully with Fig. 4.

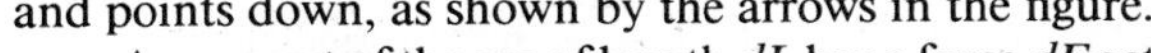

Sample Problem 6 A straight, horizontal stretch of copper wire carries a current $i = 28$ A. What are the magnitude and direction of the magnetic field needed to "float" the wire, that is, to balance its weight? Its linear density is 46.6 g/m.

Figure 23 shows the arrangement. For a length L of wire we have (see Eq. 23)

$$mg = LiB,$$

or

$$B = \frac{(m/L)g}{i} = \frac{(46.6 \times 10^{-3}\ \text{kg/m})(9.8\ \text{m/s}^2)}{28\ \text{A}}$$
$$= 1.6 \times 10^{-2}\ \text{T}\ (= 160\ \text{gauss}). \qquad \text{(Answer)}$$

This is about 160 times the strength of the earth's magnetic field.

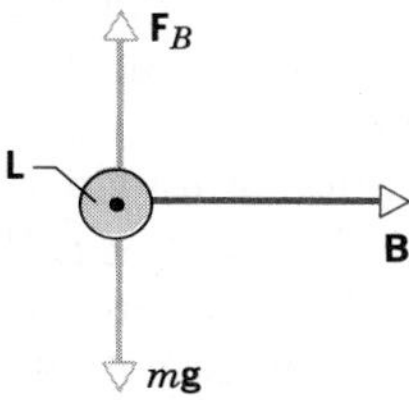

Figure 23 Sample Problem 6. A wire (shown in cross section) can be made to "float" in a magnetic field. The current in the wire emerges from the figure and the magnetic field points to the right.

Sample Problem 7 Figure 24 shows a wire segment, placed in a uniform magnetic field **B** that points out of the plane of the figure. If the segment carries a current i, what resultant magnetic force **F** acts on it?

The force that acts on each straight section, from Eq. 23, has the magnitude

$$F_1 = F_3 = iLB$$

and points down, as shown by the arrows in the figure.

A segment of the arc of length dL has a force dF acting on it, whose magnitude is given by

$$dF = iB\ dL = iB(R\ d\theta)$$

and whose direction is radially toward O, the center of the arc. Note that only the downward component of this force element is effective. The horizontal component is canceled by an oppositely directed horizontal component associated with a symmetrically located segment on the opposite side of the arc.

Thus, the total force on the central arc points down and is given by

$$F_2 = \int_0^\pi dF \sin\theta = \int_0^\pi (iBR\ d\theta)\sin\theta$$
$$= iBR \int_0^\pi \sin\theta\ d\theta = 2iBR.$$

The resultant force on the entire wire is then

$$F = F_1 + F_2 + F_3 = iLB + 2iBR + iLB$$
$$= 2iB(L + R). \qquad \text{(Answer)}$$

Note that this force is just the same as the force that would act on a straight wire of length $2(L + R)$. This would be true no matter what the shape of the central segment, shown as a semicircle in Fig. 24. Can you convince yourself that this is so?

30-8 Torque on a Current Loop

Much of the world's work is done by electric motors. The forces that do this work are the magnetic forces that we studied in the previous section; that is, they are the forces that a magnetic field exerts on a wire that carries a current.

Figure 25 shows a simple motor, in which the two

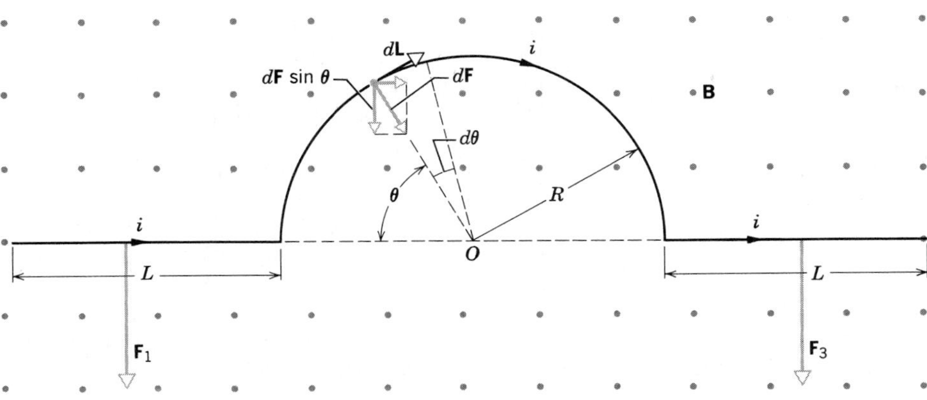

Figure 24 Sample Problem 7. A wire segment carrying a current i is immersed in a magnetic field. The resultant force on the wire is directed downward.

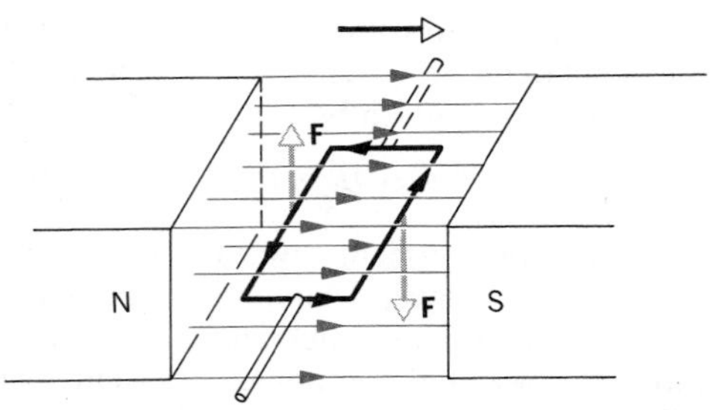

Figure 25 The rudiments of an electric motor. A rectangular coil, carrying a current and free to rotate about a fixed axis, is placed in a magnetic field. A commutator (not shown) reverses the direction of the current every half revolution so that the magnetic torque always acts in the same direction.

magnetic forces combine to exert a torque on a current loop, tending to rotate it about the central axis. Although we omit many essential details of how an electric motor works, it seems clear that the action of a magnetic field in exerting a torque on a current loop is at the heart of it.

Figure 26 shows two views of a rectangular loop of sides a and b, carrying a current i and immersed in a uniform magnetic field **B**. We place it in the field so that its long sides, labeled 1 and 3, are at right angles to the field direction. Wires to lead the current into and out of the loop are needed but, for simplicity, we do not show them.

Imagine now that the loop is held at rest, so that **n**, a vector normal to its plane, makes an angle θ with the field direction, as Fig. 26*b* shows. Note that we have defined the direction of **n** by a right-hand rule. That is, (1) curl the fingers of your right hand around the coil in the direction of the current; (2) your extended thumb then points in the direction of **n**. What net force and what net torque act on the loop in this position?

The net force is the vector sum of the forces acting on each of the four sides of the loop. For side 2 the vector **L** in Eq. 24 points in the direction of the current and has the magnitude b. The angle between **L** and **B** for side 2 (see Fig. 26*b*) is $90° - \theta$. Thus, the magnitude of the force acting on this side is

$$F_2 = ibB \sin(90° - \theta) = ibB \cos \theta. \quad (26)$$

You can show that the force $\mathbf{F}_4$ acting on side 4 has the same magnitude as $\mathbf{F}_2$ but points in the opposite direction. Thus, $\mathbf{F}_2$ and $\mathbf{F}_4$, taken together, cancel out exactly. Their net force is zero and, because they have the same line of action, so is their net torque.

The situation is different for sides 1 and 3. Here the common magnitude of $\mathbf{F}_1$ and $\mathbf{F}_3$ is iaB and they point in opposite directions so that they do not tend to move the coil bodily. However, as Fig. 26*b* shows, these two forces do *not* share the same line of action so they *do* tend to turn the coil. There is a net torque.

The magnitude τ' of the torque due to forces $\mathbf{F}_1$ and $\mathbf{F}_3$ is (see Fig. 26*b*)

$$\tau' = (iaB)(b/2)(\sin \theta) + (iaB)(b/2)(\sin \theta) = iaB \sin \theta.$$

This torque acts on every turn of the coil. If there are N turns, the total torque is

$$\tau = N\tau' = NiabB \sin \theta = (NiA)B \sin \theta. \quad (27)$$

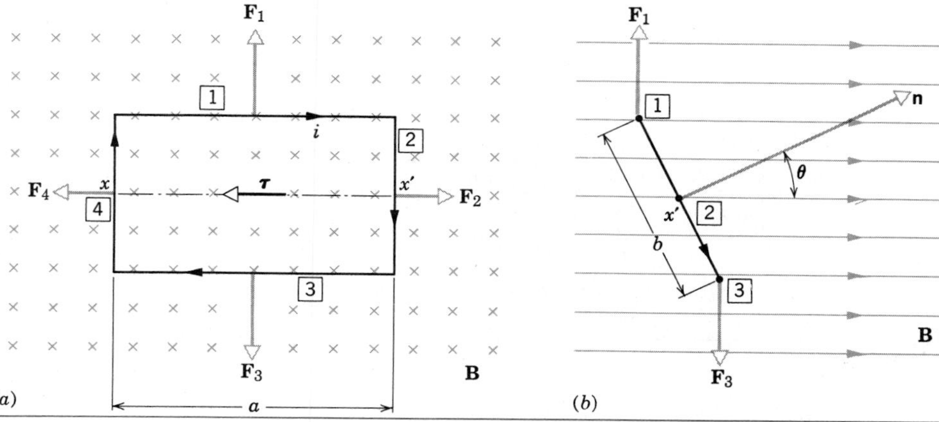

Figure 26 A rectangular coil carrying a current i is placed in a uniform magnetic field. A torque acts to align the normal vector **n** with the direction of the field.

in which A (= ab) is the area of the coil. The quantities in parentheses (= NiA) are grouped together because they are all properties of the coil: its number of turns, its area, and the current it carries. We can show (see Problem 57) that this equation holds for all plane loops, no matter what their shape.

Instead of watching the motion of the coil, it is simpler to watch the vector **n**, which is normal to its plane. Equation 27 tells us that a current-carrying coil placed in a magnetic field will tend to rotate so that this normal vector points in the field direction. This is just what a compass needle does.

Sample Problem 8 Analog voltmeters and ammeters, in which the reading is displayed by the deflection of a pointer over a scale, work by measuring the torque exerted by a magnetic field on a current loop. Figure 27 shows the rudiments of a *galvanometer*, on which both analog ammeters and analog voltmeters are based. The coil is 2.1 cm high and 1.2 cm wide; it has 250 turns and is mounted so that it can rotate about a vertical axis in a uniform radial magnetic field with $B =$ 0.23 T. A spring Sp provides a countertorque that balances the magnetic torque, resulting in a steady angular deflection ϕ corresponding to a given steady current i in the coil. If a current of 100 μA produces an angular deflection of 28°, what must be the torsional constant of the spring (see Eq. 14–20)?

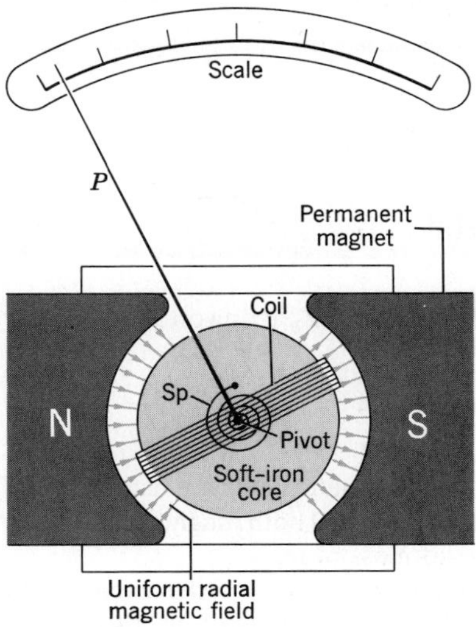

Figure 27 Sample Problem 8. The rudiments of a galvanometer. Depending on the external circuit, this device can be wired up as either a voltmeter or an ammeter.

Setting the magnetic torque equal to the spring torque (see Eq. 27) yields

$$\tau = NiAB \sin \theta = \kappa\phi, \tag{28}$$

in which ϕ is the angular deflection of the pointer and A $(= 2.52 \times 10^{-4}\ \text{m}^2)$ is the area of the coil. Note that the normal to the plane of the coil (that is, the pointer) is always at right angles to the (radial) magnetic field so that $\theta = 90°$ for all pointer positions.

Solving Eq. 28 for κ, we find

$$\kappa = \frac{NiAB \sin \theta}{\phi}$$

$$= \frac{(250)(100 \times 10^{-6}\ \text{A})(2.52 \times 10^{-4}\ \text{m}^2)(0.23\ \text{T})(\sin 90°)}{28°}$$

$$= 5.2 \times 10^{-8}\ \text{N}\cdot\text{m/degree}. \qquad \text{(Answer)}$$

Many modern ammeters and voltmeters are of the digital, direct-reading type and operate in a way that does not involve a moving coil.

30-9 A Magnetic Dipole

In physics, we like to identify the main features of a problem, ignoring details that do not matter. In this spirit we describe the current loop of the preceding section by a single vector $\boldsymbol{\mu}$, its *magnetic dipole moment*. We take the direction of $\boldsymbol{\mu}$ to be that of the normal vector **n** to the plane of the loop, as in Fig. 26*b*. We take as the magnitude of $\boldsymbol{\mu}$ the quantity NiA. Thus, we can rewrite Eq. 27 as

$$\tau = \mu B \sin \theta, \tag{29}$$

in which θ is the angle between the vectors $\boldsymbol{\mu}$ and **B**.

We can generalize this to the vector relation

$$\boldsymbol{\tau} = \boldsymbol{\mu} \times \mathbf{B}, \tag{30}$$

which reminds us very much of the corresponding equation for the torque exerted by an *electric* field on an *electric* dipole, namely (see Eq. 25 of Chapter 24),

$$\boldsymbol{\tau} = \mathbf{p} \times \mathbf{E}.$$

In each case the torque exerted by the external field—be it magnetic or electric—is equal to the vector product of the corresponding dipole moment and the field vector.

If a magnetic field exerts a torque on a magnetic dipole, then work must be done to change the orientation of the dipole. The magnetic dipole must have a *magnetic potential energy* that depends on its orientation in the external field. For electric dipoles we have shown (see Eq. 28 of Chapter 24) that

$$U(\theta) = -\mathbf{p} \cdot \mathbf{E}.$$

In strict analogy, we can write for the magnetic case

$$U(\theta) = -\boldsymbol{\mu} \cdot \mathbf{B}. \tag{31}$$

Thus, a magnetic dipole has its lowest energy $(= -\mu B \cos 0° = -\mu B)$ when it is lined up with the magnetic field. It has its greatest energy $(= -\mu B \cos 180° = +\mu B)$ when it points in a direction opposite to the field. The difference in energy between these two positions is

$$\begin{aligned}\Delta U &= U(180°) - U(0°) \\ &= (-\mu B \cos 180°) - (-\mu B \cos 0°) \\ &= (+\mu B) - (-\mu B) = 2\mu B.\end{aligned} \tag{32}$$

This much work must be done by an external agent to turn a magnetic dipole through 180°, starting from its lined-up position.

So far, we have identified a magnetic dipole as a current loop. However, a simple bar magnet is also a magnetic dipole. So is a rotating sphere of charge. The earth itself is a magnetic dipole. Finally, most subatomic particles, including the electron, the proton, and the neutron, have magnetic dipole moments. As we shall see, all these quantities can be viewed—in some sense or other—as current loops. Here, for comparison, are some approximate magnetic dipole moments:

A proton	1.4×10^{-26} J/T
An electron	9.3×10^{-24} J/T
The coil of Sample Problem 9	5.4×10^{-6} J/T
A small bar magnet	5 J/T
The earth	8.0×10^{22} J/T

Sample Problem 9 (a) What is the magnetic dipole moment of the coil of Sample Problem 8, assuming that it carries a current of 85 μA?

The *magnitude* of the magnetic dipole moment of the coil, whose area A is 2.52×10^{-4} m², is

$$\begin{aligned}\mu &= NiA \\ &= (250)(85 \times 10^{-6}\ \text{A})(2.52 \times 10^{-4}\ \text{m}^2) \\ &= 5.36 \times 10^{-6}\ \text{A}\cdot\text{m}^2 = 5.36 \times 10^{-6}\ \text{J/T}. \quad \text{(Answer)}\end{aligned}$$

Show that these two sets of units are identical. Note that the second set of units follows logically from Eq. 31.

The *direction* of $\boldsymbol{\mu}$, as inspection of Fig. 27 shows, is that of the pointer. You can verify this by showing that, if we assume $\boldsymbol{\mu}$ to be in the pointer direction, the torque predicted by Eq. 30 is such that it would indeed move the pointer clockwise across the scale.

(b) The magnetic dipole moment of the coil is lined up with an external magnetic field whose strength is 0.85 T. How much work is required to turn the coil end for end?

The work is equal to the increase in potential energy, which is

$$\begin{aligned}W = \Delta U &= 2\mu B = 2(5.36 \times 10^{-6}\ \text{J/T})(0.85\ \text{T}) \\ &= 9.1 \times 10^{-6}\ \text{J} = 9.1\ \mu\text{J}. \quad \text{(Answer)}\end{aligned}$$

This is about equal to the work needed to lift an aspirin tablet through a vertical height of about 3 mm.

REVIEW AND SUMMARY

Magnetic Field B

The magnetic field **B** is defined in terms of the sideways force $\mathbf{F}_B$ acting on a test particle with charge q and moving with velocity **v**,

$$\mathbf{F}_B = q\mathbf{v} \times \mathbf{B} \tag{6}$$

study Figs. 4 through 7 and Sample Problem 1. The SI unit for **B** is the *tesla* (abbr. T) where 1 T = 1 N/(A·m) = 10^4 gauss. Some representative magnetic fields are listed in Table 1.

***e/m* for the Electron**

J. J. Thomson, in his 1897 discovery of the electron, used both magnetic and electric fields (see Section 30–3 and Fig. 9) to determine its charge-to-mass ratio.

The Hall Effect

When a conducting strip of thickness t carrying a current i is placed in a magnetic field **B**, some charge carriers (with charge e) build up on the sides of the conductor, as illustrated in Fig. 10. A potential difference V builds up across the strip. The polarity of V gives the sign of the charge carriers; the density of charge carriers may be calculated from

$$n = \frac{Bi}{Vte}. \tag{15}$$

Sample Problem 2 illustrates the magnitudes of the quantities involved.

A Charged Particle Circulating in a Magnetic Field

A charged particle with mass m and charge q moving with velocity $\mathbf{v}$ perpendicular to a magnetic field $\mathbf{B}$ will travel in a circle of radius

$$r = \frac{mv}{qB} \quad \text{(radius).} \tag{16}$$

Its frequency of revolution in the field (the *cyclotron frequency*) is

$$\nu = \frac{\omega}{2\pi} = \frac{1}{T} = \frac{qB}{2\pi m} \quad \text{(frequency, period).} \qquad [18,19]$$

See Figs. 11 through 14 and Sample Problems 3 and 4.

Cyclotrons and Synchrotrons

A cyclotron is a particle accelerator that uses a magnetic field to hold a charged particle in a circular orbit so that a modest accelerating potential may act on it repeatedly, resulting in high energies. Because the moving particle gets out of step with the oscillator as its speed approaches that of light, there is an upper limit to the energy attainable with the cyclotron. A synchrotron avoids this difficulty. Here both B and the oscillator frequency ν_{osc} are programmed to change cyclically so that the particle can not only go to high energies but can do so at a constant orbital radius; this allows the use of a ring magnet rather than a solid magnet, at great saving in cost. See Section 30–6, especially Figs. 15 through 19 and Sample Problem 5, for an introduction to some exciting applications in modern physics.

Magnetic Force on a Current

A straight wire carrying a current i in a uniform magnetic field experiences a sideways force

$$\mathbf{F}_B = i\mathbf{L} \times \mathbf{B}. \tag{24}$$

The sideways force acting on a current element $id\mathbf{L}$ in a magnetic field is

$$d\mathbf{F} = id\mathbf{L} \times \mathbf{B}. \tag{25}$$

The direction of the length element $d\mathbf{L}$ is that of the current density vector $\mathbf{J}$. See Figs. 20 through 22 and Sample Problems 6 and 7.

Torque on a Current Loop

A current loop (area A, current i with N turns) in a uniform magnetic field $\mathbf{B}$ will experience a torque $\boldsymbol{\tau}$ given by

$$\boldsymbol{\tau} = \boldsymbol{\mu} \times \mathbf{B}. \tag{30}$$

Here $\boldsymbol{\mu}$ is the *magnetic dipole moment* with magnitude $\mu = NiA$ and a direction perpendicular to the plane of the loop in the right-hand-rule direction. This torque is the operating principle in electric motors and analog voltmeters and ammeters. See Figs. 25 through 27 and Sample Problems 8 and 9. Bar magnets, molecules, atoms, basic particles (electrons, protons, neutrons, etc.) all have magnetic dipole properties.

Orientation Energy of a Magnetic Dipole

The potential energy of orientation of a magnetic dipole in a magnetic field is

$$U(\theta) = -\boldsymbol{\mu} \cdot \mathbf{B}. \tag{31}$$

See Sample Problem 9.

QUESTIONS

1. Of the three vectors in the equation $\mathbf{F}_B = q\mathbf{v} \times \mathbf{B}$, which pairs are always at right angles? Which may have any angle between them?

2. Why do we not simply define the direction of the magnetic field $\mathbf{B}$ to be the direction of the magnetic force that acts on a moving charge?

3. Imagine that you are sitting in a room with your back to one wall and that an electron beam, traveling horizontally from the back wall toward the front wall, is deflected to your right. What is the direction of the uniform magnetic field that exists in the room?

4. How could we rule out that the forces between two magnets are electrostatic forces?

5. If an electron is not deflected in passing through a certain region of space, can we be sure that there is no magnetic field in that region?

6. If a moving electron is deflected sideways in passing through a certain region of space, can we be sure that a magnetic field exists in that region?

7. A beam of electrons can be deflected either by an electric field or by a magnetic field. Is one method better than the other? . . . in any sense easier?

8. A charged particle passes through a magnetic field and is deflected. This means that a force acted on it and changed its momentum. Where there is a force, there must be a reaction force. On what object does it act?

9. Imagine the room in which you are seated to be filled with a uniform magnetic field with **B** pointing vertically downward. At the center of the room two electrons are suddenly projected horizontally with the same initial speed but in opposite directions. (*a*) Describe their motions. (*b*) Describe their motions if one particle is an electron and one a positron, that is, a positively charged electron. (The electrons will gradually slow down as they collide with molecules of the air in the room.)

10. In Fig. 6 why are the low-energy electron tracks spirals? That is, why does the radius of curvature change in the constant magnetic field in which the chamber is immersed?

11. What are the primary functions of (*a*) the electric field and (*b*) the magnetic field in the cyclotron?

12. What central fact makes the operation of a conventional cyclotron possible? Ignore relativistic considerations.

13. A bare copper wire emerges from one wall of a room, crosses the room, and disappears into the opposite wall. You are told that there is a steady current in the wire. How can you find its direction? Describe as many ways as you can think of. You may use any reasonable piece of equipment, but you may not cut the wire.

14. In Section 7 we state that a magnetic field **B** exerts a sideways force on the conduction electrons in, say, a copper wire carrying a current i. We have tacitly assumed that this same force acts on the conductor itself. Are there some missing steps in this argument? If so, supply them.

15. A current in a magnetic field experiences a force. Therefore, it should be possible to pump conducting liquids by sending a current through the liquid (in an appropriate direction) and letting it pass through a magnetic field. Design such a pump. This principle is used to pump liquid sodium (a conductor, but highly corrosive) in some nuclear reactors, where it is used as a coolant. What advantages would such a pump have?

16. An airplane is flying west in level flight over Massachusetts, where the earth's magnetic field is directed downward below the horizontal in a northerly direction. As a result of the magnetic force on the free electrons in its wings, one of its wingtips will have more electrons than the other. Which one (right or left) is it? Will the answer be different if the plane is flying east?

17. A conductor, even though it is carrying a current, has zero net charge. Why, then, does a magnetic field exert a force on it?

18. You wish to modify a galvanometer (see Sample Problem 8) to make it into (*a*) an ammeter and (*b*) a voltmeter. What do you need to do in each case?

19. A rectangular current loop is in an arbitrary orientation in an external magnetic field. How much work is required to rotate the loop about an axis perpendicular to its plane?

20. Equation 30 ($\boldsymbol{\tau} = \boldsymbol{\mu} \times \mathbf{B}$) shows that there is no torque on a current loop in an external magnetic field if the angle between the axis of the loop and the field is (*a*) 0° or (*b*) 180°. Discuss the nature of the equilibrium (that is, is it stable, neutral, or unstable?) for these two positions.

21. In Sample Problem 9 we showed that the work required to turn a current loop end-for-end in an external magnetic field is $2\mu B$. Does this result hold no matter what the original orientation of the loop was?

22. Imagine that the room in which you are seated is filled with a uniform magnetic field with **B** pointing vertically upward. A circular loop of wire has its plane horizontal. For what direction of current in the loop, as viewed from above, will the loop be in stable equilibrium with respect to forces and torques of magnetic origin?

23. The torque exerted by a magnetic field on a magnetic dipole can be used to measure the strength of that magnetic field. For an accurate measurement, does it matter whether the dipole moment is small or not? Recall that, in the case of measurement of an electric field, the test charge was to be as small as possible so as not to disturb the source of the field.

24. You are given a smooth sphere the size of a ping-pong ball and told that it contains a magnetic dipole. What experiments would you carry out to find the magnitude and the direction of its magnetic dipole moment?

EXERCISES AND PROBLEMS

Section 30-2 The Definition of B

1E. Express magnetic field B and magnetic flux Φ_B in terms of the dimensions M, L, T, and Q (mass, length, time, and charge).

2E. Four particles follow the paths shown in Fig. 28 as they pass through the magnetic field there. What can one conclude about the charge of each particle?

3E. An electron in a TV camera tube is moving at 7.2 ×

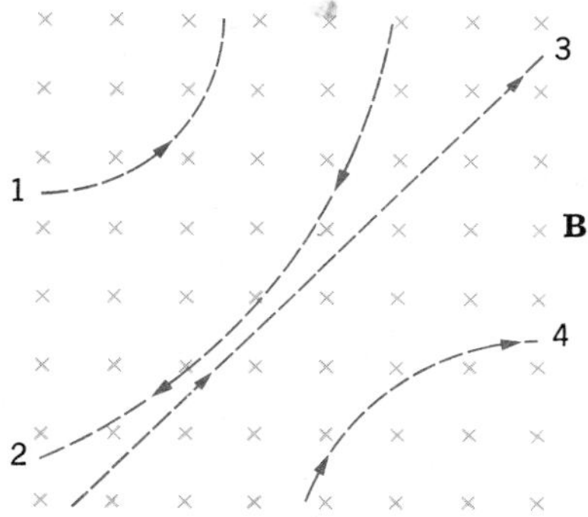

Figure 28 Exercise 2.

10^6 m/s in a magnetic field of strength 83 mT. (*a*) Without knowing the direction of the field, what could be the greatest and least magnitudes of the force the electron could feel due to the field? (*b*) At one point the acceleration of the electron is 4.9×10^{14} m/s^2. What is the angle between the electron's velocity and the magnetic field?

4E. A proton traveling at 23° with respect to a magnetic field of strength 2.6 mT experiences a magnetic force of 6.5×10^{-17} N. Calculate (*a*) the speed and (*b*) the kinetic energy in eV of the proton.

5P. An electron has a velocity given in m/s by $\mathbf{v} = 2.0 \times 10^6\mathbf{i} + 3.0 \times 10^6\mathbf{j}$. It enters a magnetic field given in T by $\mathbf{B} = 0.030\mathbf{i} - 0.15\mathbf{j}$. (*a*) Find the magnitude and direction of the force on the electron. (*b*) Repeat your calculation for a proton having the same velocity.

6P. An electron in a uniform magnetic field has a velocity $\mathbf{v} = 40\mathbf{i} + 35\mathbf{j}$ km/s. It experiences a force $\mathbf{F} = -4.2\mathbf{i} + 4.8\mathbf{j}$ fN. If $B_x = 0$, calculate the magnetic field.

7P. The electrons in the beam of a television tube have an energy of 12 keV. The tube is oriented so that the electrons move horizontally from magnetic south to magnetic north. The vertical component of the earth's magnetic field points down and has a magnitude of 55 μT. (*a*) In what direction will the beam deflect? (*b*) What is the acceleration of a given electron due to the magnetic field? (*c*) How far will the beam deflect in moving 20 cm through the television tube?

8P*. An electron has an initial velocity $12\mathbf{j} + 15\mathbf{k}$ km/s and a constant acceleration of $\mathbf{i}(2.0 \times 10^{12}$ m/s$^2)$ in a region in which uniform electric and magnetic fields are present. If $\mathbf{B} = 400\mathbf{i}$ μT, find the electric field $\mathbf{E}$.

Section 30-3 Discovering the Electron

9E. A typical cathode-ray oscilloscope employs a cathode-ray tube in which electric fields are used for both horizontal and vertical deflections of the electron beam, but that is otherwise similar to the tube shown in Fig. 30–12. Figure 29 shows the face of such a tube. The solid straight line results when the electron beam is repeatedly swept left to right by a time-varying electric field. If a uniform magnetic field is applied perpendicularly inward through the face of the tube, one might expect the horizontal line to be shifted or tilted. Which of the four dashed lines in the figure will be the line that will result?

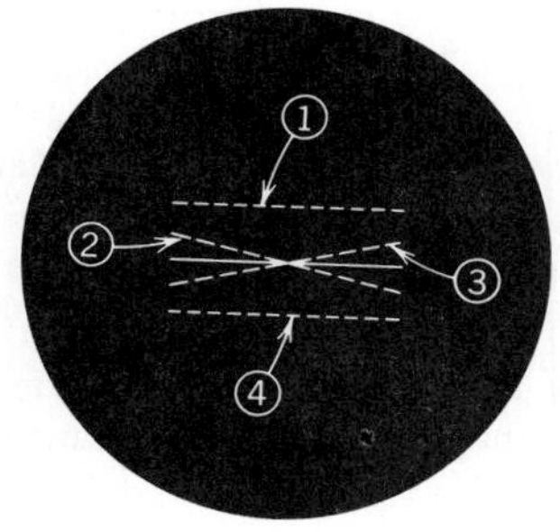

Figure 29 Exercise 9.

10E. A 10-keV electron moving horizontally enters a region of space in which there is a downward-directed electric field of magnitude 10 kV/m. (*a*) What are the magnitude and direction of the (smallest) magnetic field that will allow the electron to continue to move horizontally? Ignore the gravitational force, which is rather small. (*b*) Is it possible for a proton to pass through this combination of fields undeflected? If so, under what circumstances?

11E. An electric field of 1.5 kV/m and a magnetic field of 0.40 T act on a moving electron to produce no force. (*a*) Calculate the minimum electron speed v. (*b*) Draw the vectors **E**, **B**, and **v**.

12P. An electron is accelerated through a potential difference of 1.0 kV and directed into a region between two parallel plates separated by 20 mm with a potential difference of 100 V between them. If the electron enters moving perpendicular to the electric field between the plates, what magnetic field is necessary perpendicular to both the electron path and the electric field so that the electron travels in a straight line?

13P. An ion source is producing ions of ^{6}Li (mass = 6.0 u) each carrying a single positive elementary charge ($+e$). The ions are accelerated by a potential difference of 10 kV and pass horizontally into a region in which there is a vertical magnetic field $B = 1.2$ T. Calculate the strength of the smallest electric field to be set up over the same region that will allow the ^{6}Li ions to pass through undeflected.

Section 30-4 The Hall Effect

14E. Figure 30 shows the cross section of a sample carrying a current directed out of the page. (*a*) Which pair of the four terminals (a, b, c, d) should be used to measure the Hall voltage if the magnetic field is in the $+x$ direction and the charge

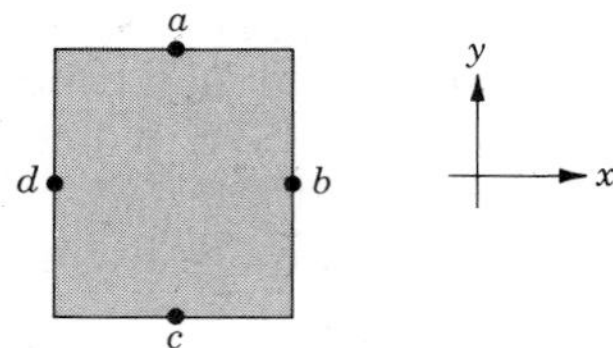

Figure 30 Exercise 14.

carriers are negative? What is the expected polarity of this voltage? (*b*) Repeat if the magnetic field is in the $-y$ direction and the charge carriers are positive. (*c*) Discuss the situation if the magnetic field is in the $+z$ direction.

15E. Show that, in terms of the Hall electric field E and the current density J, the number of charge carriers per unit volume is given by

$$n = \frac{JB}{eE}.$$

16P. In a Hall-effect experiment, a current of 3.0 A lengthwise in a conductor 1.0 cm wide, 4.0 cm long, and 10 μm thick produced a tranverse Hall voltage (across the width) of 10 μV when a magnetic field of 1.5 T is passed perpendicularly through the thin conductor. From these data, find (*a*) the drift velocity of the charge carriers and (*b*) the number density of charge carriers. (*c*) Show on a diagram the polarity of the Hall voltage with a given current and magnetic field direction, assuming the charge carriers are (negative) electrons.

17P. (*a*) Show that the ratio of the Hall electric field E to the electric field E_C responsible for the current is

$$\frac{E}{E_C} = \frac{B}{ne\rho},$$

where ρ is the resistivity of the material. (*b*) Compute the ratio numerically for Sample Problem 2. See Table 1 in Chapter 28.

18P. A metal strip 6.5 cm long, 0.85 cm wide, and 0.76 mm thick moves with constant velocity v through a magnetic field $B = 1.2$ mT perpendicular to the strip, as shown in Fig. 31. A potential difference of 3.9 μV is measured between points x and y across the strip. Calculate the speed v.

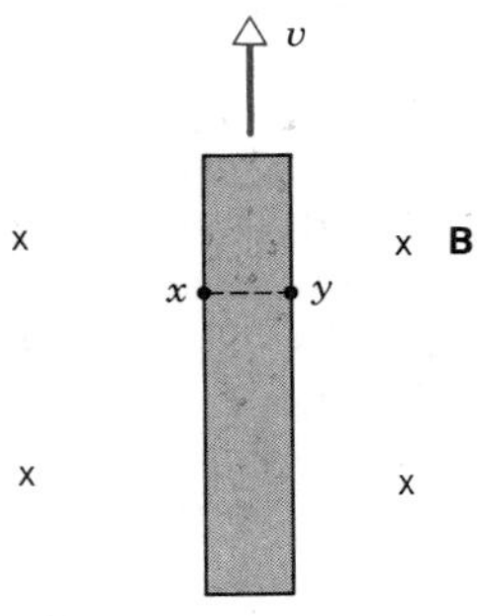

Figure 31 Problem 18.

Section 30-5 A Circulating Charge

19E. Magnetic fields are often used to bend a beam of electrons in physics experiments. What uniform magnetic field, applied perpendicular to a beam of electrons moving at 1.3×10^6 m/s, is required to make the electrons travel in a circular arc of radius 0.35 m?

20E. (*a*) In a magnetic field with $B = 0.50$ T, for what path radius will an electron circulate at 0.10 the speed of light? (*b*) What will be its kinetic energy in eV? Ignore the small relativistic effects.

21E. What uniform magnetic field must be set up in space to permit a proton of speed 1.0×10^7 m/s to move in a circle the size of the earth's equator?

22E. A 1.2-keV electron is circulating in a plane at right angles to a uniform magnetic field. The orbit radius is 25 cm. Calculate (*a*) the speed of the electron, (*b*) the magnetic field, (*c*) the frequency of revolution, and (*d*) the period of the motion.

23E. An electron is accelerated from rest by a potential difference of 350 V. It then enters a uniform magnetic field of magnitude 200 mT, its velocity being at right angles to this field. Calculate (*a*) the speed of the electron and (*b*) the radius of its path in the magnetic field.

24E. *Time-of-flight spectrometer.* S. A. Goudsmit devised a method for measuring accurately the masses of heavy ions by timing their period of revolution in a known magnetic field. A singly charged ion of iodine makes 7.00 rev in a field of 45.0 mT in1.29 ms. Calculate its mass, in atomic mass units. Actually, the mass measurements are carried out to much greater accuracy than these approximate data suggest.

25E. An alpha particle ($q = +2e$, $m = 4.0$ u) travels in a circular path of radius 4.5 cm in a magnetic field with $B = 1.2$ T. Calculate (*a*) its speed, (*b*) its period of revolution, (*c*) its kinetic energy in eV, and (*d*) the potential difference through which it would have to be accelerated to achieve this energy.

26E. (*a*) Find the frequency of revolution of an electron with an energy of 100 eV in the earth's magnetic field of 35 μT. (*b*) Calculate the radius of the path of this electron if its velocity is perpendicular to the magnetic field.

27E. A beam of electrons whose kinetic energy is K emerges from a thin-foil "window" at the end of an accelerator tube. There is a metal plate a distance d from this window and at right angles to the direction of the emerging beam. See Fig. 32. Show that we can prevent the beam from hitting the plate if we apply a magnetic field B such that

$$B \geq \sqrt{\frac{2mK}{e^2 d^2}},$$

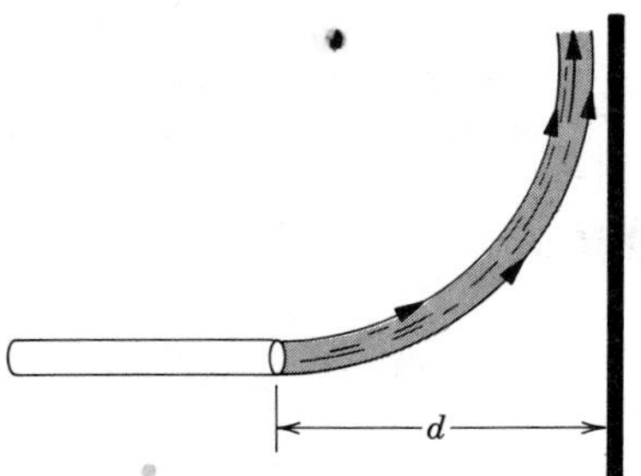

Figure 32 Exercise 27.

in which m and e are the electron mass and charge. How should **B** be oriented?

28P. In a nuclear experiment a 1.0-MeV proton moves in a uniform magnetic field in a circular path. What energy must (*a*) an alpha particle and (*b*) a deuteron have if they are to circulate in the same orbit? (Recall that for an alpha particle $q = +2e$, $m = 4.0$ u.)

29P. A proton, a deuteron, and an alpha particle, accelerated through the same potential difference, enter a region of uniform magnetic field, moving at right angles to **B**. (*a*) Compare their kinetic energies. If the radius of the proton's circular path is 10 cm, what are the radii of (*b*) the deuteron and (*c*) the alpha-particle paths?

30P. A proton, a deuteron, and an alpha particle with the same kinetic energies enter a region of uniform magnetic field, moving at right angles to **B**. Compare the radii of their circular paths.

31P. *Mass spectrometer.* Figure 33 shows an arrangement used to measure the masses of ions. An ion of mass m and charge $+q$ is produced essentially at rest in source S, a chamber in which a gas discharge is taking place. The ion is accelerated by potential difference V and allowed to enter a magnetic field **B**. In the field it moves in a semicircle, striking a photographic plate at distance x from the entry slit. Show that the ion mass m is given by

$$m = \frac{B^2 q}{8V} x^2.$$

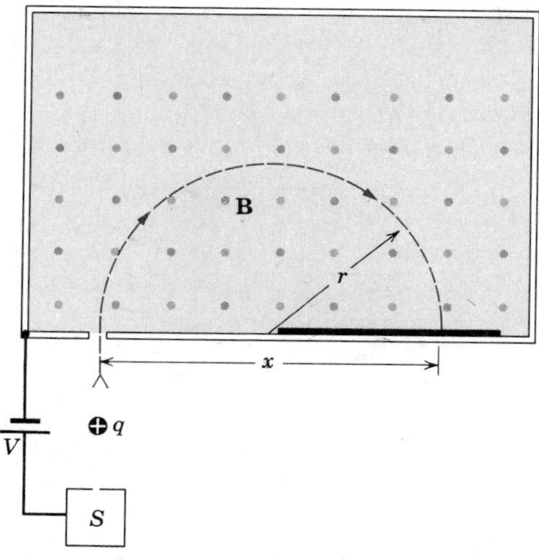

Figure 33 Problems 31, 32 and 33.

32P. Two types of singly-ionized atoms having the same charge q and mass differing by a small amount Δm are introduced into the mass spectrometer described in Problem 31. (*a*) Calculate the difference in mass in terms of V, q, m (of either), B, and the distance Δx between the spots on the photographic plate. (*b*) Calculate Δx for a beam of singly ionized chlorine atoms of masses 35 and 37 u if $V = 7.3$ kV and $B = 0.50$ T.

33P. In a mass spectrometer (see Problem 31) used for commercial purposes, uranium ions of mass 3.92×10^{-25} kg and charge 3.2×10^{-19} C are separated from related species. The ions are first accelerated through a potential difference of 100 kV and then pass into a magnetic field, where they are bent in a path of radius 1.0 m. After traveling through 180°, they are collected in a cup after passing through a slit of width 1.0 mm and a height of 1.0 cm. (*a*) What is the magnitude of the (perpendicular) magnetic field in the separator? If the machine is designed to separate out 100 mg of material per hour, calculate (*b*) the current of the desired ions in the machine and (*c*) the thermal energy dissipated in the cup in one hour.

34P. Bainbridge's mass spectrometer, shown in Fig. 34, separates ions having the same velocity. The ions, after entering through slits S_1 and S_2, pass through a velocity selector composed of an electric field produced by the charged plates P and P', and a magnetic field **B** perpendicular to the electric field and the ion path. Those ions that pass undeviated through the crossed **E** and **B** fields enter into a region where a second magnetic field **B′** exists, and are bent into circular paths. A photographic plate registers their arrival. Show that $q/m = E/(rBB')$, where r is the radius of the circular orbit.

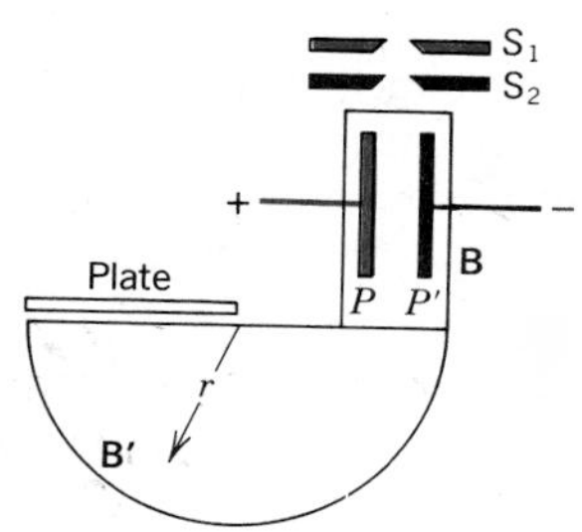

Figure 34 Problem 34.

35P. A 2.0-keV positron is projected into a uniform magnetic field **B** of 0.10 T with its velocity vector making an angle of 89° with **B**. Find (*a*) the period, (*b*) the pitch *p*, and (*c*) the radius *r* of the helical path. See Fig. 12*b*.

36P. A neutral particle is at rest in a uniform magnetic field of magnitude *B*. At time $t = 0$ it decays into two charged particles each of mass *m*. (*a*) If the charge of one of the particles is $+q$, what is the charge of the other? (*b*) The two particles move off in separate paths both of which lie in the plane perpendicular to **B**. At a later time the particles collide. Express the time from decay until collision in terms of *m*, *B*, and *q*.

37P. (*a*) What speed would a proton need to circle the earth at the equator, if the earth's magnetic field is everywhere horizontal there and directed along longitudinal lines? Relativistic effects must be taken into account. Take the magnitude of the earth's magnetic field to be 41 μT at the equator. (*Hint:* Replace the momentum mv in Eq. 16 with the relativistic momentum given in Eq. 20 of Chapter 9.) (*b*) Draw the velocity and magnetic field vectors corresponding to this situation.

Section 30-6 Cyclotrons and Synchrotrons

38E. In a certain cyclotron a proton moves in a circle of radius 0.50 m. The magnitude of the magnetic field is 1.2 T. (*a*) What is the cyclotron frequency? (*b*) What is the kinetic energy of the proton, in eV?

39E. A physicist is designing a cyclotron to accelerate protons to one-tenth the speed of light. The magnet used will produce a field of 1.4 T. Calculate (*a*) the radius of the cyclotron and (*b*) the corresponding oscillator frequency. Relativity considerations are not significant.

40P. The cyclotron of Sample Problem 5 was normally adjusted to accelerate deuterons. (*a*) What energy of protons could it produce, using the same oscillator frequency as that used for deuterons? (*b*) What magnetic field would be required? (*c*) What energy of protons could be produced if the magnetic field was left at the value used for deuterons? (*d*) What oscillator frequency would then be required? (*e*) Answer the same questions for alpha particles, instead of protons. (For an alpha particle, $q = +2e$, $m = 4.0$ u.)

41P. A deuteron in a cyclotron is moving in a magnetic field with $B = 1.5$ T and an orbit radius of 50 cm. Because of a grazing collision with a target, the deuteron breaks up, with a negligible loss of kinetic energy, into a proton and a neutron. Discuss the subsequent motions of each. Assume that the deuteron energy is shared equally by the proton and neutron at breakup.

42P. Estimate the total path length traversed by a deuteron in the cyclotron of Sample Problem 5 during the acceleration process. Assume an accelerating potential between the dees of 80 kV.

Section 30-7 The Magnetic Force on a Current

43E. Figure 35 shows a magnet and a straight wire in which electrons are flowing out of the page at right angles to it. In which case will there be a force on the wire that points toward the top of the page?

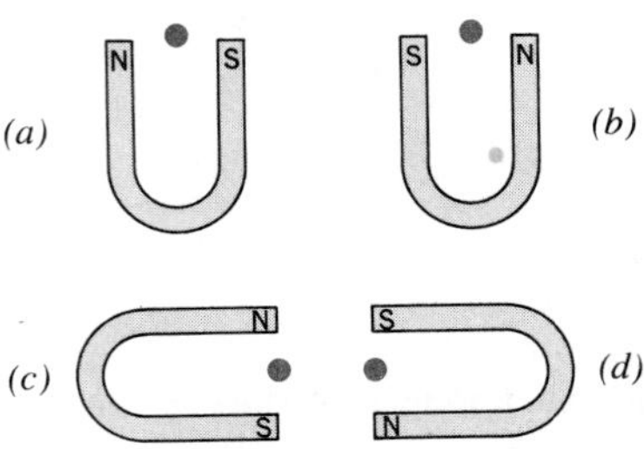

Figure 35 Exercise 43.

44E. A horizontal conductor in a power line carries a current of 5000 A from south to north. The earth's magnetic field (60 μT) is directed toward the north and is inclined downward at 70° to the horizontal. Find the magnitude and direction of the magnetic force on 100 m of the conductor due to the earth's field.

45E. A wire 1.8 m long carries a current of 13 A and makes an angle of 35° with a uniform magnetic field $B = 1.5$ T. Calculate the magnetic force on the wire.

46P. A wire of 62 cm length and mass 13 g is suspended by a pair of flexible leads in a magnetic field of 0.44 T. What are the magnitude and direction of the current required to remove the tension in the supporting leads? See Fig. 36.

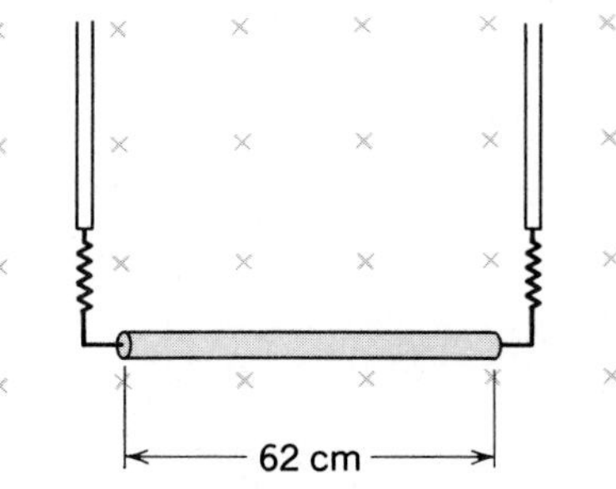

Figure 36 Problem 46.

47P. A wire 50 cm long lying along the *x* axis carries a current of 0.50 A in the positive *x* direction. A magnetic field is present that is given in T by $\mathbf{B} = 0.0030\mathbf{j} + 0.010\mathbf{k}$. Find the components of the force on the wire.

48P. A metal wire of mass *m* slides without friction on two

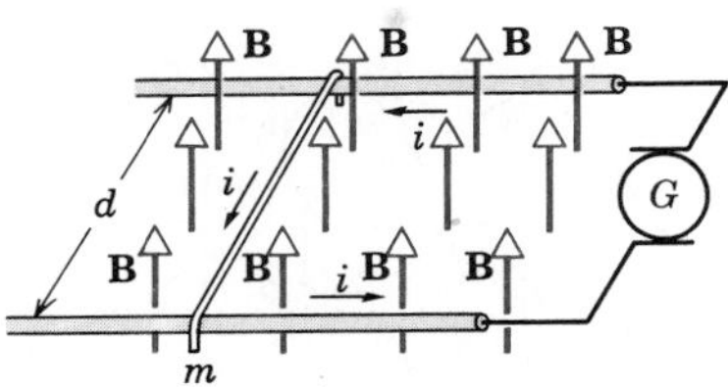

Figure 37 Problem 48.

horizontal rails spaced a distance d apart, as in Fig. 37. The track lies in a vertical uniform magnetic field **B**. A constant current i flows from generator G along one rail, across the wire, and back down the other rail. Find the velocity (speed and direction) of the wire as a function of time, assuming it to be at rest at $t = 0$.

49P. Figure 38 shows a wire of arbitrary shape carrying a current i between points a and b. The wire lies in a plane at right angles to a uniform magnetic field **B**. Prove that the force on the wire is the same as that on a straight wire carrying a current i directly from a to b. (*Hint:* Replace the wire by a series of "steps" parallel and perpendicular to the straight line joining a and b.)

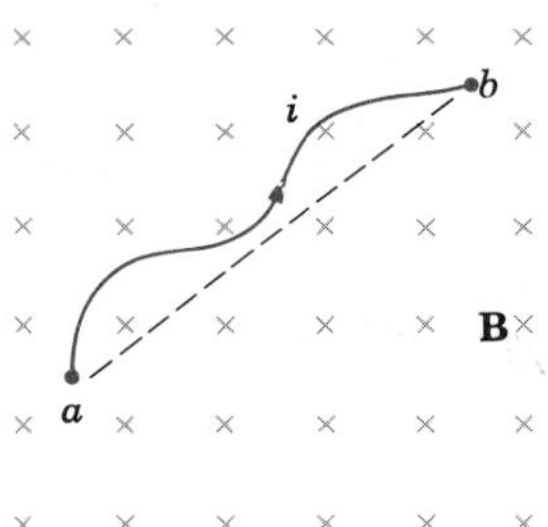

Figure 38 Problem 49.

50P. A long, rigid conductor, lying along the x axis, carries a current of 5.0 A in the $-x$ direction. A magnetic field **B** is present, given by $\mathbf{B} = 3\mathbf{i} + 8x^2\mathbf{j}$, with x in meters and **B** in mT. Calculate the force on the 2.0-m segment of the conductor that lies between $x = 1.0$ m and $x = 3.0$ m.

51P. Consider the possibility of a new design for an electric train. The engine is driven by the force due to the vertical component of the earth's magnetic field on a conducting axle. Current is passed down one rail, into a conducting wheel, through the axle, through another conducting wheel, and then back to the source via the other rail. (*a*) What current is needed to provide a modest 10-kN force? Take the vertical component of the earth's field to be 10 μT and the length of the axle to be 3.0 m. (*b*) How much power would be lost for each ohm of resistance in the rails? (*c*) Is such a train totally unrealistic or just marginally unrealistic?

52P. A 1.0-kg copper rod rests on two horizontal rails 1.0 m apart and carries a current of 50 A from one rail to the other. The coefficient of static friction is 0.60. What is the smallest magnetic field (not necessarily vertical) that would cause the bar to slide?

Section 30-8 Torque on a Current Loop

53E. A single-turn current loop, carrying a current of 4.0 A, is in the shape of a right triangle with sides 50 cm, 120 cm, and 130 cm. The loop is in a uniform magnetic field of magnitude 75 mT whose direction is parallel to the current in the 130-cm side of the loop. (*a*) Find the magnetic force on each of the three sides of the loop. (*b*) Show that the total magnetic force on the loop is zero.

54E. Figure 39 shows a rectangular, 20-turn loop of wire, 10 cm by 5.0 cm. It carries a current of 0.10 A and is hinged at one side. It is mounted with its plane at an angle of 30° to the direction of a uniform magnetic field of 0.50 T. Calculate the torque about the hinge line acting on the loop.

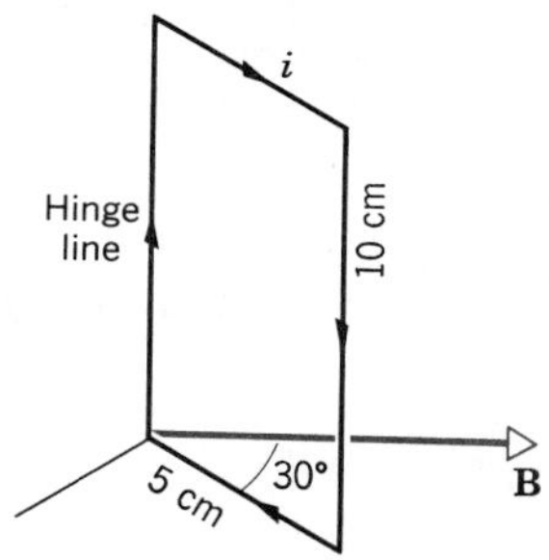

Figure 39 Exercise 54.

55E. A stationary, circular wall clock has a face with a radius of 15 cm. Six turns of wire are wound around its perimeter; the wire carries a current 2.0 A in the clockwise direction. The clock is located where there is a constant, uniform external magnetic field of 70 mT (but the clock still keeps perfect time). At exactly 1:00 p.m., the hour hand of the clock points in the direction of the external magnetic field. (*a*) After how many minutes will the minute hand point in the direction of the torque on the winding due to the magnetic field? (*b*) What is the magnitude of this torque?

56P. A length L of wire carries a current i. Show that if the wire is formed into a circular coil, the maximum torque in a given magnetic field is developed when the coil has one turn only and the maximum torque has the magnitude

$$\tau = \frac{1}{4\pi} L^2 iB.$$

57P. Prove that the relation $\tau = NiAB \sin \theta$ holds for closed loops of arbitrary shape and not only for rectangular loops as in Fig. 26. (*Hint:* Replace the loop of arbitrary shape by an assembly of adjacent long, thin, approximately rectangular, loops that are nearly equivalent to it as far as the distribution of current is concerned.)

58P. A closed loop of wire carries a current i. The loop is in a uniform magnetic field **B**. Show that the total magnetic force on the loop is zero. Does your proof also hold for a nonuniform magnetic field?

59P. Figure 40 shows a wire ring of radius a at right angles to the general direction of a radially-symmetric diverging magnetic field. The magnetic field at the ring is everywhere of the same magnitude B, and its direction at the ring is everywhere at an angle θ with a normal to the plane of the ring. The twisted lead wires have no effect on the problem. Find the magnitude and direction of the force the field exerts on the ring if the ring carries a current i as shown in the figure.

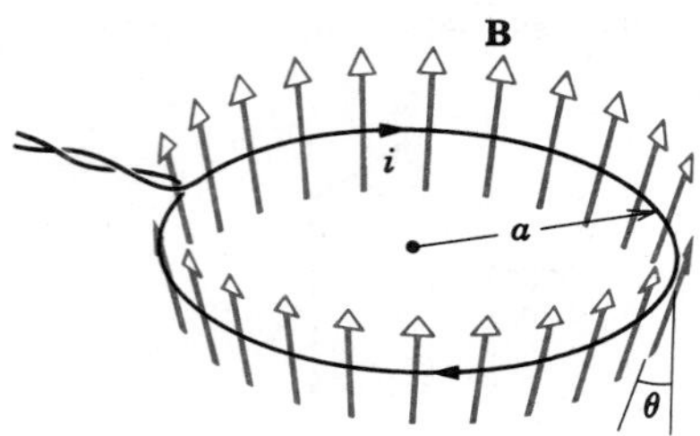

Figure 40 Problem 59.

60P. A certain galvanometer has a resistance of 75.3 Ω; its needle experiences a full-scale deflection when a current of 1.62 mA passes through its coil. (*a*) Determine the value of the auxiliary resistance required to convert the galvanometer into a voltmeter that reads 1.00 V at full-scale deflection. How is it to be connected? (*b*) Determine the value of the auxiliary resistance required to convert the galvanometer into an ammeter that reads 50.0 mA at full-scale deflection. How is it to be connected?

61P. Figure 41 shows a wooden cylinder with a mass $m = 0.25$ kg and a length $L = 0.10$ m, with $N = 10$ turns of wire wrapped around it longitudinally, so that the plane of the wire loop contains the axis of the cylinder. What is the least current through the loop that will prevent the cylinder from rolling down a plane inclined at an angle θ to the horizontal, in the presence of a vertical, uniform magnetic field of 0.50 T, if the plane of the windings is parallel to the inclined plane?

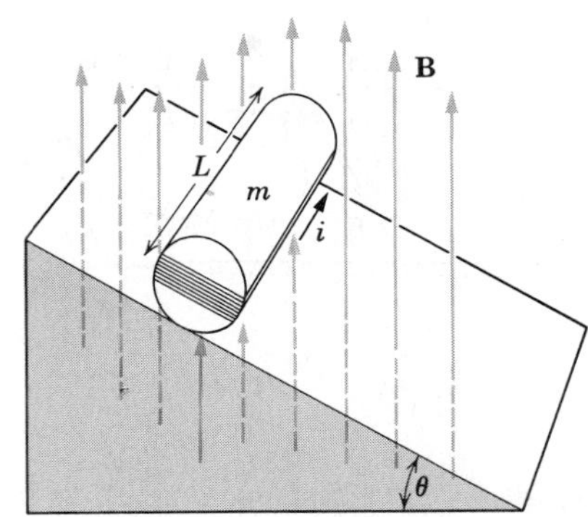

Figure 41 Problem 61.

Section 30-9 A Magnetic Dipole

62E. A circular coil of 160 turns has a radius of 1.9 cm. (*a*) Calculate the current that results in a magnetic moment of 2.3 $A \cdot m^2$. (*b*) Find the maximum torque that the coil, carrying this current, can experience in a uniform 35-mT magnetic field.

63E. The magnetic dipole moment of the earth is 8.0×10^{22} J/T. Assume that this is produced by charges flowing in the molten outer core of the earth. If the radius of the circular path is 3500 km, calculate the required current.

64E. A circular wire loop whose radius is 15 cm carries a current of 2.6 A. It is placed so that the normal to its plane makes an angle of 41° with a uniform magnetic field of 12 T. (*a*) Calculate the magnetic dipole moment of the loop. (*b*) What torque acts on the loop?

65E. A single-turn current loop, carrying a current of 5.0 A, is in the shape of a right triangle with sides 30, 40, and 50 cm. The loop is in a uniform magnetic field of magnitude 80 mT whose direction is parallel to the current in the 50-cm side of the loop. Find the magnitude of (*a*) the magnetic dipole moment of the loop and (*b*) the torque on the loop.

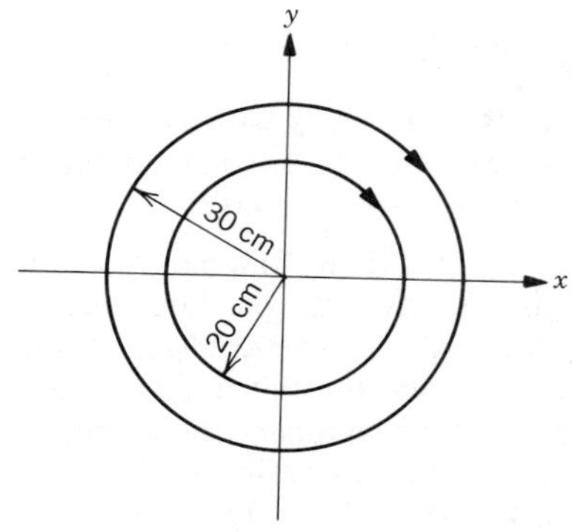

Figure 42 Exercise 66.

66E. Two concentric circular loops (radii 20 cm and 30 cm) in the xy plane each carry a clockwise current of 7.0 A, as shown in Fig. 42. (a) Find the net magnetic moment of this system. (b) Repeat if the current in the inner loop is reversed.

67P. A circular loop of wire having a radius of 8.0 cm carries a current of 0.20 A. A unit vector parallel to the dipole moment $\boldsymbol{\mu}$ of the loop is given by $0.60\mathbf{i} - 0.80\mathbf{j}$. If the loop is located in a magnetic field given in T by $\mathbf{B} = 0.25\mathbf{i} + 0.30\mathbf{k}$, find ($a$) the magnitude and direction of the torque on the loop and (b) the magnetic potential energy of the loop.

68P. Figure 43 shows a current loop $ABCDEFA$ carrying a current $i = 5.0$ A. The sides of the loop are parallel to the coordinate axes, with $AB = 20$ cm, $BC = 30$ cm, and $FA = 10$ cm. Calculate the magnitude and direction of the magnetic dipole moment of this loop. (*Hint:* Imagine equal and opposite currents i in the line segment AD, then treat the two rectangular loops $ABCDA$ and $ADEFA$.)

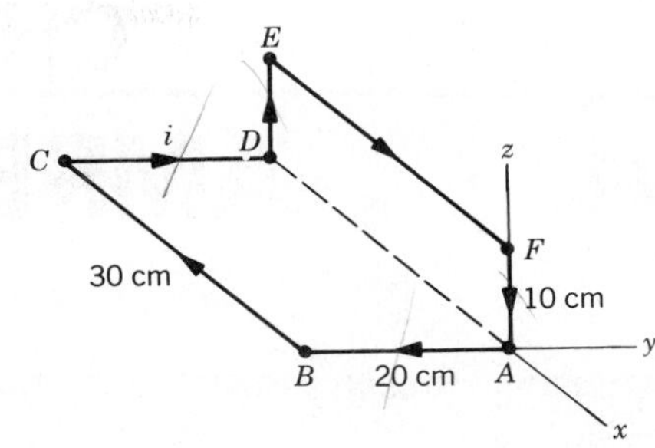

Figure 43 Problem 68.

CHAPTER 31

AMPERE'S LAW

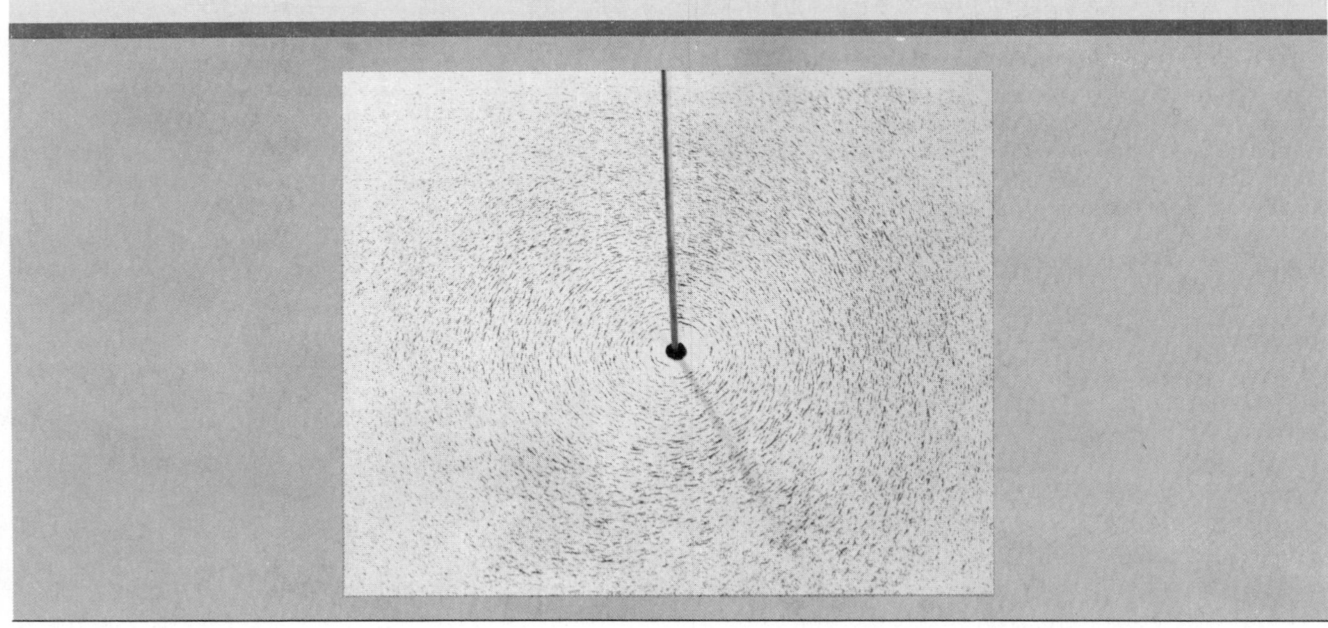

In 1820, the Danish physicist Hans Christian Oersted discovered that a current in a wire causes a nearby compass needle to deflect. In modern language, we say that a current in a wire sets up a magnetic field *in the surrounding space, as the iron filings in this photo attest. Oersted's discovery was a real breakthrough in a deliberate search for a connection between the then separate sciences of electricity and magnetism.*

31-1 Current and the Magnetic Field

A basic fact of *electrostatics* is that two charges exert forces on each other. Earlier, we wrote

$$\text{charge} \rightleftharpoons \text{electric field} \rightleftharpoons \text{charge} \qquad (1)$$

in which we introduced the electric field **E** as an intermediary in this interaction. Equation 1 suggests that (1) charges generate electric fields and (2) electric fields exert forces on charges.

A basic fact of *magnetism* is that two parallel wires carrying currents also exert forces on each other. By analogy with Eq. 1, we can write

$$\text{current} \rightleftharpoons \text{magnetic field} \rightleftharpoons \text{current} \qquad (2)$$

in which we introduce the magnetic field **B** as an intermediary. Equation 2 suggests that (1) currents generate magnetic fields and (2) magnetic fields exert forces on currents. We dealt with the second part of this interaction in Chapter 30. We deal with the first part in this chapter.

31-2 Calculating the Magnetic Field

The central question of this chapter is

> *How can you calculate the magnetic field that a given distribution of currents sets up in the surrounding space?*

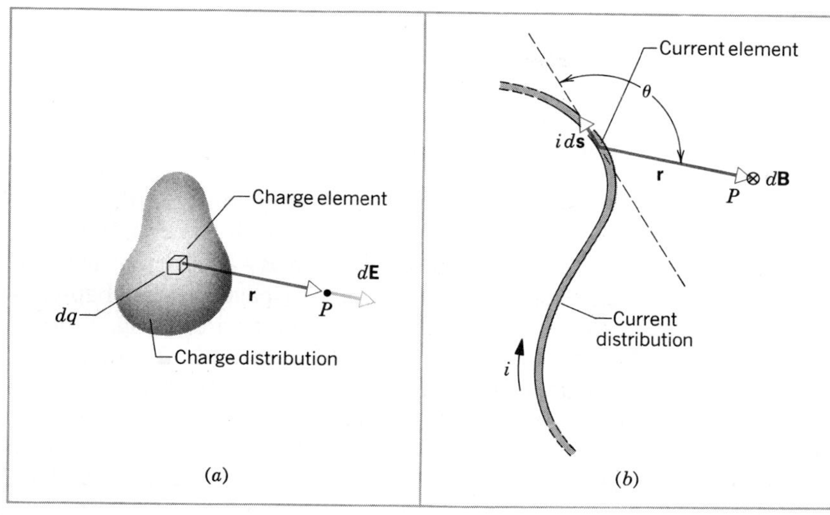

Figure 1 (*a*) A charge element dq establishes a differential electric field element $d\mathbf{E}$ at point P. (*b*) A current element $i\,ds$ establishes a differential magnetic field $d\mathbf{B}$ at point P. The symbol ⊗ (the tail of an arrow) shows that the element $d\mathbf{B}$ points *into* the page.

Let us recall the equivalent central question in electrostatics:

How can you calculate the electric field that a given distribution of charges sets up in the surrounding space?

We learned to do this in Chapter 24, for static charge distributions such as a uniform sphere, line, ring, or disk. Our approach was to divide the charge distribution into charge elements dq as in Fig. 1*a*. We then calculated the field $d\mathbf{E}$ set up by a given charge element at an arbitrary field point P. Finally, we calculated $\mathbf{E}$ at point P by integrating $d\mathbf{E}$ over the entire charge distribution.

The *magnitude* of $d\mathbf{E}$ in such calculations is given by

$$dE = \left(\frac{1}{4\pi\epsilon_0}\right)\frac{dq}{r^2} \qquad (3)$$

in which r is the distance from the charge element to point P. Equation 3 is essentially Coulomb's law, the quantity in the parentheses being a familiar constant. For a positive element of charge, the *direction* of $d\mathbf{E}$ is that of $\mathbf{r}$, where $\mathbf{r}$ is the vector pointing *from* the charge element dq *to* the field point P.

We can express both the magnitude and the direction of $d\mathbf{E}$ by writing Eq. 3 in vector form, as

$$d\mathbf{E} = \left(\frac{1}{4\pi\epsilon_0}\right)\frac{dq}{r^3}\,\mathbf{r} \quad \text{(Coulomb's law)}, \qquad (4)$$

which shows formally that, for a positive element of charge, the direction of $d\mathbf{E}$ is the direction of $\mathbf{r}$. It may seem that Coulomb's law has suddenly become an inverse cube law—rather than an inverse square law—but that is not the case. The exponent 3 in the denominator is needed because we have added a factor of magnitude r in the numerator; Eq. 4 is still an inverse square law.

In the magnetic case, we proceed by analogy. Figure 1*b* shows a wire of arbitrary shape carrying a current i. What is the magnetic field $\mathbf{B}$ at an arbitrary field point P near this wire? We first break up the wire into differential current elements $i\,d\mathbf{s}$, corresponding to the charge elements dq of Fig. 1*a*. Here the vector $d\mathbf{s}$ is a differential element of length, pointing tangent to the wire in the direction of the current. We note at once a complexity in our analogy to the electrostatic case: The differential charge element dq is a *scalar* but the differential current element $i\,d\mathbf{s}$ is a *vector.*

The *magnitude* of the magnetic field contribution set up at point P by a given current element is

$$dB = \frac{\mu_0}{4\pi}\frac{i\,ds\sin\theta}{r^2}. \qquad (5)$$

Here μ_0 is a constant, called the *permeability constant,* whose value is exact, by definition,

$$\mu_0 = 4\pi \times 10^{-7}\ \mathrm{T\cdot m/A} \approx 1.26 \times 10^{-6}\ \mathrm{T\cdot m/A}. \qquad (6)$$

This constant plays a role in magnetic problems much

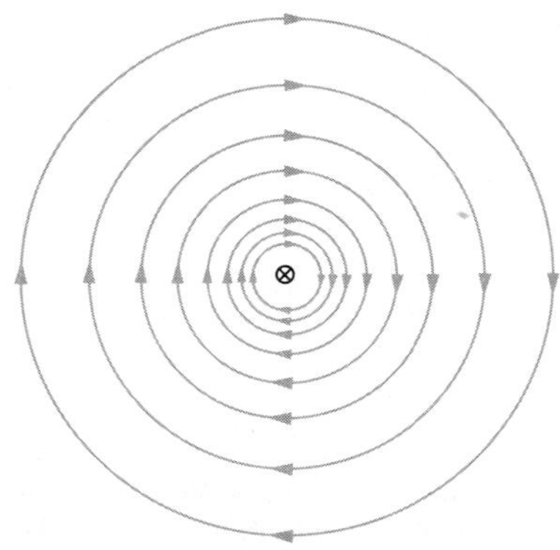

Figure 2 The lines of the magnetic field for a current i in a long straight wire are concentric circles. Their direction is given by a right-hand rule. Note that the current is directed into the page.

like the role that the permittivity constant ϵ_0 plays in electrostatic problems.

The *direction* of $d\mathbf{B}$ in Fig. 1*b* is that of the vector $d\mathbf{s} \times \mathbf{r}$, where $\mathbf{r}$ is a vector that points *from* the current element *to* the point P at which you wish to know the field. The symbol ⊗ in Fig. 1*b* (representing the tail of an arrow) shows that $d\mathbf{B}$ at point P is directed into the plane of the page at right angles. We can write the expression for $d\mathbf{B}$ in vector form as

$$\boxed{d\mathbf{B} = \left(\frac{\mu_0}{4\pi}\right)\frac{i\, d\mathbf{s} \times \mathbf{r}}{r^3}} \quad \text{(Biot–Savart law).} \quad (7)$$

Here again, Eq. 7 remains an inverse square law because a factor of magnitude r in the numerator cancels one of the factors of r in the denominator. Equation 7, which we can call Coulomb's law for magnetism, is more often called *the law of Biot and Savart.** It contains within its structure information about both the magnitude and the direction of the differential field element $d\mathbf{B}$.

Equation 4 is our basic tool for calculating the electric field set up by a given distribution of charge. In the same way, Eq. 7 is our basic tool for calculating the magnetic field set up at various points by a given distribution of current.

A Long Straight Wire. The simplest problem in magnetism is that of a long straight wire carrying a current i. Below, we shall use the law of Biot and Savart to prove that the magnitude of the magnetic field at a perpendicular distance r from such a wire is given by

$$\boxed{B(r) = \frac{\mu_0 i}{2\pi r}} \quad \text{(long straight wire).} \quad (8)$$

The magnitude of $\mathbf{B}$ depends only on the current and on radial distance r from the wire. We shall show in our derivation that the lines of $\mathbf{B}$—consistent with the picture shown on the opening page of this chapter—form concentric circles around the wire, as Fig. 2 shows. The increase in spacing of the lines in Fig. 2 with increasing distance from the wire represents the $1/r$ decrease in the magnitude of $\mathbf{B}$ predicted by Eq. 8.

An All-Purpose Right-Hand Rule. Here is a simple right-hand rule for finding the direction of the magnetic field set up by a current in a long wire:

Grasp the wire in your right hand with your extended thumb pointing in the direction of the current. Your fingers will then naturally curl around in the direction of the magnetic field lines.

This rule is a specific application of a more general right-hand rule that we shall find useful. In magnetism we shall find many situations with cylindrical symmetry in which there is what we can call a "curly element" (in this case, the circular magnetic field lines) and a "straight element" (in this case, the current in the wire). Your right hand also has a "curly element" (your fingers) and a "straight element" (your extended thumb.) By matching your right hand to the physical situation, you can find the directions in which various physical quantities point. We shall clarify this general rule by applying it to new situations as they arise.

Proof of Equation 8. Figure 3, which is just like Fig. 1*b* except that the wire is straight, illustrates the problem. The differential magnetic field set up at point P by the current element $i\, d\mathbf{s}$ is given in magnitude by Eq. 5, or

$$dB = \frac{\mu_0}{4\pi}\frac{i\, ds \sin\theta}{r^2}. \quad (9)$$

The direction of $d\mathbf{B}$ in Fig. 3 is that of the vector $d\mathbf{s} \times \mathbf{r}$, namely, into the plane of the figure at right angles.

Note that $d\mathbf{B}$ at point P has this same direction for every current element into which the wire can be divided. Thus, to find the total magnetic field $\mathbf{B}$ at point P, we integrate Eq. 9, obtaining

$$B = \int dB = \frac{\mu_0 i}{4\pi}\int_{s=-\infty}^{s=+\infty}\frac{\sin\theta\, ds}{r^2}. \quad (10)$$

* Rhymes with "Leo and bazaar."

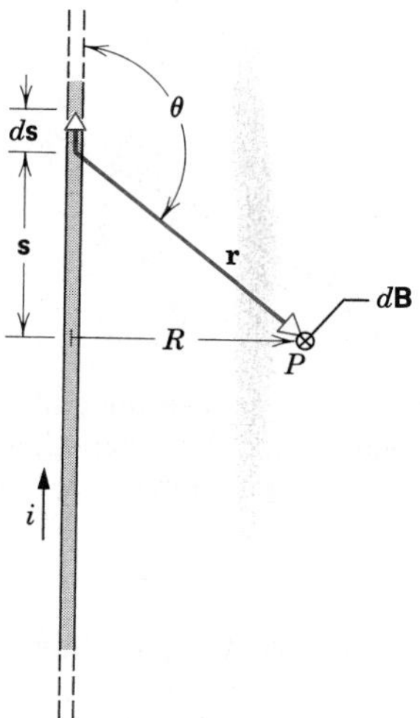

Figure 3 Calculating the magnetic field set up by a current i in a long straight wire. The field $d\mathbf{B}$ associated with the current element $i\,ds$ points into the page, as shown.

The variables θ, s, and r in this equation are not independent, being related by

$$\sin\theta = \sin(\pi - \theta) = \frac{R}{\sqrt{s^2 + R^2}}$$

and

$$r = \sqrt{s^2 + R^2}.$$

With these substitutions, Eq. 10 becomes

$$B = \frac{\mu_0 i}{4\pi}\int_{-\infty}^{+\infty}\frac{R}{(s^2+R^2)^{3/2}}\,ds = \frac{\mu_0 i}{4\pi R}\left|\frac{s}{(s^2+R^2)^{1/2}}\right|_{-\infty}^{+\infty}$$

$$= \frac{\mu_0}{2\pi}\frac{i}{R}.$$

With a small change in notation, we have Eq. 8, the relation we set out to prove.

Sample Problem 1 The magnetic field a distance $r = 2.3$ cm from the axis of a long straight wire is 13 mT. What is the current in the wire? The earth's magnetic field has a strength of only about 0.1 mT so that we may ignore its influence in this problem.

Solving Eq. 8 for i, we obtain

$$i = \frac{2\pi Br}{\mu_0} = \frac{(2\pi)(13\times10^{-3}\ \text{T})(2.3\times10^{-2}\ \text{m})}{4\pi\times10^{-7}\ \text{T}\cdot\text{m/A}}$$

$$= 1500\ \text{A}. \qquad \text{(Answer)}$$

31-3 The Magnetic Force on a Wire Carrying a Current

We recall from Section 30-7 that a section of a long straight wire of length L, carrying a current i and placed in a uniform magnetic field, experiences a sideways deflecting force given by

$$\mathbf{F} = i\,\mathbf{L}\times\mathbf{B}_{\text{ext}}. \qquad (11)$$

We introduce the subscript in $\mathbf{B}_{\text{ext}}$ to remind ourselves that the magnetic field in this force equation must be the field set up by some external agent, an electromagnet of the type shown in Fig. 3 of Chapter 30 being one possibility.

In particular, the *external* field that appears in Eq. 11 must be distinguished carefully from the *intrinsic* field $\mathbf{B}_{\text{intr}}$ that is set up by the current in the wire itself. In Fig. 2, for example, the field shown is the intrinsic field resulting from the current in the wire. There is no external field in that figure so that, according to Eq. 11, no magnetic deflecting force acts on the wire. In the corresponding electrostatic case, the electric field set up by a point charge (which we could have called $\mathbf{E}_{\text{intr}}$) exerted no electric force on that charge.

Figure 4 shows the lines of the *resultant* magnetic field $\mathbf{B}$ associated with a current in a wire that is oriented at right angles to a uniform external magnetic field $\mathbf{B}_{\text{ext}}$. At any point, the resultant field $\mathbf{B}$ will be the vector sum

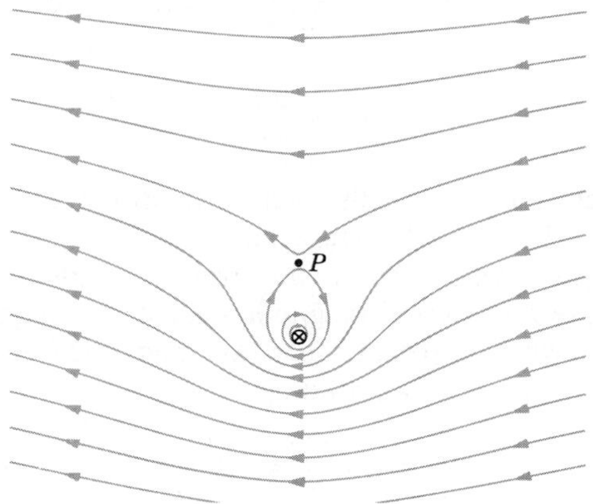

Figure 4 A long straight wire carrying a current i into the page is immersed in a uniform external magnetic field $\mathbf{B}_{\text{ext}}$. The field lines shown represent the resultant field formed by combining vectorially at each point the uniform external field and the intrinsic field associated with the current in the wire.

of $\mathbf{B}_{ext}$ and $\mathbf{B}_{intr}$, or

$$\mathbf{B} = \mathbf{B}_{ext} + \mathbf{B}_{intr}. \tag{12}$$

These two fields tend to cancel each other above the wire and to reinforce each other below it. At point P in Fig. 4, $\mathbf{B}_{ext}$ and $\mathbf{B}_{intr}$ cancel exactly. Very near the wire, $\mathbf{B}_{intr}$ dominates and the field lines are closely represented by concentric circles, like those of Fig. 2. Far from the wire, $\mathbf{B}_{ext}$ dominates and the field lines are represented closely by uniformly spaced parallel lines.

Michael Faraday, who originated the concept of lines of force, endowed them with more reality than we currently give them. He imagined that, like stretched rubber bands, they represent the site of mechanical forces. Using Faraday's analogy, can you not readily believe that the wire in Fig. 4 will be deflected upward? Verify that it will, using Eq. 11.

31-4 Two Parallel Conductors

Two long parallel wires carrying currents exert forces on each other. Figure 5 shows two such wires, separated by a distance d and carrying currents i_a and i_b. Let us analyze the forces that these wires exert on each other, in terms of Eq. 2:

$$\text{current} \rightleftharpoons \text{magnetic field} \rightleftharpoons \text{current}.$$

Wire a in Fig. 5 produces a magnetic field $\mathbf{B}_a$ at all points. The magnitude of $\mathbf{B}_a$ at the site of wire b is, from Eq. 8,

$$B_a = \frac{\mu_0 i_a}{2\pi d}. \tag{13}$$

The right-hand rule tells us that the direction of $\mathbf{B}_a$ at wire b is down, as the figure shows.

Wire b, which carries a current i_b, finds itself immersed in an external magnetic field $\mathbf{B}_a$. A length L of this wire will experience a sideways magnetic force given by Eq. 11, whose magnitude is

$$F_{ba} = i_b L B_a = \frac{\mu_0 L i_b i_a}{2\pi d}. \tag{14}$$

The rule for vector products tells us that $\mathbf{F}_{ba}$ lies in the plane of the wires and points to the left in Fig. 5.

We could equally well have computed the force on wire a by determining the magnetic field that wire b produces at the site of wire a. For parallel currents, this force would point to the right in Fig. 5, which means that

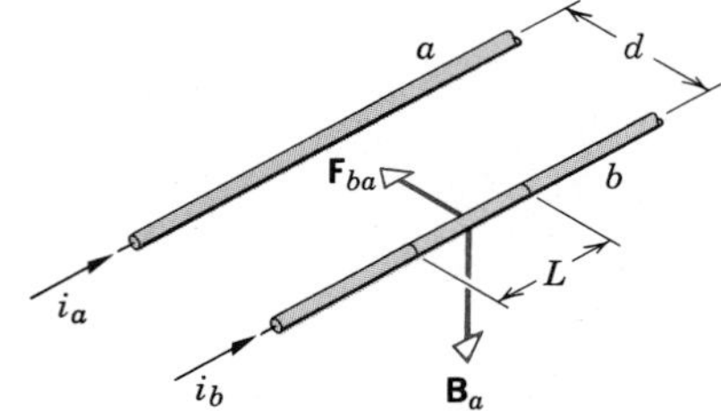

Figure 5 Two parallel wires carrying currents in the same direction attract each other. The magnetic field at wire b set up by the current in wire a is shown.

the two wires would attract each other. Note that (1) the forces on the two wires form an action–reaction pair and (2) the *external* field in which either wire finds itself is the *intrinsic* field of the other wire.

You should be able to show that, for antiparallel currents, the two wires repel each other. The rule is:

> *Parallel currents attract and antiparallel currents repel.*

This rule is opposite to the rule for the forces between charges in this sense: Although like (parallel) currents attract each other, like (same sign) charges repel each other.

The force acting between currents in parallel wires is the basis for the definition of the ampere, which is one of the seven SI base units.* The definition, adopted in 1946, is:

> *The ampere is that constant current which, if maintained in two straight parallel conductors of infinite length, of negligible circular cross section, and placed 1 meter apart in vacuum, would produce on each of these conductors a force equal to 2×10^{-7} newtons per meter of length.*

In practice, multiturn coils of carefully controlled geometries are substituted for the "conductors of infinite length" of the definition.

Sample Problem 2 Show that Eq. 14 is consistent with the definition of the ampere given above.

Let us put $i_a = i_b = 1$ A and $d = 1$ m in that equation. We

* See Appendix A for a list of the SI base units, along with their definitions.

find

$$\frac{F}{L}=\frac{\mu_0 i_a i_b}{2\pi d}=\frac{(4\pi\times10^{-7}\ \mathrm{T\cdot m/A})(1\ \mathrm{A})(1\ \mathrm{A})}{(2\pi)(1\ \mathrm{m})}$$

$$=2\times10^{-7}\ \mathrm{T\cdot A}=2\times10^{-7}\ \mathrm{N/m}. \qquad \text{(Answer)}$$

The unit transformation is helped by a study of the units in Eq. 14, in which we see that

$$1\ \mathrm{N}=1\ \mathrm{T\cdot A\cdot m},$$

or $1\ \mathrm{T\cdot A}=1\ \mathrm{N/m}$.

Sample Problem 3 Two long parallel wires a distance $2d$ apart carry equal currents i in opposite directions, as shown in Fig. 6*a*. Derive an expression for $B(x)$, the magnitude of the resultant magnetic field for points at a distance x from the point midway between the wires, as shown.

Study of the figure and the use of the right-hand rule show that the fields set up by the currents in the individual wires point in the same direction for all points between the wires. From Eq. 8 we then have

$$B(x)=B_a(x)+B_b(x)=\frac{\mu_0 i}{2\pi(d+x)}+\frac{\mu_0 i}{2\pi(d-x)}$$

$$=\frac{\mu_0 i d}{\pi(d^2-x^2)}. \qquad \text{(Answer)} \quad (15)$$

Inspection of this relation shows that (1) $B(x)$ is symmetrical about the midpoint ($x=0$); (2) $B(x)$ has its minimum value ($=\mu_0 i/\pi d$) at this point; and (3) $B(x)\to\infty$ as $x\to\pm d$. At these locations, the point P in Fig. 6*a* is within the wires on their axes. Our derivation of Eq. 8, however, is valid only for points outside the wires so that Eq. 15 above holds only up to the surface of the wires.*

Figure 6*b* shows a plot of Eq. 15 for $i=25$ A and $2d=50$ mm. We leave it as an exercise to show that Eq. 15 holds also for points beyond the wires, that is, for points with $|x|>d$.

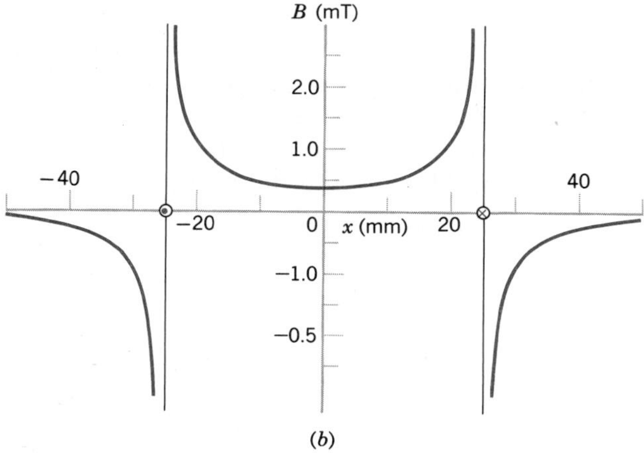

Figure 6 Sample Problem 3. (*a*) Two parallel wires carry currents of the same magnitude in opposite directions. At points between the wires, such as *P*, the magnetic fields for the separate currents point in the same direction. (*b*) A plot of $B(x)$ for $i=25$ A and a wire separation of 50 mm.

Sample Problem 4 Figure 7*a* shows two long parallel wires carrying currents i_1 and i_2 in the directions shown. What are the magnitude and the direction of the resultant magnetic field at point P? Assume the following values: $i_1=15$ A, $i_2=32$ A, and $d=5.3$ cm.

Figure 7*b* shows the individual magnetic fields $\mathbf{B}_1$ and $\mathbf{B}_2$ set up by the currents i_1 and i_2, respectively. Verify that their

* Actually, for points on the axes of the long straight wires of Fig. 6, symmetry considerations lead us to conclude that the intrinsic magnetic field must be zero. This conclusion follows because there is no preferred direction for the field at such points; see Fig. 11.

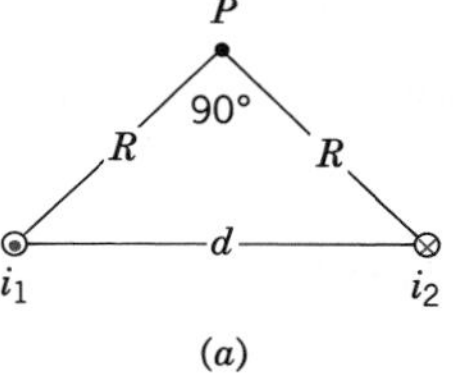

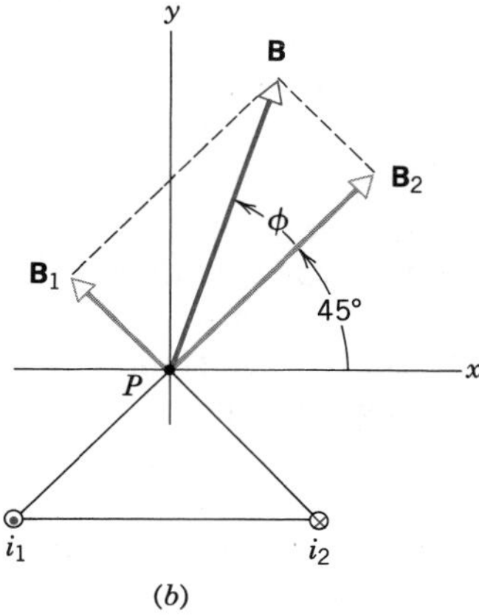

Figure 7 Sample Problem 4. (*a*) Two wires carry currents in opposite directions. What is the magnetic field at *P*? (*b*) The separate fields combine vectorially to yield the resultant field **B**.

directions are correctly given by the right-hand rule. The magnitudes of these fields are given by Eq. 8 as

$$B_1 = \frac{\mu_0 i_1}{2\pi R} = \frac{\mu_0 i_1}{2\pi(d/\sqrt{2})} = \frac{\sqrt{2}\mu_0}{2\pi d} i_1$$

and

$$B_2 = \frac{\mu_0 i_1}{2\pi R} = \frac{\mu_0 i_2}{2\pi(d/\sqrt{2})} = \frac{\sqrt{2}\mu_0}{2\pi d} i_2,$$

in which we have replaced R by its equal, $d/\sqrt{2}$.

The magnitude of the resultant magnetic field **B** is

$$B = \sqrt{B_1^2 + B_2^2} = \frac{\sqrt{2}\mu_0}{2\pi d}\sqrt{i_1^2 + i_2^2}$$

$$= \frac{(\sqrt{2})(4\pi \times 10^{-7}\ \text{T}\cdot\text{m/A})\sqrt{(15\ \text{A})^2 + (32\ \text{A})^2}}{(2\pi)(5.3 \times 10^{-2}\ \text{m})}$$

$$= 1.89 \times 10^{-4}\ \text{T} \approx 190\ \mu\text{T}. \qquad \text{(Answer)}$$

The angle ϕ between **B** and $\mathbf{B}_2$ in Fig. 7*b* follows from

$$\phi = \tan^{-1}\frac{B_1}{B_2} = \tan^{-1}\frac{i_1}{i_2}$$

$$= \tan^{-1}\frac{15\ \text{A}}{32\ \text{A}} = 25°.$$

The angle between **B** and the x axis is then

$$\phi + 45° = 25° + 45° = 70°. \qquad \text{(Answer)}$$

31-5 Ampere's Law

In electrostatics, we can use Coulomb's law—that workhorse of electrostatics—to calculate the electric field caused by any charge distribution. For complex distributions, we may have to resort to a computer but we can always get a numerical answer to any accuracy we wish. However, when we draw together the laws of electromagnetism in Table 2 of Chapter 37 (Maxwell's equations), we do not represent the field of electrostatics by Coulomb's law but rather by Gauss' law. In electrostatics, where the charges are stationary or only slowly moving, these two laws are equivalent. However, Gauss' law is more compatible in form with the other equations of electromagnetism than is Coulomb's law and allows us to solve electric field problems of appropriately high symmetry with ease and elegance.*

The situation in magnetism is similar. We can calculate the magnetic field caused by any current distribution, using the law of Biot and Savart—the magnetic equivalent of Coulomb's law. Again, in difficult cases, we may have to resort to a numerical calculation, using a computer. However, if we turn to Table 2 in Chapter 37 and examine the collected equations of electromagnetism (Maxwell's equations), we do not find the law of Biot and Savart among them. In its place we find *Ampere's law,* first advanced by Andre Marie Ampère (1775–1836) for whom the SI unit of current is named. Both Ampere's law and the law of Biot and Savart are relations between a current distribution and the magnetic field that it generates. Ampere's law, however, has a simplicity and form that makes it more compatible with the other equations of electromagnetism and—in the spirit of Gauss' law—allows us to solve magnetic field problems of appropriately high symmetry with ease and elegance.

Our plan is to display Ampere's law:

$$\oint \mathbf{B} \cdot d\mathbf{s} = \mu_0 i \qquad \text{(Ampere's law)} \qquad (16)$$

and then to become familiar with it by using it. Ampere's law is applied to an arbitrary closed loop, called an *Amperian loop;* the circle on the integral sign indicates that the quantity $\mathbf{B} \cdot d\mathbf{s}$ is to be integrated around that (closed) loop. The current i in Eq. 16 is the net current enclosed by the loop. Speaking loosely, Ampere's law relates the distribution of the magnetic field at points on the loop to the current that passes through the loop.

Let us examine Ampere's law by seeing how to apply it in the situation of Fig. 8. The figure shows the cross sections of three long straight wires that pierce the plane of the page at right angles to it. The wires carry currents i_1, i_2, and i_3 in the directions shown. The arbi-

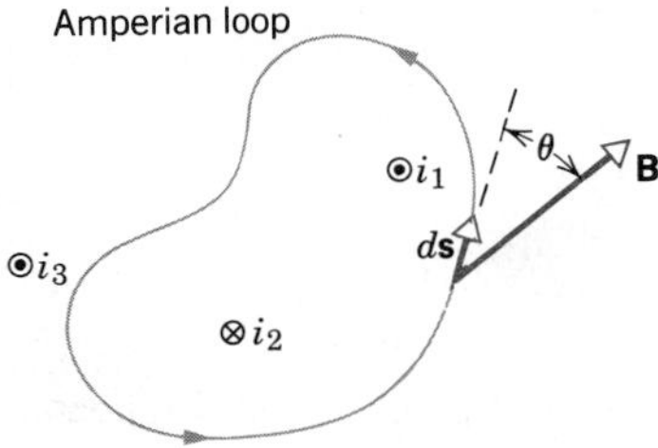

Figure 8 Ampere's law is applied to an arbitrary Amperian loop that encloses two long straight wires but excludes a third wire. Note the directions of the currents.

* Gauss' law is also more general than Coulomb's law in situations involving electric fields set up by rapidly moving charges.

trary Amperian loop to which we intend to apply Ampere's law lies entirely in the plane of the figure and threads its way among the wires, enclosing two of them but excluding the third.

We divide the Amperian loop of Fig. 8 into differential line segments of length $d\mathbf{s}$, one of which we show. At this line element, the magnetic field will have a particular value $\mathbf{B}$, also shown in the figure. Because of symmetry, $\mathbf{B}$ must lie in the plane of the figure, making an angle θ with the extended direction of the line element $d\mathbf{s}$.

The quantity $\mathbf{B} \cdot d\mathbf{s}$ on the left side of Eq. 16 is a scalar product and has the value $B \cos\theta\, ds$. The integral on the left side of Eq. 16 then becomes

$$\oint \mathbf{B} \cdot d\mathbf{s} = \oint B \cos\theta\, ds.$$

This *line integral,* as integrals of this type are called, instructs us to go around the Amperian loop of Fig. 8, adding (that is, integrating) the quantity $B \cos\theta\, ds$ as we go. We choose, arbitrarily, to traverse the loop in a counterclockwise sense.

So much for the left side of Eq. 16. The term i on the right side is the *net* current encircled by the loop, the currents being added algebraically. For the special case of Fig. 8, we have

$$i = i_1 - i_2.$$

For a counterclockwise traversal of the loop, currents pointing out of the loop (out of the page in Fig. 8) are taken as positive, those pointing inward being negative.* Note that i_3 in Fig. 8 is not included in calculating i; it lies outside the loop and is not encircled by it.

Applying Ampere's law (Eq. 16) to the situation of Fig. 8 then gives us

$$\oint B \cos\theta\, ds = \mu_0 (i_1 - i_2).$$

This result, even though we can push it no further, demonstrates the power and elegance of Ampere's law. Because the symmetry is not high enough, we cannot explicitly evaluate the closed line integral on the left side of this equation for the arbitrary Amperian loop shown. However, from the right hand side of the equation we know what that value must be. It depends *only* on the net current passing through the surface having the Amperian loop as a boundary. Note the similarity to Gauss's law. There the integral of $\mathbf{E}$ over any closed surface depends *only* on the net charge enclosed by that surface. These two laws are part of Nature's table of integrals.

We turn now to the simpler and familiar case of a single long straight wire carrying a current i, which *does* have enough symmetry so that we can use Ampere's law to find the magnetic field, $\mathbf{B}$. As Fig. 9 shows, we take our Amperian loop to be a concentric circle of radius r. This choice permits us to take full advantage of the cylindrical symmetry of the problem. Because of this symmetry, we conclude that $\mathbf{B}$ has the same magnitude, B, at every point on the circular Amperian loop. It remains to decide whether $\mathbf{B}$ is everywhere *tangent* to that loop or *perpendicular* to it, these two possibilities being equally symmetrical. We can rely on a simple compass experiment to show that the former possibility is the case; that is, $\mathbf{B}$ is tangent to the loop. (Can you show that radial lines of $\mathbf{B}$ are not consistent with Ampere's law?)

We assume further that the direction of $\mathbf{B}$ is that given by the right-hand rule of Section 31–2. Thus, $\mathbf{B}$ and $d\mathbf{s}$ point in the same direction, the angle θ between them being zero. The left-hand side of Ampere's law then becomes

$$\oint \mathbf{B} \cdot d\mathbf{s} = \oint B \cos\theta\, ds = B \oint ds = B(2\pi r).$$

Note that $\oint ds$ above is simply the circumference of the circular loop, which is $2\pi r$. The right side of Ampere's law is simply $\mu_0 i$ so that

$$B(2\pi r) = \mu_0 i$$

or

$$B = \frac{\mu_0 i}{2\pi r}. \tag{17}$$

Amperian loop
i
B
$(\theta = 0)$
$d\mathbf{s}$

Figure 9 Using Ampere's law to find the magnetic field set up by a current i in a long straight wire. The Amperian loop is a concentric circle that lies outside the wire.

* Our right-hand rule applies here, in this form: If the fingers of your right hand (the curly element) represent the direction of traversal around the Amperian loop, then your extended right thumb (the straight element) represents the positive direction for currents encircled by the loop.

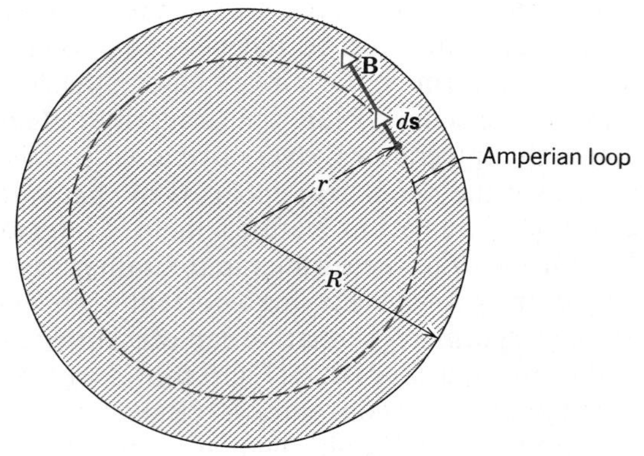

Figure 10 Using Ampere's law to find the magnetic field set up by a current i in a long straight wire of circular cross section. The Amperian loop is drawn inside the wire. The current is uniformly distributed over the cross section of the wire and emerges from the page.

This is precisely Eq. 8, which we derived earlier—with considerably more effort—using the law of Biot and Savart.

Figure 10 shows another situation in which we can usefully apply Ampere's law. It shows a cross section of a long straight wire of radius R, carrying a current i_0 uniformly distributed over the cross section of the wire. What magnetic field does this wire set up, both for points outside the wire and for points inside the wire?

For outside points, for which $r > R$, the answer is given by Eq. 8. Figure 10 shows an Amperian loop of radius r that is suitable for considering inside points, that is, points for which $r < R$. Symmetry suggests that $\mathbf{B}$ is tangent to the loop, as shown. Ampere's law

$$\oint \mathbf{B} \cdot d\mathbf{s} = \mu_0 i$$

becomes

$$(B)(2\pi r) = \mu_0 i = \mu_0 i_0 \left(\frac{\pi r^2}{\pi R^2}\right).$$

Note that the current i that appears in Ampere's law is not the total current i_0 in the wire but only that fraction of the total current that is enclosed by the Amperian loop.

Solving for B and dropping the subscript on the current, we find

$$B = \left(\frac{\mu_0 i}{2\pi R^2}\right) r, \tag{18}$$

which shows that, within the wire, B is proportional to r, starting from a value of zero at the center of the wire.

At the surface of the wire ($r = R$), Eq. 18 reduces to the same expression found by putting $r = R$ in Eq. 8 ($B = \mu_0 i / 2\pi R$). That is, the expressions for the magnetic field outside the wire and inside the wire yield the same result at the surface of the wire.

Sample Problem 5 A long straight wire of radius $R =$ 1.5 mm carries a steady current i_0 of 32 A. (a) What is the magnetic field at the surface of the wire?

Equations 8 and 18 each apply. From the former, we have

$$B = \frac{\mu_0 i}{2\pi r} = \frac{(4\pi \times 10^{-7}\ \mathrm{T \cdot m/A})(32\ \mathrm{A})}{(2\pi)(1.5 \times 10^{-3}\ \mathrm{m})}$$
$$= 4.27 \times 10^{-3}\ \mathrm{T} \approx 4.3\ \mathrm{mT}. \quad \text{(Answer)}$$

(b) What is the magnetic field at $r = 1.2$ mm?

Such points lie inside the wire so that Eq. 18 applies. We have

$$B = \frac{\mu_0 i r}{2\pi R^2} = \frac{(4\pi \times 10^{-7}\ \mathrm{T \cdot m/A})(32\ \mathrm{A})(1.2 \times 10^{-3}\ \mathrm{m})}{(2\pi)(1.5 \times 10^{-3}\ \mathrm{m})^2}$$
$$= 3.41 \times 10^{-3}\ \mathrm{T} \approx 3.4\ \mathrm{mT}. \quad \text{(Answer)}$$

Figure 11 is a plot of the magnetic field, both inside and outside the wire. Note that it reaches its maximum value at the surface of the wire.

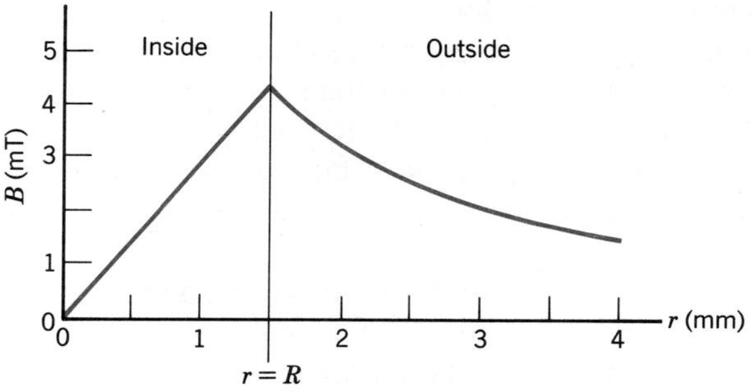

Figure 11 The magnetic field for the conductor of Fig. 10 and Sample Problem 5, both inside and outside the wire. The maximum field occurs at the surface of the wire.

31-6 Solenoids and Toroids

We now turn our attention to another problem of high symmetry in which Ampere's law will prove useful. It is the magnetic field set up by the current in a long tightly

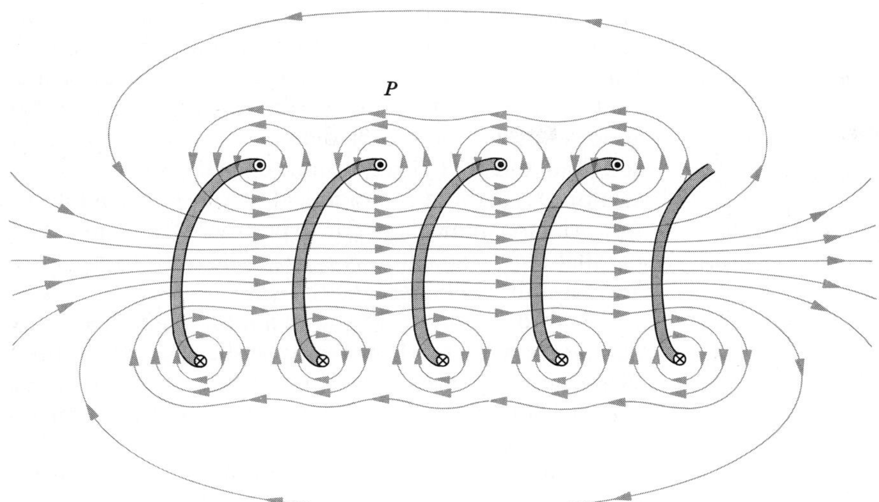

Figure 12 A section of a "stretched-out" solenoid, showing the magnetic field lines. Only the back portions of the separated windings are shown.

wound helical coil. Such a coil is called a *solenoid.* We assume here that its length is much greater than its diameter.

Figure 12 shows a section of a "stretched-out" solenoid. The solenoid field is the vector sum of the fields set up by the individual turns. For points close to these turns, the observer is not aware that the wire is bent into an arc. The wire behaves magnetically almost like a long straight wire and the lines of **B** associated with each single turn are almost concentric circles. Figure 12 suggests that the field tends to cancel between the turns. It also suggests that, at points inside the solenoid and reasonably far from the wires, **B** is parallel to the solenoid axis. In the limiting case of tightly packed square wires, the solenoid becomes essentially a cylindrical current sheet and the requirements of symmetry make the statement just given rigorously true.

For points such as *P* in Fig. 12, the field set up by the upper part of the solenoid turns (marked ⊙) points to the left and tends to cancel the field set up by the lower part of the turns (marked ⊗), which points to the right. As the solenoid approaches the configuration of an infinitely long cylindrical current sheet, the magnetic field **B** outside the solenoid approaches zero. Taking the external field to be zero is an excellent assumption for a real solenoid if its length is much greater than its diameter and if we consider points near the central region of the solenoid, far from its ends. The direction of the magnetic field along the solenoid axis follows from our all-purpose right-hand rule, interpreted this way: Grasp the solenoid with your right hand so that your fingers (the curly element) follow the direction of the current in the windings. Your extended right thumb (the straight element) will then point in the direction of the axial magnetic field.

Figure 13 shows the lines of **B** for a real solenoid. The spacing of the lines of **B** in the central plane shows that the internal field is fairly strong and uniform over its

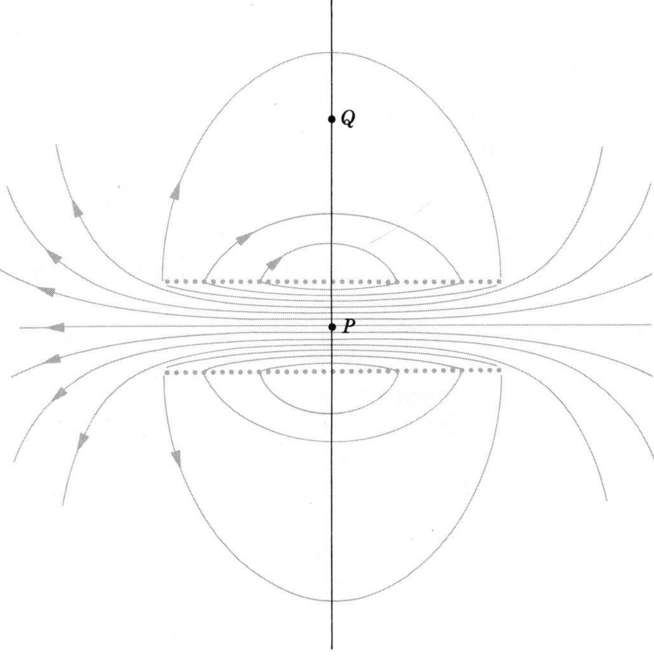

Figure 13 Magnetic field lines for a solenoid of finite length. Note that the field is strong and uniform at internal points such as *P* but is relatively weak for external points such as *Q*.

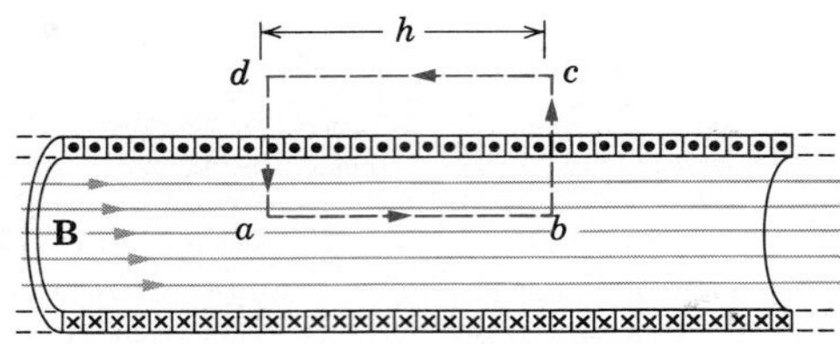

Figure 14 An application of Ampere's law to a section of a long idealized solenoid. The Amperian loop is the rectangle *abcd.*

cross section and that the external field is relatively weak.

Let us apply Ampere's law

$$\oint \mathbf{B} \cdot d\mathbf{s} = \mu_0 i \tag{19}$$

to the rectangular Amperian loop *abcd* in the ideal (infinite) solenoid of Fig. 14. We write the integral $\oint \mathbf{B} \cdot d\mathbf{s}$ as the sum of four integrals, one for each path segment:

$$\oint \mathbf{B} \cdot d\mathbf{s} = \int_a^b \mathbf{B} \cdot d\mathbf{s} + \int_b^c \mathbf{B} \cdot d\mathbf{s} + \int_c^d \mathbf{B} \cdot d\mathbf{s} + \int_d^a \mathbf{B} \cdot d\mathbf{s}. \tag{20}$$

The first integral on the right of Eq. 20 is Bh, where B is the magnitude of $\mathbf{B}$ inside the solenoid and h is the arbitrary length of the path from a to b. Note that path ab, though parallel to the solenoid axis, is some distance r from it; it will turn out that B inside the solenoid is constant over its cross section and thus independent of r.

The second and the fourth integrals on the right of Eq. 20 are zero because for every element of these paths $\mathbf{B}$ is at right angles to the path (or is zero) and thus $\mathbf{B} \cdot d\mathbf{s}$ is zero. The third integral, which includes the part of the rectangle that lies outside the solenoid, is zero because $B = 0$ for all external points. Thus, $\oint \mathbf{B} \cdot d\mathbf{s}$ for the entire rectangular path has the value Bh.

The net current i enclosed by the rectangular Amperian loop in Fig. 14 is not the same as the current i_0 in the solenoid windings because the windings pass more than once through this loop. Let n be the number of turns per unit length of the solenoid; then

$$i = i_0(nh).$$

Ampere's law (Eq. 19) then becomes

$$Bh = \mu_0 i_0 nh$$

or

$$\boxed{B = \mu_0 i_0 n} \quad \text{(ideal solenoid).} \tag{21}$$

Although we derived Eq. 21 for an infinitely long solenoid, it holds quite well for actual solenoids if we apply it only at internal points near the solenoid center. Equation 21 is consistent with the experimental fact that B does not depend on the diameter or the length of the solenoid and that B is constant over the solenoid cross section. A solenoid is a practical way to set up a known uniform magnetic field for experimentation, just as a parallel-plate capacitor is a practical way to set up a known uniform electric field.

A Toroid. Figure 15 shows a *toroid,* which we may describe as a solenoid bent into the shape of a doughnut. What magnetic field is set up at its interior points? We can find out from Ampere's law and from certain considerations of symmetry.

From symmetry, the lines of $\mathbf{B}$ form concentric circles inside the toroid, as shown in the figure. Let us choose a concentric circle of radius r as an Amperian loop and traverse it in the clockwise direction. Ampere's law (see Eq. 19) yields

$$(B)(2\pi r) = \mu_0 i_0 N,$$

where i_0 is the current in the toroid windings (and is positive) and N is the total number of turns. This gives

$$\boxed{B = \frac{\mu_0 i_0 N}{2\pi} \frac{1}{r}} \quad \text{(toroid).} \tag{22}$$

In contrast to the solenoid, B is not constant over the cross section of a toroid. It is easy to show, from Ampere's law, that $B = 0$ for points outside an ideal toroid.

Close inspection of Eq. 22 will justify our earlier statement that "a toroid is a solenoid bent into the shape

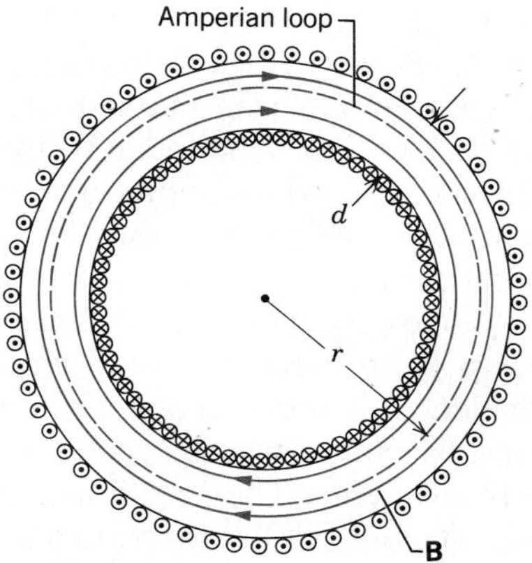

Figure 15 A toroid. The internal magnetic field can be found by applying Ampere's law to the Amperian loop shown.

of a doughnut." The denominator in Eq. 22, which is $2\pi r$, is the central circumference of the toroid and $N/2\pi r$ is just n, the number of turns per unit length. With this substitution, Eq. 22 reduces to $B = \mu_0 i_0 n$, the equation for the magnetic field in the central region of a solenoid.

The direction of the magnetic field within a toroid follows from our all-purpose right-hand rule: Grasp the toroid with the fingers of your right hand curling in the direction of the current in the windings; your extended right thumb points in the direction of the magnetic field.

Toroids form the central feature of the *tokamak*, a device showing promise as the basis for a fusion power reactor. We discuss its mode of operation in Chapter 48 of the extended version of this book.

Sample Problem 6 A solenoid has a length $L = 1.23$ m and an inner diameter $d = 3.55$ cm. It has five layers of windings of 850 turns each and carries a current $i_0 = 5.57$ A. What is B at its center?

From Eq. 21

$$B = \mu_0 i_0 n = (4\pi \times 10^{-7}\ \text{T}\cdot\text{m/A})(5.57\ \text{A})\left(\frac{5 \times 850\ \text{turns}}{1.23\ \text{m}}\right)$$

$$= 2.42 \times 10^{-2}\ \text{T} = 24.2\ \text{mT}. \qquad \text{(Answer)}$$

Note that Eq. 21 applies even if the solenoid has more than one layer of windings because the diameter of the windings does not enter into the equation.

31-7 A Current Loop as a Magnetic Dipole

So far we have studied the magnetic field set up by a long straight wire, a solenoid, and a toroid. We turn our attention here to the field set up by a single current loop. We learned in Section 30–9 that such a loop behaves like a magnetic dipole in that, if we place it in an external magnetic field, a torque τ given by

$$\boldsymbol{\tau} = \boldsymbol{\mu} \times \mathbf{B} \qquad (23)$$

acts on it. Here $\boldsymbol{\mu}$ is the magnetic dipole moment of the loop, given in magnitude by NiA where N is the number of turns, i is the current in the loop, and A is the area enclosed by the loop.

The direction of $\boldsymbol{\mu}$ is given by our all-purpose right-hand rule: Grasp the loop so that the fingers of your right hand curl around the loop in the direction of the current; your extended thumb then points in the direction of the dipole moment $\boldsymbol{\mu}$, which happens also to be the direction of the magnetic field on the axis of the loop.

We turn now to the other aspect of the current loop as a magnetic dipole: What magnetic field does it set up in the surrounding space? The problem does not have enough symmetry to make Ampere's law useful so that we must turn to the law of Biot and Savart. We consider only points on the axis of the loop, which we take to be a z axis. We shall show below that the solution is

$$B(z) = \frac{\mu_0 i R^2}{2(R^2 + z^2)^{3/2}} \qquad (24)$$

in which R is the radius of the circular loop and z is the distance of the point in question from the center of the loop.

For axial points far from the loop, we have $z \gg R$ in Eq. 24. With that approximation, this equation reduces to

$$B(z) \approx \frac{\mu_0 i R^2}{2z^3}.$$

Recalling that πR^2 is the area A of the loop and extending our result to include a loop of N turns, we can write this equation as

$$B(z) = \frac{\mu_0}{2\pi}\frac{NiA}{z^3}$$

or, since $\mathbf{B}$ and $\boldsymbol{\mu}$ have the same direction,

$$\boxed{\mathbf{B}(z) = \frac{\mu_0}{2\pi}\frac{\boldsymbol{\mu}}{z^3}} \qquad \text{(current loop)}, \quad (25)$$

where $\boldsymbol{\mu}$ is the magnetic dipole moment of the current loop.

Thus, we have two ways in which we can regard a current loop as a magnetic dipole: It experiences a torque when we place it in an external magnetic field; it generates its own intrinsic magnetic field given, for distant points along the axis, by Eq. 25.

Equation 25 reminds us of the result of Eq. 11 of Chapter 24 for the *electric* field on the axis of an *electric* dipole; namely,

$$\mathbf{E}(z) = \frac{1}{2\pi\epsilon_0}\frac{\mathbf{p}}{z^3}$$

in which $\mathbf{p}$ is the electric dipole moment. Table 1 is a summary of the properties of electric and magnetic dipoles as we have developed them so far. The symmetry between the two sets of equations is striking.

Proof of Equation 24. Figure 16 shows a circular loop of radius R carrying a current i. Consider a point P on the axis of the loop, a distance z from its plane. Let us apply

Table 1 Some Dipole Equations

Property	*Dipole Type*	*Relation*	*Equation Number*
Torque in an external field	Electric	$\tau = \mathbf{p} \times \mathbf{E}$	24–25
	Magnetic	$\tau = \boldsymbol{\mu} \times \mathbf{B}$	30–30
Energy in an external field	Electric	$U = -\mathbf{p} \cdot \mathbf{E}$	24–28
	Magnetic	$U = -\boldsymbol{\mu} \cdot \mathbf{B}$	30–31
Field at distant axial points	Electric	$\mathbf{E}(z) = \dfrac{1}{2\pi\epsilon_0} \dfrac{\mathbf{p}}{z^3}$	24–11
	Magnetic	$\mathbf{B}(z) = \dfrac{\mu_0}{2\pi} \dfrac{\boldsymbol{\mu}}{z^3}$	31–25

the law of Biot and Savart to a current element located at the left side of the loop. The vector $d\mathbf{s}$ for this element points out of the page at right angles. The angle θ between $d\mathbf{s}$ and the vector $\mathbf{r}$ in Fig. 16 is 90° and the plane formed by these two vectors is at right angles to the plane of the figure. From the law of Biot and Savart, the differential field $d\mathbf{B}$ set up by this current element is at right angles to this plane and thus lies in the plane of the figure and at right angles to $\mathbf{r}$, as Fig. 16 shows.

Let us resolve $d\mathbf{B}$ into two components, one, $dB_\parallel$, along the axis of the loop and another, $dB_\perp$, at right angles to this axis. Only $dB_\parallel$ contributes to the total magnetic field B at point P. This follows because, from symmetry, the axial direction is the only preferred direction, so that the vector sum of all magnetic field components at right angles to the axis must add up to zero. This leaves only the axial components and we have

$$B = \int dB_\parallel,$$

where the integral is a simple scalar integration.

For the element shown in Fig. 16, the law of Biot and Savart (Eq. 5) gives

$$dB = \left(\frac{\mu_0}{4\pi}\right) \frac{i\, ds \sin 90^\circ}{r^2}.$$

We also have

$$dB_\parallel = dB \cos \alpha.$$

Combining these two relations, we have

$$dB_\parallel = \frac{\mu_0 i \cos \alpha\, ds}{4\pi r^2}. \tag{26}$$

Figure 16 shows that r and α are not independent but are related to each other. Let us express each in terms of the variable z, the distance of point P from the center of the loop. The relations are

$$\cos \alpha = \frac{R}{r} = \frac{R}{\sqrt{R^2 + z^2}} \tag{27}$$

and

$$r = \sqrt{R^2 + z^2}. \tag{28}$$

Substituting Eqs. 27 and 28 into Eq. 26, we find

$$dB = \frac{\mu_0 i R}{4\pi(R^2 + z^2)^{3/2}}\, ds.$$

Note that i, R, and z have the same values for all current elements. Integrating this equation and noting that $\int ds$ is simply the circumference of the loop, we find

$$B = \int dB = \frac{\mu_0 i R}{4\pi(R^2 + z^2)^{3/2}} \int ds$$

or

$$B(z) = \frac{\mu_0 i R^2}{2(R^2 + z^2)^{3/2}},$$

which is Eq. 24, the relation we sought to prove.

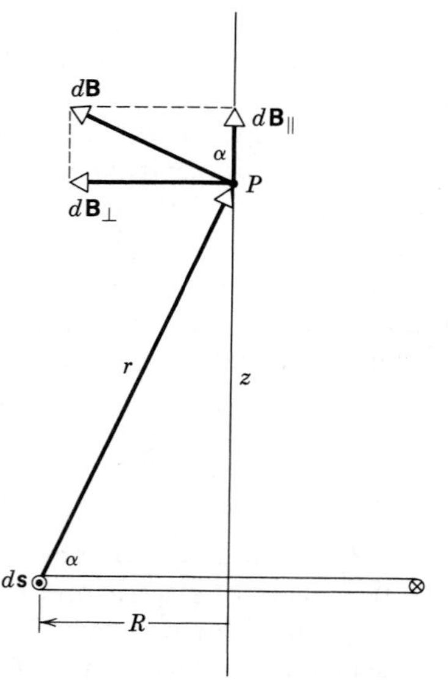

Figure 16 A current loop. We use the law of Biot and Savart to find the magnetic field at axial points.

REVIEW AND SUMMARY

The Biot-Savart Law

The magnetic field set up by a current-carrying conductor can be found from the *Biot-Savart law.* This asserts that the contribution $d\mathbf{B}$ to the field set up by a current element $i\, d\mathbf{s}$ at a point distant $\mathbf{r}$ from the current element has a magnitude given by

$$d\mathbf{B} = \left(\frac{\mu_0}{4\pi}\right)\frac{i\, d\mathbf{s} \times \mathbf{r}}{r^3} \quad \text{(Biot-Savart law).} \qquad [7]$$

μ_0

Study Fig. 3 carefully. The quantity μ_0, called the permeability constant, has the value $4\pi \times 10^{-7}\ \text{T}\cdot\text{m/A} \approx 1.26 \times 10^{-6}\ \text{T}\cdot\text{m/A}$.

A Long, Straight Wire

For a *long straight wire* carrying a current i the Biot-Savart law gives, for the magnetic field at a distance r from the wire,

$$B(r) = \frac{\mu_0 i}{2\pi r} \quad \text{(long straight wire);} \qquad [8]$$

see Sample Problem 1. Figure 2 shows the concentric field lines. The right-hand rule described in Section 31–2 is helpful in relating the directions of $\mathbf{B}$ and $i\, d\mathbf{s}$ in this and other magnetic problems.

$\mathbf{B}_{ext}$, $\mathbf{B}_{intr}$, and Their Resultant

If a wire carrying a current is placed in an external magnetic field $\mathbf{B}_{ext}$, a force given by Eq. 30–24 will act on it. $\mathbf{B}_{ext}$ must not be confused with $\mathbf{B}_{intr}$, the intrinsic magnetic field set up by the current in the wire itself. Figure 4 shows field lines representing the vector sum of $\mathbf{B}_{ext}$ and $\mathbf{B}_{intr}$ for a particular case.

Parallel wires carrying currents in the same (opposite) direction attract (repel) each other. For a length L of the wire the magnitude of the force on either wire is

The Force Between Parallel Wires

$$F_{ba} = i_b L B_a = \frac{\mu_0 L i_a i_b}{2\pi d}, \qquad [14]$$

where d is the wire separation. The ampere is defined from this relation. See Sample Problems 2–4.

For current distributions of high symmetry *Ampere's law,*

Ampere's Law

$$\oint \mathbf{B} \cdot d\mathbf{s} = \mu_0 i \quad \text{(Ampere's law)} \qquad [16]$$

can be used in placed of the Biot-Savart law to calculate the magnetic field. One first chooses a closed *Amperian loop* around which to apply the law. Study the fairly general case of Fig. 8 to learn how to evaluate the line integral $\oint \mathbf{B} \cdot d\mathbf{s}$ and to find i. See Sample Problem 5.

Using Ampere's law we show that B inside a *long solenoid,* carrying current i_0, at points near its center is given by

A Solenoid

$$B = \mu_0 i_0 n \quad \text{(ideal solenoid)} \qquad [21]$$

where n is the number of turns per unit length; see Sample Problem 6. Also, B inside a *toroid* is

A Toroid

$$B = \frac{\mu_0 i_0 N}{2\pi}\frac{1}{r} \quad \text{(toroid);} \qquad [22]$$

See Fig. 15 and Sample Problem 6.

The magnetic field produced by a *current loop* (a *magnetic dipole*) at a distance z along its axis is parallel to the axis and has magnitude

Field of a Magnetic Dipole

$$\mathbf{B}(z) = \frac{\mu_0}{2\pi}\frac{\boldsymbol{\mu}}{z^3} \quad \text{(current loop),} \qquad [25]$$

$\boldsymbol{\mu}$ being the dipole moment of the loop. Table 1 summarizes and compares the impressively symmetric properties of electric and magnetic dipoles.

QUESTIONS

1. A beam of 20-MeV protons emerges from a cyclotron. Do these particles cause a magnetic field?

2. Discuss analogies and differences between Coulomb's law and the Biot-Savart law.

3. Consider a magnetic field line. Is the magnitude of **B** constant or variable along such a line? Can you give an example of each case?

4. In electronics, wires that carry equal but opposite currents are often twisted together to reduce their magnetic effect at distant points. Why is this effective?

5. Drifting electrons constitute the current in a wire and a magnetic field is associated with this current. What current and magnetic field would be measured by an observer moving along with the drifting electrons?

6. Consider two charges, first (*a*) of the same sign and then (*b*) of opposite signs, that are moving along separated parallel paths with the same velocity. Compare the directions of the mutual electric and magnetic forces in each case.

7. Is there any way to set up a magnetic field other than by causing charges to move?

8. Like currents attract and unlike currents repel. Like stationary charges repel and unlike stationary charges attract. Is this a paradox?

9. Figure 17 shows four vertical wires carrying equal currents in the same direction. What is the direction of the force on the left-hand wire, caused by the currents in the other three wires?

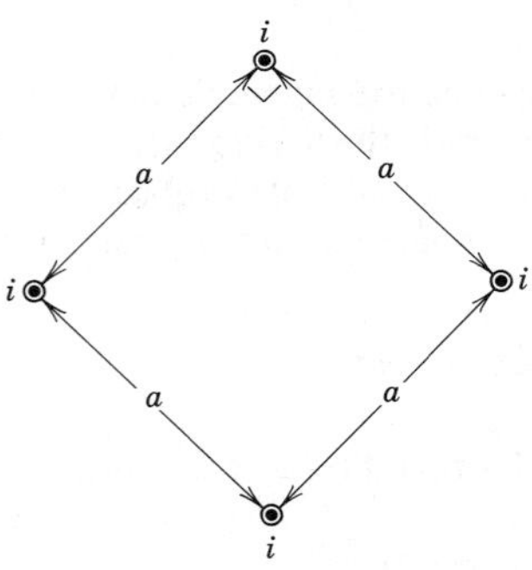

Figure 17 Question 9.

10. Two long parallel conductors carry equal currents i in the same direction. Sketch roughly the resultant lines of **B** due to the action of both currents. Does your figure suggest an attraction between the wires?

11. A current is sent through a vertical spring from whose lower end a weight is hanging; what will happen?

12. Two long straight wires pass near one another at right angles. If the wires are free to move, describe what happens when currents are sent through both of them.

13. Two fixed wires cross each other perpendicularly so that they do not actually touch but are close to each other, as shown in Fig. 18. Equal currents i exist in each wire in the directions indicated. In what region(s) will there be some points of zero net magnetic field?

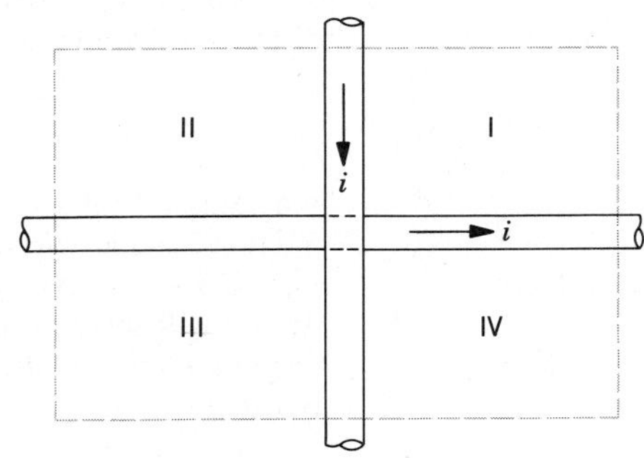

Figure 18 Question 13.

14. A messy loop of limp wire is placed on a smooth table and anchored at points a and b as shown in Fig. 19. If a current i is now passed through the wire, will it try to form a circular loop or will it try to bunch up further?

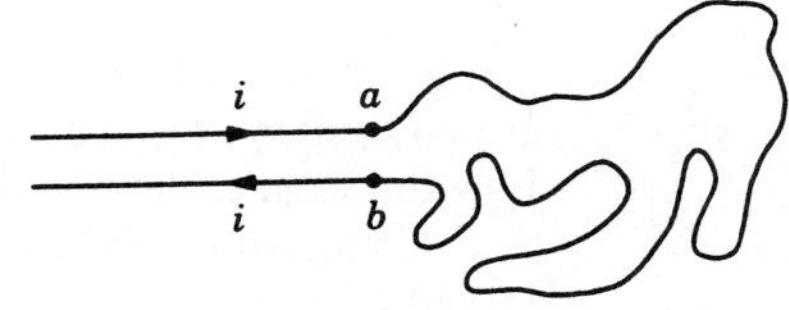

Figure 19 Question 14.

15. Can the path of integration around which we apply Ampere's law pass through a conductor?

16. Suppose we set up a path of integration around a cable that contains 12 wires with different currents (some in opposite directions) in each wire. How do we calculate i in Ampere's law in such a case?

17. Apply Ampere's law qualitatively to the three paths shown in Fig. 20.

18. Discuss analogies and differences between Gauss's law and Ampere's law.

19. A steady longitudinal uniform current is set up in a long

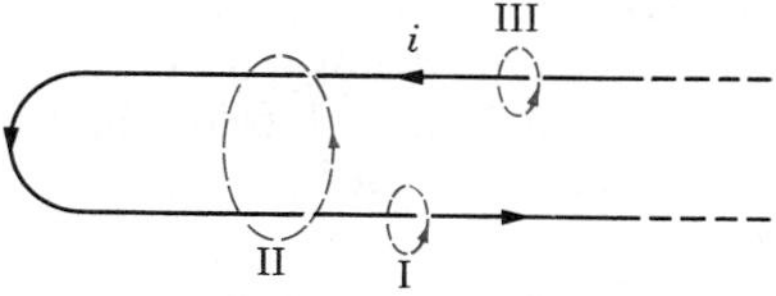

Figure 20 Question 17.

copper tube. Is there a magnetic field (*a*) inside and/or (*b*) outside the tube?

20. A long straight wire of radius R carries a steady current i. How does the magnetic field generated by this current depend on R? Consider points both outside and inside the wire.

21. A long straight wire carries a constant current i. What does Ampere's law require for (*a*) a loop that encloses the wire but is not circular, (*b*) a loop that does not enclose the wire, and (*c*) a loop that encloses the wire but does not all lie in one plane?

22. Two long solenoids are nested on the same axis, as Fig. 21 shows. They carry identical currents but in opposite directions. If there is no magnetic field inside the inner solenoid, what can you say about n, the number of turns per unit length, for the two solenoids? Which one, if either, has the larger value?

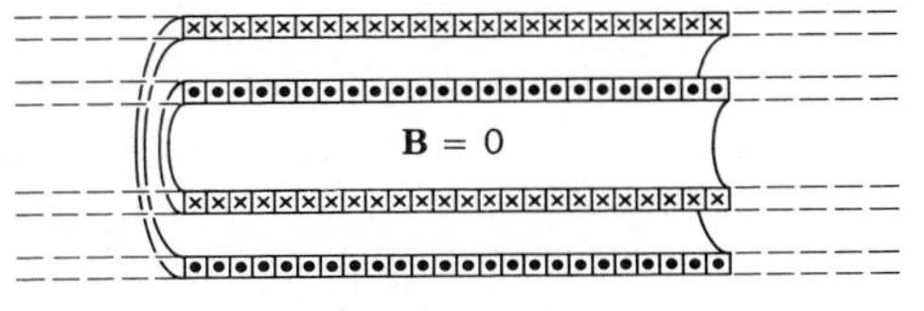

Figure 21 Question 22.

23. If the windings on a solenoid are considered to be helical, rather than being equivalent to a cylindrical current sheet, then the magnetic field outside is not zero strictly but is a circling field, just the same as the field of a straight wire. Explain.

24. The magnetic field at the center of a circular current loop has the value $\mu_0 i/2R$; see Eq. 24. However, the *electric* field at the center of a ring of charge is *zero*. Why this difference?

25. A steady current is set up in a cubical network of resistive wires, as in Fig. 22. Use symmetry arguments to show that the magnetic field at the center of the cube is zero.

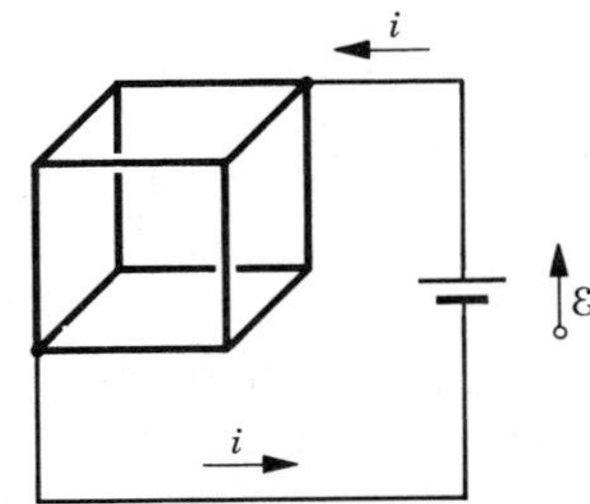

Figure 22 Question 25.

26. Does Eq. 21 ($B = \mu_0 i_0 n$) hold for a solenoid of square cross section?

27. Give details of three ways in which you can measure the magnetic field **B** at a point P, a perpendicular distance r from a long straight wire carrying a constant current i. Base them on: (*a*) projecting a particle of charge q through point P with velocity **v**, parallel to the wire; (*b*) measuring the force per unit length exerted on a second wire, parallel to the first wire and carrying a current i'; (*c*) measuring the torque exerted on a small magnetic dipole located a perpendicular distance r from the wire.

28. How might you measure the magnetic dipole moment of a compass needle?

29. A circular loop of wire lies on the floor of the room in which you are sitting. It carries a constant current i in a clockwise direction, as viewed from above. What is the direction of the magnetic dipole moment of this current loop?

30. Is **B** uniform for all points within a circular loop of wire carrying a current? Explain.

EXERCISES AND PROBLEMS

Section 31-2 Calculating the Magnetic Field

1E. A #10 bare copper wire (2.6 mm in diameter) can carry a current of 50 A without overheating. For this current, what is the magnetic field at the surface of the wire?

2E. The magnitude of the magnetic field 88 cm from the axis of a long straight wire is 7.3 μT. Calculate the current in the wire.

3E. A surveyor is using a magnetic compass 20 ft below a power line in which there is a steady current of 100 A. Will this interfere seriously with the compass reading? The horizontal component of the earth's magnetic field at the site is 20 μT (= 0.20 gauss).

4E. The 25-kV electron gun in a TV tube fires an electron beam 0.22 mm in diameter at the screen, 5.6×10^{14} electrons arriving each second. Calculate the magnetic field produced by the beam at a point 1.5 mm from the axis of the beam.

5E. Figure 23 shows a 3.0-cm segment of wire, centered at the origin, carrying a current of 2.0 A in the $+y$ direction. (Of course this segment must be part of some complete circuit.) To calculate the **B** field at a point P several meters from the origin

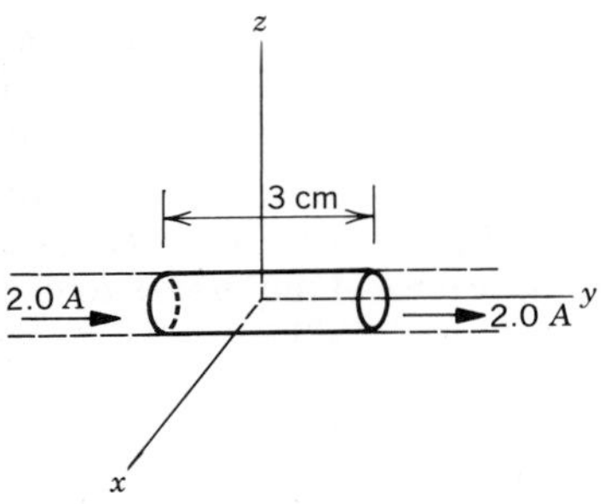

Figure 23 Exercise 5.

one may use the Biot-Savart law in the form $B = (\mu_0/4\pi)i\ \Delta s \sin\theta/r^2$ with $\Delta s = 3.0$ cm. This is because r and θ are essentially constant over the segment of wire. Calculate **B** (magnitude and direction) at the following (x, y, z) locations: (*a*) (0, 0, 5 m), (*b*) (0, 6 m, 0), (*c*) (7 m, 7 m, 0), (*d*) (−3 m, −4 m, 0).

6E. A long wire carrying a current of 100 A is placed in a uniform external magnetic field of 5.0 mT (= 50 gauss). The wire is at right angles to this magnetic field. Locate the points at which the resultant magnetic field is zero.

7E. At a position in the Philippines the earth's magnetic field of 39 μT is horizontal and due north. The net field is zero exactly 8.0 cm above a long straight horizontal wire that carries a constant current. (*a*) What is the magnitude of the current and (*b*) what is the direction of the current?

8E. A positive point charge of magnitude q is a distance d from a long straight wire carrying a current i and is traveling with speed v perpendicular to the wire. What are the direction and magnitude of the force acting on it if the charge is moving (*a*) toward, or (*b*) away from the wire?

9E. A long straight wire carries a current of 50 A. An electron, traveling at 1.0×10^7 m/s, is 5.0 cm from the wire. What force acts on the electron if the electron velocity is directed (*a*) toward the wire, (*b*) parallel to the wire, and (*c*) at right angles to the directions defined by (*a*) and (*b*)?

10E. A straight conductor carrying a current i is split into identical semicircular turns as shown in Fig. 24. What is the magnetic field at the center C of the circular loop so formed?

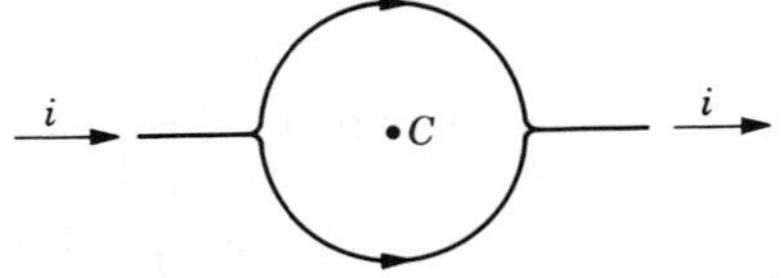

Figure 24 Exercise 10.

11E. Two infinitely long wires carry equal currents, i. Each follows a 90° arc on the circumference of the same circle of radius R in the configuration shown in Fig. 25. Show, without doing a detailed calculation, that **B** at the center of the circle is the same as the **B** field a distance R below an infinite straight wire carrying a current i to the left.

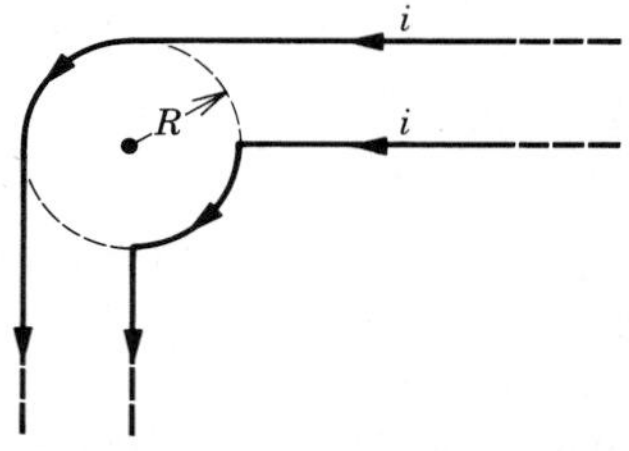

Figure 25 Exercise 11.

12P. Use the Biot-Savart law to calculate the magnetic field **B** at C, the common center of the semicircular arcs AD and HJ, of radii R_2 and R_1, respectively, forming part of the circuit $ADJHA$ carrying current i, as shown in Fig. 26.

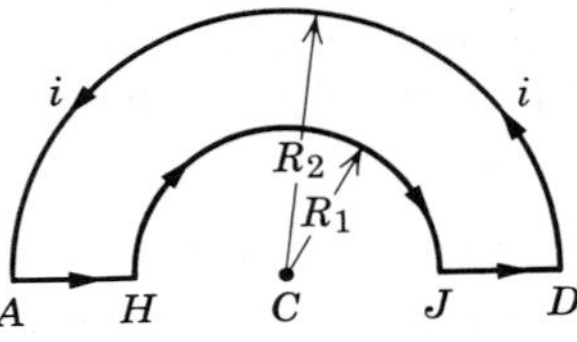

Figure 26 Problem 12.

13P. A long hairpin is formed by bending a piece of wire as shown in Fig. 27. If the wire carries a 10-A current, (*a*) what are the direction and magnitude of **B** at point a? (*b*) At point b? Take $R = 5.0$ mm.

Figure 27 Problem 13.

14P. A wire carrying current i has the configuration shown in Fig. 28. Two semi-infinite straight sections, each tangent to the

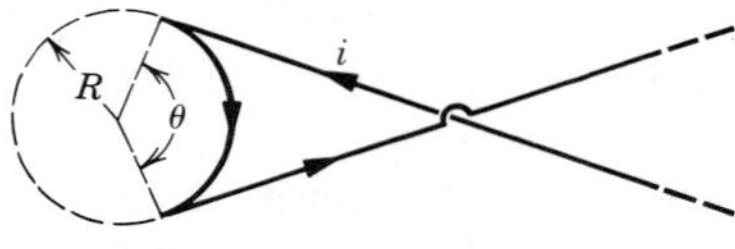

Figure 28 Problem 14.

same circle, are connected by a circular arc, of angle θ, along the circumference of the circle, with all sections lying in the same plane. What must θ be in order for B to be zero at the center of the circle?

15P. Consider the circuit of Fig. 29. The curved segments are arcs of circles of radii a and b. The straight segments are along the radii. Find the magnetic field **B** at P, assuming a current i in the circuit.

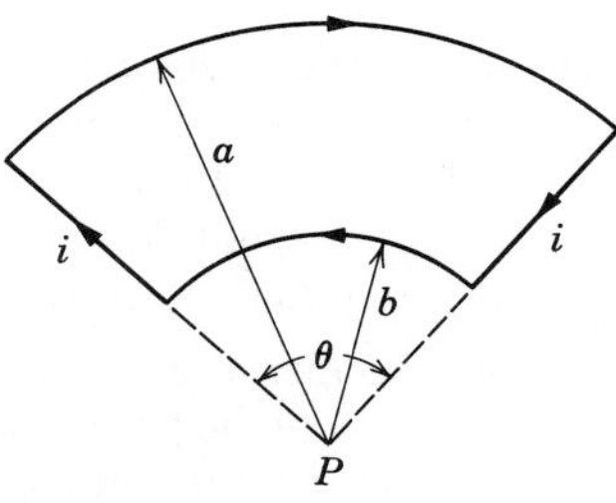

Figure 29 Problem 15.

16P. The wire shown in Fig. 30 carries a current i. What is the magnetic field **B** at the center C of the semicircle arising from (a) each straight segment of length L, (b) the semicircular segment of radius R, and (c) the entire wire?

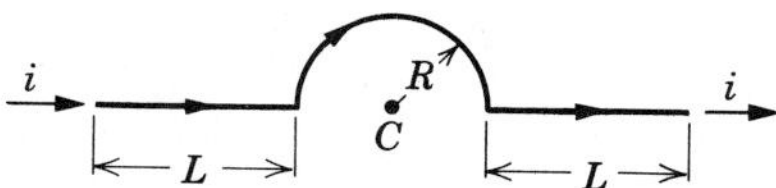

Figure 30 Problem 16.

17P. A straight wire segment of length L carries a current i. Show that the magnetic field **B** associated with this segment, at a distance R from the segment along a perpendicular bisector (see Fig. 31), is given in magnitude by

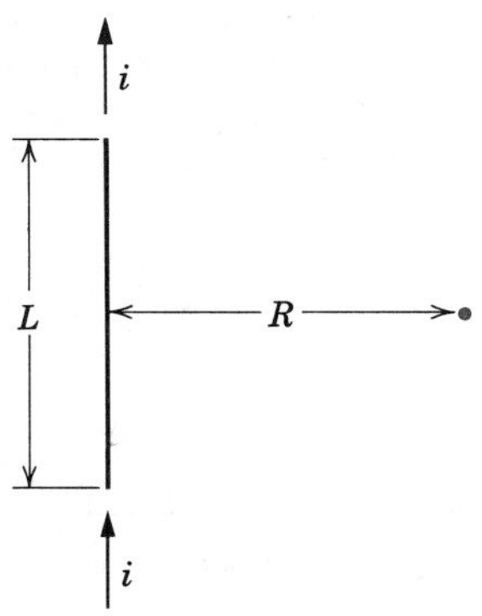

Figure 31 Problem 17.

$$B = \frac{\mu_0 i}{2\pi R} \frac{L}{(L^2 + 4R^2)^{1/2}}.$$

Show that this expression reduces to an expected result as $L \to \infty$.

18P. A square loop of wire of edge a carries a current i. Show that the value of B at the center is given by

$$B = \frac{2\sqrt{2}\mu_0 i}{\pi a}.$$

(*Hint:* See Problem 17.)

19P. Show that B at the center of a rectangular loop of wire of length L and width W, carrying a current i, is given by

$$B = \frac{2\mu_0 i}{\pi} \frac{(L^2 + W^2)^{1/2}}{LW}.$$

Show that this reduces to a result consistent with Sample Problem 3 for $L \gg W$.

20P. A square loop of wire of edge a carries a current i. Show that B for a point on the axis of the loop and a distance x from its center is given by

$$B(x) = \frac{4\mu_0 i a^2}{\pi(4x^2 + a^2)(4x^2 + 2a^2)^{1/2}}.$$

Show that the result is consistent with the result of Problem 18.

21P. You are given a length L of wire in which a current i may be established. The wire may be formed into a circle or a square. Show that the square yields the greater value for B at the central point.

22P. A straight section of wire of length L carries a current i. Show that the magnetic field associated with this segment at P a perpendicular distance D from one end of the wire (see Fig. 32) is given by

$$B = \frac{\mu_0 i}{4\pi D} \frac{L}{(L^2 + D^2)^{1/2}}.$$

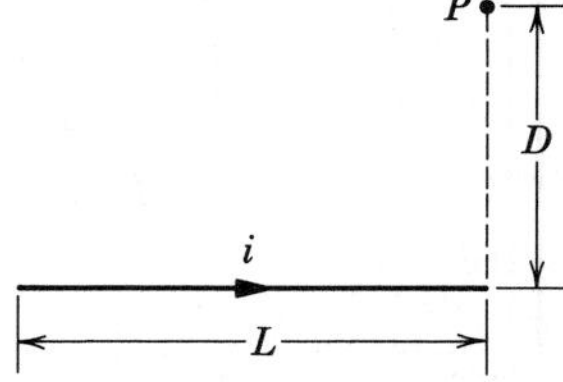

Figure 32 Problem 22.

23P. A current i flows in a straight wire segment of length a, as in Fig. 33. Show that the magnetic field at point Q in that figure is zero and that the field at P is given in magnitude by

$$B = \frac{\sqrt{2}\mu_0 i}{8\pi a}.$$

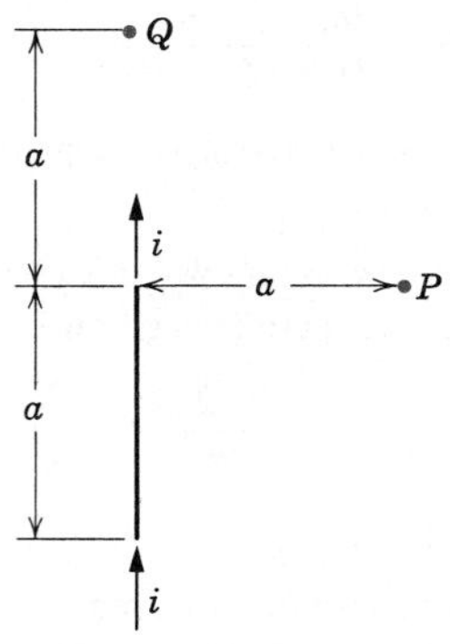

Figure 33 Problem 23.

24P. Find the magnetic field **B** at point P in Fig. 34. See Problem 23.

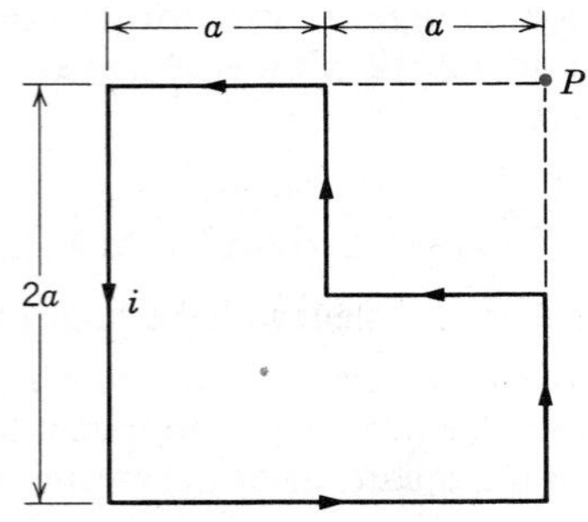

Figure 34 Problem 24.

25P. Calculate **B** at point P in Fig. 35. Assume that $i = 10$ A and $a = 8.0$ cm.

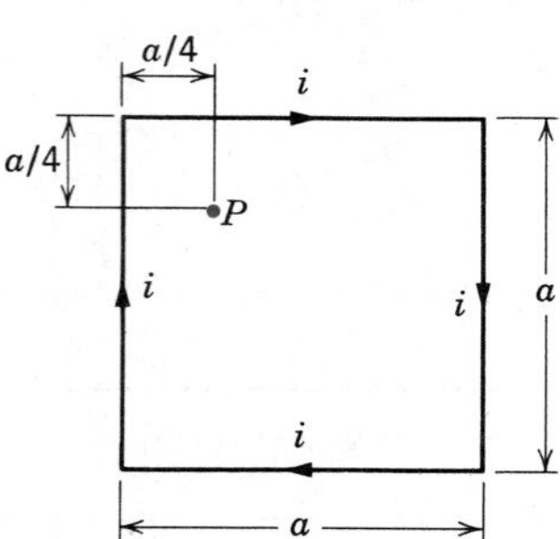

Figure 35 Problem 25.

26P. Figure 36 shows a cross section of a long, thin ribbon of width w that is carrying a uniformly-distributed total current i into the page. Calculate the magnitude and direction of the magnetic field, **B**, at a point P in the plane of the ribbon at a distance d from its edge. (*Hint:* Imagine the ribbon to be constructed from many long, thin, parallel wires.)

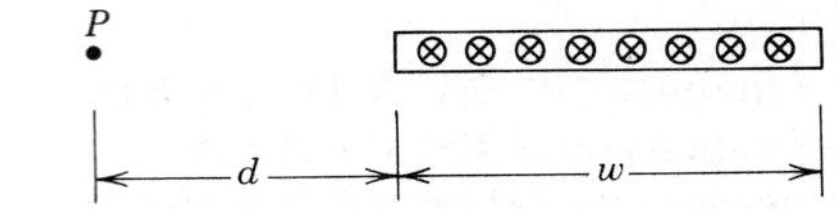

Figure 36 Problem 26.

Section 31-4 Two Parallel Conductors

27E. Two long parallel wires are 8.0 cm apart. What equal currents must flow in the wires if the magnetic field halfway between them is to have a magnitude of 300 μT? Consider both (*a*) parallel and (*b*) antiparallel currents.

28E. Two long straight parallel wires, separated by 0.75 cm, are perpendicular to the plane of the page as shown in Fig. 37. Wire W_1 carries a current of 6.5 A into the page. What must be the current (magnitude and direction) in wire W_2 for the resultant magnetic field at point P to be zero?

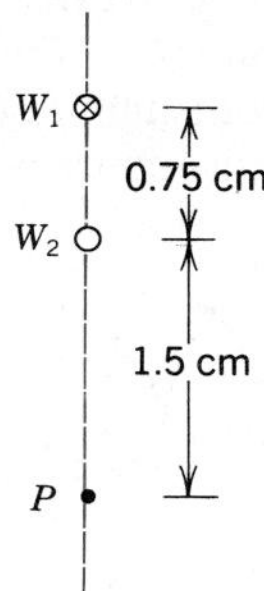

Figure 37 Exercise 28.

29E. Two long parallel wires a distance d apart carry currents of i and $3i$ in the same direction. Locate the point or points at which their magnetic fields cancel.

30E. Figure 38 shows five long parallel wires in the xy plane. Each wire carries a current $i = 3.0$ A in the positive x direction. The separation between adjacent wires is $d = 8.0$ cm. Find the magnetic force per meter exerted on each of these five wires.

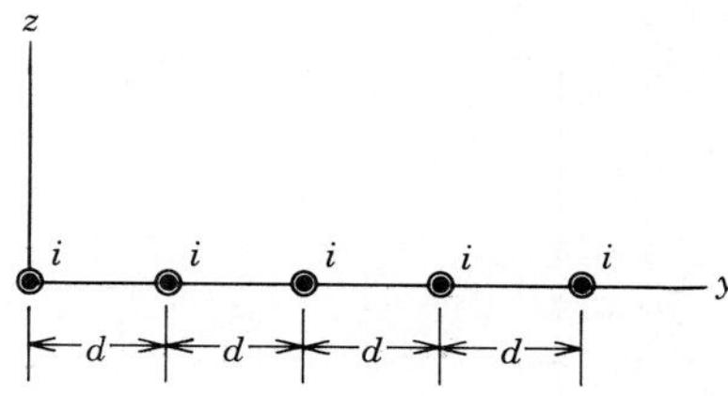

Figure 38 Exercise 30.

31E. For the wires in Sample Problem 3, show that Eq. 15 holds for points beyond the wires, that is, for points with $|x| > d$.

32E. Two long straight parallel wires 10 cm apart each carry a current of 100 A. Figure 39 shows a cross section, with the wires running perpendicular to the page and point P lying on the perpendicular bisector of d. Find the magnitude and direction of the magnetic field at P when the current in the left-hand wire is out of the page and the current in the right-hand wire is (*a*) out of the page and (*b*) into the page.

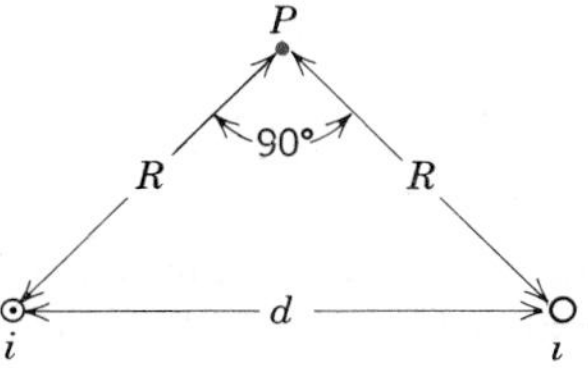

Figure 39 Exercise 32.

33P. In Fig. 6*a* assume that both currents are in the same direction, out of the plane of the figure. Show that the magnetic field in the plane defined by the wires is

$$B(x) = \frac{\mu_0 i x}{\pi(x^2 - d^2)}.$$

Assume $i = 10$ A and $d = 2.0$ cm in Fig. 6*a* and plot $B(x)$ for the range -2 cm $< x < +2$ cm. Assume that the wire diameters are negligible.

34P. Four long copper wires are parallel to each other, their cross section forming a square 20 cm on edge. A 20-A current is set up in each wire in the direction shown in Fig. 40. What are the magnitude and direction of **B** at the center of the square?

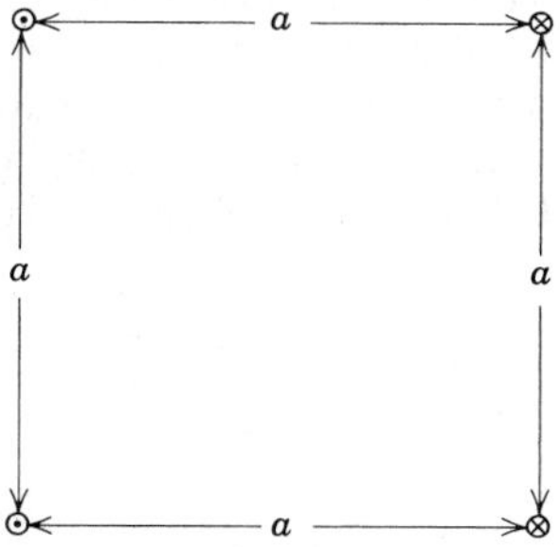

Figure 40 Problems 34, 35, and 36.

35P. Suppose, in Fig. 40, that the currents are all out of the page. What is the force per unit length (magnitude and direction) on any one wire? In the case of parallel motion of charged particles in a plasma, this is known as the pinch effect.

36P. In Fig. 40 what is the force per unit length acting on the lower left wire, in magnitude and direction?

37P. Two long wires a distance d apart carry equal antiparallel currents i, as in Fig. 41. (*a*) Show that **B** at point P, which is equidistant from the wires, is given by

$$B = \frac{2\mu_0 i d}{\pi(4R^2 + d^2)}.$$

(*b*) In what direction does **B** point?

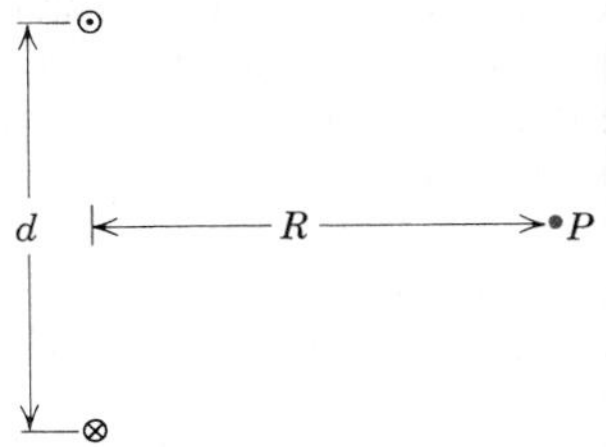

Figure 41 Problem 37.

38P. Figure 42 shows a long wire carrying a current of 30 A. The rectangular loop carries a current of 20 A. Calculate the resultant force acting on the loop. Assume that $a = 1.0$ cm, $b = 8.0$ cm, and $L = 30$ cm.

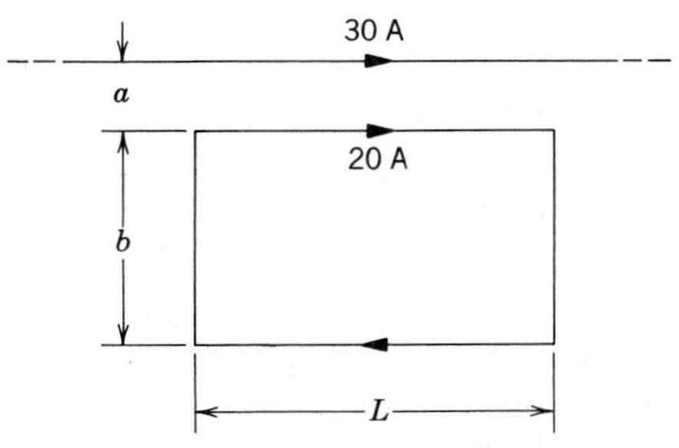

Figure 42 Problem 38.

39P. Figure 43 shows an idealized schematic of an "electromagnetic rail gun," designed to fire projectiles at speeds up to 10 km/s. (The feasibility of these devices as defenses against ballistic missiles is being studied.) The projectile P sits between and in contact with two parallel rails along which it can slide. A generator G provides a current that flows up one rail, across the projectile, and back down the other rail. (*a*) Let w = distance between the rails, r the radius of the rails (presumed circular) and i the current. Show that the force on the projectile is to the right and given approximately by

$$F = \frac{1}{2}\,(i^2\mu_0/\pi)\ln\left(\frac{w + r}{r}\right).$$

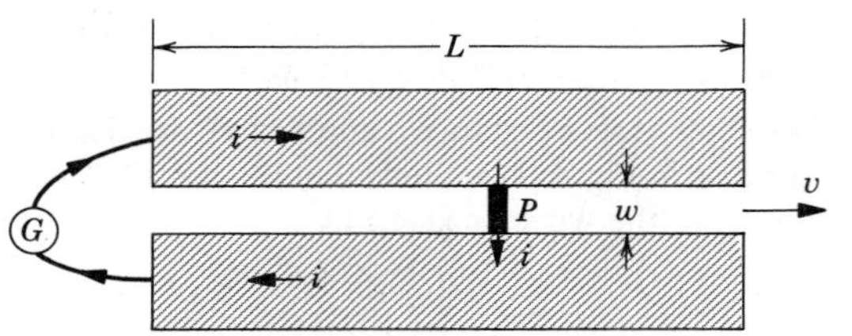

Figure 43 Problem 39.

(*b*) If the projectile (in this case a test slug) starts from the left end of the rail at rest, find the speed v at which it is expelled at the right. Assume that $i = 450$ kA, $w = 12$ mm, $r = 6.7$ cm, $L = 4.0$ m, and that the mass of the slug is $m = 10$ g.

Section 31-5 Ampere's Law

40E. Each of the indicated eight conductors in Fig. 44 carries 2.0 A of current into or out of the page. Two paths are indicated for the line integral $\oint \mathbf{B} \cdot d\mathbf{s}$. What is the value of the integral for (*a*) the dotted path? (*b*) the dashed path?

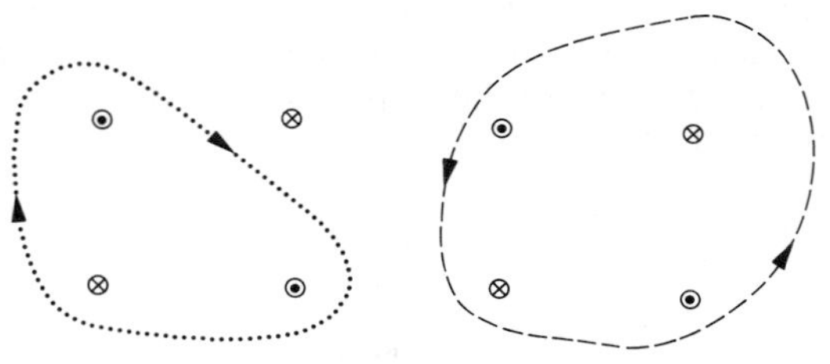

Figure 44 Exercise 40.

41E. Eight wires cut the page perpendicularly at the points shown in Fig. 45. A wire labeled with the integer k ($k = 1, 2, \ldots, 8$) bears the current ki_0. For those with odd k, the current is out of the page; for those with even k it is into the page. Evaluate $\oint \mathbf{B} \cdot d\mathbf{s}$ along the closed path shown in the direction shown.

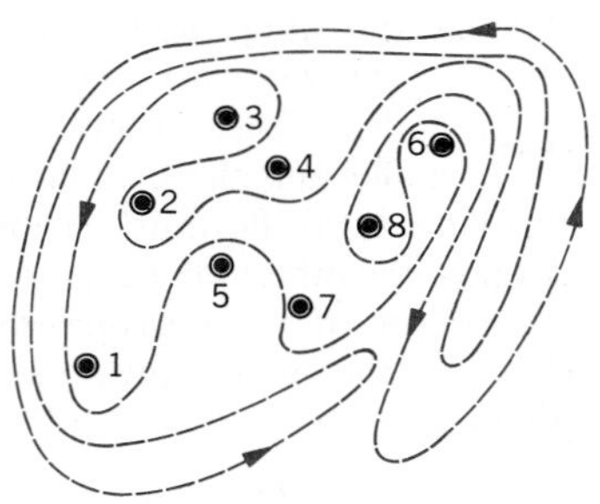

Figure 45 Exercise 41.

42E. Figure 46 shows a cross section of a long cylindrical conductor of radius a, carrying a uniformly distributed current i. Assume $a = 2.0$ cm and $i = 100$ A and plot $B(r)$ over the range $0 < r < 6.0$ cm.

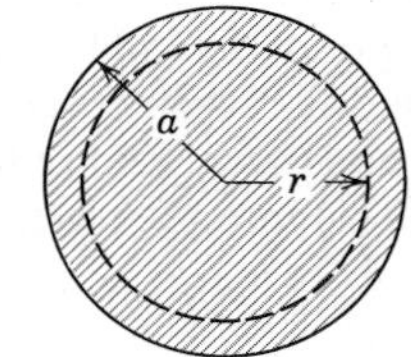

Figure 46 Exercise 42.

43E. In a certain region there is a uniform current density of 15 A/m² in the positive z direction. What is the value of $\oint \mathbf{B} \cdot d\mathbf{s}$ when the line integral is taken along the four straight-line segments from $(4d, 0, 0)$ to $(4d, 3d, 0)$ to $(0, 0, 0)$ to $(4d, 0, 0)$, where $d = 20$ cm?

44P. Two square conducting loops carry currents of 5.0 A and 3.0 A as shown in Fig. 47. What is the value of the line integral $\oint \mathbf{B} \cdot d\mathbf{s}$ for each of the two closed paths shown?

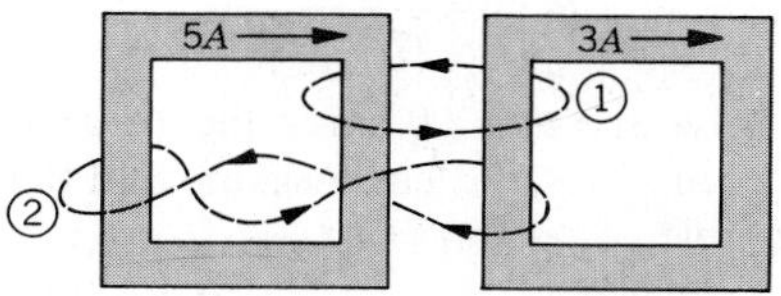

Figure 47 Problem 44.

45P. Show that a uniform magnetic field **B** cannot drop abruptly to zero as one moves at right angles to it, as suggested by the horizontal arrow through point a in Fig. 48. (*Hint:* Apply Ampere's law to the rectangular path shown by the dashed lines.) In actual magnets "fringing" of the lines of **B** always occurs, which means that **B** approaches zero in a grad-

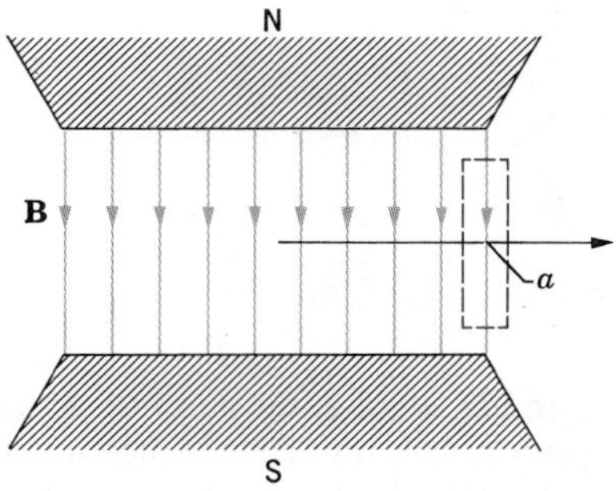

Figure 48 Problem 45.

ual manner. Modify the **B** lines in the figure to indicate a more realistic situation.

46P. Figure 49 shows a cross section of a hollow cylindrical conductor of radii a and b, carrying a uniformly distributed current i. (a) Show that $B(r)$ for the range $b < r < a$ is given by

$$B(r) = \frac{\mu_0 i}{2\pi(a^2 - b^2)}\left(\frac{r^2 - b^2}{r}\right).$$

(b) Test this formula for the special cases of $r = a$, $r = b$, and $b = 0$. (c) Assume $a = 2.0$ cm, $b = 1.8$ cm, and $i = 100$ A and plot $B(r)$ for the range $0 < r < 6$ cm.

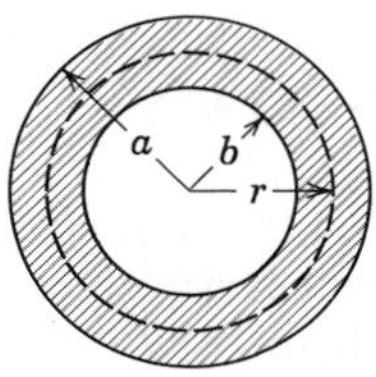

Figure 49 Problem 46.

47P. Figure 50 shows a cross section of a long conductor of a type called a coaxial cable. Its radii (a, b, c) are shown in the figure. Equal but opposite currents i exist in the two conductors. Derive expressions for $B(r)$ in the ranges (a) $r < c$, (b) $c < r < b$, (c) $b < r < a$, and (d) $r > a$. (e) Test these expressions for all the special cases that occur to you. (f) Assume $a = 2.0$ cm, $b = 1.8$ cm, $c = 0.40$ cm, and $i = 120$ A and plot $B(r)$ over the range $0 < r < 3$ cm.

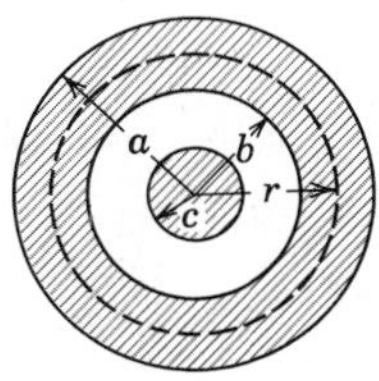

Figure 50 Problem 47.

48P. The current density inside a long, solid, cylindrical wire of radius a is in the direction of the axis and varies linearly with radial distance r from the axis according to $J = J_0 r/a$. Find the magnetic field inside the wire.

49P. A long circular pipe, with an outside radius of R, carries a (uniformly distributed) current of i_0 (into the paper as shown in Fig. 51). A wire runs parallel to the pipe at a distance of $3R$ from center to center. Calculate the magnitude and direction of the current in the wire that would cause the resultant magnetic field at the point P to have the same magnitude, but the opposite direction, as the resultant field at the center of the pipe.

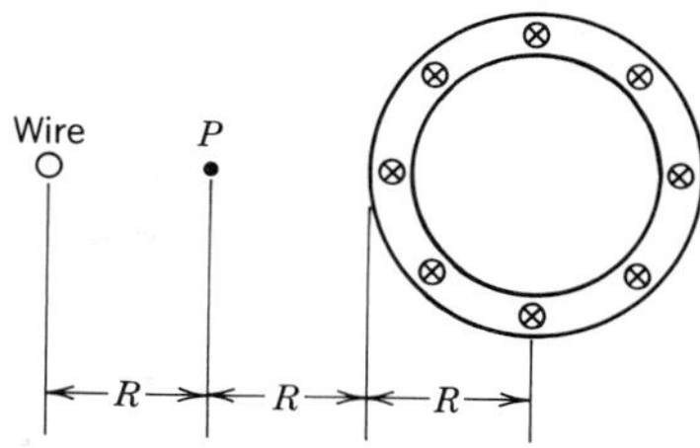

Figure 51 Problem 49.

50P. Figure 52 shows a cross section of a long cylindrical conductor of radius a containing a long cylindrical hole of radius b. The axes of the two cylinders are parallel and are a distance d apart. A current i is uniformly distributed over the cross-hatched area in the figure. (a) Use superposition ideas to show that the magnetic field at the center of the hole is

$$B = \frac{\mu_0 i d}{2\pi(a^2 - b^2)}.$$

(b) Discuss the two special cases $b = 0$ and $d = 0$. (c) Can you use Ampere's law to show that the magnetic field in the hole is uniform? (*Hint:* Regard the cylindrical hole as filled with two equal currents moving in opposite directions, thus canceling each other. Assume that each of these currents has the same current density as that in the actual conductor. Thus we superimpose the fields due to two complete cylinders of current, of radii a and b, each cylinder having the same current density.)

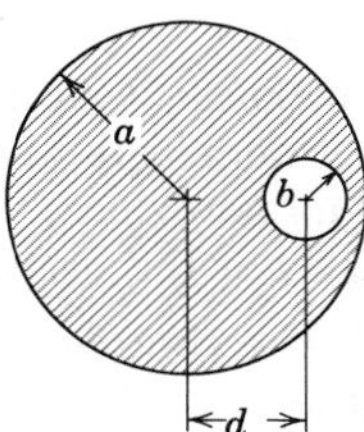

Figure 52 Problem 50.

51P. Figure 53 shows a cross section of an infinite conducting sheet with a current per unit x-length λ emerging from the page at right angles. (a) Use the right-hand rule and symmetry arguments to convince yourself that the magnetic field **B** is constant for all points P above the sheet (and for all points P' below it) and is directed as shown. (b) Use Ampere's law to prove that $B = \frac{1}{2}\mu_0\lambda$.

52P*. In a certain region there is a magnetic field given in mT by $\mathbf{B} = 3\mathbf{i} + 8(x^2/d^2)\mathbf{j}$ where x is the x-coordinate distance in

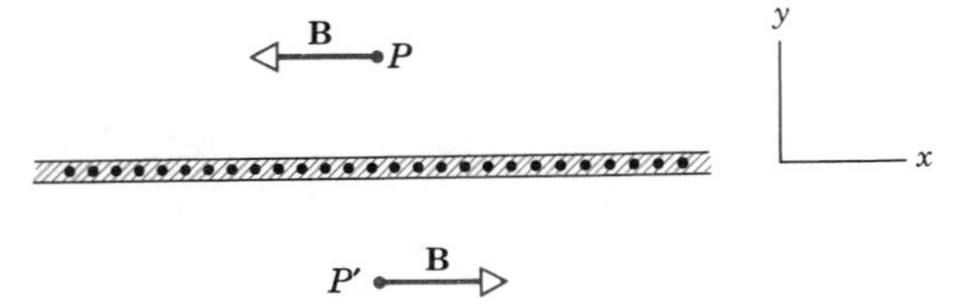

Figure 53 Problem 51.

meters and d is a constant, with units of length. Some current must be flowing in the region to cause the specified **B** field. (*a*) Evaluate the integral $\oint \mathbf{B} \cdot d\mathbf{s}$ along the straight path from $(d, 0, 0)$ to $(d, d, 0)$. (*b*) Let $d = 0.5$ m in the expression for **B** and apply Ampere's law to determine what current flows through a square of side length d ($= 0.5$ m) that lies in the first quadrant of the xy plane with one corner at the origin. (*c*) Is this current in the **k** or $-$**k** direction?

Section 31-6 Solenoids and Toroids

53E. A solenoid 95 cm long has a radius of 2.0 cm, a winding of 1200 turns and carries a current of 3.6 A. Calculate the strength of the magnetic field inside the solenoid.

54E. A 200-turn solenoid having a length of 25 cm and diameter of 10 cm carries a current of 0.30 A. Calculate the magnitude of the magnetic field **B** near the center of the solenoid.

55E. A solenoid 1.3 m long and 2.6 cm in diameter carries a current of 18 A. The magnetic field inside the solenoid is 23 mT. Find the length of the wire forming the solenoid.

56E. A toroid having a square cross section, 5.0 cm on edge, and an inner radius of 15 cm has 500 turns and carries a current of 0.80 A. What is the magnetic field inside the toroid at (*a*) the inner radius and (*b*) the outer radius of the toroid?

57E. Show that if the thickness of a toroid is very small compared to its radius of curvature (very skinny toroid), then Eq. 21 for the field inside a toroid reduces to Eq. 20 for the field inside a solenoid. Explain why this result is expected.

58P. Treat a solenoid as a continuous cylindrical current sheet, whose current per unit length, measured parallel to the cylinder axis, is λ. B inside a solenoid is given by Eq. 21 ($B = \mu_0 i_0 n$). Show that this can also be written as $B = \mu_0 \lambda$. This is the value of the *change* in **B** that you encounter as you move from inside the solenoid to outside, through the solenoid wall. Show that this same change occurs as you move through an infinite plane current sheet such as that of Fig. 53 (see Problem 51). Does this equality surprise you?

59P. In Section 6 we showed that the magnetic field at any radius r inside a toroid is given by

$$B = \frac{\mu_0 i_0 N}{2\pi r}.$$

Show that as you move from a point just inside a toroid to a point just outside the magnitude of the *change* in **B** that you encounter—at any radius r—is just $\mu_0 \lambda$. Here λ is the current per unit length for any circumference ($= 2\pi r$) of the toroid. Compare the similar result found in Problem 58. Is the equality surprising?

60P. A long solenoid with 10 turns/cm and a radius of 7.0 cm carries a current of 20 mA. A current of 6.0 A flows in a straight conductor along the axis of the solenoid. (*a*) At what radial distance from the axis will the direction of the resulting **B** field be at 45° from the axial direction? (*b*) What is the magnitude of the magnetic field?

61P. A long solenoid has 100 turns per centimeter and carries current i. An electron moves within the solenoid in a circle of radius 2.30 cm perpendicular to the solenoid axis. The speed of the electron is $0.046c$ (c = speed of light). Find the current i in the solenoid.

62P. An interesting (and frustrating) effect occurs when one attempts to confine a collection of electrons and positive ions (a plasma) in the magnetic field of a toroid. Particles whose motion is perpendicular to the **B** field will not execute circular paths because the field strength varies with radial distance from the axis of the toroid. This effect, which is shown (exaggerated) in Fig. 54, causes particles of opposite sign to drift in opposite directions parallel to the axis of the toroid. (*a*) What is the sign of the charge on the particle whose path is sketched in the figure? (*b*) If the particle path has a radius of curvature of 11 cm when its radial distance from the axis of the toroid is 125 cm, what will be the radius of curvature when the particle is 110 cm from the axis?

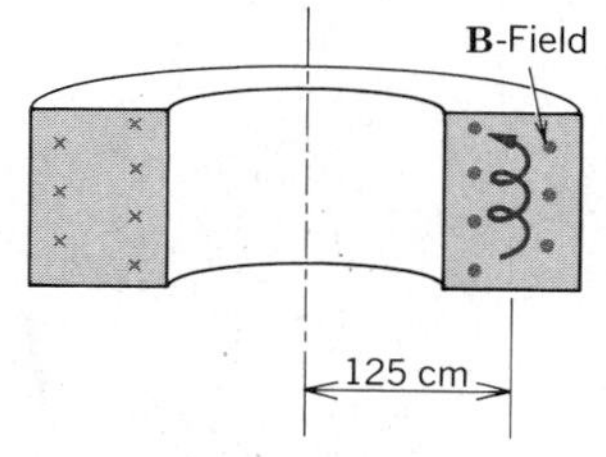

Figure 54 Problem 62.

Section 31-7 A Current Loop as a Magnetic Dipole

63E. What is the magnetic dipole moment $\boldsymbol{\mu}$ of the solenoid described in Exercise 54?

64E. Figure 55*a* shows a length of wire carrying a current i and bent into a circular coil of one turn. In Fig. 55*b* the same length of wire has been bent more sharply, to give a double loop of smaller radius. (*a*) If B_a and B_b are magnitudes of the magnetic fields at the centers of the two loops, what is the ratio B_b/B_a? (*b*) What is the ratio of their dipole moments, μ_b/μ_a?

65E. Figure 56 shows an arrangement known as a Helmholtz coil: It consists of two circular coaxial coils each of N turns and

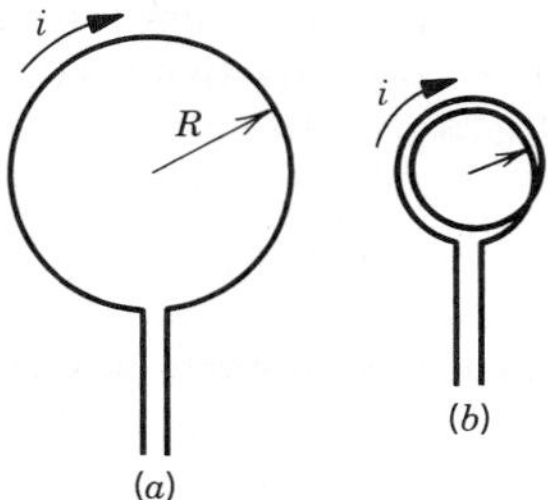

Figure 55 Exercise 64.

radius R, separated by a distance R. They carry equal currents i in the same direction. Find the magnetic field at P, midway between the coils.

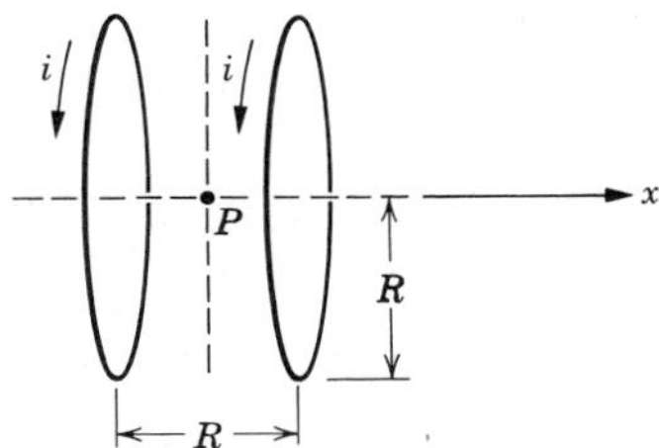

Figure 56 Exercise 65; Problems 69 and 70.

66E. A student makes an electromagnet by winding 300 turns of wire around a wooden cylinder of diameter 5.0 cm. The coil is connected to a battery producing a current of 4.0 A in the wire. (*a*) What is the magnetic moment of this device? (*b*) At what axial distance $R \gg d$ will the magnetic field of this dipole be 5.0 μT (approximately one-tenth that of the earth's magnetic field)?

67E. The magnetic field $B(x)$ for various points on the axis of a square current loop of side a is given in Problem 20. (*a*) Show that the axial field for this loop for $x \gg a$ is that of a magnetic dipole (see Eq. 25). (*b*) What is the magnetic dipole moment of this loop?

68P. You are given a closed circuit with radii a and b, as shown in Fig. 57, carrying a current i. (*a*) What are the magnitude and direction of **B** at point P? (*b*) Find the magnetic dipole moment of the circuit.

69P. Two 300-turn coils each carry a current i. They are arranged a distance apart equal to their radius, as in Fig. 56. For $R = 5.0$ cm and $i = 50$ A, plot B as a function of distance x along the common axis over the range $x = -5$ cm to $x = +5$ cm, taking $x = 0$ at the midpoint P. (Such coils provide an especially uniform field B near point P.) (*Hint:* See Eq. 24.)

70P. In Exercise 65 (Fig. 56) let the separation of the coils be

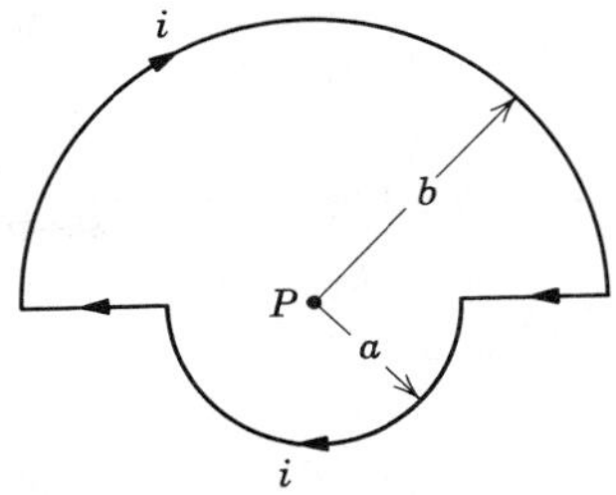

Figure 57 Problem 68.

a variable s (not necessarily equal to the coil radius R). (*a*) Show that the first derivative of the magnetic field (dB/dx) vanishes at the midpoint P regardless of the value of s. Why would you expect this to be true from symmetry? (*b*) Show that the second derivative of the magnetic field (d^2B/dx^2) also vanishes at P provided $s = R$. This accounts for the uniformity of B near P for this particular coil separation.

71P. A circular loop of radius 12 cm carries a current of 15 A. A second loop of radius 0.82 cm, having 50 turns and a current of 1.3 A is at the center of the first loop. (*a*) What magnetic field **B** does the large loop set up at its center? (*b*) What torque acts on the small loop? Assume that the planes of the two loops are at right angles and that the magnetic field due to the large loop is essentially uniform throughout the volume occupied by the small loop.

72P. A conductor carries a current of 6.0 A along the closed path *abcdefgha* involving 8 of the 12 edges of a cube of side 10 cm as shown in Fig. 58. (*a*) Why can one regard this as the superposition of three square loops: *bcfgb, abgha,* and *cdefc*? (*b*) Find the magnetic dipole moment $\boldsymbol{\mu}$ (magnitude and direction) of this current. (*c*) Calculate **B** at the points $(x, y, z) = (0, 5\text{ m}, 0)$ and $(5\text{ m}, 0, 0)$.

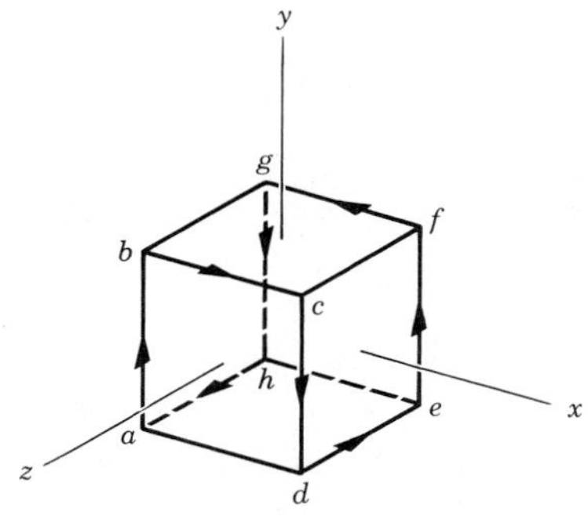

Figure 58 Problem 72.

73P. (*a*) A long wire is bent into the shape shown in Fig. 59, without cross contact at P. The radius of the circular section is

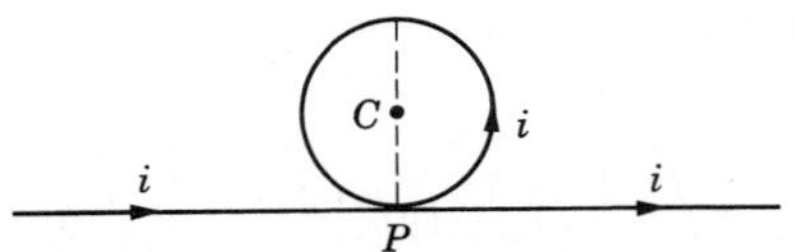

Figure 59 Problem 73.

R. Determine the magnitude and direction of $\mathbf{B}$ at the center C of the circular portion when the current i is as indicated. (*b*) The circular part of the wire is rotated without distortion about its (dashed) diameter perpendicular to the straight portion of the wire. The magnetic moment associated with the circular loop is now in the direction of the current in the straight part of the wire. Determine $\mathbf{B}$ at C in this case.

74P*. A thin plastic disk of radius R has a charge q uniformly distributed over its surface. If the disk is rotated at an angular frequency ω about its axis, show that (*a*) the magnetic field at the center of the disk is

$$B = \frac{\mu_0 \omega q}{2\pi R}$$

and (*b*) the magnetic dipole moment of the disk is

$$\mu = \frac{\omega q R^2}{4}.$$

(*Hint:* The rotating disk is equivalent to an array of current loops.)

75P*. Derive the solenoid equation (Eq. 21) starting from the expression for the field on the axis of a circular loop (Eq. 24). (*Hint:* Subdivide the solenoid into a series of current loops of infinitesimal thickness and integrate. See Fig. 16.)

CHAPTER 32

FARADAY'S LAW OF INDUCTION

Faraday's law of induction is closer than you may think. Behind these slots are copper wires that may snake out of your house and over the countryside for hundreds of miles. If you could follow them, you would almost certainly find yourself at an electric generator, *in which a potential difference is established between the wires by Faraday's law. On your way from the wall plug to the generator, you would have to portage around several* transformers, *in which Faraday's law is again at work, stepping up or down the potential difference between the wires.*

32-1 Two Symmetries

If you put a closed conducting loop in an external magnetic field and send a current through it, a torque will act on the loop tending to cause it to rotate.* We summarize this as follows:

$$\text{current} \rightarrow \text{torque} \quad \text{(loop in magnetic field).} \quad (1)$$

Equation 1 is the principle of the electric motor.

Symmetry—on which we lean so much in physics—compels us to ask: "What if we try it the other way around? Suppose that we put a closed conducting loop in an external magnetic field and rotate it by exerting a torque on it from some external source. Will an electric current appear in the loop?" That is, can we write

$$\text{torque} \rightarrow \text{current} \quad \text{(loop in magnetic field)?} \quad (2)$$

This does indeed happen! It is the principle of the electric generator. The law that governs the appearance of this current is called *Faraday's law of induction.* Physicists have learned to be guided in thinking about nature by such pleasing symmetries as those displayed by Eqs. 1

* See Section 30-8.

and 2. Symmetry is beauty, in physics as well as in art.

There is another symmetry—and a curious one—at the human level associated with the law of induction. This law was discovered in 1831 by Michael Faraday and also, independently and at about the same time, by the American physicist Joseph Henry. The self-educated Faraday was apprenticed at age 14 to a London bookbinder. He wrote: "There were plenty of books there and I read them." Henry was apprenticed at age 13 to a watchmaker in Albany, New York.

In later years, Faraday was appointed Director of the Royal Institution in London, whose founding was due in large part to an American, Benjamin Thomson (Count Rumford). Henry, on the other hand, became Secretary (that is, Director) of the Smithsonian Institution in Washington, DC, which was founded by an endowment from an Englishman, James Smithson.

Even though Faraday published his results first, which gave him priority of discovery, the SI unit of inductance (see Chapter 33) is called the *henry* (abbr. H). On the other hand, the SI unit of capacitance, as we have seen, is called the *farad* (abbr. F). In Chapter 35 we shall study the *electromagnetic oscillator,* in which energy is shuttled back and forth between inductive and capacitive forms, just as it is between kinetic and potential forms in an oscillating pendulum. It is pleasant to see the names of these two gifted scientists so closely interwoven in the operation of the electromagnetic oscillator, a device made possible by their joint discovery of electromagnetic induction.

32-2 Two Experiments

Two simple tabletop experiments will guide us to an understanding of Faraday's law of induction.

First Experiment. Figure 1 shows the terminals of a wire loop connected to a sensitive galvanometer G that can detect the presence of a current in the loop. Normally, we would not expect this meter to deflect because there is no battery in the circuit. However, if you push a bar magnet toward the loop a curious thing happens. While the magnet is moving (and *only* while it is moving) the meter deflects, showing that a current has been set up in the coil. Furthermore, the faster you move the magnet, the greater the deflection. When you stop moving the magnet, the deflection stops and the needle of the meter returns to zero. If you move the magnet *away* from the loop, the meter again deflects while the magnet

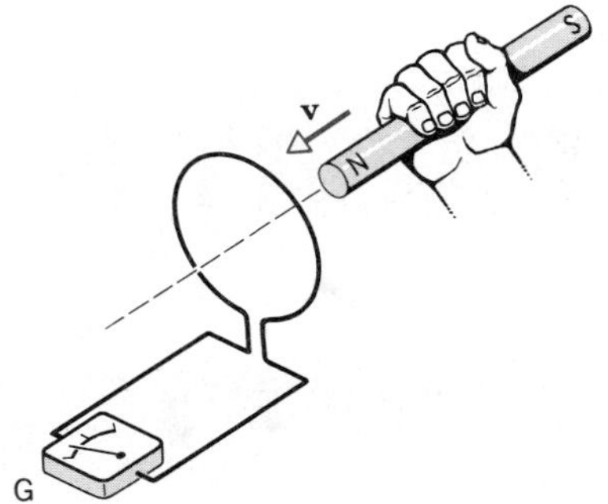

Figure 1 Galvanometer G deflects when the magnet is moving with respect to the coil.

is moving, but in the opposite direction, which tells us that the current in the loop is in the opposite direction.

If you reverse the magnet end for end, so that the south pole (rather than the north pole) faces the loop, the experiments work just as before except that the directions of the deflections are reversed. Further tests would convince you that what matters is the relative motion of the magnet and the loop. It makes no difference whether you move the loop toward the magnet or the magnet toward the loop.

The current that appears in the loop in this experiment is called an *induced current* and we say that it is set up by an *induced electromotive force.* Such induced electromotive forces play an important part in our daily lives. The chances are good that the lights in the room in which you are reading this book are operated by an induced electromotive force produced in a commercial electric generator.

Second Experiment. Here we use the apparatus of Fig. 2. The loops are close to each other but remain at rest and have no direct electrical contact. If you close switch

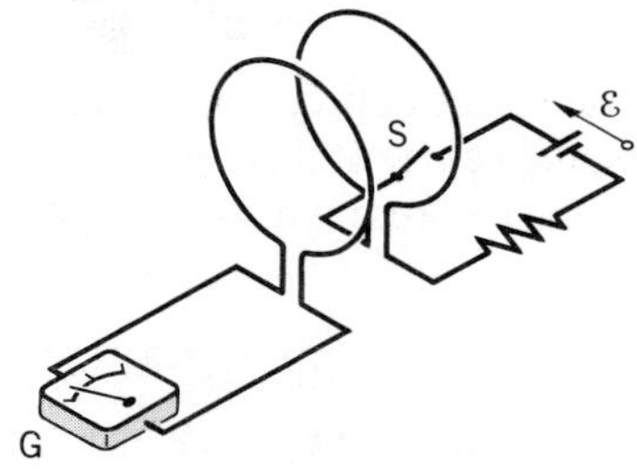

Figure 2 Galvanometer G deflects momentarily when switch S is closed or opened. No physical motion of the coils is involved.

S, thus setting up a current in the right-hand loop, the meter in the left-hand loop deflects momentarily and then returns to zero. If you then open the switch, thus interrupting this current, the meter again deflects but in the opposite direction. None of the apparatus is physically moving in this experiment.

Only when the current in the right-hand loop is rising or falling does an induced electromotive force appear in the left-hand loop. When there is a steady current in the right-hand loop, there is no induced electromotive force, no matter how large that steady current may be.

In thinking about these two experiments, we are led to conclude the following:

> *An induced electromotive force appears only when something is changing. In a static situation, in which no physical objects are moving and the currents are steady, there is no induced electromotive force. The key word is* change.

32-3 Faraday's Law of Induction

Faraday had the insight to see the common feature of the two experiments described in the preceding section. In language that he might have used:

> *An induced emf appears in the left-hand loop of Figs. 1 and 2 only when the number of magnetic field lines that pass through that loop is changing.*

The actual *number* of lines of the magnetic field that passes through the left-hand loop at any moment is of no concern; it is the *rate at which this number is changing* that determines the induced electromotive force.

In the experiment of Fig. 1, the lines originate in the bar magnet and the number of lines passing through the left-hand loop increased when you brought the magnet closer and decreased when you pulled the magnet away.

In the experiment of Fig. 2, the magnetic lines are associated with the current in the right-hand loop. The number of lines through the left-hand loop increased (from zero) when you closed switch S and decreased (back to zero) when you then opened this switch.

A Quantitative Treatment. Consider a surface—which may or may not be plane—bounded by a closed loop. We represent the number of magnetic lines that pass through that surface by the *magnetic flux* Φ_B for that surface, where Φ_B is defined from*

$$\boxed{\Phi_B = \oint \mathbf{B} \cdot d\mathbf{A}} \quad \text{(magnetic flux defined).} \quad (3)$$

Here $d\mathbf{A}$ is a differential element of surface area and the integration is to be carried out over the entire surface.

If the magnetic field has a constant magnitude B and is everywhere at right angles to a plane surface of area A, Eq. 3 reduces to

$$\boxed{\Phi_B = BA} \quad \text{(special case of Eq. 3),} \quad (4)$$

in which Φ_B is the absolute value of the flux.

From Eqs. 3 and 4 we see that the SI unit for the magnetic flux is the tesla·meter², to which we give the name *weber* (abbr. Wb). That is,

$$1 \text{ weber} = 1 \text{ Wb} = 1 \text{ T}\cdot\text{m}^2. \quad (5)$$

Having established the definition of magnetic flux, we are now ready to state Faraday's law of induction in a quantitative way:

> *The induced emf in a circuit is equal (except for a change in sign) to the rate at which the magnetic flux through that circuit is changing with time.*

In equation form this law becomes

$$\boxed{\mathcal{E} = -\frac{d\Phi_B}{dt}} \quad \text{(Faraday's law).} \quad (6)$$

If the rate of change of flux is in webers per second, the induced emf will be in volts. The minus sign has to do with the direction of the induced emf, as we explain at the end of this section.

If we change the magnetic flux through a coil of N turns, an induced emf appears in every turn and these emfs—like those of batteries connected in series—are to be added. If the coil is so tightly wound that each turn can be said to occupy the same region of space, the flux

* This definition of the flux of the magnetic field **B** is just like the definition of the flux of the electric field **E** given in Section 25-4 (which you may wish to review) or the flux of the current density **J** given by Eq. 5 of Chapter 28. Incidentally, we may add magnetic flux to our growing list of physical quantities that are quantized when examined on a fine enough scale; see Essay 12, *Superconductivity*, by Peter Lindenfeld for some details.

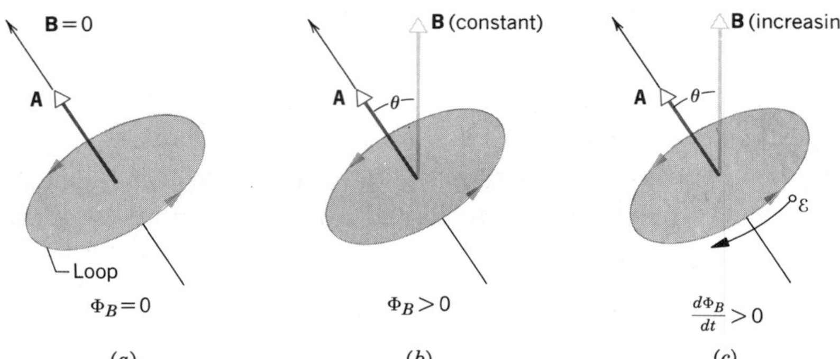

Figure 3 (*a*) A flat surface of area A bounded by a conducting loop. The positive direction of circulation around the loop is related to the vector **A** by the right-hand rule. (*b*) A steady magnetic field now pierces the loop; the magnetic flux through the loop is positive. (*c*) The magnitude of the magnetic field increases with time. The induced emf is in the negative direction of circulation, as shown.

Φ_B through each turn will be the same. Such coils are described as *close packed.* The flux through each turn is also the same for (ideal) solenoids and toroids (see Section 31–6). The induced electromotive force in all such arrangements is then

$$\boxed{\mathcal{E} = -N\frac{d\Phi_B}{dt}} \quad \text{(Faraday's law).} \tag{7}$$

A Word About Signs (Optional). The minus signs in Eqs. 6 and 7 guide us in finding the *direction* of the induced emf in a particular situation. Although we shall rely on Lenz's law (see Section 32–4) to give us this information, a well-trained physicist should also be able to deduce this direction in a formal manner directly from Faraday's law. Let us do so now.

We start with Eq. 3, the definition of the magnetic flux. Flux, which is a scalar, can be either positive or negative, a matter we have not yet addressed. In defining the flux of the *electric field* in connection with Gauss' law (see Section 25–4), we dealt only with closed surfaces and defined the direction of the area element $d\mathbf{A}$ uniquely, as the outward direction. In defining the flux of the magnetic field, however, we deal with a surface that is not closed and we must make an arbitrary decision as to which direction from that surface we take as the positive direction for the area element vector $d\mathbf{A}$.

In Fig. 3*a*, for example, we consider a plane surface of area A bounded by a loop and we arbitrarily take the direction shown in the figure as the direction of the area vector **A**. Our all-purpose right-hand rule then tells us that the direction around the loop indicated by the arrows will be our positive reference direction.

In Fig. 3*b*, we show a (constant) magnetic field **B** enclosed by the loop, the direction of **B** making an angle θ with the area vector **A**, where $\theta < 90°$. From Eq. 3 then, the flux Φ_B enclosed by the loop is positive; that is, $\Phi_B > 0$. If now we allow the strength of the magnetic field to *increase* with time, as in Fig. 3*c*, the flux through the loop will also increase with time. That is, $d\Phi_B/dt$ is also a *positive* quantity. From Faraday's law (Eq. 6), the minus sign requires that the induced emf be *negative,* which corresponds to the direction shown for the emf arrow in Fig. 3*c*. For practice, you might work out the direction of the induced emf assuming that **B** is decreasing in magnitude or that **B** points in the opposite direction to that shown in Figs. 3*b* and 3*c* and is either increasing or decreasing in magnitude.

Sample Problem 1 The long solenoid S of Fig. 4 has 220 turns/cm and carries a current $i = 1.5$ A; its diameter d is 3.2 cm. At its center we place a 130-turn close-packed coil C of diameter 2.1 cm. The current in the solenoid is reduced to zero and then increased to 1.5 A in the other direction at a steady rate over a period of 50 ms. What is the absolute value (that is, the magnitude without regard for sign) of the induced emf that appears in the central coil while the current in the solenoid is being changed?

The induced emf follows from Faraday's law (Eq. 7), in which we ignore the minus sign because we are concerned only with the absolute value of the emf. Thus,

$$\mathcal{E} = \frac{N\,\Delta\Phi_B}{\Delta t}$$

in which N is the number of turns in the inner coil C.

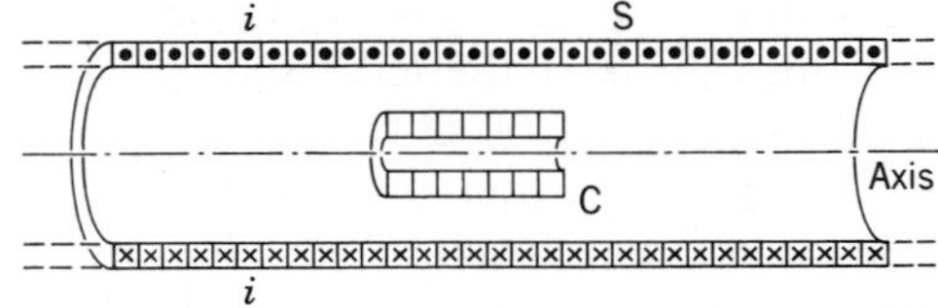

Figure 4 Sample Problem 1. A coil C is located inside a solenoid S. When the current in the solenoid is changed, an induced emf appears in the coil.

The absolute value of the initial flux through each turn of this coil is given from Eq. 4 as

$$\Phi_B = BA.$$

The magnetic field B at the center of the solenoid is given by Eq. 21 of Chapter 31 or

$$\begin{aligned} B &= \mu_0 in \\ &= (4\pi \times 10^{-7}\ \text{T}\cdot\text{m/A})(1.5\ \text{A})(220\ \text{turns/cm})(100\ \text{cm/m}) \\ &= 4.15 \times 10^{-2}\ \text{T}. \end{aligned}$$

The area of the central coil (not of the solenoid) is given by $\frac{1}{4}\pi d_C^2$, where d_C is the diameter of coil C, and works out to be $3.46 \times 10^{-4}\ \text{m}^2$. The absolute value of the initial flux through each turn of the coil is then

$$\begin{aligned} \Phi_B &= (4.15 \times 10^{-2}\ \text{T})(3.46 \times 10^{-4}\ \text{m}^2) \\ &= 1.44 \times 10^{-5}\ \text{Wb} = 14.4\ \mu\text{Wb}. \end{aligned}$$

The flux changes sign as the current is changed so that the absolute value of the *change* in flux $\Delta\Phi_B$ for each turn of the central coil is thus $2 \times 14.4\ \mu$Wb or $28.8\ \mu$Wb. This change occurs in 50 ms, giving for the magnitude of the induced emf,

$$\begin{aligned} \mathcal{E} &= \frac{N\,\Delta\Phi_B}{\Delta t} = \frac{(130\ \text{turns})(28.8 \times 10^{-6}\ \text{Wb})}{50 \times 10^{-3}\ \text{s}} \\ &= 7.5 \times 10^{-2}\ \text{V} = 75\ \text{mV}. \qquad \text{(Answer)} \end{aligned}$$

We shall explain in the next section how to use Lenz's law to find the *direction* of the induced emf.

32-4 Lenz's Law

If you drop a hammer from the top of a ladder, you do not need a rule for signs to tell you whether the hammer will move toward the center of the earth or in the opposite direction. If asked how you know that the hammer will fall, your best answer is: "It always has in the past." If pressed for a more formal answer, you might say: "If it falls, gravitational potential energy decreases and kinetic energy increases, which is acceptable. If the hammer were to rise, however, its potential energy and its kinetic energy would *both* increase, a violation of the conservation of energy principle." In this section, we want to apply this kind of reasoning to find the direction in which an induced emf will act. That is, it will act in a direction that is consistent with the conservation of energy principle.

In 1834, just 3 years after Faraday put forward his law of induction, Heinrich Friedrich Lenz gave us this rule (known as *Lenz's law*) for determining the direction of an induced current in a closed conducting loop:

An induced current in a closed conducting loop will appear in such a direction that it opposes the change that produced it.

The minus sign in Faraday's law carries with it this symbolic notion of opposition.

Lenz's law refers to induced *currents* and not to induced emfs, which means that we can apply it directly only to closed conducting loops. If the loop is not closed, however, we can usually think in terms of what would happen if it were closed and in this way find the direction of the induced emf.

To understand Lenz's law, let us apply it to a specific case, namely, the first of Faraday's experiments, shown in Fig. 1. We interpret Lenz's law as applied to this experiment in two different but equivalent ways.

First Interpretation. A current loop, like a bar magnet, has a north pole and a south pole. In each case the north pole is the region *from which* the magnetic lines emerge. If the loop is to oppose the motion of the magnet toward it, the face of the loop toward the magnet must become a north pole; see Fig. 5. The two north poles—one of the current loop and one of the magnet—will then repel each other. The right-hand rule applied to the wire of the loop shows that to produce a north pole on the face of the loop toward the magnet, the current must be as shown. Specifically, the current will be counterclockwise as we sight along the magnet toward the loop.

In the language of Lenz's law, the pushing of the magnet is the "change" that produces the induced current and that current acts to oppose the "push." If you pull the magnet away from the coil, the induced current will oppose the "pull" by creating a south pole on the face of the loop toward the magnet. This would require that the direction of the induced current be reversed.

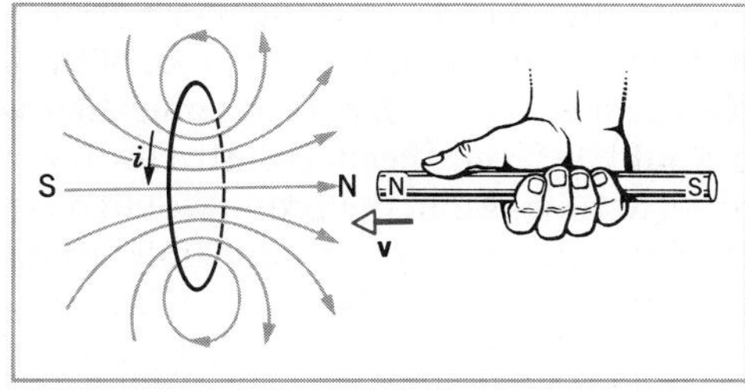

Figure 5 Lenz's law at work. If you push the magnet toward the loop, the induced current points as shown, setting up a magnetic field that opposes the motion of the magnet.

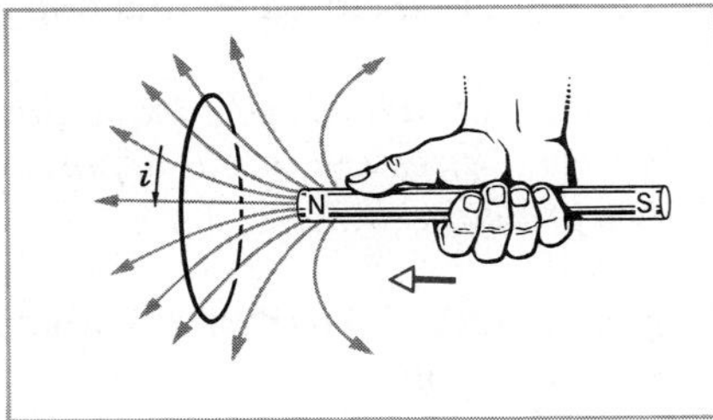

Figure 6 Lenz's law at work. If you push the magnet toward the loop, you increase the magnetic flux through the loop. The induced current in the loop sets up a magnetic field that opposes this increase in flux.

Whether you push or pull the magnet, the motion will always be opposed.

Second Interpretation. Let us now apply Lenz's law to the experiment of Fig. 1 in a different way. Figure 6 shows the lines of **B** for the bar magnet.* From this point of view the "change" referred to in Lenz's law is the increase in Φ_B through the loop caused by bringing the magnet closer. The induced current opposes this change by setting up a field of its own that opposes this increase. Thus, the magnetic field caused by the induced current must point from left to right through the plane of the loop of Fig. 6, in agreement with our earlier conclusion.

The induced magnetic field does not intrinsically oppose the *magnetic field* of the magnet; it opposes the *change* in this field, which in this case is the increase in magnetic flux through the loop. If you withdraw the magnet, you reduce Φ_B through the loop. The induced magnetic field will now oppose this decrease in Φ_B (that is, the change) by *reinforcing* the magnetic field. In each case the induced field opposes the change that gives rise to it.

Lenz's Law and Energy Conservation. Think what would happen if Lenz's law were turned the other way around, that is, if the induced current acted to *aid* the change that produced it. That would mean, for example, that a *south* pole would appear on the face of the loop in Fig. 5 as you pushed the *north* pole of a magnet toward it.

You would then only need to push a resting magnet slightly to get it moving and the action would be self-perpetuating. The magnet would accelerate toward the loop, gaining kinetic energy as it did so. At the same time thermal energy would appear in the loop. This would indeed be a something-for-nothing situation! Needless to say, it does not happen. Lenz's law is no more than a statement of the principle of conservation of energy in a form suitable for use in circuits in which there are induced currents.

Whether you push the magnet toward the loop in Fig. 1 or pull it away from the loop, you will always experience a resisting force and will thus have to do work. From the conservation of energy principle, this work must be exactly equal to the thermal energy that appears in the coil because these are the only two energy transfers that take place in this isolated system. The faster you move the magnet, the more rapidly you do work, and the greater the rate of production of thermal energy in the coil. If you cut the loop and then do the experiment, there will be no induced current, no thermal energy, no resisting force on the magnet, and no work required to move it. There will still be an emf in the loop but, like a battery in an open circuit, it will not set up a current.

* There are two magnetic fields in this experiment—one associated with the current in the loop and one associated with the bar magnet. You must always be certain with which one you are dealing.

32-5 Induction: A Quantitative Study

Figure 7 shows a rectangular loop of wire of width L, one end of which is in a uniform external magnetic field that is directed into the plane of the loop at right angles. This field may be produced, for example, by a large electromagnet. The dashed lines in Fig. 7 show the assumed limits of the magnetic field, the fringing of the field at its edges being neglected. You are asked to pull this loop to the right at a constant speed v.

The setup of Fig. 7 does not differ in any essential way from the setup of Fig. 6. In each case a magnet and a conducting loop are in relative motion; in each case the flux of the field through the loop is changing with time. It is true that in Fig. 6 the flux is changing because **B** is changing and in Fig. 7 the flux is changing because the effective loop area is changing but that difference is not essential. The important difference between the two arrangements for our purposes is that the arrangement of Fig. 7 is easier to calculate. Let us then calculate the rate at which you do mechanical work as you pull steadily on the loop in Fig. 7.

The Mechanical Work. If we know both the force F that you must exert and the speed v with which you pull the loop, we can calculate the rate at which you must do

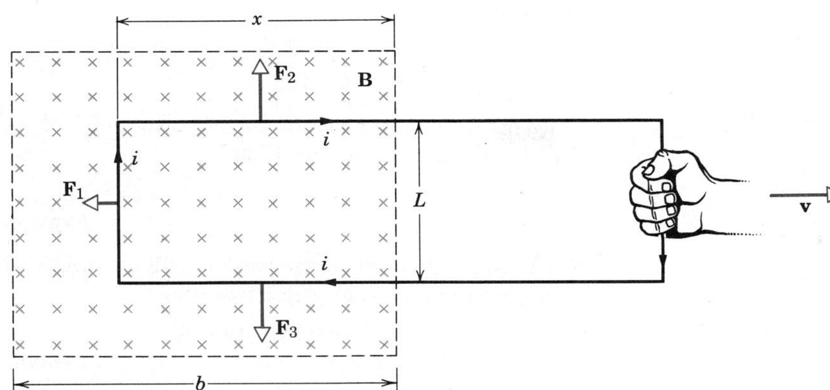

Figure 7 You pull a closed conducting loop out of a magnetic field at constant speed. While the loop is moving, a clockwise induced current appears in the loop. Thermal energy appears in the loop at a rate equal to the rate at which mechanical work is done on the loop.

work. The rate is

$$P = Fv. \tag{8}$$

Because the current-carrying loop is immersed in an external magnetic field, forces $\mathbf{F}_1$, $\mathbf{F}_2$, and $\mathbf{F}_3$ will act on three of the conductors that make up the loop, as Fig. 7 shows. Their magnitudes and directions follow from Eq. 24 of Chapter 30, or

$$\mathbf{F} = i\,\mathbf{L} \times \mathbf{B}.$$

Because $\mathbf{F}_2$ and $\mathbf{F}_3$ are equal and opposite, they cancel. $\mathbf{F}_1$ is the force that opposes your effort to move the loop. $\mathbf{F}$ is then given in magnitude by

$$F = F_1 = iLB \sin 90° = iLB. \tag{9}$$

The induced current in the loop is given by

$$i = \frac{\mathcal{E}}{R}, \tag{10}$$

where R is the loop resistance.

From Lenz's law, this current (and thus also $\mathcal{E}$) must be clockwise in Fig. 7. By circulating with this sense, it sets up a magnetic field within the loop that opposes the "change," that is, the decrease in Φ_B caused by pulling the loop out of the field.

We can now use Faraday's law to find the induced emf in the loop; from Eq. 7 (neglecting the minus sign because we are only interested in the absolute value of the emf) we have

$$\mathcal{E} = \frac{d\Phi_B}{dt}.$$

The absolute value of the flux ϕ_B enclosed by the loop in Fig. 7 is

$$\Phi_B = BLx,$$

where Lx is the area of that part of the loop in which the magnetic field is not zero. The emf is then

$$\mathcal{E} = \frac{d}{dt} BLx = BL\frac{dx}{dt} = BLv,$$

in which we have replaced dx/dt by v, the speed at which you are pulling the loop out of the field. Note that if v is constant—which we assume—the induced emf will also be constant.

From Eq. 10, the induced current in the loop is

$$i = \frac{\mathcal{E}}{R} = \frac{BLv}{R}. \tag{11}$$

The force that you would have to exert is

$$F = iLB = \frac{B^2L^2v}{R} \tag{12}$$

and the rate that you would do work is then, from Eq. 8,

$$P = Fv = \frac{B^2L^2v^2}{R} \quad \text{(rate of doing work)}. \tag{13}$$

The Thermal Energy. Now let us find the rate at which thermal energy appears in the loop as you pull it along at constant speed. We calculate it from Eq. 21 of Chapter 28 or

$$P = i^2R. \tag{14}$$

Substituting for i from Eq. 11, we find

$$P = \left(\frac{BLv}{R}\right)^2 R = \frac{B^2L^2v^2}{R} \quad \text{(thermal energy rate)}, \tag{15}$$

which is exactly equal to the rate at which you are doing work on the loop; see Eq. 13. Thus, the work that you do in pulling the loop through the magnetic field appears as

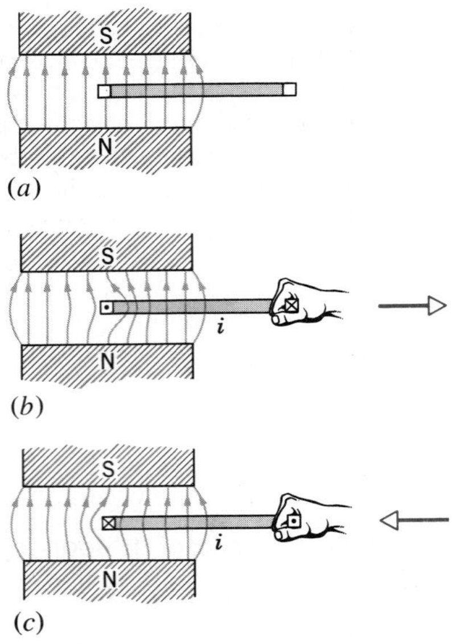

Figure 8 The pattern of the lines of force strongly suggests that any attempt to move the closed conducting loop, in either direction, will give rise to an opposing force. (Recall that field lines act somewhat like stretched rubber bands.)

thermal energy in the loop, manifesting itself as a small increase in the temperature of the loop.

Figure 8 shows a side view of the loop in the magnetic field. In Fig. 8*a* the loop is stationary; in Fig. 8*b* you are pulling it to the right; in Fig. 8*c* you are pushing it to the left. The magnetic field lines in these figures represent the *resultant* magnetic field produced by adding vectorially the external field of the electromagnet and the self-field—if any—set up by the induced current in the loop. The patterns of these lines suggest convincingly that, no matter which way you move the loop, you experience a resisting force.

Sample Problem 2 Suppose that the "loop" in Fig. 7 is actually a tightly wound coil of 85 turns, made of copper wire. Suppose further that $L = 13$ cm, $B = 1.5$ T, $R = 6.2\ \Omega$, and $v = 18$ cm/s. (a) What induced emf appears in the coil?

The induced emf appears in every turn of the coil so that, from Eq. 11, its absolute value is

$$\mathcal{E} = NBLv = (85\ \text{turns})(1.5\ \text{T})(0.13\ \text{m})(0.18\ \text{m/s})$$
$$= 2.98\ \text{V} \approx 3.0\ \text{V}. \qquad \text{(Answer)}$$

(b) What is the induced current?

We have

$$i = \frac{\mathcal{E}}{R} = \frac{2.98\ \text{V}}{6.2\ \Omega} = 0.48\ \text{A}. \qquad \text{(Answer)}$$

(c) What force must you exert on the coil to pull it along? The force is given from Eq. 12 as

$$F = NiLB = (85\ \text{turns})(0.48\ \text{A})(0.13\ \text{m})(1.5\ \text{T})$$
$$= 8.0\ \text{N}\ (=1.8\ \text{lb}). \qquad \text{(Answer)}$$

(d) At what rate must you do work to pull the coil along? The power you must exert follows from

$$P = Fv = (8.0\ \text{N})(0.18\ \text{m/s})$$
$$= 1.4\ \text{W}. \qquad \text{(Answer)}$$

Thermal energy appears in the loop at this same rate.

Sample Problem 3 Figure 9*a* shows a rectangular loop of resistance R, width L, and length b being pulled at constant speed v through a region of depth d in which a uniform magnetic field **B** is set up by an electromagnet. (a) Plot the flux Φ_B through the loop as a function of the position x of the right side of the loop. Assume that $L = 40$ mm, $b = 10$ cm, $d = 15$ cm, $R = 1.6\ \Omega$, $B = 2.0$ T, and $v = 1.0$ m/s.

The flux is zero when the loop is not in the field; it is BLb ($=8$ mWb) when the loop is entirely in the field; it is BLx when the loop is entering the field and $BL[b-(x-d)]$ when the loop is leaving the field. These conclusions, which you should verify, are plotted in Fig. 9*b*.

(b) Plot the induced emf as a function of the position of the loop.

The induced emf (see Eq. 6) is given by $-d\Phi_B/dt$, which we can write as

$$\mathcal{E} = -\frac{d\Phi_B}{dt} = -\frac{d\Phi_B}{dx}\frac{dx}{dt} = -\frac{d\Phi_B}{dx}v,$$

where $d\phi_B/dx$ is the slope of the curve of Fig. 9*b*. The emf is plotted as a function of x in Fig. 9*c*.

Lenz's law, from the arguments that we have given earlier, shows that when the loop is entering the field, the emf acts counterclockwise in Fig. 9*a*; when the loop is leaving the field, the emf is clockwise in that figure. There is no emf when the loop is either entirely out of the field or entirely in it because, in these two situations, the flux through the loop is not changing with time.

(c) Plot the rate of production of thermal energy in the loop as a function of the position of the loop.

This is given by $P = \mathcal{E}^2/R$. It may be calculated by squaring the ordinate of the curve of Fig. 9*c* and dividing by R. The result is plotted in Fig. 9*d*.

In practice, the external magnetic field **B** cannot drop sharply to zero at its boundary but must approach zero

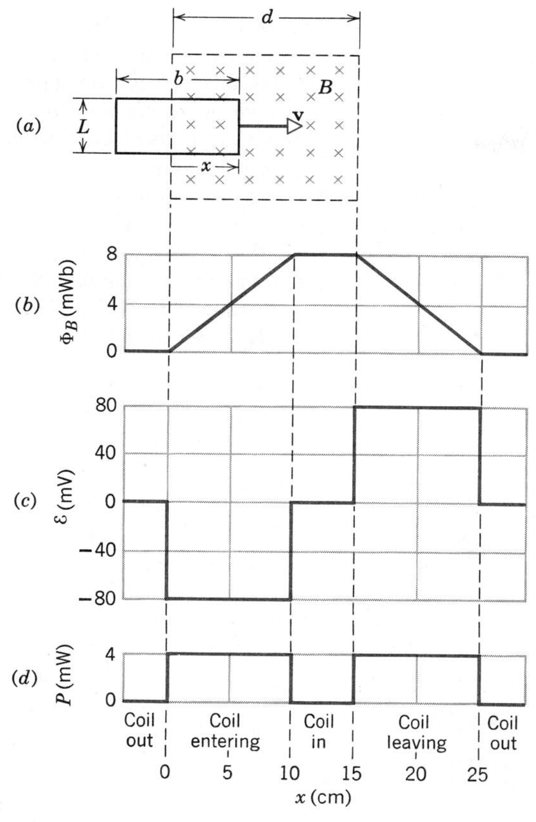

Figure 9 Sample Problem 3. (*a*) A closed conducting loop is pulled at constant speed completely through a magnetic field. (*b*) The flux through the loop for various positions of the right side of the loop. (*c*) The induced emf for various positions. (*d*) The rate at which thermal energy appears in the loop for various positions.

smoothly. The result will be a rounding of the corners of the curves plotted in Fig. 9. What changes would occur in these curves if the loop were cut, so that it no longer formed a closed conducting path?

32-6 Induced Electric Fields

Let us place a copper ring of radius r in a uniform external magnetic field, as in Fig. 10a. The field—neglecting fringing—fills a cylindrical volume of radius R. Suppose that you increase the strength of this field at a steady rate, perhaps by increasing—in an appropriate way—the current in the windings of the electromagnet that sets up the field. The magnetic flux through the ring will then change at a steady rate and—by Faraday's law—an induced emf and thus an induced current will appear in the ring. From Lenz's law you can deduce that the direction of the induced current is counterclockwise in Fig. 10a.

If there is a current in the copper ring, an electric field must be present at various points within the ring, set up by the changing magnetic flux. This *induced electric field* **E** is just as real as an electric field set up by static charges; each field, no matter what its source, will exert a force $q_0\mathbf{E}$ on a test charge. By this line of reasoning, we are lead to a useful and informative restatement of Faraday's law of induction:

A changing magnetic field produces an electric field.

The striking feature of this statement is that the electric fields are induced even if there is no copper ring.

To fix these ideas, consider Fig. 10*b*, which is just like Fig. 10*a* except the copper ring has been replaced by a hypothetical circular path of radius *r*. We assume as before that the magnetic field **B** is increasing in magnitude at the same constant rate dB/dt.

The circular path in Fig. 10*b* encloses a magnetic flux Φ_B. Because this flux is changing with time, an induced emf given by $\mathcal{E} = -d\Phi_B/dt$ will appear around the path. The electric fields induced at various points around the circular path must—from symmetry—be tangent to the circle, as Fig. 10*b* shows.* Thus, the electric lines of force set up by the changing magnetic field are in this case a set of concentric circles, as in Fig. 10*c*.

As long as the magnetic field is increasing with time, the electric field represented by the circular lines of force in Fig. 10*c* will be present. If the magnetic field remains constant with time, there will be no induced electric field and thus no lines of force. If the magnetic field is *decreasing* with time (at the same rate), the lines of force will be as they are in Fig. 10*c* but will point in the opposite direction. All this is what we have in mind when we say: "A changing magnetic field produces an electric field."

A Reformulation of Faraday's Law. Consider a test charge q_0 moving around the circular path of Fig. 10*b*.

* Arguments of symmetry would also permit the lines of **E** around the circular path to be *radial,* rather than tangential. However, such radial lines would imply that there are free charges distributed symmetrically about the axis of symmetry; there are no such charges.

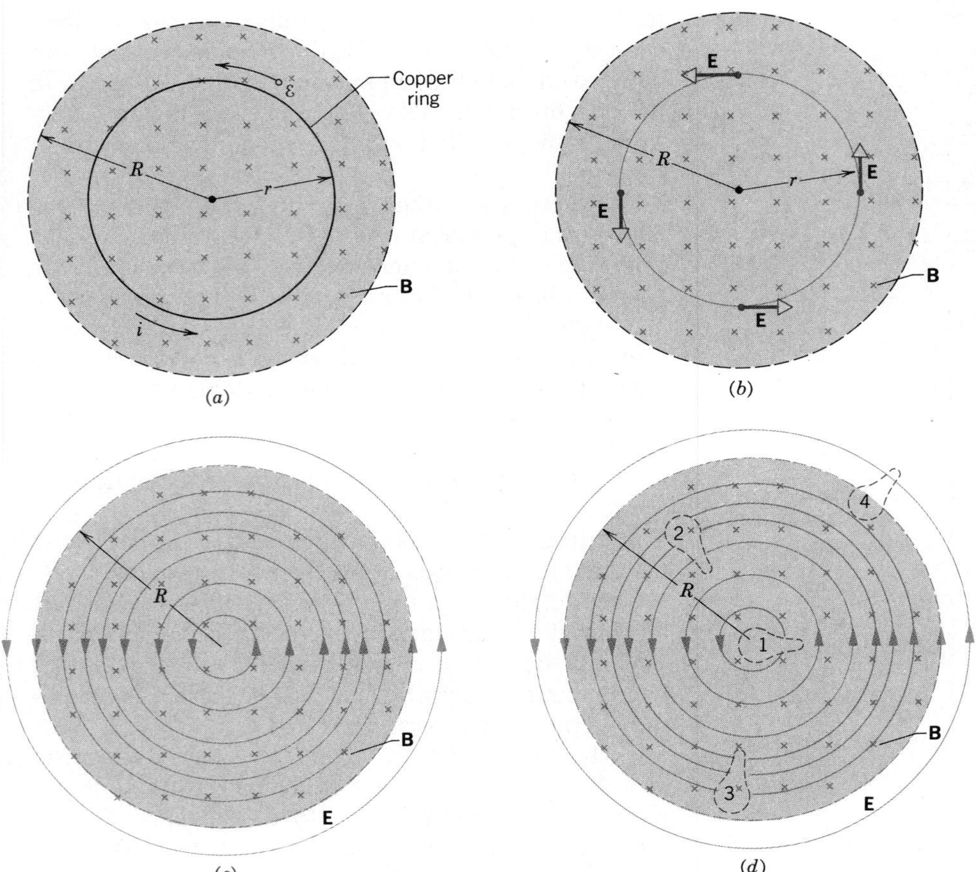

Figure 10 (*a*) If the magnetic field increases at a steady rate, a constant induced current appears, as shown, in the copper ring of radius r. (*b*) Induced electric fields appear at various points even when the ring is removed. (*c*) The complete picture of the induced electric fields, displayed as lines of force. (*d*) Four similar closed paths, around which an induced emf may—or may not—appear.

The work W done on it in one revolution by the electric field is, in terms of our definition of an emf, simply $\mathcal{E}q_0$. From another point of view, the work is $(q_0E)(2\pi r)$, where q_0E is the magnitude of the force acting on the test charge and $2\pi r$ is the distance over which that force acts. Setting these two expressions for W equal to each other and canceling q_0, we find

$$\mathcal{E} = E2\pi r. \tag{16}$$

In a more general case than that of Fig. 10*b*, we can write

$$\mathcal{E} = \oint \mathbf{E} \cdot d\mathbf{s}. \tag{17}$$

This integral reduces at once to Eq. 16 if we evaluate it for the special case of Fig. 10*b*.

If we combine Eq. 17 with Eq. 6 ($\mathcal{E} = -d\Phi_B/dt$), we can write Faraday's law of induction as

$$\boxed{\oint \mathbf{E} \cdot d\mathbf{s} = -\frac{d\Phi_B}{dt}} \quad \text{(Faraday's law).} \tag{18}$$

This is the form in which Faraday's law is displayed in Table 2 of Chapter 37, this table being our summary of Maxwell's equations of electromagnetism. Equation 18 is again what we have in mind when we say: "A changing magnetic field produces an electric field." The changing magnetic field appears on the right side of this equation, the electric field on the left.

Faraday's law in the form of Eq. 18 can be applied to *any* closed path that can be drawn in a changing magnetic field. Figure 10*d*, for example, shows four such paths, all having the same shape and area but located in different positions in the changing field. For paths 1 and 2, the induced emf is the same because these paths lie entirely in the magnetic field and thus have the same value of $d\Phi_B/dt$. Note that, even though the emf $\mathcal{E}$ ($= \oint \mathbf{E} \cdot d\mathbf{s}$) is the same for these two paths, the distribution of the electric field vectors around these paths is different, as indicated by the pattern of electric lines of force. For path 3 the induced emf is smaller because Φ_B (and hence $d\Phi_B/dt$) is smaller, and for path 4 the induced emf is zero, even though the electric field is not zero at any point on the path.

A New Look at Electric Potential. The induced electric fields that are set up by the induction process are not associated with static charges but with a changing magnetic flux. Although electric fields set up in either way exert forces on test charges, there is an important difference between them. The simplest evidence of this difference is that the lines of force associated with induced electric fields form closed loops, as in Fig. 10*c*. Lines associated with static charges never do so but must start on positive charges and end on negative charges.

In a more formal sense, we can state the difference between electric fields set up by induction and those set up by static charges in these words:

> *Electric potential has meaning only for electric fields set up by static charges; it has no meaning for electric fields set up by induction.*

We can understand this qualitatively by considering what happens to a test charge that makes a single journey around the circular path in Fig. 10*b*. It starts at a certain point and, after it returns to that same point, has experienced an emf $\mathcal{E}$ of, let us say, 5 V. Its potential should have increased by this amount. This is impossible, however, because otherwise the same point in space would have two different values of potential. We can only conclude that potential has no meaning for electric fields that are set up by changing magnetic fields.

We can take a more formal look by recalling Eq. 8 of Chapter 26, which defines the potential difference between two points i and f:

$$V_f - V_i = \frac{W_{if}}{q_0} = -\int_i^f \mathbf{E} \cdot d\mathbf{s}. \tag{19}$$

In Chapter 26 we had not yet encountered Faraday's law of induction so that the electric fields involved in the derivation of Eq. 19 were those due to static charges. If i and f in Eq. 19 are the same point, the path connecting them is a closed loop, V_i and V_f are identical, and Eq. 19 reduces to

$$\oint \mathbf{E} \cdot d\mathbf{s} = 0. \tag{20}$$

However, when a changing magnetic flux is present, this integral is *not* zero but is $-d\Phi_B/dt$, as Eq. 18 asserts. Again, we conclude that electric potential has no meaning for electric fields associated with induction.

Sample Problem 4 In Fig. 10*b*, assume that $R = 8.5$ cm and that $dB/dt = 0.13$ T/s. (a) What is the magnitude of the electric field **E** for $r = 5.2$ cm?

From Faraday's law (Eq. 18) we have

$$(E)(2\pi r) = -\frac{d\Phi_B}{dt}.$$

We note that $r < R$. The flux Φ_B through a closed path of radius r is then

$$\Phi_B = B(\pi r^2).$$

so that

$$(E)(2\pi r) = -(\pi r^2)\frac{dB}{dt}.$$

Solving for E and dropping the minus sign, we find

$$E = \tfrac{1}{2}(dB/dt)\, r. \tag{21}$$

Note that the induced electric field E depends on dB/dt but not on B. For $r = 5.2$ cm, we have, for the magnitude of **E**,

$$E = \tfrac{1}{2}(dB/dt)r = (\tfrac{1}{2})(0.13\text{ T/s})(5.2 \times 10^{-2}\text{ m})$$
$$= 0.0034\text{ V/m} = 3.4\text{ mV/m}. \quad \text{(Answer)}$$

(b) What is the magnitude of the induced electric field for $r = 12.5$ cm?

In this case we have $r > R$ so that the entire flux of the magnet passes through the circular path. Thus,

$$\Phi_B = B(\pi R^2).$$

From Faraday's law (Eq. 18) we then find

$$(E)(2\pi r) = \frac{d\Phi_B}{dt} = -(\pi R^2)\frac{dB}{dt}.$$

Solving for E and again dropping the minus sign, we find

$$E = \tfrac{1}{2}(dB/dt)R^2\frac{1}{r}. \tag{22}$$

Interestingly, an electric field is induced in this case even at points that are well outside the (changing) magnetic field, an

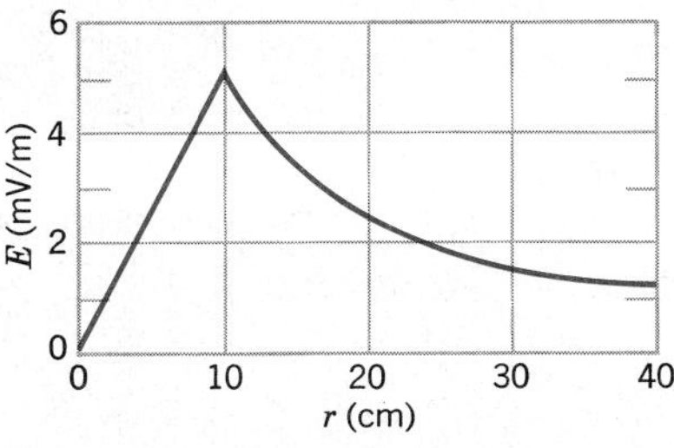

Figure 11 A plot of the induced electric field $E(r)$ for the conditions of Sample Problem 4.

important result that makes transformers possible. For $r = 12.5$ cm, Eq. 22 gives

$$E = \frac{(\frac{1}{2})(0.13\ \text{T/s})(8.5 \times 10^{-2}\ \text{m})^2}{12.5 \times 10^{-2}\ \text{m}}$$

$$= 3.8 \times 10^{-3}\ \text{V/m} = 3.8\ \text{mV/m}. \quad \text{(Answer)}$$

Equations 21 and 22 yield the same result, as they must, for $r = R$. Figure 11 shows a plot of $E(r)$ based on these two equations.

32-7 The Betatron

The betatron is a device used to accelerate electrons to high energies by allowing them to be acted on by induced electric fields. Although the betatron is not widely used today, we describe it because it is a perfect example of the reality of these induced fields.

Figure 12 shows a cross section of a betatron in a plane that contains its vertical symmetry axis. The (time-varying) magnetic field that is shown there has several functions: (1) It guides the electrons in a circular path. (2) The changing magnetic flux generates an electric field that accelerates the electrons in this path. (3) It keeps the radius of the electron orbit essentially constant during the acceleration process. (4) It injects the electrons into the orbit initially and extracts them from the orbit after they have reached their full energy. (5) It provides a restoring force that resists any tendency for the electrons to stray away from their orbit, either vertically or radially. It is remarkable that it is possible to do all these things by proper shaping and control of the magnetic field. Although the concept of the betatron had been proposed by others, it was Don W. Kerst, at the University of Illinois, who first succeeded in 1941 in providing a magnetic field that performed all these functions in a working betatron.

The object marked D in Fig. 12 is an evacuated ceramic "doughnut," inside of which the electrons circulate and are accelerated. Their orbit is a circle of constant radius R, its plane being at right angles to the page. In the figure, the electrons are circulating counterclockwise as viewed from above. Thus, we see them emerging from the figure on the left ⊙ and entering it on the right ⊗.

The alternating magnetic field shown in Fig. 12 is produced by means of an alternating current in coils (not shown) around the iron pole pieces. The field B_{orb} at the orbit position serves to guide the electrons in their orbit. The field in the area enclosed by the orbit, whose average value is $B_{av}(=2B_{orb})$, contributes to the central flux Φ_B. It is the time variation of this central flux that induces the electric field that acts on the electrons and accelerates them.

Sample Problem 5 In a 100-MeV betatron built at the General Electric Company, the orbit radius R is 84 cm. The magnetic field in the region enclosed by the orbit rises periodically (60 times per second) from zero to a maximum average value

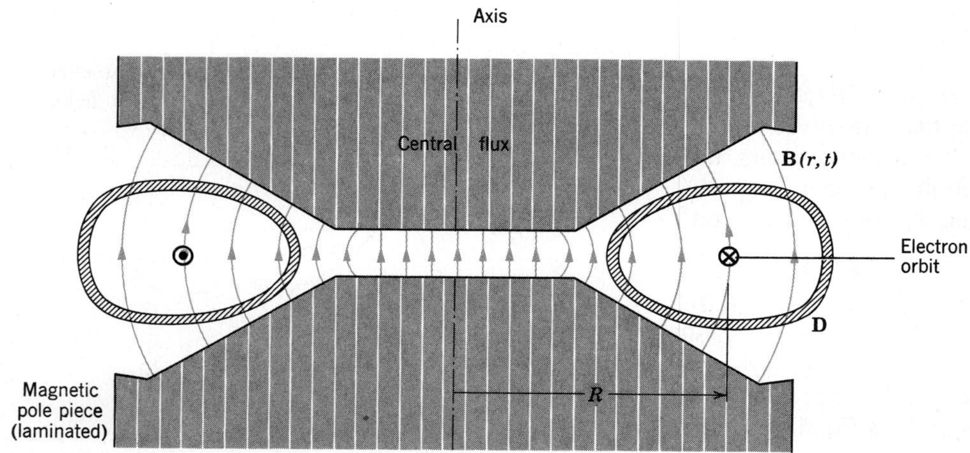

Figure 12 A cross section of a betatron, showing the orbit of the accelerating electrons and a "snapshot" of the time-varying magnetic field at a certain moment during the acceleration cycle.

$B_{av,m} = 0.80$ T in an accelerating interval of one fourth of a period, or 4.2 ms. (a) How much energy does the electron gain in one average trip around its orbit in this changing flux?

The central flux rises during the accelerating interval from zero to a maximum of

$$\Phi_m = (B_{av,m})(\pi R^2) = (0.80\ \text{T})(\pi)(0.84\ \text{m})^2 = 1.8\ \text{Wb}.$$

The average value of $d\Phi_B/dt$ during the accelerating interval is then

$$\left(\frac{d\Phi_B}{dt}\right)_{av} = \frac{1.8\ \text{Wb}}{4.2 \times 10^{-3}\ \text{s}} = 430\ \text{Wb/s}.$$

From Faraday's law (Eq. 6) this is also the average emf in volts. Thus, the electron increases its energy by an average of 430 eV per revolution in this changing flux. To achieve its full final energy of 100 MeV, it has to make about 230,000 revolutions in its orbit, a total path length of about 1200 km.

(b) What is the *average* speed of an electron during its acceleration cycle?

The length of the acceleration cycle is given as 4.2 ms and the path length is calculated above to be 1200 km. The average speed is then

$$\bar{v} = \frac{1200 \times 10^3\ \text{m}}{4.2 \times 10^{-3}\ \text{s}} = 2.86 \times 10^8\ \text{m/s}. \quad \text{(Answer)}$$

This is 95% of the speed of light. The actual speed of the fully accelerated electron, when it has reached its final energy of 100 MeV, can be shown to be 99.9987% of the speed of light.

32-8 Physics: A Human Enterprise (Optional)

The world honors some of its citizens by naming things after them. On the endless list are countries (Bolivia), states (Pennsylvania), cities (Seattle), universities (Harvard), mountains (Pike's Peak), automobiles (Chevrolet), birds (Bachman's warbler), plants (poinsettia), minerals (dolomite), and food containers (Mason jars and Tupperware).

It is much the same in physics. Among the ways in which we honor some of those who have contributed to our discipline is attaching their name to something. Because of his wide range of interests and the remarkable standard of his achievements, Michael Faraday has been especially honored in this way:

1. The Farad. As we have seen, the unit of capacitance is named in honor of his early discoveries in this field.

2. Faraday's Law of Induction. This entire chapter deals with this important law, of which Faraday was the co-discoverer.

3. The Faraday. This physical constant, named to recognize Faraday's extensive work in electrochemistry, is the charge possessed by 1 mole of single-ionized atoms. Its value is 96,580.35 C.

4. The Faraday Effect. This effect, which we do not discuss in this book, deals with the rotation of the plane of polarization of polarized light (see Section 38-7) in a solution of organic materials such as sugar.

Most SI units are named after an individual closely associated with the branch of physics with which the unit deals. There are—among others—the newton (force; Isaac Newton), the ampere (current; Andre Marie Ampère), the coulomb (charge; Charles August Coulomb), the joule (energy; James Prescott Joule), the hertz (frequency; Heinrich Hertz), the watt (power; James Watt), the kelvin (temperature; Lord Kelvin), and the henry (inductance; Joseph Henry).

Units are only one of many things that are named after physicists. Among items mentioned in this book we can list—proceeding alphabetically—Archimedes' principle, Bernoulli's equation, the Carnot cycle, the Doppler shift, the Eötvös experiment, the Foucault pendulum, Gauss' law, the Hall effect, the Ives-Stillwell experiment, Joule's law, the Kelvin scale, Lenz's law, the Mach number, Newton's rings, Ohm's law, the Planck constant, Rayleigh's criterion, Schrödinger's equation, Torricelli's law, the Van de Graaff accelerator, Wien's law, Young's modulus, and the Zeeman effect.

This practice is a constant reminder that physics is a human enterprise; that behind every discovery and every new insight, there are the minds and hands of people not unlike ourselves. It is of course true that many physicists—equally meritorious—do not happen to be recognized in this particular way.

You may wish to read about the lives and works of physicists for whom readable and authoritative biographies exist. Table 1 lists some of these, starting (arbitrarily) with Michael Faraday.

It has often been said that physics is a young person's game and that, if one has not made some contribution of merit by age 30, there is little hope for the future! Although this is far from true, it *is* true that many physicists did make major contributions in the full flush of youth, Newton at 24 and Einstein at 26 being examples that stand out. As a more useful chronological guide than birth and death date, we list in Table 1 the year in which the physicist in question became 30.

Table 1 Biographical Information About Some Physicists[a]

Person	Year Turned 30	Reference
Michael Faraday	1821	*Michael Faraday,* by L. Pearce Williams, Basic Books, New York, 1964
Marie Curie	1897	*Madame Curie,* by Eve Curie, Doubleday, Doran & Company, New York, 1937
Albert Einstein	1909	*Subtle is the Lord . . . ,* by Abraham Pais, Clarendon Press, New York, 1982 *Einstein,* by Jeremy Bernstein, The Viking Press, New York, 1973.
H. G. J. Moseley	—[b]	*H. G. J. Moseley,* by J. L. Heilbron, University of California Press, Berkeley, 1974
Niels Bohr	1915	*Niels Bohr,* by Ruth Moore, Alfred A. Knopf, New York, 1966
I. I. Rabi	1928	*Rabi: American Physicist,* by John S. Rigden, Basic Books, New York, 1987
Enrico Fermi	1931	*Atoms in the Family,* by Laura Fermi, The University of Chicago Press, Chicago, 1954
Ernest Lawrence Robert Oppenheimer	1931 1934	*Lawrence and Oppenheimer,* by Nuel Pharr Davis, Simon and Schuster, New York, 1968
Hans A. Bethe	1936	*Hans Bethe: Prophet of Energy,* by Jeremy Bernstein, Basic Books, New York, 1980
Luis W. Alvarez	1941	*Alvarez: Adventures of a Physicist,* by Luis W. Alvarez, Basic Books, New York, 1987
Richard Feynman	1948	*Surely You're Joking, Mr. Feynman,* by Richard P. Feynman, Bantam Books, New York, 1986
Jeremy Bernstein	1959	*The Life it Brings: One Physicist's Beginnings,* by Jeremy Bernstein, Ticknor & Fields, New York, 1987

[a] For fascinating accounts of the life and work of many prominent physicists, read:

The Physicists—The History of a Scientific Community in Modern America, by Daniel J. Kevles, Vintage Books, New York, 1979

The Second Creation—Makers of the Revolution in Twentieth-Century Physics, by Robert P. Crease and Charles C. Mann, Macmillan Publishing, New York, 1986

From Falling Bodies to Radio Waves—Classical Physicists and Their Discoveries, by Emilio Segre, W. H. Freeman, New York, 1984

From X-Rays to Quarks—Modern Physicists and Their Discoveries, by Emilio Segre, W. H. Freeman, New York, 1984

[b] Moseley, who laid the experimental basis for the concept of atomic number, never reached 30. He was killed by a sniper's bullet in 1915, during World War I.

REVIEW AND SUMMARY

Definition of Magnetic Flux

The flux $\mathbf{\Phi}_B$ for a given surface immersed in a magnetic field **B** is defined from

$$\Phi_B = \int \mathbf{B} \cdot d\mathbf{A} \qquad [3]$$

where the integral is taken over the surface. Review Section 25-4, which deals with the corresponding flux Φ_E for a surface immersed in an electric field **E**. The SI unit of magnetic flux is the weber, where 1 Wb = 1 T·m^2.

Faraday's law of induction states that if Φ_B for a surface bounded by a closed loop changes with time an emf given by

Faraday's Law of Induction

$$\mathcal{E} = -N\frac{d\Phi_B}{dt} \quad \text{(Faraday's law)} \qquad [7]$$

appears in the loop. If the loop is a conductor, a current will be set up. The two experiments of Section 32–2 and Sample Problem 1 illustrate this law.

Lenz's Law

Lenz's law specifies the direction of the current induced in a closed conducting loop by a changing magnetic flux. The law states: *An induced current in a closed conducting loop will appear in such a direction that it opposes the change that produced it.* Lenz's law is a consequence of the conservation of energy principle. Section 32–5, for example, shows that work is needed to pull a closed conducting loop out of a magnetic field and that this energy is accounted for as thermal energy of the loop material. If the direction of the induced current were opposite to that predicted by Lenz's law, work would be done *by* the loop and energy would not be conserved.

Conservation of Energy

An induced emf will be present even if the loop through which the magnetic flux is changing is not a physical conductor but an imaginary line. The changing flux induces an electric field **E** at every point of such a loop, the emf being related to **E** by

Emf and the Induced Electric Field

$$\mathcal{E} = \oint \mathbf{E} \cdot d\mathbf{s}. \qquad [17]$$

The integral is taken around the loop. Combining Eqs. 7 and 17 lets us write Faraday's law in its most general form, or

Faraday's Law of Induction

$$\oint \mathbf{E} \cdot d\mathbf{s} = -\frac{d\Phi_B}{dt} \quad \text{(Faraday's law).} \qquad [18]$$

The essence of this law is that *a changing magnetic field ($d\Phi_B/dt$) induces an electric field* (**E**). Sample Problem 4 illustrates the law in this form.

Induced Electric Fields

Electric fields induced by changing magnetic fields differ from those associated with static charges in that their lines of force form closed loops, as in Fig. 10. Electric potential has no meaning for such induced electric fields; the forces associated with them are nonconservative.

The Betatron

The betatron (Fig. 11) uses induced electric fields to accelerate electrons to high energy. Sample Problem 5 shows how to use Faraday's law of induction to calculate the average energy gain per revolution of the orbit for a particular machine.

QUESTIONS

1. Are induced emfs and currents different in any way from emfs and currents provided by a battery connected to a conducting loop?

2. Is the size of the voltage induced in a coil through which a magnet moves affected by the strength of the magnet? If so, explain how.

3. Explain in your own words the difference between a magnetic field **B** and the flux of a magnetic field Φ_B. Are they vectors or scalars? In what units may each be expressed? How are these units related? Are either or both (or neither) properties of a given point in space?

4. Can a charged particle at rest be set in motion by the action of a magnetic field? If not, why not? If so, how?

5. You drop a bar magnet along the axis of a long copper tube. Describe the motion of the magnet and the energy interchanges involved. Neglect air resistance.

6. You are playing with a metal loop, moving it back and forth in a magnetic field, as in Fig. 8. How can you tell, without detailed inspection, whether or not the loop has a narrow saw cut across it, rendering it nonconducting?

7. Figure 13 shows an inclined wooden track that passes, for part of its length, through a strong magnetic field. You roll a copper disk down the track. Describe the motion of the disk as it rolls from the top of the track to the bottom.

8. Figure 14 shows a copper ring, hung from a ceiling by two

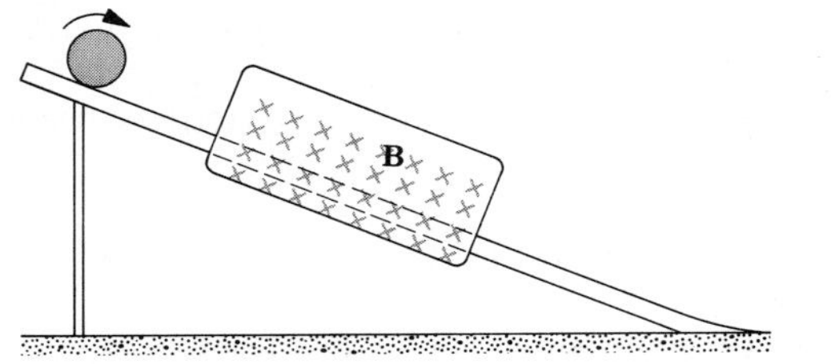

Figure 13 Question 7.

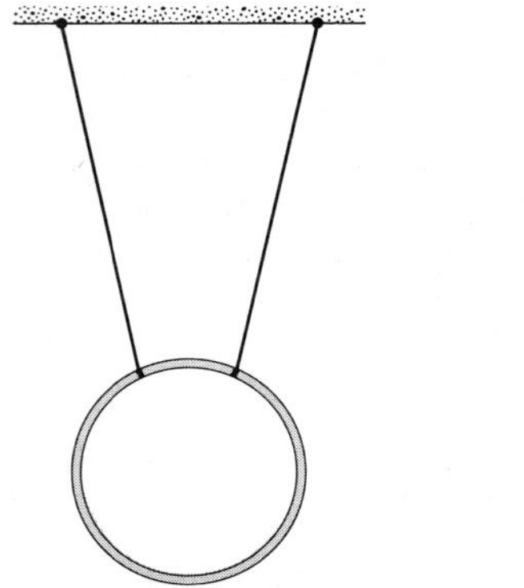

Figure 14 Question 8.

threads. Describe in detail how you might most effectively use a bar magnet to get this ring to swing back and forth.

9. Is an emf induced in a long solenoid by a bar magnet that moves inside it along the solenoid axis? Explain your answer.

10. Two conducting loops face each other a distance d apart (Fig. 15). An observer sights along their common axis from left to right. If a clockwise current i is suddenly established in the larger loop, by a battery not shown, (*a*) what is the direction of the induced current in the smaller loop? (*b*) What is the direction of the force (if any) that acts on the smaller loop?

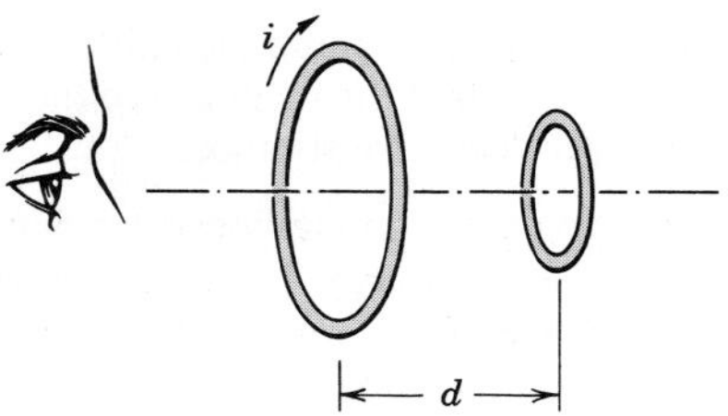

Figure 15 Question 10.

11. What is the direction of the induced emf in coil Y of Fig. 16 (*a*) when coil Y is moved toward coil X? (*b*) When the

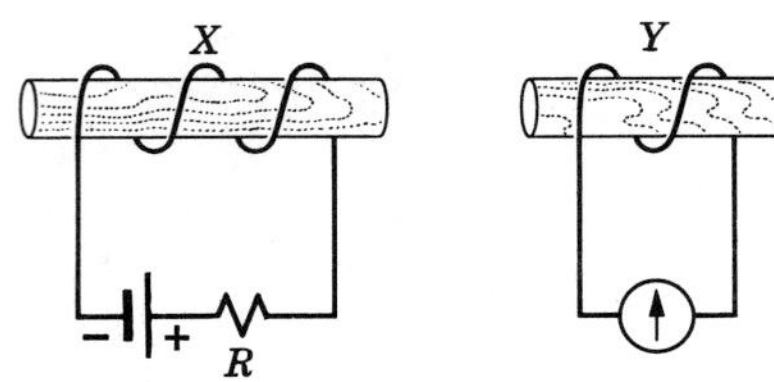

Figure 16 Question 11.

current in coil X is decreased, without any change in the relative positions of the coils?

12. The north pole of the magnet is moved away from a copper ring, as in Fig. 17. In the part of the ring farthest from the reader, which way does the current point?

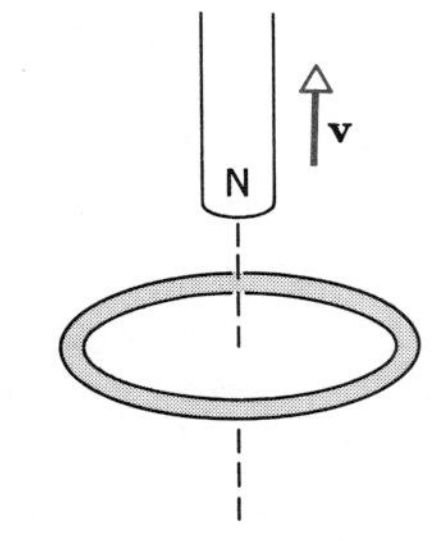

Figure 17 Question 12.

13. A circular loop moves with constant velocity through regions where uniform magnetic fields of the same magnitude are directed into or out of the plane of the page, as indicated in Fig. 18. At which of the seven indicated positions will the emf be (*a*) clockwise? (*b*) counterclockwise? (*c*) zero?

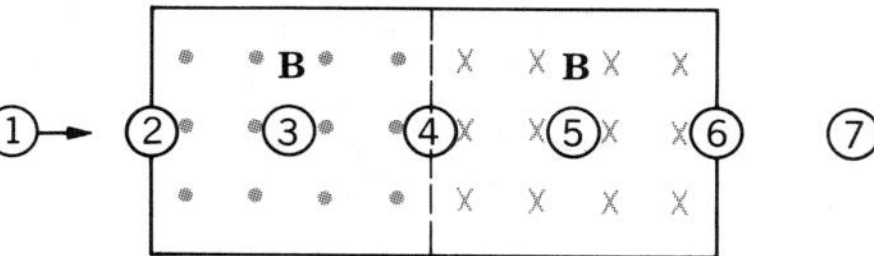

Figure 18 Question 13.

14. A short solenoid carrying a steady current is moving toward a conducting loop as in Fig. 19. What is the direction of the induced current in the loop as one sights toward it as shown?

15. The resistance R in the left-hand circuit of Fig. 20 is being increased at a steady rate. What is the direction of the induced current in the right-hand circuit?

16. What is the direction of the induced current through resistor R in Fig. 21 (*a*) immediately after switch S is closed, (*b*)

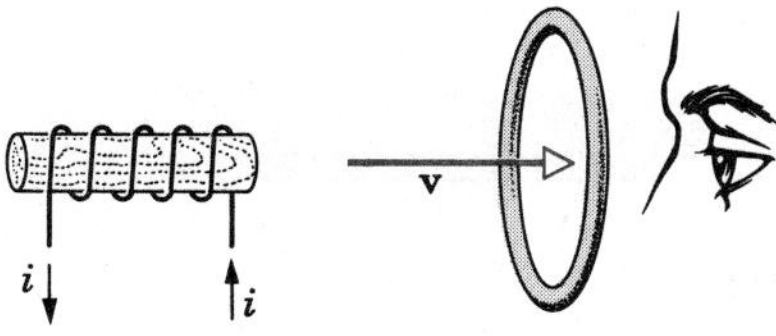

Figure 19 Question 14.

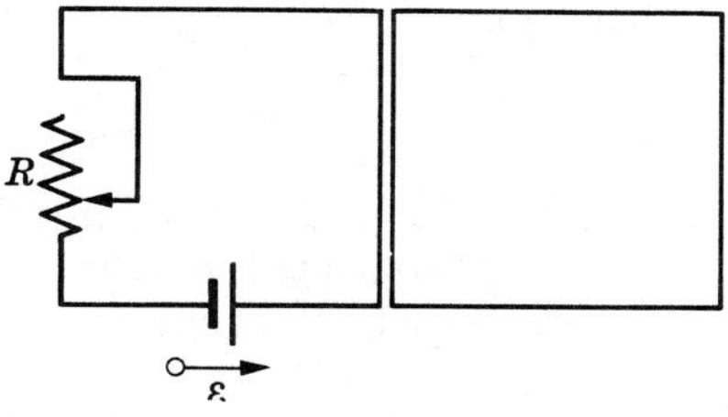

Figure 20 Question 15.

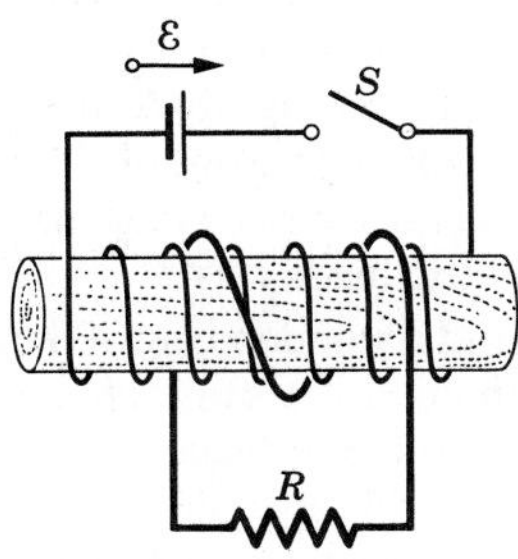

Figure 21 Question 16.

some time after switch S was closed, and (*c*) immediately after switch S is opened? (*d*) When switch S is held closed, from which end of the longer coil do field lines emerge? This is the effective north pole of the coil. (*e*) How do the conduction electrons in the coil containing R know about the flux within the long coil? What really gets them moving?

17. In Faraday's law of induction, does the induced emf depend on the resistance in the circuit? If so, how?

18. Suppose that the direction of induced emfs was governed by what we can call the Antilenz law: The induced current will appear in such a direction that it aids the change that produced it. Design a machine based on this law that would make a lot of money for you. (Alas, the Antilenz law is *false*.)

19. The loop of wire shown in Fig. 22 rotates with constant angular speed about the x axis. A uniform magnetic field **B**, whose direction is that of the positive y axis is present. For what portions of the rotation is the induced current in the loop (*a*) from P to Q, (*b*) from Q to P, (*c*) zero? (*d*) Repeat if the direction of rotation is reversed from that shown in the figure.

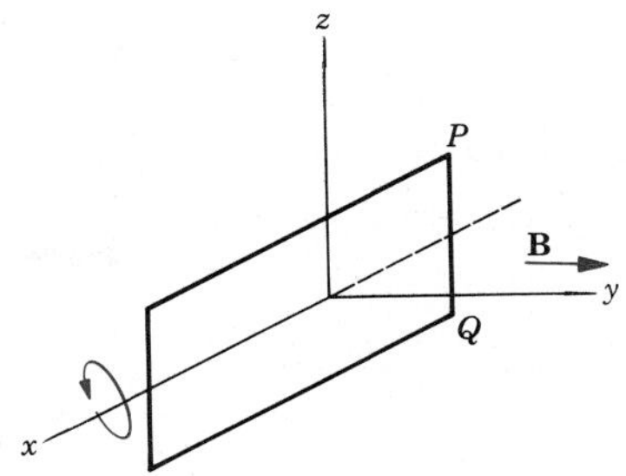

Figure 22 Question 19.

20. In Fig. 23 the straight movable wire segment is moving to the right with a constant velocity **v**. An induced current appears in the direction shown. What is the direction of the uniform magnetic field (assumed constant and perpendicular to the page) in region A?

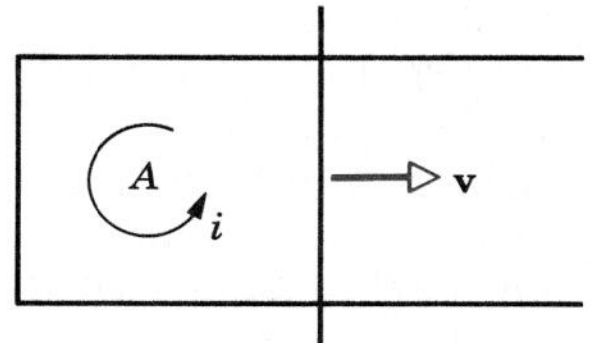

Figure 23 Question 20.

21. A conducting loop, shown in Fig. 24, is removed from the permanent magnet by pulling it vertically upward. (*a*) What is the direction of the induced current? (*b*) Is a force required to remove the loop? (*c*) Does the total amount of thermal energy produced in removing the loop depend on the time taken to remove it?

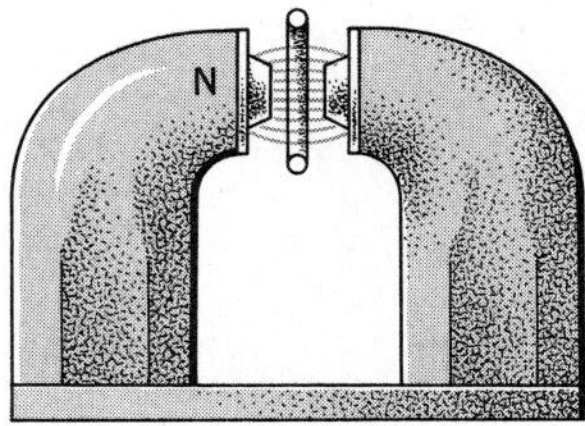

Figure 24 Question 21.

22. A plane closed loop is placed in a uniform magnetic field. In what ways can the loop be moved without inducing an emf? Consider motions of both translation and rotation.

23. *Eddy currents.* A sheet of copper is placed in a magnetic field as shown in Fig. 25. If we attempt to pull it out of the field

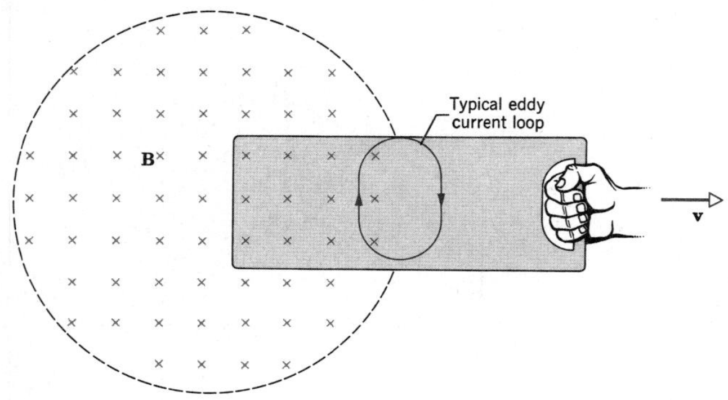

Figure 25 Question 23.

or push it further in, a resisting force automatically appears. Explain its origin. (*Hint:* Currents, called eddy currents, are induced in the sheet in such a way as to oppose the motion.)

24. *Magnetic damping.* A strip of copper is mounted as a pendulum about O in Fig. 26. It is free to swing through a magnetic field normal to the page. If the strip has slots cut in it as shown, it can swing freely through the field. If a strip without slots is substituted, the vibratory motion is strongly damped. Explain the observations. (*Hint:* Use Lenz's law; consider the paths that the charge carriers in the strip must follow if they are to oppose the motion.)

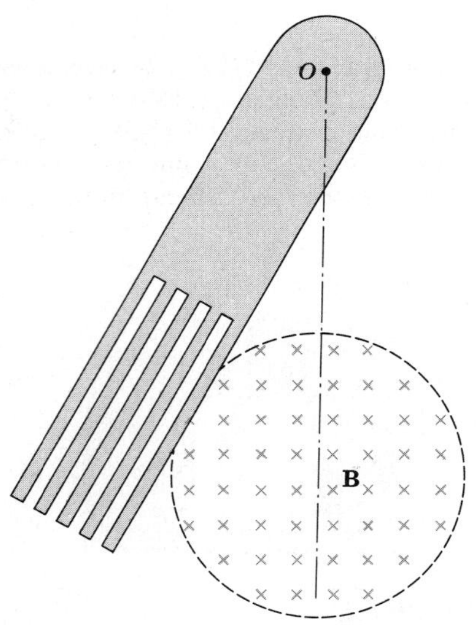

Figure 26 Question 24.

25. *Electromagnetic shielding.* Consider a conducting sheet lying in a plane perpendicular to a magnetic field **B**, as shown in Fig. 27. (*a*) If **B** suddenly changes, the full change in **B** is not immediately detected at points near P. Explain. (*b*) If the resistivity of the sheet is zero, the change is not ever detected at P. Explain. (*c*) If **B** changes periodically at high frequency and the conductor is made of a material of low resistivity, the region near P is almost completely shielded from the changes in flux. Explain. (*d*) Why is such a conductor not useful as a shield from static magnetic fields?

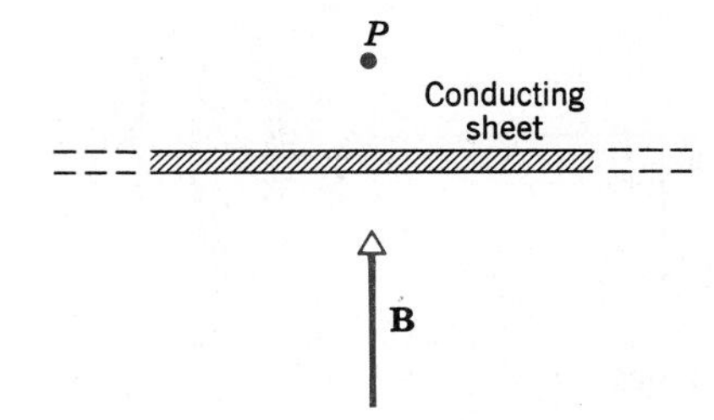

Figure 27 Question 25.

26. (*a*) In Fig. 10*b*, need the circle of radius r be a conducting loop in order that **E** and $\mathcal{E}$ be present? (*b*) If the circle of radius r were not concentric (moved slightly to the left, say), would $\mathcal{E}$ change? Would the configuration of **E** around the circle change? (*c*) For a concentric circle of radius r, with $r > R$, does an emf exist? Do electric fields exist?

27. A copper ring and a wooden ring of the same dimensions are placed so that there is the same changing magnetic flux through each. Compare the induced electric fields in the two rings.

28. An airliner is cruising in level flight over Alaska, where the earth's magnetic field has a large downward component. Which of its wingtips (right or left) has more electrons than the other?

29. In Fig. 10*d* how can the induced emfs around paths 1 and 2 be identical? The induced electric fields are much weaker near path 1 than near path 2, as the spacing of the lines of force shows. See also Fig. 11.

30. Show that, in the betatron of Fig. 12, the directions of the lines of **B** are correctly drawn to be consistent with the direction of circulation shown for the electrons.

31. In the betatron of Fig. 12 you want to increase the orbit radius by suddenly imposing an additional central flux $\Delta\Phi_B$ (set up by suddenly establishing a current in an auxilliary coil not shown). Should the lines of **B** associated with this flux increment be in the same direction as the lines shown in the figure or in the opposite direction? Assume that the magnetic field at the orbit position remains relatively unchanged by this flux increment.

32. In the betatron of Fig. 12, why is the iron core of the magnet made of laminated sheets rather than of solid metal as for the cyclotron of Section 6 of Chapter 30? (*Hint:* Consider the implications of Questions 24 and 25.)

EXERCISES AND PROBLEMS

Section 32-3 Faraday's Law of Induction

1E. At a certain location in the northern hemisphere, the earth's magnetic field has a magnitude of 42 μT and points downward at 57° to the vertical. Calculate the flux through a horizontal surface of area 2.5 m^2; see Fig. 28.

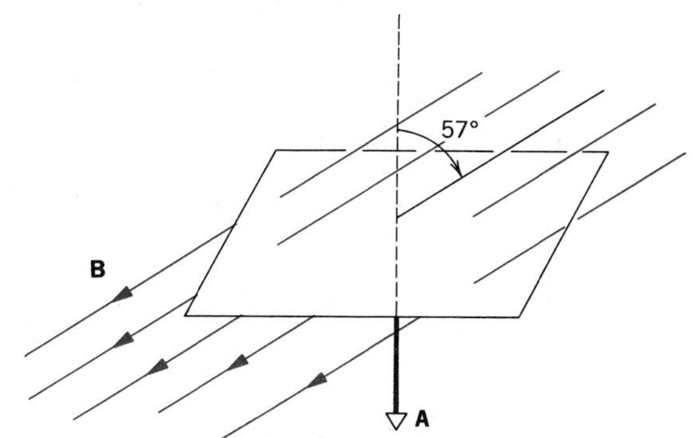

Figure 28 Exercise 1.

2E. A small loop of area A is inside of, and has its axis in the same direction as, a long solenoid of n turns per unit length and current i. If $i = i_0 \sin \omega t$, find the emf in the loop.

3E. A circular UHF television antenna has a diameter of 11 cm. The magnetic field of a TV signal is normal to the plane of the loop and, at one instant of time, its magnitude is changing at the rate 0.16 T/s. The field is uniform. What is the emf in the antenna?

4E. A uniform magnetic field **B** is perpendicular to the plane of a circular wire loop of radius r. The magnitude of the field varies with time according to $B = B_0 e^{-t/\tau}$ where B_0 and τ are constants. Find the emf in the loop as a function of time.

5E. In Fig. 29 the magnetic flux through the loop shown increases according to the relation

$$\Phi_B = 6t^2 + 7t,$$

where Φ_B is in milliwebers and t is in seconds. (a) What is the magnitude of the emf induced in the loop when $t = 2.0$ s? (b) What is the direction of the current through R?

6E. The magnetic field through a one-turn loop of wire 12 cm in radius and 8.5 Ω in resistance changes with time as shown in Fig. 30. Calculate the emf in the loop as a function of time. Consider the time intervals (a) $t = 0$ to $t = 2$ s; (b) $t = 2$ s to $t = 4$ s; (c) $t = 4$ s to $t = 6$ s. The (uniform) magnetic field is at right angles to the plane of the loop.

7E. A loop antenna of area A and resistance R is perpendicular to a uniform magnetic field **B**. The field drops linearly to zero in a time interval Δt. Find an expression for the total thermal energy dissipated in the loop.

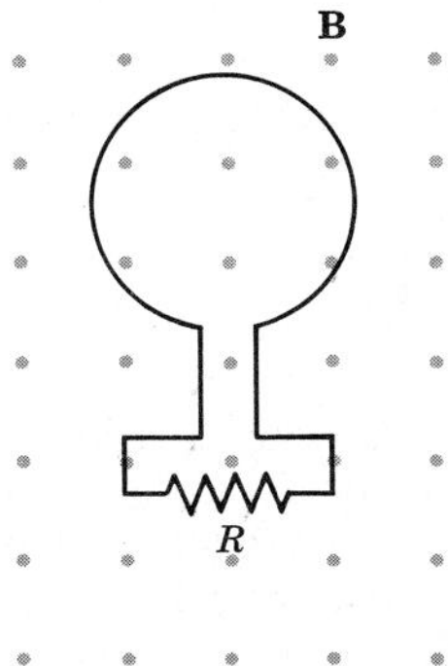

Figure 29 Exercise 5 and Problem 17.

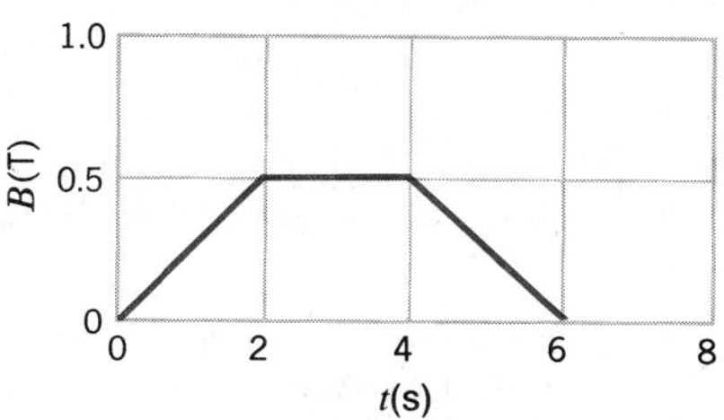

Figure 30 Exercise 6.

8E. A uniform magnetic field is normal to the plane of a circular loop 10 cm in diameter made of copper wire (diameter = 2.5 mm). (a) Calculate the resistance of the wire. (See Table 1 in Chapter 28.) (b) At what rate must the magnetic field change with time if an induced current of 10 A is to appear in the loop?

9P. The current in the solenoid of Sample Problem 1 changes, not as in that Sample Problem, but according to $i = 3.0t + 1.0t^2$, where i is in amperes and t is given in seconds. (a) Plot the induced emf in the coil from $t = 0$ to $t = 4$ s. (b) The resistance of the coil is 0.15 Ω. What is the current in the coil at $t = 2.0$ s?

10P. In Fig. 31 a 120-turn coil of radius 1.8 cm and resistance 5.3 Ω is placed outside a solenoid like that of Sample Problem 1. If the current in the solenoid is changed as in that Sample Problem, (a) what current appears in the loop while the solenoid current is being changed? (b) How do the conduction electrons in the loop "get the message" from the solenoid that they should move to establish a current? After all, the magnetic flux is entirely confined to the interior of the solenoid.

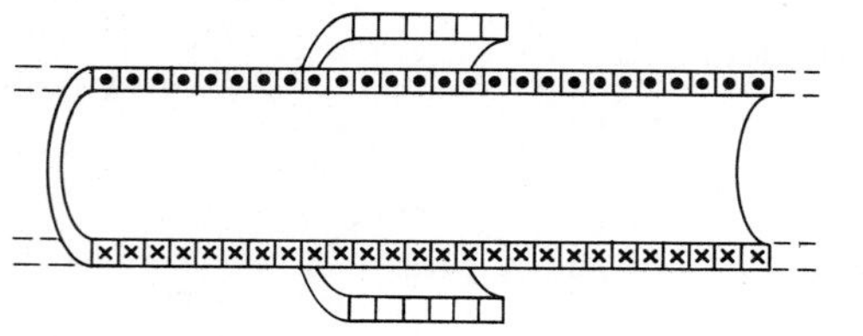

Figure 31 Problem 10.

11P. A long solenoid with a radius of 25 mm has 100 turns/cm. A single loop of wire of radius 5.0 cm is placed around the solenoid, the axis of the loop and the solenoid coinciding. The current in the solenoid is reduced from 1.0 A to 0.50 A at a uniform rate over a time interval of 10 ms. What emf appears in the loop?

12P. Derive an expression for the flux through a toroid of N turns carrying a current i. Assume that the windings have a rectangular cross section of inner radius a, outer radius b, and height h.

13P. A toroid having a 5.0-cm square cross section and an inside radius of 15 cm has 500 turns of wire and carries a current of 0.80 A. What is the magnetic flux through the cross section?

14P. You are given 50 cm of copper wire (diameter = 1.0 mm). It is formed into a circular loop and placed at right angles to a uniform magnetic field that is increasing with time at the constant rate of 10 mT/s. At what rate is thermal energy generated in the loop?

15P. A closed loop of wire consists of a pair of equal semicircles, radius 3.7 cm, lying in mutually perpendicular planes. The loop was formed by folding a circular loop along a diameter until the two halves became perpendicular. A uniform magnetic field **B** of magnitude 76 mT is directed perpendicular to the fold diameter and makes equal angles (= 45°) with the planes of the semicircles as shown in Fig. 32. The magnetic field is reduced at a uniform rate to zero during a time interval 4.5 ms. Determine the magnitude of the induced emf and the direction of the induced current in the loop during this interval.

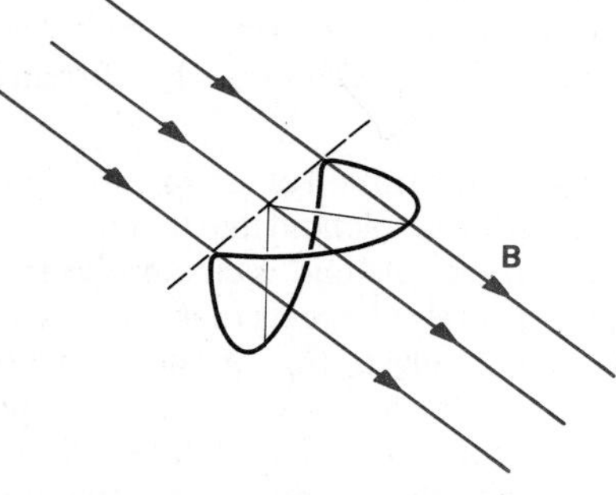

Figure 32 Problem 15.

16P. Figure 33 shows two parallel loops of wire having a common axis. The smaller loop (radius r) is above the larger loop (radius R), by a distance $x \gg R$. Consequently the magnetic field, due to the current i in the larger loop, is nearly constant throughout the smaller loop. Suppose that x is increasing at the constant rate $dx/dt = v$. (*a*) Determine the magnetic flux across the area bounded by the smaller loop as a function of x. (*b*) Compute the emf generated in the smaller loop. (*c*) Determine the direction of the induced current flowing in the smaller loop. (*Hint:* See Equation 25 in Chapter 31.)

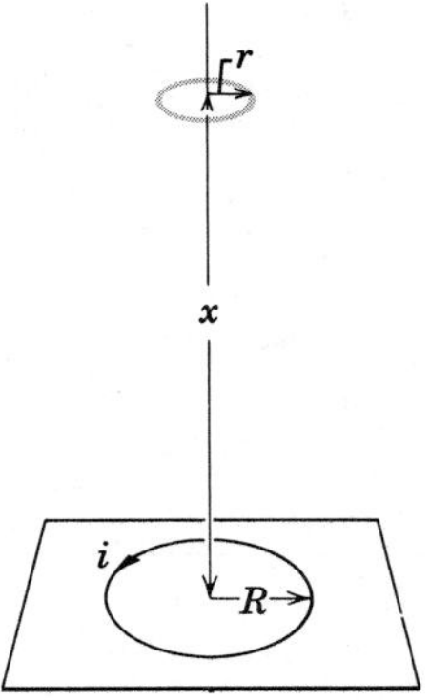

Figure 33 Problem 16.

17P. In Fig. 29 let the flux for the loop be $\Phi_B(0)$ at time $t = 0$. Then let the magnetic field **B** vary in a continuous but unspecified way, in both magnitude and direction, so that at time t the flux is represented by $\Phi_B(t)$. (*a*) Show that the net charge $q(t)$ that has passed through resistor R in time t is

$$q(t) = \frac{1}{R}\,[\Phi_B(0) - \Phi_B(t)],$$

independent of the way **B** has changed. (*b*) If $\Phi_B(t) = \Phi_B(0)$ in a particular case we have $q(t) = 0$. Is the induced current necessarily zero throughout the interval $0 \rightarrow t$?

18P. A hundred turns of insulated copper wire are wrapped around a wooden cylindrical core of cross-sectional area 1.2×10^{-3} m^2. The two terminals are connected to a resistor. The total resistance in the circuit is 13 Ω. If an externally applied uniform longitudinal magnetic field in the core changes from 1.6 T in one direction to 1.6 T in the opposite direction, how much charge flows through the circuit? (*Hint:* See Problem 17.)

19P. A square wire loop with 2.0-m sides is perpendicular to a uniform magnetic field, with half the area of the loop in the field, as shown in Fig. 34. The loop contains a 20-V battery with negligible internal resistance. If the magnitude of the field varies with time according to $B = 0.042 - 0.87t$, with B in tesla and t in seconds, (*a*) what is the total emf in the circuit? (*b*) What is the direction of the current through the battery?

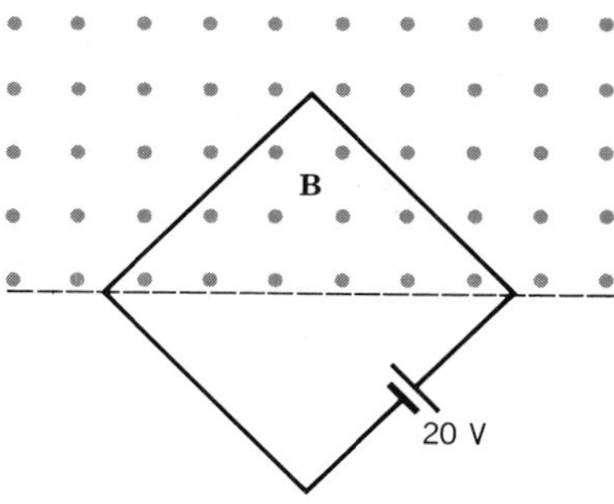

Figure 34 Problem 19.

20P. A wire is bent into three circular segments of radius $r = 10$ cm as shown in Fig. 35. Each segment is a quadrant of a circle, ab lying in the xy plane, bc lying in the yz plane, and ca lying in the zx plane. (a) If a uniform magnetic field **B** points in the positive x direction, what is the magnitude of the emf developed in the wire when B increases at the rate of 3.0 mT/s? (b) What is the direction of the current in the segment bc?

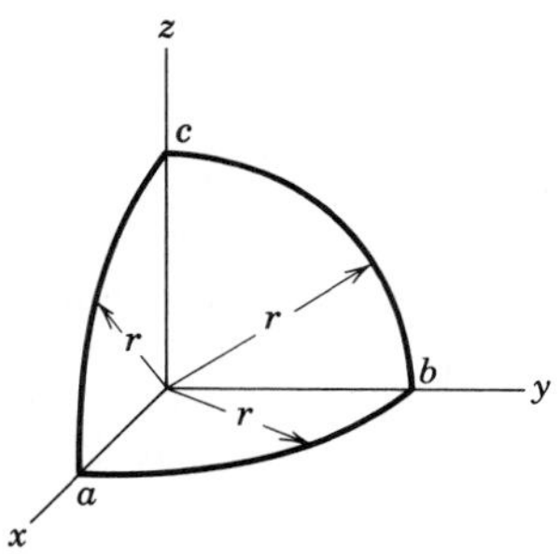

Figure 35 Problem 20.

21P. Two long, parallel copper wires (diameter = 2.5 mm) carry currents of 10 A in opposite directions. (a) If their centers are 20 mm apart, calculate the flux per meter of wire that exists in the space between the axes of the wires. (b) What fraction of this flux lies inside the wires? (c) Repeat the calculation of (a) for parallel currents.

Section 32-5 Induction: A Quantitative Study

22E. An automobile having a radio antenna 1.1 m long travels at 90 km/h in a region where the earth's magnetic field is 55 μT. Is an emf induced in the antenna? If so, what is its maximum possible value?

23E. A spaceship 12 m wide moves at a speed of 2.4×10^7 m/s through a weak interstellar magnetic field of 0.36 nT. Regard the ship as a metal rod 12 m long and assume it moves perpendicular to the field. Calculate the emf generated across the width of the ship.

24E. A circular loop of wire 10 cm in diameter is placed with its normal making an angle of 30° with the direction of a uniform 0.50-T magnetic field. The loop is "wobbled" so that its normal rotates in a cone about the field direction at the constant rate of 100 rev/min; the angle between the normal and the field direction (= 30°) remains unchanged during the process. What emf appears in the loop?

25E. A metal rod moves with constant velocity along two parallel metal rails, connected with a strip of metal at one end, as shown in Fig. 36. A magnetic field $B = 0.35$ T points out of the page. (a) If the rails are separated by 25 cm and the speed of the rod is 55 cm/s, what emf is generated? (b) If the rod has a resistance of 18 Ω and the rails have negligible resistance, what is the current in the rod?

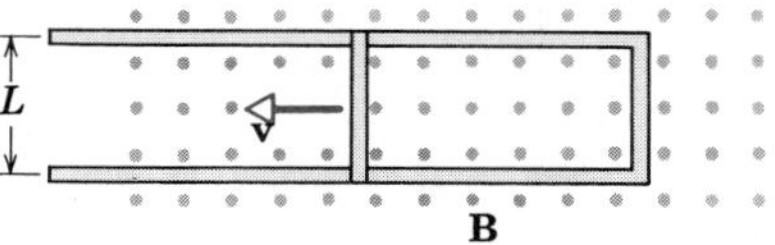

Figure 36 Exercises 25 and 26.

26E. Figure 36 shows a conducting rod of length L being pulled along horizontal, frictionless, conducting rails at a constant velocity **v**. A uniform vertical magnetic field **B** fills the region in which the rod moves. Assume that $L = 10$ cm, $v = 5.0$ m/s, and $B = 1.2$ T. (a) What is the induced emf in the rod? (b) What is the current in the conducting loop? Assume that the resistance of the rod is 0.40 Ω and that the resistance of the rails is negligibly small. (c) At what rate is thermal energy being generated in the rod? (d) What force must be applied to the rod by an external agent to maintain its motion? (e) At what rate does this external agent do work on the rod? Compare this answer with the answer to (c).

27E. In Fig. 37 a conducting rod of mass m and length L slides without friction on two long horizontal rails. A uniform vertical magnetic field **B** fills the region in which the rod is free to move. The generator G supplies a constant current i that flows down one rail, across the rod, and back to the generator along the other rail. Find the velocity of the rod as a function of time, assuming it to be at rest at $t = 0$.

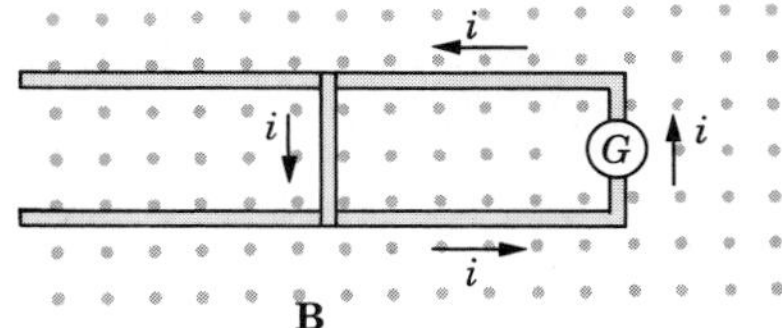

Figure 37 Exercise 27 and Problem 34.

28P. A circular loop made of a stretched conducting elastic material has a 12 cm radius. It is placed with its plane at right angles to a uniform 0.80-T magnetic field. When released, the radius of the loop starts to shrink at an instantaneous rate of 75 cm/s. What emf is induced in the loop at that instant?

29P. Two straight conducting rails form a right angle where their ends are joined. A conducting bar in contact with the rails starts at the vertex at time $t = 0$ and moves with a constant velocity of 5.2 m/s to the right, as shown in Fig. 38. A 0.35 T magnetic field points out of the page. Calculate (*a*) the flux through the triangle formed by the rails and bar at $t = 3.0$ s and (*b*) the emf around the triangle at that time. (*c*) In what manner does the emf around the triangle vary with time?

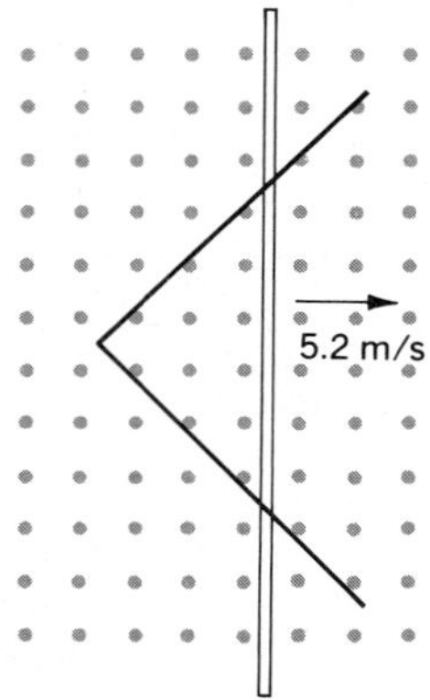

Figure 38 Problem 29.

30P. A stiff wire bent into a semicircle of radius a is rotated with a frequency ν in a uniform magnetic field, as suggested in Fig. 39. What are (*a*) the frequency and (*b*) the amplitude of the emf induced in the loop?

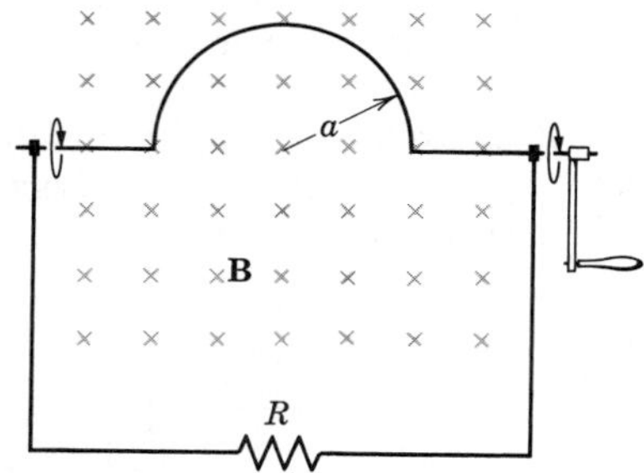

Figure 39 Problem 30.

31P. *Alternating current generator.* A rectangular loop of N turns and of length a and width b is rotated at a frequency ν in a uniform magnetic field **B**, as in Fig. 40. (*a*) Show that an induced emf given by

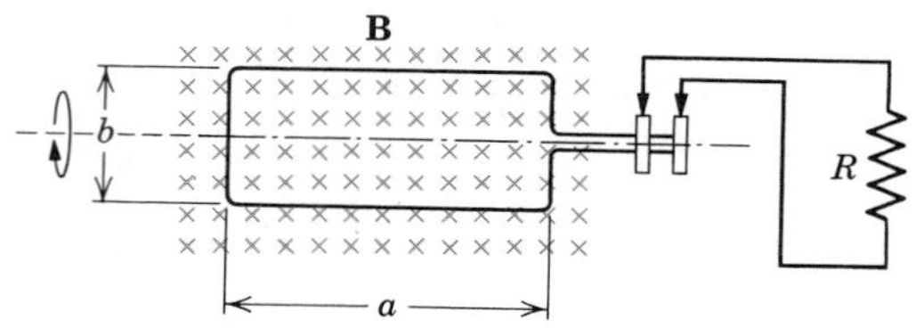

Figure 40 Problem 31.

$$\mathcal{E} = 2\pi\nu NabB \sin 2\pi\nu t = \mathcal{E}_0 \sin 2\pi\nu t$$

appears in the loop. This is the principle of the commercial alternating-current generator. (*b*) Design a loop that will produce an emf with $\mathcal{E}_0 = 150$ V when rotated at 60 rev/s in a magnetic field of 0.50 T.

32P. A generator consists of 100 turns of wire formed into a rectangular loop 50 cm by 30 cm, placed entirely in a uniform magnetic field with magnitude $B = 3.5$ T. What is the maximum value of the emf produced when the loop is spun at 1000 revolutions per minute about an axis perpendicular to **B**?

33P. Calculate the average power supplied by the generator of Problem 31*b* if it is connected to a circuit of 42-Ω resistance.

34P. In Exercise 27 (see Fig. 37) the constant-current generator G is replaced by a battery that supplies a constant emf $\mathcal{E}$. (*a*) Show that the velocity of the rod now approaches a constant terminal value **v** and give its magnitude and direction. (*b*) What is the current in the rod when this terminal velocity is reached? (*c*) Analyze both this situation and that of Exercise 27 from the point of view of energy transfers.

35P. At a certain place, the earth's magnetic field has magnitude $B = 0.59$ gauss and is inclined downwards at an angle of 70° to the horizontal. A flat horizontal circular coil of wire with a radius of 10 cm has 1000 turns and a total resistance of 85 Ω. It is connected to a galvanometer with 140 Ω resistance. The coil is flipped through a half revolution about a diameter, so it is again horizontal. How much charge flows through the galvanometer during the flip? (*Hint:* See Problem 17.)

36P. Figure 41 shows a rod of length L caused to move at constant speed v along horizontal conducting rails. In this case the magnetic field in which the rod moves is not uniform but is provided by a current i in a long parallel wire. Assume that $v = 5.0$ m/s, $a = 10$ mm, $L = 10$ cm, and $i = 100$ A. (*a*) Calculate the emf induced in the rod. (*b*) What is the current in the conducting loop? Assume that the resistance of the rod is 0.40 Ω and that the resistance of the rails is negligible. (*c*) At what rate is thermal energy being generated in the rod? (*d*) What force must be applied to the rod by an external agent to maintain its motion? (*e*) At what rate does this external agent do work on the rod? Compare this answer to (*c*).

37P. For the situation shown in Fig. 42, $a = 12$ cm, $b =$

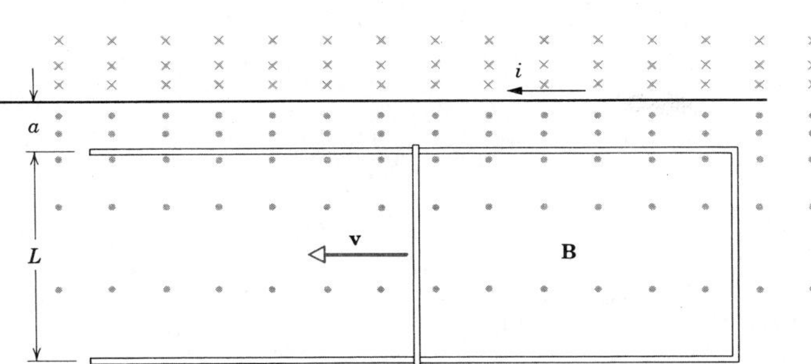

Figure 41 Problem 36.

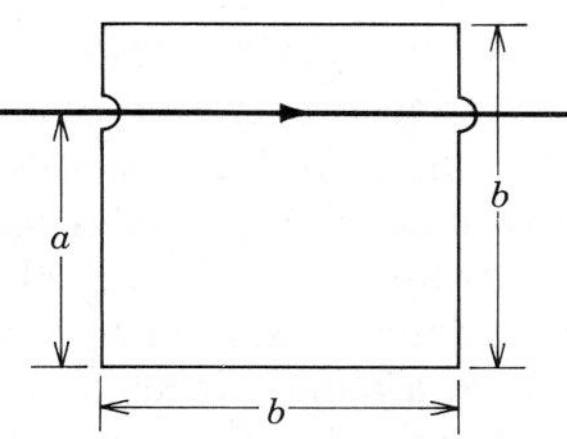

Figure 42 Problem 37.

16 cm. The current in the long straight wire is given by $i = 4.5t^2 - 10t$, where i is in amperes and t is in seconds. (*a*) Find the emf in the square loop at $t = 3.0$ s. (*b*) What is the direction of the induced current in the loop?

38P. In Fig. 43, the square has sides of length 2.0 cm. A magnetic field points out of the page; its magnitude is given by $B = 4t^2y$, where B is in tesla, t is in seconds, and y is in meters. Determine the emf around the square at $t = 2.5$ s and give its direction.

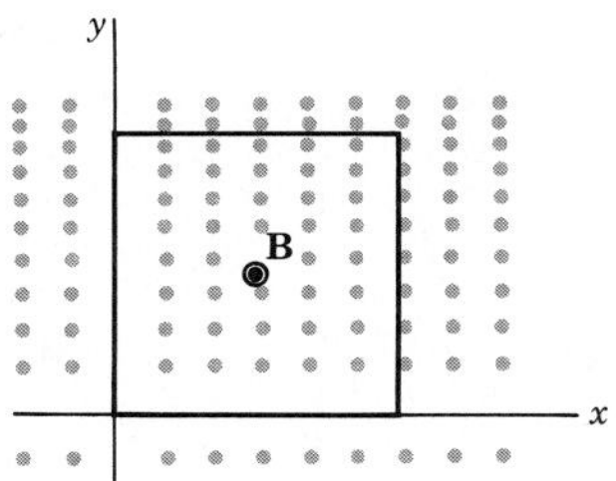

Figure 43 Problem 38.

39P. A rectangular loop of wire with length a, width b, and resistance R is placed near an infinitely long wire carrying current i, as shown in Fig. 44. The distance from the long wire to the center of the loop is r. Find (*a*) the magnitude of the magnetic flux through the loop and (*b*) the current in the loop as it moves away from the long wire with speed v.

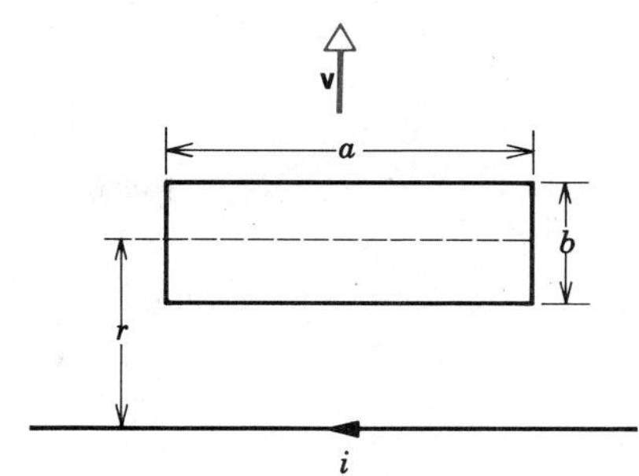

Figure 44 Problem 39.

40P. Figure 45 shows a "homopolar generator," a device with a solid conducting disk as rotor. This machine can produce a greater emf than wire loop rotors, since they can spin at a much higher angular speed before centrifugal forces disrupt the rotor. (*a*) Show that the emf produced is given by

$$\mathcal{E} = \pi \nu B R^2$$

where ν is the spin frequency, R the rotor radius, and B the uniform magnetic field perpendicular to the rotor. (*b*) What torque must be provided by the motor spinning the rotor when the output current is i?

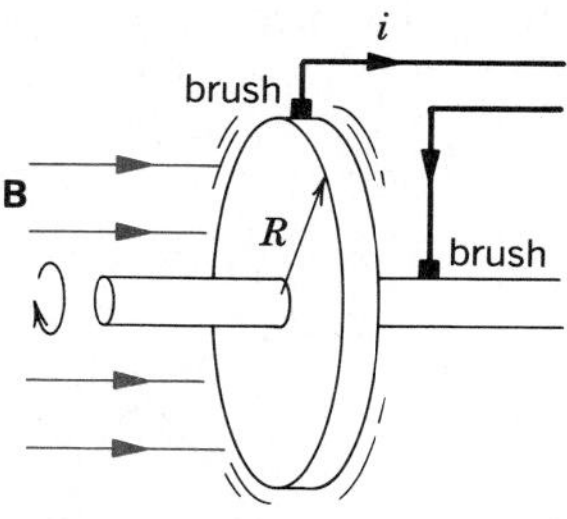

Figure 45 Problem 40.

41P. A rod with length l, mass m, and resistance R slides without friction down parallel conducting rails of negligible resistance, as in Fig. 46. The rails are connected together at the bottom as shown, forming a conducting loop with the rod as the top member. The plane of the rails makes an angle θ with the horizontal and a uniform vertical magnetic field **B** exists throughout the region. (*a*) Show that the rod acquires a steady-state terminal velocity whose magnitude is

$$v = \frac{mgR}{B^2l^2}\frac{\sin\theta}{\cos^2\theta}.$$

(*b*) Show that the rate at which thermal energy is being generated in the rod is equal to the rate at which the rod is losing

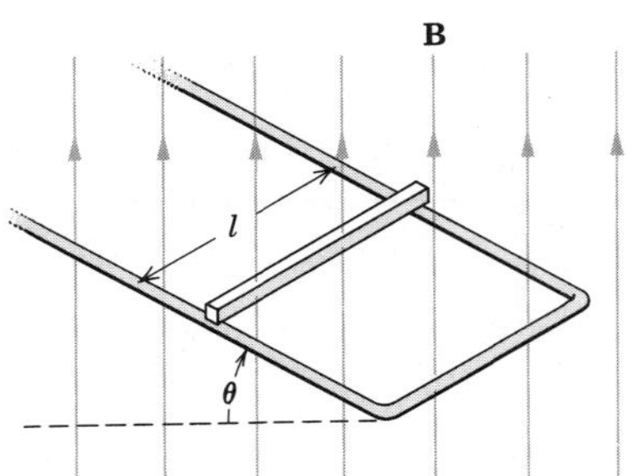

Figure 46 Problem 41.

gravitational potential energy. (*c*) Discuss the situation if **B** were directed down instead of up.

42P*. A wire whose cross-sectional area is 1.2 mm² and whose resistivity is $1.7 \times 10^{-8}\ \Omega \cdot \text{m}$ is bent into a circular arc of radius $r = 24$ cm as shown in Fig. 47. An additional straight length of this wire, *OP*, is free to pivot about *O* and makes sliding contact with the arc at *P*. Finally, another straight length of this wire, *OQ*, completes the circuit. The entire arrangement is located in a magnetic field $B = 0.15$ T directed out of the plane of the figure. The straight wire *OP* starts from rest with $\theta = 0$ and has a constant angular acceleration of 12 rad/s². (*a*) Find the resistance of the loop *OPQO* as a function of θ. (*b*) Find the magnetic flux through the loop as a function of θ. (*c*) For what value of θ is the induced current in the loop a maximum? (*d*) What is the maximum value of the induced current in the loop?

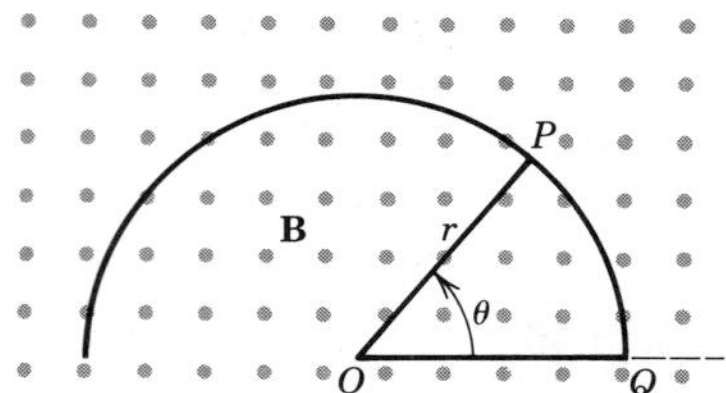

Figure 47 Problem 42.

Section 32-6 Induced Electric Fields

43E. A long solenoid has a diameter of 12 cm. When a current *i* is passed through its windings, a uniform magnetic field $B = 30$ mT is produced in its interior. By decreasing *i*, the field is caused to decrease at the rate 6.5 mT/s. Calculate the magnitude of the induced electric field (*a*) 2.2 cm and (*b*) 8.2 cm from the axis of the solenoid.

44E. Figure 48 shows two circular regions R_1 and R_2 with radii $r_1 = 20$ cm and $r_2 = 30$ cm, respectively. In R_1 there is a uniform magnetic field $B_1 = 50$ mT into the page and in R_2 there is a uniform magnetic field $B_2 = 75$ mT out of the page (ignore any fringing of these fields). Both fields are decreasing at the rate 8.5 mT/s. Calculate the integral $\oint \mathbf{E} \cdot d\mathbf{s}$ for each of the three indicated paths.

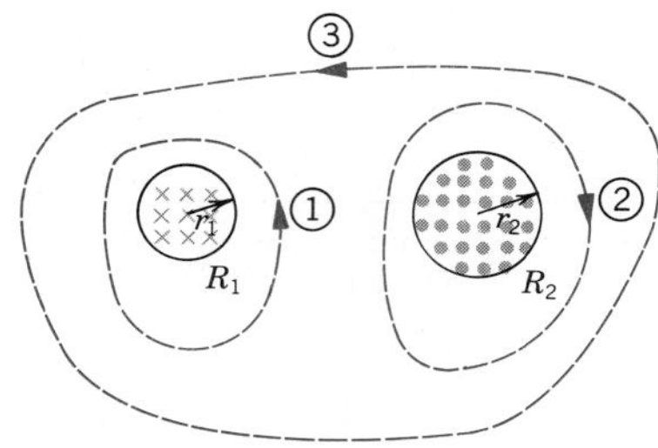

Figure 48 Exercise 44.

45P. Early in 1981 the Francis Bitter National Magnet Laboratory at M.I.T. commenced operation of a 3.3-cm diameter cylindrical magnet, which produces a 30 T field, then the world's largest steady-state field. The field can be varied sinusoidally between the limits of 29.6 T and 30 T at a frequency of 15 Hz. When this is done, what is the maximum value of the induced electric field at a radial distance of 1.6 cm from the axis? This magnet is described in *Physics Today,* August 1984. (*Hint:* See Sample Problem 4.)

46P. Figure 49 shows a uniform magnetic field **B** confined to a cylindrical volume of radius *R*. **B** is decreasing in magnitude at a constant rate of 10 mT/s. What is the instantaneous acceleration (direction and magnitude) experienced by an electron placed at *a*, at *b*, and at *c*? Assume $r = 5.0$ cm. (The necessary fringing of the field beyond *R* will not change your answer as long as there is axial symmetry about the perpendicular axis through *b*.)

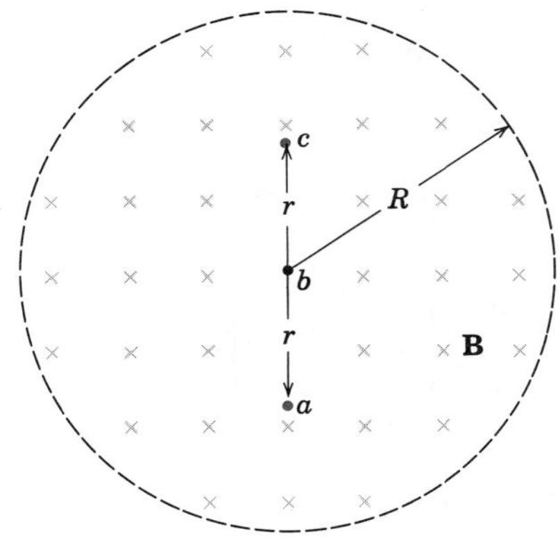

Figure 49 Problem 46.

47P. Prove that the electric field **E** in a charged parallel-plate capacitor cannot drop abruptly to zero as one moves at right angles to it, as suggested by the arrow in Fig. 50 (see point *a*). In

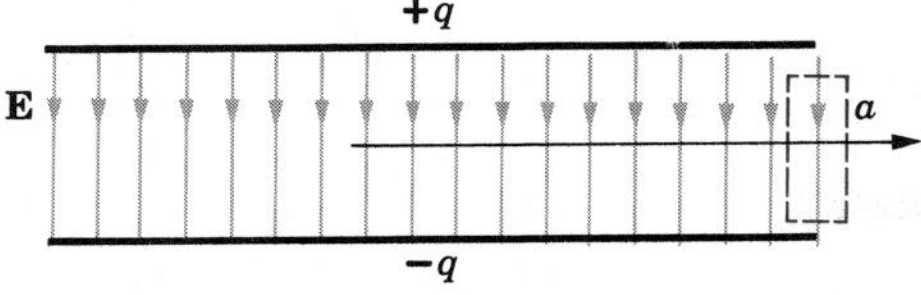

Figure 50 Problem 47.

actual capacitors fringing of the lines of force always occurs, which means that **E** approaches zero in a continuous and gradual way; compare with Problem 45, Chapter 31. (*Hint:* Apply Faraday's law to the rectangular path shown by the dashed lines.)

Section 32-7 The Betatron

48E. Figure 51*a* shows a top view of the electron orbit in a betatron. Electrons are accelerated in a circular orbit in the *xy* plane and then withdrawn to strike the target *T*. The magnetic field **B** is along the *z* axis (the positive *z* axis is out of the page). The magnetic field B_z along this axis varies sinusoidally as shown in Fig. 51*b*. Recall that the magnetic field must (*i*) guide the electrons in their circular path, and (*ii*) generate the electric field that accelerates the electrons. Which quarter cycle(s) in Fig. 51*b* are suitable (*a*) according to (*i*), (*b*) according to (*ii*), (*c*) for operation of the betatron?

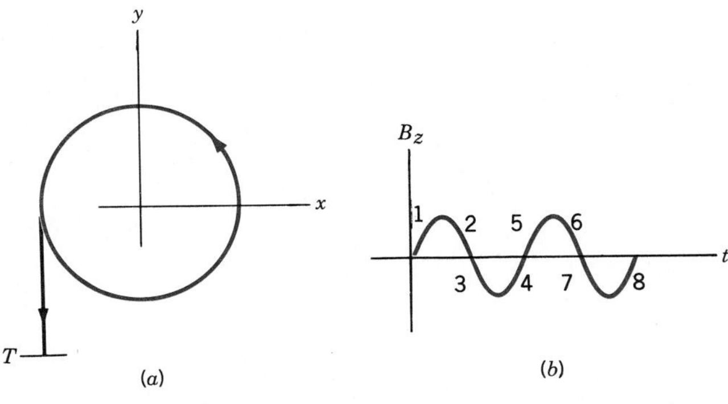

Figure 51 Problem 48.

49E. In a certain betatron the radius of the electron orbit is $r = 32$ cm and the magnetic field at this radius is given by $B_{orb} = (0.28) \sin 120\pi t$, where t in seconds gives B_{orb} in tesla. (*a*) Calculate the induced electric field felt by the electrons at $t = 0$. (*b*) Find the acceleration of the electrons at this instant. Ignore relativistic effects.

50P. Some measurements of the maximum magnetic field as a function of radius for a betatron are:

r, cm	*B*, tesla	*r*, cm	*B*, tesla
0	0.950	81.2	0.409
10.2	0.950	83.7	0.400
68.2	0.950	88.9	0.381
73.2	0.528	91.4	0.372
75.2	0.451	93.5	0.360
77.3	0.428	95.5	0.340

Show by graphical analysis that the relation $\overline{B} = 2B_{orb}$ mentioned in Section 32-7 as essential to betatron operation is satisfied at the orbit radius, $R = 84$ cm. (*Hint:* Note that

$$\overline{B} = \frac{1}{\pi R^2} \int_0^R B(r) 2\pi r \, dr$$

and evaluate the integral graphically.)

ESSAY 12
SUPERCONDUCTIVITY

PETER LINDENFELD

RUTGERS, THE STATE UNIVERSITY OF NEW JERSEY

Introduction

In 1911, when H. Kamerlingh Onnes and his assistant Gilles Holst at the University of Leiden in the Netherlands cooled some mercury to a temperature lower than anyone ever had before, they could not believe that it had no electrical resistance at all (Figs. 1 and 2). It seemed that some accidental short circuit was preventing a proper measurement. Only after they tried again and actually saw the resistance come and go on heating and cooling through the "transition temperature" did they realize that something totally unexpected was happening. This was the first in a string of surprises that superconductivity has provided for us since that time.

It turned out that the loss of resistance is just part of what makes superconductors interesting. They also have magnetic properties unlike those of any other materials. On the one hand they can generate extremely high magnetic fields, on the other hand they allow the measurement, control, and utilization of weaker magnetic fields than

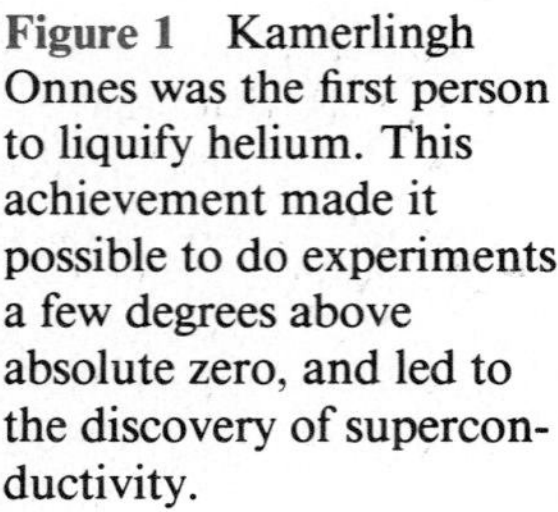

Figure 1 Kamerlingh Onnes was the first person to liquify helium. This achievement made it possible to do experiments a few degrees above absolute zero, and led to the discovery of superconductivity.

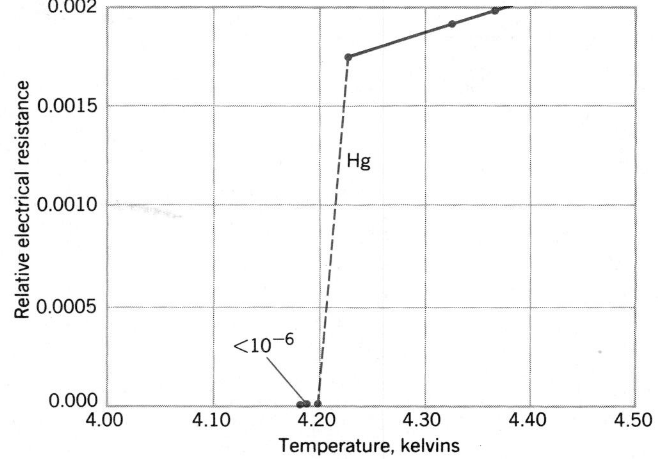

Figure 2 This graph was used to describe the discovery of superconductivity. The horizontal axis is the temperature of a sample of mercury in kelvins. The vertical axis is the electrical resistance of the sample in arbitrary units.

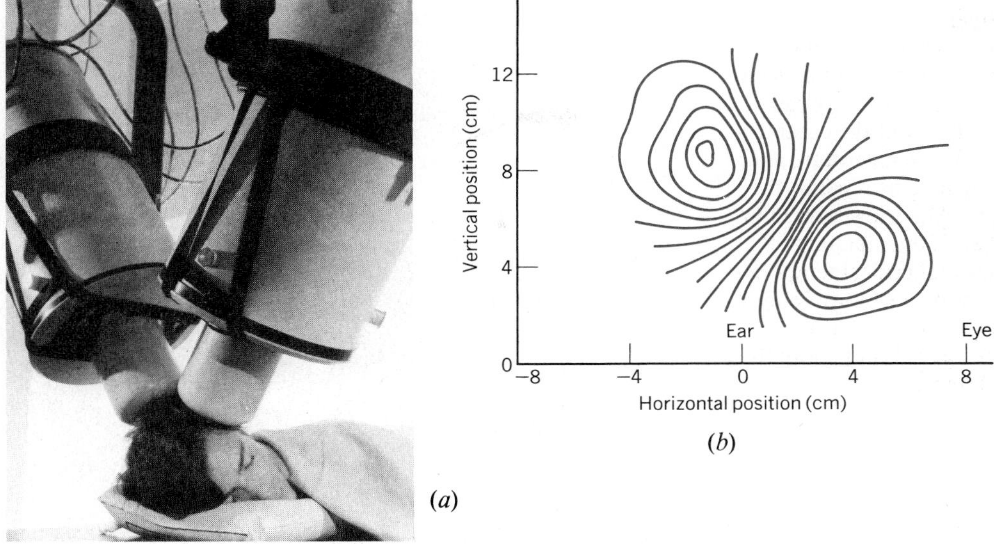

Figure 3 (*a*) Superconducting quantum interference devices immersed in liquid helium detect the magnetic fields generated by brain activity. (*b*) The graph shows the magnetic field generated when a person hears a 600-Hz tone. Adjacent magnetic field lines differ by 10 fT or 10^{-14} T. This experiment makes it possible to determine which part of the brain participates in hearing.

had been thought possible. The reason that their marvellous properties were not more fully exploited is, of course, that until 1986 they were known to occur only at temperatures below about 20 K, or 20 Celsius degrees above absolute zero.

In spite of the considerable cost and complication of the refrigeration, both the high-field and the low-field applications of superconductivity were being developed more and more even before superconductivity was found to exist also at much higher temperatures. At one end of the scale are the superconducting magnets of the accelerator at the Fermi National Laboratory, which accelerates elementary particles to the highest energies, and at the other end the "SQUIDS" that record the magnetic fields generated by the human brain (Fig. 3). In early 1987 the science of superconductivity was revolutionized by the discovery of materials that become superconducting above 90 K.

Magnets and Large-Scale Applications

An electromagnet is a device with zero efficiency. It may be expensive to build and to operate, it may be an essential part of a generator or accelerator, but if, as is usually the case, its function is to produce a constant magnetic field, all the energy supplied to it is dissipated as heat energy. This fact is most evident when we consider permanent magnets, which generate magnetic fields without any energy input. Each electron is a little permanent magnet, and the only reason that we are not more often aware of the magnetic properties of matter is that the magnetic fields of different electrons usually cancel because of their different orientations.

In iron and other "magnetic" materials the electrons can be lined up, but even at best they produce fields of only about 2 T (tesla). For larger fields current-carrying coils are invariably used. Ordinarily a significant amount of energy must be supplied and is wasted as heat at the rate of I^2R watts (where I is the current and R the resistance). More or less elaborate systems must be used to carry the heat away. To make R zero by using superconducting coils is clearly a huge advantage.

Kamerlingh Onnes was immediately aware of the possibilities of generating large fields without useless heat, but saw his hopes dashed with the first experiments. In the

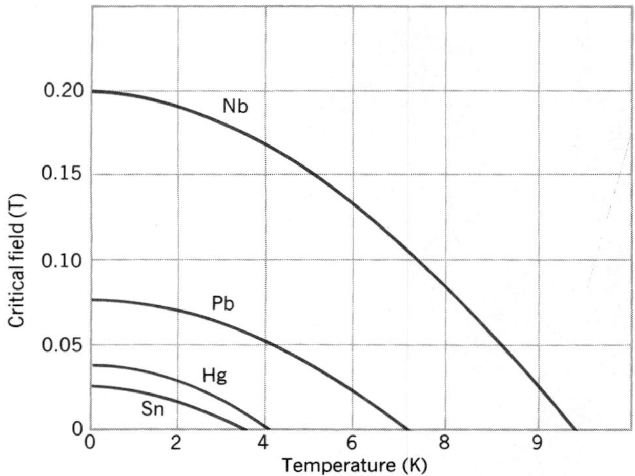

Figure 4 Each line represents the "critical field curve" of an element. Superconductivity can exist only at the combinations of temperature and magnetic field below the line. Alloys discovered in the 1950s extended the range of critical temperatures by a factor of about three and the range of critical fields by more than a hundred.

superconductors known at that time superconductivity is destroyed by magnetic fields less than $\frac{1}{10}$ T (Fig. 4).

Today we know that magnetism and superconductivity are natural enemies. Macroscopic magnetic properties depend on electrons that are lined up parallel to one another, while superconductivity requires pairs of electrons with their spins in opposite directions so that their magnetic fields cancel. A magnetic field causes a torque on an electron that tends to line it up with the field. It therefore acts to break up the superconducting pairs and so to destroy superconductivity.

Figure 5 A record speed of 321 miles per hour was set in 1979 by a magnetically levitated train in Japan. This train floats without touching the ground as a result of the repulsion between its superconducting magnets and the magnetic field that they induce in the tracks.

With hindsight we can see that there were times when the discovery of high-field superconductivity was tantalizingly close, but we cannot be too surprised that it was not until the 1950s, about 40 years after the discovery of superconductivity, that materials and processes were developed that allowed the first superconducting high-field magnets to be made.

Today almost every large physics laboratory and many small ones have superconducting magnets. They are used in the most powerful particle accelerators, and the realization of magnetic fusion devices for energy generation is expected to depend on them. But there are also applications that were not even imagined before the advent of superconductivity. One of these is the development of magnetically levitated trains or "electromagnetic flight" as it is sometimes more imaginatively called. The flight here is only a very small distance above the ground, just enough to keep car and rails from touching, so that friction between the two is eliminated, and with it a major obstacle to the achievement of higher speeds (Fig. 5).

Forces on Magnets and Superconductors

Put a bar magnet, an electron, or a current-carrying coil in a uniform magnetic field. Each of these objects has a north pole and a south pole. The north pole experiences a force in the direction of the magnetic field, the south pole a force of equal magnitude in the opposite direction. There is no net force. Unless the object is already lined up with the field there will, however, be a torque tending to make the north–south axis parallel to the field. (Although we are using the language of poles, the description in terms of forces on currents is entirely equivalent and leads to the same result.)

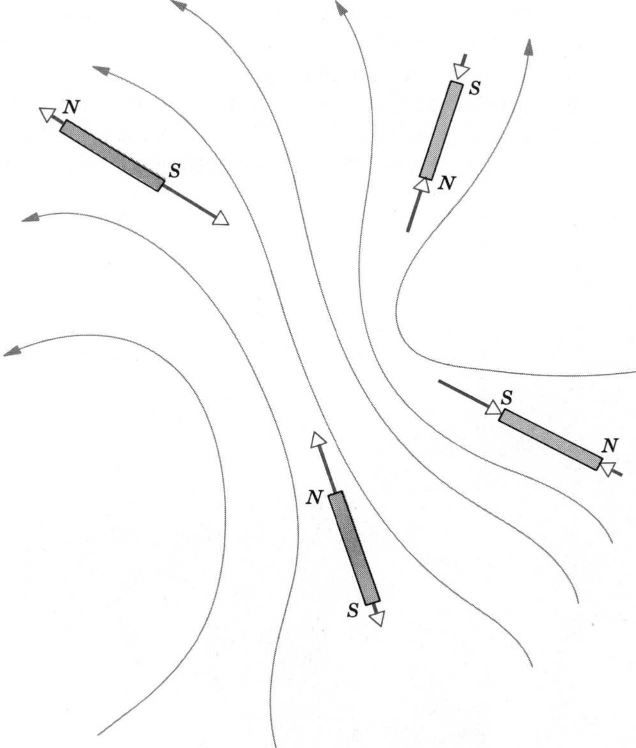

Figure 6 The forces in a magnetic field on permanent magnets (grey) and on superconductors (colored).

The situation is quite different if the field is not uniform. Suppose, for example, that the field lines converge, indicating that the field becomes stronger where the lines are closer together. There will be a torque as before, but when the north pole points along the field it will now be in a stronger field than the south pole. The net result is that the whole object (magnet, coil, or electron) is pulled toward the strongest part of the field. If the field lines diverge it is the south pole that is in the stronger field and the object will be pulled back, again toward the stronger field. A small magnet has a field that is strongest close to it, and two such magnets will attract (Fig. 6).

In a superconductor the magnetization is in the direction opposite to that of the external magnetic field. This is called diamagnetism. We can see how it arises by considering a cylinder or a closed coil made of perfectly conducting wire, but without a current source, and, at least initially, without any current. We do not have a magnetized object to start with, but currents, and hence magnetization, may be induced in accordance with Faraday's law.

If we take the coil to a region where there is a magnetic field, the increase in magnetic flux through the coil will produce an induced emf, and with it an induced current and an induced magnetic field. In accord with Lenz's law the induced field will be in a direction opposite to the change in the flux through the coil, in this case opposite to the direction of the increasing external field through the superconducting coil.

The induced emf disappears as soon as the coil comes to rest and the field through it stops increasing. In a coil made of normal wire the current and its magnetic field will then also cease. In a perfectly conducting coil, however, or on the surface of a cylinder made of perfectly conducting material, the current will continue to flow even without an emf, since the resistance is zero and no energy input is required.

In an external converging field it is now the opposite pole, the south pole, which is in the stronger field. The result is that the net force is toward the weakest part of the field. In the field of another coil or magnet the diamagnet is repelled!

For a superconductor we have to go a step further. Experiment shows that a superconductor is always diamagnetic. When a superconductor is cooled from above to below T_c in a magnetic field there is no change in the external flux, so that Faraday's

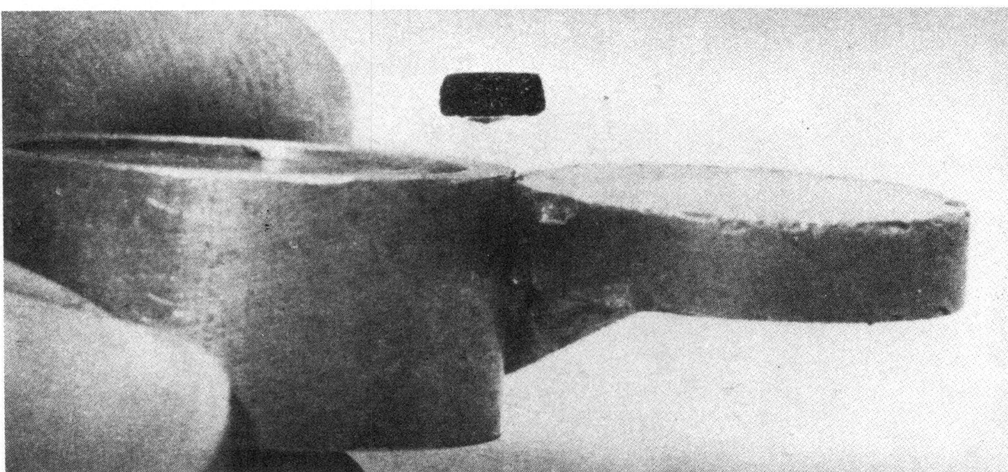

Figure 7 A sample of superconducting yttrium barium copper oxide ($T_c = 95$ K) floating above a permanent magnet.

law would not predict induced magnetization. Yet the superconductor becomes diamagnetic. This is called the Meissner effect, first demonstrated by Meissner and Ochsenfeld in 1933. It shows that the properties of a superconductor cannot simply be described by saying that it is a perfect conductor. The diamagnetism is illustrated in Figure 7, which shows the repulsion of a superconductor by a permanent magnet.

Flux Quantization and Small-Scale Applications

The use of superconductors to detect extremely small magnetic fields depends on two phenomena, flux quantization and "Josephson tunneling," named after Brian Josephson, who predicted the effect while he was still a student at Cambridge University in England.

Just as electric charge is quantized and occurs only in multiples of the electronic charge e (1.6×10^{-19} C) so is the magnetic flux through a superconducting loop. The flux quantum is equal to $h/2e$ (where h is Planck's constant), equal to about 2×10^{-15} $\mathrm{T \cdot m^2}$. This tiny amount of flux and even small fractions of it can be detected by means of the Josephson effect.

What Josephson showed is that the pairs of electrons in a superconducting current can move ("tunnel") through a thin insulating barrier. A loop of superconducting material with such a barrier or "tunnel junction" can still be superconducting. Suppose now that we try to increase the magnetic flux through the loop. Since the flux is quantized it cannot increase continuously. Instead the current through the loop will change so as to keep the flux constant. The junction, however, cannot support more than a very small current. When this amount is reached, the junction ceases momentarily to be superconducting and allows the flux through the loop to change discontinuously. Under the right conditions the change can be made equal to just one flux quantum in one step. By counting the number of steps, the flux through the coil, and therefore also the magnetic field itself, can be determined with great precision.

A loop that contains one or more Josephson junctions for flux detection or measurement is called a "superconducting quantum interference device" or SQUID (Fig. 8).

The discontinuous, stepwise change of flux is also the basis for the use of Josephson junction devices as memory and processing elements in digital computers. Because of their superconductivity and resulting lack of heat dissipation, they can be packed very closely together.

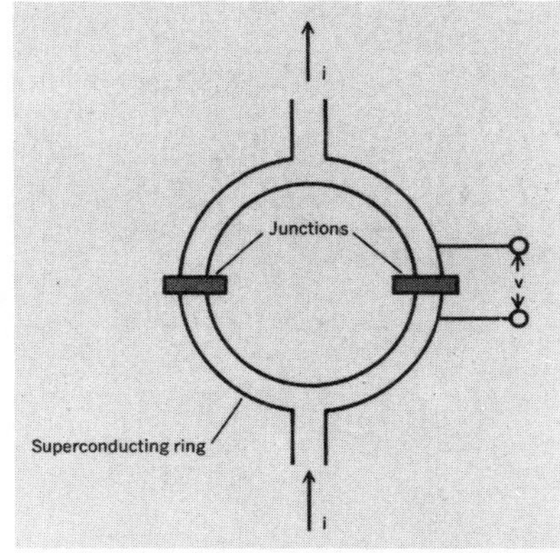

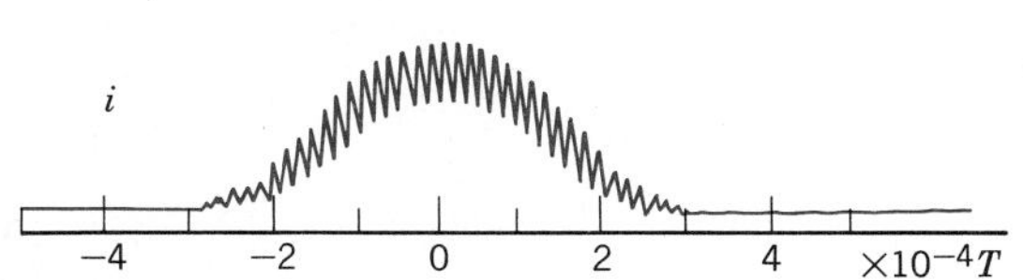

Figure 8 (*a*) Schematic diagram of a two-junction SQUID. (*b*) Data from a circuit containing a two-junction SQUID. The graph shows the variation of the current as a function of magnetic field. Each cycle represents a change in flux by one flux quantum.

Materials

After the discovery in 1911 of superconductivity in mercury, other elements in the same region of the periodic table of elements were also found to be superconductors. Tin, indium, lead, and thallium led the way, with transition temperatures (T_c) ranging from 2.4 K for thallium to 7.2 K for lead. All were discovered in Kamerlingh Onnes' laboratory, which was then the only place where liquid helium, and with it these low temperatures, were available. In 1923 Toronto joined the exclusive club, and Berlin in 1925, where Walther Meissner and his co-workers soon discovered superconductivity in a different part of the periodic table, among the "transition elements," including niobium, which remains, with its T_c of 9.2 K, the element with the highest transition temperature. Further progress was made when it was realized that metallic compounds could have even higher values. By 1940 the record holder was NbC with a T_c of 10.1 K, in 1954 it was Nb_3Sn with T_c equal to 18 K. In 1973 Nb_3Ge, with a T_c of 23.2 K was in first place, and remained there for 14 years until the dramatic discoveries of 1986 and 1987 of superconductivity in ceramic oxides, first demonstrated by Müller and Bednorz at about 35K in lanthanum-barium-copper oxide, and later by Chu and his collaborators near 91 K in yttrium-barium-copper oxide (Fig. 9).

Together with the advances in T_c came increases in the magnetic fields where superconductivity can persist, and which can therefore be generated by superconducting magnets. For about 40 years the compounds of niobium held undisputed first place in both areas. In 1987 the ceramic oxides became the focus of intense research toward further improvements and new applications.

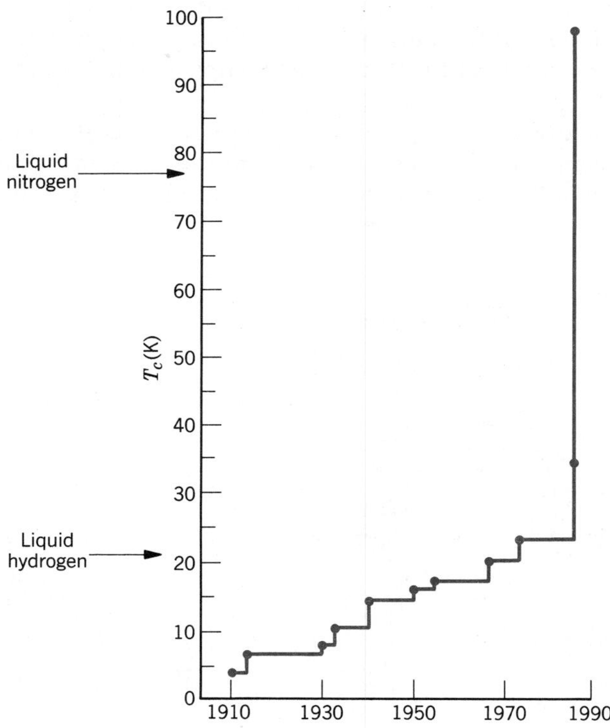

Figure 9 The highest superconducting transition temperatures (in kelvins) from 1911 to 1987.

Theory

Until 1957 superconductivity eluded almost all fundamental understanding. Much had been learned empirically but there was hardly even the beginning of a theory based on atomic and electronic properties. The frustration among physicists is illustrated by the statement of Felix Bloch, a pioneer in the theory of conductivity, when he said "superconductivity is impossible."

The reason for the difficulty is easy enough to understand. It was necessary to find how electrons can cooperate more fully than they normally do, in spite of their mutual repulsion as described by Coulomb's law. The subtle mechanism was finally discovered by Bardeen, Cooper, and Schrieffer ("BCS"). Their theory describes the behavior of superconductors so well and in so much detail that it was widely thought that except for following up some loose ends there would be little further interest in the subject among physicists. It was not the first time that a field of inquiry had been declared to be in terminal decline when in fact it was entering a time of rebirth full of new and unexpected discoveries.

The BCS mechanism depends on the fact that the negatively charged electrons are moving through a lattice of positively-charged ions. The ions are attracted toward the path of an electron. Another electron that comes along before the ions return to their equilibrium positions finds itself in an environment distorted by the passage of the first electron and in this way, indirectly, the two electrons interact.

The BCS theory allows us to understand the delicate balance between opposing tendencies that, under the right conditions, can lead to superconductivity. Its quantitative details describe the properties of the superconductors known at that time very fully, including the parameters that affect the transition temperature. Yet in the 30 years following its development there were only minor discoveries of new superconductors, and almost no increase in their transition temperatures. It began to be suspected that there might be physical laws that prevented higher values of T_c, and papers were written to show why improvements were likely to be impossible.

This era ended abruptly in 1986 with the discovery of a whole new class of materials with much higher transition temperatures. Once again nature showed that she holds unsuspected secrets, which she is, however, willing, sparingly, to uncover to diligent and imaginative search.

CHAPTER 33

INDUCTANCE

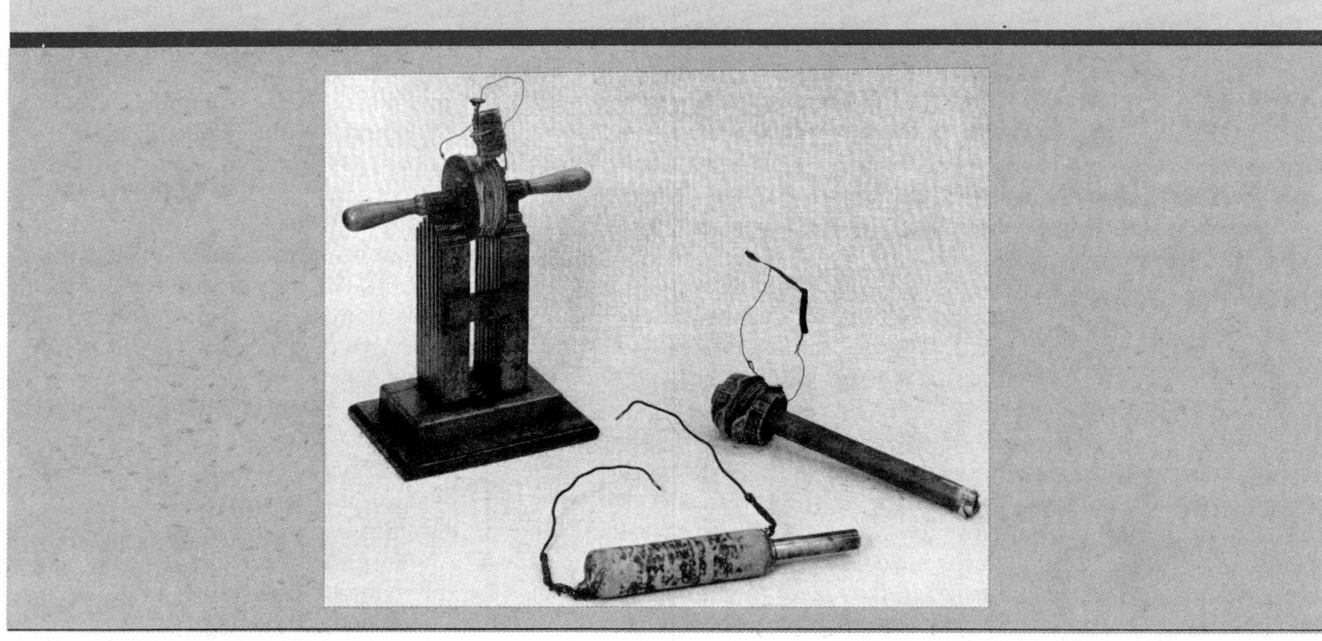

The figure shows the modest kind of apparatus with which Michael Faraday discovered the law of induction. It must be remembered, however, that in those early days such amenities as insulated wire were not commercially available. It is said that Faraday insulated his wires by wrapping them with strips from one of his wife's petticoats. There are not many fields of physics today in which significant new results can be obtained with such simple apparatus.

33-1 Capacitors and Inductors

We found in Chapter 27 that a *capacitor* (symbol ⊣⊢; see Fig. 1*a*) is an arrangement that we can conveniently use to set up a known *electric* field in a given region of space. We took the parallel plate arrangement as a convenient prototype.

Symmetrically, we can define an *inductor* (symbol —ʘʘʘʘʘʘ—; see Fig. 1*b*) as an arrangement that we can conveniently use to set up a known *magnetic* field in a specified region. We take a long solenoid (more specifically, a short length near the center of a long solenoid) as a convenient prototype.

We can express the connection between capacitors and inductors in symbolic form:

inductor : magnetic field :: capacitor : electric field.

Physicists delight in such symmetries, parallelisms and equivalencies. Apart from their inherent aesthetic appeal, as a practical matter we are able to learn new things by leaning on old knowledge. For example, we shall learn that energy can be stored in the magnetic field of an inductor just as it can in the electric field of a capacitor.

Furthermore, we have seen that, if we connect a battery to a capacitor and a resistor in series, the circuit does not come to its final equilibrium state at once but

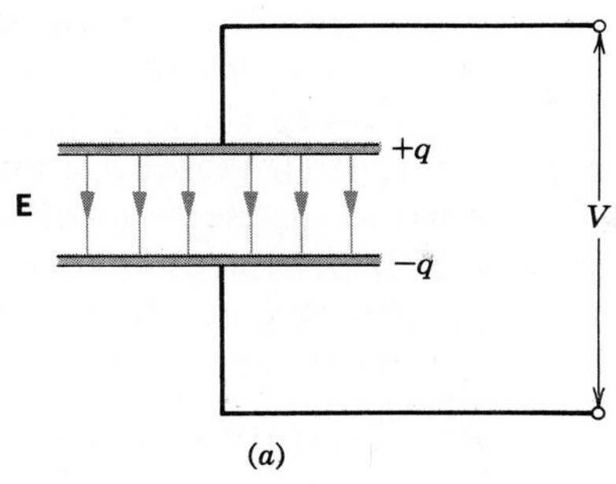

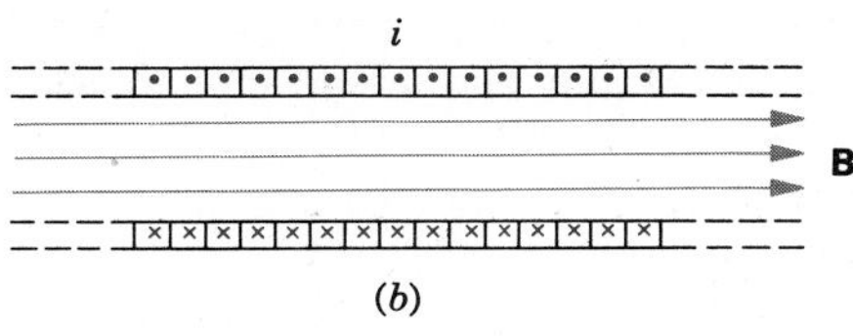

Figure 1 (a) A *capacitor,* of the parallel-plate geometry, displaying its associated *electric* field. (b) An *inductor,* (the central portion of a long solenoid), displaying its associated *magnetic* field.

approaches it exponentially. We shall learn in this chapter that the same thing is true if we connect a battery to an inductor and a resistor in series.

Let us start by defining the *inductance* of an inductor.

33-2 Inductance

A *capacitor* is an arrangement such that, if you place a charge q on its plates, a potential difference V appears across them, the *capacitance* C being given by

$$C = \frac{q}{V} \quad \text{(capacitance defined).} \tag{1}$$

The SI unit of capacitance is, as we have seen, the *farad,* named after Michael Faraday.

An *inductor* is an arrangement such that, if you establish a current i in its windings, a magnetic flux Φ links each winding, the *inductance* L being given by

$$\boxed{L = \frac{N\Phi}{i}} \quad \text{(inductance defined),} \tag{2}$$

in which N is the number of turns. The product $N\Phi$ is called the number of *flux linkages.*

The SI unit of magnetic flux is the tesla·meter² so that the SI unit of inductance is the T·m²/A. We call this the *henry* (abbr. H), after the American physicist Joseph Henry, the co-discoverer of the law of induction and a contemporary of Faraday. Thus

$$1 \text{ henry} = 1 \text{ H} = 1 \text{ T}\cdot\text{m}^2/\text{A}. \tag{3}$$

Throughout this chapter we assume that all inductors, no matter what their geometrical arrangement, have no magnetic materials such as iron in their vicinity.

The Inductance of a Solenoid. Consider a long solenoid of cross-sectional area A. What is the inductance per unit length near its center?

To use the defining equation for inductance (Eq. 2) we must calculate the number of flux linkages set up by a given current in the solenoid windings. Consider a length l near the center of this solenoid. The number of flux linkages for this section of the solenoid is

$$N\Phi = (nl)(BA)$$

in which n is the number of turns per unit length of the solenoid and B is the magnetic field within the solenoid.

B is given by Eq. 21 of Chapter 31

$$B = \mu_0 n i$$

so that, from Eq. 2,

$$L = \frac{N\Phi}{i} = \frac{(nl)(B)(A)}{i} = \frac{(nl)(\mu_0 n i)(A)}{i} = \mu_0 n^2 l A. \tag{4}$$

Thus the inductance per unit length for a long solenoid near its center is then

$$\boxed{L/l = \mu_0 n^2 A} \quad \text{(solenoid).} \tag{5}$$

The inductance—like the capacitance—depends only on geometrical factors. The dependence on the square of the number of turns per unit length is to be expected. If you triple n you not only triple the number of turns (N) but you also triple the flux (Φ) through each turn, a factor of nine for the flux linkages and hence (see Eq. 2) for the inductance.

If the solenoid is very much longer than its radius, then Eq. 4 gives its inductance to a good approximation. This approximation neglects the spreading of the magnetic field lines near the ends of the solenoid, just as the parallel-plate capacitor formula ($C = \varepsilon_0 A/d$) neglects the fringing of the electric field lines near the edges of the capacitor plates.

Inductance of a Toroid. Figure 2 shows a toroid of N turns and of rectangular cross section with the dimensions indicated. What is its inductance?

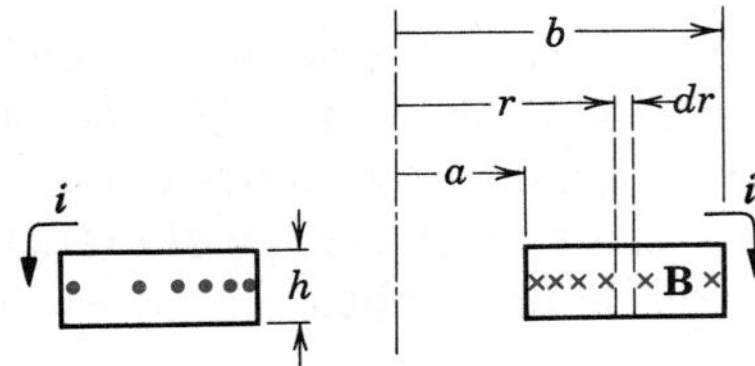

Figure 2 A cross section of a toroid, showing the current in the windings and the associated magnetic field.

Once more, to use the defining equation for inductance (Eq. 2), we must calculate the number of flux linkages that are set up by a given current. To do so, we must know how the magnetic field within the toroid depends on the current in its windings. We solved this problem in Chapter 31, where we learned that the magnetic field, which is not uniform over the cross section of a toroid, is given by Eq. 22 of that chapter, or

$$B = \frac{\mu_0 i N}{2\pi r}, \tag{6}$$

in which i is the current in the toroid windings.*

The flux Φ over the toroid cross section must be found by integration. If $h\,dr$ is the area of the elementary strip shown between the dashed lines in Fig. 2, we have, using Eq. 6

$$\Phi = \int \mathbf{B}\cdot d\mathbf{A} = \int_a^b (B)(h\,dr) = \int_a^b \frac{\mu_0 i N}{2\pi r}\,h\,dr$$
$$= \frac{\mu_0 i N h}{2\pi}\int_a^b \frac{dr}{r} = \frac{\mu_0 i N h}{2\pi}\ln\frac{b}{a}.$$

The inductance then follows from Eq. 2, its defining equation,

$$L = \frac{N\Phi}{i} = \frac{\mu_0 i N^2 h}{2\pi i}\ln\frac{b}{a},$$

or

$$\boxed{L = \frac{\mu_0 N^2 h}{2\pi}\ln\frac{b}{a} \quad \text{(toroid).}} \tag{7}$$

Note again that the inductance depends only on geometrical factors and that the number of turns enters as a square.

* Equation 6 holds no matter what the shape or dimensions of the toroid cross section.

Recall (see Section 27–3) that a capacitance can always be written as the *permittivity constant* ϵ_0 times a quantity with the dimensions of a length; thus ϵ_0 can be expressed in farads per meter. We see from Eq. 7 that an inductance can be written as the *permeability constant* μ_0 times a quantity with the dimensions of a length. This means that the permeability constant μ_0 can be expressed in henrys per meter, or

$$\mu_0 = 4\pi \times 10^{-7}\ \mathrm{T\cdot m/A} = 4\pi \times 10^{-7}\ \mathrm{H/m}. \tag{8}$$

Sample Problem 1 The toroid of Fig. 2 has $N = 1250$ turns, $a = 52$ mm, $b = 95$ mm, and $h = 13$ mm. What is its inductance?

From Eq. 7

$$L = \frac{\mu_0 N^2 h}{2\pi}\ln\frac{b}{a}$$
$$= \frac{(4\pi \times 10^{-7}\ \mathrm{H/m})(1250)^2(13 \times 10^{-3}\ \mathrm{m})}{2\pi}\ln\frac{95\ \mathrm{mm}}{52\ \mathrm{mm}}$$
$$= 2.45 \times 10^{-3}\ \mathrm{H} = 2.45\ \mathrm{mH}. \quad \text{(Answer)}$$

33–3 Self-Induction

If two coils—which we can now call inductors—are near each other, a current i in one coil will set up a magnetic flux Φ through the second coil. We learned in Chapter 32 that, if we change this flux by changing the current, an induced emf will appear in the second coil according to Faraday's law; see Fig. 2 of Chapter 32.

An induced emf also appears in a coil if we change the current in that same coil; see Fig. 3.

This is called *self-induction* and the electromotive force produced is called a *self-induced emf.* It obeys Faraday's law of induction just as other induced emfs do.

For any inductor, Eq. 2 tells us that

$$N\Phi = Li. \tag{9}$$

Faraday's law tells us

$$\mathcal{E} = -\frac{d(N\Phi)}{dt}. \tag{10}$$

By combining Eqs. 9 and 10 we can write, for the self-in-

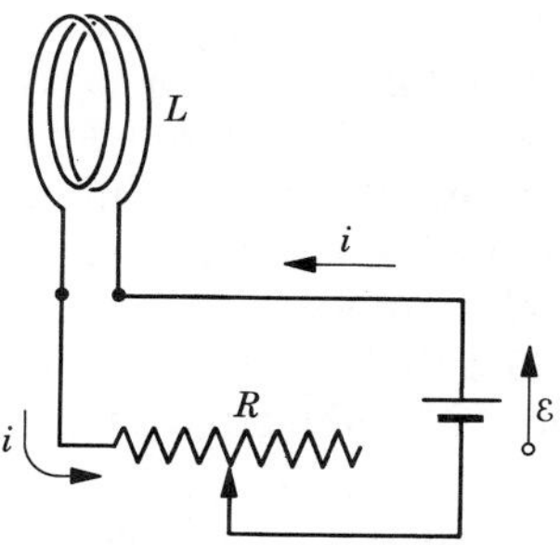

Figure 3 If the current in the coil L is changed, by varying the contact position on resistor R, a self-induced emf will appear in the coil.

duced emf

$$\mathcal{E} = -L\frac{di}{dt} \quad \text{(self-induced emf).} \tag{11}$$

Thus in any inductor (such as a coil, a solenoid, or a toroid) a self-induced emf appears whenever the current changes with time. The magnitude of the current itself has no influence on the induced emf; only the rate of change of the current counts.

You can find the *direction* of a self-induced emf from Lenz's law. The minus sign in Eq. 11 represents the fact that—as the law states—the self-induced emf acts to oppose the change that brings it about.

Suppose that, as in Fig. 4*a*, you set up a current i in a coil and arrange to have it increase with time at a rate di/dt. In the language of Lenz's law this increase in the current is the "change" that the self-induction must oppose. To do so a self-induced emf must appear in the coil, pointing—as the figure shows—so as to oppose the increase in the current. If you cause the current to decrease with time, as in Fig. 4*b*, the self-induced emf must point in a direction that tends to oppose the decrease in the current, as the figure shows.

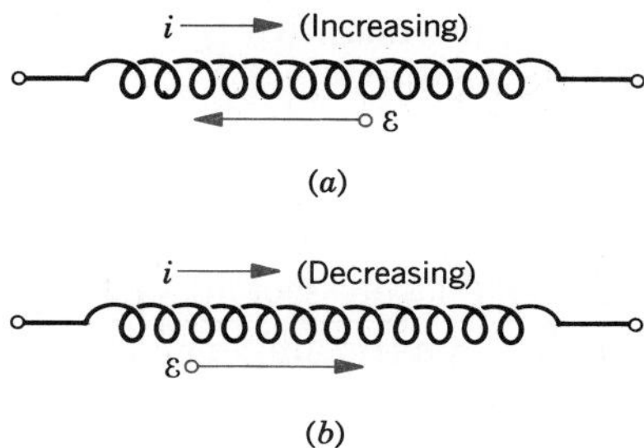

Figure 4 (a) If the current i is increasing, the induced emf appears in a direction such that it opposes the increase. (b) If the current i is decreasing, the induced emf appears in a direction such that it opposes the decrease.

33-4 An LR Circuit

In Section 29-8 we saw that if you suddenly introduce an emf $\mathcal{E}$ into a single-loop circuit containing a resistor R and a capacitor C, the charge does not build up immediately to its final equilibrium value $C\mathcal{E}$ but approaches it in an exponential fashion described by Eq. 29 of Chapter 29, or

$$q = C\mathcal{E}(1 - e^{-t/\tau_C}). \tag{12}$$

The rate at which the charge builds up is determined by the capacitive time constant τ_C, defined from

$$\tau_C = RC. \tag{13}$$

If you suddenly remove the emf from this same circuit, the charge does not immediately fall to zero but approaches zero in an exponential fashion, described by Eq. 34 of Chapter 29, or

$$q = q_0 e^{-t/\tau_C}. \tag{14}$$

The same time constant τ_C describes the fall of the charge as well as its rise.

An analogous slowing of the rise (or fall) of the current occurs if we introduce an emf $\mathcal{E}$ into (or remove it from) a single-loop circuit containing a resistor R and an inductor L. With the switch S in Fig. 5 closed on a, for example, the current in the resistor starts to rise. If the inductor were not present, the current would rise rapidly to a steady value $\mathcal{E}/R$. Because of the inductor, however, a self-induced emf, which we label $\mathcal{E}_L$, appears in the circuit and, from Lenz's law, this emf opposes the rise of current, which means that it opposes the battery emf $\mathcal{E}$ in

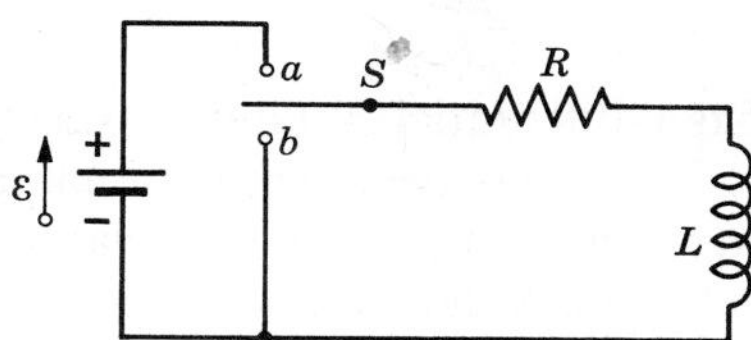

Figure 5 An LR circuit. When switch S is closed on a, the current rises and approaches a limiting value of $\mathcal{E}/R$.

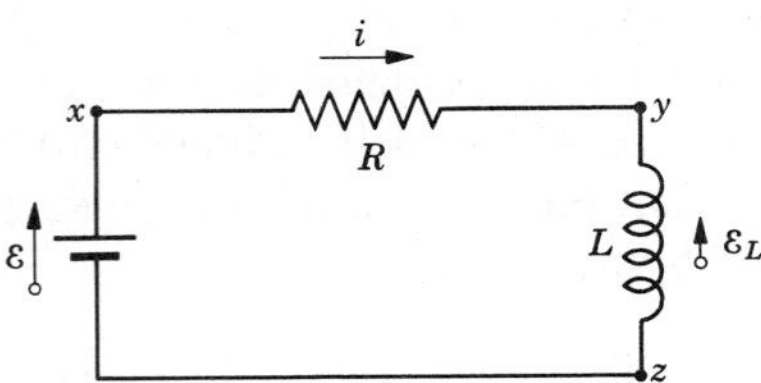

Figure 6 The circuit of Fig. 5 with the switch closed on a. Apply the loop theorem clockwise, starting at x.

polarity. Thus the resistor responds to the difference between two emfs, a constant one $\mathcal{E}$ due to the battery and a variable one $\mathcal{E}_L$ $(= -L\,di/dt)$ due to self-induction. As long as this second emf is present, the current in the resistor will be less than $\mathcal{E}/R$.

As time goes on, the rate at which the current increases becomes less rapid and the magnitude of the self-induced emf, which is proportional to di/dt, becomes smaller. Thus the current in the circuit approaches $\mathcal{E}/R$ asymptotically.

Now let us analyze the situation quantitatively. With the switch S in Fig. 5 thrown to a, the circuit is equivalent to that of Fig. 6. Let us apply the loop theorem, starting at x in this figure and going clockwise around the loop. For the direction of current shown, x will be higher in potential than y, which means that we encounter a drop in potential of $-iR$ as we traverse the resistor. Point y is higher in potential than point z because, for an increasing current, the induced emf will oppose the rise of the current by pointing as shown. Thus, as we traverse the inductor from y to z, we encounter a drop in potential of $-L(di/dt)$. We encounter a rise in potential of $+\mathcal{E}$ in traversing the battery from z to x. The loop theorem thus gives

$$-iR - L\frac{di}{dt} + \mathcal{E} = 0$$

or

$$\boxed{iR + L\frac{di}{dt} = \mathcal{E}} \qquad (LR \text{ circuit}). \qquad (15)$$

Equation 15 is a differential equation involving the variable i and its first derivative di/dt. We seek the function $i(t)$ such that when it and its first derivative are substituted in Eq. 15, the equation is satisfied and the initial condition $i(0) = 0$ is also satisfied.

Although there are formal rules for solving various classes of differential equations (and Eq. 15 can, in fact, be easily solved by direct integration, after rearrangement), we often find it simpler to guess at the solution, guided by physical reasoning and by previous experience. We can test any proposed solution by substituting it in the differential equation and seeing whether this equation reduces to an identity. In this case, we will be guided by the fact that our solution should be closely analogous to Eq. 12 for the build-up of charge in an RC circuit.

The solution to Eq. 15 which satisfies the initial condition is, we then claim,

$$i = \frac{\mathcal{E}}{R}(1 - e^{-Rt/L}). \qquad (16)$$

To test this solution by substitution, we find the derivative di/dt, which is

$$\frac{di}{dt} = \frac{\mathcal{E}}{L}e^{-Rt/L}. \qquad (17)$$

Substituting i and di/dt into Eq. 15 leads to an identity, as you can easily check. Thus Eq. 16 is indeed a solution of Eq. 15.

We can rewrite Eq. 16 as

$$\boxed{i = \frac{\mathcal{E}}{R}(1 - e^{-t/\tau_L}), \qquad \text{(rise of current)}} \qquad (18)$$

in which τ_L, the *inductive time constant,* is given by

$$\boxed{\tau_L = L/R} \qquad \text{(time constant)}. \qquad (19)$$

Figure 7 shows how the potential difference V_R across the resistor $(= iR)$ and V_L across the inductor $(= L\,di/dt)$ vary with time for particular values of $\mathcal{E}$, L, and R. Compare this figure carefully with the corresponding figure for an RC circuit (Fig. 16 of Chapter 29).

To show that the quantity τ_L $(= L/R)$ has the dimensions of time we put

$$\frac{1\text{ henry}}{\text{ohm}} = \frac{1\text{ henry}}{\text{ohm}}\left(\frac{1\text{ volt}\cdot\text{second}}{\text{henry}\cdot\text{ampere}}\right)\left(\frac{1\text{ ohm}\cdot\text{ampere}}{\text{volt}}\right) = 1\text{ second}.$$

The first quantity in parentheses is a conversion factor based on Eq. 11. The second conversion factor is based on the relation $V = iR$.

The physical significance of the time constant fol-

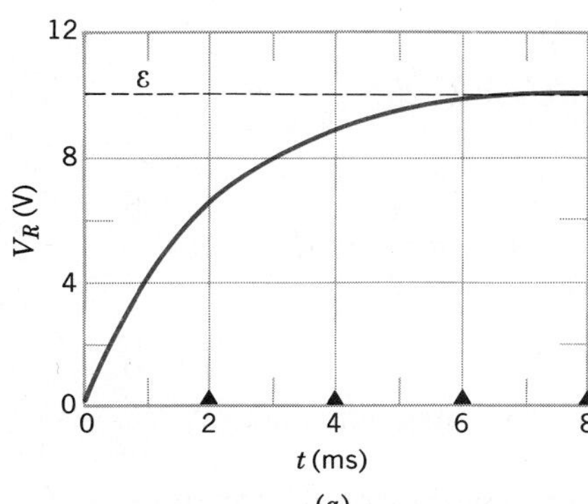

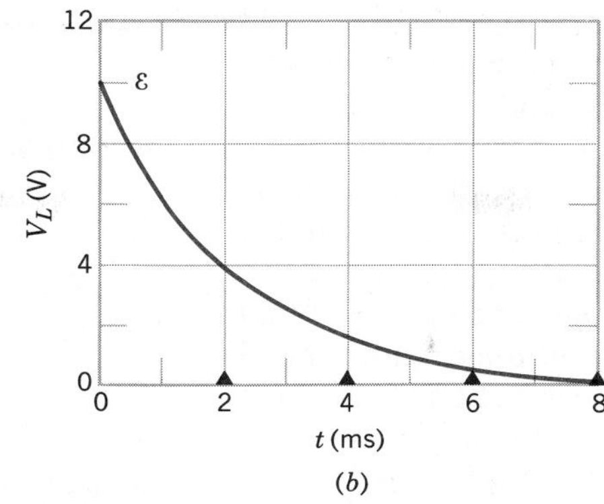

Figure 7 The variation with time of (a) V_R, the potential difference across the resistor in the circuit of Fig. 6 and (b) V_L, the potential difference across the inductor in that circuit. The triangles represent successive inductive time constants ($t = \tau_L$, $2\tau_L$, $3\tau_L$, . . .). The figure is plotted for $R = 2000\,\Omega$, $L = 4.0\,\text{H}$, and $\mathcal{E} = 10\,\text{V}$.

lows from Eq. 18. If we put $t = \tau_L = L/R$ in this equation, it reduces to

$$i = \frac{\mathcal{E}}{R}(1 - e^{-1}) = 0.63\,\frac{\mathcal{E}}{R}.$$

Thus the time constant τ_L is that time at which the current in the circuit will reach within $1/e$ (about 37%) of its final equilibrium value (see Fig. 7).

If the switch S in Fig. 5, having been left in position a long enough time for the equilibrium current $\mathcal{E}/R$ to be established, is thrown to b, the effect is to remove the battery from the circuit.* The differential equation that governs the subsequent decay of the current in the circuit can be found by putting $\mathcal{E} = 0$ in Eq. 15, or

$$L\frac{di}{dt} + iR = 0.$$

You can show by the test of substitution that the solution of this differential equation which satisfies the initial condition $i(0) = i_0 = \dfrac{\mathcal{E}}{R}$ is

$$\boxed{i = \frac{\mathcal{E}}{R}e^{-t/\tau_L} = i_0 e^{-t/\tau_L}} \qquad \text{(decay of current).} \qquad (20)$$

We see that both the rise in current (Eq. 18) and the decay of the current (Eq. 20) in an LR circuit are governed by the same inductive time constant, τ_L.

We have used i_0 in Eq. 20 to represent the current at time $t = 0$, which in our case happened to be $\mathcal{E}/R$ but could have been any initial value. The time dependence of the decaying current is identical to that of the decaying potential difference, plotted in Fig. 7b.

Sample Problem 2 A solenoid has an inductance of 53 mH and a resistance of 0.37 Ω. If it is connected to a battery, how long will it take for the current to reach one-half its final equilibrium value?

The equilibrium value of the current is reached as $t \to \infty$; from Eq. 16 it is $\mathcal{E}/R$. If the current has half this value at a particular time t_0, this equation becomes

$$\frac{1}{2}\frac{\mathcal{E}}{R} = \frac{\mathcal{E}}{R}(1 - e^{-t_0/\tau_L}).$$

Solving for t_0 by rearranging and taking the (natural) logarithm of each side, we find

$$\begin{aligned} t_0 &= \tau_L \ln 2 \\ &= \ln 2\,\frac{L}{R} = \ln 2\,\frac{53 \times 10^{-3}\,\text{H}}{0.37\,\Omega} \\ &= 0.10\,\text{s} = 100\,\text{ms}. \end{aligned} \qquad \text{(Answer)}$$

33-5 Energy and the Magnetic Field

When we lift a stone we do work, which we can get back by lowering the stone. It is convenient to think of the

* The connection to b must be made before the connection to a is broken. A switch that does this is called a *make-before-break* switch.

energy being stored temporarily in the gravitational field of the earth and the lifted stone and being withdrawn from this field when we lower the stone.

When we pull two unlike charges apart we like to say that the resulting electric potential energy is stored in the electric field of the charges. We can get it back from the field by letting the charges move closer together again.

In the same way we can consider energy to be stored in a magnetic field. For example, two long, rigid, parallel wires carrying current in the same direction attract each other, and we must do work to pull them apart. We can get this stored energy back at any time by letting the wires move back to their original positions.

To derive a quantitative expression for the energy stored in the magnetic field, consider Fig. 6, which shows a source of emf $\mathcal{E}$ connected to a resistor R and an inductor L. Equation 15

$$\mathcal{E} = iR + L\frac{di}{dt}, \qquad (21)$$

is the differential equation that describes the growth of current in this circuit. We stress that this equation follows immediately from the loop theorem and that the loop theorem in turn is an expression of the principle of conservation of energy for single-loop circuits. If we multiply each side of Eq. 21 by i, we obtain

$$\mathcal{E}i = i^2R + Li\frac{di}{dt}, \qquad (22)$$

which has the following physical interpretation in terms of work and energy:

1. If a charge dq passes through the seat of emf $\mathcal{E}$ in Fig. 6 in time dt, the seat does work on it in amount $\mathcal{E}\,dq$. The rate of doing work is $(\mathcal{E}\,dq)/dt$, or $\mathcal{E}i$. Thus the left term in Eq. 22 is the rate at which the seat of emf delivers energy to the circuit.
2. The second term in Eq. 22 is the rate at which energy appears as thermal energy in the resistor.
3. Energy that does not appear as thermal energy must, by our hypothesis, be stored in the magnetic field. Since Eq. 22 represents a statement of the conservation of energy for LR circuits, the last term must represent the rate dU_B/dt at which energy is stored in the magnetic field, or

$$\frac{dU_B}{dt} = Li\frac{di}{dt}. \qquad (23)$$

We can write this as

$$dU_B = Li\,di.$$

Integrating yields

$$U_B = \int dU_B = \int_0^i Li\,di$$

or

$$\boxed{U_B = \tfrac{1}{2}Li^2} \quad \text{(magnetic energy)}, \qquad (24)$$

which represents the total energy stored by an inductor L carrying a current i.

We can compare this relation with the expression for the energy associated with a capacitor C carrying a charge q, namely,

$$U_E = \frac{q^2}{2C}. \qquad (25)$$

Here the energy is stored in an electric field. In each case the expression for the stored energy was derived by setting it equal to the work that must be done to set up the field.

Sample Problem 3 A coil has an inductance of 53 mH and a resistance of 0.35 Ω. (a) If a 12-V emf is applied, how much energy is stored in the magnetic field after the current has built up to its equilibrium value?

The stored energy is given by Eq. 24

$$U_B = \tfrac{1}{2}Li^2.$$

To find the equilibrium stored energy, we must substitute the equilibrium current in this expression. From Eq. 16 the equilibrium current is

$$i_\infty = \frac{\mathcal{E}}{R} = \frac{12\ \text{V}}{0.35\ \Omega} = 34.3\ \text{A}.$$

The substitution yields

$$U_{B\infty} = \tfrac{1}{2}Li_\infty^2 = (\tfrac{1}{2})(53 \times 10^{-3}\ \text{H})(34.3\ \text{A})^2$$
$$= 31\ \text{J}. \qquad \text{(Answer)}$$

(b) After how many time constants will half of this equilibrium energy be stored in the magnetic field?

We are asked: At what time will the relation

$$U_B = \tfrac{1}{2}U_{B\infty}$$

be satisfied? Equation 24 allows us to rewrite this as

$$\tfrac{1}{2}Li^2 = (\tfrac{1}{2})\tfrac{1}{2}Li_\infty^2$$

or

$$i = (1/\sqrt{2})i_\infty.$$

But i is given by Eq. 16 and i_∞ (see above) is $\mathcal{E}/R$, so that

$$\frac{\mathcal{E}}{R}(1 - e^{-t/\tau_L}) = \frac{\mathcal{E}}{\sqrt{2}R}.$$

This can be written as

$$e^{-t/\tau_L} = 1 - 1/\sqrt{2} = 0.293,$$

which yields

$$\frac{t}{\tau_L} = -\ln 0.293 = 1.23$$

or

$$t = 1.23\ \tau_L. \qquad \text{(Answer)}$$

Thus the stored energy will reach half of its equilibrium value after 1.23 time constants.

Sample Problem 4 A 3.56-H inductor is placed in series with a 12.8-Ω resistor, an emf of 3.24 V being suddenly applied to the combination. At 0.278 s (which is one inductive time constant) after the contact is made, (a) what is the rate P at which energy is being delivered by the battery?

The rate, P, is available from $P = \mathcal{E}i$. The current is given by Eq. 18, or

$$i = \frac{\mathcal{E}}{R}(1 - e^{-t/\tau_L}),$$

which, after one time constant, becomes

$$i = \frac{3.24\ \text{V}}{12.8\ \Omega}(1 - e^{-1}) = 0.1600\ \text{A}.$$

The rate at which the battery delivers energy is then

$$P = \mathcal{E}i = (3.24\ \text{V})(0.1600\ \text{A}) = 0.5184\ \text{W} \approx 518\ \text{mW}. \qquad \text{(Answer)}$$

(b) At what rate P_R does energy appear as thermal energy in the resistor?

This is given by

$$P_R = i^2R = (0.1600\ \text{A})^2(12.8\ \Omega) = 0.3277\ \text{W} \approx 328\ \text{mW}. \qquad \text{(Answer)}$$

(c) At what rate P_B is energy being stored in the magnetic field?

This is given by Eq. 23, which requires that we know di/dt. Differentiating Eq. 18 yields

$$\frac{di}{dt} = \frac{\mathcal{E}}{R}\frac{R}{L}(e^{-t/\tau_L}) = \frac{\mathcal{E}}{L}e^{-t/\tau_L}.$$

After one time constant we have

$$\frac{di}{dt} = \frac{3.24\ \text{V}}{3.56\ \text{H}}e^{-1} = 0.3348\ \text{A/s}.$$

From Eq. 23 the desired rate is then

$$P_B = \frac{dU_B}{dt} = Li\frac{di}{dt}$$
$$= (3.56\ \text{H})(0.1600\ \text{A})(0.3348\ \text{A/s})$$
$$= 0.1907\ \text{W} \approx 191\ \text{mW}. \qquad \text{(Answer)}$$

Note that, as required by energy conservation,

$$P = P_R + P_B,$$

or

$$P = 0.3277\ \text{W} + 0.1907\ \text{W} = 0.5184\ \text{W} \approx 518\ \text{mW}.$$

33-6 Energy Density and the Magnetic Field

So far we have dealt with the energy U stored in the magnetic field of a specific current-carrying inductor. Here we turn our attention to the magnetic field itself—regardless of its source—and seek an expression for the *energy density* u_B, the energy per unit volume stored at any point in the field.

Consider a length l near the center of a long solenoid; Al is the volume associated with this length. The stored energy must lie entirely within this volume because the magnetic field outside such a solenoid is essentially zero. Moreover, the stored energy must be uniformly distributed throughout the volume of the solenoid because the magnetic field is uniform everywhere inside.

Thus, we can write

$$u_B = \frac{U_B}{Al}$$

or, since

$$U_B = \tfrac{1}{2}Li^2,$$

we have

$$u_B = \frac{(L/l)i^2}{2A}.$$

We can substitute for L/l for the solenoid from Eq. 5, obtaining

$$u_B = \tfrac{1}{2}\mu_0 n^2 i^2.$$

From Eq. 21 of Chapter 31 ($B = \mu_0 in$) we can write this finally as

$$\boxed{u_B = \frac{B^2}{2\mu_0}} \qquad \text{(magnetic energy density).} \qquad (26)$$

Table 1 Some Corresponding Electrical and Magnetic Quantities

Definition	$C = q/V$	$L = N\Phi/i$
Dimensions	$C = \epsilon_0 \times$ a length	$L = \mu_0 \times$ a length
Constants	$\epsilon_0 = 8.85$ pF/m	$\mu_0 = 1.26$ μH/m
Energy storage	$U_C = \frac{1}{2}CV^2 = \frac{q^2}{2C}$	$U_L = \frac{1}{2}Li^2 = \frac{(N\Phi)^2}{2L}$
Energy density	$u_E = (\epsilon_0/2)E^2$	$u_B = (1/2\mu_0)B^2$
Time constant	$\tau_C = RC$	$\tau_L = L/R$

This equation gives the energy density stored at any point where the magnetic field is B. Even though we derived Eq. 26 by considering a special case, the solenoid, it remains true for all magnetic field configurations, no matter how they are generated. Equation 26 is to be compared with Eq. 23 of Chapter 27, namely

$$u_E = \tfrac{1}{2}\epsilon_0 E^2, \tag{27}$$

which gives the energy density (in a vacuum) at any point in an electric field. Note that both u_B and u_E are proportional to the square of the appropriate field quantity, B or E.

The solenoid plays a role in relation to magnetic fields similar to the role the parallel-plate capacitor plays with respect to electric fields. In each case we have a simple device that can be used for setting up a uniform field throughout a well-defined region of space and for deducing, in a simple way, some properties of these fields. Table 1 compares some corresponding electrical and magnetic quantities.

Sample Problem 5 A long coaxial cable (Fig. 8) consists of two concentric cylinders with radii a and b. Its central conductor carries a steady current i, the outer conductor providing the return path. (a) Calculate the energy stored in the magnetic field between the conductors for a length l of such a cable.

Consider a volume dV consisting of a cylindrical shell whose radii are r and $r + dr$ and whose length is l. The energy dU contained within this shell is

$$dU = u_B\, dV.$$

The energy density, from Eq. 26, is

$$u_B = \frac{B^2}{2\mu_0}.$$

In the space between the two conductors Ampere's law,

$$\int \mathbf{B}\cdot d\mathbf{s} = \mu_0 i,$$

leads to

$$(B)(2\pi r) = \mu_0 i,$$

or

$$B = \frac{\mu_0 i}{2\pi r}.$$

The energy density is then

$$u_B = \frac{1}{2\mu_0}\left(\frac{\mu_0 i}{2\pi r}\right)^2 = \frac{\mu_0 i^2}{8\pi^2 r^2}.$$

The volume, dV, of the shell is $(2\pi r l)(dr)$ so the energy dU contained within the shell is

$$dU = \frac{\mu_0 i^2}{8\pi^2 r^2}(2\pi r l)(dr) = \frac{\mu_0 i^2 l}{4\pi}\frac{dr}{r}.$$

The total energy follows by integrating this expression over the volume between the cylinders.

$$U = \int dU = \frac{\mu_0 i^2 l}{4\pi}\int_a^b \frac{dr}{r} = \frac{\mu_0 i^2 l}{4\pi}\ln\frac{b}{a}. \tag{28}$$

No energy is stored outside the outer conductor or inside the hollow inner conductor because the magnetic field is zero there, as you can show from Ampere's law. Note that we could also have arrived at this result by calculating the inductance of a length l of the cable, using Eq. 2, then used Eq. 24 to find the stored energy.

(b) What is the stored energy per unit length for $a = 1.2$ mm, $b = 3.5$ mm, and $i = 2.7$ A?

From Eq. 28 we have

$$U/l = \frac{\mu_0 i^2}{4\pi}\ln\frac{b}{a} = \frac{(4\pi \times 10^{-7}\ \text{H/m})(2.7\ \text{A})^2}{4\pi}\ln\frac{3.5\ \text{mm}}{1.2\ \text{mm}}$$
$$= 7.8 \times 10^{-7}\ \text{J/m} = 780\ \text{nJ/m}. \quad \text{(Answer)}$$

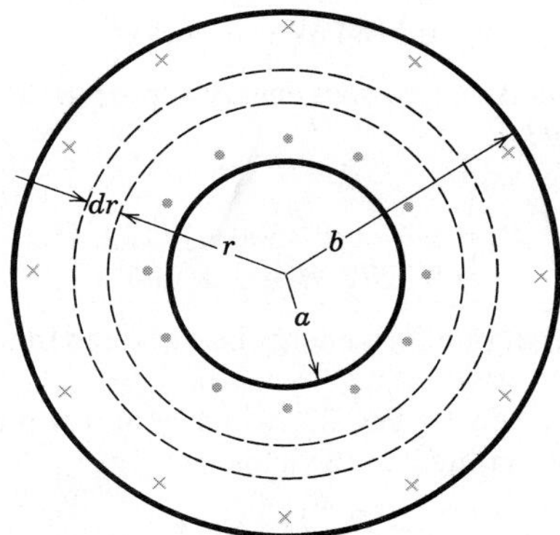

Figure 8 Sample Problem 5. A cross section of a long coaxial cable, of inner radius a and outer radius b. The dots and crosses represent equal but opposite currents in the two thin concentric conductors.

Sample Problem 6 Compare the energy required to set up, in a cube 10 cm on edge (a) a uniform electric field of 100 kV/m and (b) a uniform magnetic field of 1.0 T. Both these fields would be judged reasonably large but readily available in the laboratory.

(a) In the electric case we have, where V_0 is the volume of the cube,

$$\begin{aligned} U_E &= u_E V_0 = \tfrac{1}{2}\epsilon_0 E^2 V_0 \\ &= (\tfrac{1}{2})(8.85 \times 10^{-12}\ \text{F/m})(10^5\ \text{V/m})^2(0.1\ \text{m})^3 \\ &= 4.4 \times 10^{-5}\ \text{J} = 44\ \mu\text{J}. \qquad \text{(Answer)} \end{aligned}$$

(b) In the magnetic case, we have

$$\begin{aligned} U_B &= u_B V_0 = \frac{B^2}{2\mu_0} V_0 = \frac{(1.0\ \text{T})^2(0.1\ \text{m})^3}{(2)(4\pi \times 10^{-7}\ \text{T}\cdot\text{m/A})} \\ &= 398\ \text{J} \approx 400\ \text{J}. \qquad \text{(Answer)} \end{aligned}$$

In terms of fields normally available in the laboratory, much larger amounts of energy can be stored in a magnetic field than in an electric one, the ratio being about 10^7 in this example. Conversely, much more energy is required to set up a magnetic field of "reasonable" laboratory magnitude than is required to set up an electric field of similarly reasonable magnitude.

33-7 Mutual Induction (Optional)

In this section we return to the case of two interacting coils, which we first discussed in Section 32-2, and we treat it in a somewhat more formal manner. In Fig. 2 of that chapter, we saw that if two coils are close together, a steady current i in one coil will set up a magnetic flux Φ linking the other coil. If we change i with time, an emf $\mathcal{E}$ given by Faraday's law appears in the second coil; we called this process *induction.* We could better have called it *mutual induction,* to suggest the mutual interaction of the two coils and to distinguish it from *self-induction,* in which only one coil is involved.

Let us look a little more quantitatively at mutual induction. Figure 9*a* shows two circular close-packed coils near each other and sharing a common axis.* There is a steady current i_1 in coil 1, set up by the battery in the external circuit. This current produces a magnetic field suggested by the lines of B_1 in the figure. Coil 2 is connected to a sensitive galvanometer G but contains no battery; a magnetic flux Φ_{21} (the flux through coil 2 associated with the current in coil 1) is linked by its N_2 turns.

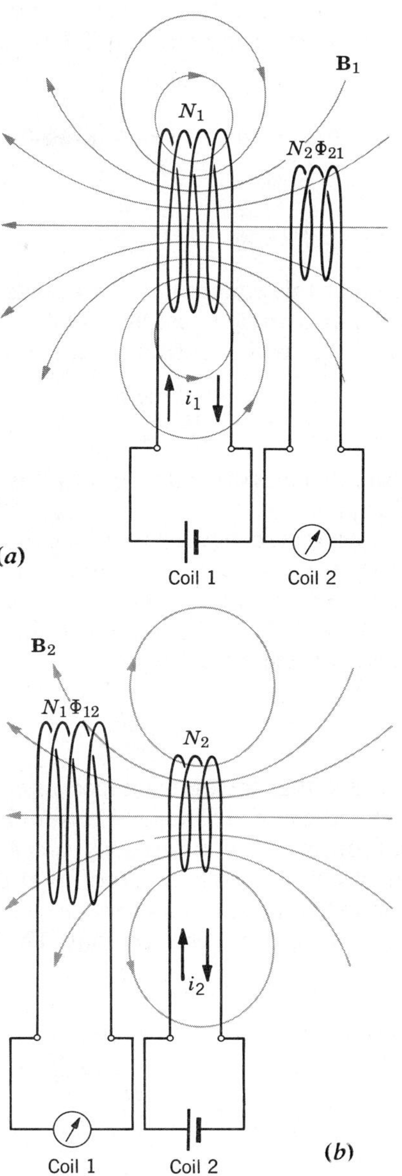

Figure 9 Mutual inductance. (a) If the current in coil 1 changes, an emf will be induced in coil 2. (b) If the current in coil 2 changes, an emf will be induced in coil 1.

We define the mutual inductance M_{21} of coil 2 with respect to coil 1 as

$$M_{21} = \frac{N_2\Phi_{21}}{i_1}. \qquad (29)$$

* For ease of representation, the two coils in Fig. 9 are not actually drawn as close-packed, which requires that every turn of a given coil enclose the same flux.

Compare this with Eq. 2 ($L = N\Phi/i$), the definition of (self) inductance. We can recast Eq. 29 as

$$M_{21}i_1 = N_2\Phi_{21}.$$

If, by external means, we cause i_1 to vary with time, we have

$$M_{21}\frac{di_1}{dt} = N_2\frac{d\Phi_{21}}{dt}.$$

The right side of this equation, from Faraday's law, is, apart from a change in sign, just the emf $\mathcal{E}_2$ appearing in coil 2 due to the changing current in coil 1, or

$$\mathcal{E}_2 = -M_{21}\frac{di_1}{dt}, \tag{30}$$

which you should compare with Eq. 11 ($\mathcal{E} = -L\,di/dt$) for self-inductance.

Let us now interchange the roles of coils 1 and 2, as in Fig. 9*b*. That is, we set up a current i_2 in coil 2, by means of a battery, and this produces a magnetic flux Φ_{12} that links coil 1. If we change i_2 with time, we have, by the same argument given above,

$$\mathcal{E}_1 = -M_{12}\frac{di_2}{dt}. \tag{31}$$

Thus we see that the emf induced in either coil is proportional to the rate of change of current in the other coil. The proportionality constants M_{21} and M_{12} seem to be different. We assert, without proof, that they are in fact the same so that no subscripts are needed. This conclusion is true but is in no way obvious. Thus, we have

$$M_{21} = M_{12} = M \tag{32}$$

and we can rewrite Eqs. 30 and 31 as

$$\boxed{\mathcal{E}_2 = -M\,di_1/dt} \tag{33}$$

and

$$\boxed{\mathcal{E}_1 = -M\,di_2/dt.} \tag{34}$$

The induction is indeed mutual. The SI unit for M (as for L) is the henry.

Sample Problem 7 Figure 10 shows two circular close-packed coils, the smaller (radius R_2) being coaxial with the larger (radius R_1) and in the same plane. (a) Derive an expression for the coefficient of mutual inductance M for this arrangement of these two coils, assuming that $R_1 \gg R_2$.

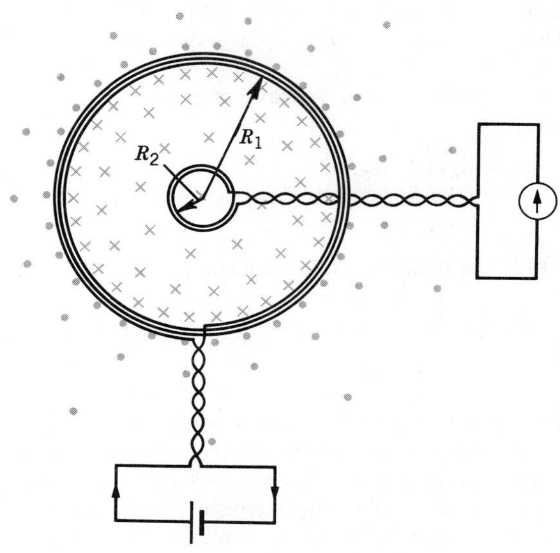

Figure 10 Sample Problem 7. A small coil is placed at the center of a large coil. What is their mutual inductance?

As the figure suggests we imagine that we establish a current i_1 in the larger coil and we note the magnetic field B_1 that it sets up. The value of B_1 at the center of this coil is (see Eq. 24 of Chapter 31, with $z = 0$)

$$B_1 = \frac{\mu_0 i_1 N_1}{2R_1}.$$

Because we have assumed that $R_1 \gg R_2$, we may take B_1 to be the magnetic field at all points within the boundary of the smaller coil. The flux linkages for the smaller coil are then

$$N_2\Phi_{21} = N_2(B_1)(\pi R_2^2) = \frac{\pi\mu_0 N_1 N_2 R_2^2 i_1}{2R_1}.$$

From Eq. 29 we then have

$$M = \frac{N_2\Phi_{21}}{i_1} = \frac{\pi\mu_0 N_1 N_2 R_2^2}{2R_1}. \quad \text{(Answer)}$$

(b) What is the value of M for $N_1 = N_2 = 1200$ turns, $R_2 = 1.1$ cm and $R_1 = 15$ cm.

The above equation yields

$$M = \frac{(\pi)(4\pi \times 10^{-7}\ \text{H/m})(1200)(1200)(0.011\ \text{m})^2}{(2)(0.15\ \text{m})}$$

$$= 2.29 \times 10^{-3}\ \text{H} \approx 2.3\ \text{mH}. \quad \text{(Answer)}$$

Consider the situation if we reverse the roles of the two coils in Fig. 10, that is, if we set up a current i_2 in the smaller coil and try to calculate M from Eq. 29

$$M = \frac{N_1 \Phi_{12}}{i_2}.$$

The calculation of Φ_{12} (the flux of the smaller coil's magnetic field encompassed by the larger coil) is not simple. If we were to calculate it numerically using a computer, we would find exactly 2.3 mH, as above! This should make us appreciate the fact that Eq. 32 ($M_{12} = M_{21} = M$) is not obvious.

REVIEW AND SUMMARY

An Inductor

An *inductor* is an arrangement that can be used to set up a known magnetic field in a specified region. If a current i is established through each of the N windings of an inductor, a magnetic flux Φ links those windings. The *inductance L* of the inductor is

Inductance

$$L = \frac{N\Phi}{i} \quad \text{(inductance defined).} \qquad [2]$$

The SI unit of inductance is the *henry* (abbr. H), with

$$1 \text{ henry} = 1 \text{ H} = 1 \text{ T} \cdot \text{m}^2/\text{A}. \qquad [3]$$

The inductance per unit length of a long solenoid near its center is

Solenoid Inductance

$$L/l = \mu_0 n^2 A \quad \text{(solenoid)} \qquad [5]$$

and the inductance of a rectangular toroid (height h, inner and outer radii a and b, N total turns) is

Toroid Inductance

$$L = \frac{\mu_0 N^2 h}{2\pi} \ln \frac{b}{a} \quad \text{(toroid);} \qquad [7]$$

see Sample Problem 1.

If a current i in a coil changes with time, an emf described by Faraday's law is induced in the coil itself. This self-induced emf is given by

Self-induction

$$\mathcal{E} = -L \frac{di}{dt} \quad \text{(self-induced emf).} \qquad [11]$$

The direction of $\mathcal{E}$ is found from Lenz's law; the self-induced emf acts to oppose the change that brings it about.

If a constant emf $\mathcal{E}$ is introduced into a single-loop circuit containing a resistor R and an inductor L (see Fig. 6), the current rises to an equilibrium value of $\mathcal{E}/R$ according to

Series *LR* Circuits

$$i = \frac{\mathcal{E}}{R}(1 - e^{-t/\tau_L}) \quad \text{(rise of current).} \qquad [18]$$

Here τ_L ($= L/R$) governs the rate of rise of the current and is called the inductive time constant of the circuit. Figure 7, which shows the potential differences across the resistor and the inductor while the current is rising, deserves careful study; see also Sample Problem 2. The decay of the current when the source of constant emf is removed is given by

$$i = i_0 e^{-t/\tau_L} \quad \text{(decay of current).} \qquad [20]$$

By applying the conservation-of-energy principle to the rise of current in the LR circuit of Fig. 6 we deduce that, if an inductor L carries a current i, an energy given by

Storage of Energy by an Inductor

$$U_B = \tfrac{1}{2} L i^2 \quad \text{(magnetic energy)} \qquad [24]$$

can be viewed as stored in its magnetic field. Sample Problems 3 and 4 analyze the energy transfers in an LR circuit.

Applying Eq. 24 to a section of a long solenoid leads us to deduce the general result that, if B is the magnetic field at any point, the density of stored magnetic energy at that point is

Magnetic Field Energy

$$u_B = \frac{B^2}{2\mu_0} \quad \text{(magnetic energy density).} \tag{26}$$

Sample Problems 5 and 6 deal with magnetic energy. Table 1 displays some interesting and useful analogies between electrical and magnetic quantities.

If two coils or similar conductors are near each other (see Fig. 9) a changing current in either coil can induce an emf in the other. This mutual induction phenomenon is described by

Mutual Induction

$$\mathcal{E}_2 = -M\, di_1/dt \quad \text{and} \quad \mathcal{E}_1 = -M\, di_2/dt. \qquad [33,34]$$

where M (measured in henries) is the coefficient of mutual inductance for the coil arrangement; see Sample Problem 7.

QUESTIONS

1. Explain how a long straight wire can show self-induction effects. How would you go about looking for them?

2. If the flux passing through each turn of a coil is the same, the inductance of the coil may be computed from $L = N\Phi_B/i$ (Eq. 2). How might one compute L for a coil for which this assumption is not valid?

3. Show that the dimensions of the two expressions for L, $N\Phi_B/i$ (Eq. 2) and $\mathcal{E}/(di/dt)$ (Eq. 11), are the same.

4. You want to wind a coil so that it has resistance but essentially no inductance. How would you do it?

5. Is the inductance per unit length for a solenoid near its center the same as, less than, or greater than the inductance per unit length near its ends? Justify your answer.

6. Explain why the self-inductance of a coaxial cable is expected to increase when the radius of the outer conductor is increased, the radius of the inner conductor remaining fixed.

7. A steady current is set up in a coil with a very large inductive time constant. When the current is interrupted with a switch, a heavy arc tends to appear at the switch blades. Explain why. (*Note:* Interrupting currents in highly inductive circuits can be destructive and dangerous.)

8. Suppose that you connect an ideal (that is, essentially resistanceless) coil across an ideal (again, essentially resistanceless) battery. You might think that, because there is no resistance in the circuit, the current would jump at once to a very large value. On the other hand, you might think that, because the inductive time constant ($= L/R$) is extremely large, the current would rise very slowly, if at all. What actually happens?

9. In an LR circuit like that of Fig. 6, can the self-induced emf ever be larger than the battery emf?

10. In an LR circuit like that of Fig. 6, is the current in the resistor always the same as the current in the inductor?

11. In the circuit of Fig. 5 the self-induced emf is a maximum at the instant the switch is closed on a. How can this be—there is no current in the inductor at this instant?

12. The switch in Fig. 5, having been closed on a for a "long" time, is thrown to b. What happens to the energy that is stored in the inductor?

13. A coil has a (measured) inductance L and a (measured) resistance R. Is its inductive time constant necessarily given by $\tau_L = L/R$? Bear in mind that we derived that equation (see Fig. 5) for a situation in which the inductive and resistive elements are physically separated. Discuss.

14. Figure 7*a* in this chapter and Figure 16*b* in Chapter 29, are plots of $V_R(t)$ for, respectively, an LR circuit and an RC circuit. Why are these two curves so different? Account for each in terms of physical processes going on in the appropriate circuit.

15. Two solenoids, A and B, have the same diameter and length and contain only one layer of copper windings, with adjacent turns touching, insulation thickness being negligible. Solenoid A contains many turns of fine wire and solenoid B contains fewer turns of heavier wire. (*a*) Which solenoid has the larger self-inductance? (*b*) Which solenoid has the larger inductive time constant? Justify your answers.

16. Can you make an argument based on the manipulation of bar magnets to suggest that energy may be stored in a magnetic field?

17. Draw all the formal analogies that you can think of between a parallel-plate capacitor (for electric fields) and a long solenoid (for magnetic fields).

18. In each of the following operations energy is expended. Some of this energy is returnable (can be reconverted) into electrical energy that can be made to do useful work and some becomes unavailable for useful work or is wasted in other ways. In which case will there be the *least* percentage of returnable electrical energy? (*a*) Charging a capacitor; (*b*) charging a storage battery; (*c*) sending a current through a resistor; (*d*) setting up a magnetic field; (*e*) moving a conductor in a magnetic field.

19. The current in a solenoid is reversed. What changes does this make in the magnetic field **B** and the energy density u at various points along the solenoid axis?

20. Commercial devices such as motors and generators that are involved in the transformation of energy between electrical and mechanical forms involve magnetic rather than electrostatic fields. Why should this be so?

21. A heavy current is passed, clockwise, through both coils of a standard mutual inductance shown in Fig. 11. Q is the horizontal midpoint of the large coil whose ends are P and S. The midpoint of the small coil starts a distance x from R. Describe its subsequent motion.

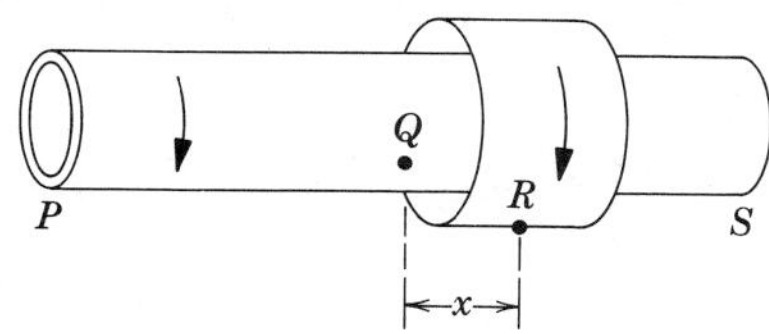

Figure 11 Question 21.

22. In a case of mutual induction, such as in Fig. 9, is self-induction also present? Discuss.

23. You are given two similar flat circular coils of N turns each. The centers of the coils are maintained at a fixed distance apart. For what orientation will their mutual inductance M be the greatest? For what orientation will it be the least?

24. A flat circular coil is placed completely outside a long solenoid, near its center, the axes of the coil and the solenoid being parallel. Describe the mutual induction, if any, of this combination.

25. A circular coil of N turns surrounds a long solenoid. Is the mutual inductance greater when the coil is near the center of the solenoid or when it is near one end? Justify your answer.

26. A coil, connected to an ac household power outlet, carries an alternating current. Explain why the amplitude of the current in the coil might change if you place a conducting metal object near it. Design a metal detector based on this principle.

27. A long cylinder is wound from left to right with one layer of wire, giving it n turns per unit length with a self-inductance of L_1, as in Fig. 12*a*. If the winding is now continued, in the same *sense* but returning from right to left, as in Fig. 12*b*, so as to give a second layer also of n turns per unit length, then what is the value of the self-inductance? Explain.

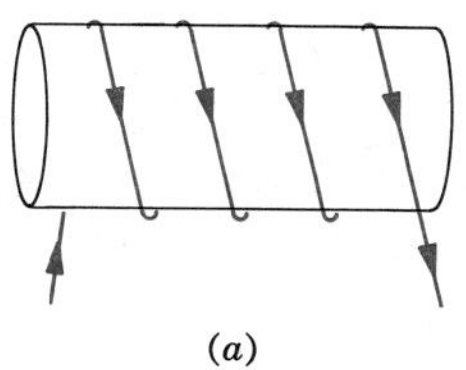

(*a*)

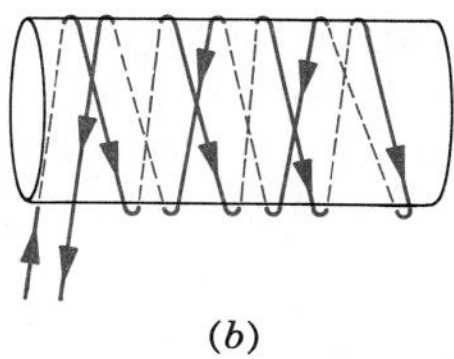

(*b*)

Figure 12 Question 27.

EXERCISES AND PROBLEMS

Section 33-2 Inductance

1E. The inductance of a close-packed coil of 400 turns is 8.0 mH. Calculate the magnetic flux through the coil when the current is 5.0 mA.

2E. A circular coil has a 10-cm radius and consists of 30 closely-wound turns of wire. An externally-produced magnetic field of 2.6 mT is perpendicular to the coil. (*a*) If no current is in the coil, what is the flux linkage? (*b*) When the current in the coil is 3.8 A in a certain direction, the net flux through the coil is found to vanish. What is the inductance of the coil?

3E. A solenoid is wound with a single layer of insulated copper wire (diameter, 2.5 mm). It is 4.0 cm in diameter and 2.0 m long. (*a*) How many windings are on the solenoid? (*b*) What is the inductance per meter for the solenoid near its center? Assume that adjacent wires touch and that insulation thickness is negligible.

4P. A long thin solenoid can be bent into a ring to form a toroid. Show that if the solenoid is long and thin enough, the equation for the inductance of a toroid (Eq. 7) is equivalent to that for a solenoid of the appropriate length (Eq. 5).

5P. *Inductors in series.* Two inductors L_1 and L_2 are connected in series and are separated by a large distance. (*a*) Show that the equivalent inductance is given by

$$L_{eq} = L_1 + L_2.$$

(*b*) Why must their separation be large for this relationship to hold? (*c*) What is the generalization of (*a*) for N inductors in series?

6P. *Inductors in parallel.* Two inductors L_1 and L_2 are connected in parallel and separated by a large distance. (*a*) Show that the equivalent inductance is given by

$$\frac{1}{L_{eq}} = \frac{1}{L_1} + \frac{1}{L_2}.$$

(*b*) Why must their separation be large for this relationship to hold? (*c*) What is the generalization of (*a*) for N inductors in parallel?

7P. A wide copper strip of width W is bent into a piece of slender tubing of radius R with two plane extensions, as shown in Fig. 13. A current i flows through the strip, distributed uniformly over its width. In this way a "one-turn solenoid" has been formed. (*a*) Derive an expression for the magnitude of the magnetic field **B** in the tubular part (far away from the edges). (*Hint:* Assume that the field outside this one-turn solenoid is negligibly small.) (*b*) Find also the inductance of this one-turn solenoid, neglecting the two plane extensions.

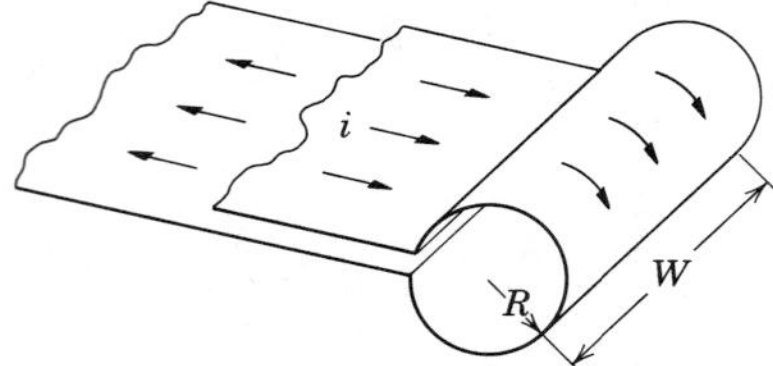

Figure 13 Problem 7.

8P. Two long parallel wires, each of radius a, whose centers are a distance d apart carry equal currents in opposite directions. Show that, neglecting the flux within the wires themselves, the inductance of a length l of such a pair of wires is given by

$$L = \frac{\mu_0 l}{\pi} \ln \frac{d-a}{a}.$$

See Sample Problem 3, Chapter 31. (*Hint:* Calculate the flux through a rectangle of which the wires form two opposite sides.)

Section 33-3 Self-Induction

9E. At a given instant the current and the induced emf in an inductor are as indicated in Fig. 14. (*a*) Is the current increasing or decreasing? (*b*) The emf is 17 V and the rate of change of the current is 25 kA/s; what is the value of the inductance?

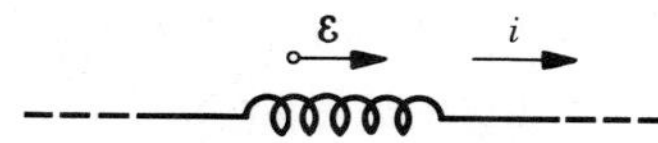

Figure 14 Exercise 9.

10E. A 12-H inductor carries a steady current of 2.0 A. How can a 60-V self-induced emf be made to appear in the inductor?

11E. A long cylindrical solenoid with 100 turns per cm has a radius of 1.6 cm. Assume the magnetic field it produces is parallel to its axis and is uniform in its interior. (*a*) What is its inductance per meter of length? (*b*) If the current changes at the rate 13 A/s, what emf is induced per meter?

12E. The inductance of a closely wound N-turn coil is such that an emf of 3.0 mV is induced when the current changes at the rate 5.0 A/s. A steady current of 8.0 A produces a magnetic flux of 40 μWb through each turn. (*a*) Calculate the inductance of the coil. (*b*) How many turns does the coil have?

13P. The current i through a 4.6-H inductor varies with time t as shown on the graph of Fig. 15. The inductor has a resistance of 12 Ω. Calculate the induced emf during the time intervals (*a*) $t = 0$ to $t = 2$ ms, (*b*) $t = 2$ ms to $t = 5$ ms, (*c*) $t = 5$ ms to $t = 6$ ms. (Ignore the behavior at the ends of the intervals.)

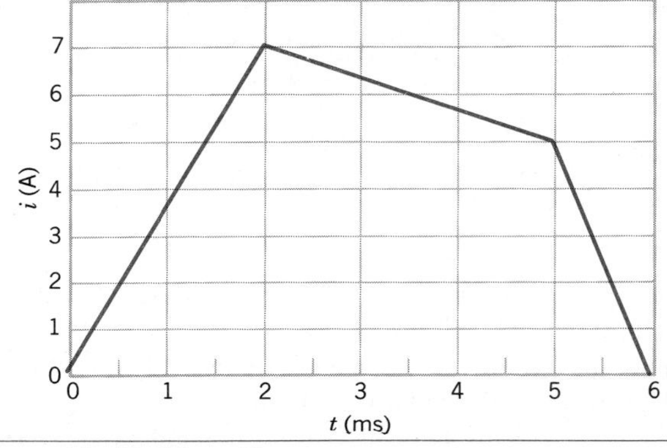

Figure 15 Problem 13.

Section 33-4 An LR Circuit

14E. The current in an LR circuit builds up to one-third of its steady-state value in 5.0 s. Calculate the inductive time constant.

15E. How many "time constants" must we wait for the current in an LR circuit to build up to within 0.10% of its equilibrium value?

16E. The current in an LR circuit drops from 1.0 A to 10 mA in the first second following removal of the battery from the circuit. If L is 10 H, find the resistance R in the circuit.

17E. How long would it take, following the removal of the battery, for the potential difference across the resistor in an LR circuit ($L = 2.0$ H, $R = 3.0$ Ω) to decay to 10% of its initial value?

18E. (*a*) Consider the LR circuit of Fig. 5. In terms of the battery emf $\mathcal{E}$, what is the self-induced emf $\mathcal{E}_L$ when the switch has just been closed on *a*? (*b*) What is $\mathcal{E}_L$ after two time constants? (*c*) After how many time constants will $\mathcal{E}_L$ be just one half of the battery emf $\mathcal{E}$?

19E. A solenoid having an inductance of 6.3 μH is connected in series with a 1.2-kΩ resistor. (*a*) If a 14-V battery is switched across the pair, how long will it take for the current through the resistor to reach 80% of its final value? (*b*) What is the current through the resistor after one time constant?

20E. The flux linkage through a certain coil of 0.75-Ω resistance is 26 mWb when there is a current of 5.5 A in it. (*a*) Calculate the inductance of the coil. (*b*) If a 6.0-V battery is suddenly connected across the coil, how long will it take for the current to rise from 0 to 2.5 A?

21P. Suppose the emf of the battery in the circuit of Fig. 6 varies with time t so the current is given by $i(t) = 3.0 + 5.0t$, where i is in amperes and t is in seconds. Take $R = 4.0\ \Omega$, $L = 6.0$ H, and find an expression for the battery emf as a function of time. (*Hint:* Apply the loop theorem.)

22P. At $t = 0$ a battery is connected to an inductor and resistor connected in series. The table below gives the measured potential difference, in volts, across the inductor as a function of time, in ms, following the connection of the battery. Deduce (*a*) the emf of the battery and (*b*) the time constant of the circuit.

t (ms)	V_L (V)	t (ms)	V_L (V)
1.0	18.2	5.0	5.98
2.0	13.8	6.0	4.53
3.0	10.4	7.0	3.43
4.0	7.90	8.0	2.60

23P. A 45-V potential difference is suddenly applied to a coil with $L = 50$ mH and $R = 180\ \Omega$. At what rate is the current increasing after 1.2 ms?

24P. A wooden toroidal core with a square cross section has an inner radius of 10 cm and an outer radius of 12 cm. It is wound with one layer of wire (diameter, 1.0 mm; 'resistance' 0.02 Ω/m). What are (*a*) the inductance and (*b*) the inductive time constant? Ignore the thickness of the insulation.

25P. In Fig. 16, $\mathcal{E} = 100$ V, $R_1 = 10\ \Omega$, $R_2 = 20\ \Omega$, $R_3 = 30\ \Omega$, and $L = 2.0$ H. Find the values of i_1 and i_2 (*a*) immediately after switch S is closed; (*b*) a long time later; (*c*) immediately after switch S is opened again; (*d*) a long time later.

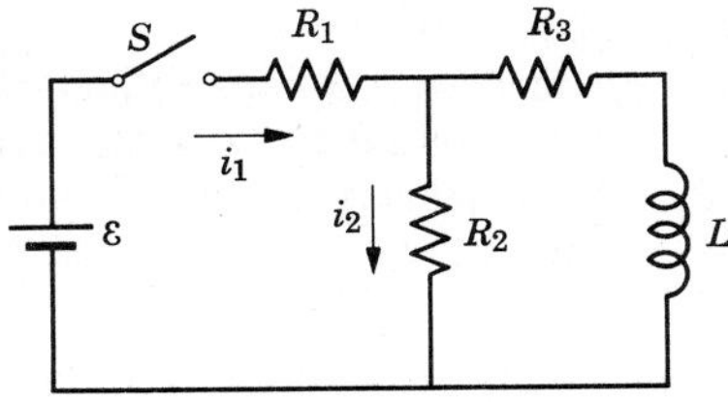

Figure 16 Problem 25.

26P. In the circuit shown in Fig. 17, $\mathcal{E} = 10$ V, $R_1 = 5.0\ \Omega$, $R_2 = 10\ \Omega$, and $L = 5.0$ H. For the two separate conditions (I) switch S just closed and (II) switch S closed for a long time, calculate (*a*) the current i_1 through R_1, (*b*) the current i_2 through R_2, (*c*) the current i through the switch, (*d*) the potential difference across R_2, (*e*) the potential difference across L, and (*f*) di_2/dt.

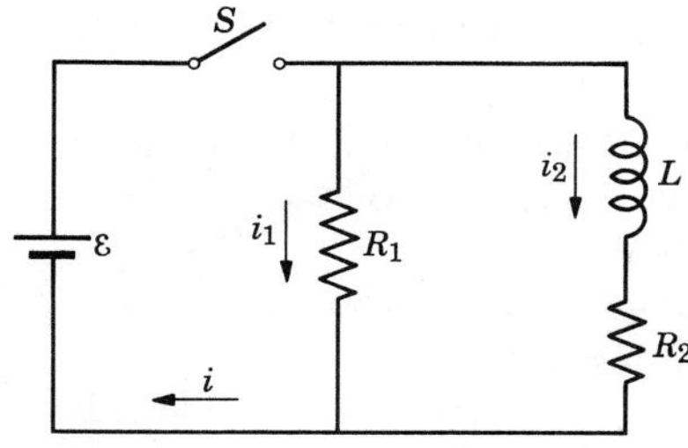

Figure 17 Problem 26.

27P. In Fig. 18 the component in the upper branch is an ideal 3-A fuse. It has zero resistance as long as the current through it remains less than 3 A. If the current reaches 3 A, it "blows" and thereafter it has infinite resistance. Switch S is closed at time $t = 0$. (*a*) When does the fuse blow? (*b*) Sketch a graph of the current i through the inductor as a function of time. Mark the time at which the fuse blows.

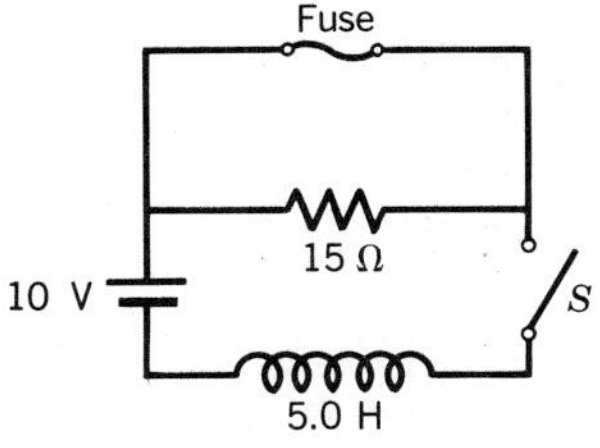

Figure 18 Problem 27.

28P*. For the circuit shown in Fig. 19, switch S is closed at time $t = 0$. Thereafter the constant current source, by varying its emf, maintains a constant current i out of its upper terminal. (*a*) Derive an expression for the current through the inductor as a function of time. (*b*) Show that the current through the

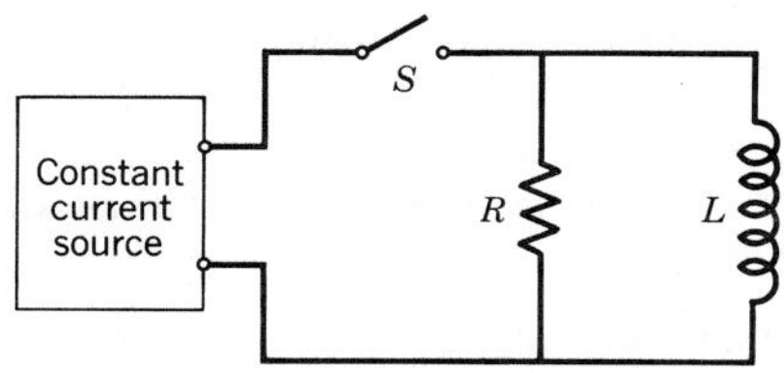

Figure 19 Problem 28.

resistor equals the current through the inductor at time $t = (L/R) \ln 2$.

Section 33-5 Energy and the Magnetic Field

29E. The magnetic energy stored in a certain inductor is 25 mJ when the current is 60 mA. (*a*) Calculate the inductance. (*b*) What current is required for the magnetic energy to be four times as much?

30E. A 90-mH toroidal inductor encloses a volume of 0.020 m^3. If the average energy density in the toroid is 70 J/m^3, what is the current?

31E. Consider the circuit of Fig. 6. In terms of the time constant, at what instant after the battery is connected will the energy stored in the magnetic field of the inductor be half its steady-state value?

32E. A coil with an inductance of 2.0 H and a resistance of 10 Ω is suddenly connected to a resistanceless battery with $\mathcal{E} = 100$ V. (*a*) What is the equilibrium current? (*b*) How much energy is stored in the magnetic field when this current exists in the coil?

33E. A coil with an inductance of 2.0 H and a resistance of 10 Ω is suddenly connected to a resistanceless battery with $\mathcal{E} = 100$ V. At 0.10 s after the connection is made, what are the rates at which (*a*) energy is being stored in the magnetic field, (*b*) thermal energy is appearing, and (*c*) energy is being delivered by the battery?

34P. Suppose that the inductive time constant for the circuit of Fig. 6 is 37 ms and the current in the circuit is zero at time $t = 0$. At what time does the rate at which energy is dissipated in the resistor equal the rate at which energy is being stored in the inductor?

35P. A coil is connected in series with a 10-kΩ resistor. When a 50-V battery is applied to the two, the current reaches a value of 2.0 mA after 5.0 ms. (*a*) Find the inductance of the coil. (*b*) How much energy is stored in the coil at this same moment?

36P. For the circuit of Fig. 6, assume that $\mathcal{E} = 10$ V, $R = 6.7$ Ω, and $L = 5.5$ H. The battery is connected at time $t = 0$. (*a*) How much energy is delivered by the battery during the first 2.0 s? (*b*) How much of this energy is stored in the magnetic field of the inductor? (*c*) How much has been dissipated in the resistor?

37P. (*a*) Find an expression for the energy density as a function of the radial distance for the toroid of Sample Problem 1. (*b*) Integrating the energy density over the volume of the toroid, calculate the total energy stored in the field of the toroid; assume $i = 0.50$ A. (*c*) Using Eq. 18 in this chapter evaluate the energy stored in the toroid directly from the inductance and compare with (*b*).

38P. A solenoid, with length 80 cm and radius 5.0 cm, consists of 3000 turns distributed uniformly over its length. Its total resistance is 10 Ω. At the instant 5.0 ms after it is connected to a 12-V battery, (*a*) how much energy is stored in its magnetic field and (*b*) how much energy has been supplied by the battery up to that time? (Neglect end effects.)

39P. Prove that, after switch *S* in Fig. 5 is thrown from *a* to *b*, all the energy stored in the inductor ultimately appears as thermal energy in the resistor.

Section 33-6 Energy Density and the Magnetic Field

40E. A solenoid 85 cm long has a cross sectional area of 17 cm^2. There are 950 turns of wire carrying a current of 6.6 A. (*a*) Calculate the magnetic field energy density inside the solenoid. (*b*) Find the total energy stored in the magnetic field inside the solenoid. (Neglect end effects)

41E. What must be the magnitude of a uniform electric field if it is to have the same energy density as that possessed by a 0.50-T magnetic field?

42E. The magnetic field in the interstellar space of our galaxy has a magnitude of about 10^{-10} T. How much energy is stored in this field in a cube 10 light-years on edge? (For scale, note that the nearest star is 4.3 light-years distant and the radius of our galaxy is about 8×10^4 light-years.)

43E. Use the result of Sample Problem 5 to obtain an expression for the inductance of a length *l* of the coaxial cable.

44E. A circular loop of wire 50 mm in radius carries a current of 100 A. (*a*) Find the magnetic field strength at the center of the loop. (*b*) Calculate the energy density at the center of the loop.

45P. A length of copper wire carries a current of 10 A, uniformly distributed. Calculate (*a*) the magnetic energy density and (*b*) the electric energy density at the surface of the wire. The wire diameter is 2.5 mm and its resistance per unit length is 3.3 Ω/km.

46P. (*a*) What is the magnetic energy density of the earth's magnetic field of 50 μT? (*b*) Assuming this to be relatively constant over distances small compared with the earth's radius and neglecting the variations near the magnetic poles, how much energy would be stored in a shell between the earth's surface and 16 km above the surface?

Section 33-7 Mutual Induction

47E. Two coils are at fixed locations. When coil 1 has no current and the current in coil 2 increases at the rate 15 A/s, the emf in coil 1 is 25 mV. (*a*) What is their mutual inductance? (*b*) When coil 2 has no current and coil 1 has a current of 3.6 A, what is the flux linkage in coil 2?

48E. Coil 1 has $L_1 = 25$ mH and $N_1 = 100$ turns. Coil 2 has $L_2 = 40$ mH and $N_2 = 200$ turns. The coils are rigidly positioned with respect to each other, their coefficient of mutual inductance *M* being 3.0 mH. A 6.0-mA current in coil 1 is changing at the rate of 4.0 A/s. (*a*) What flux Φ_{12} links coil 1 and what self-induced emf appears there? (*b*) What flux Φ_{21} links coil 2 and what mutually induced emf appears there?

49P. In Fig. 20 two coils are connected as shown. The coils separately have inductances L_1 and L_2. The coefficient of mutual inductance is M. (*a*) Show that this combination can be replaced by a single coil of equivalent inductance given by

$$L_{eq} = L_1 + L_2 + 2M.$$

(*b*) How could the coils in Fig. 20 be reconnected to yield an equivalent inductance of

$$L_{eq} = L_1 + L_2 - 2M?$$

Note that this problem is an extension of Problem 5, in which the requirement that the coils be far apart is removed.

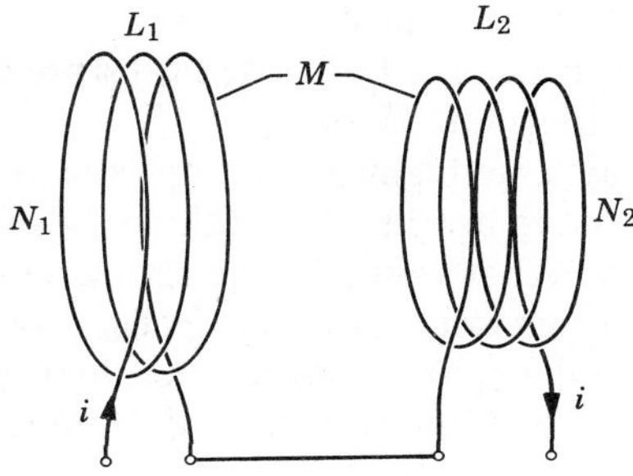

Figure 20 Problem 49.

50P. A coil C of N turns is placed around a long solenoid S of radius R and n turns per unit length, as in Fig. 21. Show that the coefficient of mutual inductance for the coil-solenoid combination is given by

$$M = \mu_0 \pi R^2 n N.$$

Explain why M does not depend on the shape, size, or possible lack of close-packing of the coil.

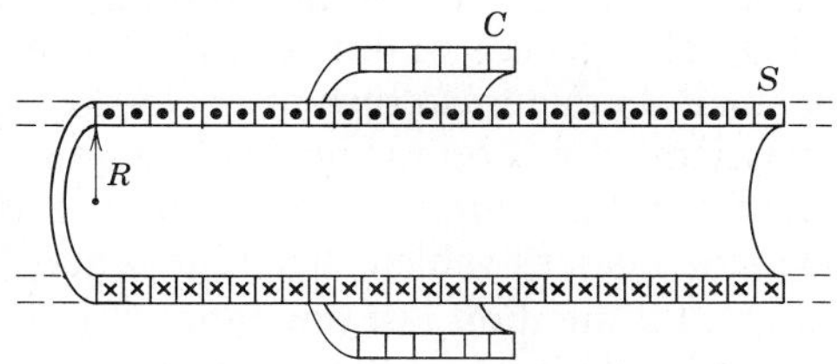

Figure 21 Problem 50.

51P. Figure 22 shows a coil of N_2 turns linked as shown to a toroid of N_1 turns. The inner toroid radius is a, its outer radius is b, and its height is h, as in Fig. 22. Show that the coefficient of mutual inductance M for the toroid-coil combination is

$$M = \frac{\mu_0 N_1 N_2 h}{2\pi} \ln \frac{b}{a}.$$

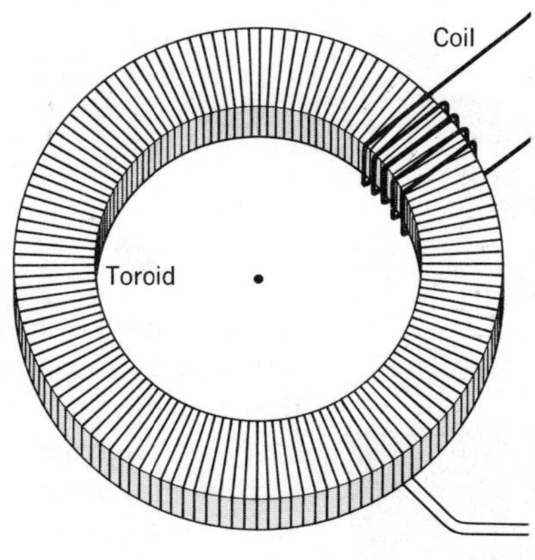

Figure 22 Problem 51.

52P. Figure 23 shows, in cross section, two coaxial solenoids. Show that the coefficient of mutual inductance M for a length l of this solenoid-solenoid combination is given by

$$M = \pi R_1^2 l \mu_0 n_1 n_2,$$

in which n_1 and n_2 are the respective numbers of turns per unit length and R_1 is the radius of the inner solenoid. Why does M depend on R_1 and not on R_2?

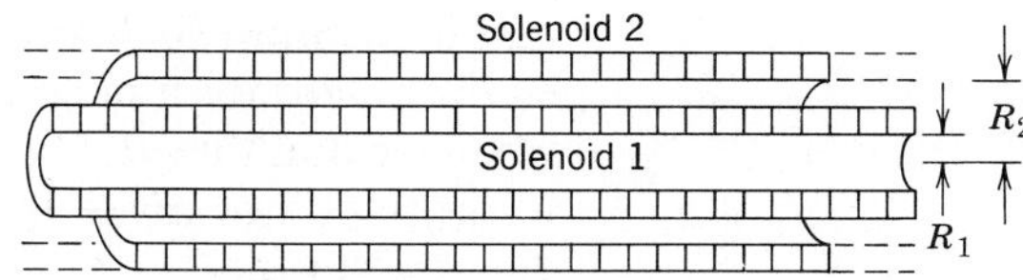

Figure 23 Problem 52.

53P. A rectangular loop of N close-packed turns is positioned near a long straight wire as in Fig. 24. (*a*) What is the coefficient of mutual inductance M for the loop-wire combination? (*b*) Evaluate M for $N = 100$, $a = 1.0$ cm, $b = 8.0$ cm, and $l = 30$ cm.

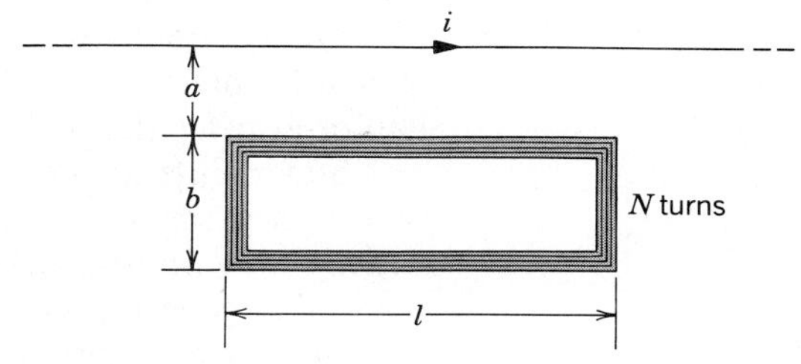

Figure 24 Problem 53.

ESSAY 13
MAGNETIC FLIGHT

GERARD K. O'NEILL
O'NEILL COMMUNICATIONS, INC.

Imagine traveling from New York to Orlando, Florida, in half an hour, at a cruising speed of a thousand meters per second (2300 miles per hour)! And imagine making that journey in the greatest comfort, in a seat with plenty of legroom, with unlimited carry-on baggage as close as you like it. If that's hard enough to believe, imagine further that your vehicle makes not a sound as it goes, is unseen by anyone outside it, and uses less energy in getting you there than the chemical energy in a liter of gasoline. Your imagination can take a break now, because all of those things are within the range of engineering based on the physics of magnetic flight.

As we know all too well by observing any large highway, or any railway or airport, all of our present transport systems move people only at the price of noise pollution, and by using a great deal of energy. What would be the characteristics of an ideal inter-city transportation system? Most of us would list high speed, economy, and comfort. But we'd also list attributes to make the system easy on our environment: silence, avoiding using valuable land area, and freedom from chemical pollution. Magnetic flight can achieve all of these.

By "magnetic flight" we mean the support of a vehicle not by wheels on a road or track, nor by the aerodynamic forces of flight through the air, but by the strong lift provided by the interaction of an electric current with a magnetic field. We provide our lifting force in that manner, with no physical contact and no need for an aerodynamic medium such as air. Using that method we can see at once that the ideal environment for a magnetic-flight vehicle is a vacuum. Why move through the air when we don't need it, and when it only causes drag? And if we don't need lifting wings, why not make the vehicle long and slim, and let it fly in vacuum in a small-diameter tunnel?

We've become quite used to the idea that high speed can be achieved only in aircraft, and that to reach that speed at a price we can afford we must resign ourselves to traveling with only a tiny amount of personal space, hunched in minimal seats while we stare at our knees. In magnetic flight all such constraints vanish. Traveling in vacuum, a vehicle indeed should be of small diameter, so that its enclosing tunnel need not be too large. But there is no aerodynamic limit on the vehicle's length. Before looking into the physics of magnetic flight, but simply knowing that it exists, we can draw up a first design for our "magnetic floater" vehicle. See Figs. 1 and 2.

As we check the physics of magnetic flight, I'll mention just one more of its unique characteristics: it has no particular speed limit, such as the speed of sound. Your magnetic-flight trip from New York to Orlando, or a similar trip from Boston to San Francisco in just over an hour, would cost no more in energy if carried out at half or at twice the speed of 1000 m/s chosen for our examples.

Statements such as these suggest ultra-high technology, and would seem to require new scientific breakthroughs. Not at all. The basic facts of magnetic flight depend only on Maxwell's four equations, the fundamental equations of electromagnetics that govern everything from the operation of an electric toaster to the transmission of starlight. And Maxwell's equations were completed by 1865, more than a century ago. Indeed, recent breakthroughs in high-temperature superconductors will make "floater" systems a little easier to engineer, and will make them a little cheaper

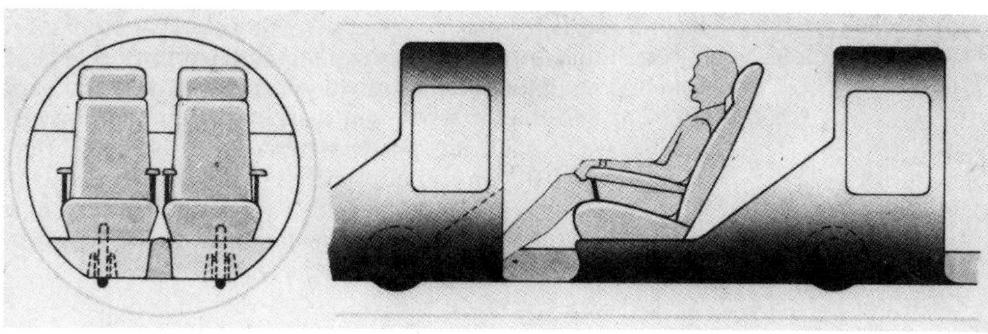

Figure 1 Side and front views of the modular carriage of a floater car. Each module has reclining seats for two passengers, and has a baggage compartment.

Figure 2 Boarding area for a floater high-speed underground transport system. The passenger-seating "carriage" of the floater vehicle rolls out of the air-lock at the far end into the central fenced area. The fences then sink into the floor to permit passengers to step on or off the carriage conveniently. When all passengers have boarded, the carriage returns through the air-lock into the enclosing shell of the vehicle, which maintains normal air pressure during flight.

to operate—but without them all of the foregoing statements could still come true within a few years.

We can measure the lifting force on which magnetic flight depends by an experiment in a basic geometry: imagine two circles of current-carrying wire, of equal diameter, spaced apart by a short distance. Each current produces a magnetic field at the other. It is the interaction of the upper current's magnetic field with the lower current that lifts the lower coil.

Let's do the experiment. We use two circular wire loops of diameter one meter. Each wire turns out to weigh 4.5 newtons (about one pound). When spaced so that the wire centers are 0.065 meter apart, we find that the lower wire can be supported when we pass a current of 788 amperes through each wire. Notice that the currents have to circulate in the same direction; if we connect the wires for currents in opposite directions, the wires repel rather than attract each other. By the way, it makes no difference whether we use single wires carrying 788 amperes, or coils of 788 turns, carrying one ampere in each individual turn. Only the product of amperes and turns of wire matters.

It's always well to check theory against experiment, and vice versa. In this case our theory comes from Maxwell's equations. In one of its simplest applications, it says that when we space two very long straight wires a distance s apart (in meters), and pass currents through the wires, the force between the wires, per meter of length, will be

$$\text{force/meter} = \left(\frac{\mu_0}{2\pi}\right)\left(\frac{1}{s}\right) \text{(current in first wire)(current in second wire).}$$

In metric units of amperes, meters, and newtons, that is

$$\text{force/meter} = \left(\frac{2 \times 10^{-7}}{s}\right) \text{(current in first wire)(current in second wire).}$$

That's a clue to the force between currents in other geometries: it tells us that the forces between currents are stronger when the currents are close to each other. With that clue, rather than using a more complicated formula for the exact force between two circular currents, let's approximate our geometry by observing that most of the force between the two circular coils must be coming from portions of the currents that are close to each other. In the circular coils, a current section in the lower coil must be affected most by the current section just above it, and the distant part of the current in the upper coil must not be contributing much to the force on that current section.

The geometry of two long straight wires is plain enough that a physicist or engineer could calculate the formula above, even without using a reference book. The exact formula for two circles of wire, though, is another story: to calculate that requires more advanced mathematics and a lot more work. Knowing that, let's be bold and bend those long straight wires into circles, and persist in using the formula above to find the force between them. Of course our answer won't be precisely right, because we've made an approximation by our boldness. But if we come anywhere near right, it will confirm our understanding and assure us that we haven't lost a factor of 10 or $\frac{1}{10}$ in our work. Let's try it. Bending the long wires into circles, the total length of each circle is pi times the diameter, or 3.14 meters. The long-wire formula gives us a force per meter of

$$\left(\frac{2 \times 10^{-7}}{.065}\right)(788)^2 = 1.9 \text{ newtons/meter of length.}$$

We multiply that force per meter by the 3.14 meters of total length, and find

$$(1.9 \text{ newtons/meter})(3.14 \text{ meters}) = 6.0 \text{ newtons.}$$

Not bad! Recall that the weight of the lower wire was 4.5 newton, which also had to equal the magnetic force on the wire when it lifted free. We got within 33% of the experimental result, while having made a pretty outrageous approximation to arrive at our estimate. That process of making a "quick and dirty" check calculation as a first step before working out a problem in complete and accurate detail is exactly the way that experienced physicists and engineers do their work. The quick check confirms basic understanding and makes sure that careless mistakes in factors of 10 haven't been made. Only when that's done—and it doesn't take long—does it make sense to take the time to work out detailed formulas and arrive at precise answers.

Just as in the case of an iron nail attracted to a permanent magnet, the attraction of the two currents in our example is unstable. Once it moves, the nail flies to and sticks to the magnet. In the same way the lower wire of Fig. 3 jumps upward to the fixed wire when the current reaches 788 amperes. But our experiment lacks vital elements of a practical magnetic flight system: a sensing mechanism to monitor the height of the wire, and control circuits to reduce the upper current when the lower wire is too high, or to increase the upper current when the lower wire is too low. With those additional elements our experiment can be made stable. The same principles are used in modern aircraft. Those aircraft are made stable by the addition of elements to sense roll, pitch, and yaw, directing motors to drive the aircraft control surfaces.

In West Germany, attractive magnetic lift has been carried to a late stage of experimental development, with the support of the West German government. Electronic systems for sensing and control of the currents have been made so reliable that they have been certified for full-size systems carrying passengers. Over many trips during a West German transport fair held several years ago, more than 40,000 people were carried in an attractive lift vehicle. I rode in a similar vehicle built by JAL, Japan Air Lines, at a site near Tokyo in the early 1980s. Attractive magnetic lift has strong points in its favor: it is so efficient that lifting a mass of several tons requires no more power than one can draw from a car battery. The corresponding currents are low enough that the system can be built out of ordinary copper, aluminum, and steel, rather than needing superconductors. Because the control of position is electronic,

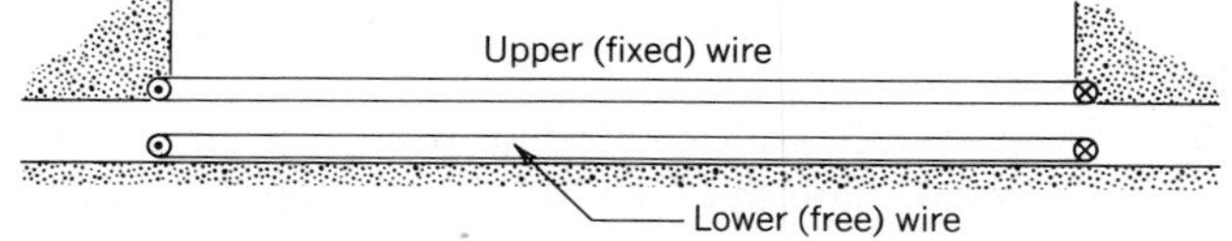

Figure 3 Experiment on magnetic lift. Each wire is a circle of diameter one meter.

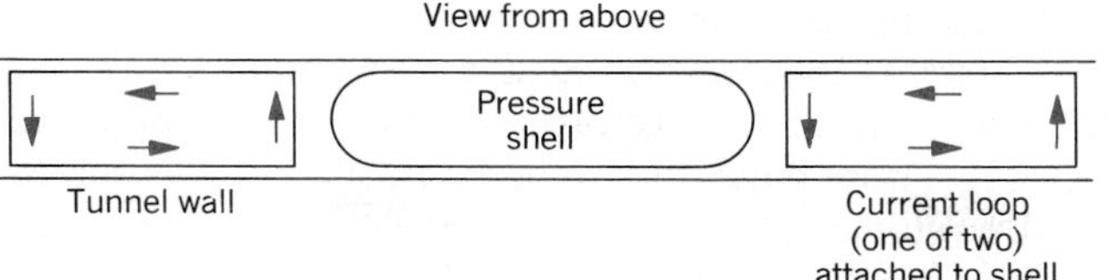

Figure 4 Essentials of a floater vehicle. The pressure shell encloses the passenger carriage of Fig. 1. The current loops function like the free wire of Fig. 3, providing lift.

control circuits using the memory of a computer can compensate for irregularities in the track. The quality of the ride can therefore be quite smooth.

But there is an alternative, which can be demonstrated by inverting our experiment, placing the fixed wire below and the floating wire above it. In that case we must also reverse one of the currents. Then, even without control circuits, the experimental setup is stable in vertical motion. That alternative, called "repulsive magnetic lift," has been developed by the Japanese National Railways to the point of a full-size two-car train, running on a test track several kilometers long. Many passengers have taken demonstration rides on that train.

Armed with the knowledge of the fundamental formulas of magnetic lift, we can move further toward the full design of our "floater" vehicle. We'll need a passenger compartment, a long slim cylinder, enclosed to hold normal atmospheric pressure inside. That's much like the fuselage of an airliner. Instead of wings, we'll have long rectangular loops of wire ahead of and behind the passenger compartment. They're the equivalents of the lower wire in our experiment. Because the vehicle will fly in vacuum, it doesn't need an aerodynamic shape. Our floater can be as angular and lumpy as a spacecraft that never needs to enter an atmosphere (Fig. 4).

We have two basic choices in providing the stationary current—the equivalent of the current in the upper wire—to make the magnetic field that will lift the lower current. One way is by induction: the rapid motion of the floater's currents could induce currents in a passive sheet of metal, a "guideway," below the vehicle, and the magnetic field of those induced currents would repel and so lift the moving current. That is repulsive magnetic lift, which requires wider spacings and strong fixed supercurrents in the vehicle.

Alternatively we can use attractive magnetic lift. That way, our "guideway" can consist of a series of pairs of short, nearly rectangular loops of wire, so arranged that in each section of the tunnel we can choose to divide the supporting current unequally between the left and right loops. The division will be determined by a central computer, based on the exact measured trajectories of the vehicles. The system is adaptive, maintaining the path of the vehicles straight even though the tunnel itself may have moved a little out of line over the years. To save power, we need only turn on the supporting currents in a particular section of the tunnel when the vehicle is approaching. As soon as it's past, we can turn off the currents again (Fig. 5).

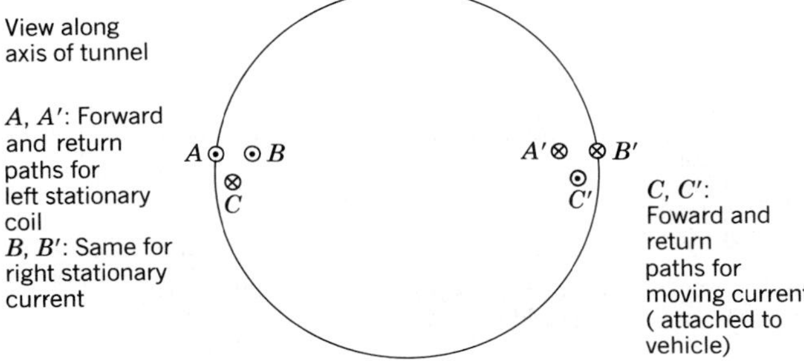

Figure 5 View along axis of tunnel, of current paths for stationary and moving coils.

There is another alternative that also uses attractive magnetic lift. We can replace the fixed currents by a simple, passive steel guideway, as is done in the West German system. The vehicle can have ordinary currents in copper or aluminum (not superconductors). Those currents can be controlled by a computer to rise or fall as needed to keep the vehicle on a smooth trajectory, even though the guideway is irregular because of movement of the ground since its construction.

Now for the matter of acceleration and deceleration. We accomplish them in the same way as in our experiment: our vehicle can have circular current-loops centered on the tunnel, and we can install controllable current-loops in the tunnel walls (Fig. 6). We can pull on our vehicle as it approaches one of the loops, and push on it, with a current in the opposite direction, when it is past. We reverse the currents to provide deceleration as the vehicle approaches its destination. Notice how simple our vehicle is now. Its currents are low enough that they can be sustained by batteries, and the vehicle has almost no moving parts: only those of an air-conditioning system to keep the air fresh and at the right temperature. The trip times are so short that meal services aren't required, and as the entire floater system is controlled by a computer, monitored at a single control center, there doesn't have to be a flight crew. There can be entertainment systems, probably in the form of flat-screen television sets at the front of each compartment, and stereo programs as we are used to on airplane flights.

When the additional physics is worked out, one finds that a floater system uses very little energy. In cruising flight, the drag force is about $\frac{1}{100}$ of the weight of the vehicle. (By contrast, the figure for the best jet aircraft is about $\frac{1}{14}$.) Energy is fed into kinetic form as the floater is accelerated to its high speed, but most of that energy is recovered and returned to the floater system electrical power grid during the deceleration process. The system runs as a motor during acceleration, and as a generator during deceleration.

Two other kinds of power loss must be considered in the design of working magnetic flight systems: power for the ordinary currents in copper or aluminum, and power to maintain the moderate vacuum needed in the tunnels. As the numbers for the coil sizes work out, the power lost to the coil currents can be made small compared to the drag losses. The power for pumps to maintain the vacuum is also small, because the tunnels can be made relatively leak-tight. As an illustration, it is common in pipelines the size of the magnetic flight tunnels to have no leaks over hundreds of kilometers length that are large enough for measurable amounts of oil or natural gas to leak out.

With all of the system design elements in place, we can now calculate trip times and the amount of energy required for a journey. We've chosen a cruising speed of 1000 m/s so that the curves in the underground tunnel don't have to be excessively long. From the formula for acceleration in circular motion,

$$a = \frac{v^2}{r}$$

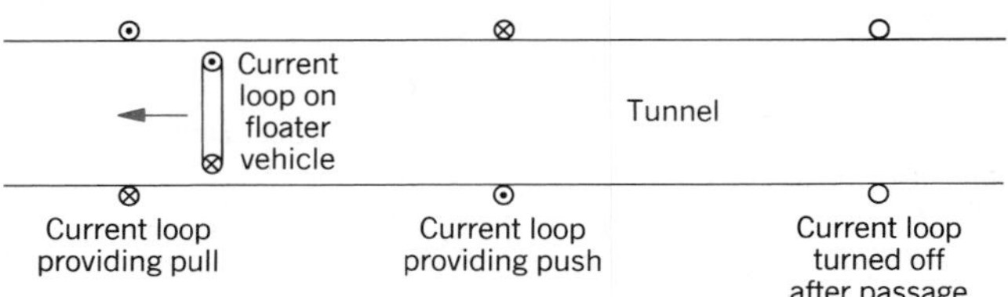

Figure 6 Side view of tunnel, with current loops providing forward acceleration.

with a the acceleration in meters per second squared, v the velocity in meters per second, and r the radius of the circle in meters, one finds that limiting the transverse acceleration to two tenths of the free-fall acceleration g requires curves with radii of at least 500 kilometers (just over 300 miles). Commercial passengers in airliners are often subjected to much more than that, but two tenths of g is a standard figure for railroads. Military pilots flying supersonic aircraft at 1000 m/s experience accelerations according to the same formula. But they are willing to take accelerations 50 times higher than those of railway passengers (10 g) in order to turn in radii of about 10 kilometers.

To be gentle to our passengers, we'll limit the forward acceleration and deceleration to 3 m/s^2, about 0.3 g. To reach our cruising speed of 1000 m/s from a standing start will require, using the formula

$$\text{velocity} = (\text{acceleration})(\text{time}),$$

a time just under six minutes. The deceleration time will be the same. During the acceleration and deceleration phases the average speed of the vehicle will be half of its cruising speed, or 500 m/s. The distance covered during acceleration will be

$$\begin{aligned}(\text{distance}) &= (\text{average velocity})(\text{time}) \\ &= (500\ \text{m/s})(333\ \text{s}) = 1.7 \times 10^5\ \text{m}\end{aligned}$$

For the New York to Orlando trip, the distance is 1507 km. Subtracting 170 km at each end for acceleration and deceleration, we're left with 1170 km of cruising flight. At our cruising speed of 1000 m/s that will take 1200 s, or 20 min. The entire trip will therefore require just about 30 min.

The energy per passenger required for the journey can be calculated from the fact that the drag on the vehicle will be about $\frac{1}{100}$ of its weight. The number of seats in the vehicle doesn't matter; it could be as few as 25 or as many as 300. Suppose each passenger has an average mass of 77 kilograms (170 pounds), and that the average passenger brings 23 kilograms of baggage (50 pounds). That brings the total mass for each passenger to a round number of 100 kilograms. Assuming that the vehicle mass is 100 kilograms per passenger, then we have 200 kilograms total, for a weight of

$$\begin{aligned}F = ma &= (200\ \text{kg})(9.8\ \text{m/s}^2) \\ &= 1960\ \text{newtons per passenger.}\end{aligned}$$

Over a total distance of 1507 km, the drag force will cost us, in energy,

$$\begin{aligned}\text{energy} &= (\text{force})(\text{distance}) = (\text{weight}/100) \times \text{distance} \\ &= (1960\ \text{N})(1{,}507{,}000\ \text{m})(\tfrac{1}{100}) \\ &= 29.5\ \text{million joules per passenger.}\end{aligned}$$

That is the energy released by burning about one liter of gasoline. The kinetic energy of motion is

$$\begin{aligned}\text{kinetic energy} &= (\tfrac{1}{2})(\text{mass})(\text{velocity squared}) = \tfrac{1}{2}mv^2 \\ &= (\tfrac{1}{2})(200\ \text{kilogram})(1000\ \text{m/s})^2 \\ &= 106\ \text{million joules per passenger.}\end{aligned}$$

Interestingly enough, that's about 3.6 times higher than the energy per passenger lost by the vehicle to drag during the entire journey. It tells us that if we accelerated to our full cruising speed in the first 170 km out of the New York terminal, as assumed above, we could coast all the way to Orlando!

Given the advantages of "floater" magnetic flight systems, it will be only a question of time before such systems are built.

CHAPTER 34

MAGNETISM AND MATTER

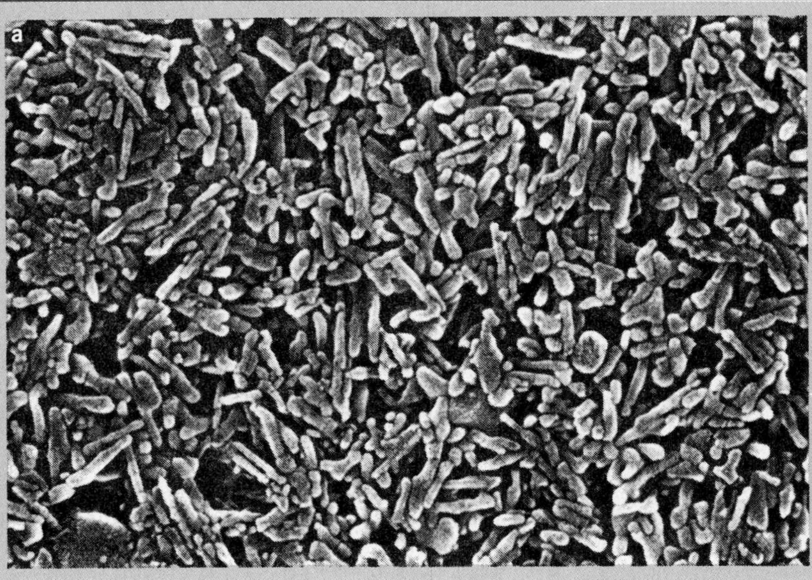

The figure shows an electron micrograph of the needle-like magnetic particles in a floppy disk. Current hard disk drives can store information magnetically at densities exceeding one million bits per cm². If scaled up to aircraft size, the problem of controlling the position of the read-record head of a such a drive is equivalent to that of flying a jumbo jet at 500 mi/h at 0.1 in. above the ground. The market for devices in which information is stored magnetically is estimated at 35 billion dollars per year.

34–1 Magnets

The most familiar magnets may well be the small decorative devices used to fasten notes to the refrigerator door. Magnets play a much larger part in our daily lives than that, however. Even if we restrict ourselves solely to the household environment, we can find magnets or magnetic materials in the motors of our electric appliances, in TVs and VCRs; in doorbells and thermostats; in relays, circuit breakers, and the electric utility meter; in loudspeakers, Walkman headsets, floppy disks, and aquarium pumps. Magnets or magnetic materials are in the ink on checks and on dollar bills; in answering machines, telephones, tape decks, credit cards, audio cassettes, cupboard doors, and portable chess sets. Can you think of any household application that we have left out? We have ignored entirely the many uses of magnets and of magnetic materials in industry and in scientific research.

We start our study of magnetic materials by looking at Fig. 1, which shows iron filings sprinkled on a sheet of cardboard under which there is a short bar magnet. The pattern of the filings suggests that the magnet has two

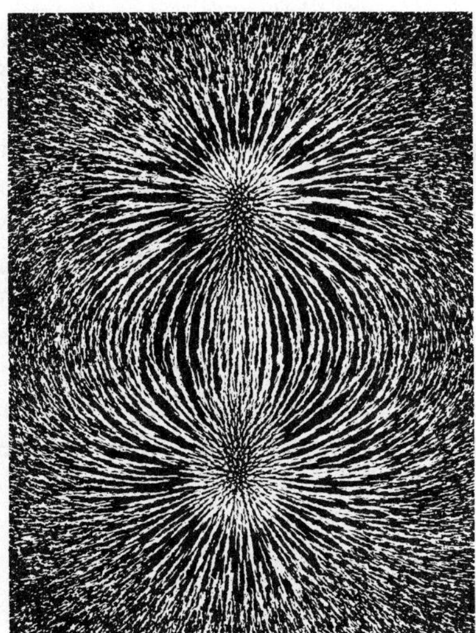

Figure 1 A bar magnet is a magnetic dipole. The iron filings suggest the magnetic lines of force.

poles, similar to the positive and the negative charges of an electric dipole. We conventionally label the magnetic poles as *north* and *south.* The north magnetic pole is the one from which the magnetic field lines emerge from the magnet into the surrounding space; these field lines then reenter the magnet at its south magnetic pole.

All attempts to isolate these poles fail. If we break the magnet, as in Fig. 2, we end up with three smaller magnets, each with its north and south pole. We can push this breakup of the magnet as far as its constituent atoms and electrons and we still fail to find anything that we can call an isolated magnetic pole, or a *magnetic monopole,* as we have come to call it. We conclude:

> *The simplest magnetic structure that can exist in nature is the* magnetic dipole. *There are no magnetic monopoles, that is, there are no magnetic structures analogous to isolated electric charges.*

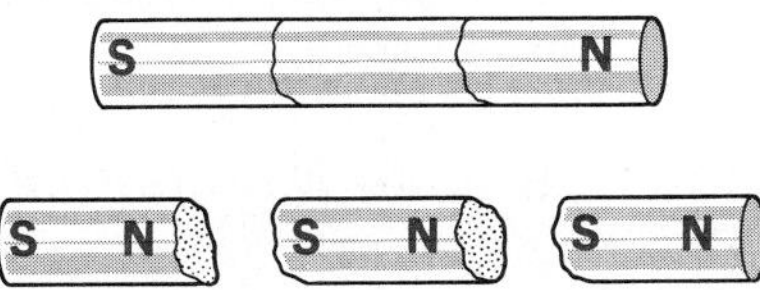

Figure 2 If you break a magnet, each fragment becomes a separate magnet, with its own north and south pole.

The fundamental magnetic dipole in nature—the one responsible for the magnetic properties of bulk matter—is that associated with the electron.

34-2 Magnetism and the Electron

Electrons can generate magnetism in three ways.

1. The Magnetism of Moving Charges. By streaming through an evacuated space or drifting through a conducting wire, electrons—like other charged particles—can set up an external magnetic field. We have studied fields produced in this way in earlier chapters and will not pursue this aspect of magnetism further here.

2. Magnetism and Spin. An isolated electron can be viewed classically as a tiny spinning negative charge, with an intrinsic *spin angular momentum S.* Associated with this spin angular momentum is an intrinsic *spin magnetic moment* μ_s. The magnitude of the spin angular momentum, as predicted by quantum theory and as measured in the laboratory, is

$$S = \frac{h}{4\pi} = 5.2729 \times 10^{-35}\ \text{J}\cdot\text{s}.$$

Here h is the Planck constant, the central constant of quantum physics.

When we deal with magnetism at the level of electrons and atoms, we find it convenient to adopt a non-SI unit for measuring magnetic moments. It is called the *Bohr magneton* (symbol μ_B) and is defined in terms of three fundamental constants of nature as

$$\boxed{1\ \mu_B = \frac{eh}{4\pi m} = 9.27 \times 10^{-24}\ \text{J/T}} \quad \text{(the Bohr magneton)}, \quad (1)$$

in which e is the elementary charge, and m is the electron mass. Expressed in these units, the magnitude of the spin magnetic moment of the electron is*

$$\mu_s = 1\ \mu_B. \quad (2)$$

Figure 3 shows the associated electric and magnetic

* The quantum theory of the electron (called *quantum electrodynamics,* or QED) predicts that the 1 in Eq. 2 should be replaced by 1.001 159 652 193. This remarkable prediction has been verified by equally remarkable experiments. However, for our purposes, 1 will do nicely.

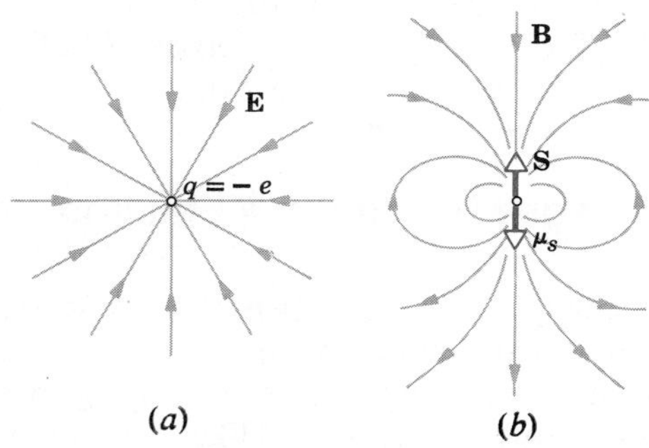

Figure 3 The lines of (a) the electric field and (b) the magnetic field for an isolated electron.

field lines for an isolated resting electron. The lines of the electric field point radially inward, as we expect for a particle with a negative charge. The magnetic lines, however, are those of a magnetic dipole. Note that the direction of the intrinsic spin magnetic moment vector for the electron (labeled μ_s) is opposite to that for the intrinsic spin angular momentum vector (labeled **S**). This is just what we would expect classically—as you should be sure to verify—for a spinning *negative* charge. Table 1 summarizes the properties of the free electron.

3. Magnetism and Orbital Motion. Electrons attached to atoms exist in states that have an intrinsic *orbital angular momentum* L_{orb}, corresponding classically to the motion of the electron in an orbit around the nucleus of the atom. These orbiting electrons are equivalent to tiny current loops and have an *orbital magnetic moment* μ_{orb} associated with them. Like essentially all physical properties examined at the atomic level, the orbital magnetic moment of the electron is quantized, being restricted to integral multiples of the Bohr magneton.

With so many possibilities for magnetism, you may well ask, "Every solid contains electrons; why isn't *everything* magnetic? Why can only magnetized iron and a few other substances pick up nails?" There are two answers.

Our first answer is that, in most cases, the magnetic moments of the electrons in a solid combine so as to cancel each other in their external effects. It is only when atoms contain unpaired electrons and special circumstances permit the large-scale alignment of their dipole moments that the familiar external effects are possible.

Our second answer is that, in a sense, everything *is* magnetic. When we speak popularly of magnetism, we almost always mean *ferromagnetism,* the familiar strong magnetism of the bar magnet or the compass needle. As we shall see, however, there are other kinds of magnetism—whose magnetic forces are too feeble to detect with our fingertips—that occur rather generally in matter.

Table 1 Some Properties of the Electron[a]

Mass	m	9.1094×10^{-31} kg
Charge	$-e$	-1.6022×10^{-19} C
Angular momentum	S	5.2729×10^{-35} J·s
Magnetic moment	μ_s	9.2848×10^{-24} J/T

[a] Although these quantities are known to considerably higher precision, we display only 5 significant figures here.

Sample Problem 1 Devise a method for measuring the magnetic dipole moment μ for a bar magnet.

(a) Place the magnet in a uniform external magnetic field **B**, with μ making an angle θ with **B**. A torque τ will act on the magnet, given by Eq. 30 of Chapter 30, or

$$\boldsymbol{\tau} = \boldsymbol{\mu} \times \mathbf{B}. \tag{3}$$

The magnitude of this torque is

$$\tau = \mu B \sin \theta. \tag{4}$$

Clearly we can find μ if we measure τ, B, and θ.

(b) A second technique is to suspend the magnet from its center of mass and to allow it to oscillate about its stable equilibrium position in the external field B. For small oscillations, $\sin \theta$ can be replaced by θ and Eq. 4 becomes

$$\tau = -(\mu B)\theta = -\kappa\theta, \tag{5}$$

where κ is a constant. We have inserted the minus sign to show that τ is a *restoring torque,* acting always in the opposite direction to the angular displacement θ.

Since τ is proportional to θ, the condition for simple angular harmonic motion is met. The period of oscillation T is given by Eq. 21 of Chapter 14.

$$T = 2\pi \sqrt{\frac{I}{\kappa}} = 2\pi \sqrt{\frac{I}{\mu B}}. \tag{6}$$

in which I is the rotational inertia. With this equation we can find μ from the measured quantities T, B, and I.

34-3 Orbital Angular Momentum and Magnetism

Here we derive the relation between the orbital magnetic moment of an electron and its orbital angular momentum.

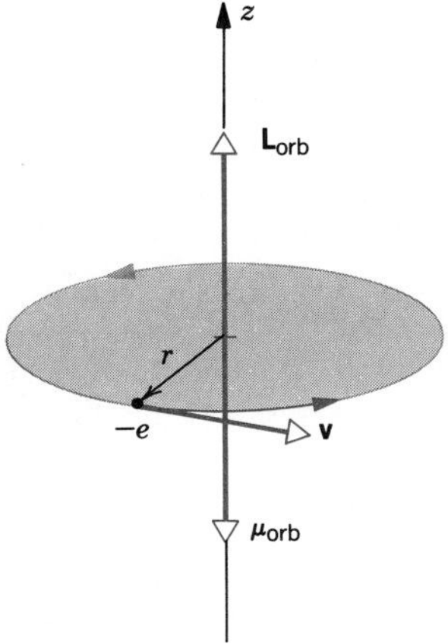

Figure 4 A classical representation of an electron circulating in an orbit of radius r with constant speed v. The orbital angular momentum vector and the orbital magnetic moment vector are both shown. Note that—because the electron carries a negative charge—these vectors point in opposite directions.

Figure 4 shows an electron moving in a circular orbit of radius r with speed v. The circulating electron is equivalent to a single-turn current loop. The magnetic moment of a current loop (see Eq. 27 of Chapter 30) is

$$\mu_{orb} = iA. \tag{7}$$

The current, i, is the amount of charge that passes any point in the orbit divided by the time required. The charge involved is e and the time, T, for one revolution is

$$T = \frac{2\pi r}{v} \tag{8}$$

so that

$$i = \frac{e}{(2\pi r/v)}.$$

The area of the single-turn loop (orbit) is $A = \pi r^2$, giving

$$\mu_{orb} = \frac{e}{(2\pi r/v)}\pi r^2 = \tfrac{1}{2}evr. \tag{9}$$

The angular momentum, on the other hand, is

$$L_{orb} = mvr. \tag{10}$$

Combining Eqs. 9 and 10 and generalizing to a vector formulation, we find

$$\boxed{\boldsymbol{\mu}_{orb} = -\frac{e}{2m}\mathbf{L}_{orb}} \quad \text{(orbital motion)}, \tag{11}$$

as the relation between the orbital angular momentum $\mathbf{L}_{orb}$ and orbital magnetic moment $\boldsymbol{\mu}_{orb}$. The minus sign arises because the orbiting electron carries a negative charge. Convince yourself that this requires that the orbital angular momentum vector and the orbital magnetic moment vector point in opposite directions.

Quantum physics tells us that the orbital angular momentum of an electron is quantized and that its smallest (non-zero) value is $h/2\pi$. Substituting this for L_{orb} in Eq. 11 yields, considering magnitudes only,

$$\mu_{orb} = \frac{e}{2m}\frac{h}{2\pi} = \frac{eh}{4\pi m}. \tag{12}$$

We recognize this quantity (see Eq. 1) as the *Bohr magneton,* the unit in which magnetic moments are expressed at the level of atoms and electrons. Physically then:

A Bohr magneton is equal to the orbital magnetic moment of an electron circulating in an orbit with the smallest allowed (non-zero) value of orbital angular momentum.

Sample Problem 2 What value for the Bohr magneton may be found from Eq. 12, its defining equation?

We have, from this equation,

$$\mu_B = \frac{eh}{4\pi m} = \frac{(1.60 \times 10^{-19}\ \text{C})(6.63 \times 10^{-34}\ \text{J}\cdot\text{s})}{(4\pi)(9.11 \times 10^{-31}\ \text{kg})}$$

$$= 9.27 \times 10^{-24}\ \text{C}\cdot\text{J}\cdot\text{s/kg}$$
$$= 9.27 \times 10^{-24}\ \text{J/T}. \quad \text{(Answer)}$$

The unit transformation follows from the relation $F = iLB$, which shows that $1\ \text{T} = 1\ \text{N/A}\cdot\text{m} = 1\ \text{kg/C}\cdot\text{s}$.

34-4 Gauss' Law for Magnetism

Gauss' law for magnetism, which is one of the basic equations of electromagnetism (see Table 2 of Chapter 37), is a formal way of stating that conclusion that has been forced on us by the facts of magnetism, namely, that isolated magnetic poles do not exist. This equation asserts that the magnetic flux Φ_B through any closed Gaussian surface must be zero, or

$$\boxed{\Phi_B = \oint \mathbf{B}\cdot d\mathbf{A} = 0} \quad \text{(Gauss' law for magnetism)}. \tag{13}$$

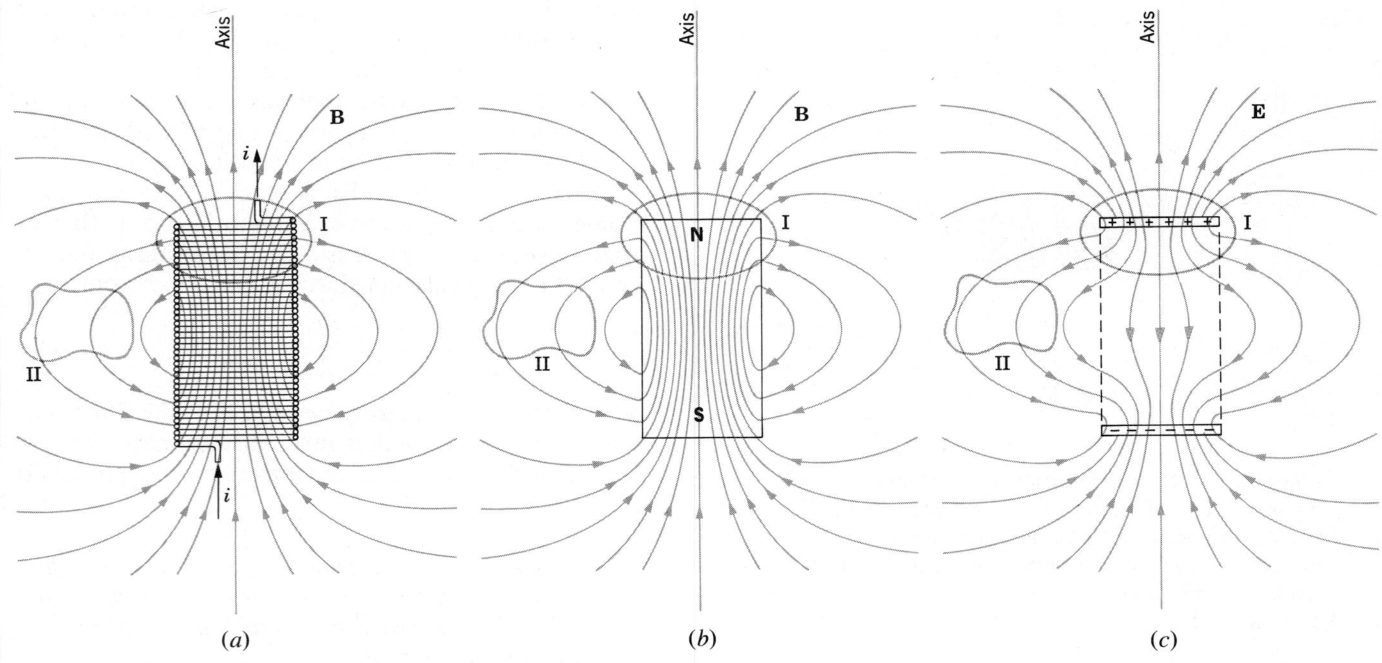

Figure 5 The lines of the magnetic field for (a) a short solenoid and (b) a short bar magnet. In each case the top end of the structure is a north magnetic pole. (c) The lines of the electric field for two charged disks. At large distances, all three fields resemble those for a dipole. The curves labeled I or II represent closed Gaussian surfaces.

We contrast this equation with Gauss' law for electricity, which is

$$\Phi_E = \epsilon_0 \oint \mathbf{E} \cdot d\mathbf{A} = q$$

(Gauss' law for electricity). (14)

In both of these laws, the integral is to be taken over the entire closed Gaussian surface. The fact that a zero appears at the right of Eq. 13, but not at the right of Eq. 14, means that in magnetism there is to be no counterpart to the free charge q in electricity.

Figure 5*a* suggests a Gaussian surface, marked I, enclosing one end of a short solenoid. Such a solenoid, as we have seen, sets up a magnetic dipole field at large distances; the end of the solenoid enclosed by surface I behaves, for such distant points, like a north magnetic pole. Note that the lines of the magnetic field **B** enter the Gaussian surface inside the solenoid and leave it outside the solenoid. No lines originate or terminate inside this surface; in other words, there are no *sources* or *sinks* of **B**; in still other words, there are no free magnetic poles. Thus the total flux Φ_B for surface I in Fig. 5*a* is zero, as Gauss' law for magnetism (Eq. 13) requires.

We also have $\Phi_B = 0$ for surface II in Fig. 5*a* and indeed for any closed surface that can be drawn in this figure. The situation is just the same if we replace the short solenoid by a short bar magnet, as in Fig. 5*b*. Here too $\Phi_B = 0$ for any closed surface that we can draw.

Figure 5*c* shows a close electrostatic analog to the two magnetic dipoles that we have been discussing. It consists of two oppositely charged circular disks facing each other as shown. The electric field **E** set up at distant points by this arrangement is also that of a dipole. In this case there is a net (outward) flux of the field lines for the Gaussian surface marked I; there *is* a source of the field, namely, the positive charges enclosed by the surface. (The negative charges of the other disk constitute a *sink* of the electric field.) Of course, for a Gaussian surface such as that marked II in Fig. 5*c*, we have $\Phi_E = 0$ because this surface happens to enclose no charge.

34-5 The Magnetism of the Earth

William Gilbert, physician in residence to Queen Elizabeth I and author of *De Magnete,* the first systematic survey of magnetic phenomena, knew in 1600 that the

Figure 6 Sir William Gilbert showing Queen Elizabeth I a spherical magnet whose magnetic field represents that of the earth. In those early days of bold seafaring exploration and uncertain navigation, the discovery that the earth was a huge magnet was a practical discovery of prime importance.

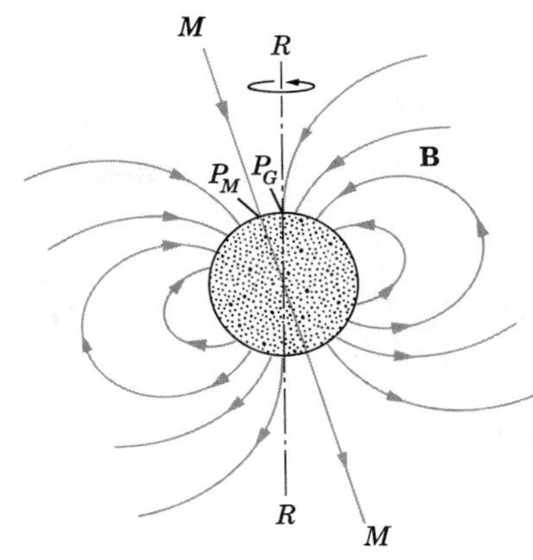

Figure 7 The earth's magnetic field represented as a dipole field. Its axis *MM* makes an angle of 11.5° with the earth's rotational axis *RR*. P_G and P_M are, respectively, the earth's geographic north pole and its magnetic north pole. The latter is actually the south pole of the earth's internal magnetic dipole.

earth was a huge magnet; see Fig. 6. As Fig. 7 suggests, the earth's magnetic field for points near its surface can be represented by a magnetic dipole located near the center of the earth.* To a good approximation, the earth's magnetic dipole moment has a magnitude of 8.0×10^{22} J/T; the dipole axis, shown as *MM* in Fig. 7, makes an angle of 11.5° with the earth's rotation axis, shown as *RR* in that figure. *MM* intersects the earth's surface at two points, defining what cartographers call the *north magnetic pole* (in northwest Greenland) and the *south magnetic pole* (in Antarctica). In general, lines of **B** for the earth's field emerge from the earth's surface in the southern hemisphere and reenter it in the northern hemisphere; thus the magnetic pole of the earth's dipole, deep inside the earth in the nothern hemisphere, is actually a south magnetic pole, just like the bottom end of the bar magnet of Fig. 5*b*.

Because of its practical applications in navigation, communication, and prospecting, the earth's magnetic field at its surface has been studied extensively for many years. The quantities of interest, as for any vector field, are the magnitude and direction of the field at different locations on the earth's surface and in the surrounding space. Field directions near the earth's surface are conveniently specified, with reference to the earth itself, in terms of the field declination (the angle between true geographic north and the horizontal component of the field) and the inclination (the angle between a horizontal plane and the field direction). There are a variety of commercially-available magnetometers used to measure these quantities with high precision, but rough measurements can be made with a compass and a dip meter. The compass is just a magnet mounted horizontally, so that it can rotate freely about a vertical axis. The angle between its direction and true geographic north is the field declination. A dip meter is a similar magnet mounted for free rotation in a vertical plane. When aligned with its plane of rotation parallel to the compass direction, the angle between the needle and the horizontal is the inclination. The strength of the field can be estimated by measuring the frequency of the oscillation of either instrument after being displaced from its equilibrium position. The calculations are the same as we described earlier in Sample Problem 1.

As an example, the north pole of a compass needle in Tucson, Arizona, pointed about 13° east of geographic north (the declination) in 1964. The magnitude of the horizontal component of the earth's field was 26 μT (= 0.26 gauss), and the north end of a dip needle pointed downward making an angle of about 59° (the inclination) with a horizontal plane. As we expect from Fig. 7, lines of B are *entering* the earth's surface at this point.

At any point on the earth's surface the observed magnetic field may differ appreciably from the idealized,

* The magnetic field shown in Fig. 7 is idealized, suggesting what the field would be like for an isolated dipole, uninfluenced by the effects of the solar wind; see Fig. 9.

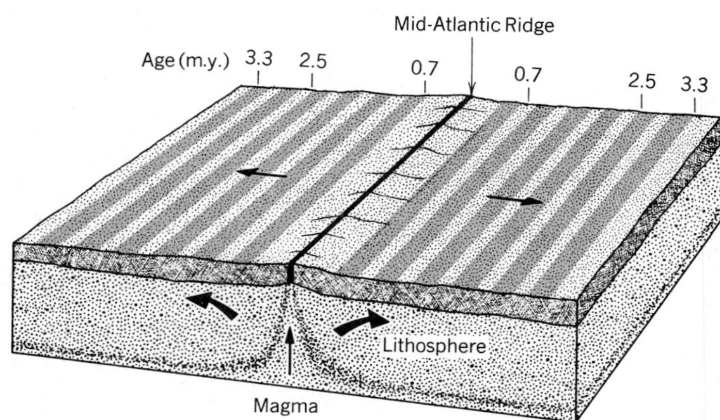

Figure 8 A magnetic profile of the sea floor on either side of the Mid-Atlantic Ridge. Points at which the earth's magnetic field has reversed its polarity are indicated. The sea floor, extruded from the ridge and spreading out as part of the continental drift system, displays a record of the past magnetic history of the earth's core.

best-fit, dipole field, in both magnitude and direction. At any location this observed field varies with time, by measurable amounts over a period of a few years and by substantial amounts over, say, 100 years. Thus between 1580 and 1820 the direction of the compass needle at London changed by 35°.

In spite of these local variations, the best-fit dipole field changes only slowly over such time periods, which are after all very short compared to the age of the earth. The variation of the earth's field over much longer intervals can be studied by measuring the weak intrinsic magnetism of the ocean floor on either side of the Mid-Atlantic Ridge. These deposits are formed by molten magma that oozes from the ridge. The magma solidifies and spreads out laterally at the rate of a few centimeters per year. The weak magnetism of the solidified magma preserves a "frozen-in" record of the earth's magnetic field at the time of solidification and thus allows us to study the direction and magnitude of the earth's magnetic field in the distant past; see Fig. 8. Such studies tell us, among other things, that the earth's field completely changes its direction (reverses its polarity) every million years or so.

There is at present no satisfactory detailed explanation for the origin of the earth's magnetic field. It seems certain that it arises in some way from current loops induced in the liquid and highly conducting outer region of the earth's core. The precise mode of action of this internal geomagnetic "dynamo" and the source of energy needed to keep it operating remain matters of continuing research interest.

Our moon, having no molten core, has no magnetic field. Most of the other planets in our solar system, Mercury and Jupiter among them, have magnetic fields. So do the sun and many other stars. In particular, neutron stars are thought to have magnetic fields of many millions of teslas in strength. There is also a magnetic field associated with our home galaxy. This field is weak (≈ 2 pT) but also important because of the immense volume it occupies.

At distances above the earth of the order of a few radii the earth's dipole field becomes modified in a major way by the action of the *solar wind,* which consists of streams of charged particles that constantly pour out of the sun. Figure 9 shows the resultant magnetic field of the earth at such large distances. A long *magnetotail* stretches outward away from the sun, extending for many thousands of earth diameters. The study of the

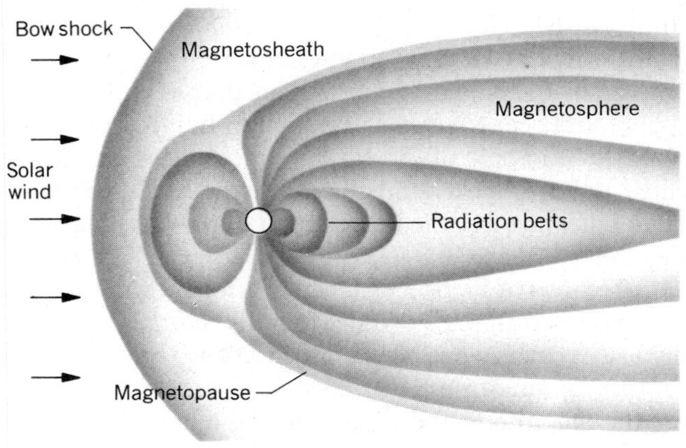

Figure 9 The earth's magnetic field, a composite of the earth's intrinsic dipole field as severely modified by the solar wind. The long *magnetotail* stretches out for several thousand earth diameters in a direction downstream from the solar wind.

magnetic field configurations of the earth and of our sister planets is high on the priority list of the space exploration programs of several countries.

Sample Problem 3 From data given earlier in this section find (a) the vertical component B_v of the earth's magnetic field and (b) the magnitude of the resultant magnetic field B at Tucson.

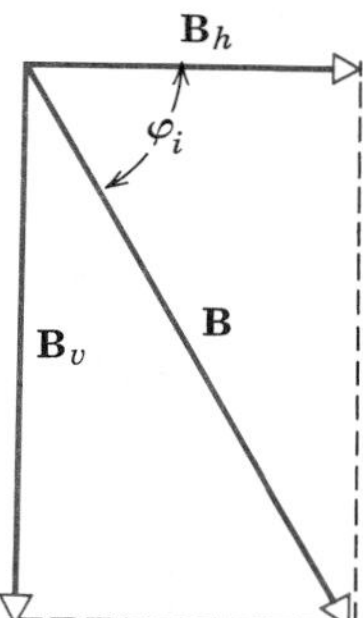

Figure 10 Sample Problem 3. The earth's magnetic field, along with its horizontal and vertical components, at Tucson, Arizona.

Figure 10 shows the situation. We have

(a)
$$\begin{aligned} B_v &= B_h \tan \phi_i \\ &= (26\ \mu\text{T})(\tan 59°) \\ &= 43\ \mu\text{T} = 0.43 \text{ gauss.} \qquad \text{(Answer)} \end{aligned}$$

and

(b)
$$\begin{aligned} B &= \sqrt{B_h^2 + B_v^2} \\ &= \sqrt{(26\ \mu\text{T})^2 + (43\ \mu\text{T})^2} \\ &= 50\ \mu\text{T} = 0.50 \text{ gauss.} \qquad \text{(Answer)} \end{aligned}$$

Note that the magnetic declination at Tucson plays no role in this problem.

34-6 Paramagnetism

Magnetism as we know it in our daily experience is an important but special branch of the subject called *ferromagnetism;* we discuss this in Section 34-8. Here we discuss a weaker and thus less familiar form of magnetism called *paramagnetism.*

For most atoms and ions, the magnetic effects of the electrons, including both their spins and orbital motions, exactly cancel so that the atom or ion is not magnetic. This is true for the rare gases such as neon and for ions such as Cu^+, which make up ordinary copper.* For other atoms or ions the magnetic effects of the electrons do not cancel, so that the atom as a whole has a magnetic dipole moment $\boldsymbol{\mu}$. Examples are found among the transition elements, such as Mn^{++}; the rare earths, such as Gd^{+++}, and the actinide elements, such as U^{++++}.

If we place a sample of N atoms, each of which has a magnetic dipole moment $\boldsymbol{\mu}$, in a magnetic field, the elementary atomic dipoles tend to line up with the field. This tendency to align is called *paramagnetism.* For perfect alignment, the sample as a whole would have a magnetic dipole moment of $N\mu$. However, the aligning process is seriously disturbed by thermal agitation. The importance of thermal agitation may be measured by comparing two energies: One ($=\frac{3}{2}$ kT) is the mean translational kinetic energy of an atom at temperature T. The other ($=2\ \mu B$) is the difference in energy between an atom whose dipole moment is lined up with the magnetic field and one pointing in the opposite direction. As Sample Problem 4 shows, the effect of the collisions at ordinary temperatures and fields is very important. The sample acquires a magnetic moment when placed in an external magnetic field, but this moment is usually very much smaller than the maximum possible moment $N\mu$.

We can express the extent to which a given specimen of a material is magnetized by dividing its measured magnetic moment by its volume. This vector quantity, the magnetic moment per unit volume, is called the *magnetization* **M** of the sample.

In 1895 Pierre Curie discovered experimentally that the magnetization **M** of a paramagnetic specimen is directly proportional to **B**, the effective magnetic field in which the specimen is placed, and inversely proportional to the kelvin temperature T. In equation form

$$\boxed{M = C\left(\frac{B}{T}\right)} \quad \text{(Curie's law)}, \qquad (15)$$

in which C is a constant. This equation is known as *Curie's law* and the constant C is called *the Curie constant.* The law is physically reasonable in that increasing B tends to align the elementary dipoles in the specimen,

* Cu^+ stands for a neutral copper atom from which one electron has been removed, leaving a net charge for the copper ion of $+e$. Similarly, Mn^{++} stands for a neutral manganese atom from which two electrons have been removed, etc.

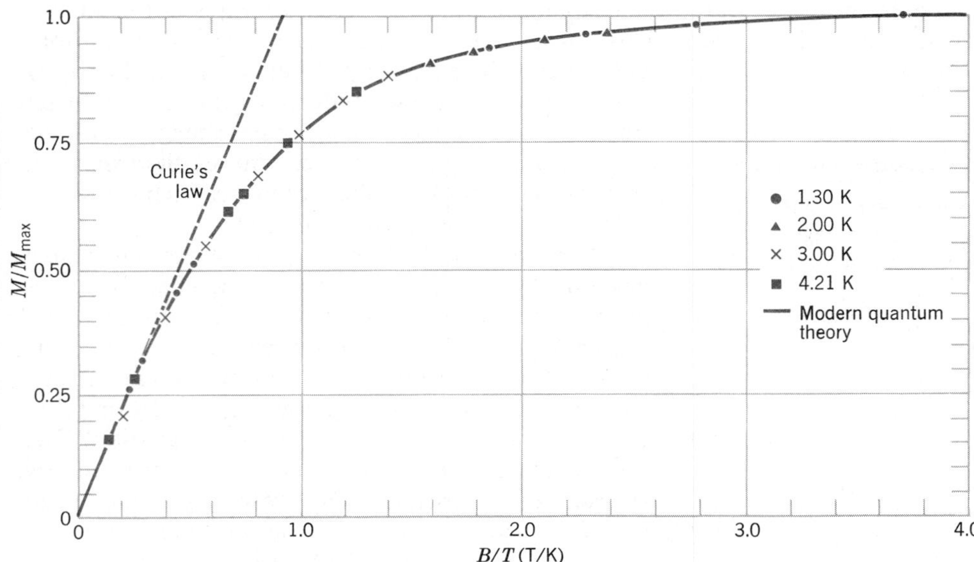

Figure 11 The ratio M to M_{max} for a paramagnetic salt, measured as a function of magnetic field for various low temperatures. After W. E. Henry.

that is, to increase M, whereas increasing T tends to interfere with this alignment, that is, to decrease M. Curie's law is well verified experimentally, provided that the ratio B/T does not become too large.

M cannot increase without limit, as Curie's law implies, but must approach a value M_{max} ($= \mu N/V$) corresponding to the complete alignment of the N dipoles contained in the volume V of the specimen. Figure 11 shows this saturation effect for a sample of potassium chromium sulfate. The chromium ions are responsible for all the paramagnetism of this salt, all its other elements being paramagnetically inert.

It is not easy to achieve anything like total alignment in a paramagnetic sample. Even at a temperature as low as 1.3 K, the magnetic field required to achieve 99.5% saturation in the material to which Fig. 11 refers, is about 5 T (= 50,000 gauss). The curve that passes through the experimental points in Fig. 11 is calculated from quantum physics; we see that it is in excellent agreement with experiment.

Sample Problem 4 A paramagnetic gas, whose atoms have a magnetic dipole moment of 1.0 Bohr magnetons, is placed in an external magnetic field of magnitude 1.5 T. At room temperature ($T = 300$ K), calculate and compare U_T, the mean kinetic energy of translation ($= \frac{3}{2}$ kT), and U_B the magnetic energy ($= 2\ \mu$B).

We have

$$U_T = (\tfrac{3}{2})\,\text{kT} = (\tfrac{3}{2})(1.38 \times 10^{-23}\ \text{J/K})(300\ \text{K}) = 6.2 \times 10^{-21}\ \text{J} = 0.039\ \text{eV}. \qquad \text{(Answer)}$$

Using Eq. 1,

$$U_B = 2\ \mu B = (2)(9.27 \times 10^{-24}\ \text{J/T})(1.5\ \text{T}) = 2.8 \times 10^{-23}\ \text{J} = 0.00017\ \text{eV}. \qquad \text{(Answer)}$$

Because U_T equals about 230 U_B, we see that energy exchanges in collisions can interfere seriously with the alignment of the dipoles with the external field.

34-7 Diamagnetism (Optional)

Perhaps the earliest observation in electrostatics is that uncharged bits of paper will be *attracted* to a charged rod if placed in the nonuniform *electric* field near the tip of the rod; see Fig. 14 of Chapter 27. The molecules of the paper do not have intrinsic electric dipole moments and we account for the attraction in terms of dipole moments that are induced in the paper by the action of the external electric field.

A parallel effect occurs in magnetism. Some materials, called *diamagnetic,* do not have intrinsic magnetic dipoles (that is, they are not paramagnetic), but dipole moments may be induced in them by the action of an external magnetic field. If a sample of such a material is placed in the nonuniform magnetic field near one pole of a strong magnet, a (very weak) force will act on the sample. In contrast to the electric case, however, the sample will not be attracted toward the pole of the magnet but *repelled.*

We trace this difference in behavior between the electric and the magnetic cases to the fact that induced electric dipoles point in the same direction as the external electric field but induced magnetic dipoles point in the *opposite* direction to the external magnetic field.

Diamagnetism is a manifestation of Faraday's law of induction acting on the atomic electrons, whose motions—on a classical picture—are equivalent to tiny current loops.* The fact that the induced magnetic moment is *opposite* to the direction of the inducing magnetic field may be viewed as a consequence of Lenz's law, acting on the atomic scale.†

Diamagnetism is present in all atoms, but if these atoms happen to have intrinsic magnetic dipole moments, the diamagnetic effects are masked by the far stronger paramagnetic or ferromagnetic behavior.

34-8 Ferromagnetism

When we speak of magnetism in everyday conversation we almost certainly have in mind an image of a bar magnet picking up nails or a magnetic disk clinging to a refrigerator. We almost certainly do not have in mind the relatively weak paramagnetic or diamagnetic effects discussed in the two preceding sections.

It turns out that for iron and several other elements (most notably cobalt, nickel, gadolinium and dysprosium) and for many alloys of these and other elements a special interaction, called *exchange coupling,* serves to align the atomic dipoles in rigid parallelism, in spite of the randomizing tendency of the thermal motions of the atoms. This phenomenon, called *ferromagnetism,* is a purely quantum effect and cannot be explained in terms of classical physics.

If the temperature is raised above a certain critical value, called the *Curie temperature,* the exchange coupling ceases to be effective and most such materials become simply paramagnetic. For iron the Curie temperature is 1043 K (= 770°C). Ferromagnetism is evidently a property not only of the individual atom or ion but also of the interaction of each atom or ion with its neighbors in the crystal lattice of the solid.

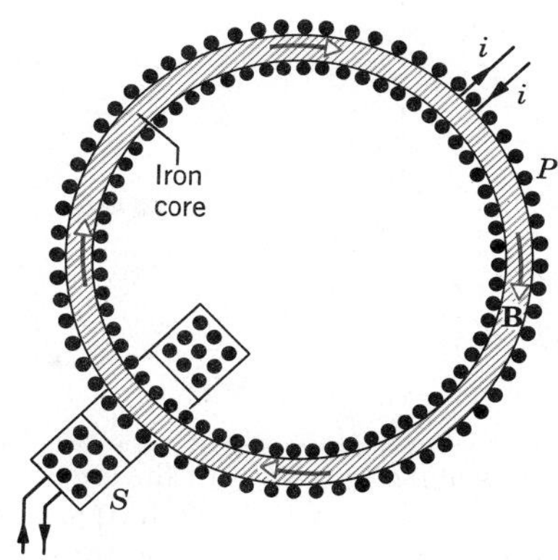

Figure 12 A specimen of iron whose ferromagnetic properties are being studied is formed into a Rowland ring. Primary coil P is used to magnetize the ring. Secondary coil S is used to measure the total magnetic field within the specimen.

To study the magnetization of a ferromagnetic material such as iron, it is convenient to form it into a toroidal or doughnut-shaped ring, as in Fig. 12. Primary coil P, containing n turns per unit length, is wrapped around it. This coil is essentially a long solenoid bent into a circle; if it carries a current i_p, we can find the magnetic field within the toroidal space from Eq. 21 of Chapter 31:

$$B_0 = \mu_0 n i_p. \tag{16}$$

Note carefully that B_0 is the field that would be present within the toroid if the iron core were not in place. The arrangement of Fig. 12 is called a *Rowland ring,* after the physicist H. A. Rowland (1848–1901), who devised it.

The actual magnetic field B in the toroidal space of the Rowland ring of Fig. 12, with the iron core in place, is greater than B_0, commonly by a large factor. We can write

$$B = B_0 + B_M, \tag{17}$$

where B_M is the contribution of the iron core to the total magnetic field B. B_M is associated with the alignment of the elementary atomic dipoles in the iron and is proportional to the magnetization M of the iron, that is, to its

* Interestingly, diamagnetism was discovered (and so named) by Michael Faraday, another of his many accomplishments.

† We do not push the semiclassical explanation of this very weak effect further. For a more complete treatment, see *Electricity and Magnetism,* by Edward M. Purcell, McGraw-Hill Book Company, New York 1985, 2nd ed., Section 11.5.

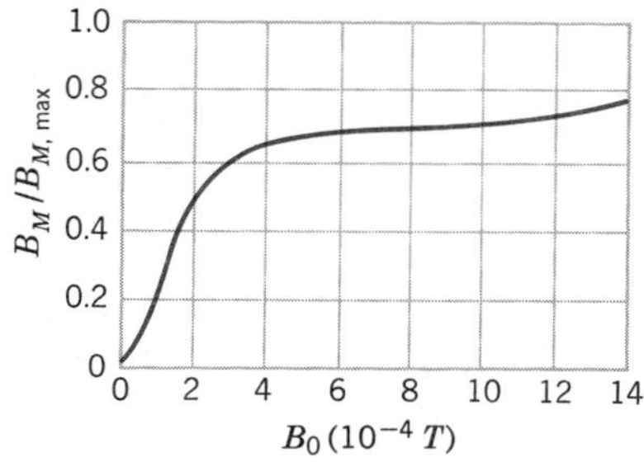

Figure 13 A *magnetization curve* for the core material of the Rowland ring of Fig. 12. On the vertical axis, 1.0 corresponds to complete alignment of the atomic dipoles within the specimen.

magnetic moment per unit volume. B_M has a certain maximum value, $B_{M,\text{max}}$, corresponding to complete alignment of the atomic dipoles.

It is instructive to plot, for any ferromagnetic specimen that can be formed into a Rowland ring, the ratio of B_M to $B_{M,\text{max}}$ as a function of B_0. Such a *magnetization curve,* as it is called, is displayed in Fig. 13. It is similar to the magnetization curve of Fig. 11 for a paramagnetic substance. Both are measures of the extent to which an applied magnetic field can succeed in aligning the elementary dipoles that make up the material in question.

For the ferromagnetic core material to which Fig. 13 refers, the alignment of the dipoles is about 75% complete for $B_0 \approx 1 \times 10^{-3}$ T. If B_0 were increased to 1 T, the fractional saturation of the specimen would increase to about 99.7%. Such a large value of B_0, if plotted in Fig. 13, would be about 95 ft to the right of the origin on the horizontal axis; complete saturation of the dipoles is approached only with considerable difficulty.

The use of iron in transformers, electromagnets, and other devices greatly increases the strength of the magnetic field that can be generated by a given current in a given set of windings. That is, very often, $B_M \gg B_0$ in Eq. 17. However, when B_0 is very large the presence of iron offers no advantage because of the saturation effect suggested in Fig. 13. When generating magnetic fields greater than this saturation limit, it is often simpler to abandon the use of iron and to rely on the "brute force" application of very large currents.

The magnetization curve for paramagnetism (Fig. 11) is explained in terms of the mutually opposing tendencies of alignment with the external field and of randomization because of the temperature motions. In ferromagnetism, however, we have assumed that adjacent atomic dipoles are locked in rigid parallelism. Why, then, does the magnetic moment of the specimen not reach its saturation value for very low—even zero—values of B_0? Why is not every iron nail a strong permanent magnet?

To understand this consider first a specimen of a ferromagnetic material such as iron that is in the form of a single crystal. That is, the arrangement of atoms that make it up—its crystal lattice—extends with unbroken regularity throughout the volume of the specimen. Such a crystal will, in its normal unmagnetized state, be made up of a number of *magnetic domains.* These are regions of the crystal throughout which the alignment of the atomic dipoles is essentially perfect. For the crystal as a whole, however, the domains are so oriented that they largely cancel each other as far as their external magnetic effects are concerned.

Figure 14 is a magnified photograph of such an assembly of domains in a single crystal of nickel. It was made by sprinkling a colloidal suspension of finely powdered iron oxide on a properly etched surface of the crystal. The domain boundaries, which are thin regions in which the alignment of the elementary dipoles changes from a certain orientation in one domain to a quite different orientation in the other, are the sites of intense, but highly localized and nonuniform magnetic

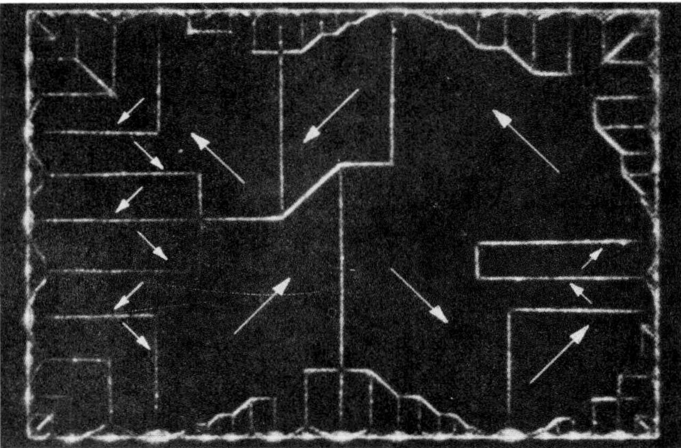

Figure 14 Domain patterns for a single crystal of nickel. The white lines show the boundaries of the domains. The arrows show the orientation of the magnetic dipoles within the domains.

fields. The suspended colloidal particles are attracted to these regions. Although the atomic dipoles in the individual domains are completely aligned, the crystal as a whole may have a very small resultant magnetic moment.

Actually a piece of iron as we ordinarily find it—an iron nail, say—is not a single crystal but is an assembly of many tiny crystals, randomly arranged; we call it a *polycrystalline solid.* Each tiny crystal, however, has its array of domains, just as in Fig. 14. Let us magnetize such a specimen by placing it in an external magnetic field of gradually increasing strength. Two effects take place, both contributing to the magnetization curve of Fig. 14. One is a growth in size of the domains that are favorably oriented at the expense of those that are not. Second, the orientation of the dipoles within a domain may swing around as a unit, becoming closer to the field direction.

Measuring B_M. It is possible to find B_M for a ferromagnetic specimen by measuring B, calculating B_0 from Eq. 16, and then subtracting, using Eq. 17:

$$B_M = B - B_0. \qquad (18)$$

B can be measured by setting up an induced current in secondary coil S of Fig. 12. With the iron core initially unmagnetized, set up a steady current i_p in the primary coil P. The magnetic flux linking secondary coil S (of resistance R_s and N_s turns) will rise from zero to a value BA in a time Δt, A being the cross-sectional area of the iron core.

From Faraday's law of induction an emf whose average value is

$$\mathcal{E}_s = -N_s \frac{d\Phi}{dt} = -\frac{N_s BA}{\Delta t}$$

will appear in the secondary coil. The average current in this coil during the time that the flux is changing will then be

$$i_s = \frac{\mathcal{E}_s}{R_s} = \frac{N_s BA}{R_s \Delta t}.$$

We have then for B

$$B = \frac{(i_s \Delta t) R_s}{N_s A} = \frac{\Delta q\, R_s}{N_s A},$$

in which $\Delta q\ (= i_s \Delta t)$ is the charge that passes through the secondary coil during the time that the magnetic field

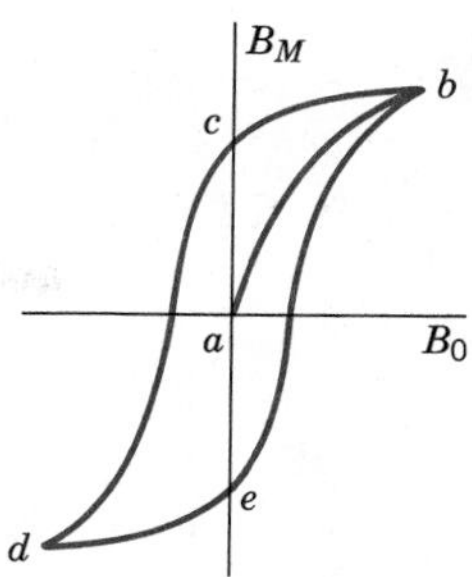

Figure 15 A magnetization curve (*ab*) for a ferromagnetic specimen and an associated hysteresis loop (*bcdeb*).

in the iron core is changing.* Thus a measurement of Δq (using an instrument called a ballistic galvanometer or in some other way) yields a measurement of the magnetic field B within the iron core.

Hysteresis. Magnetization curves for ferromagnetic materials do not retrace themselves as we increase and then decrease the external magnetic field B_0. Figure 15 shows the following operations with a Rowland ring: (1) starting with the iron unmagnetized (point *a*), increase the toroid current until $B_0\ (= \mu_0 ni)$ has the value corresponding to point *b*; (2) reduce the current in the toroid winding back to zero (point *c*); (3) reverse the toroid current and increase it in magnitude until point *d* is reached; (4) reduce the current to zero again (point *e*); (5) reverse the current once more until point *b* is reached again.

The lack of retraceability shown in Fig. 15 is called *hysteresis.* Note that at points *c* and *e* the iron core is magnetized, even though there is no current in the toroid windings; this is the familiar phenomenon of permanent magnetism.

Hysteresis can be understood on the basis of the magnetic domain concept. Evidently the motions of the domain boundaries and the reorientations of the domain directions are not totally reversible. When the applied magnetic field B_0 is increased and then decreased back to its initial value the domains do not return completely to their original configuration but retain some "memory" of the initial increase. The "memory" of magnetic materials is essential for the magnetic storage of information, as on cassette tapes or computer disks.

* Although the rate of increase of the flux will not be strictly linear, and the time Δt will not be well-defined, the charge Δq is a quantity that can be precisely measured.

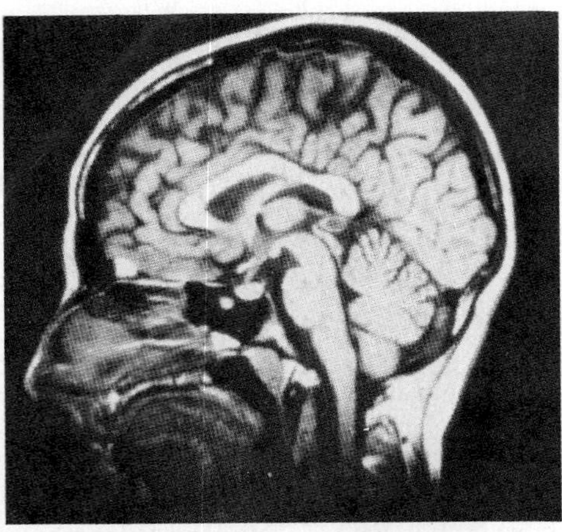

Figure 16 A cross section of a human head, taken by magnetic resonance imaging (MRI) techniques. It shows detail not visible on x-ray images and involves no radiation health risk to the patient.

34-9 Nuclear Magnetism—An Aside

The nuclei of many atoms are also magnetic dipoles. Their magnitudes tend to be about a thousand times smaller than those associated with the atomic electrons so that nuclear magnetism does not contribute in a measurable way to the gross magnetic properties of solids. It is nevertheless a subject of vital interest, both for what it can tell us about the internal structures of atoms and nuclei and for its practical applications. Figure 16, for example, shows a cross section of a human head taken by recently developed magnetic resonance imaging (MRI) techniques. For another example, much information about the structure of our galaxy and our universe comes to us from observations made with radio telescopes. Radiation at a wavelength of 21 cm, emitted from hydrogen atoms when the relative orientation of their nuclear and their electronic magnetic dipole moments changes, has proven particularly useful.

REVIEW AND SUMMARY

Magnetic Poles and Dipoles

There is no convincing evidence that magnetic monopoles (the magnetic equivalent of free electric charges) exist; see Fig. 2. The simplest sources of the magnetic field are magnetic dipoles, associated either with the orbital motions of electrons in atoms (see Fig. 4) or with the intrinsic spin motions (see Fig. 3*b*) of electrons, protons, and many other such particles. The magnetic dipole moments associated with electron motion are measured in terms of the *Bohr magneton* μ_B, where

The Bohr Magneton

$$1\ \mu_B = \frac{eh}{4\pi m} = 9.27 \times 10^{-24}\ \text{J/T} \quad \text{(the Bohr magneton).} \qquad [1]$$

The dipole moment associated with intrinsic electron spin is almost exactly $-1\ \mu_B$, the minus sign indicating that the magnetic dipole moment vector is opposite the direction of the spin angular momentum vector. The dipole moment associated with electron orbital motion is $\mu_{\text{orb}} = -\frac{e}{2m}\mathbf{L}_{\text{orb}}$; see Sample Problem 2.

Gauss' Law for Magnetism

Gauss' law for magnetism,

$$\Phi_B = \oint \mathbf{B} \cdot d\mathbf{A} = 0 \quad \text{(Gauss' law for magnetism),} \qquad [13]$$

states that the magnetic flux through any closed Gaussian surface must be zero. See Fig. 5 for illustrations and for a comparison with Gauss' law for electrostatics. Equation 13 is simply a formal statement of the observation that there are no magnetic monopoles.

The Earth's Magnetic Field

The earth's magnetic field is approximately that of a dipole, with a dipole moment of 8.0×10^{22} J/T, near the earth's center. The dipole moment makes an angle of 11.5° with the earth's rotation axis, lines of **B** emerging from the earth's southern hemisphere. The field is thought to originate in current loops induced in the earth's liquid outer core by a mechanism not yet fully understood.

The direction of the local magnetic field at any point is given by its angle of *declination* (in a horizontal plane) from true north and its angle of *inclination* (in a vertical plane) from the horizontal; see Sample Problem 3.

Paramagnetism

In the nature of their response to an external magnetic field materials may be broadly grouped as diamagnetic, paramagnetic, or ferromagnetic. Paramagnetic materials, which are (weakly) attracted by a magnetic pole, have intrinsic magnetic dipole moments which tend to line up with an external magnetic field, thus enhancing the field. This tendency is interfered with (see Sample Problem 4) by thermal agitation. The *magnetization* M of a specimen, which is its magnetic moment per unit volume, is given approximately by Curie's law, or

$$M = C\frac{B}{T} \quad \text{(Curie's law).} \qquad [15]$$

For strong enough fields or low enough temperatures this law breaks down as the atomic dipoles approach complete alignment and produce a maximum magnetization $M_{\text{max}} = \mu N/V$; see the *magnetization curve* of Fig. 11.

Diamagnetism

Diamagnetic materials are (weakly) repelled by the pole of a strong magnet. The atoms of such materials do not have intrinsic magnetic dipole moments. A dipole moment may be induced, however, by an external magnetic field, its direction being opposite to that of the field.

Ferromagnetism

In ferromagnetic materials such as iron a quantum interaction between neighboring atoms locks the atomic dipoles in rigid parallelism in spite of the disordering tendency of thermal agitation. This interaction abruptly disappears at a well-defined Curie temperature, above which the material becomes simply paramagnetic.

Magnetization Curves

Figure 12 shows how a known external magnetic field, B_0 can be applied to a ring-shaped ferromagnetic specimen. The resultant magnetic field can be measured by an induction method and a magnetization curve such as that of Fig. 13 can be plotted. The course of this curve as B_0 increases corresponds to the growth of favorably oriented magnetic domains in the material; see Fig. 14 and the related text.

Hysteresis

As Fig. 15 shows, ferromagnetic magnetization curves do not retrace themselves, a phenomenon called hysteresis. Some alignment of dipoles remains even when the external magnetic field is completely removed; the result is the familiar "permanent" magnet.

Nuclear Magnetism

The nuclei of many atoms are magnetic dipoles, a fact of some importance in studying nuclear structure. This property also allows us to develop ingenious strategies for studying the location and density of particular atoms; magnetic resonance imaging (MRI) in medicine and mapping the density of hydrogen in the universe are important examples.

QUESTIONS

1. Two iron bars are identical in appearance. One is a magnet and one is not. How can you tell them apart? You are not permitted to suspend either bar as a compass needle or to use any other apparatus.

2. Two iron bars always attract, no matter the combination in which their ends are brought near each other. Can you conclude that one of the bars must be unmagnetized?

3. How can you determine the polarity of an unlabeled magnet?

4. The neutron, which has no charge, has a magnetic dipole moment. Is this possible on the basis of classical electromagnetism, or does this evidence alone indicate that classical electromagnetism has broken down?

5. Must all permanent magnets have identifiable north and south poles? Consider geometries other than the bar or horseshoe magnet.

6. A certain short iron rod is found, by test, to have a north pole at each end. You sprinkle iron filings over the rod. Where (in the simplest case) will they cling? Make a rough sketch of what the lines of **B** must look like, both inside and outside the rod.

7. Starting with A and B in the positions and orientations shown in Fig. 17, with A fixed but B free to rotate, what happens (*a*) if A is an electric dipole and B is a magnetic dipole; (*b*) if A and B are both magnetic dipoles; (*c*) if A and B are both electric dipoles? Answer the same questions if B is fixed and A is free to rotate.

8. Cosmic rays are charged particles that strike our atmo-

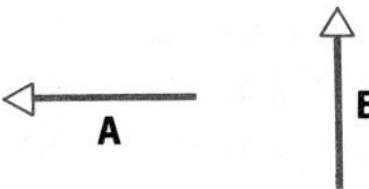

Figure 17 Question 7.

sphere from some external source. We find that more low-energy cosmic rays reach the earth near the north and south magnetic poles than at the (magnetic) equator. Why is this so?

9. How might the magnetic dipole moment of the earth be measured?

10. Give three reasons for believing that the flux Φ_B of the earth's magnetic field is greater through the boundaries of Alaska than through those of Texas.

11. You are a manufacturer of compasses. (*a*) Describe ways in which you might magnetize the needles. (*b*) The end of the needle that points north is usually painted a characteristic color. Without suspending the needle in the earth's field, how might you find out which end of the needle to paint? (*c*) Is the painted end a north or a south magnetic pole?

12. Would you expect the magnetization at saturation for a paramagnetic substance to be very much different from that for a saturated ferromagnetic substance of about the same size? Why or why not?

13. The magnetization induced in a given diamagnetic sphere by a given external magnetic field does not vary with temperature, in sharp contrast to the situation in paramagnetism. Explain this behavior in terms of the description that we have given of the origin of diamagnetism.

14. Explain why a magnet attracts an unmagnetized iron object such as a nail.

15. Does any net force or torque act on (*a*) an unmagnetized iron bar or (*b*) a permanent bar magnet when placed in a uniform magnetic field?

16. A nail is placed at rest on a smooth tabletop near a strong magnet. It is released and attracted to the magnet. What is the source of the kinetic energy it has just before it strikes the magnet?

17. Superconductors are said to be perfectly diamagnetic. Explain.

18. Compare the magnetization curves for a paramagnetic substance (see Fig. 11) and for a ferromagnetic substance (see Fig. 13). What would a similar curve for a diamagnetic substance look like?

19. Why do iron filings line up with a magnetic field, as in Fig. 1? After all, they are not intrinsically magnetized.

20. The earth's magnetic field can be represented closely by that of a magnetic dipole located at or near the center of the earth. The earth's magnetic poles can be thought of as (*a*) the points where the axis of this dipole passes through the earth's surface or as (*b*) the points on the earth's surface where a dip needle would point vertically. Are these necessarily the same points?

21. A "friend" borrows your favorite compass and paints the entire needle red. When you discover this you are lost in a cave and have with you two flashlights, a few meters of wire, and (of course) this book. How might you discover which end of your compass needle is the north-seeking end?

22. A Rowland ring is supplied with a constant current. What happens to the magnetic induction in the ring if a small slot is cut out of the ring, leaving an air gap?

23. How can you magnetize an iron bar if the earth is the only magnet around?

24. How would you go about shielding a certain volume of space from constant external magnetic fields? If you think it can't be done, explain why.

EXERCISES AND PROBLEMS

Section 34-2 Magnetism and the Electron

1E. Using the values of (spin) angular momentum S and (spin) magnetic moment μ_s given in Table 1 for the free electron, show that

$$\mu_s = \frac{e}{m} S.$$

Verify that the units and dimensions are consistent. This result is a prediction of a relativistic theory of the electron advanced by P. A. M. Dirac in 1928.

2P. Figure 18 shows four arrangements of pairs of small compass needles, set up in a space in which there is no external magnetic field. Identify the equilibrium in each case as stable or unstable. For each pair consider only the torque acting on one needle due to the magnetic field set up by the other. Explain your answers.

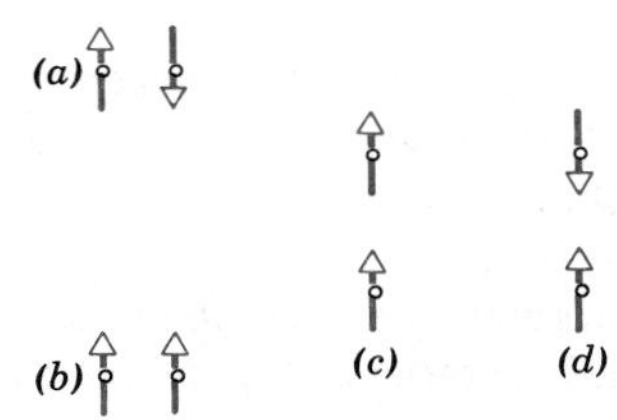

Figure 18 Problem 2.

3P. A simple bar magnet hangs from a string as in Fig. 19. A uniform magnetic field **B** directed horizontally is then established. Sketch the resulting orientation of the string and the magnet.

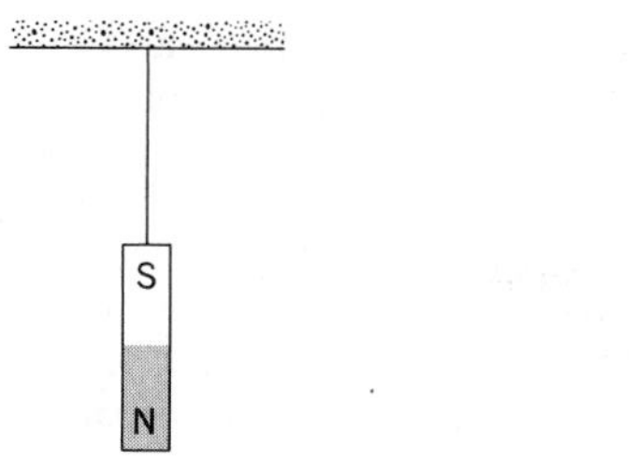

Figure 19 Problem 3.

Section 34-3 Orbital Angular Momentum and Magnetism

4E. In the lowest energy state of the hydrogen atom the most probable distance between the single orbiting electron and the central proton is 5.2×10^{-11} m. Calculate (*a*) the electric field and (*b*) the magnetic field set up by the proton at this distance, measured along the proton's axis of spin. The charge and magnetic moment of the proton are $+1.6 \times 10^{-19}$ C and 1.4×10^{-26} J/T, respectively.

5P. A charge q is distributed uniformly around a thin ring of radius r. The ring is rotating about an axis through its center and perpendicular to its plane at an angular speed ω. (*a*) Show that the magnetic moment due to the rotating charge is

$$\mu = \tfrac{1}{2} q \omega r^2.$$

(*b*) What is the direction of this magnetic moment if the charge is positive?

Section 34-4 Gauss' Law for Magnetism

6E. Imagine rolling a sheet of paper into a cylinder and placing a bar magnet near its end as shown in Fig. 20. (*a*) Sketch the **B** field lines as they cross the paper cylinder. (*b*) What can you say about the sign of $\mathbf{B} \cdot d\mathbf{A}$ for every $d\mathbf{A}$ of this paper cylinder? (*c*) Does this contradict Gauss' law for magnetism? Explain.

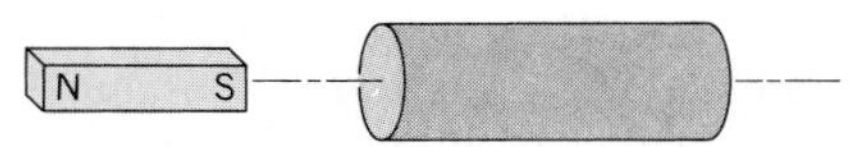

Figure 20 Exercise 6.

7E. The magnetic flux through each of five faces of a die (singular of "dice") is given by $\Phi_B = \pm N$ Wb, where $N (= 1$ to $5)$ is the number of spots on the face. The flux is positive (outward) for N even and negative (inward) for N odd. What is the flux through the sixth face of the die?

8P. A Gaussian surface in the shape of a right circular cylinder has a radius of 12 cm and a length of 80 cm. Through one end there is an inward magnetic flux of 25 μWb. At the other end there is a uniform magnetic field of 1.6 mT, normal to the surface and directed outward. What is the net magnetic flux through the curved surface?

9P*. Two wires, parallel to the z axis and a distance $4r$ apart, carry equal currents i in opposite directions, as shown in Fig. 21. A circular cylinder of radius r and length L has its axis on the z axis, midway between the wires. Use Gauss' law for magnetism to calculate the net outward magnetic flux through the half of the cylindrical surface above the x axis. (*Hint:* Find the flux through that portion of the xz plane that is within the cylinder.)

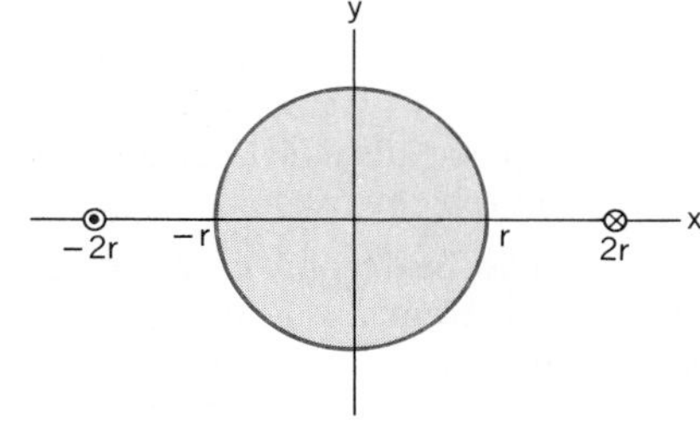

Figure 21 Problem 9.

Section 34-5 The Magnetism of the Earth

10E. In New Hampshire the average horizontal component of the earth's magnetic field in 1912 was 16 μT and the average inclination or "dip" was 73°. What was the corresponding magnitude of the earth's magnetic field?

11E. In Sample Problem 3 the vertical component of the earth's magnetic field in Tucson, Arizona, was found to be 43 μT. Assume this is the average value for all of Arizona, which has an area of 295,000 square kilometers, and calculate the net magnetic flux through the rest of the earth's surface (the entire surface excluding Arizona). Is the flux outward or inward?

12E. The earth has a magnetic dipole moment of 8.0×10^{22} J/T. (*a*) What current would have to be set up in a single turn of wire going around the earth at its magnetic equator if we wished to set up such a dipole? (*b*) Could such an arrangement be used to cancel out the earth's magnetism at points in space well above the earth's surface? (*c*) On the earth's surface?

13P. The magnetic field of the earth can be approximated as a dipole magnetic field, with horizontal and vertical components, at a point a distance r from the earth's center, given by

$$B_h = \frac{\mu_0 \mu}{4\pi r^3} \cos \lambda_m, \quad B_v = \frac{\mu_0 \mu}{2\pi r^3} \sin \lambda_m,$$

where λ_m is the *magnetic latitude* (latitude measured from the magnetic equator toward the north or south magnetic pole). The magnetic dipole moment $\mu = 8.0 \times 10^{22}$ A·m². (*a*) Show that the strength at latitude λ_m is given by

$$B = \frac{\mu_0 \mu}{4\pi r^3} \sqrt{1 + 3 \sin^2 \lambda_m}.$$

(*b*) Show that the inclination ϕ_i of the magnetic field is related to the magnetic latitude λ_m by

$$\tan \phi_i = 2 \tan \lambda_m.$$

14P. Use the results displayed in Problem 13 to predict the value of the earth's magnetic field, magnitude, and inclination at (*a*) the magnetic equator; (*b*) a point at magnetic latitude 60°; (*c*) the north magnetic pole.

15P. Find the altitude above the earth's surface where the earth's magnetic field has a magnitude one-half the surface value at the same magnetic latitude. (Use the dipole field approximation given in Problem 13.)

16P. Using the dipole field approximation to the earth's magnetic field (see Problem 13), calculate the maximum strength of the magnetic field at the core-mantle boundary, which is 2900 km below the earth's surface.

17P. Use the results displayed in Problem 13 to calculate the magnitude and inclination angle of the earth's magnetic field at the north geographic pole. (*Hint:* The angle between the magnetic axis and the rotational axis of the earth is 11.5°.) Why do the calculated values probably not agree with the measured values?

Section 34-6 Paramagnetism

18E. A 0.50-T magnetic field is applied to a paramagnetic gas whose atoms have an intrinsic magnetic dipole moment of 1.0×10^{-23} J/T. At what temperature will the mean kinetic energy of translation of the gas atoms be equal to the energy required to reverse such a dipole end for end in this magnetic field?

19E. A cylindrical rod magnet has a length of 5.0 cm and a diameter of 1.0 cm. It has a uniform magnetization of 5.3×10^3 A/m. What is its magnetic dipole moment?

20E. A paramagnetic substance is (weakly) attracted to a pole of a magnet. Figure 22 shows a model of this phenomenon. The "paramagnetic substance" is a current loop L, which is placed on the axis of a bar magnet nearer to its north pole than its south pole. Because of the torque $\boldsymbol{\tau} = \boldsymbol{\mu} \times \mathbf{B}$ exerted on the loop by the **B** field of the bar magnet, the magnetic dipole moment $\boldsymbol{\mu}$ of the loop will align itself to be parallel to **B**. (*a*) Make a sketch showing the **B** field lines due to the bar magnet. (*b*) Show the direction of the current i in the loop. (*c*) Using $d\mathbf{F} = i\, d\mathbf{s} \times \mathbf{B}$ show from (*a*) and (*b*) that the net force on L is toward the north pole of the bar magnet.

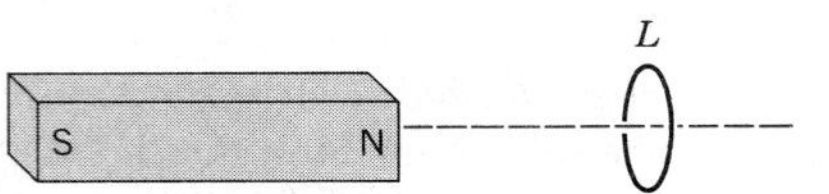

Figure 22 Exercises 20 and 26.

21P. The paramagnetic salt to which the magnetization curve of Fig. 11 applies is to be tested to see whether it obeys Curie's law. The sample is placed in a 0.50-T magnetic field that remains constant throughout the experiment. The magnetization M is then measured at temperatures ranging from 10 to 300 K. Would it be found that Curie's law is valid under these conditions?

22P. A sample of the paramagnetic salt to which the magnetization curve of Fig. 11 applies is held at room temperature (300 K). At what applied magnetic field would the degree of magnetic saturation of the sample be (*a*) 50 percent? (*b*) 90 percent? (*c*) Are these fields attainable in the laboratory?

23P. A sample of the paramagnetic salt to which the magnetization curve of Fig. 11 applies is immersed in a magnetic field of 2.0 T. At what temperature would the degree of magnetic saturation of the sample be (*a*) 50 percent? (*b*) 90 percent?

24P. An electron with kinetic energy K_e travels in a circular path that is perpendicular to a uniform magnetic field, subject only to the force of the field. (*a*) Show that the magnetic dipole moment due to its orbital motion has magnitude $\mu = K_e/B$ and that it is in the direction opposite to that of **B**. (*b*) What is the magnitude and direction of the magnetic dipole moment of a positive ion with kinetic energy K_i under the same circumstances? (*c*) An ionized gas consists of 5.3×10^{21} electrons/m^3 and the same number of ions/m^3. Take the average electron kinetic energy to be 6.2×10^{-20} J and the average ion kinetic energy to be 7.6×10^{-21} J. Calculate the magnetization of the gas for a magnetic field of 1.2 T.

25P. Consider a solid containing N atoms per unit volume, each atom having a magnetic dipole moment $\boldsymbol{\mu}$. Suppose the direction of $\boldsymbol{\mu}$ can be only parallel or antiparallel to an externally applied magnetic field **B** (this will be the case if $\boldsymbol{\mu}$ is due to the spin of a single electron). According to statistical mechanics, it can be shown that the probability of an atom being in a state with energy U is proportional to $e^{-U/(kT)}$ where T is the temperature and k is Boltzmann's constant. Thus, since $U = -\boldsymbol{\mu} \cdot \mathbf{B}$, the fraction of atoms whose dipole moment is parallel to **B** is proportional to $e^{(\mu B)/(kT)}$ and the fraction of atoms whose dipole moment is antiparallel to **B** is proportional to $e^{-(\mu B)/(kT)}$. (*a*) Show that the magnetization of this solid is $M = N\mu \tanh(\mu B/kT)$. Here tanh is the hyperbolic tangent function: $\tanh(x) = (e^x - e^{-x})/(e^x + e^{-x})$. (*b*) Show that (*a*) reduces to $M = N\mu^2 B/kT$ for $\mu B \ll kT$. (*c*) Show that (*a*) reduces to $M = N\mu$ for $\mu B \gg kT$. (*d*) Show that (*b*) and (*c*) agree qualitatively with Fig. 11.

Section 34-7 Diamagnetism

26E. A diamagnetic substance is (weakly) repelled by a pole of a magnet. Figure 22 shows a model of this phenomenon. The "diamagnetic substance" is a current loop L that is placed on the axis of a bar magnet nearer to its north pole than its south pole. Because the substance is diamagnetic the magnetic moment $\boldsymbol{\mu}$ of the loop will align itself to be antiparallel to the **B** field of the bar magnet. (*a*) Make a sketch showing the **B** field lines due to the bar magnet. (*b*) Show the direction of the current i in the loop. (*c*) Using $d\mathbf{F} = i\, d\mathbf{s} \times \mathbf{B}$, show from (*a*) and (*b*) that the net force on L is away from the north pole of the bar magnet.

27P. Analyze qualitatively the appearance of induced magnetic dipole moments in diamagnetism from the point of view of Faraday's law of induction. [*Hint:* See Fig. 10*b*, Chapter 32. Also, note that for orbiting electrons the inductive effects (any change in speed) persist after the magnetic field has stopped changing; they vanish only when the field is removed.]

Section 34-8 Ferromagnetism

28E. Measurements in mines and boreholes indicate that the temperature in the earth increases with depth at the average rate of 30°C/km. Assuming a surface temperature of 10°C, at what depth does iron cease to be ferromagnetic? (The Curie temperature of iron varies very little with pressure.)

29E. The exchange coupling mentioned in Section 8 as being responsible for ferromagnetism is *not* the mutual magnetic interaction energy between two elementary magnetic dipoles. To show this calculate (*a*) the magnetic field a distance of 10 nm away along the dipole axis from an atom with magnetic dipole moment 1.5×10^{-23} J/T (cobalt), and (*b*) the minimum energy required to turn a second identical dipole end for end in this field. Compare with the results of Sample Problem 4. What do you conclude?

30E. The saturation magnetization of the ferromagnetic metal nickel is 4.7×10^5 A/m. Calculate the magnetic moment of a single nickel atom. (The density of nickel is 8.90 g/cm^3 and its atomic mass is 58.71.)

31E. The dipole moment associated with an atom of iron in an iron bar is 2.1×10^{-23} J/T. Assume that all the atoms in the bar, which is 5.0 cm long and has a cross-sectional area of 1.0 cm^2, have their dipole moments aligned. (*a*) What is the dipole moment of the bar? (*b*) What torque must be exerted to hold this magnet at right angles to an external field of 1.5 T? The density of iron is 7.9 g/cm^3.

32P. The magnetic dipole moment of the earth is 8.0×10^{22} J/T. (*a*) If the origin of this magnetism were a magnetized iron sphere at the center of the earth, what would be its radius? (*b*) What fraction of the volume of the earth would such a sphere occupy? Assume complete alignment of the dipoles. The density of the earth's inner core is 14 g/cm^3. The magnetic dipole moment of an iron atom is 2.1×10^{-23} J/T. (*Note:* The earth's inner core is in fact thought to be in both liquid and solid form and partly iron, but a permanent magnet as the source of the earth's magnetism has been ruled out by several considerations. For one, the temperature is certainly above the Curie point.)

33P. Figure 23 shows the apparatus used in a lecture demonstration of para- and diamagnetism. A sample of the magnetic material is suspended by a string ($L = 2$ m) in a region ($d = 2$ cm) between the two poles of a powerful electromagnet. Pole P_1 is sharply pointed and pole P_2 is rounded as indicated. Any deflection of the string from the vertical is visible to the audience by means of an optical projection system (not shown). (*a*) First a bismuth (highly diamagnetic) sample is used. When the electromagnet is turned on, the sample is observed to deflect slightly (about 1 mm) toward one of the poles. What is the direction of this deflection? (*b*) Next an aluminum (paramagnetic, conducting) sample is used. When the electromagnet is turned on, the sample is observed to deflect strongly (about 1 cm) toward one pole for about a second and then deflect moderately (a few mm) toward the other pole. Explain and indicate the direction of these deflections. (*Hint:* Note that the sample is a conductor.) (*c*) What would happen if a ferromagnetic sample were used?

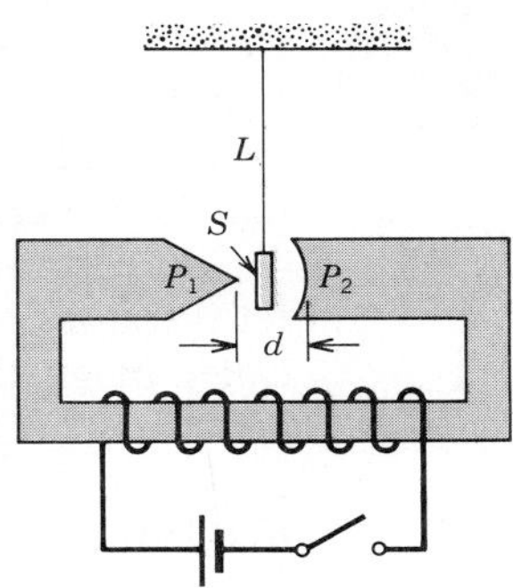

Figure 23 Problem 33.

34P. A Rowland ring is formed of ferromagnetic material. It is circular in cross section, with an inner radius of 5.0 cm and an outer radius of 6.0 cm and is wound with 400 turns of wire. (*a*) What current must be set up in the windings to attain a toroidal field $B_0 = 0.20$ mT? (*b*) A secondary coil wound around the toroid has 50 turns and has a resistance of 8.0 Ω. If, for this value of B_0, we have $B_M = 800B_0$, how much charge moves through the secondary coil when the current in the toroid windings is turned on?

ESSAY 14
MAGNETISM AND LIFE

CHARLES P. BEAN
RENSSELAER
POLYTECHNIC
INSTITUTE

Introduction

Owing to the apparent mystery of magnetic forces, people of the past, and even today, have looked for effects of the magnetic field on human and other animal life. In many cases they have believed they have found such effects but never in a way that could be replicated. Recently, however, a young scientist has discovered a completely replicable effect of the earth's magnetic field on a class of living organisms—the magnetotactic bacteria. This essay tells you about that discovery and some of its consequences. Before telling that story, it is necessary to give some historical and scientific perspective.

We said that magnetism was mysterious. Take two permanent magnets and have them approach one another. In one orientation, they attract one another through empty space and in another repel. We have all felt the strangeness of this effect. For the five-year-old Einstein, the observation of the deflection of a compass needle caused him first to think about fields of force and what they might be. Our present understanding, due largely to Einstein's thoughts, is that there exists only one field, the electromagnetic field, and that our perception depends on our motion with respect to that field. For instance, what we call a magnetic field is caused by the motion of electric charges with respect to us. One consequence is that, since the relative velocities of motion are usually much less than the speed of light, magnetic forces are usually much less than electrical forces. Electrical forces, for instance, hold atoms and solids together. Magnetic forces provide usually only a small fraction of the total binding.

In common terms, when we speak of a magnet we mean the type that sticks on refrigerator doors—the permanent magnet. In nature, it principally exists in one form called magnetite with the chemical formula, Fe_3O_4. More informatively, it can be written as $FeO \cdot Fe_2O_3$ to show that each molecule has one ferrous (Fe^{++}) ion and two ferric (Fe^{+++} ions). It is an unremarkable-looking black stone (lodestone), but one that led to the great age of geographic discovery in the twelfth to sixteenth centuries as well as to many aspects of modern science.

The details of the discovery of the properties of magnetite are lost in antiquity. The fact that magnetite can attract iron was known to the Chinese over 2000 years ago and knowledge of that interaction either spread to the Mediterranean or was independently discovered there in classical times. The crucial discovery that magnetite tends to align itself in the earth's field was, apparently, a discovery of the Chinese. (Curiously enough, they called it a "south pointer.") Further it was found by experiment that needles of iron could be magnetized by stroking them with lodestone or, alternatively, by heating the needles red hot and allowing them to cool in the earth's magnetic field. A magnetized needle suspended in water or on a string was found to point roughly north and south. With this discovery, ships could safely sail the open ocean even if the stars were invisible. For the first time a ship could follow a charted course of exploration and return to its home port.

This essay shows that a large class of bacteria, some billions of years ago, developed a magnetic guidance system that guided them in their movements. Thus one of man's great discoveries is now known to have been antedated by the simplest organism on earth.

Magnetic Navigation by Bacteria

Bacteria are single-cell organisms that live everywhere. They flourish both inside and on the surface of our bodies. They can be found in hot springs at 85 °C and at the bottom of the ocean. Typically, they are a few micrometers in size and so can only be seen with a microscope. Owing to the resolution limit imposed by the wavelength of visible light (about 0.4 μm in water), their detailed interior structure cannot be seen with the optical microscope. An electron microscope is required to see the finer points of their structure.

Despite their small size, bacteria show great powers of adaptation to local conditions and a wide range of behavior. For instance, a large class of bacteria can function best in conditions under which oxygen is not present. It is thought that these arose early in the development of the earth before the evolution of plants and the consequent release of oxygen to the atmosphere. (Oxygen is a waste product of photosynthesis just as carbon dioxide is of our metabolism.) Now these so-called anaerobic bacteria can be found today in many aquatic environments with low oxygen levels. Decaying animal and vegetable matter provide these conditions. Typically the waste product of anaerobic metabolism is methane — also known as marsh gas. Most such bacteria can swim using one or more appendages — called flagellae when they are long and few and pili when mutiple and short. In addition they have receptors that can sense chemicals and dissolved gases in the water and so can swim toward food and away from poisons. For the anaerobic bacterium oxygen is such a poison.

In 1975, Richard Blakemore made a remarkable discovery. At the time he was a graduate student in microbiology at the University of Massachusetts working at the Woods Hole Oceanographic Institute on Cape Cod. His area of research concerned the role of anaerobic bacteria in the ecology of muds and swamps. He took some mud from a nearby saltwater pond, mixed it with a little seawater, put a drop on a microscope slide and observed it at high power. Doubtless this had been done tens of thousands of times before, dating back to the days of Pasteur in the 1850s. But Blakemore noticed what no one had recorded ever noticing earlier. In some drops the bacteria swam to one side of the drop. Were they swimming away or toward the room light? He covered the microscope and its internal light with a box. He turned the microscope around. He moved the microscope to another room. In each case the bacteria continued to swim in the same geographic direction — north. Biological and chemical laboratories are well equipped with small permanent magnets coated with plastic. In conjunction with a larger external rotating magnet they are used to stir and mix solutions. Blakemore picked up such a stirrer and brought it near the drop. In one orientation, the stirrer bar did not affect the motion, in the opposite orientation the direction of motion of the bacteria was reversed! This observation, when the stirrer magnet was tested with a compass, showed that almost all the bacteria swam in the direction of the north-seeking end of the magnetic compass.

Here was a discovery that was unique in the history of magnetism and biology. It showed a direct and repeatable influence of the earth's magnetic field on a living organism. Blakemore and co-workers quickly showed that life was not essential to the orientation of bacteria. Killed bacteria would rotate to follow an imposed magnetic field but, since they were dead, would not migrate in that field.

The Mechanism of Orientation

Each discovery in science leads to more questions. In this case, they were: what is the physical mechanism of the effect and what are its biological implications? For the first

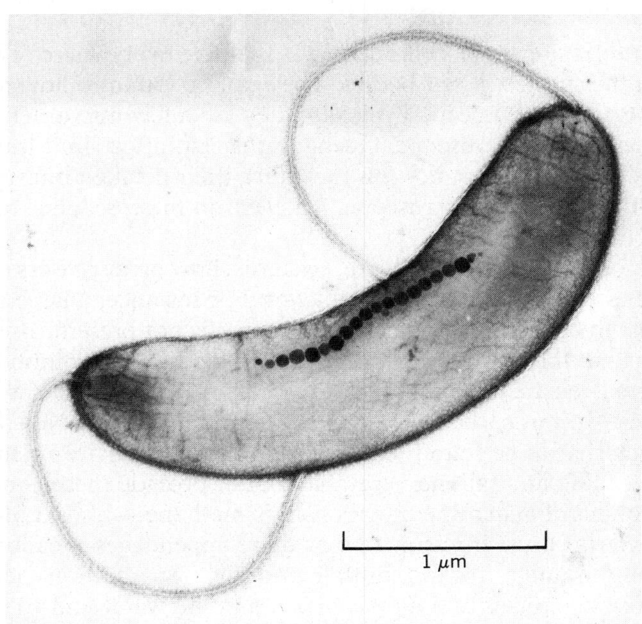

Figure 1 A magnetotactic bacterium as seen in a transmission electron micrograph. The unusual feature of the internal structure of this type of bacterium is a chain of electron dense particles roughly 50 nm in diameter. These are composed of magnetite (Fe_3O_4) and are each a fully magnetized permanent magnet. The chain functions as a compass that orients the bacterium in the same direction as the lines of force of the earth's magnetic field. The species shown here has a flagellum at each end. It is capable of swimming forward or backward. Many species have flagellae only at one end. (R. P. Blakemore and R. B. Frankel, *Scientific American,* December 1981.)

question, Blakemore reasoned that the simplest mechanism was to have each bacterium be, in itself, a small compass that would rotate in a magnetic field. An electron micrograph of a typical bacterium (Fig. 1) showed a novel alignment of electron-dense particles within a typical bacterium that responded to a magnetic field. Could they be the compass? In discussing this conjecture with Edward M. Purcell of Harvard, the latter suggested that these were probably single-domain magnetic particles of some iron-containing substance and, if so, one consequence would be that a short pulse of a high oppositely-directed magnetic field would reverse the magnetization of each bacterium before it could rotate to accommodate itself to the new field. An experiment, in conjunction with Adrianus J. Kalmijn of the Woods Hole Oceanographic Institute, showed just such an effect. After such a pulse, the bacteria were not south-seeking.

Using the physics of magnetism developed in Chapter 34, one can estimate how much material would be necessary to orient a bacterium in the earth's magnetic field. Sample Problem 4 shows that an atom with a magnetic moment of one Bohr magneton is not significantly aligned at room temperature in a field of 1.5 T. To have significant alignment the magnetic energy μB must be comparable or greater than the energy associated with thermal disorder, kT. (The factors of 2 and $\frac{3}{2}$ used in Sample Problem 4 are not necessary in this order of magnitude calculation.) This requirement says that

$$\mu B > kT \quad \text{or} \quad \mu > kT/B$$

where μ is the total magnetic moment of the bacterium, k is Boltzmann's constant (1.38×10^{-23} J/K), T is the absolute temperature (300 K), and B is the earth's magnetic field ($\sim 5 \times 10^{-5}$ T). Using these figures

$$\mu > (1.38 \times 10^{-23})(3 \times 10^{2})/(5 \times 10^{-5}) = 8.3 \times 10^{-17} \text{ J/T}.$$

If magnetite is fully magnetized, it has a magnetic moment per unit volume of 5×10^5 J/Tm^3. (This value corresponds to 4 Bohr magnetons per molecule of Fe_3O_4, the moment of Fe^{+2}. The magnetic moments of the Fe^{+3} ions exactly cancel one another.) Since the total magnetic moment, μ, is given by the product of the magnetic moment per unit volume and the volume, V, we derive that the volume, V, of fully magnetized material needed for significant alignment is

$$V > (8.3 \times 10^{-17})/(5 \times 10^5) = 1.7 \times 10^{-22}\ \text{m}^3$$

These lower limits can be compared with the volume estimated from the electron micrograph shown in Fig. 1. We see that the chain of particles has 20 members of diameter ~ 50 nm. Assuming them to be spheres, their total volume is $20\pi(50 \times 10^{-9})^3/6 = 13 \times 10^{-22}$ m^3 or about eight times the minimum calculated above. This implies, of course, a magnetic moment about eight times that for which $\mu B = kT$. Consequently, the bacteria are well aligned in the earth's magnetic field.

Another method of calculating the magnetic moment of the material comes from an ingenious experiment by Adrianus Kalmijn. He observes, using a microscope, a single live bacterium swimming along a field direction. For low fields the bacterium is buffeted by the thermal motions of the water molecules and its average forward velocity is low. For higher fields the motion of the bacterium closely follows the field lines. Kalmijn notes the time it takes the bacterium to move between two fixed lines in the optical field of the microscope, then reverses the magnetic field. The bacterium makes a U-turn and goes in the opposite direction. He has watched hundreds of round trips and averaged times to obtain an average velocity. A specimen result is shown in Fig. 2. The interpretation of this result is that the bacterium swims always at 117 μm/s but forward motion only results from the component of velocity parallel to the field direction. At low fields the bacterium swims in many directions with only slow progress along the field direction. At high fields the bacterium swims only parallel to the field. Conceptually this is identical to the problem of calculating the average moment of a paramagnet such as the chromium salt shown in Fig. 12 of Chapter 34. The experimental points can be fitted with two parameters, the swimming velocity when fully aligned and the magnetic moment of the bacterium. The solid curve of the figure is a theoretical curve with a maximum speed of 117 μm/s and a

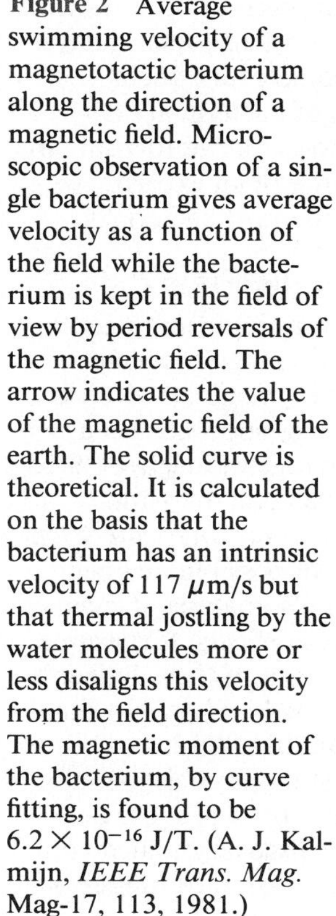

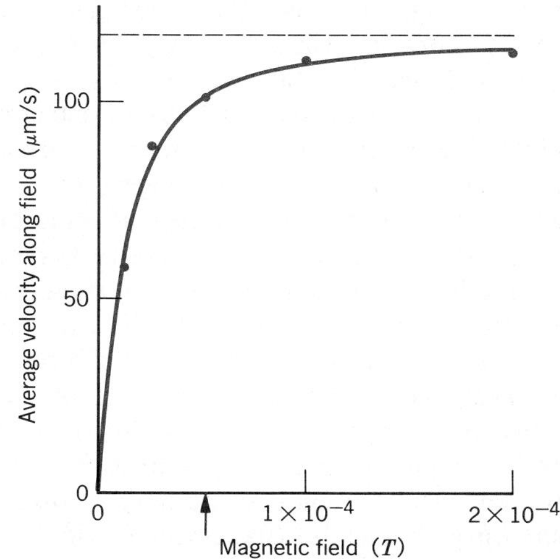

Figure 2 Average swimming velocity of a magnetotactic bacterium along the direction of a magnetic field. Microscopic observation of a single bacterium gives average velocity as a function of the field while the bacterium is kept in the field of view by period reversals of the magnetic field. The arrow indicates the value of the magnetic field of the earth. The solid curve is theoretical. It is calculated on the basis that the bacterium has an intrinsic velocity of 117 μm/s but that thermal jostling by the water molecules more or less disaligns this velocity from the field direction. The magnetic moment of the bacterium, by curve fitting, is found to be 6.2×10^{-16} J/T. (A. J. Kalmijn, *IEEE Trans. Mag.* Mag-17, 113, 1981.)

magnetic moment of 6.2×10^{-16} J/T. For the saturation magnetization of 5×10^5 J/T·m^3 given earlier, this corresponds to about 12×10^{-22} m^3 of Fe_3O_4. This value is very close to that derived from examination of the photograph in Fig. 1 although it is not the same bacterium. Thus two lines of evidence are consistent with the hypothesis that the mechanism of alignment is simply that of a compass.

The Consequences of Orientation in the Earth's Magnetic Field

In biology, one often looks for an advantage gained by an organism with a special sense. For instance, bats can both emit and detect ultrasound. By use of this faculty they can both fly in the dark and locate prey, such as moths. In the case of the magnetotactic bacteria, one can identify a clear advantage. Figure 8 of Chapter 34 shows the earth's magnetic field represented as a dipole. In the Northern hemisphere, there is a downward component. Bacteria swimming along the field line would go down. Thus if anaerobic bacteria were stirred from their usual environment, they would be aided in their return to the mud. One consequence of this concept is that bacteria in the Southern hemisphere would have south-seeking magnetic moments. An expedition by Blakemore and collaborators to New Zealand showed exactly that. At the equator, one finds both polarities but the advantage is not completely clear since there the bacteria are constrained to move in horizontal lines.

In each population of bacteria there may be a few bacteria per thousand of the "wrong" polarity. Investigators have taken a sample of mud and water and put it in a field whose vertical component is reversed. The vast majority of the bacteria swim to the surface and the consequent oxygen-rich environment. In that environment their metabolism and reproductive capacity are lowered. The former aberrant bacteria who now go to the mud are favored. After a matter of eight weeks almost all bacteria have reversed polarity. We know no mechanism whereby bacteria could rotate their internal magnetic particles. Most probably the new population of bacteria are descendents of the few aberrant bacteria. If so, the experiment shows, in microcosm, an evolutionary adaptation to environmental change. (Can you think of an experiment to prove whether the new population truly descends from the aberrant bacteria?)

Do Other Organisms Respond to the Earth's Magnetic Field?

Just as mariners use a magnetic compass to guide them, migratory birds and nectar-seeking bees could use a magnetic sense. Over the years many investigators have explored this possibility. They have attached magnets and dummy magnets to birds and claimed an altered behavior for the case of magnets. The suggestion has been made that pigeons not only detect direction but can detect a change in magnetic field strength of 2 parts in 10^4. Bees are thought by some to use a magnetic map and convey, by dances, directions to prospective foragers. (J. L. Gould in *American Scientist,* May–June 1980 gives a good review of the understanding of possible magnetic sensitivity of birds and bees.) Contrary to the case of Blakemore's analysis of magnetotactic bacteria, no one has found a mechanism of these possible effects. If it were a local compass, it would have to be connected to the nervous system of these higher organisms rather than act as a passive torque as in the case of the bacteria. No such connection has been found. Indeed no one has been able to condition a bird or bee to be attracted or repelled by a magnetic field. Consequently, most biophysicists do not believe that the case for magnetic sensitivity of birds and bees has been proven. But the question is still open. Another Blakemore may yet make a discovery that revolutionizes this field as thoroughly as did Blakemore's original observation of magnetotactic bacteria.

CHAPTER 35

ELECTROMAGNETIC OSCILLATIONS

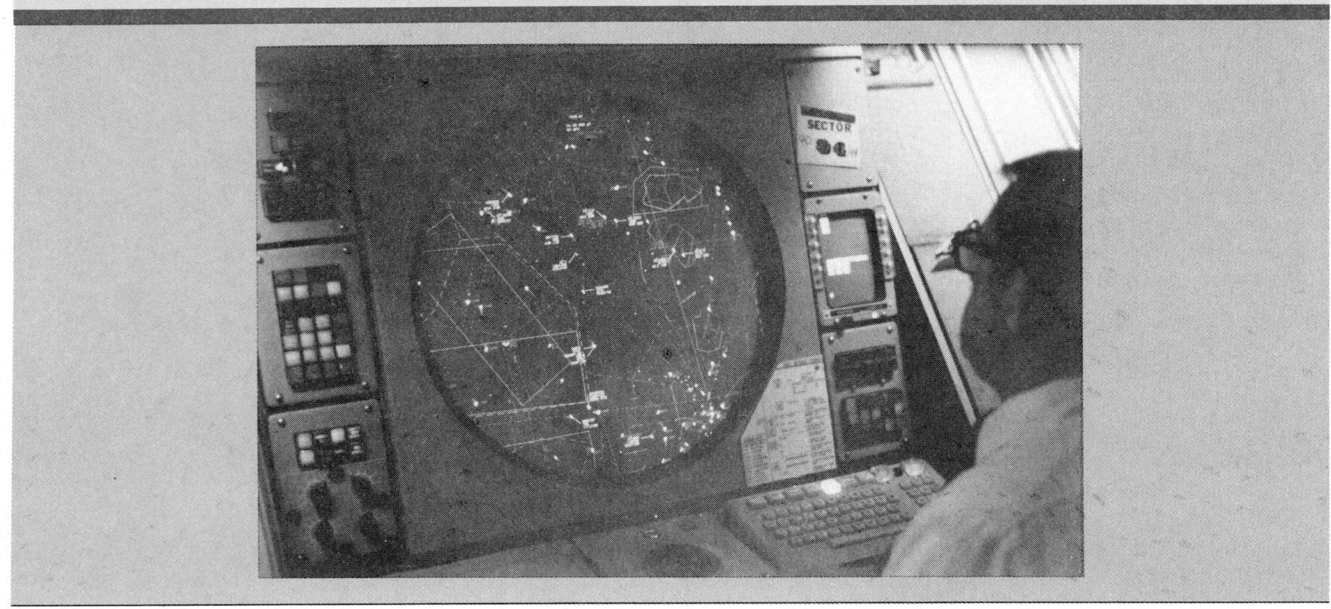

The radar beams that make modern air travel possible are generated in the electromagnetic oscillator *of a ground-based radar transmitter. When the beam strikes the aircraft it is detected by a receiver containing another* electromagnetic oscillator *tuned to the same frequency. The detected radar beam triggers a transponder beacon (still another* electromagnetic oscillator*) which transmits coded identity and altitude information, all of which is detected by a tuned receiver at the ground antenna and displayed on the air traffic controller's video terminal. Radio communication with the aircraft involves two other* electromagnetic oscillators, *one in the transmitter and another in the receiver.*

35-1 New Physics — Old Mathematics

In this chapter we describe how electric charge q varies with time in a circuit made up of an inductor L, a capacitor C, and a resistor R. From another point of view, we are going to describe how energy shuttles back and forth between the magnetic field of the inductor and the electric field of the capacitor, being gradually dissipated — as the oscillations proceed — as thermal energy in the resistor.

We have studied all these things before, in another context. In Chapter 14 we saw how displacement x varies with time in a mechanical oscillating system made up of a block of mass m, a spring k, and a viscous or frictional element such as a dash pot; see Fig. 17 of Chapter 14. We also studied how energy shuttles back and forth between the kinetic energy of the oscillating mass and the potential energy of the spring, being gradually dissipated — as the oscillations proceed — as thermal energy.

The parallel between the two idealized systems is exact and the controlling differential equations are identical. Thus, we need to learn no new mathematics; we can simply change the symbols and give our full attention to the physics of the situation.

35-2 *LC* Oscillations: Qualitative

Of the three circuit elements, resistance R, capacitance C, and inductance L, we have so far studied the combinations RC (Section 29-8) and RL (Section 33-4). In these two kinds of circuits we found that the charge, current, and potential difference both grow and decay exponentially. The time scale of the growth or decay is given by the *time constant* τ, which is either capacitive or inductive as the case may be.

We now examine the remaining two-element circuit, LC. We shall see that in this case the charge, current, and potential difference vary not exponentially (with time constant τ) but *sinusoidally* (with angular frequency ω). In other words, the circuit *oscillates.* Let us find out what is going on in such a circuit, from a physical point of view.

Assume that initially the capacitor C in Fig. 1a carries a charge q and the current i in the inductor is zero.* At this instant the energy stored in the electric field of the capacitor is given by Eq. 21 of Chapter 27:

$$U_E = \frac{q^2}{2C}. \tag{1}$$

The energy stored in the magnetic field of the inductor, given by

$$U_B = \tfrac{1}{2}Li^2, \tag{2}$$

is zero because the current is zero.

The capacitor now starts to discharge through the inductor, positive charge carriers moving counterclockwise, as shown in Fig. 1b. This means that a current i, given by dq/dt and pointing down in the inductor, is

* When we deal with sinusoidally oscillating electrical quantities such as charge, current, or potential difference, we represent their instantaneous values by small letters (q, i, and v) and their oscillation amplitudes by capital letters (Q, I, and V).

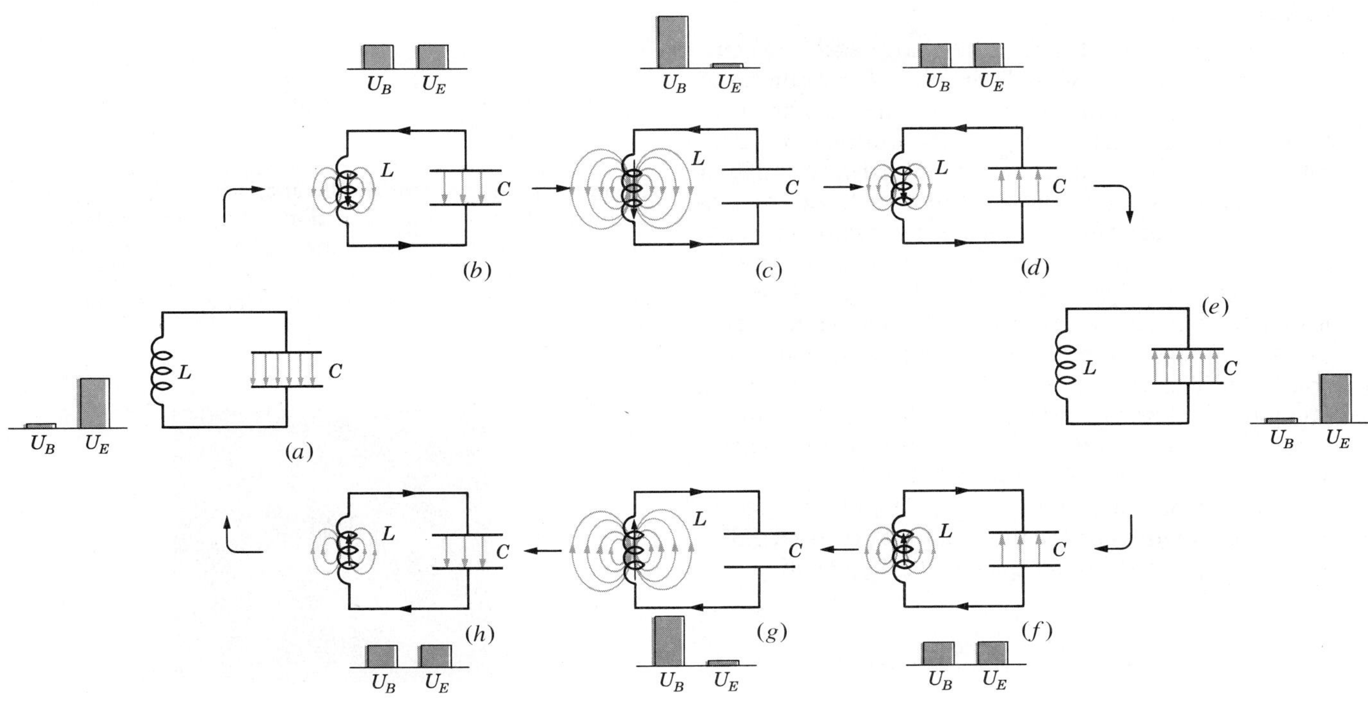

Figure 1 Eight stages in a single cycle of oscillation of a resistanceless LC circuit. The bar graphs by each figure show the stored magnetic and electric energies. The vertical arrows on the inductor axis show the current.

established. As q decreases, the energy stored in the electric field in the capacitor also decreases. This energy is transferred to the magnetic field that appears around the inductor because of the current i that is building up there. Thus, the electric field decreases, the magnetic field builds up, and energy is transferred from the former to the latter.

At a time corresponding to Fig. 1*c*, all the charge on the capacitor will have disappeared. The electric field in the capacitor will be zero, the energy stored there having been transferred entirely to the magnetic field of the inductor. According to Eq. 2, there must then be a current—and indeed one of maximum value—in the inductor. Note that even though q equals zero, the current (which is the rate at which q is changing with time) is not zero at this time.

The large current in the inductor in Fig. 1*c* continues to transport positive charge from the top plate of the capacitor to the bottom plate, as shown in Fig. 1*d*; energy now flows from the inductor back to the capacitor as the electric field builds up again. Eventually, the energy will have been transferred completely back to the capacitor, as in Fig. 1*e*. The situation of Fig. 1*e* is like the initial situation, except that the capacitor is charged oppositely.

The capacitor will start to discharge again, the current now being clockwise, as in Fig. 1*f*. Reasoning as before, we see that the circuit eventually returns to its initial situation and that the process continues at a definite frequency ν to which corresponds a definite angular frequency ω $(= 2\pi\nu)$. Once started, such *LC* oscillations (in the ideal case that we are describing, in which the circuit contains no resistance) continue indefinitely, energy being shuttled back and forth between the electric field in the capacitor and the magnetic field in the inductor. Any configuration in Fig. 1 can be set up as an initial condition. The oscillations will then continue from that point, proceeding clockwise around the figure. Compare these oscillations carefully with those of the block–spring system described in Fig. 6 of Chapter 8.

To find the charge q as a function of time, we can use a voltmeter to measure the time-varying potential difference v_C that exists across capacitor C. The relation

$$v_C = \left(\frac{1}{C}\right) q$$

shows that v_C is proportional to q. To measure the current, we can insert a small resistance R in the circuit and measure the time-varying potential difference v_R across it. This is proportional to i through the relation

$$v_R = Ri.$$

We assume here that R is so small that its effect on the behavior of the circuit is negligible. Both q and i, or more correctly v_C and v_R which are proportional to them, can be displayed on the screen of a cathode ray oscilloscope, as suggested by Fig. 2.

In an actual *LC* circuit, the oscillations will not continue indefinitely because there is always some resistance present that will drain energy from the electric and magnetic fields and dissipate it as thermal energy; the circuit will become warmer. The oscillations, once started, will die away as Fig. 3 suggests. Compare this figure with Fig. 18 of Chapter 14, which shows the decay of the mechanical oscillations of a block–spring system caused by frictional damping.

It is possible to sustain the electromagnetic oscillations if you arrange to supply, automatically and periodi-

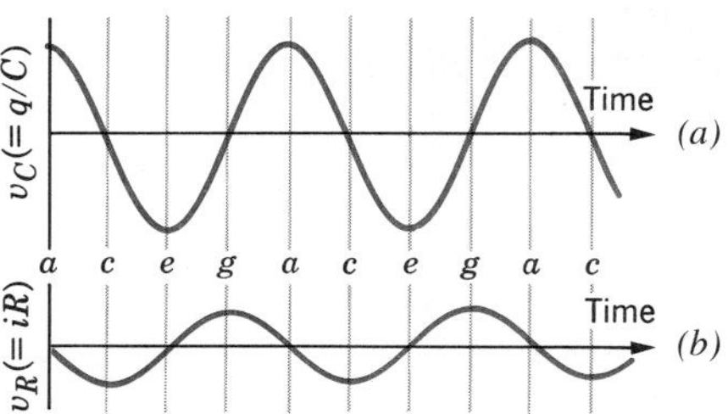

Figure 2 (*a*) The potential difference across the capacitor of the circuit of Fig. 1 as a function of time. This quantity is proportional to the charge on the capacitor. (*b*) A quantity proportional to the current in the circuit of Fig. 1. The letters indicate the corresponding oscillation stages marked in Fig. 1.

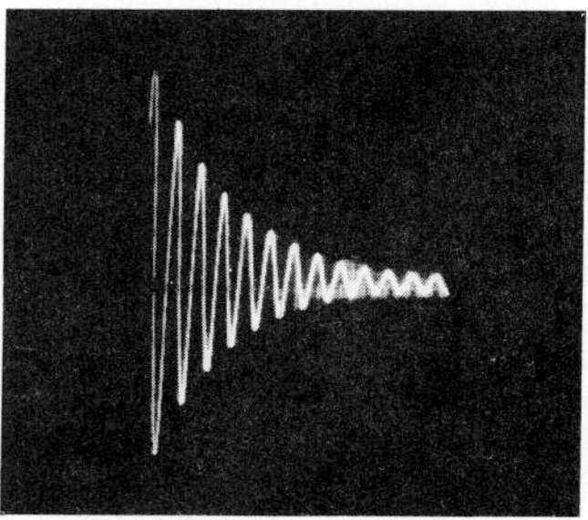

Figure 3 A photograph of an oscilloscope trace showing how the oscillations in an *LCR* circuit die away because energy is dissipated in the resistor in thermal form.

cally (once a cycle, say), enough energy from an outside source to compensate for that dissipated as thermal energy. We are reminded of a clock escapement, which is a device for feeding energy from a spring or a falling weight into an oscillating pendulum, thus compensating for frictional losses that would otherwise cause the oscillations to die away. *LC* oscillators, whose frequency may be varied between specified limits, are commercially available as packaged units over a wide range of frequencies, extending from low audiofrequencies (lower than 10 Hz) to microwave frequencies (higher than 10 GHz).

Sample Problem 1 A 1.5-μF capacitor is charged to 57 V. The charging battery is then disconnected and a 12-mH coil is connected across the capacitor, so that *LC* oscillations occur. What is the maximum current in the coil? Assume that the circuit contains no resistance.

From the conservation of energy principle, the maximum stored energy in the capacitor must equal the maximum stored energy in the inductor. This leads, from Eqs. 1 and 2, to

$$\frac{Q^2}{2C} = \tfrac{1}{2}LI^2,$$

where I is the maximum current and Q is the maximum charge. Note that the maximum current and the maximum charge do not occur at the same time but one-fourth of a cycle apart; see Figs. 1 and 2. Solving for I and substituting CV for Q, we find

$$I = V\sqrt{\frac{C}{L}} = (57\ \text{V})\sqrt{\frac{1.5 \times 10^{-6}\ \text{F}}{12 \times 10^{-3}\ \text{H}}}$$

$$= 0.637\ \text{A} \approx 640\ \text{mA}. \qquad \text{(Answer)}$$

Sample Problem 2 A 1.5-mH inductor in an *LC* circuit stores a maximum energy of 17 μJ. What is the peak current I?

When the current has its maximum value, the energy is all stored in the inductor, none being at that instant in the capacitor. If we solve Eq. 2 ($U_B = \tfrac{1}{2}LI^2$) for I we find

$$I = \sqrt{\frac{2U_B}{L}} = \sqrt{\frac{(2)(17 \times 10^{-6}\ \text{J})}{1.5 \times 10^{-3}\ \text{H}}}$$

$$= 0.151\ \text{A} \approx 150\ \text{mA}. \qquad \text{(Answer)}$$

35-3 Developing the Mechanical Analogy

Let us look a little closer at the analogy between the *LC* system of Fig. 1 and the block–spring system of Fig. 6 of Chapter 8. In the oscillating block–spring system, two kinds of energy occur. One is potential energy of the compressed or extended spring; the other is kinetic energy of the moving block. These are given by the familiar formulas in the first column of Table 1. The table suggests that a capacitor is in some mathematical way like a spring and an inductor is like a mass and that certain electromagnetic quantities "correspond" to certain mechanical ones, namely,

Table 1 The Energy in Two Oscillating Systems Compared

Mechanical System (Figure 8-6)		*Electromagnetic System* (Figure 1)	
Element	Energy	Element	Energy
Spring	Potential, $\frac{1}{2}kx^2$	Capacitor	Electric, $\frac{1}{2}(1/C)q^2$
Block	Kinetic, $\frac{1}{2}mv^2$	Inductor	Magnetic, $\frac{1}{2}Li^2$
$v = dx/dt$		$i = dq/dt$	

q corresponds to x,
i corresponds to v,
C corresponds to $1/k$,
L corresponds to m.

Comparison of Fig. 1 with Fig. 6 of Chapter 8 shows how close the correspondence is. Note how v and i correspond in the two figures; also x and q. Note too how in each case the energy alternates between two forms, magnetic and electric for the *LC* system, and kinetic and potential for the block–spring system.

In Section 14-3 we saw that the natural angular frequency of oscillation of a (frictionless) block–spring system is

$$\omega = \sqrt{\frac{k}{m}} \quad \text{(block–spring system).} \qquad (3)$$

The method of correspondences suggests that to find the natural frequency for a (resistanceless) *LC* circuit, k should be replaced by $1/C$ and m by L, obtaining

$$\omega = \frac{1}{\sqrt{LC}} \quad (LC\ \text{system}). \qquad (4)$$

This result is indeed correct, as we show in the next section.

35-4 *LC* Oscillations: Quantitative

Here we want to show explicitly that Eq. 4 for the angular frequency of the *LC* oscillations is correct. At the same time, we want to examine even more closely the analogy between the *LC* oscillations and the block-spring oscillations. We start by extending somewhat our earlier treatment of the mechanical block-spring oscillator.

The Block-Spring Oscillator. We analyzed block-spring oscillations in Chapter 14 in terms of energy transfers and did not—at that early stage—derive the fundamental differential equation that governs that oscillatory motion. We do so now.

We can write, for the mechanical energy of a block-spring oscillator at any instant,*

$$U = U_m + U_k = \tfrac{1}{2}mv^2 + \tfrac{1}{2}kx^2, \qquad (5)$$

where U_m and U_k are, respectively, the kinetic energy of the moving block and the potential energy of the stretched or compressed spring. If there is no friction—which we assume—the total energy U remains constant with time, even through v and x vary. In more formal language, $dU/dt = 0$. This leads to

$$\frac{dU}{dt} = \frac{d}{dt}\left(\tfrac{1}{2}mv^2 + \tfrac{1}{2}kx^2\right) = mv\frac{dv}{dt} + kx\frac{dx}{dt} = 0. \qquad (6)$$

But $dx/dt = v$ and $dv/dt = d^2x/dt^2$. With these substitutions, Eq. 6 becomes

$$\boxed{m\frac{d^2x}{dt^2} + kx = 0} \quad \text{(block-spring oscillations)}. \qquad (7)$$

Equation 7 is the fundamental *differential equation* that governs the frictionless block-spring oscillations. It involves the displacement x and its second derivation with respect to time.

The general solution to Eq. 7, that is, the function $x(t)$ that describes the block-spring oscillations, is (as we have seen)

$$\boxed{x = X\cos(\omega t + \phi)} \quad \text{(displacement)}, \qquad (8)$$

in which X is the amplitude of the mechanical oscillations.

* Earlier, we used E to represent the mechanical energy; we make small changes in notation here to conform to the usage of the parallel electrical situations.

The *LC* Oscillator. Now let us analyze the oscillations of a (resistanceless) *LC* circuit, proceeding in exactly the same way as we have just done for the block-spring oscillator. The total energy U present at any instant in an oscillating *LC* circuit is given by

$$U = U_B + U_E = \tfrac{1}{2}Li^2 + \frac{q^2}{2C},$$

which expresses the fact that at any arbitrary time the energy is stored partly in the magnetic field in the inductor and partly in the electric field in the capacitor. Since we have assumed the circuit resistance to be zero, there is no energy transfer to thermal energy and U remains constant with time, even though i and q vary. In more formal language, dU/dt must be zero. This leads to

$$\frac{dU}{dt} = \frac{d}{dt}\left(\tfrac{1}{2}Li^2 + \frac{q^2}{2C}\right) = Li\frac{di}{dt} + \frac{q}{C}\frac{dq}{dt} = 0. \qquad (9)$$

Now, $dq/dt = i$ and $di/dt = d^2q/dt^2$. With these substitutions, Eq. 9 becomes

$$\boxed{L\frac{d^2q}{dt^2} + \frac{1}{C}q = 0} \quad \text{(}LC\text{ oscillations)}. \qquad (10)$$

This is the *differential equation* that describes the oscillations of a (resistanceless) *LC* circuit. A careful comparison of Eq. 10 with Eq. 7 shows that the two equations are exactly of the same mathematical form, differing only in the symbols used.

Since the differential equations are mathematically identical, their solutions must also be mathematically identical. Because q corresponds to x, we can write the general solution of Eq. 10, giving q as a function of time, by analogy to Eq. 8 as

$$\boxed{q = Q\cos(\omega t + \phi)} \quad \text{(charge)}, \qquad (11)$$

where Q is the amplitude of the charge variations and ω is the angular frequency of the electromagnetic oscillations.

We can test whether Eq. 11 is indeed a solution of Eq. 10 by substituting it and its second derivative in that equation. To find the second derivative, we write

$$\frac{dq}{dt} = i = -\omega Q\sin(\omega t + \phi)$$

and

$$\frac{d^2q}{dt^2} = -\omega^2 Q\cos(\omega t + \phi).$$

Substituting q and d^2q/dt^2 into Eq. 10, we have

$$-L\omega^2 Q\cos(\omega t+\phi)+\frac{1}{C}Q\cos(\omega t+\phi)=0.$$

Canceling $Q\cos(\omega t+\phi)$ and rearranging lead to

$$\omega=\frac{1}{\sqrt{LC}}. \qquad (12)$$

Thus, if ω is given the constant value $1/\sqrt{LC}$, Eq. 11 is indeed a solution of Eq. 10. Equation 12 is exactly the same expression for ω given by Eq. 4, which we arrived at by the method of correspondences.

The phase constant ϕ in Eq. 10 is determined by the conditions that prevail at $t=0$. If we put $\phi=0$, for example, Eq. 11 requires that, at $t=0$, $q=Q$ and i $(=dq/dt)=0$; these are the initial conditions represented by Fig. 1*a*.

The stored electric energy in the *LC* circuit, using Eq. 1, is

$$U_E=\frac{q^2}{2C}=\frac{Q^2}{2C}\cos^2(\omega t+\phi) \qquad (13)$$

and the magnetic energy, using Eq. 2, is

$$U_B=\tfrac{1}{2}Li^2=\tfrac{1}{2}L\omega^2Q^2\sin^2(\omega t+\phi).$$

Substituting the expression for ω (see Eq. 4) into this last equation, we have

$$U_B=\frac{Q^2}{2C}\sin^2(\omega t+\phi). \qquad (14)$$

Figure 4 shows plots of $U_E(t)$ and $U_B(t)$ for the case of $\phi=0$. Note that (1) the maximum values of U_E and U_B are the same $(=Q^2/2C)$; (2) at any instant the sum of U_E and U_B is a constant $(=Q^2/2C)$; and (3) when U_E has its maximum value, U_B is zero and conversely. Compare this discussion with that given in Section 14–4 for the energy transfers in a block–spring system.

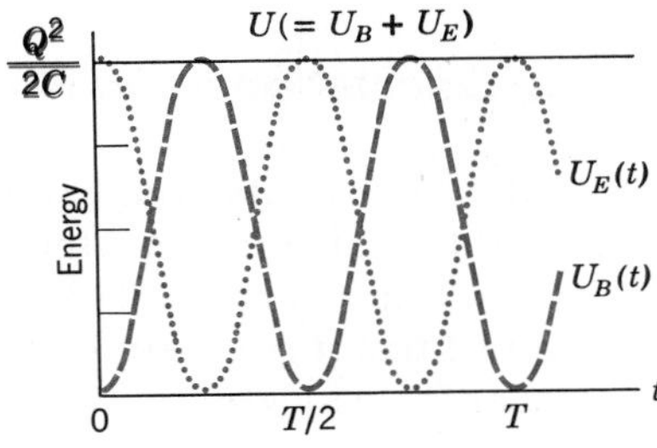

Figure 4 The stored magnetic energy (---) and the stored electric energy (···) in the circuit of Fig. 1 as a function of time. Note that their sum remains constant. T is the period of oscillation.

Sample Problem 3 (a) In an oscillating *LC* circuit, what value of charge, expressed in terms of the maximum charge Q, is present on the capacitor when the energy is shared equally between the electric and the magnetic field? Assume that $L=12$ mH and $C=1.7\ \mu$F.

The problem requires that $U_E=\frac{1}{2}U_{E,m}$. The instantaneous and maximum stored energy in the capacitor are, respectively,

$$U_E=\frac{q^2}{2C} \quad\text{and}\quad U_{E,m}=\frac{Q^2}{2C},$$

so that

$$\frac{q^2}{2C}=\frac{1}{2}\frac{Q^2}{2C}$$

or

$$q=\frac{1}{\sqrt{2}}Q=0.707\ Q. \qquad \text{(Answer)}$$

(b) How much time is required for this condition to arise, assuming the capacitor to be fully charged initially?

We write, putting $\phi=0$ in Eq. 11 and using the result just found,

$$\frac{q}{Q}=\frac{Q\cos\omega t}{Q}=\frac{1}{\sqrt{2}} \quad\text{or}\quad \omega t=45^\circ=\frac{\pi}{4}\text{ rad},$$

which corresponds to $\frac{1}{8}$ of one period of oscillation. The angular frequency ω is found from Eq. 4, or

$$\omega=\frac{1}{\sqrt{LC}}=\frac{1}{\sqrt{(12\times10^{-3}\text{ H})(1.7\times10^{-6}\text{ F})}}$$
$$=7.00\times10^3\text{ rad/s}.$$

The time t is then

$$t=\frac{\pi/4\text{ rad}}{\omega}=\frac{\pi}{(4)(7.00\times10^3\text{ rad/s})}$$
$$=1.12\times10^{-4}\text{ s}\approx110\ \mu\text{s}. \qquad \text{(Answer)}$$

Convince yourself that the frequency ν and the period T of the oscillation are ~1.1 kHz and ~900 μs, respectively.

35-5 Damped *LC* Oscillations

If resistance R is present in an *LC* circuit, the total electromagetic energy U is no longer constant but decreases with time as it is transformed steadily to thermal energy in the resistor. As we shall see, the analogy with the damped block–spring oscillator of Section 14–8 is exact. As before, we have

$$U = U_B + U_E = \tfrac{1}{2}Li^2 + \frac{q^2}{2C}. \qquad (15)$$

U is no longer constant but rather

$$\frac{dU}{dt} = -i^2R, \qquad (16)$$

the minus sign signifying that the stored energy U decreases with time, being converted to thermal energy at the rate i^2R. Differentiating Eq. 15 and combining the result with Eq. 16, we have

$$Li\frac{di}{dt} + \frac{q}{C}\frac{dq}{dt} = -i^2R.$$

Substituting dq/dt for i and d^2q/dt^2 for di/dt and dividing by i, we obtain

$$L\frac{d^2q}{dt^2} + R\frac{dq}{dt} + \frac{1}{C}q = 0 \qquad (LCR \text{ circuit}), \quad (17)$$

which is the differential equation that describes the damped LC oscillations. If we put $R = 0$, this equation reduces—as it must—to Eq. 10, which is the differential equation that described the undamped LC oscillations.

We state without proof that the general solution of Eq. 17 can be written in the form

$$q = Qe^{-Rt/2L}\cos(\omega't + \phi), \qquad (18)$$

in which

$$\omega' = \sqrt{\omega^2 - (R/2L)^2}.$$

Equation 18 is the exact equivalent of Eq. 31 of Chapter 14, the equation for the displacement as a function of time in damped simple harmonic motion.

Equation 18, which can be described as a cosine function with an amplitude that decreases exponentially with time, is the equation of the decay curve of Fig. 3. The frequency ω' is strictly less than the frequency ω $(= 1/\sqrt{LC})$ of the undamped oscillations but here we shall consider only those cases in which the resistance R is so small that we can put $\omega' = \omega$ with negligible error. Recall that we made a similar assumption of small damping of block–spring oscillations in Section 14–8.

Sample Problem 4 A circuit has $L = 12$ mH, $C = 1.6\ \mu$F, and $R = 1.5\ \Omega$. (a) After what time t will the amplitude of the charge oscillations drop to one-half of its initial value?

This will occur when the amplitude factor $e^{-Rt/2L}$ in Eq. 18 has the value $\frac{1}{2}$, or

$$e^{-Rt/2L} = \tfrac{1}{2}.$$

Taking the logarithm of each side gives us

$$(-Rt/2L)(\ln e) = \ln 1 - \ln 2.$$

But $\ln e = 1$ and $\ln 1 = 0$, so that

$$t = \frac{2L}{R}\ln 2 = \frac{(2)(12 \times 10^{-3}\ \text{H})(\ln 2)}{1.5\ \Omega}$$
$$= 0.0111\ \text{s} \approx 11\ \text{ms}. \qquad \text{(Answer)}$$

(b) To how many periods of oscillation does this correspond?

The number of oscillations is the elapsed time divided by the period, which is related to the angular frequency ω by $T = 2\pi/\omega$. The angular frequency is

$$\omega = \frac{1}{\sqrt{LC}} = \frac{1}{\sqrt{(12 \times 10^{-3}\ \text{H})(1.6 \times 10^{-6}\ \text{F})}}$$
$$= 7216\ \text{rad/s} \approx 7200\ \text{rad/s}.$$

The period of oscillation is then

$$T = \frac{2\pi}{\omega} = \frac{2\pi}{7216\ \text{rad/s}} = 8.707 \times 10^{-4}\ \text{s}$$

The elapsed time, expressed in terms of the period of oscillation, is then

$$\frac{t}{T} = \frac{0.0111\ \text{s}}{8.707 \times 10^{-4}\ \text{s}} \approx 13 \qquad \text{(Answer)}$$

Thus, the amplitude drops to one-half after about 13 cycles of oscillation. By comparison, the damping in this example is less severe than that shown in Fig. 3, where the amplitude drops to one-half in about three cycles.

35-6 Forced Oscillations and Resonance

We have discussed the free oscillations of an LC circuit and also the damped oscillations, in which a resistive element R is present. If the damping is small enough—and we have assumed that it is—both kinds of oscillation have an angular frequency given by

$$\omega_0 = \frac{1}{\sqrt{LC}} \qquad \text{(natural frequency)}, \quad (19)$$

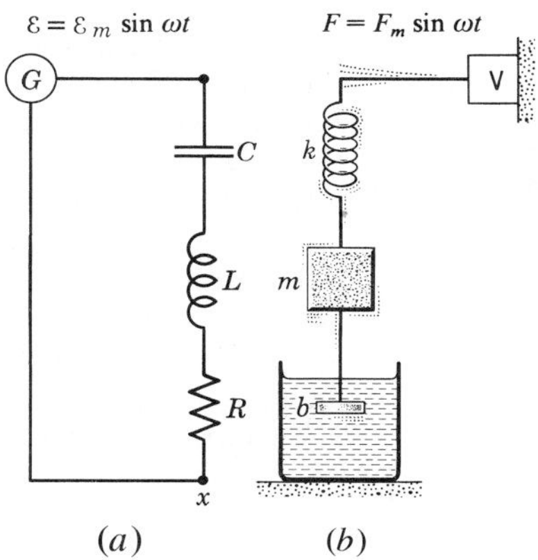

Figure 5 Forced oscillations at angular frequency ω in (*a*) an electromagnetic oscillating system and (*b*) a corresponding mechanical oscillating systems. Corresponding elements in the two systems are drawn opposite each other.

which we call the *natural frequency* of the oscillating system. Although earlier we identified this simply as ω, we relabel it here as ω_0.

Suppose now that we impress a time-varying emf given by

$$\mathcal{E} = \mathcal{E}_m \sin \omega t \tag{20}$$

on the system by an external generator. Here ω, which can be varied at will, is the frequency imposed by this external source. We describe such oscillations as *forced.* When the emf described by Eq. 20 is first applied, there will be certain time-varying transient currents in the circuit. Our interest, however, is in the sinusoidal currents that exist in the circuit after these start-up transients have died away.

> *Whatever the (constant) natural frequency ω_0 may be, the oscillations of charge, current, or potential difference in the circuit must occur at the external or driving frequency ω.*

Figure 5 compares the electromagnetic oscillating system with a corresponding mechanical system. A vibrator V, which imposes an external alternating force, corresponds to generator G, which imposes an external alternating emf. Other quantities "correspond" as before. Note incidentally that, although we use the same pictorial symbol for a spring and an inductor, they are not corresponding elements. In the appropriate differential equations, a spring is described mathematically like a capacitor and an inductor like a massive body.

We shall solve the problem of the forced oscillations of an *LCR* circuit exactly in Chapter 36, which deals with alternating currents. Here we content ourselves with giving the solution and examining some graphical results.

The electrical variable of most interest in the circuit of Fig. 5*a* is the current and we assert that this may be written

$$i = I \sin(\omega t - \phi). \tag{21}$$

The current amplitude I in Eq. 21 is a measure of the response of the circuit of Fig. 5*a* to the impressed emf. It is reasonable to suppose, from experience (in pushing swings, for example), that I will be larger the closer the driving frequency ω is to the natural frequency ω_0 of the system. In other words, we expect that a plot of I versus ω will exhibit a maximum when

$$\boxed{\omega = \omega_0} \quad \text{(resonance)}, \tag{22}$$

which we call the *resonance condition.*

Figure 6 shows three plots of I as a function of the ratio ω/ω_0, each plot corresponding to a different value of the resistance R. We see that each of these resonance peaks does indeed have a maximum value when the resonance condition of Eq. 22 is satisfied. Note that as R is decreased, the resonance peak becomes sharper, as shown by the three horizontal arrows drawn at the half-maximum level of each curve.

Figure 6 suggests the common experience of tuning a radio set. In turning the tuning knob, we are adjusting the natural frequency ω_0 of an internal *LC* circuit to match the frequency ω of the signal transmitted by the antenna of the broadcasting station; we are looking for resonance. In a metropolitan area, where there are many signals whose frequencies are often close together, sharpness of tuning becomes important.

Figure 6 finds a counterpart in Fig. 20 of Chapter 14, which shows resonance peaks for the forced oscillations of a mechanical oscillator such as that of Fig. 5*b*. In this case also, the maximum response occurs when $\omega = \omega_0$ and the resonance peaks become sharper as the damping factor (the coefficient b) is reduced. A careful observer will note that the curves of Fig. 6 and of Fig. 20 of Chapter 14 do not exactly "correspond." The former is a plot of current amplitude, the mechanical quantity that corresponds to the current being the velocity. The latter figure, however, is not a plot of velocity amplitude

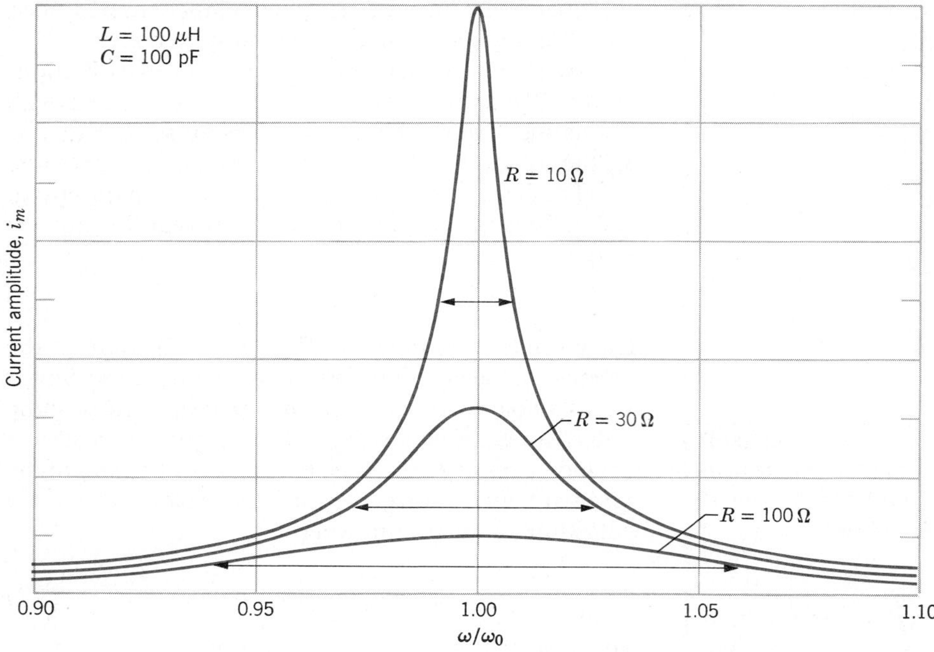

Figure 6 Resonance curves for forced oscillations in the circuit of Fig. 5*a*. The values of L and C are the same for all three curves but the values of R are different, as marked on the curves. The horizontal arrows on each curve measure its width at the half-maximum level, a measure of the sharpness of the resonance. Note that the current amplitude is a maximum in each case at resonance ($\omega/\omega_0 = 1$).

versus frequency but of displacement amplitude versus frequency. Nevertheless, both sets of curves illustrate the resonance phenomenon.

35-7 Other Oscillators: A Taste of Electronics (Optional)

We must not leave the impression that all oscillators are based on LC circuits like that of Fig. 1.

Crystal Oscillators. When you see the word quartz on your wristwatch or wall clock it means that the device is based on a *quartz crystal oscillator.* Quartz has the interesting property that, if you cut it into a thin wafer and squeeze it, equal but opposite charges appear on the opposing surfaces of the wafer; conversely, if you apply a potential difference between opposing faces of the wafer, the dimensions of the wafer change.

This property of *piezoelectricity** gives a convenient coupling between the mechanical oscillations of the crystal, which occur at a very sharply defined frequency, and the electrical properties of the circuit of which the crystal is a part. Quartz crystal oscillators are used in situations—such as watches and radio transmitters—where stability of the oscillator frequency is a fundamental concern. Piezoelectric crystals are used in other applications, in which a mechanical movement must be converted into an electrical signal—as in a phonograph pickup arm—or an electrical signal into a mechanical movement—as in a sonar transmitter.

Feedback Oscillators. When you push the buttons on your telephone handset, the tones that you hear are generated by another kind of oscillator, based on resistive and capacitive elements alone, with no inductance involved. Such oscillators are useful in microelectronics because—although it is easy enough to embody a resistor or a capacitor in a microelectronic chip—it is more difficult to embody an inductor.

Figure 7 suggests schematically how such oscillators work. Figure 7*a* shows a simple amplifier, which has the property that it amplifies an input signal v_i by a factor g to generate an output signal v_o or

$$g = \frac{v_o}{v_i} \quad \text{(Fig. 7}a\text{)}.$$

The dimensionless quantity g, called the *voltage gain* of the amplifier, is greater than unity.

Figure 7*b* shows how this basic amplifier can be turned into an oscillator by the mechanism of *positive*

* The word *piezo-* comes from the Greek "to push." Piezoelectricity was discovered in 1880 by the brothers Pierre and Jacques Curie when they were, respectively, 21 and 24 years old.

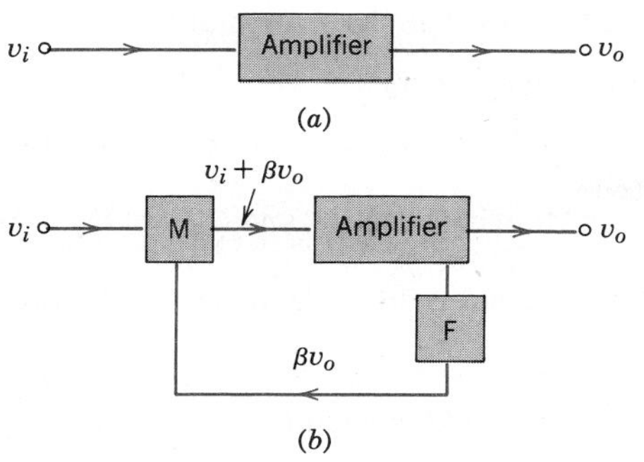

Figure 7 (*a*) A simple amplifier, showing the input and the output signals. (*b*) The amplifier turned into an oscillator by positive feedback. The feedback circuit F sends a fraction β of the output signal back to the input, where it is added to the input signal in mixer M.

feedback. A fraction βv_o of the output signal (where $\beta < 1$) is sampled by the feedback circuitry (box F) and fed back to the input, where it is mixed (in box M) with the original input signal in such a way as to reinforce that signal. If it is desired that the circuit oscillate at a specific frequency, the feedback circuit must be so designed that the fed-back signal joins the input signal after a delay corresponding to one period of oscillation at that frequency. In that way, the input signal will be reinforced in the proper phase and energy will be added to the input circuit to compensate for resistive losses.

In Fig. 7*b*, we can write the gain of the *amplifier alone* as

$$g = \frac{v_o}{v_i + \beta v_o} \quad \text{(amplifier alone).} \tag{23}$$

However, the effective gain g' of the *overall circuit* of Fig. 7*b* is

$$g' = \frac{v_o}{v_i} \quad \text{(overall circuit).} \tag{24}$$

Combining Eqs. 23 and 24 leads to

$$\boxed{g' = \frac{g}{1 - \beta g},} \tag{25}$$

which shows that g' can be much greater than g. In fact, g' can, in principle, be made infinitely great if the feedback loop is designed so that $\beta g = 1$. In practice,* this means that the oscillator will not require an input signal to cause it to oscillate; the slightest random electrical disturbance will start it off. It will then oscillate at the resonant frequency determined by the design of the frequency-dependent feedback circuit.

* In practice, an infinite gain cannot be achieved because, in all amplifiers, the intrinsic gain g decreases as the magnitude of the input signal increases.

REVIEW AND SUMMARY

LC Energy Transfers

In a (resistanceless) oscillating *LC* circuit, energy may be stored in the capacitor or in the inductor, according to

$$U_E = \frac{Q^2}{2C} \quad \text{and} \quad U_B = \frac{1}{2}Li^2. \tag{1,2}$$

The total energy $E\,(= U_E + U_B)$ remains constant as energy oscillates back and forth between these elements, as in Fig. 1; see Sample Problems 1 and 2.

A Mechanical Analogy

Energy transfers in an oscillating circuit are analogous to those in an oscillating mass-spring system. The correspondences outlined in Table 1 allow us to predict that the angular frequency of the *LC* oscillations will be given by

$$\omega = \frac{1}{\sqrt{LC}} \quad (LC\ \text{circuit}). \tag{4}$$

The conservation of energy principle leads to

A Quantitative Solution

$$L\frac{d^2q}{dt^2} + \frac{q}{C} = 0 \quad (LC\ \text{oscillations}) \tag{10}$$

as the differential equation of free resistanceless LC oscillations. The solution is

$$q = Q\cos(\omega t + \phi) \quad \text{(charge)} \qquad [17]$$

with ω given by Eq. 4. The charge amplitude Q and the phase constant ϕ are fixed by the initial conditions of the system. See Sample Problem 3.

$U_E(t)$ and $U_B(t)$

Given $q(t)$, and its derivative $i(t)$, we can find the energies $U_E(t)$ and $U_B(t)$ of Fig. 1. See Eqs. 13 and 14 and Fig. 4.

Damped Oscillations

For damped LC oscillations a dissipative element R is present and the conservation of energy principle shows the differential equation to be

$$L\frac{d^2q}{dt^2} + R\frac{dq}{dt} + \frac{1}{C}q = 0 \quad (LCR \text{ circuit}). \qquad [17]$$

Its solution is

$$q = Qe^{-Rt/2L}\cos(\omega' t + \phi). \qquad [18]$$

We consider only low damping situations, in which ω' may be set equal to the ω of Eq. 4. See Sample Problem 4.

Forced Oscillations

An LCR circuit such as that of Fig. 5*a* may be set into *forced oscillation* at angular frequency ω by an impressed emf such as

$$\mathcal{E} = \mathcal{E}_m \sin \omega t. \qquad [20]$$

Here we relabel the (constant) *natural frequency* $(= 1/\sqrt{LC})$ of the resonant system as ω_0, reserving ω for the (variable) frequency of the impressed emf. The current in the circuit is

$$i = I\sin(\omega t - \phi). \qquad [21]$$

Resonance

The current amplitude I has a maximum value when $\omega = \omega_0$, a condition called *resonance*. Figure 6 shows that resonance peaks becomes sharper as the circuit resistance R is decreased. Review the analogy to forced mechanical oscillations (see Fig. 5*b*).

Quartz Oscillators

Quartz crystal oscillators depend on the *piezoelectric* property of quartz; stresses resulting from mechanical oscillation cause alternating electric potentials that may then be used as part of electrical timing circuits of various kinds.

Feedback Oscillators

Feedback oscillators are amplifier-feedback combinations with overall gains near +1. The circuit oscillates at a resonant frequency determined by the properties of the feedback circuit.

QUESTIONS

1. Why doesn't the LC circuit of Fig. 1 simply stop oscillating when the capacitor has been completely discharged?

2. How might you start an LC circuit into oscillation with its initial condition being represented by Fig. 1*c*? Devise a switching scheme to bring this about.

3. The lower curve (*b*) in Figure 2 is proportional to the derivative of the upper curve (*a*). Explain why.

4. In an oscillating LC circuit, assumed resistanceless, what determines (*a*) the frequency and (*b*) the amplitude of the oscillations?

5. In connection with Figs. 1*c* and *g*, explain how there can be a current in the inductor even though there is no charge on the capacitor.

6. In Fig. 1, what changes are required if the oscillations are to proceed counterclockwise around the figure?

7. In Fig. 1, what phase constants ϕ in Eq. 11 would permit the eight circuit situations shown to serve in turn as initial conditions?

8. What constructional difficulties would you encounter if you tried to build an LC circuit of the type shown in Fig. 1 to oscillate (*a*) at 0.01 Hz, or (*b*) at 10^{10} Hz?

9. Two inductors L_1 and L_2 and two capacitors C_1 and C_2 can be connected in series according to the arrangement in Fig. 8*a* or Fig. 8*b*. Are the frequencies of the two oscillating circuits equal? Consider the cases (*i*) $C_1 = C_2$, $L_1 = L_2$; (*ii*) $C_1 \neq C_2$, $L_1 \neq L_2$.

10. In the mechanical analogy to the oscillating LC circuit, what mechanical quantity corresponds to potential difference?

11. Discuss the assertion that the resonance curves of Fig. 6 and Chapter 14, Fig. 20 cannot truly be compared because the

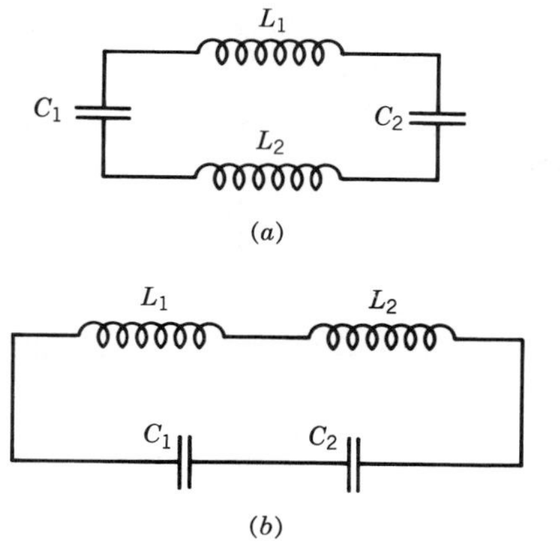

Figure 8 Question 9.

former is a plot of the current amplitude (I) and the latter of the displacement amplitude (x_m). Are these "corresponding" quantities? Does it make any difference if our purpose is only to exhibit the resonance phenomenon?

12. In comparing an electromagnetic oscillating system to a mechanical oscillating system, to what mechanical properties are the following electromagnetic properties analogous: capacitance, resistance, charge, electric field energy, magnetic field energy, inductance, current?

13. Two identical springs are joined and connected to an object with mass m, the arrangement being free to oscillate on a horizontal frictionless surface as in Fig. 9. Sketch the electromagnetic analog of this mechanical oscillating system.

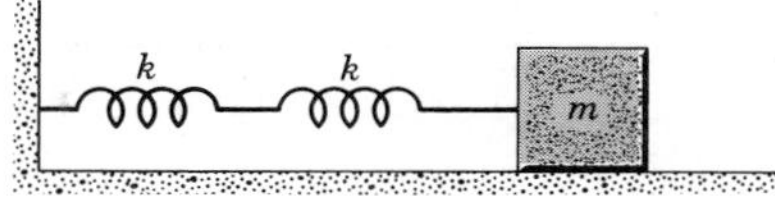

Figure 9 Question 13.

14. Explain why it is not possible to have (*a*) a real *LC* circuit without resistance, (*b*) a real inductor without inherent capacitance, or (*c*) a real capacitor without inherent inductance. Discuss the practical validity of the *LC* circuit of Fig. 1, in which each of the above realities is ignored.

15. All practical *LC* circuits have to contain some resistance. However, one can buy a packaged audio oscillator in which the output maintains a constant amplitude indefinitely and does not decay, as it does in Fig. 3. How can this happen?

16. What would a resonance curve for $R = 0$ look like if plotted in Fig. 6?

17. What is the difference between free, damped, and forced oscillating circuits?

EXERCISES AND PROBLEMS

Section 35-2 *LC* Oscillations—Qualitative

1E. What is the capacitance of an *LC* circuit if the maximum charge on the capacitor is 1.6 μC and the total energy is 140 μJ?

2E. 1.5-mH inductor in an *LC* circuit stores a maximum energy of 10 μJ. What is the peak current?

3E. In an oscillating *LC* circuit $L = 1.1$ mH and $C = 4.0$ μF. The maximum charge on C is 3.0 μC. Find the maximum current.

4E. An *LC* circuit consists of a 75-mH inductor and a 3.6-μF capacitor. If the maximum charge on the capacitor is 2.9 μC, (*a*) what is the total energy in the circuit and (*b*) what is the maximum current?

5E. For a certain *LC* circuit the total energy is converted from electrical energy in the capacitor to magnetic energy in the inductor in 1.5 μs. (*a*) What is the period of oscillation? (*b*) What is the frequency of oscillation? (*c*) How long after the magnetic energy is a maximum will it be a maximum again?

Section 35-3 Developing the Mechanical Analogy

6E. A 0.50-kg body oscillates on a spring that, when extended 2.0 mm from equilibrium, has a restoring force of 8.0 N. (*a*) Calculate the angular frequency of oscillation. (*b*) What is its period of oscillation? (*c*) What is the capacitance of the analogous *LC* system if L is chosen to be 5.0 H?

7P. The energy in an *LC* circuit containing a 1.25-H inductor is 5.70 μJ. The maximum charge on the capacitor is 175 μC. Find (*a*) the mass, (*b*) the spring constant, (*c*) the maximum displacement, and (*d*) the maximum speed for the analogous mechanical system.

Section 35-4 *LC* Oscillations—Quantitative

8E. *LC* oscillators have been used in circuits connected to loudspeakers to create some of the sounds of "electronic music." What inductance must be used with a 6.7-μF capacitor to produce a frequency of 10 kHz, about the middle of the audible range of frequencies?

9E. What capacitance would you connect across a 1.3-mH inductor to make the resulting oscillator resonate at 3.5 kHz?

10E. In an *LC* circuit with $L = 50$ mH and $C = 4.0$ μF, the current is initially a maximum. How long will it take before the capacitor is fully charged for the first time?

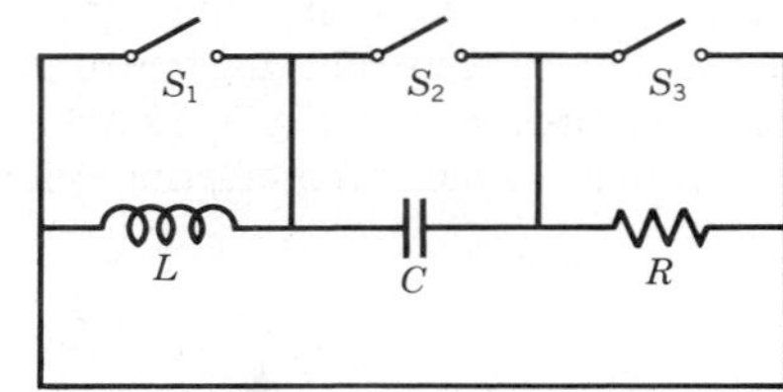

Figure 10 Exercise 11.

11E. Consider the circuit shown in Fig. 10. With switch S_1 closed and the other two switches open, the circuit has a time constant τ_C (Section 8, Chapter 29). With switch S_2 closed and the other two switches open, the circuit has a time constant τ_L (Section 4, Chapter 33). With switch S_3 closed and the other two switches open, the circuit oscillates with a period T. Show that $T = 2\pi\sqrt{\tau_C \tau_L}$.

12E. Derive the differential equation for an LC circuit (Eq. 10) using the loop theorem.

13E. A single loop consists of several inductors (L_1, L_2, . . .), several capacitors (C_1, C_2, . . .), and several resistors (R_1, R_2, . . .) connected in series as shown, for example, in Fig. 11*a*. Show that, regardless of the sequence of these circuit elements in the loop, the behavior of this circuit is identical to that of a simple damped LC circuit shown in Fig. 11*b*. (*Hint:* Consider the loop equation.)

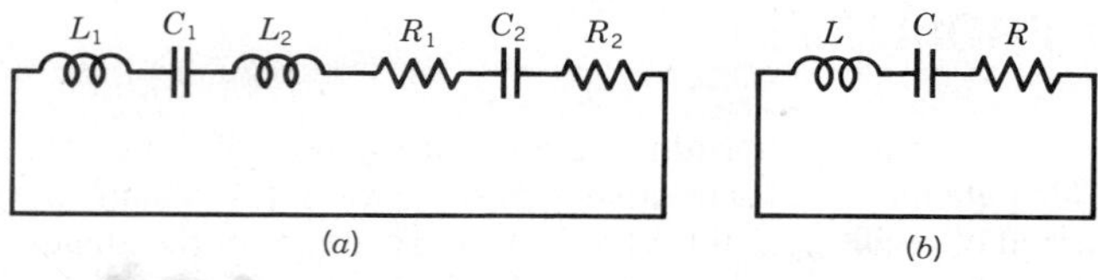

Figure 11 Exercise 13.

14P. An oscillating LC circuit consisting of a 1.0-nF capacitor and a 3.0-mH coil has a peak voltage of 3.0 V. (*a*) What is the maximum charge on the capacitor? (*b*) What is the peak current through the circuit? (*c*) What is the maximum energy stored in the magnetic field of the coil?

15P. An LC circuit has an inductance of 3.0 mH and a capacitance of 10 μF. (*a*) Calculate the angular frequency and (*b*) the period of the oscillation. (*c*) At time $t = 0$ the capacitor is charged to 200 μC, and the current is zero. Sketch roughly the charge on the capacitor as a function of time.

16P. In an LC circuit in which $C = 4.0\ \mu$F the maximum potential difference across the capacitor during the oscillations is 1.5 V and the maximum current through the inductor is 50 mA. (*a*) What is the inductance L? (*b*) What is the frequency of the oscillations? (*c*) How much time does it take for the charge on the capacitor to rise from zero to its maximum value?

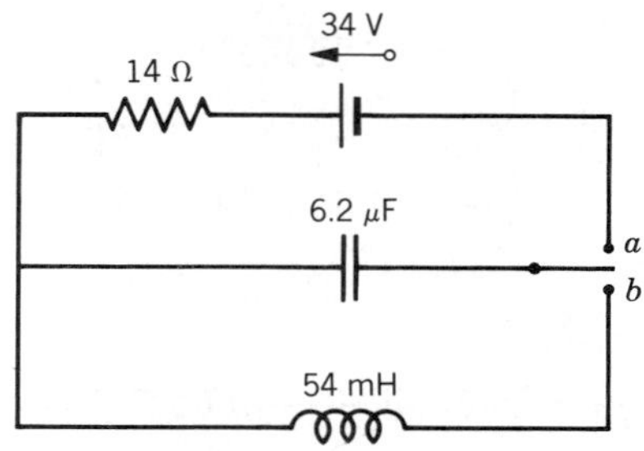

Figure 12 Problem 17.

17P. In the circuit shown in Fig. 12 the switch has been in position *a* for a long time. It is now thrown to *b*. (*a*) Calculate the frequency of the resulting oscillating current. (*b*) What will be the amplitude of the current oscillations?

18P. You are given a 10-mH inductor and two capacitors, of 5.0-μF and 2.0-μF capacitance. List the resonant frequencies that can be generated by connecting these elements in various combinations.

19P. An LC circuit oscillates at 10.4 kHz. (*a*) If the capacitance is 340 μF, what is the inductance? (*b*) If the maximum current is 7.20 mA, what is the total energy in the circuit? (*c*) Calculate the maximum charge on the capacitor.

20P. (*a*) In an oscillating LC circuit, in terms of the maximum charge on the capacitor, what value of charge is present when the energy in the electric field is one-half that in the magnetic field? (*b*) What fraction of a period must elapse following the time the capacitor is fully charged for this condition to arise?

21P. At some instant in an oscillating LC circuit, three-fourths of the total energy is stored in the magnetic field of the inductor. (*a*) In terms of the maximum charge on the capacitor, what is the charge on the capacitor at this instant? (*b*) In terms of the maximum current in the inductor, what is the current in the inductor at this instant?

22P. An inductor is connected across a capacitor whose capacitance can be varied by turning a knob. We wish to make the frequency of the LC oscillations vary linearly with the angle of rotation of the knob, going from 2×10^5 Hz to 4×10^5 Hz as the knob turns through 180°. If $L = 1.0$ mH, plot C as a function of angle for the 180° rotation.

23P. A variable capacitor with a range from 10 to 365 pF is used with a coil to form a variable-frequency LC circuit to tune the input to a radio. (*a*) What ratio of maximum to minimum frequencies may be tuned with such a capacitor? (*b*) If this capacitor is to tune from 0.54 to 1.60 MHz, the ratio computed in (*a*) is too large. By adding a capacitor in parallel to the variable capacitor this range may be adjusted. How large should this capacitor be and what inductance should be chosen in order to tune the desired range of frequencies?

24P. In an LC circuit $L = 25$ mH and $C = 7.8$ μF. At time $t = 0$ the current is 9.2 mA, the charge on the capacitor is 3.8 μC, and the capacitor is charging. (*a*) What is the total energy in the circuit? (*b*) What is the maximum charge on the capacitor? (*c*) What is the maximum current? (*d*) If the charge on the capacitor is given by $q = Q\cos(\omega t + \phi)$, what is the phase angle ϕ? (*e*) Suppose the data are the same, except that the capacitor is discharging at $t = 0$. What then is the phase angle ϕ?

25P. In an oscillating LC circuit $L = 3.0$ mH and $C = 2.7$ μF. At $t = 0$ the charge on the capacitor is zero and the current is 2.0 A. (*a*) What is the maximum charge that will appear on the capacitor? (*b*) In terms of the period T of oscillation, how much time will elapse after $t = 0$ until the energy stored in the capacitor will be increasing at its greatest rate? (*c*) What is this greatest rate at which energy flows into the capacitor?

26P. In an LC circuit with $C = 64$ μF the current as a function of time is given by $i = (1.6)\sin(2500t + 0.68)$, where t is in seconds, i in amperes, and the phase angle in radians. (*a*) How soon, after $t = 0$, will the current reach its maximum value? (*b*) Calculate the inductance. (*c*) Find the total energy in the circuit.

27P. The resonant frequency of a series circuit containing inductance L_1 and capacitance C_1 is ω_0. A second series circuit, containing inductance L_2 and capacitance C_2, has the same resonant frequency. In terms of ω_0, what is the resonant frequency of a series circuit containing all four of these elements? Neglect resistance. (*Hint:* Use the formulas for equivalent capacitance and equivalent inductance.)

28P. Three identical inductors L and two identical capacitors C are connected in a two-loop circuit as shown in Fig. 13. (*a*) Suppose the currents are as shown in Fig. 13*a*. What is the current in the middle inductor? Write down the loop equations

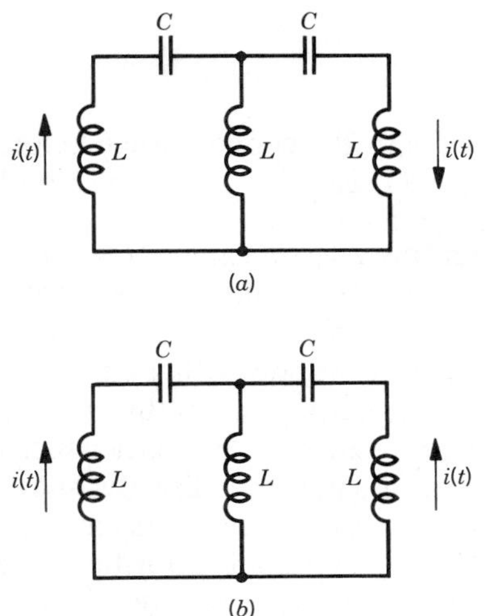

Figure 13 Problem 28.

and show that they are satisfied provided that the current oscillates with angular frequency $\omega = 1/\sqrt{LC}$. (*b*) Now suppose the currents are as shown in Fig. 13*b*. What is the current in the middle inductor? Write down the loop equations and show that they are satisfied provided the current oscillates with angular frequency $\omega = 1/\sqrt{3LC}$. (*c*) In view of the fact that the circuit can oscillate at two different frequencies, show that it is not possible to replace this two-loop circuit by an equivalent single-loop LC circuit.

29P*. In Fig. 14 the 900-μF capacitor is initially charged to 100 V and the 100-μF capacitor is uncharged. Describe in detail how one might charge the 100-μF capacitor to 300 V by manipulating switches S_1 and S_2.

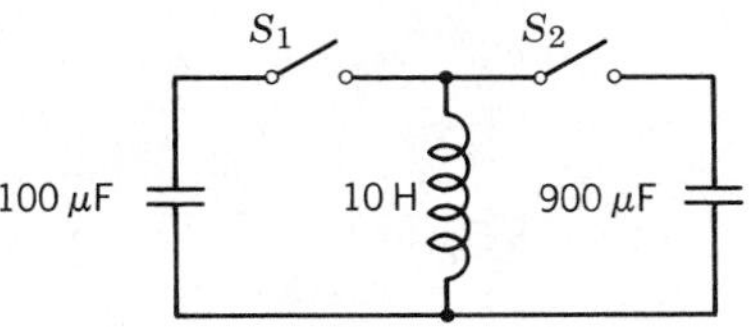

Figure 14 Problem 29.

Section 35-5 Damped *LC* Oscillations

30E. What resistance R should be connected to an inductor $L = 220$ mH and capacitor $C = 12$ μF in series in order that the maximum charge on the capacitor decay to 99% of its initial value in 50 cycles?

31E. Consider a damped LC circuit. (*a*) Show that the damping term $e^{-Rt/2L}$ (which involves L but not C) can be rewritten in the more symmetric manner (involving both L and C) as $e^{-\pi R\sqrt{C/L}(t/T)}$. Here T is the period of oscillation (neglecting resistance). (*b*) Using (*a*), show that the SI unit of $\sqrt{L/C}$ is "ohm." (*c*) Using (*a*), show that the condition that the fractional energy loss per cycle be small is $R \ll \sqrt{L/C}$.

32P. In a damped LC circuit, find the time required for the maximum energy present in the capacitor during one oscillation to fall to one-half of its initial value. Assume $q = Q$ at $t = 0$.

33P. A single loop circuit consists of a 7.2-Ω resistor, a 12-H inductor, and a 3.2-μF capacitor. Initially the capacitor has a charge of 6.2 μC and the current is zero. Calculate the charge on the capacitor N complete cycles later for $N = 5, 10$, and 100.

34P. (*a*) By direct substitution of Eq. 18 into Eq. 17, show that $\omega' = \sqrt{(1/LC) - (R/2L)^2}$. (*b*) By what fraction does the frequency of oscillation shift when the resistance is increased from 0 to 100 Ω in a circuit with $L = 4.4$ H and $C = 7.3$ μF?

35P*. In a damped LC circuit show that the fraction of the energy lost per cycle of oscillation, $\Delta U/U$, is given to a close approximation by $2\pi R/\omega L$. The quantity $\omega L/R$ is often called the Q of the circuit (for "quality"). A "high-Q" circuit has low resistance and a low fractional energy loss per cycle ($= 2\pi/Q$).

CHAPTER 36

ALTERNATING CURRENTS

It is one thing to invent the electric light bulb and quite another to devise an electric power distribution system so that the benefits of electric light can be enjoyed on a city-wide basis. Thomas Edison, who did the former, did not succeed so well in the latter because of his commitment to direct-current distribution systems. Charles Proteus Steinmetz, a gifted engineer and mathematician, favored alternating current systems in the fierce debate on this subject that raged at the end of the last century. His side won.

36-1 Why Alternating Current?

Most homes and offices in this country are wired for *alternating current,* that is, for currents whose value varies with time in a sinusoidal fashion, changing direction (most commonly) 120 times per second. At first sight this may seem a strange arrangement. We have seen that the drift speed of the conduction electrons in household wiring may typically be 4×10^{-5} m/s.* If we now reverse their direction every $\frac{1}{120}$ s, such electrons could move only about 3×10^{-7} m in one half cycle. At this rate, a typical electron could drift past no more than about ten atoms in the copper lattice before it was required to reverse itself. How, you may wonder, can the electron ever get anywhere?

Although this question may be worrisome, it is a needless concern. The conduction electrons do not have to "get anywhere." When we say that the current in a wire is one ampere we mean that charge carriers pass through any plane cutting across that wire at the rate of one coulomb per second. The speed at which the carriers cross that plane does not enter directly; one ampere may correspond to a lot of charge carriers moving very slowly or to a few moving very rapidly. Furthermore, the signal to the electrons to reverse directions—which originates in the alternating emf provided by the generator—is

* See Sample Problem 3, Chapter 28.

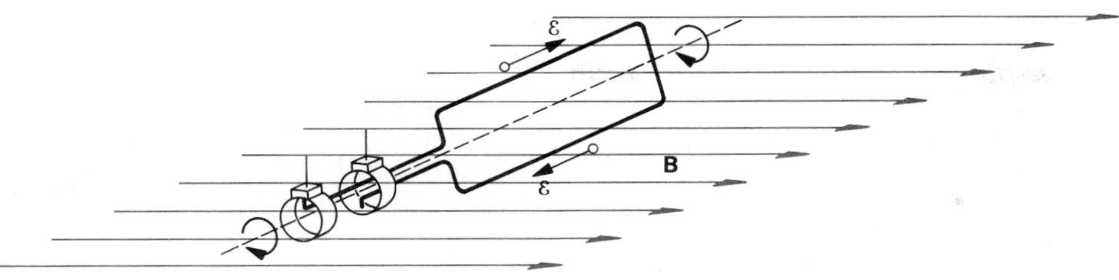

Figure 1 The basic principle of an alternating-current generator is a conducting loop rotated in an external magnetic field. In practice, the alternating emf induced in the loop of many turns of wire is made accessible by means of slip rings attached to the rotating shaft and each connected to one end of the wire.

propagated along the conductor at a speed close to that of light. All electrons, no matter where they are located, get their reversal instructions at about the same instant. Finally, we note that many devices, such as electric light bulbs or toasters, do not care in what direction the electrons are moving as long as they are indeed moving and thus delivering energy to the conductor in thermal form.

The basic advantage of alternating currents is this: *As the current alternates, so does the magnetic field that surrounds the conductor.* This makes possible the use of Faraday's law of induction which, among other things, means that we can step up or step down the magnitude of an alternating potential difference at will, using a transformer, as we shall show later in this chapter. Finally, alternating currents are more readily adaptable to rotating machinery such as generators or motors than are direct currents. If, for example, we rotate a coil in an external magnetic field, as in Fig. 1, the emf induced in the coil is an alternating emf. To extract an alternating potential difference from the coil and then to transform it into a potential difference of constant magnitude and polarity to supply a direct current power distribution system is a challenging engineering problem.

Alternating emfs and the alternating currents generated by them are central—not only to power distribution systems as we have discussed above—but also to radio, television, satellite communications systems, computer systems, and much else that helps to fashion our modern lifestyle.

36-2 Our Plan for this Chapter

In Section 35-6 we saw that if an alternating emf given by

$$\mathcal{E} = \mathcal{E}_m \sin \omega t \qquad (1)$$

is applied to a circuit such as that of Fig. 2, an alternating current given by

$$i = I \sin(\omega t - \phi) \qquad (2)$$

is established in the circuit.* Just as in direct current circuits, the alternating current i in the circuit of Fig. 2 has the same value at any given instant in all parts of the (single-loop) circuit. Furthermore, the angular frequency ω of the current in Eq. 2 is necessarily the same as the angular frequency of the generator that appears in Eq. 1.

The basic characteristics of the alternating emf supplied by the generator are its amplitude $\mathcal{E}_m$ and its angular frequency ω. The basic characteristics of the circuit of Fig. 2 are the resistance R, the capacitance C, and the inductance L. The basic characteristics of the alternating current given by Eq. 2 are its amplitude I and its phase constant ϕ. Our aim in this Chapter can be expressed as follows:

Given	*Find*
Generator: $\mathcal{E}_m$ and ω	I and ϕ
Circuit: R, C, and L	

Rather than attempt to find I and ϕ by solving the differential equation that applies to the circuit of Fig. 2 we will use a geometrical method, introducing the concept of *phasors*.

* Throughout this chapter, small letters, such as i, will represent instantaneous, time-varying quantities and capital letters, such as I, will represent the corresponding amplitudes.

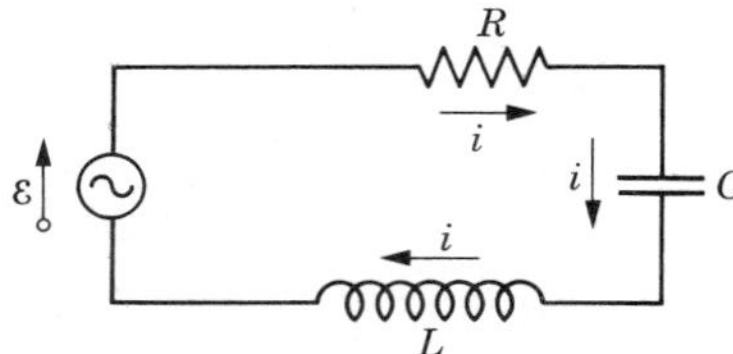

Figure 2 A single-loop circuit containing a resistor, a capacitor, and an inductor. A generator sets up an alternating emf that establishes an alternating current.

36-3 Three Simple Circuits

Let us first simplify the problem suggested by Fig. 2 by considering three simpler circuits, each containing the alternating current generator and only one other element, R, C, or L. We start with R.

A Resistive Circuit. Figure 3a shows a circuit containing a resistive element only, acted on by the alternating emf of Eq. 1. By the loop theorem the alternating potential difference across the resistor is

$$v_R = V_R \sin \omega t. \tag{3}$$

From the definition of resistance we can also write

$$i_R = \frac{v_R}{R} = \frac{V_R}{R} \sin \omega t = I_R \sin \omega t. \tag{4}$$

By comparison with Eq. 2 we see that, in this case of a purely resistive load, the phase constant $\phi = 0°$. From Eq. 4 we also see that the voltage amplitude and the current amplitude are related by

$$\boxed{V_R = I_R R} \quad \text{(resistor)}. \tag{5}$$

Although we developed this relation for the circuit of Fig. 3a, it applies to an individual resistor in any alternating current circuit whatever, no matter how complex.

Comparison of Eqs. 3 and 4 shows that the time-varying quantities v_R and i_R are in phase, which means that their corresponding maxima occur at the same time. Figure 3b, a plot of $v_R(t)$ and $i_R(t)$, illustrates this.

Figure 3c shows a useful geometrical way of looking at the same situation, the method of *phasors*. The two open arrows—the phasors—rotate counterclockwise about the origin with an angular frequency ω. The *length* of a phasor is proportional to the *amplitude* of the alternating quantity involved, that is, to V_R and to I_R. The *projection* of a phasor on the *vertical* axis is proportional to the *instantaneous value* of this alternating quantity,

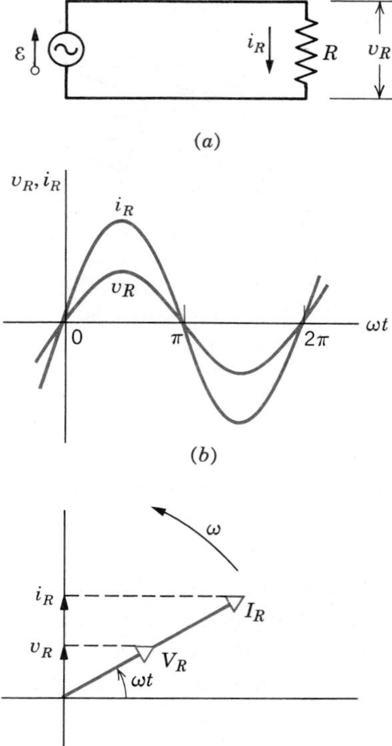

Figure 3 (a) A resistor is connected across an alternating-current generator. (b) The current and the potential difference across the resistor are in phase. (c) A phasor diagram shows the same thing.

that is, to v_R and to i_R. That v_R and i_R are in phase follows from the fact that their phasors lie along the same line in Fig. 3c. Follow the rotation of the phasors in this figure and convince yourself that it completely and correctly describes Eqs. 3 and 4.

A Capacitive Circuit. Figure 4a shows a circuit containing a capacitor only, acted on by the alternating emf of Eq. 1. From the loop theorem the potential difference across the capacitor is

$$v_C = V_C \sin \omega t. \tag{6}$$

From the definition of capacitance we can also write

$$q_C = C v_C = (C V_C) \sin \omega t. \tag{7}$$

Our concern, however, is with the current rather than with the charge. Thus we differentiate Eq. 7 to find

$$i_C = \frac{dq_C}{dt} = \omega C V_C \cos \omega t. \tag{8}$$

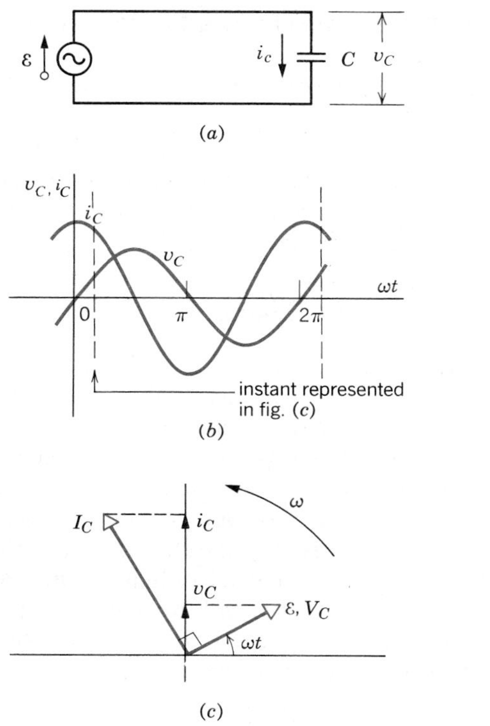

Figure 4 (*a*) A capacitor is connected across an alternating-current generator. (*b*) The potential difference across the capacitor lags the current by 90°. (*c*) A phasor diagram shows the same thing.

We now recast Eq. 8 in two ways. First, for reasons of symmetry of notation, we introduce the quantity X_C, called the *capacitive reactance* of the capacitor at the frequency in question. It is defined from

$$X_C = \frac{1}{\omega C} \quad \text{(capacitive reactance)}, \tag{9}$$

and we see that its value depends not only on the value of the capacitance but also on the frequency at which the capacitor is used. We know from the definition of the capacitive time constant ($\tau = RC$) that the SI unit for C can be expressed as second/ohm. Applying this knowledge to Eq. 9 shows that the SI unit of X_C is the *ohm*, just as for resistance R.

As our second modification of Eq. 8, we replace $\cos \omega t$ by a phase-shifted sine, namely

$$\cos \omega t = \sin(\omega t + 90°).$$

You can easily verify this identity by expanding the right-hand side above according to the formula for $\sin(\alpha + \beta)$ listed in Appendix G.

With these two modifications, Eq. 8 becomes

$$i_C = \left(\frac{V_C}{X_C}\right) \sin(\omega t + 90°) = I_C \sin(\omega t + 90°). \tag{10}$$

A comparison with Eq. 2 shows that, in this case of a purely capacitive load, the phase constant $\phi = -90°$. We see also from Eq. 10 that the voltage amplitude and the current amplitude are related by

$$V_C = I_C X_C \quad \text{(capacitor)}. \tag{11}$$

Although we developed this relation for the specific circuit of Fig. 4*a*, it holds for an individual capacitor in any alternating current circuit, no matter how complex.

Comparison of Eqs. 6 and 10, or inspection of Fig. 4*b*, shows that the quantities v_C and i_C are 90°, or one-quarter cycle, out of phase. Furthermore, we see that i_C *leads* v_C, which means that, if you monitored the current i_C and the potential difference v_C in the circuit of Fig. 4*a*, you would find that i_C reached its maximum *before* v_C did, by one-quarter cycle.

This relation between i_C and v_C is illustrated with equal clarity from the phasor diagram of Fig. 4*c*. As the phasors representing these two quantities rotate counterclockwise, we see that the phasor labeled I_C does indeed lead that labeled V_C, and by an angle of 90°. That is, the phasor I_C coincides with the vertical axis one-quarter cycle before the phasor V_C does. Be sure to convince yourself that the phasor diagram of Fig. 4*c* is consistent with Eqs. 6 and 10.

An Inductive Circuit. Figure 5*a* shows a circuit containing an inductive element only, acted on by the alternating emf of Eq. 1. From the loop theorem, we can write

$$v_L = V_L \sin \omega t. \tag{12}$$

From the definition of inductance we can also write

$$v_L = L \frac{di_L}{dt}. \tag{13}$$

If we combine these two equations we have

$$\frac{di_L}{dt} = \left(\frac{V_L}{L}\right) \sin \omega t. \tag{14}$$

Our concern, however, is with the current rather than with its time derivative. We find it by integrating Eq. 14,

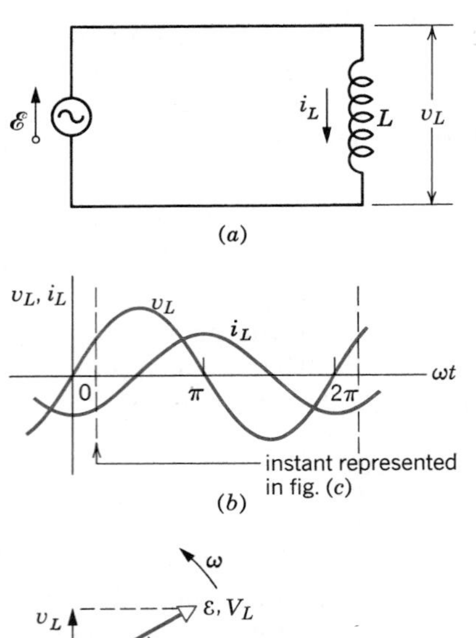

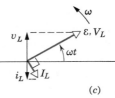

Figure 5 (*a*) An inductor is connected across an alternating-current generator. (*b*) The potential difference across the inductor leads the current by 90°. (*c*) A phasor diagram shows the same thing.

or

$$i_L = \int di_L = \frac{V_L}{L} \int_0^t \sin \omega t \, dt$$

$$= -\left(\frac{V_L}{\omega L}\right) \cos \omega t. \qquad (15)$$

Again, we recast this equation in two ways. First, once more for reasons of symmetry of notation, we introduce the quantity X_L, called the *inductive reactance* of the inductor at the frequency in question. It is defined from

$$\boxed{X_L = \omega L} \quad \text{(inductive reactance).} \qquad (16)$$

Analysis shows that the SI units of X_L is the *ohm,* just as it is for X_C and for R. Second, we replace $-\cos \omega t$ in Eq. 15 by a phase-shifted sine, namely

$$-\cos \omega t = \sin(\omega t - 90°).$$

You can verify this identity by expanding its right-hand side according to the formula for $\sin(\alpha - \beta)$ given in Appendix G.

With these two changes, Eq. 15 becomes

$$i_L = \left(\frac{V_L}{X_L}\right) \sin(\omega t - 90°) = I_L \sin(\omega t - 90°). \qquad (17)$$

By comparison with Eq. 2 we see that, in this case of a purely inductive load, the phase constant $\phi = +90°$. From Eq. 17 we also see that the current amplitude and the voltage amplitude are related by

$$\boxed{V_L = I_L X_L} \quad \text{(inductor).} \qquad (18)$$

Although we developed Eq. 18 for the specific circuit of Fig. 5*a*, it holds for an individual inductor in any alternating current circuit, no matter how complex.

Comparison of Eqs. 12 and 17, or inspection of Fig. 5*b*, shows that the quantities i_L and v_L are 90° out of phase. In this case, however, i_L *lags* v_L. That is, if you monitored the current i_L and the potential difference v_L in the circuit of Fig. 5*a*, you would find that i_L reached its maximum value *after* v_L did, by one-quarter cycle.

The phasor diagram of Fig. 5*c* also contains this information. As the phasors rotate in the figure, we see that the phasor labeled I_L does indeed lag that labeled V_L, and by an angle of 90°. Be sure to convince yourself that Fig. 5*c* does indeed represent Eqs. 12 and 17.

Table 1 summarizes what we have learned about the relations between the current *i* and the voltage *v* for each of the three kinds of circuit elements.

Sample Problem 1 In Fig. 4*a* let $C = 15\ \mu\text{F}$, $\nu = 60$ Hz, and $\mathscr{E}_m = V_C = 36$ V. (a) Find the capacitive reactance X_C.

From Eq. 9 we have

$$X_C = \frac{1}{\omega C} = \frac{1}{2\pi\nu C}$$

$$= \frac{1}{(2\pi)(60\text{ Hz})(15 \times 10^{-6}\text{ F})} = 177\ \Omega. \qquad \text{(Answer)}$$

Understand that a capacitive reactance, although measured in ohms, is not a resistance. It is what the definition of Eq. 9 says it is.

(b) Find the current amplitude I_C in this circuit.

From Eq. 11 (see also Table 1) we have

$$I_C = \frac{V_C}{X_C} = \frac{36\text{ V}}{177\ \Omega} = 0.203\text{ A} \approx 200\text{ mA}. \qquad \text{(Answer)}$$

We see that, although a reactance is not a resistance, the capacitive reactance plays the same role for a capacitor that the resistance does for a resistor. Note also that, if you doubled the

Table 1 Phase and Amplitude Relations for Alternating Currents and Voltages

Circuit Element	*Symbol*	*Impedance**	*Phase of the Current*	*Phase Angle ϕ*	*Amplitude Relation*
Resistor	R	R	In phase with v_R	0°	$V_R = I R$
Capacitor	C	X_C	Leads v_C by 90°	−90°	$V_C = I X_C$
Inductor	L	X_L	Lags v_L by 90°	+90°	$V_L = I X_L$

Many students have remembered the phase relations from:

"ELI the ICE man"

Here L and C stand for inductance and capacitance; E stands for voltage and I for current. Thus in an inductive circuit (ELI) the current (I) lags the voltage (E).

* As we shall see, *impedance* is a general term that includes both resistance and reactance.

frequency, the capacitive reactance would drop to half its value and the current amplitude would double. We can understand this physically: To get the same value of V_C you must deliver the same charge to the capacitor ($V_C = q/C$). If the frequency doubles, then you have only half the time to deliver this charge so that the maximum current must double. To sum up: For capacitors, the higher the frequency, the lower the reactance.

Sample Problem 2 In Fig. 5*a* let $L = 230$ mH, $\nu = 60$ Hz, and $\mathcal{E}_m = V_L = 36$ V. (a) Find the inductive reactance X_L.

(a) From Eq. 16

$$X_L = \omega L = 2\pi\nu L = (2\pi)(60 \text{ Hz})(230 \times 10^{-3} \text{ H}) = 87\ \Omega. \quad \text{(Answer)}$$

(b) Find the current amplitude I_L in the circuit.

From Eq. 18

$$I_L = \frac{V_L}{X_L} = \frac{36 \text{ V}}{87\ \Omega} = 0.414 \text{ A} \approx 410 \text{ mA}. \quad \text{(Answer)}$$

Note that, if you doubled the frequency, the inductive reactance would double and the current amplitude would be cut in half. We can also understand this physically: To get the same value of V_L, you must change the current at the same rate ($V_L = L\, di/dt$). If the frequency doubles, you cut the time of change in half so that the maximum current is also cut in half. To sum up: For inductors, the higher the frequency, the higher the reactance.

36-4 The Series *LCR* Circuit

We are now ready to consider the full problem posed by Fig. 2, in which the applied alternating emf is (see Eq. 1)

$$\mathcal{E} = \mathcal{E}_m \sin \omega t \quad \text{(applied emf)} \tag{19}$$

and the resulting alternating current is (see Eq. 2)

$$i = I \sin(\omega t - \phi) \quad \text{(the alternating current).} \tag{20}$$

Our task, we recall, is to find the current amplitude I and the phase constant ϕ. As a starting point we apply the loop theorem to the circuit of Fig. 2, obtaining

$$\mathcal{E} = v_R + v_C + v_L. \tag{21}$$

This relation among these four time-varying quantities remains true at all instants of time.

Consider now the phasor diagram of Fig. 6*a*. It shows, at an arbitrary instant of time, the as yet unknown alternating current. Its maximum value I, its phase ($\omega t - \phi$), and its instantaneous value i are all shown. Recall that, although the potential differences across the circuit elements in Fig. 2 all vary in time with different phases, the current i is common to all elements; there is only one current.

From the phase information summarized in Table 1 we can then draw Fig. 6*b*, in which we show also the three phasors representing the voltages across the three circuit elements R, C, and L at that same instant. As we have learned, the current is in phase with v_R, it leads v_C by 90°, and it lags behind v_L by 90°.

Note that the algebraic sum of the projections of the phasors V_R, V_C, and V_L on the vertical axis is just the right-hand side of Eq. 21. This sum of projections must then equal the left-hand side of that equation, which is $\mathcal{E}$, the projection of the phasor $\mathcal{E}_m$.

In vector operations the (algebraic) sum of the pro-

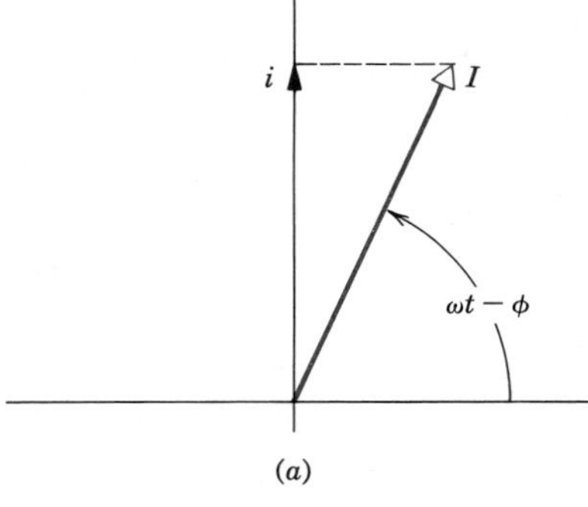

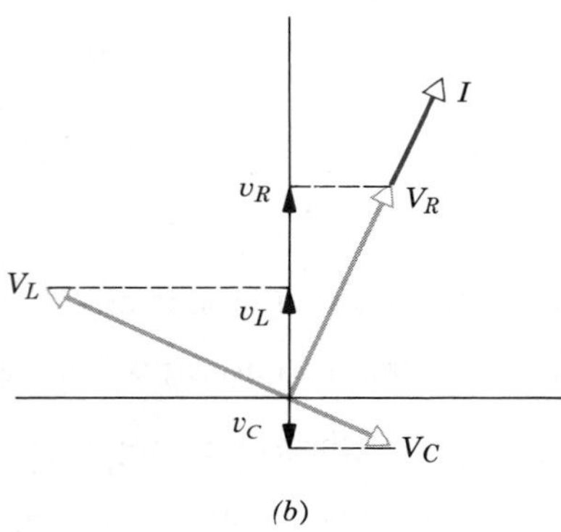

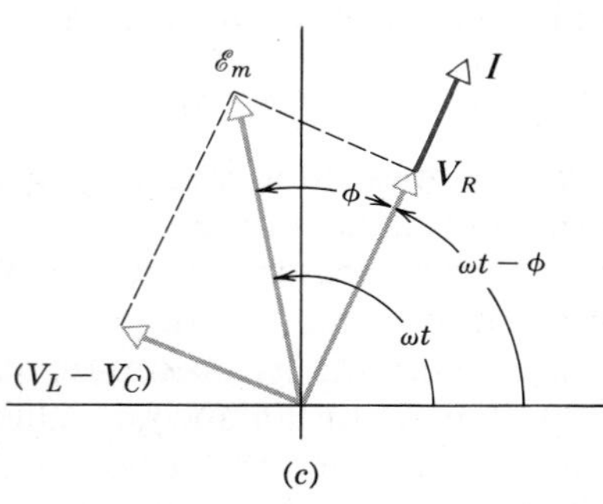

Figure 6 (*a*) A phasor representing the alternating current in the *LCR* circuit of Fig. 2. The amplitude *I*, the instantaneous value *i*, and the phase $(\omega t - \phi)$ are all shown. (*b*) Phasors representing the alternating potential differences across the resistor, the capacitor, and the inductor are added. Note their phase differences with respect to the alternating current. (*c*) A phasor representing the alternating emf is added.

jections of a set of vectors on a given axis is equal to the projection on that axis of the (vector) sum of those vectors. It follows that the phasor $\mathcal{E}_m$ is equal to the (vector) sum of the phasors V_R, V_C, and V_L, as is shown in Fig. 6*c*. The figure also shows the angle ϕ, the phase difference between the current and applied emf that appears in Eqs. 1 and 2 (and also in Eqs. 19 and 20).

In Fig. 6*c* we have formed the phasor difference $V_L - V_C$ and we note that it is at right angles to V_R. We see from the figure that

$$\mathcal{E}_m^2 = V_R^2 + (V_L - V_C)^2.$$

From the amplitude information displayed in Table 1 we can write this as

$$\mathcal{E}_m^2 = (IR)^2 + (IX_L - IX_C)^2,$$

which we can rearrange in the form

$$I = \frac{\mathcal{E}_m}{\sqrt{R^2 + (X_L - X_C)^2}}. \tag{22}$$

The denominator in Eq. 21 is called the *impedance* Z of the circuit for the frequency in question; that is

$$\boxed{Z = \sqrt{R^2 + (X_L - X_C)^2}} \quad \text{(impedance defined)}. \tag{23}$$

We can then write Eq. 22 as

$$I = \frac{\mathcal{E}_m}{Z}. \tag{24}$$

If we substitute for X_C and X_L from Eqs. 9 and 16 we can write Eq. 22 more explicitly as

$$\boxed{I = \frac{\mathcal{E}_m}{\sqrt{R^2 + (\omega L - 1/\omega C)^2}}} \quad \text{(current amplitude)}. \tag{25}$$

This relation is half of the solution of the problem we set for ourselves in Section 36–2. It is an expression for the current amplitude I in Eq. 2 in terms of $\mathcal{E}_m$, ω, R, C, and L.

Equation 25 is essentially the equation of the resonance curves of Fig. 6 of Chapter 35. Inspection of Eq. 25 shows that the maximum value of I occurs when

$$\frac{1}{\omega C} = \omega L \quad \text{or} \quad \omega = \frac{1}{\sqrt{LC}}. \quad \text{(resonance)}$$

This is precisely the resonance condition that we discussed in connection with Fig. 6 of Chapter 35. The value of I at resonance is, as we see from Eq. 25, just $\mathcal{E}_m/R$. Again this agrees with Fig. 6 of Chapter 35, in which we see that the resonance maximum increases as the circuit resistance decreases.

The Phase Constant. It remains to find an equivalent expression for the phase constant ϕ in Eq. 2. From Fig. 6c and Table 1 we can write

$$\tan\phi = \frac{V_L - V_C}{V_R} = \frac{IX_L - IX_C}{IR}$$

or

$$\tan\phi = \frac{X_L - X_C}{R} \quad \text{(phase constant).} \quad (26)$$

Thus, we have now solved the second half of our problem; we have expressed ϕ in terms of ω, R, C, and L. Note that $\mathscr{E}_m$ is not involved in this case.

We drew Fig. 6c arbitrarily with $X_L > X_C$, that is, we assumed the circuit of Fig. 1 to be more inductive than capacitive. Study of Eq. 25 shows that—as far as the amplitude of the current is concerned—the quantity $X_L - X_C$ is squared so that it does not matter whether $X_L > X_C$ or $X_L < X_C$. However, in using Eq. 26 to compute the phase of the current with respect to the applied emf, it does matter. In Fig. 6c, the phase constant is positive, as befits an inductive load (see Table 1).

Two Limiting Cases. In one limiting case we can put $R = X_L = 0$ in Eqs. 25 and 26, leading to $I = \mathscr{E}_m/X_C$ and $\tan\phi = -\infty$. The physical interpretation of this is that the circuit is now purely capacitive, as in Fig. 4a, and the phase constant ϕ is $-90°$, as in Fig. 4b,c.

In a second limiting case we can put $R = X_C = 0$ in Eqs. 25 and 26, leading to $I = \mathscr{E}_m/X_L$ and $\tan\phi = +\infty$. This corresponds to a purely inductive circuit, as in Fig. 5a and the phase constant ϕ is $+90°$, as in Figs. 5b,c.

We note that the currents described by Eqs. 20, 22, and 26 is the *steady-state* current that occurs after the alternating emf has been applied for some time. When the emf is first applied to a circuit, there is also a *transient current* whose duration is determined by the effective circuit time constants $\tau_L = L/R$ and $\tau_C = RC$. This transient current can be quite substantial and can, for example, destroy a motor on start-up if it is not properly allowed for in the circuit design.

Sample Problem 3 In Fig. 2 let $R = 160$ ohm, $C = 15\ \mu$F, $L = 230$ mH, $\nu = 60$ Hz, and $\mathscr{E}_m = 36$ V. (a) Find the capacitive reactance X_C.

$X_C = 177\ \Omega$, as in Sample Problem 1. (Answer)

(b) Find the inductive reactance X_L.

$X_L = 87\ \Omega$, as in Sample Problem 2. (Answer)

Note that $X_C > X_L$ so that the circuit is more capacitive than inductive.

(c) Find the impedance Z for the circuit.

From Eq. 23,

$$Z = \sqrt{R^2 + (X_L - X_C)^2}$$
$$= \sqrt{(160\ \Omega)^2 + (87\ \Omega - 177\ \Omega)^2} = 184\ \Omega. \quad \text{(Answer)}$$

(d) Find the current amplitude I.

From Eq. 24,

$$I = \frac{\mathscr{E}_m}{Z} = \frac{36\ \text{V}}{184\ \Omega} = 0.196\ \text{A} \approx 200\ \text{mA}. \quad \text{(Answer)}$$

(e) Find the phase constant ϕ in Eq. 2 (or Eq. 20)

From Eq. 26 we have

$$\tan\phi = \frac{X_L - X_C}{R} = \frac{87\ \Omega - 177\ \Omega}{160\ \Omega} = -0.563$$

Thus we have

$$\phi = \tan^{-1}(-0.563) = -29.4°. \quad \text{(Answer)}$$

As Table 1 suggests, a negative phase constant is appropriate for a capacitive load.

36-5 Power in Alternating-Current Circuits

In the *LCR* circuit of Fig. 2 the source of energy is the alternating-current generator. Of the energy that it provides some is stored in the electric field in the capacitor, some in the magnetic field in the inductor, and some is dissipated as thermal energy in the resistor. In steady-state operation—which we assume—the average energy stored in the capacitor and in the inductor remains constant. The net flow of energy is thus from the generator to the resistor, as in Fig. 7, where it is transformed from electromagnetic to thermal form.

The instantaneous rate at which energy is transformed in the resistor can be written, with the help of Eq. 2, as

$$P = i^2R = [I\sin(\omega t - \phi)]^2R$$
$$= I^2R[\sin(\omega t - \phi)]^2. \quad (27)$$

The *average* rate at which power is transferred to the resistor, however, which is our real concern, is the average of Eq. 27 over time. Figure 8b shows that the average value of $\sin^2\theta$ over one complete cycle, where θ is any angular variable, is just one-half. Note in Fig. 8b how the (shaded) parts of the curve that lie above the horizontal

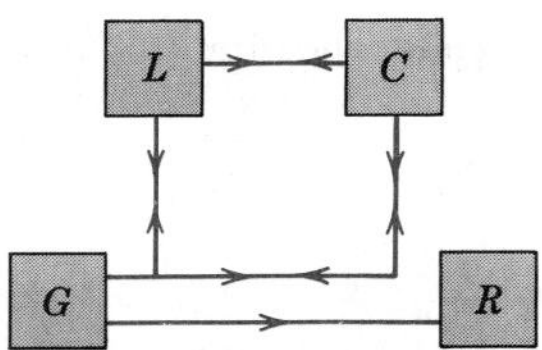

Figure 7 The flow of energy in the *LCR* circuit of Fig. 2. All energy comes from generator *G*. During each half cycle the capacitor *C* and the inductor *L* receive energy and give out as much as they have received. Some of this energy (and, at the resonant frequency, all of it) is shuttled back and forth between the capacitor and the inductor. Energy continues to flow from the generator to resistor *R*, where it appears in thermal form.

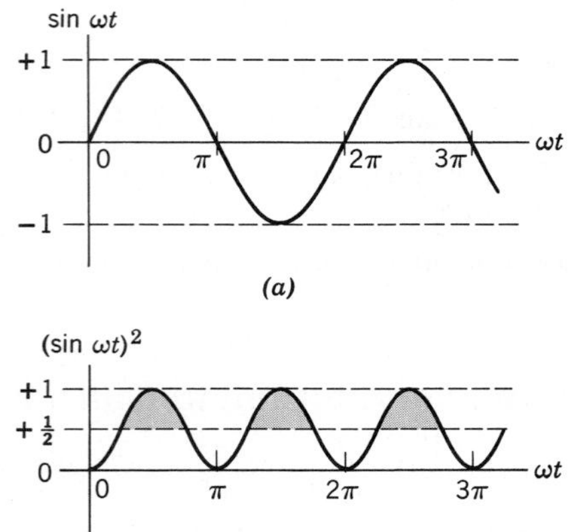

Figure 8 (*a*) A plot of sin θ versus θ. Its average value over one cycle is zero. (*b*) A plot of $\sin^2\theta$ versus θ. Its average value over one cycle is one-half. The angle θ represents any angular quantity.

line marked $\frac{1}{2}$ just fill in the empty spaces below that line. Thus we can write Eq. 27 as

$$P_{av} = \tfrac{1}{2}I^2R = (I/\sqrt{2})^2R. \quad (28)$$

We choose to call $I/(\sqrt{2})$ the *root-mean-square,* or *rms,* value of the current i and we write, in place of Eq. 28,

$$\boxed{P_{av} = I_{rms}{}^2R} \quad \text{(average power).} \quad (29)$$

The rms notation is appropriate. First we *square* the current I, obtaining I^2. Then we average it (that is, we find its *mean* value), obtaining $\frac{1}{2}I^2$, as in Fig. 8*b*. Finally we take its *square root,* obtaining $I/\sqrt{2}$, which we relabel as I_{rms}.

Equation 29 looks much like Eq. 21 of Chapter 28; the message is that if we use rms quantities for I, for V, and for $\mathcal{E}$, the average power dissipation, which is all that usually matters, will be the same for alternating-current circuits as for direct-current circuits with a constant emf.

Alternating-current instruments, such as ammeters and voltmeters, are usually calibrated to read I_{rms}, V_{rms}, and $\mathcal{E}_{rms}$. Thus, if you plug an alternating-current voltmeter into a household electric outlet and it reads 120 V, that is an rms voltage. The *maximum* value of the potential difference at the outlet is $\sqrt{2} \times (120\text{ V})$ or 170 V. The sole reason for using rms values in alternating-current circuits is to let us use the familiar direct-current power relationships of Section 28–7. We summarize the relations between the maximum and the rms values for the three variables of interest:

$$I_{rms} = I/\sqrt{2}, \quad V_{rms} = V/\sqrt{2}, \quad \text{and} \quad \mathcal{E}_{rms} = \mathcal{E}_m/\sqrt{2}. \quad (30)$$

Because the proportionality factor in Eq. 30 ($= 1/\sqrt{2}$) is the same for all three variables, we can write the important Eqs. 24 and 25 as

$$I_{rms} = \frac{\mathcal{E}_{rms}}{Z} = \frac{\mathcal{E}_{rms}}{\sqrt{R^2 + (X_L - X_c)^2}}, \quad (31)$$

and, indeed, this is the form that we almost always use.

We can recast Eq. 29 in a useful equivalent way by combining it with the relationship $I_{rms} = \mathcal{E}_{rms}/Z$ (see Eq. 31) or

$$P_{av} = \frac{\mathcal{E}_{rms}}{Z} I_{rms}R = \mathcal{E}_{rms}I_{rms}(R/Z). \quad (32)$$

From Fig. 6*b* and Table 1, however, we see that R/Z is just the cosine of the phase constant ϕ. Thus

$$\cos\phi = \frac{V_R}{\mathcal{E}_m} = \frac{IR}{IZ} = \frac{R}{Z}.$$

Equation 32 then becomes

$$\boxed{P_{av} = \mathcal{E}_{rms}I_{rms}\cos\phi} \quad \text{(average power),} \quad (33)$$

in which $\cos\phi$ is called the *power factor.* Since $\cos\phi = \cos(-\phi)$, Eq. 33 is independent of whether the phase constant ϕ is positive or negative.

If we wish to deliver maximum power to a resistor in an *LCR* circuit, we should keep the power factor $\cos\phi$ as close to unity as possible. This is equivalent to keeping the phase constant ϕ in Eq. 2 as close to zero as possible. If, for example, a load is highly inductive, it can be made

Table 2 The Average Power Transferred from the Generator for Three Special Cases

Circuit Element	Impedance Z	Phase Constant ϕ	Power Factor $\cos\phi$	Average Power P_{av}
R	R	Zero	1	$\mathcal{E}_{rms}I_{rms}$
C	X_C	$-90°$	Zero	Zero
L	X_L	$+90°$	Zero	Zero

less so by adding more capacitance to the circuit, thus reducing the phase constant and increasing the power factor in Eq. 33. Power utility companies place capacitors throughout their transmission system to bring this about.

Table 2 shows that Eq. 33 gives reasonable results in the three special cases in which the single circuit element present is a resistor (as in Fig. 3*a*), a capacitor (as in Fig. 4*a*), or an inductor (as in Fig. 5*a*).

Sample Problem 4 Consider again the circuit of Fig. 2, using the same parameters that we used in Sample Problem 3, namely: $R = 160\ \Omega$, $C = 15\ \mu\text{F}$, $L = 230$ mH, $\nu = 60$ Hz, and $\mathcal{E}_m = 36$ V. (a) Find the rms electromotive force, $\mathcal{E}_{rms}$.

$$\mathcal{E}_{rms} = \mathcal{E}_m/\sqrt{2} = 36\ \text{V}/\sqrt{2} = 25.5\ \text{V} \approx 26\ \text{V}. \quad \text{(Answer)}$$

(b) Find the rms current, I_{rms}.

In Sample Problem 3 we see that $I = 0.196$ A. We then have

$$I_{rms} = I/\sqrt{2} = 0.196\ \text{A}/\sqrt{2} = 0.139\ \text{A} \approx 0.14\ \text{A}. \quad \text{(Answer)}$$

(c) Find the power factor, $\cos\phi$.

In Sample Problem 3 we found that the phase constant ϕ was $-29.4°$. Thus

$$\text{Power factor} = \cos(-29.4°) = 0.871 \approx 0.87. \quad \text{(Answer)}$$

(d) Find the average power P_{av} dissipated in the resistor. From Eq. 29 we have

$$P_{av} = I_{rms}^2 R = (0.139\ \text{A})^2(160\ \Omega) = 3.1\ \text{W}. \quad \text{(Answer)}$$

Alternatively, Eq. 33 yields

$$P_{av} = \mathcal{E}_{rms}I_{rms}\cos\phi = (25.5\ \text{V})(0.139\ \text{A})(0.871) = 3.1\ \text{W}, \quad \text{(Answer)}$$

in full agreement. Note that, to get agreement of these results to two significant figures, we had to use three significant figures for the currents and voltages. These precautions against numerical rounding errors should not detract from the fact the Eqs. 29 and 33 are identically equal.

36-6 The Transformer

For alternating-current circuits, the average power dissipation in a resistive load is given by Eq. 33*

$$P_{av} = IV. \tag{34}$$

This means that, for a given power requirement, we have a range of choices, from a relatively large current I and a relatively small potential difference V or just the reverse, provided only that their product remains constant.

In electric power distribution systems it is desirable, both for reasons of safety and for efficient design of equipment, to deal with relatively low voltages at both the generating end (the electric power plant) and the receiving end (the home or factory). Nobody wants an electric toaster or a child's electric train to operate at, say, 10 kV.

On the other hand, in the transmission of electric energy from the generating plant to the consumer, we want the lowest practical current (and thus the largest practical potential difference) so as to minimize the I^2R ohmic losses in the transmission line. $V = 500$ kV is not uncommon. Thus, there is a fundamental mismatch between the requirements for efficient transmission on the one hand and efficient and safe generation and consumption on the other.

We need a device that can, as design considerations require, raise or lower the potential difference in a circuit, keeping the product current × voltage essentially constant. The *transformer* of Fig. 9 is such a device. It has no moving parts, operates by Faraday's law of induction, and has no direct-current counterpart of equivalent simplicity.

In Fig. 9 two coils are shown wound around a soft iron core. The primary winding, of N_p turns, is connected to an alternating-current generator whose emf $\mathcal{E}$ is given by

$$\mathcal{E} = \mathcal{E}_m \sin \omega t. \tag{35}$$

The secondary winding, of N_s turns, is an open circuit as long as switch S is open, which we assume for the present. Thus there is no secondary current. We assume further that the resistances of the primary and secondary

* In this Section, we assume a purely resistive load, so that $\phi = 0$ (and thus $\cos\phi$, which is the power factor, is equal to unity) in Eq. 33. Furthermore, we follow conventional practice and drop the subscripts identifying rms quantities. Practicing engineers and scientists assume that all time-varying current and voltages are reported as their rms values; that is what the meters read.

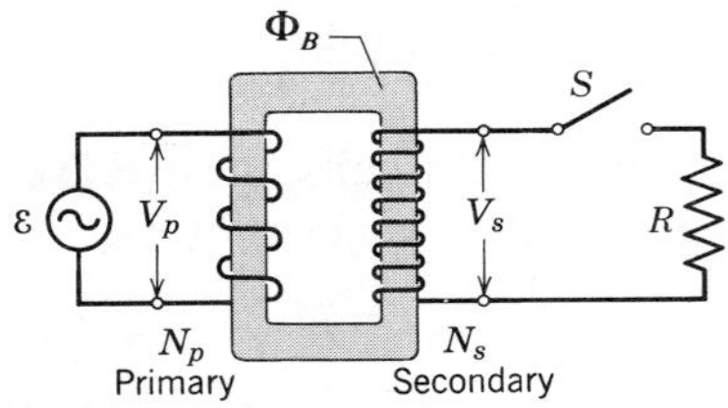

Figure 9 An ideal transformer, showing two coils wound on a soft iron core.

windings and also the magnetic hysteresis losses in the iron core are negligible. Well-designed, high-capacity transformers can have energy losses as low as one percent so that our assumption of an ideal transformer is not unreasonable.

For the above conditions the primary winding is a pure inductance; compare Fig. 5*a*. Thus the (very small) primary current, called the magnetizing current I_{mag}, lags the primary potential difference V_p by 90°; the power factor (= cos ϕ in Eq. 33) is zero and thus no power is delivered from the generator to the transformer.

However, the small alternating primary current I_{mag} induces an alternating magnetic flux Φ_B in the iron core and we assume that all this flux links the turns of the secondary windings. From Faraday's law of induction the induced emf per turn $\mathcal{E}_{\text{turn}}$ is the same for both the primary and secondary windings. Thus, assuming that the symbols represent rms values, we can write

$$\mathcal{E}_{\text{turn}} = \frac{d\Phi_B}{dt} = \frac{V_p}{N_p} = \frac{V_s}{N_s}$$

or

$$V_s = V_p(N_s/N_p) \quad \text{(transformation of voltage).} \quad (36)$$

If $N_s > N_p$, we speak of a *step-up transformer;* if $N_s < N_p$, we speak of a *step-down transformer.*

In all of the above we have assumed an open-circuit secondary so that no power is transmitted through the transformer. Now let us close switch S in Fig. 9, thus connecting the secondary winding to a resistive load R. In general the load would also contain inductive and capacitive elements, but we confine ourselves to this special case.

Several things happen when we close switch S. (1) An alternating current I_s appears in the secondary circuit, with a corresponding power dissipation I_s^2R $(= V_s^2/R)$ in the resistive load. (2) This current induces its own alternating magnetic flux in the iron core and this flux induces (from Faraday's law and Lenz's law) an opposing emf in the primary windings.* (3) V_p, however, cannot change in response to this opposing emf because it must always equal the emf that is provided by the generator; closing switch S cannot change this fact. (4) To ensure this, a new alternating current I_p must appear in the primary circuit, its magnitude and phase constant being just that needed to cancel the opposing emf generated in the primary windings by I_s.

Rather than analyze the above rather complex process in detail, we take advantage of the overall view provided by the conservation of energy principle. For an ideal transformer with a resistive load this tells us that

$$I_pV_p = I_sV_s.$$

Because Eq. 36 holds whether or not the secondary circuit of Fig. 9 is closed, we then have

$$I_s = I_p(N_p/N_s) \quad \text{(transformation of currents)} \quad (37)$$

as the transformation relation for currents.

Finally, knowing that $I_s = V_s/R$, we can use Eqs. 36 and 37 to obtain

$$I_p = \frac{V_p}{(N_p/N_s)^2R},$$

which tells us that, from the point of view of the primary circuit, the equivalent resistance of the load is not R but

$$R_{\text{eq}} = (N_p/N_s)^2R \quad \text{(transformation of resistances).} \quad (38)$$

Equation 38 suggests still another function for the transformer. We have seen that, for maximum transfer of energy from a seat of emf to a resistive load, the resistance of the generator and the resistance of the load must be equal.† The same relation holds for ac circuits except that the *impedance* (rather than the resistance) of the generator must be matched to that of the load. It often happens—as when we wish to connect a speaker to an amplifier—that this condition is far from met, the amplifier being of high impedance and the speaker of low impedance. We can match the impedances of the two devices by coupling them through a transformer with a suitable turns ratio.

* In Chapter 33 we neglected the magnetic effects resulting from induced currents. Here, however, the magnetic effects of current induced in the secondary winding is not only not small, it is essential to the operation of the transformer!

† See Problem 22 of Chapter 29.

Sample Problem 5 A transformer on a utility pole operates at $V_p = 8.5$ kV on the primary side and supplies electric energy to a number of nearby houses at $V_s = 120$ V, both quantities being rms values. Assume an ideal transformer, a resistive load, and a power factor of unity. (a) What is the turns ratio N_p/N_s of this step-down transformer?

From Eq. 36 we have

$$\frac{N_p}{N_s} = \frac{V_p}{V_s} = \frac{8.5 \times 10^3 \text{ V}}{120 \text{ V}} = 70.8 \approx 71. \qquad \text{(Answer)}$$

(b) The rate of average energy consumption in the houses served by the transformer at a given time is 78 kW. What are the rms currents in the primary and secondary windings of the transformer?

From Eq. 33 we have (with $\cos \phi = 1$)

$$I_p = \frac{P_{\text{av}}}{V_p} = \frac{78 \times 10^3 \text{ W}}{8.5 \times 10^3 \text{ V}} = 9.18 \text{ A} \approx 9.2 \text{ A} \qquad \text{(Answer)}$$

and

$$I_s = \frac{P_{\text{av}}}{V_s} = \frac{78 \times 10^3 \text{ W}}{120 \text{ V}} = 650 \text{ A}. \qquad \text{(Answer)}$$

(c) What is the equivalent resistive load in the secondary circuit?

Here we have

$$R_s = \frac{V_s}{I_s} = \frac{120 \text{ V}}{650 \text{ A}} = 0.185\ \Omega \approx 0.19\ \Omega. \qquad \text{(Answer)}$$

(d) What is the equivalent resistive load in the primary circuit?

Here we have

$$R_p = \frac{V_p}{I_p} = \frac{8.5 \times 10^3 \text{ V}}{9.18 \text{ A}} = 926\ \Omega \approx 930\ \Omega. \qquad \text{(Answer)}$$

We can verify this from Eq. 38, which we write as

$$R_p = (N_p/N_s)^2 R_s = (70.8)^2(0.185\ \Omega)$$
$$= 925\ \Omega \approx 930\ \Omega. \qquad \text{(Answer)}$$

Except for rounding errors, the two results are in full agreement.

REVIEW AND SUMMARY

Current Amplitude and Phase

The basic problem in alternating-current analysis is to find expressions for the current amplitude I and the phase angle ϕ in

$$i = I \sin(\omega t - \phi) \qquad [2]$$

when an emf given by $\mathcal{E} = \mathcal{E}_m \sin \omega t$ is applied to a circuit; we use the series *LCR* circuit of Fig. 2 as an example. The *phase constant* ϕ is an angle that indicates the extent to which the current lags the applied emf.

Single Circuit Elements

The alternating potential difference across a resistor has amplitude $V = IR$; the current is in phase with the potential difference. For a *capacitor,* $V = IX_C$, $X_C = 1/\omega C$ being the *capacitive reactance;* current leads potential difference by 90°. For an *inductor,* $V = IX_L$, $X_L = \omega L$ being the *inductive reactance;* current lags potential difference by 90°. These results are summarized in Table 1.

Phasors

Phasors provide a powerful mathematical tool for representing alternating currents and voltages (and other phase-related quantities). We represent the maximum value (amplitude) of any alternating quantity by an open arrow rotating counterclockwise about the origin at the angular frequency ω. The projection of this arrow on the vertical axis gives the instantaneous value. The voltage and currents of simple R, C, and L elements are represented by phasors in Figs. 3, 4, and 5; these should be studied carefully. See also Sample Problems 1 and 2.

Series *LCR* Circuit

The phasor diagrams Fig. 6*b*, drawn with the help of Table 1, shows the relationship of the potential drops to the currents for the three elements in the series *LCR* circuit of Fig. 2. The loop theorem (see Eq. 21) allows construction of the circuit emf phasor, as in Fig. 6*c*. Derivations based on that figure yield

Current Amplitude

$$I = \frac{\mathcal{E}_m}{\sqrt{R^2 + (X_L - X_C)^2}} = \frac{\mathcal{E}_m}{\sqrt{R^2 + (\omega L - 1/\omega C)^2}} \quad \text{(current amplitude)} \qquad [22,25]$$

and

Phase

$$\tan \phi = \frac{X_L - X_C}{R} \quad \text{(phase constant).} \qquad [26]$$

Introducing the impedance,

Impedance

$$Z = \sqrt{R^2 + (X_L - X_C)^2} \quad \text{(impedance),} \qquad [23]$$

allows us to write Eq. 22 as $I = \mathcal{E}_m/Z$. See Sample Problem 3 for a summary of this analysis.

Resonance

Equation 25 is the equation of the *resonance curves* of Fig. 6 in Chapter 35. Peak current amplitude occurs when $X_C = X_L$ (the resonance condition). It has the value $\mathcal{E}_m/R$; the phase angle ϕ is zero at resonance.

In the series *LCR* circuit of Fig. 2, the *average power* output P_{av} of the generator is delivered to the resistor, where it appears as thermal energy:

Power

$$P_{\text{av}} = (I_{\text{rms}})^2 R = \mathcal{E}_{\text{rms}} I_{\text{rms}} \cos \phi \quad \text{(average power).} \qquad [29,33]$$

Rms Notation

Here "rms" stands for *root-mean-square;* rms quantities are related to maximum quantities by $I_{\text{rms}} = I/\sqrt{2}$ and $\mathcal{E}_{\text{rms}} = \mathcal{E}_m/\sqrt{2}$. Alternating-current voltmeters and ammeters have their scales adjusted to read rms values. The term $\cos \phi$ above is called the *power factor.* Table 2 examines P_{av} for the three special cases of Figs. 3, 4, and 5. See Sample Problem 4.

Transformers

A *transformer* (assumed "ideal"; see Fig. 9) is a soft-iron yoke on which are wound a primary coil of N_p turns and a secondary coil of N_s turns. If the primary is connected to an alternating current generator, the primary and secondary voltages are related by

$$V_s = V_p(N_s/N_p) \quad \text{(transformation of voltage).} \qquad [36]$$

The currents are related by

$$I_s = I_p(N_p/N_s) \quad \text{(transformation of currents)} \qquad [37]$$

and the effective resistance of the circuit is

$$R_{\text{eq}} = (N_p/N_s)^2 \quad \text{(transformation of resistances).} \qquad [38]$$

See Sample Problem 5.

QUESTIONS

1. In the relation $\omega = 2\pi\nu$ when using SI units we measure ω in radians per second and ν in hertz or cycles per second. The radian is a measure of angle. What connection do angles have with alternating currents?

2. If the output of an ac generator such as that in Fig. 1 is connected to an *LCR* circuit such as that of Fig. 2, what is the ultimate source of the power dissipated in the resistor?

3. Why would power distribution systems be less effective without alternating emfs?

4. In the circuit of Fig. 2, why is it safe to assume that (*a*) the alternating current of Eq. 2 has the same angular frequency ω as the alternating emf of Eq. 1, and (*b*) that the phase angle ϕ in Eq. 2 does not vary with time? What would happen if either of these (true) statements were false?

5. How does a phasor differ from a vector? We know, for example, that emfs, potential differences, and currents are not vectors. How then can we justify constructions such as Fig. 6?

6. Would any of the discussion of Section 3 be invalid if the phasor diagrams were to rotate in the clockwise direction, rather than the counterclockwise direction that we assumed?

7. Suppose that, in a series *LCR* circuit, the frequency of the applied voltage is changed continuously from a very low value to a very high value. How does the phase constant change?

8. Does it seem intuitively reasonable that the capacitive reactance ($= 1/\omega C$) should vary inversely with the angular frequency, whereas the inductive reactance ($= \omega L$) varies directly with this quantity?

9. During World War II, at a large research laboratory in this country, an alternating current generator was located a mile or so from the laboratory building it served. A technician increased the speed of the generator to compensate for what he called "the loss of frequency along the transmission line" connecting the generator with the laboratory building. Comment on this procedure.

10. Discuss in your own words what it means to say that an alternating current "leads" or "lags" an alternating emf.

11. If, as we stated in Section 4, a given circuit is "more inductive than capacitive," that is, that $X_L > X_C$, (*a*) does this mean, for a fixed angular frequency, that L is relatively "large" and C is relatively "small," or L and C are both relatively "large"? (*b*) For fixed values of L and C does this mean that ω is relatively "large" or relatively "small"?

12. How could you determine, in a series LCR circuit, whether the circuit frequency is above or below resonance?

13. What is wrong with this statement: "If $X_L > X_C$, then we must have $L > 1/C$"?

14. How, if at all, must Kirchoff's rules (the loop and junction theorems) for direct current circuits be modified when applied to alternating current circuits?

15. Do the loop theorem and the junction theorem apply to multiloop ac circuits as well as to multiloop dc circuits?

16. In Sample Problem 4 what would be the effect on P_{av} if you increased (*a*) R, (*b*) C, and (*c*) L? How would ϕ in Eq. 33 change in these three cases?

17. Do commercial power station engineers like to have a low power factor or a high one, or does it make any difference to them? Between what values can the power factor range? What determines the power factor; is it characteristic of the generator, of the transmission line, of the circuit to which the transmission line is connected, or of some combination of these?

18. Can the instantaneous power delivered by a source of alternating current ever be negative? Can the power factor ever be negative? If so, explain the meaning of these negative values.

19. In a series LCR circuit the emf is leading the current for a particular frequency of operation. You now lower the frequency slightly. Does the total impedance of the circuit increase, decrease, or stay the same?

20. If you know the power factor (= cos ϕ in Eq. 33) for a given LCR circuit, can you tell whether or not the applied alternating emf is leading or lagging the current? If so, how? If not, why not?

21. What is the permissible range of values of the phase angle ϕ in Eq. 2? Of the power factor in Eq. 33?

22. Why is it useful to use the rms notation for alternating currents and voltages?

23. You want to reduce your electric bill. Do you hope for a small or a large power factor or does it make any difference? If it does, is there anything that you can do about it? Discuss.

24. In Eq. 33 is ϕ the phase angle between $\mathcal{E}(t)$ and $i(t)$ or between $\mathcal{E}_{rms}$ and i_{rms}? Explain.

25. A doorbell transformer is designed fór a primary rms input of 120 V and a secondary rms output of 6 V. What would happen if the primary and secondary connections were accidentally interchanged during installation? Would you have to wait for someone to push the doorbell to find out? Discuss.

26. You are given a transformer enclosed in a wooden box, its primary and secondary terminals being available at two opposite faces of the box. How could you find its turns ratio without opening the box?

27. In the transformer of Fig. 9, with the secondary on open circuit, what is the phase relationship between (*a*) the impressed emf and the primary current, (*b*) the impressed emf and the magnetic field in the transformer core, and (*c*) the primary current and the magnetic field in the tranformer core?

28. What are some applications of a step-up transformer? a step-down transformer?

29. What determines which winding of a transformer is the primary and which the secondary? Can a transformer have a single primary and two secondaries? a single secondary and two primaries?

30. Instead of the 120-V, 60-Hz current typical of the United States, Europeans use 240-V, 50 Hz alternating currents. While on vacation in Europe, you would like to use some of your American appliances, such as a clock, an electric razor, and a hair dryer. Can you do so simply by plugging in a 2:1 step-up transformer? Explain why this apparently simple step may or may not suffice.

EXERCISES AND PROBLEMS

Section 36-3 Three Simple Circuits

1E. Let Eq. 1 describe the effective emf available at an ordinary 60-Hz ac outlet. What angular frequency ω does this correspond to? How does the utility company establish this frequency?

2E. A 1.5-μF capacitor is connected as in Fig. 4*a* to an ac generator with $\mathcal{E}_m = 30$ V. What is the amplitude of the resulting alternating current if the frequency of the emf is (*a*) 1.0 kHz, (*b*) 8.0 kHz?

3E. A 5.0-mH inductor is connected as in Fig. 5*a* to an ac generator with $\mathcal{E}_m = 30$ V. What is the amplitude of the resulting alternating current if the frequency of the emf is (*a*) 1.0 kHz, (*b*) 8.0 kHz?

4E. A 50-Ω resistor is connected as in Fig. 3*a* to an ac generator with $\mathcal{E}_m = 30$ V. What is the amplitude of the resulting alternating current if the frequency of the emf is (*a*) 1.0 kHz, (*b*) 8.0 kHz?

5E. A 45-mH inductor has a reactance of 1.3 kΩ. (*a*) What is the frequency? (*b*) What is the capacitance of a capacitor with the same reactance at that frequency? (*c*) If the frequency is doubled, what are the reactances of the inductor and capacitor?

6E. A 1.5-μF capacitor has a capacitive reactance of 12 Ω. (*a*) What must be the frequency? (*b*) What will be the capacitive reactance if the frequency is doubled?

7E. (*a*) At what frequency would a 6.0-mH inductor and a 10-μF capacitor have the same reactance? (*b*) What would this reactance be? (*c*) Show that this frequency would be equal to the natural frequency of free LC oscillations.

8P. The output of an ac generator is $\mathcal{E} = \mathcal{E}_m \sin \omega t$, with $\mathcal{E}_m = 25$ V and $\omega = 377$ rad/s. It is connected to a 12.7-H inductor. (*a*) What is the maximum value of the current? (*b*) When the current is a maximum, what is the emf of the generator? (*c*) When the emf of the generator is -12.5 V and increasing in magnitude, what is the current? (*d*) For the conditions of part (*c*), is the generator supplying energy to or taking energy from the rest of the circuit?

9P. The ac generator of Problem 8 is connected to a 4.15-μF capacitor. (*a*) What is the maximum value of the current? (*b*) When the current is a maximum, what is the emf of the generator? (*c*) When the emf of the generator is -12.5 V and increasing in magnitude, what is the current? (*d*) For the conditions of part (*c*), is the generator supplying energy to or taking energy from the rest of the circuit?

10P. The output of an ac generator is given by $\mathcal{E} = \mathcal{E}_m \sin(\omega t - \pi/4)$, where $\mathcal{E}_m = 30$ V and $\omega = 350$ rad/s. The current is given by $i(t) = I \sin(\omega t - 3\pi/4)$, where $I = 620$ mA. (*a*) At what time, after $t = 0$, does the generator emf first reach a maximum? (*b*) At what time, after $t = 0$, does the current first reach a maximum? (*c*) The circuit contains a single element other than the generator. Is it a capacitor, and inductor, or a resistor? Justify your answer. (*d*) What is the value of the capacitance, inductance, or resistance, as the case may be?

11P. The output of an ac generator is given by $\mathcal{E} = \mathcal{E}_m \sin(\omega t - \pi/4)$, where $\mathcal{E}_m = 30$ V and $\omega = 350$ rad/s. The current is given by $i(t) = I \sin(\omega t + \pi/4)$, where $I = 620$ mA. (*a*) At what time, after $t = 0$, does the generator emf first reach a maximum? (*b*) At what time, after $t = 0$, does the current first reach a maximum? (*c*) The circuit contains a single element other than the generator. Is it a capacitor, an inductor, or a resistor? Justify your answer. (*d*) What is the value of the capacitance, inductance, or resistance, as the case may be?

12P. *Three-phase power transmission.* A three-phase generator G produces electrical power that is transmitted by means of three wires as shown in Fig. 10. The potentials (relative to a common reference level) of these wires are $V_1 = A \sin(\omega t)$, $V_2 = A \sin(\omega t - 120°)$, and $V_3 = A \sin(\omega t - 240°)$. Some industrial equipment (e.g., motors) has three terminals and is designed to be connected directly to these three wires. To use a more conventional two-terminal device (e.g., a light bulb), one connects it to any two of the three wires. Show that the potential difference between *any two* of the wires (*i*) oscillates sinusoidally with angular frequency ω, and (*ii*) has amplitude $A\sqrt{3}$.

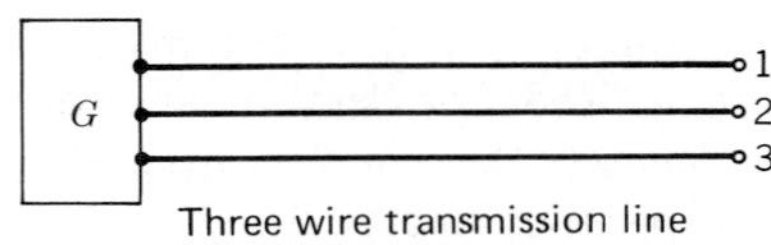

Figure 10 Problem 12.

Section 36-4 The Series *LCR* Circuit

13E. (*a*) Recalculate all the quantities asked for in Sample Problem 3 if the capacitor is removed from the circuit, all other parameters in that sample problem remaining unchanged. (*b*) Draw to scale a phasor diagram like that of Fig. 6*c* for this new situation.

14E. (*a*) Recalculate all the quantities asked for in Sample Problem 3 if the inductor is removed from the circuit, all other parameters in that sample problem remaining unchanged. (*b*) Draw to scale a phasor diagram like that of Fig. 6*c* for this new situation.

15E. (*a*) Recalculate all the quantities asked for in Sample Problem 3 for $C = 70$ μF, the other parameters in that sample problem remaining unchanged. (*b*) Draw to scale a phasor diagram like that of Fig. 6*c* for this new situation and compare the two diagrams closely.

16E. Consider the resonance curves of Fig. 6, Chapter 35. (*a*) Show that for frequencies above resonance the circuit is predominantly inductive and for frequencies below resonance it is predominantly capacitive. (*b*) How does the circuit behave at resonance? (*c*) Sketch a phasor diagram like that of Fig. 6*c* for conditions at a frequency higher than resonance, at resonance, and lower than resonance.

17P. Verify mathematically that the following geometrical construction correctly gives both the impedance Z and the phase constant ϕ. Referring to Fig. 11, (*i*) draw an arrow in the $+y$ direction of magnitude X_C, (*ii*) draw an arrow in the $-y$ direction of magnitude X_L, (*iii*) draw an arrow of magnitude R in the $+x$ direction. Then the magnitude of the "resultant" of these arrows is Z and the angle (measured below the $+x$ axis) of this resultant is ϕ.

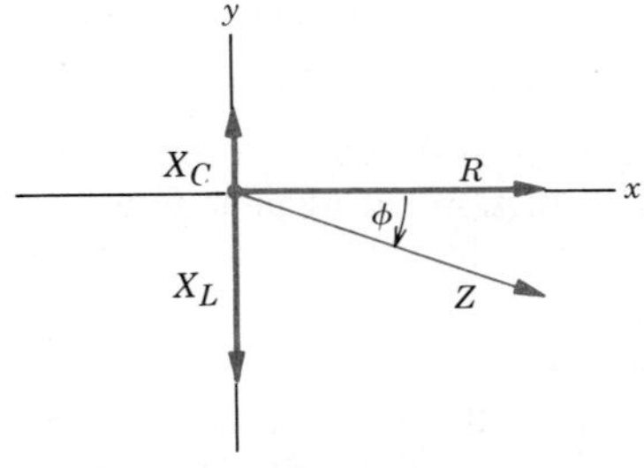

Figure 11 Problem 17.

18P. Can the amplitude of the voltage across an inductor be greater than the amplitude of the generator emf in an *LCR* circuit? Consider an *LCR* circuit with $\mathcal{E}_m = 10$ V, $R = 10\ \Omega$, $L = 1.0\ H$, and $C = 1.0\ \mu$F. Find the amplitude of the voltage across the inductor at resonance.

19P. A coil of self-inductance 88 mH and unknown resistance and a 0.94-μF capacitor are connected in series with an oscillator of frequency 930 Hz. If the phase angle between the applied voltage and current is 75°, what is the resistance of the coil?

20P. When the generator emf in Sample Problem 3 is a maximum, what is the voltage across (*a*) the generator, (*b*) the resistor, (*c*) the capacitor, and (*d*) the inductor? (*e*) By summing these with appropriate signs, verify that the loop rule is satisfied.

21P. An *LCR* circuit such as that of Fig. 2 has $R = 5.0\ \Omega$, $C = 20\ \mu$F, $L = 1.0$ H, and $\mathcal{E}_m = 30$ V. (*a*) At what angular frequency ω_0 will the current have its maximum value, as in the resonance curves of Fig. 6, Chapter 35? (*b*) What is this maximum value? (*c*) At what two angular frequencies ω_1 and ω_2 will the current amplitude have one-half of this maximum value? (*d*) What is the fractional half-width [$= (\omega_1 - \omega_2)/\omega_0$] of the resonance curve?

22P. For a certain *LCR* circuit the maximum generator emf is 125 V and the maximum current is 3.20 A. If the current leads the generator emf by 0.982 rad, (*a*) what is the impedance and (*b*) what is the resistance of the circuit? (*c*) Is the circuit predominantly capacitive or inductive?

23P. In a certain *LCR* circuit, operating at 60 Hz, the maximum voltage across the inductor is twice the maximum voltage across the resistor, while the maximum voltage across the capacitor is the same as the maximum voltage across the resistor. (*a*) By what phase angle does the current lag the generator emf? (*b*) If the maximum generator emf is 30 V, what should be the resistance of the circuit to obtain a maximum current of 300 mA?

24P. The circuit of Sample Problem 3, to which the phasor diagram of Fig. 6*c* corresponds, is not in resonance. (*a*) How can you tell? (*b*) What capacitor would you combine in parallel with the capacitor already in the circuit to bring resonance about? (*c*) What would the current amplitude then be?

25P. A resistor-inductor-capacitor combination, R_1, L_1, C_1 has a resonant frequency that is just the same as that of a different combination, R_2, L_2, C_2. You now connect the two combinations in series. Show that this new circuit also has the same resonant frequency as the separate individual circuits.

26P. A high-impedance ac voltmeter is connected in turn across the inductor, the capacitor, and the resistor in a series circuit having an ac source of 100 V (rms) and gives the same reading in volts in each case. What is this reading?

27P. Show that the fractional half-width of a resonance curve is given to a close approximation by

$$\frac{\Delta\omega}{\omega} = \sqrt{\frac{3C}{L}}\,R,$$

in which ω is the angular frequency at resonance and $\Delta\omega$ is the width of the resonance curve at half-amplitude. Note that $\Delta\omega/\omega$ decreases with *R*, as Fig. 6, Chapter 35, shows. Use this formula to check the answer to part (*d*) of Problem 21.

28P*. The ac generator in Fig. 12 supplies 120 V (rms) at 60 Hz. With the switch open as in the diagram, the resulting current leads the generator emf by 20°. With the switch in position 1 the current lags the generator emf by 10°. When the switch is in position 2 the rms current is 2.0 A. Find the values of *R*, *L*, and *C*.

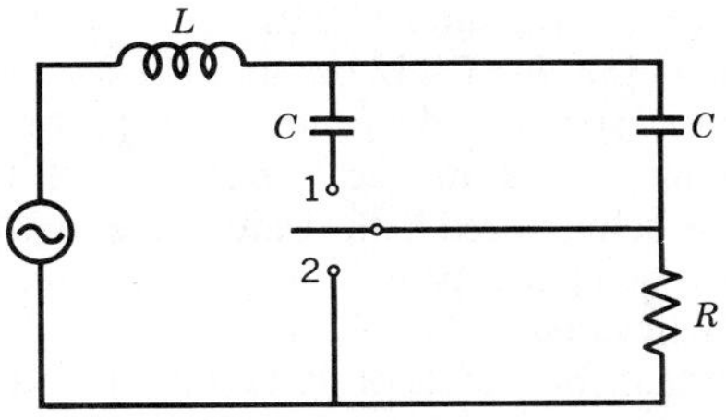

Figure 12 Problem 28.

Section 36-5 Power in Alternating Current Circuits

29E. What is the maximum value of an ac voltage whose rms value is 100 volts?

30E. What direct current will produce the same amount of heat, in a particular resistor, as an alternating current that has a maximum value of 2.6 A?

31E. Calculate the average power dissipated in the circuits of Exercises 3, 4, 13, and 14.

32E. Show that the average power delivered to the circuit of Fig. 2 can also be written as

$$P_{av} = \mathcal{E}_{rms}^2 R/Z^2.$$

Show that this expression gives reasonable results for a purely resistive circuit, for an *LCR* circuit at resonance, for a purely capacitive circuit, and for a purely inductive circuit.

33E. An electric motor connected to a 120-V, 60-Hz power outlet does mechanical work at the rate of 0.10 hp (1 hp = 746 W). If it draws an rms current of 0.65 A, what is its effective resistance, in terms of power transfer? Would this be the same as the resistance of its coils, as measured with an ohmmeter with the motor disconnected from the power outlet?

34E. An air conditioner connected to a 120-V rms ac line is equivalent to a 12-Ω resistance and a 1.3-Ω inductive reactance in series. (*a*) Calculate the impedance of the air conditioner. (*b*) Find the average power supplied to the appliance.

35E. An electric motor has an effective resistance of 32 Ω and an inductive reactance of 45 Ω when working under load. The rms voltage across the alternating source is 420 V. Calculate the rms current.

36P. Show mathematically, rather than graphically as in Fig. 8*b*, that the average value of $\sin^2(\omega t - \phi)$ over an integral number of quarter-cycles is one-half.

37P. For an *LCR* circuit show that in one cycle with period T (*a*) the energy stored in the capacitor does not change; (*b*) the energy stored in the inductor does not change; (*c*) the generator supplies energy $(\frac{1}{2}T)\mathcal{E}_m I \cos\phi$; and (*d*) the resistor dissipates energy $(\frac{1}{2}T)RI^2$. (*e*) Show that the quantities found in (c) and (*d*) are equal.

38P. In an *LCR* circuit, $R = 16.0\ \Omega$, $C = 31.2\ \mu\text{F}$, $L = 9.20$ mH, and $\mathcal{E} = \mathcal{E}_m \sin \omega t$, with $\mathcal{E}_m = 45$ V, and $\omega = 3000$ rad/s. For time $t = 0.442$ ms find (*a*) the rate at which energy is being supplied by the generator, (*b*) the rate at which energy is being stored in the capacitor, (*c*) the rate at which energy is being stored in the inductor, and (*d*) the rate at which energy is being dissipated in the resistor. (*e*) What is the meaning of a negative result for any of parts (*a*), (*b*), and (*c*)? (*f*) Show that the results of parts (*b*), (*c*), and (*d*) sum to the result of part (*a*).

39P. In Fig. 13 show that the power dissipated in the resistor R is a maximum when $R = r$, in which r is the internal resistance of the ac generator. In the text we have tacitly assumed, up to this point, that $r = 0$.

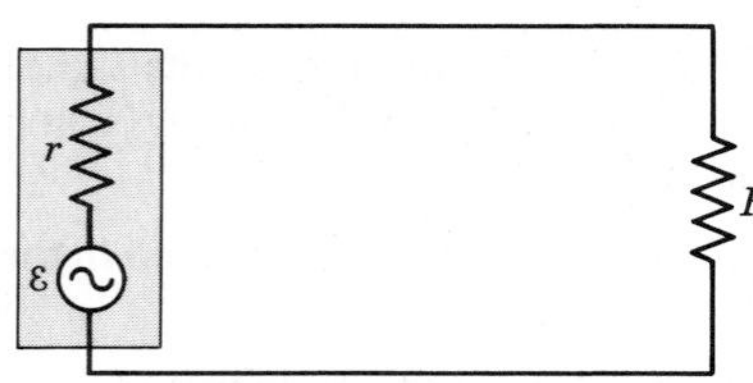

Figure 13 Problems 39 and 48.

40P. Fig. 14 shows an ac generator connected to a "black box" through a pair of terminals. The box contains an *LCR* circuit, possibly even a multiloop circuit, whose elements and arrangements we do not know. Measurements outside the box reveal that

$$\mathcal{E}(t) = (75\ V) \sin \omega t$$

and

$$i(t) = (1.2\ \text{A}) \sin(\omega t + 42°).$$

(*a*) What is the power factor? (*b*) Does the current lead or lag the emf? (*c*) Is the circuit in the box largely inductive or largely capacitive in nature? (*d*) Is the circuit in the box in resonance? (*e*) Must there be a capacitor in the box? an inductor? a resistor? (*f*) What power is delivered to the box by the generator? (*g*) Why don't you need to know the angular frequency ω to answer all these questions?

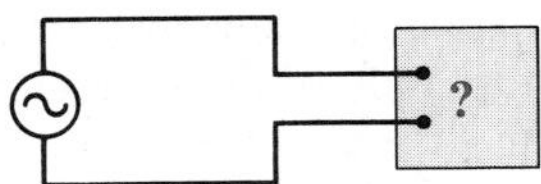

Figure 14 Problem 40.

41P. In an *LCR* circuit such as that of Fig. 2 assume that $R = 5.0\ \Omega$, $L = 60$ mH, $\nu = 60$ Hz, and $\mathcal{E}_m = 30$ V. For what values of the capacitance would the average power dissipated in the resistor be (*a*) a maximum and (*b*) a minimum? (*c*) What are these maximum and minimum powers? (*d*) The corresponding phase angles? (*e*) The corresponding power factors?

42P. A typical "light dimmer" used to dim the stage lights in a theater consists of a variable inductor L connected in series with the light bulb B as shown in Fig. 15. The power supply is 120 V (rms) at 60 Hz; the light bulb is marked "120 V, 1000 W." (*a*) What maximum inductance L is required if the power in the light bulb is to be varied by a factor of five? Assume that the resistance of the light bulb is independent of its temperature. (*b*) Could one use a variable resistor instead of an inductor? If so, what maximum resistance is required? Why isn't this done?

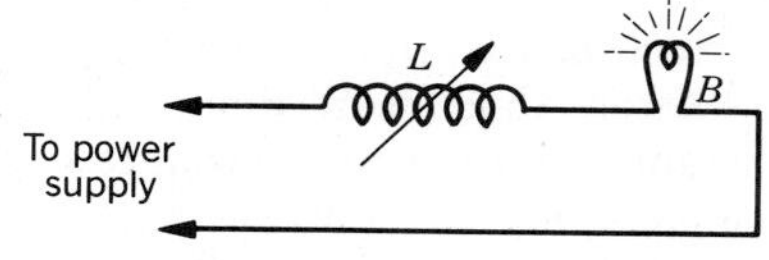

Figure 15 Problem 42.

43P. In Fig. 16, $R = 15\ \Omega$, $C = 4.7\ \mu\text{F}$, and $L = 25$ mH. The generator provides a sinusoidal voltage of 75 V (rms) and frequency $\nu = 550$ Hz. (*a*) Calculate the rms current amplitude. (*b*) Find the rms voltages V_{ab}, V_{bc}, V_{cd}, V_{bd}, V_{ad}. (*c*) What average power is dissipated by each of the three circuit elements?

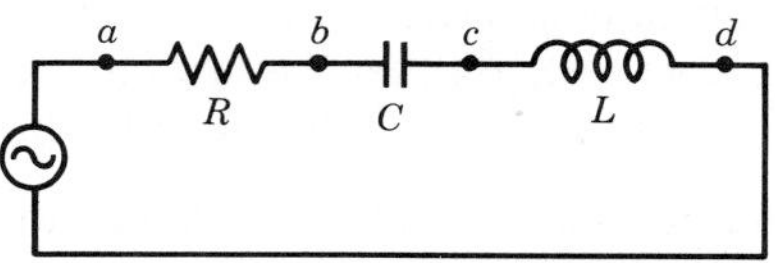

Figure 16 Problem 43.

Section 36-6 The Transformer

44E. A generator supplies 100 V to the primary coil of a transformer of 50 turns. If the secondary coil has 500 turns, what is the secondary voltage?

45E. A transformer has 500 primary turns and 10 secondary turns. (*a*) If V_P for the primary is 120 V (rms), what is V_S for the secondary, assumed on open circuit? (*b*) If the secondary is now connected to a resistive load of 15 Ω, what are the currents in the primary and secondary windings?

46E. Figure 17 shows an "autotransformer." It consists of a single coil (with an iron core). Three "taps" are provided. Between taps T_1 and T_2 there are 200 turns and between taps T_2 and T_3 there are 800 turns. Any two taps can be considered the "primary terminals" and any two taps can be considered the "secondary terminals." List all the ratios by which the primary voltage may be changed to a secondary voltage.

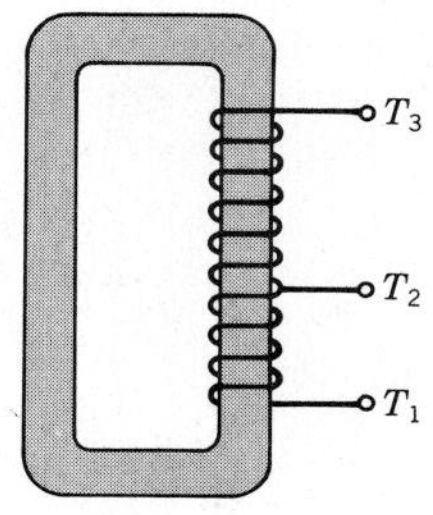

Figure 17 Exercise 46.

47P. An ac generator delivers power to a resistive load in a remote factory over a two-cable transmission line. At the factory a step-down transformer reduces the voltage from its (rms) transmisson value V_t to a much lower value, safe and convenient for use in the factory. The transmission line resistance is 0.30 Ω/cable and the power delivered by the generator is 250kW. Calculate the voltage drop along the transmission line and the power dissipated in the line as thermal energy. Assume (*a*) V_t = 80 kV, (*b*) V_t = 8.0 KV, and (*c*) V_t = 0.80 kV and comment on the acceptability of each choice.

48P. *Impedance matching.* In Fig. 13 let the rectangular box on the left represent the (high-impedance) output of an audio amplifier, with $r = 1000\ \Omega$. Let $R = 10\ \Omega$ represent the (low-impedance) coil of a loudspeaker. In Problem 39 we learned that for maximum transfer of power to the load R we must have $R = r$ and that is not true in this case. However, we also learned, in Section 6, that a transformer can be used to "transform" resistances, making them behave electrically as if they were larger or smaller than they actually are. Sketch the primary and secondary coils of a transformer to be introduced between the "amplifier" and the "speaker" in Fig. 13 to "match the impedances." What must be the turns ratio?

CHAPTER 37

MAXWELL'S EQUATIONS

In days of old, knights wore heraldic symbols on their shields to give information about themselves to the outside world. Today, that role has been taken over by messages on T-shirts and bumper stickers. The message here is: "If you know that these are Maxwell's equations, we share certain insights." The form in which the equations are displayed in this case (differential form) is not the same as the form (integral form) in which we display them in this book. Nevertheless, the content of the equations is identical.

37-1 Pulling Things Together

Fourteen chapters ago, when we began our venture into electromagnetism, we promised that, at some stage, we would assemble what we had learned into a single set of equations, called Maxwell's equations. The time has come to do so. Our plan is to assemble these equations in a preliminary way, at which time we shall discover—using arguments of symmetry—that there is an important term missing in one of them. We then develop this term and, finally, display the equations in their complete form.

All equations of physics that serve, as these do, to correlate experiments in a vast area and to predict new results have a certain beauty about them that can be appreciated, by those who understand them, on an aesthetic level. This is true for Newton's laws of motion, for the laws of thermodynamics, for the theory of relativity, and for the theories of quantum physics.

As for Maxwell's equations, the physicist Ludwig Boltzmann (quoting a line from Goethe) wrote: "Was it a God who wrote these lines. . . ." In more recent times J. R. Pierce, in a book chapter entitled "Maxwell's Wonderful Equations" writes: "To anyone who is motivated by anything beyond the most narrowly practical, it is worthwhile to understand Maxwell's equations simply for the good of his soul." The scope of these equations has been well summarized by the remark that Maxwell's equations account for the facts that a compass needle points north, that light bends when it enters water, and that your car starts when you turn the ignition key. These equations are the basis for the operation of all such electromagnetic and optical devices as electric motors, telescopes, cyclotrons, eyeglasses, television transmitters and receivers, telephones, electromagnets, radar, and microwave ovens.

James Clerk Maxwell, who was born in the same year that Faraday discovered the law of induction, died at age 48 in 1879, the year that Einstein was born.* One is reminded that Newton was born in the year that Galileo, his illustrious predecessor, died. Maxwell spent much of his short but highly productive life providing a theoretical basis for the experimental discoveries of Faraday. It is fair to say that Einstein was led to his theory of relativity by means of his close scrutiny of Maxwell's equations. Einstein, a great admirer of Maxwell, once wrote of him, "Imagine his feelings when the differential equations he had formulated proved to him that electromagnetic fields spread in the form of polarized waves and with the speed of light!"

37-2 Maxwell's Equations: A Tentative Listing

When we studied classical mechanics and thermodynamics, our aim was to identify the smallest, most compact set of equations or laws that would define the subject as completely as possible. In mechanics we found this in Newton's three laws of motion and in the associated force laws, such as Newton's law of gravitation. In thermodynamics we found it in the three laws that we have described in Chapters 19, 20, and 22.

* See "James Clerk Maxwell," by James R. Newman, *Scientific American,* June 1955.

We have now reached the point in our studies of electromagnetism at which we can begin to assemble its basic equations. Table 1 shows a tentative set of them, pulled together from earlier sections of this book. As you examine this short list, it may occur to you that you have encountered many more than four equations during the course of the last 14 chapters! That is certainly true but most of those equations—the expression for the electric field strength on the axis of an electric dipole, for example—apply to special situations and are not basic, in the sense that they are derived from more fundamental equations. You may still ask: "What about Coulomb's law and the law of Biot and Savart? We certainly treated them as fundamental equations." However, as we pointed out earlier, these laws may be viewed as fundamental only for stationary or slowly moving charges. We generalize them in Table 1 by Eq. I (for Coulomb's law) and Eq. IV (for the law of Biot and Savart.) Equations I and IV hold for rapidly time varying as well as for static or near-static situations.

The missing term to which we referred above will prove to be no trifling correction but will round out the complete description of electromagnetism and, beyond this, will establish optics as an integral part of electromagnetism. In particular, it will allow us to prove—as we shall do in the following chapter—that the speed of light c in free space is related to purely electric and magnetic quantities by

$$\boxed{c = \frac{1}{\sqrt{\epsilon_0 \mu_0}}} \quad \text{(the speed of light).} \qquad (1)$$

It will also lead us to the concept of the electromagnetic spectrum, which lies behind the experimental discovery of radio waves.

We have seen how the principle of symmetry permeates physics and how it has often led to new insights or discoveries. For example: If body A attracts body B with a force $\mathbf{F}$, then perhaps body B attracts body A with a

Table 1 The Basic Equations of Electromagnetism: A Tentative List

Number	*Name*	*Equation*	*Reference*
I	Gauss' law for electricity	$\oint \mathbf{E} \cdot d\mathbf{A} = q/\epsilon_0$	25-9
II	Gauss' law for magnetism	$\oint \mathbf{B} \cdot d\mathbf{A} = 0$	34-13
III	Faraday's law of induction	$\oint \mathbf{E} \cdot d\mathbf{s} = -d\Phi_B/dt$	32-18
IV	Ampere's law	$\oint \mathbf{B} \cdot d\mathbf{s} = \mu_0 i$	31-16

force $-\mathbf{F}$ (it does). For another example: If there is a negative electron, there may well be a positive electron (there is).

Let us examine Table 1 from this point of view. First, we say that when we are dealing with symmetry considerations alone (that is, not making quantitative calculations) we can ignore the quantities ϵ_0 and μ_0. These constants result from our choice of unit systems and play no role in arguments of symmetry. In fact, the equations displayed on the T-shirt in the photo at the beginning of this chapter are expressed in a unit system in which $\epsilon_0 = \mu_0 = 1$.

With this in mind, we see that the left sides of the equations in Table 1 are completely symmetrical, in pairs. Equations I and II are surface integrals of **E** and **B**, respectively, over closed surfaces. Equations III and IV are line integrals of **E** and **B**, respectively, around closed loops. (Note in passing that, if we had substituted Coulomb's law for Eq. I and the law of Biot and Savart for Eq. IV, these appealing symmetries would be entirely missing.)

The right sides of these equations, however, do not seem symmetrical at all. We identify two kinds of asymmetry, which we discuss separately.

The First Asymmetry. This asymmetry deals with the apparent fact that although there are isolated centers of charge (electrons and protons, say) isolated centers of magnetism (magnetic monopoles, see Section 34–1) do not seem to exist in nature. That is how we interpret the fact that a q occurs on the right side of Eq. I but no corresponding magnetic quantity appears on the right side of Eq. II. In the same way, the term $\mu_0 i$ $(= \mu_0\, dq/dt)$ appears on the right of Eq. IV but no similar term (a current of magnetic monopoles) appears on the right of Eq. III.

These considerations of symmetry—coupled with the detailed predictions of certain preliminary theories of the nature of elementary particles and forces—have motivated physicists to search for the magnetic monopole in great earnest and in many ways; none has been found. It may yet be found. It is as though nature were hinting and guiding physicists in their explorations.

The Second Asymmetry. This one sticks out like a sore thumb. On the right side of Eq. III (Faraday's law of induction), we find the term $-d\Phi_B/dt$ and we interpret this law loosely by saying:

If you change a magnetic field ($d\Phi_B/dt$), you produce an electric field ($\oint \mathbf{E} \cdot d\mathbf{s}$).

We learned this in Section 32–2 where we showed that if you shove a bar magnet through a closed conducting loop, you do indeed induce an electric field, and thus a current, in that loop.

From the principle of symmetry, we are entitled to suspect that the symmetrical relation holds, that is:

If you change an electric field ($d\Phi_E/dt$), you produce a magnetic field ($\oint \mathbf{B} \cdot d\mathbf{s}$).

This conclusion from symmetry arguments proves to be correct when submitted to the test of laboratory experiment. This supposition supplies us with the important "missing" term in Eq. IV in Table 1; we develop this idea fully in the following section.

37–3 Induced Magnetic Fields

Here we discuss in detail the evidence for the supposition of the previous section, namely: "A changing electric field induces a magnetic field." Although we shall be guided by considerations of symmetry alone, we shall also point to direct experimental verification.

Figure 1*a* shows a uniform electric field **E** filling a cylindrical region of space. It might be produced by a circular parallel-plate capacitor, as suggested in Fig. 1*b*. We assume that E is increasing at a steady rate dE/dt, which means that charge must be supplied to the capacitor plates at a steady rate; to supply this charge requires a steady current i into the positive plate and an equal steady current i out of the negative plate.

Direct experiment shows that a magnetic field is set up by this changing electric field both inside the changing electric field and in the space outside that field. Figure 1*a* shows **B** for four selected points. This figure suggests a beautiful example of the symmetry of nature. A changing magnetic field induces an electric field (Faraday's law); now we see that a changing electric field induces a magnetic field.

To describe this new effect quantitatively, we are guided by analogy with Faraday's law of induction,

$$\oint \mathbf{E} \cdot d\mathbf{s} = -\frac{d\Phi_B}{dt} \quad \text{(Faraday's law of induction)}, \quad (2)$$

which asserts that an electric field (left side) is produced

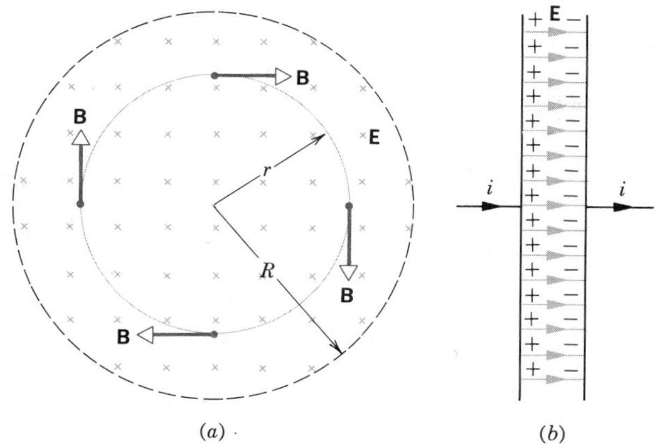

Figure 1 (*a*) A uniform electric field **E**, which is increasing in magnitude, fills a cylindrical region of space. The magnetic fields induced by this changing electric field are shown for four points on a circle of arbitrary radius *r*. (*b*) Such a changing electric field might be produced by charging a parallel-plate capacitor, as shown in this side view.

by a changing magnetic field (right side). For the symmetrical counterpart we might well venture to write

$$\oint \mathbf{B} \cdot d\mathbf{s} = -\frac{d\Phi_E}{dt} \quad \text{(not correct).} \tag{3}$$

This is certainly symmetrical with Eq. 2, but there are two things wrong with it. The first is that experiment requires that we replace the minus sign by a plus sign. This in itself is a kind of symmetry and, in any case, nature requires it.

The second difficulty with Eq. 3 is a formal one, based on the fact that we are committed to SI units. In this unit system, Eq. 3 is not dimensionally correct; to make it correct, we must insert a factor $\mu_0\epsilon_0$ on the right-hand side. Thus, the (correct) symmetrical counterpart of Eq. 2, which we may well call Maxwell's law of induction, is

$$\boxed{\oint \mathbf{B} \cdot d\mathbf{s} = +\mu_0\epsilon_0 \frac{d\Phi_E}{dt}} \quad \text{(Maxwell's law of induction).} \tag{4}$$

Note that in Eq. 2 the "induction" refers to the induced electric field; in Eq. 4 it refers to the induced magnetic field.

Figure 2, in which we compare Fig. 1*a* and Fig. 10*b* of Chapter 32, displays these electromagnetic symmetries more clearly. Figure 2*a* shows a magnetic field produced by a changing electric field; Fig. 2*b* shows the converse. In each figure the appropriate flux, Φ_E or Φ_B, is increasing. However, experiment requires that the lines of **B** in Fig. 2*a* be clockwise whereas those of **E** in Fig. 2*b* must be counterclockwise. This is why Eqs. 2 and 4 differ by a minus sign.

In Section 31-2 we saw that a magnetic field can also be set up by a current in a wire. We described this quantitatively by Ampere's law, which we now recognize

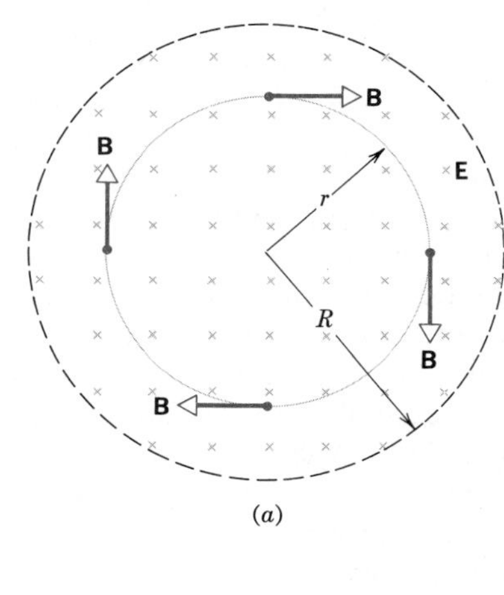

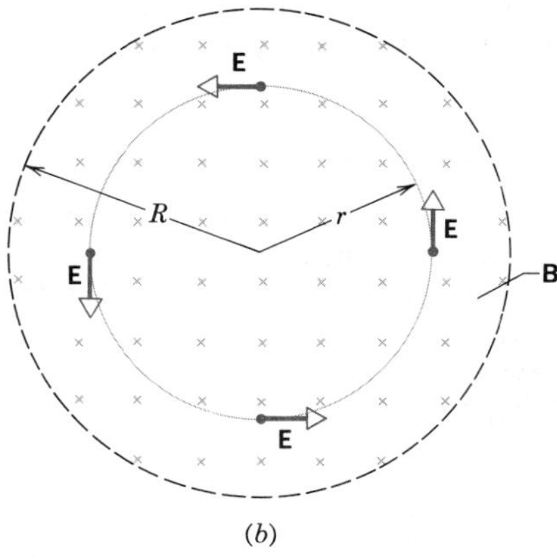

Figure 2 (*a*) The same as Fig. 1*a*; a changing electric field induces a magnetic field. (*b*) The same as Fig. 10*b* of Chapter 32; a changing magnetic field induces an electric field. In each case the fields marked by the crosses are increasing in magnitude. Note that the induced fields in these two cases point in opposite directions.

as being incomplete in form. Thus,

$$\oint \mathbf{B} \cdot d\mathbf{s} = \mu_0 i \quad \text{(Ampere's law—incomplete)}, \quad (5)$$

in which i is the conduction current passing through the Amperian loop around which the line integral is taken.

Thus, there are at least two ways of setting up a magnetic field: (1) by a changing electric field (Eq. 4) and (2) by a current (Eq. 5). In general, we must allow for both possibilities. By combining Eqs. 4 and 5, we arrive at the law in its complete form,

$$\boxed{\oint \mathbf{B} \cdot d\mathbf{s} = +\mu_0\epsilon_0 \frac{d\Phi_E}{dt} + \mu_0 i}$$

$$\text{(Ampere–Maxwell law)}. \quad (6)$$

Maxwell is responsible for this important generalization of Ampere's law. It is a central and vital contribution, as we have pointed out earlier.

In Chapter 31 we assumed that no changing electric fields were present so that the term $d\Phi_E/dt$ in Eq. 6 was zero. In the discussion leading to Eq. 3, we assumed that there were no conduction currents in the space containing the electric field. Thus, the term i in Eq. 6 was zero. We see now that each of these situations is a special case.

We cannot claim to have derived Eq. 6 from deeper principles. Although our symmetry arguments should have made this equation at least reasonable, it basically must stand or fall on whether or not its predictions agree with experiment. As we shall see in Chapter 38, this agreement is complete and impressive.

Sample Problem 1 A parallel-plate capacitor with circular plates is being charged as in Fig. 1*b*. (a) Derive an expression for the induced magnetic field at various radii r for the case of $r < R$.

There are no conduction currents between the plates so that $i = 0$ in Eq. 6, leaving

$$\oint \mathbf{B} \cdot d\mathbf{s} = \mu_0\epsilon_0 \frac{d\Phi_E}{dt}. \quad (7)$$

We can write this equation, for the case of $r < R$ in Fig. 1*a*, as

$$(B)(2\pi r) = \mu_0\epsilon_0 \frac{d}{dt}[(E)(\pi r^2)] = \mu_0\epsilon\pi r^2 \frac{dE}{dt}.$$

Solving for B, we find

$$B = \tfrac{1}{2}\mu_0\epsilon_0 r \frac{dE}{dt} \quad (r < R). \quad \text{(Answer)}$$

We see that $B = 0$ at the center of the capacitor, where $r = 0$, and that B increases linearly with r out to the edge of the circular capacitor plates.

(b) Evaluate B for $r = R = 55$ mm and for $dE/dt = 1.5 \times 10^{12}$ V/m·s.

From the expression just derived, we have

$$\begin{aligned} B &= (\tfrac{1}{2})(4\pi \times 10^{-7}\ \text{T·m/A})(8.85 \times 10^{-12}\ \text{C}^2/\text{N·m}^2) \\ &\quad \times (55 \times 10^{-3}\ \text{m})(1.5 \times 10^{12}\ \text{V/m·s}) \\ &= 4.59 \times 10^{-7}\ \text{T} \approx 460\ \text{nT}. \quad \text{(Answer)} \end{aligned}$$

Be sure to check the unit cancellations.

(c) Derive an expression for the induced magnetic field in this example for the case of $r > R$.

For points with $r > R$, the electric field E is equal to zero, so that Eq. 7 becomes

$$(B)(2\pi r) = \mu_0\epsilon_0 \frac{d}{dt}[(E)(\pi R^2)] = \mu_0\epsilon_0\pi R^2 \frac{dE}{dt}.$$

Solving for B, we find

$$B = \frac{\mu_0\epsilon_0 R^2}{2r}\frac{dE}{dt} \quad (r > R). \quad \text{(Answer)}$$

Note that the two expressions for B that we have derived yield the same result—as we expect—for $r = R$. Furthermore, the value of B at $r = R$, which we calculated in (b) above, is the maximum value that occurs for any value of r.

The induced magnetic field calculated in (b) above is so small that it can scarcely be measured with simple apparatus. This is in sharp contrast to induced electric fields (Faraday's law), which can be demonstrated easily. This experimental difference is in part due to the fact that induced emfs can easily be multiplied by using a coil of many turns. No technique of comparable simplicity exists for induced magnetic fields. In experiments involving oscillations at very high frequencies, dE/dt above can be very large, resulting in significantly larger values of the induced magnetic field. In any case, the experiment of this Sample Problem has been done and the presence of the induced magnetic fields verified quantitatively.

37-4 Displacement Current

If you look closely at the right side of Eq. 6, you will see that the term $\epsilon_0\, d\Phi_E/dt$ must have the dimensions of a current. Even though no motion of charge is involved, there are advantages in giving this term the name *displacement current* and representing it by the symbol i_d.*

* The word *displacement* was introduced for historical reasons that need not concern us here.

That is,

$$i_d = \epsilon_0 \frac{d\Phi_E}{dt} \quad \text{(displacement current).} \tag{8}$$

Thus, we can say that a magnetic field can be set up either by a *conduction current* i or by a *displacement current* i_d and we can rewrite Eq. 6 as

$$\oint \mathbf{B} \cdot d\mathbf{s} = \mu_0(i_d + i) \quad \text{(Ampere–Maxwell law).} \tag{9}$$

By generalizing the definition of current in this way, we can hold on to the notion that current is continuous, a principle established for steady conduction current in Section 28-2. In Fig. 1*b*, for example, a (conduction) current i enters the positive plate and leaves the negative plate. The conduction current is not continuous across the capacitor gap because no charge is transported across this gap. However, the displacement current i_d in the gap will prove to be exactly i, thus retaining the concept of the continuity of current.

To calculate the displacement current, recall that E in the gap of the capacitor of Fig. 1*b* is given by Eq. 4 of Chapter 27, or

$$E = \frac{q}{\epsilon_0 A},$$

in which q is the charge on the positive capacitor plate and A is the plate area. Differentiation gives

$$\frac{dE}{dt} = \frac{1}{\epsilon_0 A}\frac{dq}{dt} = \frac{1}{\epsilon_0 A} i,$$

so that we can write for the conduction current

$$i = \epsilon_0 A \frac{dE}{dt} \quad \text{(conduction current).} \tag{10}$$

The displacement current, defined by Eq. 8, is

$$i_d = \epsilon_0 \frac{d\Phi_E}{dt} = \epsilon_0 \frac{d(EA)}{dt}$$

or

$$i_d = \epsilon_0 A \frac{dE}{dt} \quad \text{(displacement current).} \tag{11}$$

As Eqs. 10 and 11 show, the conduction current in the connecting wires of the capacitor of Fig. 1*b* and the displacement current in the gap between the plates have the same value. When the capacitor is fully charged, up to the value of the applied emf, the current in the connecting wires drops to zero. The electric field between the plates assumes a steady value so that $dE/dt = 0$, which means that the displacement current also drops to zero.

The displacement current i_d, given by Eq. 11, has a direction as well as a magnitude. The direction of the conduction current i is that of the conduction current density vector $\mathbf{J}$. Similarly, the direction of the displacement current i_d is that of the displacement current density vector $\mathbf{J}_d$ which—as we see from Eq. 11—is just $\epsilon_0(d\mathbf{E}/dt)$. The right-hand rule applied to $\mathbf{J}_d$ gives the direction of the associated magnetic field, just as it does for the conduction current density $\mathbf{J}$.

Sample Problem 2 What is the displacement current for the situation of Sample Problem 1?

From Eq. 8, the definition of displacement current,

$$\begin{aligned} i_d &= \epsilon_0 \frac{d\Phi_E}{dt} = \epsilon_0 \frac{d}{dt}[(E)(\pi R^2)] = \epsilon_0 \pi R^2 \frac{dE}{dt} \\ &= (8.85 \times 10^{-12}\ \mathrm{C^2/N\cdot m^2})(\pi)(55 \times 10^{-3}\ \mathrm{m})^2 \\ &\quad \times (1.5 \times 10^{12}\ \mathrm{V/m\cdot s}) \\ &= 0.126\ \mathrm{A} \approx 130\ \mathrm{mA}. \qquad \text{(Answer)} \end{aligned}$$

You may be asking yourself: "This is a reasonably large current. Yet in Sample Problem 1(b) we saw that it produced a magnetic field of only 460 nT at the edge of the capacitor plates. Why such a very small value of the magnetic field?" It is true that a conduction current of 130 mA in a thin wire would produce a much larger magnetic field at the surface of the wire, easily detectable by a compass needle.

The difference is *not* caused by the fact that one current is a conduction current and the other is a displacement current. Under the same conditions, both kinds of current are equally effective in generating a magnetic field. The difference arises because the conduction current, in this case, is confined to a thin wire but the displacement current is spread out over an area equal to the surface area of the capacitor plates. Thus, the capacitor behaves like a "fat wire" of radius 55 mm, carrying a (displacement) current of 130 mA. Its largest magnetic effect, which occurs at the capacitor edge, is much smaller than would be the case at the surface of a thin wire.

37-5 Maxwell's Equations

Equation 6 completes our presentation of the basic equations of electromagnetism, called Maxwell's equa-

Table 2 Maxwell's Equations[a]

Number	Name	Equation	Describes	Reference
I	Gauss' law for electricity	$\oint \mathbf{E} \cdot d\mathbf{A} = q/\epsilon_0$	Charge and the electric field	Chapter 25
II	Gauss' law for magnetism	$\oint \mathbf{B} \cdot d\mathbf{A} = 0$	The magnetic field	Chapter 34
III	Faraday's law	$\oint \mathbf{E} \cdot d\mathbf{s} = -\frac{d\Phi_B}{dt}$	An electric field produced by a changing magnetic field	Chapter 32
IV	Ampere–Maxwell law	$\oint \mathbf{B} \cdot d\mathbf{s} = \mu_0\epsilon_0 \frac{d\Phi_E}{dt} + \mu_0 i$	A magnetic field produced by a changing electric field or by a current or both	Chapters 31 and 37

[a] Written on the assumption that no dielectric or magnetic materials are present.

tions. We display them in Table 2, which rounds out the preliminary listing of Table 1 by supplying the missing term in Eq. IV of that table. These equations are, of course, not purely theoretical speculations but were developed to explain certain crucial laboratory experiments and observations. We list some of these in Table 3.

Maxwell described his theory of electromagnetism in a lengthy *Treatise on Electricity and Magnetism,* published in 1873, just 6 years before his death. The *Treatise* makes difficult reading and, as a matter of fact, does not contain Maxwell's equations in the form in which we have presented them (nor in the form shown in the photo at the head of this chapter.) It fell to Oliver Heaviside (1850–1925), described as "an unemployed, largely self-educated former telegrapher" who "set out in the 1870's to master electromagnetic theory" to cast Maxwell's theory into the form of the four equations that we know today.

Maxwell's ideas were slow to be adopted by the generation of practical men, trained in rule-of-thumb methods, who were his contemporaries.* Figure 3, first published in 1888, suggests the nature of the controversy. It shows W. H. Preece ("Experience"), the head of the British telegraph service and a leading "practical man," in triumph over Oliver Lodge ("Experiment"), a prominent supporter of Maxwellian ideas. The triumph

Figure 3 An 1888 cartoon suggesting the opposition of those trained in rule-of-thumb methods to the introduction of Maxwellian ideas. The triumph of the "practical men" (as they called themselves) slowly declined as it became more and more apparent that it was Maxwell's equations that were "practical."

* See "Practice vs. Theory—The British Electrical Debate, 1888–1891," by Bruce J. Hung, *Isis,* September 1983.

Table 3 Some Crucial Electromagnetic Experiments

1. Like charges repel and unlike charges attract, with a force that varies as the inverse square of their separation.
2. A charge placed on an insulated conductor moves entirely to its outer surface.
3. It has not been possible to verify the existence of a magnetic monopole.
4. A bar magnet, thrust through a closed coil, will set up a current in that coil.
5. If the current in a coil changes, a current will be set up in a second closed coil, placed nearby.
6. A current in a wire can influence a compass needle.
7. Parallel wires carrying currents in the same direction attract each other.
8. The speed of light can be calculated from purely electric and magnetic measurements.

was short-lived; the year in which the cartoon appeared was the year in which Heinrich Hertz, guided by Maxwell's theory, discovered radio waves.

We suggested in Section 37–2 that Maxwell's equations (as they appear in Table 2) bear the same relation to electromagnetism that Newton's laws of motion do to mechanics. There is, however, an important difference. Einstein presented his special theory of relativity in 1905, roughly 200 years after Newton's laws appeared and about 40 years after Maxwell's equations. As it turns out, Newton's laws had to be drastically modified in cases in which the relative speeds approached that of light. However, no changes whatever were required in Maxwell's equations; they are totally consistent with the special theory of relativity. In fact, Einstein's theory grew out of his deep and careful thinking about the electromagnetic equations of James Clerk Maxwell.

REVIEW AND SUMMARY

In Table 1 we summarize the basic equations of electromagnetism as they have been presented in earlier chapters. In studying them for symmetry we come to see that, to make Ampere's law symmetrical with Faraday's law, we must write it as

Maxwell's Extension of Ampere's Law

$$\oint \mathbf{B} \cdot d\mathbf{s} = \mu_0 \epsilon_0 \frac{d\Phi_E}{dt} + \mu_0 i \quad \text{(Ampere-Maxwell law).} \qquad [6]$$

The new first term on the right states that *a changing electric field ($d\Phi_E/dt$) generates a magnetic field* ($\oint \mathbf{B} \cdot d\mathbf{s}$). It is the symmetrical counterpart of Faraday's law: *A changing magnetic field ($d\Phi_B/dt$) generates an electric field* ($\oint \mathbf{E} \cdot d\mathbf{s}$). See Sample Problem 1.

We define displacement current as

Displacement Current

$$i_d = \epsilon_0 \frac{d\Phi_E}{dt} \quad \text{(displacement current).} \qquad [8]$$

Equation 6 then becomes

$$\oint \mathbf{B} \cdot d\mathbf{s} = \mu_0 (i_d + i) \quad \text{(Ampere–Maxwell law).} \qquad [9]$$

We thus retain the notion of continuity of current (conduction current + displacement current). Sample Problem 2 illustrates this for the charging capacitor of Fig. 1 and Sample Problem 1. Displacement current involves a changing electric field and *not* a transfer of charge.

Maxwell's Equations

Maxwell's equations, displayed in Table 2, summarize all of electromagnetism and form its foundation. They will be used later and warrant careful study.

QUESTIONS

1. In your own words explain why Faraday's law of induction (see Table 2) can be interpreted by saying, "A changing magnetic field generates an electric field."

2. If a uniform flux Φ_E through a plane circular ring decreases with time, is the induced magnetic field (as viewed along the direction of **E**) clockwise or counterclockwise?

3. Why is it so easy to show that "a changing magnetic field produces an electric field" but so hard to show in a simple way that "a changing electric field produces a magnetic field"?

4. In Fig. 1*a* consider a circle with $r > R$. How can a magnetic field be induced around this circle, as Sample Problem 1 shows? After all, there is no electric field at the location of this circle and $dE/dt = 0$ here.

5. In Fig. 1*a*, **E** is into the figure and is increasing in magnitude. Find the direction of **B** if, instead, (*a*) **E** is into the figure and decreasing, (*b*) **E** is out of the figure and increasing, (*c*) **E** is out of the figure and decreasing, and (*d*) **E** remains constant.

6. In Fig. 1*c*, Chapter 35, a displacement current is needed to maintain continuity of current in the capacitor. How can one exist, considering that there is no charge on the capacitor?

7. In Figs. 1*a*,*b* what is the direction of the displacement current i_d? In this same figure, can you find a rule relating the direction (*a*) of **B** and **E**? (*b*) of **B** and $d\mathbf{E}/dt$?

8. What advantages are there in calling the term $\epsilon_0 d\Phi_E/dt$ in Eq. IV, Table 2 a displacement current?

9. Can a displacement current be measured with an ammeter? Explain.

10. Why are the magnetic effects of conduction currents in wires so easy to detect but the magnetic effects of displacement currents in capacitors so hard to detect?

11. In Table 2 there are three kinds of apparent lack of symmetry in Maxwell's equations. (*a*) The quantities ϵ_0 and/or μ_0 appear in I and IV but not in II and III. (*b*) There is a minus sign in III but no minus sign in IV. (*c*) There are missing "magnetic pole terms" in II and III. Which of these represent genuine lack of symmetry? If magnetic monopoles were discovered, how would you rewrite these equations to include them? (*Hint:* Let p be the magnetic pole strength.)

EXERCISES AND PROBLEMS

Section 37-2 Maxwell's Equations—A Tentative Listing

1E. Verify the numerical value of the speed of light from Eq. 1 and show that the equation is dimensionally correct. (See Appendix B.)

2E. (*a*) Show that $\sqrt{\mu_0/\epsilon_0} = 377\ \Omega$ (called the "impedance of free space"). (*b*) Show that the angular frequency of ordinary 60 Hz ac is 377 rad/s. (*c*) Compare (*a*) with (*b*). Do you think that this coincidence is the reason that 60 Hz was originally chosen as the frequency for ac generators? Recall that, in Europe, 50 Hz is used.

Section 37-3 Induced Magnetic Fields

3E. For the situation of Sample Problem 1, where is the induced magnetic field equal to one-half of its maximum value?

4P. Suppose that a circular-plate capacitor has a radius R of 30 mm and a plate separation of 5.0 mm. A sinusoidal potential difference with a maximum value of 150 V and a frequency of 60 Hz is applied between the plates. Find $B_m(R)$, the maximum value of the induced magnetic field at $r = R$.

5P. For the conditions of Problem 4, plot $B_m(r)$ for the range $0 < r < 10$ cm.

Section 37-4 Displacement Current

6E. Prove that the displacement current in a parallel-plate capacitor can be written as

$$i_d = C\frac{dV}{dt}.$$

7E. You are given a 1.0-μF parallel-plate capacitor. How would you establish an (instantaneous) displacement current of 1.0 A in the space between its plates?

8E. In Sample Problem 1 show that the *displacement current density* J_d is given, for $r < R$, by

$$J_d = \epsilon_0 \frac{dE}{dt}.$$

9E. Fig. 4 shows the plates P_1 and P_2 of a circular parallel-plate capacitor of radius R. They are connected as shown to long straight wires in which a constant conduction current i exists. A_1, A_2, and A_3 are hypothetical circles of radius r, two of them outside the capacitor and one between the plates. Show that the magnetic field at the circumference of each of these circles is given by

$$B = \frac{\mu_0 i}{2\pi r}.$$

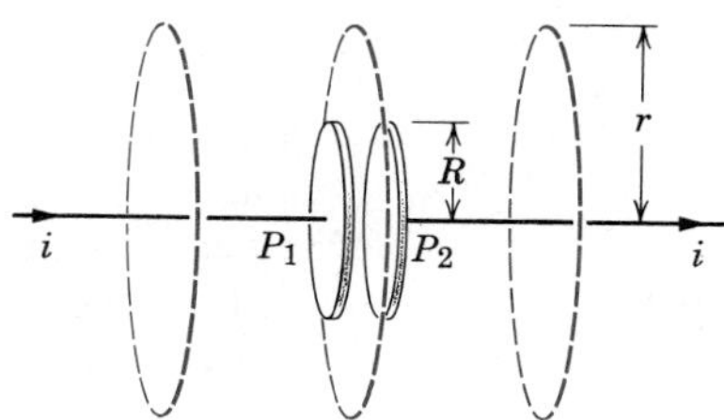

Figure 4 Exercise 9.

10P. In Sample Problem 1 show that the expressions derived for $B(r)$ can be written as

$$B(r) = \frac{\mu_0 i_d}{2\pi r} \quad (r \geq R)$$

and

$$B(r) = \frac{\mu_0 i_d r}{2\pi R^2} \quad (r \leq R).$$

Note that these expressions are of just the same form as those derived in Chapter 31 except that the conduction current i has been replaced by the displacement current i_d.

11P. A parallel-plate capacitor with circular plates 20 cm in diameter is being charged as in Fig. 1*b*. The displacement current density throughout the region is uniform, into the paper in the diagram, and has a value of 20 A/m². (*a*) Calculate the magnetic field B at a distance $r = 50$ mm from the axis of symmetry of the region. (*b*) Calculate dE/dt in this region.

12P. A uniform electric field collapses to zero from an initial strength of 6.0×10^5 N/C in a time of 15 μs in the manner shown in Fig. 5. Calculate the displacement current, through a 1.6 m² region perpendicular to the field, during each of the time intervals (*a*), (*b*), and (*c*) shown on the graph. (Ignore the behavior at the ends of the intervals.)

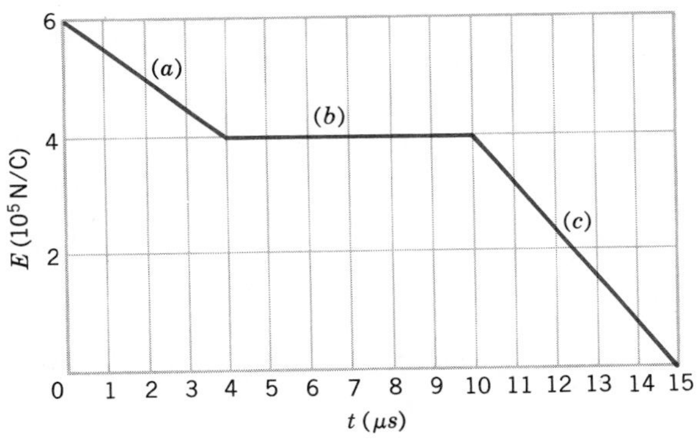

Figure 5 Problem 12.

13P. A parallel-plate capacitor has square plates 1.0 m on a side as in Fig. 6. There is a charging current of 2.0 A flowing into (and out of) the capacitor. (*a*) What is the displacement current through the region between the plates? (*b*) What is dE/dt in this region? (*c*) What is the displacement current through the square dashed path between the plates? (*d*) What is $\oint \mathbf{B} \cdot d\mathbf{s}$ around this square dashed path?

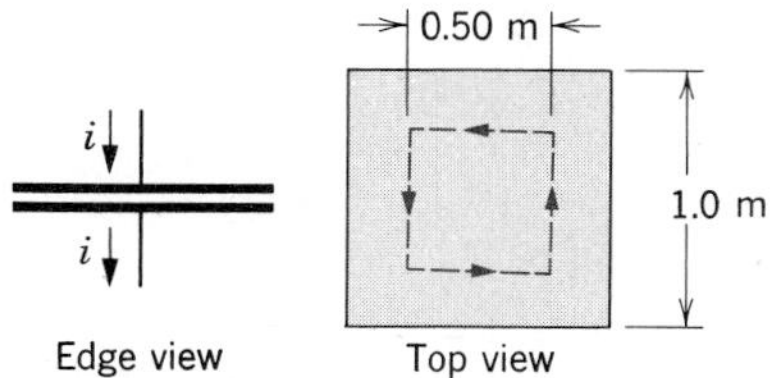

Figure 6 Problem 13.

14P. In 1929 M. R. Van Cauwenberghe succeeded in measuring directly, for the first time, the displacement current i_d between the plates of a parallel-plate capacitor to which an alternating potential difference was applied, as suggested by Fig. 1. He used circular plates whose effective radius was 40 cm and whose capacitance was 100 pF. The applied potential difference had a maximum value V_m of 174 kV at a frequency of 50 Hz. (*a*) What maximum displacement current was present between the plates? (*b*) Why was the applied potential difference chosen to be as high as it is? (The delicacy of these measurements is such that they were only performed in a direct manner more than 60 years after Maxwell enunciated the concept of displacement current! The experiment is described in *Journal de Physique,* No. 8, 1929.)

15P. The capacitor in Fig. 7 consisting of two circular plates with radius $R = 18$ cm is connected to a source of emf $\mathcal{E} = \mathcal{E}_m \sin \omega t$, where $\mathcal{E}_m = 220$ V and $\omega = 130$ rad/s. The maximum value of the displacement current is $i_d = 7.6$ μA. Neglect fringing of the electric field at the edges of the plates. (*a*) What is the maximum value of the current i? (*b*) What is the maximum value of $d\Phi_E/dt$, where Φ_E is the electric flux through the region between the plates? (*c*) What is the separation d between the plates? (*d*) Find the maximum value of the magnitude of **B** between the plates at a distance $r = 11$ cm from the center.

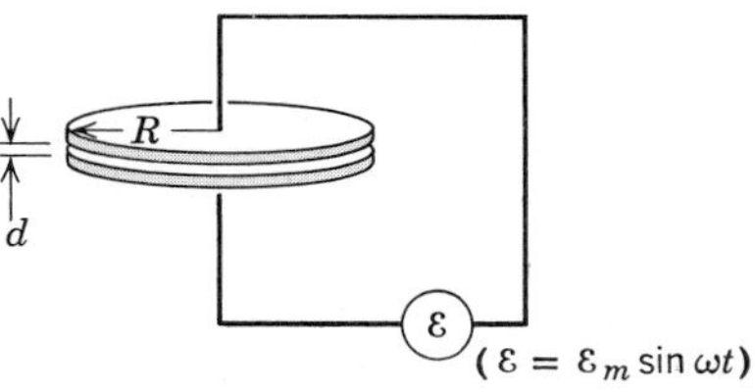

Figure 7 Problem 15.

Section 37-5 Maxwell's Equations

16E. Which of Maxwell's equations (Table 2) is most closely associated with each of the crucial experiments listed in Table 3?

17P. *A self-consistency property of two of the Maxwell equations* (numbers III and IV in Table 2). Two adjacent closed paths *abefa* and *bcdeb* share the common edge *bc* as shown in Fig. 8. (*a*) We may apply $\oint \mathbf{E} \cdot d\mathbf{s} = -d\Phi_B/dt$ (III) to each of

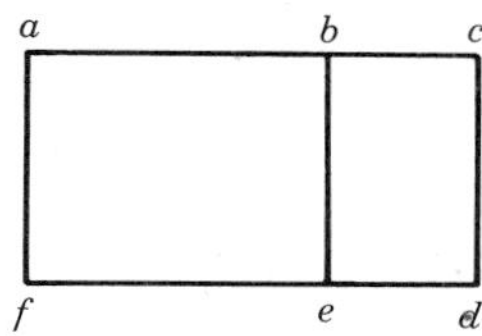

Figure 8 Problem 17.

these two closed paths separately. Show that, from this alone, Eq. III is *automatically* satisfied for the composite closed path *abcdefa*. (*b*) Repeat using Eq. IV.

18P. *A self-consistency property of two of the Maxwell equations* (numbers I and II in Table 2). Two adjacent closed parallelepipeds share a common face as shown in Fig. 9. (*a*) We may apply $\oint \mathbf{E} \cdot d\mathbf{A} = q/\epsilon_0$ (I) to each of these two closed surfaces separately. Show that, from this alone, Eq. I is *automatically* satisfied for the composite closed surface. (*b*) Repeat using Eq. II.

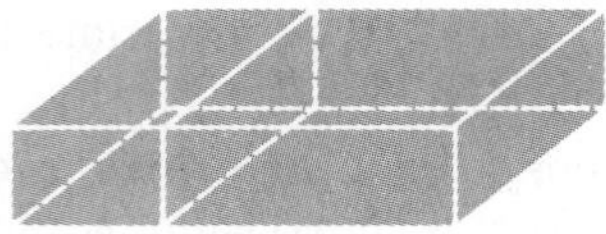

Figure 9 Problem 18.

19P. Maxwell's equations as displayed in Table 2 are written on the assumption that no dielectric materials are present. How should the equations be written if this restriction is removed?

20P. A long cylindrical conducting rod with radius R is centered on the x axis as shown in Fig. 10. A narrow saw cut is made in the rod at $x = b$. A conduction current i, increasing with time and given by $i = \alpha t$, flows toward the right in the rod; α is a (positive) proportionality constant. At $t = 0$ there is no charge on the cut faces near $x = b$. (*a*) Find the magnitude of the charge on these faces, as a function of time. (*b*) Use Eq. I in Table 2 to find E in the gap as a function of time. (*c*) Sketch the lines of $\mathbf{B}$ for $r < R$, where r is the distance from the x axis. (*d*) Use Eq. IV in Table 2 to find $B(r)$ in the gap for $r < R$. (*e*) Compare the above answer with $B(r)$ in the rod for $r < R$.

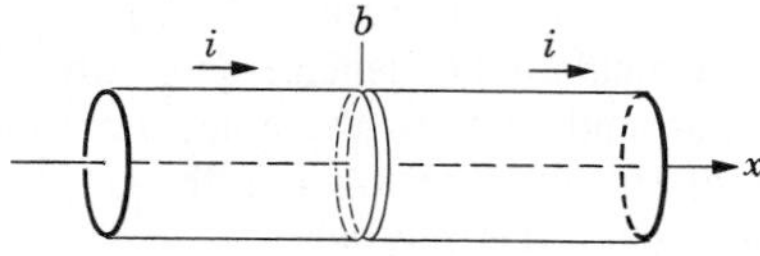

Figure 10 Problem 20.

CHAPTER 38
ELECTROMAGNETIC WAVES

A rainbow over Woolsthorpe Manor, Newton's birthplace. (Newton's analysis of the rainbow in his book Opticks *begins, "This Bow never appears, but where it rains in the Sunshine . . . ") Maxwell and his wondrous equations have taught us to view a beam of sunlight as a configuration of electric and magnetic fields, traveling through space. The sunlight is here spread out for us in the rainbow by wavelength, displaying a tiny segment of the vast electromagnetic spectrum.*

38-1 "Maxwell's Rainbow"

Maxwell's crowning achievement was to show that optics, the study of visible light, is a branch of electromagnetism and that a beam of light is a traveling configuration of electric and magnetic fields. This chapter, the last in our study of strictly electric and magnetic phenomena, is intended to round out that study and to form a bridge to the subject of optics.

In Maxwell's day, visible light and the adjoining infrared and ultraviolet radiations were the only electromagnetic radiations known. Spurred on by Maxwell's predictions, however, Heinrich Hertz discovered what we now call radio waves and verified that they move through the laboratory at the same speed as visible light. It is for this acheivement that we commemorate Hertz by using his name as the SI unit of frequency.

As Fig. 1 shows, we now know an entire spectrum of electromagnetic waves, referred to by one imaginative writer as "Maxwell's rainbow." Think of the extent to which we are bathed in electromagnetic radiation from various regions of the spectrum. The sun, whose radia-

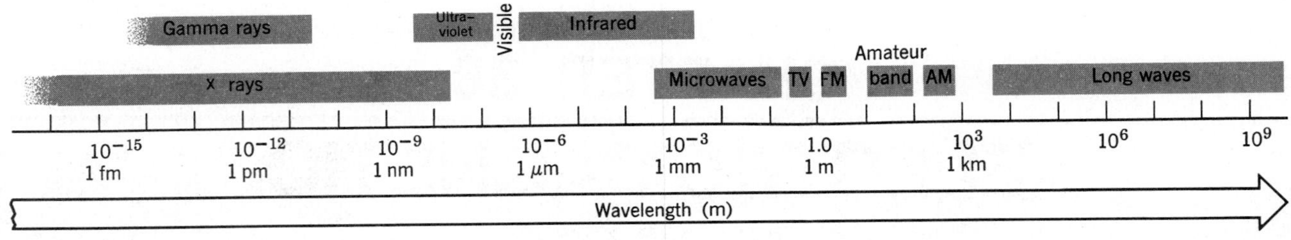

Figure 1 The electromagnetic spectrum. See "The Allocation of the Radio Spectrum," by Charles Lee Jackson, *Scientific American,* February 1980, for a description of the complex process of allocating frequencies in the range from ~ 10 kHz to ~ 300 GHz.

tions define the environment to which we as a species have evolved and adapted, is the dominant source. We are also crisscrossed by radio and TV signals. Microwaves from radar systems and from telephone relay systems may reach us. There are electromagnetic waves from light bulbs, from the heated engine blocks of automobiles, from x-ray machines, from lightning flashes, and from radioactive materials buried in the earth. Beyond this, radiation reaches us from stars and other objects in our galaxy and from other galaxies. We are even exposed, however weakly, to radiation (wavelength ≈ 2 mm) from the primeval fireball, thought by many to be associated with the creation of our universe. Electromagnetic waves also travel in the other direction. Television signals, transmitted from Earth since about 1950, have now taken news about us to whatever technically sophisticated inhabitants there may be on whatever planets may encircle the nearest 400 or so stars.

The wavelength scale in Fig. 1 (and similarly the corresponding frequency scale) is drawn so that each scale marker represents a change in wavelength (and correspondingly in frequency) by a factor of 10. The scale is open-ended, the wavelengths of electromagnetic waves having no inherent upper or lower bounds.

Certain regions of the electromagnetic spectrum in Fig. 1 are identified by familiar labels, *x rays* and *microwaves* being examples. These labels denote roughly defined wavelength ranges within which certain kinds of sources and detectors of the radiations in question are in common use. Other regions of Fig. 1, of which those labeled *TV* and *AM* are examples, represent specific wavelength bands assigned by law for certain commercial or other purposes. Only the most familiar of such allocated bands are shown.

There are no gaps in the electromagnetic spectrum. For example, we can produce electromagnetic radiation with wavelengths in the millimeter range either by microwave techniques (microwave oscillators) or by infrared techniques (heated sources). We stress that all electromagnetic waves, no matter where they lie in the spectrum, travel through free space with the same speed c.

The visible region of the spectrum is of course of particular interest to us. Figure 2 shows the relative sensitivity of the eye of an assumed standard human observer to radiations of various wavelength. The center of the visible region is about 555 nm; light of this wavelength produces the sensation that we call yellow-green.

The limits of the visible spectrum are not well defined because the eye sensitivity curve approaches the axis asymptotically at both long and short wavelengths. If we take the limits, arbitrarily, as the wavelengths at which the eye sensitivity has dropped to 1% of its maximum value, these limits are about 430 and 690 nm, less than a factor of two in wavelength. The eye can detect radiation beyond these limits if it is intense enough. In many experiments in physics we use photographic plates

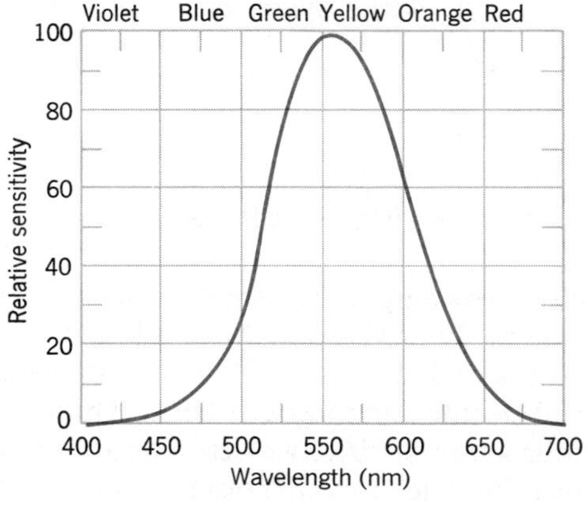

Figure 2 The relative sensitivity of the human eye at different wavelengths.

or light-sensitive electronic detectors in place of the human eye.

38-2 Generating an Electromagnetic Wave

Let us see how electromagnetic waves are generated. Some radiations such as x rays, gamma rays, and visible light come from sources that are of atomic or nuclear size, where quantum physics rules. To simplify matters, we restrict ourselves here to that region of the spectrum ($\lambda \approx 1$ m) in which the source of the radiation (a shortwave radio antenna, say) is both macroscopic and of manageable dimensions.

Figure 3 shows, in broad outline, a generator of such waves. At its heart is an *LC oscillator,* which establishes an angular frequency ω ($= 1/\sqrt{LC}$). Charges and currents in this circuit vary sinusoidally at this frequency, as depicted in Fig. 1 of Chapter 35. An external source—possibly a battery—supplies the energy needed to compensate both for thermal losses in the circuit and for energy carried away by the radiated electromagnetic wave.

The LC oscillator of Fig. 3 is transformer-coupled to a *transmission line,* which might be a coaxial cable among other possibilities, that feeds the *antenna.* The two branches of the antenna alternate sinusoidally in potential at the angular frequency ω set by the oscillator, causing charges to surge back and forth along the antenna axis. The effect is that of an electric dipole whose electric dipole moment varies sinusoidally with time.

Figure 4 shows the electromagnetic wave generated by the accelerating charges in the oscillating electric dipole antenna. The electric and magnetic field lines, which form a figure of revolution about the dipole axis, travel away from the antenna with speed c. The intensity

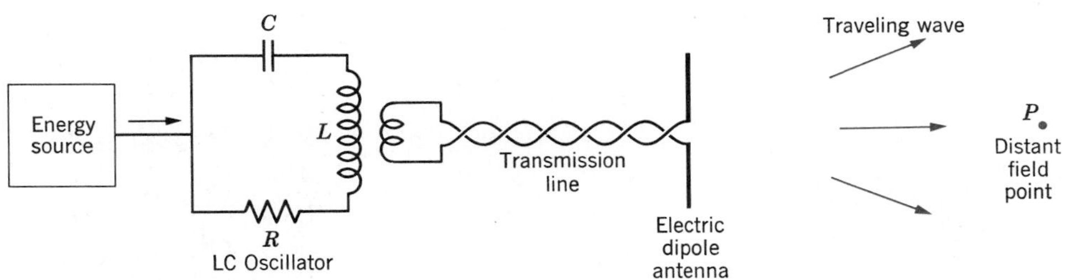

Figure 3 An arrangement for generating a traveling electromagnetic wave in the shortwave radio region of the spectrum. P is a distant point at which an observer can monitor the wave.

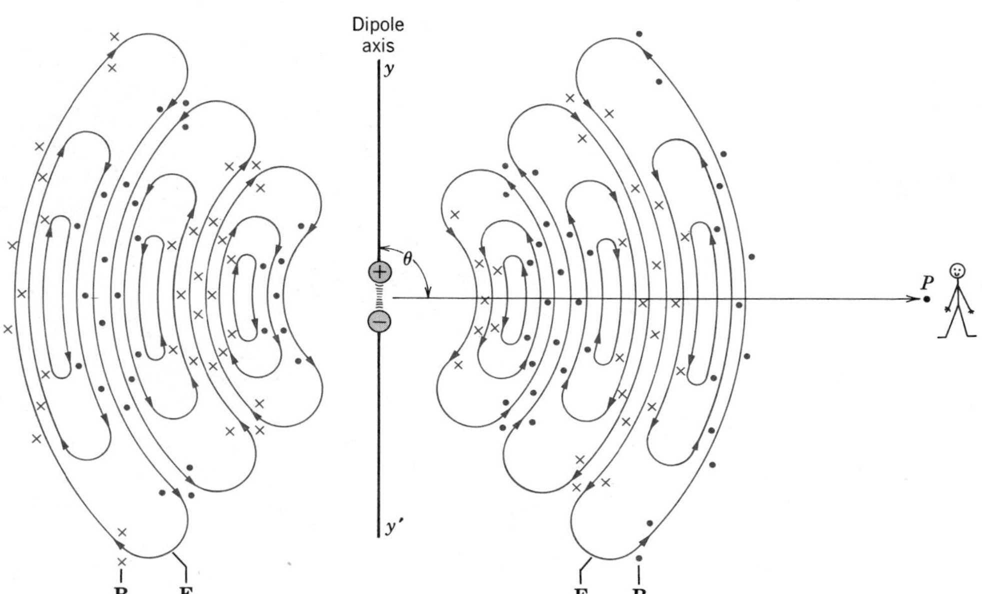

Figure 4 A close-up view of the oscillating dipole antenna of Fig. 3, showing the electric and magnetic field lines associated with the radiated electromagnetic wave. The dots and crosses, as usual, represent field lines emerging from and entering into the plane of the figure. The field pattern close to the antenna (the *near field*) is more complicated and is not shown.

of the traveling wave in any direction is proportional to $\sin^2 \theta$, being zero in the direction of the axis of the dipole ($\theta = 0$ or $180°$) and a maximum in the equatorial plane of the dipole ($\theta = 90°$). The electromagnetic field pattern close to the antenna (the *near field*) is complicated and is not shown in Fig. 4. The field pattern that *is* shown in that figure (the *radiation field*) holds for all radial distances from the antenna such that $r \gg \lambda$.

38-3 The Traveling Electromagnetic Wave — Qualitative

Consider now an observer stationed at some distant fixed point P, far enough from the antenna of Fig. 3 so that the wave fronts sweeping past him would be essentially plane. How would such an observer describe the varying electric and magnetic field patterns that constitute the wave?

Figure 5 suggests how the electric field **E** and the magnetic field **B** change with time as the wave sweeps past our stationary observer. Note that **E** and **B** are perpendicular to each other and to the direction of propagation of the wave and that they are in phase. That is, both wave components reach their maxima at the same time.

We postulate that our observer would quantify his observations by writing down, for the magnitudes of the electric and magnetic field vectors,

$$\boxed{E = E_m \sin(kx - \omega t)} \quad \text{(electric component)} \quad (1)$$

and

$$\boxed{B = B_m \sin(kx - \omega t)} \quad \text{(magnetic component).} \quad (2)$$

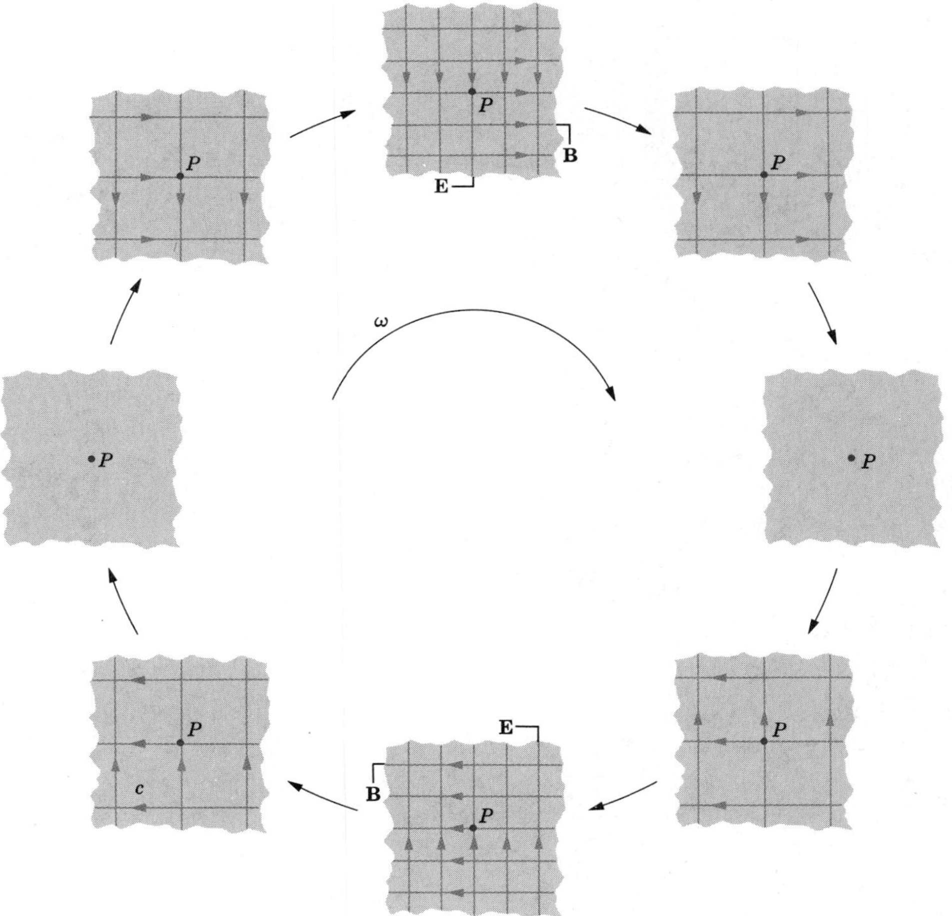

Figure 5 The electromagnetic wave moves directly toward an observer stationed at a remote field point, such as P in Fig. 4. The "patches" show the electric and magnetic field patterns that the observer, facing the oncoming wave, would measure during one period of the wave as the wave sweeps past his position.

Here x is the distance (measured from any convenient origin) in the direction in which the wave is traveling. The speed of the wave, which we represent by the symbol c, is ω/k; see Eq. 14 of Chapter 17.

Equation 1 describes a time-varying electric field; from Maxwell's law of induction, it should be associated with a magnetic field. On the other hand, Eq. 2 describes a time-varying magnetic field; from Faraday's law of induction, it should be associated with an electric field. As a matter of fact, the magnetic field associated with Eq. 1 is precisely the field described by Eq. 2. Furthermore, the electric field associated with Eq. 2 is precisely the field described by Eq. 1.

What a neat arrangement! The two wave components—the electric and the magnetic—feed on each other. The spatial variation of each is associated with the time variation of the other, the entire pattern forming a consistent whole, traveling through space with speed c and completely described by Maxwell's equations.

Let us write down these equations as they apply to an electromagnetic wave traveling through free space. In free space there are no charges ($q = 0$) and no conduction currents ($i = 0$). Equations identified as III and IV in Table 2 of Chapter 37 then become

$$\oint \mathbf{E} \cdot d\mathbf{s} = -\frac{d\Phi_{\mathrm{B}}}{dt}$$

(Faraday's law of induction) (3)

and

$$\oint \mathbf{B} \cdot d\mathbf{s} = \mu_0 \epsilon_0 \frac{d\Phi_{\mathrm{E}}}{dt}$$

(Maxwell's law of induction). (4)

Our question now is:

Is our description of the electromagnetic wave, summarized by Eqs. 1 and 2, consistent with Maxwell's equations as represented by Eqs. 3 and 4?

The answer, which we shall prove in the next section is: Yes, provided that

$$\frac{E_m}{B_m} = c \quad \text{(magnitude ratio),} \qquad (5)$$

where $c\,(=\omega/k)$ is the speed of the waves, given by

$$c = \frac{1}{\sqrt{\mu_0 \epsilon_0}} \quad \text{(the wave speed).} \qquad (6)$$

The requirement expressed by Eq. 5 is not surprising. If the electric and the magnetic components of the wave are as intricately intertwined as we have described, their wave amplitudes E_m and B_m cannot be independent of each other.

Equation 6 is the basis for Maxwell's claim that his electromagnetic equations apply to light and that optics is a branch of electromagnetism. When the electromagnetic quantities μ_0 and ϵ_0 are substituted into Eq. 6, the speed that results is indeed the speed of light.

A Special Note on Eq. 6 (Optional). The quantities μ_0 and ϵ_0 in Eq. 6 are characteristics of the SI unit system. In Maxwell's day that unit system did not yet exist and Eq. 6 would have appeared in a different but totally equivalent form that need not concern us here. What is important is that Maxwell predicted that certain experiments—purely electrical and magnetic in character and in which light beams played no essential role—would yield a speed equal to the speed of light. Maxwell's own words convey his excitement:

> "The velocity of transverse undulations in our hypothetical medium, calculated from the electromagnetic experiments of MM Kohlrausch and Weber, agrees so exactly with the velocity of light calculated from the optical experiments of M Fizeau, that we can scarcely avoid the inference that *light consists in the transverse undulations of the same medium which is the source of the electric and magnetic phenomena.*"

The language reflects the insights and the vocabulary of an earlier day, but it is still clear that the speed of light has been measured by strictly electrical and magnetic experiments. The emphasis is Maxwell's.

After the SI unit system was introduced, Maxwell's prediction was cast in the form of Eq. 6. From the beginning, μ_0 in that equation was given an arbitrarily assigned value, exact by definition:

$$\mu_0 = 4\pi \times 10^{-7}\ \mathrm{H/m} \quad \text{(exact by definition).} \qquad (7)$$

The quantity ϵ_0 can be measured in the laboratory, the most common method being to measure the capacitance of a parallel-plate capacitor of known plate area A and plate separation d and to compute ϵ_0 from the relation

A Sin(kx - ωt)

$C = \epsilon_0 A/d$. One can then calculate the speed of light c from Eq. 6 and compare it with the measured value.

Since 1983, however, as part of the redefinition of the meter, the speed of light has been given an assigned value, exact and by definition, of

$$c = 299{,}792{,}458 \text{ m/s} \quad \text{(exact by definition).} \tag{8}$$

Today, our confidence in Maxwell's theory is so great that we now *assume* Eq. 6 to be correct and we use it to assign an exact value to ϵ_0:

$$\epsilon_0 = 1/c^2 \mu_0 \quad \text{(exact by definition).} \tag{9}$$

Thus, since 1983, because of our redefinition of the meter and of our confidence in Maxwell's prediction, measuring the speed of light and verifying Eq. 6 have lost their significance as possible experiments.

38-4 The Traveling Electromagnetic Wave — Quantitative (Optional)

In this Section, we show that Eqs. 5 and 6 do indeed follow from Maxwell's equations. Figure 6 shows a three-dimensional "snapshot" of a plane wave traveling in the x direction and displaying the instantaneous value of **E** and **B** whose magnitudes are given by Eqs. 1 and 2. The lines of **E** are chosen to be parallel to the y axis and those of **B** to the z axis.

Figure 7 shows two sections through the three-dimensonal diagram of Fig. 6. In Fig. 7*a* the plane of the page is the xy plane and in Fig. 7*b* it is the xz plane. Note that, consistent with Eqs. 1 and 2, **E** and **B** are in phase, that is, at any point through which the wave is moving they reach their maximum values at the same time. We divide our proof into two parts, dealing separately with the induced electric field and the induced magnetic field.

The Induced Electric Field. The shaded rectangle of dimensions dx and h in Fig. 7*a* is fixed at a particular point P on the x axis. As the wave passes over it, the magnetic flux Φ_B through the rectangle will change and — according to Faraday's law of induction — induced electric fields should appear around the rectangle. These induced electric fields are, in fact, simply the electric component of the traveling electromagnetic wave.

Let us apply Lenz's law. The flux Φ_B for the shaded rectangle of Fig. 7*a* is decreasing with time because the wave is moving through the rectangle to the right and a region of weaker magnetic fields is moving into the rectangle. The induced electric field will act to oppose this change, which means that if we imagine the boundary of the rectangle to be a conducting loop, a counterclockwise induced current would appear in it. This current would produce a field of **B** that, within the rectangle, would point out of the page, thus opposing the decrease in Φ_B. There is, of course, no conducting loop, but the net induced electric field does indeed act counterclockwise around the rectangle because $E + dE$, the field magnitude at the right edge of the rectangle, is greater than E, the field magnitude at the left edge. Thus, the electric field configuration is entirely consistent with the concept that it is induced by the changing magnetic field.

For a more detailed analysis let us apply Eq. 3, Faraday's law of induction,

$$\oint \mathbf{E} \cdot d\mathbf{s} = -\frac{d\Phi_B}{dt} \quad \text{(Faraday's law of induction),} \tag{10}$$

going counterclockwise around the shaded rectangle of Fig. 7*a*. There is no contribution to the integral from the top or bottom of the rectangle because **E** and $d\mathbf{s}$ are at right angles here. The integral then becomes

$$\oint \mathbf{E} \cdot d\mathbf{s} = [(E + dE)(h) - [(E)(h)] = h\, dE. \tag{11}$$

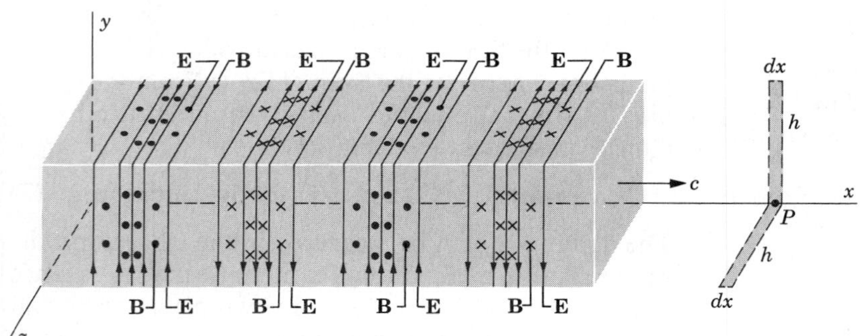

Figure 6 Another view of the plane electromagnetic wave traveling directly toward an observer stationed at point P in Fig. 4. The lines, dots, and crosses represent the electric and magnetic components of the wave. The shaded rectangles at P refer to Fig. 7.

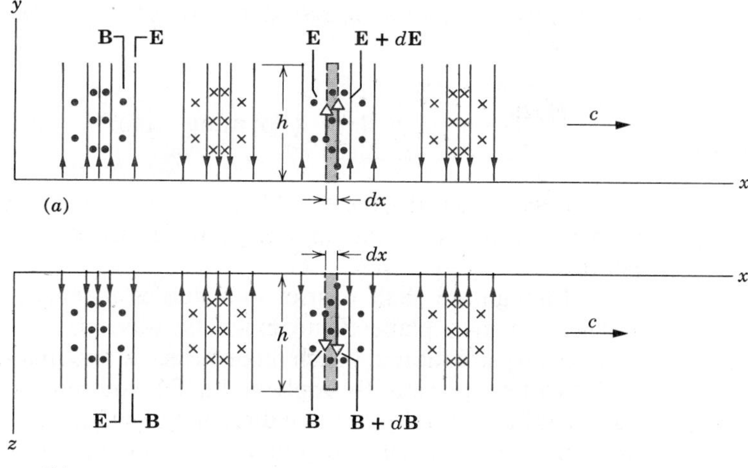

Figure 7 (*a*) The wave of Fig. 6 viewed in the *xy* plane. As the wave sweeps past, the magnetic flux through the shaded rectangle changes, inducing an electric field. (*b*) The wave of Fig. 6 viewed in the *xz* plane. As the wave sweeps past, the electric flux through the shaded rectangle changes, inducing a magnetic field.

The flux Φ_B through this rectangle is

$$\Phi_B = (B)(h\,dx), \tag{12}$$

where B is the (average) magnitude of B at the rectangular strip and $h\,dx$ is the area of the strip. Differentiating Eq. 12 gives

$$\frac{d\Phi_B}{dt} = h\,dx\,\frac{dB}{dt}. \tag{13}$$

If we substitute Eqs. 11 and 13 into Faraday's law (Eq. 10), we find

$$h\,dE = -h\,dx\,\frac{dB}{dt}$$

or

$$\frac{dE}{dx} = -\frac{dB}{dt}. \tag{14}$$

Actually, both B and E are functions of *two* variables, x and t. In evaluating dE/dx we assume that t is constant because Fig. 7*a* is an "instantaneous snapshot." Also, in evaluating dB/dt we assume that x is constant since what we want is the time rate of change of B at a particular place, the point P in Fig. 7*a*. The derivatives under these circumstances are called *partial derivatives,* and a special notation is used for them. In this notation Eq. 14 becomes

$$\frac{\partial E}{\partial x} = -\frac{\partial B}{\partial t}. \tag{15}$$

The minus sign in his equation is appropriate and necessary because, although E is increasing with x at the site of the shaded rectangle in Fig. 7*a*, B is decreasing with t. Since $E(x, t)$ and $B(x, t)$ are known (see Eqs. 1 and 2), Eq. 15 reduces to

$$kE_m \cos(kx - \omega t) = \omega B_m \cos(kx - \omega t).$$

In Section 17–5 we saw that the ratio ω/k for a traveling wave is just its speed, which we choose here to call c. The above equation then becomes

$$\frac{E_m}{B_m} = c \quad \text{(magnitude ratio)}, \tag{16}$$

which is just Eq. 5. We have accomplished the first part of our task.

The Induced Magnetic Field. We now turn to Fig. 7*b*, in which the flux Φ_E for the shaded rectangle is decreasing with time as the wave moves through it. Let us apply Eq. 4, Maxwell's law of induction,

$$\oint \mathbf{B} \cdot d\mathbf{s} = \epsilon_0 \mu_0 \frac{d\Phi_E}{dt} \quad \text{(Maxwell's law of induction)}. \tag{17}$$

The changing flux Φ_E will induce a magnetic field at points around the periphery of the rectangle. This induced magnetic field is simply the magnetic component of the electromagnetic wave. Thus, the electric and the magnetic components of the wave are intimately connected, each depending on the time rate of change of the other.

The integral in Eq. 17, evaluated by proceeding

counterclockwise around the shaded rectangle of Fig. 7*b*, is

$$\oint \mathbf{B} \cdot d\mathbf{s} = [-(B + dB)(h) + [(B)(h)] = -h\, dB, \quad (18)$$

where B is the magnitude of B at the left edge of the strip and $B + dB$ is its magnitude at the right edge.

The flux Φ_E through the rectangle of Fig. 7*b* is

$$\Phi_E = (E)(h\, dx). \quad (19)$$

Differentiating Eq. 19 gives

$$\frac{d\Phi_E}{dt} = h\, dx \frac{dE}{dt}. \quad (20)$$

If we substitute Eqs. 18 and 20 into Maxwell's law of induction (Eq. 17), we find

$$-h\, dB = \epsilon_0 \mu_0 \left(h\, dx \frac{dE}{dt} \right)$$

or, changing to partial derivatives as we did before,

$$-\frac{\partial B}{\partial x} = \epsilon_0 \mu_0 \frac{\partial E}{\partial t}. \quad (21)$$

Again, the minus sign in this equaton is necessary because, although B is increasing with x at point P in the shaded rectangle in Fig. 7*b*, E is decreasing with t.

Evaluating Eq. 21 by using Eqs. 1 and 2 leads to

$$-kB_m \cos(kx - \omega t) = -\epsilon_0 \mu_0 \omega E_m \cos(kx - \omega t),$$

which we can write as

$$\frac{E_m}{B_m} = \frac{1}{\epsilon_0 \mu_0 (\omega/k)} = \frac{1}{\epsilon_0 \mu_0 c}. \quad (22)$$

Combining Eqs. 16 and 22 leads at once to

$$c = \frac{1}{\sqrt{\epsilon_0 \mu_0}} \quad \text{(the wave speed)}, \quad (23)$$

which is exactly Eq. 6. We have completed our proof that Eqs. 1 and 2, describing a traveling electromagnetic wave, are consistent with Maxwell's equations.

38-5 Energy Transport and the Poynting Vector

All sunbathers know that an electromagnetic wave can transport energy and can deliver it to a body on which it falls. The rate of energy transport per unit area (that is to say, the intensity) in such a wave is described by a vector **S**, called the *Poynting vector* after John Henry Poynting (1852–1914), who first discussed its properties. **S** is defined from

$$\mathbf{S} = \frac{1}{\mu_0} \mathbf{E} \times \mathbf{B} \quad \text{(Poynting vector)} \quad (24)$$

its SI units being watts/meter². The direction of **S** at any point gives the direction of energy transport at that point.

Although we shall restrict our studies to energy transported by a plane electromagnetic wave, Eq. 24 holds in any situation in which related electric and magnetic fields are present. By applying Eq. 24, you can use the Poynting vector to trace the direction of energy flow from an alternating current generator, through a transformer, along a transmission line, through another transformer, and into your electric toaster.* Along the transmission line, the vector **S** has a large component parallel to the line to represent the energy flow from the generator to your house. It will also have a small component pointing directly into the conductors forming the line, to account for the i^2R resistive losses in the line.

Let us test that Eq. 24 gives a correct direction for **S** by applying it to the plane electromagnetic wave shown in Fig. 6. Note that at places where the lines of **E** point in the direction of increasing y, the lines of **B** point in the direction of increasing z. At other places, where the lines of **E** point in the direction of decreasing y, the lines of **B** point in the direction of decreasing z. In both cases, however, the product $\mathbf{E} \times \mathbf{B}$ points in the direction of increasing x, that is, the direction in which the wave is traveling. We conclude that—as far as predicting the correct direction of energy transport in the plane wave of Fig. 6 is concerned—Eq. 24 gives the right answer.

The magnitude of S for the traveling electromagnetic wave of Fig. 6 is given by†

$$S = \frac{1}{\mu_0} EB, \quad (25)$$

in which S, E, and B are instantaneous values. E and B are so closely coupled to each other that we only need to deal with one of them; we choose E, largely because most instruments for detecting electromagnetic waves deal

* See, for example, "The Poynting Vector Field and Energy Flow Within a Transformer," by F. Herrmann and Bruno Schmid, *American Journal of Physics,* June 1986.

† **E** and **B** are at right angles to each other, so that the magnitude of $\mathbf{E} \times \mathbf{B}$ is $EB \sin 90° = EB$.

with the electric rather than the magnetic component of the wave.

Equation 5 ($E_m/B_m = c$) expresses the relation between the amplitudes of these two wave components. From Eqs. 1 and 2, we see that this same relation must hold between the instantaneous values, that is

$$\frac{E_m}{B_m} = \frac{E}{B} = c. \tag{26}$$

Thus we can replace B in Eq. 25 by E/c, obtaining

$$S = \frac{1}{c\mu_0} E^2 \quad \text{(energy flow; plane wave)} \tag{27}$$

as the Poynting equation for the special case of a plane electromagnetic wave. We shall show below that this equation can also be derived from information we already have about energy storage in electric and magnetic fields.

In practice, we are more interested in the average value of the time-varying quantity S, which is the *intensity* I of the wave. From Eq. 27 and Eq. 1 we can then write

$$I = \overline{S} = \frac{1}{c\mu_0}\overline{E^2} = \frac{1}{c\mu_0} E_m^2\, \overline{\sin^2 (kx - \omega t)}. \tag{28}$$

We have many times used the fact that the average value of the square of the sine of any angle (over an integral number of half-waves) is one-half. Furthermore, $E_m = \sqrt{2}\, E_{rms}$. With these two substitutions, Eq. 28 becomes

$$I = \overline{S} = \frac{1}{c\mu_0} E_{rms}^2 \quad \text{(energy flow; plane wave).} \tag{29}$$

This useful equation follows directly from Eq. 27, whose proof we now present.

Proof of Eq. 27 (Optional). Figure 8 shows a traveling plane wave, along with a thin "box" of thickness dx and face area A. The box, a mathematical construction, is fixed with respect to the axes and the wave moves through it. At any instant the energy stored in the box is

$$dU = dU_E + dU_B = (u_E + u_B)(A\, dx)$$

where u_E and u_B are, respectively, the electric and the magnetic energy densities and $A\, dx$ is the volume of the box. From Eq. 23 of Chapter 27 and Eq. 26 of Chapter 33, we can write this as

$$dU = \left(\frac{1}{2}\epsilon_0 E^2 + \frac{1}{2\mu_0} B^2\right)(A\, dx). \tag{30}$$

We can use Eq. 6 to eliminate ϵ_0 and can (again) replace B by E/c. The result is

$$dU = \left(\frac{E^2}{c^2\mu_0}\right)(A\, dx). \tag{31}$$

Although we here write dU in terms of E, we must remember that the energy stored in an electromagnetic wave is evenly distributed between the electric and the magnetic field components of the wave.

This energy dU will pass through the right face of the box of Fig. 8 in a time dt equal to dx/c. Thus, the energy per unit area per unit time, which is S, is given by

$$S = \frac{dU}{A\, dt} = \frac{(E^2/c^2\mu_0)(A\, dx)}{A\, (dx/c)} = \frac{1}{c\mu_0} E^2, \tag{32}$$

which (see Eq. 27) is just what we set out to prove.

Sample Problem 1 An observer is 1.8 m from a point light source whose power output P is 250 W. Calculate the rms values of the electric and the magnetic fields at the position of the observer. Assume that the source radiates uniformly in all directions.

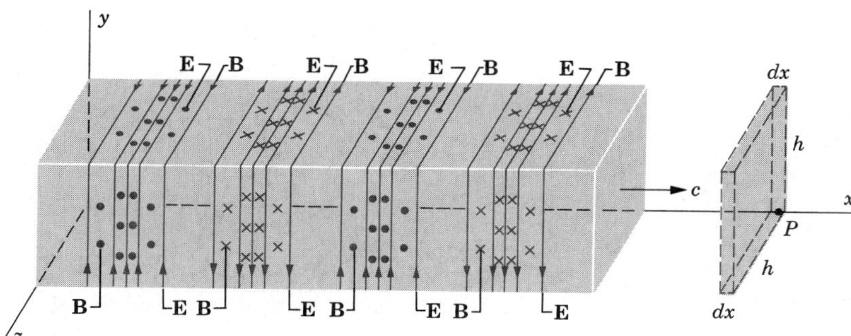

Figure 8 The wave of Fig. 6 once more. It transports energy through a hypothetical rectangular box held fixed at position P. Note that, at all points in the wave, the vector $\mathbf{E} \times \mathbf{B}$ points in the direction in which the wave is moving.

The intensity of the light at a distance r from the source is given by

$$I = \frac{P}{4\pi r^2},$$

where $4\pi r^2$ is the area of a sphere of radius r centered on the source. The intensity is also given by Eq. 29, so that

$$I = \frac{P}{4\pi r^2} = \frac{1}{c\mu_0} E_{\text{rms}}^2.$$

The rms electric field is then

$$E_{\text{rms}} = \sqrt{\frac{Pc\mu_0}{4\pi r^2}}$$
$$= \sqrt{\frac{(250\ \text{W})(3.00 \times 10^8\ \text{m/s})(4\pi \times 10^{-7}\ \text{H/m})}{(4\pi)(1.8\ \text{m})^2}}$$
$$= 48.1\ \text{V/m} \approx 48\ \text{V/m}. \qquad \text{(Answer)}$$

The rms value of the magnetic field follows from Eq. 5 and is

$$B_{\text{rms}} = \frac{E_{\text{rms}}}{c} = \frac{48.1\ \text{V/m}}{3.00 \times 10^8\ \text{m/s}}$$
$$= 1.6 \times 10^{-7}\ \text{T}. \qquad \text{(Answer)}$$

Note that E_{rms} (= 48 V/m) is appreciable as judged by ordinary laboratory standards but that B_{rms} (= 0.0016 gauss) is quite small. This helps to explain why most instruments used for the detection and measurement of electromagnetic waves respond to the electric component of the wave. It is wrong, however, to say that the electric component of an electromagnetic wave is "stronger" than the magnetic component. You cannot compare quantities that are measured in different units. As we have seen, the electric and the magnetic components are on an absolutely equal basis as far as the propagation of the wave is concerned, their average energies—which *can* be compared—being exactly equal.

38-6 Radiation Pressure

We all know that electromagnetic waves transport energy. Perhaps you also know that such waves can transport linear momentum. That is, it is possible to exert a pressure (a *radiation pressure*) on an object by shining a light on it. Such forces must be small in relation to forces of our daily experience because, for example, we do not feel a recoil force when we turn on a flashlight.

Let a parallel beam of radiation, light for example, fall on an object for a time Δt and be entirely absorbed by the object. Maxwell showed that, if an energy U is absorbed during this interval, the magnitude of the momentum p delivered to the object is given by

$$\boxed{p = \frac{U}{c}} \quad \text{(total absorption)}, \qquad (33)$$

where c is the speed of light. The direction of $\mathbf{p}$ is the direction of the incident beam. If the energy U is entirely reflected, the magnitude of the momentum delivered at normal incidence will be twice that given above, or

$$\boxed{p = \frac{2U}{c}} \quad \text{(total reflection)}. \qquad (34)$$

In the same way, twice as much momentum is delivered to an object when a perfectly elastic tennis ball is bounced from it as when it is struck by a perfectly inelastic ball (a lump of putty, say) of the same mass and velocity. If the light energy U is partly reflected and partly absorbed, the delivered momentum will lie between U/c and $2U/c$.

The first measurement of radiation pressure was made in 1901–1903 by Nichols and Hull at Dartmouth College and by Lebedev in Russia, about 30 years after the existence of such effects had been predicted theoretically by Maxwell. Nichols and Hull measured radiation pressures and verified Eqs. 33 and 34, using a torsion balance technique. They allowed light to fall on mirror M as in Fig. 9; the radiation pressure caused the balance arm to turn through a measured angle θ, twisting the torsion fiber F. Assuming a suitable calibration for their torsion fiber, the experimenters could calculate a nu-

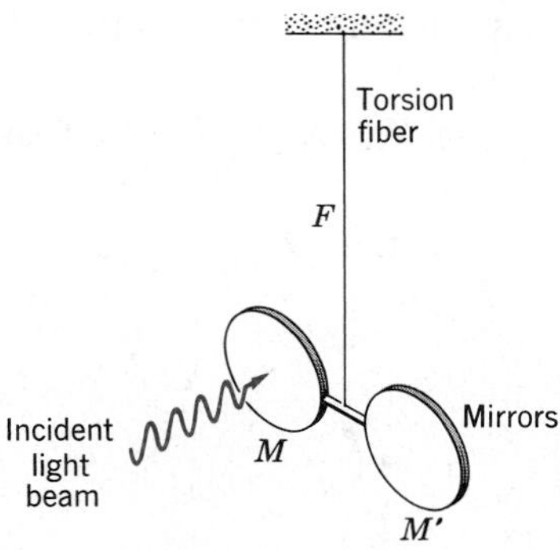

Figure 9 The arrangement of Nichols and Hull for measuring radiation pressure. Many details, essential to the success of this delicate experiment, are omitted.

merical value for this pressure. Nichols and Hull measured the intensity of their light beam by allowing it to fall on a blackened metal disk of known absorptivity and by measuring the temperature rise of this disk. In a particular run these experimenters measured a radiation pressure of 7.01 $\mu N/m^2$; the predicted value was 7.05 $\mu N/m^2$, in excellent agreement. Assuming a mirror area of 1 cm^2, this represents a force on the mirror of only 7×10^{-10} N, a remarkably small force.

The development of laser technology has permitted the achievement of radiation pressures much higher than those discussed so far in this section. This comes about because a beam of laser light—unlike a beam of light from a small lamp filament—can be focused to a tiny spot only a few wavelengths in diameter. This permits the delivery of very large energy fluxes to small objects placed at the focal spot.*

Sample Problem 2 A beam of light with an energy flux S of 12 W/cm^2 falls perpendicularly on a perfectly reflecting plane mirror of 1.5-cm^2 area. What force acts on the mirror?

From Newton's second law, the average force on the mirror is given by

$$F = \frac{\Delta p}{\Delta t},$$

where Δp is the momentum transferred to the mirror in time Δt. From Eq. 34 we have

$$\Delta p = \frac{2\,\Delta U}{c} = \frac{2SA\,\Delta t}{c}.$$

We have then for the force

$$F = \left(\frac{2S}{c}\right) A = \frac{(2)(12 \times 10^4\ \text{W/m}^2)}{3.00 \times 10^8\ \text{m/s}} (1.5 \times 10^{-4}\ \text{m}^2)$$

$$= 1.2 \times 10^{-7}\ \text{N}. \qquad \text{(Answer)}$$

This is a very small force, about equal to the weight of a very small grain of table salt. Note that the quantity $2S/c$, set in the large parentheses above, is just the radiation pressure.

38-7 Polarization

Sharp-eyed travelers will notice that most TV antennas on British rooftops are vertical but in the United States they are horizontal. The difference comes about because the British elect to transmit their TV signals so that the electric field vectors oscillate in a vertical plane. In this country—with equal arbitrariness—we have chosen to have them oscillate in a horizontal plane.

The wave characteristic that enters here is its *polarization.* The transverse electromagnetic wave of Fig. 6 is said to be *polarized* (more specifically, *plane polarized*) in the y direction, which means that the alternating electric field vectors are parallel to this direction for all points in the wave. (The magnetic field vectors are parallel to the z direction, but in dealing with polarization questions we focus our attention on the electric field, to which most detectors of electromagnetic radiation are sensitive.)

Figure 10, which is simply another representation of the traveling electromagnetic wave of Fig. 6, is drawn to stress the polarization feature. The wave in this figure is traveling in the x direction and is polarized in the y direction. The plane defined by the direction of propagation (the x axis) and the direction of polarization (the y axis) is called the *plane of vibration.*

Electromagnetic waves in the radio and microwave range readily exhibit polarization. Such a wave, generated by the surging of charge up and down in the dipole that forms the transmitting antenna of Fig. 11, has (for points along the horizontal axis) an electric field vector **E** that is parallel to the dipole axis. The plane of vibration of the transmitted wave is the plane of the figure. See also Fig. 12*a*, which shows a polarized beam emerging at right angles from the plane of the figure.

When the polarized wave of Fig. 11 falls on a second dipole connected to a microwave receiver, the alternating electric component of the wave will cause electrons to surge back and forth in the receiving antenna, producing a reading on the detector. If we turn the receiving antenna through 90° about the direction of propagation, the detector reading drops to zero. In this orientation the electric field vector is not able to cause charge to move along the dipole axis because it points at right angles to this axis. Check this out with your rabbit-ears TV antenna.

In radio and microwave sources the elementary radiators, which are the electrons surging back and forth in the transmitting antenna, act in unison; we say that they form a *coherent* source. In common light sources, however, such as the sun or a fluorescent lamp, the elementary radiators, which are the atoms that make up the source, act independently. Because of this difference the light propagated from such sources in a given direction

* See "The Pressure of Laser Light," by Arthur Ashkin, *Scientific American,* February 1972.

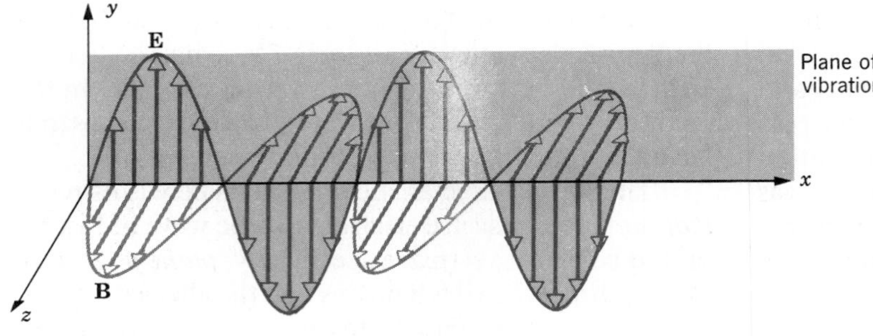

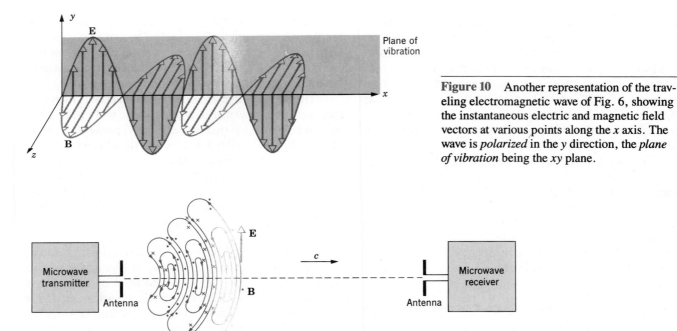

Figure 10 Another representation of the traveling electromagnetic wave of Fig. 6, showing the instantaneous electric and magnetic field vectors at various points along the x axis. The wave is *polarized* in the y direction, the *plane of vibration* being the xy plane.

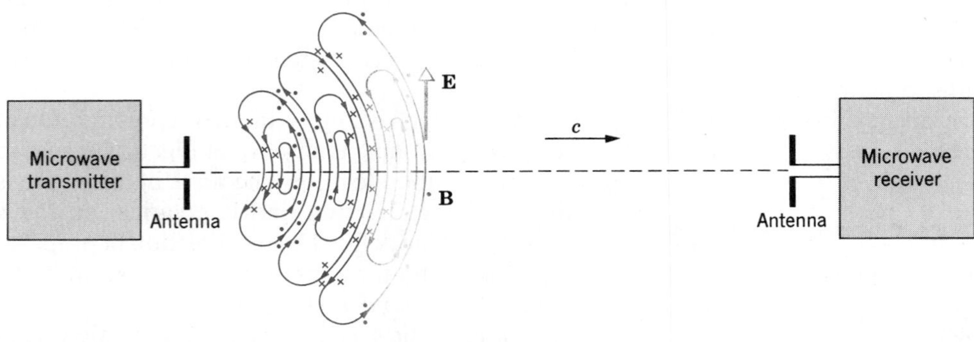

Figure 11 The electromagnetic wave generated by the transmitter is polarized, the plane of vibration being the plane of the page. The receiver can detect this wave with maximum effectiveness if (as shown) its receiving antenna lies in this plane. If the receiving antenna were to be turned at right angles to this plane, no signal would be detected.

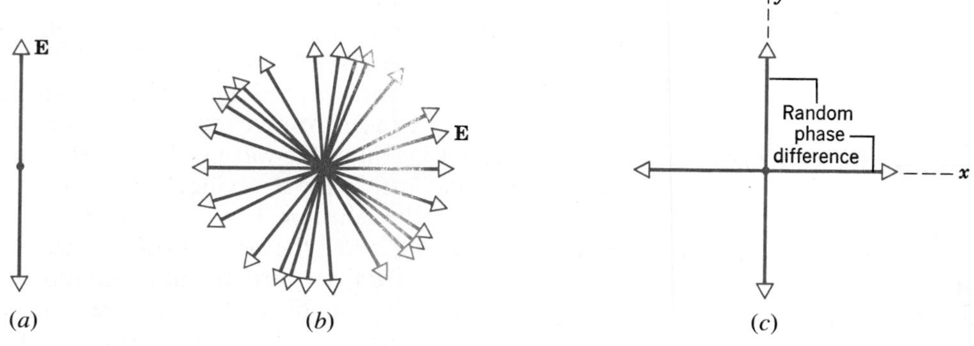

Figure 12 (*a*) A polarized wave moving out of the plane of the page, showing only the electric field vector. (*b*) An unpolarized wave, such as might be emitted by a light bulb. It is made up of a random array of polarized wave trains. (*c*) A second way of looking at an unpolarized wave. It can be viewed as the superposition of two plane polarized waves at right angles to each other but with a random phase difference between them.

consists of many independent wavetrains whose planes of vibration are randomly oriented about the direction of propagation, as in Fig. 12*b*. Such light is *unpolarized.* Furthermore, the random orientation of the planes of vibration conceals the transverse nature of the waves. To study this transverse nature, we must find a way to unscramble the different planes of vibration.

Figure 13 shows unpolarized light falling on a sheet

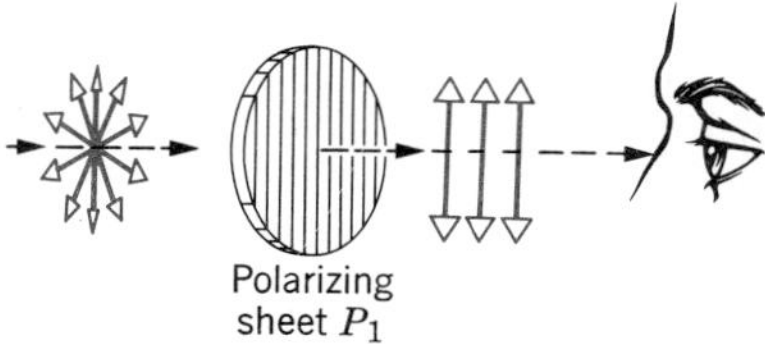

Figure 13 Unpolarized light (see Fig. 12*b*) is polarized by the action of a single polarizing sheet.

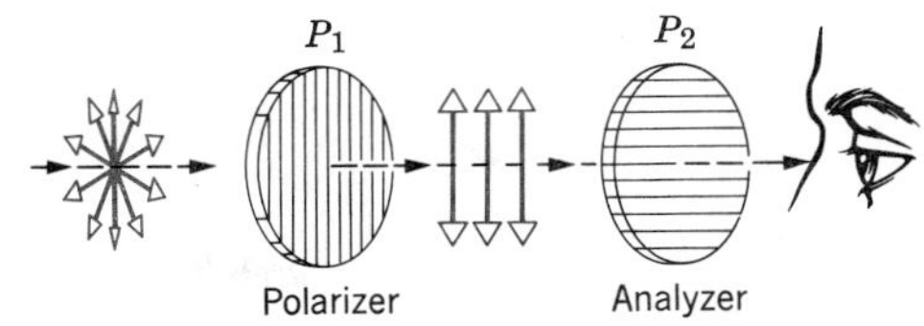

Figure 15 Unpolarized light is not transmitted by crossed polarizing sheets.

of commercial polarizing material called *Polaroid.* There exists in the sheet a certain characteristic polarizing direction, shown by the parallel lines.

> *The sheet will transmit only those wavetrain components whose electric vectors vibrate parallel to the polarizing direction and will absorb those that vibrate at right angles to this direction.*

The light emerging from a polarizing sheet will be polarized. The polarizing direction of the sheet is established during the manufacturing process by embedding certain long-chain molecules in a flexible plastic sheet and then stretching the sheet so that the molecules are aligned parallel to each other. Such a sheet absorbs radiation polarized in a direction parallel to the long molecules; radiation perpendicular to them passes through.

In Fig. 14 the polarizing sheet or *polarizer* lies in the plane of the page and the direction of propagation is into the page. The arrow **E** shows the plane of vibration of a randomly selected wavetrain falling on the sheet. It can be broken down into two vector components, E_x (of magnitude $E \sin\theta$) and E_y (of magnitude $E\cos\theta$). Only E_y will be transmitted; E_x will be absorbed within the sheet.

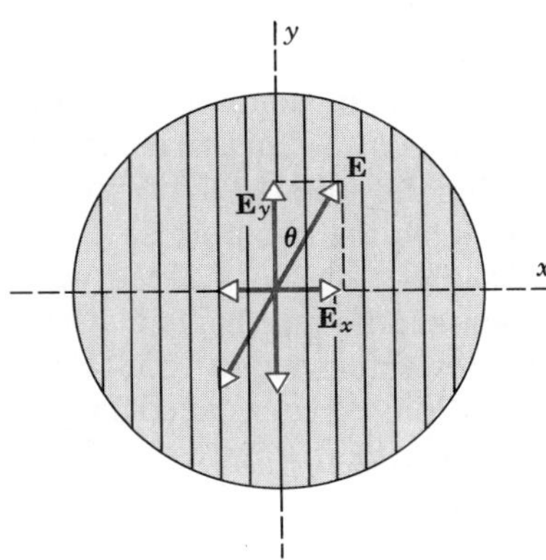

Figure 14 Another view of Fig. 13, showing a wavetrain represented by **E** falling on the polarizing sheet. Only the component E_y will be transmitted, the component E_x being absorbed within the sheet.

Let us place a second polarizing sheet P_2 (usually called, when so used, an *analyzer*) as in Fig. 15. If we rotate P_2 about the direction of propagation, there are two positions, 180° apart, at which the transmitted light intensity is almost zero; these are the positions in which the polarizing directions of P_1 and P_2 are at right angles.

If the amplitude of the polarized light falling on P_2 is E_m, the amplitude of the light that emerges is $E_m \cos\theta$, where θ is the angle between the polarizing directions of P_1 and P_2. Recalling that the intensity of the light beam is proportional to the square of the amplitude, we see that the transmitted intensity I varies with θ according to

$$\boxed{I = I_m \cos^2\theta} \quad \text{(law of Malus)} \qquad (35)$$

in which I_m is the maximum value of the transmitted intensity. This maximum occurs when the polarizing directions of P_1 and P_2 are parallel, that is, when $\theta = 0$ or 180°. Figure 16*a*, in which two overlapping polarizing sheets are in the parallel position ($\theta = 0$ or 180° in Eq. 35), shows that the light transmitted through the region of overlap has its maximum value. In Fig. 16*b* one or the other of the sheets has been rotated through 90° so that θ in Eq. 35 has the value 90° or 270°; the light transmitted through the region of overlap is now a minimum.

We shall see in Chapter 39 that light can also be polarized by being reflected at a certain critical angle from a sheet of glass or other dielectric material. This fact was discovered by chance in 1809 by Etienne Louis Malus, as he was gazing at the reflection of the setting sun in the Luxembourg Palace in Paris. Malus was not gazing through a polarizing sheet—such sheets were not invented until many years later—but through a crystal of calcite, which (like many other naturally-occurring crystals) also has polarizing properties. Equation 35, developed by Malus as a follow-up to his chance observation, is called the *law of Malus*.

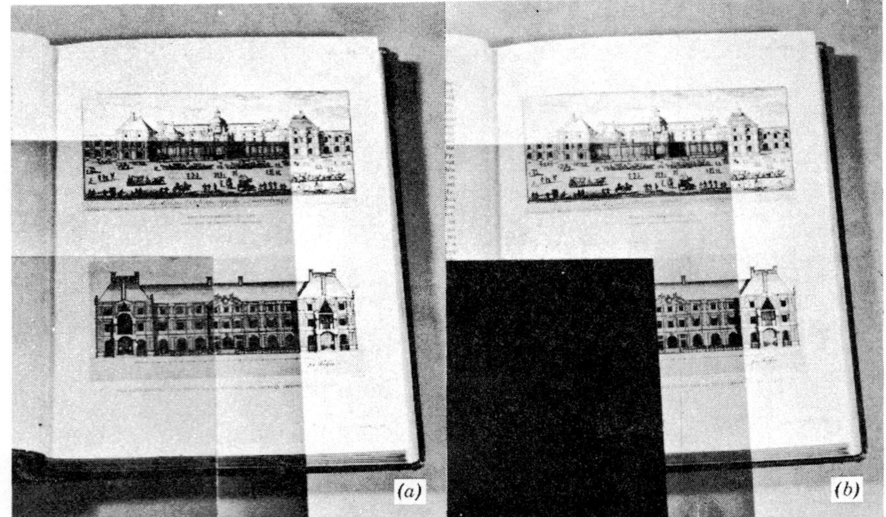

Figure 16 Two sheets of Polaroid are laid over a drawing of the Luxembourg Palace in Paris. In (*a*) the axes of the sheets are parallel, so that light is transmitted through both sheets. In (*b*) one of the sheets has been turned through 90° so that the condition of Fig. 15 prevails and no light passes through. (Malus discovered the phenomenon of polarization by reflection while looking through a polarizing crystal at sunlight reflected from the windows of this building.)

We can find out a lot about an object by examining the polarization state of the light emitted by or scattered from that object. For example, from studies of the polarization of light reflected from grains of cosmic dust present in our home galaxy, we can tell that they have been oriented in the weak galactic magnetic field ($\sim 10^{-8}$ T) so that their long dimension is parallel to this field. Polarization studies have suggested that Saturn's rings consist of ice crystals. We can find the size and shape of virus particles by the polarization of ultraviolet light scattered from them. We can learn a lot about the structure of atoms and nuclei from polarization studies of their emitted radiations. Thus, we have a useful research technique for structures ranging in size from a galaxy ($\sim 10^{+20}$ m) to a nucleus ($\sim 10^{-14}$ m).

Sample Problem 3 Two polarizing sheets have their polarizing directions parallel so that the intensity I_m of the transmitted light is a maximum. Through what angle must either sheet be turned if the intensity is to drop by one-half?

From Eq. 35, since $I = \frac{1}{2}I_m$, we have

$$\tfrac{1}{2}I_m = I_m \cos^2\theta$$

or

$$\theta = \cos^{-1} \pm \frac{1}{\sqrt{2}} = \pm 45° \text{ and } \pm 135°. \qquad \text{(Answer)}$$

The same effect is obtained no matter which sheet is rotated or in which direction.

38-8 The Speed of Electromagnetic Waves

The speed of electromagnetic radiation in free space—which we shall call the speed of light—is central not only to Maxwell's theory of electromagnetism but also to Einstein's theory of relativity. Its value has been measured so often and so precisely that, in a sense, we can say that it has been "measured to death." We have in mind the fact that the speed of light has been assigned an exact value by definition as part of the 1983 redefinition of the meter, namely

$$c = 299{,}792{,}458 \text{ m/s} \quad \text{(exactly).}$$

Thus, the long history of measuring the speed of light is over. Today, if you send a light beam from one point to another (in free space) and measure its transit time, you are not measuring the speed of light; you are measuring the distance between the two points.

An Historical Perspective. Galileo was perhaps the first to try to measure the speed of light. In his chief work, *Two New Sciences,* published in 1638, is a conversation among three fictitious persons called Salviati, Simplicio, and Sagredo (who evidently represents Galileo himself). Here is part of what they say about the speed of light:

Simplicio: Everyday experience shows that the propagation of light is instantaneous; for when we see a piece of artillery fired, at a great distance, the flash reaches our eyes without lapse of time; but the sound reaches the ear only after a noticeable interval.

Sagredo: Well, Simplicio, the only thing I am able to infer from this familiar bit of experience is that sound, in reaching our ear, travels more slowly than light; it does not inform me whether the coming of the light is instantaneous or whether, although extremely rapid, it still occupies time.

Sagredo then describes a possible method for measuring the speed of light. He and an assistant stand facing each other some distance apart, at night. Each carries a lantern, which can be covered or uncovered at will. Galileo started the experiment by uncovering his lantern. When the light reached the assistant he uncovered his own lantern, whose light was then seen by Galileo. Galileo tried to measure the time between the instant at which he uncovered his own lantern and the instant at which the light from his assistant's lantern reached him. The method fails because human reaction times are far too slow.

Then followed various astronomical methods and then methods involving the timing of light pulses produced by shining light through a toothed wheel or reflecting it from a rotating mirror. The most recent (and perhaps the last) measurement of the speed of light involved measuring the frequency of light trapped as standing waves in a laser cavity of known length and computing c from the relation $c = \lambda\nu$. Table 3 of Chapter 23 points out some landmarks in the history of the measurement of the speed of light.

Einstein and the Speed of Light. When we say that the speed of sound in dry air at 0°C is 331.7 m/s, we imply a reference frame fixed with respect to the air mass through which the sound wave is moving. When we say that the speed of electromagnetic waves in free space is 299,792,458 m/s, what reference frame are we talking about? It cannot be the medium through which the light wave travels because, in contrast to sound, no medium is required.

Physicists of the nineteenth century, influenced as they then were by an analogy between light waves and sound waves or other purely mechanical disturbances, did not accept the idea of a wave requiring no medium. They postulated the existence of an ether, which was a tenuous substance that filled all space and served as a medium of transmission for light.

Although it proved useful for many years, the ether concept did not survive the test of experiment. In particular, careful attempts to measure the speed of the earth through the ether always gave the result of zero. Physicists were not willing to believe that the earth was permanently at rest in the ether and that all other bodies in the universe were in motion through it.

In 1905 Einstein resolved the dilemma by making a bold postulate to which we have already referred in Section 17–7 which you may wish to reread at this time. If a number of observers are moving (at uniform velocity) with respect to each other and to a light source and if each observer measures the speed of the light emerging from the source, they will all measure the same value. In formal language:

The speed of light in free space has the same value c in all directions and in all inertial reference frames.

This is a fundamental assumption of Einstein's theory of relativity, which we develop in detail in Chapter 42. The postulate does away with the need for an ether by asserting that the speed of light is the same in all reference frames; none is singled out as fundamental. The theory of relativity, derived from this postulate, has been tested many times, and agreement with the predictions of theory has always emerged. These agreements, extending over half a century, lend strong support to Einstein's basic postulate about light propagation.

REVIEW AND SUMMARY

The Electromagnetic Spectrum

Maxwell's equations predict the existence of a spectrum of electromagnetic waves (see Fig. 1) that travel through free space with a common speed c. Figure 4 shows the spacial distribution of fields produced by the dipole shortwave-radio antenna of Fig. 3. The wave is propagated because the electric and magnetic fields are generated by the alternating magnetic and electric fields, respectively, in accord with Faraday's law and the Ampere–Maxwell law.

The Fields

The electric and magnetic fields of the wave have the forms

$$E = E_m \sin(kx - \omega t) \quad \text{and} \quad B = B_m \sin(kx - \omega t). \qquad [1,2]$$

Magnitude Ratio and Wave Speed

By applying Maxwell's equations we can show that

$$\frac{E_m}{B_m} = c = \frac{1}{\sqrt{\epsilon_0 \mu_0}}. \qquad [16,23]$$

Energy Flow

The Poynting vector, defined from

$$\mathbf{S} = \frac{1}{\mu_0}\mathbf{E} \times \mathbf{B} \quad \text{(Poynting vector),} \qquad [24]$$

gives the energy flux (W/m^2) in an electromagnetic wave. The intensity of the wave (the average of S) is

Intensity

$$I = \overline{S} = \frac{1}{c\mu_0} E^2_{rms} \quad \text{(energy flow; plane wave).} \qquad [29]$$

See Sample Problem 1.

Radiation Pressure

Electromagnetic radiation falling on a surface exerts a radiation pressure on it. Given the energy flux S falling normally on a plane mirror, Sample Problem 2 shows how to calculate the radiation pressure $p = 2S/c$.

Polarization

An electromagnetic wave from an antenna like that of Fig. 3 is polarized; its electric field vectors are parallel. The direction of the electric field **E** is called the direction of polarization; the plane containing **E** and the direction of propagation is called the plane of vibration. See Figs. 10, 11, and 12*a*.

Unpolarized Light

Light from an "ordinary" source such as the sun is unpolarized; its energy is emitted in wavetrains of finite length whose orientation about the propagation direction is random; see Fig. 12*b*.

Commercial Polarizing Sheets

A sheet of Polaroid only transmits light with electric field components parallel to a given direction, strongly absorbing the others; the emerging light is polarized. The intensity of polarized light passing through a polarizing sheet is reduced from I_m to I, where

Law of Malus

$$I = I_m \cos^2\theta \quad \text{(law of Malus).} \qquad [35]$$

Here θ (see Fig. 14) is the angle between the plane of vibration and the polarizing direction of the sheet; see Figs. 15 and 16 and Sample Problem 3.

Einstein's Postulate

Einstein postulated that no medium is required for the propagation of light and that the speed of light in free space has the same value c in all directions and in all inertial reference frames. This experimentally-verified prediction leads directly to the theory of relativity.

QUESTIONS

1. Electromagnetic waves reach us from the farthest depths of space. From the information they carry, can we tell what the universe is like at the present moment? At any selected time in the past?

2. If you are asked on an examination what fraction of the electromagnetic spectrum lies in the visible range, what would you reply?

3. Why are danger signals in red when the eye is most sensitive to yellow-green?

4. Comment on this definition of the limits of the spectrum of visible light, given by a physiologist: "The limits of the visible spectrum occur when the eye is no better adapted than any other organ of the body to serve as a detector."

5. How might an eye-sensitivity curve like that of Fig. 2 be measured?

6. In connection with Fig. 2, (*a*) do you think it possible that the wavelength of maximum sensitivity could vary if the intensity of the light is changed? (*b*) What might the curve of Fig. 2 look like for a color-blind person who could not, for example, distinguish red from green?

7. What feature of light corresponds to loudness in sound?

8. List several ways in which radio waves differ from visible light waves. In what ways are they the same?

9. Suppose that human eyes were insensitive to visible light but were very sensitive to infrared light. What environmental changes would be needed if you were to (*a*) walk down a long corridor and (*b*) drive a car? Could the phenomenon of color exist? How would traffic lights have to be modified?

10. Speaking loosely we can say that the electric and the magnetic components of a traveling electromagnetic wave "feed on each other." What does this mean?

11. "Displacement currents are present in a traveling electromagnetic wave and we may associate the magnetic field component of the wave with these currents." Is this statement true? Discuss it in detail.

12. H. G. Wells, in his novel *The Invisible Man,* described a concoction developed by a "mad scientist" that would render the person who drank it invisible. Give arguments to prove that a truly invisible man would be blind.

13. Can an electromagnetic wave be deflected by a magnetic field? By an electric field?

14. How does a microwave oven cook food? You can boil water in a plastic bag in such an oven. How can this happen?

15. Why is Maxwell's modification of Ampere's law (that is, the term $\epsilon_0 d\Phi_E/dt$ in Table 2, Chapter 37) needed to understand the propagation of electromagnetic waves?

16. If you were to calculate the Poynting vector for various points in and around a transformer, what would you expect the field pattern of these vectors to look like? Assume that an alternating potential difference has been applied to the primary windings and that a resistive load is connected across the secondary.

17. Can an object absorb light energy without having linear momentum transferred to it? If so, give an example. If not, explain why.

18. When you turn on a flashlight does it experience any force associated with the emission of the light?

19. We associated energy and linear momentum with electromagnetic waves. Is angular momentum present also?

20. What is the relation, if any, between the intensity I of an electromagnetic wave and the magnitude S of its Poynting vector?

21. As you recline in a beach chair in the sun, why are you so conscious of the thermal energy delivered to you but totally unresponsive to the linear momentum delivered from the same source? Is it true that when you catch a hard-pitched baseball you are conscious of the energy delivered but not of the momentum?

22. As we normally experience them, radio waves are almost always polarized and visible light is almost always unpolarized. Why should this be so?

23. Sample Problem 3 shows that, when the angle between the two polarizing directions is turned from 0° to 45°, the intensity of the transmitted beam drops to one half its initial value. What happens to this "missing" energy?

24. You are given a number of polarizing sheets. Explain how you would use them to rotate the plane of polarization of a plane polarized wave through any given angle. How could you do it with the least energy loss?

25. What determines the desirable length and orientation of the rabbit ears antenna on a portable TV set?

26. Why do sunglasses made of polarizing materials have a marked advantage over those that simply depend on absorption effects? What disadvantages might they have?

27. Why aren't sound waves polarized?

28. Unpolarized light falls on two polarizing sheets so oriented that no light is transmitted. If a third polarizing sheet is placed between them, can light be transmitted? If so, explain how.

29. Find a way to identify the polarizing direction of a sheet of Polaroid. No marks appear on the sheet.

30. When observing a clear sky through a polarizing sheet, you find that the intensity varies on rotating the sheet. This does not happen when viewing a cloud through the sheet. Why?

31. How could Galileo have tested experimentally that reaction times were an overwhelming source of error in his attempt to measure the speed of light, described in Section 8?

32. Can you think of any "everyday" observation (that is, without experimental apparatus) to show that the speed of light is not infinite? Think of lightning flashes, possible discrepancies between the predicted time of sunrise and the observed time, radio communications between earth and astronauts in orbiting space ships, and so on.

33. Atoms are mostly empty space. However, the speed of light passing through a transparent solid made up of such atoms is often considerably less than the speed of light in free space. How can this be?

34. In a vacuum, does the speed of light depend on (*a*) the wavelength, (*b*) the frequency, (*c*) the intensity, (*d*) the state of polarization, (*e*) the speed of the source, or (*f*) the speed of the observer?

EXERCISES AND PROBLEMS

Section 38-1 "Maxwell's Rainbow"

1E. Project Seafarer was an ambitious program to construct an enormous antenna, buried underground on a site about 4000 square miles in area. Its purpose was to transmit signals to submarines while they were deeply submerged. If the effective wavelength was 10^4 earth radii; what would be (*a*) the frequency and (*b*) the period of the radiations emitted? Ordinarily electromagnetic radiations do not penetrate very far into conductors such as sea water. Can you think of any reason why such ELF (extremely low frequency) radiations should penetrate more effectively? Think of the limiting case of zero frequency. (Why not transmit signals at zero frequency?)

2E. (*a*) How long does it take a radio signal to travel 150 km from a transmitter to a receiving antenna? (*b*) We see a full moon by reflected sunlight. How much earlier did the light that enters our eye leave the sun? The earth–moon and the earth–

sun distances are 3.8×10^5 km and 1.5×10^8 km. (*c*) What is the round-trip travel time for light between Earth and a spaceship orbiting Saturn, 1.3×10^9 km distant? (*d*) The Crab nebula, which is about 6500 light-years distant, is thought to be the result of a supernova explosion recorded by Chinese astronomers in A.D. 1054. In approximately what year did the explosion actually occur?

3E. (*a*) The wavelength of the most energetic x rays produced when electrons accelerated to 18 GeV in the Stanford Linear Accelerator slam into a solid target is 0.067 fm. What is the frequency of these x rays? (*b*) A VLF (very low frequency) radio wave has a frequency of only 30 Hz. What is its wavelength?

4E. (*a*) At what wavelengths does the eye of a standard observer have half its maximum sensitivity? (*b*) What are the wavelength, the frequency, and the period of the light for which the eye is the most sensitive?

5E. In the electromagnetic spectrum displayed in Fig. 1, the wavelength scale markers represent successive powers of 10 and are evenly spaced. Show that the frequency markers must also be evenly spaced, with the same interval.

6P. The radiation from a certain HeNe laser, although centered on 632.8 nm, has a finite "linewidth" of 0.010 nm. Calculate the linewidth in frequency units.

7P. One method for measuring the speed of light, based on observations by Roemer in 1676, consisted in observing the apparent times of revolution of one of the moons of Jupiter. The true period of revolution is 42.5 h. (*a*) Taking into account the finite speed of light, how would you expect the apparent time of revolution to alter as the earth moves in its orbit from point *x* to point *y* in Fig. 17? (*b*) What observations would be needed to compute the speed of light? Neglect the motion of Jupiter in its orbit. Fig. 17 is not drawn to scale.

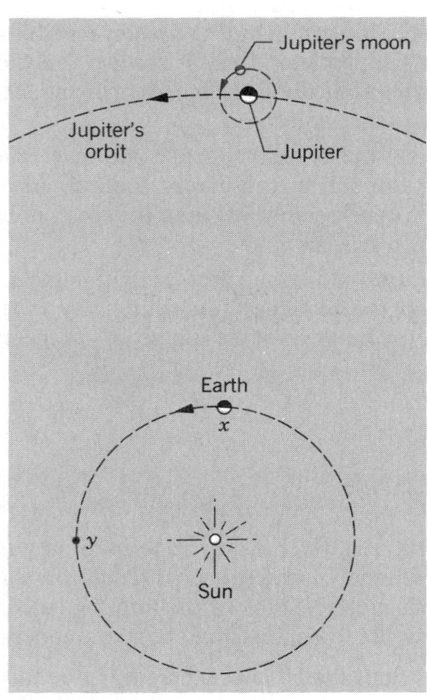

Figure 17 Problem 7.

Section 38-2 Generating an Electromagnetic Wave

8E. Find the angle from the axis of an oscillating dipole at which the intensity of the radiation field is one-half its value in the equatorial plane.

9E. What inductance is required with a 17-pF capacitor in order to construct an oscillator capable of generating 550-nm (i.e., visible) electromagnetic waves? Comment on your answer.

10P. Figure 18 shows an *LC* oscillator connected by a transmission line to an antenna of a so-called *magnetic* dipole type. Compare with Fig. 3, which shows a similar arrangement but with an *electric* dipole type of antenna. (*a*) What is the basis for the names of these two antenna types? (*b*) Draw figures corresponding to Figs. 4 and 5 to describe the electromagnetic wave that sweeps past the observer at point *P* in Fig. 18.

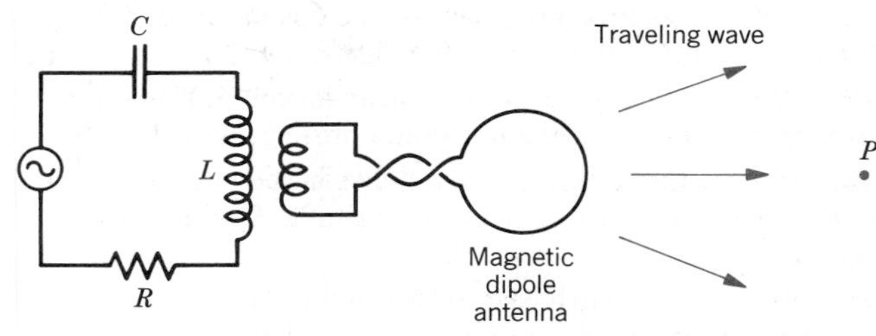

Figure 18 Problem 10.

Section 38-4 The Traveling Electromagnetic Wave—Quantitative

11E. A plane electromagnetic wave has a maximum electric field of 3.2×10^{-4} V/m. Find the maximum magnetic field.

12E. The electric field associated with a plane electromagnetic wave is given by $E_x = 0$; $E_y = 0$; $E_z = 2 \cos[\pi \times 10^{15}(t - x/c)]$, with $c = 3.0 \times 10^8$ m/s and all quantities in SI units. The wave is propagating in the $+x$ direction. Write expressions for the components of the magnetic field of the wave.

13P. Start from Eqs. 15 and 21 and show that $E(x, t)$ and $B(x, t)$, the electric and magnetic field components of a plane traveling electromagnetic wave, must satisfy the "wave equations"

$$\frac{\partial^2 E}{\partial t^2} = c^2 \frac{\partial^2 E}{\partial x^2}$$

and

$$\frac{\partial^2 B}{\partial t^2} = c^2 \frac{\partial^2 B}{\partial x^2}.$$

14P. (*a*) Show that Eqs. 1 and 2 satisfy the wave equations displayed in Problem 13. (*b*) Show that any expressions of the form

$$E = E_m f(kx \pm \omega t)$$

and

$$B = B_m f(kx \pm \omega t),$$

where $f(kx \pm \omega t)$ denotes an arbitrary function, also satisfy these wave equations.

Section 38-5 Energy Transport and the Poynting Vector

15E. Show, by finding the direction of the Poynting vector **S**, that the directions of the electric and magnetic fields at all points in Figs. 4, 5, 6, and 7 are consistent at all times with the assumed directions of propagation.

16E. Currently-operating Nd:glass lasers can provide 100 TW of power in 1.0-ns pulses at a wavelength of 0.26 μm. How much energy is contained in a single pulse?

17E. Our closest stellar neighbor, α-Centauri, is 4.3 light-years away. It has been suggested that TV programs from our planet have reached this star and may have been viewed by the hypothetical inhabitants of a hypothetical planet orbiting this star. A TV station on earth has a power output of 1.0 MW. What is the intensity of its signal at α-Centauri?

18E. An electromagnetic wave is traveling in the negative y direction. At a particular position and time, the electric field is along the positive z axis and has a magnitude of 100 V/m. What are the direction and magnitude of the magnetic field at that position and at that time?

19E. The earth's mean radius is 6.37×10^6 m and the mean earth–sun distance is 1.5×10^8 km. What fraction of the radiation emitted by the sun is intercepted by the disc of the earth?

20E. The radiation emitted by a laser is not exactly a parallel beam; rather, the beam spreads out in the form of a cone with circular cross section. The angle θ of the cone, see Fig. 19, is called the *full-angle beam divergence*. An argon laser, radiating at 514.5 nm, is aimed at the moon in a ranging experiment. If the beam has a full-angle beam divergence of 0.88 μrad, what area on the moon's surface is illuminated by the laser?

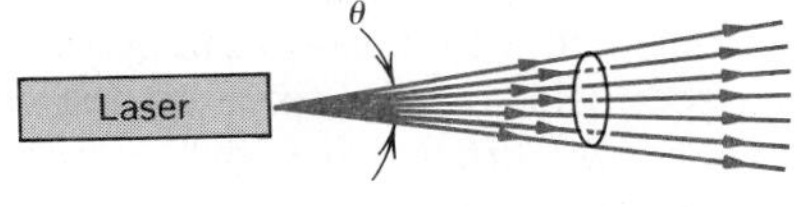

Figure 19 Exercise 20.

21E. The intensity of direct solar radiation that was unabsorbed by the atmosphere on a particular summer day is 100 W/m^2. How close would you have to stand to a 1.0-kW electric heater to feel the same intensity? Assume that the heater radiates uniformly in all directions.

22E. Show that in a plane traveling electromagnetic wave the average intensity, that is, the average rate of energy transport per unit area, is given by

$$\bar{S} = \frac{E_m^2}{2\mu_0 c} = \frac{cB_m^2}{2\mu_0}.$$

23E. What is the average intensity of a plane traveling electromagnetic wave if B_m, the maximum value of its magnetic field component, is 1.0×10^{-4} T (= 1.0 gauss)? (*Hint:* See Exercise 22.)

24E. In a plane radio wave the maximum value of the electric field component is 5.0 V/m. Calculate (*a*) the maximum value of the magnetic field component and (*b*) the wave intensity. (*Hint:* See Exercise 22.)

25P. You walk 150 m directly toward a street lamp and find that the intensity increases to 1.5 times the intensity at your original position. How far from the lamp were you first standing? (The lamp radiates uniformly in all directions.)

26P. Prove that, for any point in an electromagnetic wave such as that of Fig. 6, the time-averaged density of the energy stored in the electric field equals that of the energy stored in the magnetic field.

27P. Sunlight strikes the earth, outside its atmosphere, with an intensity of 1.4 kW/m^2. Calculate E_m and B_m for sunlight, assuming it to be a plane wave.

28P. The maximum electric field at a distance of 10 m from a point light source is 2.0 V/m. Calculate (*a*) the maximum value of the magnetic field, (*b*) the average intensity, and (*c*) the power output of the source.

29P. A cube of edge a has its edges parallel to the x, y, and z axes of a rectangular coordinate system. A uniform electric field **E** is parallel to the y axis and a uniform magnetic field **B** is parallel to the x axis. Calculate (*a*) the rate at which, according to the Poynting vector point of view, energy may be said to pass through each face of the cube and (*b*) the net rate at which the energy stored in the cube may be said to change.

30P. Frank D. Drake, an active investigator in the SETI (Search for Extra-Terrestrial Intelligence) program, has said that the large radio telescope in Arecibo, Puerto Rico, ". . . can detect a signal which lays down on the entire surface of the earth a power of only one picowatt." See Fig. 20. (*a*) What is the power actually received by the Arecibo antenna for such a signal? The antenna diameter is 1000 ft. (*b*) What would be the power output of a source at the center of our galaxy that could provide such a signal? The galactic center is 2.2×10^4 ly away. Take the source as radiating uniformly in all directions.

Figure 20 Problem 30.

31P. A HeNe laser, radiating at 632.8 nm, has a power output of 3.0 mW and a full-angle beam divergence (see Exercise 20) of 0.17 mrad. (*a*) What is the intensity of the beam 40 m from the laser? (*b*) What would be the power output of an isotropic source that provides this same intensity at the same distance?

32P. An airplane flying at a distance of 10 km from a radio transmitter receives a signal of power 10 $\mu W/m^2$. Calculate (*a*) the amplitude of the electric field at the airplane due to this signal; (*b*) the amplitude of the magnetic field at the airplane; (*c*) the total power radiated by the transmitter, assuming the transmitter to radiate uniformly in all directions.

33P. During a test, a NATO surveillance radar system, operating at 12 GHz with 180 kW of output power, attempts to detect an incoming "enemy" aircraft at 90 km. The target aircraft is designed to have a very small effective area for reflection of radar waves of 0.22 m^2. Assume that the radar beam spreads out isotropically into the forward hemisphere both upon transmission and reflection and ignore absorption in the atmosphere. For the reflected beam as received back at the radar site, calculate (*a*) the intensity, (*b*) maximum value of the electric field vector, and (*c*) the rms value of the magnetic field.

34P. A copper wire (diameter, 2.5 mm; resistance, 1.0 Ω per 300 m) carries a current of 25 A. Calculate (*a*) **E**, (*b*) **B**, and (*c*) **S** for a point on the surface of the wire.

35P. Figure 21 shows a cylindrical resistor of length l, radius a, and resistivity ρ, carrying a current i. (*a*) Show that the Poynting vector **S** at the surface of the resistor is everywhere directed normal to the surface, as shown. (*b*) Show that the rate P at which energy flows into the resistor through its cylindrical surface, calculated by integrating the Poynting vector over this surface, is equal to the rate at which thermal energy is produced; that is,

$$\int \mathbf{S} \cdot d\mathbf{A} = i^2 R,$$

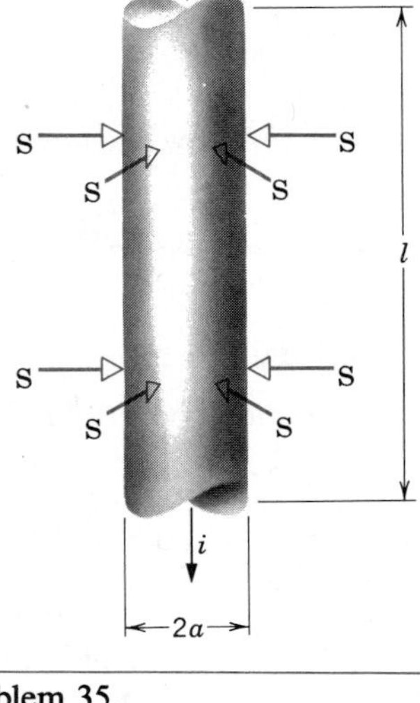

Figure 21 Problem 35.

where $d\mathbf{A}$ is an element of area of the cylindrical surface. This suggests that, according to the Poynting vector point of view, the energy that appears in a resistor as thermal energy does not enter it through the connecting wires but through the space around the wires and the resistor.

Section 38-6 Radiation Pressure

36E. Show (*a*) that the force F exerted by a laser beam of intensity I on a perfectly reflecting object of area A normal to the beam is given by $F = 2IA/c$, and (*b*) that the pressure $P = 2I/c$.

37E. High-power lasers are used to compress gas plasmas by radiation pressure. The reflectivity of a plasma is unity if the electron density is high enough. A laser generating pulses of radiation of peak power 1.5×10^3 MW is focused onto 1.0 mm^2 of high-electron-density plasma. Find the pressure exerted on the plasma.

38E. The average intensity of the solar radiation that falls normally on a surface just outside the earth's atmosphere is 1.4 kW/m^2. (*a*) What radiation pressure is exerted on this surface, assuming complete absorption? (*b*) How does this pressure compare with the earth's sea-level atmospheric pressure, which is 1.0×10^5 N/m^2?

39E. Radiation from the sun striking the earth has an intensity of 1.4 kW/m^2. (*a*) Assuming that the earth behaves like a flat disk at right angles to the sun's rays and that all the incident energy is absorbed, calculate the force on the earth due to radiation pressure. (*b*) Compare it with the force due to the sun's gravitational attraction.

40E. What is the radiation pressure 1.5 m away from a

500-W light bulb? Assume that the surface on which the pressure is exerted faces the bulb and is perfectly absorbing and that the bulb radiates uniformly in all directions.

41P. A plane electromagnetic wave, with wavelength 3.0 m, travels in free space in the $+x$ direction with its electric vector **E**, of amplitude 300 V/m, directed along the y axis. (*a*) What is the frequency ν of the wave? (*b*) What is the direction and amplitude of the magnetic field associated with the wave? (*c*) If $E = E_m \sin(kx - \omega t)$, what are the values of k and ω? (*d*) What is the time-averaged rate of energy flow in W/m² associated with this wave? (*e*) If the wave falls upon a perfectly absorbing sheet of area 2.0 m², at what rate would momentum be delivered to the sheet and what is the radiation pressure exerted on the sheet?

42P. A helium-neon laser of the type often found in physics laboratories has a beam power output of 5.0 mW at a wavelength of 633 nm. The beam is focused by a lens to a circular spot whose effective diameter may be taken to be 2.0 wavelengths. Calculate (*a*) the intensity of the focused beam, (*b*) the radiation pressure exerted on a tiny perfectly-absorbing sphere whose diameter is that of the focal spot, (*c*) the force exerted on this sphere, and (*d*) the acceleration imparted to it. Assume a sphere density of 5.0×10^3 kg/m³.

43P. A laser has a power output of 4.6 W and a beam diameter of 2.6 mm. If it is aimed vertically upward, what is the height H of a perfectly reflecting cylinder that can be made to "hover" by the radiation pressure exerted by the beam? Assume that the density of the cylinder is 1.2 g/cm³. See Fig. 22.

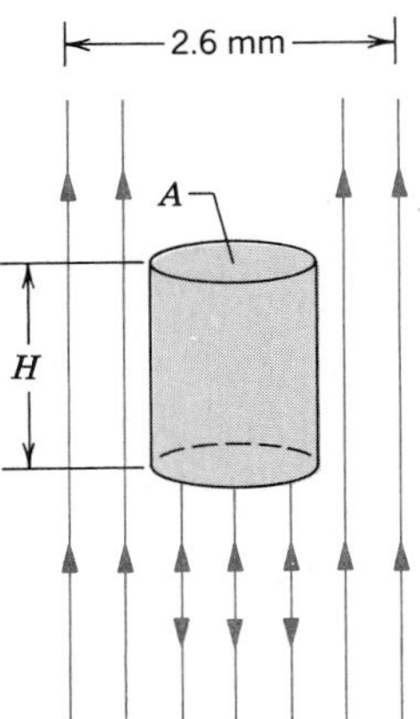

Figure 22 Problem 43.

44P. Radiation of intensity I is normally incident on an object that absorbs a fraction f of it and reflects the rest. What is the radiation pressure?

45P. Prove, for a plane wave at normal incidence on a plane surface, that the radiation pressure on the surface is equal to the energy density in the beam outside the surface. This relation holds no matter what fraction of the incident energy is reflected.

46P. Prove, for a stream of bullets striking a plane surface at normal incidence, that the "pressure" is twice the kinetic energy density in the stream above the surface; assume that the bullets are completely absorbed by the surface. Constrast this with the behavior of light; see Problem 45.

47P. A small spaceship whose mass, with occupant, is 1.5×10^3 kg is drifting in outer space, where no gravitational field exists. If the astronaut turns on a 10-kW laser beam, what speed would the ship attain in one day because of the reaction force associated with the momentum carried away by the beam?

48P. It has been proposed that a spaceship might be propelled in the solar system by radiation pressure, using a large sail made of foil. How large must the sail be if the radiation force is to be equal in magnitude to the sun's gravitational attraction? Assume that the mass of the ship + sail is 1500 kg, that the sail is perfectly reflecting, and that the sail is oriented at right angles to the sun's rays. See Appendix C for needed data.

49P. A particle in the solar system is under the combined influence of the sun's gravitational attraction and the radiation force due to the sun's rays. Assume that the particle is a sphere of density 1.0×10^3 kg/m³ and that all of the incident light is absorbed. (*a*) Show that all particles with radius less than some critical radius, R_0, will be blown out of the solar system. (*b*) Calculate R_0. Note that R_0 does not depend on the distance from the particle to the sun.

Section 38-7 Polarization

50E. The magnetic field equations for an electromagnetic wave in free space are $B_x = B \sin(ky + \omega t)$, $B_y = B_z = 0$. (*a*) What is the direction of propagation? (*b*) Write the electric field equations. (*c*) Is the wave polarized? If so, in what direction?

51E. A beam of unpolarized light of intensity 0.01 W/m² falls at normal incidence upon a polarizing sheet. (*a*) Find the maximum value of the electric field of the transmitted beam. (*b*) What is the radiation pressure exerted on the polarizing sheet?

52E. Unpolarized light falls on two polarizing sheets placed one on top of the other. What must be the angle between the characteristic directions of the sheets if the intensity of the transmitted light is one-third the intensity of the incident beam? Assume that each polarizing sheet is ideal, that is, that it reduces the intensity of unpolarized light by exactly 50%.

53E. Three polarizing plates are stacked. The first and third are crossed; the one between has its axis at 45° to the axes of the other two. What fraction of the intensity of an incident unpolarized beam is transmitted by the stack?

54P. An unpolarized beam of light is incident on a group of four polarizing sheets that are lined up so that the characteristic direction of each is rotated by 30° clockwise with respect to the preceding sheet. What fraction of the incident intensity is transmitted?

55P. A beam of polarized light strikes two polarizing sheets. The characteristic direction of the second is 90° with respect to the incident light. The characteristic direction of the first is at angle θ with respect to the incident light. Find angle θ for a transmitted beam intensity that is 0.10 times the incident beam intensity.

56P. A beam of light is plane polarized in the vertical direction. The beam falls at normal incidence on a polarizing sheet with its polarizing direction at 70° to the vertical. The transmitted beam falls, also at normal incidence, on a second polarizing sheet with its polarizing direction horizontal. If the intensity of the original beam is 43 W/m^2, what is the intensity of the beam transmitted by the second sheet?

57P. Suppose that in Problem 56 the incident beam was unpolarized. What now is the intensity of the beam transmitted by the second sheet?

58P. A beam of light is a mixture of polarized light and unpolarized light. When it is sent through a Polaroid sheet, we find that the transmitted intensity can be varied by a factor of five depending on the orientation of the Polaroid. Find the relative intensities of these two components of the incident beam.

59P. It is desired to rotate the plane of vibration of a beam of polarized light by 90°. (*a*) How might this be done using only Polaroid sheets? (*b*) How many sheets are required in order for the total intensity loss to be less than 40%?

60P. At a beach the light is generally partially polarized. At a particular beach on a particular day near sundown the horizontal component of the electric field vector is 2.3 times the vertical component. A standing sunbather puts on polaroid sunglasses; the glasses suppress the horizontal field component. (*a*) What fraction of the light energy received before the glasses were put on now reaches the eyes? (*b*) The sunbather, still wearing the glasses, lies on his side. What fraction of the light energy received before the glasses were put on reaches the eyes now?

ESSAY 15
PHYSICS AND TOYS

RAYMOND C. TURNER
CLEMSON UNIVERSITY

The basic principles of physics can often be demonstrated with ordinary toys. By understanding how these toys work, you can better understand the world around you.

Toys are often used to illustrate various physical principles of mechanics. You are probably already aware that the motion of *weebles* that wobble but don't fall down and other balancing toys can be explained using the concepts of center of mass, torque, and equilibrium.[1] Similarly, a *water rocket* can be used to demonstrate conservation of momentum, while a *hot wheels* race car and track can be analyzed by using the law of conservation of energy.[2] This article will be limited, however, to discussing several toys that deal with magnetism and light.

You have probably seen a *flicker light* in a toy store or a gift shop. What makes its filament vibrate? An example of such a lamp is shown in Fig. 1. Its operation can be analyzed by considering the Lorentz magnetic force on a current element, $d\mathbf{F} = i\, d\mathbf{s} \times \mathbf{B}$. This type of lamp has a small permanent magnet mounted near the lamp's filament. When there is a current in the filament, the magnetic field of the magnet exerts a force on the filament, pushing it in a direction perpendicular to both the field and the current. Since the lamp is operated from the usual household 60 Hz ac line, the current changes direction 120 times each second, and this changes the direction of the force at the same rate. This causes the filament to vibrate back and forth, and the light to appear to flicker. But how can this explanation be verified? Usually the magnet in a flicker light is mounted inside the glass bulb, but the particular lamp shown in the figure has its magnet attached to the ring on the outside of the bulb. This lets you move the magnet relative to the filament, thus changing the magnetic field. The farther you move the magnet from the lamp, the smaller the magnetic field at the filament, and the smaller the force on the filament. You can further verify this explanation by operating the lamp from a dc voltage source. In this case there is a force in only one direction, and the filament is simply pushed one way instead of vibrating.

Figure 1 Photograph of the flicker light.

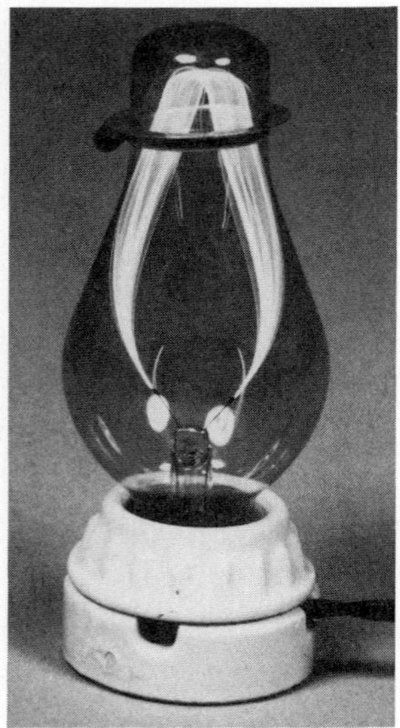

A very clever magnetic toy is the *seal and ball,* which is shown in Fig. 2.[3] When the toy is placed on a flat table and the seal is pushed near the ball, the ball begins to rotate. As long as the seal is kept near the ball, the ball continues to rotate. What is clever about the toy is that this is done with just two permanent magnets, one in the seal and one in the ball. You can understand the operation of the toy by considering the forces and torques that the magnets exert on each other. The magnet in the ball is cylindrical in shape with its magnetic moment $\boldsymbol{\mu}$ along the cylinder axis. The magnet is positioned below the center of the ball, so that the ball's center of mass is on its vertical axis, but below its geometric center, and its magnetic moment is vertical. The magnet in the seal has its magnetic moment at an angle of about 45° to the vertical. By separately considering the horizontal and vertical components of the seal's magnetic moment, you can understand the motion of the ball. The horizontal component

Figure 2 Photograph of the seal and ball.

sets up a horizontal magnetic field **B** at the ball. This will exert a torque $\boldsymbol{\tau} = \boldsymbol{\mu} \times \mathbf{B}$ on the ball, which will tip the ball slightly as indicated in Fig. 3. This tipping will move the center of mass of the ball slightly off the vertical axis (exaggerated in the figure). The vertical component of the seal's magnetic moment is essentially parallel to the ball's magnetic moment. This is the same as having two magnets side by side with their north poles together and their south poles together, which gives a repulsive force **F** between the magnets. This means that the ball will be pushed in a direction away from the seal. This magnetic force will exert a torque about the vertical axis, which, coupled with the frictional force of the table on the ball, will cause the ball to rotate. In short, the horizontal component of the seal's magnetic moment exerts a magnetic torque on the ball that causes the ball to tip slightly about a horizontal axis, and the magnetic repulsive force then exerts a torque causing the ball to rotate about the vertical axis. You can now also understand the source of energy that is necessary if the ball is to gain kinetic energy. The seal exerts a repulsive force on the ball, so the ball must in turn exert a repulsive force on the seal. In order for you to push the seal toward the ball you must do work against this repulsive force. This work done by you then increases the kinetic energy of the ball. By understanding magnetic forces and torques you can understand the operation of this clever toy.

A group of toys often seen in gift shops are shown in Fig. 4. These are the *rolling dolphin,* the *space circle* and the *mystery top.*[4] All of these are dynamic toys with one feature in common; they all keep moving. The dolphins roll, the circle swings, and the top spins. How do they work? Why do they keep moving? They all operate on the

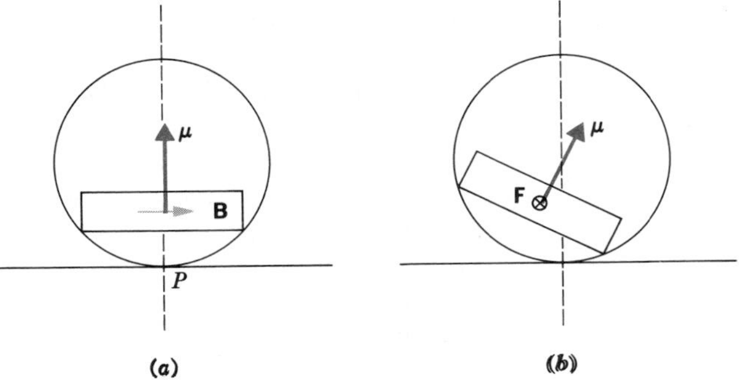

Figure 3 View of the ball facing the seal. The rectangle is the side view of the flat cylindrical magnet. (*a*) The horizontal component **B** of the magnetic field from the seal exerts a clockwise torque on the magnetic moment $\boldsymbol{\mu}$, causing the ball to tip about point *P*. (*b*) The component of the magnetic field of the seal that is parallel to $\boldsymbol{\mu}$ will exert a repulsive force **F**. This will cause the ball to rotate about the vertical axis.

Figure 4 Photograph of the space circle, mystery top, and rolling dolphin.

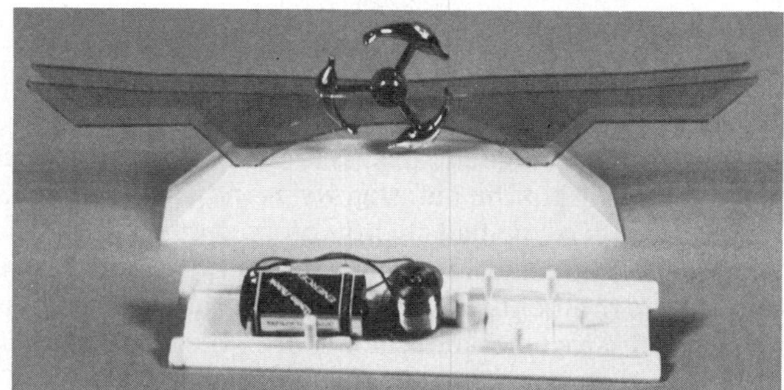

Figure 5 Photograph of the rolling dolphin open.

same principle, and you can get a clue to this mechanism by opening the base of one and looking inside. The *rolling dolphin* with its base open is shown in Fig. 5. You find that there is a coil of wire (actually, two concentric coils), a transistor, and a battery. In addition, you find that each dolphin has a small magnet in it. The battery provides the energy to keep the dolphins rolling, but you must understand Faraday's law and Lenz's law in order to understand the operation of the toy. As the magnet in one of the dolphins passes near the coil, it induces a voltage in the coil as predicted by Faraday's law. This voltage is applied to a transistor circuit in such a way as to turn on a second circuit consisting of the other coil and the battery. This second coil is wound in the opposite direction from the first. According to Lenz's law, the voltage induced in the first coil is such as to oppose the motion of the dolphin. The current in the second coil is then in a direction to enhance the motion. The dolphin is given a small pull if it is coming in toward the coil, or a small push if it is moving away; and it doesn't matter whether the dolphin is coming in from the right or the left. While the other two toys shown in the figure look quite different from the *rolling dolphin,* they operate in exactly the same way. Each toy contains a moving magnet, which, as predicted by Faraday's law, induces a voltage in a coil, and this voltage turns on a second circuit. The current in this second circuit is such as to increase the speed of the moving magnet.

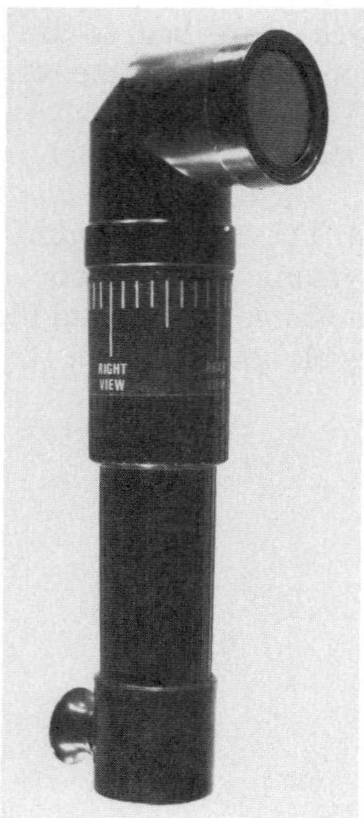

Figure 6 Photograph of the rotating periscope.

A toy periscope can be used to demonstrate reflection of light and image formation with flat mirrors. A particularly interesting version is a *rotating periscope* shown in Fig. 6. The top of the periscope can be rotated so that you can use it to look behind you. But when you do this, you find that the image is inverted. Why is this if the image is normal when you are looking straight ahead? Examination of the periscope shows that it contains two mirrors each at a 45° angle to the vertical. Light from an object will reflect down from the top mirror, and then out from the bottom mirror. An

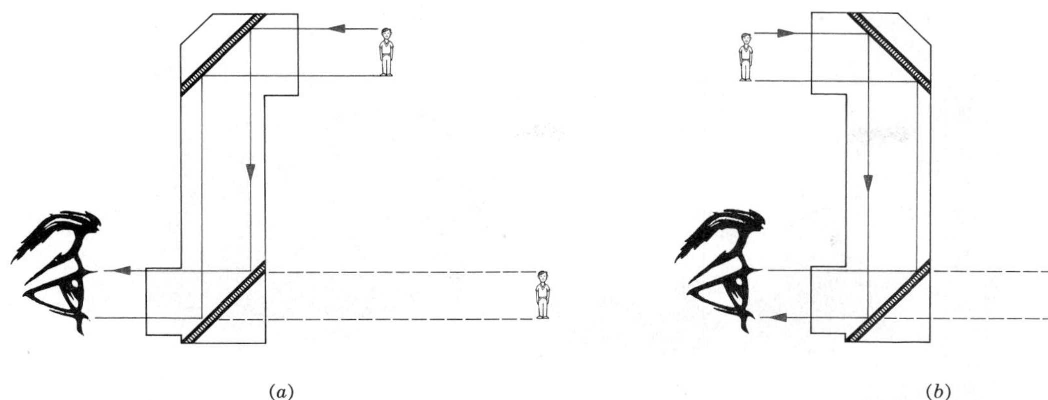

Figure 7 Image formation with the periscope. (*a*) The image is upright when the periscope is in its normal position. (*b*) The image is inverted when one periscope mirror is rotated 180°.

upright image will normally be observed, as sketched in Fig. 7*a*. If the top mirror is reversed, however, then the reflected light rays form an inverted image as shown in Fig. 7*b*. By using the law of reflection and tracing light rays, you can readily see why the image is upright in a normal periscope, but is inverted when one mirror is turned 180°. You can ask another question. What would you observe if the top mirror is rotated 90° so that you are looking to the side? You decide the answer to that question.

The *radiometer* shown in Fig. 8 has been sold in novelty shops for many years, but its operation is probably a mystery to most people.[5] It consists of four vanes, black on one side and white on the other, that are free to rotate about a vertical axis inside a glass housing. If the radiometer is placed in a beam of sunlight, or near a reasonably bright light bulb, the vanes begin to rotate rapidly. A possible cause of the rotation is radiation pressure from the light. Assume that the light is incident normal to the surface of the vanes as sketched in Fig. 9. The light striking the black side of the vane would be absorbed, transferring the light's total momentum to the vane; light incident on the white side of the vane would be reflected, reversing the light's momentum and transferring twice as much momentum to the vane. The net momentum transfer would cause the vanes to rotate, with the black side of the vanes leading and the white side following. But if you observe the rotation, you will see that the white side of the vanes is always leading. The problem is that the bulb does not have a good

Figure 8 Photograph of the radiometer.

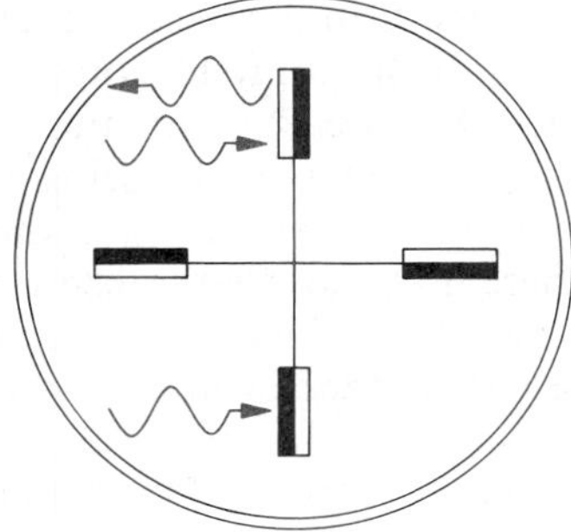

Figure 9 Drawing of light incident on black and white radiometer vanes.

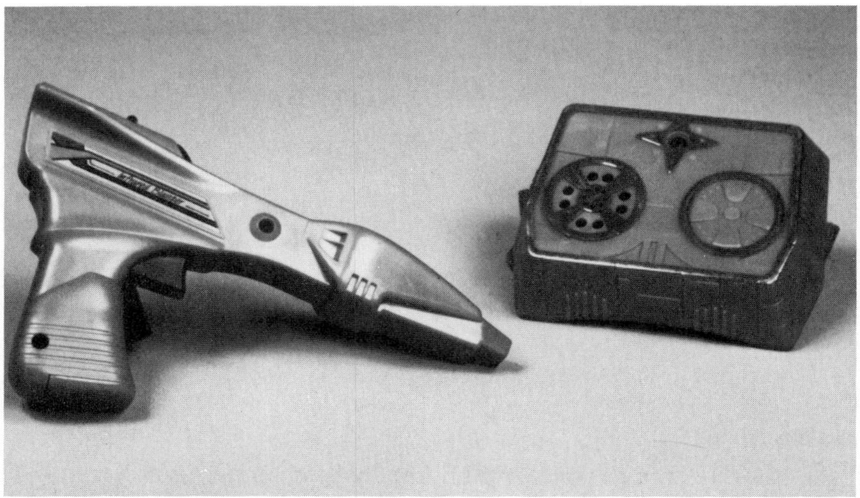

Figure 10 Photograph of the "laser gun" and target.

vacuum in it, and the air in the bulb plays an important role in the motion. Since the black side of the vane absorbs the light's energy, while the white side reflects it, the black side of the vane becomes hotter than the white side. This temperature difference causes the vanes to move due both to a circulation of the air in the bulb and to a momentum transfer. Air molecules that come in contact with the vane will leave the vane with an energy that is dependent on the vane's temperature. Molecules leaving the hotter black vane will on the average have a larger energy and thus a larger momentum than those leaving the cooler white vane. The recoil momentum of the vane will be larger for the hotter black side than for the cooler white side. As a result, the vane will gain a net momentum, rotating in such a way that the white side is always leading. This is still a momentum transfer that helps cause the vanes to rotate, but it is not the momentum of the light that does it.

A relatively new toy that is popular now is sometimes called *laser tag.* As shown in Fig. 10, it consists of a "laser gun" that emits an invisible beam of electromagnetic radiation and a sensor that detects the beam and indicates if the wearer has been "tagged." What are the properties of this electromagnetic radiation? There are any number of questions that you might ask about the radiation, and then answer by experimentation. Does the beam travel in straight lines, or will it bend around corners? Can the beam be reflected? If it can, what materials are reflecting, and does it obey the law of reflection? Can it be refracted and if it can, can it be focused with a lens? What is the maximum distance at which the beam can be detected, and can this range be extended by using a lens? Are there materials that will transmit this radiation, but not visible light; that is, can the beam go through some kinds of walls? Are there materials that transmit visible light, but not this radiation, so you can build a transparent shield? You can undoubtedly think of many more questions that can be asked, and then can be investigated experimentally. You may be able to guess the answers to some of the above questions by knowing that the beam of electromagnetic radiation from the "laser gun" is, in fact, infrared light.

These are only a few of the many toys that can be used to illustrate basic physical principles. Only your imagination and ingenuity limit you in your application of the fundamental laws of physics to ordinary objects, even toys. Science can be fun!

REFERENCES [1] "Toys in Physics Teaching: Balancing Man," by Raymond C. Turner, *American Journal of Physics,* January 1987. [2] "Hot Wheels Physics," by Stanley J. Briggs, *The Physics Teacher,* May 1970. [3] "A Physics Toy: Magnetic Seal and Ball," by Raymond C. Turner, *The Physics Teacher,* December 1987. [4] "How Things Work: A Spinning Top," by H. Richard Crane, *The Physics Teacher,* February 1984. [5] "Running Crooke's Radiometer Backwards," by Frank S. Crawford, *American Journal of Physics,* November 1985.

CHAPTER 39
GEOMETRICAL OPTICS

We are so accustomed to mirrors that we forget what a marvelous thing it is to be able to form an image of ourselves in such a simple way. This three-dimensional phantom — rigidly coupled to our own movements — looks back at us from the solid space behind the mirror where we know that only darkness exists. The Hall of Mirrors at Glacier Park in Lucerne recaptures some of the magic.

39-1 Geometrical Optics

If you are attending an outdoor concert and somebody stands up in front of you, you can still hear the music but you can no longer see the stage. Why this difference in behavior between sound waves and light waves? We trace it to the fact that the wavelength of sound (about 1 m) is about the same size as the "obstacle" but the wavelength of light (about 500 nm or 5×10^{-7} m) is very much smaller.

This experience illustrates that, under some circumstance, waves behave to a good approximation as if they traveled in straight lines, being blocked by barriers and casting sharp shadows. We need only to agree not to place in the path of the wave any obstacle or aperture, such as a mirror, lens, slit, or baffle, unless its dimensions are much larger than the wavelength. For light waves this useful special case of wave behavior is called *geometrical optics;* it is the subject of this chapter.

39-2 Reflection and Refraction

Figure 1*a* shows a beam of light falling on a plane glass surface. Part of the light is reflected from the surface and part of it is transmitted into the glass. Note that the

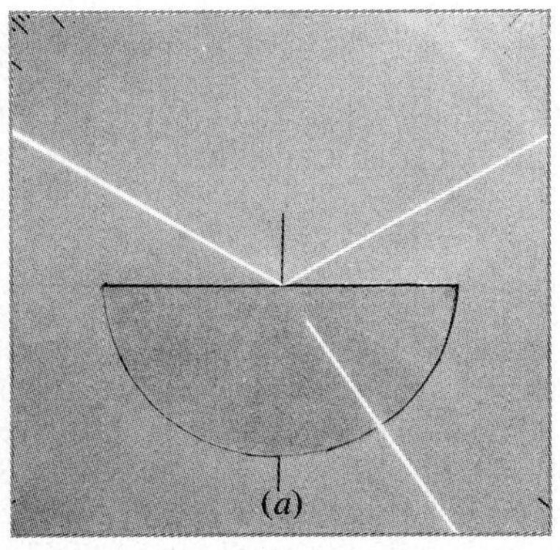

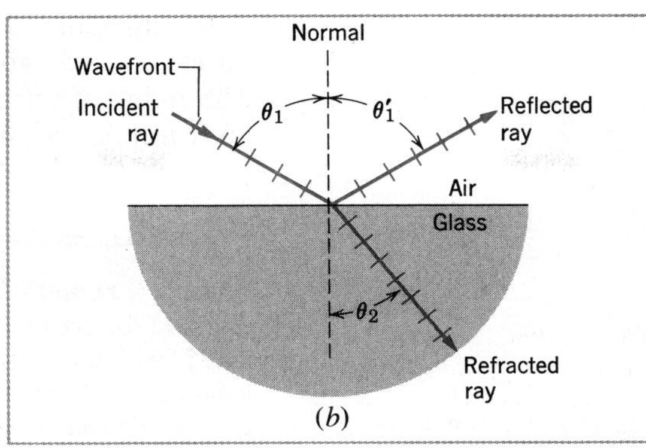

Figure 1 (*a*) A photograph showing the reflection and the refraction of an incident light beam as it falls on a plane glass surface. (*b*) A representation using rays. The angles of incidence (θ_1), of reflection (θ_1'), and of refraction (θ_2) are marked. Note that each angle is measured from the surface normal to the appropriate ray.

transmitted beam is bent or *refracted* as it passes through the surface.*

Let us define some useful quantities. In Fig. 1*b* we represent the incident beam and also the reflected and the refracted beams by *rays,* which are lines drawn at right angles to the wavefronts. The *angle of incidence* θ_1, the *angle of reflection* θ_1', and the *angle of refraction* θ_2 are also shown in the figure. Note that each of these angles is measured between the normal to the surface and the appropriate ray. The plane that contains both the incident ray and a line normal to the surface is called the *plane of incidence.* (In Fig. 1*b*, the plane of incidence is the plane of the printed page.)

Experiment shows that reflection and refraction are governed by these laws:

The law of reflection: The reflected ray lies in the plane of incidence and

$$\boxed{\theta_1' = \theta_1} \quad \text{(reflection).} \qquad (1)$$

The law of refraction: The refracted ray lies in the plane of incidence and

$$\boxed{n_1 \sin \theta_1 = n_2 \sin \theta_2} \quad \text{(refraction).} \qquad (2)$$

Here n_1 is a dimensionless constant called the *index of refraction* of medium 1 and n_2 is the index of refraction of medium 2. Eq. 2 is called Snell's law. As we shall see in Section 40-2, the index of refraction of a substance is c/v, where c is the speed of light in free space and v is its speed in the substance in question.

Table 1 shows the indices of refraction for some common substances; note that $n \approx 1$ for air. It is the high index of refraction of diamond ($n = 2.42$, versus ~ 1.5 for a typical glass) that accounts for its sparkle. As Fig. 2 shows, the index of refraction for a given substance varies with wavelength.

We give here two examples of clever applications of the laws of reflection and refraction, chosen from a very long list.

Reflection: A Neat Application. A plane mirror will reflect a beam of light back toward its source if the mirror is positioned exactly right. A *corner reflector,* which is three mirrors fastened together as an inside corner of a cube, has the property that it will reflect a beam directly backward *no matter how the reflector is oriented.*

Luminous "cat's eyes," imbedded in the roadway, work on this principle. In the Physics Building at the University of Washington, three mirrors are mounted in a corner of the ceiling in a stairwell so that you can see

Table 1 Some Indices of Refraction[a]

Medium	*Index*	*Medium*	*Index*
Vacuum (exactly)	1.00000	Typical crown glass	1.52
Air (STP)	1.00029	Sodium chloride	1.54
Water (20°C)	1.33	Polystyrene	1.55
Acetone	1.36	Carbon disulfide	1.63
Ethyl alcohol	1.36	Heavy flint glass	1.65
Sugar solution (30%)	1.38	Sapphire	1.77
Fused quartz	1.46	Heaviest flint glass	1.89
Sugar solution (80%)	1.49	Diamond	2.42

[a] For a wavelength of 589 nm (yellow sodium light).

* *Refracted* comes from the Latin for broken, as in fractured leg. If you dip a pencil on a slant part way into a fishbowl, the pencil seems to be "broken."

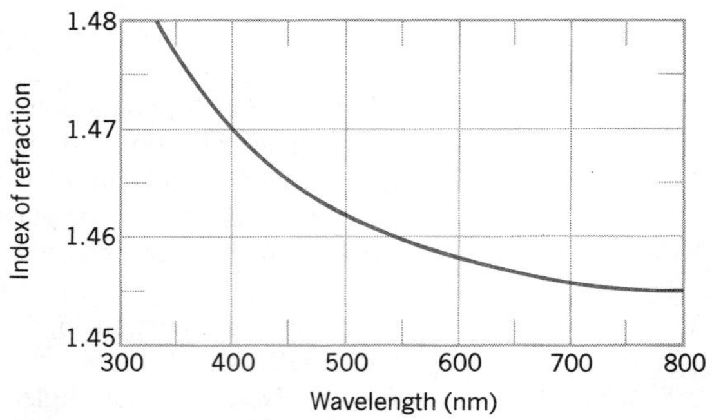

Figure 2 The index of refraction as a function of wavelength for fused quartz with respect to air. Shorter wavelengths, corresponding to higher indices of refraction, are bent through larger angles on entering quartz.

your own image from every step. Figure 3 shows a laser geodynamic satellite (LAGEOS) studded with 426 corner reflectors. Range measurements made by bouncing laser signals from it permit measurement of continental drift to within two centimeters. Finally, there is a corner reflector on the moon, from which we can bounce signals at will.

Refraction: A Neat Application. Investigators often want to trace glass fragments found at the scene of an accident so as to identify a hit-and-run motorist. The index of refraction of such fragments can be found by immersing a fragment in a transparent liquid whose index of refraction is particularly temperature-sensitive. The temperature of the liquid is then adjusted until the glass fragment disappears visually, a signal that there is no longer any refraction at the glass–liquid interface. The index of refraction of the glass is the same as the known index of the liquid at that temperature.

Figure 3 A LAser GEOdynamic Satellite (LAGEOS), launched into a 3600-mi high orbit in 1976. Corner reflectors permit precise location of its position by means of laser-pulse ranging. The satellite is 24 in. in diameter.

Sample Problem 1 An incident beam falls on the plane polished surface of a block of fused quartz, making an angle of 31.25° with the normal. This beam contains light of two wavelengths, 404.7 nm and 508.6 nm. The indices of refraction for quartz at these wavelengths are 1.4697 and 1.4619, respectively; the index of refraction for air may be taken as 1.0003 at both wavelengths. What is the angle between the two refracted rays?

From Eq. 2 we have, for the 404.7-nm ray,

$$n_1 \sin \theta_1 = n_2 \sin \theta_2$$

or

$$(1.0003) \sin 31.25° = (1.4697) \sin \theta_2,$$

which leads to

$$\theta_2 = \sin^{-1}\left(\sin 31.25° \frac{1.0003}{1.4697}\right) = 20.6761°.$$

In the same way we have, for the 508.6-nm ray,

$$\theta_2' = \sin^{-1}\left(\sin 31.25° \frac{1.0003}{1.4619}\right) = 20.7915°.$$

The angle $\Delta\theta$ between the rays is

$$\Delta\theta = 20.7915° - 20.6761° = 0.1154° \approx 6.9 \text{ min of arc.} \qquad \text{(Answer)}$$

The component with the shorter wavelength, having the larger index of refraction, has the smaller angle of refraction and is thus bent through the larger angle.

39-3 Total Internal Reflection

Figure 4 shows rays from a point source in glass falling on a glass–air interface. As the angle of incidence θ is increased, we reach a situation (see ray *e*) at which the refracted ray points along the surface, the angle of refraction being 90°. For angles of incidence larger than this *critical angle* θ_c, there is no refracted ray, and we speak of *total internal reflection.*

We find the critical angle by putting $\theta_2 = 90°$ in the law of refraction (see Eq. 2):

$$n_1 \sin \theta_c = n_2 \sin 90°,$$

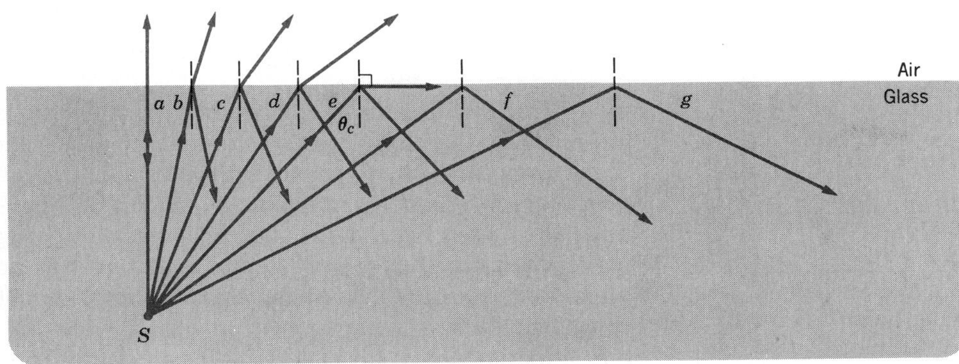

Figure 4 Total internal reflection of light from a point source S occurs for all angles of incidence greater than the critical angle θ_c. At the critical angle, the refracted ray points along the air–glass interface.

or

$$\boxed{\theta_c = \sin^{-1} \frac{n_2}{n_1}} \quad \text{(critical angle).} \tag{3}$$

The sine of an angle cannot exceed unity so that we must have $n_2 < n_1$. This tells us that total internal reflection cannot occur when the incident light is in the medium of lower index of refraction. The word *total* means just that; the reflection occurs with no loss of intensity. In ordinary reflection from a mirror—by way of contrast—there is an intensity loss of about 4%.

Total internal reflection makes possible the *colonofiberscope* and other fiber optical devices by means of which physicians can visually inspect many internal body sites; see Fig. 5. In these devices, a bundle of fibers transmits an image that can be inspected visually outside the body.

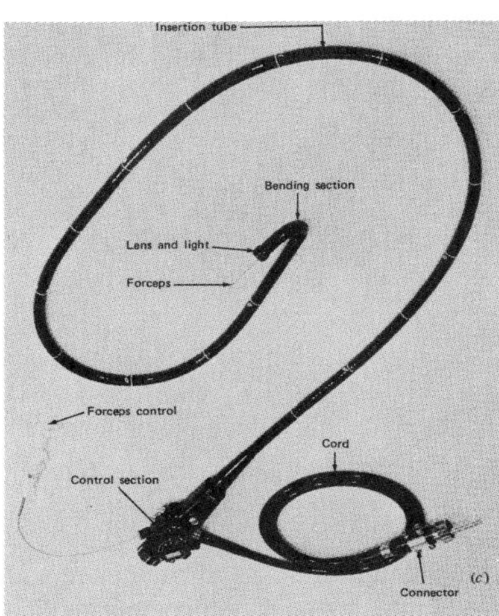

Figure 5 A *colonofiberscope,* used to examine the intestinal tract.

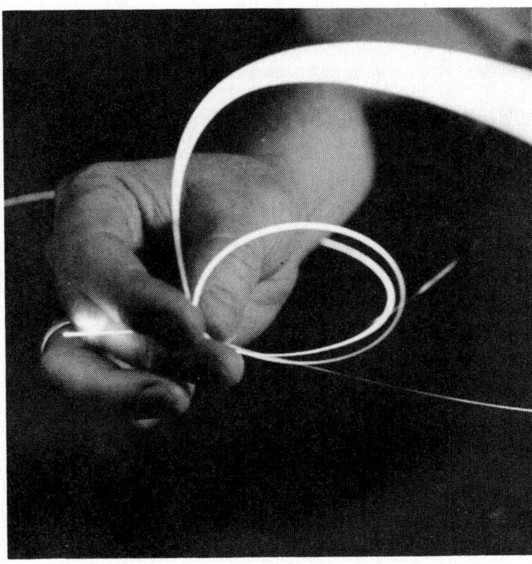

Figure 6 Light is transmitted by total internal reflection through an optical fiber of the kind used in lightwave communication systems.

One rapidly growing application of total internal reflection is lightwave communication over an optical fiber network that now spans the country and will soon span the Atlantic Ocean; see Fig. 6. As Fig. 7 shows, the fiber consists of a central core that is graded smoothly into an outer cladding layer of a material of lower index of refraction. Only those rays that are internally reflected can be propagated along the cable. To reduce attenuation of the signal as it passes along the cable, materials of extreme purity have been developed. If sea water were as transparent as the glass from which optical fibers are made, it would be possible to see the sea bottom by reflected sunlight at a depth of several miles.

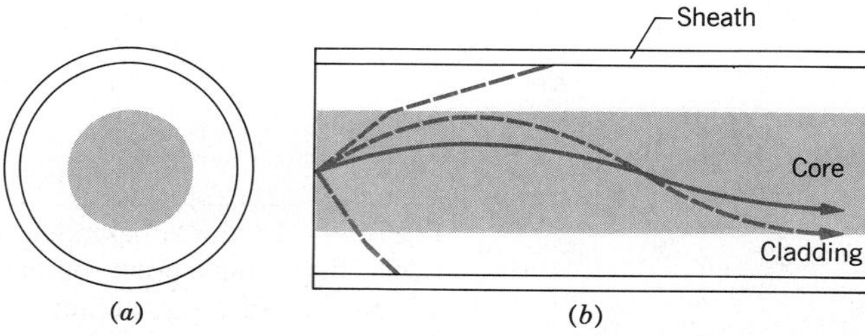

Figure 7 (*a*) An optical fiber shown in cross section. The fiber diameter is about that of a human hair. (*b*) A transverse view, showing propagation by total internal reflection. The glass core, the glass-graded cladding (of lower index than the core) and the protective sheath are shown.

Sample Problem 2 Figure 8 shows a triangular prism of glass in air, a ray incident normal to one face being totally reflected. If θ_1 is 45°, what can you say about the index of refraction of the glass?

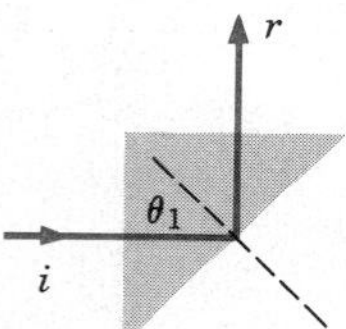

Figure 8 Sample Problem 2. The incident ray is totally internally reflected at the glass–air interface.

The angle θ_1 must be equal to or greater than the critical angle θ_c, where θ_c is given by Eq. 3:

$$\theta_c = \sin^{-1}\frac{n_2}{n_1} = \sin^{-1}\frac{1}{n}$$

in which, for all practical purposes, the index of refraction of air ($= n_2$) is set equal to unity. Since total internal reflection occurs, then θ_c must be less than 45° and

$$n > \frac{1}{\sin 45°} = 1.41. \qquad \text{(Answer)}$$

If the index of refraction of the glass is less than 1.41, total internal reflection will not occur.

39-4 Polarization by Reflection

If you rotate a polarizing sunglass lens in front of one eye, you can reduce or eliminate the glare from sunlight reflected from water or from any glossy surface. Such reflected light is fully or partially polarized by the process of reflection from the surface.

Figure 9 shows an unpolarized beam falling on a glass surface. The electric field vector for each wavetrain in the beam can be resolved into a *perpendicular component* (perpendicular to the plane of incidence), represented by dots in Fig. 9, and a *parallel component* (lying in the plane of incidence), represented by arrows. For the unpolarized incident light, these two components are of equal amplitude.

Experimentally, for glass or other dielectric materials, there is a particular angle of incidence, called the *Brewster angle* θ_B, at which the reflection for the parallel component is zero. This means that the beam reflected from the glass for this angle of incidence is completely polarized, with its plane of vibration at right angles to the plane of Fig. 9.

Since the parallel component of a ray incident at the Brewster angle is not reflected, it must be fully transmitted. Just as total internal reflection permits lossless reflection, the phenomenon of Fig. 9 permits lossless transmission (of the parallel component). In a gas laser, light must be reflected back and forth through the laser cavity by external mirrors, perhaps several hundred times. If the ends of the laser cavity are sealed by *Brewster windows* (that is, windows tilted at the Brewster angle) beam losses by reflection on successive passage through the windows can be prevented.

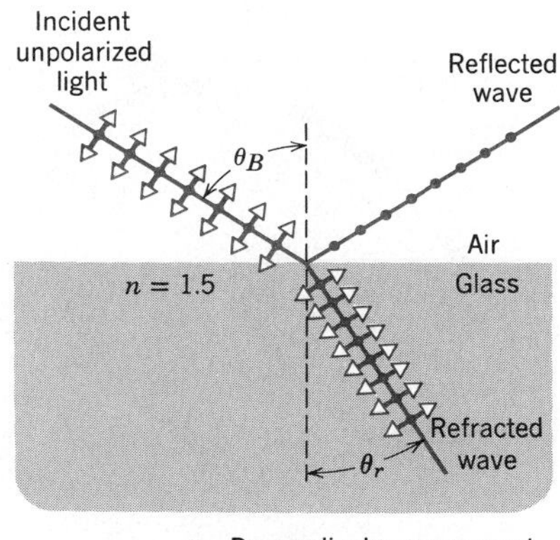

Figure 9 For a particular angle of incidence, called the *Brewster angle* θ_B, the parallel component of the incident ray is transmitted without loss. As a consequence, the reflected ray contains no parallel component and is thus completely polarized. The transmitted ray is partially polarized.

Brewster's Law. At the Brewster angle we find experimentally that the reflected and the refracted beams are at right angles, or, equivalently (Fig. 9),

$$\theta_B + \theta_r = 90°.$$

From the law of refraction we have

$$n_1 \sin \theta_B = n_2 \sin \theta_r.$$

Combining these equations leads to

$$n_1 \sin \theta_B = n_2 \sin (90° - \theta_B) = n_2 \cos \theta_B,$$

or

$$\theta_B = \tan^{-1} \frac{n_2}{n_1},$$

where the incident ray is in medium one and the refracted ray in medium two. We can write this as

$$\boxed{\theta_B = \tan^{-1} n} \quad \text{(Brewster's law)}, \qquad (4)$$

where $n(= n_2/n_1)$ is the index of refraction of medium two with respect to medium one. Equation 4 is known as *Brewster's law* after Sir David Brewster, who deduced it empirically in 1812.*

Sample Problem 3 We wish to use a plate of glass ($n = 1.57$) as a polarizer. (a) What is the Brewster angle? That is, at what angle of incidence will the reflected beam be fully polarized?

From Eq. 4

$$\theta_B = \tan^{-1} n = \tan^{-1} 1.57 = 57.5°. \qquad \text{(Answer)}$$

(b) What angle of refraction corresponds to this angle of incidence?

Since $\theta_B + \theta_r = 90°$ we have

$$\theta_r = 90° - \theta_B = 90° - 57.5° = 32.5°. \qquad \text{(Answer)}$$

39-5 A Plane Mirror

Perhaps our simplest optical experience is looking in a mirror. Figure 10 shows a point source of light O, which we will call the *object,* placed at a distance o in front of a plane mirror. The light that falls on the mirror is represented by rays emanating from O. At the point in Fig. 10 at which each ray from O strikes the mirror, we construct a reflected ray. If we extended the reflected rays backward, they intersect in a point I that we call the *image* of the object O. The image is as far behind the mirror as the object is in front of it. This fact can easily be brought home if you photograph yourself in a mirror with a sonar autofocus camera; the sound waves will focus the camera on the mirror surface and your image, which is well behind that surface, will be out of focus.

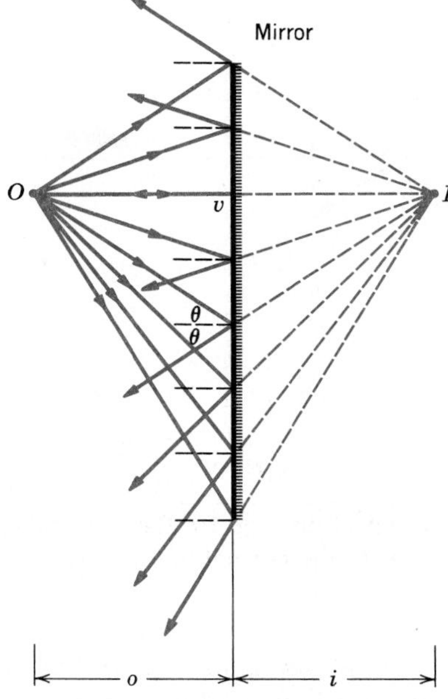

Figure 10 A point object forms a virtual image in a plane mirror. The rays *appear* to diverge from I but actually no light is present at this point.

When we look into the mirror, the reflected rays really seem to be coming from the image point, although we know that they do not. We call such an image a *virtual* image, meaning that light does not actually pass though it. We know from daily experience how "real" such a virtual image appears to be and how definite is its location in the space behind the mirror, even though this space may, in fact, be occupied by a brick wall.

Figure 11 shows two rays selected from the bundle of rays in Fig. 10. One strikes the mirror at v, along a perpendicular line. The other strikes it at an arbitrary point a, making an angle of incidence θ with the normal at that point. Elementary geometry shows that the right triangles $aOva$ and $aIva$ are congruent and thus

$$\boxed{i = -o} \quad \text{(plane mirror)}. \qquad (5)$$

* Brewster was also the inventor of the *kaleidoscope,* a word from the Greek meaning: "seeing beautiful shapes." See Essay 16.

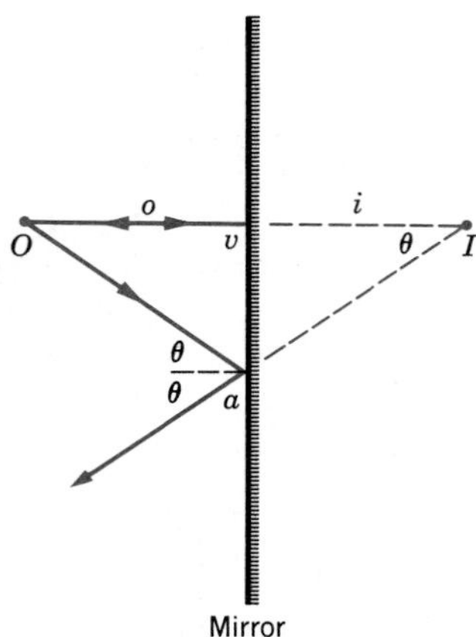

Figure 11 Two rays from Fig. 10. Ray Oa makes an arbitrary angle θ with the normal to the mirror surface.

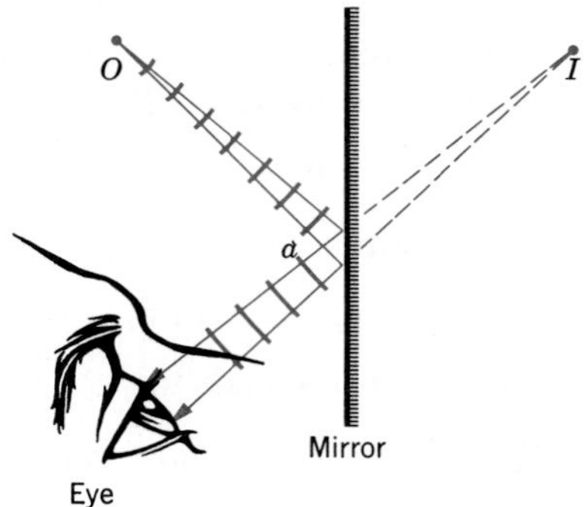

Figure 12 A pencil of rays from O enters the eye after reflection at the mirror. Only a small portion of the mirror near a is effective. The small arcs represent portions of spherical wavefronts. The eye "thinks" that the light is coming from I.

We introduce the minus sign to signal that the image is virtual, a point we will expand upon below.

Only rays that lie fairly close together can enter the eye after reflection at a mirror. For the eye position shown in Fig. 12, only a small patch of the mirror near point a (a patch smaller than the pupil of our eye) is used in forming the image. You might experiment with a mirror, closing one eye and looking at the image of a small object such as the tip of a pencil. Then move your fingertip over the mirror surface until you cannot see the image. Only that small portion of the mirror under your fingertip was used to bring the image to you.

Image Reversal. As Fig. 13a shows, the image of a left hand is a right hand, and it is often said that a mirror reverses right and left. Perceptive students then often ask: Why then does the mirror not also reverse up and down?

Figure 13b sheds some light on the matter. Of the three arrows that form the object, the images of the two that lie in a plane parallel to the mirror are just like their objects. The image of the arrow that points toward the mirror, however, is reversed front to back. Thus, it is more accurate to say that a mirror reverses front to back than that it reverses right to left. The transformation of a left hand into a right hand is accomplished, in a sense, by transferring the front of the hand to the back of the hand directly through the hand.

Sample Problem 4 Kareem Abdul-Jabbar is 7′2″ (= 218 cm) tall. How tall must a vertical mirror be if he is to be able to see his entire length?

In Fig. 14, the top of Kareem's head (h), his eyes (e), and

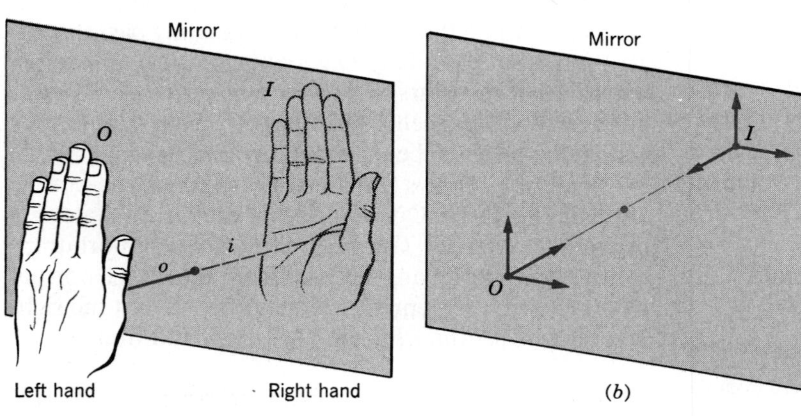

Figure 13 (a) The object O is a left hand; its image I is a right hand. (b) Study of a reflected three-arrow object shows that a mirror interchanges front and back, rather than right and left.

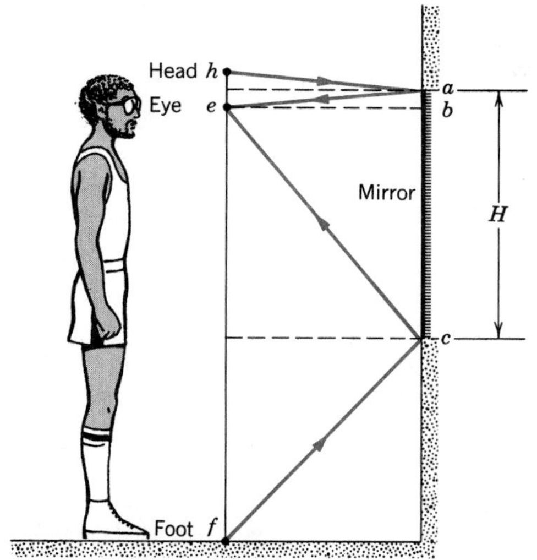

Figure 14 Sample Problem 4. A "full-length mirror" need only be half your height.

the bottoms of his feet (f) are marked by dots. The figure shows the paths followed by rays that leave his head and his feet and enter his eyes, reflecting from the mirror at points a and c, respectively. The mirror need occupy only the vertical distance H between those points.

We see that

$$ab = \tfrac{1}{2}he \quad \text{and} \quad bc = \tfrac{1}{2}ef.$$

Thus,

$$H = ac = ab + bc = \tfrac{1}{2}(he + ef)$$
$$= (\tfrac{1}{2})(218\text{ cm}) = 109\text{ cm}. \qquad \text{(Answer)}$$

The mirror need be no taller than half the height of the person. If you have a full-length mirror available, you might experiment by taping newspaper over those portions of the mirror that do not contribute to your image. You will find that what you have left is just half of your height. Mirrors that extend below point c just allow you to look at an image of the floor. Your horizontal distance from the mirror makes no difference.

39-6 Spherical Mirrors

In the last section we discussed image formation in a plane mirror. Now we will see what happens to the image if the surface of the mirror is curved. In particular, we will consider a spherical mirror, which is simply a small section of the surface of a sphere. A plane mirror is in fact a spherical mirror with an infinitely large radius of curvature.

Let us start with a plane mirror (Fig. 15a) and make it *concave* ("caved in") toward the observer but with r, its *radius of curvature,* still very large, as in Fig. 15b. Note that C, its *center of curvature,* is on the left in Fig. 15b. Compared with a plane mirror, two things happen. First, the image moves farther behind the mirror (that is, i takes on a larger negative value). Second, the size of the image increases. For a plane mirror, the image is exactly the same size as the object. For this curved mirror, the image is *larger* than the object. This is the principle of the makeup mirror or the shaving mirror, which are slightly concave and magnify the face. The relatively narrow angle of divergence of the reflected rays in Fig. 15b suggests that such a mirror possesses a narrower field of view than does a plane mirror; the image of your face is bigger but you cannot see as much of it.

If we curve the plane mirror to make it *convex* toward the observer, as in Fig. 15c, the image moves closer to the mirror and shrinks. Such convex mirrors are used as right-hand side-view mirrors in automobiles and as surveillance mirrors in buses and supermarkets. The wide angle of divergence of the reflected rays in Fig. 15c suggests that such a mirror possesses a wider field of view than does a plane mirror; the driver can see the whole interior of the bus. Note that the images are virtual for all three cases shown in Fig. 15.

As Fig. 16 shows, we can also use a spherical mirror to bring incident light to a focus. For a concave mirror (Fig. 16a) the focus is *real;* you can ignite a piece of paper by focusing sunlight on it in this way. Reflecting telescopes also use this principle to bring distant starlight to a focus. For a convex mirror (Fig. 16b) the focus is *virtual.* The point F in each of these figures is called the *focal point* and the distance f is the *focal length.* As we shall prove in Section 39-11, there is a simple relation between the distance o of the object from the mirror, the distance i of the image from the mirror, and the focal length f of the mirror. It is

$$\boxed{\frac{1}{o} + \frac{1}{i} = \frac{1}{f}} \quad \text{(spherical mirror),} \qquad (6)$$

in which f is related to r, the radius of curvature of the mirror, by

$$\boxed{f = \tfrac{1}{2}r} \quad \text{(spherical mirror).} \qquad (7)$$

We can combine these relations and write the mirror equation also in the form

$$\boxed{\frac{1}{o} + \frac{1}{i} = \frac{2}{r}} \quad \text{(spherical mirror).} \qquad (8)$$

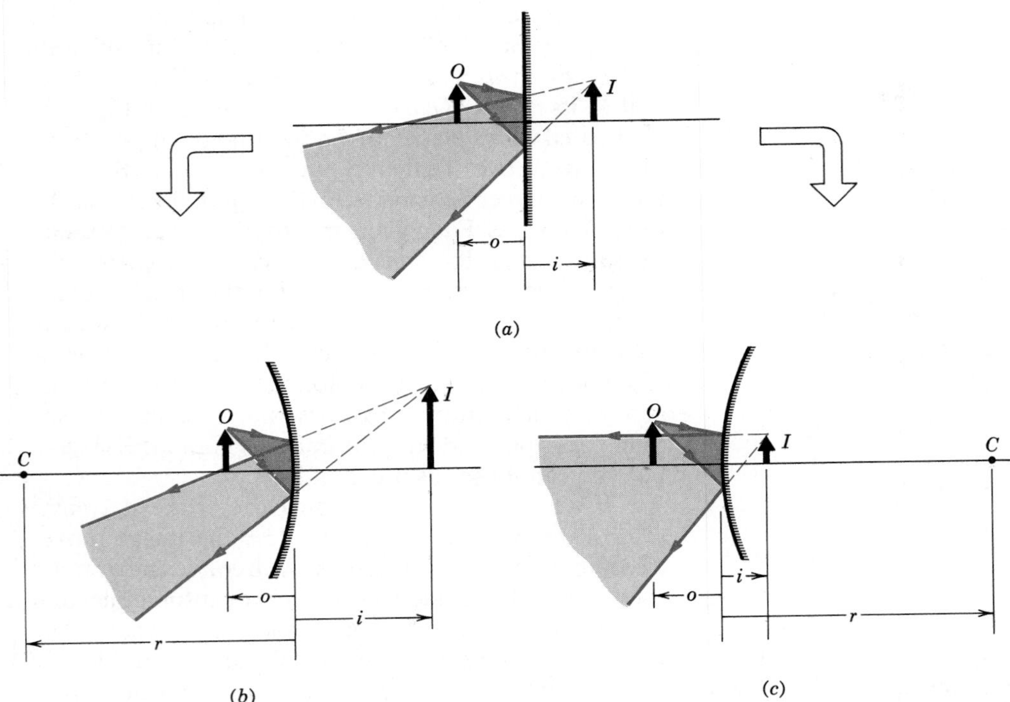

Figure 15 (*a*) An object forms a virtual image in a plane mirror. (*b*) If the mirror is bent so that it becomes *concave,* the image moves farther away and becomes larger. (*c*) If the plane mirror is bent so that it becomes *convex,* the image moves closer and becomes smaller.

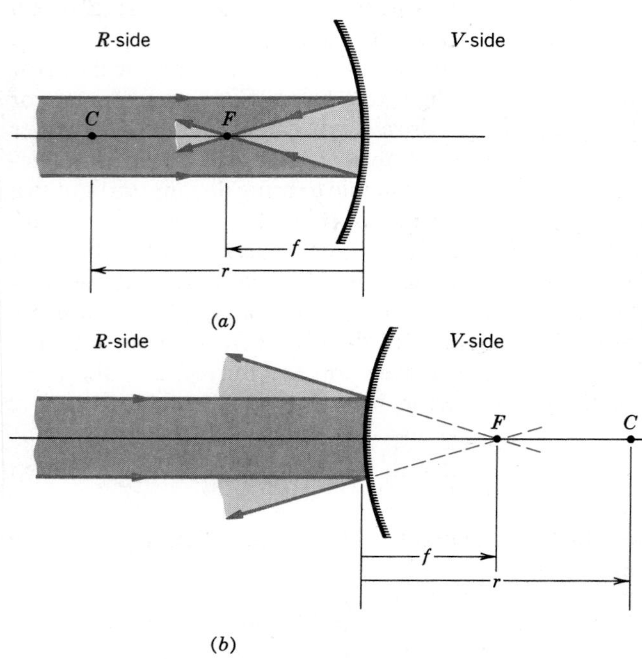

Figure 16 (*a*) In a concave mirror, incident parallel light is brought to a real focus at *F*, on the *R*-side of the mirror. (*b*) In a convex mirror, incident parallel light is made to seem to diverge from a virtual focus at *F*, on the *V*-side of the mirror.

We can test Eq. 8 by letting r become infinitely large, thus describing a plane mirror. If we do so, Eq. 8 reduces to $i = -o$, which is exactly the relation (see Eq. 5) that we gave earlier for such a mirror. We can test Eq. 6 by letting o become infinitely large, thus describing parallel incident light from an infinitely distant object as in Fig. 16. Equation 6 then reduces to $i = f$, showing that the image does indeed appear at the focal point.

In Fig. 17 the three figures show what happens as the object is moved from a point close to a concave mirror (Fig. 17*a*) to the focal point (Fig. 17*b*), and finally to a position beyond the focal point (Fig. 17*c*). Note that the image is *virtual* in the first figure but *real* in the other two. The image is *upright* in the first two figures but *inverted* in the third. In studying this figure it is well to follow the image position as the object moves in a continuous fashion from a point close to the mirror out to infinity. (Where is the image when the object is at the center of curvature?)

We shall show in Section 39–7 that the *lateral magnification m* of a mirror is given by

$$\boxed{m = -\frac{i}{o}} \quad \text{(lateral magnification).} \qquad (9)$$

For a plane mirror, for which $i = -o$ (see Eq. 5), we have $m = +1$. The magnification of 1 means that the image is the same size as the object. The plus sign, as we shall see, means that the image is *upright;* a minus sign would signify an *inverted* image.

Equations 6, 7, 8, and 9 hold for all mirrors, whether plane, concave, or convex. In using these equations, however, we must be careful about the signs of the quantities o, i, r, f, and m that enter into them.

If we set it up properly, our rule for signs will apply not only to mirrors but also to spherical refracting surfaces and to lenses. We start by calling the front of the mirror—where only real images can be formed—the *R*-side (*R* for real) of the mirror. Similarly, we call the back of the mirror—where only virtual images can be formed—the *V*-side (*V* for virtual) of the mirror. Our rule for signs is: Associate *positive* with *Real, R-side,* and *upRight;* associate *negative* with *Virtual, V-side,* and *inVerted.* In particular:

o is positive if the object is real,*
i is positive if the image is real (that is, if I is on the *R*-side),
r is positive if C is on the *R*-side,
f is positive if the focus is real (that is, if F is on the *R*-side),
m is positive if the image is upRight.

We will clarify the use of these rules in the sample problems.†

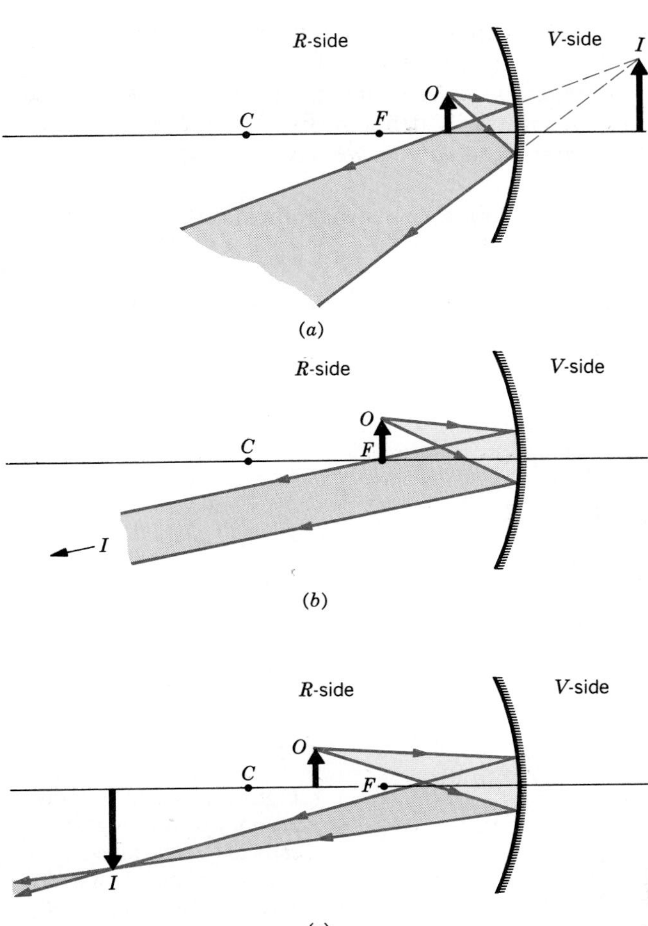

Figure 17 An object is moved at a constant slow speed from a position near the surface of a concave mirror out to infinity. How does its image move? This figure shows three stages of the motion.

Sample Problem 5 A convex mirror (see Fig. 15*c*) has a radius of curvature of 22 cm. (a) If an object is placed 14 cm away from the mirror, where is its image?

As Fig. 15*c* shows, the center of curvature for this mirror is on the *V*-side of the mirror. According to our rule for signs, the radius of curvature r is negative, as it is for all convex mirrors. We have then, from Eq. 8

$$\frac{1}{o} + \frac{1}{i} = \frac{2}{r}$$

or

$$\frac{1}{+14\text{ cm}} + \frac{1}{i} = \frac{2}{-22\text{ cm}},$$

which yields

$$i = -6.2\text{ cm}, \qquad \text{(Answer)}$$

in agreement with the graphical prediction. The negative sign for i reminds us that the image is on the *V*-side of the mirror and thus is virtual.

(b) What is the magnification in this case?

From Eq. 9 we have

$$m = -\frac{i}{o} = -\frac{-6.2\text{ cm}}{+14\text{ cm}} = +0.44. \qquad \text{(Answer)}$$

---

* Although objects, like images, can be virtual, we consider here only the familiar case of real objects.

† You may wish to use Eq. 9 to check on the following limerick, composed by Robert G. Greenler of the Optical Society of America

To a politician who would reveal
An Image with public appeal
Said his optical friend
"In my view I contend
If it's Upright it's Virtual, not Real!"

Thus the image is smaller than the object and, according to our rule for signs, it is upright. This is in full qualitative agreement with Fig. 15*c*.

39-7 Ray Tracing

Figures 18*a* and 18*b* show an object O in front of a concave mirror. We can locate an image of any off-axis point graphically by tracing any two of four special rays. Thus:

1. A ray that strikes the mirror parallel to its axis passes through the focal point (ray 1 in Fig. 18*a*).
2. A ray that strikes the mirror after passing through the focal point emerges parallel to the axis (ray 2 in Fig. 18*a*).
3. A ray that strikes the mirror after passing through the center of curvature C returns along itself (ray 3 in Fig. 18*b*).
4. A ray that strikes the vertex of the mirror will be reflected at an equal angle with the axis of the mirror (ray 4 in Fig. 18*b*).

The four rays are displayed on two separated plots to minimize clutter. Our descriptions of the rays need to be modified only slightly to apply to convex mirrors, as in Figs. 18*c* and 18*d*.

The Magnification. Here we wish to derive Eq. 9 ($m = -i/o$), the expression for the magnification of an object reflected in a mirror. Consider ray 4 in Fig. 18*b*. It is drawn to be reflected at the vertex v, making equal angles with the axis of the mirror at that point.

For the two similar right triangles in the figure we can write

$$\frac{Ib}{Oa} = \frac{vb}{va}.$$

The quantity on the left (apart from a question of sign) is the *lateral magnification* m of the mirror. Since we want to represent an *inverted* image by a *negative* magnification, we arbitrarily define m for this case as $-(Ib/Oa)$. Since $vb = i$ and $va = o$, we have at once

$$m = -\frac{i}{o} \quad \text{(magnification)}, \qquad (10)$$

which is Eq. 9, the relation we set out to prove.

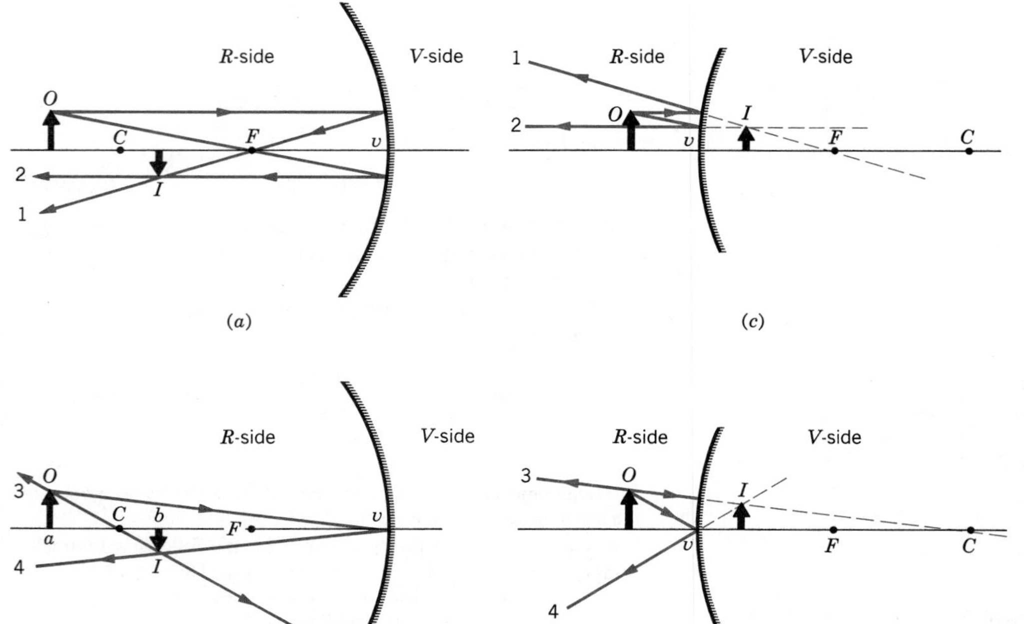

Figure 18 (*a,b*) Four rays that may be drawn graphically to find the image of an object in a concave mirror. Note that the image is real, inverted, and smaller than the object. (*c,d*) Four similar rays for the case of a convex mirror. Note that the image is virtual, upright, and smaller than the object.

39-8 Spherical Refracting Surfaces

In Fig. 19*a* light rays from a point object *O* fall on a convex spherical refracting surface of radius of curvature *r*. The surface separates two media, the index of refraction of the medium containing the incident light being n_1 and that on the other side of the surface being n_2. After refraction at the surface, the rays join to form a real image *I*.

As we shall prove in Section 39-11, the image distance *i* is related to the object distance *o*, the radius of curvature *r*, and the two indices of refraction, by

$$\frac{n_1}{o} + \frac{n_2}{i} = \frac{n_2 - n_1}{r} \quad \text{(single surface).} \quad (11)$$

This equation is quite general and holds whether the refracting surface is convex (Fig. 19*a*) or concave (Fig. 19*b*) and also for the case in which $n_2 < n_1$ (Fig. 19*c*).

The rule for signs for a single spherical refracting surface is the same as for spherical mirrors. There is a difference, however; for mirrors the *R*-side (where real images are formed) is the side toward which the incident light is *reflected.* For refracting surfaces, the *R*-side is the side toward which the incident light is *transmitted.* Figure 20 clarifies this distinction.

Sample Problem 6 Locate the image for the geometry shown in Fig. 19*a*, assuming the radius of curvature *r* to be 11 cm, $n_1 = 1$, and n_2 to be 1.9. Let the object be 19 cm to the

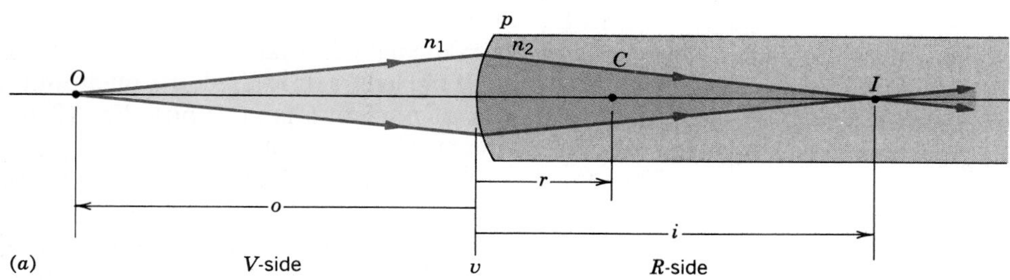

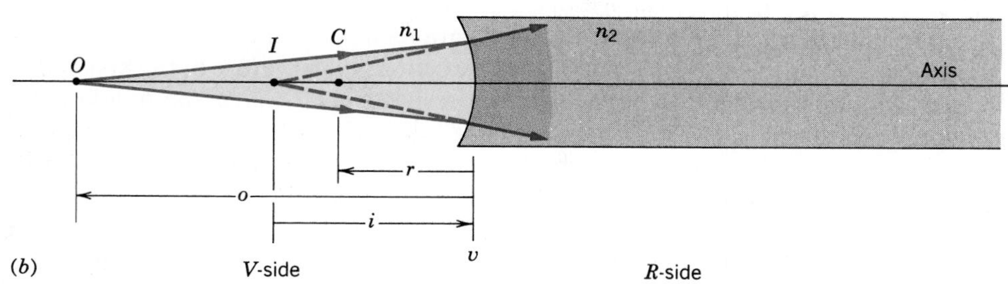

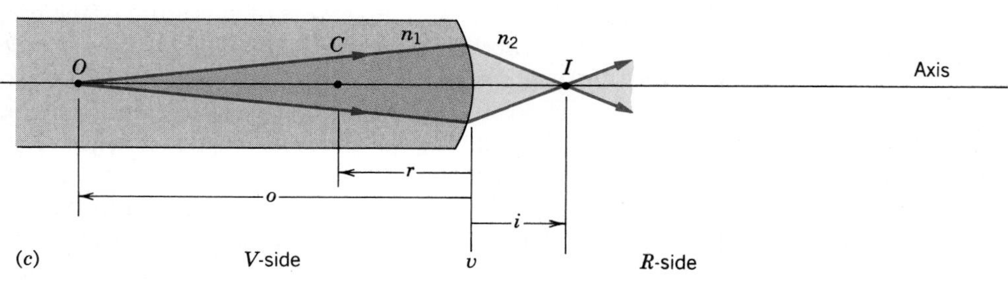

Figure 19 (*a*) A real image is formed by refraction at a convex spherical boundary between two media; in this case $n_2 > n_1$. (*b*) A virtual image is formed by refraction at a concave spherical boundary between two media; as in (*a*), $n_2 > n_1$. (*c*) The same as (*b*) except that $n_2 < n_1$.

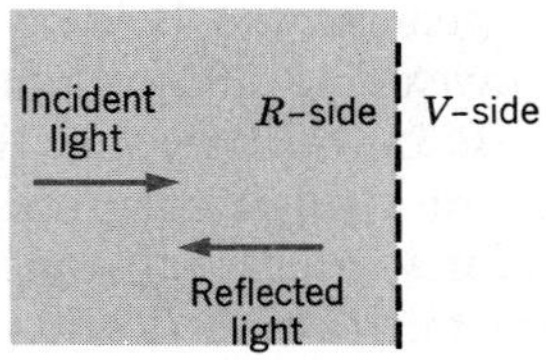

(spherical mirror)

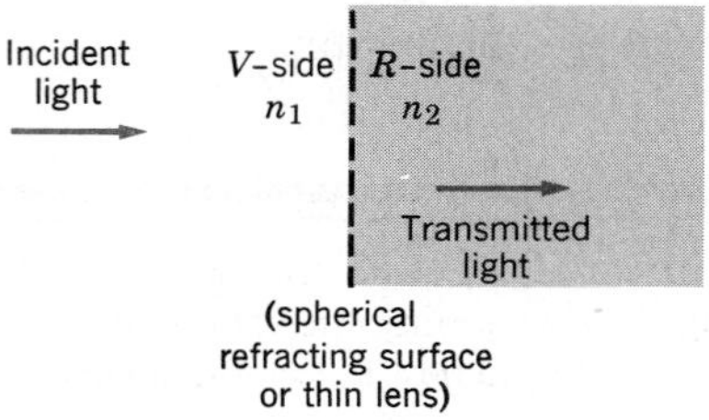

(spherical refracting surface or thin lens)

Figure 20 Real images are formed on the same side as the incident light for mirrors but on the opposite side for refracting surfaces and for lenses.

left of the vertex v.

From Eq. 11,

$$\frac{n_1}{o} + \frac{n_2}{i} = \frac{n_2 - n_1}{r},$$

we have

$$\frac{1}{+19 \text{ cm}} + \frac{1.9}{i} = \frac{1.9 - 1}{+11 \text{ cm}}.$$

Notice that r is positive because the center of curvature C of the surface in Fig. 19a lies on the R-side. If we solve the above equation for i, we find

$$i = 65 \text{ cm}. \qquad \text{(Answer)}$$

This result agrees with Fig. 19a and is consistent with the sign conventions. The light actually passes through the image point I so the image is real, as indicated by the positive sign that we found for i. Remember also that n_1 (= 1 in this case) always refers to the medium on the side of the surface from which the light comes.

39-9 Thin Lenses

In most refraction situations there is more than one refracting surface. This is true even for a spectacle lens, the light passing from air into glass and then from glass into air. We consider here only the special case of a *thin lens,* that is, a lens in which the thickness of the lens is small compared to the object distance, the image distance, or either of the two radii of curvature. For such a lens—as we shall prove in Section 39-11—these quantities are related by

$$\boxed{\frac{1}{o} + \frac{1}{i} = \frac{1}{f}} \qquad \text{(thin lens)} \qquad (12)$$

in which the focal length f of the lens is given by

$$\boxed{\frac{1}{f} = (n - 1)\left(\frac{1}{r_1} - \frac{1}{r_2}\right)} \qquad \text{(thin lens).} \qquad (13)$$

Note that Eq. 12 is the same equation that we used for spherical mirrors. Equation 13 is often called the *lens makers equation;* it relates the focal length of the lens to the index of refraction n of the lens material and the radii of curvature of the two surfaces.

In Eq. 13, r_1 is the radius of curvature of the lens surface on which the light first falls and r_2 is that of the second surface. If the lens is immersed in a medium for which the index of refraction is not unity, Eq. 13 still holds; simply replace n in that formula by $n_{\text{lens}}/n_{\text{medium}}$. The same rules for signs apply to lenses as to mirrors and spherical refracting surfaces.

A thin lens has *two* focal points, symmetrically placed on either side of the lens; we must be careful to distinguish between them.

Figure 21a shows the formation of a real inverted image in a converging lens, that is, in a lens that would cause incident parallel light to converge to a real focus. C_1 and C_2 are the centers of curvature of the first and the second surfaces, respectively. C_1 is on the R-side so that r_1 is positive but C_2 is on the V-side so that r_2 is negative. For such a (converging) lens you can show (using Eq. 13) that the focal length f is positive. A converging lens is thicker at the center than at the edges.

Figure 21b shows the formation of a virtual, upright image in a *diverging* lens, that is, in a lens that would cause incident parallel light to diverge from a virtual focus. C_1 and C_2 are the centers of curvature of the first and the second surfaces, respectively. C_1 is now on the V-side so that r_1 is negative and C_2 is now on the R-side so that r_2 is positive. For such a (diverging) lens you can show (using Eq. 13) that the focal length f is negative. A diverging lens is thinner at the center than at the edges.

Graded Index Lenses. All rays that pass through a lens from a real point object to a real point image must contain the same number of wavelengths, even though

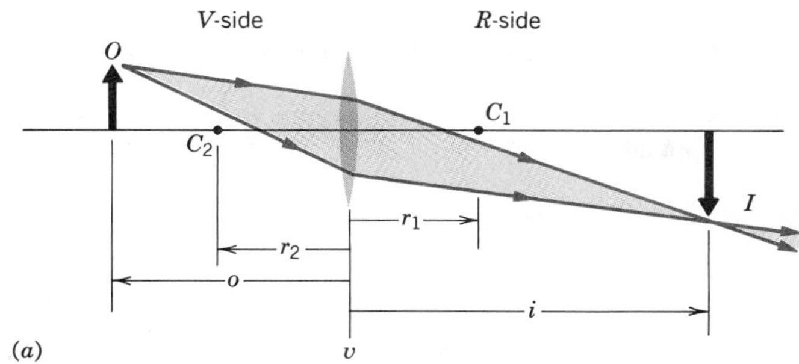

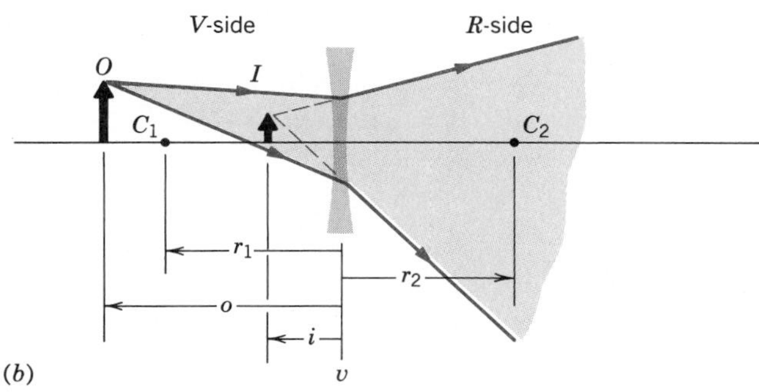

Figure 21 (*a*) A real, inverted image is formed by a converging lens. Such a lens has a positive focal length and is thicker in the center than at the edges. (*b*) A virtual, erect image is formed by a diverging lens. Such a lens has a negative focal length and is thinner at the center than at the edges.

their geometrical path lengths are necessarily different. In lens of the type we have been discussing, this is accomplished by causing the thickness of the lens to vary with radial distance from the central axis.

The same effect can be accomplished with a flat lens, if the index of refraction is caused to vary appropriately with axial distance. Figure 22 shows such a *graded index lens,* designed for use in a photo-copying machine.

Ray Tracing. We can locate the image of an off-axis point graphically as we show in Fig. 23 for a converging lens. With small changes, the construction can easily be adapted to diverging lenses. Thus:

1. A ray parallel to the axis and falling on the lens passes through the focal point F_2 (ray x in Fig. 23*a*).

2. A ray falling on a lens after passing through the focal point F_1 will emerge from the lens parallel to the axis (ray y).

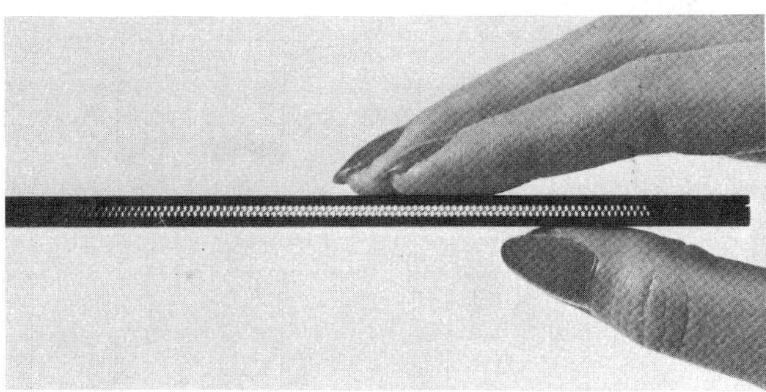

Figure 22 A side-view of a graded index lens. The lens is flat and focuses the light because its index of refraction (not its thickness) decreases in a controlled way with radial distance from the lens axis.

3. A ray falling on the lens at its center will pass through undeflected. There is no deflection because the lens, near its center, behaves like a thin piece of glass with parallel sides. The direction of the light rays is not changed and the sideways displacement can be neglected because the lens thickness has been assumed to be negligible (ray z).

Sample Problem 7 (a) The lens of Fig. 21*a* has radii of curvature of magnitude 42 cm and is made of glass with $n = 1.65$. Compute its focal length.

Since C_1 lies on the *R*-side of the lens in Fig. 21*a*, r_1 is positive ($= +42$ cm). Since C_2 lies on the *V*-side, r_2 is negative ($= -42$ cm). Substituting in Eq. 13 yields

$$\frac{1}{f} = (n-1)\left(\frac{1}{r_1} - \frac{1}{r_2}\right) = (1.65 - 1)\left(\frac{1}{+42\text{ cm}} - \frac{1}{-42\text{ cm}}\right)$$

or

$$f = +32\text{ cm}. \qquad \text{(Answer)}$$

A positive focal length indicates that, in agreement with what we have said above, parallel incident light converges after refraction to form a real focus.

(b) Find the focal length for the lens of Fig. 21*b*, again assuming 42-cm radii and $n = 1.65$.

In Fig. 21*b*, C_1 lies on the *V*-side of the lens so that r_1 is negative ($= -42$ cm). Since r_2 is positive ($= +42$ cm), Eq. 13 yields

$$\frac{1}{f} = (n-1)\left(\frac{1}{r_1} - \frac{1}{r_2}\right) = (1.65 - 1)\left(\frac{1}{-42\text{ cm}} - \frac{1}{+42\text{ cm}}\right)$$

or

$$f = -32\text{ cm}. \qquad \text{(Answer)}$$

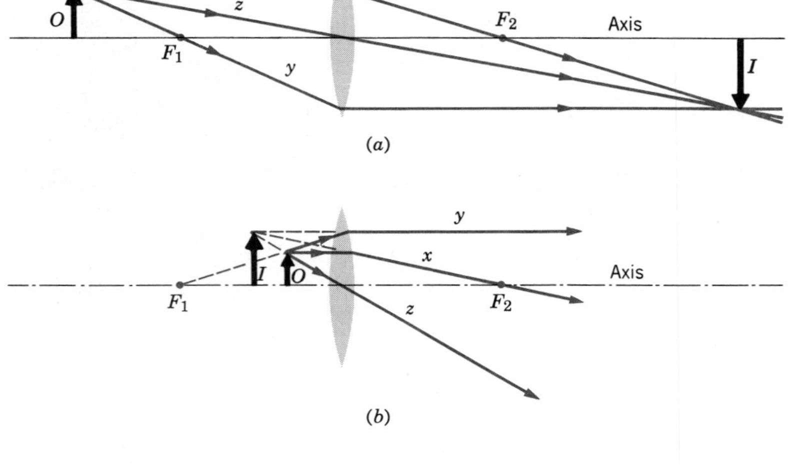

(b)

(c)

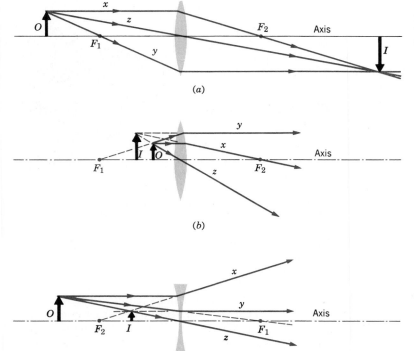

Figure 23 Three special rays allow us to locate an image formed by a thin lens.

A negative focal length indicates that, in agreement with what we have said above, incident light diverges after refraction to form a virtual image.

Sample Problem 8 A converging thin lens like that of Fig. 21*a* has a focal length of +24 cm. An object is placed 9.5 cm from the lens. Note that this is between the focal point and the lens. (a) Where is the image?

From Eq. 12

$$\frac{1}{o}+\frac{1}{i}=\frac{1}{f}$$

we have

$$\frac{1}{+9.5\text{ cm}}+\frac{1}{i}=\frac{1}{+24\text{ cm}},$$

which yields

$$i=-16\text{ cm}. \qquad \text{(Answer)}$$

The minus sign means that the image is on the *V*-side of the lens and is thus virtual.

(b) What is the lateral magnification of the lens for this object?

The lateral magnification is given by Eq. 9, which holds for thin lenses as well as for mirrors:

$$m=-\frac{i}{o}=-\frac{-16\text{ cm}}{+9.5\text{ cm}}=+1.7. \qquad \text{(Answer)}$$

The plus sign signifies an upright image.

39-10 Optical Instruments

The human eye is a remarkably effective organ, but its range can be extended in many ways by optical instruments such as spectacles, simple magnifiers, motion picture projectors, cameras (including TV cameras), microscopes, and telescopes. In many cases these devices extend the scope of our vision beyond the visible range; satellite-borne infrared cameras and x-ray microscopes are examples.

In almost all cases of modern sophisticated optical instruments, the mirror and thin lens formulas hold only as approximations. In typical laboratory microscopes the lens can by no means be considered "thin." In most optical instruments lenses are compound, that is, they are made of several components, the interfaces rarely being exactly spherical. Figure 24, for example, shows the components of a typical zoom lens, commonly used in TV cameras to provide a 20-to-one range in focal lengths.

In what follows we describe three optical instruments, assuming, for simplicity of illustration only, that the thin lens formula applies.

Simple Magnifier. The normal human eye can focus a sharp image of an object on the retina if the object *O* is located anywhere from infinity to a certain point called

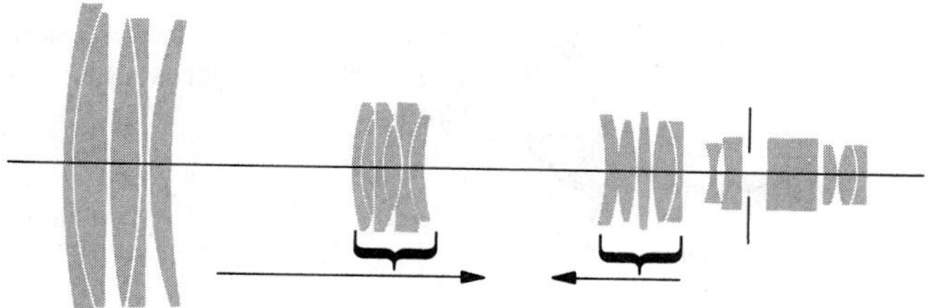

Figure 24 The components of a zoom lens in a TV camera. The central sections of the lens system move as shown. None of these lenses is "thin" and the paraxial approximation is not imposed. This is the real world of high-performance geometrical optics.

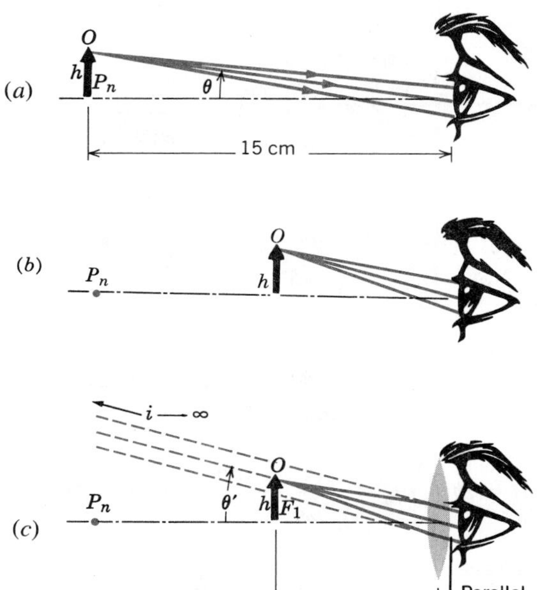

Figure 25 (*a*) An object of height *h* is placed at the near point of a human eye. (*b*) The object is moved closer but now the observer cannot bring it into focus. (*c*) A converging lens is placed close to the eye so that the rays from the object appear to come from an infinite distance and the eye can easily focus on them. A "magnifier" simply permits you to bring the object closer to your eye. Not to scale.

the *near point* P_n. If you move the object closer than the near point, the perceived retinal image becomes fuzzy. The location of the near point normally varies with age. We have all heard about people who claim not to need glasses but who read their newspapers at arm's length; their near points are receding! Find your own near point by moving this page closer to your eyes, considered separately, until you reach a position at which the image begins to become indistinct. In what follows, we take the near point to be 15 cm from the eye, a typical value for 20-year-olds.

Figure 25*a* shows an object O placed at the near point P_n. The size of the perceived image on the retina is determined by the angle θ. One way to make the object seem larger is to move it closer to your eye, as in Fig. 25*b*. The image on your retina is now larger, but the object is now so close that the eye cannot bring it into focus. We can give the eye some help by inserting a converging lens (of focal length f) just in front of the eye as in Fig. 25*c*. The eye now perceives an image at infinity, rays from which are comfortably focused by the eye.

The angle of the image rays is now θ', where $\theta' > \theta$. The *angular magnification* m_θ (not to be confused with the lateral magnification m) can be found from

$$m_\theta = \theta'/\theta$$

This is just the ratio of the sizes of the two images on the eye's retina. From Fig. 25,

$$\theta \approx h/15\text{ cm} \quad \text{and} \quad \theta' \approx h/f,$$

so that

$$\boxed{m_\theta \approx \frac{15\text{ cm}}{f}} \quad \text{(simple magnifier).} \tag{14}$$

Lens aberrations limit the angular magnifications for a single lens to something less than 10. This is enough, however, for stamp collectors and for actors portraying Sherlock Holmes.

Compound Microscope. Figure 26 shows a thin lens version of a compound microscope, used for viewing small objects that are very close to the objective lens of the instrument. The object O, of height h, is placed just outside the first focal point F_1 of the objective lens, whose focal length is f_{ob}. A real, inverted image I of height h' is formed by the objective, the lateral magnification being given by Eq. 9, or

$$m = -\frac{h'}{h} = -\frac{s\tan\theta}{f_{ob}\tan\theta} = -\frac{s}{f_{ob}}. \tag{15}$$

As usual, the minus sign indicates an inverted image.

The distance s (called the *tube length*) is so chosen that the image I falls on the first focal point F_1' of the eyepiece, which than acts as a simple magnifier as described above. Parallel rays enter the eye and a final image I' forms at infinity. The final magnification M is given by the product of the linear magnification m for

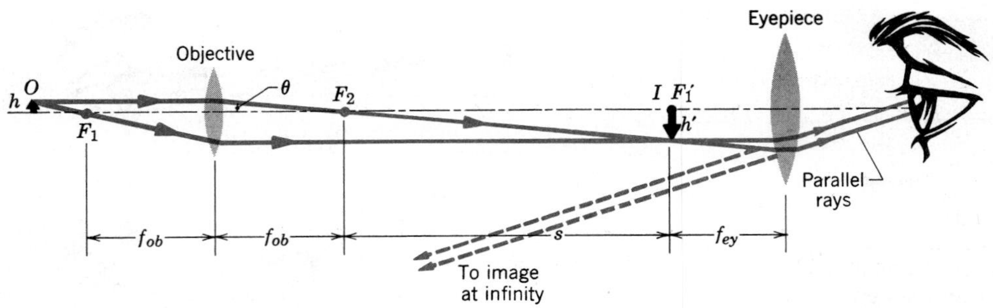

Figure 26 A thin-lens version of a compound microscope. Not to scale.

the objective lens, given by Eq. 15, and the angular magnification of the eyepiece, given by Eq. 14, or

$$M = m \times m_\theta = -\frac{s}{f_{ob}}\frac{15\text{ cm}}{f_{ey}} \quad \text{(microscope).} \quad (16)$$

Refracting Telescope. Like microscopes, telescopes come in a large variety of forms. The form we describe here is the simple refracting telescope that consists of an objective lens and an eyepiece, both represented in Fig. 27 by thin lenses, although in practice, as for microscopes, they will each be compound lens systems.

At first glance it may seem that the lens arrangements for telescopes and for microscopes are similar. However, telescopes are designed to view large objects, such as galaxies, stars, and planets, at large distances, whereas microscopes are designed for just the opposite purpose. Note also that in Fig. 27 the second focal point of the objective F_2 coincides with the first focal point of the eyepiece F_1', but in Fig. 26 these points are separated by the tube length s.

In Fig. 27 parallel rays from a distant object strike the objective lens, making an angle θ with the telescope axis and forming a real, inverted image at the common focal point F_2, F_1'. This image acts as an object for the eyepiece and a (still inverted) virtual image is formed at infinity. The rays defining the image make an angle θ_{ey} with the telescope axis.

The angular magnification m_θ of the telescope is θ_{ey}/θ_{ob}. For paraxial rays (rays close to the axis) we can write $\theta_{ob} = h'/f_{ob}$ and $\theta_{ey} = -h'/f_{ey}$ or

$$m_\theta = -\frac{f_{ob}}{f_{ey}} \quad \text{(telescope).} \quad (17)$$

Magnification is only one of the design factors of an astronomical telescope and is indeed easily achieved (How?). A good telescope needs *light-gathering power*, which determines how bright the image is. This is important when viewing faint objects such as distant galaxies and is accomplished by making the objective lens diameter as large as possible. *Field of view* is another important parameter. An instrument designed for galactic observation (narrow field of view) must be quite different from one designed for the observation of meteors (wide field of view). The telescope designer must also take account of lens and mirror aberrations including *spherical aberration* (that is, lenses and mirrors with truly spherical surfaces do not form sharp images) and *chromatic aberration* (that is, for simple lenses the focal length varies with wavelength so that fuzzy images are formed, displaying unnatural colors). There is also *resolving power*, which describes the ability of any optical

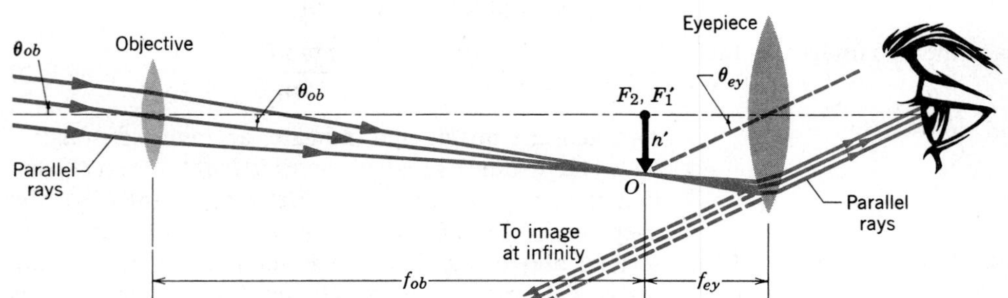

Figure 27 A thin-lens version of a refracting telescope. Not to scale.

instrument to distinguish between two objects (stars, say) whose angular separation is small. This by no means exhausts the design parameters of astronomical telescopes. We could also make a similar listing for any high-performance optical instrument.

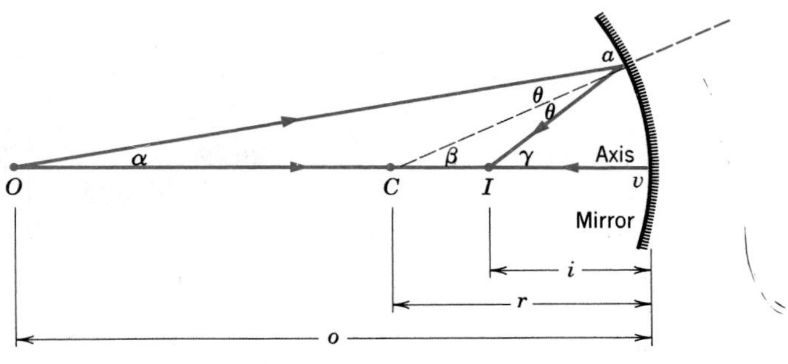

Figure 28 A point object O forms a real point image I after reflection from a concave spherical mirror.

39-11 Three Proofs (Optional)

The Spherical Mirror Formula (Eq. 8). Figure 28 shows a point object O placed on the axis of a concave spherical mirror beyond its center of curvature C. A ray from O that makes an angle α with the axis intersects the axis at I after reflection from the mirror at a. A ray that leaves O along the axis will be reflected back along itself at v and will also pass through I. Thus, I is the image of O; it is a *real* image because light actually passes through it. Let us find the image distance i in Fig. 28.

A useful theorem is that the exterior angle of a triangle is equal to the sum of the two opposite interior angles. Applying this to triangles $OaCO$ and $OaIO$ in Fig. 28 yields

$$\beta = \alpha + \theta \quad \text{and} \quad \gamma = \alpha + 2\theta.$$

If we eliminate θ between these two equations, we find

$$\alpha + \gamma = 2\beta. \tag{18}$$

We can write angles α, β, and γ, in radian measure, as

$$\alpha \approx \frac{av}{vO} = \frac{av}{o},$$
$$\beta = \frac{av}{vC} = \frac{av}{r}, \tag{19}$$
$$\gamma \approx \frac{av}{vI} = \frac{av}{i}.$$

Only the equation for β is exact, because the center of curvature of arc av is at C and not at O or I. However, the equations for α and γ are approximately correct if these angles are small enough. Substituting Eqs. 19 into Eq. 18, using Eq. 7 to replace r by $2f$ and canceling av leads exactly to Eq. 8, the relation that we set out to prove.

The Refracting Surface Formula (Eq. 11). The incident ray in Fig. 29 that falls on point a is refracted there according to

$$n_1 \sin \theta_1 = n_2 \sin \theta_2.$$

If α is small, θ_1 and θ_2 will also be small and we can replace the sines of these angles by the angles themselves. Thus, the above equation becomes

$$n_1 \theta_1 \approx n_2 \theta_2. \tag{20}$$

We again use the theorem that the exterior angle of a triangle is equal to the sum of the two opposite interior angles. Applying this to triangles $COaC$ and $ICaI$ yields

$$\theta_1 = \alpha + \beta \quad \text{and} \quad \beta = \theta_2 + \gamma. \tag{21}$$

If we eliminate θ_1 and θ_2 from Eqs. 20 and 21, we find

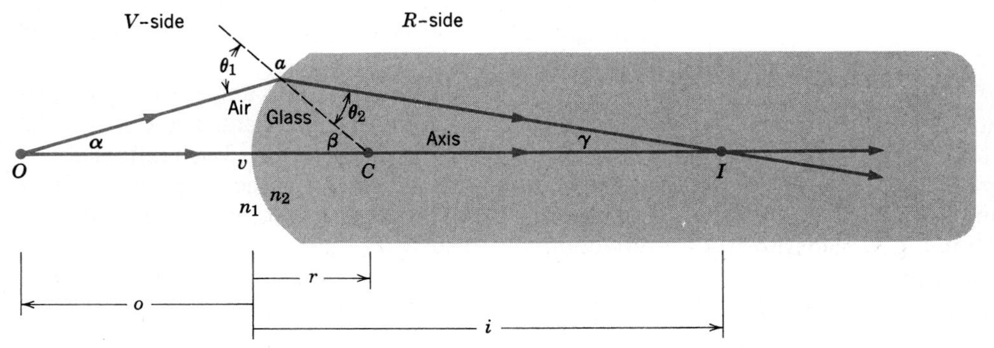

Figure 29 A point object O forms a real point image I after refraction at a spherical convex surface between two media.

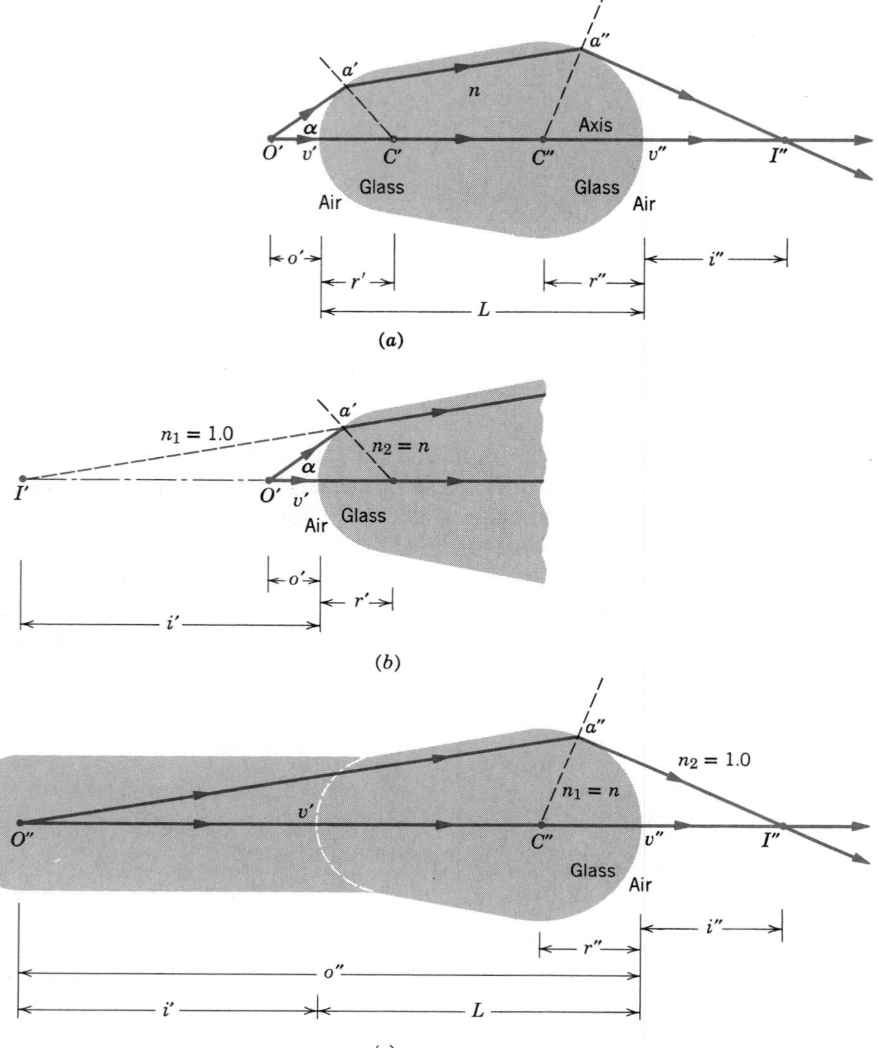

Figure 30 (*a*) Two rays from O' form a real image at I'' after being refracted at two spherical surfaces, the first surface being converging and the second diverging. (*b*) The first surface and (*c*) the second surface, shown separately. The vertical scale has been exaggerated for clarity.

$$n_1\,\alpha + n_2\,\gamma = (n_2 - n_1)\beta. \tag{22}$$

In radian measure the angles α, β, and γ are

$$\alpha \approx \frac{av}{o}; \quad \beta = \frac{av}{r}; \quad \gamma \approx \frac{av}{i}. \tag{23}$$

Only the second of these equations is exact. The other two are approximate because I and O are not the centers of circles of which av is an arc. However, for α small enough we can make the inaccuracies in Eqs. 23 as small as we wish. Substituting Eqs. 23 into Eq. 22 leads directly to Eq. 11, the relation we set out to prove.

The Thin Lens Formulas (Eqs. 12 and 13). Our plan is to consider each lens surface separately, using the image formed by the first surface as an object for the second.

Figure 30*a* shows such a thick glass "lens" of length L whose surfaces are ground to radii r' and r''. A point object O' is placed near the left surface as shown. A ray leaving O' along the axis is not deflected on entering or leaving the lens.

A second ray leaving O' at an angle α with the axis strikes the surface at point a', is refracted, and strikes the second surface at point a''. The ray is again refracted and crosses the axis at I'', which, being the intersection of two rays from O'', is the image of point O', formed after refraction at two surfaces.

Figure 30*b* shows the first surface, which forms a virtual image of O' at I'. To locate I', we use Eq. 11,

$$\frac{n_1}{o} + \frac{n_2}{i} = \frac{n_2 - n_1}{r}.$$

Putting $n_1 = 1$ and $n_2 = n$ and bearing in mind that the image distance is negative (that is, $i = -i'$ in Fig. 30*b*), we obtain

$$\frac{1}{o'} - \frac{n}{i'} = \frac{n-1}{r'}. \qquad (24)$$

In this equation i' will be a positive number because we have arbitrarily introduced the minus sign appropriate to a virtual image.

Figure 30*c* shows the second surface. Unless an observer at point a'' were aware of the existence of the first surface, he would think that the light striking that point originated at point I' in Fig. 30*b* and that the region to the left of the surface was filled with glass. Thus, the (virtual) image I' formed by the first surface serves as a real object O'' for the second surface. The distance of this object from the second surface is

$$o = i' + L. \qquad (25)$$

In applying Eq. 11 to the second surface, we insert $n_1 = n$ and $n_2 = 1$ because the object behaves as if it were imbedded in glass. If we use Eq. 25, Eq. 11 becomes

$$\frac{n}{i' + L} + \frac{1}{i''} = \frac{1-n}{r'}. \qquad (26)$$

Let us now assume that the thickness L of the "lens" in Fig. 30*a* is so small that we can neglect it in comparison with other linear quantities in this figure (such as o', i', o'', i'', r', and r''). In all that follows we make this *thin-lens approximation.* Putting $L = 0$ in Eq. 26 leads to

$$\frac{n}{i'} + \frac{1}{i''} = -\frac{n-1}{r''}. \qquad (27)$$

Adding Eqs. 24 and 27 leads to

$$\frac{1}{o'} + \frac{1}{i''} = (n-1)\left(\frac{1}{r'} - \frac{1}{r''}\right).$$

Finally, calling the original object distance simply o and the final image distance simply i leads to

$$\frac{1}{o} + \frac{1}{i} = (n-1)\left(\frac{1}{r'} - \frac{1}{r''}\right) \qquad (28)$$

which, with a small change in notation, are Eqs. 12 and 13, the relations we set out to prove.

REVIEW AND SUMMARY

Geometrical Optics

Light is most accurately described as an electromagnetic wave, whose speed and other properties are derivable from Maxwell's equations. *Geometrical optics* is an approximate treatment in which the waves can be represented by straight-line rays; it is valid if the waves do not encounter obstacles comparable in size to the wavelength of the radiation.

Reflection and Refraction

When a light ray falls on a boundary between two transparent media a *reflected* and a *refracted* ray generally appear. Both rays remain in the plane of incidence. The *angle of reflection* is equal to the angle of incidence and the *angle of refraction* is related to the angle of incidence by

$$n_1 \sin\theta_1 = n_2 \sin\theta_2 \quad \text{(refraction)}. \qquad [1]$$

See Fig. 1 and Sample Problem 1.

A wave encountering a boundary for which a transmitted wave would have a higher speed will experience *total internal reflection* (see Fig. 4 and Sample Problem 2) if its angle of incidence is at least equal to θ_c, where

Total Internal Reflection

$$\theta_c = \sin^{-1}\frac{n_2}{n_1} \quad \text{(critical angle)}. \qquad [3]$$

A reflected wave will be *polarized,* with its **E** vector perpendicular to the plane of incidence, if it strikes a boundary at the *polarizing angle* θ_B, where

Polarization by Reflection

$$\theta_B = \tan^{-1}(n_2/n_1) \quad \text{(Brewster's law)}. \qquad [4]$$

See Fig. 9 and Sample Problem 3.

Rays diverging from a point object O can recombine to form an (approximately) point image I if they encounter a spherical mirror, a spherical refracting surface, or a thin lens. For rays sufficiently close to the axis we have the following (in which o is the *object distance* and i is the *image distance*):

1. *Spherical mirror:*

Mirrors

$$\frac{1}{o}+\frac{1}{i}=\frac{1}{f}=\frac{2}{r} \quad \text{(spherical mirror).} \qquad [6,8]$$

See Figs. 15–17 and Sample Problem 6. A *plane mirror* is a special case for which $r \to \infty$, yielding $o = -i$; see Figs. 10–14 and Sample Problem 4.

2. *Spherical refracting surface:*

Refracting Surface

$$\frac{n_1}{o}+\frac{n_2}{i}=\frac{n_2-n_1}{r} \quad \text{(single surface).} \qquad [11]$$

See Fig. 19 and Sample Problem 6.

3. *Thin lens:*

Thin Lens

$$\frac{1}{o}+\frac{1}{i}=\frac{1}{f}=(n-1)\left(\frac{1}{r_1}-\frac{1}{r_2}\right) \quad \text{(thin lenses).} \qquad [12,13]$$

See Fig. 21 and Sample Problems 7 and 8.

The Sign Conventions

The rule for signs is: Associate *positive* with *Real, R-side,* and *upRight;* associate *negative* with *Virtual, V-side,* and *inVerted.* The R-side for mirrors is the side toward which the incident light is *reflected;* for refraction it is the side toward which the incident light is *transmitted.* See Fig. 20.

Images of extended objects may be found graphically by ray tracing; see Fig. 18 for mirrors and Fig. 23 for thin lenses. The *lateral magnification* in these two cases is given by

Lateral Magnification

$$m = -i/o, \qquad [9]$$

a positive value of m corresponding to an erect image.

Optical Instruments

Three simplified treatments of optical instruments are given:

1. The *simple magnifier* (Fig. 25). The angular magnification is given by

$$m_\theta = \frac{(15\text{ cm})}{f} \quad \text{(simple magnifier).} \qquad [14]$$

2. The *compound microscope* (Fig. 26). The angular magnification is given by

$$M = m \times m_\theta = \frac{s}{f_{ob}}\frac{(15\text{ cm})}{f_{ey}} \quad \text{(microscope).} \qquad [16]$$

3. The *refracting telescope* (Fig. 27). The overall angular magnification is given by

$$m_\theta = -\frac{f_{ob}}{f_{ey}} \quad \text{(telescope).} \qquad [17]$$

QUESTIONS

1. Describe what your immediate environment would be like if all objects were totally absorbing. Sitting in a chair in a room, could you see anything? If a cat entered the room could you see it?

2. Can you think of a simple test or observation to prove that the law of reflection is the same for all wavelengths, under conditions in which geometrical optics prevail?

3. A street light, viewed by reflection across a body of water in which there are ripples, appears very elongated. Explain.

4. Shortwave broadcasts from Europe are heard in the United States even though the path is not a straight line. Explain how.

5. The travel time of signals from satellites to receiving stations on the earth varies with the frequency of the signal. Why?

6. By what percent does the speed of blue light in fused quartz differ from that of red light?

7. Can (*a*) reflection phenomena or (*b*) refraction phenomena be used to determine the wavelength of light?

8. Describe and explain what a fish sees as it looks in various directions above its "horizon."

9. What is a plausible explanation for the observation that a street appears darker when wet than when dry?

10. Light, traveling through vacuum from a distant stationary source, strikes your eye. If the source starts to move rapidly toward you, how does this affect (*a*) the wavelength, (*b*) the frequency, and (*c*) the speed of the light?

11. Design a periscope, taking advantage of total internal reflection. What are the advantages compared with silvered mirrors?

12. For a plane mirror, what is the focal length? The magnification?

13. At night, in a lighted room, you blow a smoke ring toward a window pane. If you focus your eyes on the ring as it approaches the pane it will seem to go right through the glass into the darkness beyond. What is the explanation of this illusion?

14. How do your eyes adjust for seeing objects at different distances from you?

15. A machinist whose eyesight is failing finds that he can read his micrometer scale more easily if he squints at it through a tiny aperture formed by coiling his index finger into the base of his thumb. Although less bright, the image is sharper than when he looks at the scale directly. What is the explanation?

16. In driving a car you sometimes see vehicles such as ambulances with letters printed on them in such a way that they read in the normal fashion when you look at them through the rear-view mirror. Print your name so that it may be so read.

17. Brewster's law determines the polarizing angle on reflection from a material such as glass; see Fig. 9. A plausible interpretation for zero reflection of the parallel component at that angle is that the charges in the glass are caused to oscillate parallel to the reflected ray by this component and produce no radiation in this direction. Explain this and comment on the plausibility.

18. Explain how polarization by reflection could occur if the light is incident on the interface from the side with the higher index of refraction (glass to air, for example).

19. We all know that when we look into a mirror right and left are reversed. Our right hand will seem to be a left hand; if we part our hair on the left it will seem to be parted on the right, etc. Can you think of a system of mirrors that would let us see ourselves as others see us? If so draw it and prove your point by sketching some typical rays.

20. Devise a system of plane mirrors that will let you see the back of your head. Trace the rays to prove your point.

21. Can a virtual image be photographed by exposing a film at the location of the image? Explain.

22. If converging rays fall on a plane mirror, is the image virtual?

23. It is a bright sunny day and you want to create a rainbow in your back yard, using your garden hose. Exactly how do you go about it? Incidentally, why can't you walk under, or go to the end of, a rainbow?

24. Is it possible, by using one or more prisms, to recombine into white light the color spectrum formed when white light passes through a single prism? If yes, explain how.

25. How does atmospheric refraction affect the apparent time of sunset?

26. Stars twinkle but planets don't. Why?

27. In many city buses a convex mirror is suspended over the door, in full view of the driver. Why not a plane or a concave mirror?

28. What approximations were made in deriving the mirror equation (Eq. 8):

$$\frac{1}{o}+\frac{1}{i}=\frac{2}{r}?$$

29. Under what conditions will a spherical mirror, which may be concave or convex, form (*a*) a real image, (*b*) an inverted image, and (*c*) an image smaller than the object?

30. Can a virtual image be projected onto a screen? Photographed? If you put a piece of paper at the site of a virtual image (assuming a high-intensity light beam) will it ignite after sufficient exposure?

31. You are looking at a dog through a glass window pane. Where is the image of the dog? Is it real or virtual? Is it upright or inverted? What is the magnification? (*Hint:* Think of the window pane as the limiting case of a thin lens in which the radii of curvature have been allowed to become infinitely large.)

32. In some cars the right side mirror bears the notation: "Objects in the mirror are closer than they appear." What feature of the mirror requires this warning? What advantage does the mirror have to compensate for this disadvantage? Do cars viewed in this mirror seem to be moving faster or slower than they would be if viewed in a plane mirror?

33. We have all seen TV pictures of a baseball game shot from a camera located somewhere behind second base. The pitcher and the batter are about 60 ft apart but they look much closer on the TV screen. Why are images viewed through a telephoto lens foreshortened in this way?

34. An unsymmetrical thin lens forms an image of a point object on its axis. Is the image location changed if the lens is reversed?

35. Why has a lens two focal points and a mirror only one?

36. Under what conditions will a thin lens, which may be converging or diverging, form (*a*) a real image, (*b*) an inverted image, and (*c*) an image smaller than the object?

37. A diver wants to use an air-filled plastic bag as a converging lens for underwater visibility. Sketch a suitable cross section for the bag.

38. What approximations were made in deriving the thin

lens equation (Eq. 12):

$$\frac{1}{o}+\frac{1}{i}=\frac{1}{f}?$$

39. A concave mirror and a converging lens have the same focal length in air. Do they have the same focal length when immersed in water? If not, which has the greater focal length?

40. Under what conditions will a thin lens have a lateral magnification (*a*) of -1 and (*b*) of $+1$?

41. How does the focal length of a thin glass lens for blue light compare with that for red light, assuming the lens is (*a*) diverging and (*b*) converging?

42. Does the focal length of a lens depend on the medium in which the lens is immersed? Is it possible for a given lens to act as a converging lens in one medium and a diverging lens in another medium?

43. Are the following statements true for a glass lens in air? (*a*) A lens that is thicker at the center than at the edges is a converging lens for parallel light. (*b*) A lens that is thicker at the edges than at the center is a diverging lens for parallel light.

44. Under what conditions would the lateral magnification ($m = -i/o$) for lenses and mirrors become infinite? Is there any practical significance to such a condition?

45. Is the focal length of a spherical mirror affected by the medium in which it is immersed? of a thin lens? Why the difference, if any?

46. Why is the magnification of a simple magnifier (see the derivation leading to Eq. 14) defined in terms of angles rather than image/object size?

47. Ordinary spectacles do not magnify but a simple magnifier does. What, then, is the function of spectacles?

48. The "*f-number*" of a camera lens (see Problem 88) is its focal length divided by its aperture (effective diameter). Why is this useful to know in photography? How can the *f*-number of the lens be changed? How is exposure time related to *f*-number?

49. Does it matter whether (*a*) an astronomical telescope, (*b*) a compound microscope, (*c*) a simple magnifier, (*d*) a camera, including a TV camera, or (*e*) a projector, including a slide projector and a motion picture projector, produces upright or inverted images? What about real or virtual images?

50. The unaided human eye produces a real but inverted image on the retina. (*a*) Why then don't we perceive objects such as people and trees as upside down? (*b*) We don't, of course, but suppose that we wore special glasses so that we did. If you then turned this book upside down, could you read this question with the same facility that you do now?

51. Which of the following: a converging lens, a diverging lens, a concave mirror, a convex mirror, a plane mirror, is used (*a*) as a magnifying glass? (*b*) as the reflector in the lamphouse of a slide projector? (*c*) as the objective of a reflecting telescope? (*d*) in a kaleidoscope? (*e*) as the eyepiece of opera glasses? (*f*) to obtain a wider angle rear view from the driver's seat in a car?

52. In William Golding's *Lord of the Flies* the character Piggy uses his glasses to focus the sun's rays and kindle a fire. Later, the boys abuse Piggy and break his glasses. He is unable to identify them at close range because he is nearsighted. Find the flaw in this narrative. (*Boston Globe,* December 17, 1985, Letters)

53. Explain the function of the objective lens of a microscope; why use an objective lens at all? Why not just use a very powerful simple magnifier?

54. Why do astronomers use optical telescopes in looking at the sky? After all, the stars are so far away that they still appear to be points of light, without any detail discernible.

55. A watchmaker uses diverging eyeglasses for driving, no glasses for reading, and converging glasses in his occupational work. Is he nearsighted or farsighted? Explain. (See Problem 87.)

EXERCISES AND PROBLEMS

Section 39-2 Reflection and Refraction

1E. In Fig. 31 find the angles (*a*) θ_1 and (*b*) θ_2.

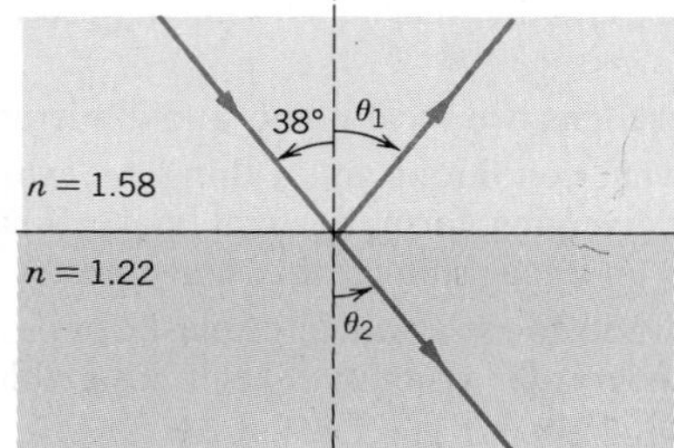

Figure 31 Exercise 1.

2E. Light in vacuum is incident on the surface of a glass slab. In the vacuum the beam makes an angle of 32° with the normal to the surface, while in the glass it makes an angle of 21° with the normal. What is the index of refraction of the glass?

3E. The speed of yellow sodium light in a certain liquid is measured to be 1.92×10^8 m/s. What is the index of refraction of this liquid, with respect to air, for sodium light?

4E. What is the speed in fused quartz of light of wavelength 550 nm? (See Fig. 2.)

5E. *Cerenkov radiation.* When an electron moves through a medium at a speed exceeding the speed of light in that medium, it radiates electromagnetic energy (the Cerenkov effect). What minimum speed must an electron have in a liquid of refractive index 1.54 in order to radiate?

6E. A laser beam travels along the axis of a straight section of pipeline, one mile long. The pipe normally contains air at standard temperature and pressure, but it may also be evacuated. In which case would the travel time for the beam be greater and by how much?

7E. When the rectangular metal tank in Fig. 32 is filled to the top with an unknown liquid, an observer with eyes level with the top of the tank can just see the corner E. Find the index of refraction of the liquid.

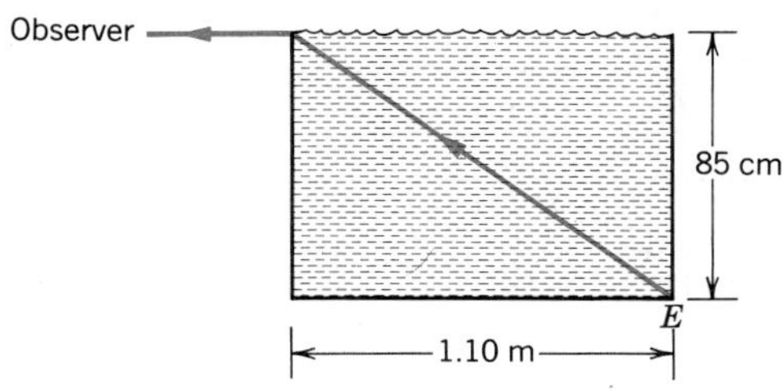

Figure 32 Exercise 7.

8E. Claudius Ptolemy (c. A.D. 150) gave the following measured values for the angle of incidence θ_1 and the angle of refraction θ_2 for a light beam passing from air to water:

θ_1	θ_2	θ_1	θ_2
10°	8°	50°	35°
20°	15°30′	60°	40°30′
30°	22°30′	70°	45°30′
40°	29°	80°	50°

Are these data consistent with the law of refraction? If so, what index of refraction results? These data are interesting as perhaps the oldest recorded physical measurements.

9P. A bottom-weighted 2.0-m-long vertical pole extends from the bottom of a swimming pool to a point 0.5 m above the water. Sunlight is incident at 55° above the horizon. What is the length of the shadow of the pole on the level bottom of the pool?

10P. Prove that a ray of light incident on the surface of a sheet of plate glass of thickness t emerges from the opposite face parallel to its initial direction but displaced sideways, as in Fig. 33. Show that, for small angles of incidence θ, this displacement is given by

$$x = t\theta \frac{n-1}{n}$$

where n is the index of refraction and θ is measured in radians.

11P. Ocean waves moving at a speed of 4.0 m/s are approaching a beach at an angle of 30° to the normal, as shown in Fig. 34. Suppose the water depth changes abruptly and the wave speed drops to 3.0 m/s. Close to the beach, what is the angle θ between the direction of wave motion and the normal? (Assume the same law of refraction as for light.) Explain why most waves come in normal to a shore even though at large distances they approach at a variety of angles.

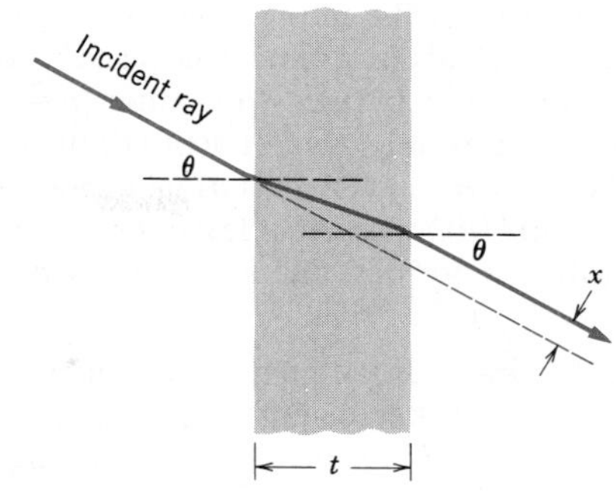

Figure 33 Problem 10.

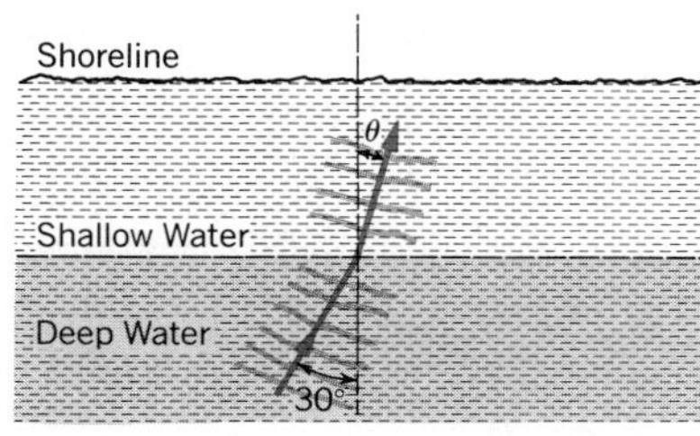

Figure 34 Problem 11.

12P. A 60° prism is made of fused quartz. A ray of light falls on one face, making an angle of 35° with the normal. Trace the ray through the prism graphically with some care, showing the paths traversed by rays representing (*a*) blue light, (*b*) yellow-green light, and (*c*) red light. (See Fig. 2.)

13P. A penny lies at the bottom of a pool with depth d and index of refraction n, as shown in Fig. 35. Show that light rays that are close to the normal appear to come from a point $d_a = d/n$ below the surface. This distance is the apparent depth of the pool.

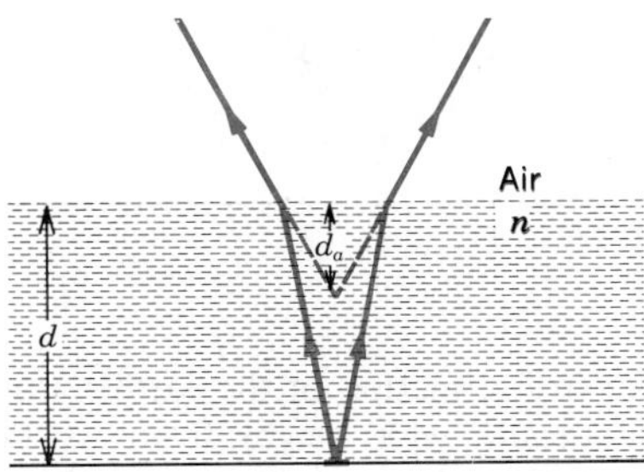

Figure 35 Problem 13.

14P. As an example of the importance of the paraxial ray assumption, consider this problem. You place a coin at the bottom of a swimming pool filled with water ($n = 1.33$) to a depth of 2.4 m. What is the apparent depth of the coin below the surface when viewed (*a*) at near normal incidence (that is, by paraxial rays) and (*b*) by rays that leave the coin making an angle of 30° with the normal (that is, definitely not paraxial rays)?

15P. Figure 36 shows a small light bulb suspended 250 cm above the surface of the water in a swimming pool. The water is 200 cm deep and the bottom of the pool is a large mirror. Where is the image of the light bulb? Consider only paraxial rays near normal incidence.

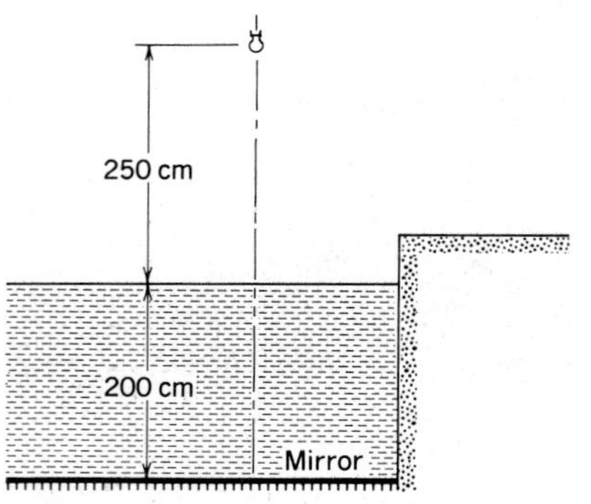

Figure 36 Problem 15.

16P. A layer of water ($n = 1.33$) 20 mm thick floats on carbon tetrachloride ($n = 1.46$) 40 mm thick. How far below the water surface, viewed at normal incidence, does the bottom of the tank seem to be?

17P. Two perpendicular mirrors form the sides of a vessel filled with water, as shown in Fig. 37. A light ray is incident from above, normal to the water surface. (*a*) Show that the emerging ray is parallel to the incident ray. Assume that there are two reflections at the mirror surfaces. (*b*) Repeat the analysis for the case of oblique incidence, the ray lying in the plane of the figure.

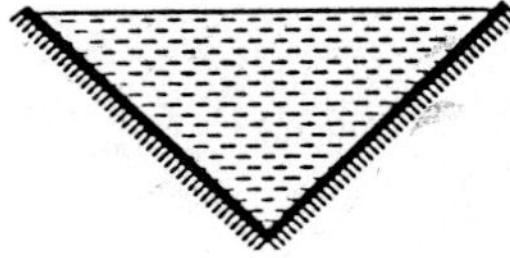

Figure 37 Problem 17.

18P. The index of refraction of the earth's atmosphere decreases monotonically with height from its surface value (about 1.00029) to the value in space (about 1.00000) at the top of the atmosphere. This continuous (or graded) variation can

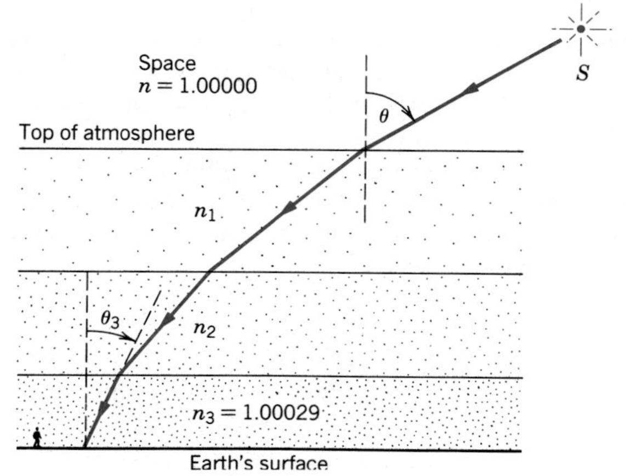

Figure 38 Problem 18.

be approximated by considering the atmosphere to be composed of three (or more) plane parallel layers in each of which the index of refraction is constant. Thus, in Fig. 38, $n_3 > n_2 > n_1 > 1.00000$. Consider a ray of light from a star *S* that strikes the top of the atmosphere at an angle θ with the vertical. (*a*) Show that the apparent direction θ_3 of the star with the vertical as seen by an observer at the earth's surface is given from

$$\sin \theta_3 = \left(\frac{1}{n_3}\right) \sin \theta.$$

(*Hint:* Apply the law of refraction to successive pairs of layers of the atmosphere; ignore the curvature of the earth.) (*b*) Calculate the shift in position of a star observed to be 20° from the vertical. (The very small effects due to atmospheric refraction can be most important; for example, they must be taken into account in using navigation satellites to obtain accurate fixes of position on the earth.)

19P. An incident ray falls on one face of a glass prism in air. The angle of incidence θ is chosen so that the emerging ray also makes an angle θ with the normal to the other face, the ray passing symmetrically through the prism; see Fig. 39. Show

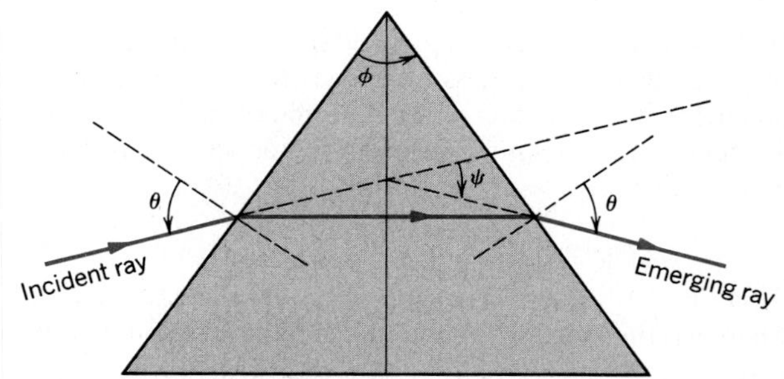

Figure 39 Problem 19.

that the index of refraction n of the glass prism is given by

$$n = \frac{\sin \frac{1}{2}(\psi + \phi)}{\sin \frac{1}{2}\phi},$$

where ϕ is the vertex angle of the prism between the two faces and ψ is the deviation angle through which the light beam has been turned in passing through the prism. (Under these conditions the deviation angle ψ has the minimum value, with respect to rotation of the prism, and is called the *angle of minimum deviation.*)

20P. A ray of light goes through an equilateral prism in the position of minimum deviation. The total deviation is 30°. What is the index of refraction of the prism? See Problem 19.

Section 39-3 Total Internal Reflection

21E. The refractive index of benzene is 1.8. What is the critical angle of incidence for a light ray traveling in benzene toward a plane layer of air above it?

22E. A light ray falls on a square glass slab as in Fig. 40. What must be the minimum index of refraction of the glass if total internal reflection occurs at the vertical face?

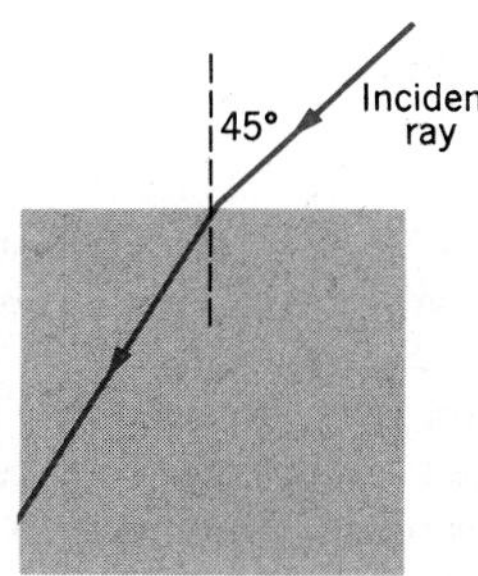

Figure 40 Exercise 22.

23E. A ray of light is incident normally on the face ab of a glass prism ($n = 1.52$), as shown in Fig. 41. (*a*) Assuming that the prism is immersed in air, find the largest value for the angle ϕ so that the ray is totally reflected at face ac. (*b*) Find ϕ if the prism is immersed in water.

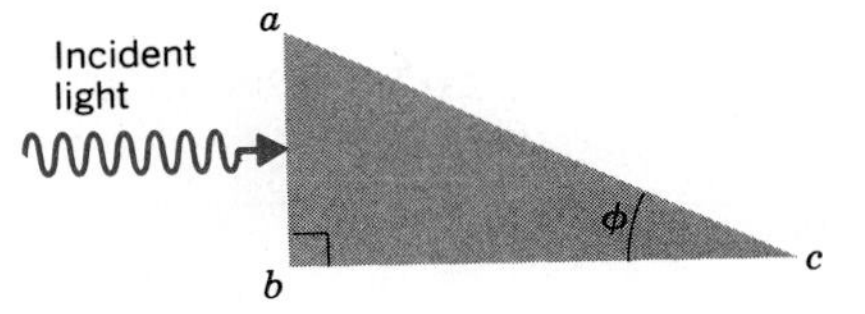

Figure 41 Exercise 23.

24E. A fish is 2.00 m below the surface of a smooth lake. At what angle above the horizontal must it look to see the light from a small fire burning at the water's edge 100 m away? Take the index of refraction for water to be 1.33.

25E. A point source is 80 cm below the surface of a body of water. Find the diameter of the largest circle at the surface through which light can emerge from the water.

26P. A glass cube has a small spot at its center. (*a*) What parts of the cube face must be covered to prevent the spot from being seen, no matter what the direction of viewing? (*b*) What fraction of the cube surface must be so covered? Assume a cube edge of 10 mm and an index of refraction of 1.5. (Neglect the subsequent behavior of an internally reflected ray.)

27P. A plane wave of white light traveling in fused quartz strikes a plane surface of the quartz, making an angle of incidence θ. Is it possible for the internally reflected beam to appear (*a*) bluish or (*b*) reddish? (*c*) Roughly what value of θ must be used? (*Hint:* White light will appear bluish if wavelengths corresponding to red are removed from the spectrum.)

28P. A given monochromatic light ray, initially in air, strikes a 90° prism at P (see Fig. 42) and is refracted there and at Q to such an extent that it just grazes the right-hand prism surface at Q. (*a*) Determine the index of refraction of the prism for this wavelength in terms of the angle of incidence θ_1 that gives rise to this situation. (*b*) Give a numerical upper bound for the index of refraction of the prism. Show, by ray diagrams, what happens if the angle of incidence at P is (*c*) slightly greater than θ_1 or (*d*) is slightly less than θ_1.

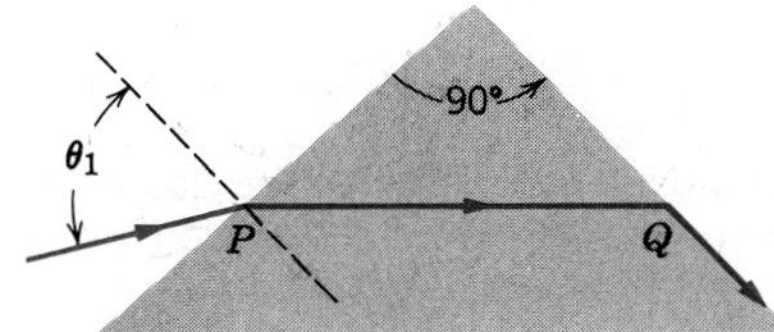

Figure 42 Problem 28.

29P. A glass prism with an apex angle of 60° has $n = 1.60$. (*a*) What is the smallest angle of incidence for which a ray can enter one face of the prism and emerge from the other? (*b*) What angle of incidence would be required for the ray to pass through the prism symmetrically? See Problem 19.

30P. A point source of light is placed a distance h below the surface of a large deep lake. (*a*) Show that the fraction f of the light energy that escapes directly from the water surface is independent of h and is given by

$$f = \frac{1}{2}(1 - \sqrt{1 - 1/n^2})$$

where n is the index of refraction of water. (*Note:* Absorption within the water and reflection at the surface (except where it is total) have been neglected.) (*b*) Evaluate this fraction for $n = 1.33$.

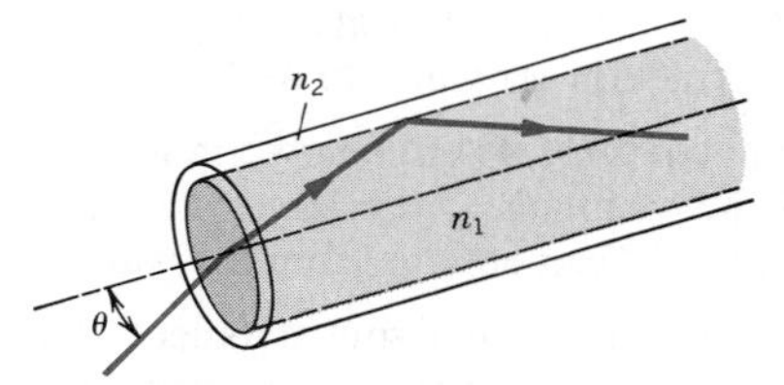

Figure 43 Problem 31.

31P. An optical fiber consists of a glass core (index of refraction n_1) surrounded by a coating (index of refraction $n_2 < n_1$). Suppose a beam of light enters the fiber from air at an angle θ with the fiber axis as shown in Fig. 43. (*a*) Show that the greatest possible value of θ for which a ray can be propagated down the fiber is given by $\theta = \sin^{-1}\sqrt{n_1^2 - n_2^2}$. (*b*) Assume the glass and coating indices of refraction are 1.58 and 1.53, respectively, and calculate the value of this angle.

32P. In an optical fiber (see preceding problem), different rays travel different paths along the fiber, leading to different travel times. This causes a light pulse to spread out as it travels along the fiber, resulting in information loss. The delay time should be minimized in designing a fiber. Consider a ray that travels a distance L along a fiber axis and another that is reflected, at the critical angle, as it travels to the same destination as the first. (*a*) Show that the difference Δt in the times of arrival is given by

$$\Delta t = \frac{L}{c}\frac{n_1}{n_2}(n_1 - n_2),$$

where n_1 is the index of refraction of the glass core and n_2 is the index of refraction of the fiber coating. (*b*) Evaluate Δt for the fiber of Problem 31, with $L = 300$ m.

33P*. Sound waves generated in the earth by the detonation of a small amount of explosive obey the same laws of reflection, refraction, and total internal reflection as do light rays. Detectors, set up on the surface in a straight line from the detonation point S (see Fig. 44) detect the arrival of the sound waves. Suppose that a layer of soil in which the sound speed is v_1 covers solid bedrock in which the sound speed is v_2; suppose also, as is often the case, that $v_2 > v_1$. Waves arrive at a detector by two routes: (*i*) a direct surface wave; (*ii*) a wave striking the interface of soil and bedrock at the critical angle for total internal reflection; this wave travels along the boundary at speed v_2, generating waves that return to the surface, leaving the interface at the same angle as that of incidence. (Waves simply reflected from the interface are not considered.) (*a*) Show that the travel time of these critically reflected waves is given by

$$t_c = \frac{2D\sqrt{v_2^2 - v_1^2}}{v_1 v_2} + \frac{x}{v_2},$$

where D is the thickness of the upper layer and x the distance from detonation S to the detector. (*b*) Show that beyond a certain distance x^*, the critically reflected wave arrives before the direct wave, and that

$$x^* = 2D\sqrt{\frac{1+n}{1-n}}, \quad n = v_1/v_2$$

and, therefore, by determining x^* the thickness D of the upper layer is determined. This method is widely employed in determining the suitability of land areas for construction purposes, to trace subsurface water-bearing zones, etc., and is called *seismic surveying.*

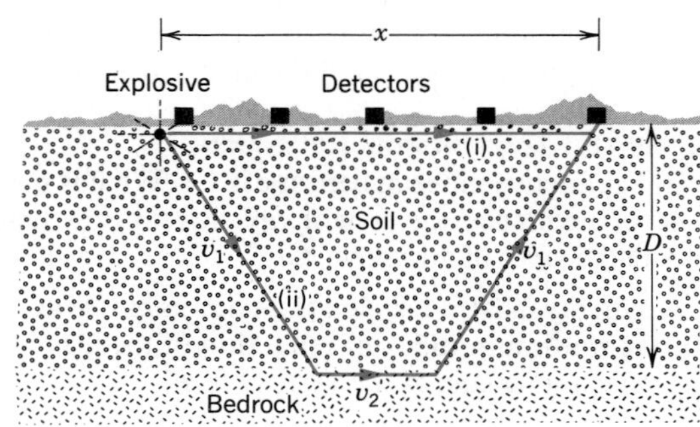

Figure 44 Problem 33.

34P*. A seismic surveying team is attempting to determine the depth to a horizontal bedrock layer. Detectors are placed 200 m apart. The data, shown graphically in Fig. 45, shows the time from the explosion to the arrival of the first sound waves at each detector plotted against the distance to the detector. Apply the theory of Problem 33 and find (*a*) the sound speeds v_1 and v_2, and (*b*) the depth to the bedrock layer.

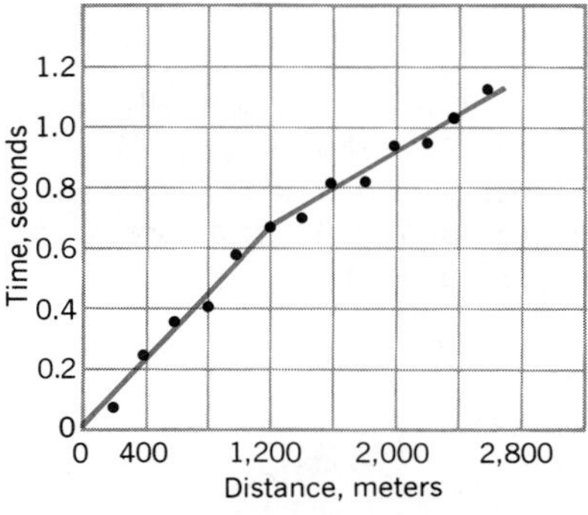

Figure 45 Problem 34.

Section 39-4 Polarization by Reflection

35E. (*a*) At what angle of incidence will the light reflected from water be completely polarized? (*b*) Does this angle depend on the wavelength of the light?

36E. Light traveling in water of refractive index 1.33 is incident upon a plate of glass of refractive index 1.53. At what angle of incidence is the reflected light completely plane polarized?

37E. Calculate the range of polarizing angles for white light incident on fused quartz. Assume that the wavelength limits are 400 and 700 nm and use the dispersion curve of Fig. 2.

38P. When red light in vacuum is incident at the polarizing angle on a certain glass slab, the angle of refraction is 32°. What are (*a*) the index of refraction of the glass and (*b*) the polarizing angle?

Section 39-5 A Plane Mirror

39E. A small object is 10 cm in front of a plane mirror. If you stand behind the object, 30 cm from the mirror, and look at its image, for what distance must you focus your eyes?

40E. Suppose you wished to photograph an object seen in a plane mirror. If the object is 5.0 m to your right and 1.0 m closer to the plane of the mirror than you, for what distance must you focus the lens of your camera, which you are holding 4.3 m from the mirror?

41E. You are standing in front of a large plane mirror, contemplating your image. If you move toward the mirror at speed v, at what speed does your image move toward you? Report this speed both (*a*) in your own reference frame and (*b*) in the reference frame of the room in which the mirror is at rest.

42E. Figure 46 shows an incident ray striking plane mirrors MM' and $M'M''$. Find the angle between the incoming ray i and the outgoing ray r'. The two mirrors are at right angles.

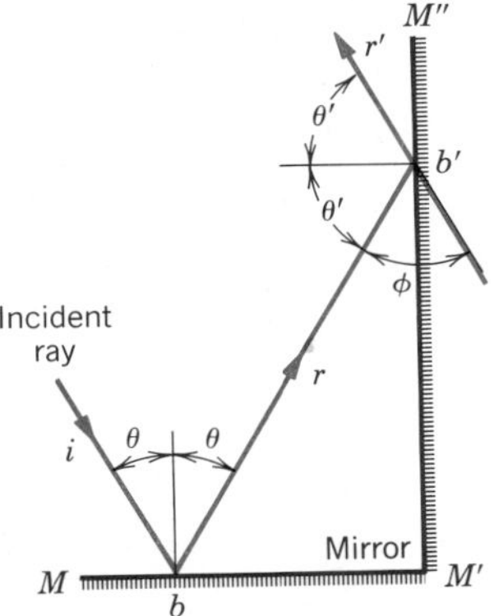

Figure 46 Problem 42.

43P. Figure 47 shows (top view) that Bernie B is walking directly toward the center of a vertical mirror. How close to the mirror will he be when Sarah S is just able to see him? Take $d = 3.0$ m.

44P. Prove that if a plane mirror is rotated through an angle α, the reflected beam is rotated through an angle 2α. Show that this result is reasonable for $\alpha = 45°$.

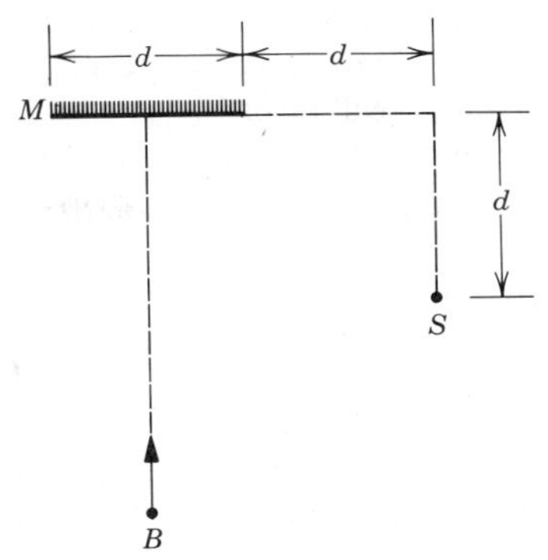

Figure 47 Problem 43.

45P. In Fig. 12 you rotate the mirror 30° counterclockwise about its bottom edge, leaving the point object O in place. Is the image point displaced? If so, where is it? Can the eye still see the image without being moved? Sketch a figure showing the new situation.

46P. A small object O is placed one-third of the way between two parallel plane mirrors as in Fig. 48. Trace appropriate bundles of rays for viewing the four images that lie closest to the object.

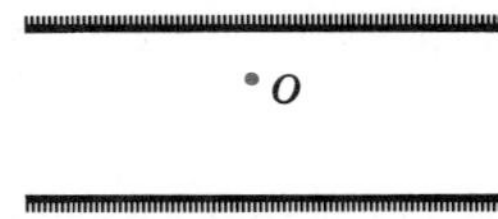

Figure 48 Problem 46.

47P. Two plane mirrors make an angle of 90° with each other. What is the largest number of images of an object placed between them that can be seen by a properly placed eye? The object need not lie on the mirror bisector.

48P. A point object is 10 cm away from a plane mirror while the eye of an observer (pupil diameter 5.0 mm) is 20 cm away. Assuming both the eye and the point to be on the same line perpendicular to the surface, find the area of the mirror used in observing the reflection of the point.

49P. You put a point source of light S a distance d in front of

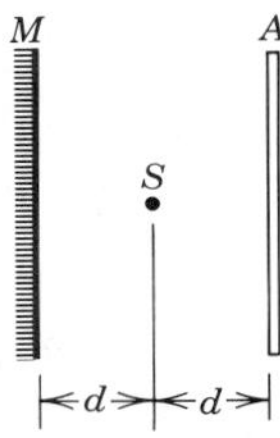

Figure 49 Problem 49.

a screen A. How is the intensity at the center of the screen changed if you put a mirror M a distance d behind the source, as in Fig. 49? (*Hint:* Recall from Chapter 38 the variation of intensity with distance from a point source of light.)

50P. Figure 50 shows an idealized submarine periscope (submarine not shown). The periscope consists of two parallel plane mirrors set at 45° to the periscope axis. An object (arrow) is sighted at a distance D from the scope, as shown. Due to the action of the mirrors, describe the image as seen by the submarine officer peering into the scope. Specifically, is the image (*a*) real or virtual; (*b*) upright or inverted; (*c*) magnified; if so, by how much? (*d*) Find the distance of the image from the bottom mirror.

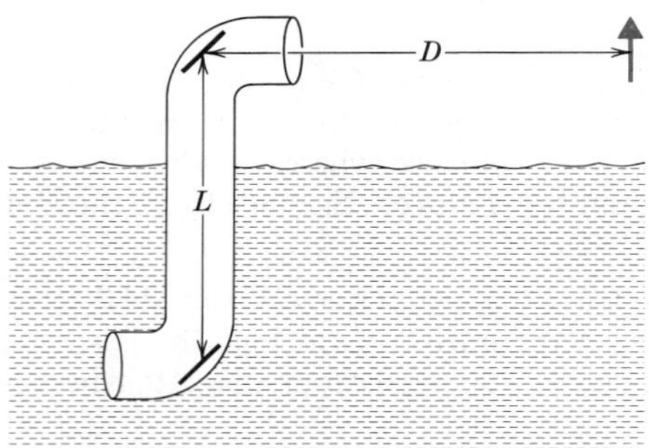

Figure 50 Problem 50.

51P. Solve Problem 47 if the angle between the mirrors is (*a*) 45°, (*b*) 60°, (*c*) 120°, the object always being placed on the bisector of the mirrors.

52P*. A *corner reflector,* much used in optical, microwave, and other applications, consists of three plane mirrors fastened together as the corner of a cube. It has the property that an incident ray is returned, after three reflections, with its direction exactly reversed. Prove this result.

Section 39-6 Spherical Mirrors

53E. A concave shaving mirror has a radius of curvature of 35 cm. It is positioned so that the image of a man's face is 2.5 times the size of his face. How far is the mirror from the man's face?

54E. For clarity, the rays in figures like Fig. 15*b* are not drawn paraxial enough for Eq. 8 to hold with great accuracy. With a ruler, measure r and o in this figure and calculate, from Eq. 8, the predicted value of i. Compare this with the measured value of i.

55P. Fill in the table below, each column of which refers to a spherical mirror and a real object. Check your results by ray-tracing. Distances are in centimeters; if a number has no plus or minus sign in front of it, find the correct sign.

56P. A short linear object of length L lies on the axis of a spherical mirror, a distance o from the mirror. (*a*) Show that its image will have a length L' where

$$L' = L\left(\frac{f}{o-f}\right)^2.$$

(*b*) Show that the *longitudinal magnification* m' $(= L'/L)$ is equal to m^2 where m is the lateral magnification.

57P. (*a*) A luminous point is moving at speed v_0 toward a spherical mirror, along its axis. Show that the speed at which the image of this point object is moving is given by

$$v_I = -\left(\frac{r}{2o-r}\right)^2 v_0.$$

(*Hint:* Start from Eq. 8.) (*b*) Assume that the mirror is concave, with $r = 15$ cm (and thus $f = 7.5$ cm) and that $v_0 = 5.0$ cm/s. Find the speed of the image if the object is far outside the focal point ($o = 30$ cm). (*c*) If it is close to the focal point ($o =$ 8.0 cm). (*d*) If it is very close to the mirror ($o = 0.1$ cm).

Section 39-8 Spherical Refracting Surfaces

58P. A parallel beam of light from a laser falls on a solid transparent sphere of index of refraction n, as shown in Fig. 51.

Table for Problem 55

	a	*b*	*c*	*d*	*e*	*f*	*g*	*h*
Type	*Concave*						*Convex*	
f (cm)	20		+20			20		
r (cm)					−40		40	
i (cm)					−10		4	
o (cm)	+10	+10	+30	+60				+24
m		+1		−0.5		+0.10		0.50
Real image?		no						
Upright image?								no

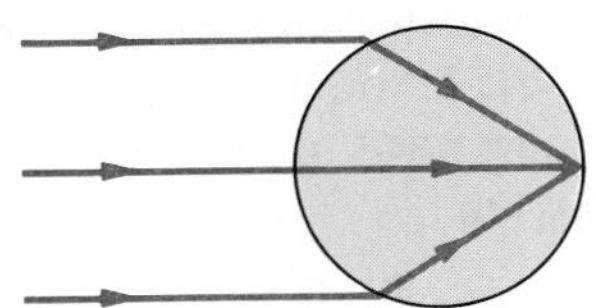

Figure 51 Problem 58.

(*a*) Show that the beam cannot be brought to a focus at the back of the sphere unless the beam width is small compared with the radius of the sphere. (*b*) If the condition in (*a*) is satisfied, what is the index of refraction of the sphere? (*c*) What index of refraction, if any, will focus the beam at the center of the sphere?

59P. Fill in the table below, each column of which refers to a spherical surface separating two media with different indices of refraction. Distances are measured in centimeters. The object is real in all cases. Draw a figure for each situation and construct the appropriate rays graphically. Assume a point object.

60P. A narrow parallel incident beam falls on a solid glass sphere at normal incidence. Locate the image in terms of the index of refraction n and the sphere radius r.

Section 39-9 Thin Lenses

61E. An object is 20 cm to the left of a thin diverging lens having a 30 cm focal length. Where is the image formed? Obtain the image position both by calculation and also from a ray diagram.

62E. Two converging lenses, with focal lengths f_1 and f_2, are positioned a distance $f_1 + f_2$ apart, as shown in Fig. 52. Arrangements like this are called *beam expanders* and are often used to increase the diameters of light beams from lasers. (*a*) If W_1 is the incident beam width, show that the width of the emerging beam is $W_2 = (f_2/f_1)W_1$. (*b*) Show how a combination of one diverging and one converging lens can also be arranged as a beam expander. Incident rays parallel to the axis should exit parallel to the axis.

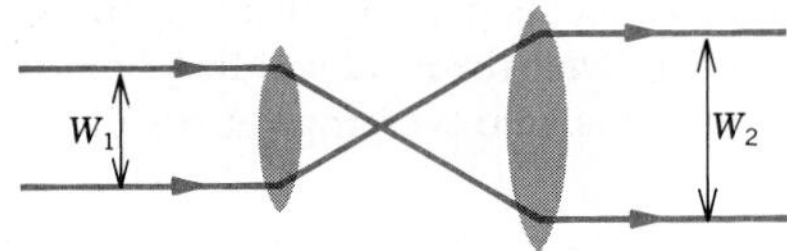

Figure 52 Exercise 62.

63E. Calculate the ratio of the intensity of the beam emerging from the beam expander of Exercise 62 to the intensity of the laser beam.

64E. A double-convex lens is to be made of glass with an index of refraction of 1.5. One surface is to have twice the radius of curvature of the other and the focal length is to be 60 mm. What are the radii?

65E. You focus an image of the sun on a screen, using a thin lens whose focal length is 20 cm. What is the diameter of the image? (See Appendix C for needed data on the sun.)

66E. A lens is made of glass having an index of refraction of 1.5. One side of the lens is flat and the other convex with a radius of curvature of 20 cm. (*a*) Find the focal length of the lens. (*b*) If an object is placed 40 cm to the left of the lens, where will the image be located?

67E. Using the lensmaker's formula (Eq. 13), decide which of the thin lenses in Fig. 53 are converging and which diverging for parallel incident light.

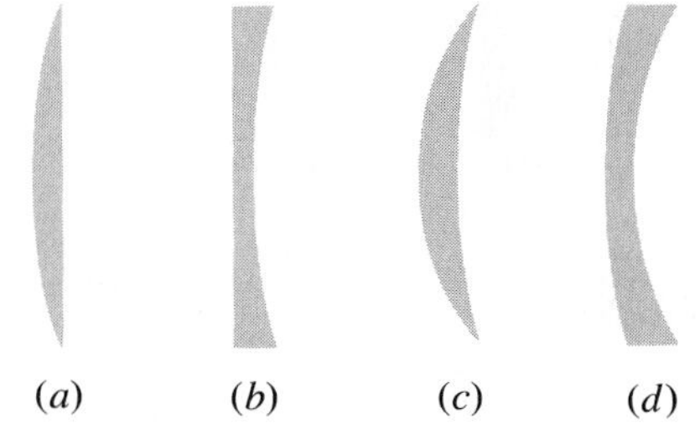

Figure 53 Exercise 67.

Table for Problem 59

	a	*b*	*c*	*d*	*e*	*f*	*g*	*h*
n_1	1.0	1.0	1.0	1.0	1.5	1.5	1.5	1.5
n_2	1.5	1.5	1.5		1.0	1.0	1.0	
o (cm)	+10	+10		+20	+10		+70	+100
i (cm)		−13	+600	−20	−6	−7.5		+600
r (cm)	+30		+30	−20		−30	+30	−30
Real image?								

68E. Show that the focal length f for a thin lens whose index of refraction is n and which is immersed in a fluid whose index of refraction is n' is given by

$$\frac{1}{f}=\frac{n-n'}{n'}\left(\frac{1}{r'}-\frac{1}{r''}\right).$$

69E. A movie camera with a lens of focal length 75 mm takes a picture of a 180-cm-high person standing 27 m away. What is the height of the image of the person on the film?

70P. You have a supply of flat glass disks ($n = 1.5$) and a lens-grinding machine that can be set to grind radii of curvature of either 40 cm or 60 cm. You are asked to prepare a set of six lenses like those shown in Fig. 54. What will be the focal length of each lens? Will the lens form a real or a virtual image of the sun? (*Note:* Where you have a choice of radii of curvature, select the smaller one.)

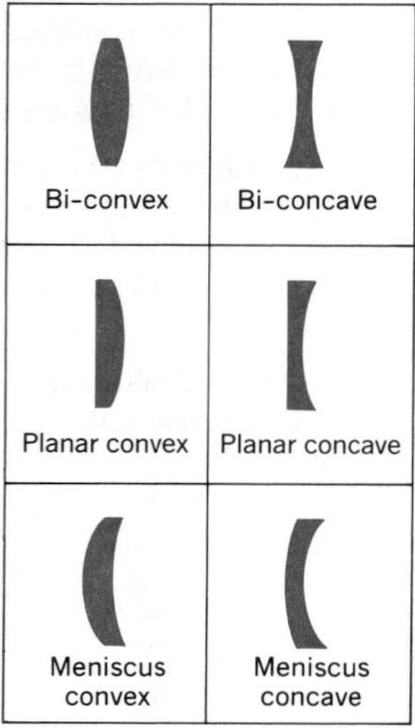

Figure 54 Problem 70.

71P. The formula

$$\frac{1}{o}+\frac{1}{i}=\frac{1}{f}$$

is called the *Gaussian* form of the thin lens formula. Another form of this formula, the *Newtonian* form, is obtained by considering the distance x from the object to the first focal point and the distance x' from the second focal point to the image. Show that

$$xx'=f^2.$$

72P. To the extent possible, fill in the table below, each column of which refers to a thin lens. Distances are in centimeters; if a number (except in row n) has no plus sign or minus sign in front of it, find the correct sign. Draw a figure for each situation and construct the appropriate rays graphically. The object is real in all cases.

73P. A converging lens with a focal length of +20 cm is located 10 cm to the left of a diverging lens having a focal length of −15 cm. If a real object is located 40 cm to the left of the first lens, locate and describe completely the image formed.

74P. An object is placed 1.0 m in front of a converging lens, of focal length 0.50 m, which is 2.0 m in front of a plane mirror. (*a*) Where is the final image, measured from the lens, that would be seen by an eye looking toward the mirror through the lens? (*b*) Is the final image real or virtual? (*c*) Is the final image upright or inverted? (*d*) What is the lateral magnification?

75P. An upright object is placed a distance in front of a converging lens equal to twice the focal length f_1 of the lens. On the other side of the lens is a converging mirror of focal length f_2

Table for Problem 72

	a	*b*	*c*	*d*	*e*	*f*	*g*	*h*	*i*
Type	converging								
f (cm)	10	+10	10	10					
r_1 (cm)					+30	−30	−30		
r_2 (cm)					−30	+30	−60		
i (cm)									
o (cm)	+20	+5	+5	+5	+10	+10	+10	+10	+10
n					1.5	1.5	1.5		
m			>1	<1				0.5	0.5
Real image?									yes
Upright image?								yes	

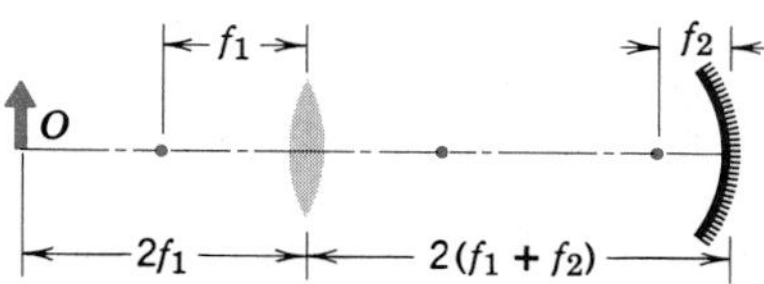

Figure 55 Problem 75.

separated from the lens by a distance $2(f_1 + f_2)$; see Fig. 55. (*a*) Find the location, nature, and relative size of the final image, as seen by an eye looking toward the mirror through the lens. (*b*) Draw the appropriate ray diagram.

76P. An illuminated arrow forms a real inverted image of itself at a distance $d = 40$ cm, measured along the optic axis of a lens; see Fig. 56. The image is just half the size of the object. (*a*) What kind of lens must be used to produce this image? (*b*) How far from the object must the lens be placed? (*c*) What is the focal length of the lens?

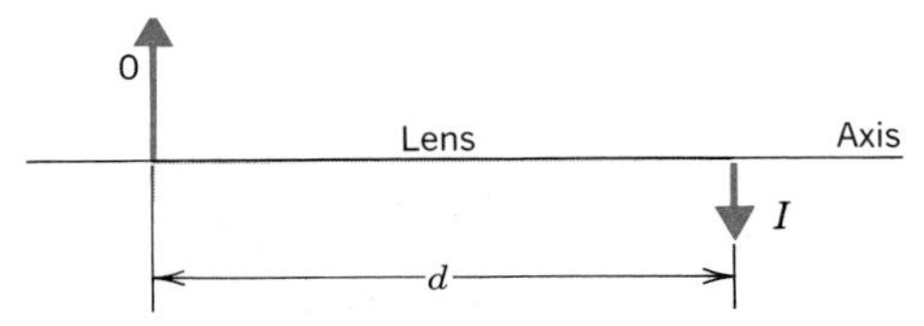

Figure 56 Problem 76.

77P. An object is 20 cm to the left of a lens with a focal length of +10 cm. A second lens of focal length +12.5 cm is 30 cm to the right of the first lens. (*a*) Using the image formed by the first lens as the object for the second, find the location and relative size of the final image. (*b*) Verify your conclusions by drawing the lens system to scale and constructing a ray diagram. (*c*) Describe the final image.

78P. Two thin lenses of focal lengths f_1 and f_2 are in contact. Show that they are equivalent to a single thin lens with a focal length given by

$$f = \frac{f_1 f_2}{f_1 + f_2}.$$

79P. The *power* P of a lens is defined by $P = 1/f$, where f is the focal length. The unit of power is the *diopter,* where 1 diopter = 1/meter. (*a*) Why is this a reasonable definition to use for lens power? (*b*) Show that the net power of two lenses in contact is given by $P = P_1 + P_2$, where P_1 and P_2 are the powers of the separate lenses. (Hint: See Problem 78.)

80P. An illuminated slide is mounted 44 cm from a screen. How far from the slide must a lens of focal length 11 cm be placed in order to focus an image on the screen?

81P. Show that the distance between a real object and its real image formed by a thin converging lens is always greater than or equal to four times the focal length of the lens.

82P. A luminous object and a screen are a fixed distance D apart. (*a*) Show that a converging lens of focal length f will form a real image on the screen for two positions that are separated by

$$d = \sqrt{D(D - 4f)}.$$

(*b*) Show that the ratio of the two image sizes for these two positions is

$$\left(\frac{D - d}{D + d}\right)^2.$$

Section 39-10 Optical Instruments

83E. A microscope of the type shown in Fig. 26 has a focal length for the objective lens of 4.0 cm and for the eyepiece lens of 8.0 cm. The distance between the lenses is 25 cm. (*a*) What is the distance s in Fig. 26? (*b*) To reproduce the conditions of Fig. 26 how far beyond F_1 in that figure should the object be placed? (*c*) What is the lateral magnification m of the objective? (*d*) What is the angular magnification m_θ of the eyepiece? (*e*) What is the overall magnification M of the microscope?

84E. The magnifying power of an astronomical telescope in normal adjustment is 36, and the diameter of the objective lens is 75 mm. What is the minimum diameter of the eyepiece required to collect all the light entering the objective from a distant point source on the axis of the instrument?

85P. In connection with Fig. 25*c*, (*a*) show that if the object O is moved from the first focal point F_1 toward the eye, the image moves in from infinity and the angle θ' (and thus the angular magnification m_θ) is increased. (*b*) If you continue this process, at what image location will m_θ have its maximum usable value? (*c*) Show that the maximum usable value of m_θ is 1 + (15 cm)/f. (*d*) Show that in this situation the angular magnification is equal to the linear magnification.

86P. *The eye—the basic optical instrument:* Fig. 57*a* suggests a normal human eye. Parallel rays entering a relaxed eye gazing at infinity produce a real, inverted image on the retina. The eye thus acts as a converging lens. Most of the refraction occurs at the outer surface of the eye, the *cornea.* Assume a focal length f for the eye of 2.50 cm. In Fig. 57*b* the object is moved in to a distance $o = 40$ cm from the eye. To form an image on the retina the effective focal length of the eye must be reduced to f'. This is done by the action of the ciliary muscles that change the shape of the lens and thus the effective focal length of the eye. (*a*) Find f' from the above data. (*b*) Would the effective radii of curvature of the lens become larger or smaller in the transition from Fig. 57*a* to 57*b*? (In the figure the structure of the eye is only roughly suggested and Fig. 57*b* is not to scale.)

87P. In an eye that is *farsighted* the eye focuses parallel rays so that the image would form behind the retina, as in Fig. 58*a*. In an eye that is *nearsighted* the image is formed in front of the

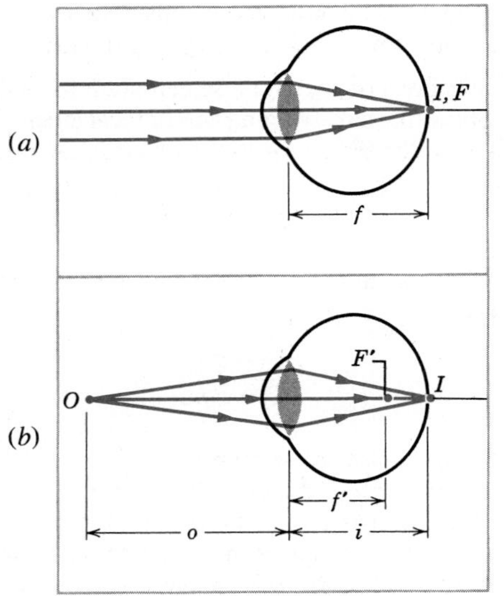

Figure 57 Problem 86.

retina, as in Fig. 58*b*. (*a*) How would you design a corrective lens for each eye defect? Make a ray diagram for each case. (*b*) If you need spectacles only for reading, are you nearsighted or farsighted? (*c*) What is the function of bifocal spectacles, in which the upper parts and lower parts have different focal lengths?

88P. *The camera:* Figure 59 shows an idealized camera focused on an object at infinity. A real, inverted image I is formed on the film, the image distance i being equal to the (fixed) focal length f(= 5.0 cm, say) of the lens system. In Fig. 59*b* the object O is closer to the camera, the object distance o being, say, 100 cm. To focus an image I on the film, we must extend the lens away from the camera (why?). (*a*) Find i' in Fig. 59*b*. (*b*) By how much must the lens be moved? Note that the camera differs from the eye (see Problem 86) in this respect. In the camera, f remains constant and the image distance i must be adjusted by moving the lens. For the eye the image distance i remains constant and the focal length f is adjusted by distorting the lens. Compare Fig. 57 and Fig. 59 carefully.

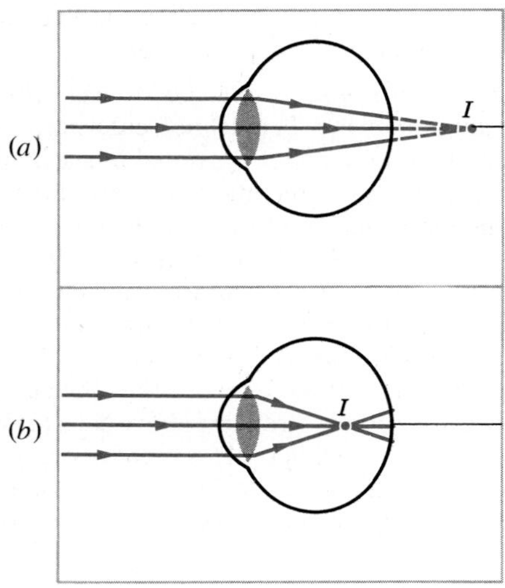

Figure 58 Problem 87.

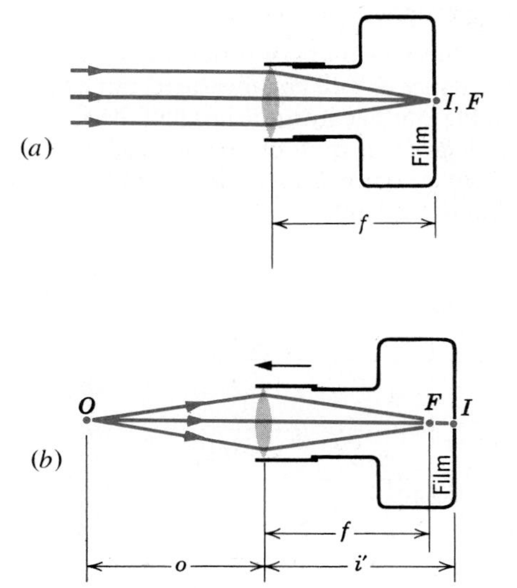

Figure 59 Problem 88.

89P. *The reflecting telescope:* Isaac Newton, having convinced himself (erroneously as it turned out) that chromatic aberration was an inherent property of refracting telescopes, invented the reflecting telescope, shown schematically in Fig. 60. He presented his second model of this telescope, which has a magnifying power of 38, to the Royal Society, which still has it. In Fig. 60 incident light falls, closely parallel to the telescope

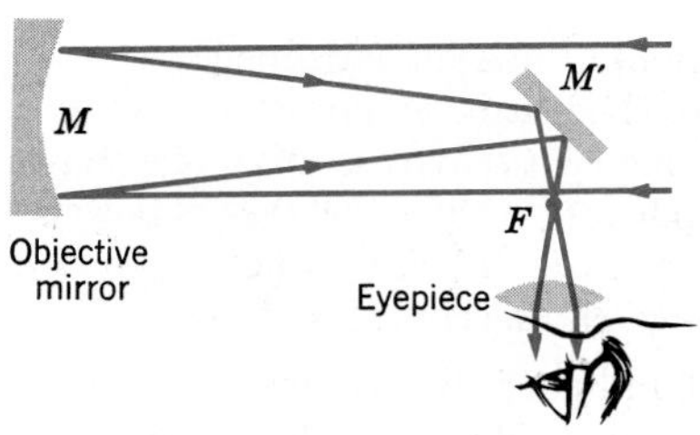

Figure 60 Problem 89.

axis, on the objective mirror M. After reflection from small mirror M' (the figure is not to scale), the rays form a real, inverted image in the focal plane through F. This image is then viewed through an eyepiece. (*a*) Show that the angular magnification m_θ is also given by Eq. 17, or

$$m_\theta = -f_{ob}/f_{ey}$$

where f_{ob} is the focal length of the objective mirror and f_{ey} that of the eyepiece. (*b*) The 200-in. mirror in the reflecting telescope at Mt. Palomar in California has a focal length of 16.8 m. Estimate the size of the image formed in the focal plane of this mirror when the object is a meter stick 2.0 km away. Assume parallel incident rays. (*c*) The mirror of a different reflecting astronomical telescope has an effective radius of curvature ("effective" because such mirrors are ground to a parabolic rather than a spherical shape, to eliminate spherical aberration defects) of 10 m. To give an angular magnification of 200, what must be the focal length of the eyepiece?

90P. In a compound microscope, the object is 10 mm from the objective lens. The lenses are 300 mm apart and the intermediate image is 50 mm from the eyepiece. What magnification is produced?

ESSAY 16
KALEIDOSCOPES

JEARL WALKER
CLEVELAND STATE UNIVERSITY

Modern kaleidoscopes offer brightly colored displays with subtle symmetries. I do not mean the inexpensive toys that yield a few murky images. Instead, I refer to the kaleidoscopes that are now an art form. They contain not only fine-quality mirrors that yield hundreds of clear images but also lenses, filters, and other devices that alter and color the images, sometimes creating vivid illusions. What makes a kaleidoscope display pleasing images and how are the images created? Were it a matter of only simple reflections, kaleidoscopes would be largely identical. Such is not the case, for modern kaleidoscopes come in wide variety.

Before describing the novel kaleidoscopes, I shall explore how simple arrangements of mirrors give multiple images of objects. Mount a vertical mirror on a table and place a small object, for example, a coin, in front of it. When you peer down into the mirror you see the coin and also its virtual image that seems to lie on the tabletop within the mirror, as distant from the lower edge of the mirror as the coin is. One mirror yields one virtual image of the coin.

Next, add another identical mirror, with its vertical edge running alongside a vertical edge of the first mirror. Adjust the angle between the mirror planes to be 60°. When you look down into the mirrors you again see the coin directly but now you also see five virtual images of it in pie-slice sectors that are clustered around the vertex at which the mirrors meet on the table. The composite display has a sixfold symmetry.

One way to locate the images is by means of "virtual mirrors." Note that the lower edges of the mirrors show up in the display as four virtual images. The images can be considered to be mirrors that reflect images just as real mirrors do. Figure 1 shows where the various images seem to lie. To eliminate the complexity of perspective, the plan is an overhead view, although such a viewing direction would scarcely allow you to see any of the reflected images.

The direct view a is reflected by mirror A to form sector b. The reflection amounts mathematically to a simple rotation. Imagine rotating sector a around a hinge lying

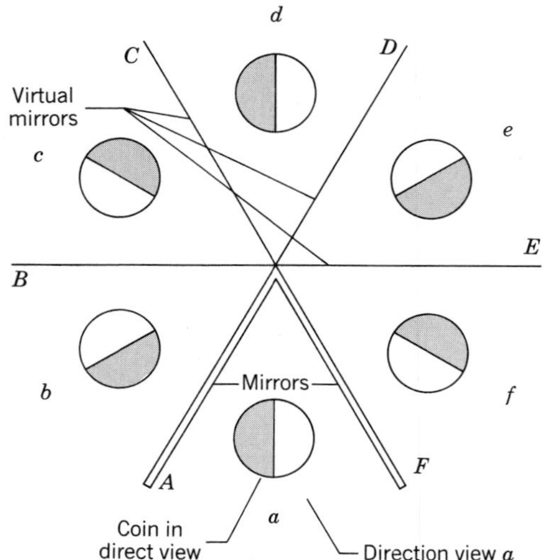

Figure 1 A sixfold cluster around a 60° vertex.

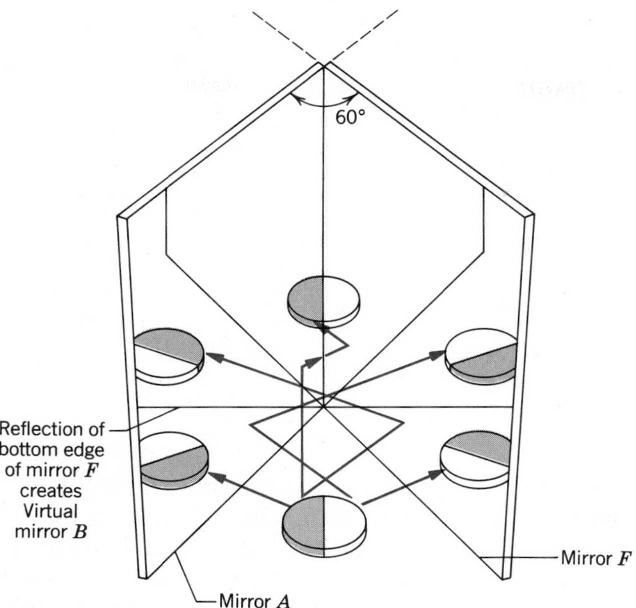

Figure 2 Reflections from the mirrors.

along A. The sector leaves the plane of the sketch and then returns to it, forming sector b.

The image of mirror F in sector b is labeled B. It is a virtual mirror in the sense that it too can serve as a hinge. This time mentally rotate sector b around a hinge along B. The sector leaves the plane of the sketch and then returns to it to form sector c. Continue the procedure. Rotate sector c around the virtual mirror labeled C so that it forms the rearmost sector d. The procedure works just as well in the counterclockwise direction, with sector a generating sector f, followed by f generating e, and finally e generating d. Note that sector d is the same whether you consider a clockwise or counterclockwise generation of the sectors. Such an arrangement is said to be unambiguous. If you change your perspective into the mirrors, the contents of sector d (as well as the other sectors) are unchanged.

The images may be produced mathematically by mental rotations of the sectors, but they are physically produced by reflections of light rays from the real mirrors as shown in Fig. 2. The image of the coin in sector b results from a reflection of rays by mirror A: A ray leaves the coin and then reflects from mirror A to you. You mentally extrapolate the ray backward, concluding that the light originated where you perceive the coin to lie in sector b. The image in sector f also depends on a single reflection of light but the images in sectors c and e require an additional reflection. For example, the image in sector c involves light that leaves the coin, reflects from mirror F and then from mirror A.

Sector d is more complex, because it depends on three reflections of the light. For example, light might leave the coin, reflect from A, then F, and finally A again. Instead, the sequence might be F, A, and then F. Which sequence is important depends on your perspective into the mirror system. If you face toward mirror A, you intercept light undergoing the reflection sequence of A, F, and A. Note the pattern in the number of reflections in the cluster. The farther from the direct view a sector is, the greater the number of reflections must be.

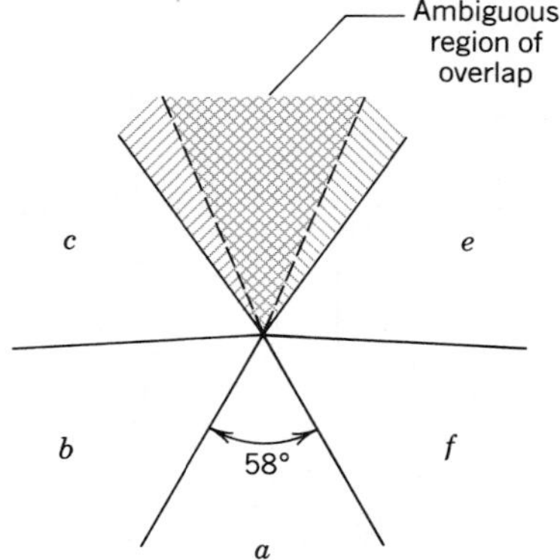

Figure 3 Effect of decreasing the angle between the mirrors.

This cluster of six sectors is common to inexpensive kaleidoscopes having two mirrors angled at 60°. The mirrors lie in a tube. At the viewing end you peer through a small opening at the center. At the other end lie small objects, often colored bits of plastic. By looking at the vertex formed by the mirrors at the far end, you see a cluster of six sectors, one of which is the direct view. If you turn the kaleidoscope, the colored bits of plastic fall into a new arrangement. Part of the kaleidoscope's charm is that the sixfold symmetry remains even when the bits of plastic fall chaotically into place.

An inexpensive kaleidoscope usually has mirrors that consist of a glass layer with a reflecting metallic coating on the rear surface. With such mirrors, light weakly reflects from the front surface of the glass and then more strongly reflects from the metal coating. The resulting double image has fuzzy edges. If the light undergoes several reflections from the mirrors, the image becomes even fuzzier. Better-quality kaleidoscopes come with "front surface mirrors" that have the metallic coating on the front of the glass, thus eliminating the double reflection and maintaining sharp edges on the images.

Return to the tabletop experiment. If you decrease the angle between the mirrors, the sectors narrow in angle while sector *d* begins to separate into two sectors that overlap at the rear of the display as shown in Fig. 3. The overlap is ambiguous: Its contents depend on which mirror you look into. If you face primarily toward mirror *A*, you see one type of scene at the rear of the display, whereas if you face toward mirror *F*, you see quite a different scene there. Moreover, the overlap of sectors at the rear leaves at least one of them smaller than the other sectors. The ambiguity and the unequal angular width of the sectors at the rear spoil the symmetry.

Are there any other angles between the mirrors that yield unambiguous sectors and offer unmarred symmetry? Yes, there are several, each an even divisor of 360°. For example, an angle of 45° gives an eightfold symmetry, with the rearmost sector requiring four reflections from the mirrors. The largest angle that gives unambiguous images is 180°, which is the same as saying that a single mirror can be mounted in the kaleidoscope tube. However, the twofold symmetry that results is dull. More interesting is the fourfold symmetry that is created when the angle is 90°.

Return to the tabletop arrangement with the mirror angle at 60°. Add a third identical mirror so that the edges at either end of the mirror system form an equilateral triangle. Look down into the assembly with a coin lying between the mirrors. The addition of the third mirror dramatically increases the number of images, with images extending in all directions in the plane of the tabletop as far as you can see. In princi-

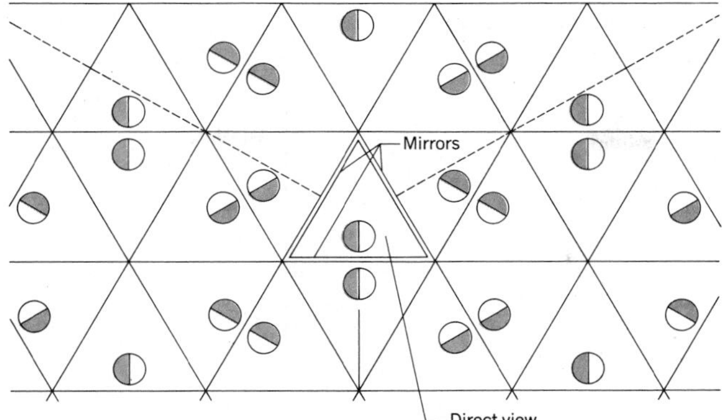

Figure 4 A portion of the image field produced by three mirrors.

ple, the number of images is infinite. However, the quality of the mirrors limits how far from the direct view you can see distinct images.

Part of the image plane is indicated in Fig. 4. Note that each vertex of the equilateral triangle of the direct view has a cluster of sectors with sixfold symmetry. The plane is completely covered with such triangles and clusters. This display is typical of what you see in a kaleidoscope where the ends of the mirrors form an equilateral triangle. The display is unambiguous. When you change your angle of view into the mirror system, the contents of the sectors are unchanged. If the direct view contains a variety of brightly colored objects, the image field can be dazzling. You see a repetition of patterns while also seeing the sixfold symmetry at each vertex.

The triangular pieces of the image field are created by reflections of light from the mirrors. At the vertices of the direct view, the number of reflections required is the same as when only two mirrors are in place. Sectors more distant from the direct view require more reflections. Figure 5 indicates the number of reflections required for some of the sectors.

Most three-mirror kaleidoscopes are based on the equilateral triangle design. What happens when the angles between the mirrors are changed? Suppose, for example, the ends of the mirrors form an isosceles triangle with angles of 50°, 65°, and 65°. Since the angles are not even divisors of 360°, many of the sectors are ambiguous, changing in content and angular size as you vary your angle of view into the kaleidoscope.

The most expensive kaleidoscope I own, costing well over $1000, contains

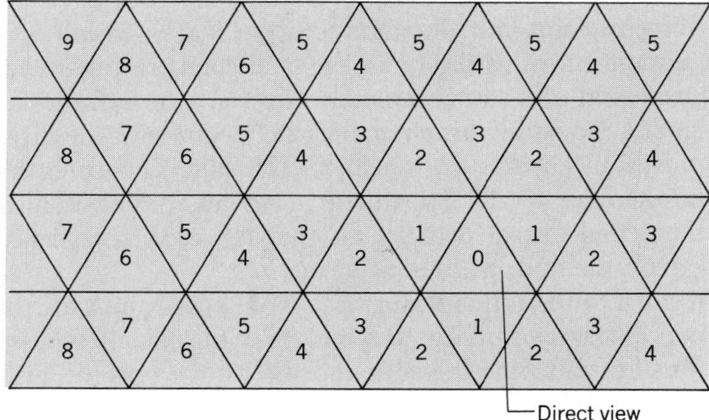

Figure 5 The number in an image triangle is the number of reflections involved in producing the image that appears to be in that triangle. The triangles continue out to infinity in all directions in the image plane.

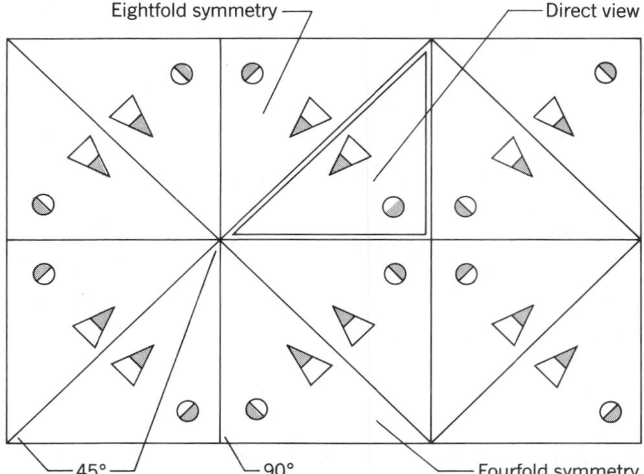

Figure 6 A mirror system giving two types of symmetry.

mirrors that form an isosceles triangle with angles of 22.5°, 78.75°, and 78.75°. At first glance, the image field produced by the instrument is enchanting because the 22.5° angle generates a 16-fold symmetry around one corner of the direct view. However, when I look at the rest of the image field, I see incomplete reflections that ruin any sense of symmetry.

Are there any mirror arrangements other than an equilateral triangle that give completely unambiguous image fields and full symmetry? Yes, there are, one arrangement being long known. Four mirrors forming a rectangle give unambiguous images with fourfold symmetry. I found two more solutions that were even more delightful because they simultaneously create more than one type of symmetry. One solution is a right triangle with two 45° angles (see Fig. 6). This mirror system gives eightfold symmetry around the 45° vertices and fourfold symmetry around the 90° vertices throughout the image field.

The other solution is the best, because it offers three types of symmetry. It is a right triangle with angles of 60° and 30° (see Fig. 7). The 90° vertices have fourfold symmetry, the 60° vertices have sixfold symmetry, and the 30° vertices have 12-fold symmetry. Soon after I published this design, two kaleidoscope makers incorporated it in their instruments. Even though I had calculated and mapped the image field, I was still stunned by the beauty of the images when I finally had the opportunity to peer into a kaleidoscope with the design.

Are there more solutions? You might try your hand at designing kaleidoscopes with three or more mirrors. One solution that is initially promising is a hexagonal arrangement of mirrors. The image field can certainly be filled with hexagons much like hexagonal tiles can completely cover a bathroom floor. However, the images are ambiguous. To show this characteristic, begin with the direct view that is hexagonal as shown in Fig. 8. Add a coin near the left side. Then mentally rotate the direct view around the edge on the left (mirror A) to get a reflected hexagon b. Then rotate b around virtual mirror B to get another reflected hexagon c. The image of the coin ends up on the bottom right of hexagon c.

Return to the initial hexagon a. Rotate it around mirror C to get another reflected hexagon c as shown in Fig. 9. Note that c fits exactly where it did with the clockwise rotations but this time the coin ends up on the bottom left. Thus, the hexagonal

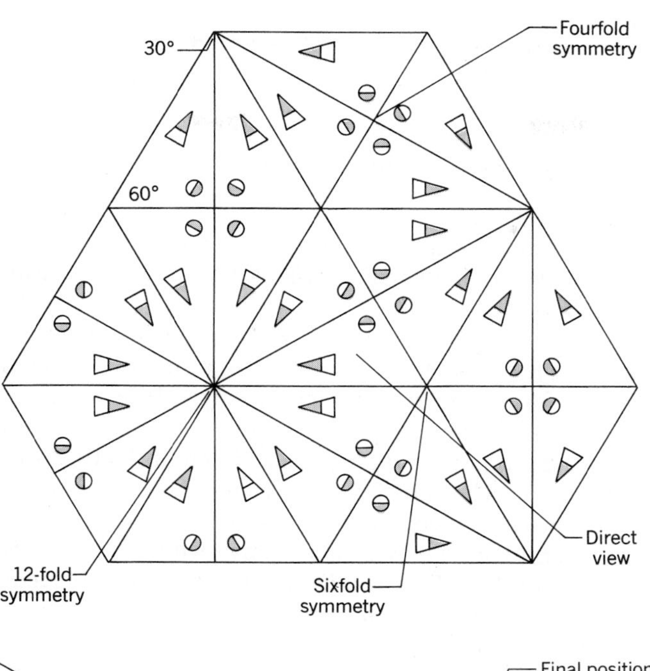

Figure 7 A mirror system giving three types of symmetry.

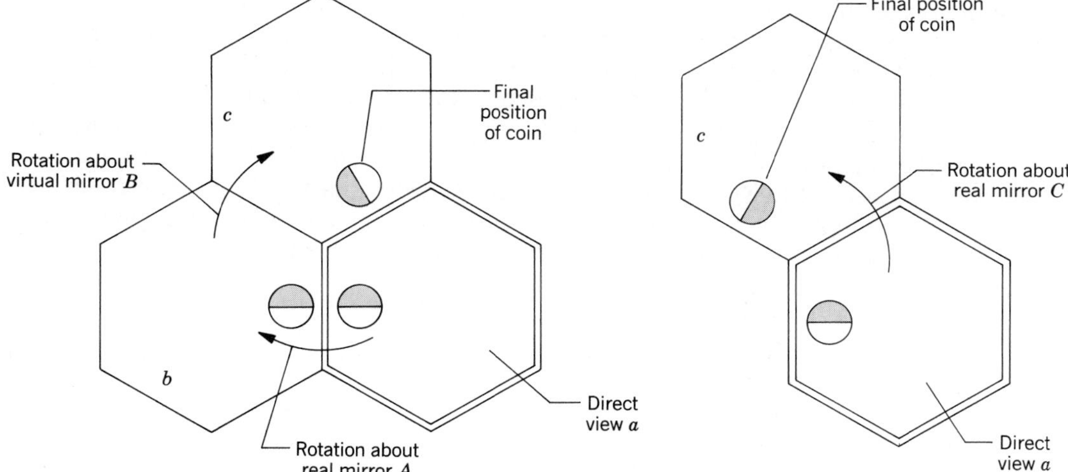

Figure 8 Clockwise reflection of a hexagon.

Figure 9 Counterclockwise reflection of a hexagon.

system is ambiguous. When you look into a kaleidoscope with a hexagonal mirror system, your angle of view determines what you see in *c*. For some angles you might even see two coins in *c*.

Many kaleidoscopes are open at the far end. When you hold them toward a colorful or moving scene, you see a mosaic of slightly different images. If the far end is equipped with a convex lens that compresses the external scene, the mosaic is even more striking. The pieces of the mosaic differ because they depend on the angle at which light rays enter the far end of the tube. The pieces that are near the direct view involve rays that strike a mirror with an incident angle that is close to 90° and then reflect once or twice before reaching you as shown in Fig. 10. These pieces are similar to the direct scene. The pieces far from the direct scene require rays that strike a

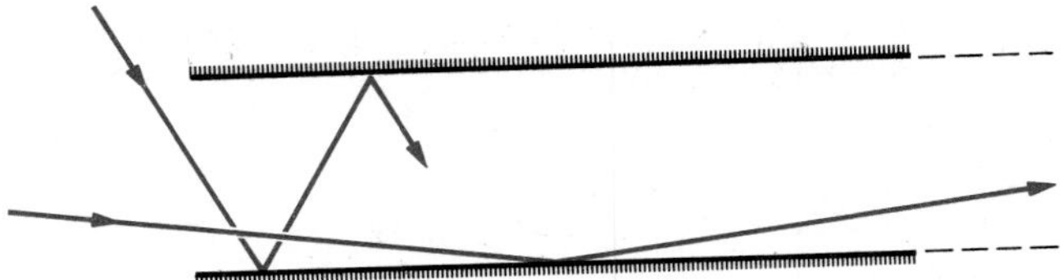

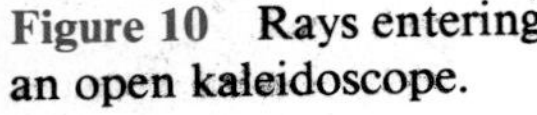

Figure 10 Rays entering an open kaleidoscope.

mirror with a much smaller incident angle and then reflect many times before reaching you. These rays originate from parts of the scenery well away from what you see in the direct view.

One of my favorite kaleidoscopes has slanted mirrors. At the far end of the instrument, the edges of the mirrors form a small equilateral triangle. The edges at the viewing end form a larger equilateral triangle. When you look into the instrument you see what appears to be a geodesic sphere that consists of many equilateral triangles and that seems to float in empty space. Each of the triangular pieces contains a slightly different perspective of the scenery toward which the kaleidoscope is pointed.

You can best understand how the geodesic sphere is created by looking down into a single mirror held on a table. Begin with the mirror vertical. The image of the table you see in the mirror is a flat continuation of the table you see directly. Now tilt the mirror away from you. The image of the table tilts down, rotating around its juncture with the table you see directly. In the kaleidoscope, the slanted mirrors tilt the images so that they seem to be glued to the side of a sphere. The images that are farther from the direct view require more reflections from the mirrors, which increases the tilt of the image and creates the illusion of there being a sphere in front of you.

A giant version of this type of kaleidoscope was once planned for Disney World by Stephen Hines, an optical engineer and designer. Spectators were to ride through the kaleidoscope toward the floating sphere. The sight would have been spectacular. However, considerations of expense reduced the kaleidoscope to one that is viewed only at the end. Still, with mirrors that are nearly 3 m long and with a triangular opening at the viewing end that is over a meter long on each edge, the scaled-down kaleidoscope is quite a sight.

CHAPTER 40

INTERFERENCE

*The pyramid-shaped PAVE PAWS radar on Cape Cod sweeps each of its two fan-shaped microwave beams back and forth through an angle of ±60° every microsecond, searching for submarine-launched missiles. The two beams are generated by the constructive interference of microwave signals from the thousands of antenna elements—shown above—that cover two faces of the pyramid. The pyramid is as immobile as its Egyptian counterparts, the sweeping being done by periodically and electronically varying the phases of the microwave signals fed to these elements.**

40-1 Interference

Sunlight, as the rainbow shows us, is a composite of all the colors of the visible spectrum. The colors reveal themselves in the rainbow because the incident wavelengths are bent through different angles as they pass through raindrops that form the bow. However, the striking colors of soap bubbles, oil slicks, peacock feathers, and the throats of hummingbirds are not produced by *refraction* but by constructive and destructive *interference* of light reflected from the upper and lower surfaces of a thin film. The interfering waves combine either to enhance or to suppress certain colors in the spectrum of the incident sunlight.

This selective enhancement or suppression of selected wavelengths has many applications. When light falls on an ordinary glass surface, for example, about 4% of the incident energy is reflected, thus weakening the transmitted beam by that amount. This unwanted loss of light can be a real problem in optical systems with many

* The coverage is ±120° in azimuth and a range of 3000 miles; see "Phased-Array Radars," by Eli Brookner, *Scientific American*, February 1985.

components. A thin transparent film, deposited on the optical surface, can largely suppress the reflected light (and thus enhance the transmitted light) by destructive interference. The bluish cast of your camera lens shows the presence of such a coating.

Sometimes we wish to enhance—rather than reduce—the reflectivity of a surface and this too can be done with interference coatings. In fact, an interference stack of a number of films, with differing thicknesses and indices of refraction, can be designed to give almost any desired wavelength profile for the reflected or the transmitted light.* For example, windows can be provided with coatings that have high reflectivity in the infrared, thus admitting the visible component of sunlight but reflecting its infrared or heating component.

The interference of light is not restricted to light reflected from the two surfaces of a thin film. Such phenomena can occur in any situation in which light from a single source is divided into two sub-beams that recombine after following paths of different lengths. To understand interference we must go beyond the restrictions of geometric optics and employ the full power of wave optics. In fact, the existence of interference phenomena—as we shall see—is perhaps our most convincing evidence that light is a wave.

40-2 Light as a Wave

The first person to advance a convincing wave theory for light was the Dutch physicist Christian Huygens, in 1678. While much less comprehensive than the later electromagnetic theory of Maxwell, it was simpler mathematically and remains useful today. Its great merit is that it can account for the laws of reflection and refraction in terms of a wave picture and that it gives physical meaning to the index of refraction.

Huygens' wave theory is based on a geometrical construction that allows us to tell where a given wavefront will be at any time in the future if we know its present position. This construction is based on *Huygens' principle,* which is:

> *All points on a wave front serve as point sources of spherical secondary wavelets. After a time t, the new position of the wave front will be the surface of tangency to these secondary wavelets.*

* See "Optical Interference Coatings," by Philip Baumeister and Gerald Pincus, *Scientific American,* December 1970.

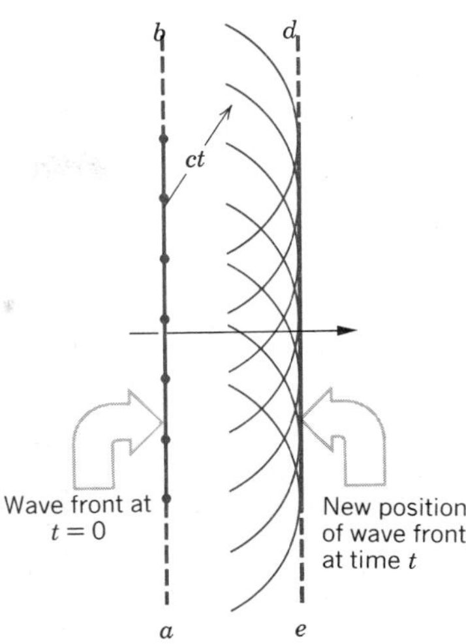

Figure 1 The propagation of a plane wave in free space, as portrayed by Huygens' principle.

Here is a simple example: Given a wave front, *ab* in Fig. 1, in a plane wave in free space, where will the wave front be a time t later? We let several points on this plane (see dots) serve as centers for secondary spherical wavelets. In a time t, the radius of these spherical waves is ct, where c is the speed of light in free space. We represent the plane of tangency to these spheres at time t by *de*. As we expect, it is parallel to plane *ab* and a perpendicular distance ct from it. Thus, plane wave fronts are propagated as planes and with speed c. Note that the Huygens method involves a three-dimensional construction and that Fig. 1 is the intersection of this construction with the plane of the page.

The Law of Refraction. We now use Huygens' principle to derive the law of refraction. Figure 2 shows four stages in the refraction of three wave fronts at a plane interface between air (medium 1) and glass (medium 2). We choose the wave fronts in the incident beam to be separated, arbitrarily, by λ_1, the wavelength in medium 1. Let the speed of light in air be v_1 and that in glass be v_2. We assume that $v_2 < v_1$, which happens to be true.

The angle θ_1 in Fig. 2*a* is the angle between the wave front and the surface; this is the same as the angle between the *normal* to the wave front (that is, the incident ray) and the *normal* to the surface; thus, θ_1 is the angle of incidence. The time $(= \lambda_1/v_1)$ for a Huygens wavelet to expand from point e in Fig. 2*b* to include point c will equal the time $(= \lambda_2/v_2)$ for a wavelet in the glass to expand at the reduced speed v_2 from h to include e'.

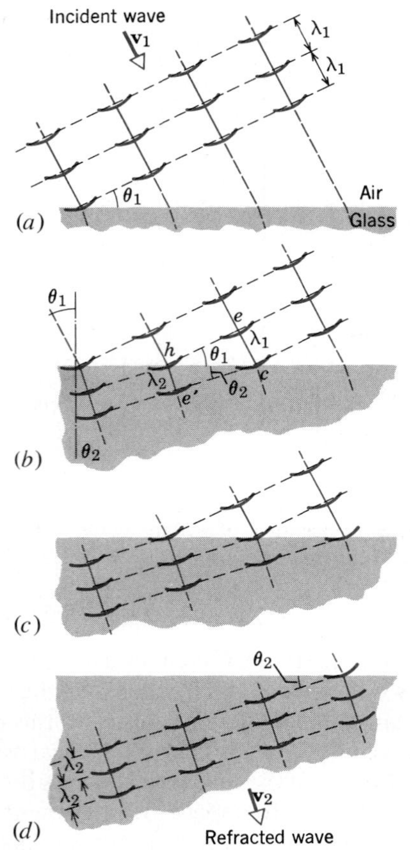

Figure 2 The refraction of a plane wave at a plane surface, as portrayed by Huygens' principle. Note that the wavelength in glass is smaller than that in air. For simplicity, the reflected wave is not shown.

Equating these times, we have

$$\frac{\lambda_1}{\lambda_2} = \frac{v_1}{v_2}, \tag{1}$$

which shows that the wavelength is proportional to the speed in the medium.

The refracted wave front must be tangent to an arc of radius λ_2 centered on h. Since c lies on the new wave front, the tangent must pass through this point also. Note that θ_2, the angle between the refracted wave front and the surface, is equal to the angle of refraction.

For the right triangles hce and hce' we may write

$$\sin\theta_1 = \frac{\lambda_1}{hc} \quad \text{(for } hce\text{)}$$

and

$$\sin\theta_2 = \frac{\lambda_2}{hc} \quad \text{(for } hce'\text{)}.$$

Dividing these two equations and using Eq. 1, we find

$$\frac{\sin\theta_1}{\sin\theta_2} = \frac{\lambda_1}{\lambda_2} = \frac{v_1}{v_2}. \tag{2}$$

We can define an *index of refraction* for each medium as the ratio of the speed of light in free space to the speed of light in the medium. Thus,

$$\boxed{n = \frac{c}{v}.} \tag{3}$$

In particular, for our two media, we have

$$n_1 = \frac{c}{v_1} \quad \text{and} \quad n_2 = \frac{c}{v_2}. \tag{4}$$

If we combine Eqs. 2 and 4 we find

$$\frac{\sin\theta_1}{\sin\theta_2} = \frac{c/v_2}{c/v_1} = \frac{n_2}{n_1} \tag{5}$$

or

$$n_1 \sin\theta_1 = n_2 \sin\theta_2 \quad \text{(law of refraction)}, \tag{6}$$

which is the law of refraction.

Counting Waves. When two waves from the same source rejoin after following separate paths, we are less concerned with the separate geometric lengths of those paths than with the number of waves each path contains. It is the difference in this number of waves between the two paths that determines the phase difference between the waves and thus the intensity of the light. If one or both paths are through media with different indices of refraction, we must take into account the fact that the wavelength of light depends on the index of refraction.

From Eq. 1 we can write, for the wavelength λ_n in a given medium in terms of the wavelength λ in a vacuum,

$$\lambda_n = \lambda \frac{v}{c} \tag{7}$$

or

$$\boxed{\lambda_n = \frac{\lambda}{n},} \tag{8}$$

where n is the index of refraction of the medium with

respect to a vacuum. Thus, as Fig. 2 makes clear, the higher the index of refraction, the shorter the wavelength of the light.

If rays diverging from a point source form a real image after passing through a converging lens, the rays travel over paths of different *geometric* length but all paths contain the same number of waves; otherwise the rays would not be in phase when they recombine at the image point. It is, in fact, the function of the lens—taking advantage of the fact that the wavelength in glass is less than that in air—to cause the rays to follow paths that contain the same number of waves. For a converging lens a short path, directly along the lens axis, passes through the largest thickness of glass. A path that passes through the edge of the lens passes through a shorter thickness of glass, which just compensates for its greater geometric length.

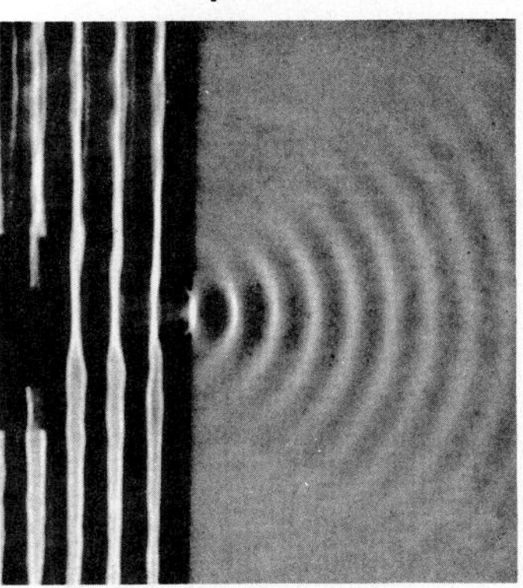

Figure 3 The diffraction of water waves in a ripple tank. Waves moving from left to right flare out through an opening in a barrier into the space beyond. Note that the opening is about the same size as the wavelength.

40-3 Diffraction

To understand the interference of two combining waves, we must first understand the central features of the diffraction of waves, a subject that we explore much more fully in Chapter 41. If a wave falls on a barrier that has an opening of dimensions similar to the wavelength, the wave will flair out into the region beyond. This phenomenon, called *diffraction,* is in the spirit of the spreading out of the wavelets in the Huygens construction of Fig. 1. Figure 3 shows the diffraction of water waves in a shallow ripple tank. These waves were produced by tapping the water surface periodically with a flat stick.

Figure 4*a* shows the situation schematically for an incident plane wave of wavelength λ falling on a slit of width $a = 6.0\ \lambda$. The light clearly flares out on the far side of the slit. Figures 4*b* ($a = 3.0\ \lambda$) and 4*c* ($a = 1.5\ \lambda$) illustrate the main feature of diffraction: The narrower the slit, the greater the diffraction.

We see here the limitations of geometric optics, whose central feature is a ray following a linear path. If we try to form such a ray physically by allowing light to fall on a narrow slit, or a series of narrow slits, we are

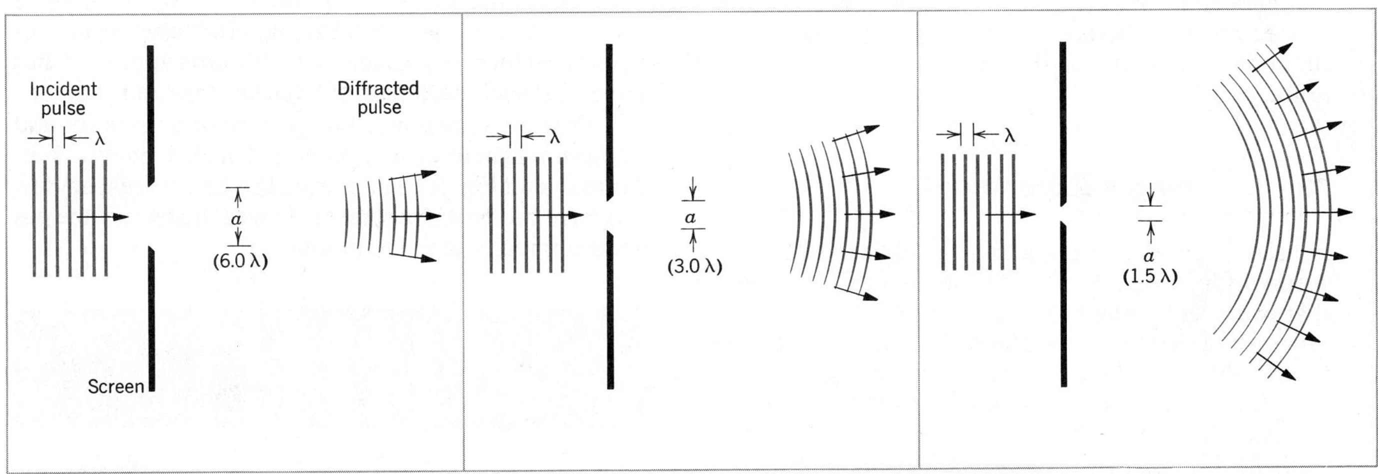

Figure 4 Diffraction represented schematically. For a given wavelength λ, the diffraction is more pronounced the smaller the slit width a. The figures show the cases for $a = 6.0\ \lambda$, $a = 3.0\ \lambda$, and $a = 1.5\ \lambda$.

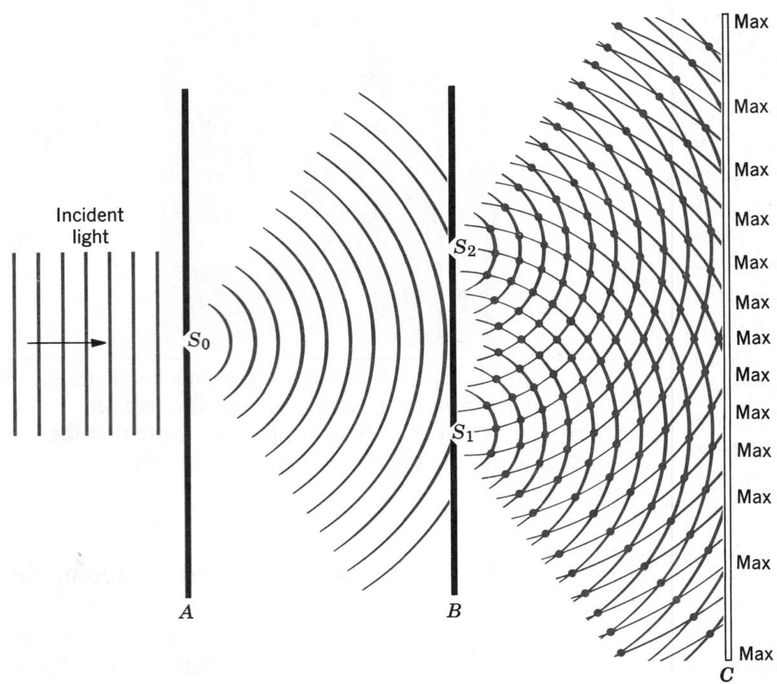

Figure 5 In Thomas Young's interference experiment, light diffracted from pinhole S_0 falls on pinholes S_1 and S_2 in screen B. Light diffracted from these two pinholes overlaps in the region between screens B and C, producing an interference pattern on screen C.

foiled at every turn by diffraction. Indeed, the more we try to define the ray physically (that is, the narrower we make the slits), the greater is the spreading out by diffraction. We said in Chapter 39 that geometric optics only holds if we place no barriers, slits, or other apertures in the path of the light beam if their dimensions are comparable to or smaller than the wavelength of light. We see now that this is equivalent to saying that geometric optics holds only to the extent that we can neglect diffraction.

40-4 Young's Experiment

In 1801, Thomas Young first established the wave theory of light on a firm experimental basis by showing that two overlapping light waves can *interfere* with each other. His experiment was especially convincing because he was able to deduce the wavelength of light from his experiments, the first measurement of this important quantity. Young's value for the average wavelength of sunlight, 570 nm in modern units, is remarkably close to the modern value of 555 nm.

Young allowed sunlight to fall on a pinhole S_0 punched in a screen A. As represented in Fig. 5, the emerging light spreads out by diffraction and falls on pinholes S_1 and S_2 punched into screen B. Diffraction occurs again at the two pinholes and two overlapping spherical waves expanded into the space to the right of screen B, where they can interfere with each other. The lines of dots extending radially outward from the slits S_1 and S_2 show the directions in which the waves from these two slits are mutually reinforcing, producing regions of maximum intensity on screen C. Minima appear on this screen between each pair of adjacent maxima.

Figure 6 shows an actual pattern of maximum and minimum light intensity, generated in an apparatus similar to that of Fig. 5. In preparing this figure, long narrow slits have been used in screens A and B instead of the pin holes originally used by Young.

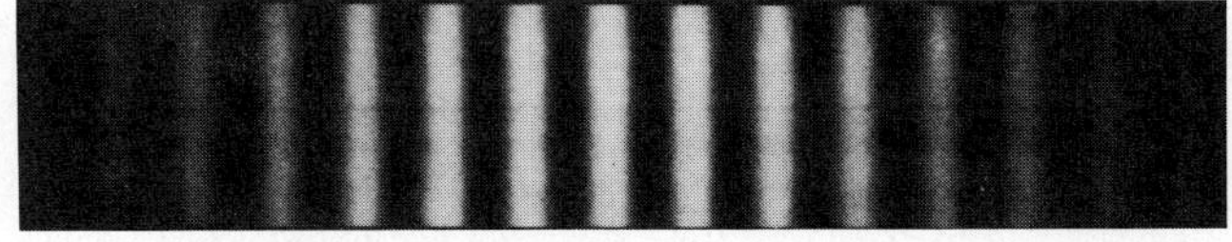

Figure 6 An interference pattern produced by the arrangement shown in Fig. 5, with narrow slits substituted for the pinholes. The alternating maxima and minima are called *interference fringes.*

Figure 7 shows an interference pattern set up by overlapping water waves in a ripple tank. The waves are generated by two spheres connected to the same mechanical vibrator and oscillating up and down through the water surface. The two spheres of Fig. 7 are to be compared to the pinholes S_1 and S_2 of Fig. 5.

Let us analyze Young's experiment qualitatively, assuming that the incident light consists of a single wavelength only. In Fig. 8, P is an arbitrary point on screen C, which is a distance D from the screen containing the two narrow slits, S_1 and S_2. Let θ be the angle between cP and the horizontal line cO. Let b be a point on the line PS_1 such that the distances PS_2 and Pb are equal. The waves emitted at the slits, S_1 and S_2, are initially in phase with each other. We can see in the figure though that the wave emitted at the slits, S_1 and S_2 are initially in phase with wave emitted at S_2. The difference in the path lengths of these two waves is simply S_1b. The nature of the interference of the two waves at P depends entirely on the length of the path difference. For example, if the path difference S_1b contains an integral number of wavelengths ($S_1b = m\lambda$, where m is an integer), the two waves will be in phase at P and will interfere constructively, producing a maximum in the intensity of light on the screen at that point.

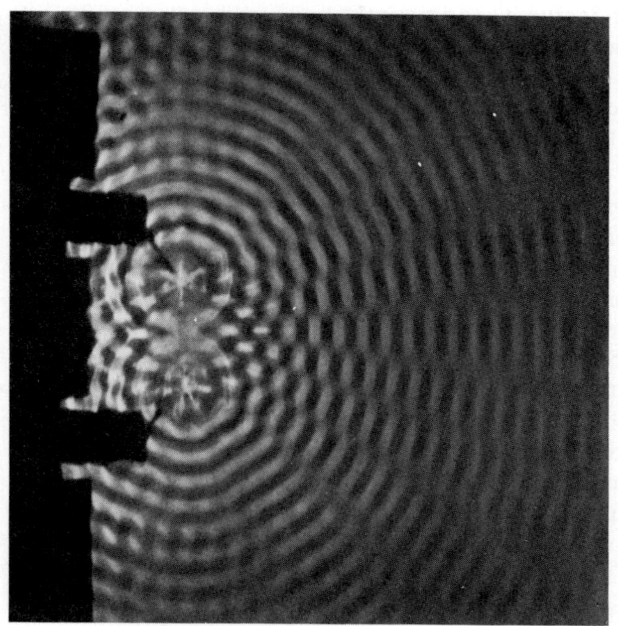

Figure 7 An interference pattern produced by water waves in a ripple tank. Compare with the region between screens B and C in Fig. 5.

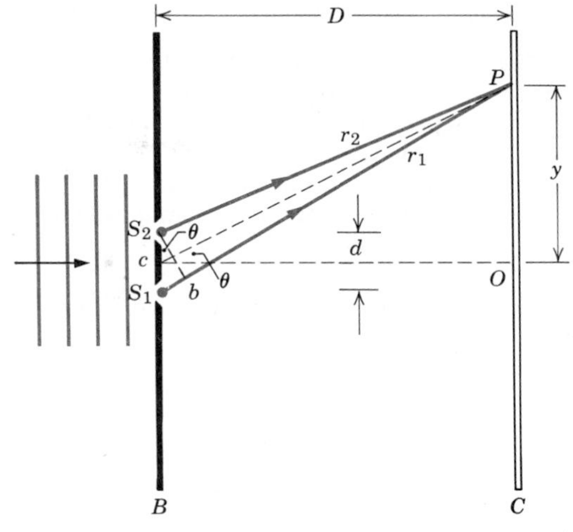

Figure 8 Waves from slits S_1 and S_2 combine at P, an arbitrary point on the screen. The angle θ serves as a convenient locator for P. Actually, $D \gg d$, the figure being distorted for clarity.

Now for the quantitative part. Let us assume that the slit spacing d is very much smaller than the distance D between the two screens (the ratio d/D in Fig. 8 has been greatly exaggerated for clarity). This means that the lines S_1P and S_2P are nearly parallel to each other.* The triangle S_1S_2b is then a right triangle, and the angle between S_1S_2 and bS_2 is equal to θ. Thus, the length of the side S_1b in this triangle is equal to $d \sin \theta$. This is the path difference of the two waves so that the conditions for a maximum intensity at P is

$$d \sin \theta = m\lambda, \qquad m = 0, 1, 2, \ldots \qquad \text{(maxima)}. \quad (9)$$

For each maximum above O in Fig. 8 there is a symmetrically located maximum below O. There is a central maximum, described by $m = 0$.

For a minimum at P, S_1b ($= d \sin \theta$) must contain a half-integral number of wavelengths, or

$$d \sin \theta = (m + \tfrac{1}{2})\lambda, \qquad m = 0, 1, 2, \ldots \qquad \text{(minima)}. \quad (10)$$

* We often put a lens behind the two slits, the screen being in the focal plane of the lens. Under these conditions the rays focussed at P are strictly parallel on striking the lens and we can drop the requirement that $d \ll D$. The lens may in practice be the lens and cornea of the eye, the screen being the retina.

Sample Problem 1 What is the linear distance on screen C in Fig. 8 between adjacent maxima? The wavelength λ is 546 nm; the slit separation d is 0.12 mm, and the slit–screen separation D is 55 cm.

We assume from the start that the angle θ in Fig. 8 will be small enough to permit us to use the approximations

$$\sin\theta \approx \tan\theta \approx \theta,$$

in which θ is to be expressed in radian measure. From Fig. 8 we see that

$$\tan\theta \approx \theta = \frac{y}{D}.$$

From Eq. 9 we have

$$\sin\theta \approx \theta = \frac{m\lambda}{d}.$$

If we equate these two expressions for θ and solve for y, we find

$$y_m = \frac{m\lambda D}{d}. \qquad (11)$$

For the adjacent fringe, we have

$$y_{m+1} = \frac{(m+1)\lambda D}{d}. \qquad (12)$$

We find the fringe separation by subtracting Eq. 11 from Eq. 12, or

$$\Delta y = y_{m+1} - y_m = \frac{\lambda D}{d}$$

$$= \frac{(546 \times 10^{-9}\text{ m})(55 \times 10^{-2}\text{ m})}{0.12 \times 10^{-3}\text{ m}}$$

$$= 2.50 \times 10^{-3}\text{ m} \approx 2.5\text{ mm}. \qquad \text{(Answer)}$$

As long as d and θ in Fig. 8 are small, the separation of the interference fringes is independent of m; that is, the fringes are evenly spaced. If the incident light contains more than one wavelength, the separate interference patterns, which will have different fringe spacings, will be superimposed.

40-5 Coherence

If we are to have well-defined interference fringes on screen C in Fig. 5, the light waves that travel from S_1 and S_2 to any point on this screen must have a well-defined phase difference ϕ that remains constant with time. That will be the case for the experiment of Fig. 5 because the two slits S_1 and S_2 are illuminated by light diffracted from the same source slit S_0. The two beams emerging from slit S_1 and S_2 are said to be *completely coherent,* because their phase difference does not change with time. Coherent beams in the radiofrequency region of the electromagnetic spectrum can be formed by connecting two separate antennas to the same oscillator.

Dark fringes on screen C in Fig. 5 correspond to places at which the phase difference ϕ of the waves arriving from the two slits is π, 3π, 5π, ... Bright fringes are places at which this phase difference is 0, 2π, 4π, 6π, ...

Imagine, however, that the slits S_1 and S_2 in Fig. 5 are replaced by two completely independent light sources, such as two fine incandescent wires placed side by side. No interference fringes would appear on screen C but only a relatively uniform illumination. We can understand this if we assume that for completely independent light sources the phase difference between the two beams arriving at any point on screen C in Fig. 5 will vary with time in a random way. At a certain instant conditions may be right for cancellation and a short time later they may be right for reinforcement. The eye, however, cannot follow these rapid variations and sees only a uniform illumination. The intensity at any point on screen C is equal to the sum of the intensities that each source S_1 and S_2 produces separately at that point. Under these conditions the two beams emerging from S_1 and S_2 are said to be *completely incoherent.* Incoherent beams in the radiofrequency part of the electromagnetic spectrum are formed by connecting two antennas to completely independent oscillators.

For completely coherent light beams we (1) combine the amplitudes vectorially, taking the (constant) phase difference properly into account, and then (2) square this resultant amplitude to obtain a quantity proportional to the resultant intensity. For completely incoherent light beams, on the other hand, we (1) square the individual amplitudes to obtain quantities proportional to the individual intensities and then (2) add the individual intensities to obtain the resultant intensity. This procedure is in agreement with the experimental fact that for completely independent light sources the resultant intensity at every point is always greater than the intensity produced at that point by either light source acting alone.

In common sources of visible light, such as incandescent wires or an electric discharge passing through a gas, the fundamental light emission processes occur in individual atoms and these atoms do not act together in a cooperative (that is, coherent) way. The act of light emission by a single atom in a typical case takes about 10^{-8} s

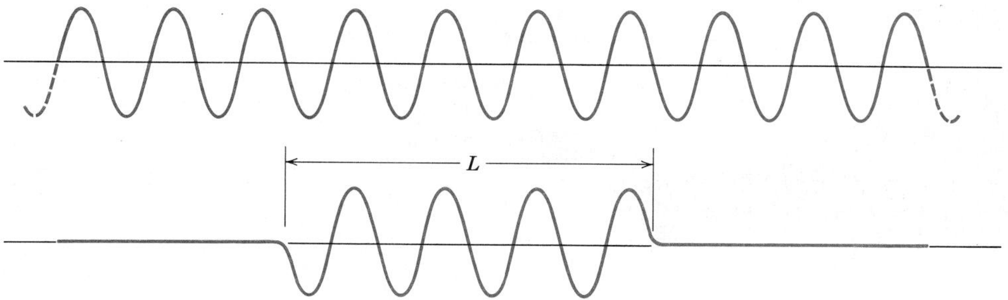

Figure 9 (*a*) A section of a pure sine wave, with no beginning and no ending. (*b*) A *wavetrain,* of finite length *L*. Light from ordinary incandescent sources is emitted as wavetrains, originating with individual atoms.

and the emitted light is properly described as a *wavetrain* (Fig. 9*b*) rather than as a wave (Fig. 9*a*). For emission times such as these, the wavetrains are a few meters long.

Interference effects from ordinary light sources may be produced by putting a narrow slit (S_0 in Fig. 5) directly in front of the source. This ensures that, by diffraction, the same families of wavetrains fall on slits S_1 and S_2. The diffracted beams emerging from S_1 and S_2 are thus coherent with respect to each other. When the phase of the light emitted from S_0 changes, this change is transmitted simultaneously to S_1 and S_2. Thus, at any point on screen *C*, a constant phase difference is maintained between the beams from these two slits and a stationary interference pattern occurs.

The lack of coherence of the light from common sources such as the sun or incandescent lamp filaments is due to the fact that the emitting atoms in such sources act independently rather than cooperatively. Since 1960 it has been possible to make light sources in which the atoms do act cooperatively, the emitted light being highly coherent. Such devices are the familiar *lasers,* a coined word derived from: *l*ight *a*mplification through the *s*timulated *e*mission of *r*adiation. Laser light (which we shall describe more fully in Chapter 45 of the extended version of this book) is not only highly coherent but also extremely monochromatic, highly directional, and focusable to a spot whose dimensions are of the order of magnitude of one wavelength. This "light fantastic," as it has been called, has a bewildering variety of applications, among which are telephone communication over optical fibers, spotwelding detached retinas, hydrogen fusion research, measuring continental drift, and holography, a technique in which three-dimensional images can be produced from information stored in a two-dimensional matrix.

40-6 Intensity in Double-Slit Interference

Equations 9 and 10 tell us how to locate the maxima and the minima of the double-slit interference fringes on screen *C* of Fig. 8 as a function of the angle θ in that figure. Note that θ is our position locator on the screen, every screen point *P* being associated with a definite value of θ. Here we wish to derive an expression for the intensity *I* of the fringes as a function of θ.

Let us assume that the electric field components of the light waves arriving at point *P* in Fig. 8 from the two slits vary with time as

$$E_1 = E_0 \sin \omega t \qquad (13)$$

and

$$E_2 = E_0 \sin(\omega t + \phi), \qquad (14)$$

where ω is the angular frequency of the waves and ϕ is the phase difference between them. We shall show below that these two waves will combine at *P* to produce an intensity of illumination *I* given by

$$\boxed{I = 4I_0 \cos^2 \tfrac{1}{2}\phi,} \qquad (15)$$

where

$$\boxed{\phi = \frac{2\pi d}{\lambda} \sin \theta.} \qquad (16)$$

In Eq. 15, I_0 is the intensity on the screen associated with light from one of the two slits, the other slit being temporarily covered. We assume that the slits are so narrow in comparison to the wavelength that this single-slit intensity is essentially uniform over the region of the screen in which we wish to examine the fringes.

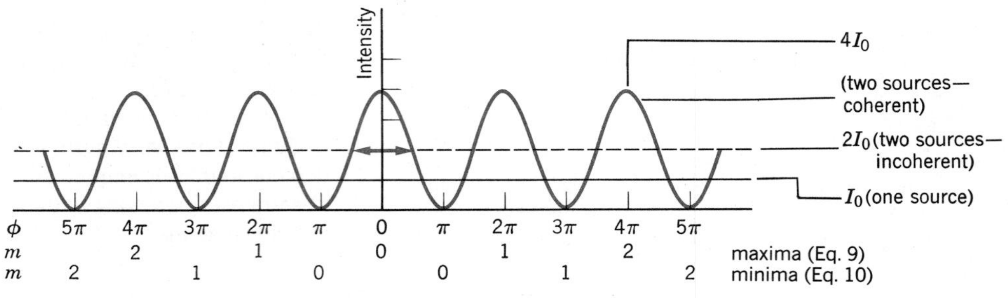

Figure 10 A plot of Eq. 15, showing the intensity of a double-slit interference pattern as a function of the phase difference between the waves from the two slits. I_0 is the (uniform) intensity that would appear on the screen if one slit were covered. The average intensity of the fringe pattern is $2I_0$ and the *maximum* intensity is $4I_0$.

Equations 15 and 16, which together tell us how the intensity I of the fringe pattern varies with the angle θ in Fig. 8, necessarily contain information about the location of the maxima and the minima. Let us see if we can extract it.

Study of Eq. 15 shows that intensity maxima will occur when

$$\tfrac{1}{2}\phi = m\pi, \qquad m = 0, 1, 2, \ldots . \tag{17}$$

If we put this result into Eq. 16 we find

$$2m\pi = \frac{2d\pi}{\lambda}\sin\theta$$

or

$$d\sin\theta = m\lambda, \qquad m = 0, 1, 2, \ldots \quad \text{(maxima)}, \tag{18}$$

which is exactly Eq. 9, the expression that we derived earlier for the location of the maxima.

The minima in the fringe pattern occur when

$$\tfrac{1}{2}\phi = (m + \tfrac{1}{2})\pi, \qquad m = 0, 1, 2, \ldots .$$

If we combine this relation with Eq. 16, we are led at once to

$$d\sin\theta = (m + \tfrac{1}{2})\lambda, \qquad m = 0, 1, 2, \ldots \quad \text{(minima)}, \tag{19}$$

which is just Eq. 10, the expression derived earlier for the location of the fringe minima.

Figure 10, which is a plot of Eq. 15, shows the intensity pattern for double-slit interference as a function of the phase angle ϕ. The horizontal solid line is I_0, the (uniform) intensity on the screen if one of the slits is covered up. Note from Eq. 15 that the intensity (which is always positive) varies from zero at the fringe minima to $4I_0$ at the fringe maxima.

If the two sources (slits) that illuminate the screen were *incoherent,* so that no enduring phase relation existed between them, there would be no fringe pattern and the intensity would have the uniform value $2I_0$ for all points on the screen; see the horizontal dashed line in Fig. 10.

Interference cannot create or destroy energy but merely redistributes it over the screen. Thus, the *average* intensity on the screen must be the same, whether the sources are coherent or not. This follows at once from Eq. 15; if we substitute one-half for the average value of the cosine-squared term, this equation reduces to $\bar{I} = 2I_0$. We have many times used the fact that the average value of the square of a sine or a cosine term over one or more half-cycles is one-half.

Proof of Eqs. 15 and 16. (Optional). We choose to combine E_1 and E_2, given by Eqs. 13 and 14, respectively, by the method of *phasors,* which we encountered in our studies of alternating currents in Chapter 36. The method will be especially useful later, when we want to combine a large number of wave disturbances with differing phases.

In Fig. 11*a*, we represent E_1 of Eq. 13 by a rotating phasor of amplitude E_0. In that figure, the alternating wave disturbance E_1 is simply the projection of the phasor E_0 on the vertical axis.

A second wave disturbance E_2, which has the same amplitude E_0 but a phase difference ϕ with respect to E_1 (see Eq. 14), can be represented (see Fig. 11*b*) as the projection on the vertical axis of a second phasor of magnitude E_0, which makes a fixed angle ϕ with the first phasor. As this figure shows, the sum of E_1 and E_2, which is the instantaneous amplitude of the resultant wave, is the sum of the projections of the two phasors on the vertical axis. This is revealed more clearly if we redraw the phasors, as in Fig. 11*c*, placing the foot of one arrow

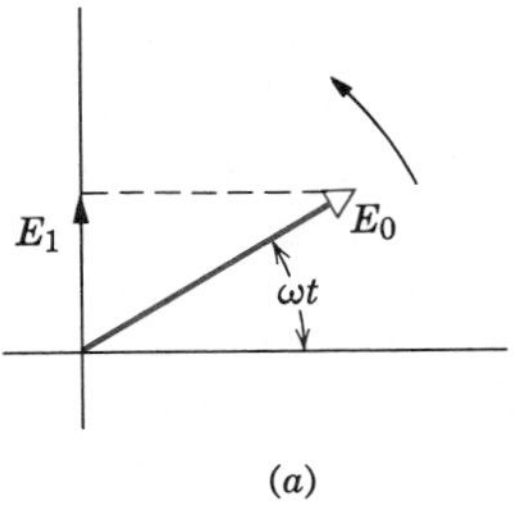

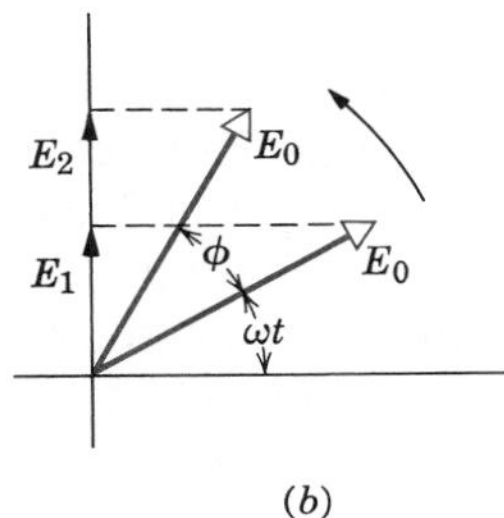

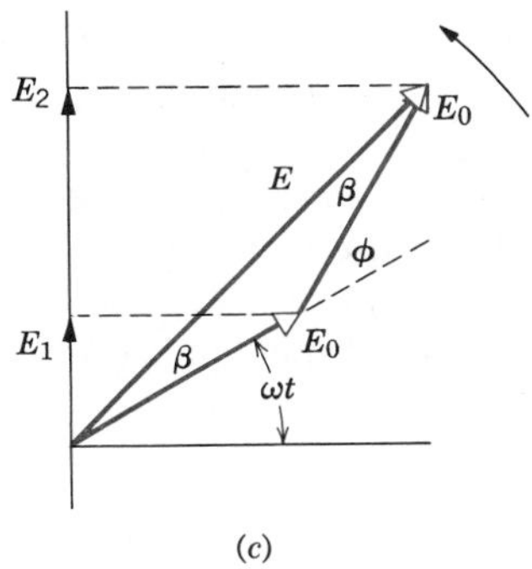

Figure 11 (*a*) A time-varying wave disturbance E_1 is represented as the projection of a rotating *phasor.* (*b*) Two phasors, with a constant phase difference ϕ between them. (*c*) Another way of drawing (*b*).

at the head of the other, maintaining the proper phase difference, and letting the whole assembly rotate counterclockwise about the origin.

In Fig. 11*c*, this sum can also be regarded as the projection on a vertical axis of a phasor of amplitude E, which makes a phase angle β with respect to the phasor that generates E_1. Note that the (algebraic) sum of the projections of the two phasors is equal to the projection of the (vector) sum of the two phasors.

In most problems in optics we are concerned only with the amplitude E of the resultant wave and not with its time variation. This is because the eye and most measuring instruments respond to the intensity of the light (that is, to the square of the amplitude) and cannot detect the rapid time variations. For light at the center of the visible spectrum ($\lambda \approx 550$ nm), the frequency $\nu (= c/\lambda)$ is about 5×10^{14} Hz. Usually, we need not consider the rotation of the phasors and can confine our attention to finding the amplitude of the resultant phasor.

Let us find the amplitude E in Fig. 11*c*. From the theorem that an exterior angle (ϕ) is equal to the sum of the two opposite interior angles ($\beta + \beta$) we see from Fig. 11*c* that $\beta = \frac{1}{2}\phi$. Thus, we have

$$E = 2(E_0 \cos \beta) = 2E_0 \cos \tfrac{1}{2}\phi. \qquad (20)$$

If we square each side of this relation we obtain

$$E^2 = 4E_0^2 \cos^2 \tfrac{1}{2}\phi. \qquad (21)$$

We learned in Section 38-5 that the intensity of a wave is proportional to the square of its amplitude. Thus, we can write for Eq. 21

$$I = 4I_0 \cos^2 \tfrac{1}{2}\phi,$$

which is Eq. 15, one of the equations that we set out to prove.

It remains to prove Eq. 16, which relates the phase difference ϕ between the waves arriving at any point P on the screen of Fig. 8 to the angle θ that serves as a locator of that point.

The phase difference ϕ in Eq. 14 is associated with a path difference S_1b in Fig. 8. If S_1b is $\frac{1}{2}\lambda$, ϕ will be π; if S_1b is λ, ϕ will be 2π, and so on. This suggests

$$\text{phase difference} = \frac{2\pi}{\lambda}\,(\text{path difference}). \qquad (22)$$

The path difference S_1b in Fig. 8 is just $d \sin \theta$ so that Eq. 22 becomes

$$\phi = \frac{2\pi d}{\lambda} \sin \theta,$$

which is just Eq. 16, the other equation that we set out to prove.

In a more general case we might want to find the resultant of a number ($n > 2$) of sinusoidally varying wave disturbances. The general procedure is as follows:

1. Construct a series of phasors representing the functions to be added. Draw them end to end, maintaining the proper phase relations between adjacent phasors.
2. Construct the vector sum of his array. Its length gives the amplitude of the resultant. The angle between it and the first phasor is the phase of the resultant with

respect to this first phasor. The projection of this phasor on the vertical axis gives the time variation of the resultant wave disturbance.

Sample Problem 2 Find graphically, using phasor methods, the resultant $E(t)$ of the following wave disturbances:

$$E_1 = E_0 \sin \omega t$$
$$E_2 = E_0 \sin(\omega t + 15°)$$
$$E_3 = E_0 \sin(\omega t + 30°)$$
$$E_4 = E_0 \sin(\omega t + 45°).$$

Figure 12 shows the assembly of four phasors that represent these functions. We find by graphical measurement with a ruler and a protractor that the amplitude E_R is 3.8 times as long as E_0 and that the resultant wave makes a phase angle ϕ_0 of 22.5° with respect to E_1. In other words,

$$E(t) = E_1 + E_2 + E_3 + E_4 = 3.8 \sin(\omega t + 22.5°). \quad \text{(Answer)}$$

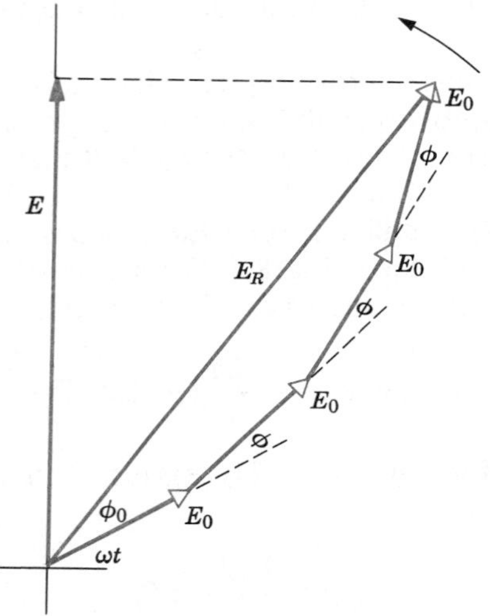

Figure 12 Sample Problem 2. Four wave disturbances added graphically, using the method of phasors. E is the amplitude of the resultant wave.

Check this result by direct trigonometric calculation or by geometric calculation from the phasor diagram of Fig. 12.

40-7 Interference from Thin Films

The colors that we see when sunlight falls on a soap bubble, an oil slick, or a ruby-throated hummingbird are caused by the interference of light waves reflected from the front and back surfaces of a thin transparent film. The film thickness is typically of the order of magnitude of the wavelength of the light involved. Thin-film technology, including the deposition of multiple-layered films, is highly developed and is widely used for the control of the reflection and/or transmission of light or radiant heat at optical surfaces.

Figure 13 shows a transparent film of uniform thickness d illuminated by monochromatic light of wavelength λ from a point source S. The eye is so positioned that a particular incident ray i from the source enters the eye, as ray r_1, having been reflected from the front surface of the film at a. The incident ray also enters the film at a as a refracted ray and is reflected from the back surface of the film at b; it then reemerges from the front surface of the film at c and also enters the eye, as ray r_2. The geometry of Fig. 13 is such that rays r_1 and r_2 are parallel. Having originated in the same point source, they are also coherent and are thus in a position to interfere. Because these two rays have traveled over paths of different lengths, have traversed different media, and—as we shall see—have suffered different kinds of reflections at a and at b, there will be a phase difference between them. The intensity perceived by the eye, as the parallel rays from the region a,c of the film enter it, is determined by this phase difference.

For near-normal incidence ($\theta_i \approx 0°$ in Fig. 13) the

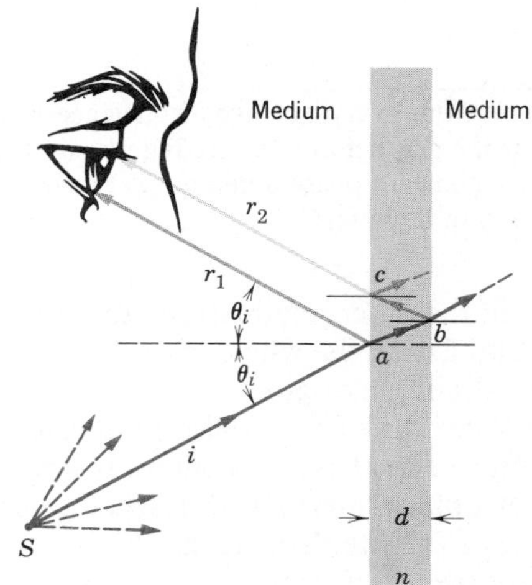

Figure 13 A thin film is viewed by light reflected from source S. Waves reflected from the front surface and the back surface enter the eye as shown, the intensity of the resultant wave being determined by the phase difference between the combining waves. The medium is assumed to be air.

geometric path difference for the two rays from S will be close to $2d$. We might expect the resultant wave reflected from the film to be an interference maximum if the distance $2d$ is an integral number of wavelengths. This statement must be modified for two reasons, which we discuss in turn.

First Reason. In counting the number of wavelengths in the path difference $2d$, we must be sure to use the wavelength in the medium λ_n and not the wavelength in air λ. These two quantities are related by (see Eq. 8)

$$\lambda_n = \frac{\lambda}{n} \tag{23}$$

in which n is the index of refraction of the film material with respect to air.

Second Reason. To bring out the second point, let us assume that the film is so thin that $2d$ is very much less than one wavelength. We might assume the phase difference between the two waves would be close to zero and we would expect such a film to appear bright on reflection. However, it appears dark. This is clear from Fig. 14, which shows a thin vertical film of soapy water, viewed by reflected monochromatic light. The action of gravity produces a wedge-shaped film, extremely thin at its top edge. To explain this and many similar phenomena, we assume that one or the other of the two rays of Fig. 13 suffers an abrupt phase change of 180° when it is reflected from the air–film interface.

Figure 14 A soapy water film on a wire loop, viewed by reflected light. The black segment at the top is not a torn film. It arises because the film, by drainage, is so thin there that there is destructive interference between waves reflected from its two surfaces.

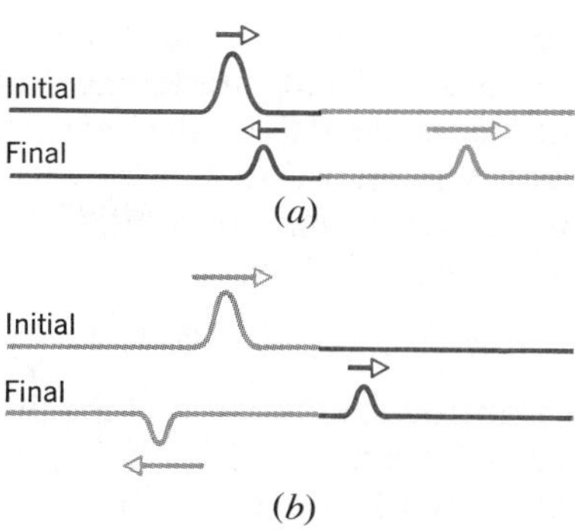

Figure 15 Phase changes on reflection at a junction between two stretched composite strings of different linear densities. The wave speed is greater in the lighter string. (*a*) The incident pulse is in the heavier string. (*b*) The incident pulse is in the lighter string.

In Section 17–12 we discussed phase changes on reflection for transverse waves in strings. To extend these ideas, consider the composite string of Fig. 15, which consists of two parts with different linear densities, stretched to the same tension. Recall from Eq. 26 of Chapter 17 that the wave speed is greater in the lighter string.

In Fig. 15*a* the incident pulse is in the heavier string and in Fig. 15*b* it is in the lighter string. In each case, the transmitted pulse suffers no change in sign as it passes through the boundary.

The behavior of the reflected pulses is different, however. When the incident pulse approaches a junction beyond which its speed will be greater (Fig. 15*a*), the reflected pulse suffers no phase change. However, when the incident pulse approaches a junction beyond which its speed is less (Fig. 15*b*), the reflected pulse is reversed in sign. That is, it undergoes a phase change of 180°.

Figure 15*a* suggests a light wave in glass approaching a surface beyond which there is a medium of lower index of refraction (which implies a larger wave speed) such as air. Figure 15*b* suggests a light wave in air approaching glass. The optical situation can be summed up as follows:

When reflection occurs from an interface beyond which the medium has a lower index of refraction, the reflected wave undergoes no phase change; when the medium beyond the interface has a higher index, there is a phase change of 180°. The transmitted wave undergoes no phase change in either case.

We are now able to take into account both factors that determine the nature of the interference, namely, the dependence of the wavelength on the index of refraction and phase changes on reflection. For the two rays of Fig. 13 to combine to give a maximum intensity, assuming normal incidence, we must have

$$2d = (m + \tfrac{1}{2})\lambda_n, \qquad m = 0, 1, 2, \ldots .$$

The term $\frac{1}{2}\lambda_n$ is introduced because of the phase change on reflection, a phase change of 180° being equivalent to half a wavelength. Substituting λ/n for λ_n yields finally

$$\boxed{2nd = (m + \tfrac{1}{2})\lambda,} \qquad m = 0, 1, 2 \ldots \quad \text{(maxima).} \quad (24)$$

The condition for a minimum intensity is

$$\boxed{2nd = m\lambda,} \qquad m = 0, 1, 2, \ldots \quad \text{(minima).} \quad (25)$$

These equations hold when the index of refraction of the film is either greater or less than the indices of the media on each side of the film. In these cases, there will be a relative phase change of 180° for reflections at the two surfaces since there will be a phase reversal at one reflection but not at the other. A water film in air and an air film in the space between two glass plates provide examples of cases to which Eqs. 24 and 25 apply.

If the film thickness is not uniform, as in Fig. 14, where the film is wedge shaped because of the action of gravity, constructive interference will occur in certain parts of the film and destructive interference will occur in others. Lines of maximum and of minimum intensity will appear, called *fringes of constant thickness.* Each fringe is the locus of points for which the film thickness d is constant. If the film is illuminated with white light rather than monochromatic light, the light reflected from various parts of the film will be modified by the various constructive or destructive interferences that occur. This accounts for the brilliant colors of soap bubbles and oil slicks.

Sample Problem 3 A water film ($n = 1.33$) in air is 320 nm thick. If it is illuminated with white light at normal incidence, what color will it appear to be in reflected light?

If we solve Eqs. 24 and 25 for λ, we find

$$\lambda = \frac{2nd}{m + \frac{1}{2}} = \frac{(2)(1.33)(320\text{ nm})}{m + \frac{1}{2}}$$

$$= \frac{851\text{ nm}}{m + \frac{1}{2}} \quad \text{(maxima)}$$

and

$$\lambda = \frac{851\text{ nm}}{m} \quad \text{(minima).}$$

Maxima and minima occur for the wavelengths shown in the following table, in which IR and UV stand for infrared and ultraviolet, respectively:

m	*0* (max)	*1* (min)	*1* (max)	*2* (min)	*2* (max)
λ (nm)	1702	851	567	426	340
Color	IR	IR	Yellow-green	Deep purple	UV

Only the maximum corresponding to $m = 1$ lies in the visible region (see Fig. 2 of Chapter 38): Light of this wavelength appears yellow-green. If white light is used to illuminate the film, the yellow-green component will be enhanced when viewed by reflection.

Sample Problem 4 Lenses are often coated with thin films of transparent substances like magnesium fluoride (MgF_2; $n = 1.38$) to reduce the reflection from the glass surface. How thick a coating is needed to produce a minimum reflection at the center of the visible spectrum ($\lambda = 550$ nm)?

We assume that the light strikes the lens at near-normal incidence, θ being exaggerated for clarity in Fig. 16. We seek the film thickness that will bring about destructive interference between rays r_1 and r_2.

Equation 25 does *not* apply in this case because a phase change of 180° is now associated with *each* ray and not with

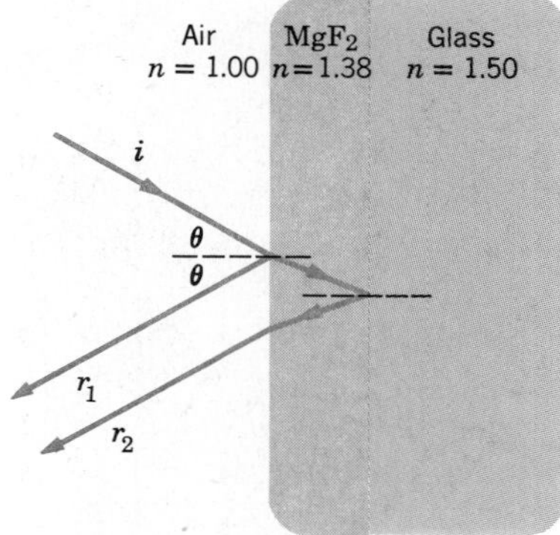

Figure 16 Sample Problem 4. Unwanted reflections from glass can be suppressed (at a chosen wavelength) by coating the glass with a thin transparent film of magnesium fluoride of properly chosen thickness.

only one ray as in Fig. 13. At both the front and rear surfaces of the MgF_2 film in Fig. 16 the reflection is from a medium of greater index of refraction so that there is no *net* change in phase produced by the two reflections.

This means that, to find the thickness of the thinnest possible film, the path difference between the two rays (which is $2d$) must equal one-half of a wavelength of light in the medium ($= \frac{1}{2}\lambda_n = \lambda/2n$). Thus,

$$2d = \frac{\lambda}{2n}$$

or

$$d = \frac{\lambda}{4n} = \frac{550 \text{ nm}}{(4)(1.38)} = 100 \text{ nm}. \qquad \text{(Answer)}$$

Note that this answer is independent of the index of refraction of the glass (or of the air, for that matter) as long as $n_{glass} > 1.38$.

40-8 Michelson's Interferometer

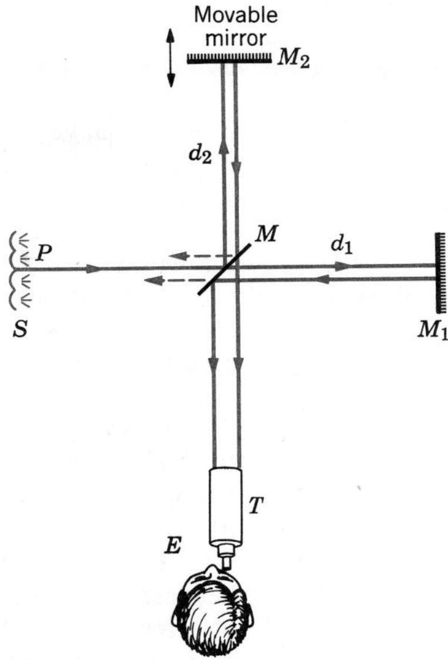

Figure 17 Michelson's interterometer, showing the path of a ray originating at point P of an extended source S. The ray splits, the two subrays following different paths and then recombining. The fringe pattern can be changed by changing the length of one path by moving mirror M_2.

An *interferometer* is a device that can be used to measure lengths or changes in length with great accuracy by means of interference fringes. We describe the form originally devised and built by A. A. Michelson in 1881.* Consider light that leaves point P on extended source S (Fig. 17) and falls on a *beam splitter* M. This is a mirror with the property that it transmits half the incident light, reflecting the rest; in the figure we have assumed, for convenience, that this mirror possesses negligible thickness. At M the light divides into two waves. One proceeds by transmision toward mirror M_1; the other proceeds by reflection toward M_2. The waves are reflected at each of these mirrors and are sent back along their directions of incidence, each wave eventually entering the eye. Since the waves are coherent, being derived from the same point on the source, they will interfere.

The path difference for the two waves when they recombine is $2d_2 - 2d_1$ and anything that changes this path difference will cause a change in the relative phase of the two waves as they enter the eye. As an example, if mirror M_2 is moved by a distance $\frac{1}{2}\lambda$, the path difference changes by λ and the observer will see the fringe pattern shift by one fringe. If the mirrors M_1 and M_2 are exactly perpendicular to each other, the effect is that of light from an extended source S falling on a uniformly thick slab of air, between glass, whose thickness is equal to $d_2 - d_1$.

By such techniques the lengths of objects can be expressed in terms of the wavelength of light. In Michelson's day, the standard of length—the meter—was chosen by international agreement to be the distance between two fine scratches on a certain metal bar preserved at Sèvres, near Paris. Michelson was able to show, using his interferometer, that the standard meter was equivalent to 1,553,163.5 wavelengths of a certain monochromatic red light emitted from a light source containing cadmium. For this careful measurement, Michelson received the 1907 Nobel prize in physics. His work laid the foundation for the eventual abandonment (in 1961) of the meter bar as a standard of length and for the redefinition of the meter in terms of the wavelength of light. In 1983, as we have seen, even this wavelength standard was not precise enough to meet the growing requirements of science and technology and was replaced by a new standard based on a defined value for the speed of light.

* See "Michelson: America's First Nobel Prize Winner in Science," by R. S. Shankland, *The Physics Teacher,* January 1977.

REVIEW AND SUMMARY

Huygens' Principle

The three-dimensional transmission of waves, including light, may often be predicted by *Huygens' principle,* which states: All points on a wave front can serve as point sources of spherical secondary wavelets; see Fig. 1.

The *law of refraction* can be derived from Huygens' principle by presuming that the index of refraction of any medium is $n = c/v$, v being the speed of light in the medium; see Fig. 2.

Geometrical Optics and Diffraction

Attempts to isolate a ray by forcing light through a narrow slit fail because of *diffraction,* the flaring out of the light into the geometrical shadow of the slit; see Figs. 3 and 4. If such slits are present, the approximations of geometrical optics (Chapter 39) fail and the full treatment of wave optics must be used.

Interference

In *Young's double-slit interference experiment* light from a slit in screen A of Fig. 5 flares out (by diffraction) and falls on the two slits in screen B. The light from these slits also flares out in the region beyond B and interference occurs between the two overlapping wavelets. A fringe pattern like that of Fig. 6 is formed on screen C. Figure 7 shows the same effect in water waves.

The light intensity at any point on screen C of Fig. 5 depends in part on the path difference between two rays drawn to that point from the two slits. If this difference is an integral number of wavelengths, the waves will interfere constructively and an intensity maximum results. If it is a half-integral multiple, there is destructive interference and an intensity minimum. Analysis shows that the conditions for maximum and minimum intensity are

Intensity in Two-slit Interference

$$d \sin \theta = m\lambda \qquad m = 0, 1, 2 \ldots \quad \text{(maxima)} \tag{9}$$

$$d \sin \theta = (m + \tfrac{1}{2})\lambda \quad m = 0, 1, 2 \ldots \quad \text{(minima).} \tag{10}$$

See Sample Problem 1.

Coherence and Incoherence

If two overlapping light waves are to interfere at a given point, the phase difference between them must remain constant with time; the waves must be *coherent.* Very long wavetrains must be involved; see Fig. 9. When two coherent waves overlap, the resulting intensity may be found by the phasor method, as in Fig. 11. In this method the amplitude E_θ of the electric field vector of the resultant wave is calculated, taking the phase difference between the two combining waves properly into account. The intensity I_θ of the resultant wave is then taken to be proportional to E_θ^2. As applied to Young's double-slit experiment we have, for the interference of two beams with intensity I_0:

Intensity in Young's Experiment

$$I_\theta = 4I_0 \cos^2(\tfrac{1}{2}\phi), \quad \text{where} \quad \phi = \left(\frac{2\pi d}{\lambda}\right) \sin \theta. \tag{15,16}$$

Equations 9 and 10, which identify the positions of the fringe maxima and minima, are contained within this relation. Figure 11 should be studied closely. Sample Problem 2 shows how to apply the phasor strategy to analyze four-slit interference.

When light falls on a *thin transparent film,* the light waves reflected from the front and the rear surfaces, as in Fig. 13, interfere. For near-normal incidence the conditions for a maximum or a minimum intensity of the light reflected from the film are

Thin-film Interference

$$2nd = (m + \tfrac{1}{2})\lambda \quad m = 0, 1, 2 \ldots \quad \text{(maxima)} \tag{24}$$

$$2nd = m\lambda \qquad m = 0, 1, 2 \ldots \quad \text{(minima).} \tag{25}$$

Phase Change on Reflection

Here the medium in which the film is immersed is assumed to be the same on each side, n is the index of the film material with respect to this medium, and λ is the wavelength in vacuum. The formula also assumes that one of the interfering waves has undergone a "hard" reflection (that is, from a higher index medium; compare Fig. 15*b*) and the other a "soft" reflection (that is, from a lower index medium; compare Fig. 15*a*). Hard reflections involve a phase change of 180°; soft reflections do not. Sample Problems 3 and 4 discuss cases in which some assumptions of the above formula do not hold and the problem must be solved from first principles.

The Michelson Interferometer

In *Michelson's interferometer* (Fig. 17) a light beam is split into two sub-beams which, after traversing paths of different lengths, are recombined so that they interfere and form a fringe pattern. By varying the path length of one of the sub-beams distances can be accurately expressed in terms of wavelengths of light. Thus, in Fig. 17, as mirror M_2 is moved through the distance in question, one simply counts the number of fringes (wavelengths) that pass through the field of view of the observer.

QUESTIONS

1. Light has (*a*) a wavelength, (*b*) a frequency, and (*c*) a speed. Which, if any, of these quantities remains unchanged when light passes from a vacuum into a slab of glass?

2. Why is it plausible that the wavelength of light should change when the light passes from air into glass but that its frequency should not?

3. The speed and wavelength of, say, red light that we see in air is reduced when the light passes into water. Would that light then appear to be another color—blue, perhaps—if you viewed it from under the water surface?

4. Would you expect sound waves to obey the laws of reflection and of refraction obeyed by light waves? Does Huygens' principle apply to sound waves in air? If Huygens' principle predicts the laws of reflection and refraction, why is it necessary or desirable to view light as an electromagnetic wave, with all its attendant complexity?

5. In Young's double-slit interference experiment, using a monochromatic laboratory light source, why is screen A in Fig. 5 necessary?

6. What changes occur in the pattern of interference fringes if the apparatus of Fig. 8 is placed under water?

7. Do interference effects occur for sound waves? Recall that sound is a longitudinal wave and that light is a transverse wave.

8. If interference between light waves of different frequencies is possible, one should observe light beats, just as one obtains sound beats from two sources of sound with slightly different frequencies. Discuss how one might experimentally look for this possibility.

9. Why are parallel slits preferable to the pinholes that Young used in demonstrating interference?

10. Is coherence important in reflection and refraction?

11. If your source of light is a laser beam, you do not need the equivalent of screen A in Fig. 5. Why?

12. Describe the pattern of light intensity on screen C in Fig. 8 if one slit is covered with a red filter and the other with a blue filter, the incident light being white.

13. If one slit in Fig. 8 is covered, what change would occur in the intensity of light in the center of the screen?

14. We are all bathed continuously in electromagnetic radiation, from the sun, from radio and TV signals, from the stars and other celestial objects. Why don't these waves interfere with each other?

15. Is polarization or interference a better test for identifying waves? Do they give the same information?

16. What causes the fluttering of a TV picture when an airplane flys overhead?

17. Is it possible to have coherence between light sources emitting light of different wavelengths?

18. Each slit in Fig. 8 is covered with a sheet of Polaroid, the polarizing directions of the two slits being at right angles. What is the pattern of light intensity on screen C? (The incident light is unpolarized.)

19. Suppose that the film coating in Fig. 16 had a refractive index greater than that of the glass. Could it still be nonreflecting? If so, what difference would it make?

20. What are the requirements for maximum intensity when viewing a thin film by *transmitted* light?

21. Why does a film (soap bubble, oil slick, etc.) have to be "thin" to display interference effects? Or does it? How thin is "thin"?

22. Why do coated lenses (see Sample Problem 4) look purple by reflected light?

23. Ordinary store windows or home windows reflect light from both their interior and exterior plane surfaces. Why then don't we see interference effects?

24. A person wets his eyeglasses to clean them. As the water evaporates he notices that for a short time the glasses become markedly less reflecting. Explain why.

25. A lens is coated to reduce reflection, as in Sample Problem 4. What happens to the energy that had previously been reflected? Is it absorbed by the coating?

26. Consider the following objects that produce colors when exposed to sunlight: (1) soap bubbles, (2) rose petals, (3) the inner surface of an oyster shell, (4) thin oil slicks, (5) nonreflecting coatings on camera lenses, and (6) peacock tail feathers. The colors displayed by all but one of these are purely interference phenomena, no pigments being involved. Which one is the exception? Why do the others seem to be "colored"?

27. An automobile directs its headlights onto the side of a barn. Why are interference fringes not produced in the region in which light from the two beams overlaps?

28. A soap film on a wire loop held in air appears black at its thinnest portion when viewed by reflected light. On the other hand, a thin oil film floating on water appears bright at its thinnest portion when similarly viewed from the air above. Explain these phenomena.

29. If the pathlength to the movable mirror in Michelson's interferometer (see Fig. 17) greatly exceeds that to the fixed mirror (say by more than a meter) the fringes begin to disappear. Explain why. Lasers greatly extend this range. Why?

30. How would you construct an acoustical Michelson interferometer to measure sound wavelengths? Discuss differences from the optical interferometer.

EXERCISES AND PROBLEMS

Section 40-2 Light as a Wave

1E. The wavelength of yellow sodium light in air is 589 nm. (*a*) What is its frequency? (*b*) What is its wavelength in glass whose index of refraction is 1.52? (*c*) From the results of (*a*) and (*b*) find its speed in this glass.

2E. How much faster, in meters per second, does light travel in sapphire than in diamond? See Table 1, Chapter 39.

3E. Derive the law of reflection using Huygens' principle.

4P. Light of free-space wavelength 600 nm travels 1.6 μm in a medium of index of refraction 1.5. Find (*a*) the wavelength in the medium, and (*b*) the phase difference after moving that distance, with respect to light traveling the same distance in free space.

5P. One end of a stick is dragged through water at a speed v, which is greater than the speed u of water waves. Applying Huygens' construction to the water waves, show that a conical wavefront is set up and that its half-angle α is given by

$$\sin \alpha = u/v.$$

This is familiar as the bow wave of a ship and the shock wave caused by an object moving through air with a speed exceeding that of sound, as in Fig. 17, Chapter 18.

Section 40-4 Young's Experiment

6E. Monochromatic green light, wavelength = 550 nm, illuminates two parallel narrow slits 7.7 μm apart. Calculate the angular deviation of the third order, $m = 3$, bright fringe (*a*) in radians, and (*b*) in degrees.

7E. What is the phase difference between the waves from the two slits arriving at the mth dark fringe in a Young's double-slit experiment?

8E. In an experiment using Young's arrangement to demonstrate the interference of light, the separation d of the two narrow slits is doubled. In order to maintain the same spacing of the fringes, how must the distance D of the screen from the slits be altered? (The wavelength of the light remains unchanged.)

9E. Young's experiment is performed with blue-green light of wavelength 500 nm. The slits are 1.2 mm apart and the screen is 5.4 m from the slits. How far apart are the bright fringes?

10E. Find the slit separation of a double-slit arrangement that will produce interference fringes 0.018 rad apart on a distant screen. Assume sodium light ($\lambda = 589$ nm).

11E. A double-slit arrangement produces interference fringes for sodium light ($\lambda = 589$ nm) that are 0.0035 rad apart. For what wavelength would the angular separation be 10% greater?

12E. In a double-slit arrangement the slits are separated by a distance equal to 100 times the wavelength of the light passing through the slits. (*a*) What is the angular separation in radians between the central maximum and an adjacent maximum? (*b*) What is the linear distance between these maxima if the screen is at a distance of 50 cm from the slits?

13E. In a double-slit experiment, $\lambda = 546$ nm, $d =$ 0.10 mm, and $D = 20$ cm. What is the linear distance between the fifth maximum and seventh minimum from the central maximum?

14E. A double-slit arrangement produces interference fringes for sodium light ($\lambda = 589$ nm) that are 0.20° apart. What is the angular fringe separation if the entire arrangement is immersed in water ($n = 1.33$)?

15P. In a double-slit experiment the distance between slits is 5.0 mm and the slits are 1.0 m from the screen. Two interference patterns can be seen on the screen, one due to light with wavelength 480 nm and the other due to light with wavelength 600 nm. What is the separation on the screen between the third-order interference fringes of the two different patterns?

16P. In Young's interference experiment in a large ripple tank (see Fig. 7) the coherent vibrating sources are placed 120 mm apart. The distance between maxima 2.00 m away is 180 mm. If the speed of ripples is 25 cm/s, calculate the frequency of the vibrators.

17P. If the distance between the first and tenth minima of a double-slit pattern is 18 mm and the slits are separated by 0.15 mm with the screen 50 cm from the slits, what is the wavelength of the light used?

18P. As shown in Fig. 18, A and B are two identical radiators of waves that are in phase and of the same wavelength λ. The radiators are separated by distance 3.0λ. Find the largest distance from A, along the line Ax, for which destructive interference occurs. Express this in terms of λ.

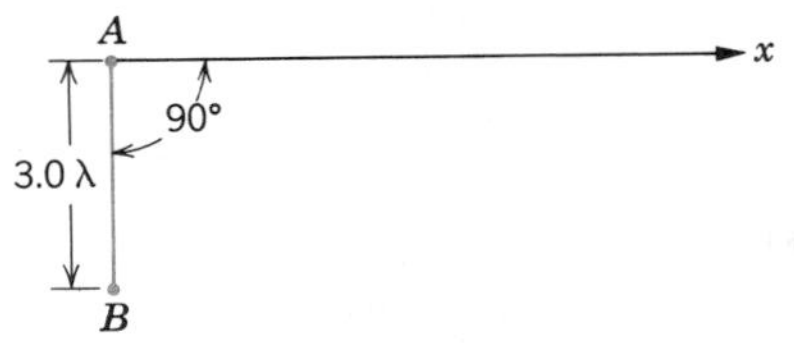

Figure 18 Problem 18.

19P. A thin flake of mica ($n = 1.58$) is used to cover one slit of a double-slit arrangement. The central point on the screen is occupied by what used to be the seventh bright fringe. If $\lambda = 550$ nm, what is the thickness of the mica?

20P. Sketch the interference pattern expected from using two pinholes, rather than narrow slits.

21P. In Fig. 19, the source emits monochromatic light of wavelength λ. S is a narrow slit in an otherwise opaque screen I. A plane mirror, whose surface includes the axis of the lens shown, is located a distance h below S. Screen II is at the focal plane of the lens. (*a*) Find the condition for maximum and minimum brightness of fringes on screen II in terms of the usual angle θ, the wavelength, and the distance h. (*b*) Show that fringes appear only in region A (above the axis of the lens), but not in region B (below the axis of the lens) or in both regions A and B. (*Hint:* Consider the image of S formed by the mirror.) This arrangement for obtaining a double-slit interference pattern from a single slit is called *Lloyd's mirror.*

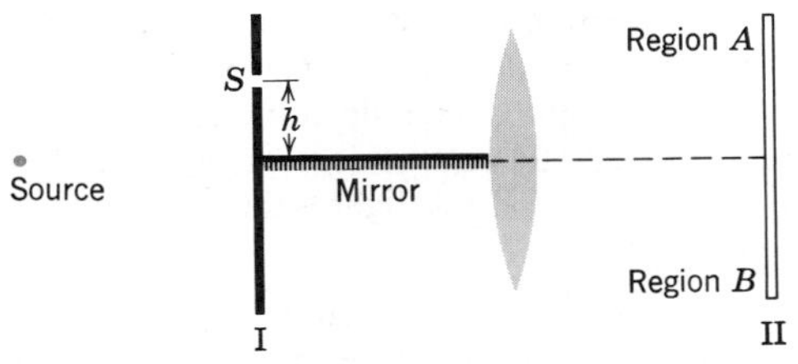

Figure 19 Problem 21.

22P. Two coherent radio point sources separated by 2.0 m are radiating in phase with $\lambda = 0.50$ m. A detector moved in a circular path around the two sources in a plane containing them will show how many maxima?

23P. Interference fringes are produced using white light with a double-slit arrangement. A piece of parallel-sided mica of refractive index 1.6 is placed in front of one of the slits, as a result of which the center of the fringe system moves to the left a distance subsequently shown to accommodate 30 dark bands when light of wavelength 480 nm is used. What is the thickness of the mica?

24P. In the front of a lecture hall, a coherent beam of monochromatic light from a helium–neon laser ($\lambda = 632.8$ nm) illuminates a double slit. From there it travels a distance $d = 20$ m to a mirror at the back of the hall, and returns the same distance to a screen. (*a*) In order that the distance between interference maxima be 10 cm what should be the distance between the two slits? (*b*) State what you will see if the lecturer slips a thin sheet of cellophane over one of the slits. The path through the cellophane contains 2.5 more waves than a path through air of the same geometric thickness.

25P. One slit of a double-slit arrangement is covered by a thin glass plate of refractive index 1.4, and the other by a thin glass plate of refractive index 1.7. The point on the screen where the central maximum fell before the glass plates were inserted is now occupied by what had been the $m = 5$ bright fringe before. Assume $\lambda = 480$ nm and that the plates have the same thickness t and find the value of t.

26P. Sodium light ($\lambda = 589$ nm) falls on a double slit of separation $d = 2.0$ mm. The slit-screen distance $D = 40$ mm. What percent error is made by using Eq. 9 to locate the tenth bright fringe on the screen?

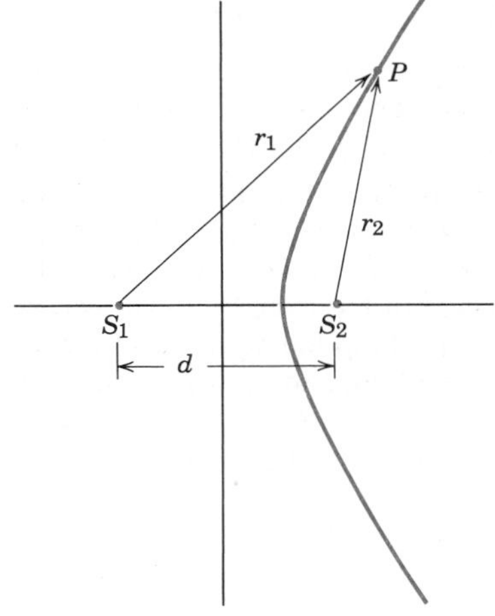

Figure 20 Problem 27.

27P. Two point sources, S_1 and S_2 in Fig. 20, emit coherent waves. Show that curves, such as that given, over which the phase difference for rays r_1 and r_2 is a constant, are hyperbolas. (*Hint:* A constant phase difference implies a constant difference in length between r_1 and r_2.) The OMEGA system of sea navigation relies on this principle. S_1 and S_2 are phase-locked transmitters. The ship's navigator notes the received phase difference on an oscilloscope and locates the ship on a hyperbola. Reception of signals from a third transmitter is needed to determine the position on that hyperbola.

Section 40-5 Coherence

28E. The *coherence length* of a wavetrain is the distance over which the phase constant is the same. (*a*) If an individual atom emits coherent light for 1×10^{-8} s, what is the coherence length of the wavetrain? (*b*) Suppose this wavetrain is separated into two parts with a partially reflecting mirror and later reunited after one beam travels 5 m and the other 10 m. Do the waves produce interference fringes observable by a human eye?

Section 40-6 Intensity in Double-slit Interference

29E. Add the following quantities graphically, using the phasor method (see Sample Problem 2):

$$y_1 = 10 \sin \omega t,$$
$$y_2 = 15 \sin(\omega t + 30°),$$
$$y_3 = 5 \sin(\omega t - 45°).$$

30E. Source A of long-range radio waves leads source B by 90°. The distance r_A to a detector is greater than the distance r_B by 100 m. What is the phase difference at the detector? Both sources have a wavelength of 400 m.

31E. Light of wavelength 600 nm is incident normally on two parallel narrow slits separated by 0.60 mm. Sketch the intensity pattern observed on a distant screen as a function of angle θ for the range of values $0 \le \theta \le 0.0040$ radians.

32P. Two waves of the same frequency have amplitudes 1 and 2, respectively. They interfere at a point where their phase difference is 60°. What is the resultant amplitude?

33P. S_1 and S_2 in Fig. 21 are effective point sources of radiation, excited by the same oscillator. They are coherent and in phase with each other. Placed 4.0 m apart, they emit equal amounts of power in the form of 1.0-m wavelength electromagnetic waves. (*a*) Find the positions of the first (that is, the nearest), the second, and the third maxima of the received signal, as the detector is moved out along Ox. (*b*) Is the intensity at the nearest minimum equal to zero? Justify your answer.

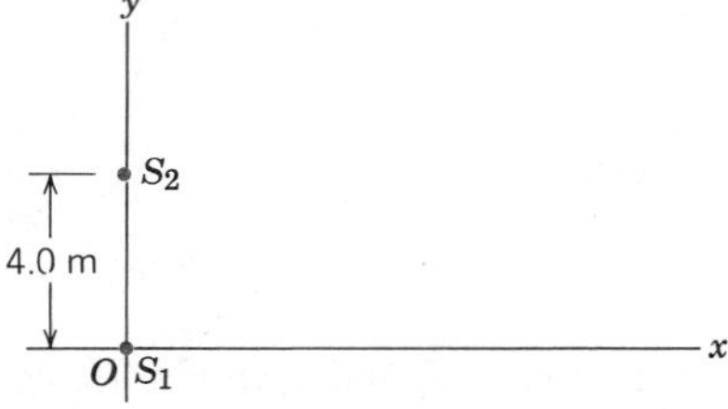

Figure 21 Problem 33.

34P. Find the sum of the following quantities (*a*) graphically, using phasors, and (*b*) using trigonometry:

$$y_1 = 10 \sin \omega t.$$
$$y_2 = 8 \sin(\omega t + 30°).$$

35P. Show that the half-width $\Delta\theta$ of the double-slit interference fringes (see horizontal arrow in Fig. 10) is given by

$$\Delta\theta = \frac{\lambda}{2d}$$

if θ is small enough so that $\sin \theta \approx \theta$.

36P*. One of the slits of a double-slit system is wider than the other, so that the amplitude of the light reaching the central part of the screen from one slit, acting alone, is twice that from the other slit, acting alone. Derive an expression for the intensity I in terms of θ, corresponding to Eqs. 15 and 16.

Section 40-7 Interference from Thin Films

37E. A thin coating of refractive index 1.25 is applied to a glass camera lens to minimize the intensity of the light reflected from the lens. In terms of λ, the wavelength in air of the incident light, what is the smallest thickness of the coating that is needed?

38E. We wish to coat a flat slab of glass ($n = 1.50$) with a transparent material ($n = 1.25$) so that light of wavelength 600 nm (in vacuum) incident normally is not reflected. What minimum thickness could the coating have?

39E. A thin film in air is 0.41 μm thick and is illuminated by white light normal to its surface. Its index of refraction is 1.5. What wavelengths within the visible spectrum will be intensified in the reflected beam?

40E. A disabled tanker leaks kerosene ($n = 1.2$) into the Persian Gulf, creating a large slick on top of the water ($n = 1.3$). (*a*) If you are looking straight down from an airplane onto a region of the slick where its thickness is 460 nm, for which wavelength(s) of visible light is the reflection the greatest? (*b*) If you are scuba-diving directly under this same region of the slick, for which wavelength(*s*) of visible light is the transmitted intensity the strongest?

41E. Light of wavelength 585 nm is incident normally on a thin soapy film ($n = 1.33$) suspended in air. If the film is 0.00121 mm thick, determine whether it appears bright or dark when observed from a point near the light source.

42E. A lens is coated with a thin transparent film to minimize reflection of the red component of white light. The index of refraction of the film is 1.30 and that of the lens is 1.65. What minimum thickness of film is needed? The wavelength of the light in air is 680 nm.

43E. In costume jewelry, rhinestones (made of glass with $n = 1.50$) are often coated with silicon monoxide ($n = 2.0$) to make them more reflective. How thick should the coating be to achieve strong reflection for 560-nm light, incident normally?

44E. A soap film ($n = 1.33$) is illuminated by light of wavelength 624 nm in air, incident normally. What is (*a*) the smallest thickness and (*b*) the second smallest thickness that results in no reflection?

45P. A plane wave of monochromatic light falls normally on a uniformly thin film of oil that covers a glass plate. The wavelength of the source can be varied continuously. Complete destructive interference of the reflected light is observed for wavelengths of 500 and 700 nm and for no wavelengths between them. If the index of refraction of the oil is 1.3 and that of the glass is 1.5, find the thickness of the oil film.

46P. White light reflected at perpendicular incidence from a soap film has, in the visible spectrum, an interference maximum at 600 nm and a minimum at 450 nm with no minimum in between. If $n = 1.33$ for the film, what is the film thickness, assumed uniform?

47P. A sheet of glass having an index of refraction of 1.40 is to be coated with a film of material having a refractive index of 1.55 such that green light (wavelength = 525 nm) is preferentially transmitted. (*a*) What is the minimum thickness of the film that will achieve the result? (*b*) Why are other parts of the visible spectrum not also preferentially transmitted? (*c*) Will the transmission of any colors be sharply reduced?

48P. A plane monochromatic light wave in air falls at normal incidence on a thin film of oil that covers a glass plate. The wavelength of the source may be varied continuously. Complete destructive interference in the reflected beam is observed for wavelengths of 500 and 700 nm and for no wavelength in between. The index of refraction of glass is 1.5. Show that the index of refraction of the oil must be less than 1.5.

49P. A thin film of acetone (index of refraction = 1.25) is coating a thick glass plate (index of refraction = 1.50). Plane light waves of variable wavelengths are incident normal to the film. When one views the reflected wave, it is noted that complete destructive interference occurs at 600 nm and constructive interference at 700 nm. Calculate the thickness of the acetone film.

50P. An oil drop ($n = 1.20$) floats on a water ($n = 1.33$) surface and is observed from above by reflected light (see Fig. 22). (*a*) Will the outer (thinnest) regions of the drop correspond to a bright or a dark region? (*b*) Approximately how thick is the oil film where one observes the third blue region from the outside of the drop? (*c*) Why do the colors gradually disappear as the oil thickness becomes larger?

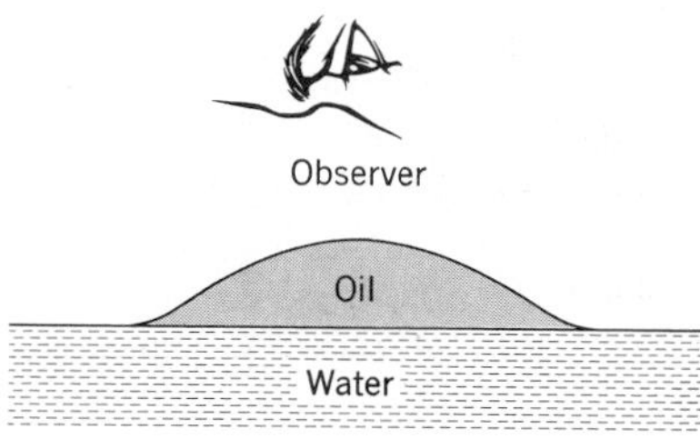

Figure 22 Problem 50.

51P. From a medium of index of refraction n_1, monochromatic light of wavelength λ falls normally on a thin film of uniform thickness and index of refraction n_2. The transmitted light travels in a medium with index of refraction n_3. Find expressions for the minimum film thickness (in terms of λ and the indices of refraction) for the following cases: (*a*) $n_1 < n_2 > n_3$—minimum reflected light; (*b*) $n_1 < n_2 > n_3$—maximum transmitted light; (*c*) $n_1 < n_2 < n_3$—minimum reflected light; (*d*) $n_1 < n_2 < n_3$—maximum transmitted light; and (*e*) $n_1 < n_2 < n_3$—maximum reflected light.

52P. In Sample Problem 4 assume that there is zero reflection for light of wavelength 550 nm at normal incidence. Calculate the factor by which the reflection is diminished by the coating at 450 and at 650 nm.

53P. A broad source of light ($\lambda = 680$ nm) illuminates normally two glass plates 120 mm long that touch at one end and are separated by a wire 0.048 mm in diameter at the other end (Fig. 23). How many bright fringes appear over the 120 mm distance?

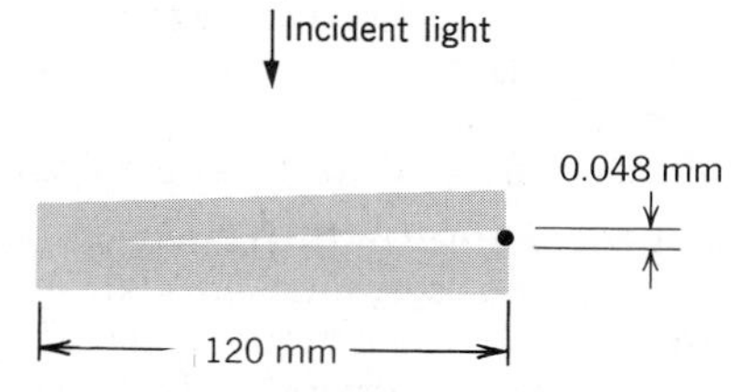

Figure 23 Problems 53 and 54.

54P. In Fig. 23 white light is incident from above. (*a*) Observed from above, why does the region near the edge, where the two glass plates touch, appear black? (*b*) For what part of the visible spectrum does destructive interference next occur? (*c*) What color does an observer see where this destructive interference occurs?

55P. A perfectly flat piece of glass ($n = 1.5$) is placed over a perfectly flat piece of black plastic ($n = 1.2$) as shown in Fig. 24*a*. They touch at *A*. Light of wavelength 600 nm is incident

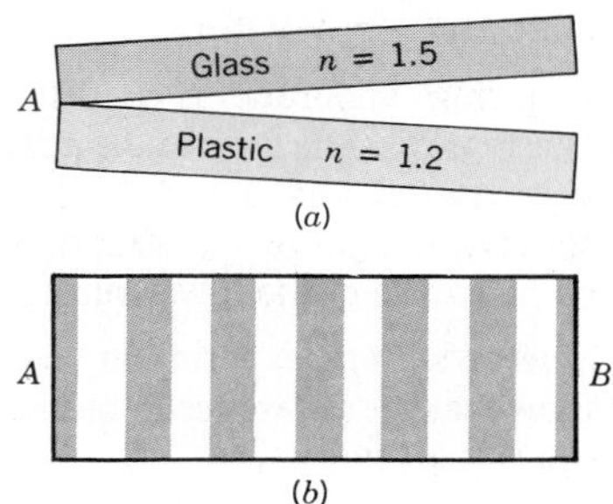

Figure 24 Problem 55.

normally from above. The location of the dark fringes in the reflected light is shown on the sketch of Fig. 24*b*. (*a*) How thick is the space between the glass and the plastic at *B*? (*b*) Water ($n = 1.33$) seeps into the region between the glass and plastic. How many dark fringes are seen when all the air has been displaced by water? (The straightness and equal spacing of the fringes is an accurate test of the flatness of the glass.)

56P. A transparent liquid of refractive index 4/3 is allowed to displace the air from an air wedge that is formed from two glass plates touching each other along one edge. What happens, as a result, to the spacing of the dark lines caused by interference of the monochromatic light reflected back?

57P. Light of wavelength 630 nm is incident normally on a thin wedge-shaped film with index of refraction 1.5. There are 10 bright and nine dark fringes over the length of film. By how much does the film thickness change over this length?

58P. Two pieces of plate glass are held together in such a way that the air space between them forms a very thin wedge. Light of wavelength 480 nm strikes the upper surface perpendicularly and is reflected from the lower surface of the top glass and the upper surface of the bottom glass, thereby producing a series of interference fringes. How much thicker is the air wedge at the sixteenth fringe than it is at the sixth?

59P. In an air wedge formed by two plane glass plates, touching each other along one edge, there are 4001 dark lines observed when viewed by reflected monochromatic light. When the air between the plates is evacuated, only 4000 such lines are observed. Calculate the index of refraction of the air from this data.

60P. *Newton's rings.* Figure 25 shows a lens with radius of curvature *R* resting on an accurately plane glass plate and illuminated from above by light with wavelength λ. Figure 26 shows that circular interference fringes (Newton's rings) appear, associated with the variable thickness *d* of the air film between the lens and the plate. Find the radii *r* of the circular interference maxima assuming that $r/R \ll 1$.

61P. In a Newton's rings (see Problem 60) experiment the radius of curvature *R* of the lens is 5.0 m and its diameter is 20 mm. (*a*) How many rings are produced? (*b*) How many rings would be seen if the arrangement were immersed in water ($n = 1.33$)? Assume that $\lambda = 589$ nm.

62P. A Newton's rings apparatus is used to determine the radius of a lens (see Fig. 25). The radii of the *n*th and $(n + 20)$th bright rings are measured and found to be 0.162 cm and 0.368 cm, respectively, in light of wavelength 546 nm. Calculate the radius of curvature of the lower surface of the lens.

63P. In the Newton's rings experiment, use the result of Problem 60 to show that the difference in radius between adjacent rings (maxima) is given by

$$\Delta r = r_{m+1} - r_m \approx \tfrac{1}{2}\sqrt{\lambda R/m},$$

assuming $m \gg 1$.

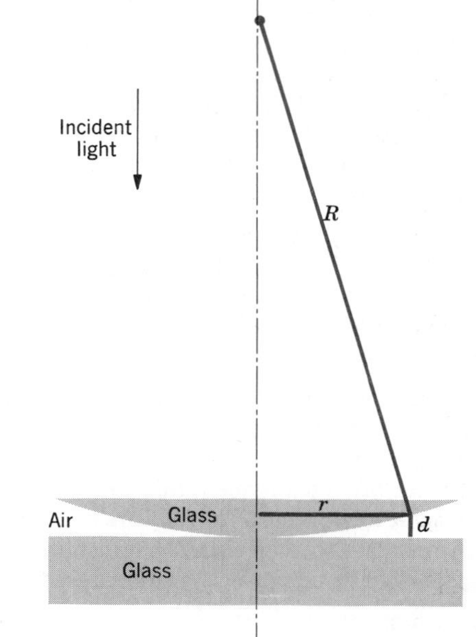

Figure 25 Problems 60–64.

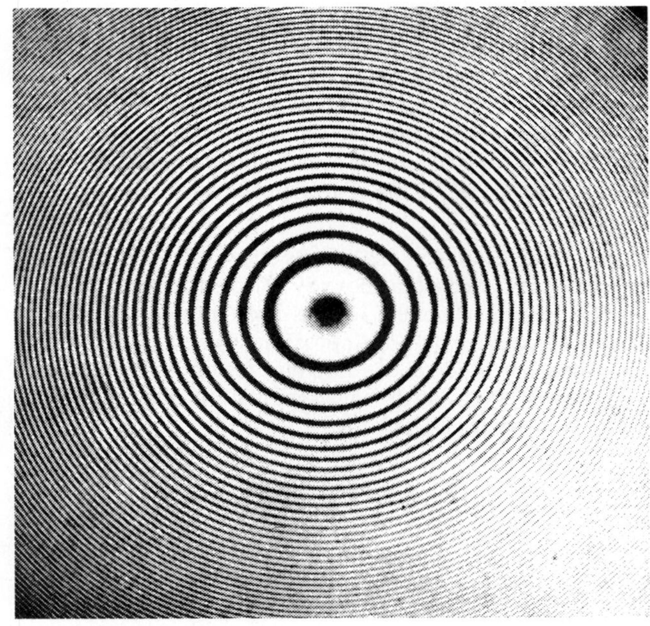

Figure 26 Problems 60–64.

64P. In the Newton's rings experiment, use the result of Problem 63 to show that the *area* between adjacent rings (maxima) is given by

$$A = \pi\lambda R,$$

assuming $m \gg 1$. Note that this area is independent of m.

Section 40-8 Michelson's Interferometer

65E. If mirror M_2 in Michelson's interferometer is moved through 0.233 mm, 792 fringes are counted with a light meter. What is the wavelength of the light?

66E. A thin film with $n = 1.40$ for light of wavelength 589 nm is placed in one arm of a Michelson interferometer. If a shift of 7.0 fringes occurs, what is the film thickness?

67P. An airtight chamber 5.0 cm long with glass windows is placed in one arm of a Michelson interferometer as indicated in Fig. 27. Light of wavelength $\lambda = 500$ nm is used. The air is slowly evacuated from the chamber using a vacuum pump. While the air is being removed, 60 fringes are observed to pass through the view. From these data, find the index of refraction of air at atmospheric pressure.

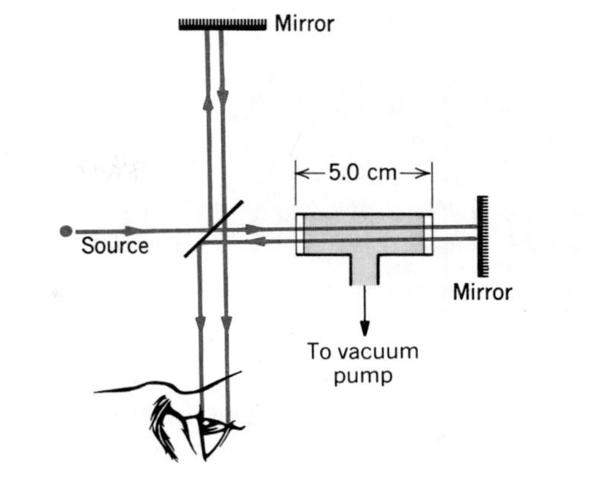

Figure 27 Problem 67.

68P. Write an expression for the intensity observed in Michelson's interferometer (Fig. 17) as a function of the position of the movable mirror. Measure the position of the mirror from the point at which $d_1 = d_2$.

ESSAY 17
LIGHTWAVE COMMUNICATIONS USING OPTICAL FIBERS

SUZANNE R. NAGEL
AT&T BELL LABORATORIES

A revolutionary new technology—lightwave communications—is transforming the communications networks of the world. Huge quantities of information—voice signals, video and digital data—can be rapidly and efficiently transmitted from one place to another by using an ever expanding grid of optical fibers. These hair-thin strands of glass carry information over very long distances in the form of pulses of light. Why is communicating with light so significant and how do these optical fiber "lightguides" work? Let's briefly explore the answer to these questions.

A basic communications system consists of a transmitter (signal source), in which information is encoded, a transmission medium (signal carrier) and a receiver (signal detector) that decodes or reconstructs the original information. Most modern communication systems are "digital" because of the excellent transmission quality that can be achieved. In a simple digital communication system, information is encoded into binary digits consisting of zeros or ones. Before we examine lightwave communications in particular, let's explore how digital encoding is done for one simple voice signal. When we talk, our voice has a maximum frequency of about 4000 hertz (cycles/sec). If this voice signal is sampled as a function of time, it can be accurately represented using a digital code of zeros and ones. Figure 1 is a simple representation of how this is done using a technique called pulse code modulation (PCM). A very small portion of a voice signal is represented in this figure. The voice signal is sampled as shown by the dots. At any sampling time, a binary amplitude code is assigned (right hand axis) based on the relative amplitude of the signal which ranges from 0 to 7 (left-hand axis). Thus, in this example, a relative amplitude of 1 is represented by 001, 7 by 111, etc. The resulting digital-pulse-coded signal is shown in the lower portion of the figure. Your voice signal can thus be represented by a sequence of 0's

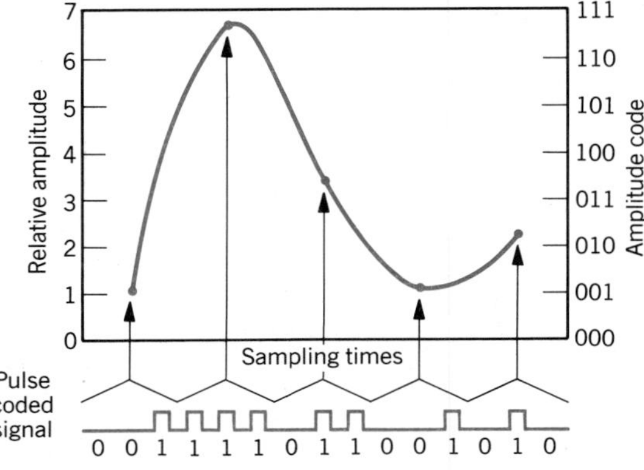

Figure 1 Pulse code modulation technique. The relative amplitide of an analogue signal such as a voice is sampled at a specific sampling rate, and assigned a binary amplitude code. A binary-pulse-coded signal results which can be used for generating a transmission signal.

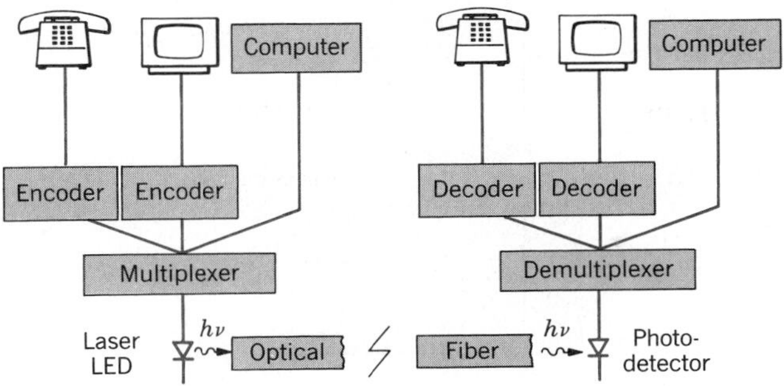

Figure 2 Schematic representation of a lightwave telecommunication system. All information is encoded in a binary data stream of zero's and one's and joined together through a multiplexer. The signal from the multiplexer is used to turn a laser or light emitting diode (LED) on and off at a given transmission rate. The light generated in this manner (represented by $h\nu$) is transmitted over an optical fiber where the output signal falls on a photodetector. The electrical signal generated at the photodetector is fed into a demultiplexer which separates the various signals and routes them to their final destination.

and 1's. To represent accurately a voice signal in digital form, roughly 64,000 pulse coded signal bits per second are needed. In other words, your voice, as you speak, is accurately represented by a sequence of 64,000 zeros and ones per second! Each zero or one is a "bit" of information.

Why can a lightwave communication system transmit so much digital information relative to conventional communication systems? Because the rate at which information can be transmitted is directly related to the signal frequency. Light has a frequency in the range of $10^{14} - 10^{15}$ hertz, compared to radio frequencies of $\sim 10^6$ hertz and microwave frequencies of $10^8 - 10^{10}$ hertz. Therefore, a transmission system that operates at the frequency of light can theoretically transmit information at a higher rate than systems that operate at radio or microwave frequencies. The digital transmission rate is defined as the number of bits/second that are transmitted.

A simple lightwave telecommunication system is shown in Figure 2. The information that is transmitted can be telephone voice signals, video signals, or digital data from a computer. Voice and video signals are encoded into a binary sequence of zeros and ones. All of these signals are blended together into a single very high-data-rate stream in the multiplexer unit. For simplicity let us only consider blending or multiplexing voice signals together for transmission. Each voice signal requires 6.4×10^4 bits/sec. If the data rate of the system is 1 Gbit/sec (1×10^9 bits/sec), the number of voice channels that can be multiplexed together is approximately 15,000! (1×10^9 divided by 6.4×10^4.) How is this actually done? In the lightwave transmitter, each "one" corresponds to an electrical pulse and each "zero" corresponds to the absence of an electrical pulse. These electrical pulses are used to turn a light source on and off very rapidly, much like turning a light switch on and off. The light source can be a laser or a light emitting diode (LED). Thus, in the transmitter in a lightwave communication system, information is blended together into a very high data rate sequence of electrical pulses, which in turn are used to turn a light source on and off very rapidly: all binary encoded information is thus transformed into a timed sequence of flashes of light for transmission.

The next important part of the lightwave communication system is the transmission medium. Although in principle these flashes of light could be transmitted through the open atmosphere, much the same as radio signals are, it would be very difficult to build a practical telecommunications system in this way. Instead, glass fiber "lightguides" are used to carry the light from the transmitter to the third part of the lightwave system, the receiver. In the receiver, each pulse of light is detected by a photodetector. As each pulse of light arrives at the photodetector, a pulse of electrical

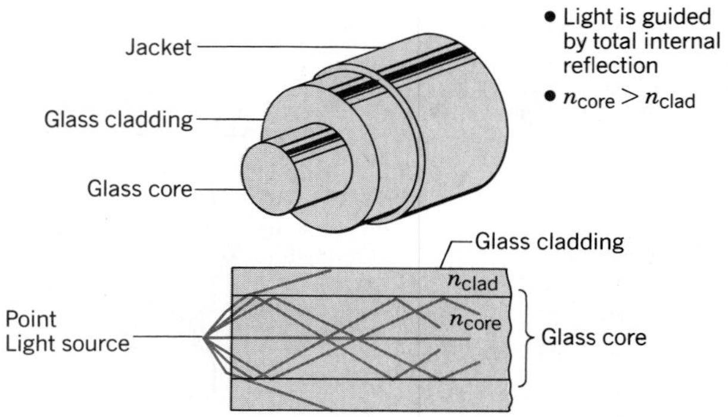

Figure 3 Simple light-guiding structure for an optical transmission medium. The core and cladding index of refraction are represented by n_{core} and n_{clad}, respectively.

current is produced. In this way, the optical pulses are converted back to electrical pulses. The receiver also has a "demultiplexer" which separates the signals and converts them back into voice, video and computer data.

Although this is only a very basic description of how a lightwave communication system operates, it shows the basic approach: information is converted into pulses of light that are transmitted over some distance through an optical fiber, then converted back into information.

Now we can examine in a little more detail how an optical fiber is used to transmit information in the form of pulses of light. The first important property of a fiber is that it be able to guide light from one place to another. The basic principle that an optical fiber uses is "total internal reflection," shown for a lightguide structure in Figure 3. A fiber structure consists of a central core of material that has a higher refractive index than the surrounding material, called the "cladding." Remember, the refractive index is the ratio of the speed of light in a perfect vacuum relative to its speed through the material. A fiber made from glass will usually also have a plastic jacket on the outside to protect the glass from mechanical abrasion and other environmental effects. A light source such as a laser or LED is placed close to the fiber core. The light source gives off or emits a "cone" of light that is coupled into the core of the fiber. For this light to be guided, it must meet the requirement for total internal reflection. As is shown in the diagram, some angles of light are guided in "zig-zag" paths in the fiber while others are not. If the angle of light from the light source is too large, the light gets refracted into the cladding instead of reflected within the core. Although the details of light guidance are far more complex than this simple description, the basic principle of total internal reflection is useful for understanding how light is guided within the core of the optical fiber.

While this simple type of fiber is very useful for guiding light over short distances (meters), more specialized fiber structures are required to guide light over the long distances used in telecommunication systems. Let us examine what determines how closely spaced in time individual pulses can be coupled into fibers. Remember, we want to carry as many pulses per second as possible through the fiber. Figure 4 shows three different types of fiber. First, let us look at the simple fiber shown at the top. It is called a "multimode" fiber because many angles of light are guided and it is said to have a step index because it has a core of constant refractive index surrounded by a cladding of lower refractive index. The refractive index profile shown depicts the relative value of the index as a function of position across the fiber cross section. A pulse of light from the source is coupled into the fiber. Many beams, or angles of

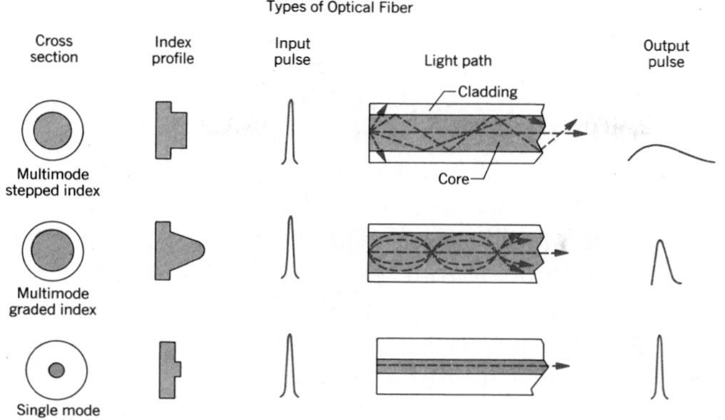

Figure 4 Types of optical fiber.

light, all traveling at the same speed, are guided in the core. But the beam that travels down the center has a shorter distance to travel than those beams that have to zig-zag back and forth. As a result, the "narrow" pulse of light that was initially coupled into the fiber is considerably broadened after traveling many kilometers through the fiber, due to the different relative distances each beam has to travel. It is this effect that limits how closely together the input pulses can be spaced and can be detected without overlapping at the output end.

Two types of fibers are used in lightwave systems to overcome the limits of this pulse broadening effect. One type of fiber is shown in the center of Figure 4 and is called a multimode graded index fiber. Note that the core of the fiber has a gradually changing refractive index profile. This type of profile results in the guided angles of light being bent in gentle periodic paths as they travel through the fiber. How does this help? The beam that travels down the center of the fiber still has the shortest distance to travel, but it travels the slowest since it is in the highest refractive index part of the fiber. A beam that travels along the periodic path spends most of its time in a lower index portion of the fiber, so it travels at a higher speed! Thus, the longer the distance a given beam has to travel, the faster on average it can be made to travel. By very carefully controlling the exact shape of the refractive index profile in the fiber, all angles of guided light can be made to travel through the fiber in roughly the same amount of time! This thus decreases the amount of spreading of the output pulse and extends the distance over which pulses of light can be transmitted before they start to overlap. The third type of fiber depicted on the bottom of Figure 4 completely eliminates the pulse spreading due to different angles of light being guided. This type of fiber is called a single mode fiber, and only the on-axis beam of light is guided. This is achieved by using very small refractive index differences between the core and cladding, and very small cores. This type of fiber can carry the data at the very highest bit rates.

Although the details of light guidance and pulse broadening are far more complicated than this simple description, you can begin to understand how the fiber is used to transmit data at high rates. Lasers or LED's are turned on and off very rapidly in a specific sequence corresponding to ones and zeros; each pulse is individually transmitted through the fiber; and at the output end of the fiber the arriving pulse of light is detected. The broadening of the pulse of light in the fiber as it travels plays an important role in determining the maximum rate at which information can be transmitted, as well as the distance over which information can be transmitted.

A second important property of the fiber is its optical attenuation, since this also determines how far the signal can be transmitted and can still be intense enough to be detected. When light travels through a material, it is not perfectly conducted, and the intensity of the signal decreases with distance. The light traveling through a fiber is diminished by both absorption and scattering. Very simply, absorption results in a fraction of the intensity of the light being transferred to the material itself instead of being transmitted. In contrast, scattering results in the light being diffused or deflected in many different directions. Therefore, some of the light is scattered out from the core rather than being transmitted, causing a decrease in the light signal intensity. In glass fibers made from very high purity, low scattering silica glasses, the amount of scattering and absorption is extremely low at wavelengths of light in the near-infrared. These wavelengths are just a little longer than the wavelengths of the visible spectrum. Remember, visible light ranges from 380 nm (violet) to 700 nm (red). The most common transmission wavelength is 1300 nm (approximately 2×10^{14} hertz). In silica glass fibers, more than 95 percent of the light is transmitted over a distance of one kilometer. This very high transparency allows light to be transmitted over distances of 20 to 200 kilometers and still be intense enough for the signal to be detected by a photodetector device in the receiver.

While this discussion of lightwave communication and the optical fiber transmission medium has been simple and qualitative, the interested student will find a wealth of reference material available to explore this emerging area of technology in greater depth. A few advanced references are listed at the end of this discussion.

At this time, the quest to utilize the frequency potential of communication systems based on light continues, and many advances are still being made. In the years 1987 and 1988, telecommunication systems are being installed that operate at 1.7 Gbits/sec (1.7×10^9 bits/sec), corresponding to ~25,000 voices being carried over a fiber that is roughly the size of a human hair! Impressive as this may seem, it is still several orders magnitude below the theoretical capacity. In the coming years, these fiber-based lightwave communications systems will increasingly be used to carry voice, video, and computer data across the country and the world as well as to and from homes and businesses.

REFERENCES FOR FURTHER READINGS *Optical Fiber Telecommunications,* S. E. Miller and A. Chynoweth (eds.), Academic Press, 1979, 705 pp. J. E. Midwinter, *Optical Fibers for Transmission,* John Wiley & Sons, Inc. 1979, 410 pp. A. W. Snyder and J. D. Love, *Optical Waveguide Theory,* Chapman and Hall, 1983, 734 pp. *Optical Fiber Communications,* Vol. 1, Fiber Fabrication, T. Li, (ed.) Academic Press, 1985, 363 pp. D. J. Morris, *Pulse Code Formats for Fiber Optical Data Communication,* Marcel Decker, Inc., 1983, 217 pp.

CHAPTER 41

DIFFRACTION

About 160,000 years ago, a star in a nearby galaxy blew up. In February of 1987 the light from this supernova reached Earth and was first identified by Ian Shelton, a Canadian astronomer working at the Las Campanas observatory in Chile. The cross that is such a feature of this photo has nothing to do with the explosion but is formed by the diffraction of light by the internal supporting structure of the telescope. We shall see in this chapter that the ability of all optical instruments to form sharp images of point sources is limited by the diffraction of light.

41-1 Diffraction and the Wave Theory of Light

In Chapter 40 we defined diffraction rather loosely as the flaring out of light as it emerges from a narrow slit. Its main feature is: for a given wavelength, the narrower the slit the more the flaring out. Figure 1 shows a narrow slit and the diffraction pattern formed when monochromatic light from a distant source falls on it. The incident light certainly "flares out" into the geometric shadow of the slit, but there is more to it than that. There is a broad and intense central maximum — which was our principal concern in Chapter 40 — but there are also secondary maxima and minima. Figure 2 shows the diffraction pattern formed when light from a distant point source falls on a razor blade.

If geometrical optics prevailed, light would form a sharp shadow of the slit edges, as in Fig. 1*b*. Figure 1*c* is what we actually see. The assumption of geometrical optics — that light travels in straight lines — though useful in many situations, is only an approximation.

Today, diffraction finds a ready explanation in terms of the wave theory of light. This theory, advanced by Huygens and used by Young to explain double-slit interference, was very slow in being adopted, largely because it ran counter to the theory of Newton, who held

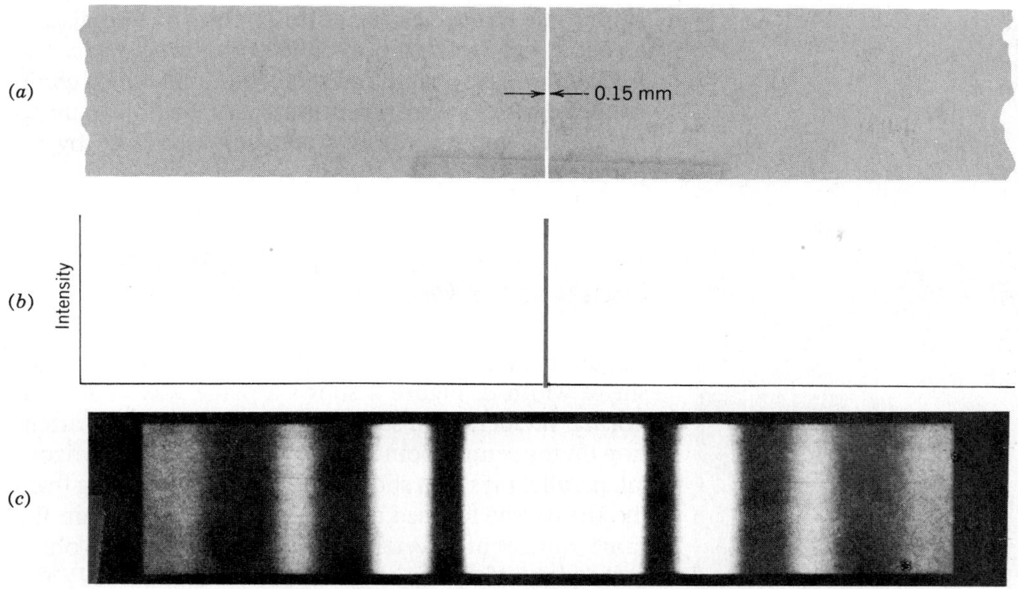

Figure 1 (*a*) A slit in an otherwise opaque screen. The slit width (0.15 mm) has been somewhat exaggerated for clarity. (*b*) The pattern you would expect on a diffusing screen if light *truly* traveled in straight lines. (*c*) The pattern that actually occurs. The light flares out substantially beyond the slit. Note the broad central diffraction maximum, the regions of zero intensity, and the secondary diffraction maxima.

the view that light was a stream of particles.

Support for Newton's views prevailed in French scientific circles. Enter Augustin Fresnel,* a young military engineer who followed his passion for optics largely in the spare time that he could wrench from his military duties. Fresnel believed in the wave theory of light and submitted memoirs to the French Academy of Sciences describing his experiments and his wave-theory explanations of them.

In 1819, the Academy, dominated by the supporters of Newton† and thinking to challenge the wave point of view, organized a prize competition for an essay on the subject of diffraction. Fresnel won. The Newtonians, however, were neither converted nor silenced. One of them, Poisson, pointed out the "strange result" that, if Fresnel's theories were correct, a bright spot should appear in the center of the shadow of a sphere. The prize

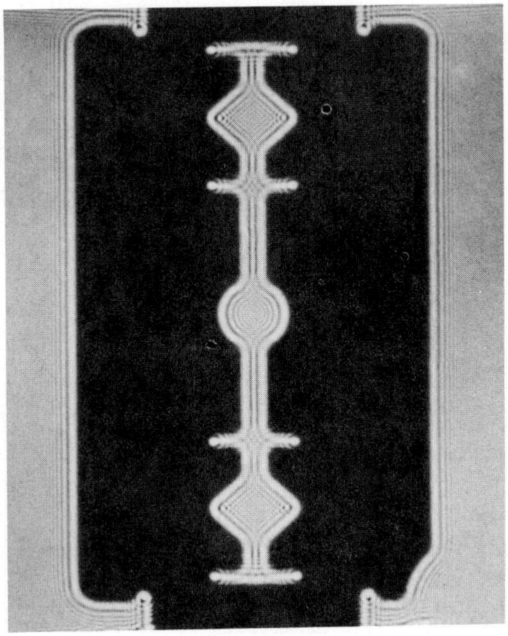

Figure 2 The diffraction pattern of a razor blade in monochromatic light. Note the fringes of alternating maximum and minimum intensity.

* Pronounced Fra-nel′.

† In 1819, Newton's great work *Opticks* had been in print for 115 years. For an appreciation of Newton's views on light, from today's perspective, see I. Bernard Cohen's preface to the 1952 reprinting of this work (Dover Publications).

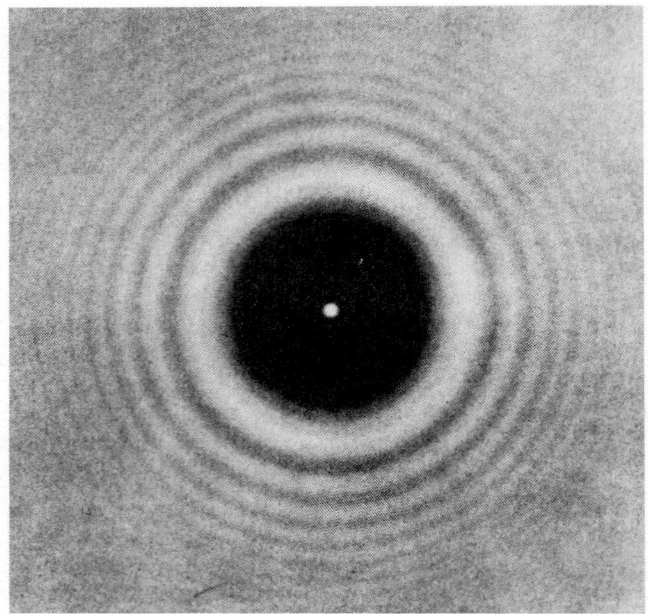

Figure 3 The diffraction pattern of a disk. Note the concentric diffraction rings and—especially—Fresnel's bright spot in the center of the pattern. If geometrical optics prevailed, the center of the pattern (being the most effectively screened) is the last place you would expect to find a bright spot.

committee arranged a test of this prediction and discovered (see Fig. 3) that the predicted *Fresnel bright spot,* as we call it today, was indeed there! Nothing builds confidence in a theory so much as having one of its unexpected and counterintuitive predictions verified by experiment.

41-2 Diffraction from a Single Slit—Locating the Minima

Let us consider how light is diffracted by a long narrow slit of width a. Figure 4 shows a plane wave falling at normal incidence on such a slit. Let us focus our attention on the central point P_0 of screen C. A set of horizontal, parallel rays (not shown in the figure) emerging from the slit will be focused at P_0. These rays all contain the same number of wavelengths and, since they are in phase at the plane of the slit, they will still be in phase at P_0 and the central point of the diffraction pattern that appears on screen C has a maximum intensity.

We now consider another point on the screen. Light rays that reach P_1 in Fig. 4 leave the slit at an angle θ as shown. The ray xP_1 determines θ because it passes undeflected through the center of the lens. Ray r_1 originates at the top of the slit and ray r_2 at its center. If θ is chosen so that the distance bb' in the figure is one-half a wave-

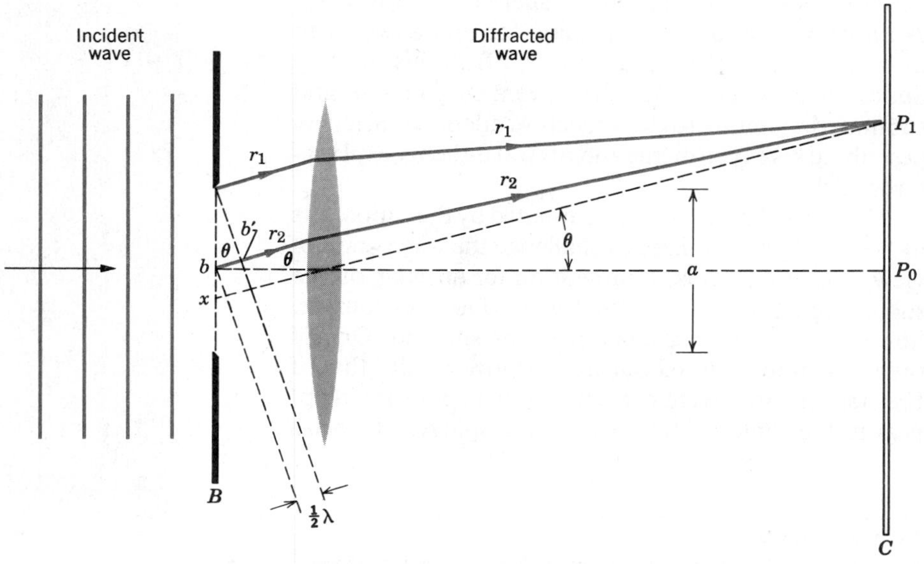

Figure 4 Conditions at the first minimum of the diffraction pattern. Angle θ is such that the distance bb' is one-half of a wavelength.

Figure 5 A marine radar antenna. You can tell from its dimensions that its transmitted beam is fan-shaped, being broad in the vertical plane and narrow in the horizontal plane.

length, r_1 and r_2 will be out of phase and will produce no effect at P_1. In fact, every ray from the upper half of the slit will be canceled by a ray from the lower half, originating at a point $a/2$ below the first ray. The point P_1, the first minimum of the diffraction pattern, will have zero intensity.

The condition shown in Fig. 4 is

$$\frac{a}{2}\sin\theta = \frac{\lambda}{2},$$

or

$$a\sin\theta = \lambda \quad \text{(first minimum)}. \tag{1}$$

Equation 1 shows that the central maximum becomes *wider* as the slit becomes *narrower.* That is, for a fixed wavelength, θ increases as a decreases, just as Fig. 4 of Chapter 40 suggests. For $a = \lambda$ we have $\theta = 90°$, so that the central maximum fills the entire forward hemisphere.

Equation 1 helps us to understand why the antennas of marine radars (see Fig. 5) are much longer in the horizontal dimension than in the vertical dimension. The microwave beam from such radars is fan shaped, being very narrow in the horizontal plane (hence the long horizontal dimension) and wide in the vertical plane (hence the short vertical dimension).

In Fig. 6 the slit is divided into four equal zones, with a ray shown leaving the top of each zone. Let θ be chosen so that the distance bb' is one-half a wavelength. Rays r_1 and r_2 will then cancel at P_2. Rays r_3 and r_4 will also be half a wavelength out of phase and will also cancel. Consider four other rays, emerging from the slit a given distance below the four rays above. The two rays below r_1 and r_2 will exactly cancel, as will the two rays below r_3 and r_4. We can proceed across the entire slit and conclude again that no light reaches P_2; we have located a second point of zero intensity.

The condition described (see Fig. 6) requires that

$$\frac{a}{4}\sin\theta = \frac{\lambda}{2},$$

or

$$a\sin\theta = 2\lambda \quad \text{(second minimum)}. \tag{2}$$

By extension of Eqs. 1 and 2, the general formula for the minima in the diffraction pattern on screen C can be seen to be

$$\boxed{a\sin\theta = m\lambda \quad m = 1, 2, 3, \ldots} \quad \text{(minima)}. \tag{3}$$

There is a maximum approximately halfway between each adjacent pair of minima. There are also minima and maxima on the bottom half of the screen, the pattern being symmetrical about P_0 in Fig. 6.

Sample Problem 1 A slit of width a is illuminated by white light. For what value of a will the first minimum for red light ($\lambda = 650$ nm) fall at $\theta = 15°$?

At the first minimum, $m = 1$ in Eq. 3. Solving for a, we then find

$$a = \frac{m\lambda}{\sin\theta} = \frac{(1)(650 \text{ nm})}{\sin 15°}$$
$$= 2511 \text{ nm} \approx 2.5\ \mu\text{m}. \quad \text{(Answer)}$$

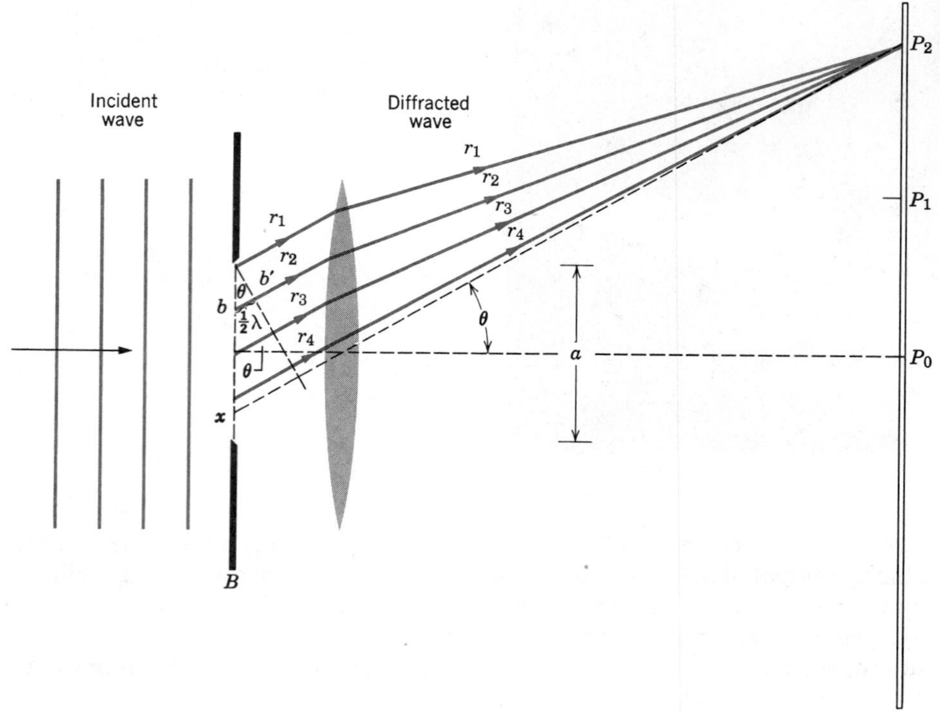

Figure 6 Conditions at the second minimum of the diffraction pattern of a single slit. The path difference bb' is one-half a wavelength.

For the incident light to flare out that much ($\pm 15°$) the slit must be very fine indeed, amounting to about 4 times the wavelength. Note that a fine human hair may only be about 100 μm in diameter.

Sample Problem 2 In Sample Problem 1, what is the wavelength λ' of the light whose first diffraction maximum (not counting the central maximum) falls at 15°, thus coinciding with the first minimum for red light?

This maximum is about halfway between the first and second minima. We can find it without too much error by putting $m = 1.5$ in Eq. 3, or*

$$a \sin \theta = 1.5\, \lambda'.$$

From Sample Problem 1, however,

$$a \sin \theta = \lambda.$$

Dividing gives

$$\lambda' = \frac{\lambda}{1.5} = \frac{650 \text{ nm}}{1.5} = 433 \text{ nm}. \qquad \text{(Answer)}$$

Light of this wavelength is violet. The second maximum for light of wavelength 430 nm will always coincide with the first minimum for light of wavelength 650 nm, no matter what the slit width. If the slit is relatively narrow, the angle θ at which this overlap occurs will be relatively large, and conversely.

* $m = 1.43$ turns out to be a much better value but $m = 1.5$ will do for our purpose.

41-3 Diffraction: A Closer Look

Figure 7 shows the generalized diffraction situation. The curved surfaces on the left represent wavefronts of the incident light falling on the diffracting object B which, in the case we examined in Section 41-2, was a narrow slit. In Fig. 7 we show the diffracting object as an opaque screen containing an aperture of arbitrary shape. C in Fig. 7 is a screen or photographic film that receives the light that passes through or around the diffracting object.

We can calculate the pattern of light intensity on screen C by subdividing the wavefront into elementary areas dA, each of which becomes a source of an expanding Huygens wavelet. The light intensity at an arbitrary point P is found by superimposing the wave disturbances (that is, the **E**-vectors) caused by the wavelets reaching P from all these elementary sources. Instead of a screen with a hole in it, the diffracting object could equally well

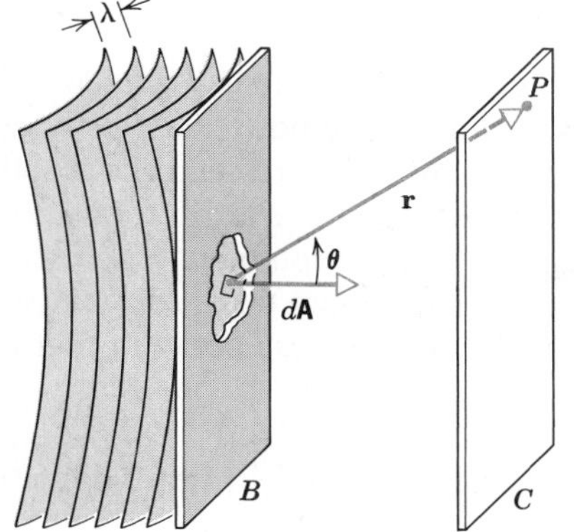

Figure 7 The elements of diffraction. Coherent wavefronts of light fall on screen B, which contains an aperture of arbitrary shape. The diffraction pattern is formed on diffusing screen C.

be a wire mesh, a screen with a single slit (Fig. 1), or a razor blade (Fig. 2).

The wave disturbances reaching P differ in amplitude and in phase because the elementary radiators are at varying distances from P. Diffraction calculations—simple in principle—may become difficult in practice. The calculation must be repeated for every point on screen C at which we wish to know the light intensity.

Figure 8*a* shows light from source S falling on a slit in screen B and forming a diffraction pattern on screen C. The rays falling on the slit and emerging from it are not parallel, which makes the calculation of the light intensity for various points on screen C more difficult.

A simplification results if source S and screen C are

Figure 8 (*a*) Light from point source S illuminates a slit in screen B. The intensity at point P on screen C depends on the relative phases of the light received from the various parts of the slit. (*b*) If source S and screen C are moved to a large distance from the slit, both the incident and the emergent light at B are closely parallel and the diffraction calculations are much simplified. (*c*) Rather than move the source and the screen, you can put each in the focal plane of a lens, as shown; the light entering and emerging from the slit is parallel, as in (*b*).

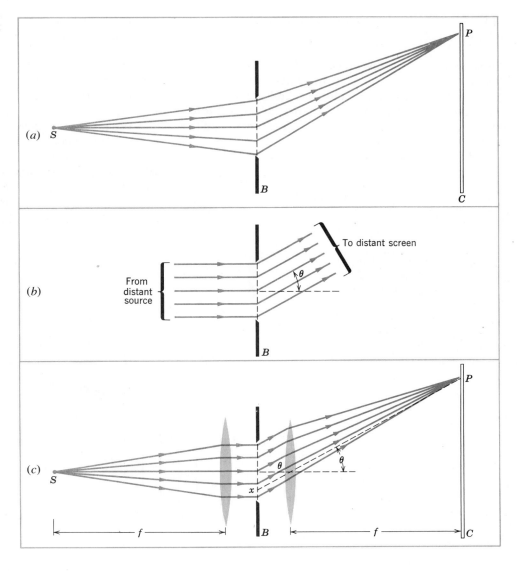

moved to a large distance from the diffraction aperture, as in Fig. 8*b*. The wavefronts arriving at the diffracting aperture from the distant source *S* are now planes, and the rays associated with these wavefronts are parallel to each other. This condition can be established in the laboratory by using two converging lenses, as in Fig. 8*c*. The first of these converts the diverging wave from the source into a plane wave. The second lens causes plane waves leaving the diffracting aperture to converge to point *P*. All rays that illuminate *P* will leave the diffracting aperture parallel to the dashed line *Px* drawn from *P* through the center of this second (thin) lens. In this chapter, we shall assume that the conditions described in Figs. 8*b* and 8*c* prevail.*

41-4 Diffraction from a Single Slit—Qualitative

In Section 41-2 we saw how to find the positions of the maxima and the minima in the single-slit diffraction pattern. Now we turn to a more general problem: Find an expression for the intensity I of the pattern as a function of position on the screen, that is, as a function of the angle θ in Fig. 6.

Figure 9 shows a slit of width a divided into N parallel strips of width Δx. Each strip acts as a radiator of Huygens' wavelets and produces a characteristic wave disturbance at point P. We may take the amplitudes ΔE_0 of the wave disturbances at P from the various strips as equal if θ in Fig. 9 is not too large.

The wave disturbances from adjacent strips have a constant phase difference $\Delta\phi$ between them at P given by

$$\text{phase difference} = \left(\frac{2\pi}{\lambda}\right)(\text{path difference})$$

or

$$\Delta\phi = \left(\frac{2\pi}{\lambda}\right)(\Delta x \sin\theta), \tag{4}$$

where $\Delta x \sin\theta$, as the figure insert shows, is the path difference for rays originating at the top edges of adjacent strips. Thus at point P, N vectors with the same amplitude ΔE_0, the same wavelength λ, and the same phase difference $\Delta\phi$ between adjacent members, combine to produce a resultant disturbance. We ask: What is the amplitude E_θ of the resultant wave disturbance?

We find the answer by representing the individual wave disturbances ΔE_0 by phasors and calculating the resultant phasor amplitude, just as we did in Section 40-6. We do so qualitatively, reserving a quantitative treatment for the following section.

At the center of the diffraction pattern, θ equals zero, and the phase shift between adjacent strips (see Eq.

* The general case of Fig. 8*a* is called *Fresnel diffraction;* the special case of Fig. 8*c* is called *Fraunhofer diffraction.*

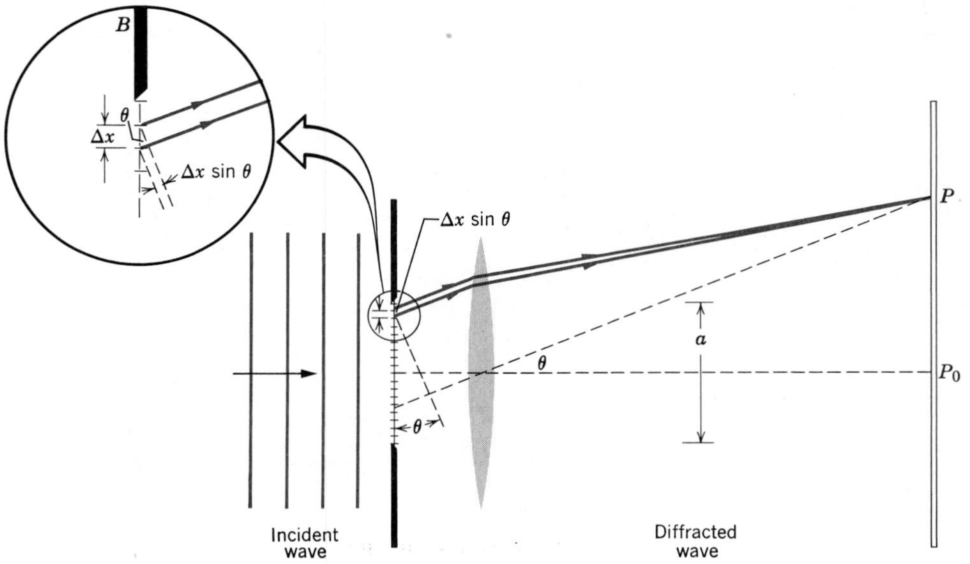

Figure 9 A slit of width a is divided into N strips of width Δx. The insert shows conditions at the second strip more clearly. In the differential limit, the slit is divided into an infinite number of strips of differential width dx.

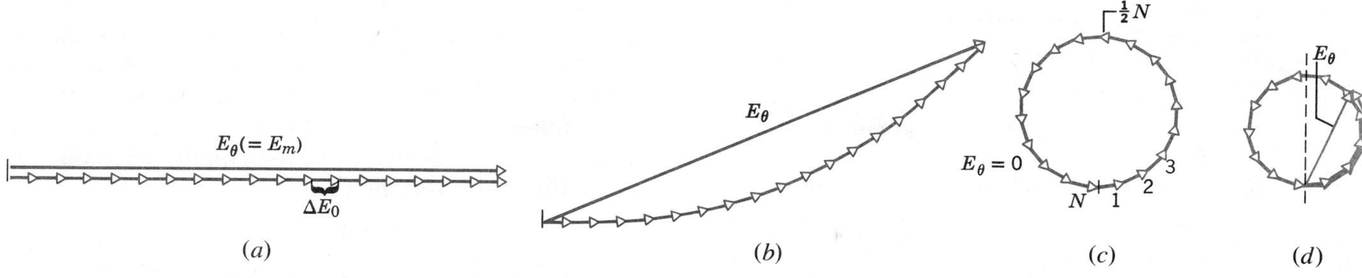

Figure 10 Phasors in single-slit diffractions. Conditions at (*a*) the central maximum, (*b*) a direction θ slightly removed from the central maximum, (*c*) the first minimum, and (*d*) the first maximum beyond the central maximum. The figure corresponds to $N = 18$ in Fig. 9.

4) is also zero. As Fig. 10*a* shows, the phasor arrows in this case are laid end to end and the amplitude of the resultant has its maximum value E_m. This corresponds to the center of the central maximum.

As we move to a value of θ other than zero, $\Delta\phi$ assumes a definite nonzero value (again see Eq. 4), and the array of arrows is now as shown in Fig. 10*b*. The resultant amplitude E_θ is less than before. Note that the length of the "arc" of small arrows is the same for both figures and indeed for all figures in this series. As θ increases further, a situation is reached (Fig. 10*c*) in which the chain of arrows curls around through 360°, the tip of the last arrow touching the foot of the first arrow. This corresponds to $E_\theta = 0$, that is, to the first minimum. For this condition the ray from the top of the slit (marked 1 in Fig. 10*c*) is 180° out of phase with the ray from the center of the slit (marked $\frac{1}{2}N$ in Fig. 10*c*). These phase relations are consistent with Fig. 4, which also represents the first minimum.

As θ increases further, the phase shift continues to increase, and the chain of arrows coils around through an angular distance greater than 360°, as in Fig. 10*d*, which corresponds to the first maximum beyond the central maximum. The intensity of this maximum is much smaller than that of the central maximum. In making this comparison, recall that the arrows marked E_θ in Fig. 10*a* correspond to the *amplitudes* of the wave disturbance and not to the *intensities*. The amplitudes must be squared to obtain the corresponding relative intensities.

41-5 Diffraction from a Single Slit—Quantitative

Equation 3 tells us how to locate the minima of the single-slit diffraction pattern on screen C of Fig. 9 as a function of the angle θ in that figure. Note that θ is our position locator, every screen point P being associated with a definite value of θ. Here we wish to derive an expression for the intensity I of the pattern as a function of θ.

We state, and shall prove below, that the intensity is given by

$$I = I_m\left(\frac{\sin\alpha}{\alpha}\right)^2 \tag{5}$$

where

$$\alpha = \tfrac{1}{2}\phi = \left(\frac{\pi a}{\lambda}\right)\sin\theta. \tag{6}$$

We can here view α as a convenient transfer angle, serving as a connection between θ of Fig. 9 and I of Eq. 5. In Eq. 5, I_m is the maximum value of the intensity, which occurs at the center of the diffraction pattern, corresponding to $\theta = 0$. Figure 11 shows plots of the intensity of a single-slit diffraction pattern, calculated from Eqs. 5 and 6. Note that, as the slit width is decreased from 10λ (in Fig. 11*c*) to 5λ (in Fig. 11*b*) to λ (in Fig. 11*a*), the central maximum becomes correspondingly broader.

Equations 5 and 6, which together tell us how the intensity of the diffraction pattern varies with the angle θ in Fig. 11, necessarily contain information about the location of the intensity minima. Let us see if we can extract it.

Study of Eq. 5 shows that intensity minima will occur when

$$\alpha = m\pi \quad m = 1, 2, 3, \ldots \tag{7}$$

If we put this result into Eq. 6 we find

$$m\pi = \frac{\pi a}{\lambda}\sin\theta,$$

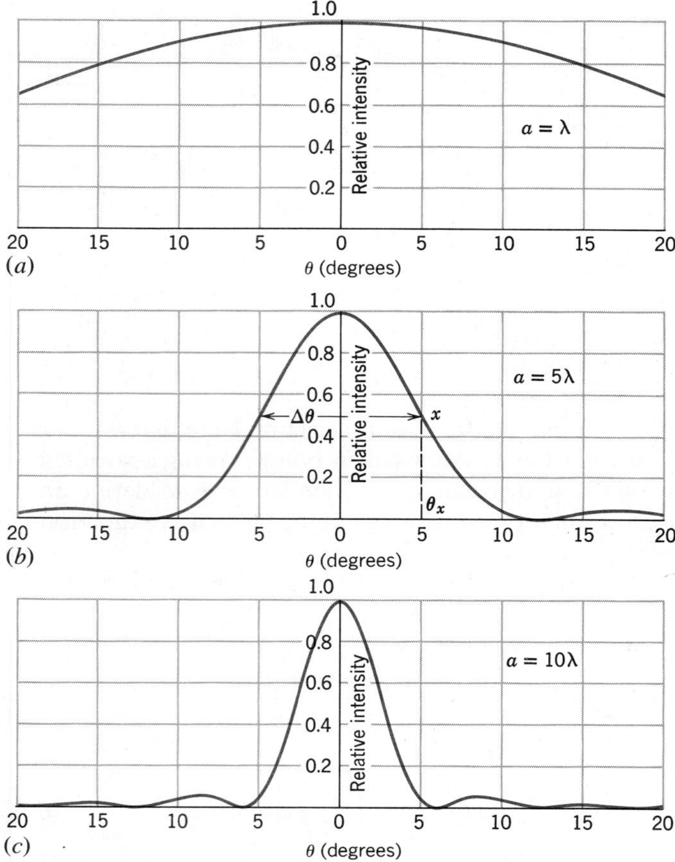

Figure 11 The relative intensity in single-slit diffraction for three different values of the ratio a/λ. The wider the slit, the narrower is the central diffraction peak.

or

$$a \sin \theta = m\lambda \quad m = 1, 2, 3, \ldots \quad \text{(minima)} \quad (8)$$

which is exactly Eq. 3, the expression that we derived earlier for the location of the minima.

Proof of Eq. 5. The arc of small arrows in Fig. 12 shows the phasors representing, in amplitude and phase, the wave disturbances that reach an arbitrary point P on the screen of Fig. 9, corresponding to a particular angle θ. The resultant amplitude at P is E_θ. If we divide the slit of Fig. 9 into infinitesimal strips of width dx, the arc of arrows in Fig. 12 approaches the arc of a circle, its radius R being indicated in that figure. The length of the arc is E_m, the amplitude at the center of the diffraction pattern, for at the center of the pattern the wave disturbances are all in phase and this "arc" becomes a straight line as in Fig. 10*a*.

The angle ϕ in the lower part of Fig. 12 is the difference in phase between the infinitesimal vectors at the left and the right ends of the arc E_m. This means that ϕ is the phase difference between rays from the top and the bottom of the slit of Fig. 6. From geometry we see that ϕ is also the angle between the two radii marked R in Fig. 12. From this figure we can write

$$E_\theta = 2R \sin \tfrac{1}{2}\phi.$$

In radian measure ϕ (from the figure) is

$$\phi = \frac{E_m}{R}.$$

Combining yields

$$E_\theta = \left(\frac{E_m}{\frac{1}{2}\phi}\right) \sin \frac{1}{2}\phi.$$

We have several times made use of the fact that the intensity of a wave is proportional to the square of its amplitude. If we do so again with the above equation, substituting α for $\frac{1}{2}\phi$, we are led to an expression for the intensity as a function of θ that we can write as

$$I = I_m \left(\frac{\sin \alpha}{\alpha}\right)^2.$$

This is exactly Eq. 5, one of the two equations we set out to prove.

It remains to relate the angle α to the angle θ. The phase difference ϕ between rays from the top and the

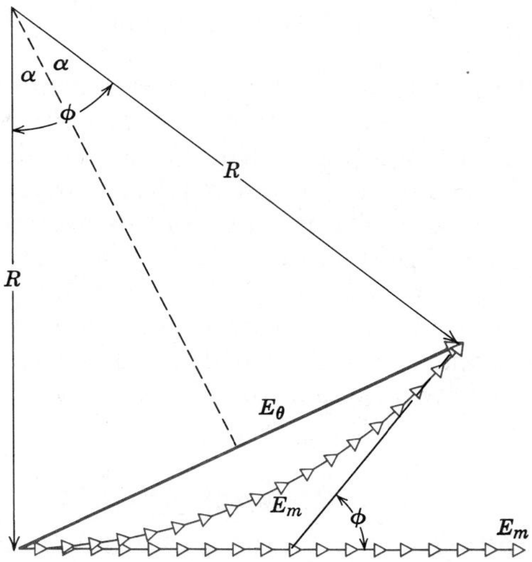

Figure 12 A construction used to calculate the intensity in single-slit diffraction. The situation corresponds to that of Fig. 10*b*.

bottom of the slit in Fig. 9 is related to a path difference for those rays by Eq. 4, or

$$\phi = \left(\frac{2\pi}{\lambda}\right)(a \sin \theta).$$

But $\phi = 2\alpha$, so that this equation readily reduces to Eq. 6.

Sample Problem 3 Find the intensities of the secondary maxima in the single-slit diffraction pattern of Fig. 1, measured relative to the intensity of the central maximum.

The secondary maxima lie approximately halfway between the minima, which are given by Eq. 7 ($\alpha = m\pi$). The maxima are then given (approximately) by

$$\alpha = (m + \tfrac{1}{2})\pi \quad m = 1, 2, 3, \ldots$$

If we substitute this result into Eq. 5 we obtain

$$\frac{I}{I_m} = \left(\frac{\sin \alpha}{\alpha}\right)^2 = \left(\frac{\sin(m + \frac{1}{2})\pi}{(m + \frac{1}{2})\pi}\right)^2 \quad m = 1, 2, 3, \ldots$$

The first secondary maximum occurs for $m = 1$, its relative intensity being

$$\frac{I}{I_m} = \left(\frac{\sin(1 + \frac{1}{2})\pi}{(1 + \frac{1}{2})\pi}\right)^2 = \left(\frac{\sin 1.5\pi}{1.5\pi}\right)^2 = \left(\frac{\sin 270^\circ}{1.5\pi}\right)^2$$

$$= 4.50 \times 10^{-2} \approx 4.5\%. \qquad \text{(Answer)}$$

For $m = 2$ and $m = 3$ we find that $I/I_m = 1.6\%$ and 0.83%, respectively.

The successive maxima decrease rapidly in intensity. The pattern of Fig. 1 has been deliberately overexposed to reveal these faint secondary maxima.

41-6 Diffraction from a Circular Aperture

Here we consider diffraction by a circular aperture of diameter d, the aperture constituting the boundary of a converging lens.

Figure 13 shows the image of a distant point source of light (a star, for instance) formed on a photographic film placed in the focal plane of a converging lens. It is not a point, as the (approximate) geometrical optics treatment suggests, but a circular disk surrounded by several progressively fainter secondary rings. Comparison with Fig. 1 leaves little doubt that we are dealing with a diffraction phenomenon in which, however, the aperture is a circle rather than a rectangular slit. The ratio d/λ, where d is the diameter of the lens (or of a circular aperture placed in front of the lens), determines the scale of the diffraction pattern, just as the ratio a/λ does for a slit.

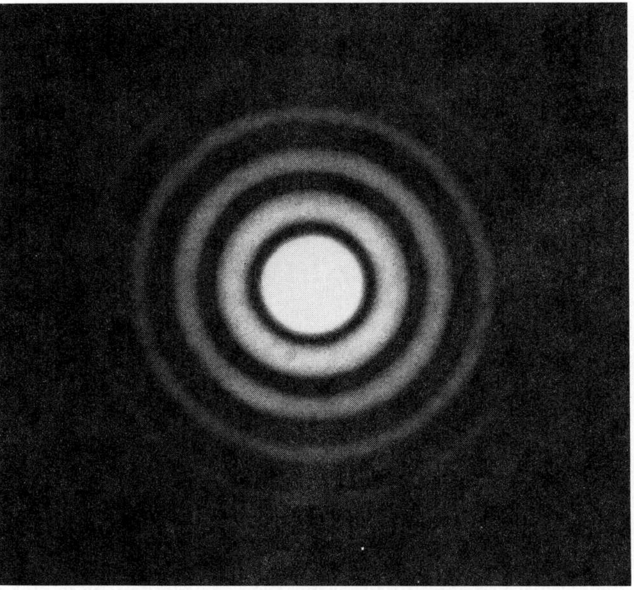

Figure 13 The diffraction pattern of a circular aperture. Note the central maximum and the circular secondary maxima. The figure has been deliberately overexposed to bring out these secondary maxima, which are much less intense than the central disk.

Analysis shows that the first minimum for the diffraction pattern of a circular aperture of diameter d is given by

$$\sin \theta = 1.22 \frac{\lambda}{d} \quad \text{(1st minimum; circular aperture).} \qquad (9)$$

This is to be compared with Eq. 1, or

$$\sin \theta = \frac{\lambda}{a} \quad \text{(1st minimum; single slit),} \qquad (10)$$

which locates the first minimum for a long narrow slit of width a. The factor 1.22 emerges from the mathematical analysis when we integrate over the elementary radiators into which the circular aperture may be divided.

Resolving Power. The fact that lens images are diffraction patterns is important when we wish to distinguish two distant point objects whose angular separation is small. Figure 14 shows the visual appearances and the corresponding intensity patterns for two distant point objects (stars, say) with small angular separations. In

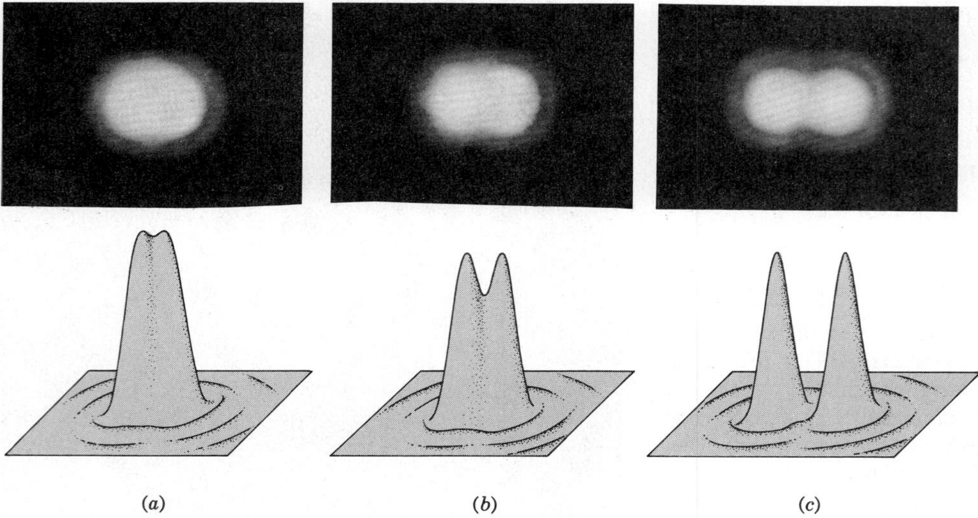

Figure 14 The images of two point sources (stars) are formed by a converging lens. In (*a*) the sources are so close together that they can scarcely be distinguished. In (*b*) they can be marginally distinguished and in (*c*) they are clearly distinguished. The criterion in (*b*) (Rayleigh's criterion) is that the central maximum of one diffraction pattern coincide with the first minimum of the other. Computer-generated profiles of the intensities are shown below the images.

Figure 14*a* the objects are not resolved; that is, they cannot be distinguished from a single point object. In Fig. 14*b* they are barely resolved and in Fig. 14*c* they are fully resolved.

In Fig. 14*b* the angular separation of the two point sources is such that the maximum of the diffraction pattern of one source falls on the first minimum of the diffraction pattern of the other, a condition called *Rayleigh's criterion* for resolvability. From Eq. 9, two objects that are barely resolvable by this criterion must have an angular separation θ_R of

$$\theta_R = \sin^{-1} \frac{1.22\lambda}{d}.$$

Since the angles involved are small, we can replace $\sin \theta_R$ by θ_R expressed in radians:

$$\boxed{\theta_R = 1.22 \frac{\lambda}{d}} \quad \text{(Rayleigh's criterion).} \quad (11)$$

If the angular separation θ between the objects is greater than θ_R, we can resolve the two objects; if it is significantly less, we cannot. The objects must be of comparable brightness for Rayleigh's criterion to be useful.

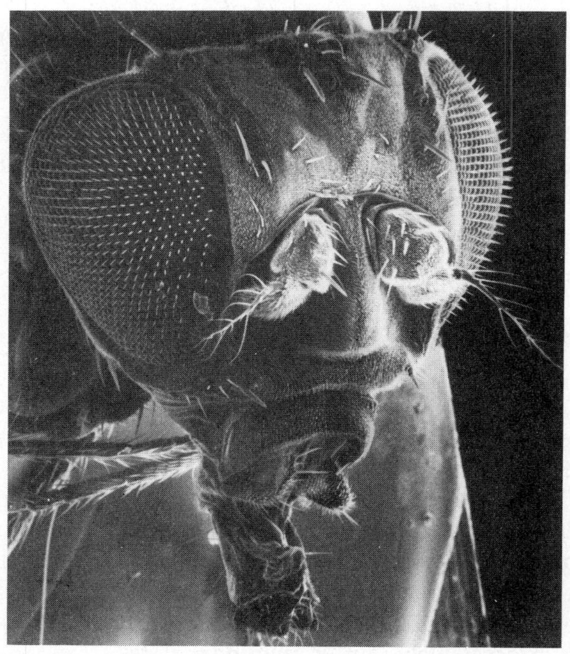

Figure 15 A head-on view of a midge, suggesting the fine resolution that can be obtained with the small wavelengths that are possible in an electron microscope.

When we wish to use a lens to resolve objects of small angular separation, it is desirable to make the central disk of the diffraction pattern as small as possible. This can be done (see Eq. 11) by increasing the lens diameter or by using a shorter wavelength.

To reduce diffraction effects in microscopes we often use ultraviolet light, which, because of its shorter wavelength, permits finer detail to be examined than would be possible for the same microscope operated with visible light. In Chapter 44 of the extended version of this text, we show that beams of electrons behave like waves under some circumstances. In the electron microscope such beams may have an effective wavelength as much as 10^5 times shorter than the wavelength of visible light. This permits the detailed examination of tiny structures that would be hopelessly blurred by diffraction if viewed with an optical microscope; see Fig. 15.

Sample Problem 4 A converging lens 32 mm in diameter has a focal length f of 24 cm. (a) What angular separation must two distant point objects have to satisfy Rayleigh's criterion? Assume that $\lambda = 550$ nm.

From Eq. 11

$$\theta_R = 1.22\frac{\lambda}{d} = \frac{(1.22)(550 \times 10^{-9}\text{ m})}{32 \times 10^{-3}\text{ m}}$$
$$= 2.10 \times 10^{-5}\text{ rad} = 4.3\text{ arc seconds.} \quad \text{(Answer)}$$

(b) How far apart are the centers of the diffraction patterns in the focal plane of the lens?

The linear separation is

$$\Delta x = f\theta = (0.24\text{ m})(2.10 \times 10^{-5}\text{ rad})$$
$$= 5.0\ \mu\text{m}. \quad \text{(Answer)}$$

This is about 9 wavelengths of the light employed.

41-7 Diffraction from a Double Slit (Optional)

In Young's double-slit experiment (Section 40-4) we assumed that the slits were arbitrarily narrow, that is, that $a \ll \lambda$. For such narrow slits, the central part of the screen on which the light falls is uniformly illuminated by the diffracted waves from each slit. When such waves interfere, they produce interference fringes of uniform intensity.

In practice, for visible light, the condition $a \ll \lambda$ is usually not met. For such relatively wide slits, the intensity of the interference fringes formed on the screen will *not* be uniform. Instead, the intensity of the fringes will be governed by an intensity envelope that is the diffraction pattern of a single slit.

Figure 16*a*, for example, suggests the double-slit fringe pattern that would occur if the slits were infinitely narrow. Figure 16*b* shows the diffraction pattern of an actual slit; the broad central maximum and one weaker secondary maximum (at $\pm 17°$) are shown. Figure 16*c*, which is found by multiplying Figs. 16*a* and 16*b*, shows the resulting interference pattern. We see that the positions of the fringes remain unchanged but their intensities are indeed governed by the single-slit diffraction pattern.

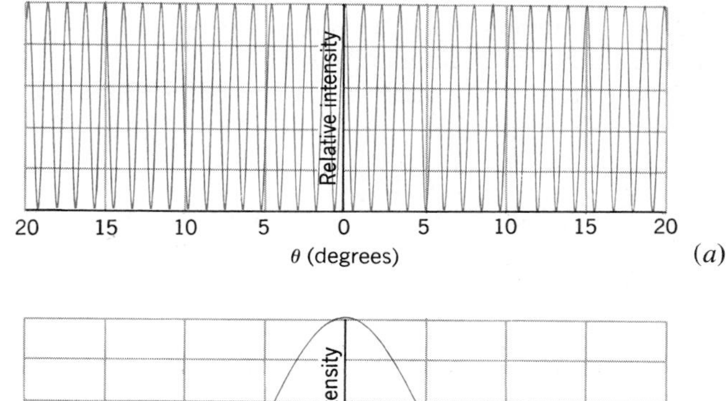

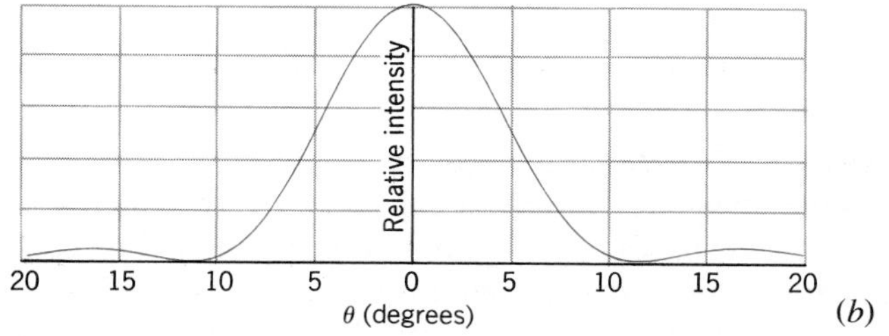

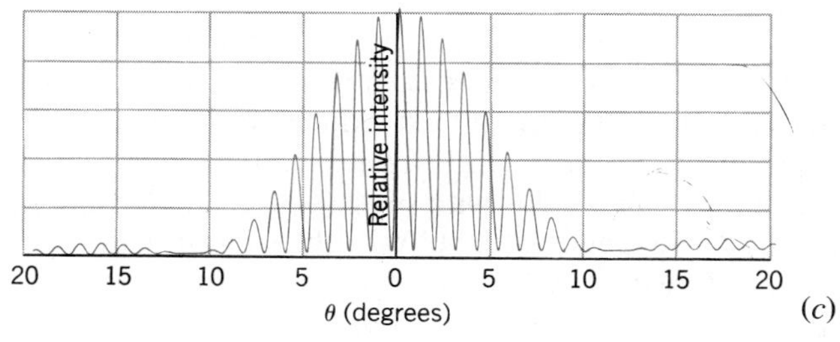

Figure 16 (*a*) The uniform fringe pattern to be expected from a double slit with vanishingly narrow slits. (*b*) The diffraction pattern of a typical slit of width *a* (not vanishingly narrow). (*c*) The fringe pattern formed by two of these slits. The pattern is equivalent to the pattern in (*a*) multiplied point by point by the pattern in (*b*).

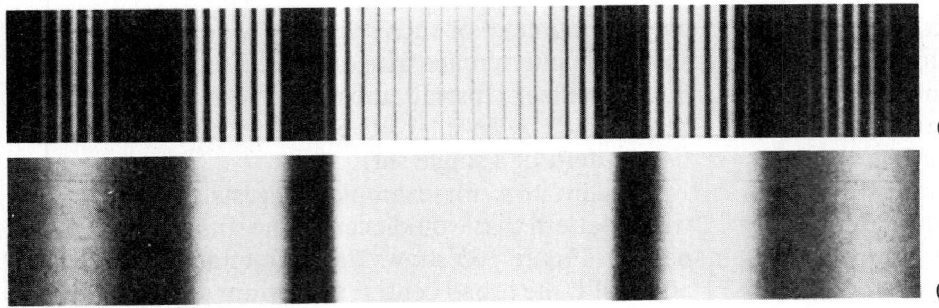

Figure 17 (*a*) Interference fringes for a double-slit system; compare Fig. 16*c*. (*b*) The diffraction pattern of a single slit; compare Fig. 16(*b*).

Figure 17*a* shows an actual double-slit interference–diffraction pattern. If one slit is covered, the single-slit diffraction pattern of Fig. 17*b* results. Note that Fig. 17*a* corresponds to Fig. 16*c* and Fig. 17*b* to Fig. 16*b*. In making these comparisons, bear in mind that Fig. 17 has been deliberately overexposed to bring out the faint secondary maxima and that two secondary maxima (rather than one) are shown.

The intensity of double-slit interference pattern is given by

$$I = I_m(\cos\beta)^2\left(\frac{\sin\alpha}{\alpha}\right)^2 \quad \text{(double slit)} \quad (12)$$

in which

$$\beta = \left(\frac{\pi d}{\lambda}\right)\sin\theta \quad \text{and} \quad \alpha = \left(\frac{\pi a}{\lambda}\right)\sin\theta. \quad (13)$$

Here d is the distance between the centers of the slits and a is the slit width. A study of Eq. 12 should convince you that it is the product of the interference pattern for a pair of narrow slits with slit separation d (see Eqs. 15 and 16 of Chapter 40) and the diffraction pattern for a single slit of width a (see Eqs. 5 and 6).

Let us look at Eqs. 12 and 13 more closely. If we put $a = 0$ in Eq. 13, for example, then $\alpha = 0$ and $\sin\alpha/\alpha = 1$. Equation 12 then reduces, as it must, to an equation describing the interference pattern for a pair of vanishingly narrow slits with slit separation d. Similarly, putting $d = 0$ is equivalent physically to causing the two slits to merge into a single slit of width a. Putting $d = 0$ in Eq. 13 yields $\beta = 0$ and $\cos^2\beta = 1$. In this case Eq. 12 reduces, as it must, to an equation describing the diffraction pattern for a single slit of width a.

The double-slit pattern described by Eqs. 12 and 13 and displayed in Fig. 17*a* combines interference and diffraction in an intimate way. Both are superposition effects and depend on adding wave disturbances at a given point, taking phase differences properly into account. If the waves to be combined originate from a finite (and usually small) number of elementary coherent sources—as in Young's double-slit experiment—we call the process *interference*. If the waves to be combined originate by subdividing a wavefront into infinitesimal coherent sources of differential size—as we did in Fig. 6—we call the process *diffraction*. This distinction between interference and diffraction (which is somewhat arbitrary and not always adhered to) is convenient, but we should not lose sight of the fact that both are superposition effects and that often both are present simultaneously, as in Fig. 17*a*.

Sample Problem 5 In a double-slit experiment, the distance D of the screen from the slits is 52 cm, the wavelength λ is 480 nm, the slit separation d is 0.12 mm, and the slit width a is 0.025 mm. (a) What is the spacing between adjacent fringes?

The intensity pattern is given by Eq. 12, the fringe spacing being determined by the interference factor $\cos^2\beta$. From Sample Problem 1, Chapter 40, we have

$$\Delta y = \frac{\lambda D}{d}.$$

Substituting yields

$$\Delta y = \frac{(480\times10^{-9}\ \text{m})(52\times10^{-2}\ \text{m})}{0.12\times10^{-3}\ \text{m}}$$

$$= 2.080\times10^{-3}\ \text{m} \approx 2.1\ \text{mm}. \qquad \text{(Answer)}$$

(b) What is the distance from the central maximum to the first minimum of the fringe envelope?

The angular position of the first minimum follows from Eq. 1, or

$$\sin\theta = \frac{\lambda}{a} = \frac{480\times10^{-9}\ \text{m}}{25\times10^{-6}\ \text{m}} = 0.0192.$$

This is so small that, with little error, we can put $\sin\theta \approx$

$\tan\theta \approx \theta$. Thus

$$y = D\tan\theta \approx D\theta = (52 \times 10^{-2}\ \text{m})(0.0192)$$
$$= 9.98 \times 10^{-3}\ \text{m} \approx 10\ \text{mm}. \qquad \text{(Answer)}$$

You can show that there are about 9 fringes in the central peak of the diffraction envelope.

Sample Problem 6 What requirements must be met for the central maximum of the envelope of the double-slit interference pattern to contain exactly 11 fringes?

The required condition will be met if the sixth minimum of the interference factor ($\cos^2\beta$ in Eq. 12) coincides with the first minimum of the diffraction factor $(\sin\alpha/\alpha)^2$ in Eq. 12.

The sixth minimum of the interference factor occurs when

$$\beta = (11/2)\pi$$

in Eq. 12. The first minimum in the diffraction term occurs for

$$\alpha = \pi$$

in Eq. 12. Dividing (see Eq. 13) yields

$$\frac{\beta}{\alpha} = \frac{d}{a} = \frac{11}{2}.$$

This condition depends only on the ratio of the slit separation d to the slit width a and not at all on the wavelength. For long waves the pattern will be broader than for short waves, but there will always be 11 fringes in the central peak of the envelope.

41-8 Multiple Slits

A logical extension of Young's double-slit interference experiment is to increase the number of slits from two to a larger number N. An arrangement like that of Fig. 18, usually involving many more slits—as many as 10^3/mm is not uncommon—is called a *diffraction grating*. As for a double slit, the intensity pattern that results when monochromatic light falls on a grating consists of a series of interference fringes.

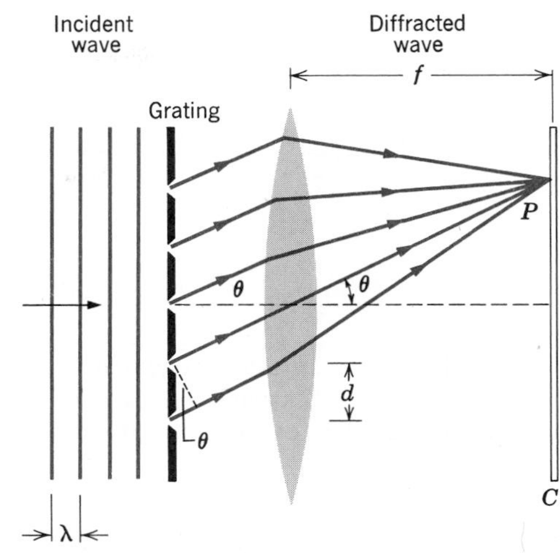

Figure 18 An idealized diffraction grating containing 5 slits. The figure is distorted for clarity.

Figure 19 compares the intensity patterns for $N = 2$ and for $N = 5$, showing only the central maximum of the single-slit diffraction envelope. We see two important changes that occur when we increase the number of slits from two to five. (1) The fringes become narrower. (2) Faint secondary maxima (three in this case) appear between the fringes. As N increases, perhaps to 10^4 for a useful grating, the fringes become very sharp indeed and the secondary maxima, while increasing in number, become so reduced in intensity as to be negligible in their effects; we shall ignore them in what follows.

Figure 19 (*a*) A diffraction pattern for a two-slit diffraction "grating." (*b*) The same for a five-slit grating. Note that the fringes are sharper and that secondary maxima of low intensity make their appearance. The pattern for the five-slit grating was deliberately overexposed to bring out these faint secondary maxima.

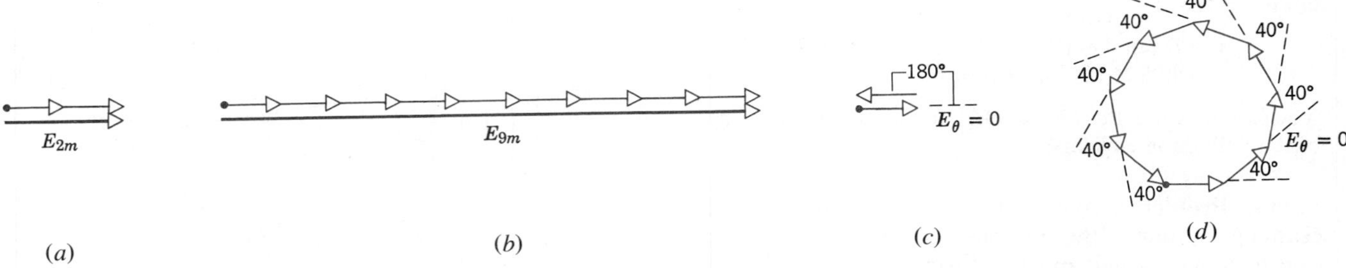

Figure 20 Drawings (*a*) and (*b*) show conditions at the central maximum for a two-slit and a nine-slit grating, respectively. Drawings (*c*) and (*d*) show conditions at the minimum of zero intensity that lies on either side of this central maximum.

Positions of the Fringes. A sharply defined fringe (or *line*) will occur when $d \sin \theta$, which is the path difference between rays from adjacent slits in Fig. 18, is equal to an integral number of wavelengths, or

$$\boxed{d \sin \theta = m\lambda} \quad m = 0, 1, 2, \ldots \quad \text{(maxima)}. \tag{14}$$

Here m is called the *order number* of the line in question, $m = 0$ corresponding to the central line. This equation is identical with Eq. 9 of Chapter 40, which locates the intensity maxima for a double slit. The locations of the diffracted lines are thus determined only by the ratio λ/d and are independent of N.

Width of the Fringes. Here we seek to understand how the fringes narrow into sharp lines as N is increased. We use a graphical argument, based on phasors. Figures 20*a* and 20*b* show conditions at the central maximum for a two-slit and a nine-slit grating. The small arrows represent the amplitudes of the wave disturbances arriving at the center of the screen.

Consider the angle $\Delta\theta$, corresponding to the position of zero intensity that lies on either side of the central maximum. Figures 20*c* and 20*d* show the phasors at this point. The phase difference between waves from adjacent slits, which is zero at the central maximum, must increase by an amount $\Delta\phi$ chosen so that the array of the phasors just closes on itself, yielding zero resultant intensity. For $N = 2$, $\Delta\phi = 2\pi/2$ $(= 180°)$; for $N = 9$, $\Delta\phi = 2\pi/9$ $(= 40°)$. In the general case of N slits, the phase difference at the first intensity minimum is given by

$$\Delta\phi = \frac{2\pi}{N}. \tag{15}$$

We see that, as N increases, the phase difference between adjacent slits that corresponds to the first intensity minimum becomes smaller and smaller. This phase difference for adjacent waves corresponds to a path difference ΔL given by Eq. 22 of Chapter 40, or

$$\text{path difference} = \left(\frac{\lambda}{2\pi}\right) \text{(phase difference)}$$

or, from Eq. 15,

$$\Delta L = \left(\frac{\lambda}{2\pi}\right)(\Delta\phi) = \left(\frac{\lambda}{2\pi}\right)\left(\frac{2\pi}{N}\right) = \frac{\lambda}{N}. \tag{16}$$

From Fig. 18, however, the path difference ΔL at the first minimum is also given by $d \sin \Delta\theta$, so that we can write

$$d \sin \Delta\theta = \frac{\lambda}{N},$$

or (because $\Delta\theta$ is very small for a sharp maximum)

$$\Delta\theta = \frac{\lambda}{Nd} \quad \text{(central maximum)}. \tag{17}$$

We interpret $\Delta\theta$ as the *line width* of the central maximum. Equation 17 shows specifically that, if λ and d are fixed, the line width $\Delta\theta$ of the central maximum will decrease as we increase N. That is, the central maximum becomes sharper as we add more slits to the grating. (Comparison of Eq. 17 with Eq. 3 suggests that, as far as the location of its first diffraction minimum is concerned, the grating behaves like a single slit of width Nd.)

Equation 17 gives the line width for the central line only. We state without proof that the general expression for the width of lines of other orders, that occur at an angle θ, is

$$\boxed{\Delta\theta = \frac{\lambda}{Nd \cos \theta}} \quad \text{(line width)} \tag{18}$$

in which θ defines the direction in which the line in question occurs. (Note that, in this case, the grating behaves like a single slit whose width in the direction defined by θ is $Nd \cos \theta$, the *projected* width of the grating.)

41-9 Diffraction Gratings

The grating spacing d for a typical grating that contains 12,000 "slits" distributed over a 1-in. width is 25.4 mm/12,000 or 2100 nm. Gratings are widely used to measure wavelengths and to study the structure and intensity of spectral lines.

Gratings are made by ruling equally spaced parallel grooves on a glass or a metal plate, using a diamond cutting point whose motion is automatically controlled by an elaborate ruling engine. Gratings ruled on metal are called *reflection gratings* because the interference effects are viewed in reflected rather than in transmitted light. Once such a master grating has been prepared, replicas can be formed by pouring a liquid plastic on the grating, allowing it to harden, and stripping it off. The stripped plastic, fastened to a flat piece of glass or other backing, forms a good grating.

Figure 21 shows a simple grating spectroscope, used for viewing the spectrum of a light source, assumed to emit a number of discrete wavelengths, or spectral lines. The light from source S is focused by lens L_1 on a slit S_1 placed in the focal plane of lens L_2. The parallel light emerging from collimator C falls on grating G. Parallel rays associated with a particular maximum occurring at angle θ fall on lens L_3, being brought to a focus in a plane $F - F'$. The image formed in this plane is examined, using a magnifying lens arrangement E, called an eyepiece. A symmetrical interference pattern is formed on the other side of the central position, as shown by the dotted lines. The entire spectrum can be viewed by rotating telescope T through various angles. Instruments used for scientific research or in industry are more complex than the simple arrangement of Fig. 21. They invariably employ photographic or photoelectric recording and are called *spectrographs.*

Grating instruments can be used to make absolute measurements of wavelength, since the grating spacing d in Eq. 14 can be measured accurately with a microscope. Several spectra are normally produced in such instruments, corresponding to $m = 1, 2, 3, \ldots$ in Eq. 14; see Fig. 22. This may cause some confusion if the spectra overlap. Further, this multiplicity of spectra reduces the recorded intensity of any given spectrum line because the available energy is divided among a number of spectra. However, by controlling the shape of the grating rulings, a large fraction of the energy can be concentrated in a particular order; this is called *blazing.*

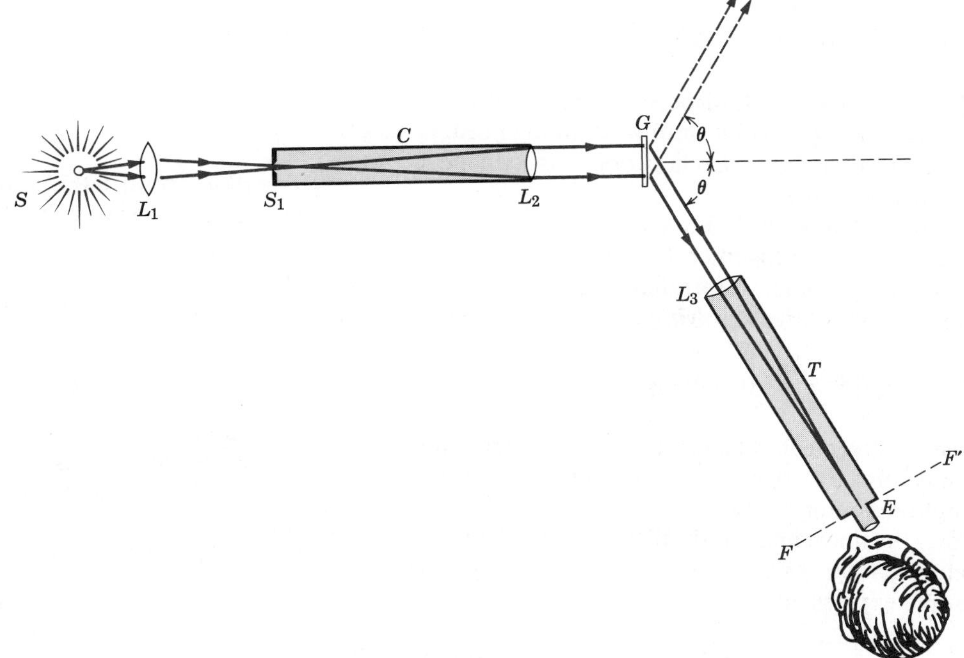

Figure 21 A simple type of grating spectroscope used to analyze the wavelengths of the light emitted by source S.

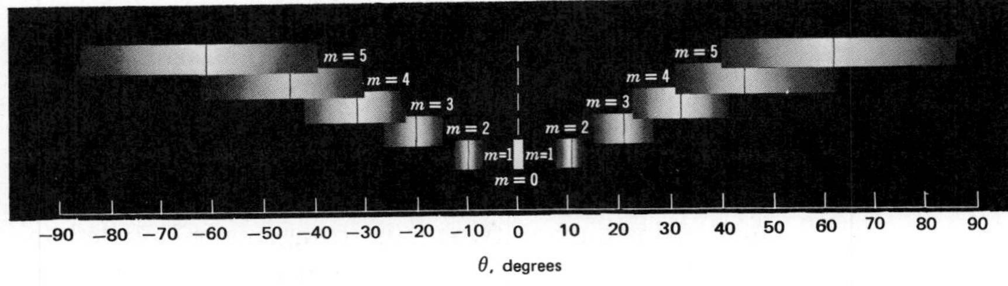

Figure 22 The spectrum of white light as viewed in an instrument like that of Fig. 21. The different orders, identified by the order number m, are shown separated vertically for clarity. As actually viewed, they overlap. The central line in each order corresponds to $\lambda = 550$ nm, the center of the visible spectrum.

41-10 Gratings: Dispersion and Resolving Power (Optional)

Dispersion. It is desirable in a grating that wavelengths that are close together be as widely separated in angle as possible. That is, we wish a grating to have a large *dispersion D*, defined from

$$D = \frac{\Delta\theta}{\Delta\lambda} \quad \text{(dispersion defined).} \tag{19}$$

Here $\Delta\theta$ is the angular separation of two lines whose wavelengths differ by $\Delta\lambda$. We show below that the dispersion of a grating is given by

$$\boxed{D = \frac{m}{d\cos\theta}} \quad \text{(dispersion of a grating).} \tag{20}$$

Thus, to achieve high dispersion, use a grating of small grating spacing (small d) and work in high orders (large m). Note that the dispersion does not depend on the number of rulings.

Resolving Power. To separate lines whose wavelengths are close together, the line widths should be as small as possible. Expressed otherwise, the grating should have a large *resolving power R*, defined from

$$R = \frac{\lambda}{\Delta\lambda} \quad \text{(resolving power defined).} \tag{21}$$

Here λ is the mean wavelength of two spectrum lines that can barely be recognized as separate and $\Delta\lambda$ is the wavelength difference between them. The smaller $\Delta\lambda$ is, the closer the lines can be and still be resolved. We shall show below that the resolving power of a grating is given by the simple expression

$$\boxed{R = Nm} \quad \text{(resolving power of a grating).} \tag{22}$$

It is to achieve high *dispersion* that grating rulings are closely spaced (d small; see Eq. 20). It is to achieve high *resolving power* that many rulings are used (N large; see Eq. 22).

Proof of Eq. 20. Let us start with Eq. 14, the expression for the angular positions of the lines in the diffraction pattern of a grating:

$$d\sin\theta = m\lambda.$$

Let us regard θ and λ as variables and take differentials of this expression. We find

$$d\cos\theta\, d\theta = m\, d\lambda.$$

For small enough angles, we can write these differentials as small differences, thus

$$d\cos\theta\,\Delta\theta = m\,\Delta\lambda \tag{23}$$

or

$$\frac{\Delta\theta}{\Delta\lambda} = \frac{m}{d\cos\theta}.$$

The ratio on the left is simply D (see Eq. 19) so that we have indeed derived Eq. 20.

Proof of Eq. 22. We start with Eq. 23, which was derived from Eq. 14, the expression for the angular position of the lines in the diffraction pattern formed by a grating. Thus

$$d\cos\theta\,\Delta\theta = m\,\Delta\lambda. \tag{24}$$

Here $\Delta\lambda$ is the small wavelength difference between the two waves and $\Delta\theta$ is the angular separation between them. If $\Delta\theta$ is to be the largest angle that will permit the two lines to be resolved, it must (by Rayleigh's criterion) be equal to the width of the line, which is given by Eq. 18, or

$$\Delta\theta = \frac{\lambda}{Nd\cos\theta} \quad \text{(line width).} \tag{25}$$

Table 1 Three Gratings[a]

Grating	*N*	*d* nm	θ	*D* °/μm	*R*
A	10,000	2540	13.3°	23.2	10,000
B	20,000	2540	13.3°	23.2	20,000
C	10,000	1370	15.5°	43.4	10,000

[a] For $\lambda = 589$ nm and $m = 1$.

If we substitute Eq. 25 into Eq. 24, we find

$$d\cos\theta = \frac{\lambda}{Nd\cos\theta} = m\,\Delta\lambda,$$

from which it readily follows that

$$R = \frac{\lambda}{\Delta\lambda} = Nm.$$

This is Eq. 22, which we set out to derive.

Dispersion and Resolving Power Compared. The resolving power of a grating must not be confused with its dispersion. Table 1 shows the characteristics of three gratings, each illuminated with light of $\lambda = 589$ nm, the diffracted light being viewed in the first order ($m = 1$ in Eq. 14). You should verify that the values of D and R are given in the table can be calculated from Eqs. 20 and 22, respectively.

For the conditions noted in Table 1, gratings A and B have the same *dispersion* and A and C have the same *resolving power.*

Figure 23 shows the intensity patterns (line shapes) that would be produced by these gratings for two lines of wavelength λ_1 and λ_2, in the vicinity of $\lambda = 589$ nm. Grating B, which has high resolving power, has narrow lines and is inherently capable of distinguishing lines that are much closer together in wavelength than those in the figure. Grating C, which was high dispersion, produces twice the angular separation between lines λ_1 and λ_2 as does grating B.

Sample Problem 7 A diffraction grating has 1.26×10^4 rulings uniformly spaced over 25.4 mm. It is illuminated at normal incidence by yellow light from a sodium vapor lamp. This light contains two closely spaced lines (the well-known sodium doublet) of wavelengths 589.00 nm and 589.59 nm. (a) At what angle will the first-order maximum occur for the first of these wavelengths?

The grating spacing d is given by

$$d = \frac{L}{N} = \frac{25.4 \times 10^{-3}\ \text{m}}{1.26 \times 10^4}$$

$$= 2.016 \times 10^{-6}\ \text{m} = 2016\ \text{nm}.$$

The first-order maximum corresponds to $m = 1$ in Eq. 14. We thus have

$$\theta = \sin^{-1}\frac{m\lambda}{d} = \sin^{-1}\left(\frac{(1)(589.00\ \text{nm})}{2016\ \text{nm}}\right)$$

$$= 16.99° \approx 17.0°. \qquad \text{(Answer)}$$

(b) What is the angular separation between these two lines (in first order)?

Here the *dispersion* of the grating comes into play. From Eq. 20, the dispersion is

$$D = \frac{m}{d\cos\theta} = \frac{1}{(2016\ \text{nm})(\cos 16.99°)}$$

$$= 5.187 \times 10^{-4}\ \text{rad/nm}.$$

From Eq. 19, the defining equation for dispersion, we have

$$\begin{aligned}\Delta\theta &= D\,\Delta\lambda \\ &= (5.187 \times 10^{-4}\ \text{rad/nm})(589.59\ \text{nm} - 589.00\ \text{nm}) \\ &= 3.06 \times 10^{-4}\ \text{rad} = 0.0175° = 1.05\ \text{arc min}. \qquad \text{(Answer)}\end{aligned}$$

As long as the grating spacing d remains fixed, this result holds no matter how many lines there are in the grating.

(c) How close in wavelength can two lines be (in first order) and still be resolved by this grating?

Here the *resolving power* of the grating comes into play. From Eq. 22, the resolving power is

$$R = Nm = (1.26 \times 10^4)(1) = 1.26 \times 10^4.$$

From Eq. 21, the defining equation for resolving power, we have,

$$\Delta\lambda = \frac{\lambda}{R} = \frac{589\ \text{nm}}{1.26 \times 10^4} = 0.047\ \text{nm}. \qquad \text{(Answer)}$$

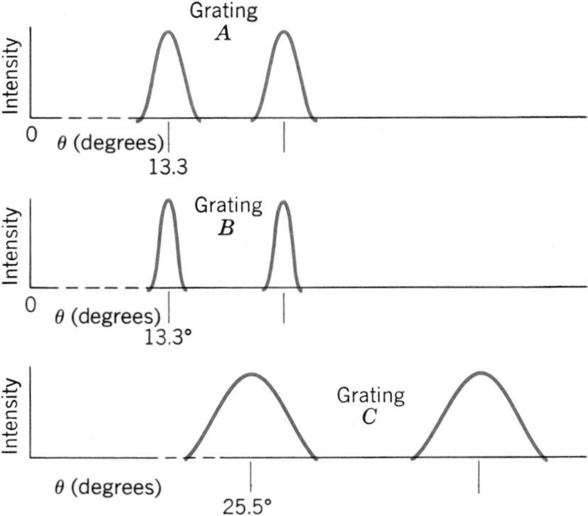

Figure 23 The intensity patterns for light of wavelengths λ_1 and λ_2 falling on the gratings of Table 1. Grating B has the highest resolving power and grating C the highest dispersion.

Thus, this grating can easily resolve the two sodium lines, which have a wavelength separation of 0.59 nm. Note that this result depends only on the number of grating rulings and is independent of d, the spacing between adjacent rulings.

(d) How many rulings can a grating have and just resolve the sodium doublet lines?

From Eq. 21, the defining equation for R, the grating must have a resolving power of

$$R = \frac{\lambda}{\Delta\lambda} = \frac{589\text{ nm}}{0.59\text{ nm}} = 998.$$

From Eq. 23, the number of rulings needed to achieve this resolving power (in first order) is

$$N = \frac{R}{m} = \frac{998}{1} = 998\text{ rulings.} \qquad \text{(Answer)}$$

Since the grating has about 13 times as many rulings as this, it can easily resolve the sodium doublet lines, as we have already shown in (c) above.

Sample Problem 8 A grating has 8200 lines uniformly spaced over 25.4 mm and is illuminated by light from a mercury vapor discharge. (a) What is the expected dispersion, in the third order, in the vicinity of the intense green line ($\lambda = 546$ nm)?

The grating spacing is given by

$$d = \frac{L}{N} = \frac{25.4 \times 10^{-3}\text{ m}}{8200} = 3.098 \times 10^{-6}\text{ m} \approx 3100\text{ nm}.$$

We must find the angle θ at which the line in question occurs. From Eq. 14, we have

$$\theta = \sin^{-1}\frac{m\lambda}{d} = \sin^{-1}\frac{(3)(546\text{ nm})}{3100\text{ nm}} = 31.9°.$$

We can now calculate the dispersion. From Eq. 20

$$D = \frac{m}{d\cos\theta} = \frac{3}{(3100\text{ nm})(\cos 31.9°)} = 1.14 \times 10^{-3}\text{ rad/nm} = 0.0653°/\text{nm} = 3.92\text{ arc min/nm.} \qquad \text{(Answer)}$$

(b) What is the resolving power of this grating in the fifth order? From Eq. 22

$$R = Nm = (8200)(5) = 4.10 \times 10^4. \qquad \text{(Answer)}$$

Thus, near $\lambda = 546$ nm and in fifth order, a wavelength difference given by (see Eq. 21)

$$\Delta\lambda = \frac{\lambda}{R} = \frac{546\text{ nm}}{4.10 \times 10^4} = 0.013\text{ nm}$$

can be resolved.

41-11 X-ray Diffraction

X rays are electromagnetic radiation whose wavelengths are of the order of 1 Å (= 10^{-10} m).* Compare this with 550 nm (= 5.5×10^{-7} m) for the center of the visible spectrum. Figure 24 shows how the x rays are produced when electrons from a heated filament F are accelerated by a potential difference V and strike a metal target T.

For such small wavelengths a standard optical diffraction grating, as normally employed, cannot be used to discriminate between different wavelengths. For $\lambda = 1$ Å (= 0.1 nm) and $d = 3000$ nm, for example, Eq. 14 shows that the first-order maximum occurs at

$$\theta = \sin^{-1}\frac{m\lambda}{d} = \sin^{-1}\frac{(1)(0.1\text{ nm})}{3000\text{ nm}} = 0.0019°.$$

This is too close to the central maximum to be practical. A grating with $d \approx \lambda$ is desirable, but, since x-ray wavelengths are about equal to atomic diameters, such gratings cannot be constructed mechanically.

In 1912 it occurred to the German physicist Max von Laue that a crystalline solid, consisting as it does of a regular array of atoms, might form a natural three-dimensional "diffraction grating" for x rays. The idea is that in a crystal, such as sodium chloride (NaCl), there is a basic unit of atoms (called the *unit cell*) that repeats itself throughout the array. In NaCl four sodium ions

* The *ångström* (1 Å = 10^{-10} m) is "a unit to be used with SI for a limited time." Its usefulness stems from the fact that atoms are about this same size.

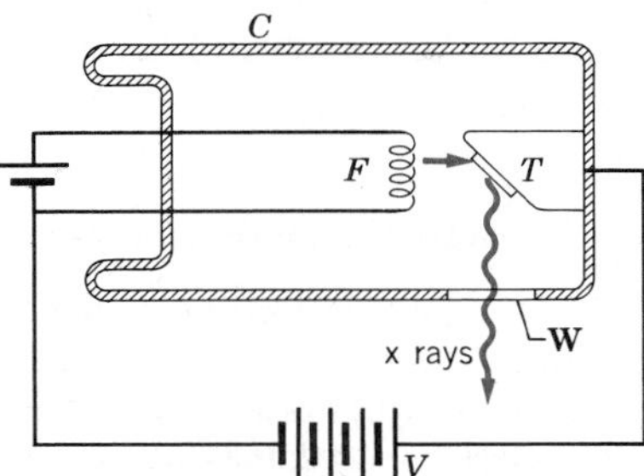

Figure 24 X rays are generated when electrons from heated filament F, accelerated through a potential difference V, strike a metal target T. W is a "window"—transparent to x rays—in the evacuated chamber C.

and four chlorine ions are associated with each unit cell. Figure 25 represents a section through a crystal of NaCl and identifies this basic unit. This crystal has a cubic structure and the unit cell is thus itself a cube, measuring a_0 on the side.

Each unit cell in the crystal acts as a diffracting center, just as a ruled groove acts as a linear diffracting center in a conventional grating. For both the conventional grating and the crystal grating, the intensity at any point outside the array is determined by the phase difference and intensity of radiation diffracted from each center to the point in question.

Figure 26*a* is a copy of Fig. 25*b* with a set of dashed lines added. These lines represent one of a large number of arbitrary families of planes that can be drawn through the crystal, each such family having a characteristic *interplanar spacing d.* Figure 26*b* shows how a strong diffracted beam can be "reflected" from such a family of

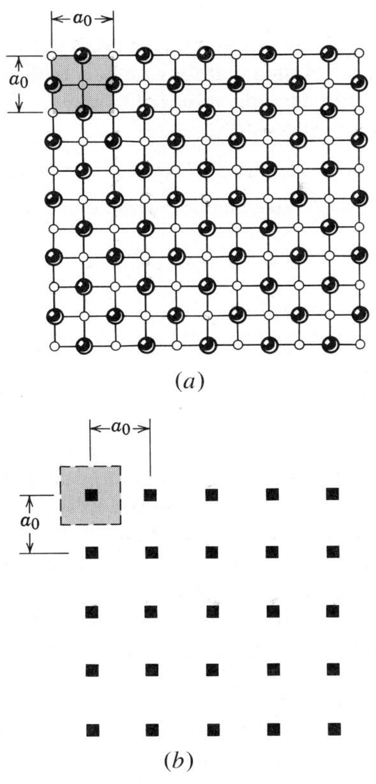

Figure 25 (*a*) A plane through a crystal of sodium chloride, showing the sodium and the chlorine ions. (*b*) The corresponding unit cells in this section, each cell being represented by a small black square.

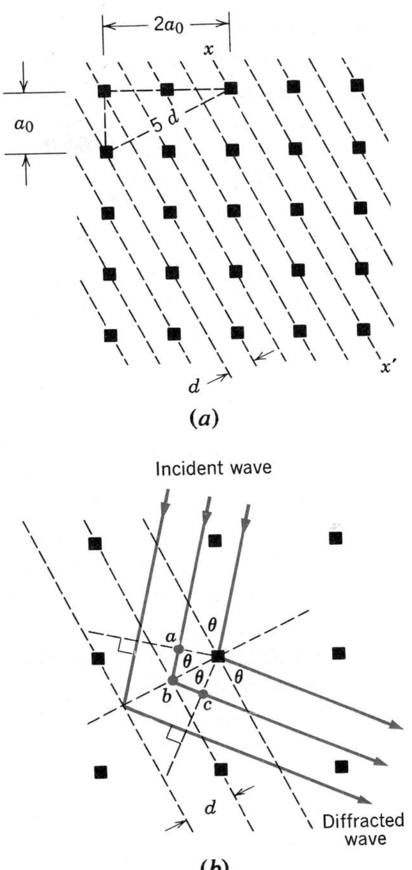

Figure 26 (*a*) A section through the NaCl cell lattice of Fig. 25*b*. The dashed sloping lines represent an arbitrary family of planes, with interplanar spacing *d*. (*b*) An incident beam falls on the family of planes shown in (*a*). A strong diffracted beam will be formed if Bragg's law is satisfied.

planes if the incident beam falls on this family in a direction that is proper for its wavelength.

If we consider only one of the planes that make up the family shown in Fig. 26*b*, mirrorlike "reflection" occurs for *any* incident angle. To have a constructive interference in the beam diffracted in the direction θ from the *entire family* of planes, the rays from the separate planes must reinforce each other. This means that the path difference for rays from adjacent planes (*abc* in Fig. 26*b*) must be an integral number of wavelengths, or

$$\boxed{2d \sin \theta = m\lambda} \qquad m = 1, 2, 3, \ldots \qquad \text{(Bragg's law).} \quad (26)$$

Each of the many families of planes has its own characteristic interplanar spacing d. Note in Fig. 26*b* that, contrary to the practice in optics, the angle θ is measured from the surface of the crystal, not from the normal to that surface. Bragg's law is named after the British physicist W. L. Bragg, who first derived it. He and his father shared the 1915 Nobel Prize for their use of x rays to study the structures of crystals.

For the planes of Fig. 26*a* you can show that the interplanar spacing d is related to the unit cell dimension a_0 by

$$d = \frac{a_0}{\sqrt{5}}. \tag{27}$$

This relation suggests how the dimensions of the unit cell can be found if the interplanar spacing is measured by x-ray diffraction methods.

X-ray diffraction is a powerful tool for studying both x-ray spectra and the arrangement of atoms in crystals. To study spectra, a particular set of crystal planes, having a known spacing d, is chosen. Diffraction from these planes diffracts different wavelengths at different angles. A detector, then, that can discriminate one angle from another can be used to determine the wavelength of radiation reaching it. On the other hand, we can study the crystal itself, using a monochromatic x-ray beam, determining not only the spacings of various crystal planes but also the structure of the unit cell. The DNA molecule, and many other equally complex structures, have been mapped by x-ray diffraction methods.

Sample Problem 9 At what angles must an x-ray beam with $\lambda = 1.10$ Å fall on the family of planes represented in Fig. 26*b* if a diffracted beam is to exist? Assume the material to be sodium chloride ($a_0 = 5.63$ Å).

The interplanar spacing d for these planes is given by Eq. 27 or

$$d = \frac{a_0}{\sqrt{5}} = \frac{5.63\ \text{Å}}{\sqrt{5}} = 2.518\ \text{Å}.$$

Equation 26 gives

$$\theta = \sin^{-1}\frac{m\lambda}{2d} = \sin^{-1}\left(\frac{(m)(1.10\ \text{Å})}{(2)(2.518\ \text{Å})}\right)$$
$$= \sin^{-1}(0.2184\, m).$$

Diffracted beams are possible for $\theta = 12.6°$ ($m = 1$), $\theta = 25.9°$ ($m = 2$), $\theta = 40.9°$ ($m = 3$), and $\theta = 60.9°$ ($m = 4$). Higher order beams cannot exist because they require that $\sin\theta > 1$.

Actually, the unit cell in cubic crystals such as NaCl has diffraction properties such that the intensity of diffracted x-ray beams corresponding to odd values of m is zero. Thus the only beams that are expected are

$$\theta = 25.9° \quad (m = 2) \quad \text{and} \quad \theta = 60.9° (m = 4). \quad \text{(Answer)}$$

REVIEW AND SUMMARY

Diffraction

Diffraction, which occurs when a wave encounters an obstacle or hole whose size is comparable to the wavelength of the wave, is convincing evidence of the wave theory of light. Using the Huygens' construction, the wave is divided at the obstruction into infinitesimal wavelets that then interfere with each other as they proceed (see Fig. 7). We treat diffraction effects that occur a large distance from the obstruction, often studied experimentally by interposing a lens so that patterns that would otherwise occur at infinite distance are focused onto a screen at the lens focal plane.

Single-slit Diffraction

Waves passing through a long narrow slit of width a produce a *single-slit diffraction pattern* with a central maximum together with minima corresponding to diffraction angles θ that satisfy

$$a \sin\theta = m\lambda \quad m = 1, 2, 3, \ldots \quad \text{(minima)}. \tag{3}$$

See Sample Problems 1 and 2. Using *phasor diagrams* to add the Huygens' wavelets, we show in Section 41–5 that the diffracted intensity for a given diffraction angle θ is

$$I = I_m\left(\frac{\sin\alpha}{\alpha}\right)^2 \quad \text{where} \quad \alpha = \frac{\pi a}{\lambda}\sin\theta. \tag{6}$$

See Sample Problem 3 for a calculation of the intensity of the secondary maxima.

Diffraction by a *circular* aperture or lens with diameter d also produces a central maximum

Circular Diffraction

with a first minimum at a diffraction angle θ given by

$$\sin \theta = 1.22 \frac{\lambda}{d} \quad \text{(1st minimum, circular aperature).} \qquad [9]$$

See Fig. 13.

Rayleigh's Criterion

Rayleigh's criterion suggests that two objects viewed through a telescope or microscope are on the verge of resolvability if the central diffraction maximum of one is at the first minimum of the other. Their angular separation must be at least

$$\theta_R = 1.22 \frac{\lambda}{d} \quad \text{(Rayleigh's criterion).} \qquad [11]$$

in which d is the diameter of the objective lens; Sample Problem 4.

Double-slit Diffraction

Waves passing through two slits, each of width a, whose centers are distance d apart, display diffraction patterns whose intensity I at various diffraction angles θ is given by

$$I = I_m \cos^2 \beta \left(\frac{\sin \alpha}{\alpha}\right)^2 \quad \text{(double slit),} \qquad [12]$$

with $\beta = (\pi d/\lambda) \sin \theta$ and α the same as for single-slit diffraction. See Sample Problems 5 and 6.

Multiple-slit Diffraction

Diffraction by N *multiple slits* results in principal maxima whenever

$$d \sin \theta = m\lambda \quad m = 0, 1, 2 \ldots \quad \text{(maxima)} \qquad [14]$$

with the angular width of the maxima given by

$$\Delta\theta = \frac{\lambda}{Nd \cos \theta} \quad \text{(line width).} \qquad [18]$$

Diffraction Gratings

A *diffraction grating* is a series of "slits" used to separate an incident wave into different wavelength components whose principal diffraction maxima are directionally dispersed by the grating. A grating is characterized by two parameters, the dispersion D and the resolving power R.

$$D = \frac{\Delta\theta}{\Delta\lambda} = \frac{m}{d \cos \theta} \quad \text{and} \quad R = \frac{\lambda}{\Delta\lambda} = Nm. \qquad [19\text{–}22]$$

See Fig. 23 and Sample Problems 7 and 8.

X-ray Diffraction

The regular array of atoms in a crystal is a three-dimensional diffraction grating for short-wavelength waves such as x rays. The atoms can be visualized as being arranged in planes with characteristic interplanar spacing d. Diffraction maxima (constructive interference) occur if the incident direction of the wave, measured from the surface of a plane of atoms, and the wavelength λ of the radiation satisfy *Bragg's law:*

Bragg's Law

$$2d \sin \theta = m\lambda \quad m = 1, 2, 3 \ldots \quad \text{(Bragg's law).} \qquad [26]$$

See Sample Problem 9.

QUESTIONS

1. Why is the diffraction of sound waves more evident in daily experience than that of light waves?

2. Sound waves can be diffracted. About what width of a single slit should you use if you wish to broaden the distribution of an incident plane sound wave of frequency 1 kHz?

3. Why aren't sound waves polarized?

4. Why do radio waves diffract around buildings, although light waves do not?

5. A loud-speaker horn, used at a rock concert, has a rectangular aperture 1 m high and 30 cm wide. Will the pattern of sound intensity be broader in the horizontal plane or in the vertical?

6. For what kind of waves could a long picket fence be considered a useful diffraction grating?

7. A particular radar antenna is designed to give accurate measurements of the altitude of an aircraft but less accurate measurements of its direction in a horizontal plane. Must the height-to-width ratio of the radar antenna be less than, equal to, or greater than unity?

8. In single-slit diffraction, what is the effect of increasing (*a*) the wavelength and (*b*) the slit width?

9. Why are the colors in the spectrum of a light source called "lines"?

10. While listening to the car radio, you may have noticed that the AM signal fades—but the FM signal doesn't—when you drive under a bridge. Could diffraction have anything to do with this?

11. What will the single-slit diffraction pattern look like if $\lambda > a$?

12. What would the pattern on a screen formed by a double slit look like if the slits did not have the same width? Would the locations of the fringes be changed?

13. The shadow of a vertical flagpole cast by the sun has clearly defined edges near its base, but less-well-defined edges near its top end. Why?

14. A crossed diffraction grating has lines ruled in two directions, at right angles to each other. Predict the pattern of light intensity on the screen if light is sent through such a grating.

15. Sunlight falls on a single slit of width 1 μm. Describe qualitatively what the resulting diffraction pattern looks like.

16. In Fig. 6, rays r_1 and r_3 are in phase; so are r_2 and r_4. Why isn't there a maximum intensity at P_2 rather than a minimum?

17. When the atmosphere is not quite clear, one may sometimes see colored circles concentric with the sun or moon, generally not more than four or five times the diameter of the sun or moon and invariably having the inner edge blue. What is the explanation of this phenomenon?

18. When we speak of diffraction by a single slit we imply that the width of the slit must be much less than its length. Suppose that, in fact, the length was equal to twice the width. Make a rough guess at what the diffraction pattern would look like.

19. In connection with Fig. 4 we stated, correctly, that the optical pathlengths from the slit to point P_0 are all the same. Why?

20. Distinguish clearly between θ, α, and ϕ in Eqs. 5 and 6.

21. If we were to redo our analysis of the properties of lenses in Chapter 39 by the methods of geometric optics but without restricting our considerations to paraxial rays and to "thin" lenses, would diffraction phenomena emerge from the analysis? Discuss.

22. The double-slit pattern of Fig. 27*a* seen with a monochromatic light source is somehow changed to the pattern of Fig. 27*b*. Consider the following possible changes in conditions: (1) The wavelength of the light was decreased. (2) The wavelength of the light was increased. (3) The width of each slit was increased. (4) The separation of the slits was increased. (5) The separation of the slits was decreased. (6) The width of each slit was decreased. Which selection(s) of the above changes would explain the alteration of the pattern?

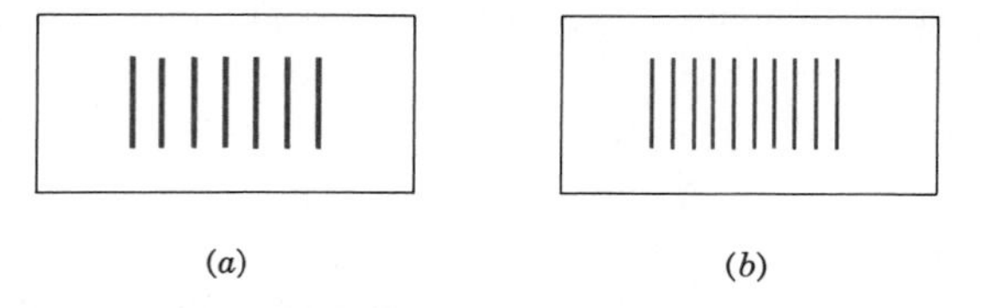

Figure 27 Question 22.

23. We have seen that diffraction limits the resolving power of optical telescopes (see Fig. 14). Does it also do so for large radio telescopes?

24. In double-slit interference patterns such as that of Fig. 17*a* we said that the interference fringes were modulated in intensity by the diffraction pattern of a single slit. Could we reverse this statement and say that the diffraction pattern of a single slit is intensity-modulated by the interference fringes? Discuss.

25. Assume that the limits of the visible spectrum are 430 and 680 nm. How would you design a grating, assuming that the incident light falls normally on it, such that the first-order spectrum barely overlaps the second-order spectrum?

26.. A glass prism can form a spectrum. Explain how. How many "orders" of spectra will a prism produce?

27. For the simple spectroscope of Fig. 21 show (*a*) that θ increases with λ for a grating and (*b*) that θ decreases with λ for a prism.

28. Explain in your own words why increasing the number of slits, *N*, in a diffraction grating sharpens the maxima. Why does decreasing the wavelength do so? Why does increasing the slit spacing, *d*, do so?

29. How much information can you discover about the structure of a diffraction grating by analyzing the spectrum it forms of a monochromatic light source? Let $\lambda = 589$ nm, for an example.

30. (*a*) Why does a diffraction grating have closely spaced rulings? (*b*) Why does it have a large number of rulings?

31. Two nearly-equal wavelengths are incident on a grating of *N* slits and are not quite resolvable. However, they become resolved if the number of slits is increased. Formulas aside, is the explanation of this that: (*a*) more light can get through the grating? (*b*) the principal maxima become more intense and hence resolvable? (*c*) the diffraction pattern is spread more and hence the wavelengths become resolved? (*d*) there are a larger number of orders? or (*e*) the principal maxima become narrower and hence resolvable?

32. How can the resolving power of a lens be increased?

33. The relation $R = Nm$ suggests that the resolving power of a given grating can be made as large as desired by choosing an arbitrarily high order of diffraction. Discuss this possibility.

34. Show that at a given wavelength and a given angle of diffraction the resolving power of a grating depends only on its width $W (= Nd)$.

35. How would you experimentally measure (*a*) the dispersion D and (*b*) the resolving power R of a grating spectrograph?

36. For a given family of planes in a crystal, can the wavelength of incident x rays be (*a*) too large or (*b*) too small to form a diffracted beam?

37. If a parallel beam of x rays of wavelength λ is allowed to fall on a randomly-oriented crystal of any material, generally no intense diffracted beams will occur. Such beams appear if (*a*) the x-ray beam consists of a continuous distribution of wavelengths rather than a single wavelength or (*b*) the specimen is not a single crystal but a finely divided powder. Explain each case.

38. Does an x-ray beam undergo refraction as it enters and leaves a crystal? Explain your answer.

EXERCISES AND PROBLEMS

Section 41-2 Diffraction from a Single Slit—Locating the Minima

1E. When monochromatic light is incident on a slit 0.022 mm wide, the first diffraction minimum is observed at an angle of 1.8° from the direction of the direct beam. What is the wavelength of the incident light?

2E. Monochromatic light of wavelength 441 nm falls on a narrow slit. On a screen 2.0 m away, the distance between the second minimum and the central maximum is 1.5 cm. (*a*) Calculate the angle of diffraction θ of the second minimum. (*b*) Find the width of the slit.

3E. Light of wavelength 633 nm is incident on a narrow slit. The angle between the first minimum on one side of the central maximum and the first minimum on the other side is 1.2°. What is the width of the slit?

4E. A single slit is illuminated by light whose wavelengths are λ_a and λ_b, so chosen that the first diffraction minimum of the λ_a component coincides with the second minimum of the λ_b component. (*a*) What relationship exists between the two wavelengths? (*b*) Do any other minima in the two patterns coincide?

5E. The distance between the first and fifth minima of a single-slit pattern is 0.35 mm with the screen 40 cm away from the slit, using light having a wavelength of 550 nm. (*a*) Calculate the diffraction angle θ of the first minimum. (*b*) Find the width of the slit.

6E. For a single slit the first minimum occurs at $\theta = 90°$, thus filling the forward hemisphere beyond the slit with light. What must be the ratio of the slit width to the wavelength for this to take place?

7E. A plane wave, wavelength = 590 nm, falls on a slit with $a = 0.40$ mm. A thin converging lens, focal length = +70 cm, is placed behind the slit and focuses the light on a screen. (*a*) How far is the screen behind the lens? (*b*) What is the linear distance on the screen from the center of the pattern to the first minimum?

8P. A slit 1.0 mm wide is illuminated by light of wavelength 589 nm. We see a diffraction pattern on a screen 3.0 m away. What is the distance between the first two diffraction minima on the same side of the central diffraction maximum?

9P. Sound waves with frequency 3000 Hz diffract out of a speaker cabinet with a 0.30-m diameter opening into a large auditorium. Where does a listener standing against a wall 100 m from the speaker have the most difficulty hearing? Assume a sound speed of 343 m/s.

10P. Manufacturers of wire (and other objects of small dimensions) sometimes use a laser to continually monitor the thickness of the product. The wire intercepts the laser beam, producing a diffraction pattern like that of a single slit of the same width as the wire diameter; see Fig. 28. Suppose a He–Ne laser, wavelength 632.8 nm, illuminates a wire, the diffraction pattern being projected onto a screen 2.6 m away. If the desired wire diameter is 1.37 mm, what would be the observed distance between the two tenth order minima on each side of the central maximum?

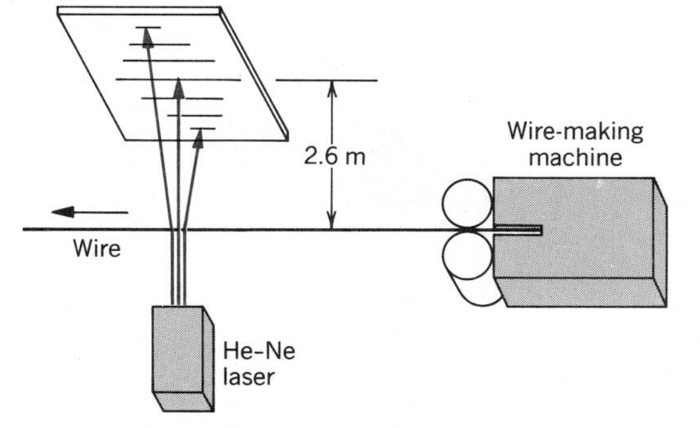

Figure 28 Problem 10.

Section 41-5 Diffraction from a Single Slit—Quantitative

11E. A 0.10-mm-wide slit is illuminated by light with a wavelength 589 nm. Consider rays for which $\theta = 30°$ in Fig. 8*b* and calculate the phase difference at the screen of Huygens' wavelets from the top and midpoint of the slit. (*Hint:* See Eq. 4.)

12E. Monochromatic light with wavelength 538 nm falls on a slit with width 0.025 mm. The distance from the slit to a screen is 3.5 m. Consider a point on the screen 1.1 cm from the central maximum. (*a*) Calculate θ. (*b*) Calculate α. (*c*) Calculate the ratio of the intensity at this point to the intensity at the central maximum.

13P. If you double the width of a single slit, the intensity of the central maximum of the diffraction pattern increases by a factor of four, even though the energy passing through the slit only doubles. Explain this quantitatively.

14P. *Babinet's principle:* A monochromatic beam of parallel light is incident on a "collimating" hole of diameter $x \gg \lambda$. Point P lies in the geometrical shadow region on a distant screen, as shown in Fig. 29*a*. Two obstacles, shown in Fig. 28*b*, are placed in turn over the collimating hole. A is an opaque circle with a hole in it and B is the "photographic negative" of A. Using superposition concepts, show that the intensity at P is identical for each of the two diffracting objects A and B. In this connection, it can be shown that the diffraction pattern of a wire is that of a slit of equal width. See "Measuring the Diameter of a Hair by Diffraction" by S.M. Curry and A.L. Schawlow, *American Journal of Physics,* May 1974.

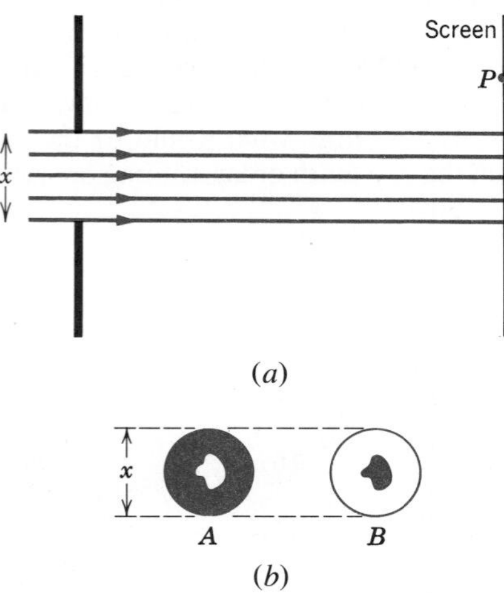

Figure 29 Problem 14.

15P. The full width at half maximum (FWHM) of the central diffraction maximum is defined as the angle between the two points in the pattern where the intensity is one-half that at the center of the pattern. (See Fig. 11*b*.) (*a*) Show that the intensity drops to one-half of the maximum value when $\sin^2 \alpha = \alpha^2/2$. (*b*) Verify that $\alpha = 1.39$ radians (about 80°) is a solution to the transcendental equation of part (*a*). (*c*) Show that the FWHM is $\Delta\theta = 2 \sin^{-1}(0.443\lambda/a)$. (*d*) Calculate the FWHM of the central maximum for slits whose widths are 1, 5, and 10 wavelengths.

16P. (*a*) Show that the values of α at which intensity maxima for single-slit diffraction occur can be found exactly by differentiating Eq. 5 with respect to α and equating to zero, obtaining the condition

$$\tan \alpha = \alpha.$$

(*b*) Find the values of α satisfying this relation by plotting graphically the curve $y = \tan \alpha$ and the straight line $y = \alpha$ and finding their intersections or by using a pocket calculator to find an appropriate value of α by trial and error. (*c*) Find the (nonintegral) values of m corresponding to successive maxima in the single-slit pattern. Note that the secondary maxima do not lie exactly halfway between minima.

17P*. Derive this expression for the intensity pattern for a three-slit "grating":

$$I = \tfrac{1}{9} I_m (1 + 4 \cos \phi + 4 \cos^2 \phi),$$

where

$$\phi = \frac{2\pi d \sin \theta}{\lambda}.$$

Assume that $a \ll \lambda$ and be guided by the derivation of the corresponding double-slit formula (Eq. 15 of Chapter 40).

Section 41-6 Diffraction from a Circular Aperture

18E. The two headlights of an approaching automobile are 1.4 m apart. At what (*a*) angular separation and (*b*) maximum distance will the eye resolve them? Assume a pupil diameter of 5.0 mm and a wavelength of 550 nm. Also assume that diffraction effects alone limit the resolution.

19E. An astronaut in a satellite claims he can just barely resolve two point sources on the earth, 160 km below him. Calculate their (*a*) angular and (*b*) linear separation, assuming ideal conditions. Take $\lambda = 540$ nm, and the pupil diameter of the astronaut's eye to be 5.0 mm.

20E. Find the separation of two points on the moon's surface that can just be resolved by the 200-in. (= 5.1-m) telescope at Mount Palomar, assuming that this distance is determined by diffraction effects. The distance from the earth to the moon is 3.8×10^5 km. Assume a wavelength of 550 nm.

21E. The wall of a large room is covered with acoustic tile in which small holes are drilled 5.0 mm from center to center. How far can a person be from such a tile and still distinguish the individual holes, assuming ideal conditions? Assume the diameter of the pupil of the observer's eye to be 4.0 mm and the wavelength to be 550 nm.

22E. The pupil of a person's eye has a diameter of 5.0 mm. What distance apart must two small objects be if, when 250 mm from the eye, their images are just resolved when they are illuminated with light of wavelength 500 nm?

23E. Under ideal conditions, estimate the linear separation of two objects on the planet Mars that can just be resolved by an observer on earth (*a*) using the naked eye, and (*b*) using the 200-in. (= 5.1 m) Mount Palomar telescope. Use the following data: distance to Mars = 8.0×10^7 km; diameter of pupil = 5.0 mm; wavelength of light = 550 nm.

24E. If Superman really had x-ray vision at 0.10 nm wavelength and a 4.0 mm pupil diameter, at what maximum altitude could he distinguish villains from heroes assuming the minimum detail required was 5.0 cm?

25E. A navy cruiser employs radar with a wavelength of 1.6 cm. The circular antenna has a diameter of 2.3 m. At a range of 6.2 km, what is the smallest distance that two speedboats can be from each other and still be resolved as two separate objects by the radar system?

26P. Nuclear-pumped x-ray lasers are seen as a possible weapon to destroy ICBM booster rockets at ranges up to 2000 km. One limitation on such a device is the spreading of the beam due to diffraction, with resulting dilution of beam intensity. Consider such a laser operating at a wavelength of 1.4 nm. The lasing element is a wire with diameter 0.20 mm. (*a*) Calculate the diameter of the central beam at the target 2000 km away. (*b*) By what factor is the beam intensity reduced in transit to target? (The laser is fired from space, so that atmospheric absorption can be ignored.)

27P. The paintings of Georges Seurat consist of closely spaced small dots ($\approx$ 2 mm in diameter) of pure pigment, as indicated in Fig. 30. The illusion of color mixing occurs because the pupils of the observer's eyes diffract light entering them. Calculate the minimum distance an observer must stand from such a painting to achieve the desired blending of color. Take the wavelength of the light to be 550 nm and the diameter of the pupil to be 1.5 mm.

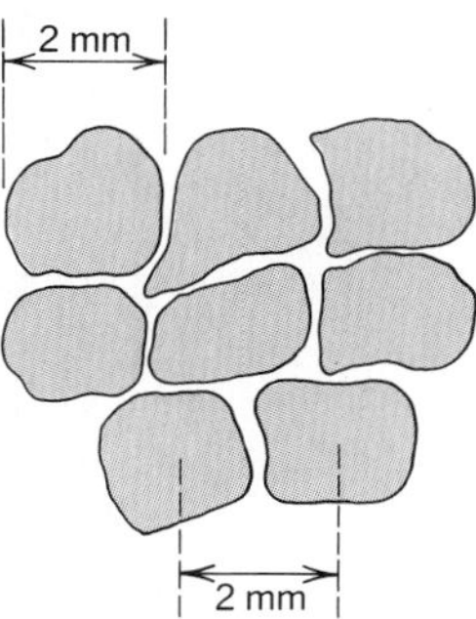

Figure 30 Problem 27.

28P. (*a*) A circular diaphragm 60 cm in diameter oscillates at a frequency of 25 kHz in an underwater source of sound used for submarine detection. Far from the source the sound intensity is distributed as a diffraction pattern for a circular hole whose diameter equals that of the diaphragm. Take the speed of sound in water to be 1450 m/s and find the angle between the normal to the diaphragm and the direction of the first minimum. (*b*) Repeat for a source having an (audible) frequency of 1.0 kHz.

29P. In June 1985 a laser beam was fired from the Air Force Optical Station on Maui, Hawaii, and reflected back from the shuttle *Discovery* as it sped by, 220 miles overhead. The diameter of the central maximum of the beam at the shuttle position was said to be 30 ft and the beam wavelength was 500 nm. What is the effective diameter of the laser aperture at the Maui ground station? (*Hint:* A laser beam spreads because of diffraction; assume a circular exit aperture.)

30P. A "spy in the sky" satellite orbiting at 160 km above the earth's surface has a lens with a focal length of 3.6 m. Its resolving power for objects on the ground is 30 cm; it could easily measure the size of an aircraft's air intake. What is the effective lens diameter, determined by diffraction consideration alone? Assume $\lambda = 550$ nm. Far more effective satellites are reported to be in operation today. See "Reconnaissance and Arms Control" by Ted Greenwood, *Scientific American,* February 1973.

31P. Millimeter-wave radar generates a narrower beam than conventional microwave radar. This makes them less vulnerable to antiradar missiles. (*a*) Calculate the angular width, from first minimum to first minimum, of the central "lobe" produced by a 220-GHz radar beam emitted by a 0.55-m diameter circular antenna. (The frequency is chosen to coincide with a low-absorption atmospheric "window.") (*b*) Calculate the same quantity for the ship's radar described in Exercise 25.

32P. (*a*) How small is the angular separation of two stars if their images are barely resolved by the Thaw refracting telescope at the Allegheny Observatory in Pittsburgh? The lens diameter is 76 cm and its focal length is 14 m. Assume $\lambda =$ 550 nm. (*b*) Find the distance between these barely resolved stars if each of them is 10 light years distant from the earth. (*c*) For the image of a single star in this telescope, find the diameter of the first dark ring in the diffraction pattern, as measured on a photographic plate placed at the focal plane. Assume that the star image structure is associated entirely with diffraction at the lens aperture and not with (small) lens "errors," etc.

33P. It can be shown that, except for $\theta = 0$, a circular obstacle produces the same diffraction pattern as a circular hole of the same diameter. Furthermore, if there are many such obstacles, such as water droplets located randomly, then the interference effects vanish leaving only the diffraction associated with a single obstacle. (*a*) Explain why one sees a "ring" around the moon on a foggy night. The ring is usually reddish in color; explain why. (*b*) Calculate the size of the water droplets in the air if the ring around the moon appears to have a diameter 1.5 times that of the moon. The angular diameter of the moon in the sky is 0.5°. (*c*) At what distance from the moon might a bluish ring be seen? Sometimes the rings are white; why? (*d*)

The color arrangement is opposite to that in a rainbow: why should this be so?

34P. In a Soviet–French experiment to monitor the moon's surface with a light beam, pulsed radiation from a ruby laser ($\lambda = 0.69\ \mu$m) was directed to the moon through a reflecting telescope with a mirror radius of 1.3 m. A reflector on the moon behaved like a circular plane mirror with radius 10 cm, reflecting the light directly back toward the telescope on earth. The reflected light was then detected by a photometer after being brought to a focus by this telescope. What fraction of the original light energy was picked up by the detector? Assume that for each direction of travel all the energy is in the central diffraction circle.

Section 41-7 Diffraction from a Double Slit

35E. Suppose that, as in Sample Problem 6, the envelope of the central peak contains 11 fringes. How many fringes lie between the first and second minima of the envelope?

36E. For $d = 2a$ in Fig. 31, how many interference fringes lie in the central diffraction envelope?

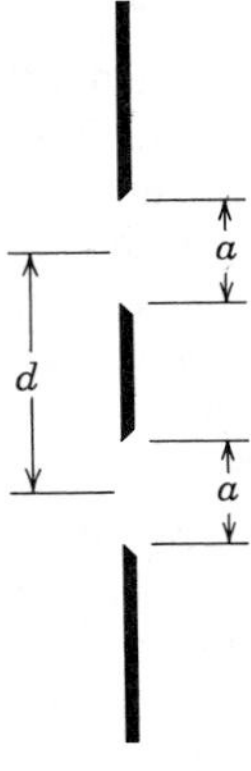

Figure 31 Exercise 36.

37P. If we put $d = a$ in Fig. 31, the two slits coalesce into a single slit of width $2a$. Show that Eq. 12 reduces to the diffraction pattern for such a slit.

38P. (*a*) Design a double-slit system in which the fourth fringe, not counting the central maximum, is missing. (*b*) What other fringes, if any, are also missing?

39P. Two slits of width a and separation d are illuminated by a coherent beam of light of wavelength λ. What is the linear separation of the interference fringes observed on a screen that is at a distance D away?

40P. (*a*) How many complete fringes appear between the first minima of the fringe envelope to either side of the central maximum for a double-slit pattern if $\lambda = 550$ nm, $d = 0.15$ mm, and $a = 0.030$ mm? (*b*) What is the ratio of the intensity of the third fringe to the side of the center to that of the central fringe?

41P. Light of wavelength 440 nm passes through a double slit, yielding the diffraction pattern of intensity I versus deflection angle θ shown in Fig. 32. Calculate (*a*) the slit width, and (*b*) the slit separation. (*c*) Verify the displayed intensities of the $m = 1$ and $m = 2$ interference fringes.

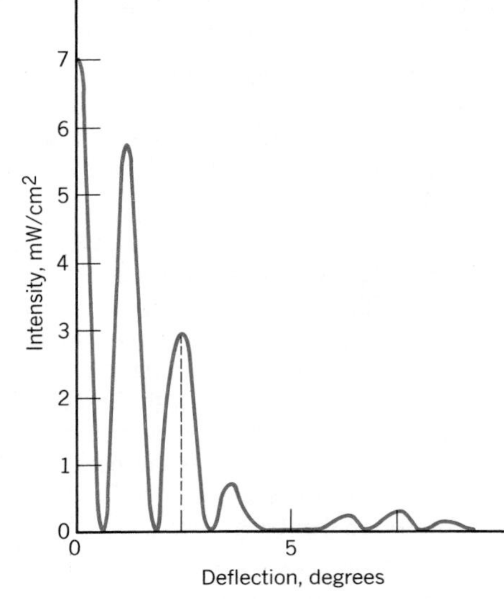

Figure 32 Problem 41.

42P. An acoustic double-slit system (slit separation d, slit width a) is driven by two loudspeakers as shown in Fig. 33. By use of a variable delay line, the phase of one of the speakers may be varied. Describe in detail what changes occur in the inten-

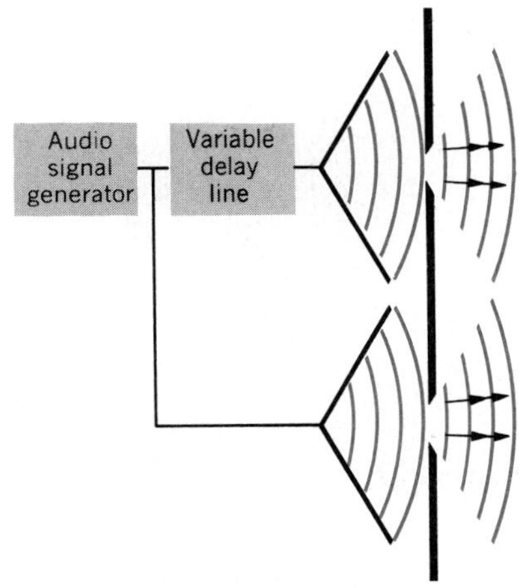

Figure 33 Problem 42.

sity pattern at large distances as this phase difference is varied from zero to 2π. Take both interference and diffraction effects into account.

Section 41-9 Diffraction Gratings

43E. A diffraction grating 20 mm wide has 6000 rulings. (*a*) Calculate the distance d between adjacent rulings. (*b*) At what angles will maximum-intensity beams occur if the incident radiation has a wavelength of 589 nm?

44E. A diffraction grating has 200 rulings/mm, and a strong diffracted beam is noted at $\theta = 30°$. (*a*) What are the possible wavelengths of the incident light? (*b*) What colors are they?

45E. A grating has 315 rulings/mm. For what wavelengths in the visible spectrum can fifth-order diffraction be observed?

46E. Given a grating with 400 lines/mm, how many orders of the entire visible spectrum (400–700 nm) can be produced?

47E. A diffraction grating 3.0 cm wide produces a deviation of 33° in the second order with light of wavelength 600 nm. What is the total number of lines on the grating?

48E. A diffraction grating exactly one centimeter wide has 10,000 parallel slits. Monochromatic light that is incident normally is deviated through 30° in first order. What is the wavelength of the light?

49P. Light of wavelength 600 nm is incident normally on a diffraction grating. Two adjacent principal maxima occur at $\sin\theta = 0.2$ and $\sin\theta = 0.3$, respectively. The fourth order is missing. (*a*) What is the separation between adjacent slits? (*b*) What is the smallest possible individual slit width? (*c*) Name all orders actually appearing on the screen with the values derived in (*a*) and (*b*).

50P. A diffraction grating is made up of slits of width 300 nm with a 900-nm separation between centers. The grating is illuminated by monochromatic plane waves, $\lambda = 600$ nm, the angle of incidence being zero. (*a*) How many diffraction maxima are there? (*b*) What is the width of the spectral lines observed in first order if the grating has 1000 slits? See Eq. 18.

51P. Assume that the limits of the visible spectrum are arbitrarily chosen as 430 and 680 nm. Calculate the number of rulings per mm of a grating that will spread the first-order spectrum through an angular range of 20°.

52P. With light from a gaseous discharge tube incident normally on a grating with a distance 1.73 μm between adjacent slit centers, a green line appears with sharp maxima at measured transmission angles $\theta = \pm 17.6°$, 37.3°, $-37.1°$, 65.2°, and $-65.0°$. Compute the wavelength of the green line that best fits the data.

53P. Assume that light is incident on a grating at an angle ψ as shown in Fig. 34. Show that the condition for a diffraction maximum is

$$d(\sin\psi + \sin\theta) = m\lambda, \quad m = 0, 1, 2, \ldots$$

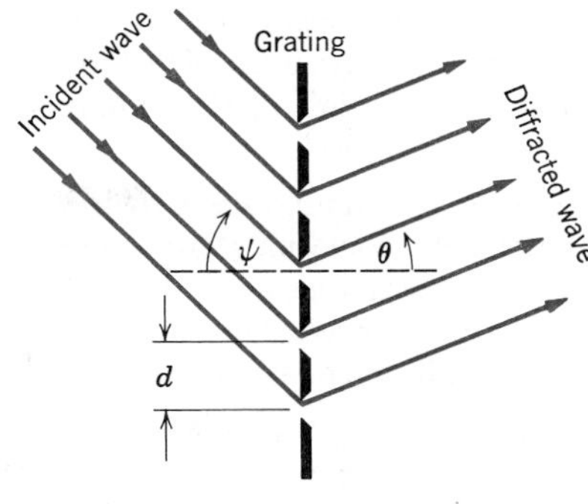

Figure 34

Only the special case $\psi = 0$ has been treated in this chapter (compare with Eq. 14).

54P. A transmission grating with $d = 1.50$ μm is illuminated at various angles of incidence by light of wavelength 600 nm. Plot as a function of angle of incidence (0 to 90°) the angular deviation of the first-order diffracted beam from the incident direction. See Problem 53.

55P. Two spectral lines have wavelengths λ and $\lambda + \Delta\lambda$, respectively, where $\Delta\lambda \ll \lambda$. Show that their angular separation $\Delta\theta$ in a grating spectrometer is given approximately by

$$\Delta\theta = \frac{\Delta\lambda}{\sqrt{(d/m)^2 - \lambda^2}},$$

where d is the separation of adjacent slit centers and m is the order at which the lines are observed. Notice that the angular separation is greater in the higher orders than in lower orders.

56P. White light (400 nm $< \lambda <$ 700 nm) is incident on a grating. Show that, no matter what the value of the grating spacing d, the second- and third-order spectra overlap.

57P. Show that in a grating with alternately transparent and opaque strips of equal width, all the even orders (except $m = 0$) are absent.

58P. A grating has 350 rulings/mm and is illuminated at normal incidence by white light. A spectrum is formed on a screen 30 cm from the grating. If a 10-mm square hole is cut in the screen, its inner edge being 50 mm from the central maximum and parallel to it, what range of wavelengths passes through the hole?

59P. Derive Eq. 18, that is, the expression for the width of lines other than the central maximum.

Section 41-10 Gratings: Dispersion and Resolving Power

60E. The D line in the spectrum of sodium is a doublet with wavelengths 589.0 nm and 589.6 nm. Calculate the minimum number of lines in a grating needed to resolve this doublet in the second order spectrum. See Sample Problem 7.

61E. A grating has 600 rulings/mm and is 5.0 mm wide. (*a*) What is the smallest wavelength interval that can be resolved in the third order at $\lambda = 500$ nm? (*b*) How many higher orders can be seen?

62E. A source containing a mixture of hydrogen and deuterium atoms emits light containing two closely spaced red colors at $\lambda = 656.3$ nm whose separation is 0.18 nm. Find the minimum number of lines needed in a diffraction grating that can resolve these lines in the first order.

63E. (*a*) How many rulings must a 4.0-cm-wide diffraction grating have to resolve the wavelengths 415.496 nm and 415.487 nm in the second order? (*b*) At what angle are the maxima found?

64E. In a particular grating the sodium doublet (see Sample Problem 7) is viewed in third order at 10° to the normal and is barely resolved. Find (*a*) the grating spacing and (*b*) the total width of the rulings.

65E. Show that the dispersion of a grating can be written as

$$D = \frac{\tan\theta}{\lambda}.$$

66E. A grating has 40,000 rulings spread over 76 mm. (*a*) What is its expected dispersion D for sodium light ($\lambda =$ 589 nm) in the first three orders? (*b*) What is its resolving power in these orders?

67P. Light containing a mixture of two wavelengths, 500 nm and 600 nm, is incident normally on a diffraction grating. It is desired (1) that the first and second principal maxima for each wavelength appear at $\theta \leq 30°$, (2) that the dispersion be as high as possible, and (3) that the third order for 600 nm be a missing order. (*a*) What should be the separation between adjacent slits? (*b*) What is the smallest possible individual slit width? (*c*) Name all orders for 600 nm that actually appear on the screen with the values derived in (*a*) and (*b*).

68P. In Problem 50, calculate the product of line width and resolving power of the grating in first order.

69P. A diffraction grating has a resolving power $R = \lambda/\Delta\lambda = Nm$. (*a*) Show that the corresponding frequency range $\Delta\nu$ that can just be resolved is given by $\Delta\nu = c/Nm\lambda$. (*b*) From Fig. 18, show that the "times of flight" of the two extreme rays differ by an amount $\Delta t = (Nd/c)\sin\theta$. (*c*) Show that $(\Delta\nu)(\Delta t) = 1$, this relation being independent of the various grating parameters. Assume $N \gg 1$.

Section 41-11 X-ray Diffraction

70E. The x-ray wavelength 0.12 nm is found to reflect in the second order from the face of a lithium fluoride crystal at a Bragg angle of 28°. Calculate the distance between adjacent crystal planes.

71E. A beam of x rays of wavelength 30 pm is incident on a calcite crystal of lattice spacing 0.30 nm. What is the smallest angle between the crystal planes and the x-ray beam that will result in constructive reflection of the x rays?

72E. Monochromatic high-energy x rays are incident on a crystal. If first-order reflection is observed at Bragg angle 3.4°, at what angle would second-order reflection be expected?

73E. An x-ray beam, containing radiation of two distinct wavelengths, is scattered from a crystal, yielding the intensity spectrum shown in Fig. 35. The interplanar spacing of the scattering planes is 0.94 nm. Determine the wavelengths of the x rays present in the beam.

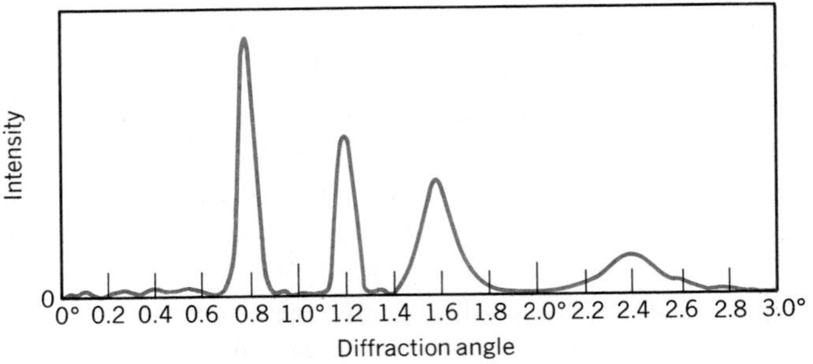

Figure 35 Problem 73.

74E. In comparing the wavelengths of two monochromatic x-ray lines, it is noted that line A gives a first-order reflection maximum at a glancing angle of 23° to the smooth face of a crystal. Line B, known to have a wavelength of 97 pm, gives a third-order reflection maximum at an angle of 60° from the same face of the same crystal. (*a*) Calculate the interplanar spacing. (*b*) Find the wavelength of line A.

75E. Monochromatic x rays are incident on a set of NaCl crystal planes whose interplanar spacing is 39.8 pm. When the beam is rotated 60° from the normal, first-order Bragg reflection is observed. What is the wavelength of the x rays?

76P. Prove that it is not possible to determine both wavelength of radiation and spacing of Bragg reflecting planes in a crystal by measuring the angles for Bragg reflection in several orders.

77P. Assume that the incident x-ray beam in Fig. 36 is not monochromatic but contains wavelengths in a band from 95 to 130 pm. Will diffracted beams, associated with the planes shown, occur? If so, what wavelengths are diffracted? Assume $d = 275$ pm.

78P. First-order Bragg scattering from a certain crystal occurs at an angle of incidence of 63.8°; see Fig. 37. The wavelength of the x rays is 0.26 nm. Assuming that the scattering is from the dashed planes shown, find the unit cell size a_0.

79P. Consider an infinite two-dimensional square lattice as in Fig. 25*b*. One interplanar spacing is obviously a_0 itself. (*a*)

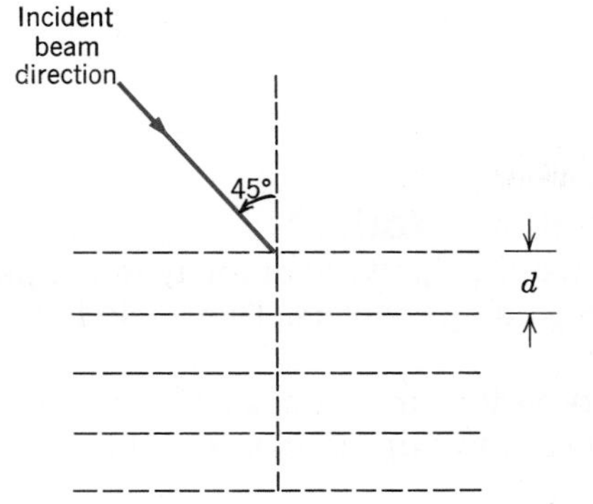

Figure 36 Problems 77 and 80.

Calculate the next five smaller interplanar spacings by sketching figures similar to Fig. 26*a*. (*b*) Show that your answers obey the general formula

$$d = a_0/\sqrt{h^2 + k^2}$$

where *h* and *k* are both relatively prime integers that have no common factor other than unity.

80P. Monochromatic x rays ($\lambda = 0.125$ nm) fall on a crystal of sodium chloride, making an angle of 45° with the reference line shown in Fig. 36. The planes shown are those of Fig. 26*a*, for which $d = 0.252$ nm. Through what angles must the crystal be turned to give a diffracted beam associated with the planes shown? Assume that the crystal is turned about an axis that is perpendicular to the plane of the page.

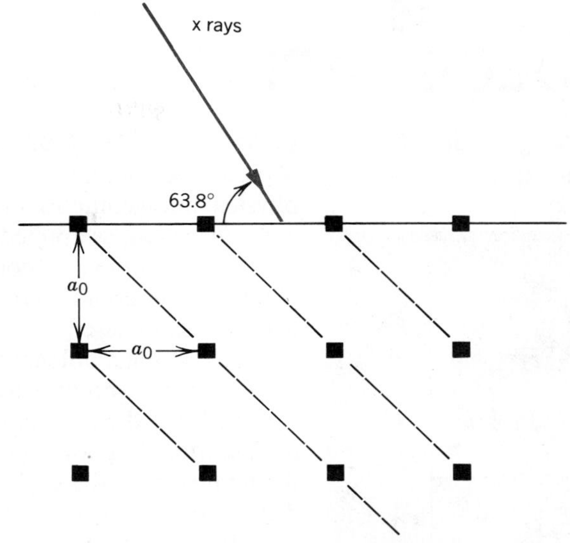

Figure 37 Problem 78.

ESSAY 18
HOLOGRAPHY

TUNG H. JEONG
LAKE FOREST COLLEGE

What Is a Hologram and How Does It "Work"?

A basic understanding of *holography*[1,2,3] requires a review of many major principles in physical and geometric optics found in preceding chapters. They include *interference* and *diffraction,* as applied to visible light.

A *hologram* is a recording on a light-sensitive medium, such as a photographic plate, of *interference patterns* formed between two or more beams of light derived from the same laser.

In making a hologram, part of the output from a *laser* is spread out by a *lens* or curved *mirror* and directed onto the plate. This is called the *reference beam (R).* The remainder of the light illuminates a three-dimensional object being recorded. The light scattered by the object toward the plate is called the *object beam (O).* Because all the light is from the same laser, the two beams are mutually *coherent* and form distinct interference patterns.

When a hologram is illuminated by R, it behaves as a complex *grating* and diffracts the light. The diffraction pattern precisely recreates the *wavefronts* that emanated from the original object.

To understand this process better, first consider the simplest of all objects, a point in space. Figure 1*a* shows two beams, situated far from the plate and interfering at 90° with respect to each other. The interference pattern is precisely the same as that from a *Young's double slit* with very wide separation. The exposed and developed hologram is a *diffraction grating* consisting of $d = \lambda$, where λ is the wavelength of the laser.

Figure 1*b* shows how to reconstruct the *wavefront* of O. Laser light from R illuminates the hologram. The diffracted light, according to the equation $m\lambda = d \sin \theta$, yields $\theta = 90°$, for $m = 1$. There is no room for higher orders. Thus all the diffracted light forms a *virtual image* of O. If R were directed backwards (called a conjugate beam R') as shown in Figure 1*c*, O is reconstructed backward also. A screen placed in the location of O' will show the *real image.*

If we replace the point O with a three-dimensional object illuminated by laser light, the new object beam O consists of a large collection of point sources representing the *scattering* centers on the object. The recording on the plate now consists of a *superposition* of gratings. When illuminated by R (or R'), a virtual (or real) image of the object can be observed.

The above recording is called a "laser transmission" hologram.[2] It has the remarkable property that any small area on it is capable of recreating a complete picture of the object.

Figure 2 represents the general interference pattern between R and O on a plane containing the sources.[4] The lines represent the locations of *maxima;* halfway between them are the *minima.* An analog using water waves can be produced in a *ripple tank.* The perpendicular bisector of RO is the *zeroth order* interference, the loci of points that have the same *optical path* to R and O.

In three-dimensional space, the pattern is a figure of revolution with RO as the axis. It is a family of hyperboloids with foci R and O.

In region A sufficiently far from R and O, the pattern is precisely that of the Young's double-slit interference as discussed before. Region B consists of waves moving in opposite directions, forming *standing waves.* The *antinodes* along the line joining R and O are separated by $(\frac{1}{2})\lambda$. Region C represents *fringes* of a *Michelson interferometer.*

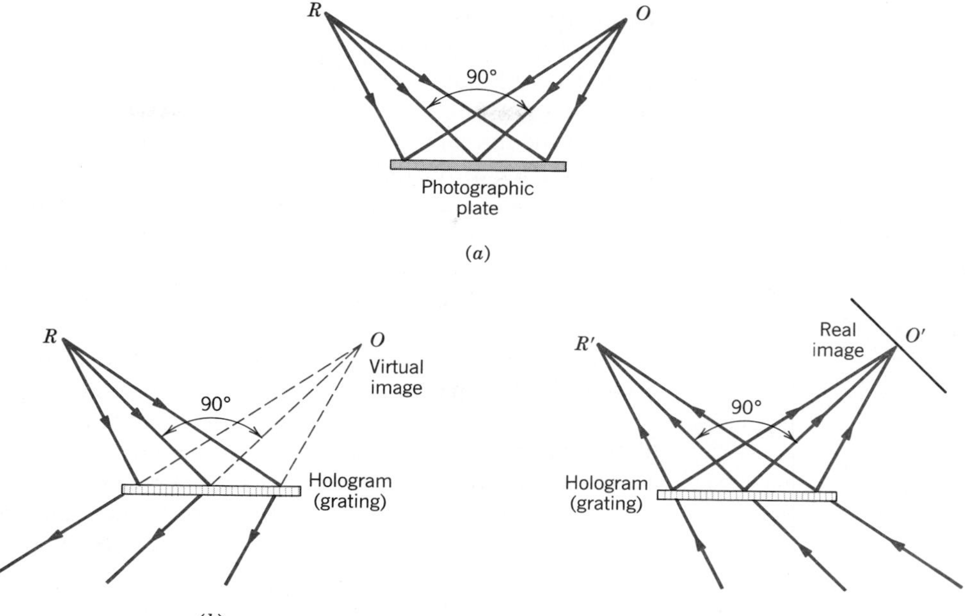

Figure 1 (*a*) Light from two widely separated coherent point sources R and O produces fine interference fringes on the photoplate, which becomes a diffraction grating upon chemical processing. (*b*) When R alone is directed at the grating, the diffracted light precisely recreates the wavefronts of 0. (*c*) If R were directed in a reverse direction, the real image of 0 can be projected on a screen.

Generally, the distance *RO* is many thousand wavelengths of light, and the patterns are microscopic, beyond visual resolution. Figure 2 is a special case in which *R* and *O* are only a few wavelengths apart for simplicity.

If a hologram is made by placing the plate in region *B*, parallel to the zeroth order of Fig. 2, it records the standing wave pattern. The photographic emulsion on the plate is generally 10 λ thick, thus it records up to 20 hyperboloidal planes. This remarkable "white light reflection" hologram[3] can be viewed with a point source of *incandescent* light from *R*. Because it is a "volume" hologram, it performs *Bragg diffraction* and selects the same wavelength λ from the white light and reconstruct the wavefront of *O*.

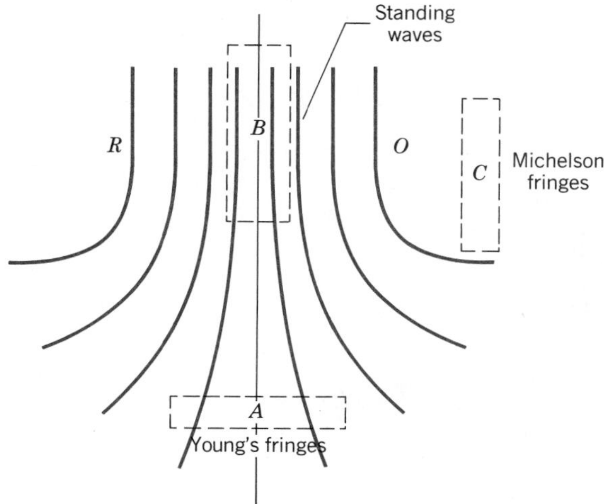

Figure 2 The general interference pattern between two point sources. It is a family of hyperbolas

Making Holograms

Figure 3 show a simple system for making laser transmission holograms.[5] The output from a 1- to 5-milliwatt HeNe laser is spread by a front-surfaced curved mirror. Some light arrives at the plate directly and serves as the reference beam, where *R* is the *focal point* of the mirror. Another part of the light illuminates the object and is scattered onto the plate as the object beam.

The plate and object are supported by a steel plate on top of an inflated rubber tube, which absorbs mechanical vibrations from beneath. All components are held down by magnets or glue. This is necessary because during the exposure, which may be several seconds long, any relative movement between the object and the plate will smear the microscopic interference patterns being recorded, resulting in failure.

In general, much more complicated optical arrangements are necessary in order to illuminate large scenes in more artistic or useful ways.

Using a ruby laser, which can emit more than one joule of light energy in less than 20 nanoseconds, a hologram can be made of deep transient scenes. Although the subjects are generally in motion, the brief exposure time allows the recording of the necessary patterns without smearing. For human and animal subjects, eye-safety rules must be followed. These include closing the eyes, wearing goggles or opaque contact lenses, or using ingenious illumination techniques.

Figure 4 is a photograph of the reconstructed wavefronts from such a hologram. The subject is the author in a laboratory. One can look around through this small hologram as if it were a window on the wall of the laboratory and see everything in full dimension. If an undiverged laser beam were directed at any small area of the hologram in a reversed direction of *R*, a real image could be projected.

Figure 5 shows a setup for recording a white-light reflection hologram. Notice that the object is in contact with the plate and on the opposite site of the mirror.[5]

Because it is a volume hologram, the separation between *Bragg planes* can be shifted by chemically swelling or shrinking the emulsion. This allows one to reconstruct the image in any color desired. Artists can produce beautiful work in multiple

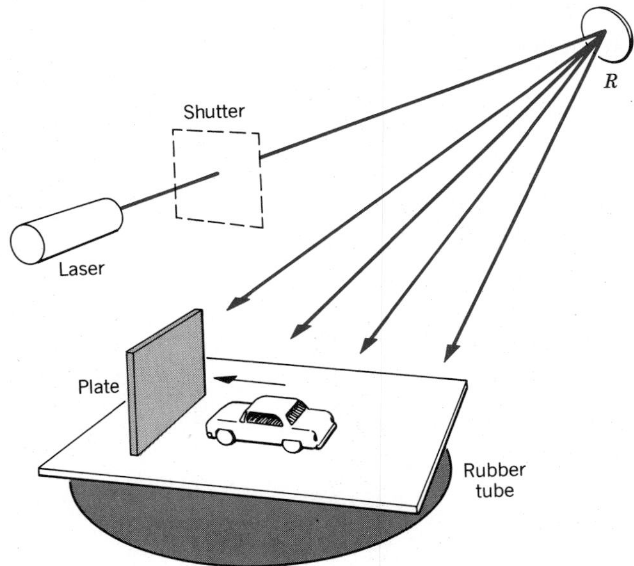

Figure 3 Simple configuration for making a transmission hologram viewable with a laser light.

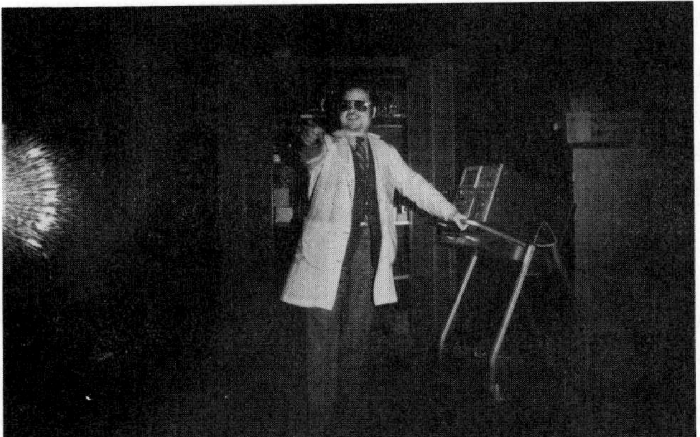

Figure 4 Deep scene image of the author from a transmission hologram made with a pulsed ruby laser.

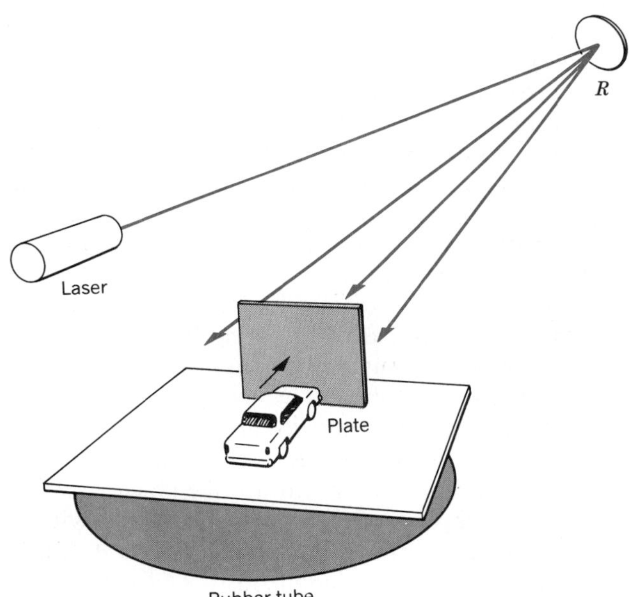

Figure 5 System for making the simplest hologram viewable with incandescent light.

color by making multiple exposures, and swelling the emulsion between exposures. Upon developing, each exposure shrinks differently and the image appears in many colors.

Other Types of Holograms

Many variations from the above two basic types are now available for artistic as well as technical applications. The main hybrid holograms are the following.

Rainbow hologram[6,7]: A white-light viewable transmission hologram that offers three-dimensionality only in the horizontal direction. The color observed depends on the vertical viewing angle. When such holograms are aluminized in the back, they can be attached to credit cards or other surfaces and serves as pseudo-reflection holograms.

Integral stereogram[8]: A hologram synthesized from a series of motion picture frames, allowing live subjects as well as outdoor scenes to be recorded without initial laser illumination. Computer-generated images can also be synthesized.

Focused image hologram[9]: A hologram made using the real image of an object such as *O*, located through the plane of the plate. The real image is usually from another hologram.

Holographic cinema[10]: By making a projection screen in the form of a large hologram that behaves as a multiple elliptical mirror, real images from a series of holograms can be projected to each viewer in an auditorium.

There are numerous other variations the reader can find in the open literature.

Other Applications

Although most holograms encountered are made for visual observations, much more work in the field is aimed at technical applications. Some major fields include "holographic interferometry," "HOE" (holographic optical elements), and "phase conjugation," an important and exotic application that will play a major role in the new field of "optronics" (optical electronics). A brief discussion of each follows.

Interferometry[11]: When two separate exposures are made, and the object undergoes minute mechanical change in between, two slightly different images will appear at reconstruction. The precise change is revealed visually through the mutual interference of the two reconstructed wavefronts. Figure 6 shows a "real-time" observation of a growing mushroom.[12] The virtual image from a hologram made a few seconds before is interfering with light from the actual mushroom, resulting in "live" fringes that allows precise interpretation. Fringes caused by the swaying motion are separated from those caused by the net growth by using additional techniques.

Holographic optical elements (HOE): Diffraction gratings can now be made holographically. Indeed, a transmission hologram of a point source nearby behaves as a *concave grating.* HOEs are used for laser scanning at store checkout counters. HUDs (head up displays) are transparent holograms on windshields of aircrafts (in the future, cars) that relay visual information to the pilot looking straight ahead. HOEs are generally more compact and less costly than their classical counterparts, and can perform optical functions otherwise impossible.

Phase conjugation[13]: When R' is directed through a hologram, O' is directed backward (time-reversed beam) toward the object. Consider a material such as barium titanate, which is capable of recording a hologram, without any chemical processing, as soon as sufficient energy from O and R fall on it. If an intense conjugate beam (the "pump" beam) is present simultaneously, O' further illuminates the actual object and

Figure 6 "Real time" observation of a live specimen through a hologram made moments before.

causes it to become brighter with time. This is a form of light amplification. The process has innumerable other novel applications and has already become a field of research by itself.

REFERENCES [1]D. Gabor, Proc. Roy. Soc. (London) **A197,** 454–487 (1949). E. N. Leith and J. Upatnieks, [2]J. Opt. Soc. Am. **54,** 1295 (1964). [3]Y. N. Denisyuk, Opt. Spectgry. (USSR) **18** (2), 152 (1965). [4]T. Jeong, Am. J. Phys. August, 714–717 (1975). [5]T. Jeong, *Laser Holography,* published by Thomas A. Edison Foundation, 21000 W. Ten Mile Road, Southfield, MI 48075. [6]S. Benton, Proc. of the Int'l. Sym. on Display Holography, Vol. I, 5–14 (1982) (Published by Holography Workshops, Lake Forest College, Lake Forest, IL 60045). [7]S. St. Cyr, Proc. of the Int'l. Sym. on Display Holography, Vol. II, 191–221 (1985). [8]W. Molteni, Proc. of the Int'l. Sym. on Display Holography, Vol. I, 15–26 (1982). [9]H. Bjelkhagen, Proc. of the Int'l. Sym. on Display Holography, Vol. I, 45–54 (1982). [10]T. Jeong, *Holography,* pub. by Integraf, Box 586, Lake Forest, IL 60045. [11]N. Abramson, *Holograms and Their Evaluations,* Academic Press, (1981). [12]T. Jeong, Pro. of S. P. I. E. Vol. 746, 16–19 (1987). [13]R. A. Fisher, *Optical Phase Conjugation,* Academic Press (1983).

CHAPTER 42

RELATIVITY*

"Everything should be as simple as possible—but not simpler."
Albert Einstein (1879–1955)

This statue of Albert Einstein, by the sculptor Robert Berks, is located on Constitution Avenue in Washington, DC, in front of the National Academy of Sciences building. It is a favorite of children, many of whom like to sit on his lap.

42–1 What Is Relativity All About?

In 1905, the 26-year-old Albert Einstein (see Fig. 1) put forward his *special theory of relativity.*† At that time, Einstein was Technical Expert (Third Class) in the Swiss Patent Office, working on physics in his spare time and in what has been termed "splendid isolation" from physicists in the academic community. During that same year, he published a second paper on relativity and two other world-class papers on entirely different subjects, for one of which he was later awarded the Nobel prize.

Relativity, an aesthetically appealing theory, is about the nature of space and time. It has survived every one of the many searching experimental tests to which it has been subjected during the last eight decades. Its status today is such that, if an experimental result is proposed that is inconsistent with relativity, physicists

† Einstein also put forward a *general theory* of relativity, in 1917. It deals with the interpretation of gravity—not as a force but as a curvature in space and time. Although we do not deal with the general theory in this chapter, we gave a preview of its central principle, the principle of equivalence, in Section 15–10. In this chapter, the word *relativity* will always refer to the *special theory,* in which gravity plays no role.

* A fuller treatment of relativity can be found in *Basic Concepts in Relativity and Early Quantum Theory,* 2nd ed., by Robert Resnick and David Halliday, John Wiley & Sons, New York, 1985.

Figure 1 Einstein in the early 1900s, at his desk in the Bern Patent Office.

everywhere would conclude that there must be something wrong with the experiment. Asked about the influence of relativity on the development of physics, one physicist replied, "Well, relativity is simply *there.*"

Relativity has a reputation, among those who have not studied it, as a difficult subject. It is not mathematical complexity that stands in the way of understanding; if you can solve a quadratic equation, you are overqualified. The difficulty lies entirely with the fact that relativity forces us to reexamine critically our ideas of space and time.

Our life experiences are restricted in that we have no direct experience with tangible objects moving faster than a tiny fraction of the speed of light. It is no wonder that our ideas of space and time, molded by this restricted experience, are also restricted. In much the same way a bacterium, spending its life in a fluid environment dominated by viscous forces, knows nothing of gravity. Our advice: Be receptive to new ideas and keep an open mind.

42-2 Our Plan

Relativity rests on two postulates, which we shall present in the next section. We ask you to accept them provisionally but uncritically and not to say to yourself, "Well, this postulate can't be true because. . . ." It is an admirable trait to question all statements in physics but

Table 1 Earlier Sections on Relativity

Section	*Chapter Title*	*Section Title*
4-9	Motion in a Plane	Relative Motion at High Speeds
7-7	Work and Energy	Kinetic Energy at High Speeds
8-9	The Conservation of Energy	Mass and Energy
10-6	Collisions	Reactions and Decay Processes
15-10	Gravity	A Closer Look at Gravity
17-7	Waves—I	The Speed of Light
38-8	Electromagnetic Waves	The Speed of Electromagnetic Waves

in this case we propose the following:

1. Accept the postulates provisionally.
2. Examine, with us, the consequences that flow from these postulates.
3. Examine, again with us, the universal agreement of these consequences with experiment.*

If you follow this course, you can master the basic ideas of relativity. You will come to see how natural it is, how it simply extends the classical view, and how much common sense it contains.

Relativity is so important for physics that we did not feel that we could postpone discussing it until so late in the book. Therefore, to provide some foretaste of the subject, we inserted a number of optional relativity-related sections at appropriate places in earlier chapters; they are listed in Table 1. Throughout this chapter, we suggest at appropriate points that you go back and read (or reread) this earlier material.

42-3 The Postulates

1. The Postulate of Relativity. *The laws of physics are the same for observers in all inertial*

* See "Modern Tests of Special Relativity," by Mark P. Haugan and Clifford M. Will, *Physics Today,* May 1987, for a summary of the scope and precision of these experimental tests and of their impressive confirmation of the predictions of relativity.

reference frames. No frame is singled out as preferred.

It was assumed by Galileo that the laws of *mechanics* were the same in all inertial reference frames. (Newton's first law of motion is one important consequence.) Einstein extended that idea to include *all* the laws of physics, including especially electromagnetism and optics. This postulate does *not* say that the measured values of all physical quantities are the same for all inertial observers; most are not. It is the *laws of physics* that relate these measurements to each other that are the same. Stated another way, one of the tests of any proposed law of physics is that it must satisfy the relativity postulate.

2. The Postulate of the Speed of Light. *The speed of light in free space has the same value c in all directions and in all inertial reference frames.*

We can also phrase this postulate to say that—as Fig. 2 whimsically suggests—there is in nature an *ultimate speed c,* the same in all directions and in all inertial reference frames. Light happens to travel at this ultimate speed, as do massless particles such as neutrinos. The ultimate speed *c* also sets a limit to which any material particle such as an electron can be accelerated.*

We first introduced this postulate in Section 17–7. We suggest that you reread that section, study Fig. 9 of that chapter carefully, and become thoroughly familiar with exactly what the second postulate says.

The Ultimate Speed. The reality of the existence of an ultimate speed for accelerated electrons is shown in a 1964 experiment of W. Bertozzi. He accelerated electrons to various measured speeds and—by an independent calorimetric method—also measured their kinetic energies. Figure 3 shows that as the force that acts on a very fast electron is increased, its measured kinetic energy increases toward very large values but its speed does not increase appreciably. Electrons have been accelerated to at least 0.999 999 999 95 times the speed of light but—close though it may be—that speed is still less than the ultimate speed *c*.

Testing the Speed of Light Postulate. If the speed of light is the same in all inertial reference frames, then the speed of light emitted by a moving source should be the same as the speed of light emitted by a source that is at rest in the laboratory. This claim has been tested directly, in an experiment of high precision. The "light source" was the *neutral pion* (symbol π^0), an unstable, short-lived particle that may be produced by collisions in a particle

* Relativity can be developed—with exactly the same results—without reference to light. See "Relativity Without Light," by N. David Mermin, *American Journal of Physics,* February 1984.

Figure 2 A whimsical illustration for an article "The Ultimate Speed Limit," by Isaac Asimov (*Saturday Review,* July 1972).

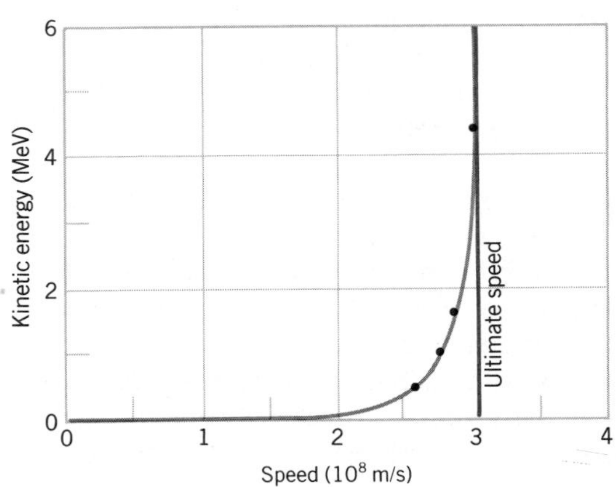

Figure 3 The dots show the measured values of the kinetic energy of an electron plotted against its measured speed. No matter how much energy you impart to an electron (or to any other particle, for that matter) you can never get its speed to equal or to exceed the ultimate limiting speed *c*. This "brick wall" can be approached as closely as you like but never reached. The curve is the prediction of relativity theory (see Eq. 37).

accelerator. It decays into two gamma rays by the process

$$\pi^0 \rightarrow \gamma + \gamma. \tag{1}$$

Gamma rays are part of the electromagnetic spectrum and obey the postulate of the speed of light, just as visible light does.

In a 1964 experiment, physicists at CERN, the European particle-physics laboratory near Geneva, generated a beam of pions moving at a speed of $0.99975c$ with respect to the laboratory. The experimenters then measured the speed of the gamma rays emitted from these very rapidly-moving sources. Their results for the speed of light were

From the moving pions: 2.9977×10^8 m/s;
From a resting source (accepted value): 2.9979×10^8 m/s.

There seems little doubt that the speed of light emitted by these pions—which were racing along at almost the speed of light—is the same as we would measure if the pions had been at rest in the laboratory.

Sample Problem 1 A 20-GeV electron, such as might be generated in the Stanford Linear Accelerator, can be shown to have a speed $v = 0.999\,999\,999\,67c$. If such an electron raced a light pulse to the nearest star outside the solar system (Proxima Centauri, 4.3 light years or 4.0×10^{16} m distant), by how much time would the light pulse win the race?

If L is the distance to the star, the difference in travel times is

$$\Delta t = \frac{L}{v} - \frac{L}{c} = L\frac{c - v}{vc}.$$

Now v is so close to c that we can put $v = c$ in the denominator of this expression (but not in the numerator!). If we do so, we find

$$\Delta t = \frac{L}{c}\left(1 - \frac{v}{c}\right) = \frac{(4.0 \times 10^{16}\text{ m})(1 - 0.999\,999\,999\,67)}{3.00 \times 10^8\text{ m/s}}$$
$$= 0.044\text{ s} = 44\text{ ms}. \qquad \text{(Answer)}$$

The 20-GeV electron certainly comes close to the ultimate speed c but it does not equal or surpass it.

42-4 Measuring an Event

An *event* is something that happens to which an observer can assign three space coordinates and one time coordinate. Among many possible events are (1) the turning on or off of a tiny light bulb, (2) the collision of two particles, (3) the passage of a pulse of light through a specified point in space, or (4) the coincidence of the hand of a clock with a marker on the rim of the clock. An observer, fixed in an inertial reference frame, may assign to event A (the turning on of a light bulb, say) the following spacetime coordinates:*

Record of Event A

Coordinate	*Value*
x	3.58 m
y	−1.29 m
z	2.77 m
t	34.5 s

A given event may be recorded by any number of observers, each in their own inertial reference frame. In general, all such observers will assign different spacetime coordinates to the same event. Note that an event does not, in any sense, "belong" to a particular inertial reference frame. An event is just something that happens and anyone may look at it and assign spacetime coordinates to it.

We need to understand in some detail how a single observer, fixed in an inertial reference frame, assigns space and time coordinates to a single event. Many of the procedures that we outline will seem totally impractical. However, they are *thought procedures,* outlining *in principle* how the measurements are to be carried out.

The Space Coordinates. We imagine the observer's coordinate system fitted with a close-packed, three-dimensional array of measuring rods, one set of rods parallel to each of the three coordinate axes. Thus, if the event is the turning on of a small light bulb, the observer need only read the three space coordinates at the location of the bulb.

The Time Coordinate. For the time coordinate, we imagine that every point of intersection of the array of measuring rods has a tiny clock, which the observer can read by the light generated by the event. Figure 4 suggests the "jungle gym" of meter rods and clocks that we have described.

The array of clocks must be synchronized properly. You may think that it is enough to assemble a set of

* Space and time are so closely linked in relativity that we describe them collectively as *spacetime,* without a hyphen.

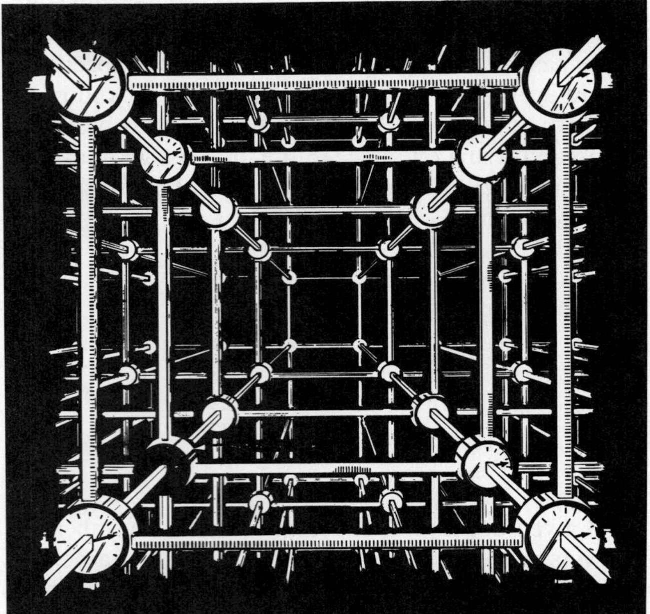

Figure 4 Suggesting the "jungle gym" of rods and clocks used by an observer in an inertial reference frame to assign spacetime coordinates to an event.

identical clocks, set them all to the same time, and then move them to their assigned positions. However, we are committed to question everything. How do we know, for example, that moving the clocks does not change their rates? (Actually, it does.) We must put the clocks in place and *then* synchronize them.

If we had a method of transmitting signals at infinite speed, synchronization would be a simple matter. However, no known signal has this property. We choose light (interpreted broadly to include the entire electromagnetic spectrum) to send out our synchronizing signals because, in free space, light travels at the highest possible speed, the limiting speed c.

Here is one of many ways that we might synchronize an array of clocks with the help of light signals: The observer enlists the help of a large number of temporary helpers, one for each clock. The observer then stands at a point selected as the origin and sends out a pulse of light when the origin clock reads $t = 0$. When the light pulse reaches each helper, that helper sets his or her clock to read $t = r/c$, where r is the distance of the helper from the origin.

All these procedures refer to a *single* observer in a *single* inertial reference frame. Other observers in other inertial reference frames must have a similar array of rods and clocks in order to assign spacetime coordinates to the events that they obseve.

42-5 Simultaneous Events

Suppose that one observer (Sam) notes that two independent events (event Red and event Blue) occur at the same time. Suppose also that another observer (Sally), who is moving at a constant velocity **v** with respect to Sam, also records these same two events. Will Sally also find that they occurred at the same time?

The answer is that in general she will not. Let us be clear about what we are saying:

> *If two observers are in relative motion, they will not, in general, agree as to whether two events are simultaneous. If one observer finds them to be simultaneous, the other will not, and conversely.*

We cannot say that one observer is right and the other wrong. The situation is completely symmetrical and there is no reason to choose one observer over the other. We conclude the following:

> *Simultaneity is not an absolute concept but a relative one, depending on the state of motion of the observer.*

Of course, if the relative speed of the observers is very much less than the speed of light, the measured departures from simultaneity become so small that they are not noticeable. Such is the case for all our experiences of daily living; this is why the relativity of simultaneity is unfamiliar.

42-6 Simultaneity: A Closer Look

Let us clarify the relativity of simultaneity by a specific example. We base our analysis directly on the postulates of relativity, no clocks or measuring rods being directly involved.

Figure 5 shows two long spaceships (the *SS Fitzgerald* and the *SS Lorentz*), which can serve as inertial reference frames for observer Sam and observer Sally. The two observers are stationed at the midpoints of their ships. The ships are separating along a common x axis, the relative velocity of *Fitzgerald* with respect to *Lorentz*

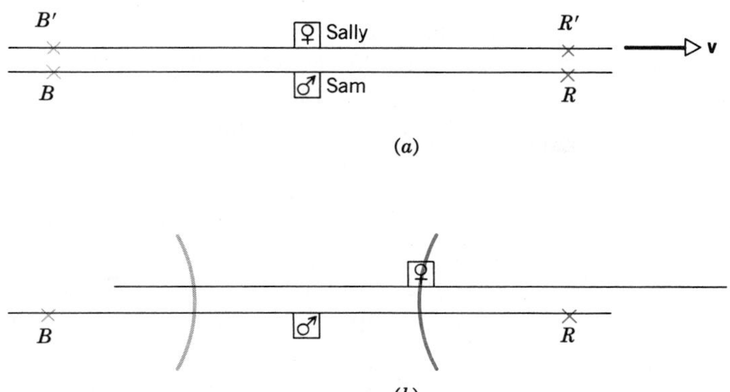

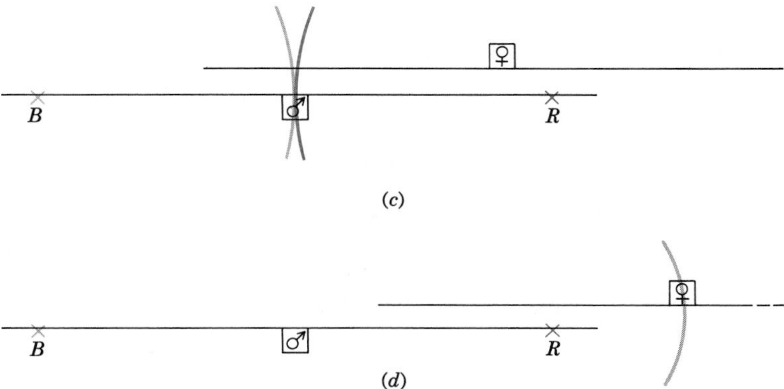

Figure 5 The two spaceships are represented by simple straight lines. The figure shows the situation from Sam's point of view, in which Sally's spaceship is moving to the right with speed v. (*a*) Light waves leave the site of the Red event (RR') and the Blue event (BB'). Successive drawings correspond to the assumption that Sam will perceive events Red and Blue to be simultaneous. (*b*) The Red wave front reaches Sally. (*c*) Both wave fronts reach Sam. (*d*) The Blue wave front reaches Sally; this last figure is not to the scale of the previous three.

being **v**. Figure 5*a* shows the ships with the two observer stations momentarily aligned opposite each other.

Two meteorites strike the ships, one setting off a red flare (event Red) and the other a blue flare (event Blue). Each event leaves a permanent mark on each ship, at positions R, R' and B, B'.

Let us suppose that the expanding wave fronts from the two events happen to reach Sam at the same time, as Fig. 5*c* shows. Let us further suppose that, after the episode, Sam finds, by measurement, that he was stationed exactly halfway between the markers B and R on his ship. He will say:

Sam: Light from event Red and event Blue reached me at the same time. From the marks on my spaceship, I find that I was standing halfway between the two sources when the light from them reached me. Therefore, event Red and event Blue are simultaneous events.

As study of Fig. 5 shows, however, the expanding wave front from event Red will reach Sally *before* the expanding wave front from event Blue does. She will say:

Sally: Light from event Red reached me before light from event Blue did. From the marks on my spaceship, I found that I too was standing halfway between the two sources. Therefore, the events were *not* simultaneous; event Red occurred first, followed by event Blue.

Their reports do not agree. Nevertheless, both observers are correct. That is what the relativity of simultaneity is all about.

It is essential to understand that there is only one wave front expanding from the site of each event and that *this wave front travels with the same speed c in both reference frames,* exactly as the speed of light postulate requires.

It *might* have happened that the meteorites struck the ships in such a way that they appeared simultaneous to Sally. Sam would then declare them not to be simultaneous. The experiences of the two observers are exactly symmetrical. (Note that we have avoided saying anything like: The meteorites struck the ships simultaneously. That would raise the question: Simultaneous in which reference frame? We always said: The meteorites struck the ships *in such a way that*. . . .)

42-7 The Relativity of Time

The relativity of simultaneity is closely related to the relativity of time. That is, if different observers measure the time interval between a given pair of events, they will in general not agree as to how long that interval is. We describe a simple case, again basing our analysis directly on the postulates of relativity.

In Fig. 6, Sally is in a train that is moving with uniform velocity **v** with respect to the station. She has an electronic clock, which she uses to measure the time Δt_0 between two events:

Event 1. The turning on of a flash bulb B.

Event 2. The arrival of the light back at its source after reflection from a mirror on the ceiling.

For the time interval between these two events, Sally finds

$$\Delta t_0 = \frac{2D}{c} \quad \text{(Sally)}, \tag{2}$$

where D is the distance between the source and the mirror. Note that for Sally these two events occur *at the same place* and she can time the interval between them with *a single clock C located at that place.* A time interval measured with a single resting clock is called a *proper time interval,* identified by the subscript zero.

Consider now how these same two events look to Sam, who is standing on the station platform as the train goes by. Because of the postulate of the speed of light, the light travels at the same speed c for Sam as for Sally. It travels a larger distance for Sam, however, namely, $2L$. The time interval measured by Sam between these two events is

$$\Delta t = \frac{2L}{c} \quad \text{(Sam)}, \tag{3}$$

where

$$L = \sqrt{(\tfrac{1}{2}v\,\Delta t)^2 + D^2}. \tag{4}$$

From Eq. 2, we can write this as

$$L = \sqrt{(\tfrac{1}{2}v\,\Delta t)^2 + (\tfrac{1}{2}c\,\Delta t_0)^2}. \tag{5}$$

If we eliminate L between Eqs. 3 and 5 and solve for Δt, we find

$$\Delta t = \frac{\Delta t_0}{\sqrt{1-(v/c)^2}}. \tag{6}$$

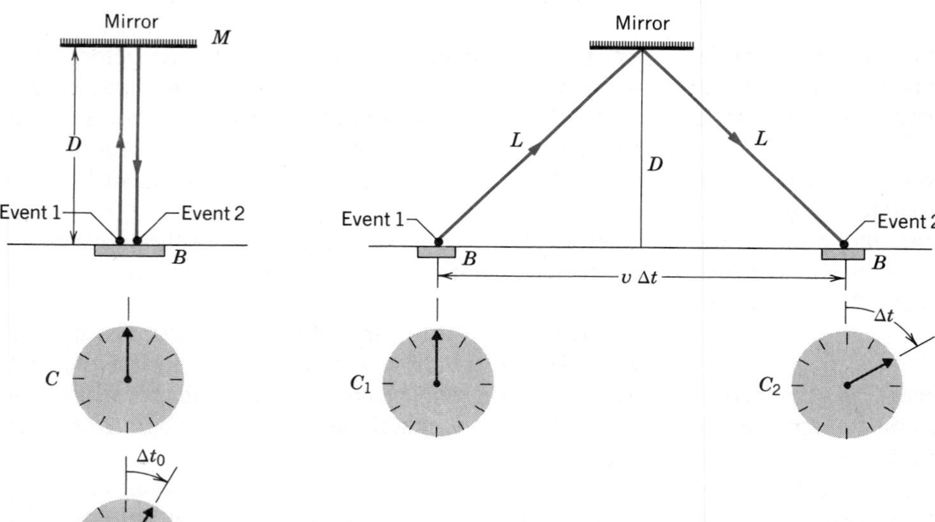

Figure 6 (*a*) Sally times the light pulse excursion using a single resting clock C, reporting a proper time Δt_0. (*b*) Sam, watching the moving train, requires two synchronized clocks, C_1 and C_2, to measure the elapsed time Δt.

Note that for Sam the two events occur *in different places* and, to measure Δt, he must use *two synchronized clocks, C_1 and C_2, located at different places in his reference frame.* The interval that he measures is *not* a proper time interval because he does not measure it with a single resting clock. For this reason, the situations for Sam and Sally (unlike their situations in Fig. 5) are *not* symmetrical.*

We can rewrite Eq. 6 as

$$\Delta t = \frac{\Delta t_0}{\sqrt{1-\beta^2}} \quad \text{(time dilation)}, \tag{7}$$

in which we have replaced the dimensionless ratio v/c by the symbol β, which we call the *speed parameter.*

Because $\beta < 1$, we always have $\Delta t > \Delta t_0$. Sam, standing on the station platform, might say:

Sam: Sally and I have identical clocks and we each measure the time interval between the same two events. Sally's reference frame is special because, for her, the events occur at the same place so that she can use a single clock to measure the time interval. For me, the events occur at different places and I must use *two* synchronized clocks, one at the location of each event. I find that—no matter how fast (or in what direction) the train is moving—my measured value for the time interval is always greater than Sally's.

This *time dilation effect*† is very real and has nothing to do with any mechanical change that takes place in a clock because of its motion. It is simply the nature of time.

The Lorentz Factor. We can also write Eq. 7 in the form

$$\Delta t = \gamma\, \Delta t_0 \quad \text{(time dilation)} \tag{8}$$

in which the dimensionless quantity γ, called the *Lorentz factor,* is given by

$$\gamma = \frac{1}{\sqrt{1-\beta^2}} \quad \text{(Lorentz factor)}. \tag{9}$$

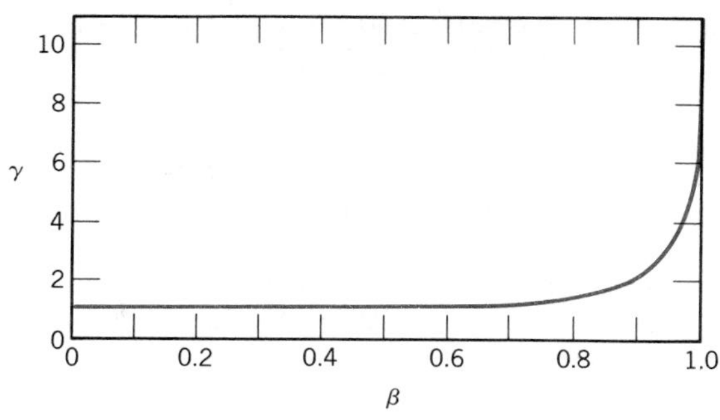

Figure 7 A plot of the Lorentz factor γ as a function of the speed parameter β ($= v/c$).

This combination of variables occurs often in relativity. Figure 7 shows how γ varies with the speed parameter β. As Table 2 shows, the Lorentz factor γ, which is always greater than 1, is our guide to the importance of relativity to the problem at hand.

A Test of Time Dilation: Microscopic Clocks. Subatomic particles called *muons* are unstable and, when at rest in the laboratory, decay with an average lifetime of 2.200 μs. This average lifetime, measured for resting muons with a single resting laboratory clock, is thus a *proper time interval* and we can label it Δt_0.

In a 1968 experiment at CERN, a beam of muons, circulating in a storage ring of 2.5-m radius, was accelerated to a speed of $0.9966c$.* The average lifetime of these muons was then measured while they were in flight at this very high speed.

The accelerated muons can serve as tiny moving clocks, to which Eq. 8, $\Delta t = \gamma\, \Delta t_0$, applies. We first calculate the Lorentz factor γ for the moving muons from Eq. 9,

$$\gamma = \frac{1}{\sqrt{1-(v/c)^2}} = \frac{1}{\sqrt{1-(0.9966)^2}} = 12.14,$$

which is substantially greater than one. Equation 8 yields

$$\Delta t = \gamma\, \Delta t_0 = (12.14)(2.200\ \mu\text{s}) = 26.7\ \mu\text{s}.$$

The measured value was 26.2 μs, in excellent agreement

* To make them symmetrical, you would have to move Sally's flash–mirror arrangement to the platform and give it to Sam. Then Sam would measure a proper time interval and Sally would measure a dilated interval.

† To dilate is to expand or stretch.

* You may object that these muons, in uniform circular motion with a centripetal acceleration of $10^{15}g$, are not in an inertial reference frame so that relativity does not apply. It can be shown, however, that it does apply in this case. Indeed, this experiment can be taken as a verification of that prediction.

Table 2 The Speed Parameter and the Lorentz Factor

Particle	*Speed Parameter* β	*Lorentz Factor* γ	*Can I use Newtonian Mechanics?*[a]
Fastest aircraft (Mach 6.72)	0.0000068	1.000000. . .	Certainly
Earth's orbital speed	0.000099	1.000000005	Yes
1-keV electron	0.063	1.0020	Yes (?)
Around the world in one second	0.13	1.009	Maybe
1-MeV electron	0.94	2.9	No
1-GeV electron	0.999 999 88	2000	Never
Electron in Sample Problem 1	0.999 999 999 67	40,000	Don't even think of it

[a] Relativity applies at *all* speeds, Newtonian mechanics only at speeds much less than the speed of light.

within the experimental error. The time dilation factor is universally accepted and indeed turned to advantage in the design of certain high-energy particle experiments. One physicist has written: "We frequently transport beams of unstable particles over long distances such that no particles would be left in the beam without the help of Einstein's factor."

A Test of Time Dilation: Macroscopic Clocks. In October 1977, Joseph Hafele and Richard Keating carried out what must have been a grueling experiment. They flew four portable atomic clocks twice around the world on commercial airlines, once in each direction. Their purpose was ". . . to test Einstein's theory of relativity with macroscopic clocks." As we have just seen, the time dilation predictions of Einstein's theory had already been confirmed on a microscopic scale but there is great comfort in seeing a confirmation made with an actual clock. Such measurements became possible only because of the very high precision of modern atomic clocks. Hafele and Keating verified the predictions of the theory to within 5–10%.*

A few years later, physicists at the University of Maryland carried out a similar experiment with improved precison. They flew an atomic clock round and round over Chesapeake Bay for flights of 15-h duration and succeeded in checking the time dilation prediction to better than 1%. Today, when atomic clocks are transported from one place to another for calibration or other purposes, the effect of time dilation caused by their motion must always be taken into account.

* Einstein's *general* theory of relativity, which predicts that the rate of a clock is influenced by gravity, also plays a role in this experiment.

The Twin Paradox. Time dilation applies not only to clocks but to all naturally occurring time intervals including pulse rates and average lifetimes. This gives rise to the twin paradox, which is no paradox at all to those who understand relativity and accept its predictons.

One twin sister embarks on a round-trip journey to a star in a very fast spaceship while the other twin remains on earth. As measured by the earth-bound twin, her sister's heart beat, respiration rate, and indeed her aging processes slow down because of the time dilation effect. The traveling twin, however, measuring her pulse rate with her wristwatch (a co-traveling clock!) is totally unaware of any such changes and judges herself to be normal in all such respects. Nevertheless, when the spaceship returns, the traveling twin will be found to be younger than the stay-at-home twin. The stay-at-home twin may in fact have died and be represented only by her aging greatgrandchildren!

Some will say that the traveling twin could argue: Can I not regard myself as stationary and my earth-bound sister as traveling? If so, she should be younger than I when we meet and not the other way around. Clearly, A cannot be younger than B and also B younger than A; that is the paradox.

The *answer to the paradox* is that the twins are *not* in symmetrical situations and the traveling twin *cannot* make the argument outlined above. If the traveling twin is to return, she must at some point turn around, which involves a very necessary and readily measurable acceleration. Put another way, the traveling twin (in the simplest case) switches inertial reference frames, from an outward bound frame to an inward bound frame. The stay-at-home twin undergoes no acceleration and remains in a single reference frame throughout. It is the traveling twin—and not the stay-at-home twin—who will be the younger when they meet again.*

Although this experiment has not (yet) been carried out with actual twins, the round-the-world experiment of Hafele and Keating, mentioned above, amounts to the same thing. Instead of twin people, they used twin clocks, the stay-at-home clock remaining at the Bureau of Standards and the traveling clock going round the world, and necessarily accelerating in the process. When the two clocks "met" at the end of the trip, the traveling clock (in terms of accumulated time) was indeed "younger" than the stay-at-home clock, by 273 ns for the westward circuit.

Sample Problem 2 At what relative speed would a moving clock appear, to a stationary observer, to run at half the rate observed by a person moving with the clock?

The person moving with the clock records a proper time Δt_0, since the clock is at rest relative to him. The person who is watching the moving clock records a dilated time Δt for that clock. That is, if $\Delta t_0 = 1$ h then $\Delta t = 2$ h and, in general, $\Delta t = 2\,\Delta t_0$. From Eq. 9 then,

$$\gamma = 2 = \frac{1}{\sqrt{1-\beta^2}}.$$

Squaring both sides yields

$$(4)(1-\beta^2) = 1$$

or

$$\beta = \sqrt{3/4} = 0.866. \qquad \text{(Answer)}$$

Thus, a clock must be traveling at about 87% of the speed of light for the time dilation factor to amount to a factor of 2. Such a speed corresponds to circling the globe at the equator 6.7 times per second.

* For a detailed account, with the arguments carefully spelled out, see *Basic Concepts,* 2nd ed., Supplementary Topic B (The Twin Paradox), by Robert Resnick and David Halliday, John Wiley & Sons, New York, 1985.

42-8 The Relativity of Length

If you want to measure the length of a rod that is at rest with respect to you, you can—at your leisure—note the positions of its end points on a long stationary scale and subtract the two readings. If the rod is moving, however, you must note the positions of the end points *simultaneously* (in your reference frame) or your measurement cannot be called a length. Figure 8 suggests the difficulty of trying to measure the length of a swimming goldfish by locating its ends at different times. Because simultaneity is relative and it enters into length measurements, you expect that length is also a relative quantity.

If the length of an object at rest in your reference frame—called its *proper length*—is L_0, the length that you would measure if the rod is moving past you (parallel to itself) at speed $v(=\beta c)$. is given by

$$L = L_0\sqrt{1-\beta^2} = \frac{L_0}{\gamma} \qquad \text{(length contraction).} \qquad (10)$$

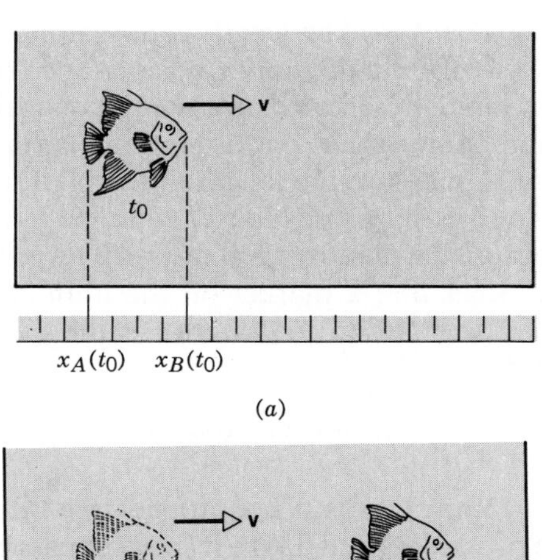

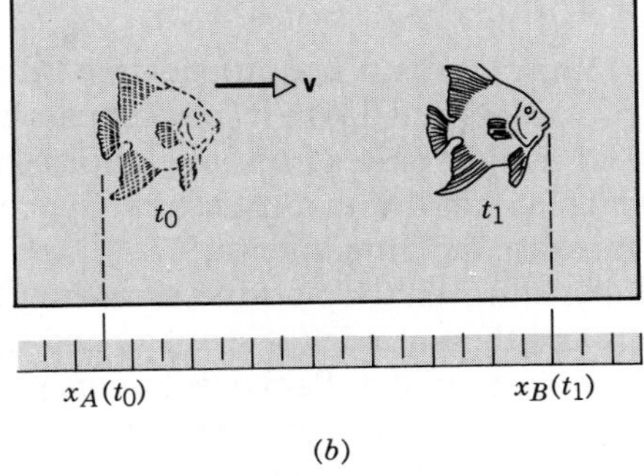

Figure 8 If you want to measure the length of your pet goldfish while it is swimming, you must mark the positions of its head and its tail simultaneously (in your reference frame), as in (*a*), rather than at arbitrary times, as in (*b*).

Because the Lorentz factor γ is always greater than unity, the length of a moving rod is always measured to be smaller than the length of the rod when it is at rest. Like the time dilation, the length contract is very real. We can summarize both phenomena as follows:

> *If two events occur at the same place in an inertial reference frame, the time interval Δt_0 between them, measured by a single resting clock, is called a proper time interval.* All other inertial observers will measure a larger value for this interval.

> *The length L_0 of a rod, measured in an inertial reference frame in which the rod is at rest, is called its proper length.* All other inertial observers will measure a shorter length.

The questions, "Does the rod *really* shrink?" and "Do the atoms in the rod *really* get pushed closer together?" are not proper questions within the framework of relativity. The length of a rod is what you measure it to be and motion affects measurements.

Proof of Eq. 10. The length contraction is a direct consequence of time dilation. Consider once more our two observers. Sally is seated on a train moving through a station and Sam is again on the station platform. They both want to measure the length of the platform. Sam, using a tape measure, finds the length to be L_0, a *proper length* because the platform is at rest with respect to him. Sam also notes that a marker on the train covers this length in a time $\Delta t = L_0/v$, where v is the speed of the train. That is,

$$L_0 = v\,\Delta t \quad \text{(Sam)}. \tag{11}$$

This time interval Δt is not a proper time interval because the two events that define it (marker passes back of platform and marker passes front of platform) occur at two different places and Sam must use two synchronized clocks to measure the time interval Δt.

For Sally, however, the platform is moving. She sees it approach and then recede (at speed v) and finds that the two events measured by Sam occur *at the same place* in her reference frame. She can time them with a single resting clock so that the interval Δt_0 that she measures is a proper time interval. To her, the length L of the platform is given by

$$L = v\,\Delta t_0 \quad \text{(Sally)}. \tag{12}$$

If we divide Eq. 12 by Eq. 11 and use Eq. 8, the time dilation equation, we have

$$\frac{L}{L_0} = \frac{v\,\Delta t_0}{v\,\Delta t} = \frac{1}{\gamma},$$

or

$$L = \frac{L_0}{\gamma}, \tag{13}$$

which is exactly Eq. 10, the length contraction equation.

Sample Problem 3 Two spaceships, each of proper length $L_0 = 230$ m, pass each other as in Fig. 9. Sally, located at point A on one of the spaceships, measures a time interval of 3.57 μs for the second ship to pass her. What is the relative speed parameter β of the two ships? Let AB be the coincidence of points A and B and AC the coincidence of points A and C.

The time interval between events AB and AC, measured by Sally using a single clock at A, is a *proper time interval* $\Delta t_0 = 3.57$ μs. The length L that Sally measures for the other ship is

$$L = v\,\Delta t_0 = \beta c\,\Delta t_0.$$

However, Sally knows that the *proper* length of the other ship is $L_0 = 230$ m, where (see Eq. 10)

$$L = \frac{L_0}{\gamma} = L_0\sqrt{1-\beta^2}.$$

Setting these two values for L equal to each other, we find

$$\beta c\,\Delta t_0 = L_0\sqrt{1-\beta^2}.$$

If we square each side of this equation and solve for β, we find, after a little algebra,

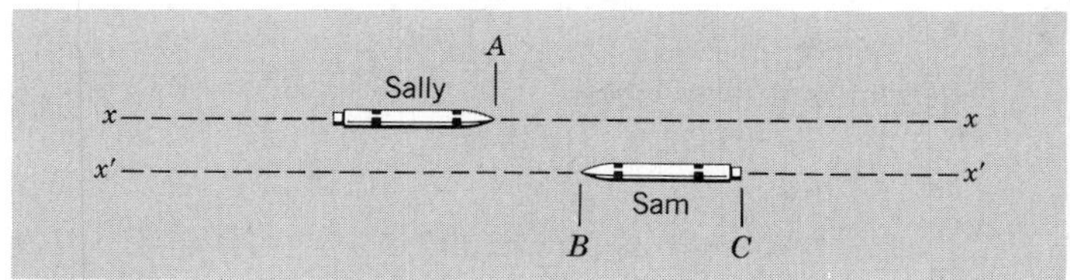

Figure 9 Sample Problem 3. Sally measures the length of Sam's spaceship as it drifts past.

$$\beta = \frac{L_0}{\sqrt{(c\,\Delta t_0)^2 + L_0^2}}$$

$$= \frac{230\text{ m}}{\sqrt{(3.00\times10^8\text{ m/s})^2(3.57\times10^{-6}\text{ s})^2 + (230\text{ m})^2}}$$

$$= 0.210. \qquad \text{(Answer)}$$

Thus, the ships are separating at about 21% of the speed of light.

42-9 The Transformation Equations

As Fig. 10 shows, observer S sees observer S' moving with speed v in the positive direction of their common xx' axis. Observer S reports spacetime coordinates x, y, z, t for an event and observer S' reports x', y', z', t' for the same event. How are these sets of numbers related?

We claim at once (although it requires proof) that the y and z coordinates, which are at right angles to the motion, are not affected by the motion. That is, we will always have $y = y'$ and $z = z'$. Our interest then reduces to the relation between x and x' and between t and t'.

The Galilean Transformation Equations. In prerelativity days, the relations we sought would be given by

$$x' = x - vt \qquad t' = t. \qquad \text{(Galilean transformation equations; valid at low speeds only).} \tag{14}$$

The first of these equations seems to follow from Fig. 10, coupled with the assumption that each observer chooses $t = t' = 0$ to represent the instant that their origins coincided.

The second of these equations would have seemed so obvious to prerelativity physicists that it would seem strange to even write it down. After all, they might have said: "Time is absolute and the same for everybody," or perhaps more likely: "Time is time!"

The fact that Eqs. 14 seem almost obviously true is a reflection of the fact that all our experience with space and time coordinates is limited to the very special case of $v \ll c$. In fact, at speeds comparable to the speed of light, each of the Galilean transformation equations fails to agree with experiment.

The Lorentz Transformation Equations.* We state without proof that the correct transformation equations, which remain valid for all speeds up to the speed of light, can be derived from the postulates of relativity. The results are

$$x' = \gamma(x - vt) \qquad t' = \gamma(t - vx/c^2). \qquad \text{(Lorentz transformation equations; valid at all speeds).} \tag{15}$$

*You may wonder why we do not call these the *Einstein transformation equations* (and why not the *Einstein factor* for γ). The great Dutch physicist H. A. Lorentz actually derived these equations before Einstein did but (as Lorentz graciously conceded) he did not take the further bold step of interpreting these equations as describing the true nature of space and time. It is this interpretation that is at the heart of relativity.

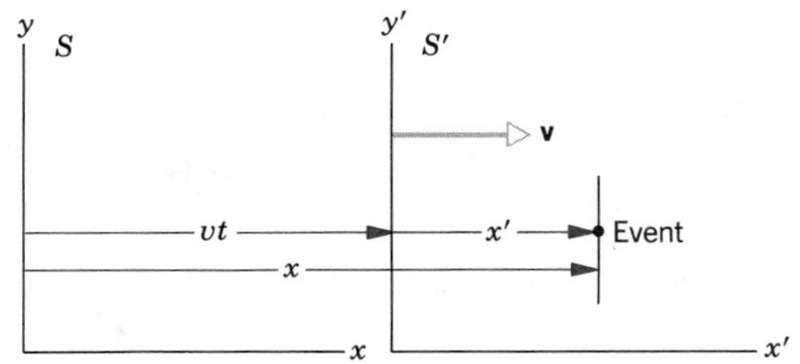

Figure 10 Two inertial reference frames share a common xx' axis. Frame S' is receding from frame S with speed v.

Note how, in the second Lorentz equation, the variable x is bound up with the determination of t'. Time and space are closely intertwined in relativity.

It is a formal requirement of relativistic equations that they should reduce to familiar classical equations if we let c approach infinity. After all, if the speed of light were infinitely great, *all* finite speeds would be "low" and classical equations would never fail. If we let $c \to \infty$ in Eqs. 15, $\gamma \to 1$ and these equations reduce—as we expect—to the Galilean equations (Eqs. 14).

Equations 15 are written in a form that is useful if we are given x and t and wish to find x' and t'. We may wish to go the other way, however. In that case we simply solve Eqs. 15 for x and t, obtaining

$$x = \gamma(x' + vt') \qquad t = \gamma(t' + vx'/c^2). \tag{16}$$

Comparison shows that, starting from either Eqs. 15 or Eqs. 16, you can find the other set by interchanging primed and unprimed quantities and reversing the sign of the relative velocity v.

Equations 15 and 16 relate the coordinates of a single event as seen by two observers. Sometimes we want to know not the coordinates of a single event but the differences between coordinates for a pair of events. That is, if we label our events 1 and 2, we may want to know

$$\Delta x = x_2 - x_1 \quad \text{and} \quad \Delta t = t_2 - t_1,$$

as seen by observer S, and

$$\Delta x' = x_2' - x_1' \quad \text{and} \quad \Delta t' = t_2' - t_1',$$

as observed by S'.

Table 3 displays the Lorentz equations in difference form, suitable for analyzing pairs of events. The equations in the table were derived by simply taking differences between the four variables displayed in Eqs. 15 and 16.

Table 3 The Lorentz Transformation Equations[a]

1. $\Delta x = \gamma(\Delta x' + v\,\Delta t')$	1′. $\Delta x' = \gamma(\Delta x - v\,\Delta t)$
2. $\Delta t = \gamma(\Delta t' + v\,\Delta x'/c^2)$	2′. $\Delta t' = \gamma(\Delta t - v\,\Delta x/c^2)$

$$\gamma = \frac{1}{\sqrt{1-(v/c)^2}} = \frac{1}{\sqrt{1-\beta^2}}$$

[a] Written for pairs of events, as difference equations.

42-10 Some Consequences of the Lorentz Equations

Here we use the transformation equations of Table 3 to affirm some of the conclusions that we reached earlier by arguments based directly on the postulates.

Simultaneity. Consider Eq. 2 of Table 3,

$$\Delta t = \gamma(\Delta t' + v\,\Delta x'/c^2) \quad \text{(a Lorentz equation).} \qquad (17)$$

If two events occur at different places in S', then $\Delta x'$ in this equation is not zero. It follows that even if the events are simultaneous in S' ($\Delta t' = 0$), they will not be simultaneous in S. The time interval in S will be $\Delta t = \gamma v\,\Delta x'/c^2$. This is in accord with the conclusion we reached in Section 42-6.

Time Dilation. Suppose now that two events occur at the same place in $S'(\Delta x' = 0)$ but at different times ($\Delta t' \neq 0$). Equation 17 then reduces to

$$\Delta t = \gamma\,\Delta t' \quad \text{(events in same place in } S'). \qquad (18)$$

This confirms time dilation. Because the two events occur at the same place in S', the time interval $\Delta t'$ between them can be measured with a single clock, located at that place. Under these conditions, the measured interval is a *proper time interval,* and we can label it Δt_0. Thus, Eq. 18 becomes

$$\Delta t = \gamma\,\Delta t_0 \quad \text{(time dilation),}$$

which is exactly Eq. 8, the time dilation equation.

Length Contraction. Consider Eq. 1′ of Table 3,

$$\Delta x' = \gamma(\Delta x - v\,\Delta t) \quad \text{(a Lorentz equation).} \qquad (19)$$

If a rod lies parallel to the xx' axis and is at rest in reference frame S', an observer can measure its length at leisure. The value $\Delta x'$ that is obtained by subtracting the coordinates of the end points of the rod will be its *proper length* L_0.

The rod is moving in frame S. Thus, Δx can only be identified as the length L of the rod if the coordinates of the end points are measured *simultaneously,* that is, if $\Delta t = 0$. If we put $\Delta x' = L_0$, $\Delta x = L$, and $\Delta t = 0$ in Eq. 19, we find

$$L = \frac{L_0}{\gamma} \quad \text{(length contraction),} \qquad (20)$$

which is exactly Eq. 10, the length contraction formula.

Sample Problem 4 In inertial frame S, a blue light flashes, followed after 5.35 μs by a red flash. The separation of the two flashes is $\Delta x = 2.45$ km, with the red flash occurring at the larger value of x. S' is moving in the direction of increasing x with a speed parameter $\beta = 0.855$. What is the distance between the two events and the time interval between them as measured in S'?

Equations 1′ and 2′ of Table 3, with v replaced by βc, are

$$\Delta x' = \gamma(\Delta x - \beta c\,\Delta t) \qquad (21)$$

and

$$\Delta t' = \gamma(\Delta t - \beta\,\Delta x/c). \qquad (22)$$

We are told that

$$\Delta x = x_R - x_B = 2.45 \text{ km} = 2450 \text{ m}$$

and

$$\Delta t = t_R - t_B = 5.35\ \mu\text{s} = 5.35 \times 10^{-6} \text{ s}$$

and that

$$\gamma = \frac{1}{\sqrt{1-\beta^2}} = \frac{1}{\sqrt{1-(0.855)^2}} = 1.928.$$

Thus, we have, from Eq. 21,

$$\begin{aligned}\Delta x' &= (1.928)[2450 \text{ m} - (0.855) \\ &\quad \times (3.00 \times 10^8 \text{ m/s})(5.35 \times 10^{-6} \text{ s})] \\ &= 2078 \text{ m} \approx 2.08 \text{ km} \qquad \text{(Answer)}\end{aligned}$$

and from Eq. 22,

$$\begin{aligned}\Delta t' &= (1.928)\left(5.35 \times 10^{-6} \text{ s} - \frac{(0.855)(2450 \text{ m})}{3.00 \times 10^8 \text{ m/s}}\right) \\ &= -3.147 \times 10^{-6} \text{ s} \approx -3.15\ \mu\text{s}. \qquad \text{(Answer)}\end{aligned}$$

We conclude that in S' the red flash is also more distant but that distance is 2.08 km (rather than 2.45 km). The negative sign in the last result tells us that in S'—contrary to what is observed in S—the red flash occurs first. The time between the flashes in S' is 3.15 μs (not 5.35 μs).

Sample Problem 5 A plane flying at a speed u travels from Seattle to Atlanta. In inertial frame S, which is fixed with respect to the ground, the plane's takeoff from Seattle and its landing in Atlanta are separate events, separated in space by a distance Δx and in time by an interval Δt. Assume that an observer S', moving with respect to observer S, measures these same two events. Is it possible for these events to be recorded reversed in sequence, that is, can an observer in S' see the plane land in Atlanta before it takes off from Seattle?

Consider Eq. 2′ of Table 3,

$$\Delta t' = \gamma(\Delta t - \beta\,\Delta x/c) \quad \text{(a Lorentz equation).}$$

The quantities Δx (the distance from Seattle to Atlanta) and Δt (the flight time) are both positive quantities. Let us find the value of β such that $\Delta t'$ would be negative. Study of the Lorentz equation above shows that we must have

$$\frac{\beta\,\Delta x}{c} > \Delta t.$$

But $\Delta x/\Delta t$ is just the speed u of the plane in S. We then have the requirement that

$$\beta > \frac{c}{u}.$$

But, since the plane cannot exceed the speed of light, $c/u > 1$ so that the requirement is

$$\beta > 1. \qquad \text{(Answer)}$$

This requirement is impossible to fulfill because it would require observer S' to be traveling at a speed greater than the speed of light. A plane cannot land before it takes off, even in special relativity! The takeoff and landing of the plane are not *independent* events because one must necessarily occur before the other, in *all* reference frames. It is never possible to reverse the sequence of events that are causally related. If A causes B, then all observers will agree that A precedes B; you cannot be born before your mother is born!

42-11 The Transformation of Velocities

Here we wish to use the Lorentz equations to compare the velocities that two observers in different inertial reference frames S and S' would measure for the same moving particle.

Suppose that the particle, moving with constant speed parallel to the xx' axis, sends out two signals as it moves. Each observer measures the spatial interval and the time interval between these two events. These four measurements are related by Eqs. 1 and 2 of Table 3, or

$$\Delta x = \gamma(\Delta x' + u\,\Delta t') \quad \text{and} \quad \Delta t = \gamma(\Delta t' + u\,\Delta x'/c^2),$$

in which we now represent by u the velocity of S' with respect to S. If we divide the first of these equations by the second, we find

$$\frac{\Delta x}{\Delta t} = \frac{\Delta x' + u\,\Delta t'}{\Delta t' + u\,\Delta x'/c^2}.$$

Dividing the numerator and the denominator of the right side by $\Delta t'$, we find

$$\frac{\Delta x}{\Delta t} = \frac{\Delta x'/\Delta t' + u}{1 + u(\Delta x'/\Delta t')/c^2}.$$

But, in the differential limit, $\Delta x/\Delta t$ is v, the velocity measured in S and $\Delta x'/\Delta t'$ is v', the velocity measured in S'. We have finally that

$$v = \frac{v' + u}{1 + uv'/c^2} \qquad \text{(relativistic velocity law)} \quad (23)$$

as the relativistic velocity transformation law. We discussed this law earlier, using a slightly different notation, in optional Section 4-9. You may wish to reread that section and, in particular, to study Sample Problems 10 and 11 of Chapter 4. Equation 23 reduces to the classical, or Galilean, velocity transformation law,

$$v = v' + u \quad \text{(classical velocity law),} \quad (24)$$

when we apply the formal test of letting $c \rightarrow \infty$.

42-12 The Doppler Effect

We discussed the Doppler effect for sound waves in air in Section 18-7. In that case, there were two Doppler formulas because the velocities of the source and of the detector with respect to the transmitting medium (air) must be considered separately. In prerelativity days, physicists believed that a medium (called the *luminiferous ether*) was needed to support the propagation of light. Thus, the two Doppler formulas for sound were taken over for light by simply substituting the speed of light c for the speed of sound.

With the advent of relativity, Einstein declared that the propagation of light requires no medium and that the only relevant velocity is the relative velocity of the source and the detector. Thus, there should be just one Doppler formula for light, not two, and it must be derived from relativity theory.

The classical and the relativistic Doppler formulas for the shifted frequency are

Prerelativity theory:
source fixed, detector receding $\qquad \nu = \nu_0(1 - \beta) \qquad (25)$

Relativity theory: source and detector separating $$\nu = \nu_0 \sqrt{\frac{1-\beta}{1+\beta}}. \tag{26}$$

Prerelativity theory: source receding, detector fixed $$\nu = \nu_0 \frac{1}{1+\beta}. \tag{27}$$

Here $\beta = v/c$, where v is the relative speed of separation of source and detector and c is the speed of light. The quantity ν_0 is the frequency (a *proper* frequency) that would be measured by an observer for whom the source is at rest. The frequency ν is that which would be measured by an observer in relative motion with respect to the source. All three formulas apply to the case in which source and detector are moving farther apart. If they are moving closer together, it is only necessary to change the sign of β in the above three equations.

To test the theory of relativity, we must be able to show by experiment that Doppler shifts for light (or other electromagnetic radiation) obey Eq. 26 and not Eqs. 25 or 27. At first sight this does not look too difficult because the equations look so different. The three equations, however, are deceptively similar at low speeds, which is the speed region in which we must conduct our tests.

If $\beta \ll 1$ and if we expand Eqs. 25, 26, and 27 as power series in β, keeping no terms higher than β^2, we find that all three equations reduce to

$$\nu \approx \nu_0(1 - \beta + C\beta^2) \tag{28}$$

The equations differ only in the numerical constant C, the coefficient of the term in β^2. The predicted values for C are

Equation	*Value of C*
25	0
26	$\frac{1}{2}$
27	1

For a source moving at low speed, β will be small and β^2 correspondingly smaller so that clever experiments are needed. In 1938, H. E. Ives and G. R. Stilwell of the Bell Telephone Laboratories devised just such an experiment. They sent the light beam back and forth through the apparatus in such a way that the term containing β in Eq. 28 effectively canceled out, leaving only the crucial term in β^2. These workers verified the prediction of relativity theory to a precision of a few percent. A repetition of the experiment in 1985 with modern techniques improved the agreement with the prediction of relativity to 1 part in 25,000 or 0.004%.

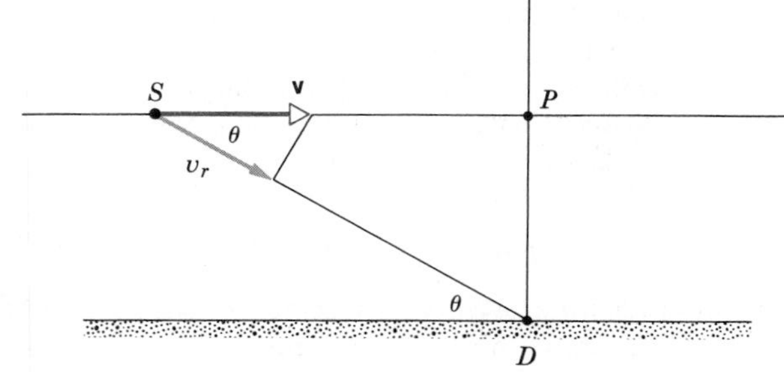

Figure 11 A light source S travels with velocity **v** past a detector at D. As the source passes through point P, the radial component of its velocity is zero. According to relativity, there should be a transverse Doppler effect at this point; classical theory predicts that there should not be.

Transverse Doppler Effect. Our entire discussion of the Doppler effect, both here and in Chapter 18, is for the case in which the source and the detector are moving directly toward each other or directly away from each other. Suppose, however, that a light source S is moving past a detector D as shown in Fig. 11. When the moving source passes through point P in Fig. 11, the radial component of its velocity is zero. The classical and the relativistic predictions for the Doppler effect in his case are

$$\text{Prerelativity theory} \quad \nu = \nu_0 \tag{29}$$

and

$$\text{Relativity theory} \quad \nu = \nu_0 \sqrt{1-\beta^2}. \tag{30}$$

That is, there is (Eq. 30) or there is not (Eq. 29) a *transverse Doppler effect.*

The transverse Doppler effect is really another test of time dilation. If we write Eq. 30 in terms of the period T of oscillation of the emitted rays instead of the frequency, we have, recalling that $T = 1/\nu$,

$$T = \frac{T_0}{\sqrt{1-\beta^2}} = \gamma T_0, \tag{31}$$

in which $T_0 (= 1/\nu_0)$ is the *proper period* of the source. As comparison with Eq. 5 shows, Eq. 31 is simply the time dilation formula. We can understand this physically if we think of the moving source as a moving clock, beating out electromagnetic oscillations at a proper frequency ν_0 and a corresponding proper period T_0. We expect to measure a longer period (that is, a smaller frequency) for such a moving clock.

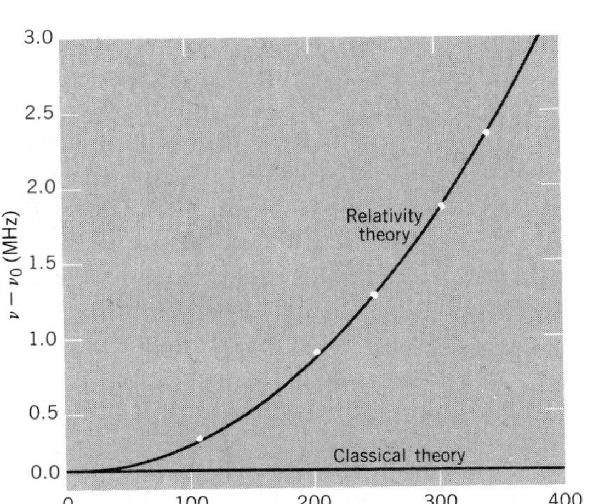

Figure 12 The results of Kundig on the transverse Doppler effect. The experiments fall very nicely on the curve predicted by relativity theory and not at all on the curve predicted by classical theory.

A Transverse Doppler Experiment. In 1963, W. Kundig obtained excellent quantitative data confirming Eq. 30 to within an experimental error of about 1%. In his experiment, a radioactive source emitting gamma rays was located on the rotor of a centrifuge. By sensitive techniques, Kundig was able to measure the shift in frequency detected by a detector placed on the rim of the centrifuge. Figure 12 shows his results, which confirm the prediction of the relativity theory beyond any question.

42-13 A New Look at Momentum

Suppose that a number of observers, each in his or her own inertial reference frame, watch an isolated collision between two particles. In classical mechanics, we have seen that—even though the observers measure different velocities for the colliding particles—they all find that the law of conservation of momentum holds. That is, they find that the momentum of the system of particles after the collision is the same as it was before the collision.

How is this situation affected by relativity? We find that, if we continue to define the momentum $\mathbf{p}$ of a particle as $m\mathbf{v}$, the product of its mass and its velocity, momentum is *not* conserved for all inertial observers. We have two choices: (1) Give up the law of conservation of momentum. (2) See if we can redefine the momentum of a particle in some new way so that the law of conservation of momentum still holds. We choose the second of these alternatives.

Consider a particle moving with constant speed v in the x direction. Classically, its momentum is

$$p = mv = m\frac{\Delta x}{\Delta t} \quad \text{(momentum—classical)}, \quad (32)$$

in which Δx is the distance covered in time Δt. As we turn our thoughts to relativity, this classical definition is suspect from the beginning because it sets an upper limit ($=mc$) on the momentum that a particle may attain. If momentum is to be useful in relativity, it must be able to attain high values, without any discernible upper limit.

It seems clear that we must generalize the classical definition of momentum. There is no way to derive a generalized definition rigorously from a restricted one. We can only make an educated guess and then test it by seeing whether, using this new definition, momentum is conserved in collisions between particles.

The definition we propose is

$$p = m\frac{\Delta x}{\Delta t_0}.$$

Here, as before, Δx is the distance covered by a moving particle as viewed by an observer watching that particle. However, Δt_0 is the time to cover that distance, measured not by the observer watching the moving particle but by an observer moving with the particle. The particle is at rest with respect to this second observer so that the time that observer measures is a proper time Δt_0.

Using the time dilation formula (Eq. 8), we can then write

$$p = m\frac{\Delta x}{\Delta t_0} = m\frac{\Delta x}{\Delta t}\frac{\Delta t}{\Delta t_0} = m\frac{\Delta x}{\Delta t}\gamma.$$

But $\Delta x/\Delta t$ is just the particle velocity v so that

$$\boxed{p = \gamma mv} \quad \text{(momentum—relativistic)}. \quad (33)$$

Note that this differs from the classical definition of Eq. 32 only by the Lorentz factor γ. Unlike the classical definition, this relativistic definition permits the momentum p to approach infinitely large values as the particle speed v approaches the speed of light as a limiting value.

We can generalize the definition of Eq. 33 to vector

form as

$$\boxed{\mathbf{p} = \gamma m \mathbf{v}} \quad \text{(momentum—relativistic).} \tag{34}$$

We introduced this definition without elaboration in Section 9–4 as a foretaste of things to come; see Eq. 20 of Chapter 9. We state without further proof that, if we adopt the definition of momentum presented in Eq. 34, we can continue to use the conservation-of-momentum principle up to the very highest particle speeds.

Relativistic Mass. We can also write Eq. 33 in the form

$$p = m'v, \tag{35}$$

in which m', called the *relativistic mass* of the particle, is given by

$$\boxed{m' = \gamma m} \quad \text{(relativistic mass).} \tag{36}$$

To distinguish between the two masses, we now refer to m as the *rest mass* of the particle. As its name implies, the rest mass of a particle is the mass measured in a reference frame in which the particle is at rest. As Eq. 35 shows, we can carry over the classical definition of momentum by simply substituting the relativistic mass for the rest mass. The relativistic mass m' increases with velocity (because the Lorentz factor γ in Eq. 36 does so), approaching an infinite value as the speed of the particle approaches the speed of light.*

42–14 A New Look at Energy

In Section 7–7 we presented, without elaboration, the following relativistic expression for the kinetic energy of a particle:

$$K = mc^2\left(\frac{1}{\sqrt{1-(v/c)^2}} - 1\right),$$

which we can now write as

$$\boxed{K = mc^2(\gamma - 1)} \quad \text{(relativistic kinetic energy).} \tag{37}$$

We showed in Section 7–7 that—unlikely as it may seem—this expression reduces to the familiar classical $K = \frac{1}{2}mv^2$ at low speeds. The derivation of Eq. 37 follows exactly the same path that we followed in deriving the classical kinetic energy expression ($K = \frac{1}{2}mv^2$) in Section 7–5 except that we use the relativistic, rather than the classical, expression for the momentum. In both the classical case and the relativistic case, we set the kinetic energy K equal to the work that must be done to accelerate the particle from rest to its observed speed.*

Let us point to some of the consequences of Eq. 37. We start by defining the *total energy E* of a particle as γmc^2. With the help of Eq. 37 we can then write

$$\boxed{\begin{aligned} E &= \gamma mc^2 \\ &= mc^2 + K \end{aligned}} \quad \begin{array}{l}\text{(total energy;}\\ \text{single particle).}\end{array} \tag{38}$$

We interpret the total energy E of the moving particle as made up of mc^2, which we call the *rest energy* of the particle, and K, its kinetic energy. Table 1 of Chapter 8 lists the rest energies of a few particles and other objects. The rest energy of an electron, for example, is 0.511 MeV and for a proton it is 938.3 MeV.

The total energy of a system of particles is

$$\begin{aligned} E &= \sum E_i = \sum (\gamma mc^2) \\ &= \sum mc^2 + \sum K \end{aligned} \quad \begin{array}{l}\text{(total energy;}\\ \text{system of particles).}\end{array} \tag{39}$$

In relativity, the *conservation of energy principle* is stated as follows:

> *For an isolated system of particles, the total energy E of the system, defined by Eq. 39, remains constant, no matter what interactions may occur among the particles.*

Thus, in any isolated reaction or decay process involving two or more particles, the total energy of the system after the interaction must remain equal to the total energy after the reaction. During the reaction, the total rest energy of the interacting particles may change but the total kinetic energy must then also change by an equal amount in the opposite direction to compensate.

We note from Eqs. 36 and 39 that the total energy E of a system of particles can also be written as $\Sigma(m'c^2)$, where m' is the relativistic mass. Thus, the conservation

* In treatments in which relativistic mass plays a central role, it is usually represented by m, the rest mass being then represented by m_0. In this book, m will always represent the rest mass.

* See the footnote reference on the opening page of this chapter for details of the proof.

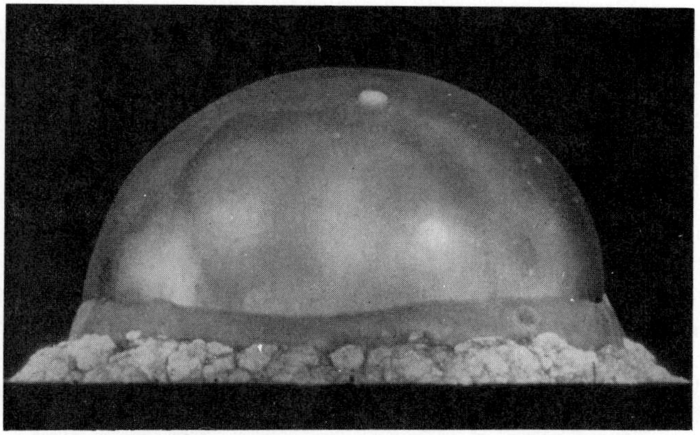

Figure 13 The fireball of a nuclear explosion, a striking example of the conversion of (rest) mass into energy.

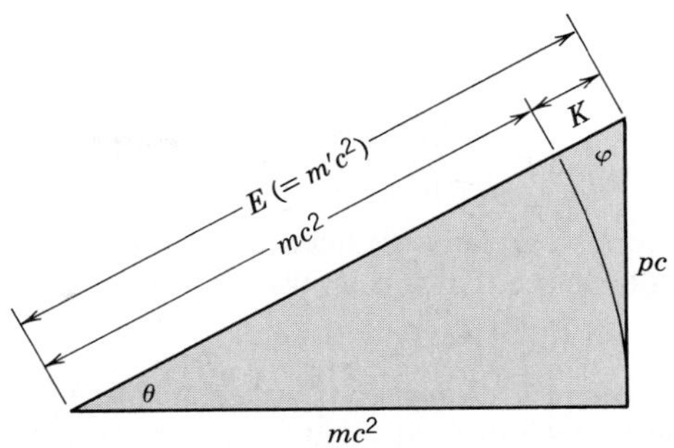

Figure 14 A useful mnemonic device for remembering the relativistic relations among the total energy E, the rest energy mc^2, the kinetic energy K, and the momentum p.

of total energy is equivalent to the conservation of relativistic mass. Therefore, although *rest* mass may not be conserved for an isolated system, *relativistic* mass is always conserved.

Considerations of this sort are at the root of Einstein's well-known $E = mc^2$ relation, which asserts that rest energy is freely convertible into other forms. All reactions—whether chemical or nuclear—in which energy is released or absorbed involve a corresponding change in the rest energy of the reactants; Fig. 13 is a particularly striking example. We discussed the $E = mc^2$ relation in detail in Section 8-9; see, in particular, Sample Problems 7, 8, and 9 in that Section.

Momentum and Kinetic Energy. In classical mechanics, the momentum p of a particle is mv and its kinetic energy K is $\frac{1}{2}mv^2$. If we eliminate v between these two expressions, we find a direct relation between the momentum and the kinetic energy:

$$p^2 = 2Km \quad \text{(classical).} \tag{40}$$

We can find a similar connection in relativity by eliminating v between the relativistic definition of momentum (Eq. 33) and the relativistic definition of kinetic energy (Eq. 37). Doing so leads, after some algebra, to

$$\boxed{(pc)^2 = K^2 + 2Kmc^2} \quad \text{(relativistic).} \tag{41}$$

With the aid of Eq. 38, we can transform Eq. 41 into a relation between the momentum p and the total energy E of a particle, or

$$\boxed{E^2 = (pc)^2 + (mc^2)^2} \quad \text{(relativistic),} \tag{42}$$

The right triangle of Fig. 14 helps to keep these useful relations in mind. You can also show that, in that triangle,

$$\sin\theta = \beta \quad \text{and} \quad \sin\phi = 1/\gamma. \tag{43}$$

Sample Problem 6 An electron has a kinetic energy K of 2.53 MeV. (a) What is its total energy E?

From Eq. 38 we have

$$E = mc^2 + K.$$

Earlier, we listed mc^2 for an electron as 0.5110 MeV so that

$$E = 0.511 \text{ MeV} + 2.53 \text{ MeV} = 3.04 \text{ MeV}. \quad \text{(Answer)}$$

(b) What is its momentum p?

From Eq. 42,

$$E^2 = (pc)^2 + (mc^2)^2,$$

we can write

$$(3.04 \text{ MeV})^2 = (pc)^2 + (0.511 \text{ MeV})^2$$

or

$$pc = \sqrt{(3.04 \text{ MeV})^2 - (0.511 \text{ MeV})^2}$$
$$= 3.00 \text{ MeV}.$$

It is customary to report momentum in particle physics in units of energy divided by c. Thus,

$$p = 3.00 \text{ MeV}/c. \quad \text{(Answer)}$$

(c) What is the electron's relativistic mass?

From Eqs. 36 and 38 we see that

$$E = \gamma mc^2 = m'c^2,$$

so that

$$m' = \frac{E}{c^2} = \frac{E}{mc^2}\,m$$

$$= \frac{3.04\text{ MeV}}{0.511\text{ MeV}}\,m = 5.95m. \qquad \text{(Answer)}$$

Thus, the electron's relativistic mass is 5.95 times its rest mass. From Eq. 36 we see that this is just the Lorentz factor ($\gamma =$ 5.95).

Sample Problem 7 A proton is accelerated to 1.08 TeV in the Fermilab accelerator. (a) What is its speed parameter β?

Let us write Eq. 37 in the form

$$K = \gamma mc^2 - mc^2.$$

If we solve for γ we find

$$\gamma = \frac{K + mc^2}{mc^2} = \frac{K}{mc^2} + 1$$

$$= \frac{1.08 \times 10^{12}\text{ eV}}{938.3 \times 10^6\text{ eV}} + 1 = 1151.0 + 1 = 1152.$$

This Lorentz factor γ (=1152) is much greater than unity, indicating that relativity is absolutely vital in this problem. This is borne out by our observation that the kinetic energy of the moving proton (1.08×10^{12} eV) is more than 1000 times greater than its rest energy (938.3×10^6 eV).

Knowing γ, it would seem a straightforward matter to solve Eq. 9 for the speed parameter β. It simplifies our work greatly, however, to realize that β will be very close indeed to unity and that we would do well to solve not for β itself but for $1 - \beta$, its departure from unity.

Equation 9 then becomes

$$\gamma = \frac{1}{\sqrt{1-\beta^2}} = \frac{1}{\sqrt{(1-\beta)(1+\beta)}} \approx \frac{1}{\sqrt{(2)(1-\beta)}}.$$

We have here taken advantage of the fact that β is so close to unity that $1 + \beta$ is very close to 2.

If we solve the above equation for $1 - \beta$, we find

$$1 - \beta = \frac{1}{2\gamma^2} = \frac{1}{(2)(1152)^2} = 3.77 \times 10^{-7}.$$

Finally, we find for β

$$\beta = 1 - (1 - \beta) = 1 - 3.77 \times 10^{-7}$$

$$= 0.999\,999\,623. \qquad \text{(Answer)}$$

If you think that proceeding by approximation did not simplify the work, try solving the problem without making any approximation. You will overload your calculator.

If you have any lingering doubts about relativity theory, try solving this problem using the classical kinetic energy formula, $K = \frac{1}{2}mv^2$. You will find a speed 48 times the speed of light!

(b) How much slower than the speed of light is this high-energy proton moving?

We can write

$$\Delta v = c - v = c(1 - v/c) = c(1 - \beta).$$

Using data from (a), we find

$$\Delta v = c(1 - \beta) = (3.00 \times 10^8\text{ m/s})(3.77 \times 10^{-7})$$

$$= 113\text{ m/s}. \qquad \text{(Answer)}$$

42-15 The Common Sense of Relativity

We have come to a point from which we can look back and think about the common sense of relativity. We must start by agreeing that relativity permeates physics to its roots and that the theory has passed all experimental tests without the emergence of the slightest flaw. It is a theory that is aesthetically pleasing, inherently simple, comprehensively consistent, of great predictive value, and highly practical. If, for example, engineers were to design a high-energy particle accelerator ignoring relativity, the accelerator can be guaranteed not to work.

Are not the two postulates of the theory reasonable? Most of us willingly embrace the first postulate, the postulate of relativity, which gives a larger scope to a concept already familiar to Galileo. Einstein's second postulate, which asserts that there is in nature an ultimate limiting speed at which signals and energy can be transmitted from point to point, seems more reasonable the longer we think about it. If it were not true, it would then be possible to transmit signals instantaneously to all parts of the universe. It is really this assumption of instantaneous action-at-a-distance (which underlies classical physics) that is unreasonable. If there *is* a limiting speed, then from the principle of relativity, should it not be the same for all observers, regardless of their state of motion?

The relativity of simultaneity, the shrinking of moving rods, and the slowing down of moving clocks may disturb the rigidly classical mind; but are not all these phenomena based on measurement, following procedures derived in a reasonable way from the postulates? If the limiting speed c were only a lot smaller (perhaps 1000 mi/h), would not all these phenomena seem to be common sense of the highest order? Is not the relativity of the effects just what we would expect? That is, if A's clock seems to B to run slow, then do we not find it comfortingly consistent that B's clock will seem to A to run slow? After all, who are A and B that we should be forced to choose between them?

Relativity broadens immeasurably our view of the world around us. In classical physics we have the separate laws of the conservation of mass and of energy. Is it not a great step forward to replace these by the single conservation law of total energy (or of relativistic mass)? There are other examples that we have not been able to explore fully. For example, in relativity space and time are linked together as spacetime and the electric field and the magnetic field are linked as aspects of a single electromagnetic field. The thoughtful student must conclude with us that, in relativity, we have the very model of a theory. We owe it all to Albert Einstein, who described well the nature of his contemplation of the world around him when he remarked: "I want to know God's thoughts . . . the rest are details." Figure 15 showed the tribute paid to Einstein by the cartoonist Herblock at the time of Einstein's death in 1955.

Figure 15 A cartoon by Herblock, published at Einstein's death in 1955.

REVIEW AND SUMMARY

The Postulates

Einstein's *special theory of relativity* is based on two postulates:

1. The law of physics are the same for observers in all inertial reference frames. No frame is singled out as preferred.
2. The speed of light in free space has the same value c in all directions and in all inertial reference frames.

The free-space speed of light c is an ultimate speed that cannot be exceeded by any entity carrying energy or information. See Sample Problem 1.

Coordinates of an Event

Three space coordinates and one time coordinate specify an *event.* One problem of special relativity is to relate these coordinates as assigned by two observers in uniform motion with respect to each other. Section 42–4 describes thought experiments by which a single observer might assign such coordinates.

Simultaneous Events

If two observers are in relative motion, they will not, in general, agree as to whether two events are simultaneous or not. If one observer finds two events at different locations to be simultaneous, the other will not, and conversely. Simultaneity is *not* an absolute concept but a relative one, depending on the motion of the observer. In Section 42–6, we show that this relativity of simultaneity is a direct consequence of the finite ultimate speed c.

Relativity of Time

If events occur at the same place in an inertial reference frame, the time interval Δt_0 is the *proper time* of the events. *All other observers will measure a larger value for this interval.* For an observer moving with speed v with respect to the original inertial frame, the measured time interval is

Time Dilation

$$\Delta t = \frac{\Delta t_0}{\sqrt{1-(v/c)^2}} = \frac{\Delta t_0}{\sqrt{1-\beta^2}} = \gamma\,\Delta t_0 \quad \text{(time dilation).} \qquad [7,8]$$

Here $\beta = v/c$ is the *speed parameter* and $\gamma = 1/\sqrt{1-\beta^2}$ is the *Lorentz factor;* see Table 2. An important consequence is that moving clocks run slow as measured by an observer at rest. See Sample Problem 2.

Relativity of Length

The length L_0 of a rod, measured in an inertial reference frame in which the rod is at rest, is called its *proper length. All other inertial observers will measure a shorter length.* For an observer moving with speed v with respect to the original inertial frame, the measured length is

Length Contraction

$$L_0\sqrt{1-\beta^2} = \frac{L_0}{\gamma} \quad \text{(length contraction).} \qquad [10]$$

See Sample Problem 3.

The Transformation Equations

The *Lorentz transformation equations* relate the coordinates of a single event as seen by two observers, S and S', with S' moving with respect to S with a velocity v in the positive x direction. The two perpendicular coordinates (y and z) are not affected. The x and t coordinates are related by

The Lorentz Transformation

$$\begin{aligned} x' &= \gamma(x - vt) \\ t' &= \gamma(t - vx/c^2) \end{aligned} \quad \text{(Lorentz transformation equations; valid at all speeds).} \qquad [15]$$

See Sample Problems 4 and 5.

Transformation of Velocities

A particle moving with speed v' in the positive x' direction in an inertial frame of reference S' that itself is moving with speed u parallel to the x direction of a second interial frame S will be measured in S as having speed

$$v = \frac{v' + u}{1 + uv'/c^2} \quad \text{(relativistic velocity law).} \qquad [23]$$

If a source with frequency ν_0 moves directly away from a detector with relative velocity $\mathbf{v}$, the frequency ν measured by the detector is

Longitudinal Doppler Effect

$$\nu = \nu_0\sqrt{\frac{1-\beta}{1+\beta}}. \qquad [26]$$

If the relative motion of the source is transverse, the Doppler formula is

Transverse Doppler Effect

$$\nu = \nu_0\sqrt{1-\beta^2}. \qquad [30]$$

This *transverse Doppler effect* is entirely a consequence of time dilation.

Momentum and Energy

Presuming the validity of generalized laws of conservation of momentum and energy, the expressions for linear momentum $\mathbf{p}$, kinetic energy K, and total energy E that must be used for particles in relativistic dynamics are

$$\mathbf{p} = \gamma m\mathbf{v} = m'\mathbf{v} \quad \text{(momentum—relativistic),} \qquad [34]$$

$$K = mc^2(\gamma - 1) \quad \text{(relativistic kinetic energy),} \qquad [37]$$

and

$$E = \gamma mc^2 = mc^2 + K = m'c^2 \quad \text{(total energy, single particle).} \qquad [38]$$

Here m is the *rest mass* of the particle and $m' = \gamma m$ is its *relativistic mass.* With these definitions, conservation of total energy for a system of particles takes the form

$$E = \sum(\gamma mc^2) = \sum mc^2 + \sum K \quad \text{(total energy, system of particles).} \qquad [39]$$

Two additional relationships, derivable from the definitions, are often useful:

$$(pc)^2 = K^2 + 2Kmc^2 \quad \text{(relativistic)} \qquad [41]$$

and

$$E^2 = (pc)^2 + (mc^2)^2 \quad \text{(relativistic).} \qquad [42]$$

See Sample Problems 6 and 7 for representative calculations.

QUESTIONS

1. How would you test a proposed reference frame to find out whether or not it is an inertial frame?

2. Give examples in which effects associated with the earth's rotation are significant enough in practice to rule out a laboratory frame as being a good enough approximation to an inertial frame.

3. The speed of light in a vacuum is a true constant of nature, independent of the wavelength of the light or the choice of an (inertial) reference frame. Is there any sense, then, in which Einstein's second postulate can be viewed as contained within the scope of his first postulate?

4. Discuss the problem that young Einstein grappled with; that is, what would be the appearance of an electromagnetic wave to a person running along with it at speed c?

5. Quasars *(quasi-stellar objects)* are the most intrinsically luminous objects in the universe. Many of them fluctuate in brightness, often on a time scale of a day or so. How can the rapidity of these brightness changes be used to estimate an upper limit to the size of these objects? (*Hint:* Separated points cannot change in a coordinated way unless information is sent from one to the other.)

6. The sweep rate of the tail of a comet can exceed the speed of light. Explain this phenomenon and show that there is no contradiction with relativity.

7. Borrowing two phrases from Herman Bondi, we can catch the spirit of Einstein's two postulates by labeling them: (1) the principle of "the irrelevance of velocity" and (2) the principle of "the uniqueness of light." In what senses are velocity irrelevant and light unique in these two statements?

8. A beam from a laser falls at right angles on a plane mirror and rebounds from it. What is the speed of the reflected beam if the mirror is (*a*) fixed in the laboratory? (*b*) Moving directly toward the laser with speed v?

9. Give an example from classical physics in which the motion of a clock affects its rate, that is, the way it runs. (The magnitude of the effect may depend on the detailed nature of the clock.)

10. Although in relativity (where motion is relative and not absolute) we find that "moving clocks run slow," this effect has nothing to do with the motion altering the way a clock works. What does it have to do with?

11. We have seen that if several observers watch two events, labeled A and B, one of them may say that event A occurred first but another may claim that it was event B that did so. What would you say to a friend who asked you which event *really did* occur first?

12. Two events occur at the same place and at the same time for one observer. Will they be simultaneous for all other observers? Will they also occur at the same place for all other observers?

13. Two observers, one at rest in S and one at rest in S', each carry a meter stick oriented parallel to their relative motion. *Each* observer finds upon measurement that the *other* observer's meter stick is shorter than his own meter stick. Does this seem like a paradox to you? Explain. (*Hint:* Compare the following situation. Harry waves goodbye to Walter who is in the rear of a station wagon driving away from Harry. Harry says that Walter gets smaller. Walter says that Harry gets smaller. Are they measuring the same thing?)

14. How does the concept of simultaneity enter into the measurement of the length of a body?

15. In relativity the time and space coordinates are intertwined and treated on a more or less equivalent basis. Are time and space fundamentally of the same nature, or is there some essential difference between them that is preserved even in relativity?

16. In the "twin paradox," explain (in terms of heartbeats, physical and mental activities, and so on) why the younger returning twin has not lived any longer than his own proper time even though his stay-at-home brother may say that he has. Hence, explain the remark: "You age according to your own proper time."

17. Can we simply substitute m' for m in classical equations to obtain the correct relativistic equations? Give examples.

18. If zero-mass particles have a speed c in one reference frame, can they be found at rest in any other frame? Can such particles have any speed other than c?

19. How many relativistic expressions can you think of in which the Lorentz factor γ enters as a simple multiplier?

20. In a given magnetic field, would a proton or an electron, traveling at the same speed, have the greater frequency of revolution? See Problem 57.

21. Is the rest mass of a stable, composite particle (a gold nucleus, for example) greater than, equal to, or less than the sum of the rest masses of its constituents? Explain.

22. "The rest mass of the electron is 0.511 MeV." Exactly what does this statement mean?

23. "The relation $E = mc^2$ is essential to the operation of a power plant based on nuclear fission but has only a negligible relevance for a fossil-fuel plant." Is this a true statement? Explain why or why not.

24. A hydroelectric plant generates electricity because water falls under gravity through a turbine, thereby turning the shaft of a generator. According to the mass–energy concept, must the appearance of energy (the electricity) be identified with a mass decrease somewhere? If so, where?

25. Exactly why is it that, kilogram for kilogram, nuclear explosions release so much more energy than do TNT explosions?

26. A hot metallic sphere cools off as it rests on the pan of a scale. If the scale were sensitive enough, would it indicate a change in rest mass? If so, woudl it be an increase or a decrease?

27. Some say that relativity complicates things. Give examples to the contrary, wherein relativity simplifies matters.

EXERCISES AND PROBLEMS

Section 42-3 The Postulates

1E. What fraction of the speed of light does each of the following speeds represent; that is, what is the speed parameter β? (*a*) A typical rate of continental drift (1 in. per year). (*b*) A typical drift speed for electrons in a current-carrying conductor (0.5 mm/s). (*c*) A highway speed limit of 55 mi/h. (*d*) The root-mean-square speed of a hydrogen molecule at room temperature. (*e*) A supersonic plane flying at Mach 2.5 (1200 km/h). (*f*) The escape speed of a projectile from the surface of the earth. (*g*) The speed of the earth in its orbit around the sun. (*h*) A typical recession speed of a distant quasar (3.0×10^4 km/s).

2E. Quite apart from effects due to the earth's rotational and orbital motions, a laboratory frame is not strictly an inertial frame because a particle placed at rest there will not, in general, remain at rest; it will fall under gravity. Often, however, events happen so quickly that we can ignore free fall and treat the frame as inertial. Consider, for example, a 1.0-MeV electron (for which $v = 0.992c$) projected horizontally into a laboratory test chamber and moving through a distance of 20 cm. (*a*) How long would it take, and (*b*) how far would the electron fall during this interval? What can you conclude about the suitability of the laboratory as an inertial frame in this case?

3P. Find the speed of a particle that takes 2 years longer than light to travel a distance of 6.0 ly.

Section 42-7 The Relativity of Time

4E. What must be the speed parameter β if the Lorentz factor γ is to be (*a*) 1.01? (*b*) 10? (*c*) 100? (*d*) 1000?

5E. The mean lifetime of muons stopped in a lead block in the laboratory is measured to be 2.2 μs. The mean lifetime of high-speed muons in a burst of cosmic rays observed from the earth is measured to be 16 μs. Find the speed of these cosmic ray muons.

6P. An unstable high-energy particle enters a detector and leaves a track 1.05 mm long before it decays. Its speed relative to the detector was $0.992c$. What is its proper lifetime? That is, how long would it have lasted before decay had it been at rest with respect to the detector?

7P. A pion is created in the higher reaches of the earth's atmosphere when an incoming high-energy cosmic-ray particle collides with an atomic nucleus. A pion so formed descends toward earth with a speed of $0.99c$. In a reference frame in which they are at rest, pions decay with a mean life of 26 ns. As measured in a frame fixed with respect to the earth, how far (on the average) will such a typical pion move through the atmosphere before it decays?

8P. You wish to make a round trip from earth in a spaceship, traveling at constant speed in a straight line for six months and then returning at the same constant speed. You wish further, on your return, to find the earth as it will be a thousand years in the future. (*a*) How fast must you travel? (*b*) Does it matter whether or not you travel in a straight line on your journey? If, for example, you traveled in a circle for one year, would you still find that a thousand years had elapsed by earth clocks when you returned?

Section 42-8 The Relativity of Length

9E. A rod lies parallel to the x axis of reference frame S, moving along this axis at a speed of $0.63c$. Its rest length is 1.7 m. What will be its measured length in frame S?

10E. The length of a spaceship is measured to be exactly half its rest length. (*a*) What is the speed of the spaceship relative to the observer's frame? (*b*) By what factor do the spaceship's clocks run slow, compared to clocks in the observer's frame?

11E. A 100-MeV electron, for which $\beta = 0.999987$, moves along the axis of an evacuated tube that has a length of 3.0 m as measured by a laboratory observer S with respect to whom the tube is at rest. An observer S' moving with the electron, however, would see this tube moving past with speed v $(= \beta c)$. What length would this observer measure for the tube?

12E. The rest radius of the earth is 6370 km and its orbital speed about the sun is 30 km/s. By how much would the earth's diameter appear to be shortened to an observer stationed so as to be able to watch the earth move past him at this speed?

13E. A spaceship of rest length 130 m drifts past a timing station at a speed of $0.74c$. (*a*) What is the length of the spaceship as measured by the timing station? (*b*) What time interval between the passage of the front and back end of the ship will the station monitor record?

14P. A space traveler takes off from earth and moves at speed $0.99c$ toward the star Vega, which is 26 ly distant. How much time will have elapsed by earth clocks (*a*) when the traveler reaches Vega? (*b*) when the earth observers receive word from him that he has arrived? (*c*) How much older will the earth observers calculate the traveler to be when he reaches Vega than he was when he started the trip?

15P. An airplane whose rest length is 40 m is moving at a uniform velocity with respect to the earth at a speed of 630 m/s. (*a*) By what fraction of its rest length will it appear to be shortened to an observer on earth? (*b*) How long would it take by earth clocks for the airplane's clock to fall behind by 1 μs? (Assume that only special relativity applies.)

16P. (*a*) Can a person, in principle, travel from earth to the galactic center (which is about 23,000 ly distant) in a normal lifetime? Explain, using either time-dilation or length-contraction arguments. (*b*) What constant speed would be needed to make the trip in 30 y (proper time)?

Section 42-10 Some Consequences of the Lorentz Equations

17E. Observer S assigns the following spacetime coordinates to an event:

$$x = 100 \text{ km}, \quad t = 200\ \mu\text{s}.$$

What are the coordinates of this event in frame S', which moves in the direction of increasing x with speed $0.95c$? Assume $x = x'$ at $t = t' = 0$.

18E. Observer S reports that an event occurred on his x axis at $x = 3.0 \times 10^8$ m at a time $t = 2.50$ s. (*a*) Observer S' is moving in the direction of increasing x at a speed of $0.40c$. What coordinates would he report for the event? (*b*) What coordinates would he report if he were moving in the direction of *decreasing* x at this same speed?

19E. Inertial frame S' moves at a speed of $0.60c$ with respect to frame S. Two events are recorded. In frame S, event 1 occurs at the origin at $t = 0$ and event 2 occurs on the x axis at $x = 3.0$ km and at $t = 4.0\ \mu$s. What times of occurrence does observer S' record for these same events? Explain the difference in the time order.

20E. An experimenter arranges to trigger two flashbulbs simultaneously, a blue flash located at the origin of his reference frame and a red flash at $x = 30$ km. A second observer, moving at a speed of $0.25c$ in the direction of increasing x, also views the flashes. (*a*) What time interval between them does she find? (*b*) Which flash does she say occurs first?

21E. In Table 3 the Lorentz transformation equations in the right-hand column can be derived from those in the left-hand column simply by (1) exchanging primed and unprimed quantities and (2) changing the sign of v. Verify this procedure by deriving one set of equations directly from the other by algebraic manipulation.

22P. A clock moves along the x axis at a speed of $0.60c$ and reads zero as it passes the origin. (*a*) Calculate the Lorentz factor. (*b*) What time does the clock read as it passes $x = 180$ m?

23P. An observer S sees a flash of red light 1200 m from his position and a flash of blue light 720 m closer to him and on the same straight line. He measures the time interval between the occurrence of the flashes to be 5.00 μs, the red flash occurring first. (*a*) What is the relative velocity **v** (magnitude and direction) of a second observer S' who would record these flashes as occurring at the same place? (*b*) From the point of view of S', which flash occurs first? (*c*) What time interval between them would S' measure?

24P. In Problem 23, observer S sees the two flashes in the same positions, but they now occur closer together in time. How close together in time can they be and still have it possible to find a frame S' in which they occur at the same place?

Section 42-11 The Transformation of Velocities

25E. A particle moves along the x' axis of frame S' with a speed of $0.40c$. Frame S' moves with a speed of $0.60c$ with respect to frame S. What is the measured speed of the particle in frame S?

26E. Frame S' moves relative to frame S at $0.62c$ in the direction of increasing x. In frame S' a particle is measured to have a velocity of $0.47c$ in the direction of increasing x'. (*a*) What is the velocity of the particle with respect to frame S? (*b*) What would be the velocity of the particle with respect to S if it moved (at $0.47c$) in the direction of *decreasing* x' in the S' frame? In each case, compare your answers with the predictions of the classical velocity transformation equation.

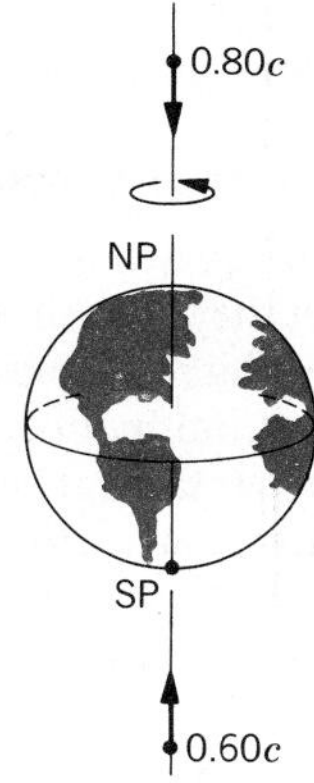

Figure 16 Exercise 27.

27E. One cosmic-ray particle approaches the earth along its axis with a velocity of $0.80c$ toward the North Pole and another, with a velocity $0.60c$, toward the South Pole. See Fig. 16. What is the relative speed of approach of one particle with respect to the other? (*Hint:* It is useful to consider the earth and one of the particles as the two inertial reference frames.)

28E. Galaxy A is reported to be receding from us with a speed of $0.35c$. Galaxy B, located in precisely the opposite direction, is also found to be receding from us at this same speed. What recessional speed would an observer on Galaxy A find (*a*) for our galaxy? (*b*) for Galaxy B?

29E. It is concluded from measurements of the red shift of the emitted light that quasar Q_1 is moving away from us at a speed of $0.80c$. Quasar Q_2, which lies in the same direction in space but is closer to us, is moving away from us at speed $0.40c$. What velocity for Q_2 would be measured by an observer on Q_1?

30P. A spaceship whose rest length is 350 m has a speed of $0.82c$ with respect to a certain reference frame. A micrometeorite, also with a speed of $0.82c$ in this frame, passes the spaceship on an antiparallel track. How long does it take this object to pass the spaceship?

31P. To circle the earth in low orbit a satellite must have a speed of about 17,000 mi/h. Suppose that two such satellites orbit the earth in opposite directions. (*a*) What is their relative speed as they pass? Evaluate using the classical Galilean velocity transformation equation. (*b*) What fractional error was made because the (correct) relativistic transformation equation was not used?

32P. A spaceship, at rest in a certain reference frame *S*, is given a speed increment of $0.50c$. It is then given a further $0.50c$ increment in this new frame, and this process is continued until its speed with respect to its original frame *S* exceeds $0.999c$. How many increments does it require?

Section 42-12 The Doppler Effect

33E. A spaceship, moving away from the earth at a speed of $0.90c$, reports back by transmitting on a frequency (measured in the spaceship frame) of 100 MHz. To what frequency must earth receivers be tuned to receive these signals?

34E. In the spectrum of quasar 3C9, some of the familiar hydrogen lines appear but they are shifted so far toward the red that their wavelengths are observed to be three times as large as that observed in the light from hydrogen atoms at rest in the laboratory. (*a*) Show that the classical Doppler equation gives a velocity of recession greater than *c*. (*b*) Assuming that the relative motion of 3C9 and the earth is entirely one of recession, find the recession speed predicted by the relativistic Doppler equation.

35E. Give the Doppler wavelength shift $\lambda - \lambda_0$, if any, for the sodium D_2 line (589.00 nm) emitted from a source moving in a circle with constant speed ($= 0.10c$) as measured by an observer fixed at the center of the circle.

36P. A spaceship is receding from the earth at a speed of $0.20c$. A light on the rear of the ship appears blue ($\lambda = 450$ nm) to passengers on the ship. What color would it appear to an observer on earth?

37P. A radar transmitter *T* is fixed to a reference frame *S′* that is moving to the right with speed *v* relative to reference frame *S* (see Fig. 17). A mechanical timer (essentially a clock) in frame *S′*, having a period τ_0 (measured in *S′*) causes transmitter *T* to emit radar pulses, which travel at the speed of light and are received by *R*, a receiver fixed in frame *S*. (*a*) What would be the period τ of the timer relative to observer *A*, who is fixed in frame *S*? (*b*) Show that the receiver *R* would observe the time interval between pulses arriving from *T*, not as τ or as τ_0, but as

$$\tau_R = \tau_0 \sqrt{\frac{c+v}{c-v}}.$$

(*c*) Explain why the observer at *R* measures a different period for the transmitter than does observer *A*, who is in the same reference frame. (*Hint:* A clock and a radar pulse are not the same.)

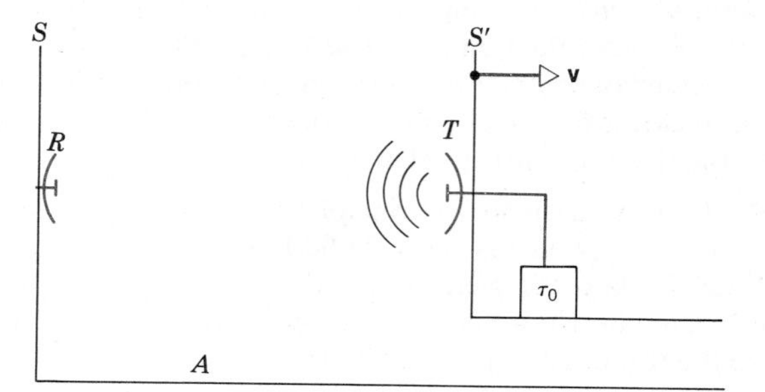

Figure 17 Problem 37.

Section 42-14 A New Look at Energy

38E. How much work must be done to increase the speed of an electron from rest (*a*) to $0.50c$? (*b*) to $0.990c$? (*c*) to $0.9990c$?

39E. An electron is moving at a speed such that it could circumnavigate the earth at the equator in one second. (*a*) What is its speed, in terms of the speed of light? (*b*) Its kinetic energy *K*? (*c*) What percent error do you make if you use the classical formula to calculate *K*?

40E. Find the speed parameter β and the Lorentz factor γ for an electron whose kinetic energy is (*a*) 1.0 keV; (*b*) 1.0 MeV; (*c*) 1.0 GeV.

41E. Find the speed parameter β and the Lorentz factor γ for a particle whose kinetic energy is 10 MeV if the particle is (*a*) an electron; (*b*) a proton; (*c*) an alpha particle.

42E. Verify the statement in Sample Problem 5, Chapter 32, that a 100-MeV electron travels at 99.9987% of the speed of light.

43E. A particle has a speed of $0.99c$ in a laboratory reference frame. What are its kinetic energy, its total energy, and its momentum if the particle is (*a*) a proton or (*b*) an electron?

44E. The United States consumed about 2.2×10^{12} kWh of electrical energy in 1979. How much matter would have to vanish to account for the generation of this energy? Does it make any difference to your answer if this energy is generated in oil-burning, nuclear, or hydroelectric plants?

45E. Quasars are thought to be the nuclei of active galaxies in the early stages of their formation. A typical quasar radiates energy at the rate of 10^{41} W. At what rate is the mass of this quasar being reduced to supply this energy? Express your answer in solar mass units per year, where one solar mass unit (smu $= 2 \times 10^{30}$ kg) is the mass of our sun.

46P. How much work must be done to increase the speed of an electron from *(a)* $0.18c$ to $0.19c$? *(b)* $0.98c$ to $0.99c$? Note that the speed increase ($= 0.01c$) is the same in each case.

47P. What is the speed of a particle *(a)* whose kinetic energy is equal to twice its rest energy? *(b)* whose total energy is equal to twice its rest energy?

48P. *(a)* What potential difference would accelerate an electron to the speed of light, according to classical physics? *(b)* With this potential difference, what speed would the electron actually attain?

49P. A particle has a momentum equal to mc. *(a)* What is its speed? *(b)* Its relativistic mass? *(c)* Its kinetic energy?

50P. What must be the momentum of a particle with rest mass m in order that its total energy be three times its rest energy?

51P. Consider the following, all moving in free space: a 2.0-eV photon, a 0.40-MeV electron, and a 10-MeV proton. *(a)* Which is moving the fastest? *(b)* The slowest? *(c)* Which has the greatest momentum? *(d)* The least? (*Note:* A photon is a light-particle of zero rest mass.)

52P. A 5-grain aspirin tablet has a mass of 320 mg. For how many miles would the energy equivalent of this mass, in the form of gasoline, power an automobile? Assume 30 mi/gal and a heat of combustion of 1.3×10^8 J/gal for the gasoline.

53P. *(a)* If the kinetic energy K and the momentum p of a particle can be measured, it should be possible to find its rest mass m and thus identify the particle. Show that

$$m = \frac{(pc)^2 - K^2}{2Kc^2}.$$

(b) Show that this expression reduces to an expected result as $u/c \to 0$, in which u is the speed of the particle. *(c)* Find the rest mass of a particle whose kinetic energy is 55.0 MeV and whose momentum is 121 MeV/c; express your answer in terms of the rest mass of the electron.

54P. In a high-energy collision of a primary cosmic-ray particle near the top of the earth's atmosphere, 120 km above sea level, a pion is created with a total energy E of 1.35×10^5 MeV, traveling vertically downward. In its proper frame this pion decays 35 ns after its creation. At what altitude above sea level does the decay occur? The rest energy of a pion is 139.6 MeV.

55P. The average lifetime of muons at rest is 2.2 μs. A laboratory measurement on the decay in flight of the muons in a beam emerging from a particle accelerator yields an average lifetime of 6.9 μs. what is: *(a)* The speed of these muons in the laboratory? *(b)* The relativistic mass (in terms of m_e, the rest mass of the electron)? *(c)* The kinetic energy? *(d)* The momentum? The rest mass of a muon is 207 times that of an electron.

56P. *(a)* How much energy is released in the explosion of a fission bomb containing 3.0 kg of fissionable material? Assume that 0.10 percent of the rest mass is converted to released energy. *(b)* What mass of TNT would have to explode to provide the same energy release? Assume that each mole of TNT liberates 3.4 MJ of energy on exploding. The molecular mass of TNT is 0.227 kg/mol. *(c)* For the same mass of explosive, how much more effective are nuclear explosions than TNT explosions? That is, compare the fractions of the rest mass that are converted to energy in each case.

57P. In Section 5 of Chapter 30 we showed that a particle of charge q and mass m moving with speed v perpendicular to a uniform magnetic field B moves in a circle of radius r given by (see Eq. 16, Chapter 30)

$$r = \frac{mv}{qB}.$$

Also, it was demonstrated that the period T of the circular motion is independent of the speed of the particle. Now, these results hold only if $v \ll c$. For particles moving faster, the radius of the circular path can be shown to be

$$r = \frac{p}{qB} = \frac{m(\gamma v)}{qB} = \frac{mv}{qB\sqrt{1-\beta^2}}.$$

This equation is valid at all speeds. Compute the radius of the path of a 10-MeV electron moving perpendicular to a uniform 2.2-T magnetic field. Use both the *(a)* "classical" and *(b)* "relativistic" formulas. *(c)* Calculate the true period of the circular motion. Is the result independent of the speed of the electron?

58P. Ionization measurements show that a particular nuclear particle carries a double charge ($= 2e$) and is moving with a speed of $0.71c$. Its measured radius of curvature in a magnetic field of 1.00 T is 6.28 m. Find the rest mass of the particle and identify it. (*Hint:* Light nuclear particles are made up of neutrons [which carry no charge] and protons [charge $= +e$], in roughly equal numbers. Take the rest mass of either of these particles to be 1.00 u. See Problem 57.)

59P. A 10-GeV proton in the cosmic radiation approaches the earth in the plane of its geomagnetic equator, in a region over which the earth's average magnetic field is 55 μT. What is the radius of its curved path in that region?

60P. A 2.50-MeV electron moves at right angles to a magnetic field in a path whose radius of curvature is 3.0 cm. What is the magnetic field B?

61P. The proton synchrotron at Fermilab accelerates protons to a kinetic energy of 500 GeV. At this energy, calculate *(a)* the Lorentz factor, *(b)* the speed parameter, and *(c)* the magnetic field at the proton orbit that has a radius of curvature of 750 m. (The proton has a rest energy of 938.3 MeV.)

CHAPTER 43

QUANTUM PHYSICS—I

Dartmouth students enjoying thermal radiation the night before the Harvard game. Studies of such radiations—under controlled laboratory conditions—introduced the Planck constant into physics, established the quantization of energy, and laid the foundations of modern quantum physics.

43-1 A New Direction

So far we have studied light—and by that word we now mean not only visible light but radiation throughout the entire electromagnetic spectrum—under the headings of reflection, refraction, polarization, interference, and diffraction. We can explain all these phenomena by treating light as an electromagnetic *wave*, governed by Maxwell's equations. The experimental support for this belief is pretty overwhelming.

We now move off in an entirely new direction and consider experiments that can only be understood by making quite a different assumption about light, namely, that it behaves like a stream of *particles*, each with a specified energy and momentum.

You may well ask, "Well, which is it, wave or particle?" These concepts are so different that it is hard to see how light can model itself after both at the same time. We will face this question squarely in Section 44-10. Meanwhile, we ask you not to worry about this puzzle for now but simply to look, with us, at the strong experimental evidence that light is particlelike. The path we are taking will open the door to the world of quantum physics and will allow us to begin to understand how atoms are constructed.

43-2 Einstein Makes a Proposal

In 1905, Einstein made the bold hypothesis—since convincingly confirmed by experiment—that light sometimes behaves as if its energy were concentrated in discrete bundles that he called *light quanta;* we now call them *photons*. He proposed that the energy of a single photon is

$$\boxed{E = h\nu} \quad \text{(photon energy)}, \qquad (1)$$

in which ν is the frequency of the light and h is the Planck constant. This constant, introduced into physics a few years earlier in another connection by Max Planck (see Section 43–5) has the value

$$h = 6.63 \times 10^{-34}\ \text{J}\cdot\text{s} = 4.14 \times 10^{-15}\ \text{eV}\cdot\text{s}. \qquad (2)$$

Photons carry not only energy but also linear momentum. We can find an expression for it by starting with Eq. 42 of Chapter 42,

$$E^2 = (pc)^2 + (mc^2)^2. \qquad (3)$$

This expression gives the relativistic relationship between the momentum p and the total energy E of a particle, such as an electron or a proton, whose rest mass is m.

We can apply Eq. 3 to a photon by putting $E = h\nu$ and $m = 0$, since a photon, traveling at the speed of light, must have zero rest mass. Equation 3 then becomes $h\nu = pc$; solving for p and using the relation $c = \lambda\nu$ leads to

$$\boxed{p = \frac{h}{\lambda}} \quad \text{(photon momentum)}, \qquad (4)$$

in which λ is the wavelength of the light and h is again the Planck constant.

Note how the wave and the photon models are intimately connected. The energy E of the *photon* is related to the frequency ν of the *wave* by Eq. 1. Similarly, the momentum p of the *photon* is related to the wavelength λ of the *wave* by Eq. 4. In each case, the factor of proportionality is the Planck constant h.

Equations 1 and 4 permit us to look at the electromagnetic spectrum in a new way. In Figure 1 of Chapter 38 we displayed this spectrum as a range of wavelengths or, equivalently, of frequencies. We can now also display it as a range of photon energies or (if we wish) of photon momenta. Table 1 shows some corresponding wavelengths, frequencies, and photon energies for selected regions of the electromagnetic spectrum.

In 1905 most physicists were quite comfortable with the wave theory of light and did not look kindly upon Einstein's photon idea. Prominent among those who were slow to believe was Max Planck, the very person who introduced the constant h into physics. In recommending Einstein for membership in the Royal Prussian Academy of Sciences in 1913, for example, Planck wrote, "that he may sometimes have missed the target in his speculations, as for example in his theory of light quanta, cannot really be held against him." It is almost commonplace that radical ideas are accepted only slowly, even by such men of genius as Planck. It was, incidentally, for his photon theory that Einstein received the Nobel Prize in physics for 1921.

Table 1 Some Corresponding Wavelengths, Frequencies, and Photon Energies

Region	*Wavelength*	*Frequency* (Hz)	*Photon Energy*
Gamma ray	50 fm	6×10^{21}	25 MeV
X ray	50 pm	6×10^{18}	25 keV
Ultra violet	100 nm	3×10^{15}	12 eV
Visible	550 nm	5×10^{14}	2 eV
Infrared	10 μm	3×10^{13}	120 meV
Microwave	1 cm	3×10^{10}	120 μeV
Radio wave	1 km	3×10^{5}	1.2 neV

Sample Problem 1 Yellow light from a sodium vapor lamp has an effective wavelength of 589 nm. What is the energy of the corresponding photons?

From Eq. 1 we have, using the relation $c = \lambda\nu$,

$$E = h\nu = \frac{hc}{\lambda}$$

$$= \frac{(4.14 \times 10^{-15}\ \text{eV}\cdot\text{s})(3.00 \times 10^{8}\ \text{m/s})}{589 \times 10^{-9}\ \text{m}}$$

$$= 2.11\ \text{eV}. \qquad \text{(Answer)}$$

This is the energy that a single electron or proton would acquire if it were accelerated through a potential difference of 2.11 V.

Sample Problem 2 During radioactive decay, a certain nucleus emits a gamma ray whose photon energy is 1.35 MeV. (a) To what wavelength does this photon correspond?

From Eq. 1 and the relation $c = \lambda\nu$ we have

$$\lambda = \frac{c}{\nu} = \frac{hc}{h\nu} = \frac{hc}{E}$$

$$= \frac{(4.14 \times 10^{-15}\ \text{eV}\cdot\text{s})(3.00 \times 10^{8}\ \text{m/s})}{1.35 \times 10^{6}\ \text{eV}}$$

$$= 9.20 \times 10^{-13}\ \text{m} = 920\ \text{fm}. \qquad \text{(Answer)}$$

(b) What is the momentum of this photon?

From Eq. 4 we can write (again using the relation $c = \lambda\nu$)

$$p = \frac{h}{\lambda} = \frac{h\nu}{\lambda\nu} = \frac{E}{c}, \tag{5}$$

in which E is the photon energy. Substituting in a straightforward way, we find

$$p = \frac{E}{c} = \frac{(1.35\text{ MeV})(1.60 \times 10^{-13}\text{ J/MeV})}{3.00 \times 10^{8}\text{ m/s}}$$
$$= 7.20 \times 10^{-22}\text{ kg}\cdot\text{m/s}. \qquad \text{(Answer)}$$

Although this answer is correct, physicists in the field of high-energy particle physics do not ordinarily express the momenta of photons (or of particles such as electrons or protons) in SI units. Instead, they express the momentum as an energy divided by the speed of light. Thus, from Eq. 5,

$$p = \frac{E}{c} = \frac{1.35\text{ MeV}}{c} = 1.35\text{ MeV}/c. \qquad \text{(Answer)}$$

One advantage of this practice is that, given the energy of a photon, you at once know its momentum, and conversely.*

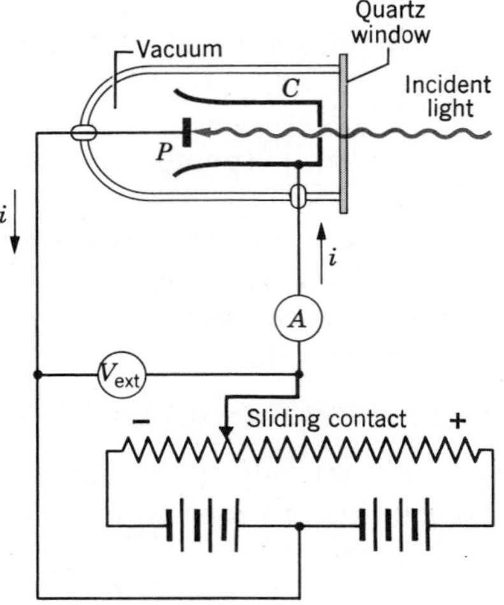

Figure 1 An apparatus used to study the photoelectric effect. The incident light falls on plate P, ejecting photoelectrons, which are collected by collector cup C. The photoelectrons move in the circuit in a direction opposite to that of the conventional current arrows.

43-3 The Photoelectric Effect

Here we consider the first of a large class of experiments that cannot be interpreted in terms of a wave model for light but that find a ready explanation if we assume that light is made up of photons.

If you shine a beam of light on a clean metal surface it can—if conditions are right—knock out electrons from that surface. Most of us are familiar with applications of this *photoelectric effect,* as it is called, in automatic door openers or security alarm systems. When the photoelectric effect is studied carefully in the laboratory, we find that the experimental results cannot be explained at all in terms of the wave model of light. However, as Einstein pointed out, the explanation of the photoelectric effect is quite straightforward if we view it as a "collision" between an incident photon and an electron within the metal.

Figure 1 shows a typical apparatus for studying the photoelectric effect. Light of frequency ν falls on a metal plate P and kicks out electrons. A suitable potential difference* V between P and collector cup C sweeps up these *photoelectrons,* displaying them as a photoelectric current in ammeter A.

The essential data are displayed in two figures. Figure 2 shows the photoelectric current i as a function of V, for two levels of illumination of light of the same wavelength. The *stopping potential* V_0 is the potential difference required to stop the fastest photoelectrons, thus bringing the photoelectric current to zero. Note that eV_0 measures the kinetic energy of the most energetic photoelectrons. That is,

$$K_m = eV_0. \tag{7}$$

The central feature of Fig. 2 is that V_0 is the same for both curves, so that *the kinetic energy of the most energetic photoelectrons is independent of the intensity of the incident light.*

Figure 3 shows the stopping potential plotted against the frequency of the incident light. We see by extrapolation that there is a sharp *cutoff frequency* ν_0,

* This same advantage holds for material particles if their total energies greatly exceed their rest energies, so that the last term in Eq. 3 can be neglected.

* V is given by

$$V = V_{\text{ext}} + V_{\text{cpd}}, \tag{6}$$

in which the first term is the reading of the voltmeter in Fig. 1 and the second is the measured *contact potential difference* (a battery effect) introduced by the fact that the plate and the collector are usually made of different materials.

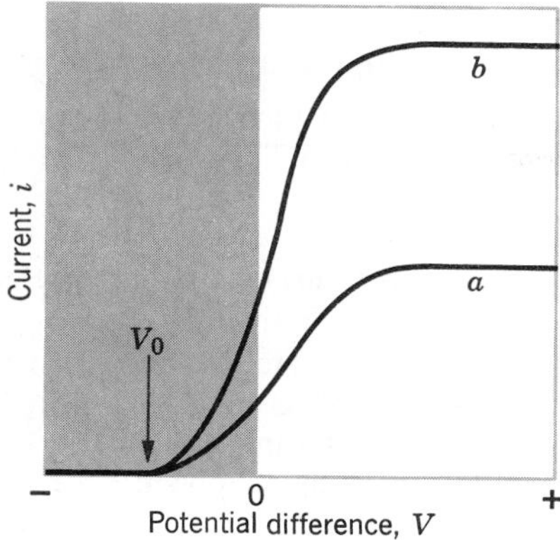

Figure 2 A plot (not to scale) of data taken with the apparatus of Fig. 1. The intensity of the incident light is twice as great for curve *b* as for curve *a*. The wavelength of the incident light remains the same for each curve.

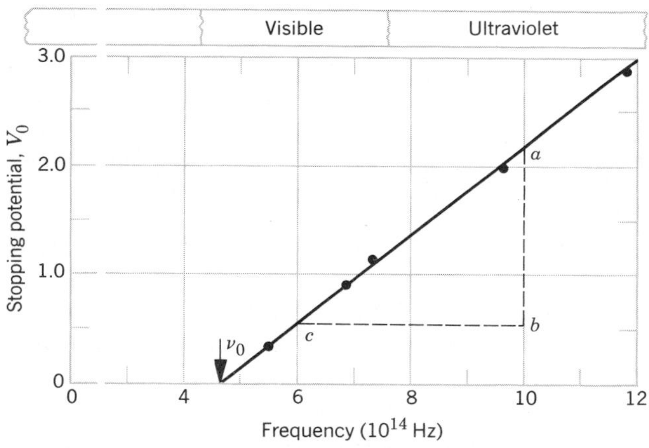

Figure 3 The stopping potential for sodium as a function of the frequency of the incident light. Data reported by R. A. Millikan in 1916.

corresponding to a stopping potential of zero. For light with frequency below this value the photoelectric effect simply does not happen.

Let us now see how the photon model succeeds—and the wave model fails—in understanding these experiments:

1. The Intensity Problem. In wave theory, when you increase the intensity of a beam of light, you increase the magnitude of the oscillating electric field vector **E**. The force that the incident beam exerts on an electron is $e\mathbf{E}$. You might then expect that the more intense the light, the more energetic would be the ejected photoelectrons. However, as Fig. 2 shows, V_0 (and thus K_m; see Eq. 7) does not depend on the light intensity; this has been tested experimentally over an intensity range of $\sim 10^7$.

Using the photon picture, however, the "intensity problem" is no problem. If we double the light intensity, we simply double the number of photons but we do not change the energy of the individual photons. Thus K_m, the maximum kinetic energy that an electron can pick up from a photon during a collision, remains unchanged.

2. The Frequency Problem. According to wave theory, the photoelectric effect should occur at *any* frequency of the incident light, provided only that the light is intense enough. However, as Fig. 3 shows, there is a characteristic cutoff frequency below which there is no photoelectric effect, *no matter how intense the light.*

Again, the "frequency problem" is no problem if we think in terms of photons. The conduction electrons are held within the metal target by a potential barrier at the metal surface. Thus, to eject a photoelectron, you must give it a certain minimum energy ϕ, called the work function *of the material. If the photon energy exceeds the work function (that is, if $h\nu > \phi$), the photoelectric effect can occur. If it does not (that is, if $h\nu < \phi$), the effect will not occur. This is exactly what Fig. 3 shows.*

3. The Time Delay Problem. In wave theory, the energy of an ejected photoelectron must be soaked up from the incident wave. The effective area from which the electron soaks up this energy cannot be much larger than the cross section of an atom. Thus, if the light is feeble enough, there should be a measurable time delay between the light striking the surface and the electron emerging from it. No such delay has ever been found. The time delay problem is that there is no time delay!

The "time delay problem" does not exist on the photon picture because the photon energy is delivered to the ejected photoelectron in a single collision event.

A Quantitative Analysis. Einstein wrote the conservation-of-energy principle for the photoelectric effect as

$$\boxed{h\nu = \phi + K_m} \quad \text{(photoelectric equation)} \quad (8)$$

where $h\nu$ is the energy of the photon. Equation 8 tells us that a photon carries an energy $h\nu$ into the surface, where it interacts with an electron. If the electron is to escape, an amount of energy ϕ (the *work function* of the material) must be provided to surmount the potential barrier that exists at the surface. The remaining energy ($= h\nu - \phi$) is equal to K_m, the *maximum* kinetic energy that the ejected electron can have.

Let us rewrite Eq. 8 by substituting for K_m from Eq. 7. After some rearrangement, we have

$$V_0 = (h/e)\nu - (\phi/e). \tag{9}$$

Thus Einstein's photon theory predicts a linear relationship between the stopping potential V_0 and the frequency ν, in complete agreement with Fig. 3. The slope of the experimental curve in that figure should be h/e, or

$$\frac{h}{e} = \frac{ab}{bc} = \frac{2.35 \text{ V} - 0.72 \text{ V}}{(10 \times 10^{14} - 6 \times 10^{14}) \text{ Hz}}$$
$$= 4.1 \times 10^{-15} \text{ V}\cdot\text{s}.$$

We can find h by multiplying by the electronic charge e, or

$$h = (4.1 \times 10^{-15} \text{ V}\cdot\text{s})(1.6 \times 10^{-19} \text{ C})$$
$$= 6.6 \times 10^{-34} \text{ J}\cdot\text{s},$$

in full agreement with the value given in Eq. 2.

Sample Problem 3 A potassium foil is a distance $r = 3.5$ m from a light source whose power output P is 1.5 W. Assuming that the incident light is a wave, calculate how long it would take for the foil to soak up enough energy ($= 1.8$ eV) to eject a photoelectron. Assume that the electron collected its energy from a circular area of the foil whose radius is 5.3×10^{-11} m.*

The target area A is $\pi(5.3 \times 10^{-11} \text{ m})^2$ or $8.8 \times 10^{-21} \text{ m}^2$. If the light source radiates uniformly in all directions, the illumination I at the foil is

$$I = \frac{P}{4\pi r^2} = \frac{1.5 \text{ W}}{(4\pi)(3.5 \text{ m})^2} = 9.7 \times 10^{-3} \text{ W/m}^2.$$

The rate at which energy falls on the target is then

$$R = IA = (9.7 \times 10^{-3} \text{ W/m}^2)(8.8 \times 10^{-21} \text{ m}^2)$$
$$= 8.5 \times 10^{-23} \text{ W}.$$

If all this incoming energy is absorbed, the time required to accumulate enough energy for the electron to escape is

$$t = \left(\frac{1.8 \text{ eV}}{8.5 \times 10^{-23} \text{ J/s}}\right)\left(\frac{1.60 \times 10^{-19} \text{ J}}{1 \text{ eV}}\right)\left(\frac{1 \text{ min}}{60 \text{ s}}\right)$$
$$= 56 \text{ min!} \qquad \text{(Answer)}$$

However, no measurable time delay whatsoever is observed.

Sample Problem 4 At what rate do photons strike the metal plate in Sample Problem 3? Assume a wavelength of 589 nm (yellow sodium light) and an area of 1 cm².

Using results from Sample Problem 3, we can express the illumination at the plate as

$$I = (9.7 \times 10^{-3} \text{ W/m}^2)(1 \text{ eV}/1.6 \times 10^{-19} \text{ J})$$
$$= 6.1 \times 10^{16} \text{ eV/m}^2\cdot\text{s}.$$

In Sample Problem 1, we found the energy of each photon to be 2.1 eV. The rate at which photons strike the plate is then

$$R = (6.1 \times 10^{16} \text{ eV/m}^2\cdot\text{s})\left(\frac{1 \text{ photon}}{2.1 \text{ eV}}\right)(10^{-4} \text{ m}^2)$$
$$= 2.9 \times 10^{12} \text{ photons/s}. \qquad \text{(Answer)}$$

Even at this low level of illumination ($\sim 1\ \mu\text{W/cm}^2$), the photon rate is very great, about 10^8 photons falling every second on a patch the size of a period on this page. Small wonder that we do not ordinarily notice the granularity of light.

Sample Problem 5 Find the work function of sodium from the data plotted in Fig. 3.

The intersection of the straight line in Fig. 3 with the frequency axis is the cutoff frequency ν_0. Putting $V_0 = 0$ and $\nu = \nu_0$ in Eq. 9 yields

$$\phi = h\nu_0 = (6.63 \times 10^{-34} \text{ J}\cdot\text{s})(4.3 \times 10^{14} \text{ Hz})$$
$$= 2.9 \times 10^{-19} \text{ J} = 1.8 \text{ eV}. \qquad \text{(Answer)}$$

We note from Eq. 9 that, to find the Planck constant h, you need only know the slope of the straight line in Fig. 3. To find the work function, you need only know the intercept.

43-4 The Compton Effect*

Here we have another experiment that can be understood readily in terms of a photon model for light but that can not be understood at all in terms of a wave model. Historically, this experiment proved to be a great

* This value, called the *Bohr radius,* is roughly equal to the radius of an average atom. It is a (non-SI) length unit useful on the scale of atomic dimensions.

* For an historical account, see "The Scattering of X-Rays as Particles," A. H. Compton, *American Journal of Physics,* December 1961.

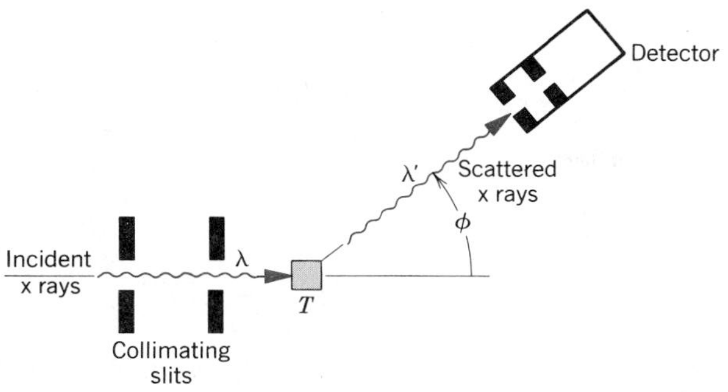

Figure 4 An apparatus used to study the Compton effect. A beam of x rays falls on a graphite target T. The x rays scattered from the target are observed at various angles to the incident direction. The detector measures both the intensity and the wavelength of these scattered x rays.

"convincer" of the reality of photons because it introduced the photon *momentum,* as well as the photon energy, into an experimental situation. Further, it showed that the photon model applies not only to visible and ultraviolet light—the domain of the photoelectric effect—but also to x rays.

In 1923, Arthur Holly Compton at Washington University in St. Louis arranged for a beam of x rays of wavelength λ to fall on a graphite target T, as in Fig. 4. He measured, as a function of wavelength, the intensity of the x rays scattered from the target in several selected directions. Figure 5 shows his results. We see that, although the incident beam contains only a single wavelength, the scattered x rays have intensity peaks at two wavelengths. One peak corresponds to the incident wavelength λ; the other has a wavelength λ' that is longer than λ by an amount $\Delta\lambda$. This *Compton shift,* as it is called, varies with the angle at which the scattered x rays are observed.

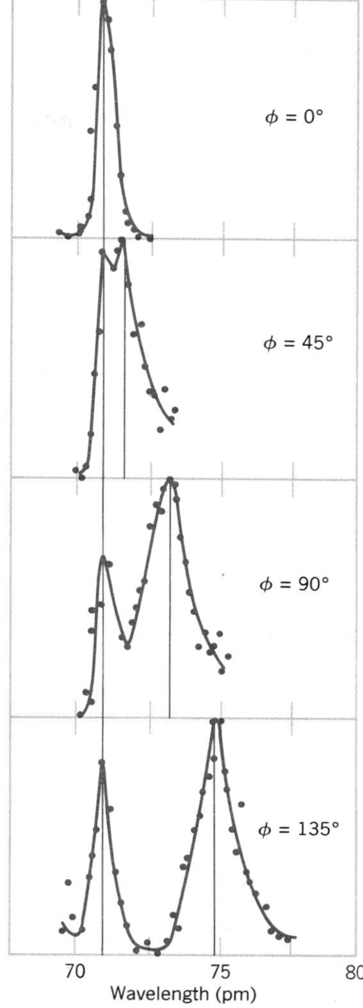

Figure 5 Compton's results, for four values of the scattering angle ϕ. (1 pm = 10^{-12} m = 100 Å.) Note that the Compton shift increases as the scattering angle increases.

The scattered peak of wavelength λ' cannot be understood at all if you think of the incident x-ray beam as a wave. On this picture, the incident wave, with frequency ν, causes the electrons in the target to oscillate at that same frequency. These oscillating electrons, like charges surging back and forth in a small transmitting antenna, radiate at this same frequency. Thus the scattered beam should have only the same frequency—and only the same wavelength—as the incident beam. It simply doesn't.

Compton viewed the incident beam as a stream of photons of energy E (= $h\nu$) and momentum p (= h/λ) and assumed that some of these photons made billiard-ball-like collisions with individual free electrons in the target. Since the electron picks up some kinetic energy in such an encounter, the scattered photon must have a lower energy E' than the incident photon. It will therefore have a lower frequency ν' and, correspondingly, a longer wavelength λ', exactly as we observe. Thus we account qualitatively for the Compton shift.

A Quantitative Analysis. Figure 6 suggests a collision between a photon and a free electron in the target. Let us apply the law of conservation of energy. Because the

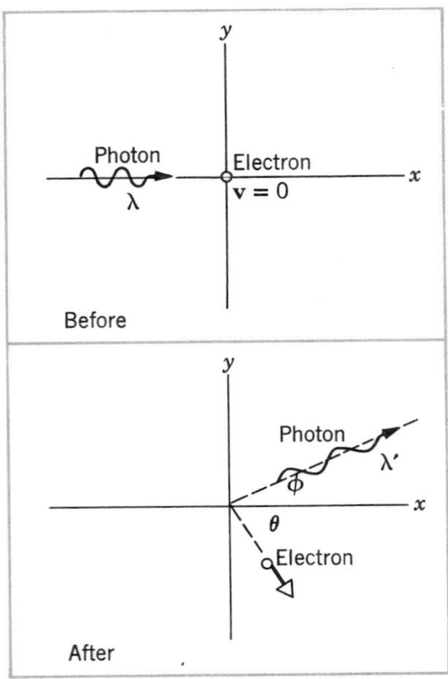

Figure 6 A photon of wavelength λ strikes a resting electron. The photon is scattered at an angle ϕ with an increased wavelength λ'. The electron moves off with a speed v at an angle θ.

electrons emerge from the collision with a speed that may be comparable to the speed of light, we must use the relativistic expression for the kinetic energy. Thus we have

$$h\nu = h\nu' + mc^2\left(\frac{1}{\sqrt{1-(v/c)^2}} - 1\right),$$

in which the second term on the right (see Section 42–14) is the kinetic energy of the recoiling electron. Substituting c/λ for ν and c/λ' for ν' leads to

$$\frac{h}{\lambda} = \frac{h}{\lambda'} + mc\left(\frac{1}{\sqrt{1-(v/c)^2}} - 1\right) \quad \text{(energy conservation).} \quad (10)$$

Now let us apply the (vector) law of conservation of momentum to the collision of Fig. 6. The momentum of a photon is given by Eq. 4 ($p = h/\lambda$). For the electron, the relativistic expression for the momentum is given by Eq. 20 of Chapter 9, or

$$p = \frac{mv}{\sqrt{1-(v/c)^2}} \quad \text{(electron momentum).} \quad (11)$$

We can then express the conservation of momentum for the photon–electron collision as

$$\frac{h}{\lambda} = \frac{h}{\lambda'}\cos\phi + \frac{mv}{\sqrt{1-(v/c)^2}}\cos\theta \quad (x\ \text{component}) \quad (12)$$

and

$$0 = \frac{h}{\lambda'}\sin\phi - \frac{mv}{\sqrt{1-(v/c)^2}}\sin\theta \quad (y\ \text{component}). \quad (13)$$

Our aim is to find $\Delta\lambda\ (= \lambda' - \lambda)$, the wavelength shift of the scattered photons. Of the five collision variables (λ, λ', v, ϕ, and θ) that appear in Eqs. 10, 12, and 13, we can eliminate two. We choose to eliminate v and θ, which deal only with the recoil electron.

Carrying out the necessary algebra (see Problem 41) leads to this simple result:

$$\boxed{\Delta\lambda = \frac{h}{mc}(1 - \cos\phi)} \quad \text{(Compton shift),} \quad (14)$$

in which the quantity h/mc has the value of 2.43×10^{-12} m or 2.43 pm. Equation 14 agrees exactly with Compton's experimental results.

Equation 14 tells us that the Compton shift depends only on the scattering angle ϕ and not on the initial photon energy. The predicted shift varies from zero (for $\phi = 0$, a grazing collision, the incident photon being scarcely deflected) to $2h/mc$ (for $\phi = 180°$, a head-on collision, the incident photon being reversed in direction).

It remains to explain the peak in Fig. 5 in which the wavelength does *not* change. This peak results from scattering from electrons that are not free—as we have assumed so far—but are tightly bound to the atoms of the target. For a carbon target, the effective mass of such electrons is that of carbon atoms, or about $22{,}000m$, m being the electron mass. If we replace m in Eq. 14 by $22{,}000m$, we see that the Compton shift for bound electrons is immeasurably small, just as we observe.

The Compton effect is responsible for the so-called *electromagnetic pulse* (EMP) caused by thermonuclear explosions high in the atmosphere. The x rays and gamma rays generated in such explosions have Compton collisions with electrons in the upper atmosphere, knocking them sharply forward. This sudden, enormous surge of charge sets up electromagnetic fields that can play havoc with unshielded electric circuits on the earth's surface. The effect was first noticed when power and communication circuits in Hawaii failed at the time of a thermonuclear airburst test in the Pacific Ocean, many miles away.

Sample Problem 5 X rays of wavelength 22 pm (photon energy = 56 keV) are scattered from a carbon target, the scattered radiation being viewed at 85° to the incident beam. (a) What is the Compton shift?

From Eq. 14 we have

$$\Delta\lambda = \frac{h}{mc}(1 - \cos\phi)$$

$$= \frac{(6.63 \times 10^{-34}\ \text{J}\cdot\text{s})(1 - \cos 85°)}{(9.11 \times 10^{-31}\ \text{kg})(3.00 \times 10^{8}\ \text{m/s})}$$

$$= 2.21 \times 10^{-12}\ \text{m} = 2.21\ \text{pm}. \qquad \text{(Answer)}$$

(b) What percentage of its initial energy does an incident x-ray photon loose?

The fractional energy loss f is

$$f = \frac{E - E'}{E} = \frac{h\nu - h\nu'}{h\nu} = \frac{(c/\lambda) - (c/\lambda')}{(c/\lambda)} = \frac{\lambda' - \lambda}{\lambda'}$$

$$= \frac{\Delta\lambda}{\lambda + \Delta\lambda}. \qquad (15)$$

Substitution yields

$$f = \frac{2.21\ \text{pm}}{22\ \text{pm} + 2.21\ \text{pm}} = 0.091 \text{ or } 9.1\%. \qquad \text{(Answer)}$$

Equation 14 reminds us that the Compton shift $\Delta\lambda$ is independent of the wavelength λ of the incident photon. Equation 15 then tells us that the shorter this incident wavelength (that is, the more energetic the incoming photon), the larger will be the fractional energy loss. The Compton effect shows up more strongly for more energetic photons.

43-5 Planck and His Constant—An Historical Aside

At any given time, there are always one or more "hot problems" that attract the attention of many of the most able physicists. The related problems of the fundamental nature of matter and of the evolution of the universe rank high on today's list. At the turn of the century, however, the problem that attracted the attention of the best and the brightest was a hot problem in more ways than one. It was that of understanding the distribution in wavelength of the radiation emitted by heated objects.

The radiation emitted by red hot pokers or bonfires depends on too many variables to be of fundamental significance. The physicists of 1900 turned instead to the study of the radiation emitted by an *ideal radiator,* that is, a radiator whose emitted radiation depends *only* on the temperature of the radiator and not on the material from which the radiator is made, the nature of its surface, or anything else.

We can make such an ideal radiator in the laboratory by forming a cavity within a body, the walls of the cavity being held at a uniform temperature. We must drill a small hole through the cavity wall so that a sample of the radiation inside the cavity can escape into the laboratory, where we can study it. Experiment shows that such *cavity radiation* has a very simple spectrum, determined indeed only by the temperature of the walls. Cavity radiation (photons in a box) helps us to understand radiation, just as the ideal gas (atoms in a box) helped us to understand matter.

Figure 7 shows a simple cavity radiator made of a thin-walled tungsten tube about a millimeter in diameter, heated to incandescence by passing a current through it. We see the bright cavity radiation emerging from a small hole in the wall. The cavity radiation is much brighter than the radiation from the outer wall of the cavity, even though the temperatures of the outer and inner walls are more or less equal.

The property of the cavity radiation that we seek to measure is its *spectral radiancy* $S(\lambda)$, defined so that $S(\lambda)\,d\lambda$ gives the radiated power per unit area of the cavity aperture that lies in the wavelength interval λ to

Figure 7 A thin-walled tungsten cylinder, heated to incandescence. Cavity radiation emerges from the small hole drilled through its wall.

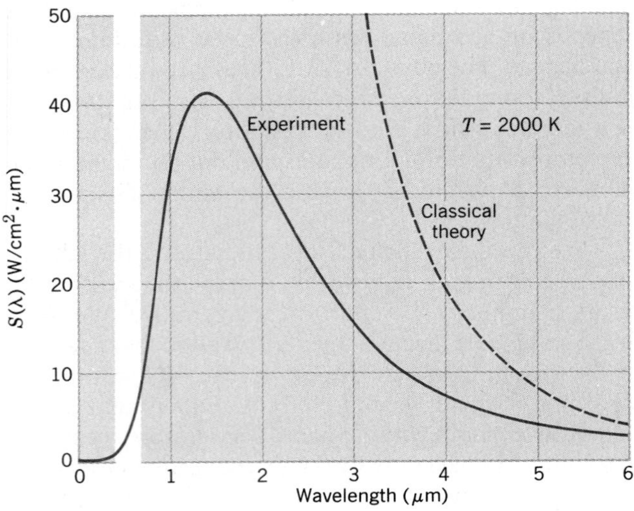

Figure 8 The solid curve shows the experimental spectral radiancy for a cavity at 2000 K. Note the failure of the classical theory. The unshaded bar is the range of visible wavelengths.

$\lambda + d\lambda$. Figure 8 shows the measured spectral radiancy for a cavity whose walls are held at 2000 K. Although such a radiator would glow brightly in a dark room, we see from the wavelength scale of the figure that only a small part of its radiated energy lies in the visible region of the spectrum. Most of it—by far—lies in the infrared. You do not have to linger too long near a bonfire to believe that it also emits plenty of energy in the warming rays of the infrared.

The prediction of classical theory for the variation of the spectral radiancy with wavelength (at a given temperature) is

$$S(\lambda) = \frac{2\pi ckT}{\lambda^4} \quad \text{(classical radiation law).} \quad (16)$$

Here c is the speed of light and k is the *Boltzmann constant,* a quantity that we met in Section 21–5; its value is

$$k = 1.38 \times 10^{-23}\ \text{J/K} = 8.62 \times 10^{-5}\ \text{eV/K}. \quad (17)$$

Equation 16 (with $T = 2000$ K) is plotted in Fig. 8. Although classical theory and experiment agree quite well at very long wavelengths (far beyond the scale shown in Fig. 8), the disagreement between theory and experiment at shorter wavelengths is total. The theoretical prediction does not even pass through a maximum. If the experiments are correct—and they are—then something is seriously wrong with the classical theory.

In 1900, Max Planck proposed a formula for the spectral radiancy that fitted the experimental data *perfectly* at all wavelengths and for all temperatures. His prediction is

$$S(\lambda) = \frac{2\pi c^2 h}{\lambda^5} \frac{1}{e^{hc/\lambda kT} - 1} \quad \text{(Planck's radiation law).} \quad (18)$$

In deriving this formula, Planck introduced the important constant h (now called the *Planck constant*) into physics for the first time. By fitting Eq. 18 to the experimental spectral radiancy data at a variety of temperatures, Planck was able to arrive at a value for his constant that agreed within a few percent with the present accepted value. Modern quantum physics began with Planck's presentation of his radiation law.

43-6 The Quantization of Energy

The assumptions that must be made to derive Planck's radiation law represent such a break with classical ideas that they were not at all clear to physicists of the day including—by his own admission—Planck himself. In 1917, however (17 years after Planck had advanced his radiation law), Einstein offered a remarkably straightforward derivation of Eq. 18 that made its underlying assumptions abundantly clear.*

The first of the assumptions underlying Eq. 18 is that the energy of the radiation in the cavity is quantized. That is, this radiation exists in the form of photons, of energy $E = h\nu$. The second assumption is that the energy of the atoms that form the cavity walls is quantized. That is:

The atoms that form the walls of the cavity can exist only in states of definite energy; states with intermediate energies are forbidden.

If you assume these quantization-of-energy principles for the radiation in the cavity and for the atoms of the cavity walls, you can derive Planck's law; if you don't, you can't.

* Einstein's derivation of Eq. 18 is given in Robert Resnick and David Halliday, *Basic Concepts in Relativity and Early Quantum Theory* (John Wiley & Sons, New York, 1985), 2nd ed., Supplementary Topic E. For an historical account of the derivation of Planck's radiation law, see Abraham Pais, *"Subtle is the Lord . . ."* (Oxford University Press, New York, 1982), Chapters 19 and 21.

We discussed energy quantization for atoms in a preliminary way in Section 8–10, which you may wish to reread at this time. Further developments showed that energy quantization is universal; it holds, not only for atoms, but for all kinds of systems—be they atoms, nuclei, molecules, or electrons in solids.

43-7 The Correspondence Principle

We have seen that the equations of relativistic mechanics reduce to those of classical Newtonian mechanics under conditions (low particle speeds) in which the classical laws are known to agree with experiment.

A similar *correspondence principle* holds in quantum physics. That is:

> *The equations of quantum physics must reduce to familiar classical laws under conditions in which the classical laws are known to agree with experiment.*

Let us explore this principle for the case of Eq. 18, Planck's radiation law. The classical radiation law (Eq. 16) is known to agree with experiment at very large wavelengths. Let us see whether Eq. 18 reduces to Eq. 16 in this limiting case. We note, however, that if we simply substitute $\lambda = \infty$ in Eq. 18, an indeterminate value for $S(\lambda)$ results. We must adopt a more subtle approach.

To simplify the algebra, we write Eq. 18 in the form

$$S = \frac{2\pi c^2 h}{\lambda^5} \frac{1}{e^x - 1}, \qquad (19)$$

in which $x = hc/\lambda kT$. The limiting case of $\lambda \to \infty$ corresponds to $x \to 0$. For small enough values of x (see Appendix G) we can approximate the series

$$e^x = 1 + x + \cdots \qquad \text{by} \quad e^x - 1 \approx x.$$

Equation 19 then becomes

$$S = \frac{2\pi c^2 h}{\lambda^5}\left(\frac{1}{x}\right) = \frac{2\pi c^2 h}{\lambda^5}\left(\frac{\lambda kT}{hc}\right) = \frac{2\pi ckT}{\lambda^4}.$$

This is exactly Eq. 16, the classical radiation law! The correspondence principle holds. Note how the Planck constant h—that sure indicator of a quantum equation—has conveniently canceled out in the process of applying this principle.

43-8 Atomic Structure

A question of ancient standing is, "What is the internal structure of an atom like?" It is appropriate that we begin to answer that question here, by studying one of the major clues to the structure of atoms, namely, the nature of the light that atoms emit.

In Fig. 8 we have an example of the light emitted by atoms when they are assembled to form the solid wall of a cavity radiator. We saw in Section 43–6 that what we can learn from the study of *this* radiation is the very important fact that the energies of the atoms that form the cavity walls are quantized. We can get no detailed information about specific atoms, however, because the cavity radiation does not depend on the nature of the atoms that make up the cavity walls.

To learn about the detailed structure of individual atoms (hydrogen, carbon, copper . . .) we must study the light that they emit (or absorb) when they stand alone, isolated from other atoms. The important feature of the light emitted by such isolated atoms is that it occurs only in a spectrum of sharply defined wavelengths or *spectral lines,* the spectrum being a characteristic signature of the atom that emits it.

Such *line spectra* are characteristic—not only of

Table 2 Some Line Spectra (see Fig. 9)

Figure Reference	*Entity*	*Wavelength Region* (m)	*Spectrum Designation*	*Mode*
a	Hg-198 nucleus	3×10^{-12}	Gamma ray	Emission
b	Mo atom	6×10^{-11}	X ray	Emission
c	Fe atom	3×10^{-7}	Ultraviolet	Emission
d	HCl molecule	3×10^{-6}	Infrared	Absorption
e	NH_3 molecule	1×10^{-2}	Microwave	Absorption
f	H_2 molecule	40	Radio	Absorption

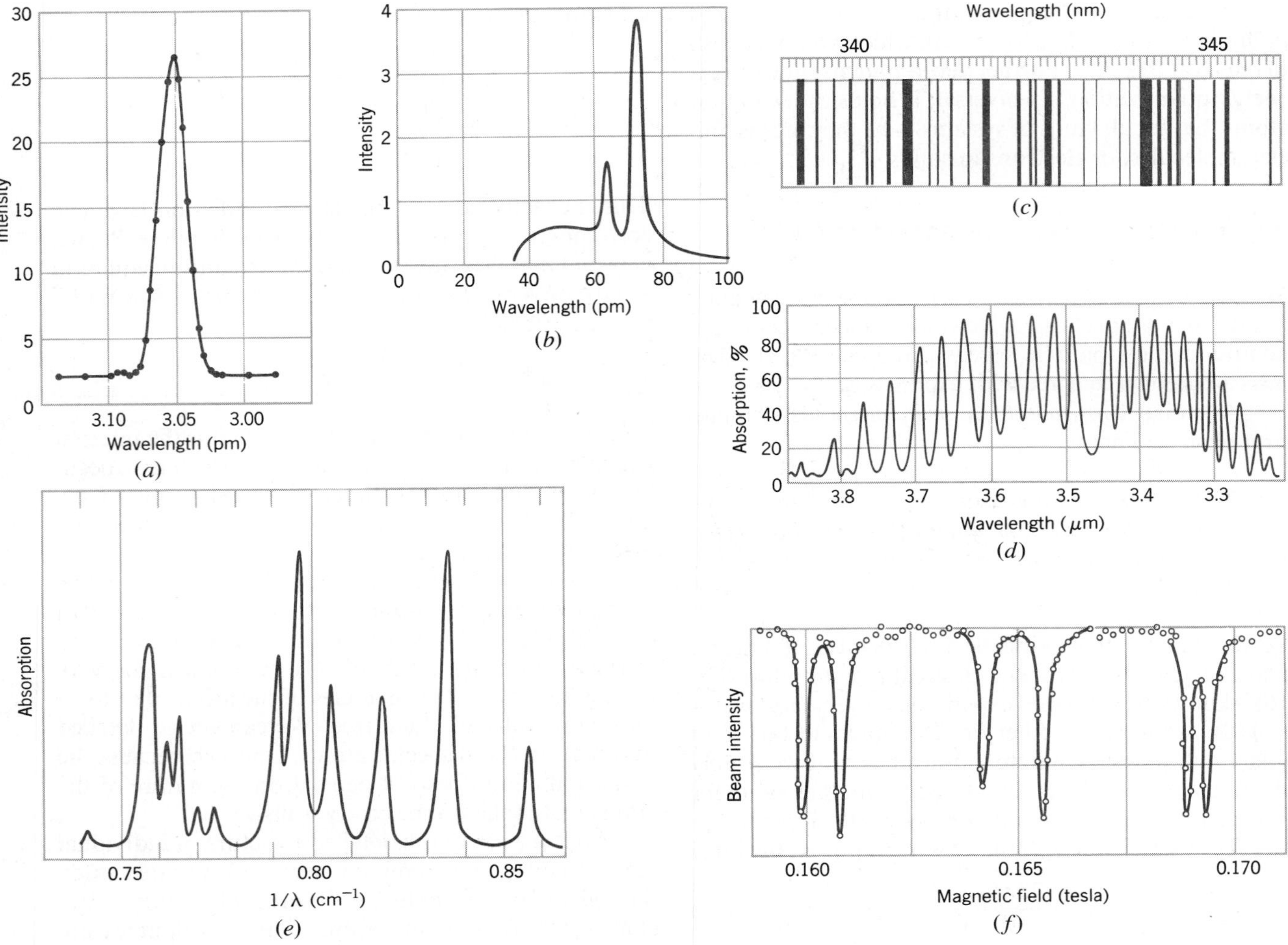

Figure 9 Selected spectral lines from various regions of the electromagnetic spectrum. See Table 2.

isolated atoms—but also of isolated molecules or atomic nuclei. Figure 9 shows some selected line spectra, associated with the emission (or absorption) or radiation by such entities. These curves, which involve wavelengths from all over the electromagnetic spectrum (see Table 2), only begin to suggest the bewildering variety of such spectra that can be measured in the laboratory.

It is our plan to concentrate on the spectrum of the hydrogen atom. Hydrogen is the simplest atom and, not surprisingly, it has the simplest spectrum. We shall trace out first the preliminary attempts by Niels Bohr to understand the structure of the hydrogen atom. In the following chapter we shall move on to a full quantum description of the hydrogen atom.

43-9 Niels Bohr and the Hydrogen Atom

The wavelengths of the lines in the spectrum of atomic hydrogen (see Fig. 10) have been known with precision for many years. They stand as a testing ground for any theory of the structure of the hydrogen atom that may be put forward.

Classical Theory. Let us first review the problems that arise when we try to solve the hydrogen-atom structure problem by the methods of classical physics. We can imagine that the electron in the hydrogen atom revolves about the central nucleus (a proton) in a circular orbit of radius r, as in Fig. 11. We can then suppose that the

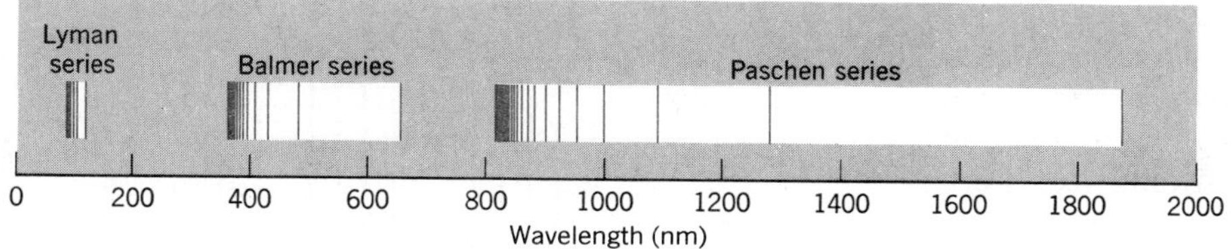

Figure 10 The spectrum of atomic hydrogen. The lines forming the Lyman, Balmer, and Paschen series are shown. Note the sharp series limits at the short-wavelength end of each series.

frequency of the radiation that the atom emits is equal to the frequency at which the electron circulates in this orbit. Classical theory predicts that such an accelerating electron will indeed radiate, *and* at its orbital frequency. However, the theory has a fatal flaw. The orbiting electron will radiate its energy completely away, emitting a continuous spectrum of radiation as it spirals in toward the nucleus. Thus the great classical theories of Newton and Maxwell stand helpless before the simplest atom. They cannot even account for the existence of the spectral lines, let alone predict their wavelengths. Indeed, they predict that atoms cannot exist!

Bohr's Theory. In 1913, just 2 years after Rutherford had put forward the idea that the atom has a nucleus, the great Danish physicist Niels Bohr (see Fig. 12) proposed a model for the hydrogen atom that not only accounted for the presence of the spectral lines but—with no adjustable parameters whatever—predicted their wavelengths to an accuracy of about 0.02%. Although Bohr's theory was successful for hydrogen, it proved less useful for more complex atoms. We now regard Bohr's theory as an inspired first step toward the more comprehensive quantum theory that followed it.

Bohr, realizing that classical physics had come to a dead end on the hydrogen atom problem, put forward two bold postulates. Both turned out to be enduring features that carry over in full force to modern quantum physics. Moreover, both turned out to be quite general, applying not only to the hydrogen atom but to atomic, molecular, and nuclear systems of all kinds. We discuss each postulate in turn.

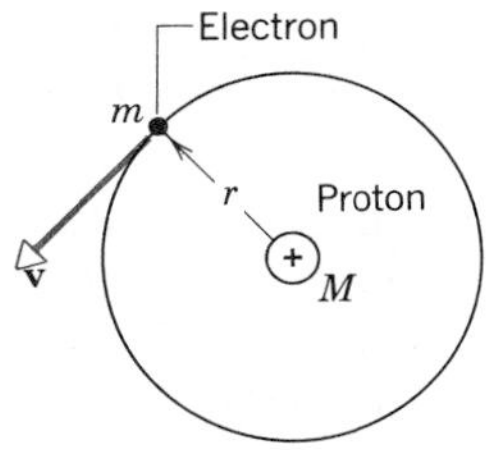

Figure 11 A classical model of the hydrogen atom, showing an electron of mass m circulating about a central nucleus of mass M. We assume that $M \gg m$.

Figure 12 Niels Bohr with Aage Bohr, one of his five sons. Both earned Nobel prizes in physics, Niels in 1922 and Aage in 1975.

1. The Postulate of Stationary States. Bohr assumed that the hydrogen atom can exist *without radiating* in any one of a discrete set of *stationary states* of fixed energy. This assumption of energy quantization flies in the face of classical theory but Bohr's attitude was, "Let's assume it anyway and see what happens." Note that this postulate says nothing at all about how we are to find the energies of these stationary states.

2. The Frequency Postulate. Bohr assumed that the hydrogen atom can emit or absorb radiation *only* when the atom changes from one of its stationary states to another. The energy of the emitted (or absorbed) photon is equal to the difference in energy between these two states. Thus, if an atom changes from an initial stage of energy E_i to a final state of (lower) energy E_f, the energy of the emitted photon is given by

$$h\nu_{if} = E_i - E_f \quad \text{(Bohr frequency condition)}, \qquad (20)$$

a relation known as the *Bohr frequency condition.* This postulate ties together neatly two new ideas (the photon hypothesis and energy quantization) with one familiar idea (the conservation of energy).

Bohr's next task was to select the stationary states by specifying their energies. Then, using Eq. 20, he could calculate the frequencies—and thus the wavelengths—of the spectrum lines. How to find the energies? Bohr actually did this in a clever way, making use of the correspondence principle. We present his result here, without proof. In Section 43-10, however, we give a semiclassical proof—also due to Bohr—that leads to the same result.

Bohr found that the energies of the stationary states of the hydrogen atom are given by

$$E = -\frac{me^4}{8\epsilon_0^2 h^2}\frac{1}{n^2} \qquad n = 1, 2, 3, \ldots, \qquad (21)$$

in which n is a *quantum number.* The negative sign tells us that the hydrogen atom states whose energies are given by this equation are bound states. That is, work must be done by an external agent to pull the atom apart. Although Bohr derived Eq. 21 in a semiclassical manner, exactly the same result follows from a rigorous derivation based on modern quantum theory.

Figure 13 shows an energy level diagram for the hydrogen atom, the energies being calculated from Eq. 21; each level is labeled with its quantum number. The

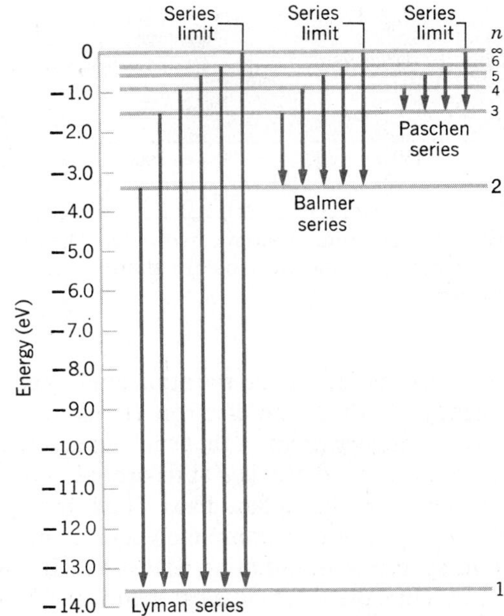

Figure 13 Energy levels and transitions in the spectrum of atomic hydrogen. Compare with Fig. 10, which displays the spectral lines.

state of lowest energy, called the *ground state,* is found by putting $n = 1$ in Eq. 21; it is easy to show that $E_1 = -13.6$ eV so that Eq. 21 can be written as

$$E = -\frac{13.6 \text{ eV}}{n^2} \qquad n = 1, 2, 3, \ldots \qquad (22)$$

Finally, we can combine Eq. 21 with the Bohr frequency condition (Eq. 20), obtaining

$$h\nu = \frac{hc}{\lambda} = \frac{me^4}{8\epsilon_0^2 h^2}\left(\frac{1}{l^2} - \frac{1}{u^2}\right). \qquad (23)$$

We can recast this, more compactly, as

$$\frac{1}{\lambda} = R\left(\frac{1}{l^2} - \frac{1}{u^2}\right). \qquad (24)$$

Here u and l are, respectively, the quantum numbers of the upper and the lower energy states involved in the transition whose wavelength is λ. The factor R, called the *Rydberg constant,* has the value

$$R = \frac{me^4}{8\epsilon_0^2 h^3 c} = 1.097 \times 10^7 \text{ m}^{-1} = 0.01097 \text{ nm}^{-1}. \qquad (25)$$

It is remarkable that Bohr's theory predicts a value for R

that agrees exceedingly well with the value of R found from experiment.

The downward pointing arrows in Fig. 13 show some transitions between pairs of levels. The transitions group themselves into several series of spectrum lines, each series having a particular level as its "home base." The Balmer series, for example, is defined by putting $l = 2$ in Eq. 24 and allowing u to take on the values 3, 4, 5, ... Each of the series has a *series limit,* corresponding to letting u approach infinity in that equation.

Bohr quite properly pushed his theory for all it was worth, taking as his goal the development of a theoretical basis for the periodic table of the elements. However, beyond hydrogen and hydrogenlike atoms such as He^+, Bohr had limited success. His "great and shining moment" was no doubt his prediction that element 72, then a blank space in the periodic table, should have chemical properties like those of zirconium. This led to the discovery of element 72 in zirconium ores and the naming of the new element *hafnium,* after an early name for Copenhagen, Bohr's home town. Word of this discovery came dramatically to Bohr in Stockholm, just hours before he was scheduled to receive the Nobel prize.

Sample Problem 6 (a) What is the wavelength of the least energetic photon in the Balmer series?

We identify the Balmer series (see Fig. 13) by putting $l = 2$ in Eq. 24. From the relation $E = h\nu$, the least energetic photon has the smallest frequency and thus the greatest wavelength. This means that we must put $u = 3$ (the smallest possible value) in Eq. 24; any higher value would yield a smaller wavelength. With these substitutions we have

$$\frac{1}{\lambda} = R\left(\frac{1}{l^2} - \frac{1}{u^2}\right)$$

$$= (0.01097\ \text{nm}^{-1})\left(\frac{1}{2^2} - \frac{1}{3^2}\right) = 1.524 \times 10^{-3}\ \text{nm}^{-1}$$

or

$$\lambda = \frac{1}{1.524 \times 10^{-3}\ \text{nm}^{-1}} = 656.3\ \text{nm}. \quad \text{(Answer)}$$

(b) What is the wavelength of the series limit for the Balmer series?

Again we put $l = 2$ in Eq. 24. To find the series limit (see Fig. 13) we let $u \to \infty$. Equation 24 then becomes

$$\frac{1}{\lambda} = (0.01097\ \text{nm}^{-1})\left(\frac{1}{2^2}\right) = 2.743 \times 10^{-3}\ \text{nm}^{-1}$$

which yields

$$\lambda = 364.6\ \text{nm}. \quad \text{(Answer)}$$

43-10 Bohr's Derivation (Optional)

Here we outline Bohr's derivation of Eq. 21, his formula for the energies of the stationary states of the hydrogen atom. As we indicated earlier, Bohr actually derived this formula in a clever way, using the correspondence principle and making no detailed assumptions about the nature of the quantum states of the hydrogen atom at low quantum numbers. Bohr also suggested an alternative derivation, which we shall follow. Although it involves ideas such as electron orbits that we no longer find useful, it nevertheless leads to a correct quantum result.

Let us apply Newton's second law ($F = ma$) to an electron revolving in a circular orbit of radius r, as in Fig. 11. Thus, applying Coulomb's law,

$$\frac{1}{4\pi\epsilon_0}\frac{(e)(e)}{r^2} = m\frac{v^2}{r}. \tag{26}$$

We can use Eq. 26 to find the kinetic energy. Thus

$$K = \tfrac{1}{2}mv^2 = \frac{e^2}{8\pi\epsilon_0 r}. \tag{27}$$

The electric potential energy is

$$U = \frac{1}{4\pi\epsilon_0}\frac{(+e)(-e)}{r} = -\frac{e^2}{4\pi\epsilon_0 r} \tag{28}$$

and the total energy is

$$E = K + U = -\frac{e^2}{8\pi\epsilon_0 r}. \tag{29}$$

This is as far as we can go with purely classical ideas. The total energy E of the stationary states depends on the orbit radius and, unless we can find out which orbit radii correspond to the stationary states, we are stuck. In modern language, we need a quantization criterion.

Having come so far, you are entitled to suppose that we are stuck at this point because the Planck constant h has not yet put in an appearance. If you think that, you are right. Bohr saw that the simplest way to introduce h was to quantize the angular momentum L of the atomic orbits, by assuming—quite arbitrarily—that L could only have values given by*

$$L = n\frac{h}{2\pi} \qquad n = 1, 2, 3, \ldots, \tag{30}$$

* The angular momentum of the electron in the hydrogen atom is indeed quantized but the true quantization rule is not quite as simple as Eq. 30; see Section 45-5. Bear in mind that Bohr's approach is semiclassical and is an intermediate step in the development of a full quantum theory.

in which n is a quantum number. Note that, although they are quite different quantities, angular momentum and the Planck constant h have the same dimensions, a fact that makes it less surprising that angular momentum should give us the simplest quantization rule.

We can use Eq. 26 to write for the angular momentum

$$L = mvr = \sqrt{\frac{me^2 r}{4\pi\epsilon_0}}. \qquad (31)$$

Combining Eqs. 30 and 31 gives us the radii of the allowed orbits, or

$$r = n^2 \frac{h^2\epsilon_0}{\pi m e^2} \qquad n = 1, 2, 3, \ldots. \qquad (32)$$

We can recast this, more compactly, as

$$r = n^2 r_B \qquad n = 1, 2, 3, \ldots, \qquad (33)$$

in which r_B can be computed from the fundamental constants in Eq. 32 to be

$$r_B = 5.292 \times 10^{-11}\ \text{m} = 52.92\ \text{pm} = 0.05292\ \text{nm}. \qquad (34)$$

The quantity r_B, called the *Bohr radius,* is equal to the radius of the Bohr orbit for hydrogen in its ground state, defined by putting $n = 1$ in Eq. 32. Even though we no longer believe in such orbits, we still use the Bohr radius as a convenient measure of distance on the atomic scale. You should be impressed that, although Bohr put nothing into his theory that says how big atoms are, a length of about the right size comes out of it.

Finally, we can substitute the quantized orbit radius r from Eq. 32 into Eq. 29, obtaining

$$E = -\frac{me^4}{8\epsilon_0^2 h^2}\frac{1}{n^2} \qquad n = 1, 2, 3 \ldots,$$

which is exactly Eq. 21, the result that we set out to prove.

REVIEW AND SUMMARY

Photons

This chapter describes some experiments that lead ultimately to the concept that light is made up of concentrated bundles of energy called photons. Einstein hypothesized that each photon has energy E and momentum p with

$$E = h\nu \quad \text{and} \quad p = h/\lambda. \qquad [1,4]$$

See Table 1 and Sample Problems 1 and 2. The Planck constant h has the value 6.63×10^{-34} J·s = 4.14×10^{-15} eV·s. This constant, although small, is not zero; this is the determining feature of modern quantum physics. The experiments are:

The Photoelectric Effect

1. The *photoelectric effect,* in which electrons are ejected from a metal surface by incident light. Einstein's equation for this effect, based on his photon hypothesis, is

$$h\nu = \phi + K_m \quad \text{(photoelectric equation)}. \qquad [8]$$

Here $h\nu$ is the energy of the photon absorbed by an electron in the metal surface. The *work function* ϕ is the energy needed to remove this electron from the metal; K_m is the maximum kinetic energy of the electron outside the surface. Study Figs. 1, 2, and 3 and Sample Problems 3, 4, and 5.

The Compton Effect

2. The *Compton effect,* in which x-ray photons, scattered from free electrons as in Fig. 4, undergo a wavelength increase $\Delta\lambda$. This *Compton shift* (see Fig. 5) is given by

$$\Delta\lambda = \frac{h}{mc}(1 - \cos\phi). \qquad [14]$$

This equation follows from applying the laws of conservation of energy and momentum to a billiard-ball-like collision between a photon and a free electron, as in Fig. 6. Study Sample Problem 5.

Cavity Radiation

3. Measurement of the distribution with wavelength of the radiation emerging from heated cavities; see Figs. 7 and 8 and Eq. 18. These studies introduced the concept of *energy quantization* and caused the Planck constant h to appear in the equations of physics for the first time.

The Correspondence Principle

The *correspondence principle* formalizes the requirement that the predictions of quantum physics must agree with the predictions of classical physics in situations where classical physics is known to be consistent with experiment.

Line Spectra

Figure 9 and Table 2 show that the absorption and emission of radiation at sharply defined wavelengths is characteristic of atoms, molecules, and nuclei. Classical physics cannot explain this phenomenon.

Bohr's Quantum Postulates

Attempts to understand line spectra in quantum terms start with Bohr's *quantum postulates,* first introduced to explain the spectrum of hydrogen atoms: (1) an atom can exist *without radiating* in any one of a discrete set of *stationary states* of fixed energy and (2) an atom can emit or absorb radiation only in transitions between stationary states, the frequencies of the spectral lines being given by

The Bohr Frequency Condition

$$h\nu_{if} = E_i - E_f \quad \text{(Bohr frequency condition).} \qquad [20]$$

Here E_i and E_f are the energies of the initial and final states involved in the transition.

To find the energies of the stationary states in hydrogen atoms, Bohr assumed that the angular momentum L of the orbiting electron can have only the discrete values given by Eq. 30. The resulting energies of the allowed states are

The Energies of the Stationary States

$$E = -\left(\frac{me^4}{8\epsilon_0^2 h^2}\right)\frac{1}{n^2} = -\frac{13.6\ \text{eV}}{n^2}, \qquad n = 1, 2, 3, \ldots . \qquad [21,22]$$

Combining Eqs. 20 and 21 leads to

The Wavelengths

$$\frac{1}{\lambda} = R\left(\frac{1}{l^2} - \frac{1}{u^2}\right) \qquad [24]$$

for the wavelengths of the lines of the hydrogen spectrum for a transition between an upper state with quantum number u to a lower state with quantum number l; $R = 0.01097\ \text{nm}^{-1}$ is the *Rydberg constant.* Study and compare Figs. 10 and 13; see Sample Problem 6.

QUESTIONS

1. Is charge quantized in physics? Is mass quantized? Describe some experiments that support your answers.

2. How can a photon energy be given by $E = h\nu$ when the very presence of the frequency ν in the formula implies that light is a wave?

3. Given that $E = h\nu$ for a photon, the Doppler shift in frequency of radiation from a receding light source would seem to indicate a reduced energy for the emitted photons. Is this in fact true? If so, what happened to the conservation of energy principle? (See "Questions Students Ask," *The Physics Teacher* December 1983, p. 616.)

4. Photon A has twice the energy of photon B. What is the ratio of the momentum of A to that of B?

5. How does a photon differ from a material particle?

6. Show that the Planck constant has the dimensions of angular momentum. Does this necessarily mean that angular momentum is a quantized quantity?

7. For quantum effects to be "everyday" phenomena in our lives, what order of magnitude would the value of h need to have? (See G. Gamow, *Mr Tompkins in Wonderland,* Cambridge University Press, Cambridge, 1957, for a delightful popularization of a world in which the physical constants c, G, and h make themselves obvious.)

8. An isolated metal plate yields photoelectrons when you first shine ultraviolet light on it but later it doesn't give up any more. Explain.

9. In Fig. 2, why doesn't the photoelectric current rise vertically to its maximum (saturation) value when the applied potential difference is just slightly more positive than V_0?

10. In the photoelectric effect, why does the existence of a cutoff frequency speak in favor of the photon theory and against the wave theory?

11. Why are photoelectric measurements so sensitive to the nature of the photoelectric surface?

12. Explain the statement that one's eyes could not detect faint starlight if light were not particlelike.

13. Consider the following procedures: (*a*) bombard a metal with electrons; (*b*) place a strong electric field near a metal; (*c*) illuminate a metal with light; (*d*) heat a metal to a high temperature. Which of the above procedures can result in the emission of electrons?

14. A certain metal plate is illuminated by light of a definite frequency. Whether or not photoelectrons are emitted as a result depends upon which of the following features? (*a*) intensity of illumination; (*b*) length of time of exposure to the light; (*c*) thermal conductivity of the plate; (*d*) area of the plate; (*e*) material of the plate.

15. Does Einstein's theory of photoelectricity, in which light is postulated to be a stream of photons, invalidate Young's double-slit interference experiment in which light is postulated to be a wave?

16. What is the direction of a Compton-scattered electron with maximum kinetic energy, compared with the direction of the incident photon?

17. In Compton scattering, why would you expect $\Delta\lambda$ to be independent of the material of which the scatterer is composed?

18. Why don't we observe a Compton effect with visible light?

19. Light from distant stars is Compton-scattered many times by free electrons in outer space before reaching us. This shifts the light toward the red. How can this shift be distinguished from the Doppler red shift due to the motion of receding stars?

20. "Pockets" formed by the coals in a coal fire seem brighter than the coals themselves. Is the temperature in such pockets appreciably higher than the surface temperature of an exposed glowing coal? Explain this common observation.

21. If we look into a cavity whose walls are maintained at a constant temperature, no details of the interior are visible. Explain.

22. We claim that all objects radiate energy by virtue of their temperature and yet we cannot see all objects in the dark. Why?

23. Only a relatively small number of Balmer lines can be observed from laboratory discharge tubes, whereas a large number are observed in the spectra of stars. Explain this in terms of the small density, high temperature, and large volume of gases in stellar atmospheres.

24. Any series of atomic hydrogen yet to be observed will probably be found to be in what region of the spectrum?

25. In Bohr's theory for the hydrogen atom, what is the implication of the fact that the potential energy is negative and is greater in magnitude than the kinetic energy? Is this a result of quantum physics or is it true classically as well?

26. Why are some lines in the hydrogen spectrum brighter than others?

27. Consider a hydrogenlike atom in which a positron (a postively charged electron) circulates about a (negatively charged) antiproton. In what ways, if any, would the emission spectrum from this "antimatter atom" differ from the spectrum of a normal hydrogen atom?

28. Radioastronomers observe lines in the hydrogen spectrum that originate in hydrogen atoms that are in states with $n = 350$ or so. Why can't hydrogen atoms in states with such high quantum numbers be produced and studied in the laboratory?

29. Can a hydrogen atom absorb a photon whose energy exceeds its binding energy (13.6 eV)?

30. List and discuss the assumptions made by Planck in connection with the cavity radiation problem, by Einstein in connection with the photoelectric effect, by Compton in connection with the Compton effect, and by Bohr in connection with the hydrogen atom problem.

31. Describe several methods that can be used to experimentally determine the value of the Planck constant h.

32. According to classical mechanics, an electron moving in an orbit should be able to do so with any angular momentum whatever. According to Bohr's theory of the hydrogen atom, however, the angular momentum is quantized according to $L = nh/2\pi$. Reconcile these two statements, using the correspondence principle.

EXERCISES AND PROBLEMS

Section 43-2 Einstein Makes a Proposal

1E. Show that the energy E of a photon (in eV) is related to its wavelength λ (in nm) by

$$E = \frac{1240}{\lambda}.$$

This result can be useful in solving many problems.

2E. The orange-colored light from a highway sodium lamp has a wavelength of 589 nm. How much energy is possessed by an individual photon from such a lamp?

3E. Consider monochromatic light falling on a photographic film. The incident photons will be recorded if they have enough energy to dissociate a AgBr molecule in the film. The minimum energy required to do this is about 0.6 eV. Find the cutoff wavelength greater than which the light will not be recorded. In what region of the spectrum does this wavelength fall?

4E. (*a*) A spectral emission line, important in radioastronomy, has a wavelength of 21 cm. What is its corresponding photon energy? (*b*) At one time the meter was defined as 1,650,763.73 wavelengths of the orange light emitted by a light source containing krypton-86 atoms. What is the corresponding photon energy of this radiation?

5E. A particular x-ray photon has a wavelength of 35 pm. Calculate the photon's (*a*) energy, (*b*) frequency, and (*c*) momentum.

6P. Under ideal conditions the normal human eye will record a visual sensation at 550 nm if incident photons are absorbed at a rate as low as 100 s^{-1}. To what power level does this correspond?

7P. What are (*a*) the frequency, (*b*) the wavelength, and (*c*) the momentum of a photon whose energy equals the rest energy of the electron?

8P. In the photon picture of radiation, show that if two parallel beams of light of different wavelengths are to have the same intensity, then the rates per unit area at which photons pass through any cross section of the beams are in the same ratio as the wavelengths.

9P. An ultraviolet light bulb, emitting at 400 nm, and an infrared light bulb, emitting at 700 nm, each are rated at 400 W. (*a*) Which bulb radiates photons at the greater rate? (*b*) How many more photons does it generate per second than does the other bulb?

10P. A satellite in earth orbit maintains a panel of solar cells of 2.6-m^2 area at right angles to the direction of the sun's rays. (*a*) At what rate does solar energy strike the panel? (*b*) At what rate do solar photons strike the panel? Assume that the solar radiation is monochromatic with a wavelength of 550 nm. (*c*) How long would it take for a "mole of photons" to strike the panel? Solar energy arrives at the rate of 1.39 kW/m^2.

11P. A special kind of light bulb emits monochromatic light at a wavelength of 630 nm. It is rated at 60 W, and is 93% efficient in converting electrical energy to light. How many photons will the bulb emit over its 730 h lifetime?

12P. *Focusing laser light.* The emerging beam from a 1.5-W argon laser ($\lambda = 515$ nm) has a diameter d of 3.0 mm. (*a*) At what rate per m^2 do photons pass through any cross section of the beam? (*b*) The beam is focused by a lens system whose effective focal length f is 2.5 mm. The focused beam forms a circular diffraction pattern whose central disk has a radius R given by $1.22\, f\lambda/d$. It can be shown that 84% of the incident power lies within this central disk, the rest falling in the fainter, concentric diffraction rings that surround the central disk. At what rate per m^2 do photons pass through the central disk of the diffraction pattern?

13P. Assume that a 100-W sodium-vapor lamp radiates its energy uniformly in all directions in the form of photons with an associated wavelength of 589 nm. (*a*) At what rate are photons emitted from the lamp? (*b*) At what distance from the lamp will the average flux of photons be 1.0 photons/($cm^2 \cdot s$)? (*c*) At what distance from the lamp will the average density of photons be 1.0 photons/cm^3? (*d*) What are the photon flux and the photon density 2.0 m from the lamp?

Section 43-3 The Photoelectric Effect

14E. You wish to pick a substance for a photocell operable with visible light. Which of the following will do (work function in parentheses): tantalum (4.2 eV); tungsten (4.5 eV); aluminum (4.2 eV); barium (2.5 eV); lithium (2.3 eV)?

15E. Satellites and spacecraft in orbit about the earth can become charged due, in part, to the loss of electrons caused by the photoelectric effect induced by sunlight on the space vehicle's outer surface. Suppose that a satellite is coated with platinum, a metal with one of the largest work functions: $\phi =$ 5.32 eV. Find the longest-wavelength photon that can eject a photoelectron from platinum. (Satellites must be designed to minimize such charging.)

16E. (*a*) The energy needed to remove an electron from metallic sodium is 2.28 eV. Does sodium show a photoelectric effect for red light, with $\lambda = 680$ nm? (*b*) What is the cutoff wavelength for photoelectric emission from sodium and to what color does this wavelength correspond?

17E. Find the maximum kinetic energy of photoelectrons if the work function of the material is 2.3 eV and the frequency of the radiation is 3.0×10^{15} Hz.

18E. Incident photons strike a sodium surface having a work function of 2.2 eV, causing photoelectric emission. When a stopping potential of 5.0 V is imposed, there is no photocurrent. What is the wavelength of the incident photons?

19E. The work function of tungsten is 4.5 eV. Calculate the speed of the fastest of the photoelectrons emitted when photons of energy 5.8 eV are incident on a sheet of tungsten.

20E. Light of wavelength 200 nm falls on an aluminum surface. In aluminum 4.2 eV are required to remove an electron. What is the kinetic energy of (*a*) the fastest and (*b*) the slowest emitted photoelectrons? (*c*) What is the stopping potential? (*d*) What is the cutoff wavelength for aluminum?

21E. (*a*) If the work function for a metal is 1.8 eV, what would be the stopping potential for light having a wavelength of 400 nm? (*b*) What would be the maximum speed of the emitted photoelectrons at the metal's surface?

22E. The wavelength of the photoelectric threshold for silver is 325 nm. Find the maximum kinetic energy of electrons ejected from a silver surface by ultraviolet light of wavelength 254 nm.

23P. The stopping potential for photoelectrons emitted from a surface illuminated by light of wavelength 491 nm is 0.71 V. When the incident wavelength is changed to a new value, the stopping potential is found to be 1.43 V. (*a*) What is this new wavelength? (*b*) What is the work function for the surface?

24P. In a photoelectric experiment in which a sodium surface is used, one finds a stopping potential of 1.85 V for a wavelength of 300 nm and a stopping potential of 0.82 V for a wavelength of 400 nm. From these data find (*a*) a value for the

Planck constant, (*b*) the work function for sodium and (*c*) the cutoff wavelength for sodium.

25P. Millikan's photoelectric data for lithium are:

Wavelength, nm	433.9	404.7	365.0	312.5	253.5
Stopping potential, V	0.55	0.73	1.09	1.67	2.57

Make a plot like Fig. 3, which is for sodium, and find (*a*) the Planck constant and (*b*) the work function for lithium.

26P. Photosensitive surfaces are not necessarily very efficient. Suppose the fractional efficiency of a cesium surface (work function 1.8 eV) is 1.0×10^{-16}, that is, one photoelectron is produced for every 10^{16} photons striking the surface. What would be the photocurrent from such a cesium surface if it is illuminated with 600-nm light from a 2.0-mW laser and all of the photoelectrons produced take part in charge flow?

27P. X rays with a wavelength of 71 pm eject photoelectrons from a gold foil, the electrons originating from deep within the gold atoms. The ejected electrons move in circular paths of radius r in a region of uniform magnetic field **B**. Experiment shows that $Br = 1.88 \times 10^{-4}$ T · m. Find (*a*) the maximum kinetic energy of the photoelectrons and (*b*) the work done in removing the electrons from the gold atoms that make up the foil.

28P*. Show, by analyzing a collision between a photon and a free electron (using relativistic mechanics), that it is impossible for a photon to give all of its energy to the free electron. In other words, the photoelectric effect cannot occur for completely free electrons; the electrons must be bound in a solid or in an atom.

Section 43-4 The Compton Effect

29E. Photons of wavelength 2.4 pm are incident on a target containing free electrons. (*a*) Find the wavelength of a photon that is scattered at 30° from the incident direction. (*b*) Do the same for a scattering angle of 120°.

30E. A 0.511-MeV gamma-ray photon is Compton-scattered from a free electron in an aluminum block. (*a*) What is the wavelength of the incident photon? (*b*) What is the wavelength of the scattered photon? (*c*) What is the energy of the scattered photon? Assume a scattering angle of 90°.

31E. An x-ray photon of wavelength 0.01 nm strikes an electron head on ($\phi = 180°$). Determine (*a*) the change in wavelength of the photon, (*b*) the change in energy of the photon, and (*c*) the kinetic energy imparted to the electron.

32E. The quantity h/mc in Eq. 14 is often called the *Compton wavelength* of the scattering particle and that equation is written as

$$\Delta\lambda = \lambda_C(1 - \cos\phi).$$

(*a*) Calculate the Compton wavelength λ_C of an electron; of a proton. (*b*) What is the energy of a photon whose wavelength is equal to the Compton wavelength of the electron? of the proton?

33E. Calculate the percent change in photon energy for a Compton collision with ϕ in Fig. 4 equal to 90° for radiation in (*a*) the microwave range, with $\lambda = 3.0$ cm, (*b*) the visible range, with $\lambda = 500$ nm, (*c*) the x-ray range, with $\lambda = 25$ pm, and (*d*) the gamma-ray range, the energy of the gamma-ray photons being 1.0 MeV. What are your conclusions about the importance of the Compton effect in these various regions of the electromagnetic spectrum, judged solely by the criterion of energy loss in a single Compton encounter? (*Hint:* See Eq. 15.)

34E. What fractional increase in wavelength leads to a 75% loss of photon energy in a Compton collision with a free electron? (*Hint:* See Eq. 15.)

35E. Find the maximum wavelength shift for a Compton collision between a photon and a free *proton.*

36P. A 6.2-keV x-ray photon falling on a carbon block is scattered by a Compton collision and its frequency is shifted by 0.010%. (*a*) Through what angle is the photon scattered? (*b*) What kinetic energy is imparted to the electron?

37P. Show that $\Delta E/E$, the fractional loss of energy of a photon during a Compton collision, is given by

$$(h\nu'/mc^2)(1 - \cos\phi).$$

38P. Through what angle must a 200-keV photon be scattered by a free electron so that it loses 10% of its energy?

39P. Show that when a photon of energy E scatters from a free electron, the maximum recoil kinetic energy of the electron is given by

$$K_{max} = \frac{E^2}{E + mc^2/2}.$$

40P. What is the maximum kinetic energy of the Compton-scattered electrons knocked out of a thin copper foil by an incident beam of 17.5-keV x rays?

41P. Carry out the algebra needed to eliminate v and θ from Eqs. 10, 12, and 13 and thus to derive Eq. 14, the equation for the Compton shift.

Section 43-5 Planck and His Constant— An Historical Aside

42E. The wavelength λ_{max} at which the spectral radiancy of a cavity radiator has its maximum value for a particular temperature T (see Fig. 8) is given by the Wien displacement law, or

$$\lambda_{max}T = \text{a constant} = 2898\ \mu\text{m} \cdot \text{K}.$$

The effective surface temperature of the sun is 5800 K. At what wavelength would you expect the sun to radiate most strongly? In what region of the spectrum is this? Why, then, does the sun appear yellow?

43E. Sensitive infrared detectors allow antiaircraft missiles to respond to the low-intensity radiation emitted by the target aircraft's airframe, and not just to the hot exhaust. This makes attack from any angle feasible. To what wavelength should a missile seeker be most sensitive if the target temperature is 290 K? Ignore atmospheric absorption. (See Exercise 42.)

44E. At what temperature is cavity radiation most visible to the human eye if the eye is most sensitive to yellow-green light, wavelength = 550 nm? (See Exercise 42.)

45E. In 1983 the Infrared Astronomical Satellite (IRAS) detected a cloud of solid particles surrounding the star Vega, radiating maximally at a wavelength of 32 μm. What is the temperature of this cloud of particles? (See Exercise 42.)

46E. Low-temperature physicists would not consider a temperature of 2.0 mK to be particularly low. (*a*) At what wavelength would a cavity whose walls were at this temperature radiate most copiously? (See Exercise 42.) (*b*) To what region of the electromagnetic spectrum would this radiation belong? (*c*) What are some of the practical difficulties of operating a cavity radiator at such a low temperature?

47E. Calculate the wavelength of maximum spectral radiancy (see Exercise 42) and identify the region of the electromagnetic spectrum to which it belongs for each of the following: (*a*) the 3.0-K microwave background radiation, a remnant of the primordial fireball; (*b*) your body, assuming a skin temperature of 20°C; (*c*) a tungsten lamp filament at 1800 K; (*d*) an exploding thermonuclear device, at an assumed fireball temperature of 10^7 K; (*e*) the universe immediately after the Big Bang, at an assumed temperature of 10^{38} K.

48P. Show that the wavelength λ_{max} at which Planck's spectral radiation law, Eq. 18, has its maximum is given by

$$\lambda_{max} = (2898\ \mu\text{m} \cdot \text{K})/T.$$

(*Hint:* Set $dS/d\lambda = 0$; an equation will be encountered whose numerical solution is 4.965.)

49P. (*a*) By integrating the Planck radiation law, Eq. 18, over all wavelengths, show that the power radiated per square meter of a cavity surface is given by

$$P = \left(\frac{2\pi^5 k^4}{15h^3c^2}\right) T^4 = \sigma T^4.$$

(*Hint:* Make a change in variables, letting $x = hc/\lambda kT$. A definite integral will be encountered that has the value

$$\int_0^\infty \frac{x^3 dx}{e^x - 1} = \frac{\pi^4}{15}.$$

(*b*) Verify that the numerical value of the constant $\sigma = 5.67 \times 10^{-8}\ \text{W/(m}^2 \cdot \text{K}^4)$.

50P. Calculate the thermal power radiated from a fireplace assuming an effective radiating surface of 0.50 m^2, and an effective temperature of 500°C. (See Problem 49.)

51P. (*a*) Show that a human body of area 1.8 m^2 and temperature 31°C radiates radiation at the rate of 872 W. (*b*) Why, then, do people not glow in the dark? (See Problem 49.)

52P. A cavity at absolute temperature T_1 radiates energy at a power level of 12 mW. At what power level does the same cavity at temperature $2T_1$ radiate? (See Problem 49.)

53P. A *thermograph* is a medical instrument used to measure radiation from the skin. For example, normal skin radiates at a temperature of about 34°C, and the skin over a tumor radiates at a slightly higher temperature. Derive an expression for the fractional difference in the radiance between adjacent areas of the skin that are at slightly different temperatures. Evaluate the expression for a temperature difference of 1°C. (See Problem 49.)

54P. An oven with an inside temperature $T_o = 227$°C is in a room having a temperature $T_r = 27$°C. There is a small opening of area 5.0 cm^2 in one side of the oven. How much net power is transferred from the oven to the room? (*Hint:* Consider both oven and room as cavities. See Problem 49.)

Section 43-9 Niels Bohr and the Hydrogen Atom

55E. An atom absorbs a photon of frequency 6.2×10^{14} Hz. By how much does the energy of the atom increase?

56E. An atom absorbs a photon having a wavelength of 375 nm and immediately emits another photon having a wavelength of 580 nm. How much net energy was absorbed by the atom in this process? Ease the computation by using the result of Exercise 1.

57E. A line in the x-ray spectrum of gold has a wavelength of 18.5 pm. The emitted x-ray photons correspond to a transition of the gold atom between two stationary states, the upper one of which has the energy −13.7 keV. What is the energy of the lower stationary state?

58E. (*a*) By direct substitution of numerical values of the fundamental constants, verify that the energy of the ground state of the hydrogen atom is −13.6 eV; see Eqs. 21 and 22. (*b*) Similarly, show that the value of the Rydberg constant R is 0.01097 nm^{-1}, as asserted in Eq. 25.

59E. Answer the questions of Sample Problem 6, but for the Lyman series.

60E. What are (*a*) the energy, (*b*) momentum, and (*c*) the wavelength of the photon that is emitted when a hydrogen atom undergoes a transition from the state $n = 3$ to $n = 1$?

61E. Using Bohr's formula, Eq. 24, calculate the three longest wavelengths of the Balmer series.

62E. Find the ratio of the shortest wavelength of the Balmer series to the shortest wavelength of the Lyman series.

63E. A hydrogen atom is excited from a state with $n = 1$ to one with $n = 4$. (*a*) Calculate the energy that must be absorbed by the atom. (*b*) Calculate and display on an energy-level diagram the different photon energies that may be emitted if the atom returns to the $n = 1$ state.

64P. How much work must be done by an external agent to pull apart a hydrogen atom if the electron initially is (*a*) in the ground state? (*b*) in the state with $n = 2$?

65P. (*a*) What are the wavelength intervals over which the Lyman, Balmer, and Paschen series extend? (Each interval extends from the longest wavelength to the series limit.) (*b*) What are the corresponding frequency intervals?

66P. Light of wavelength 486.1 nm is emitted by a hydrogen atom. (*a*) What transition of the atom is responsible for this radiation? (*b*) To what series does this radiation belong?

67P. Show, on an energy-level diagram for hydrogen, the quantum numbers corresponding to a transition in which the wavelength of the emitted photon is 121.6 nm.

68P. A hydrogen atom in a state having a *binding energy* (the energy required to remove an electron) of 0.85 eV makes a transition to a state with an *excitation energy* (the difference in energy between the state and the ground state) of 10.2 eV. (*a*) Find the energy of the emitted photon. (*b*) Show this transition on an energy-level diagram for hydrogen, labeling with the appropriate quantum numbers.

69P. Calculate the recoil speed of a hydrogen atom, assumed initially at rest, if the electron makes a transition from the $n = 4$ state directly to the ground state. (*Hint:* Apply conservation of linear momentum.)

70P. A neutron, with kinetic energy of 6.0 eV, collides with a resting hydrogen atom in its ground state. Show that this collision must be elastic (that is, energy must be conserved). (*Hint:* Show that the atom cannot be raised to a higher excitation state as a result of the collision.)

71P. From the energy-level diagram for hydrogen, explain the observation that the frequency of the second Lyman-series line is the sum of the frequencies of the first Lyman-series line and the first Balmer-series line. This is an example of the empirically discovered *Ritz combination principle.* Use the diagram to find some other valid combinations.

72P. *Radio waves from a hydrogen atom.* (*a*) Show that the smallest quantum number of adjacent levels in hydrogen between which transitions giving rise to radio waves are possible is given by $n = (2R\lambda)^{1/3}$, where λ is the wavelength of the radio wave. Note that for radio wave emission, n must be large. (*b*) If the 21-cm radio emission from interstellar hydrogen were due to such a transition (it isn't), what would be the value of n?

Section 43-10 Bohr's Derivation

73E. Verify the numerical value of r_B, given in Eq. 34, by direct computation of its equivalent expression given in Eq. 32.

74E. What is the value of the quantum number for a hydrogen atom that has an orbital radius of 0.847 nm?

75E. (*a*) If the angular momentum of the earth due to its motion around the sun were quantized according to Bohr's relation $L = nh/2\pi$, what would the quantum number be? (*b*) Could such quantization be detected if it existed?

76P. In the ground state of the hydrogen atom, according to Bohr's theory, what are (*a*) the quantum number, (*b*) the orbit radius, (*c*) the angular momentum, (*d*) the linear momentum, (*e*) the angular velocity, (*f*) the linear speed, (*g*) the force on the electron, (*h*) the acceleration of the electron, (*i*) the kinetic energy, (*j*) the potential energy, and (*k*) the total energy?

77P. How do the quantities (*b*) to (*k*) in Problem 76 vary with the quantum number?

78P. A diatomic gas molecule consists of two atoms of mass m separated by a fixed distance d rotating about an axis as indicated in Fig. 14. Assuming that its angular momentum is quantized as in the Bohr atom, determine (*a*) the possible angular velocities, and (*b*) the possible quantized rotational energies.

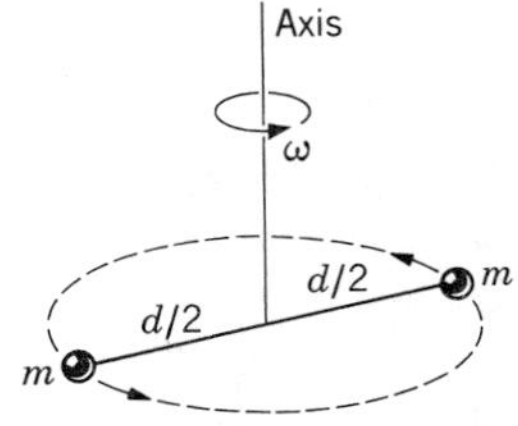

Figure 14 Problem 78.

CHAPTER 44

QUANTUM PHYSICS — II

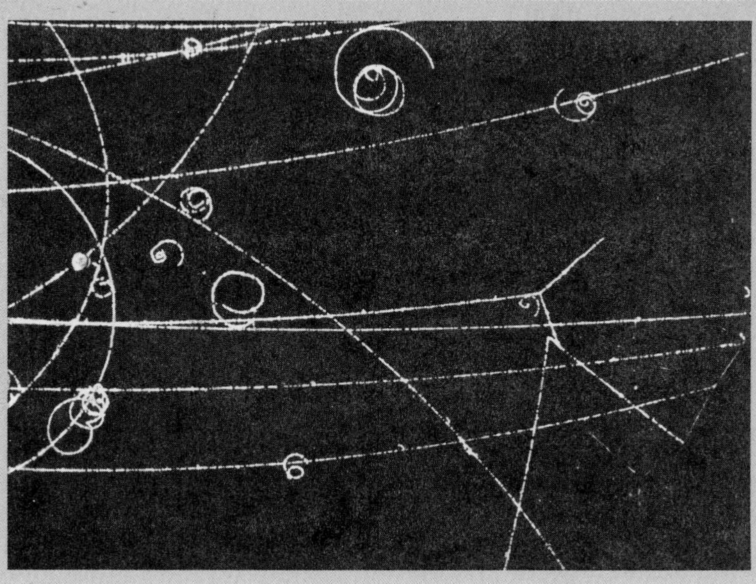

In this chapter we want to convince you that a moving particle can sometimes be represented as a wave. The tracks in the photo are strings of bubbles left in the wake of charged particles as they pass through a bubble chamber. It may seem unlikely that these tracks can be interpreted in terms of a wave picture. However, they can, and we shall explain how at the end of this chapter.

44-1 Louis Victor de Broglie Makes a Suggestion

Physicists have rarely gone wrong in banking on the underlying symmetry of nature. For example, when you learn that a changing magnetic field produces an electric field it is a good bet to guess — and it turns out to be true — that a changing electric field will produce a magnetic field. For another example: The electron has an *antiparticle,* that is, a particle of the same mass but of opposite charge. Is it not reasonable to expect that the proton should also have an antiparticle? A 5-GeV proton accelerator was built at the University of California at Berkeley to search for the antiproton. It was found.

In 1924 Louis de Broglie,* a physicist and a member of a distinguished French aristocratic family, puzzled over the fact that light seemed to have a dual wave-particle aspect but matter — at that time — seemed to be entirely particlelike. This in the face of the fact that both light and matter are forms of energy, that each can be transformed into the other and that they are both governed by the spacetime symmetries of the theory of relativity. He then began to think that matter should also have a dual character and that particles such as electrons should have wavelike properties.

* Pronounced "de Broley" by most American physicists.

If we want to describe a moving particle as a wave, our first task is to answer the question: What is its wavelength? De Broglie made the bold suggestion that the relation

$$\lambda p = h \tag{1}$$

applies both to light and to matter. If we solve Eq. 1 for p we have

$$\boxed{p = \frac{h}{\lambda}} \quad \text{(momentum of a photon),} \tag{2}$$

which we have used in connection with the Compton effect (see Section 43–4) to assign a momentum to a photon of known wavelength.

If we solve Eq. 1 for λ we have

$$\boxed{\lambda = \frac{h}{p}} \quad \text{(wavelength of a particle),} \tag{3}$$

which (said de Broglie) we can use to assign a wavelength to a particle of known momentum. A wavelength calculated from Eq. 3 is called a *de Broglie wavelength.* Look again at Eq. 1 and note the central role played by the Planck constant h in connecting the wave and the particle aspects of both light and matter.

Sample Problem 1 What is the de Broglie wavelength of an electron whose kinetic energy is 120 eV?

We can find the de Broglie wavelength from Eq. 3 if we know the momentum of the electron. For such relatively slow electrons, the momentum is related to the kinetic energy by

$$\begin{aligned} p &= \sqrt{2mK} \\ &= \sqrt{(2)(9.11 \times 10^{-31}\ \text{kg})(120\ \text{eV})(1.60 \times 10^{-19}\ \text{J/eV})} \\ &= 5.91 \times 10^{-24}\ \text{kg}\cdot\text{m/s}. \end{aligned}$$

From Eq. 3 then

$$\begin{aligned} \lambda &= \frac{h}{p} = \frac{6.63 \times 10^{-34}\ \text{J}\cdot\text{s}}{5.91 \times 10^{-24}\ \text{kg}\cdot\text{m/s}} \\ &= 1.12 \times 10^{-10}\ \text{m} = 112\ \text{pm}. \qquad \text{(Answer)} \end{aligned}$$

This is about the size of a typical atom. As we shall see, we can take advantage of this fact to verify the wave nature of such slow electrons experimentally in the laboratory.

Sample Problem 2 What is the de Broglie wavelength of a 150-g baseball traveling at 35 m/s?

From Eq. 3 we have

$$\begin{aligned} \lambda &= \frac{h}{p} = \frac{h}{mv} = \frac{6.63 \times 10^{-34}\ \text{J}\cdot\text{s}}{(150 \times 10^{-3}\ \text{kg})(35\ \text{m/s})} \\ &= 1.26 \times 10^{-34}\ \text{m}. \qquad \text{(Answer)} \end{aligned}$$

Because there is no hope of measuring such a small length, we do not speak of the wave nature of objects as large as baseballs.

44-2 Testing de Broglie's Hypothesis

If you want to prove that you are dealing with a wave a convincing thing to do is to measure the wavelength in the laboratory. That is what Thomas Young did in 1801 to establish the wave nature of visible light; it is what Max von Laue did in 1912 to establish the wave nature of x rays.

To measure a wavelength, you need two or more diffracting centers (pinholes, slits, or atoms) separated by a distance that is about the same size as the wavelength you are trying to measure. Sample Problem 2 shows at once that it is hopeless to try to measure the wavelength of a pitched baseball. You would need a pair of "slits" spaced $\sim 10^{-34}$ m apart! That is why our daily experiences with large moving objects give no clues to the wave nature of matter. Sample Problem 1, however, suggests that we *should* be able to measure the wavelength of a moving electron, using the atoms of a crystal as a three-dimensional diffraction grating.

The Davisson-Germer Experiment. Figure 1 shows the apparatus used by C. J. Davisson and L. H. Germer of what is now the AT&T Bell Laboratories to measure the de Broglie wavelength of slow electrons. In 1937 Davisson shared the Nobel prize in physics for this work, one of seven such prizes awarded (as of 1988) to scientists associated with this remarkable laboratory.

In the apparatus of Fig. 1 electrons from heated filament F are accelerated by an adjustable potential difference V. The resultant beam, made up of electrons whose kinetic energy is eV, is then allowed to fall on a crystal C which—in their experiment—was of nickel. The beam, "reflected" from the crystal surface, enters detector D and is recorded as a current I.

The experimenters then set V to a particular arbitrary value and read the detector current I for various angular settings (ϕ) of the detector. They then set V to another value and repeated the angular sweep of the detector. Figure 2 is a polar plot of $I(\phi)$ for $V = 54$ V; a strong diffracted beam at $\phi = 50°$ is evident. If the accel-

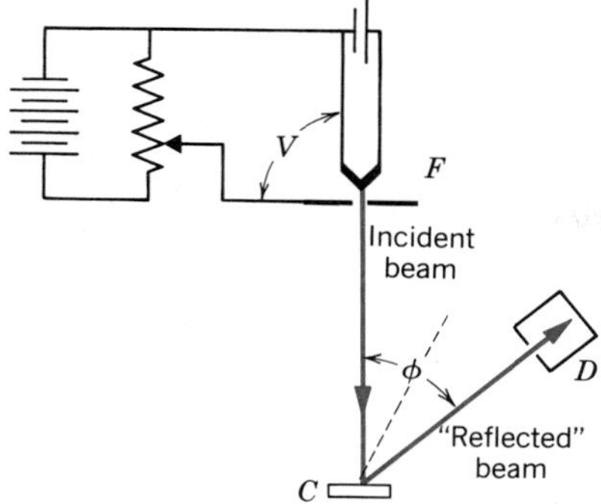

Figure 1 The apparatus of Davisson and Germer, used to demonstrate the wave nature of electrons. The electrons, emitted from heated filament *F*, are accelerated by an adjustable potential difference *V*. After reflection from crystal *C* they are recorded by detector *D*, which can be set to various angular positions ϕ.

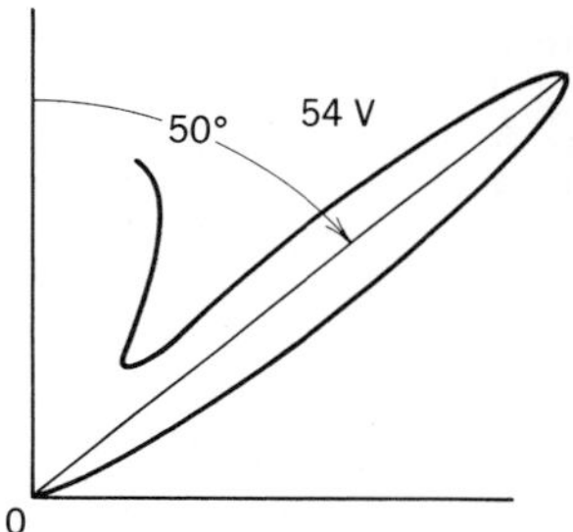

Figure 2 *V* in Fig. 1 is set to 54 V. The current *I* recorded by detector *D* in the apparatus of Fig. 1 is then plotted radially from the origin against the angle ϕ in a polar plot. The sharp diffraction maximum occurs for $\phi = 50°$. If *V* is changed from its set value of 54 V, the intensity of the diffraction maximum decreases.

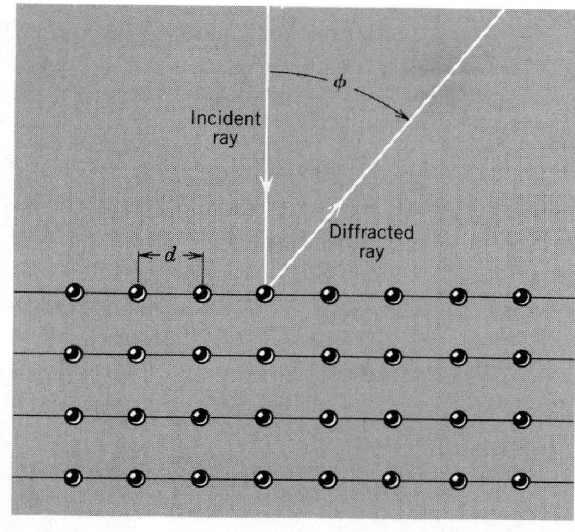

Figure 3 A schematic view of the atoms that make up crystal *C* in Fig. 1. The crystal behaves like a diffraction grating, the lines of atoms on its surface being separated by a distance *d*. For the crystal of Davisson and Germer, $d = 215$ pm.

erating potential is either increased slightly or decreased slightly, the intensity of the diffracted beam drops.

Figure 3 is a simplified representation of the nickel crystal *C* of Fig. 1. The diffracted beam is formed by Bragg reflection of the electron matter wave from a particular family of atomic planes within the crystal. However, (except for normal incidence) the electron beam bends as it enters the crystal surface and also as it leaves; thus its angle of reflection within the crystal is not the same as angle ϕ of Fig. 3, which is measured outside the crystal. When this surface bending is taken into account it can be shown* that the crystal behaves like a two-dimensional diffraction grating, its grating lines being the parallel lines of atoms lying on the crystal surface and its grating spacing being the interval marked *d* in Fig. 3. The principal maxima for such a grating must satisfy Eq. 14 of Chapter 41, or

$$d \sin \phi = m\lambda \qquad m = 0, 1, 2, 3, \ldots \tag{4}$$

For their crystal Davisson and Germer knew that $d = 215$ pm. For $m = 1$, which corresponds to a first-order diffraction peak, Eq. 4 leads to

$$\lambda = \frac{d \sin \phi}{m} = \frac{(215 \text{ pm})(\sin 50°)}{1} = 165 \text{ pm}.$$

The expected de Broglie wavelength for a 54-eV electron, calculated as in Sample Problem 1, is 167 pm, in good agreement with the measured value. De Broglie's prediction is confirmed.

G. P. Thomson's Experiment. In 1927 George P. Thomson, working at the University of Aberdeen in Scotland, independently confirmed de Broglie's equation, using a somewhat different method. As Fig. 4*a* shows, he directed a monoenergetic beam of 15-keV electrons through a thin metal target foil. The target was specifically *not* a single large crystal (as in the Davisson-Germer experiment) but was made up of a large number of tiny, randomly oriented crystallites. With this arrangement there will always, by chance, be a certain

* See *Basic Concepts in Relativity and Early Quantum Theory*, 2nd ed., Robert Resnick and David Halliday (John Wiley & Sons, New York, 1985), Supplementary Topic H.

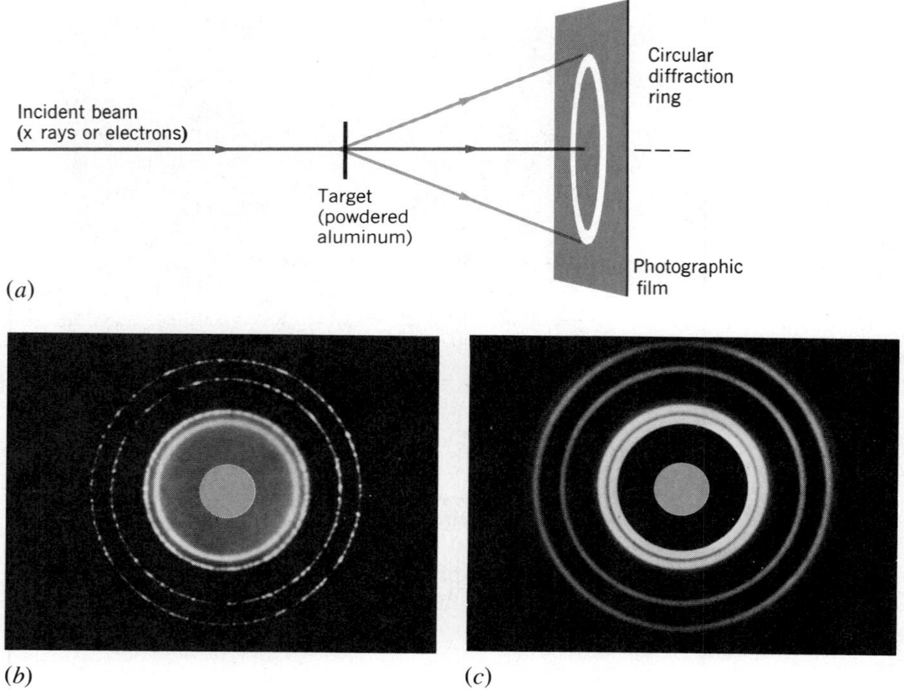

Figure 4 (*a*) An experimental arrangement used to demonstrate a diffraction pattern characteristic of the atomic arrangements in a target of powdered aluminum. (*b*) The diffraction pattern if the incident beam in (*a*) is an x-ray beam. (*c*) The diffraction pattern if the incident beam is an electron beam. Note that the main features of the two patterns in (*b*) and (*c*) are the same. The electron energy was chosen so that the de Broglie wavelength of the electrons in (*c*) matched the wavelength of the incident x rays in (*b*).

number of crystallites oriented at the proper angle to produce a diffracted beam.

If a photographic plate is placed at right angles to the incident beam, as in Fig. 4*a*, the central beam spot will be surrounded by diffraction rings. Figure 4*c* shows this pattern for a target of powdered aluminum. Figure 4*b* shows the pattern when the electron beam is replaced with an x-ray beam of the same wavelength. A simple glance at these two diffraction patterns leaves no doubt that both originated in the same way. Measurement and analysis of the patterns confirms de Broglie's hypothesis in every detail.

Thomson shared the 1937 Nobel prize with Davisson for his electron diffraction experiments. George P. Thomson was the son of J. J. Thomson, who won the Nobel prize in 1906 for his discovery of the electron and for his measurement of its charge-to-mass ratio. It has been written that:

> . . . *one may feel inclined to say that Thomson, the father, was awarded the Nobel prize for having shown that the electron is a particle, and Thomson, the son, for having shown that the electron is a wave.*

Matter Waves: Some Applications. Today the wave nature of matter is taken for granted and diffraction studies by beams of electrons or neutrons are used routinely to study the atomic structures of solids and liquids. Figure 5 shows a commercial electron diffraction apparatus, of a kind found in many chemical, physical, and metallurgical analytical laboratories.

Matter waves are a valuable supplement to x rays in studying the atomic structures of solids. Electrons, for example, are less penetrating than x rays and so are particularly useful for studying surface features. Again, x rays interact largely with the electrons in a target and for

Figure 5 The central unit of a commercial apparatus for analyzing the atomic structures of solids by low energy electron diffraction (LEED). Pumps for creating the necessary vacuum, the power supply, and the data recording instrumentation are not shown. (If electrons did not have a wavelike aspect, this equipment would not work.)

that reason it is not easy to use them to locate light atoms—particularly hydrogen—which have few electrons. Neutrons, on the other hand, interact largely with the nucleus of the atom and can be used to fill this gap. Figure 6 shows the structure of solid benzene as deduced from neutron diffraction studies. The six carbon atoms that form the familiar benzene ring are clearly there to see, as are the six hydrogen atoms that couple to them.

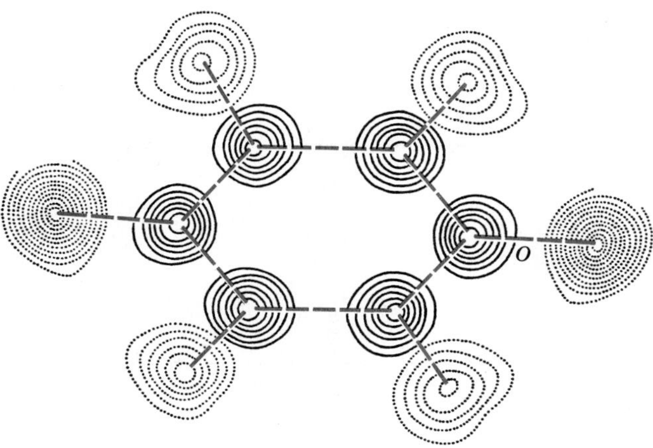

Figure 6 The atomic structure of solid benzene as deduced by neutron diffraction methods. The familiar benzene ring of six carbon atoms is clearly present, as are the hydrogen atoms coupled to them. The lines are "contour lines" that identify the electron density at various positions.

44-3 The Wave Function

When Thomas Young measured the wavelength of light in 1801 he had no idea of the nature of the beam of sunlight that fell on the two pinholes in his interference apparatus. It was more than half a century later that Maxwell filled that gap by postulating that light is a traveling configuration of electric and magnetic fields.

We are in exactly the same situation at this stage of our introduction to matter waves. We can measure the wavelength associated with an electron or a neutron in a beam of those particles but—to put it loosely—we do not know what is waving. That is, we do not know what quantity in a matter wave corresponds to the electric field in an electromagnetic wave, to the transverse displacement in a wave traveling along a stretched string, or to the local pressure variation in a sound wave traveling through an air-filled pipe.

For the time being we shall call the quantity whose variation with position and time represents the wave aspect of a moving particle simply its *wave function* and shall assign it the symbol* ψ. Later, we shall interpret the wave function in a physical way by developing the analogy:

$$\boxed{\text{light wave : photon : : matter wave : particle.}} \quad (5)$$

First, let us develop a useful theorem that applies to waves of all kinds. When we studied waves on strings, we saw that you can propagate a *traveling* wave of *any* wavelength down a stretched string of *infinite* length. However, if you deal with a stretched string of *finite* length, only *standing* waves can be set up and these occur only at a *discrete set* of wavelengths. We summarize this general experience with waves by saying:

> *Localizing the extent of a wave in space has the result that only a discrete set of wavelengths—and, correspondingly, a discrete set of frequencies—can occur. That is,* localization leads to quantization.

This theorem holds not only for waves in strings but for waves of all kinds, including electromagnetic waves and—as we shall see—matter waves.

* The lowercase symbol ψ refers only to the space-varying portion of the wave function. That is the only portion that concerns us in this chapter.

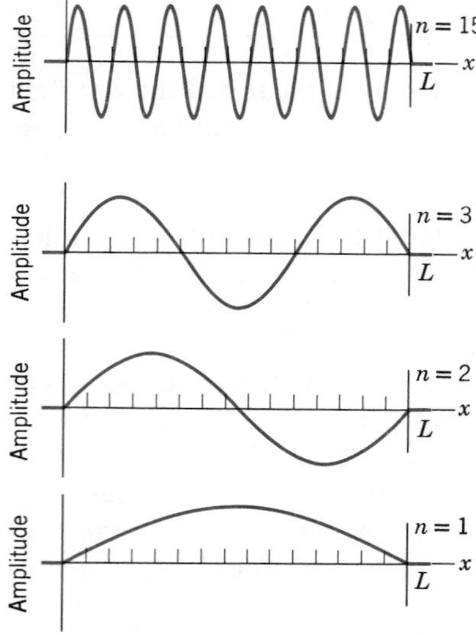

Figure 7 Four standing wave patterns for a stretched string of length L, clamped rigidly at each end. The label n appears in Eqs. 6 and 7. The pattern for $n = 1$ represents the largest possible wavelength (and the lowest possible frequency) at which the string can oscillate. We can view these patterns as *stationary states* of the vibrating string, occurring at frequencies that are *quantized* according to Eq. 7, n being a *quantum number.*

Figure 7 shows a few of the standing wave patterns that can exist when a stretched string is restricted to a finite length L, its ends secured in rigid clamps. As we learned in Section 17–12, each such pattern has a well-defined wavelength, given by

$$\lambda = \frac{2L}{n} \qquad n = 1, 2, 3, \ldots \tag{6}$$

in which the integer n, which is marked on Fig. 7, defines the oscillation mode. We shall come to call such integers *quantum numbers.* The frequencies corresponding to these wavelengths are also quantized and are given by

$$\nu = v/\lambda = (v/2L)n \qquad n = 1, 2, 3, \ldots \tag{7}$$

in which v is the wave speed.

44-4 Light Waves and Photons

We can set up one-dimensional standing electromagnetic waves, exactly like those shown in Fig. 7 for a stretched string, by trapping some radiation between two parallel, perfectly reflecting mirrors. In the visible or near-visible region, standing waves set up in the cavity of a gas laser serve nicely as an example. We can also set up such standing waves in the microwave region, using parallel copper sheets as mirrors.

For convenience in what follows, we shall deal only with the mode of oscillation that has the longest wavelength and, correspondingly, the lowest frequency. Figure 8*a* (a copy of the curve marked $n = 1$ in Fig. 7) shows a plot of the wave amplitude E_{max} as a function of position for this mode. We see that exactly half a wave fits between the walls, so that the wavelength λ is $2L$.

Figure 8*b* shows a plot of $E^2_{max}(x)$ for this same oscillation mode. In view of Eq. 23 of Chapter 27 ($u = \frac{1}{2}\epsilon_0 E^2$), we can also interpret Fig. 8*b* as a plot of the *energy density* in the standing electromagnetic wave. If we think in terms of photons, each of which carries the same energy $h\nu$, we conclude that the square of the wave amplitude at any point in a standing electromagnetic wave is proportional to the density of photons at that point, a conclusion that you can test by exploring the region between the mirrors of Fig. 8*b* with a photon probe. You would find a maximum density of photons halfway between the mirrors in Fig. 8*b* and a density approaching zero just in front of each mirror.

If the total energy in the standing wave pattern is so low that it corresponds to the energy of a single photon, you would be led to conclude:

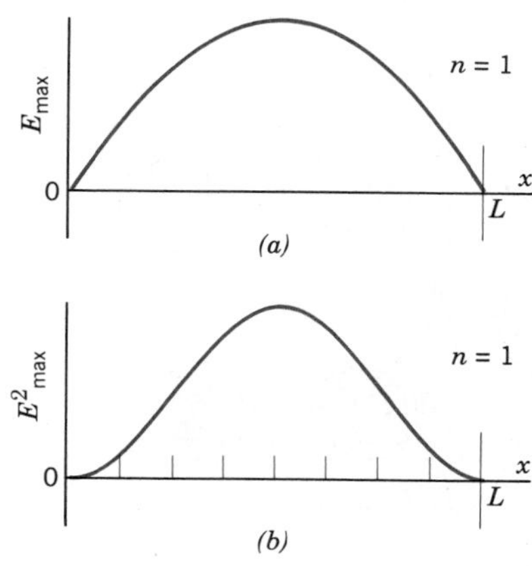

Figure 8 Light is trapped between two parallel mirrors separated by a distance L, forming a standing wave pattern. (*a*) The amplitude of oscillation for the lowest-frequency oscillation mode, corresponding to $n = 1$ in Fig. 7. (*b*) The square of this amplitude, which is proportional at any point to the density of photons at that point.

The probability of detecting a photon at any location is proportional to the square of the amplitude of the electromagnetic wave at that location.

Note that our knowledge of the photon position is inherently statistical. That is, we cannot say exactly where a photon is at a given moment; we can only speak of the relative probability that a photon will be in a certain region of space. As we shall learn, this statistical limitation is fundamental for both light and matter, that is, for both photons and particles.

44-5 Matter Waves and Electrons

In considering the relation between matter waves and particles, we will use the electron as a prototype and will be guided by the analogy between light waves and photons that we developed in the previous section.

How might we set up a standing matter wave? Recalling the localization-quantization theorem of Section 44-3, we are led to believe that we should confine the electron, by electric forces, to a certain region of space. The matter waves associated with the electron should then occur as a set of patterns of standing matter waves, each at a specified frequency.

Atoms are such electron traps. In fact, most of the electrons in the atoms that make up our planet and the living things that inhabit it have been so trapped since before the solar system was formed. It is also possible with modern technology to build up—atom layer by atom layer—solid electron traps whose dimensions are as small as a few atomic diameters. Such *quantum well structures* have many practical applications, in light-wave communication devices and as optical logic gates, for two examples.

For our purpose, however, we imagine the simple case of a hypothetical one-dimensional electron trap, in which a single electron is confined by electrical forces to move back and forth between two rigid "walls" separated by a distance L. Within the trap, we assume that the electron experiences no force. Figure 9*a* suggests the potential energy of the electron bound in a trap of this kind. We note that $U = 0$ within the trap* and that U rises rapidly to an infinitely great value at $x = 0$ and $x = L$. The "trap" of Fig. 9*a* is more formally known as an infinitely deep potential well, or an *infinite well*, which is what we shall call it from now on.

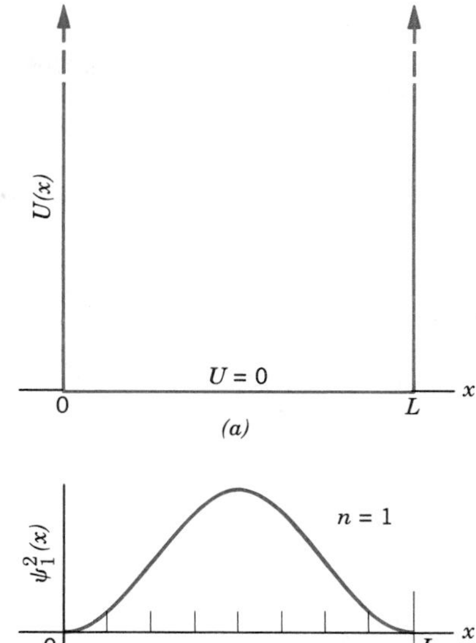

Figure 9 (*a*) An infinite well of width L. (*b*) The probability density for an electron trapped in this well; the electron is in its ground state, corresponding to $n = 1$ in Fig. 7 and Fig. 8.

Figure 9*b*—which is identical with Fig. 8*b* except that it applies to matter waves instead of to light waves—shows the square of the amplitude of the wave function of the matter wave for a single trapped electron in its $n = 1$ state. Reasoning by analogy with light waves and photons, we conclude that:

The probability of finding the electron at any given location is proportional to the square of the amplitude of the matter wave at that location.

In particular, the quantity $\psi^2(x)\,dx$ is proportional to the probability of finding the electron in the interval that lies between x and $x + dx$. For our purposes, the square of the wave function—which we call the *probability density*—is more important than the wave function itself because it tells us where the electron is likely to be. The probability that the electron will be *somewhere* in the infinite well of Fig. 9*a* is unity, which represents a certainty. Thus we have

$$\int_0^L \psi^2(x)\,dx = 1. \qquad (8)$$

* Recall that our choice of a configuration to which we assign zero potential is arbitrary; only potential differences count.

The integral in Eq. 8 is simply the area under the curve of Fig. 9*b*. We see that this area has the numerical value of unity. When a probability density obeys a relation such as Eq. 8, we say that the probability density is *normalized*.

The Energies of the Allowed States. Figure 10 shows the probability densities for four of the allowed standing matter wave patterns, corresponding to the four oscillation modes of Fig. 7. Let us find the energies of these allowed states.

Figure 9*a* shows us that the potential energy of the trapped electron is constant within the infinite well and has the value zero. Thus the total electron energy is equal to its kinetic energy, and we have

$$E = K = \frac{p^2}{2m}. \tag{9}$$

We find the momentum of the trapped electron from Eq. 2 ($p = h/\lambda$). The wavelength of an electron in a particular state is related to the quantum number n of that state by Eq. 6 ($\lambda = 2L/n$). Identifying λ as a de Broglie wavelength, we have

$$p = \frac{h}{\lambda} = \frac{hn}{2L}.$$

The total energy is then given from Eq. 9 as

$$\boxed{E_n = n^2 \frac{h^2}{8mL^2} \qquad n = 1, 2, 3, \ldots} \tag{10}$$

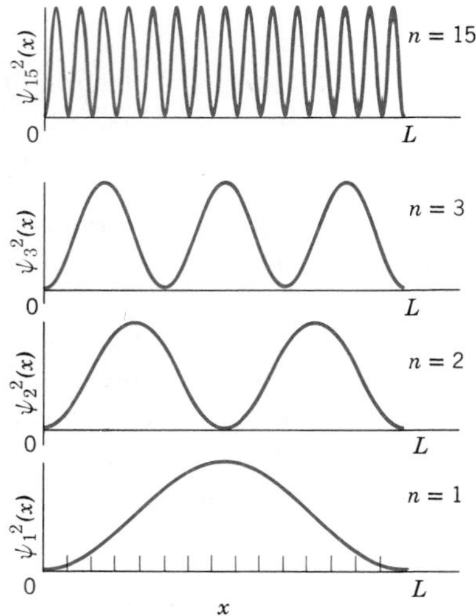

Figure 10 The probability density for four states of an electron trapped in the infinite well of Fig. 9*a*. The quantum numbers are indicated.

We see that the state with $n = 1$, whose probability density is sketched in Fig. 9*b*, is the state with the lowest total energy; we call it the *ground state.*

The Zero-Point Energy. We note that, contrary to classical expectation, *the electron cannot be at rest in its well.* We have just seen that its lowest energy is its ground state energy, corresponding to $n = 1$ in Eq. 10, or

$$E_1 = \frac{h^2}{8mL^2} \quad \text{(zero point energy).} \tag{11}$$

Even at the absolute zero of temperature, this expression for the *zero point energy* holds.

Equation 11 tells us that we can make the zero point energy as small as we like by making the well wider, that is, by increasing L. In the limit of $L \to \infty$, which corresponds to a free particle, the zero point energy approaches zero. We also see from Eq. 11 that, if we lived in a world (not ours!) in which the Planck constant were zero, there would be no such thing as a zero point energy and the electron could indeed be at rest in its well. This involvement of the Planck constant shows us that the phenomenon of a zero point energy—which turns out to be completely general—is strictly a quantum phenomenon.

Sample Problem 3 An electron is confined in an infinite well whose width L is 120 pm, about the diameter of an atom. What are the energies of the four states whose probability densities are displayed in Fig. 10?

From Eq. 10, with $n = 1$, we have

$$\begin{aligned} E_n &= n^2 \frac{h^2}{8mL^2} \\ &= (1)^2 \frac{(6.63 \times 10^{-34}\ \text{J}\cdot\text{s})^2}{(8)(9.11 \times 10^{-31}\ \text{kg})(120 \times 10^{-12}\ \text{m})^2} \\ &= 4.19 \times 10^{-18}\ \text{J} = 26.2\ \text{eV}. \qquad \text{(Answer)} \end{aligned}$$

The energies of the state with $n = 2$ is $2^2 \times 26.2$ eV or 105 eV. Similarly, the energies of the states with $n = 3$ and $n = 15$ are, respectively, 236 eV and 5900 eV.

Sample Problem 4 A 1.5-μg speck of dust moves back and forth between two rigid walls separated by 0.10 mm. It moves so slowly that it takes 120 s for the particle to cross this gap. Let us view this motion as that of a particle trapped in an infinite well. What quantum number describes the motion?

Solving Eq. 10 for n yields

$$n = \sqrt{\frac{8mEL^2}{h^2}}.$$

The energy of the particle is entirely kinetic. Noting that the speed of the particle is

$$v = \frac{0.10 \times 10^{-3}\text{ m}}{120\text{ s}} = 8.33 \times 10^{-7}\text{ m/s},$$

we find the energy to be

$$E\,(=K) = \tfrac{1}{2}mv^2 = (\tfrac{1}{2})(1.5 \times 10^{-9}\text{ kg})(8.33 \times 10^{-7}\text{ m/s})^2$$
$$= 5.2 \times 10^{-22}\text{ J}.$$

The quantum number n is then

$$n = \sqrt{\frac{8mEL^2}{h^2}}$$
$$= \sqrt{\frac{(8)(1.5 \times 10^{-9}\text{ kg})(5.2 \times 10^{-22}\text{ J})(0.10 \times 10^{-3}\text{ m})^2}{(6.63 \times 10^{-34}\text{ J}\cdot\text{s})^2}}$$
$$= 3.8 \times 10^{14}. \quad \text{(Answer)}$$

This is a very large number indeed. It is impossible to distinguish experimentally between $n = 4 \times 10^{14}$ and $(4 \times 10^{14}) + 1$. Even this tiny speck of dust is a gross macroscopic object when compared to an electron. Quantum physics and classical physics give the same answers in this problem. We are in a region governed by the correspondence principle.

44-6 The Hydrogen Atom

Let us now extend what we have learned about an electron trapped in an infinite well to the more realistic case of an electron trapped in an atom. We choose the simplest atom, hydrogen.

The hydrogen atom consists of a single electron, bound to its nucleus (a single proton) by the attractive Coulomb force. The potential energy function $U(r)$ for this system is

$$U(r) = -\frac{1}{4\pi\epsilon_0}\frac{e^2}{r}, \tag{12}$$

in which e is the magnitude of the charge of the electron and the proton and r is the distance between them. Figure 11 is a sketch of Eq. 12.

This hydrogen atom trap, unlike the (one-dimensional) infinite well of Fig. 9*a*, is three-dimensional. It has spherical symmetry, so that the potential energy depends on only one variable—the separation r between the electron and the (relatively massive) central proton. Note also that the potential energy given by Eq. 12 is negative for all values of r. This comes about because we have (arbitrarily) chosen our zero of potential to correspond to $r = \infty$. In Fig. 9*a*, on the other hand, we (arbitrarily) chose to assign $U(x) = 0$ to the region inside the well.

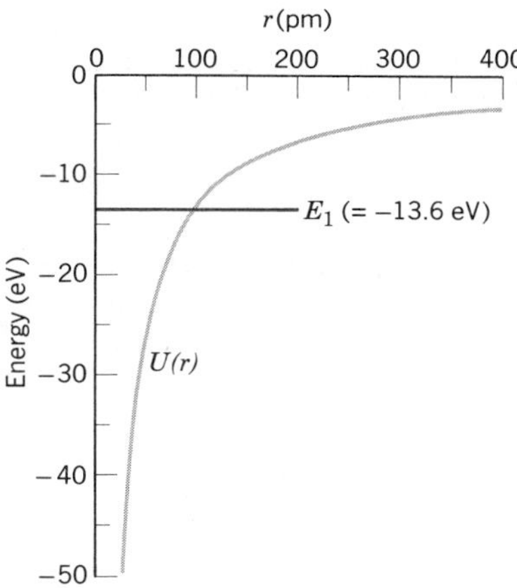

Figure 11 The potential well that governs the motion of the electron in the hydrogen atom. Compare it with the infinite well of Fig. 9*a*. The horizontal line shows the state of lowest energy (the *ground state*), corresponding to $n = 1$ in Eq. 13.

Wave mechanics* yields, for the energies of the allowed states of the hydrogen atom,

$$E_n = -\left(\frac{me^4}{8\epsilon_0^2h^2}\right)\frac{1}{n^2} \qquad n = 1, 2, 3, \ldots \tag{13}$$

Compare this relation with Eq. 10, which gives the energies for the allowed states of an electron trapped in the infinite well of Fig. 9*a*. We see that the hydrogen atom also exhibits a zero point energy, corresponding to $n = 1$ in Eq. 13; all other value of n yield higher (that is, less negative) energies.

The probability density for the ground state of the hydrogen atom is given by

$$\psi^2(r) = \frac{1}{\pi r_B^3}\,e^{-2r/r_B}, \tag{14}$$

in which r_B is the *Bohr radius,* a convenient measure of distance on the atomic scale, having the value (see Section 43-10) of

$$r_B = 5.29 \times 10^{-11}\text{ m} = 52.9\text{ pm}. \tag{15}$$

The physical meaning of Eq. 14 is that $\psi^2(r)dV$ is proportional to the probability that the electron will be found in any specified infinitesimal volume element dV.

* *Wave mechanics* is one of several equivalent forms of what we have earlier called quantum physics.

Because the probability density depends only on r, it makes sense to choose as a volume element the volume between two spherical shells whose radii are r and $r + dr$. That is, we define a volume element dV as

$$dV = (4\pi r^2)(dr). \tag{16}$$

We now define a *radial probability density* $P(r)$ such that $P(r)\,dr$ gives the probability that we will find the electron in the volume element defined by Eq. 16. Thus, from Eqs. 14 and 16,

$$P(r)\,dr = \psi^2(r)\,dV = \frac{4}{r_B^3}\,r^2\,e^{-2r/r_B}\,dr \tag{17}$$

or

$$\boxed{P(r) = \frac{4}{r_B^3}\,r^2\,e^{-2r/r_B}.} \tag{18}$$

Figure 12 shows a plot of Eq. 18. We can show (see Problem 34) that

$$\int_0^\infty P(r)\,dr = 1 \tag{19}$$

so that the area under the curve of Fig. 12 is unity; this ensures that the electron in the hydrogen atom must lie *somewhere* between the limits of zero and infinity.

In the semiclassical theory of Bohr, the electron in its ground state simply revolved in a circular orbit of radius r_B. In wave mechanics, however, we discard this mechanical picture. We think instead of the hydrogen atom as a tiny nucleus surrounded by a *probability cloud* whose value $P(r)$ at any point is given by Eq. 18. We do not ask: "Is the electron near this point?" but "What are the odds that the electron is near this point?" This probabilistic information is all that we can ever learn about the electron; as it turns out, it is also all that we ever need to know.

As Fig. 12 shows, the radial probability density—interestingly enough—has its maximum value at the classical Bohr radius. The odds are that the electron will be farther away from the nucleus than this value 68% of the time and that it will be closer the remaining 32% of the time.

It is not easy for a beginner to look at subatomic particles in this probabilistic and statistical way. The difficulty is our natural impulse to regard an electron as something like a tiny marble or a tiny jelly bean, being at a certain place at a certain time and following a well-defined path. Electrons and other subatomic particles simply do not behave in this way. In the sections that follow we will do what we can to help dispel this pervasive *jelly bean fallacy,* as we may call it.

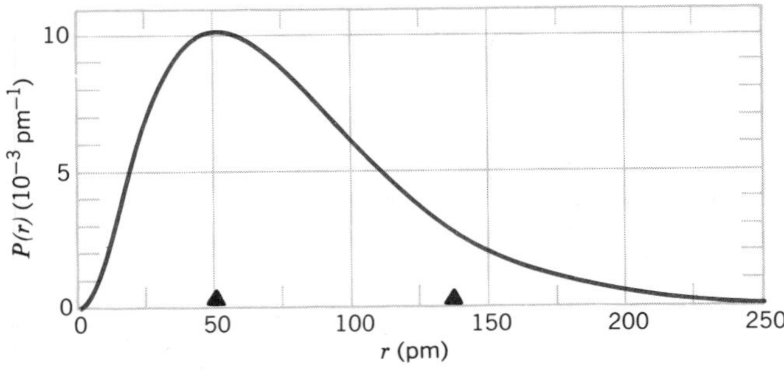

Figure 12 The radial probability density for the hydrogen atom in its ground state. Note that the electron is more likely to be found close to the Bohr radius ($r = 52.9$ pm) than to any other position. The radius of the 90% sphere is also marked, at 2.67 Bohr radii or 141 pm. The electron spends 90% of its time inside a sphere of this radius.

Sample Problem 5 Show that the maximum value of the radial probability density falls at $r = r_B$, where r_B is the Bohr radius.

The radial probability density is given by Eq. 18

$$P(r) = \frac{4}{r_B^3}\,r^2\,e^{-2r/r_B}.$$

If we differentiate with respect to r we find, using the rule for products,

$$\frac{dP}{dr} = \frac{4}{r_B^3}\,r^2\,(-2/r_B)\,e^{-2r/r_B} + \frac{4}{r_B^3}\,(2r)\,e^{-2r/r_B}$$

$$= \frac{8}{r_B^4}\,r\,(r_B - r)\,e^{-2r/r_B}.$$

At the maximum of the curve we must have $dP/dr = 0$ and, as inspection of the above shows, this does indeed occur at $r = r_B$, which is what we sought to prove. Note that we also have $dP/dr = 0$ at $r = 0$ and at $r \rightarrow \infty$. However, these conditions are quite consistent with Fig. 12 and do not correspond to maxima.

Sample Problem 6 In Problem 35 you are asked to prove that the probability $p(r)$ that the electron in the ground state of the hydrogen atom will be found inside a spherical shell of radius r is given by

$$p(r) = 1 - e^{-2x}(1 + 2x + 2x^2),$$

in which x, a dimensionless quantity, is equal to r/r_B. Find r for $p(r) = 0.90$.

We seek the radius of a sphere for which $p(r)$ is 0.90; see Fig. 12. From the foregoing expression for $p(r)$ we have

$$0.90 = 1 - e^{-2x}(1 + 2x + 2x^2)$$

or

$$10e^{-2x}(1 + 2x + 2x^2) = 1$$

and we must find the value of x that satisfies this equality. It is not possible to solve explicitly for x but a little trial and error with a pocket calculator (write a small program for it) quickly yields $x = 2.67$. This means that the radius of a sphere such that the probability that the electron will be inside 90% of the time is 2.67 Bohr radii.

44-7 Barrier Tunneling

Figure 13*a* sets the stage for an interesting quantum surprise. It shows a barrier, of height U and thickness L. An electron of total energy E approaches the barrier from the left. Classically, because $E < U$, the electron would be reflected from the barrier and would move back in the direction from which it came. In wave mechanics. however, there is a finite chance that the electron will appear on the other side of the barrier and continue its motion to the right.

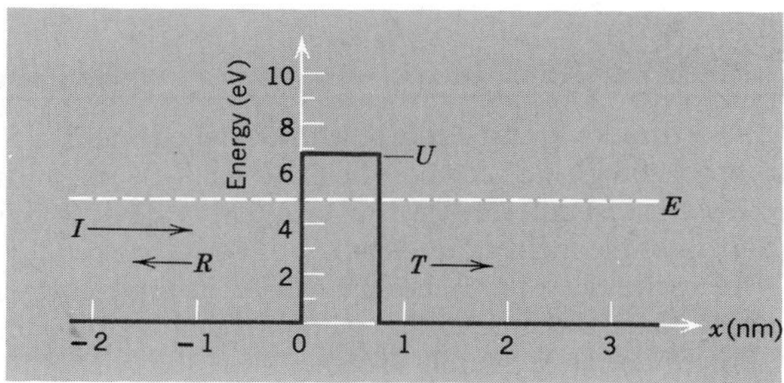

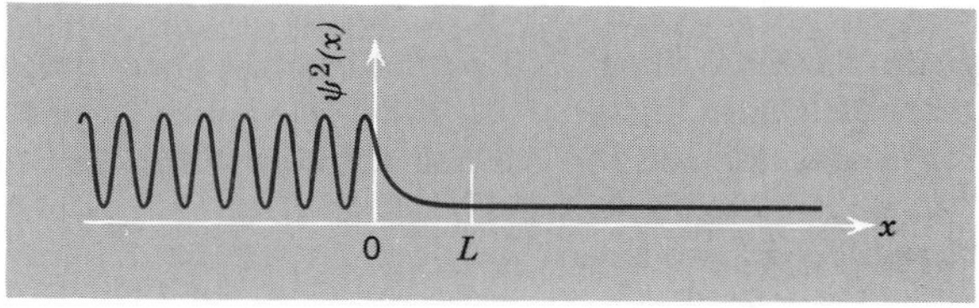

Figure 13 (*a*) An electron of total energy E approaches a potential barrier from the left. It has a probability R of being reflected from the barrier and a probability T of being transmitted through it, by the wave-mechanical barrier tunneling. (*b*) The probability density for the matter wave describing the electron in (*a*). The "fringes" on the left represent interference between the incident and the reflected matter waves.

It is as if you tossed a jelly bean at a window pane and—to your surprise—it materialized on the other side with the glass unbroken. Don't expect this to happen for jelly beans. However, electrons are not jelly beans and such *barrier tunneling,* as it is called, certainly *does* happen for electrons.

We can assign a reflection coefficient R and a transmission coefficient T to the incident electron in Fig. 13*a*, the sum of these two quantities necessarily being unity. Thus, if $T = 0.02$, of every 1000 electrons fired at the barrier, 20 (on average) will tunnel through it and 980 will be reflected.

Let us represent the electron by a matter wave. Figure 13*b* shows the appropriate probability density curve. To the left of the barrier, there is an incident matter wave moving to the right and a (somewhat less intense) reflected matter wave moving to the left. These two waves interfere, producing the fringe pattern that we see in that region of Fig. 13*b*. Within the barrier the probability density decreases exponentially. On the far side of the barrier we have only a matter wave traveling to the right, with the reduced but constant amplitude shown.

From wave mechanics, the transmission coefficient T can be shown to be given approximately (for small values of T) by

$$T = e^{-2kL}, \tag{20}$$

in which

$$k = \sqrt{\frac{8\pi^2 m(U - E)}{h^2}}. \tag{21}$$

The value of T is very sensitive to the energy of the incident particle and to the height and width of the barrier.

Barrier tunneling is a puzzle only if you cling to the jelly bean fallacy. The proper way to look at barrier tunneling is to think of it as a matter wave problem, the wave being related to the electron only in a probabilistic sense. That is, if the probability density on the far side of the barrier is not zero, there is a specific probability that you will find the electron there if you look for it.

For a simple example of barrier tunneling, consider a bare copper wire that has been cut and the two ends twisted together. It still conducts electricity readily, in spite of the fact that the wires are coated with a thin layer of copper oxide, an insulator. The electrons simply tunnel through this thin insulating barrier.

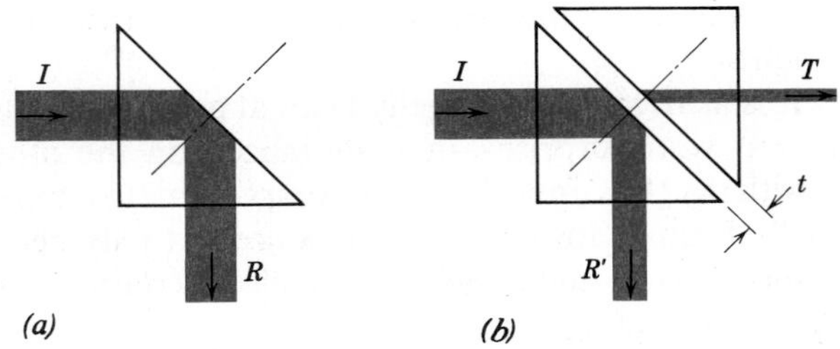

Figure 14 (*a*) A light wave is totally reflected from a glass-air interface. (*b*) If a second glass prism is held close to the first, the light wave can tunnel through the barrier associated with the air film between them. For this to happen, the thickness t of the air film must be no more than a few wavelengths of the incident light.

Among other examples, we may list the tunnel diode, in which the flow of electrons (by tunneling) through a device can be rapidly turned on or off by controlling the height of the barrier. This can be done very quickly (within 5 ps) so that the device is suitable for applications where high-speed response is critical. The 1973 Nobel prize was shared by three "tunnelers," Leo Esaki (tunneling in semiconductors), Ivar Giaever (tunneling in superconductors), and Brian Josephson (the Josephson junction, a quantum switching device based on tunneling). The 1986 Nobel prize was also awarded (to Gerd Binnig and Heinrich Rohrer) to recognize a device based on tunneling, the *scanning tunneling microscope.** In later chapters we shall learn of the importance of tunneling in understanding certain kinds of radioactive decay, nuclear fission, and nuclear fusion.

Barrier tunneling occurs, not only for matter waves, but for waves of all kinds, including water waves and light waves. Figure 14*a* shows a light wave falling on a glass–air interface at an angle of incidence such that total internal reflection occurs. When we treated this subject in Section 39–3, we assumed that the incident light did not penetrate into the air space beyond the interface. However, that treatment was based on geometric optics which, as we know, is an approximation, being a limiting case of the more general wave optics. In much the same way, Newtonian mechanics (with its ray-like trajectories) is a limiting case of the more general wave mechanics.

If we analyze total internal reflection from the wave point of view, we learn that light *does* penetrate beyond the interface, for a distance of a few wavelengths. Speaking very loosely, we can say that such a penetration is necessary because the incident wave must "feel out" the situation locally before it can "know for sure" that there *is* an interface.

In Fig. 14*b*, we place the face of a second glass prism parallel to the interface, the gap between them being no more than a few wavelengths. The incident wave can then "tunnel" through this narrow "barrier" and generate a transmitted wave T. Specialists in optics call this phenomenon *frustrated total internal reflection* (FTIR).

You can check out barrier penetration by light waves with a drinking glass partially filled with water. Tilt the glass and look down into it at the side wall, at such an angle that the light entering your eye has been totally internally reflected from the wall. The wall will look silvery when this condition holds. Then press your (moistened) thumb against the outside of the glass. You will see the ridges of your finger print because, at those points, you have interfered with the total internal reflection process, as in Fig. 14*b*. The valleys between the ridges are still far enough away from the glass that the reflection here remains total and you simply see a silvery whorl.

* See *Physics Today*, January 1987.

Sample Problem 7 An electron whose total energy E is 5.1 eV is approaching a barrier whose height U is 6.8 eV and whose thickness L is 750 pm; see Fig. 13*a*. (a) What is the de Broglie wavelength of the incident electron?

Before the electron reaches the barrier, its total energy E is entirely kinetic, the potential energy in that region being zero. Proceeding as in Sample Problem 1 we find

$$\lambda = 540 \text{ pm}. \qquad \text{(Answer)}$$

Thus, the barrier is about 750 pm/540 pm or about 1.4 de Broglie wavelengths thick.

(b) What transmission coefficient follows from Eqs. 20 and 21?

From Eq. 21, we have

$$k = \sqrt{\frac{8\pi^2 m(U-E)}{h^2}}$$

$$= \sqrt{\frac{8\pi^2(9.11\times10^{-31}\text{ kg})(6.8\text{ eV}-5.1\text{ eV})(1.60\times10^{-19}\text{ J/eV})}{(6.63\times10^{-34}\text{ J}\cdot\text{s})^2}}$$

$$= 6.67\times10^{9}\text{ m}^{-1}. \qquad \text{(Answer)}$$

The quantity $2kL$ is then

$$2kL = (2)(6.67\times10^{9}\text{ m}^{-1})(750\times10^{-12}\text{ m}) = 10.0$$

and the transmission coefficient, from Eq. 20, is

$$T = e^{-2kL} = e^{-10.0} = 45 \times 10^{-6}. \qquad \text{(Answer)}$$

Thus, of every million electrons striking the barrier, about 45 will tunnel through it.

(c) What would be the transmission coefficient if the incident particle were a proton?

Carrying out the calculation once more but with the proton mass ($= 1.67 \times 10^{-27}$ kg) substituted for the electron mass yields $T \simeq 10^{-186}$. The transmission coefficient has been very substantially reduced indeed for this more massive particle. Imagine how small it would be for a jelly bean!

44-8 Heisenberg's Uncertainty Principle

The "jelly bean fallacy" way of thinking is a natural extension of familiar experiences with objects like baseballs that we can see and touch. We have seen, however, that this model simply doesn't work at the subatomic level. If an electron were like a tiny jelly bean, we should —in principle—be able to measure both its position and its momentum at any instant, with unlimited precision.

IT CAN'T BE DONE

It is not a matter of the practical difficulties of measurement because we assume ideal measuring instruments. Nor is it as if the electron *has* an infinitely precise position and momentum but that—for some reason—nature will not let us find it out. What we are dealing with is a fundamental limitation on the concept of "particle."

Heisenberg's uncertainty principle (see Fig. 15) provides a quantitative measure of this limitation. Suppose that you try to measure both the position and the momentum of an electron constrained to move along the x axis. Let Δx be the uncertainty in your measurement of its position and Δp_x your uncertainty in the measurement of its momentum. Heisenberg's principle states that*

$$\Delta x \cdot \Delta p_x \simeq h \qquad \text{(uncertainty principle).} \qquad (22)$$

That is, if you design an experiment to pin down the position of an electron as closely as possible (by making Δx smaller), you will find that you are not able to measure its momentum very well (Δp will get bigger). If you tinker with the experiment to improve the precision of your momentum measurement, the precision of your position measurement will deteriorate. *There is nothing that you can do about it.* The product of the two uncertainties must remain fixed, this constant product being nothing other than the Planck constant. Because momentum and position are vectors, a relation like Eq. 22 holds for the y and z coordinates as well.

The uncertainty principle seems strange only if you cling to the jelly bean fallacy and think of the electron as a tiny dot. Richard Feynman, in a footnote to a published series of lectures, puts the situation with characteristic clarity and forcefulness:†

> *I would like to put the uncertainty principle in its historical place: When the revolutionary ideas of quantum physics were first coming out, people still tried to understand them in terms of old-fashioned ideas . . . [that is, the jelly bean fallacy] . . . But at a certain point the old-fashioned ideas would begin to fail, so a warning was developed that said, in effect, "Your old-fashioned ideas are no damn good . . . If you get rid of all these old-fashioned ideas and instead use the ideas that I'm explaining in these lectures . . . there is no need for an uncertainty principle!"*

Figure 15 Waiting for Heisenberg. Physics is not all serious!

* The symbol $\simeq$ is sometimes replaced by $\gtrsim$, to recognize the fact that, in practice, you can never actually do as well as the quantum limit. Also, some formulations of the principle put $h/2\pi$ or $h/4\pi$ in place of h. These small differences need not concern us here.

† Richard P. Feynman, *QED—The Strange Theory of Light and Matter* (Princeton University Press, 1985), p. 55.

What Feynman is saying, in effect, is: "Think in terms of matter waves. Throw out the notion of the electron as a tiny dot. When you want to think of electrons, do so statistically, being guided by the probability density of the matter wave."

Heisenberg's Principle—Another Formulation. Another way to formulate this principle is in terms of energy and time, both scalars. The relation is

$$\boxed{\Delta E \cdot \Delta t \simeq h} \quad \text{(uncertainty principle).} \quad (23)$$

Thus, if you try to measure the energy of a particle, allowing yourself a time interval Δt to do so, your energy measurement will be uncertain by an amount ΔE given by $h/\Delta t$. To improve the precision of your energy measurement, you must allow more time.

Another way to look at Eq. 23 is this: You can violate the law of conservation of energy by "borrowing" an energy amount ΔE *provided* that you "pay back" the borrowed energy within a time Δt given by $h/\Delta E$.

Let's apply this idea to the barrier tunneling problem of Section 44–7. If the electron in Fig. 13*a* only had an additional energy $U - E$ it could climb over the barrier in accord with classical rules. The uncertainty principle tells us that the electron can "borrow" this amount of energy if it "pays it back" in the time it would take for the electron to travel a distance equal to the barrier thickness. Thus the electron finds itself on the other side of the barrier; the energy books are balanced again and nobody is the wiser!

Sample Problem 8 A 12-eV electron can be shown to have a speed of 2.05×10^6 m/s. Assume that you can measure this speed with a precision of 1.5%. With what precision can you simultaneously measure the momentum of the electron?

The electron's momentum is

$$p = mv = (9.11 \times 10^{-31}\ \text{kg})(2.05 \times 10^6\ \text{m/s})$$
$$= 1.87 \times 10^{-24}\ \text{kg}\cdot\text{m/s}.$$

The uncertainty in momentum is 1.5% of this, or 2.80×10^{-26} kg·m/s. The uncertainty in position is then from Eq. 22,

$$\Delta x \simeq \frac{h}{\Delta p} = \frac{6.63 \times 10^{-34}\ \text{J}\cdot\text{s}}{2.80 \times 10^{-26}\ \text{kg}\cdot\text{m/s}}$$
$$= 2.4 \times 10^{-8}\ \text{m} = 24\ \text{nm}, \qquad \text{(Answer)}$$

which is about 200 atomic diameters. Given your measurement of the electron's momentum, there is simply no way to pin down its position to any greater precision than this.

Sample Problem 9 A golf ball has a mass of 45 g and a speed, which you can measure to a precision of 1.5%, of 35 m/s. What limits does the uncertainty principle place on your ability to measure the position of the golf ball?

This example is like Sample Problem 8, except that the golf ball is much more massive and much slower than the electron of that example. The same calculation yields, in this case

$$\Delta x \simeq 3 \times 10^{-32}\ \text{m}. \qquad \text{(Answer)}$$

This is about 10^{17} times smaller than the diameter of a typical atomic nucleus. Where large objects are concerned, the uncertainty principle sets no meaningful limit to the precision of measurement. All of this is in accord with the correspondence principle, which tells us that, in situations in which classical physics is known to give correct answers (that is, golf balls), the predictions of quantum physics must merge with those of classical physics.

44-9 The Uncertainty Principle—Two Case Studies

Here we explore the uncertainty principle by trying our best to beat it. We shall not succeed.

A Particle in a Box. Let us try to pin down the position of an electron by trapping it in a box from which it cannot escape and then shrinking the walls of the box, which we assume we can do without limit. We will work in one dimension so that our "box" becomes the familiar infinite well of Fig. 9*a*, its walls separated by a distance L. For the uncertainty in our position measurement we then have

$$\Delta x \simeq L \quad \text{(uncertainty in position).} \quad (24)$$

As Eq. 11 shows, if we decrease L we increase the energy (that is, the zero point energy) of the trapped electron. You might say:

> *"It is true that the energy gets bigger (and so does the momentum) but at least I know* exactly *what this larger energy is; it is given by Eq. 11 and there is no uncertainty about it."*

The flaw in your argument is that you are not taking fully into account the fact that momentum is a vector. You may know the *magnitude* of the momentum exactly but you do not know its *direction.* That is, the electron may be bouncing back and forth between its confining walls but, at any instant, you do not know whether it is moving from left to right or from right to left. On the wave picture, the standing matter wave that represents the trapped electron (see Fig. 9*b*) is made up—as all standing waves are—of two traveling waves traveling in opposite directions, each carrying momentum.

The magnitude of the uncertainty in momentum is then

$$\Delta p \simeq (+p) - (-p) = 2p.$$

The momentum of the electron is given by Eq. 2 ($p = h/\lambda$) and $\lambda = 2L$ so that

$$\Delta p = 2p = \frac{2h}{\lambda} = \frac{2h}{2L}$$

or

$$\Delta p \simeq \frac{h}{L} \quad \text{(uncertainty in momentum).} \qquad (25)$$

If we multiply Eqs. 24 and 25, we have

$$\Delta x \cdot \Delta p \simeq (L)(h/L),$$

or

$$\Delta x \cdot \Delta p \simeq \text{h},$$

which is exactly the uncertainty principle. We have failed in our attempt to pin down the position and momentum of our trapped electron. We knew we would but let's try again anyway, with a different attack.

A Particle Passing Through a Slit. Let an electron, represented by an incident matter wave, fall on a slit of width Δy in screen *A* of Fig. 16. We are going to try to pin down the *vertical* position and momentum components of the electron at the instant it passes through the slit.

If the electron gets through the slit, we know its vertical position at the instant it did so with an uncertainty Δy. By reducing the slit width, we can pin down the vertical position of the electron as closely as we like.

However, matter waves—like all waves—flare out by diffraction when they pass through a slit. Furthermore, the narrower the slit, the more they flare out. From the particle point of view, this "flaring out" means that the electron acquires a vertical component of momentum as it passes through the slit. Some electrons acquire only a little vertical momentum, others a lot, so that there is an uncertainty.

There is one particular value of the vertical momentum component that will carry the electron to the first minimum of the diffraction pattern, point *a* on screen *B* of Fig. 16. Let us take this value as a measure of the uncertainty Δp of our knowledge of the vertical momentum component of the electron.

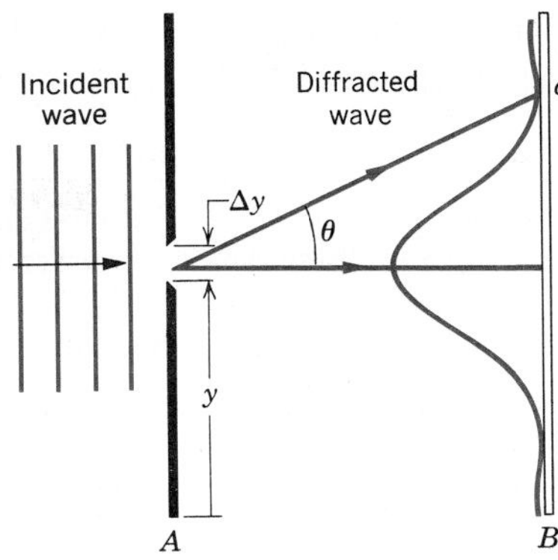

Figure 16 An arrangement for trying to beat the uncertainty principle. (It doesn't work.) An incident beam of electrons is diffracted at the slit in screen *A*, forming a typical diffraction pattern on screen *B*. The narrower the slit, the wider the pattern.

The first minimum of the diffraction pattern (see Section 41-2) occurs at an angle θ given by

$$\sin\theta = \frac{\lambda}{\Delta y}.$$

If θ is small enough, we can write replace $\sin\theta$ by θ. Also, from Eq. 3, we have $\lambda = h/p$. This equation then becomes

$$\theta \simeq \frac{h}{p\,\Delta y}, \qquad (26)$$

in which p is the horizontal momentum component. To reach the first minimum, θ must be such that

$$\theta = \frac{\Delta p}{p}. \qquad (27)$$

If we equate Eqs. 26 and 27, we find

$$\Delta y \cdot \Delta p \simeq \text{h},$$

which is once more the uncertainty principle. Foiled again!

44-10 Waves and Particles

Earlier, we promised to address the question: "How can an electron (or a photon) be wavelike under some circumstances and particlelike under others?" We now

Table 1 Selected Experiments Showing the Wave-Particle Nature of Both Light and Matter

	Matter	*Light*
Wave nature	The Davisson–Germer diffraction experiment Section 44–2	Young's double-slit interference Section 40–4
Particle nature	Millikan's oil drop experiment Section 24–8	The Compton effect Section 43–4

keep that promise. First we remind you, in Table 1, of the hard experimental evidence that both matter and light do indeed have this dual character.

Our mental images of "wave" and "particle" are drawn from our familiarity with large-scale objects such as ocean waves and tennis balls. In a way it is fortunate that we are able to extend these concepts (separately!) into the subatomic domain and to apply them to entities such as the electron, which we can neither see nor touch. We say at once, however, that no single concrete mental image, combining the features of *both* wave and particle, is possible in the quantum world. As Paul Davies, physicist and science writer, has written, "It is impossible to visualize a wave-particle, so don't try."

Niels Bohr, who not only played a major role in the development of quantum mechanics but also served as its major philosopher and interpreter, has shown a way to feel comfortable with the wave-duality problem. It is embodied in his *principle of complementarity,* which states:

> *The wave and the particle aspects of a quantum entity are both necessary for a complete description. However, both aspects cannot be revealed simultaneously in a single experiment. The aspect that is revealed is determined by the nature of the experiment being done.*

Consider a beam of light, perhaps from a laser, that passes across a laboratory table. What is the nature of the light beam? Is it a wave or a stream of particles? You cannot answer this question unless you interact with the beam in some way.

If you put a diffraction grating in the path of the beam, you reveal it as a wave. If you interpose a photoelectric apparatus such as that of Fig. 1 of Chapter 43, you will need to regard the beam as a stream of particles (photons) if you are to interpret your measurements in a satisfactory way. Try as you will, there is no single experiment that you can carry out with the beam that will require you to interpret it as a wave *and* as a particle *at the same time.* You may not like having to switch back and forth between wave and particle descriptions, depending on the experiment you are doing, but there is at least no confusion caused by overlap of the two models.

Complementarity—A Case Study. Let us see how complementarity works by trying to set up an experiment that will force nature to reveal both the wave and the particle aspects of electrons at the same time. We didn't succeed in beating the uncertainty principle and we won't succeed here either, but we'll try!

In Fig. 17 a beam of electrons falls on a double-slit arrangement in screen A and sets up a pattern of interference fringes on screen B. This is proof enough of the wave nature of the incident electrons.

Suppose now that we replace screen B with a small electron detector, designed to generate and record a "click" every time an electron hits it. We find that such clicks do indeed occur. If we move the detector up and down in Fig. 17 we can, by plotting the click rate against the detector position, trace out the pattern of interference fringes. Have we not succeeded in demonstrating both wave and particle? We see the fringes (wave) and we hear the clicks (particle)?

We have not. A mere "click" is not enough evidence that we are dealing with a particle. The concept of "particle" involves the concept of "trajectory" and a mental image of a dot following a prescribed path. As a minimum, we want to be able to know which of the two slits

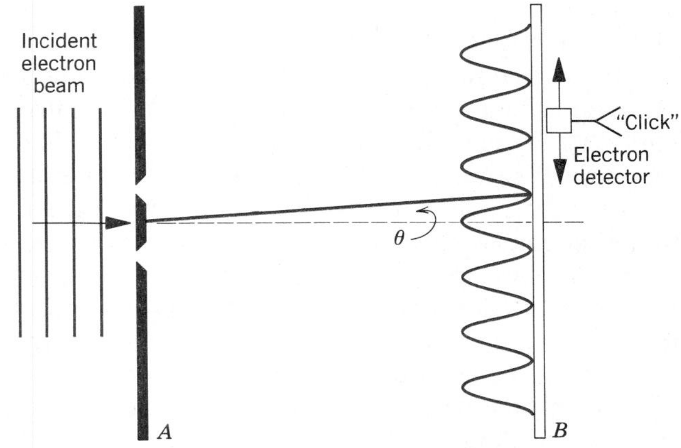

Figure 17 An arrangement for trying to prove that an incident electron beam is simultaneously both wavelike and particlelike. (It doesn't work.) You can modify the apparatus to show one aspect or the other, but not both at the same time.

in screen A the electron passed through on its way to generating a click in the detector. Can we find out?

We can, in principle, by putting a very thin detector in front of each slit, designed so that, if an electron passes through it, it will generate an electronic signal. We can then try to correlate each click, or "screen arrival signal" with a "slit passage signal," thus identifying the path of the electron involved.

If we succeed in modifying the apparatus to do this, we find a surprising thing. *The interference fringes have disappeared!* In passing through the slit detectors the electrons were affected in ways that destroyed the interference pattern. Although we have now shown the particle nature of the electron, the evidence for its wave nature has vanished.

The converse to our thought experiment is also true. If we start with an experiment that shows that electrons are particles and if we tinker with it to bring out the wave aspect, we will always find that the evidence for particles has vanished. Also, our experiment would work in precisely the same way if we substituted a light beam for the incident electron beam in Fig. 17.

A Quantum Puzzle Solved. At the beginning of this chapter we asked how tracks such as those shown in the figure at the head of this chapter, made up of tiny bubbles and so clearly suggesting the wake of a fast charged particle, can be associated with waves.

As the beginnings of an answer we look again at the thought experiment of Fig. 17, in which the pattern of fringes on screen B is neatly accounted for by the alternating constructive and destructive interference of matter wavelets radiating from each of the two slits in screen A. We can think of these as "guiding waves," their connection with the particle being that the square of their associated wave function at any point measures the probability that the particle will be found at that point. Thus, on screen B, electrons will pile up at those places where this probability is large, and they will be found in lesser abundance at those places where it is small. Figure 18, a computer simulation, shows how the fringes build up with time for a weak incident beam.

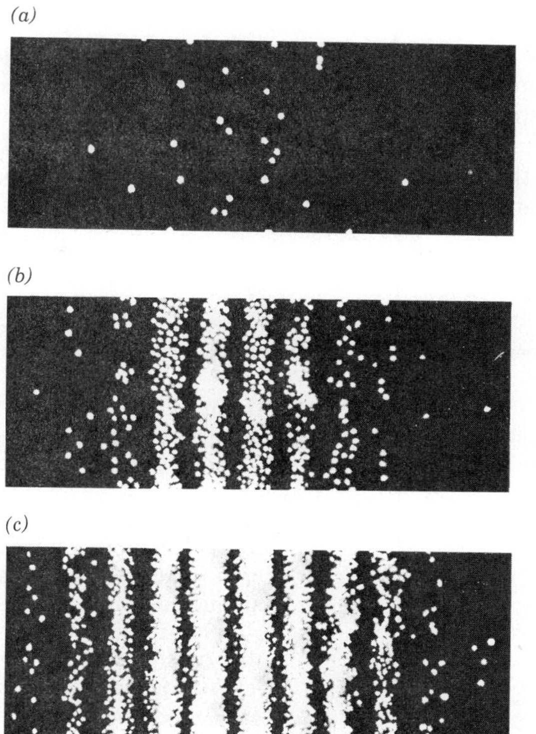

Figure 18 Photons fall on screen B of Fig. 17, gradually building up a fringe pattern. The pattern emerges even if the light intensity is so low that—on average—only one photon at a time is present in the apparatus.

These considerations apply even if the incident beam is deliberately made so weak that, by calculation, there should be—on average—*only one electron in the apparatus at any instant.* You might think that, because the single electron that chances to be in the apparatus must go through one slit or the other, the fringes must vanish; after all—you reason—the electron cannot interfere with itself and there is nothing else for it to interfere with. However, experiment shows that the fringes will *still* be formed, built up slowly as electron after electron falls on screen B. Even under these conditions the associated matter wave always passes through *both* slits and *it* is what determines where the electrons are likely to fall on screen B. The single electron *does* interfere with itself. Don't try to visualize how it does it!

With this background we are ready to answer the question of the wave nature of bubble chamber tracks. Consider Fig. 19, which shows an electron generated at point I and detected at point F. How does "it" traverse the empty space between them?

The quantum answer is that the associated matter

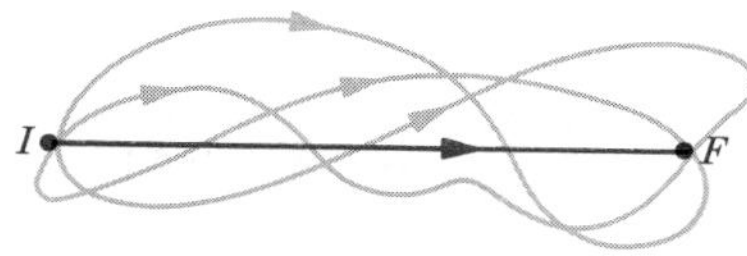

Figure 19 An electron moves from I to F. From the wave point of view, all possible paths are explored, the resultant track being the superposition of all of these paths.

wave explores all possible paths, as the figure suggests, assigning an equal probability to each. However, only for the straight line connecting the two points do the matter waves add constructively, yielding a high probability that the particle will be found there if sought. For points not on this straight line, the waves can be shown to cancel each other by destructive interference, the cancellation being more severe the more massive the particle. It is in this way that the trajectories of particles in Newtonian mechanics are related to their associated matter waves. It is on this basis that we can claim that bubble chamber tracks can be given a wave interpretation.

REVIEW AND SUMMARY

The Wave Nature of Matter

Beams of electrons and other forms of matter exhibit wave properties, including interference and diffraction, with a *de Broglie wavelength* given by

$$\lambda = h/p \quad \text{(wavelength of a particle).} \qquad [3]$$

(See Sample Problems 1 and 2.) These wave properties are most easily shown by diffraction, similar to x-ray diffraction, which occurs during reflection from atomic planes in crystals. (A moderate-energy 100-eV electron beam has a wavelength of only 120 pm, the same order of magnitude as atomic interplanar distances.) See Figures 2 and 4.

The Wave Function

Matter waves are described by a *wave function* ψ. Such waves describe particle motion in much the same way that electromagnetic waves describe photons. In particular, the probability of finding an electron at any location, called the *probability density,* is equal to the square of the wave function at that location.

A Particle Trapped Between Rigid Walls

A simple one-dimensional introduction to particle waves is the study of the motion of a particle trapped between rigid walls. We can study the relevant wave function because of its close mathematical relationship to two classical problems: the standing-wave oscillations of a short constrained string and the electromagnetic oscillations inside a cavity with perfectly reflecting walls. In this one-dimensional case, $\psi^2(x)dx$ is proportional to the probability of finding the particle in the interval between x and $x + dx$. For normalization,

$$\int_0^L \psi^2(x)dx = 1. \qquad [8]$$

Figure 9*b* shows the probability density for the lowest energy wave function. The trapped particle's energy is shown to be limited to quantized values given by

$$E_n = n^2 \frac{h^2}{8mL^2} \quad n = 1, 2, 3, \ldots . \qquad [10]$$

The lowest energy, $E_1 = h^2/8mL^2$, is the *zero-point energy,* the energy retained by the particle even at 0 K. See Sample Problems 3 and 4.

The Hydrogen Atom

The *energies* of the allowed states of an electron in a *hydrogen atom* are

$$E_n = -\left(\frac{me^4}{8\epsilon_0^2 h^2}\right)\frac{1}{n^2} \quad n = 1, 2, 3, \ldots . \qquad [13]$$

The *probability density* for the ground state (the lowest-energy state) is

$$\psi^2 = \frac{1}{\pi r_B^3} e^{-2r/r_B}, \qquad [14]$$

in which the $r_B = 5.29 \times 10^{-11}$ m is the *Bohr radius.* The *radial probability density* $P(r)$ is

$$P(r) = \frac{4}{r_B^3} r^2 e^{-2r/r_B}. \qquad [18]$$

See Fig. 12 and Sample Problems 5 and 6.

Barrier Tunneling

An electron approaching a flat potential barrier of height U and thickness L has a finite probability T (the *transmission coefficient*) of penetrating the barrier even if its kinetic energy E is less than the height of the barrier. The probability is

$$T = e^{-2kL}, \quad [20]$$

in which

$$k = \sqrt{\frac{8\pi^2 m(U-E)}{h^2}}. \quad [21]$$

See Sample Problem 7.

Heisenberg's Uncertainty Principle

The uncertainty principle suggests that the very concept of "particle" (as in "particle between rigid walls") is inherently fuzzy. In particular we cannot, even in principle, simultaneously measure both a particle's position **r** and momentum **p** with arbitrary precision. In fact, the uncertainties associated with each component of **r** and **p** must obey a relationship of the form

$$\Delta x \cdot \Delta p_x \simeq h \quad \text{(uncertainty principle)}. \quad [22]$$

The principle also applies to energy and time measurements in the form

$$\Delta E \cdot \Delta t \simeq h \quad \text{(uncertainty principle)}. \quad [23]$$

See Sample Problems 8 and 9 and study Section 44–9 carefully.

Wave–particle Duality and the Complementarity Principle

The dual wave–particle nature of both matter and radiation (see Table 1) is summarized by Bohr's principle of complementarity:

The wave and the particle aspects of a quantum entity are both necessary for a complete description. However, both aspects cannot be revealed simultaneously in a single experiment. The aspect that is revealed is determined by the nature of the experiment being done.

The thought experiment of Fig. 17 suggests the spirit of complementarity. Two-slit interference fringes show the wave nature of the incident beam, even though the detector responds to individual particles. Any attempt to demonstrate a more fundamental particle nature, perhaps by tracing such particles through the apparatus, causes the interference fringes to disappear.

QUESTIONS

1. How can the wavelength of an electron be given by $\lambda = h/p$? Doesn't the very presence of the momentum p in this formula imply that the electron is a particle?

2. In a repetition of Thomson's experiment for measuring e/m for the electron (see Section 30–3), a beam of electrons is collimated by passage through a slit. Why is the beamlike character of the emerging electrons not destroyed by diffraction of the electron wave at this slit?

3. Why is the wave nature of matter not more apparent to our daily observations?

4. Considering the wave behavior of electrons, we should expect to be able to construct an "electron microscope" using short-wavelength electrons to provide high resolution. This, indeed, has been done. (*a*) How might an electron beam be focused? (*b*) What advantages might an electron microscope have over a light microscope? (*c*) Why not make a proton microscope? a neutron microscope?

5. How many experiments can you recall that support the wave theory of light? the particle theory of light? the wave theory of matter? the particle theory of matter?

6. Is an electron a particle? Is it a wave? Explain your answer, citing relevant experimental evidence.

7. If the particles listed below all have the same energy, which has the shortest wavelength? electron; α-particle; neutron; proton.

8. What common expression can be used for the momentum of either a photon or a particle?

9. Discuss the analogy between (*a*) wave optics and geometrical optics and (*b*) wave mechanics and classical mechanics.

10. Does a photon have a de Broglie wavelength? Explain.

11. Discuss similarities and differences between a matter wave and an electromagnetic wave.

12. Can the de Broglie wavelength associated with a particle be smaller than the size of the particle? larger? Is there any relation necessarily between such quantities?

13. If, in the de Broglie formula $\lambda = h/mv$, we let $m \to \infty$, do we get the classical result for particles of matter?

14. How could Davisson and Germer be sure that the "54-eV" peak of Fig. 2 was a first-order diffraction peak, that is, that $m = 1$ in Eq. 4?

15. Do electron diffraction experiments give different information about crystals than can be obtained from x-ray diffraction experiments? from neutron diffraction experiments? Give examples.

16. A standing wave can be viewed as the superposition of two traveling waves. Can you apply this view to the problem of a particle confined between rigid walls, giving an interpretation in terms of the motion of the particle?

17. The allowed energies for a particle confined between rigid walls are given by Eq. 10. First, convince yourself that, as n increases, the energy levels become farther apart. How can this possibly be? The correspondence principle would seem to require that they move closer together as n increases, approaching a continuum.

18. How can the predictions of wave mechanics be so exact if the only information we have about the positions of the electrons in atoms is statistical?

19. In the $n = 1$ state, for a particle confined between rigid walls, what is the probability that the particle will be found in a small-length element at the surface of either wall?

20. In Fig. 10 what do you imagine the curve for $\psi^2(x)$ for $n = 100$ looks like? Convince yourself that these curves approach classical expectations as $n \to \infty$.

21. We have seen that barrier tunneling works for matter waves and for electromagnetic waves. Do you think that it also works for water waves? for sound waves?

22. Comment on the statement, "A particle can't be detected while tunneling through a barrier, so that it doesn't make sense to say that such a thing actually happens."

23. List examples of barrier tunneling occurring in nature and in man-made devices.

24. A proton and a deuteron, each having 3 MeV of energy, attempt to penetrate a rectangular potential barrier of height 10 MeV. Which particle has the higher probability of succeeding? Explain in qualitative terms.

25. A laser projects a beam of light across a laboratory table. If you put a diffraction grating in the path of the beam and observe the spectrum, you declare the beam to be a wave. If instead you put a clean metal surface in the path of the beam and observe the ejected photoelectrons, you declare this same beam to be a stream of particles (photons). What can you say about the beam if you don't put anything in its path?

26. State and discuss (*a*) the correspondence principle, (*b*) the uncertainty principle, and (*c*) the complementarity principle.

27. In Fig. 17, why would you expect the electrons from each slit to arrive at the screen over a range of positions? Shouldn't they all arrive at the same place? How does your answer relate to the complementarity principle?

28. Several groups of experimenters are trying to detect gravity waves, perhaps coming from our galactic center, by measuring small distortions in a massive object through which the hypothesized waves pass. They seek to measure displacements as small as 10^{-21} m. (The radius of a proton is $\sim 10^{-15}$ m, a million times larger!) Does the uncertainty principle put any restriction on the precision with which this measurement can be carried out?

29. Figure 10 shows that for $n = 3$ the probability function $\psi^2(x)$ for a particle confined between rigid walls is zero at two points between the walls. How can the particle ever move across these positions? (*Hint:* Consider the implications of the uncertainty principle.)

30. Why does the concept of Bohr orbits violate the uncertainty principle?

31. (*a*) Give examples of how the process of measurement disturbs the system being measured. (*b*) Can the disturbances be taken into account ahead of time by suitable calculations?

32. You measure the pressure in a tire, using a pressure gauge. The gauge, however, bleeds a little air from the tire in the process, so that the act of measuring changes the property that you are trying to measure. Is this an example of the Heisenberg uncertainty principle? Explain.

EXERCISES AND PROBLEMS

Section 44-1 Louis Victor de Broglie Makes a Suggestion

1E. A bullet of mass 40 g travels at 1000 m/s. (*a*) What wavelength can we associate with it? (*b*) Why does the wave nature of the bullet not reveal itself through diffraction effects?

2E. Using the classical relation between momentum and kinetic energy, show that the de Broglie wavelength of an electron can be written (*a*) as

$$\lambda = \frac{1.226 \text{ nm}}{\sqrt{K}}, \qquad (K \text{ in eV})$$

in which K is the kinetic energy in electron volts, or (*b*) as

$$\lambda = \sqrt{\frac{1.50}{V}},$$

where λ is in nm, and V is the accelerating potential in volts.

3E. In an ordinary color television set, electrons are accelerated through a potential difference of 25 kV. What is the de Broglie wavelength of such electrons? (*Hint:* See Exercise 2; ignore relativistic effects.)

4E. Calculate the wavelength of (*a*) a 1-keV electron, (*b*) a 1-keV photon, and (*c*) a 1-keV neutron. Use the relation of Exercise 2 for the electron.

5E. An electron and a photon each have a wavelength of 0.20 nm. Calculate their (*a*) momenta and (*b*) energies. Use the result of Exercise 2 for the electron.

6E. The wavelength of the yellow spectral emission line of sodium is 590 nm. At what kinetic energy would an electron have the same de Broglie wavelength?

7E. Thermal neutrons have an average kinetic energy $\frac{3}{2}kT$ where T may be taken to be 300 K. Such neutrons are in thermal equilibrium with normal surroundings. (*a*) What is the average energy of a thermal neutron? (*b*) What is the corresponding de Broglie wavelength?

8E. If the de Broglie wavelength of a proton is 0.10 pm, (*a*) what is the speed of the proton and (*b*) through what electric potential would the proton have to be accelerated to acquire this speed?

9P. Consider a balloon filled with (monatomic) helium gas at room temperature and pressure. (*a*) Calculate the average de Broglie wavelength of the helium atoms and the average distance between atoms under these conditions. The average kinetic energy of an atom is equal to $\frac{3}{2}kT$. (*b*) Can the molecules be treated as particles under these conditions?

10P. (*a*) A photon in free space has an energy of 1.0 eV and an electron, also in free space, has a kinetic energy of that same amount. What are their wavelengths? (*b*) Repeat for an energy of 1.0 GeV.

11P. (*a*) Photons and electrons travel in free space with wavelengths of 1.0 nm. What are the energy of the photon and the kinetic energy of the electron? (*b*) Repeat for a wavelength of 1.0 fm.

12P. Singly-charged sodium ions are accelerated through a potential difference of 300 V. (*a*) What is the momentum acquired by the ions? (*b*) Calculate their de Broglie wavelength.

13P. The 20-GeV electron accelerator at Stanford provides an electron beam of small wavelength, suitable for probing the fine details of nuclear structure by scattering experiments. What will this wavelength be and how does it compare with the size of an average nucleus? (At these energies it is sufficient to use the extreme relativistic relationship between momentum and energy, namely, $p = E/c$. This is the same relationship used for light and is justified when the kinetic energy of a particle is much greater than its rest energy, as in this case. The radius of a middle mass nucleus is about 5.0 fm.)

14P. The existence of the atomic nucleus was discovered in 1911 by Ernest Rutherford, who properly interpreted some experiments in which a beam of alpha particles was scattered from a foil of atoms such as gold. (*a*) If the alpha particles had a kinetic energy of 7.5 MeV, what was their de Broglie wavelength? (*b*) Should the wave nature of the incident alpha particles have been taken into account in interpreting these experiments? The mass of an alpha particle is 4.00 u, and its distance of closest approach to the nuclear center in these experiments was about 30 fm. (The wave nature of matter was not postulated until more than a decade after these crucial experiments were first performed.)

15P. A nonrelativistic particle is moving three times as fast as an electron. The ratio of their de Broglie wavelengths, particle to electron, is 1.813×10^{-4}. By calculating its mass, identify the particle.

16P. The highest achievable resolving power of a microscope is limited only by the wavelength used; that is, the smallest detail that can be separated is about equal to the wavelength. Suppose one wishes to "see" inside an atom. Assuming the atom to have a diameter of 100 pm this means that we wish to resolve detail of separation about 10 pm. (*a*) If an electron microscope is used, what minimum energy of electrons is needed? (*b*) If a light microscope is used, what minimum energy of photons is needed? (*c*) Which microscope seems more practical for this purpose? Why?

17P. What accelerating voltage would be required for electrons in an electron microscope to obtain the same ultimate resolving power as that which could be obtained from a gamma-ray microscope using 100-keV gamma rays? (*Hint:* See Problem 16.)

18P. (*a*) Calculate, according to the Bohr model, the speed of the electron in the ground state of the hydrogen atom. (*b*) Calculate the corresponding de Broglie wavelength. (*c*) Comparing the answers to (*a*) and (*b*), find a relation between the de Broglie wavelength λ and the radius r of the ground state Bohr orbit.

Section 44-2 Testing de Broglie's Hypothesis

19E. A potassium chloride (KCl) crystal is cut so that the layers of atomic planes parallel to its surface have an interplanar spacing of 0.314 nm. A beam of 380-eV electrons is incident normally on the crystal surface. Calculate the angles ϕ at which the detector must be positioned to record strongly diffracted beams of all orders present.

20P. In the experiment of Davisson and Germer (*a*) at what angles would the second- and third-order diffracted beams corresponding to a strong maximum in Fig. 2 occur, provided they are present? (*b*) At what angle would the first-order diffracted beam occur if the accelerating potential were changed from 54 to 60 V?

Section 44-5 Matter Waves and Electrons

21E. (*a*) A proton or (*b*) an electron is trapped in a one-dimensional box of 100 pm length. What is the minimum energy these particles can have?

22E. What must be the width of an infinite well such that the energy of an electron trapped therein in the $n = 3$ state has an energy of 4.7 eV?

23E. (*a*) Calculate the smallest allowed energy of an electron confined to an atomic nucleus (diameter about 1.4×10^{-14} m). (*b*) Compare this with the several MeV of energy binding protons and neutrons inside the nucleus; on this basis should we expect to find electrons inside nuclei?

24E. The ground state energy of an electron in an infinite well is 2.6 eV. What will the ground state energy be if the width of the well is doubled?

25E. An electron, trapped in an infinite well of width 0.25 nm, is in the ground ($n = 1$) state. How much energy must it absorb to jump up to the third excited ($n = 4$) state?

26P. (*a*) What is the separation in energy between the lowest two energy levels for a container 20 cm on a side containing argon atoms? (*b*) How does this compare with the thermal energy of the argon atoms at 300 K? (*c*) At what temperature does the thermal energy equal the spacing between these two energy levels? Assume, for simplicity, that the argon atoms are trapped in a one-dimensional well 20 cm wide. The atomic weight of argon is 39.9 g/mol.

27P. Consider a conduction electron in a cubical crystal of a conducting material. Such an electron is free to move throughout the volume of the crystal, but cannot escape to the outside. It is trapped in a three-dimensional infinite well. The electron can move in three dimensions, so that its total energy is given by (compare with Eq. 10),

$$E = \frac{h^2}{8L^2m}(n_1^2 + n_2^2 + n_3^2),$$

in which n_1, n_2, n_3 each take on the values 1, 2, ... Calculate the energies of the lowest five distinct states for a conduction electron moving in a cubical crystal of edge length $L = 0.25\ \mu$m.

28P. The wave function of a particle confined to an infinite well and in the lowest energy state is given by $\psi = A \sin(\pi x/L)$. Use the "normalization condition" expressed by Eq. 8 to show that $A = \sqrt{2/L}$.

29P. A particle is confined between rigid walls separated by a distance L. The particle is in the lowest energy state; the wavefunction for this state is given in Problem 28. Use this wavefunction to calculate the probability that the particle will be found between the points (*a*) $x = 0$ and $x = L/3$, (*b*) $x = L/3$ and $x = 2L/3$, and (*c*) $x = 2L/3$ and $x = L$.

Section 44-6 The Hydrogen Atom

30E. In Fig. 12, verify the plotted values of $P(r)$ at (*a*) $r = 0$, (*b*) $r = r_B$, and (*c*) $r = 2r_B$.

31E. In the ground state of the hydrogen atom, what is the probability that the electron will be found within a sphere whose radius is that of the first Bohr orbit? See Sample Problem 6.

32E. In the ground state of the hydrogen atom, evaluate the probability density $\psi^2(r)$ and the radial probability density $P(r)$ for the positions (*a*) $r = 0$ and (*b*) $r = r_B$. Explain what these quantities mean.

33E. Use the result of Sample Problem 6 to calculate the probability that the electron in a hydrogen atom, in the ground state, will be found between the spheres $r = r_B$ and $r = 2r_B$.

34P. For an electron in the ground state of the hydrogen atom, (*a*) verify Eq. 19 and (*b*) calculate the radius of a sphere for which the probability that the electron will be found inside the sphere equals the probability that the electron will be found outside the sphere. (*Hint:* See Sample Problem 6.)

35P. In the ground state of the hydrogen atom show that the probability $p(r)$ that the electron lies within a sphere of radius r is given by

$$p(r) = 1 - e^{-2x}(1 + 2x + 2x^2),$$

in which $x = r/r_B$, a dimensionless ratio.

36P. In atoms there is a finite, though very small, probability that, at some instant, an orbital electron will actually be found inside the nucleus. In fact, some unstable nuclei use this occasional appearance of the electron to decay by *electron capture.* Assuming that the proton itself is a sphere of radius 1.1×10^{-15} m and that the hydrogen atom electron wave function holds all the way to the proton's center, use the ground state wave function to calculate the probability that the hydrogen atom electron is inside its nucleus. (*Hint:* when $x \ll 1$, $e^{-x} \approx 1$.)

Section 44-7 Barrier Tunneling

37E. A proton and a deuteron (which has the same charge as a proton but twice the mass) are incident on a barrier of thickness 10 fm and height 10 MeV. Each particle has a kinetic energy of 3.0 MeV. Find the transmission probabilities for them.

38P. Consider a barrier such as that of Fig. 13*a*, but whose height U is 6.0 eV and whose thickness L is 0.70 nm. Calculate the energy of an incident electron such that its transmission probability is one in 1000.

39P. Suppose that an incident beam of 5.0-eV protons fell on a barrier of height 6.0 eV and thickness 0.70 nm, and at a rate equivalent to a current of 1.0 kA. How long would you have to wait—on the average—for one proton to be transmitted?

40P. Consider the barrier tunneling situation defined by Sample Problem 7. What fractional change in the transmission coefficient occurs for a 1% increase in (*a*) the barrier height, (*b*) the barrier thickness, and (*c*) the incident energy of the electron?

Section 44-8 Heisenberg's Uncertainty Principle

41E. A microscope using photons is employed to locate an electron in an atom to within a distance of 10 pm. What is the minimum uncertainty in the momentum of the electron located in this way?

42E. The uncertainty in the position of an electron is given as 50 pm, which is about the radius of the first Bohr orbit in hydrogen. What is the uncertainty in the momentum of the electron?

43E. Imagine playing baseball in a universe where Planck's constant was 0.60 J·s. What would be the uncertainty in the position of a 0.50-kg baseball moving at 20 m/s with an uncertainty of 1.0 m/s? Why would it be hard to catch such a ball?

44E. Consider an electron trapped in an infinite well whose width is 100 pm. If it is in a state with $n = 15$, what are (*a*) its energy? (*b*) the uncertainty in its momentum? (*c*) the uncertainty in its position?

45E. The lifetime of an electron in the state $n = 2$ in hydrogen is about 10^{-8} s. What is the uncertainty in the energy of the $n = 2$ state? Compare this with the energy of this state.

46P. Show that if the uncertainty in the location of a particle is equal to its de Broglie wavelength, the uncertainty in its velocity is equal to its velocity.

47P. Suppose that we wish to test the possibility that electrons in atoms move in orbits by "viewing" them with photons with sufficiently short wavelength, say 10 pm or less. (*a*) What would be the energy of such photons? (*b*) How much energy would such a photon transfer to a free electron in a head-on Compton collision? (*c*) What does this tell you about the possibility of confirming orbital motion by "viewing" an atomic electron at two or more points along its path?

CHAPTER 45

ALL ABOUT ATOMS

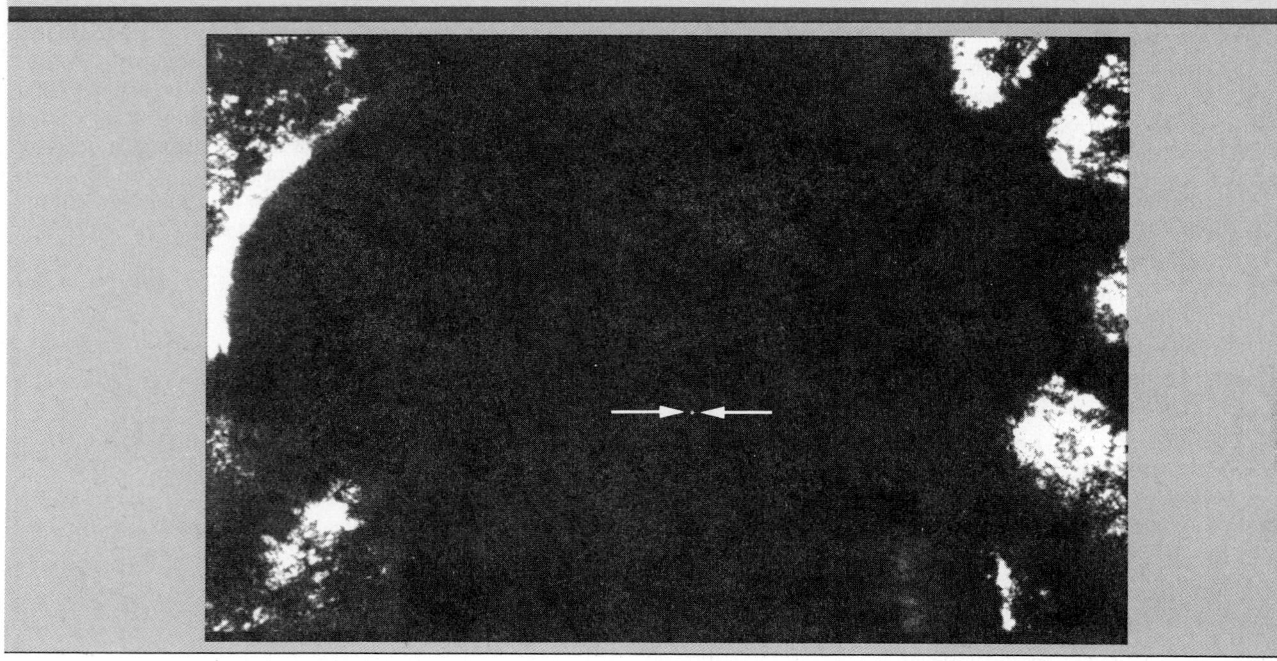

The tiny bright spot identified by the arrows is light emitted by a **single barium *ion*** *held in an electromagnetic trap at the University of Washington in Seattle. Laser stimulation causes its outermost electron to move to a higher energy level; as the electron returns to its ground level, the atom emits a photon of blue-green light. The process recurs in rapid succession, over and over again. (Atoms do not wear out!)*

45-1 Atoms and the World Around Us

What would you think if your physics or chemistry instructor told you that he did not believe in atoms? In the early years of this century, quite a few prominent scientists held just that view. Today, however, no well-informed teenager doubts that the material world around us is made up of atoms.

Why do we believe in these tiny objects that—it is often alleged—we cannot see? For one thing, with modern techniques we now *can* see individual atoms, as the photo at the head of this chapter makes clear. Figure 1, taken with a high-resolution electron microscope and Fig. 15 of Chapter 2, taken with a scanning tunneling microscope, leave little doubt. Even more to the point than these convincing pictures is the steady piling up of mountains of experimental information about atoms, all of it totally understandable in terms of modern quantum theory.

45-2 Some Properties of Atoms

Here we describe some of the properties of atoms that any theory of atomic structure must be able to explain.

Atoms Are Put Together According to a Systematic Plan. The existence of the *periodic table of the elements* (see Appendix E), with its remarkable repetitive sequences of chemical and physical properties, is evidence

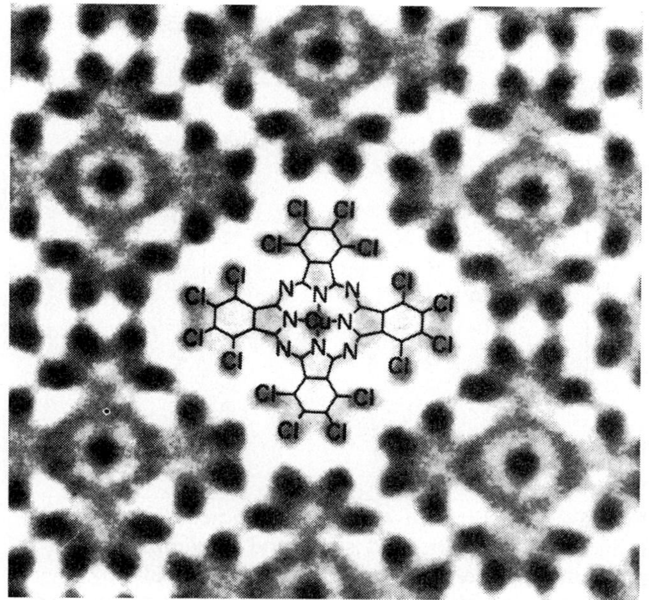

Figure 1 A photo taken with a high-resolution electron microscope of a very thin crystalline sample containing copper, chlorine, and nitrogen atoms. The copper atoms show up well at the centers of the "rosettes" formed by 16 chlorine atoms. Nitrogen atoms occupy intermediate positions.

enough that the electrons in the atoms of the elements are arranged according to a systematic plan. Figure 2 shows one simple example of such a repetitive property. It is a plot of the *ionization energy* of the elements (that is, the work required to remove a single electron from a neutral atom) as a function of the position of the element in the periodic table.

The periodic table contains six complete* horizontal periods of elements, each period starting with a highly reactive alkali metal (lithium, sodium, potassium, etc.) and ending with a chemically inert noble gas (neon, argon, krypton, etc.). The numbers of elements in these periods are:

2, 8, 8, 18, 18, and 32.

As we shall see, quantum physics predicts these numbers and leads us to a general understanding of the periodic table and thus of much of physics and nearly all of chemistry. Because the life processes that sustain us as thinking beings are (bio)chemical, we see that the influence of quantum physics in our lives runs deep.

Atoms Emit Light. Another central feature of atoms is that they emit (and absorb) light at sharply defined frequencies. Figure 3 shows an example. As we first explained in Section 8–10, atoms can exist only in certain discrete quantum states, each with its characteristic energy. An atom emits light when it transfers from one of these states to another state, of lower energy. The fre-

* The last horizontal period, starting with element 87 (francium), is incomplete.

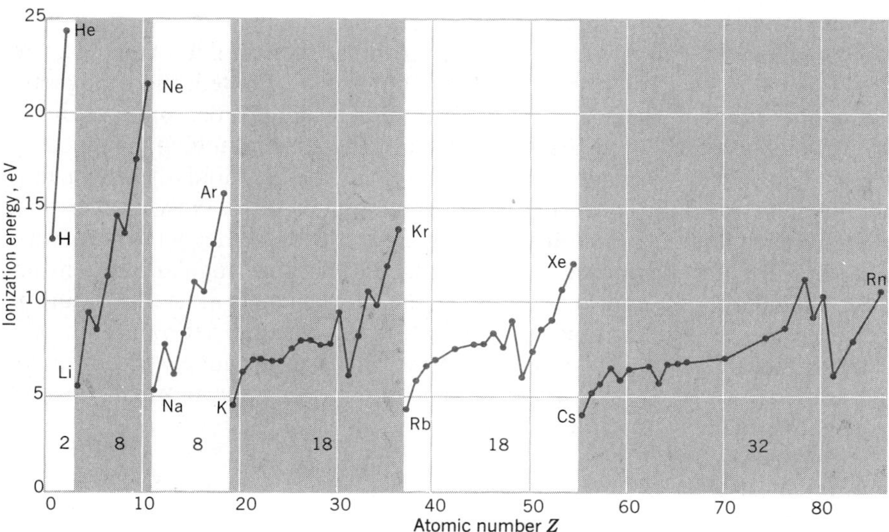

Figure 2 A plot of the ionization energy of the elements as a function of atomic number, showing the periodic repetition of properties through the six complete horizontal periods of the periodic table. The numbers of elements in these periods are indicated.

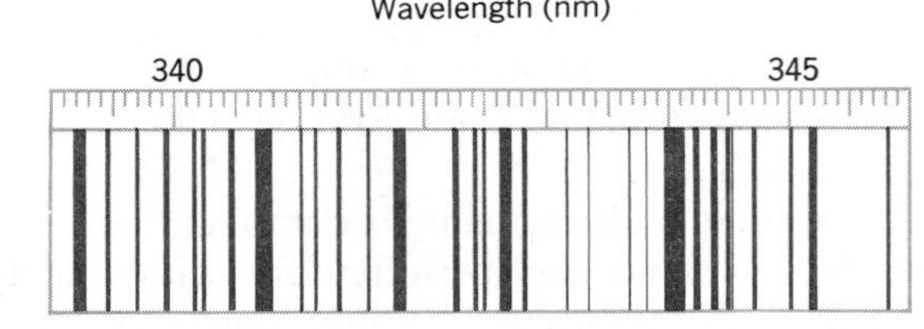

Figure 3 A small portion of the emission spectrum of iron, selected from the ultraviolet region.

quency ν of the emitted light is given by the *Bohr frequency condition* (see Eq. 33 of Chapter 8), or

$$h\nu = E_x - E_y. \tag{1}$$

Here E_x and E_y are the energies of the upper and lower states, respectively, and h is the Planck constant.

Thus, the problem of finding the frequencies of the light emitted (or absorbed) by an atom reduces to the problem of finding the energy levels for that atom. As we shall learn, the laws of quantum physics allow us—in principle at least—to calculate these energies.

Atoms Have Angular Momentum and Magnetism. Electrons in atoms behave classically like tiny current loops and have both an *orbital angular momentum* and an *orbital magnetic moment* associated with this motion. In Section 34–2 we pointed out that an electron also has an *intrinsic* angular momentum, called its *spin angular momentum.* The electron behaves classically like a spinning negative charge, thus giving rise to an intrinsic *spin magnetic moment.* Because the electron carries a negative charge, both the orbital and the spin magnetic moments are back-to-back with their corresponding angular momenta, pointing in opposite directions.

The spin and orbital angular momenta of the individual electrons in an atom combine to produce a net angular momentum for the atom as a whole. Associated with this net angular momentum is a net magnetic moment. For some atoms (neon, for example), the effects of the various electrons cancel each other so that the net angular momentum and the associated net magnet moment are zero. For many other atoms, however, the cancellation is not complete so that the atom exhibits a net angular momentum and a net magnetic moment.

The magnetism of atoms—in the special case of the ferromagnetism—is familiar to all. The *angular momentum* of atoms, however, is not so familiar. It occurred to Einstein that, if the atomic magnets in an iron bar were aligned, their associated angular momenta should also be aligned (in the opposite sense) and should

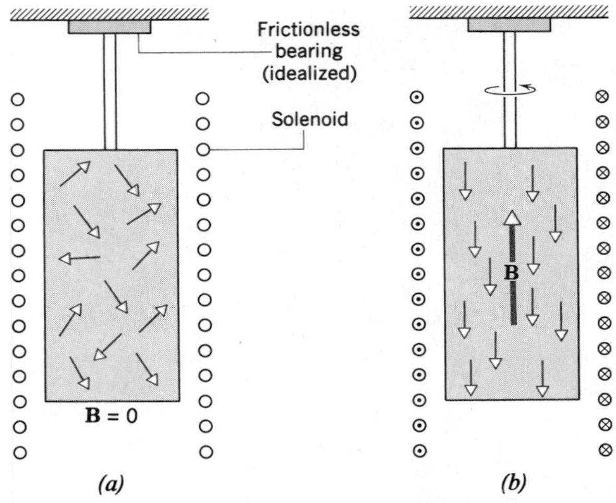

Figure 4 The Einstein–de Haas experiment (idealized.) (*a*) Initially, the magnetic field is zero and the atomic angular momentum vectors in the iron cylinder are randomly oriented, as the figure shows. The atomic magnetic moment vectors (not shown) point in the opposite direction to the atomic angular momentum vectors. (*b*) When an axial magnetic field is applied, the alignment of the magnetic moment vectors causes the atomic angular momentum vectors to line up as shown. Because the cylinder is isolated from external torques, angular momentum is conserved and the cylinder as a whole must rotate in the opposite sense, as shown.

exhibit large-scale external effects. In 1915, Einstein and W. J. de Haas carried out an ingenious experiment based on this idea.

In an ordinary iron bar such as that of Fig. 4*a*, the atomic magnets are randomly oriented, their magnetic effects cancelling at all external points. Suppose, however, that these atomic magnets are suddenly aligned by switching on a current in the solenoid shown in that figure. The angular momenta of the individual atoms, which are rigidly coupled to their magnetic moments, must also become aligned. Because angular momentum must be conserved, the bar as a whole must then rotate in the opposite sense. With this clever experiment, Einstein and de Haas demonstrated quantitatively the intimate connection between atomic angular momentum and atomic magnetism.*

* See Peter Galison, *How Experiments End* (University of Chicago Press, 1987), Chapter 2, for a detailed description of this important experiment and of the related experiments that followed it.

45-3 Schrödinger's Equation and the Hydrogen Atom

How do we use quantum theory to calculate numerical values for the atomic properties we have outlined in the previous section? In particular, how do we calculate the energies, the angular momenta, and the magnetic moments of the quantum states of an atom?

Let us start with hydrogen. We physicists love this atom because it is so simple, consisting of a single electron bonded by the electrostatic force to a single central proton. From the early days of Bohr theory (1913), through wave mechanics (1926), and into modern quantum electrodynamics (1948–49), this atom has served as a laboratory for testing, in exquisite detail, the depth of our understanding of the nature of matter.

We shall learn that it requires four quantum numbers to describe completely the quantum states of the hydrogen atom. *This same set of four numbers also serves to identify the quantum states of single electrons in multi-electron atoms.* Thus, we can carry over much of what we learn about the hydrogen atom to atoms with more than one electron.

How to proceed? For a problem in classical mechanics, we use Newton's laws. For a problem in electromagnetism, we use Maxwell's equations. For a problem in wave mechanics, we use *Schrödinger's equation,* a relation first advanced by the Austrian physicist Erwin Schrödinger in 1926.

Instead of writing down and analyzing the Schrödinger equation, we propose simply to describe how it is used.* Imagine, as in Fig. 5, a computer programmed to solve this equation. For its INPUT we insert the potential energy function that defines the problem at hand. For the hydrogen-atom problem, that function is the familiar Coulomb potential energy, given (see Eq. 22 of Chapter 26) by

$$U = -\frac{1}{4\pi\epsilon_0}\frac{e^2}{r}. \tag{2}$$

Here e is the magnitude of the charge of the electron and of the proton and r is the distance between these two particles.

When we press the RUN button (that is, when we solve the equation), the computer generates a PRINTOUT of the wave functions that define the quantized hydrogen-atom states. Printed alongside each wave function is the corresponding energy, angular momentum, and magnetic moment for the atom when it is in that state. Let us discuss these quantities in more detail.

45-4 The Energies of the Hydrogen Atom States

The Schrödinger equation has an infinite number of solutions, but most of them do not make any sense physically. In solving that equation, we program our computer to deliberately discard all solutions *except* those for which the wave function approaches zero as r in Eq. 2 approaches infinity. This is equivalent to recognizing that, beyond a certain distance, as you move away from the central proton you are less and less likely to find the electron.

> *The existence of quantized states with well-defined energies is a direct consequence of imposing this sensible requirement.*

This is another example of the *localization → quantization* principle that we have met before, first in connec-

* See Robert Eisberg and Robert Resnick, *Quantum Physics* (John Wiley & Sons, New York, 1985), 2nd ed., Chapters 5, 6, and 7, for a detailed account of the Schrödinger equation and its solutions.

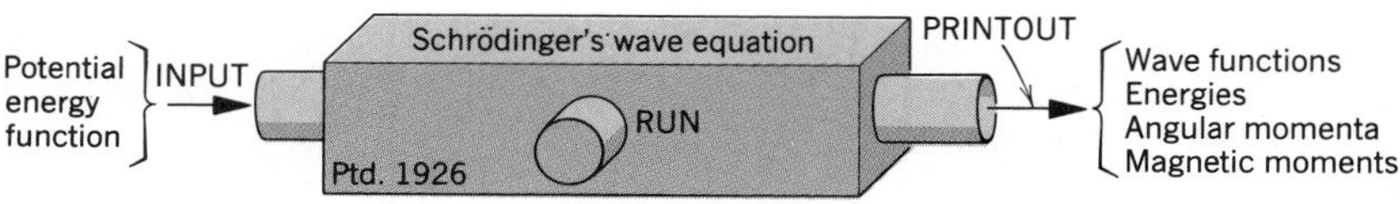

Figure 5 A computer programmed to solve Schrödinger's equation. The INPUT is the potential energy function that defines the problem, along with suitable boundary conditions suggested by the physics of the situation. The OUTPUT is the wave functions, the energies, the angular momenta, and the magnetic moments of the quantized states.

tion with waves on strings. We pointed out in Section 17–13 that a wave of *any* frequency can be propagated along a stretched string of *infinite* length but that only *discrete* frequencies of standing waves can be set up in a string of *finite* length. That is, localizing the wave quantizes the frequency. In the case of the hydrogen atom, localizing the wave function quantizes the energy.

The energies of the hydrogen-atom states are given by

$$E_n = -\frac{me^4}{8\epsilon_0^2 h^2}\frac{1}{n^2} = -\frac{13.6\text{ eV}}{n^2}, \qquad n = 1, 2, 3, \ldots . \tag{3}$$

in which the integer n is called the *principal quantum number;* it is the *first* of the four quantum numbers that we need to identify fully the allowed quantum states of the hydrogen atom.

45-5 Orbital Angular Momentum and Magnetism

Each state of the hydrogen atom has an associated orbital angular momentum **L**. We discuss first its magnitude and then its direction.

The Magnitude of **L**. In solving the Schrödinger equation we learn that the magnitude of the orbital angular momentum of the hydrogen atom states is quantized. Its allowed values are

$$L = \sqrt{\ell(\ell + 1)}\,\hbar \tag{4}$$

in which $\hbar$ (pronounced *h-bar*) is an abbreviation for $h/2\pi$ and ℓ is the *orbital quantum number;* it is the *second* of the four quantum numbers that we seek. The allowed values of ℓ depend on the value of the principal quantum number n and are

$$\ell = 0, 1, 2, \ldots (n - 1). \tag{5}$$

Thus, for $n = 1$, only $\ell = 0$ is permitted. For $n = 2$, only $\ell = 0$ and $\ell = 1$ are permitted.

The Direction of **L**. States with the same values of n and ℓ but with different wave functions correspond to different *directions* for the angular momentum vector **L**. For an isolated hydrogen atom, there is no obvious direction in space with respect to which the orientation of its angular momentum vector can be measured. To supply one, it is convenient to imagine that the atom is immersed in a weak but uniform magnetic field whose direction we may take as a z axis.

According to the rules of wave mechanics, the angular momentum vector **L** cannot make *any* angle with the z axis, but only those angles that yield a component along this axis given by

$$\boxed{L_z = m_\ell\,\hbar.} \tag{6}$$

Here m_ℓ, the *magnetic quantum number,* is restricted to the values

$$\boxed{m_\ell = 0, \pm 1, \pm 2, \ldots \pm \ell.} \tag{7}$$

The magnetic quantum number is the *third* of the four quantum numbers that we are seeking.

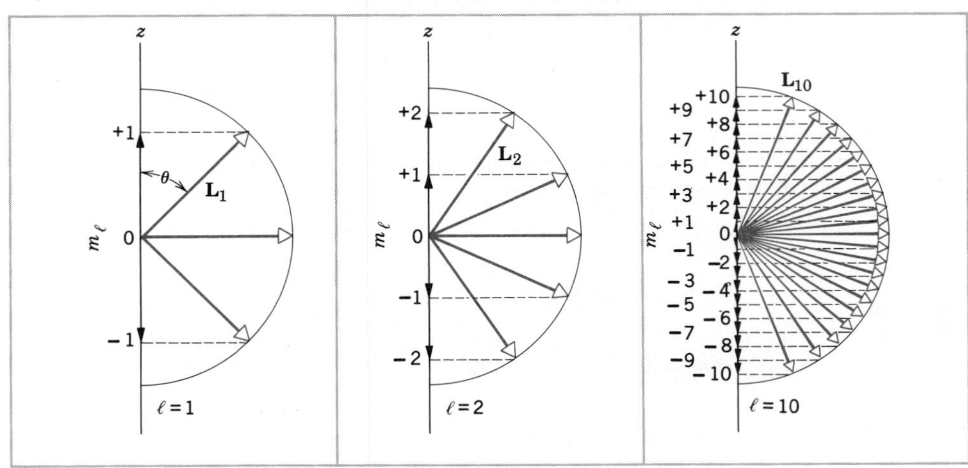

Figure 6 The allowed values of L_z, for $\ell = 1$, 2, and 10. The numbers on the z axis are the values of the magnetic quantum number m_ℓ. The figures are drawn to different scales.

Figure 6 shows the allowed values of L_z for $\ell = 1, 2,$ and 10. Note that, for a given value of ℓ, there are $2\ell + 1$ different values of m_ℓ. For $\ell = 10$, we begin to merge with the classical limit, in which the correspondence principle requires that *any* orientation of the angular momentum vector be allowed. The restriction imposed by quantum theory on the direction of the angular momentum vector is called *space quantization;* we describe an early experimental demonstration of it in Section 45-8.

A Useful Vector Model. Figure 7 suggests a classical vector model that helps us to visualize the space quantization of **L**. It shows the angular momentum vector precessing about the z direction, like a top precessing about a vertical axis in the earth's gravitational field. The projection L_z remains constant as the motion proceeds.

Heisenberg's uncertainty principle—in its angular form; compare Eq. 22 of Chapter 44—is

$$\Delta L_z \cdot \Delta\phi \simeq h, \quad (z \text{ component}) \tag{8}$$

in which ϕ is the angle of rotation about the z axis in Fig. 7. Once we have specified the magnetic quantum number, L_z is *precisely* known; that is, $\Delta L_z = 0$. Equation 8 then requires that $\Delta\phi$ must be infinitely great, which means that we have no information at all about the angular position about the z axis of the precessing angular momentum vector **L**. We know the magnitude of **L** and its projection L_z on the z axis, and *nothing else.*

Orbital Magnetic Moments. As Fig. 7 suggests, the orbital magnetic moment of an electron is rigidly coupled to its orbital angular momentum. Thus, if the orbital angular momentum vector is restricted to a discrete set of components along the z axis, then the orbital magnetic moment vector must be similarly restricted. The allowed projections of the orbital magnetic moment vector (compare Eq. 6) are

$$\boxed{\mu_{\ell,z} = -m_\ell \mu_B,} \tag{9}$$

in which μ_B, defined (see Section 34-2) from

$$\begin{aligned}\mu_B &= eh/4\pi m \\ &= 9.274 \times 10^{-24}\ \text{J/T} = 5.788 \times 10^{-5}\ \text{eV/T}\end{aligned} \tag{10}$$

is called the *Bohr magneton.* It is a convenient measure of atomic magnetism, just as $\hbar$ is a convenient measure of atomic angular momentum, and r_B (the *Bohr radius;* see Section 43-10) is a convenient measure of atomic distance. The minus sign in Eq. 9 shows that (as we expect for an orbiting negative charge) the angular momentum and the magnetic moment vectors are oppositely directed.

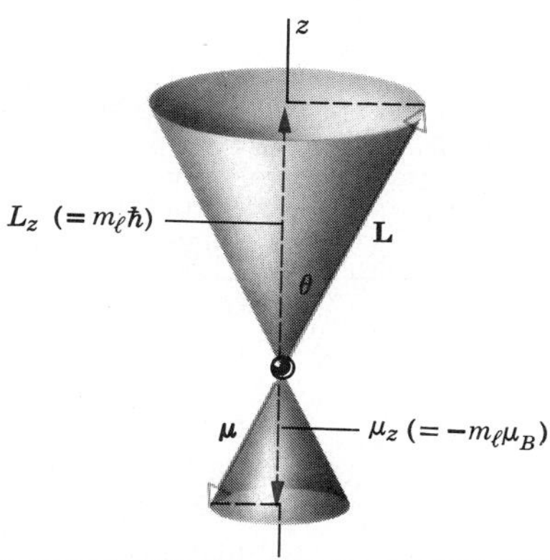

Figure 7 A vector model designed to represent the space quantization of the angular momentum and the magnetic moment vectors. Note that the magnitudes L and μ and the projections L_z, and μ_z remain constant as the vectors precess about the z axis.

Sample Problem 1 What is the minimum angle θ in Fig. 7 between the angular momentum vector and the z axis? Calculate the result for $\ell = 1, 10^2, 10^3, 10^4,$ and 10^9.

From Fig. 6 we see that the minimum angle occurs when $m_\ell = \ell$ in Eq. 6 and is

$$\begin{aligned}\theta_{\text{min}} &= \cos^{-1}\frac{L_{z,\text{max}}}{L} = \cos^{-1}\frac{\ell\hbar}{\sqrt{\ell(\ell+1)}\,\hbar} \\ &= \cos^{-1}\left(1 + \frac{1}{\ell}\right)^{-1/2}\end{aligned}$$

Substituting for ℓ in the above equation leads to:

ℓ	θ_{min}
1	45°
10^2	5.7°
10^3	1.8°
10^4	0.57°
10^9	0.0018°

For a macroscopic object like a spinning top, ℓ would be enormously larger than 10^9 and θ_{min} would be so close to zero that there would be no hope of measuring it. The correspondence principle really works!

Sample Problem 2 (a) For $n = 4$, what is the largest allowed value of ℓ?

From Eq. 5 it is

$$\ell_{max} = n - 1 \quad \text{or} \quad \ell_{max} = 3. \qquad \text{(Answer)}$$

(b) What is the magnitude of the corresponding angular momentum for $\ell = 3$?

From Eq. 4 it is

$$L = \sqrt{\ell(\ell + 1)}\,\hbar = \sqrt{(3)(3+1)}\,\hbar = 2\sqrt{3}\,\hbar. \qquad \text{(Answer)}$$

(c) How many different projections on the z axis may this angular momentum vector have?

From Eq. 7, we see that the number is

$$(2\ell + 1) = (2 \times 3 + 1) = 7. \qquad \text{(Answer)}$$

(d) What is the magnitude of the largest projected component?

This follows from Eq. 6, in which m_ℓ is given its largest value, which is ℓ. Thus

$$L_{z,max} = \ell\hbar = 3\,\hbar. \qquad \text{(Answer)}$$

(e) What is the smallest angle that the angular momentum vector can make with the z axis?

From (b) and (d) above we have

$$\theta_{min} = \cos^{-1}\frac{L_{z,max}}{L} = \cos^{-1}\frac{3\,\hbar}{2\sqrt{3}\,\hbar}$$

$$= \cos^{-1}\sqrt{3}/2 = 30°. \qquad \text{(Answer)}$$

45-6 Spin Angular Momentum and Magnetism

Whether or not it is trapped in an atom, an electron has an intrinsic angular momentum of its own. This *spin angular momentum,* as it is called, is also space quantized and can have components in the z direction given by

$$\boxed{S_z = m_s\,\hbar,} \qquad (11)$$

in which the *spin quantum number* m_s, can have only the values $+\frac{1}{2}$ and $-\frac{1}{2}$. We have used the symbol "S" for angular momentum associated with spin, to distinguish it from "L," the angular momentum associated with orbital motion. The spin quantum number is the *fourth* and last of the four quantum numbers needed to describe the hydrogen atom states. Table 1 summarizes them.

A host of experimental data requires us to assume that the corresponding spin magnetic moment of the electron can have only the values given by

$$\boxed{\mu_{s,z} = -2\,m_s\,\mu_B,} \qquad (12)$$

in which μ_B is the Bohr magneton. The factor 2 in Eq. 12 (compare Eq. 9) tells us that:

Spin orbital angular momentum is twice as effective as orbital angular momentum in generating magnetism.

This experimental result is fully supported by relativistic quantum theory.

Table 1 The Hydrogen Atom Quantum Numbers

Name	*Symbol*	*Allowed Values*	*Associated With*	*Number of Values*
Principal	n	$1, 2, 3, \ldots$	Energy	∞
Orbital	ℓ	$0, 1, 2, \ldots (n-1)$	Orbital angular momentum[a]	n
Magnetic	m_ℓ	$0, \pm 1, \pm 2, \ldots \pm \ell$	Orbital angular momentum[b]	$2\ell + 1$
Spin	m_s	$\pm\frac{1}{2}$	Spin angular momentum[c]	2

[a] Magnitude; see Eq. 4.
[b] Projected values; see Eq. 6.
[c] Projected values; see Eq. 11.

45-7 The Hydrogen-Atom Wave Functions*

To complete our discussion of the hydrogen atom, let us examine the wave functions for a few of its states. We start with the ground state, for which the quantum numbers are: $n = 1$, $\ell = 0$, and $m_\ell = 0$. The wave function for this state, as we saw in Section 44-6, depends only on r. It is reasonable that such a spherically-symmetrical "billiard-ball" state should have zero angular momentum because all directions through the center of the atom when it is in this state are completely equivalent.

The *radial probability density* for the ground state (see Eq. 18 of Chapter 44) is

$$P(r) = \left(\frac{4r^2}{r_B^3}\right) e^{-2r/r_B} \tag{13}$$

Figure 8*a* is a plot of this function. Recall that the radial probability density is defined so that $P(r)\,dr$ gives the probability that the electron will be found between shells whose radii are r and $r + dr$. Figure 8*a* shows that $P(r)$ has its maximum value for $r = r_B$.

Consider next the state with $n = 2$, $\ell = 0$, and $m_\ell = 0$. Like all states with $\ell = 0$, this state is also a spherically-symmetric, or "billiard-ball" state, its radial probability density being given by

$$P(r) = \left(\frac{r^2}{8r_B^3}\right)\left(2 - \frac{r}{r_B}\right)^2 e^{-r/r_B} \tag{14}$$

Figure 8*b* shows a plot of this function. Inspection of Eq. 14 shows that $P(r) = 0$ for $r = 2r_B$.

For $n = 2$, states with $\ell = 1$ are also permitted. There are three such states, defined by the quantum numbers:

n	ℓ	m_ℓ
2	1	+1
2	1	0
2	1	−1.

The three values of m_ℓ represent the three allowed orientations of the orbital angular momentum vector corresponding to $\ell = 1$; see Fig. 6*(a)*.

The probability densities for these three states are *not* spherically symmetric. That is, as Fig. 9 shows, the

* In this section, we omit consideration of electron spin. Its quantum number, m_s, has no effect on the wave functions and simply doubles the numbers of states defined by the quantum numbers n, ℓ, and m_ℓ.

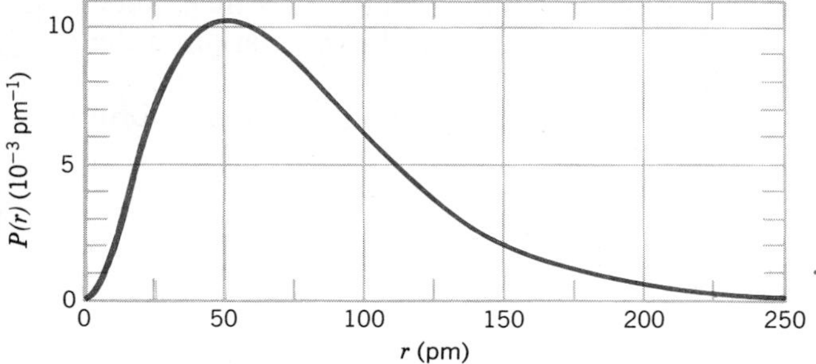

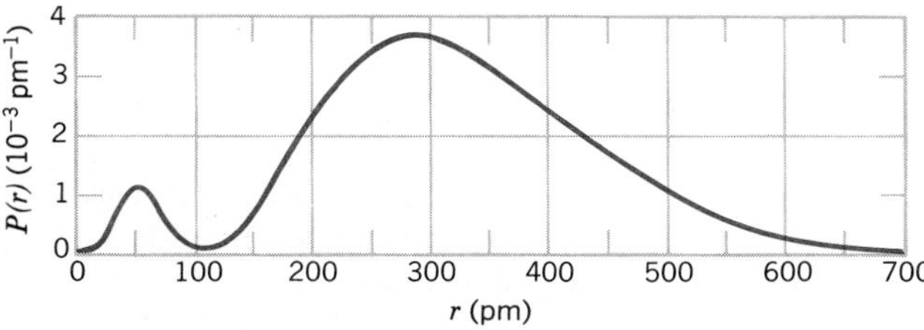

Figure 8 (*a*) The radial probability density for the ground state of the hydrogen atom, for which $n = 1$, $\ell = 0$ and $m_\ell = 0$. (*b*) The radial probability density for the state of the hydrogen atom with $n = 2$, $\ell = 0$, and $m_\ell = 0$.

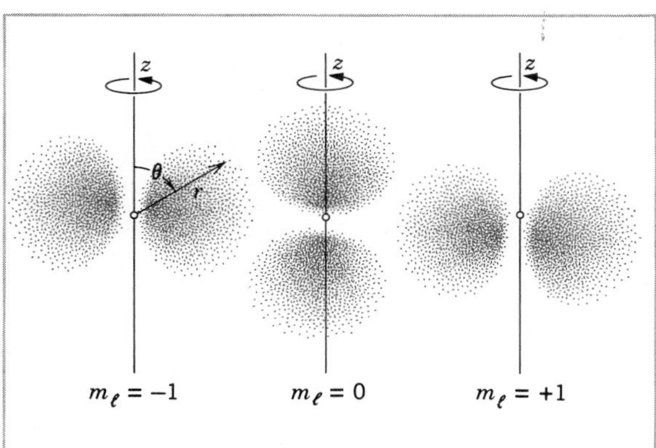

Figure 9 "Dot plots" for the three states of the hydrogen atom with $n = 2$ and $\ell = 1$. The values of m_ℓ correspond to the three allowed orientations in space of the angular momentum vector corresponding to $\ell = 1$. The patterns are symmetrical about the z axis, the density of dots at any point being proportional to the probability density at that point. (Although the probability densities in the three patterns are functions of both r and θ, the sum of all three patterns is spherically symmetrical, being a function of r alone.)

probability density at any point depends not only on the radial distance r to that point but also on the angle θ between the radial line and the z axis. The density of the dots at any point in the "dot plots" of Fig. 9 is proportional to the probability density at that point; all three plots are symmetrical if rotated about the z axis.

45-8 The Stern-Gerlach Experiment

In 1922, several years before the development of wave mechanics, space quantization was verified experimentally by Otto Stern and Walter Gerlach. Figure 10 shows their apparatus.

Silver is vaporized in an electrically-heated "oven" and silver atoms spray into the external vacuum of the apparatus from a small hole in the oven wall. The atoms (which are electrically neutral but which have a magnetic moment) are formed into a narrow beam as they pass through a collimating slit. The beam then passes between the poles of an electromagnet, finally depositing itself on a glass detector plate.

A Dipole in a Nonuniform Field. The pole faces of the magnet in Fig. 10 are shaped to make the magnetic field as *nonuniform* as possible. We digress to ask what force acts on a magnetic dipole placed such a field. Figure 11*a* shows a dipole of magnetic moment $\boldsymbol{\mu}$, making an angle θ with a *uniform* magnetic field. We can imagine the dipole to have a North and a South pole, the magnetic dipole moment vector $\boldsymbol{\mu}$ pointing (by convention) from the latter toward the former. We see that, for a uniform field, there is no net force on the dipole. The upward and downward forces on the poles are of the same magnitude and they cancel, no matter what the orientation of the dipole.

Figures 11*b,c* show the situation in a nonuniform

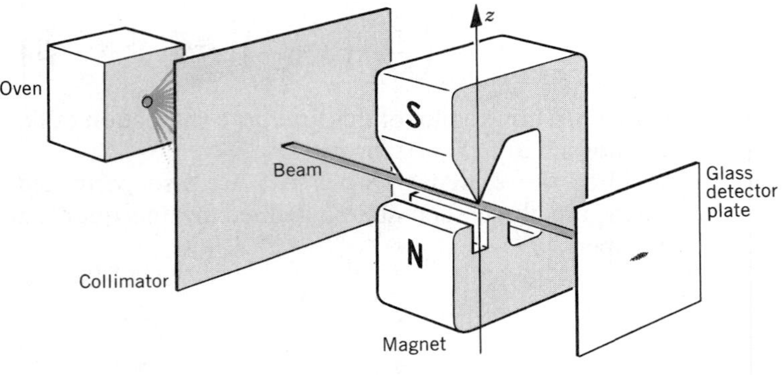

Figure 10 The apparatus of Stern and Gerlach, used to demonstrate space quantization.

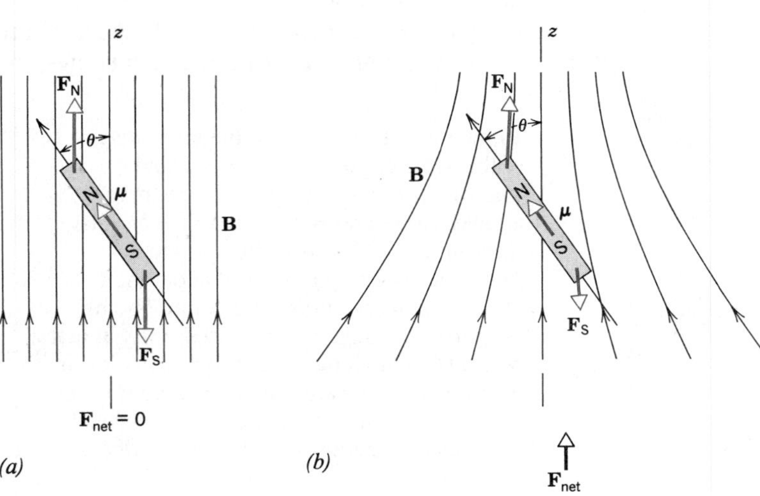

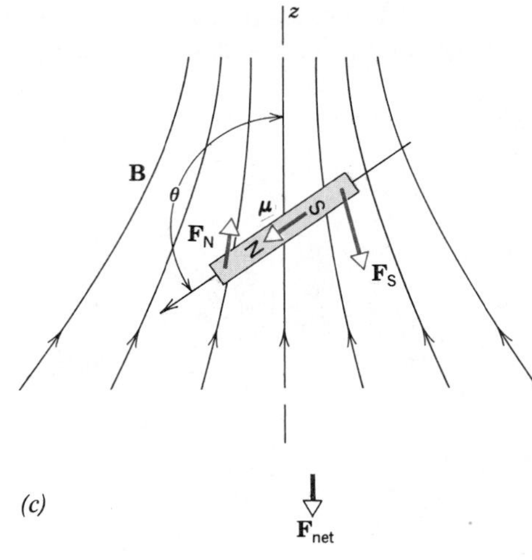

Figure 11 A magnetic dipole, represented as a small bar magnet with two poles, in (*a*) a uniform magnetic field and (*b,c*) a nonuniform field. The net force acting on the magnet is zero in (*a*), points up in (*b*) and points down in (*c*).

field. Here the upward and downward forces do *not* have the same magnitude because the two poles are immersed in fields of different strengths. In this case there *is* a net force, both its magnitude and direction depending on the orientation of the dipole, that is, on the value of θ. In Fig. 11*b* this net force is up and in Fig. 11*c* it is down. Thus, the silver atoms will be deflected as they pass through the magnet, the direction and the magnitude of the deflection depending on the orientation of their magnetic moment.

Now let us calculate the deflecting force quantitatively. The magnetic potential energy of a dipole in a magnetic field **B** is given by Eq. 31 of Chapter 30, or

$$U = -\boldsymbol{\mu} \cdot \mathbf{B} = -\mu B \cos\theta \qquad (15)$$

in which θ (see Fig. 11) is the angle between the directions of $\boldsymbol{\mu}$ and of $\boldsymbol{B}$. From Eq. 13 of Chapter 8, the net force F_z on the atom is $-(dU/dz)$ or, from Eq. 15

$$F_z = -(dU/dz) = \mu\,(dB/dz)\cos\theta. \qquad (16)$$

In Fig. 11*b,c*, B increases as z increases so that dB/dz is positive. Thus, the sign of the deflecting force F_z in Eq. 16 is determined by the angle θ. If $\theta < 90°$ (as in Fig. 11*b*), the atom will be deflected up; if $\theta > 90°$ (as in Fig. 11*c*) the deflection will be down.

The Experimental Results. When the electromagnet in Fig. 10 is turned off there will be no deflections of the atoms and the beam will form a narrow line on the detecting plate. When the electromagnet is turned on, however, strong deflecting forces come into play. Then there are two possibilities, depending upon whether space quantization exists or not. (Don't forget that the object of this experiment is to find out!) If there is no space quantization, the atomic magnetic dipoles will have a continuous distribution of angles with the direction of the magnetic field and the beam will simply broaden.

On the other hand, if space quantization *does* exist there will be only a discrete set of values for θ. This means that there will be only a discrete set of values for the deflecting force F_z in Eq. 16 and the beam will split up into a number of discrete components.

Figure 12 shows what happens. The beam does *not* broaden but splits cleanly into two subbeams.* Space quantization exists! Stern and Gerlach ended the published report of their work with the words, "We view these results as direct experimental verification of space quantization in a magnetic field." Physicists everywhere agree.

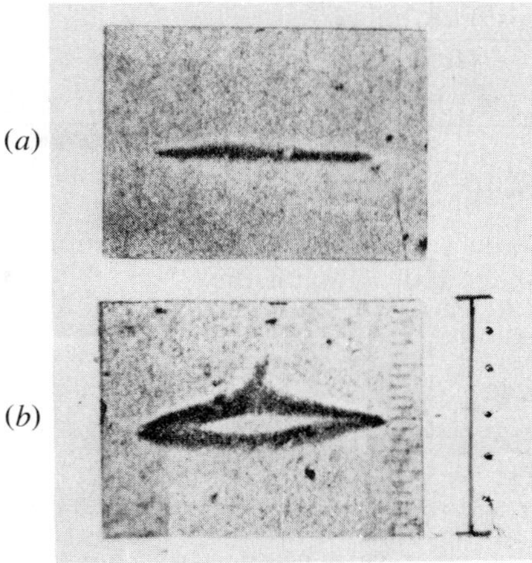

Figure 12 The results of the Stern–Gerlach experiment, showing the silver deposit on the glass detector plate of Fig. 10, with the magnetic field (*a*) turned off and (*b*) turned on. The beam has been split into two subbeams by the action of the field. The vertical bar at the right in (*b*) represents 1 mm.

Sample Problem 3 In a Stern–Gerlach experiment, the magnetic field gradient dB/dz was 1.4 T/mm along the beam path and the length w of the beam path through the magnet was 3.5 cm. The temperature of the oven in which the silver was evaporated was adjusted so that the most probable speed v for the atoms in the beam was 750 m/s. Find the separation d between the two deflected subbeams as they emerge from the magnet. The mass M of a silver atom is 1.8×10^{-25} kg and the projection of its magnetic moment on the z axis is 1.0 Bohr magneton, or 9.27×10^{-24} J/T.

The acceleration of a silver atom as it passes through the electromagnet is (see Eq. 16)

$$a = \frac{F_z}{M} = \frac{(\mu\cos\theta)(dB/dz)}{M}.$$

The vertical deflection of either subbeam as it clears the magnet is

$$\tfrac{1}{2}d = \tfrac{1}{2}at^2 = \tfrac{1}{2}\,\frac{(\mu\cos\theta)(dB/dz)}{M}\left(\frac{w}{v}\right)^2$$

so that

* The spin and the orbital angular momenta of the electrons of a silver atom cancel out *except* for the spin angular momentum of its single valence electron. This spin can have only two orientations, described by $m_s = +\frac{1}{2}$ and $m_s = -\frac{1}{2}$. Hence two subbeams, and not some other number.

$$d = \frac{(\mu \cos\theta)(dB/dz)w^2}{Mv^2}$$
$$= \frac{(9.27 \times 10^{-24}\ \mathrm{J/T})(1.4 \times 10^3\ \mathrm{T/m})(3.5 \times 10^{-2}\ \mathrm{m})^2}{(1.8 \times 10^{-25}\ \mathrm{kg})(750\ \mathrm{m/s})^2}$$
$$= 1.6 \times 10^{-4}\ \mathrm{m} = 0.16\ \mathrm{mm}. \quad \text{(Answer)}$$

This is the order of magnitude of the separation displayed in Fig. 12; note the scale in that figure.

45-9 Science, Technology and Spin— An Aside

Most discoveries in pure science turn out to have applications in technology. On November 8, 1895, for example, William Röntgen, working in his physics laboratory at the University of Wurzburg in Germany, discovered x rays. Less than three months later a skater, having fallen on the ice of the Connecticut River, had his broken arm x rayed at Dartmouth College, the first medical application of x rays in this country.

The spin of the electron was postulated in 1925 by two Dutch graduate students (George Uhlenbeck and Samuel Goudsmit), two years away from their doctorates at the University of Leiden. It took longer for spin than for x rays to make the journey from pure science to technology but it has now done so. Consider a proton, perhaps residing in a drop of water. The proton, like the electron, also has a spin of $\frac{1}{2}$ in units of $\hbar$. Its magnetic dipole moment $\boldsymbol{\mu}$ can occupy either of two quantized orientations with respect to a magnetic field $\mathbf{B}$, as Fig. 13a shows. These two positions differ in energy by $2\mu B$, this being the work required to turn a magnetic dipole end for end in a magnetic field.

If a drop of water containing our proton is subjected to an electromagnetic field that is alternating with a fre-

Figure 14 Wolfgang Pauli (Mr. Exclusion Principle) and Niels Bohr (Mr. Correspondence Principle) watching a "Tippy Top," a top that spins for a while and then turns upside down. They seem to be waiting for the "spin flip."

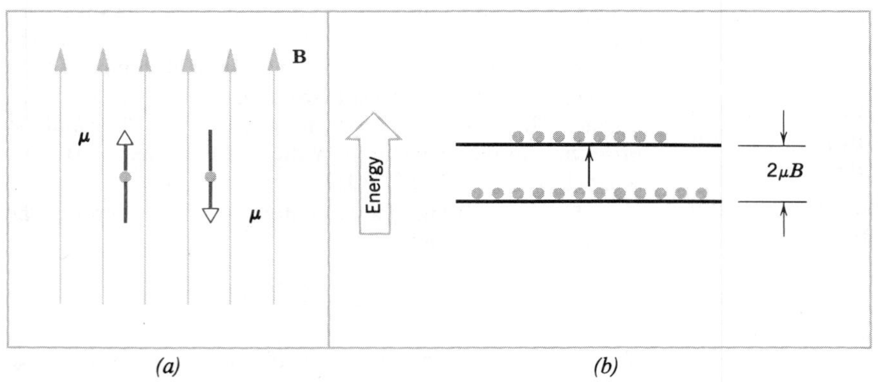

Figure 13 (a) A proton, whose spin is $\frac{1}{2}$ in units of $\hbar$, can occupy either of two quantized positions in an external magnetic field. (b) If Eq. 17 is satisfied, the protons in the sample can flip from one orientation to the other.

quency ν, transitions between the two orientations of μ_p—called *spin flips* (see Fig. 14)—may be induced. For this to happen the equation

$$h\nu = 2\mu B \tag{17}$$

must be satisfied. That is, the energy $h\nu$ of the photons associated with the alternating electromagnetic field must be just equal to the energy difference between the two spin orientations.

The spin flip may go either way (that is, from up to down or from down to up) with equal probability. However, if the water drop is in thermal equilibrium, there will be more proton spins in the lower energy state than in the higher energy state, as Fig. 13*b* suggests. This means that there will be a net *absorption* of energy from the electromagnetic field.

The usefulness of magnetic resonance studies, carried out under conditions of high resolution, lies in the fact that the factor B in Eq. 17 is *not* the imposed external magnetic field B_{ext} but rather it is that field as modified by the small local internal magnetic field B_{local} due to the electrons and nuclei in the molecule of which our proton is a part. Thus, we can rewrite Eq. 17 as

$$h\nu = 2\mu(B_{local} + B_{ext}). \tag{18}$$

In practice, it is customary to leave the frequency ν of the electromagnetic oscillator fixed and to vary B_{ext} until Eq. 18 is satisfied and an absorption peak is recorded.

Figure 15 shows a *nuclear magnetic resonance spectrum,* as it is called, for ethanol, whose formula we may write as CH_3—CH_2—OH. The various resonance peaks all represent spin flips of protons. They occur at different values of B_{ext}, however, because the local environments of the protons within the ethanol molecule are different. The spectrum of Fig. 15 is a unique signature for ethanol. The nuclear magnetic resonance method is of great value as an analytical tool in organic chemistry.

Spin technology has also been applied to medical diagnostic imaging. The protons in the various tissues of the human body find themselves in different local magnetic environments. When the body, or part of it, is immersed in a strong external magnetic field, these environmental differences can be detected by spin-flip techniques and translated by computer processing into an x-ray-like image. Figure 16, for example, shows a cross section of a human head taken by this method. The

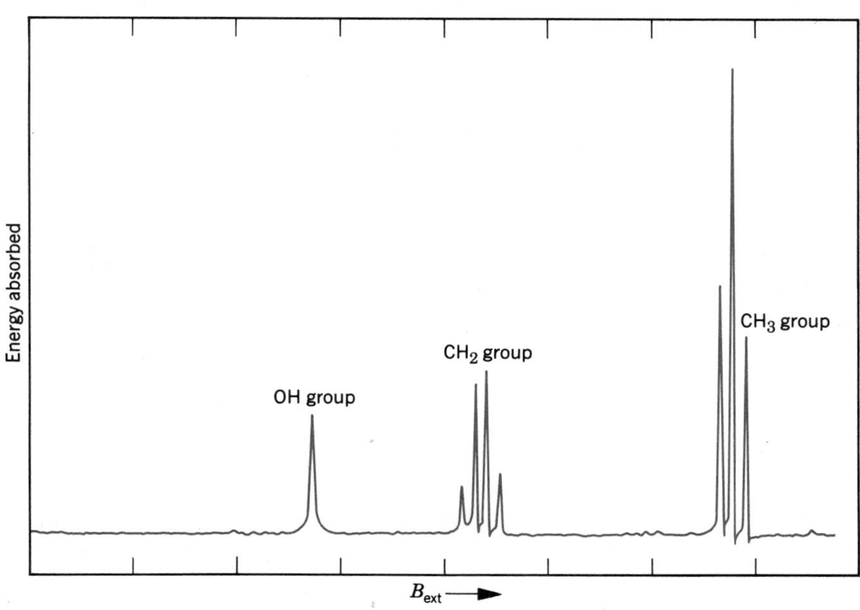

Figure 15 A nuclear magnetic resonance spectrum for ethanol. All the absorption lines are due to the spin flips of protons. The three groups of lines correspond, as indicated, to protons in the OH group, the CH_2 group, and the CH_3 group within the molecule. The entire horizontal axis encompasses considerably less than 10^{-4} T.

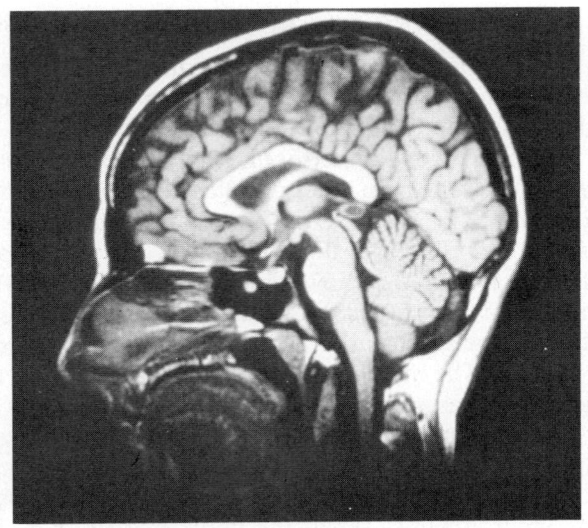

Figure 16 An image of a cross section of a human head, taken by spin-flip techniques. From the point of view of the patient, magnetic resonance imaging (MRI) is a noninvasive technique and shows anatomical detail not present on x-ray images.

method supplements x-ray imaging in many important ways. All this from a suggestion by two Dutch graduate students, more than 60 years ago!

Sample Problem 4 A drop of water is suspended in a 1.8-T magnetic field and an alternating electromagnetic field is applied, its frequency chosen so as to produce spin flips for the protons in the sample. The magnetic moment of the proton is 1.41×10^{-26} J/T. Ignore local magnetic fields within the sample. What are the frequency and the wavelength of the alternating field?

From Eq. 17 we have

$$\nu = \frac{2\mu B}{h}$$

$$= \frac{(2)(1.41 \times 10^{-26}\ \text{J/T})(1.8\ \text{T})}{6.63 \times 10^{-34}\ \text{J} \cdot \text{s}}$$

$$= 7.66 \times 10^{7}\ \text{Hz} = 76.6\ \text{MHz}. \qquad \text{(Answer)}$$

This frequency is in the VHF television band. The corresponding wavelength is

$$\lambda = \frac{c}{\nu} = \frac{3.00 \times 10^{8}\ \text{m/s}}{7.66 \times 10^{7}\ \text{Hz}} = 3.92\ \text{m}. \qquad \text{(Answer)}$$

45-10 Multi-electron Atoms and the Periodic Table

We turn now from the hydrogen atom to atoms with more than one electron, and we state without proof that:

> *The four quantum numbers listed in Table 1 that identify the states of the hydrogen atom also serve to identify the states of individual electrons in atoms with more than one electron.*

Just because states are described by the same quantum numbers does not mean that they have the same energies and wave functions; they do not. In multi-electron atoms, the potential energy associated with any given electron is determined not only by the atomic nucleus but also by the other electrons in the atom. When this is taken properly into account in solving the Schrödinger equation, it turns out that the energy of a state no longer depends only on the principal quantum number n—as in Eq. 3 for the hydrogen atom states—but also on the orbital quantum number ℓ.

When we assign electrons to states in a multi-electron atom, we must be guided by the *Pauli exclusion principle,* which asserts:

> *Only a single electron can be assigned to a given quantum state.*

If this important principle did not hold, all of the electrons in an atom would move to the state of lowest energy and the world would be a far different place.

The Orbitals. We group the electron states of a multi-electron atom into *orbitals,* each characterized by a given value of n and of ℓ. Equation 7 tells us that, for a given value of ℓ, there are $2\ell + 1$ possible states, each with a different value of the magnetic quantum number m_ℓ. Each of these states has two choices of the spin quantum number m_s, so that the total number of states in an orbital with a given value of ℓ is $2(2\ell + 1)$. For $\ell = 0$, this number is two and for $\ell = 1$, it is six. Note that the number of states in an orbital depends only on ℓ but the energy of those states depends on both n and ℓ.

Neon. This atom has 10 electrons. Two of them occupy the orbital with $n = 1$ and $\ell = 0$ (the 1,0 orbital), filling it completely. Two of the remaining eight electrons fill the orbital with $n = 2$ and $\ell = 0$ (the 2,0 orbital). The remaining six electrons fill the 2,1 orbital. Thus, the neon atom in its lowest energy state has its electrons arranged in three filled orbitals.

In a filled orbital, all possible projections of both the orbital and the spin angular momentum vectors are present in the same atom and thus cancel out for the atom as a whole. Such a filled orbital is a tightly bound unit, described by a spherically symmetric wave function. It has zero net angular momentum and, correspondingly, zero net magnetic dipole moment. This is consistent with the fact that the neon atom is chemically inert.

Sodium. Next after neon comes sodium, with 11 electrons. Ten of them form a neonlike core, leaving the remaining single electron in the 3,0 orbital. To a first approximation we can think of the sodium nucleus (whose charge is $+11e$) as partially screened by the neonlike core (charge $-10e$), leaving a reduced net central charge to govern the motion of the outer electron. The entire angular momentum and magnetic dipole moment of the sodium atom is due to this single, relatively loosely bound, valence electron. Because it is in a state with $\ell = 0$, its angular momentum and magnetic moment must be those associated with the intrinsic spin of this single electron.

The existence of a single electron in an outer, unfilled orbital is consistent with the fact that sodium is chemically active. Its valence electron is loosely bound, only 5 eV of energy being required to remove it. Contrast this with the chemically inert neon, for which 22 eV must be expended to remove an electron.

The Periodic Table. In adding electrons to a bare nucleus to form an atom, the orbitals are always filled in the order of increasing energy. For heavier atoms, however, this filling order is not always the logical sequence suggested by the quantum numbers. In krypton, for example, the 4,0 orbital lies lower in energy than the 3,2 orbital and thus the electrons in the 4,0 orbital lie deeper within the atom than do the 10 electrons in the 3,2 orbital. It is a major triumph of wave mechanics that, taking the filling order properly into account, we can account for the entire periodic table of the elements.

Sample Problem 5 Account for the numbers of elements in the six complete horizontal periods of the periodic table in terms of the populations of orbitals.

As Appendix E shows, the numbers of elements in the horizontal rows of the periodic table are 2, 8, 8, 18, 18, and 32. We have seen that the populations of the orbitals depend only on the orbital quantum number ℓ and are given by $2(2\ell + 1)$. Thus:

Orbital Quantum Number ℓ	*Population* $2(2\ell + 1)$
0	2
1	6
2	10
3	14

We can account for each horizontal period in terms of filled orbitals in this way:

Periods	*Elements in Period*	*Array of Orbitals*
1	2	2
2,3	8	$2 + 6$
4,5	18	$2 + 6 + 10$
6	32	$2 + 6 + 10 + 14$

45-11 X Rays and the Numbering of the Elements

Here we shift our attention from electrons on the outer fringes of the atom to electrons deep within the atom. We move from a region of relatively low binding energy (~ 5 eV for the valence electron of sodium, for example) to a region of higher energy (~ 70 keV for the binding energy of the innermost electron in tungsten, for example, over 10,000 times larger.) The radiations we deal with shift dramatically in wavelength, from ~ 600 nm for the yellow light from sodium to ~ 20 pm for one of the characteristic tungsten radiations. We are speaking of x rays.

Our concern here is with what these rays—whose medical, dental, and industrial usefulness is so well known—can teach us about the structure of the atoms that absorb or emit them. We focus on the work of the British physicist H. G. J. Moseley who, by x-ray methods, developed the concept of atomic number and gave physical meaning to the ordering of the elements in the periodic table.

As we saw in Section 41-11, x rays are produced when energetic electrons strike a solid target and are brought to rest in it. Figure 17 shows the wavelength spectrum of the x rays that are produced when a beam of 35-keV electrons falls on a molybdenum target. We see

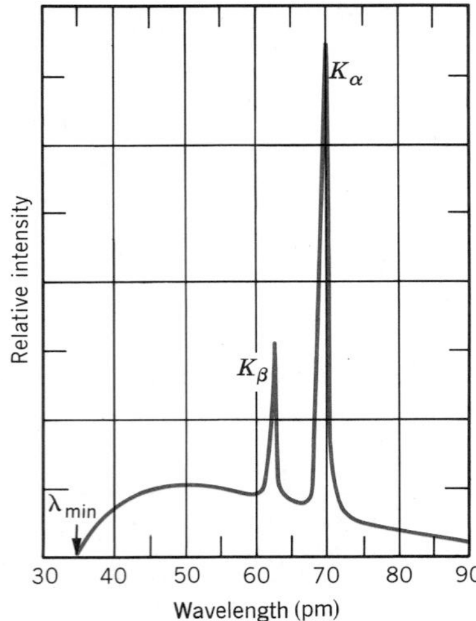

Figure 17 The distribution by wavelength of the x rays produced when 35-keV electrons strike a molybdenum target. Note the sharp peaks standing out above a continuous background. (1 pm = 10^{-12} m.)

that it consists of a broad spectrum of radiation distributed continuously in wavelength on which are superimposed x rays of sharply-defined frequency. These radiations arise in different ways, which we discuss separately.

45-12 The Continuous X-ray Spectrum

Here we examine the continuous spectrum of Fig. 17, ignoring for the time being the two prominent peaks that rise from it. If the incident electrons are accelerated through a potential difference V, their kinetic energy as they strike the target will be eV. As they are brought to rest within the target, we expect that electrons of all kinetic energies from zero to eV will be present. Consider an electron of kinetic energy K within this range that happens to pass close to the nucleus of one of the molybdenum atoms in the target, as in Fig. 18. The electron may well lose an amount of energy ΔK, which will appear as the energy of an x-ray photon that is radiated away from the site of the encounter. All electrons whose kinetic energies lie in the range from 0 to eV can undergo such *bremsstrahlung** processes and all contribute thereby to creating the continuous x-ray spectrum.

* This word means "braking radiation" in German.

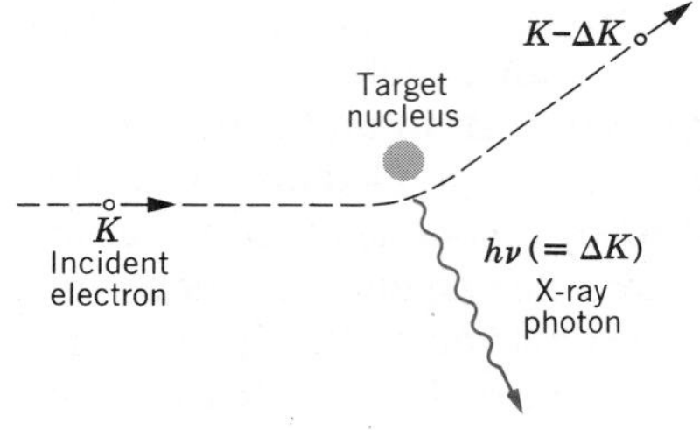

Figure 18 When an electron passes near the nucleus of a target atom, it may generate an x-ray photon, losing part of its kinetic energy in the process.

A prominent feature of the continuous spectrum of Fig. 17 is the sharply defined *cutoff wavelength* λ_{min}, below which the continuous spectrum does not exist. This minimum wavelength corresponds to an encounter in which one of the incident electrons, with initial kinetic energy eV, loses *all* of this energy in a single encounter, radiating it away as a single x-ray photon. Thus,

$$eV = h\nu = \frac{hc}{\lambda_{min}}$$

or

$$\boxed{\lambda_{min} = \frac{hc}{eV}} \quad \text{(cutoff wavelength).} \qquad (19)$$

The cutoff wavelength is totally independent of the target material. If you were to switch from a molybdenum to a copper target, for example, all features of the x-ray spectrum of Fig. 17 would change *except* the cutoff wavelength.

Sample Problem 6 A beam of 35-keV electrons falls on a molybdenum target, generating the x rays whose spectrum is shown in Fig. 17. Calculate the expected cutoff wavelength λ_{min}.

From Eq. 19 we have

$$\lambda_{min} = \frac{hc}{eV} = \frac{(4.14 \times 10^{-15}\ \text{eV} \cdot \text{s})(3.00 \times 10^{8}\ \text{m/s})}{35 \times 10^{3}\ \text{eV}}$$
$$= 3.55 \times 10^{-11}\ \text{m} = 35.5\ \text{pm}. \qquad \text{(Answer)}$$

This agrees well with the value indicated by the vertical arrow in Fig. 17. Note that our calculation contains no reference to the material of which the target is made.

45-13 The Characteristic X-ray Spectrum

We now turn our attention to the two peaks of Fig. 17, labeled K_α and K_β. These peaks, together with other peaks that appear at longer wavelengths, are characteristic of the target material and form the *characteristic x-ray spectrum* of the element in question.

Here is how *these* x-ray photons arise: (1) An energetic incoming electron strikes an atom in the target and knocks out one of its deep-lying electrons. If the electron is in the shell with $n = 1$ (called, for historical reasons, the K-shell) there remains a vacancy, or a *hole* as we shall call it, in this shell. (2) One of the outer electrons moves in to fill this hole and, in the process, the atom emits a characteristic x-ray photon. If the electron falls from the shell with $n = 2$ (called the L-shell) we have the K_α line of Fig. 17; if it falls from the shell with $n = 3$ (called the M-shell) we have the K_β line, and so on. Of course, such a transition will leave a hole in either the L or the M shell, but this will be filled in by an electron from still farther out in the atom, the atom, in the process, emitting still another characteristic x-ray spectral line.

In studying the radiations emitted by the single electron of the hydrogen atom, we found it convenient to draw an energy level diagram in which each level corresponded to a different quantum state for that single electron. We chose as our zero-energy configuration the state of the system in which the electron is at rest and is completely removed from the atom.

In studying x rays, however, we find it much more convenient to keep track of the single hole created deep in the electron cloud, rather than of the quantum states of the many electrons that remain in the atom. We choose as our zero-energy configuration the state of the system in which the hole has been completely removed from the atom, that is, the normal neutral atom in its ground state. Recall that the configuration to which we assign zero energy is quite arbitrary. Only differences in energy are important and these are the same no matter what configuration we choose to represent $E = 0$.

Figure 19 shows an x-ray energy level diagram for molybdenum, the element to which Fig. 17 refers. The base line ($E = 0$) represents, as we have said, the neutral atom in its ground state. The level marked K ($E = 20$ keV) represents the energy of the molybdenum atom with a hole in its K-shell. Similarly, the level marked L ($E = 2.7$ keV) represents the energy of the atom with a hole in its L-shell, and so on.

The transitions marked K_α and K_β in Fig. 19 show the origin of the two sharp x-ray lines in Fig. 17. The K_α line, for example, originates when an electron from the L-shell of molybdenum fills the hole in the K-shell. This corresponds to a hole moving downward on the energy level diagram of Fig. 19 from the K-level to the L-level.

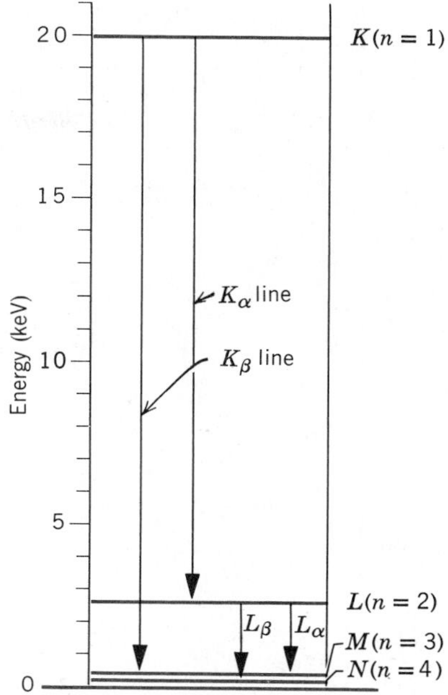

Figure 19 An atomic energy level diagram for molybdenum, showing the transitions that give rise to the characteristic x rays of that element. (All levels, except the K-level, contain a number of closely lying components, not shown here.)

Moseley and the X-ray Spectrum. In his investigation of the characteristic x-ray spectrum, Moseley (see Fig. 20) generated characteristic x rays by using as many elements as he could find—he found 38—as targets for electron bombardment in a special evacuated x-ray tube of his own design. By means of a trolley, manipulated by strings, he could put various targets in place in the path of the incident electron beam. He measured the wavelengths of the x rays by the crystal diffraction method that we described in Section 41-11.

Moseley then sought, and found, regularities in these spectra as he moved from element to element in the periodic table. In particular, he noted that if, for a given spectral line such as K_α, he plotted the square root of the line frequency against the position of the element in the periodic table, a straight line resulted. Figure 21 shows a portion of his extensive data. We shall see later why it was logical to plot the data in this way and why a straight

Figure 20 H. G. J. Moseley in his laboratory, holding some of the simple apparatus with which he established the concept of atomic number.

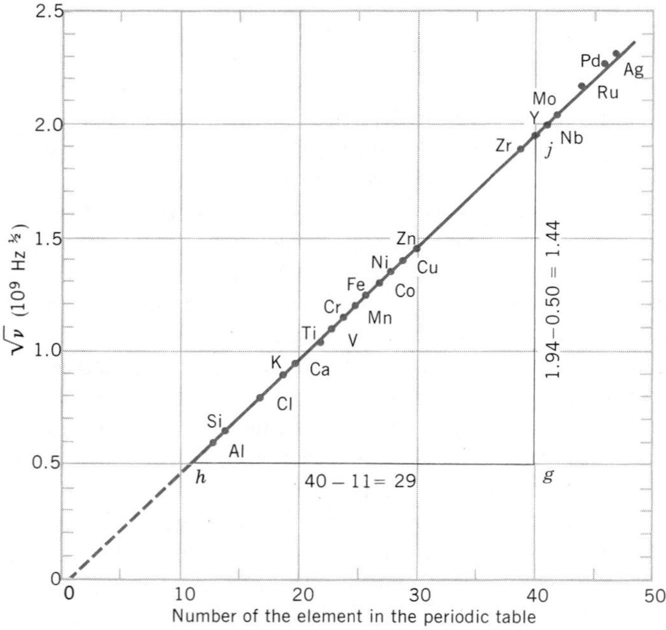

Figure 21 A Moseley plot of the K_α line of the characteristic x-ray spectrum of 21 elements. The frequency is calculated from the measured wavelength.

line is to be expected. Moseley's conclusion from the full body of his data was:

> *We have here a proof that there is in the atom a fundamental quantity, which increases by regular steps as we pass from one element to the next. This quantity can only be the charge on the central nucleus.*

Due to Moseley's work, the characteristic x-ray spectrum became the universally accepted signature of an element, permitting the solution of a number of periodic-table puzzles. Prior to that time (1913) the position of an element in the table was assigned in order of atomic *weight,* although there were several cases in which it was necessary to invert this order because of compelling chemical evidence; Moseley showed that it is the nuclear charge (that is, the atomic *number*) that is the real basis for numbering the elements.

The table had several empty squares, and a surprising number of claims for new elements had been advanced. The x-ray spectrum provided an indisputable test for such claims. The rare-earth elements, because of their similar chemical properties, had been only imperfectly sorted out. They were properly organized in short order. In more recent times, the identities of elements beyond uranium are pinned down without dispute when they are available in quantities large enough to permit a study of their x-ray spectra.

It is not hard to see why the characteristic x-ray spectrum shows such impressive regularities from element to element and the optical spectrum in the visible and near-visible region does not. The key to the identity of an element is the charge on its nucleus. Gold, for example, is what it is because its atoms have a nuclear charge of $+79e$. If it had one more nuclear charge it would not be gold, but mercury; if it had one less, it would be platinum. The K electrons, which play such a large role in the production of the characteristic x-ray spectrum, lie very close to the nucleus and are thus sensitive probes of its charge. The optical spectrum, on the other hand, involves transitions of the valence electrons. These outermost electrons are heavily screened from the nucleus by the remaining electrons of the atom and are not sensitive nuclear-charge probes.

Bohr Theory and the Moseley Plot. Moseley's experimental data are totally useful for numbering the elements, even if no theoretical basis for them could be established. Moseley went further, however, and showed that his results were consistent with Bohr's theory of atomic structure. We have seen that Bohr theory works quite well for the hydrogen atom but fails even for helium, the next element in the periodic sequence. How, then, can this theory work so well for such heavy atoms? We can glimpse some of the reasons if we focus attention on the single hole in such heavy atoms, whose transitions generate the x-ray spectrum. From this point of view, the situation is not so much different from that of the hydrogen atom, whose spectrum is produced by the transitions of its single electron. Now for the details.

Suppose that one of the two electrons in the K shell of a heavy atom is removed, leaving a hole in that shell. That hole will be filled by an electron moving inward from the L shell and the atom will emit a K_α x-ray photon in the process. The effective nuclear charge "seen" by this moving electron is not Z but something very close to $Z-1$, because the full charge of the nucleus is partially screened by the electron that has remained undisturbed in the K shell throughout this transition.

Bohr's formula for the frequency of the radiation corresponding to a transition between any two atomic levels in hydrogenlike atoms is*

$$\nu = \frac{me^4Z^2}{8\epsilon_0^2h^3}\left(\frac{1}{n_1^2} - \frac{1}{n_2^2}\right),$$

in which m is the electron mass and n_1 and n_2 are quantum numbers. For the K_α transition, it is appropriate to replace Z by $Z-1$ and to put $n_1 = 1$ and $n_2 = 2$. If we do so and then take the square root of each side, we find

$$\sqrt{\nu} = \sqrt{\frac{3me^4}{32\epsilon_0^2h^3}}\,(Z-1), \qquad (20)$$

which we can write in the form

$$\sqrt{\nu} = a(Z-1), \qquad (21)$$

where a is the constant identified by the square root sign in Eq. 20. Equation 21 is the equation of a straight line, in full agreement with the experimental data of Fig. 21. If the plot of Fig. 21 is extended to higher atomic numbers, however, it departs somewhat from a straight line. The agreement with Bohr theory is, however, surprisingly good.

Sample Problem 7 A cobalt target is bombarded with electrons and the wavelengths of its characteristic spectrum are measured. A second, fainter, characteristic spectrum is also found, due to an impurity in the target. The wavelengths of the K_α lines are 178.9 pm (cobalt) and 143.5 pm (impurity). What is the impurity?

Let us apply Eq. 21 both to cobalt and to the impurity. Putting c/λ for ν, we obtain

$$\sqrt{\frac{c}{\lambda_{\text{Co}}}} = a(Z_{\text{Co}} - 1) \quad\text{and}\quad \sqrt{\frac{c}{\lambda_x}} = a(Z_x - 1).$$

Dividing yields

$$\sqrt{\frac{\lambda_{\text{Co}}}{\lambda_x}} = \frac{Z_x - 1}{Z_{\text{Co}} - 1}.$$

Substituting gives us

$$\sqrt{\frac{178.9\text{ pm}}{143.5\text{ pm}}} = \frac{Z_x - 1}{27 - 1}.$$

Solving for the unknown, we find

$$Z_x = 30.0. \qquad \text{(Answer)}$$

A glance at the periodic table identifies the impurity as zinc.

Sample Problem 8 Calculate the constant a in Eq. 21 and compare it with the measured slope of the straight line in Fig. 21.

If we compare Eqs. 20 and 21, we can write

$$a = \sqrt{\frac{3me^4}{32\epsilon_0^2h^3}}$$

$$= \sqrt{\frac{(3)(9.11\times10^{-31}\text{ kg})(1.60\times10^{-19}\text{ C})^4}{(32)(8.85\times10^{-12}\text{ F/m})^2(6.63\times10^{-34}\text{ J}\cdot\text{s})^3}}$$

$$= 4.96\times10^{7}\text{ Hz}^{1/2}. \qquad \text{(Answer)}$$

Equation 21 shows us that a is the slope of the straight line in Fig. 21. If we measure the triangle hgj carefully, we find

$$a = \frac{gj}{hg} = \frac{(1.94 - 0.50)\times10^{9}\text{ Hz}^{1/2}}{40 - 11}$$

$$= 4.97\times10^{7}\text{ Hz}^{1/2}, \qquad \text{(Answer)}$$

in good agreement with the prediction of Bohr theory. The agreement is not nearly so good for other than K_α lines in the x-ray spectrum; here one must rely on calculations based on wave mechanics.

45-14 Lasers and Laser Light

In the late 1940s and again in the early 1960s quantum physics made two enormous contributions to technology, the transistor and the laser. The first stimulated the

* The derivation parallels that of Eq. 23 of Chapter 43, except that the nuclear charge is taken to be Ze rather than e, as for the hydrogen atom.

growth of *microelectronics,* which deals with the interaction (at the quantum level) between electrons and bulk matter. The laser is leading the way in a new field—sometimes called *photonics*—which deals with the interaction (again at the quantum level) between photons and bulk matter.

To see the importance of lasers, let us look at some of the characteristics of laser light. We shall compare it as we go along with the light emitted by such sources as a tungsten filament lamp (continuous spectrum) or a neon gas discharge tube (line spectrum). We shall see that referring to laser light as "the light fantastic" goes far beyond whimsy.

Laser Light Is Highly Monochromatic. Tungsten light, spread over a continuous spectrum, gives us no basis for comparison. The light from selected lines in a gas discharge tube, however, can have wavelengths in the visible region that are precise to about 1 part in about 10^6. The sharpness of definition of laser light can be many times greater, as much as 1 part in 10^{15}.

Laser Light Is Highly Coherent. Wave trains for laser light may be several hundred kilometers long. This means that interference fringes can be set up by combining two separate beams whose path lengths differ by such distances. The corresponding coherence length for light from a tungsten filament lamp or a gas discharge tube is typically less than 1 meter.

Laser Light Is Highly Directional. A laser beam departs from strict parallelism only because of diffraction effects, determined (see Section 41–6) by the wavelength and the diameter of the exit aperture. Light from other sources can be made into an approximately parallel beam by a lens or a mirror, but the beam divergence is much greater than from a laser. Each point on, say, a tungsten filament source forms its own separate beam, the angular divergence of the overall composite beam being determined—not by diffraction—but by the size of the filament.

Laser Light Can Be Sharply Focused. This property is related to the parallelism of the laser beam. As for star light, the size of the focused spot for a laser beam is limited only by diffraction and not by the size of the source. Energy flux densities for laser light of $\sim 10^{16}$ W/cm^2 are readily possible. An oxyacetylene flame, by contrast, has an energy flux density of only $\sim 10^3$ W/cm^2.

The smallest lasers, used for telephone communication over optical fibers,* have as their active medium a semiconducting gallium arsenide crystal about the size of a pin head. The largest lasers, used for laser fusion research and for military applications, fill a large building; see the photo on page 618. They can generate pulses of laser light of $\sim 3 \times 10^{-9}$ s duration with a power level, during the pulse, of $\sim 8 \times 10^{13}$ W. This is more than a hundred times the total electric power generating capacity of the United States.

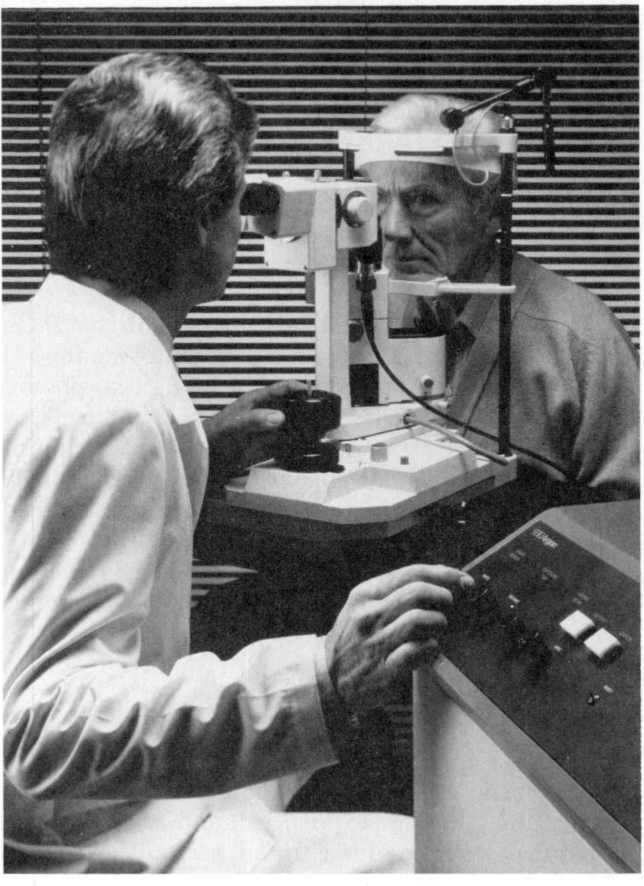

Figure 22 An ophthalmic surgeon performing an eye operation with a laser beam.

Other laser uses† include spot welding detached retinas (Fig. 22), drilling tiny holes in diamonds for drawing fine wires, cutting cloth (50 layers at a time, with no frayed edges) in the garment industry, precision surveying, precision length measurements by interferometry, the generation of holograms, and other applications too numerous to list (see Figs. 23 and 24).

* See Essay 17, "Optical Fibers and Communication," by Suzanne R. Nagel.

† See Essay 19, "Applications of Lasers," by Elsa Garmire.

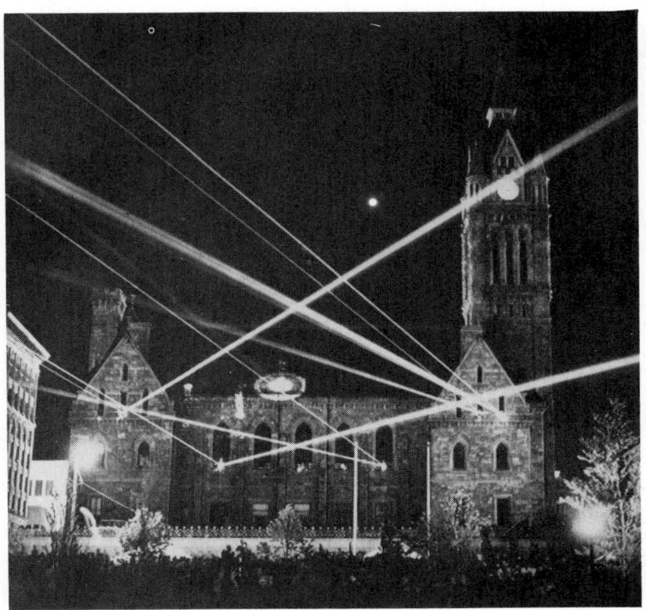

Figure 23 Laser beams in the sky at Holyoke, Massachusetts.

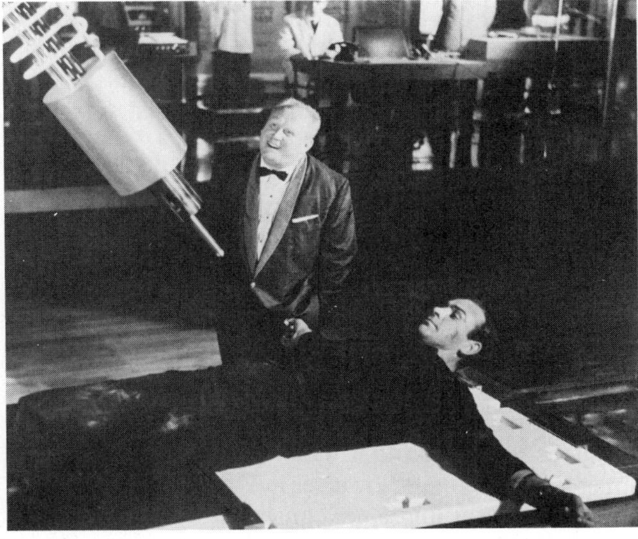

Figure 24 Goldfinger is about to use a laser beam to do some damage to James Bond.

45-15 Einstein and the Laser

The word "laser" is an acronym for **L**ight **A**mplification by the **S**timulated **E**mission of **R**adiation and we are thus not surprised to learn that *stimulated emission* is the key to its operation. Einstein introduced this concept into physics in 1917; although the world had to wait until 1960 to see an operating laser, the groundwork for its development was put in place at that earlier date.

Let us consider a single isolated atom that can exist in only one of two states, of energies E_1 and E_2. We discuss below three ways in which this atom can be caused to move from one of its two allowed states to the other.

Absorption. Figure 25*a* shows an atom initially in the lower of its two states, with energy E_1. We assume also that a continuous spectrum of radiation is present. If a photon of energy

$$h\nu = E_2 - E_1 \tag{22}$$

interacts with the atom, the photon will vanish and the atom will move to its upper energy state. We call this familiar process *absorption.*

Spontaneous Emission. In Fig. 25*b* the atom is in its upper state and no radiation is present. After a certain mean time τ, the atom moves of its own accord to the state of lower energy, emitting a photon of energy $h\nu$ in the process. We call this familiar process *spontaneous emission,* because it was not triggered by any outside influence. The light from the glowing filament in an ordinary light bulb is generated in this way.

Normally, the mean life of excited atoms before spontaneous emission occurs is $\sim 10^{-8}$ s. However, there are some states for which this mean life is much longer, perhaps as long as $\sim 10^{-3}$ s. We call such states *metastable;* they play an essential role in laser operation.

Stimulated Emission. In Fig. 25*c* the atom is again in its upper state, but this time a continuous spectrum of radiation is also present. As in absorption, a photon whose energy is given by Eq. 22 interacts with the atom. The result is that the atom moves to its lower energy state and there are now *two* photons where only one existed before.

The emitted photon in Fig. 25*c* is in every way identical to the triggering or *stimulating* photon. It has the same energy, direction, phase, and polarization. We can see how a chain reaction of similar processes could be triggered by one such *stimulated emission* event. Laser light is produced in this way.

Figure 25*c* describes the stimulated emission of a photon from a single atom. In the usual case, however, many atoms are present. Given a large number of atoms, in equilibrium at a certain temperature T, we may ask how many of them will be in level E_1 and how many in

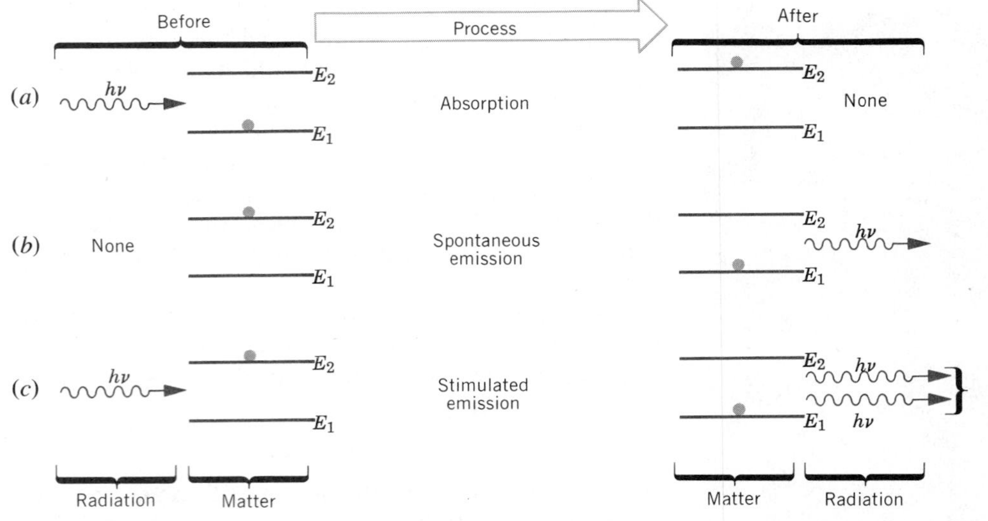

Figure 25 The interaction of matter and radiation in the processes of (*a*) absorption, (*b*) spontaneous emission, and (*c*) stimulated emission. It is the last process that accounts for the possibility of laser operation.

level E_2? Ludwig Boltzmann showed that the number n_x of atoms in any level whose energy is E_x is given by

$$n_x = Ce^{-E_x/kT}, \tag{23}$$

in which C is a constant. This seems reasonable. The quantity kT is the mean energy of agitation of an atom at temperature T and we see that, the higher the temperature, the more atoms—on long-term average—will be "bumped up" by thermal agitation (that is, atom–atom collisions) to the level E_x.

If we apply Eq. 23 to the two levels of Fig. 25 and divide, the constant C cancels out and we find, for the ratio of the number of atoms in the upper level to the number in the lower level

$$\frac{n_2}{n_1} = e^{-(E_2-E_1)/kT}. \tag{24}$$

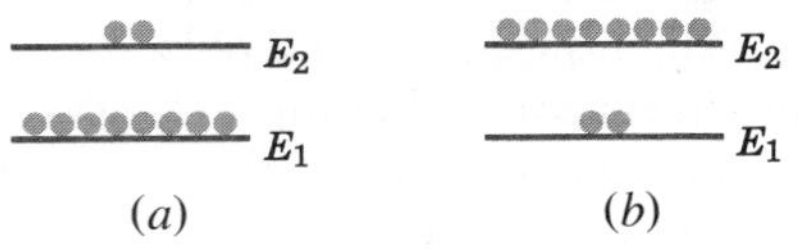

E_2 E_1 (*a*) E_2 E_1 (*b*)

Figure 26 (*a*) The normal thermal-equilibrium distribution of atoms between two states, accounted for by thermal agitation. (*b*) An inverted population, obtained by special techniques. Such an inverted population is essential for laser action.

Figure 26*a* illustrates this situation. Because $E_2 > E_1$, the ratio n_2/n_1 will always be less than unity, which means that there will always be fewer atoms in the higher energy level than in the lower. Again, this is what we would expect if the level populations are determined only by the action of thermal agitation.

If we flood the atoms of Fig. 26*a* with photons of energy $E_2 - E_1$, photons will disappear by the absorption process and will be generated by the two emission processes. However, the net effect, by sheer weight of numbers, will be absorption. To make a laser we must *generate* photons, not absorb them. The arrangement of Fig. 26*a* won't work.

To generate laser light we must have a situation in which stimulated emission dominates. The only way to do this is to fix it so that there are more atoms in the upper level than in the lower, as in Fig. 26*b*. A *population inversion* such as this is not consistent with simple thermal equilibrium and we must think of clever ways to set it up.

45-16 How a Laser Works

Of the many kinds of lasers, we describe two, the first being an optically pumped laser. The first operating laser, assembled by Theodore Maiman in 1960, used

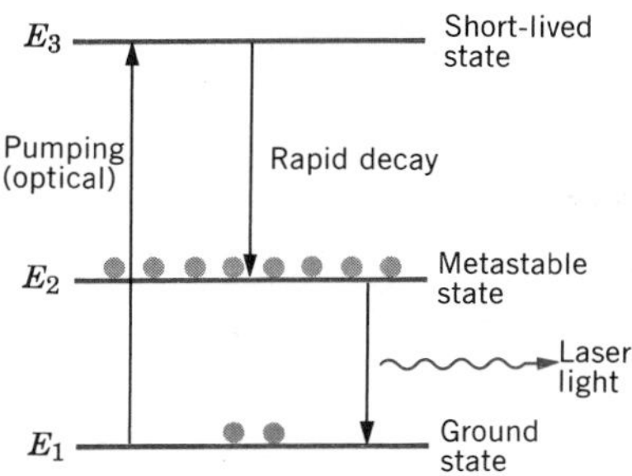

Figure 27 The basic three-level scheme for laser operation. Metastable state E_2 is more heavily populated than the ground state E_1.

crystalline ruby as the lasing material and operated in this way.

An Optically Pumped Laser. Figure 27 shows schematically how we can set up a population inversion in a lasing material so that laser action can occur. We start with essentially all of the atoms of the material in the ground state E_1. We then supply energy to the system so that many atoms are raised up to an excited state E_3. In *optical pumping,* the energy comes from an intense, continuous-spectrum light source that surrounds the lasing material.

From state E_3 many atoms decay rapidly and spontaneously to state E_2, which must be a metastable state, that is, it must have a relatively long mean life against spontaneous emission. If conditions are right, state E_2 can then become more heavily populated than state E_1 and we have our population inversion. A stray photon of the right energy can then trigger an avalanche of stimulated emission events from state E_2, and we have laser light.

The Helium–Neon Gas Laser. Figure 28 shows a type of laser commonly found in student laboratories. The glass discharge tube is filled with an 80%–20% mixture of the noble gases helium and neon, the neon being the "lasing" medium. In the gas laser, the necessary population inversion comes about—as we shall see—because of collisions between the helium and the neon atoms.

Figure 29 is a simplified version of the level schemes for these two atoms. Note that four levels, labeled E_0, E_1, E_2, and E_3, are involved, rather than three levels as in the lasing scheme of Fig. 27. Pumping is accomplished by setting up an electrically induced gas discharge in the helium–neon mixture. Electrons and ions in this discharge, frequently enough, collide with helium atoms, raising them to level E_3 in Fig. 29. This level is metastable, so that spontaneous emission to the ground state (level E_0) occurs only very slowly.

Level E_3 in helium (= 20.61 eV) is, by chance, very close to level E_2 in neon (= 20.66 eV). Thus, when a metastable helium atom and a ground-state neon atom collide, the excitation energy of the helium atom can often be transferred to the neon atom. In this way, level E_2 in Fig. 29 can become more heavily populated than level E_1 in that figure. This population inversion is maintained because (1) the metastability of level E_3 ensures a ready supply of neon atoms in level E_2 and (2) atoms in level E_1 decay rapidly (through intermediate stages not shown) to the neon ground state E_0. Stimulated emission from level E_2 to level E_1 predominates and red laser light, of wavelength 632.8 nm, is generated.

More must be done before a strong beam of laser light can be produced. Most stimulated-emission photons initially produced in the discharge tube of Fig. 28 will not happen to be parallel to the tube axis and will quickly be stopped at the tube walls. Stimulated emission photons that *are* parallel to the axis, however, can be made to move back and forth through the discharge tube many times by successive reflections from mirrors M_1 and M_2. A chain reaction thus builds up rapidly in this direction, accounting for the inherent parallelism of the laser light.

Rather than thinking in terms of the photons bouncing back and forth between the mirrors, it is perhaps more useful to think of the entire arrangement of Fig. 28 as an optical resonant cavity that, like an organ pipe for sound waves, can be tuned to sharp resonance at one (or more) wavelengths.

The mirrors M_1 and M_2 are concave, with their focal points nearly coinciding at the center of the tube. Mirror

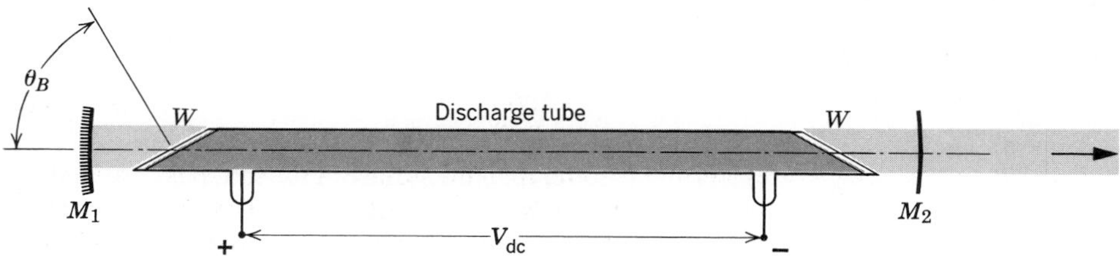

Figure 28 The elements of a helium–neon gas laser.

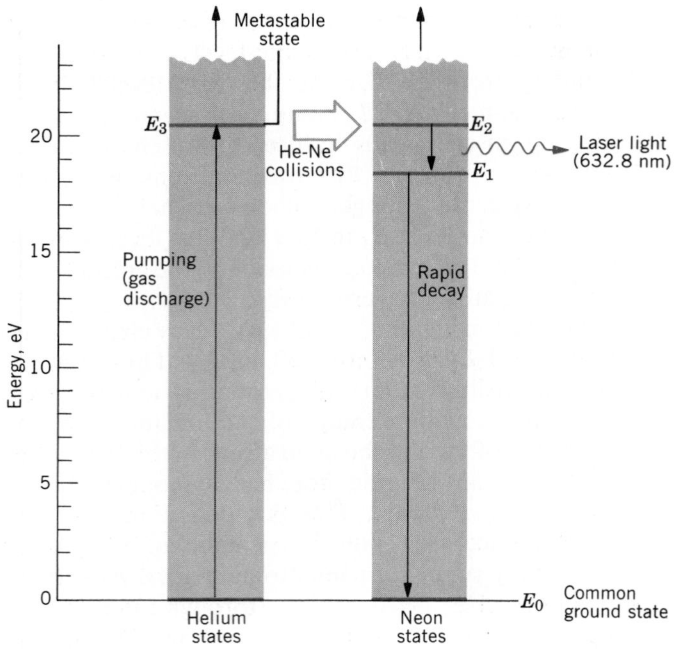

Figure 29 Four essential levels in helium and neon atoms in a He–Ne gas laser.

M_1 is coated with a dielectric film whose thickness is carefully adjusted to make the mirror as close as possible to a perfect reflector at the wavelength of the laser light. Mirror M_2, on the other hand, is coated so as to be slightly "leaky," so that a small fraction of the laser light can escape at each reflection to form a useful beam. The windows W in Fig. 28, which close the ends of the discharge tube, are slanted to the Brewster angle (see Section 39–4) to minimize loss of light by reflection.

Sample Problem 9 A three-level laser of the type shown in Fig. 27 emits laser light at a wavelength of 550 nm, near the center of the visible band. (a) If the optical-pumping mechanism is not used, what will be the equilibrium ratio of the population of the upper level (energy E_2) to that of the lower level (energy E_1)? Assume that $T = 300$ K.

From the Bohr frequency condition, the energy difference between the two levels is given by

$$E_2 - E_1 = h\nu = \frac{hc}{\lambda}$$

$$= \frac{(6.63 \times 10^{-34}\ \mathrm{J \cdot s})(3 \times 10^8\ \mathrm{m/s})}{(550 \times 10^{-9}\ \mathrm{m})(1.60 \times 10^{-19}\ \mathrm{J/eV})}$$

$$= 2.26\ \mathrm{eV}.$$

The mean energy of thermal agitation is equal to

$$kT = (8.62 \times 10^{-5}\ \mathrm{eV/K})(300\ \mathrm{K}) = 0.0259\ \mathrm{eV}.$$

From Eq. 24 we have then, for the desired ratio,

$$\frac{n_2}{n_1} = e^{-(E_2-E_1)/kT} = e^{-(2.26\ \mathrm{eV})/(0.0259\ \mathrm{eV})}$$

$$= e^{-87.26} \approx 1.3 \times 10^{-38}! \qquad \text{(Answer)}$$

This is an extremely small number. It is not unreasonable, however. An atom whose *mean* thermal agitation energy is only 0.0259 eV will not often impart an energy of 2.26 eV to another atom in a collision.

(b) At what temperature for the conditions of (a) above would the ratio n_2/n_1 be $\frac{1}{2}$?

Making this substitution in Eq. 24, taking the natural logarithm of each side, and solving for T yields

$$T = \frac{E_2 - E_1}{k\,(\ln 2)} = \frac{2.26\ \mathrm{eV}}{(8.62 \times 10^{-5}\ \mathrm{eV/K})(\ln 2)}$$

$$= 37{,}800\ \mathrm{K}. \qquad \text{(Answer)}$$

This is much hotter than the surface of the sun. It is clear that, if we are to invert the populations of these two levels, some rather special pumping or other mechanism is needed. Without population inversion, lasing is not possible.

REVIEW AND SUMMARY

Any successful theory of atomic structure must be able to explain (1) the regularities represented by the *periodic table of the elements,* (2) the details of atomic emission and absorption spectra, and (3) the fact that atoms have angular momentum and magnetic dipole moments. Modern quantum theory does this and much more.

Hydrogen Atoms

The structure of the hydrogen atom is analyzed by introducing the potential-energy function of Eq. 2 into the mathematical machinery of wave mechanics. What emerges is a set of quantized energies, angular momenta, magnetic moments and wave functions that describe the allowed states of the hydrogen atom. The energies of these states are

Hydrogen Atom Energies

$$E_n = -\frac{me^4}{8\epsilon_0 h^2}\frac{1}{n^2} = -\frac{13.6\text{ eV}}{n^2}, \qquad n = 1, 2, 3, \ldots \qquad [3]$$

with n being the *principle quantum number.* Every hydrogen atom state has an orbital angular momentum $\mathbf{L}$ whose magnitude L is fixed by the *orbital quantum number* ℓ:

Angular Momentum

$$L = \sqrt{\ell(\ell + 1)}\hbar, \qquad \ell = 0, 1, 2, \ldots (n-1). \qquad [4,5]$$

Here $\hbar = h/2\pi$ is a useful unit of angular momentum in quantum mechanics. The allowed components of $\mathbf{L}$ along the z axis depend on the *magnetic quantum number* m_ℓ and are given by

$$L_z = m_\ell \hbar, \qquad m_\ell = 0, \pm 1, \pm 2, \ldots, \pm \ell \qquad [6,7]$$

Space Quantization

The fact that only a discrete number $(= 2\ell + 1)$ of these projections is allowed is called space quantization and is represented by the vector construction of Fig. 7. The allowed projections of the magnetic moment associated with orbital angular momentum are

Orbital Magnetic Moment

$$\mu_{\ell,z} = -m_\ell(eh/4\pi m) = -m_\ell \mu_B. \qquad [9,10]$$

Here $\mu_B = 9.274 \times 10^{-24}$ J/T $= 5.788 \times 10^{-5}$ eV/T is the *Bohr magneton,* a convenient unit in which to measure atomic magnetic moments.

In addition to orbital angular momentum, each electron has a *spin angular momentum* $\mathbf{S}$ whose z component is given by

Electron Spin

$$S_z = m_s \hbar, \qquad m_s = \pm\tfrac{1}{2}. \qquad [11]$$

The magnetic moment associated with electron spin has possible projections given by

Spin Magnetic Moment

$$\mu_{s,z} = -2m_s\mu_B. \qquad [12]$$

Spin angular momentum is twice as effective as orbital angular momentum in generating magnetism. Table 1 summarizes the hydrogen atom quantum numbers; also see Sample Problems 1 and 2.

Verification of Space Quantization

The Stern-Gerlach experiment confirmed the reality of space quantization. It did so by splitting, using a nonuniform magnetic field, a beam of neutral silver atoms into two subbeams, each characterized by the orientation in space of the magnetic moments of the atoms that make it up. See Figs. 10 and 11 and Sample Problem 3.

Nuclear Magnetic Resonance provides a useful analytical technique based on the space quantization of the angular momenta of atomic nuclei, usually protons. The nuclei absorb energy from a radio-frequency beam when the frequency is just right to cause "spin flips" of the protons, that is when

Nuclear Magnetic Resonance

$$h\nu = 2\mu B. \qquad [17]$$

Since $\mathbf{B}$ near the protons is the sum of an external field plus a local field due to the atomic environment, the absorption frequency ν reveals information about chemical bonds; see Fig. 15 and Sample Problem 4. The same phenomenon is used in *Magnetic Resonance Imaging;* see Fig. 16.

Atom Building and the Periodic Table

The periodic table can be understood by assuming that electron states in multielectron atoms are to be labeled with the hydrogen-atom quantum numbers and that only one electron may be assigned to each quantum state *(the Pauli exclusion principle).* Sample Problem 5 shows how to account for the major features of the periodic table using these rules.

Fig. 17 shows the distribution by wavelength of the x rays produced when energetic electrons strike a metal target. The smooth background curve in is called the *continuous x-ray spectrum.* It arises when the electrons are decelerated in the field of a nucleus. The minimum cutoff wavelength is given by

The Continuous X-ray Spectrum

$$\lambda_{\min} = \frac{hc}{eV} \quad \text{(cutoff wavelength)}, \qquad [19]$$

where V, the only variable, is the accelerating potential of the electrons; see Sample Problem 6.

The Characteristic X-ray Spectrum

The sharp peaks in Fig. 17 constitute the *characteristic x-ray spectrum* of the target element. The peaks arise from electron transitions deep within the atom as represented in Fig. 19. The frequency of any characteristic x-ray line (the K_α line, say) varies smoothly with the atomic number of the element. This fact can be used to assign atomic numbers to the elements, by way of a Moseley plot such as that of Fig. 21. The straight-line shape of the plot can be understood in terms of one-electron-atom quantum mechanics, which predicts that the frequency ν of any x-ray transition should vary with atomic number Z according to the equation

$$\sqrt{\nu} = \sqrt{\frac{3me^4}{32\epsilon_0^2 h^3}}\,(Z-1) = a(Z-1). \qquad [20,21]$$

See Sample Problems 7 and 8.

Lasers

The special properties of laser light (see p. 1040) are directly traceable to a chain reaction of stimulated emission events in a "lasing" medium. Figure 25*c* shows the basic process. Lasing always involves a pair of levels and, for stimulated emission to be the controlling process, there are two requirements: (1) The upper level (E_2) must be more heavily populated than the lower level (E_1); see Fig. 26*b*. (2) The upper level must be metastable, that is, it must have a relatively long mean lifetime against depletion by spontaneous emission. Figures 27 and 29 show how the necessary population inversion can be brought about, by optical pumping in the first case and by gas collision excitation processes in the second. Study Sample Problem 9.

QUESTIONS

1. If Bohr's theory and wave mechanics predict the same result for the energies of the hydrogen atom states, then why do we need wave mechanics, with its greater complexity?

2. Compare Bohr's theory and wave mechanics. In what respects do they agree? In what respects do they differ?

3. In the laboratory, how would you show that an atom has angular momentum? That it has a magnetic dipole moment?

4. Why don't we observe space quantization for a spinning top?

5. How can we arrive at the conclusion that the spin magnetic quantum number m_s can have only the values $\pm\frac{1}{2}$? What kinds of experiments support this conclusion?

6. Why is the magnetic moment of the electron directed opposite to its spin angular momentum?

7. An atom in a state with zero angular momentum has spherical symmetry as far as its interaction with other atoms is concerned. It is sometimes called a "billiard-ball atom." Explain.

8. Discuss how good an analogy the rotating earth revolving about the sun is to a spinning electron moving about a proton in the hydrogen atom.

9. Angular momentum is a vector and you might expect that it would take three quantum numbers to describe it, corresponding to the three space components of a vector. Instead, in an atom, only two quantum numbers characterize the angular momentum. Explain why.

10. The angular momentum of the electron in the hydrogen atom is quantized. Why isn't the linear momentum also quantized? (*Hint:* Consider the implications of the uncertainty principle.)

11. Does the Einstein–de Haas experiment (see Fig. 4) provide any evidence that angular momentum is quantized?

12. What are the dimensions and the SI units of a hydrogen atom wave function?

13. Define and distinguish among the terms "wave function," "probability density function," and "radial probability density function."

14. What determines the number of subbeams into which a beam of neutral atoms is split in a Stern–Gerlach experiment?

15. If in a Stern–Gerlach experiment an ion beam is resolved into five component beams, then what angular momentum quantum number does each ion have?

16. In a Stern–Gerlach apparatus, is it possible to have a magnetic field configuration in which the magnetic field itself is zero along the beam path but the field gradient is not? If your answer is yes, can you design an electromagnet that will produce such a field configuration?

17. A beam of neutral silver atoms is used in a Stern–Gerlach experiment. What is the origin of both the force and the torque that act on the atom? How is the atom affected by each?

18. On what quantum numbers does the energy of an electron in (*a*) a hydrogen atom and (*b*) a vanadium atom depend?

19. The periodic table of the elements was based originally on atomic weight, rather than on atomic number, the latter concept having not yet been developed. Why were such early tables as successful as they proved to be? In other words, why is the atomic weight of an element (roughly) proportional to its atomic number?

20. How does the structure of the periodic table support the need for a fourth quantum number, corresponding to electron spin?

21. Why does it take more energy to remove an electron from neon ($Z = 10$) than from sodium ($Z = 11$)?

22. Why do the lanthanide elements (see Appendix E) have such similar chemical properties? How can we justify putting them all into a single square of the periodic table? Why is it that, in spite of their similar chemical properties, they can be so easily sorted out by measuring their characteristic x-ray spectra?

23. How would the properties of helium differ if the electron had no spin, that is, if the only operative quantum numbers were n, ℓ, and m_ℓ?

24. What is the origin of the cutoff wavelength λ_{min} of Fig. 17? Why is it an important clue to the photon nature of x rays?

25. Why do you expect the wavelengths of radiations generated by transitions deep within an atom to be shorter than those generated by transitions occurring in the outer fringes of the atom?

26. What are the characteristic x rays of an element? How can they be used to determine the atomic number of the element?

27. Compare Figs. 17 and 19. How can you be sure that the two prominent peaks in the former figure do indeed correspond numerically with the two transitions similarly labeled in the latter figure?

28. Can atomic hydrogen be caused to emit x rays? If so, describe how. If not, why not?

29. How does an x-ray energy-level diagram differ from the energy-level diagram for hydrogen? In what respects are the two diagrams similar?

30. When extended to higher atomic numbers, the Moseley plot of Fig. 21 is not a straight line but is slightly concave upwards. Does this affect the ability to assign atomic numbers to the elements?

31. Why is it that Bohr's theory, which otherwise does not work very well even for helium ($Z = 2$), gives such a good account of the characteristic x-ray spectra of the elements?

32. Why is focused laser light inherently better than focused light from a tiny incandescent lamp filament for such delicate surgical jobs as spot-welding detached retinas?

33. Laser light forms an almost parallel beam. Does the intensity of such light fall off as the inverse square of the distance from the source?

34. In what ways are laser light and starlight similar? In what ways are they different?

35. Arthur Schawlow, one of the laser pioneers, invented a typewriter eraser, based on focusing laser light on the unwanted character. What is its principle of operation?

36. In what ways do spontaneous emission and stimulated emission differ?

37. We have spontaneous emission and stimulated emission. From symmetry, why don't we also have spontaneous and stimulated absorption?

38. Why is a population inversion between two atomic levels necessary for laser action to occur?

39. What is a metastable state? What role do such states play in the operation of a laser?

40. Comment on this statement: "Other things being equal, a four-level laser scheme such as that of Fig. 29 is preferable to a three-level scheme such as that of Fig. 27 because, in the latter scheme, half of the population of atoms in level E_1 must be moved to state E_2 before a population inversion can even begin to occur."

41. Comment on this statement: "In the laser of Fig. 28, only light whose plane of polarization lies in the plane of that figure is transmitted through the right-hand window W. Therefore half of the energy potentially available to the output beam is lost." (*Hint:* Is this second statement really true? Consider what happens to photons whose effective plane of polarization is at right angles to the plane of Fig. 28. Do such photons participate fully in the stimulated emission amplification process?)

42. A beam of light emerges from an aperture in a "black box" and moves across your laboratory bench. How could you test this beam to find out the extent to which it is coherent over its cross section? How could you tell (without opening the box) whether or not the concealed light source is a laser?

43. Why is it difficult to build an x-ray laser?

EXERCISES AND PROBLEMS

Section 45-5 Orbital Angular Momentum and Magnetism

1E. Use the value of Planck's constant in Appendix B to show that, to three significant figures, $\hbar = 1.06 \times 10^{-34}$ J·s $= 6.59 \times 10^{-16}$ eV·s.

2E. Verify that $\mu_B = 9.274 \times 10^{-24}$ J/T $= 5.788 \times 10^{-5}$ eV/T, as reported in Eq. 10.

3E. Calculate the magnitude of the orbital angular momentum of an electron in a state with $\ell = 3$.

4E. (*a*) What number of possible ℓ values are associated with $n = 3$? (*b*) What number of possible m_ℓ values are associated with $\ell = 1$?

5E. If an electron in a hydrogen atom is in a state with $\ell = 5$, what is the minimum possible angle between **L** and $\mathbf{L}_z$?

6E. Write down the quantum numbers for all the hydrogen atom states for which $n = 4$ and $\ell = 3$.

7E. A hydrogen atom state is known to have the quantum number $\ell = 3$. What are the possible n, m_ℓ, and m_s quantum numbers?

8E. A hydrogen atom state has a maximum m_ℓ value of $+4$. What can you say about the rest of its quantum numbers?

9E. How many hydrogen atom states are there with $n = 5$?

10P. (*a*) Show that the magnetic moment of the electron in hydrogen, according to the Bohr model, is given by $\mu = n\mu_B$, where μ_B is the Bohr magneton and $n = 1, 2, \ldots$. (*b*) This result disagrees with the correct expression in Eq. 9. Why does the Bohr model fail to give the correct orbital magnetic moment?

11P. Calculate and tabulate, for a hydrogen atom in a state with $\ell = 3$, the allowed values of L_z, μ_z, and θ. Find also the magnitudes of **L** and $\boldsymbol{\mu}$.

12P. Estimate (*a*) the quantum number ℓ for the orbital motion of the earth around the sun and (*b*) the number of quantized orientations of the plane of the earth's orbit, according to the rules of space quantization. (*c*) Also find $\theta_{\min}$, the half-angle of the smallest cone that can be swept out by a perpendicular to the earth's orbit as the earth revolves around the sun. Discuss from the point of view of the correspondence principle.

13P. Show that Eq. 8 is a plausible version of the uncertainty principle $\Delta p \cdot \Delta x = h$. (*Hint:* Multiply by r/r; associate p with mv, and L with mvr.)

14P. Consider the relation

$$\cos\theta_{\min} = \left(1 + \frac{1}{\ell}\right)^{-1/2}$$

in Sample Problem 1 for the limiting case of large ℓ. (*a*) By expanding both sides of this equation in series form (see Appendix G), show that, to a good approximation for large ℓ,

$$\theta_{\min} \cong \frac{1}{\sqrt{\ell}},$$

where $\theta_{\min}$ is to be expressed in radian measure. (*b*) Test the validity of this approximate formula for the five values of ℓ given in Sample Problem 1. (*c*) Fill in the table in Sample Problem 1 by finding $\theta_{\min}$ for $\ell = 10^5$, 10^6, and 10^7. (*d*) What computational difficulties do you encounter if you do *not* use the above approximate formula to calculate $\theta_{\min}$ for large values of ℓ?

15P. Of the three scalar components of **L**, one, L_z, is quantized, according to Eq. 6. Show that the most that can be said about the other two components of **L** is

$$(L_x^2 + L_y^2)^{1/2} = [\ell(\ell + 1) - m_\ell^2]^{1/2}\hbar.$$

Note that these two components are not separately quantized. Show also that

$$\ell\hbar \leq (L_x^2 + L_y^2)^{1/2} \leq [\ell(\ell + 1)]^{1/2}\hbar.$$

Correlate these results with Fig. 6.

16P*. An unmagnetized iron cylinder, whose radius is 5.0 mm, hangs from a frictionless bearing inside a solenoid, so that the cylinder can rotate freely about its axis; see Fig. 4. By passing a current through the solenoid windings, a magnetic field is suddenly applied parallel to the axis, causing the magnetic dipole moments of the atoms to align themselves parallel to the field. The atomic angular momentum vectors, which are coupled back-to-back with the magnetic dipole moment vectors, also become aligned and the cylinder will start to rotate. This is the Einstein–de Haas effect (see Section 45-2). Find the period of rotation of the cylinder. Assume that each iron atom has an angular momentum of $\hbar$ and that the alignment is perfect. (*Hint:* Apply conservation of angular momentum; the answer is independent of the length of the cylinder.)

Section 45-7 The Hydrogen-Atom Wave Functions

17E. For a hydrogen atom in its ground state, what is the value, at $r = 2r_B$ of the radial probability density $P(r)$?

18E. A small sphere of radius $0.1r_B$ is located a distance r_B from the nucleus of a hydrogen atom in its ground state. What is the probability that the electron will be found inside this sphere? (Assume that ψ is constant inside the sphere.)

19E. For a hydrogen atom in its ground state, what is the probability of finding the electron between two spheres of radii $r = 1.00r_B$ and $r = 1.01r_B$?

20E. From Eq. 14 for the $n = 2$, $\ell = 0$ radial probability density, show that the corresponding wave function is

$$\psi = \frac{1}{\sqrt{32\pi r_B^3}}\left(2 - \frac{r}{r_B}\right)e^{-r/2r_B}.$$

21E. (*a*) Sketch the wavefunction for the state $n = 2$, $\ell = 0$ given in the previous Exercise. (*b*) What is the value of the wavefunction at the center of the nucleus?

22E. For a hydrogen atom in a state with $n = 2$ and $\ell = 0$, what are the values, at $r = 5r_B$, of (*a*) the wave function $\psi(r)$, (*b*) the probability density $\psi^2(r)$, and (*c*) the radial probability density $P(r)$?

23E. Using Eq. 14, show that, for the hydrogen atom state with $n = 2$ and $\ell = 0$,

$$\int_0^\infty P(r)\,dr = 1.$$

What is the physical interpretation of this result?

24P. Locate the two maxima for the radial probability density curve of Fig. 8*b*.

25P. Use the results of the previous problem to calculate the values of the radial probability density, for the state $n = 2$, $\ell = 0$, at the two maxima; compare with Fig. 8*b*.

26P. Repeat Problem 36 of Chapter 44 for an electron in the state $n = 2$, $\ell = 0$; that is, calculate the probability that the electron will be found inside the proton, radius = 1.1 fm, that constitutes the nucleus of the hydrogen atom.

27P. For a hydrogen atom in a state with $n = 2$ and $\ell = 0$, what is the probability of finding the electron between two spheres of radii $r = 5.00r_B$ and $r = 5.01r_B$?

28P. For a hydrogen atom in a state with $n = 2$ and $\ell = 0$, what is the probability of finding the electron somewhere within the smaller of the two bulges of its radial probability density function? See Fig. 8*b*.

Section 45-8 The Stern-Gerlach Experiment

29E. Calculate the two possible angles between the electron spin angular momentum vector and the magnetic field in Sample Problem 3. Bear in mind that the *orbital* angular momentum of the valence electron is zero.

30E. What is the acceleration of the silver atom as it passes through the deflecting magnet in the Stern-Gerlach experiment of Sample Problem 3?

31E. Assume that in the Stern-Gerlach experiment described for neutral silver atoms the magnetic field **B** has a magnitude of 0.50 T. (*a*) What is the energy difference between the orientations of the silver atoms in the two subbeams? (*b*) What is the frequency of the radiation that would induce a transition between these two states? (*c*) What is its wavelength, and to what part of the electromagnetic spectrum does it belong? The magnetic moment of a neutral silver atom is 1 Bohr magneton.

32P. Suppose a hydrogen atom (in its ground state) moves 80 cm in a direction perpendicular to a magnetic field that has a gradient, in the vertical direction, of 1.6×10^2 T/m. (*a*) What is the force on the atom due to the magnetic moment of the electron, which we take to be 1 Bohr magneton? (*b*) What is its vertical displacement if its speed is 1.2×10^5 m/s?

Section 45-9 Science, Technology, and Spin

33E. What is the wavelength of a photon that will induce a transition of an electron spin from parallel to antiparallel orientation in a magnetic field of magnitude 0.20 T? Assume that $\ell = 0$.

34E. The proton as well as the electron has spin $\frac{1}{2}$. In the hydrogen atom in its ground state, with $n = 1$ and $\ell = 0$, there are two energy levels, depending on whether the electron and the proton spins are in the same direction or in opposite directions. The state with the spins in the opposite direction has the higher energy. If an atom is in this state and one of the spins "flips over," the small energy difference is released as a photon of wavelength 21 cm. This spontaneous spin-flip process is very slow, the mean life for the process being about 10^7 y. However, radio astronomers observe this 21-cm radiation in deep space, where the density of hydrogen is so small that an atom can flip before being disturbed by collisions with other atoms. What is the effective magnetic field (due to the magnetic dipole moment of the proton) experienced by the electron in the emission of this 21-cm radiation?

35E. An external magnetic field of frequency 34 MHz is applied to molecules of a certain material that contains hydrogen atoms. Resonance is observed when the strength of this applied field equals 0.78 T. Calculate the strength of the local molecular field at the site of the protons undergoing spin flips.

36E. Excited sodium atoms emit two closely-spaced lines (the sodium doublet; see Fig. 30) with wave lengths 588.995 nm and 589.592 nm. (a) What is the difference in energy between the two upper energy levels? (b) This energy difference occurs because the electron's spin magnetic dipole

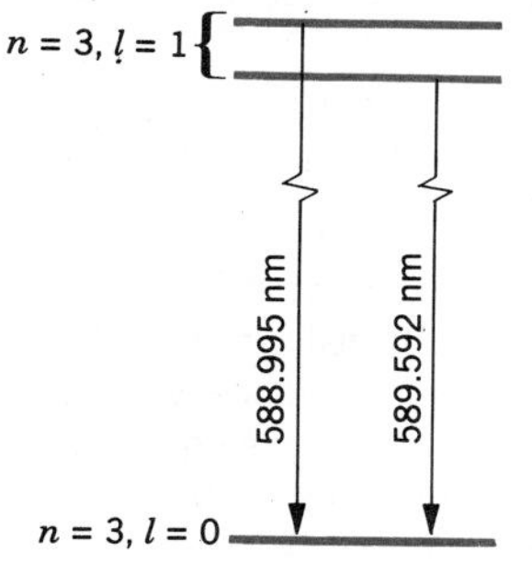

Figure 30 Exercise 36. The three energy levels that account for the two lines of the familiar sodium doublet.

moment (one Bohr magneton) can be oriented either parallel or antiparallel to the internal magnetic field associated with the electron's orbital motion. Use the result you have just calculated to find the strength of this internal magnetic field.

Section 45-10 Multi-electron Atoms and the Periodic Table

37E. Label as true or false these statements involving the quantum numbers n, ℓ, m_ℓ. (*a*) One of these orbitals cannot exist: $n = 2, \ell = 1; n = 4, \ell = 3; n = 3, \ell = 2; n = 1, \ell = 1$. (*b*) The number of values of m_ℓ that are allowed depends only on ℓ and not on n. (*c*) There are four orbitals with $n = 4$. (*d*) The smallest value of n that can go with a given ℓ is $\ell + 1$. (*e*) All states with $\ell = 0$ also have $m_\ell = 0$, regardless of the value of n. (*f*) There are n orbitals for each value of n.

38E. What are the quantum numbers n, ℓ, m_ℓ, m_s for the two electrons of the helium atom in its ground state?

39E. Two electrons in lithium ($Z = 3$) have as their quantum numbers n, ℓ, m_ℓ, m_s, the values $1, 0, 0, \pm\frac{1}{2}$. (*a*) What quantum numbers can the third electron have if the atom is to be in its ground state? (*b*) If the atom is to be in its first excited state?

40P. Suppose there are two electrons in the same system, both of which have $n = 2$ and $\ell = 1$. (*a*) If the exclusion principle did not apply, how many combinations of states are conceivably possible? (*b*) How many states does the exclusion principle forbid? Which ones are they?

41P. Show that the number of states with the same n is given by $2n^2$.

42P. If the electron had no spin, and if the Pauli exclusion principle still held, how would the periodic table be affected? In particular, which of the present elements would be noble gases?

Section 45-12 The Continuous X-ray Spectrum

43E. Show that the short-wavelength cutoff in the continuous x-ray spectrum is given by

$$\lambda_{\min} = 1240 \text{ pm}/V,$$

where V is the applied potential difference in kilovolts.

44E. Determine Planck's constant from the fact that the minimum x-ray wavelength produced by 40.0-keV electrons is 31.1 pm.

45E. What is the minimum potential difference across an x-ray tube that will produce x rays with a wavelength of 0.1 nm?

46P. A 20-keV electron is brought to rest by undergoing two successive bremsstrahlung events, thus transferring its kinetic energy into the energy of two photons. The wavelength of the second photon is 130 pm greater than the wavelength of the first photon to be emitted. (*a*) Find the energy of the electron after its first deceleration. (*b*) Calculate the wavelengths and energies of the two photons.

47P. X rays are produced in an x-ray tube by a target potential of 50 kV. If an electron makes three collisions in the target before coming to rest and loses one-half of its remaining kinetic energy on each of the first two collisions, determine the wavelengths of the resulting photons. Neglect the recoil of the heavy target atoms.

48P. Show that a moving electron cannot spontaneously change into an x-ray photon in free space. A third body (atom or nucleus) must be present. Why is it needed? (*Hint:* Examine conservation of total energy and of momentum.)

Section 45-13 The Characteristic X-ray Spectrum

49E. Electrons bombard a molybdenum target, producing both continuous and characteristic x rays as in Fig. 17. In that figure the energy of the incident electrons is 35.0 keV. If the accelerating potential applied to the x-ray tube is increased to 50.0 kV, what new values of (*a*) $\lambda_{\min}$, (*b*) λ_{K_α}, and (*c*) λ_{K_β} result?

50E. In Fig. 17, the x rays shown are produced when 35.0-keV electrons fall on a molybdenum target. If the accelerating potential is maintained at 35.0 kV but a silver target ($Z = 47$) is substituted for the molybdenum target, what values of (*a*) $\lambda_{\min}$, (*b*) λ_{K_β}, and (*c*) λ_{K_α} result? The K, L, and M atomic x-ray levels for silver (compare Fig. 19) are 25.51, 3.56, and 0.53 keV.

51E. The wavelength of the K_α line from iron is 19.3 pm. What is the energy difference between the two states of the iron atom (see Fig. 19) that give rise to this transition? What is the corresponding energy difference for the hydrogen atom? Why is the difference so much greater for iron than for hydrogen? (*Hint:* In the hydrogen atom the K shell corresponds to $n = 1$ and the L shell to $n = 2$.)

52E. From Fig. 17, calculate approximately the energy difference $E_L - E_M$ for the x-ray atomic energy levels of molybdenum. Compare with the result that may be found from Fig. 19.

53E. Calculate the ratio of the wavelengths of the K_α line for niobium (Nb) to that of gallium (Ga). Take needed data from the periodic table.

54P. Here are the K_α wavelengths of a few elements:

Ti	27.5 pm	Co	17.9 pm
V	25.0	Ni	16.6
Cr	22.9	Cu	15.4
Mn	21.0	Zn	14.3
Fe	19.3	Ga	13.4

Make a Moseley plot (see Fig. 21) and verify that its slope agrees with the value calculated in Sample Problem 8.

55P. A tungsten target ($Z = 74$) is bombarded by electrons in an x-ray tube. (*a*) What is the minimum value of the accelerating potential that will permit the production of the characteristic K_β and K_α lines of tungsten? (*b*) For this same accelerating potential, what is $\lambda_{\min}$? (*c*) What are λ_{K_β} and λ_{K_α}? The K, L, and M atomic x-ray levels for tungsten (see Fig. 19) are 69.5, 11.3, and 2.3 keV, respectively).

56P. A molybdenum target ($Z = 42$) is bombarded with 35.0-keV electrons and the x-ray spectrum of Fig. 17 results. Here $\lambda_{K_\beta} = 63$ pm and $\lambda_{K_\alpha} = 71$ pm. (*a*) What are the corresponding photon energies? (*b*) It is desired to filter these radiations through a material that will absorb the K_β line much more strongly than it will absorb the K_α line. What substance would you use? The K ionization energies for molybdenum and for four neighboring elements are as follows:

Z	40	41	42	43	44
Element	Zr	Nb	Mo	Tc	Ru
E_K (keV)	18.00	18.99	20.00	21.04	22.12

(*Hint:* A substance will selectively absorb one of two x radiations more strongly if the photons of one have enough energy to eject a K electron from the atoms of the substance but the photons of the other do not.)

57P. The binding energy of K-shell and L-shell electrons in copper are 8.979 keV and 0.951 keV, respectively. If a K_α x ray from copper is incident on a sodium chloride crystal and gives a first-order Bragg reflection at 15.9° when reflected from the alternating planes of the sodium atoms, what is the spacing between these planes?

58P. (*a*) Using Bohr's theory, estimate the ratio of energies of photons from the K_α transitions from two atoms whose atomic numbers are Z and Z'. (*b*) How much more energetic is a K_α x ray from uranium expected to be than from aluminum. (*c*) than from lithium?

59P. Determine how close the theoretical K_α x-ray energies, obtained from Eq. 20, are to the measured energies in the light elements from lithium to magnesium. To do this you must first (*a*) determine the constant in Eq. 20 to five significant figures, using data in Appendix B. Next, (*b*) calculate the percentage deviation of the theoretical from the measured energies. (*c*) Plot the deviation and comment on the trend. The measured values of the K_α x rays are:

Li	54.3 eV	O	524.9
Be	108.5	F	676.8
B	183.3	Ne	848.6
C	277	Na	1041
N	392.4	Mg	1254

(There is actually more than one K_α ray because of splitting of the L energy level, but that effect is negligible in the elements considered.)

Section 45-16 How a Laser Works

60E. A hypothetical atom has energy levels evenly spaced by 1.2 eV in energy. For a temperature of 2000 K, calculate the ratio of the number of atoms in the 13th excited state to the number in the 11th excited state.

61E. A particular (hypothetical) atom has only two atomic levels, separated in energy by 3.2 eV. In the atmosphere of a star there are 6.1×10^{13} of these atoms in the excited (upper) state per cm^3 and 2.5×10^{15} per cm^3 in the ground (lower) state. Calculate the temperature of the star's atmosphere.

62E. A population inversion for two levels is often described by assigning a negative Kelvin temperature to the system. Show that such a negative temperature would indeed correspond to an inversion. What negative temperature would describe the system of Sample Problem 9 if the population of the upper level exceeds that of the lower by 10%?

63E. A He–Ne laser emits light at a wavelength of 632.8 nm and has an output power of 2.3 mW. How many photons are emitted each minute by this laser when operating?

64E. A ruby laser emits light at wavelength 694.4 nm. If a laser pulse is emitted for 1.2×10^{-11} s and the energy release per pulse is 0.15 J, (*a*) what is the length of the pulse, and (*b*) how many photons are in each pulse?

65E. Lasers have become very small as well as very large. The active volume of a laser constructed of the semiconductor GaAlAs has a volume of only 200 (μm)^3 (smaller than a grain of sand) and yet it can continuously deliver 5.0 mW of power at 0.80-μm wavelength. Calculate the production rate of photons.

66E. It is entirely possible that techniques for modulating the frequency or amplitude of a laser beam will be developed so that such a beam can serve as a carrier for television signals, much as microwave beams do now. Assume also that laser systems will be available whose wavelengths can be precisely "tuned" to anywhere in the visible range, that is, in the range 450 nm $< \lambda <$ 650 nm. If a television channel occupies a bandwidth of 10 MHz, how many channels could be accommodated with this laser technology? Comment on the intrinsic superiority of visible light to microwaves as carriers of information.

67E. A high-powered laser beam ($\lambda = 600$ nm) with a beam diameter of 12 cm is aimed at the moon, 3.8×10^5 km distant. The spreading of the beam is caused only by diffraction effects. The angular location of the edge of the central diffraction disk (see Equation 11 in Chapter 41) is given by

$$\sin\theta = \frac{1.22\,\lambda}{d},$$

where d is the diameter of the beam aperture. What is the diameter of the central diffraction disk at the moon's surface?

68P. The active medium in a particular ruby laser ($\lambda =$ 694 nm) is a synthetic ruby crystal 6.0 cm long and 1.0 cm in diameter. The crystal is silvered at one end and—to permit the formation of an external beam—only partially silvered at the other. (*a*) Treat the crystal as an optical resonant cavity in analogy to a closed organ pipe and calculate the number of standing-wave nodes there are along the crystal axis. (*b*) By what amount $\Delta\nu$ would the beam frequency have to shift to increase this number by one? Show that $\Delta\nu$ is just the inverse of the travel time of light for one round trip back and forth along the crystal axis. (*c*) What is the corresponding fractional frequency shift $\Delta\nu/\nu$? The appropriate index of refraction is 1.75.

69P. The mirrors in the laser of Fig. 28 form a cavity in which standing waves of laser light are set up. In the vicinity of 533 nm, how far apart in wavelength are the adjacent allowed operating modes? The mirrors are 8.0 cm apart.

70P. An atom has two energy levels with a transition wavelength of 580 nm. At 300 K, 4.0×10^{20} atoms are in the lower state. (*a*) How many occupy the upper state, under conditions of thermal equilibrium? (*b*) Suppose, instead, that 7.0×10^{20} atoms are pumped into the upper state, with 4.0×10^{20} in the lower state. How much energy could be released in a single laser pulse?

71P. The beam from an argon laser ($\lambda = 515$ nm) has a diameter d of 3.0 mm and a continuous-wave power output of 5.0 W. The beam is focused onto a diffuse surface by a lens whose focal length f is 3.5 cm. A diffraction pattern such as that of Fig. 13 in Chapter 41 is formed, the radius of the central disk being given by

$$R = \frac{1.22 f\lambda}{d};$$

see Equation 11 in Chapter 41. The central disk can be shown to contain 84% of the incident power. Calculate (*a*) the radius R of the central disc, (*b*) the average power flux density in the incident beam, and (*c*) in the central disk.

72P. The use of lasers for missile defense has been proposed as part of the Strategic Defense Initiative ("Star Wars"). A beam of intensity 10^8 W/m^2 would probably burn into and destroy a hardened (nonspinning) missile in 1 s. (*a*) If the laser has 5-MW power, 3-μm wavelength and 4.0-m beam diameter (a very powerful laser indeed), would it destroy a missile at a distance of 3000 km? (*b*) If the wavelength could be changed, what maximum value would work? Use the equation for the central disk given in Exercise 67 and take the focal length to be the distance to the target.

73P. The NOVA laser at Lawrence Livermore National Laboratory is used in fusion experiments (see Chapter 48). Its beam is designed to deliver between 0.1 and 1 MJ of energy in nanoseconds to a target of solid deuterium and tritium (isotopes of hydrogen) to heat the material and increase its density. (*a*) Assume that two 150-kJ laser pulses simultaneously impact from opposite directions on a sphere of 1.0-mm radius containing 4.0 μg of fuel. If the pulses last 3.0 ns and are completely absorbed, estimate the maximum radiation pressure on the sphere. [For simplicity, assume that each pulse is uniformly spread over an area $\pi(1.0\text{ mm})^2$.] (*b*) If 1% of the energy is effective in raising the temperature, how hot will the fuel get? Assume that the fuel is entirely deuterium with an atomic weight of 2.

ESSAY 19
APPLICATIONS OF LASERS

ELSA GARMIRE
UNIVERSITY OF SOUTHERN CALIFORNIA

Introduction

LASERS—STARWARS—MAGIC LIGHT MACHINES are words that evoke dreams in children, scientists, and engineers alike. For many years since its invention in 1958, the laser has been called "the solution looking for a problem." Now, 25 years later, the laser has indeed solved many problems. In this essay I describe only a few of the thousands of applications for lasers.

Lasers emit light with the special characteristic of *coherence* (Chapter 45). That is, lasers emit light that is an electromagnetic wave with a well-defined frequency and phase. This coherence often results in a monochromatic, collimated output. The uses of lasers can be related to the properties of "coherence." For example, light with a well-defined phase can be collimated through a telescope for applications such as surveying and tracking, or it may be focused to a tiny spot, to achieve very high intensities. Furthermore, the ability to measure phase of light, in addition to intensity, provides new information. Most of the applications described below use these features. Specific uses depend on the characteristics of available lasers, which will now be outlined.

Kinds of Lasers

To emit coherent light, a laser must contain both an amplifying or "gain" medium and mirrors to provide feedback (Chapter 45). Table 1 describes some of the many lasers that have been developed to date. Gases, liquids (dye solvents), crystals (YAG—yttrium aluminum garnet), glasses, semiconductors (GaAs) all can be used in lasers. Even excimers ("molecules" composed of rare gas halogens such as ArF) and gases in outer space have been shown to exhibit laser action. Laser wavelengths range from far ultraviolet (<170 nm) to far infrared (>200 μm). Lasers operate with either a contin-

Table 1 Characteristics of Typical Lasers

Gain Medium	*Type*	*Peak Power*	*Pulse Duration*	*Wavelength*
HeNe	gas	1 mW	cw	633 nm
Argon	gas	10 W	cw	488 nm
CO_2	gas	200 W	cw	10.6 μm
CO_2 TEA	gas	5 MW	20 ns	10.6 μm
GaAs	semiconductor	10 mW	cw	840 nm
Ruby (QS)	solid-state	100 MW	10 ns	694 nm
Nd:YAG	solid-state	50 W	cw	1.06 μm
Nd:YAG (QS)	solid-state	50 MW	20 ns	1.06 μm
Nd:YAG (ML)	solid-state	2 kW	60 ps	1.06 μm
Nd:Glass	solid-state	100 TW	11 ps	1.06 μm
Rh6G (ML)	dye	10 kW	30 fs	600 nm
HF	chemical	50 MW	50 ns	3 μm
ArF	excimer	10 MW	20 ns	193 nm

uous wave (cw) output or pulsed, through techniques such as Q-switching (QS), mode-locking (ML), or transverse electrical atmospheric (TEA) discharge. Optical powers emitted by inexpensive lasers (HeNe, GaAs) are a few mW, while peak powers of pulsed lasers may be multi-gigawatts. Other characteristics such as efficiency and durability may be important for practical applications.

To create gain, excitation may be achieved by electrical discharge (gases), current injection (semiconductors), flashlamps (solid state), other lasers (dyes), or chemical reactions. Figures 1 through 4 show geometries for several laser types. Characteristics of lasers can be vastly different: costs from a few dollars to multimillion dollars; powers from microwatts to gigawatts; sizes from tenths of millimeters to tens of meters; linewidths ($\Delta f/f$) from 10^{-1} to 10^{-15}; pulse durations from 10^{-14} s to continuous wave.

Only a few lasers have found major commercial markets. The HeNe laser is the only "cheap" visible laser. The semiconductor laser is efficient, miniature, inexpensive (millions are used in Optical Disc players and fiber communication systems), but its light is invisible. The long wavelength infrared carbon dioxide (CO_2) laser, as the most efficient high-power laser, is useful for cutting and heating. The argon, excimer, and Nd:YAG lasers are used for medical applications.

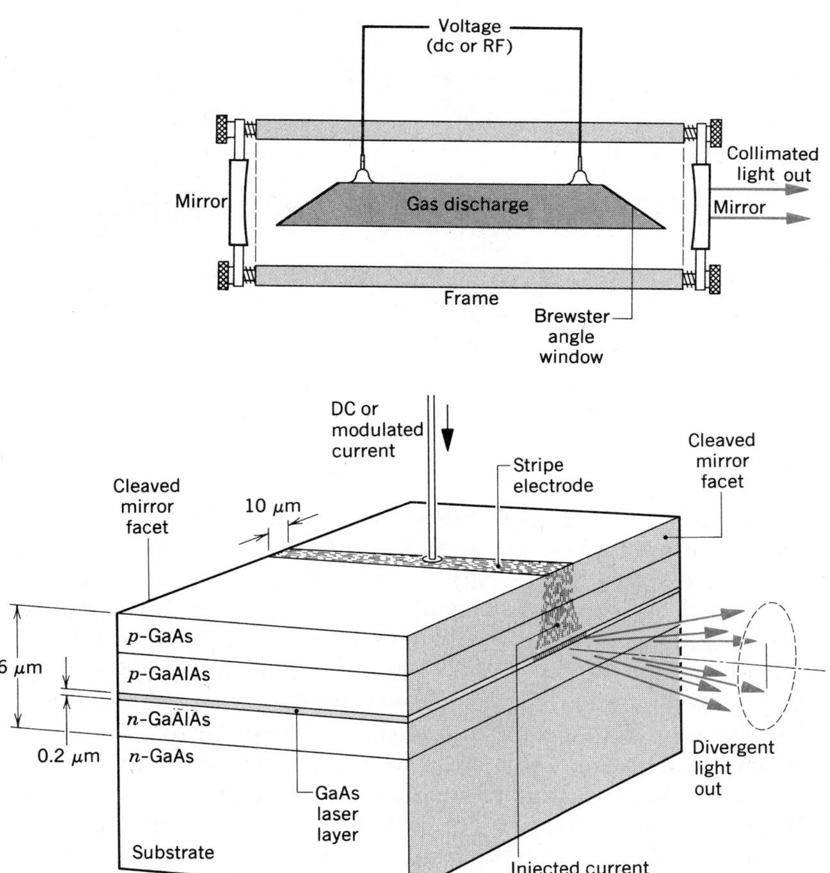

Figure 1 Gas lasers are excited by electrical discharge into a tube of very pure gas, with mirrors forming the optical cavity, which can vary from 5 cm to 5 m in length. Visible lasers contain a mixture of helium and neon, or of argon or krypton. Gas lasers can be operated cw or pulsed; CO_2 lasers are the most efficient. Carefully aligned mirrors may be external to the gas tube, as shown, or attached directly to the gas tube.

Figure 2 In a semiconductor laser, incoming current creates light at a *p-n* junction (abrupt interface between *p*-type and *n*-type material). Layers of alloy semiconductor $Ga_{0.7}Al_{0.3}As$ confine light and electrical carriers within the GaAs laser layer, forming a *double heterostructure.* Mirrors for the laser are created by cleaving end facets on the GaAs crystal 200 μm apart. Their small size, high efficiency, ease of modulation, high durability, and low cost are the primary reasons why more semiconductor lasers have been sold today than all other lasers combined.

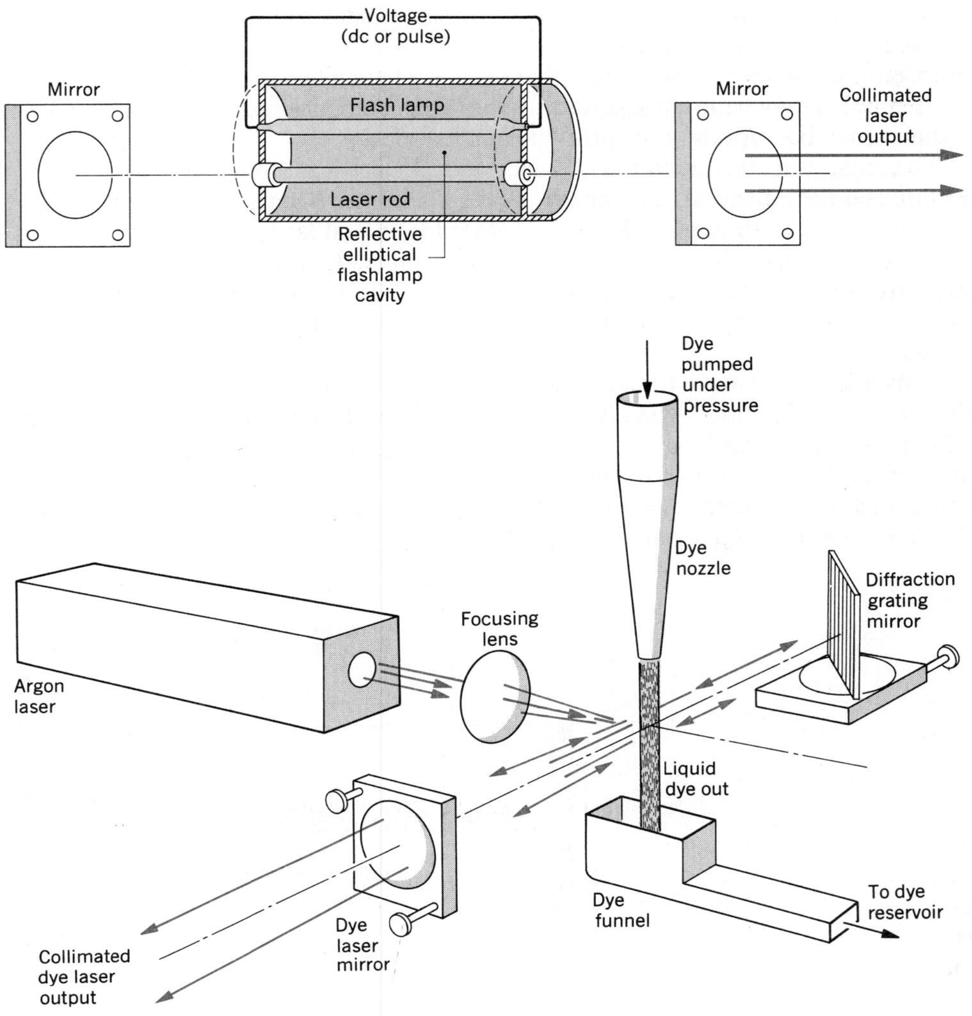

Figure 3 *Solid-state lasers* is the name given to all nonsemiconductor crystal and glass lasers. These lasers are excited by lamps, either cw or pulsed flash-lamps (as in cameras). The ruby laser, first laser ever built, has now been supplanted by a more efficient solid-state laser that uses Nd (Neodymium) in either YAG (yttrium aluminum garnet) crystals or glass. High-reflectance mirrors consist of many alternating layers of high- and low-refractive-index dielectric films.

Figure 4 Dye lasers use flowing liquids that are excited by flash lamps or other lasers. The most useful of these is Rh6G (rhodamine 6G), one of the most highly fluorescing materials known (used by the first astronauts to mark the position of their capsules when landing in the ocean). The most important advantage of dye lasers is that they are the only widely-tunable lasers. The color of a dye laser can be changed by rotating an optical element such as a diffraction grating within the laser cavity. These lasers can also be mode-locked to generate extremely short light pulses.

Cutting, Welding, and Blasting with Lasers

Phase coherence allows focusing of laser light to spots whose diameters are approximately a wavelength (10^{-4} cm). Thus, a 1-W laser can be focused to an intensity of 10^8 W/cm^2. As a graduate student in 1963, shortly after the invention of the ruby laser, I demonstrated that a pulsed ruby laser (peak power of 10^8 W), focused to a peak intensity of 10^{16} W/cm^2, can blast holes in razor blades and ionize the air! This high brightness (intensity) of lasers makes them dangerous. An unfocused 1 mW HeNe laser has a brightness equal to sunlight on a clear day (0.1 W/cm^2) and it is dangerous to stare at the beam. The more powerful lasers can cause damage very quickly. I learned the hard way that an unfocused 1-W Argon laser, with an intensity of 100 W/cm^2, can burn a hole in a dress. Students in the Center for Laser Studies are required to obey the rules for laser safety.

The high intensity of the laser can be applied to a variety of medical needs. Lasers have been commonly used to weld detached retinas in place in the eyes of many patients, including Bob Hope. Surgeons favor lasers for cutting because the light cannot infect the wound and it also cauterizes (singes to stop bleeding). By using an optical fiber to convey light into the stomach, physicians cauterize bleeding ulcers with lasers. Lasers are used to remove a variety of skin cancers as well as birthmarks. Indeed, lasers have undoubtedly saved many lives.

Focused CO_2 lasers are used to provide heat for many applications. One example is the welding of handles onto cooking pots. Here the importance of the laser lies in the fact that the copper of the pot has a high thermal conductivity (in order to distribute heat rapidly and cook the food uniformly), while the stainless steel handle has a low thermal conductivity (in order not to cook the cook). Conventional welding has great difficulty attaching metals with different thermal conductivites. The intense heat of the laser makes it possible to weld in a very short time, so that the different thermal conductivities are irrelevant.

Lasers combined with computers have made possible automated manufacturing, as for example in automated cloth-cutting. Garment sizes are programmed into a computer that drives a movable mirror, focusing the cutting beam onto the layers of cloth.

Lasers can provide localized heat to treat surfaces such as in hardening parts of automobile cam shafts. If the entire cam shaft is heated to achieve hardening, slight warping occurs, requiring larger tolerances. By contrast, the laser can heat just the cam surfaces and close tolerances can be maintained.

High-precision, tightly-focused laser beams are used for many fabrication jobs in the semiconductor wafer industry. Laser trimming of resistors removes bits of material and the resistance is increased. A generic integrated circuit chip can be customized by selectively removing some of the metallic interconnections with a Nd:YAG laser beam.

Extremely high intensities, created by pulsed lasers that store electrical (or chemical) energy and then release it in a short time, may produce a plasma hot enough to create neutrons. Laser-initiated thermonuclear reactions (laser fusion) were predicted in the 1970s to be the energy source of the future, although recent difficulties have dimmed original speculations. Several multimillion-dollar national facilities explore these ultra-high-intensity applications.

Lasers and Information Processing

Laser light may be used to communicate information. One such example is the supermarket scanner, in which a HeNe laser is scanned across a bar code on a package. The reflected light is detected as a modulated light signal, varying in time with the pattern of the bar code. The detector converts the modulated light signal into an electrical signal, which is then fed into a computer. Optical reading of information is an important application for lasers. The original information may be printed, such as in the bar code of the supermarket scanner, or may be impressed as tiny holes on a disc, as in optical disc recording.

Home entertainment systems use the optical disc to provide digital audio or video signals. Information is stored on these compact discs by punching a series of miniature holes in a reflecting layer with an intense, modulated laser beam. Within the disc player head is a semiconductor laser whose output is reflected from the disc to a detector, unless there is a hole in the reflecting layer. As the disc spins, the pattern of holes is converted to a digital signal that contains the coded audio information.

Since focused laser beams can pack information into a wavelength-sized spot, compact information storage is possible through optical memories. In fact, since a typical book is about 6 Mbits, one 12-inch disc could theoretically hold the contents of 10,000 books! First microfilm, then magnetic disks, and now the optical disc have drastically reduced the volume required for information storage.

Information can also be impressed on a light beam by modulating the laser itself. In an optical communication system this information is transmitted by the light beam to a detector that may be many kilometers away. In an optical fiber system, the output of a semiconductor laser, modulated by variation of its drive current, is focused into an optical fiber (see Essay 17). For communication to and from satellites, line-of-sight optical systems are important. In this application a Nd:YAG laser is used, together with an external modulator consisting of a crystal of $LiNbO_3$ (lithium niobate), which acts as a shutter.

More efficient external modulation can be achieved if the light is confined to optical waveguides, a concept called *integrated optics.* The ultimate dream of integrated optics is to replace complicated discrete optical systems with a single semiconductor chip that contains all the necessary optical devices for light generation, modulation, combining, processing, and detection. We are at the threshold of developing integrated optical circuits that may rival integrated electronic circuits. The term "opto-electronics" is used for circuits that combine optical and electronic components on one chip.

Lasers for Measurements

The coherence of lasers makes them ideal for a wide variety of measurements such as interferometry and spectroscopy. Monitoring small movements of the earth due to earthquakes is done with long-baseline laser interferometers. Laser radar (lidar) has provided precise measurements of astronomical distances to the moon and to satellites. The first people on the moon placed a "corner cube" (specially-constructed mirror to reflect light backwards independent of orientation) on the moon's surface. A pulsed ruby laser beam was collimated through a telescope in Texas and directed to the corner cube on the moon, where it was reflected back to the same telescope. The time delay of the return signal determined the distance to the moon to a probable error of a few centimeters.

One of the unique techniques lasers provide to the science of measurement is *holography* (see Essay 18). By interference of light between that reflected from an object and a reference beam, both the intensity and relative phase can be recorded, making possible the creation of three-dimensional images. Artistic and commercial displays, such as the covers of *National Geographic* magazine (March 1984, November 1985) and the imprints on credit cards, have allowed most laypersons to see holograms. In science and engineering, holography is used to measure distortions in objects. If the object moves even a fraction of a wavelength, the motion can be detected by holograms double-exposed, before and after the motion, resulting in bands of interference fringes observed on the image.

The monochromaticity of lasers has made possible revolutionary changes in the science of *spectroscopy,* which is the measurement of absorption or emission of light in an atomic or molecular medium. By carefully determining the characteristic wavelengths, considerable information can be obtained about the state of the atoms or molecules that emitted the light. Spectroscopy is very important in many scientific fields such as chemistry, physics, biology, and astronomy; true revolutions in understanding have been obtained by the laser's ability to reduce the linewidth in measurements by a factor of 10^{10}!

A new method of measurement uses an optical fiber to sense changes in a physical quantity. The fiber may provide a means for distributing light to a remote location, or it may provide the sensing element itself. In the former category would be a temperature sensor at the bottom of a deep well into which light is fed through a fiber. This sensor, attached to the end of the fiber, is a piece of semiconductor material whose transmission depends strongly on temperature, followed by a mirror. The amount of reflected light depends on temperature, providing a remote thermometer. The same concept can be used for a variety of applications such as the measurement of blood pressure, position (earthquake fault detection), and pollutants.

Optical fibers themselves can also be used as sensors by inserting them in one arm of an interferometer, using a long coil to increase sensitivity. For example, in our laboratory we found that a fiber coil could detect the heat of our hands at a distance of 10 cm! The two most practical fiber-coil sensors developed to date are acoustic sensors (hydrophones) and inertial rotation sensors (gyros). Considerable progress is currently being made in all types of fiberoptic sensors for medical, commercial, and military applications.

One of the most exciting developments in laser measurements uses a property that violates the monochromaticity most people associate with lasers. It is the ability to make dye lasers operate over a wide frequency region and then to create ultra-short light pulses, only 10^{-14} seconds long. Since typical electronic response times are nanoseconds, ultra-short light pulses can probe matter much more rapidly. We call the regime of $10^{-15}-10^{-13}$ s, the "femtosecond" regime.

Ultra-short pulses are created from lasers by "mode-locking," by using lasers that have many modes, and by providing a means for locking these modes together with the same phase at a given time. Because the modes oscillate at different frequencies, they rapidly lose phase coherence and effectively cancel in a time given by the inverse of their overall frequency difference, which provides the ultra-short pulse. The shorter the pulse in time, the wider the bandwidth of laser frequencies required. The shortest pulses have resulted from developing dye lasers with very wide bandwidths, covering almost the entire visible region. With these short pulses, new understanding is possible as to the physical, chemical, and electromagnetic processes occurring in the femtosecond domain.

Conclusions

Applications for lasers exist throughout our society, and new uses are discovered almost daily; we have discussed only a few of the more important ones. Considerable research is needed to bring some of these ideas to full fruition as well as to invent new applications. Lasers today are in the position of electronics in the late 1950s. Who can envisage their uses in 30 more years?

CHAPTER 46

CONDUCTION OF ELECTRICITY IN SOLIDS

The pocket calculator has made tables of logarithms obsolete, has turned slide rules into museum pieces and has put enormous calculating power at our fingertips. The device is crammed full of transistors and other semiconducting devices, none of which could have been developed without a knowledge of quantum physics. As its display shows, this calculator stores within its memory a reminder of this quantum background. Do you recognize the number?

46-1 The Properties of Solids

We have seen how well quantum theory works when we apply it to individual atoms. In this chapter we hope to show, by a single broad example, that this powerful theory works just as well when we apply it to aggregates of atoms in the form of solids.

Every solid has an enormous range of properties that we can choose to examine. Is it transparent? Can you hammer it out into a flat sheet? What kinds of waves travel through it and at what speeds? Does it have interesting magnetic properties? Is it a good heat conductor? What is its crystal structure? Does it have special surface properties? . . . The list goes on. We choose here to focus on a single question: "What are the mechanisms by which a solid conducts, or does not conduct, electricity?" As we shall see, the laws that govern electrical conduction are quantum laws.

46-2 Electrical Conductivity

In studying electrical conductivity, we choose to examine only solids whose atoms are arranged to form a periodic three-dimensional lattice. Figure 1 shows such lattice structures for carbon (in the form of diamond), for

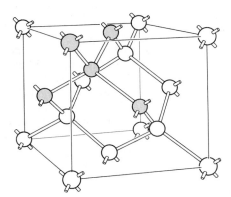

(a)

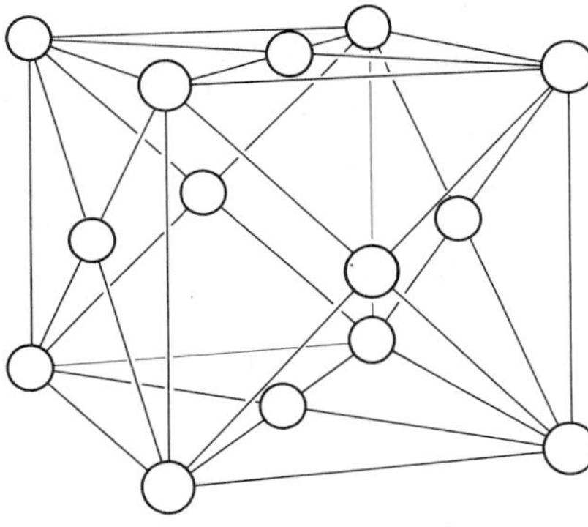

(b)

Figure 1 *(a)* The crystal structures for carbon (in the form of diamond) and for silicon, which happen to be identical. In this structure, as the darkened spheres show, each atom is bonded to four of its neighbors. The structure for carbon in the form of graphite is quite different. *(b)* The crystal structure for copper, an arrangement called *face-centered cubic.*

copper, and for silicon. We shall not consider such materials as plastic, glass, or rubber, whose atoms are not arranged in any such regular way.

The basic electrical measurement that we can make on a sample is its *electrical resistivity* ρ at room temperature; see Section 28-4. By measuring ρ at various temperatures, we can also obtain a value for α, the *temperature coefficient of resistivity.* Finally, by making Hall effect measurements (see Section 30-4) we can find a value for n, *the number of charge carriers per unit volume* in the material being tested.

Table 1 Some Electric Properties of Two Materials[a]

		Copper	*Silicon*
Type of conductor		Metal	Semiconductor
Density of charge carriers,[b] n	m^{-3}	9×10^{28}	1×10^{16}
Resistivity, ρ	$\Omega \cdot m$	2×10^{-8}	3×10^{3}
Temperature coefficient of resistivity, α	K^{-1}	$+4 \times 10^{-3}$	-70×10^{-3}

[a] All values are for room temperature.
[b] The value for the semiconductor includes both electrons and holes.

From measurements of the room temperature resistivity alone, we quickly discover that there are some materials—we call them *insulators*—that for all practical purposes do not conduct electricity at all. Diamond, for example, has a resistivity of about $10^{16}\ \Omega \cdot m$, greater than that of copper by a factor of $\sim 10^{24}$.

We can use our measured values of ρ, α and n, to divide most noninsulators into two major categories: *metals* such as copper, and *semiconductors* such as silicon. As we see from Table 1, a typical semiconductor (silicon), compared with a typical conductor (copper): (1) has far fewer charge carriers, (2) has a considerably larger resistivity, and (3) has a temperature coefficient of resistivity that is both large and negative. That is, although the resistivity of a metal *increases* with temperature, that of a semiconductor *decreases.*

We have now established an experimental basis for framing our central question about the conduction of electricity in solids. We pose it in specific terms:

What is there about diamond that makes it an insulator, about copper that makes it a metal, and about silicon that makes it a semiconductor?

As we shall see, quantum physics provides the answers.

46-3 Energy Levels in a Solid

The distance between adjacent copper atoms in solid copper is 260 pm. Consider, as in Fig. 2*a*, two copper atoms that are separated by a much greater distance than this. As Fig. 2*b* shows, each of these isolated atoms has

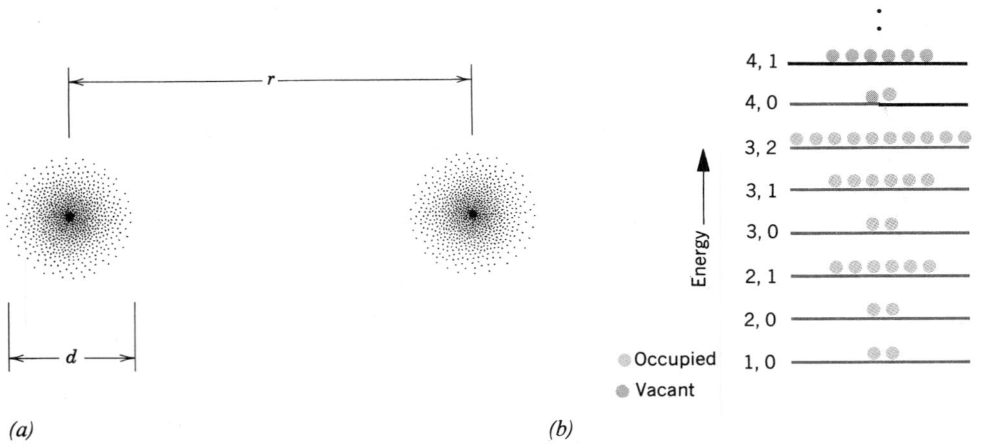

Figure 2 *(a)* Two isolated copper atoms, representing their electron clouds represented by dot-plots. *(b)* Each atom has 29 electrons distributed over a set of energy levels as shown. The levels are identified by the notation n, ℓ where n is the principal quantum number and ℓ the orbital quantum number. Each energy level contains $2(2\ell + 1)$ quantum states, defined by the quantum numbers m_ℓ and m_s. For simplicity, the levels are shown as uniformly spaced in energy.

associated with it an array of discrete quantum states, each state defined by its unique set of quantum numbers. In the ground state of the neutral copper atom, its 29 electrons occupy the 29 states of this array that are lowest in energy, each state containing but a single electron as the Pauli exclusion principle requires.

If we bring the atoms of Fig. 2*a* closer together, they will—speaking loosely—gradually begin to sense each other's presence. In the formal language of quantum physics, their wave functions will begin to overlap. This overlap will occur first for the wave functions of the valence electrons, which, because they spend most of their time in the outer regions of the electron cloud of the isolated atom, are the first to make contact.

When the wave functions overlap, we can no longer speak of two independent and isolated systems but of a single two-atom system containing 2×29 or 58 electrons. The Pauli principle requires that each of these electrons must occupy a *different* quantum state. The only way that this can happen is for each energy level of the isolated atom to split into *two* levels for the two-atom system.

We can bring up further atoms and in this way gradually construct a lattice of solid copper. If our specimen contains N atoms, each level of the isolated copper atom must be split into N levels. In this way, each *level* of the isolated atom becomes a *band of levels* in the solid. In a typical solid, an energy band is a few electron volts wide. Since N is of the order of the Avogadro number, we can see that the individual energy levels within a band are very close together indeed.

Figure 3 suggests the band structure of the levels in a hypothetical solid in which we have assumed, for simplicity, that the bands do not overlap. The gaps between the bands represent ranges of energy that no electron may possess. In much the same way, electrons in individual atoms cannot possess energies that lie between the discrete allowed levels of the atom.

Note in Fig. 3 that the bands that lie lower in energy are narrower than those that lie higher. This is because these low-energy bands correspond to levels in the isolated atom that are occupied by electrons that spend most of their time deep within the electron cloud of the atom. The wave functions of these core electrons thus do not overlap as much as do the wave functions of the outer or valence electrons and for that reason the split-

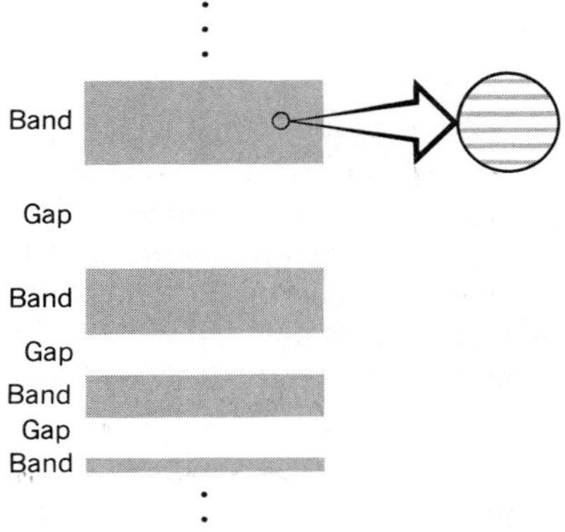

Figure 3 An idealized representation of the band-gap pattern for the energy levels of a solid. As the magnified view of the upper band suggests, each band consists of a very large number of closely lying energy levels. The levels of this idealized plot are, as yet, unoccupied by electrons.

ting of the levels—although it must occur—is not as great as it is for the levels normally occupied by the outer electrons.

Now that we have established the pattern of levels for a solid, we are ready to consider how these levels are filled with electrons. We shall see how this will lead us in convincing way to the answers to the question that we raised at the end of Section 46-2.

46-4 Insulators

The feature that defines an *insulator* is that, as Fig. 4 shows, the highest occupied level coincides with the top of a band. In addition, this band must be separated from the unoccupied band above it by a substantial energy gap E_g. For diamond, $E_g = 5.4$ eV, a value some 140 times larger than the average thermal energy of a free particle at room temperature.

By definition, an insulator is a solid through which electrons cannot flow as a directed drift current. Let us see why. If you apply an electric field **E** to an insulator, it will exert a force $-e\mathbf{E}$ on each electron. Classically, this force will cause the electron to increase the component of its velocity in the direction $-\mathbf{E}$ which, in turn, means that its kinetic energy will change. In quantum terms, if the energy of an electron changes, the electron must move to a different energy level within the solid. However, the Pauli principle prevents it from doing so because all other levels within the band into which the electron might move are already occupied. These electrons are in total grid lock. It is as if a child tries to climb a ladder on which other children are standing, one every rung; since there are no vacant rungs, no one can move.

There are plenty of vacant levels in the band above the filled band in Fig. 4 but, if an electron is to occupy one of these levels, it must somehow jump across the gap that separates the two bands. It cannot pause at a way-station within the gap because all energies in this range are strictly forbidden. In diamond, the gap is simply too wide for any detectable number of electrons to make it to the vacant band, either by the action of an external electric field or by thermal agitation.

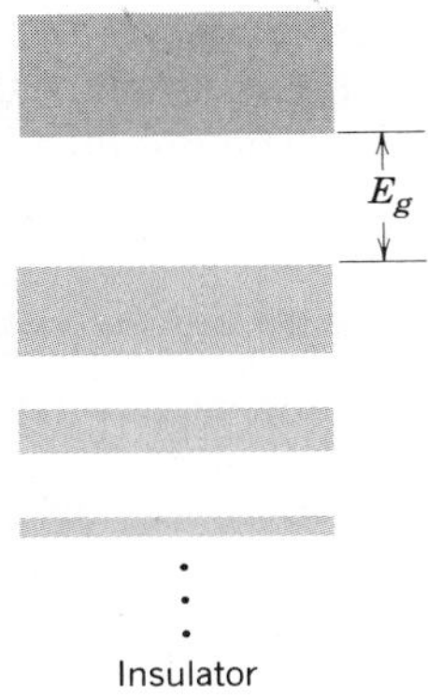

Figure 4 An idealized representation of the band-gap pattern for an insulator. Note that the highest filled level (color) lies at the top of a band and that the next highest vacant band (in black) is separated by a relatively large energy gap E_g.

46-5 Metals—Qualitative

The feature that defines a *metal* is that, as Fig. 5 shows, the highest occupied level (at the absolute zero of temperature) falls somewhere in the middle of a band. The electrons that occupy this partially filled band are the valence electrons of the atoms, which, being free to move throughout the solid, become the *conduction electrons* of the solid.

At the absolute zero of temperature, thermal agitation plays no role and all electrons occupy states of the lowest possible energy. We can assume with little error that the potential energy of the conduction electrons

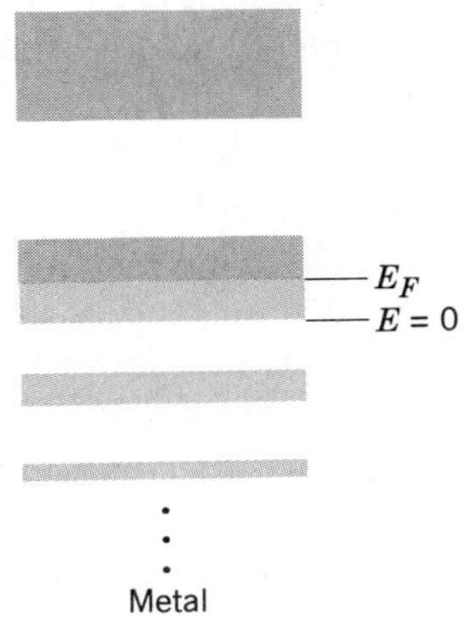

Figure 5 An idealized representation of the band-gap pattern for a metal at the absolute zero of temperature. The conduction electrons of the metal occupy the highest partially filled band. Note that vacant levels are available within the band so that these electrons can change their energies and conduction can take place. Lower lying bands are completely filled by the core electrons, that is, those electrons held close to the lattice sites and not free to move through the solid.

remains constant as they move about within the solid. If we set this constant equal to zero—as we are always permitted to do—the total energy E associated with any level is equal to the kinetic energy of the electron that occupies that level.

The level at the bottom of the partially-filled band of Fig. 5 thus corresponds to $E = 0$. The highest occupied level in this band (at the absolute zero of temperature) is called the *Fermi level* and the energy corresponding to it is called the *Fermi energy* E_F; for copper, $E_F = 7.0$ eV. The electron speed corresponding to the Fermi energy is called the *Fermi speed* v_F; for copper, $v_F = 1.6 \times 10^6$ m/s.

A glance at Fig. 5 should be enough to shatter the popular misconception that all motion ceases at the absolute zero of temperature. We see that, entirely because of Pauli's exclusion principle, the electrons are stacked up in the partially-filled band of Fig. 5 with energies that range from zero up to the Fermi energy. The *average* kinetic energy for the electrons in this band for copper is about 4.2 eV. By comparison, the average translational kinetic energy of a molecule of an ideal gas at room temperature is only 0.025 eV. The conduction electrons in a metal have plenty of energy at absolute zero!

Conditions for $T > 0$. What happens to the electron distribution of Fig. 5 as we raise the temperature above absolute zero? The short answer is that not very much happens, although the little that does is very important. It is clear that electrons in bands below the partially-filled band of Fig. 5 cannot be affected by thermal agitation; they are all grid-locked.

Only electrons close to the Fermi energy find vacant levels above them and it is only these electrons that are free to be boosted to higher levels by thermal agitation. Even at $T = 1000$ K, a temperature at which a metal sample would glow brightly in a dark room, the distribution of electrons among the available levels does not differ very much from the distribution at the absolute zero.

Let us see why. The quantity kT is a convenient measure of the energy that may be given to an electron by the random thermal jiggling of the lattice. At $T = 1000$ K, $kT = 0.086$ eV; no electron can hope to have its energy changed by more than a few times this relatively small amount by thermal agitation alone. All the "action" takes place for electrons whose energies are close to the Fermi energy. It has been said, somewhat poetically, that thermal agitation normally causes only ripples on the surface of the Fermi sea; the vast depths of that sea lie undisturbed.

Electrical Conduction in a Metal. If you apply an electric field $\mathbf{E}$ to a metal, it exerts a force $-e\mathbf{E}$ on each electron. In a metal, this force, during time Δt, causes every conduction electron to acquire a velocity increment $\Delta\mathbf{v}$ in the direction of $-\mathbf{E}$. This will require the electrons to change their energies, but there are vacant levels available so that these rearrangements can be made. To return to our previous metaphor, there are now vacant rungs on the upper half of the ladder.

The velocities of the individual conduction electrons do not increase without limit, however, because of collisions associated with the thermal vibrations of the lattice. Thus, after a certain time τ, called the *relaxation time,* the drift velocity of the conduction electrons settles down to a constant limiting value, which we associate with the constant current that is set up by the applied electric field. Note that, although *all* of the conduction electrons contribute to the current, only electrons close to the Fermi energy are able to make collisions and thus play their role in establishing the limiting value of the drift velocity. It is only these electrons that have ample vacant levels nearby into which they can move after they have experienced a scattering event.

In Section 28-6, we presented the following equation for the resistivity of a metallic conductor

$$\rho = \frac{m}{ne^2\tau} \tag{1}$$

in which m is the mass of the electron, $-e$ is its charge, and n is the number density of the charge carriers; that is, the number of conduction electrons per unit volume. Although we derived this equation on a classical basis, it holds true when the quantization of the electron energy is taken into account. The quantity τ is the relaxation time to which we referred in the preceding paragraph.

Sample Problem 1 How many conduction electrons are there in a copper cube 1 cm on edge?

In copper, there is one conduction electron per atom so that N is given by

$$N = na^3 \tag{2}$$

in which n is the number of atoms per unit volume and a (= 1 cm) is the length of the cube edge. We can find n from

$$n = \frac{N_A d}{A}$$

in which N_A is the Avogadro constant, A is the atomic weight of copper, and d is the density of copper. If we substitute values for these quantities, we find

$$n = \frac{(6.02 \times 10^{23}\ \text{atoms/mol})(8900\ \text{kg/m}^3)}{0.06357\ \text{kg/mol}}$$
$$= 8.43 \times 10^{28}\ \text{atoms/m}^3 = 8.43 \times 10^{28}\ \text{electrons/m}^3.$$

From Eq. 2 we then have

$$N = na^3 = (8.43 \times 10^{28}\ \text{electrons/m}^3)(10^{-2}\ \text{m})^3$$
$$= 8.43 \times 10^{22}\ \text{electrons.} \qquad \text{(Answer)}$$

The Pauli exclusion principle requires that each of these electrons occupy a different quantum state. These states are distributed over an energy interval of only 7.0 eV (the Fermi energy) so that the average spacing between them is very small indeed.

Not all of these states have distinct energies. Consider, for example, an electron moving along the x axis with speed v. It has the same energy as an electron moving with this same speed in any other direction in the solid but, because its motion is different, it is in a different quantum state, described by a different wave function.

Sample Problem 2 (a) What is the speed of a conduction electron in copper with a kinetic energy equal to the Fermi energy (= 7.0 eV)?

The total energy E of the conduction electrons is all kinetic and we can write, if $E = E_F$,

$$E_F = \tfrac{1}{2} m v_F^2,$$

in which v_F is the Fermi speed. Solving for v_F yields

$$v_F = \sqrt{\frac{2E_F}{m}} = \sqrt{\frac{(2)(7.0\ \text{eV})(1.6 \times 10^{-19}\ \text{J/eV})}{9.11 \times 10^{-31}\ \text{kg}}}$$
$$= 1.6 \times 10^6\ \text{m/s.} \qquad \text{(Answer)}$$

You must not confuse this speed with the *drift speed* of the conduction electrons, which is typically 10^{-5} m/s and is thus smaller by about a factor of 10^{11}. As we explained more fully in Section 28-6, the drift speed is the average speed at which electrons actually drift through a conductor when an electric field is applied; the Fermi speed is their average speed between collisions.

(b) What is the average time τ between collisions for the conduction electrons in copper? The resistivity of copper at room temperature is $1.7 \times 10^{-8}\ \Omega \cdot \text{m}$.

Solving Eq. 1 for τ yields

$$\tau = \frac{m}{ne^2\rho}$$
$$= \frac{9.11 \times 10^{-31}\ \text{kg}}{(8.43 \times 10^{28}\ \text{m}^{-3})(1.60 \times 10^{-19}\ \text{C})^2(1.7 \times 10^{-8}\ \Omega \cdot \text{m})}$$
$$= 2.5 \times 10^{-14}\ \text{s.} \qquad \text{(Answer)}$$

(c) What mean free path λ may be calculated from the results of (a) and (b) above?

We have

$$\lambda = v_F\tau = (1.6 \times 10^6\ \text{m/s})(2.5 \times 10^{-14}\ \text{s})$$
$$= 4.0 \times 10^{-8}\ \text{m} = 40\ \text{nm.} \qquad \text{(Answer)}$$

In the copper lattice the centers of neighboring atoms are 0.26 nm apart. Thus, a typical conduction electron can move a substantial distance, about 150 interatomic distances, through a copper lattice at room temperature before making a collision.

46-6 Metals—Quantitative

Now let us look at the conduction of electricity in a metal quantitatively, under several headings.

Counting the Quantum States. We start by counting the number of distinct quantum states in the partially-filled band of Fig. 5. We cannot possibly deal with this vast number of states one at a time; we must use statistical methods. Instead of asking, "What is the energy of this state?" We must ask, "How many states (per unit volume) have energies that lie in the energy range E to $E + dE$?" This number can be written as $n(E)dE$, where $n(E)$ is called the *density of states.*

If we assume that the conduction electrons move in a region of constant potential, $n(E)$ can be shown to be given by

$$n(E) = \frac{8\sqrt{2}\pi m^{3/2}}{h^3} E^{1/2} \qquad \text{(density of states).} \qquad (3)$$

Figure 6a is a plot of Eq. 3. Note that there is nothing in this equation that depends on the shape or size of the sample or on the material of which it is made.

Filling the States at $T = 0$. Equation 3 tells us how the *unoccupied* states are distributed in energy. We must now weight each energy with a factor $p(E)$, called the *probability function,* that gives us the probability that a state with that energy will actually be occupied. At the absolute zero of temperature, all states with energies less than the Fermi energy are filled and all states with energies greater than that energy are vacant. The probability function in this case must be the simple rectangle shown in Fig. 6b, in which unity corresponds to a certainty that the state will be occupied and zero to a certainty that it will *not* be occupied.

The product of these two factors gives us $n_o(E)$, the density of *occupied* states. Thus

$$n_o(E) = n(E)p(E). \qquad (4)$$

Figure 6c is a plot of this product.

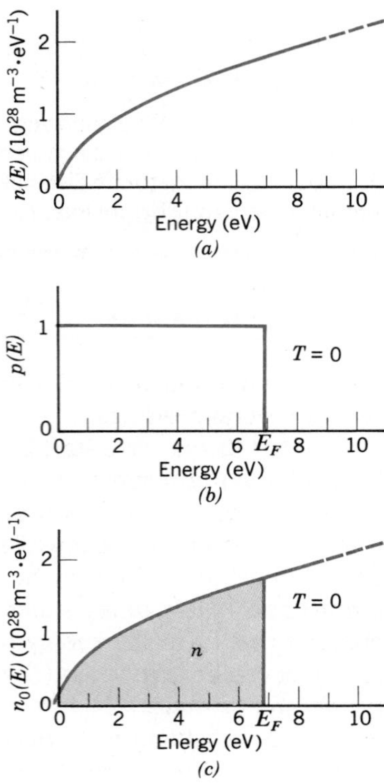

Figure 6 *(a)* The *density of states n(E)* plotted as a function of energy. *(b)* The *probability function p(E)* plotted as a function of energy at $T = 0$. *(c)* The *density of occupied states* $n_o(E)$, formed by multiplying the curves in *(a)* and *(b)*, plotted as a function of energy. Note that all states whose energies lie below the Fermi energy are occupied but all states above that energy are vacant.

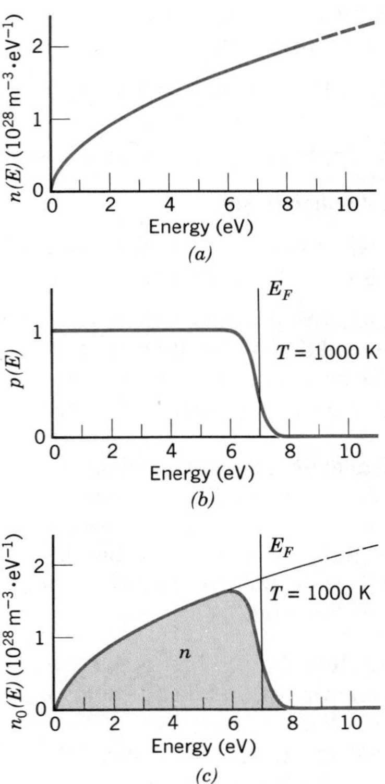

Figure 7 *(a)* The *density of states n(E)* plotted as a function of energy; this plot is the same as that of Fig. 6*a*. *(b)* The *probability function p(E)* plotted as a function of energy at $T = 1000$ K; note how little this plot differs from that of Fig. 6*b*. *(c)* The *density of occupied states* $n_o(E)$, formed by multiplying the curves in *(a)* and *(b)*, plotted as a function of energy; note how little this plot differs from that of Fig. 6*c*. (This plot is idealized in that it assumes that the conduction electrons move in a region of constant potential. Measured density of states plots do not have this simple shape.)

We can find the Fermi energy for a metal by adding up (integrating) the number of occupied states in Fig. 6*c* between $E = 0$ and $E = E_F$. The result must equal n, the number of conduction electrons per unit volume for the metal. In equation form we have

$$n = \int_0^{E_F} n(E)dE. \tag{5}$$

Note that n is represented by the shaded area in Fig. 6*c*.

Calculating the Fermi Energy. If we substitute Eq. 3 into Eq. 5, we find

$$n = \frac{8\sqrt{2}\pi m^{3/2}}{h^3}\int_0^{E_F} E^{1/2}\,dE = \left(\frac{8\sqrt{2}\pi m^{3/2}}{h^3}\right)\left(\frac{2\,E_F^{3/2}}{3}\right).$$

Solving for E_F leads to

$$E_F = \left(\frac{3}{16\sqrt{2}\pi}\right)^{2/3}\frac{h^2}{m}n^{2/3} = \frac{0.121\,h^2}{m}n^{2/3}. \tag{6}$$

Thus the Fermi energy can be calculated once n, the number of conduction electrons per unit volume, is known.

Filling the States for $T > 0$. It can be shown that the probability function for $T > 0$ is given by

$$p(E) = \frac{1}{e^{(E-E_F)/kT} + 1}$$

(probability function) (7)

in which E_F is the Fermi energy and k is the Boltzmann constant.

Note that as $T \to 0$, the exponent $(E - E_F)/kT$ in Eq. 7 approaches $-\infty$ if $E < E_F$ and $+\infty$ if $E > E_F$. In the first case we have $p(E) = 1$ and in the second $p(E) = 0$. Thus, at $T = 0$, Eq. 7 correctly yields the rectangular form shown in Fig. 6*b*. Equation 7 also shows us that the important quantity is not the energy E but rather $E - E_F$, the energy interval between E and the Fermi energy.

Figure 7*b* shows the probability function for $T = 1000$ K, calculated from Eq. 7. Note how little it differs from the rectangular form of Fig. 6*b*. Figure 7*c*, found by multiplying Figs. 7*a* and 7*b*, shows the density of occupied states for $T = 1000$ K. Note how little it differs from Fig. 6*c*, the distribution at $T = 0$.

Sample Problem 3 A cube of copper is 1 cm on an edge. How many quantum states lie in the energy interval between $E = 5.00$ eV and $E = 5.01$ eV?

These energy limits are so close together that we can say that the answer is

$$N = n(E)\Delta E\ V, \tag{8}$$

where $E = 5$ eV, $\Delta E = 0.01$ eV, and V is the volume of the cube. From Eq. 3 we have

$$\begin{aligned} n(E) &= \frac{8\sqrt{2}\pi m^{3/2}}{h^3} E^{1/2} \\ &= \frac{(8\sqrt{2}\pi)(9.11 \times 10^{-31}\text{ kg})^{3/2}(5\text{ eV})^{1/2}(1.60 \times 10^{-19}\text{ J/eV})^{1/2}}{(6.63 \times 10^{-34}\text{ J}\cdot\text{s})^3} \\ &= 9.48 \times 10^{46}\text{ m}^{-3}\text{ J}^{-1} = 1.52 \times 10^{28}\text{ m}^{-3}\text{ eV}^{-1}. \end{aligned}$$

From Eq. 8 we have, putting $V = a^3$ where a is the cube edge

$$\begin{aligned} N &= n(E)\Delta E\, a^3 \\ &= (1.52 \times 10^{28}\text{ m}^{-3}\text{ eV}^{-1})(0.01\text{ eV})(1 \times 10^{-2}\text{ m})^3 \\ &= 1.52 \times 10^{20}. \end{aligned} \qquad \text{(Answer)}$$

Sample Problem 4 (a) What is the probability that a state whose energy is 0.1 eV above the Fermi energy will be occupied? Assume a temperature of 800 K.

We can find $p(E)$ from Eq. 7. Let us first calculate the (dimensionless) exponent in that equation:

$$\frac{E - E_F}{kT} = \frac{0.1\text{ eV}}{(8.62 \times 10^{-5}\text{ eV/K})(800\text{ K})} = 1.45.$$

Inserting this exponent into Eq. 7 yields

$$p = \frac{1}{e^{1.45} + 1} = 0.19 \text{ or } 19\%. \qquad \text{(Answer)}$$

(b) What is the probability of occupancy for a state that is 0.1 eV *below* the Fermi energy?

The exponent in Eq. 7 has the same numerical value as above but is negative. Thus, from this equation

$$p = \frac{1}{e^{-1.45} + 1} = 0.81 \text{ or } 81\%. \qquad \text{(Answer)}$$

For states whose energies lie below the Fermi energy we are often more interested in the probability that the state is *not* occupied. This is, of course, just $1 - p$, or 19% in the present case. An unfilled state in an energy range in which most of the states are filled is called a *hole.* We shall see later that this is a very useful concept.

(c) What is the probability of occupancy for a state whose energy is equal to the Fermi energy?

For $E = E_F$ the exponent in Eq. 7 is zero and that equation becomes

$$p = \frac{1}{e^0 + 1} = \frac{1}{1 + 1} = 0.50 \text{ or } 50\%. \qquad \text{(Answer)}$$

This result does not depend on the temperature. We can, in fact, define the Fermi energy for a metal to be that energy for which the probability of occupancy *at any temperature* is 50%.

46-7 Semiconductors

As a comparison of Fig. 8 with Fig. 4 shows, a semiconductor is like an insulator in that its uppermost filled level (at the absolute zero of temperature) lies at the top of a band. A semiconductor differs from an insulator, however, in that the gap between this filled band and the next vacant band above it is much smaller than for an

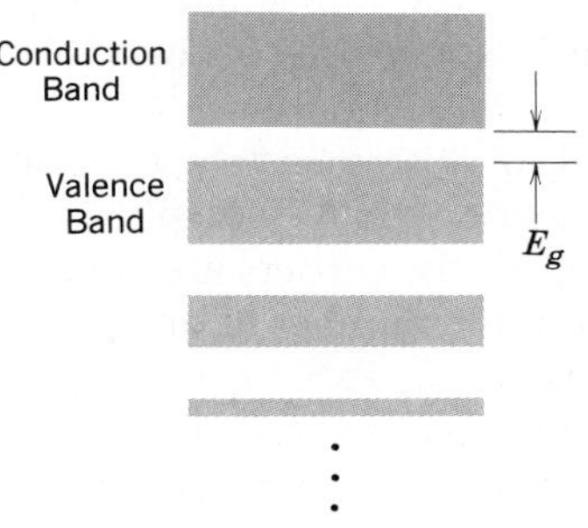

Figure 8 An idealized representation of the band-gap pattern for a semiconductor such as silicon. It resembles the pattern for an insulator (see Fig. 4) except that the gap between the valence band and the conduction band is much smaller.

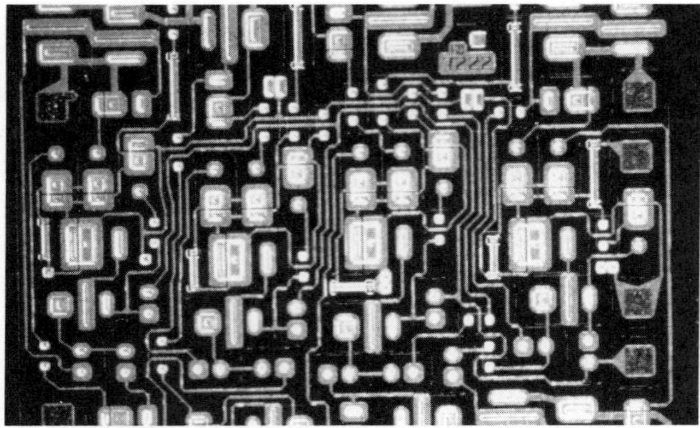

Figure 9 A photograph of an enlarged section of an integrated circuit.

insulator, so that there is a real possibility for electrons to "jump the gap" into this empty band by thermal agitation. For semiconducting materials, the highest filled band is called the *valence band* because the electrons that occupy it are the valence electrons of the isolated atom. The band above the valence band, which is vacant at $T = 0$, is called the *conduction band.*

The distinction between an insulator and a semiconductor is qualitative, depending as it does on the width of the energy gap. However, there is no doubt that diamond ($E_g = 5.4$ eV) is an insulator and silicon ($E_g =$ 1.1 eV) is a semiconductor. As it happens (see Fig. 1) these two substances have the same crystal structure.

The revolution in microelectronics that has so influenced our lives is based on semiconductors (see Fig. 9), so we would do well to learn more about them. Table 1 compares some electrical properties of silicon, our prototype semiconductor, and copper, our prototype conductor. Let us look carefully at this table, one row at a time.

The Density of Charge Carriers, n. Copper has many more charge carriers than does silicon, by a factor of $\sim 10^{13}$. For copper the carriers are the conduction electrons, present in the number of one per atom. At room temperature, to which Table 1 refers, charge carriers in silicon arise only because, at thermal equilibrium, thermal agitation has caused a certain (very small) number of electrons to be raised to the conduction band, leaving an equal number of vacant states (holes) in the valence band.

The holes in the valence band of a semiconductor also serve effectively as charge carriers because they permit a certain freedom of movement to the electrons in that band. If an electric field is set up in a semiconductor, the electrons in the valence band, being negatively charged, will drift in the direction of $-\mathbf{E}$. This causes the holes to drift in the direction of $\mathbf{E}$. That is, the holes behave like particles carrying a charge $+e$ and, in all that follows, that is exactly how we shall regard them. Conduction by holes is an important fact of life for semiconductors.

If the concept of a migrating hole seems confusing to you, think of a vacant slot in a parking lot that is otherwise filled with cars. If one of these cars moves into the slot, it fills it but creates a new vacant slot in the place it just left. This, in turn, can be filled by another car and so on. As the cars move around in this way, we can focus attention on the single migrating vacant slot as it wanders over the lot.

The Resistivity, ρ. At room temperature the resistivity of silicon is considerably higher than that of copper, by a factor of $\sim 10^{11}$. For both elements, the resistivity is determined by Eq. 1. The vast difference in resistivity between copper and silicon can be accounted for by the vast difference in n, the density of charge carriers. (The mean collision time τ will also be different for copper and for silicon but the effect of this on the resistivity is swamped by the enormous difference in n.)

The Temperature Coefficient of Resistivity, α. This quantity (see Eq. 16 of Chapter 28) is the fractional change in the resistivity per unit change or temperature, or

$$\alpha = \frac{1}{\rho}\frac{d\rho}{dT}.$$

The resistivity of copper and other metals *increases* with temperature ($d\rho/dT > 0$). This happens because collisions occur more frequently the higher the temperature, thus reducing τ in Eq. 1. For metals, the density of charge carriers n in that equation is independent of temperature.

On the other hand, the resistivity of silicon (and other semiconductors) *decreases* with temperature ($d\rho/dT < 0$). This happens because the density of charge carriers n in Eq. 1 increases rapidly with temperature. The decrease in τ mentioned before for metals also occurs for semiconductors but its effect on the resistivity is swamped by the very rapid increase of the density of charge carriers.

46-8 Doping

The versatility of semiconductors can be marvelously improved by introducing a small number of suitable replacement atoms (it seems pejorative to call them impurities) into the semiconductor lattice, a process called

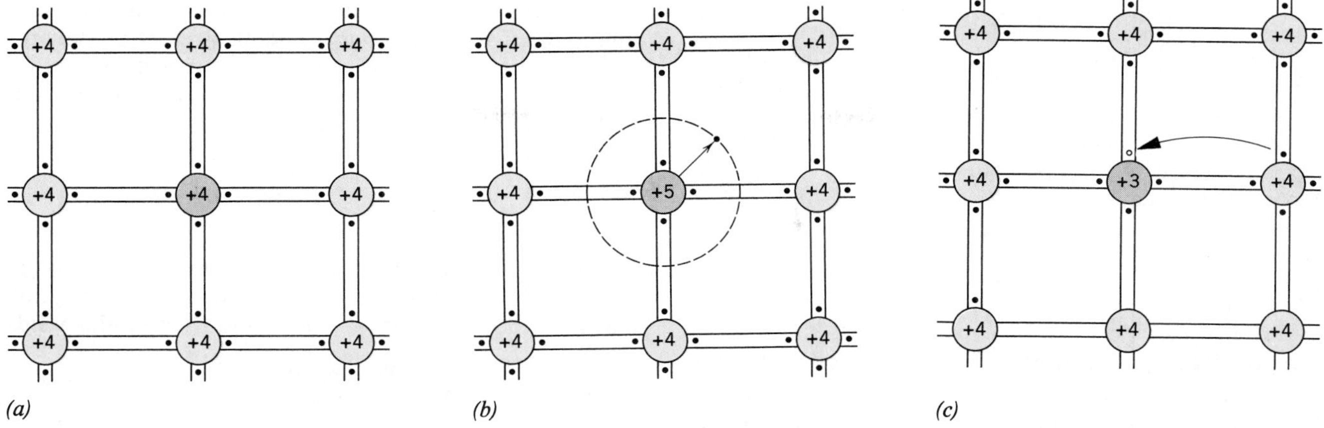

Figure 10 (*a*) A two-dimensional representation of a silicon lattice. Each silicon ion (core charge $= +4e$) is bonded to each of its four nearest neighbors by a shared two-electron bond. The dots show these valence electrons. (*b*) A phosphorus atom (valence $= +5$) is substituted for the central silicon atom, creating a donor site. (*c*) An aluminum atom (valence $= +3$) is substituted for the central silicon atom, creating an acceptor site.

doping. Essentially all practical semiconducting devices today are based on doped material. They are of two varieties, called *n-type* and *p-type;* we discuss each in turn.

n-Type Semiconductors. Figure 10*a* is a "flattened out" representation of a lattice of pure silicon; compare Fig. 1*a*. Each silicon atom forms a two-electron covalent bond with each of its four nearest neighbors, the electrons involved in the bonding making up the valence band of the sample. In Fig. 10*b* one of the silicon atoms (valence = 4) has been replaced by an atom of phosphorus (valence = 5). As Fig. 10*b* suggests, the "extra" electron is loosely bound to the phosphorus ion core because it is not involved in covalent bonds to neighboring ions. It is far easier for *this* electron to be thermally excited into the conduction band than it is for one of the silicon valence electrons to be so excited.

The phosphorus atom is called a *donor* atom because it so readily *donates* an electron to the conduction band. The "extra" electron in Fig. 10*b* can be said to lie in a localized *donor level,* as Fig. 11*a* shows. This level is separated from the bottom of the conduction band by an energy gap E_d, where $E_d \ll E_g$. By controlling the concentration of donor atoms it is possible to increase greatly the density of electrons in the conduction band.

Semiconductors doped with donor atoms are called *n-type* semiconductors, the "*n*" standing for "negative" because the negative charge carriers greatly outnumber

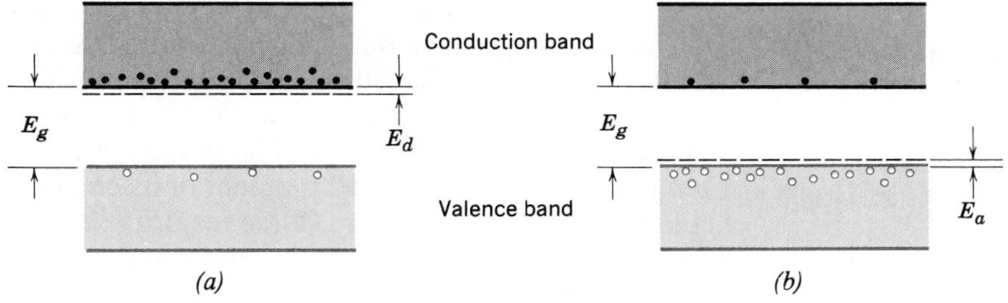

Figure 11 (*a*) An *n*-type semiconductor, showing donor levels that have contributed electrons (majority carriers) to the conduction band. The small number of holes (minority carriers) in the valence band is also shown. (*b*) A *p*-type semiconductor, showing acceptor levels that have contributed holes (majority carriers) to the valence band. The small number of electrons (minority carriers) in the conduction band is also shown.

Table 2 Properties of Two Doped Semiconductors

Matrix material	Silicon	Silicon
Dopant	Phosphorus	Aluminum
Type of dopant	Donor	Acceptor
Type of semiconductor	*n*-type	*p*-type
Dopant valence	5 (= 4 + 1)	3 (= 4 − 1)
Dopant energy gap	45 meV	57 meV
Majority carriers	Electrons	Holes
Minority carriers	Holes	Electrons
Dopant ion core charge	$+e$	$-e$

the positive charge carriers. The former, called the *majority carriers,* are the electrons in the conduction band. The latter, called the *minority carriers,* are the holes in the valence band.

p-Type Semiconductors. Figure 10*c* shows a silicon lattice in which a silicon atom (valence = 4) has been replaced by an aluminum atom (valence = 3). In this case there is a "missing" electron and it is easy for the aluminum ion core to "steal" a valence electron from a nearby silicon atom, thus creating a hole in the valence band.

The aluminum atom is called an *acceptor* atom because it so readily *accepts* an electron from the valence band. The electron so accepted moves into a localized *acceptor level,* as Fig. 11*b* shows. This level is separated from the top of the valence band by an energy gap E_a, for which $E_a \ll E_g$. By controlling the concentration of acceptor atoms it is possible to greatly increase the number of holes in the valence band.

Semiconductors doped with acceptor atoms are called *p-type* semiconductors, the "*p*" standing for "positive" because the positive charge carriers in this case greatly outnumber the negative carriers. In *p*-type semiconductors the majority carriers are the holes in the valence band and the minority carriers are the electrons in the conduction band.

Table 2 summarizes the properties of a typical *n*-type and a typical *p*-type semiconductor. Note particularly that the donor and acceptor ion cores, although they are charged, are not charge *carriers* because, at normal temperatures, they remain fixed in their lattice sites.

Sample Problem 5 The number density of conduction electrons in pure silicon at room temperature is about $10^{16}\ \mathrm{m^{-3}}$. Assume that, by doping the lattice with phosphorus, you want to increase this number by a factor of a million (10^6). What fraction of the silicon atoms must you replace by phosphorus atoms? (Assume that, at room temperature, the thermal agitation is effective enough so that essentially every phosphorus atom donates its "extra" electron to the conduction band.)

The density of the phosphorus atoms must be about $(10^{16}\ \mathrm{m^{-3}})(10^6)$ or $\sim 10^{22}\ \mathrm{m^{-3}}$. The density of silicon atoms in a pure silicon lattice may be found from

$$n_{\mathrm{Si}} = \frac{N_A d}{A},$$

in which N_A is the Avogadro constant, d is the density of silicon, and A is the atomic weight of silicon; from Appendix D we find that $d = 2330\ \mathrm{kg/m^3}$ and $A = 28.1$ g/mol. Substituting yields

$$n_{\mathrm{Si}} = \frac{(6.02 \times 10^{23}\ \mathrm{mol^{-1}})(2330\ \mathrm{kg/m^3})}{0.0281\ \mathrm{kg/mol}} = 5 \times 10^{28}\ \mathrm{m^{-3}}.$$

The ratio of these two number densities is the quantity we are looking for. Thus

$$\frac{n_{\mathrm{Si}}}{n_{\mathrm{P}}} = \frac{5 \times 10^{28}\ \mathrm{m^{-3}}}{10^{22}\ \mathrm{m^{-3}}} = 5 \times 10^6. \qquad \text{(Answer)}$$

We see that if only *one silicon atom in five million* is replaced by a phosphorus atom, the number of electrons in the conduction band will be increased by a factor of 10^6.

How can such a tiny admixture of phosphorus atoms have such a big effect? The answer is that, for pure silicon at room temperature, there were not many conduction electrons there to start with. The density of conduction electrons was $10^{16}\ \mathrm{m^{-3}}$ before doping and $10^{22}\ \mathrm{m^{-3}}$ after doping. For copper, however, the conduction electron density (see Table 1) is $\sim 10^{29}\ \mathrm{m^{-3}}$. Thus, even *after* doping, the conduction electron density of silicon remains much less than that of a typical metal such as copper.

46-9 The *p-n* Junction

Pass a hypothetical plane across a rod of a pure silicon. Dope the rod on one side of the plane with donor atoms (thus creating *n*-type material) and on the other side with acceptor atoms (thus creating *p*-type material.) You have just made a *p-n junction;* it is at the heart of essentially all semiconducting devices.* Figure 12*a* represents a *p-n* junction at the imagined moment of its creation. Let us first discuss the motions of the majority carriers, which are electrons in the *n*-type material and holes in the *p*-type material.

* In practice, to make a *p-n* junction one usually starts with, say, *n*-type material and then diffuses acceptor atoms into the solid sample at high temperature, overcompensating the donor atoms to a certain (controllable) depth below the surface.

Motions of the Majority Carriers. Electrons close to the junction plane will tend to diffuse across it (from right to left in Fig. 12), for much the same reason that gas molecules will diffuse through a permeable membrane into a vacuum beyond it. In the same way, holes will tend to diffuse across the junction plane from left to right. Both motions contribute to a *diffusion current* i_{diff}, directed from left to right as in Fig. 12*d*.

Recall that *n*-type material is studded throughout with donor ions, fixed firmly in their lattice sites. Normally, the positive charges of these ions are compensated electrically by the majority carriers, which are electrons. When an electron diffuses through the junction plane, however, it "uncovers" one of these donor ions, thus introducing a fixed positive charge in the *n*-type material. When this diffusing electron arrives on the other side of the barrier, it quickly finds a hole and combines with it,* thus neutralizing one of the positively-charged acceptor ions that are sprinkled throughout the *p*-type material, resulting in a fixed negative charge in the *p*-type material.

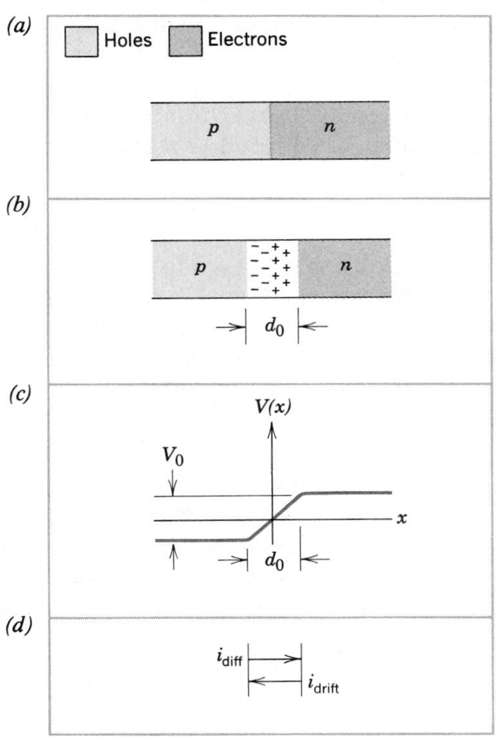

Figure 12 (*a*) A *p-n* junction at the imagined moment of its creation. Only the majority carriers are shown. (*b*) Diffusion of majority carriers across the junction plane causes a space charge of fixed donor and acceptor ions to appear. (*c*) The space charge establishes a contact potential difference V_0 across the junction plane. (*d*) In equilibrium, the diffusion of majority carriers across the junction plane is just balanced by the drift of minority carriers in the opposite direction.

Convince yourself that a hole, diffusing through the barrier from left to right, has exactly the same end result. Thus a region of fixed positive charge builds up on one side of the barrier and of fixed negative charge on the other, and the so-called *depletion zone* shown in Fig. 12*b* is created.

These fixed charges cause a *contact potential difference* to build up across the junction, as Fig. 12*c* shows. This potential difference is such that it serves as a barrier to limit further diffusion of both electrons and holes across the junction plane. An electron at the junction plane, for example, would be repelled back to its *n*-type home by the negative space charge in the *p*-type material that faces it across the plane. To complete the picture, let us turn our attention to the minority carriers.

Motions of the Minority Carriers. As Fig. 11*a* and Table 2 show, although the majority carriers in *n*-type material are electrons, there are nevertheless also a few holes, the minority carriers. Likewise in *p*-type material, although the majority carriers are holes, there are also a few conduction electrons.

Although the potential difference in Fig. 12*c* acts to retard the motions of the majority carriers—being a barrier for them—it is a downhill trip for the minority carriers, be they electrons or holes. When, by thermal agitation, an electron close to the junction plane is raised from the valence band to the conduction band of the *p*-type material in Fig. 12*a*, the contact potential difference causes it to drift steadily from left to right across the junction plane. Similarly, if a hole is created in the *n*-type material, it, too, drifts across to the other side. The space-charge region shown in Fig. 12*b* is effectively swept free of charge carriers by this process and, for that reason, we call it the *depletion* zone. The current represented by the motions of the minority carriers, called the *drift current* i_{drift}, is in the opposite direction to the diffusion current and just compensates it at equilibrium, as Fig. 12*d* shows.

Thus, at equilibrium, a *p-n* junction resting on a shelf develops a contact potential difference V_0 between

* An "electron combines with a hole" when the electron drops from the conduction band to the valence band, filling a vacancy in that band.

its ends. The diffusion current d_{diff} that moves through the junction plane from the direction p to n is just balanced by a drift current i_{drift} that moves in the opposite direction.

46-10 The Diode Rectifier

A *p-n* junction is basically a two-terminal rectifier. If you connect it across the terminals of a battery, the current in the circuit will be very much smaller for one polarity of the battery connection than for the other, as Fig. 13 shows.

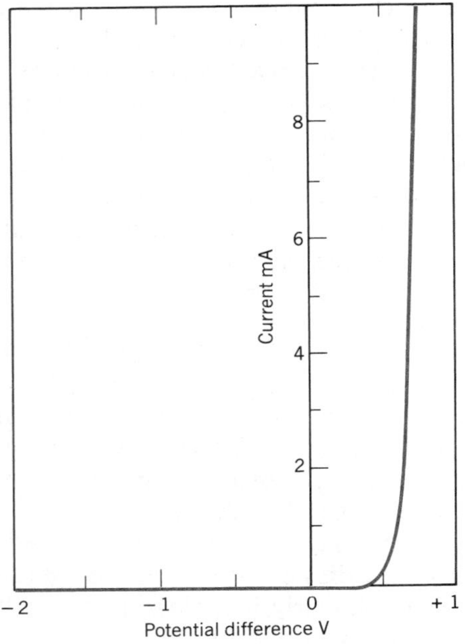

Figure 13 A current-voltage plot for a junction diode, showing that it is highly conducting in the forward direction and essentially nonconducting in the reverse direction.

Figure 14 shows one of many applications of a diode rectifier. A sine wave input potential generates a half-wave output potential, the diode rectifier acting as essentially a short circuit (a closed switch) for one polarity of the input potential and as essentially an open circuit (an open switch) for the other. An ideal diode rectifier, in fact, has only these two modes of operation. It is either ON (zero resistance) or OFF (infinite resistance.)

Figure 14 displays the conventional symbol for a diode rectifier. The arrow head corresponds to the *p*-type terminal of the device and points in the direction of "easy" conventional current flow. That is, the diode is ON when the terminal with the arrow head is (sufficiently) positive with respect to the other terminal.

Figure 15 shows details of the two connections. In Fig. 15*a*—the back-bias arrangement—the battery emf simply *adds* to the contact potential difference, thus increasing the height of the barrier that the majority carriers must surmount. Fewer of them can do so and, as a result, the diffusion current decreases markedly.

The drift current, however, senses no barrier and thus is independent of the magnitude or direction of the external potential. The nice current balance that existed at zero bias (see Fig. 12*d*) is thus upset and, as shown in Fig. 15*a*, a very small net back-current i_B appears in the circuit.

Another effect of back-bias is to widen the depletion zone, as a comparison of Figs. 12*b* and 15*a* shows. Because the depletion zone contains very few charge carriers, it is a region of high resistivity. Thus, its substantially increased width means a substantially increased resistance, consistent with the small value of the back-bias current.

Figure 15*b* shows the forward-bias connection, the positive terminal of the battery being connected to the *p*-type end of the pn junction. Here the applied emf *subtracts* from the contact potential difference, the diffusion current *rises* substantially, and a relatively *large*

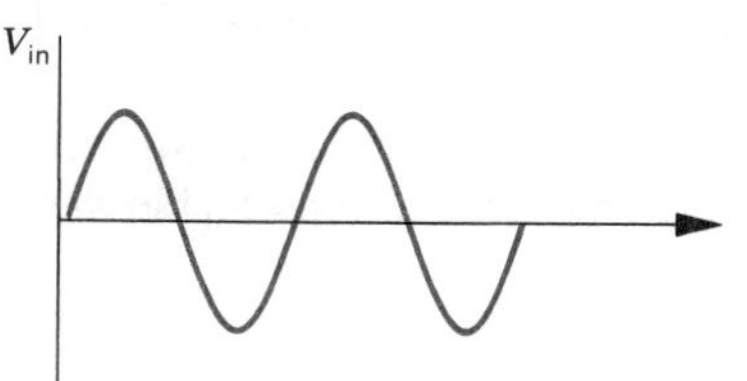

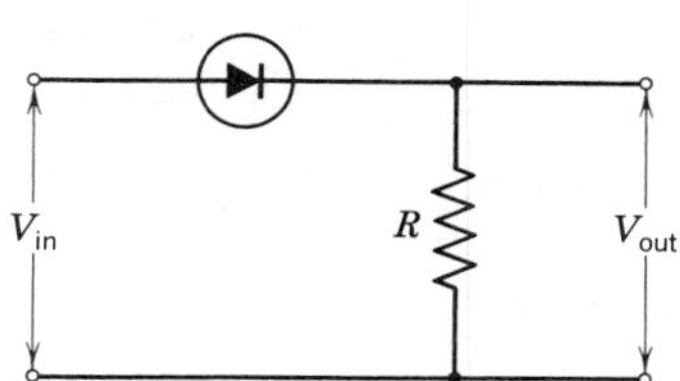

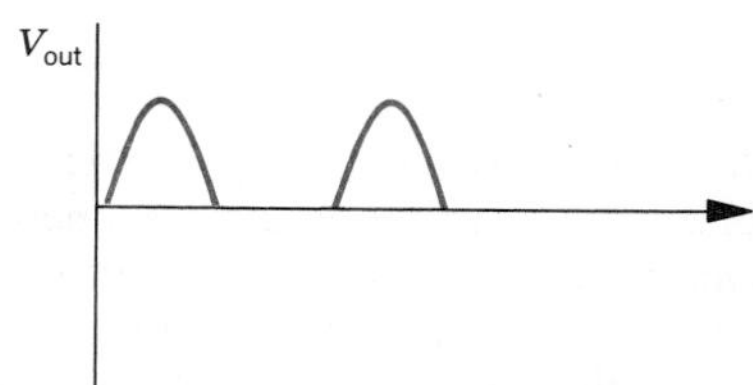

Figure 14 A *p-n* junction diode is connected as a rectifier. The action of the circuit is to pass the positive half of the input wave form but to suppress the negative half. The average potential of the input waveform is zero; that for the output wave form is positive.

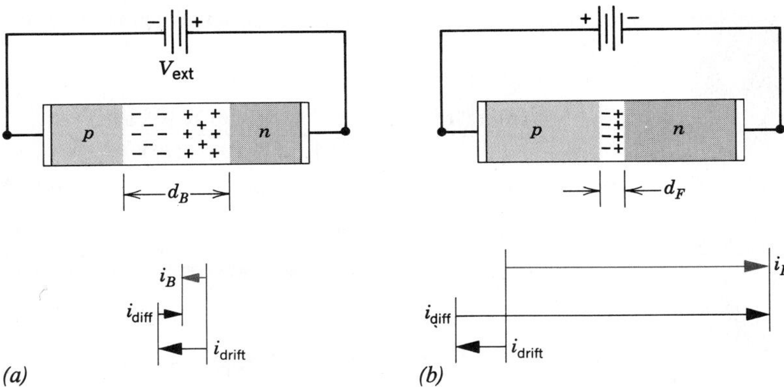

Figure 15 *(a)* The back-bias connection of a *p-n* junciton, showing the wide depletion zone and the corresponding small back current. *(b)* The forward-bias connection, showing the narrowing of the depletion zone and the large forward current.

net forward current results. The depletion zone *narrows,* its low resistance being consistent with the large forward current i_F.

46-11 The Light-Emitting Diode (LED)

We are all familiar with the brightly-colored numbers that flash at us from cash registers and gasoline pumps. In nearly all cases this light is emitted from an assembly of *p-n* junctions operating as *light-emitting diodes* (LEDs).

Figure 16*a* shows the familiar seven-segment display from which the numbers are formed. Figure 16*b* shows that each element of this display is the end of a flat plastic lens, at the other end of which is a small LED, possibly about one mm² in area. Figure 16*c* shows a typical circuit, in which the LED is forward-biased.

How can a *p-n* junction emit light? When an electron at the bottom of the conduction band of a semiconductor falls into a hole at the top of the valence band, an energy E_g is released, where E_g is the gap width. What happens to this energy? There are at least two possibilities. It might be transformed into thermal energy of the vibrating lattice and, with high probability, that is exactly what happens in a silicon-based semiconductor.

In some semiconducting materials, however, conditions are such that the emitted energy can also appear as electromagnetic radiation, the wavelength being given by

$$\lambda = \frac{c}{\nu} = \frac{c}{E_g/h} = \frac{hc}{E_g}. \tag{9}$$

Commercial LEDs designed for the visible region are usually based on a semiconducting material that is a suitably chosen gallium-arsenic-phosphorus compound. By adjusting the ratio of phosphorus to arsenic the gap width—and thus the wavelength of the emitted light—can be tailored to suit the need.

A question arises: If light is emitted when an electron falls from the conduction band to the valence band, will not light of that same wavelength be absorbed when an electron moves in the other direction, that is, from the valence band to the conduction band? It will indeed. To avoid having all of the emitted photons absorbed, it is necessary to have a great surplus of both electrons and holes present in the material, in much greater numbers than would be generated by thermal agitation in the

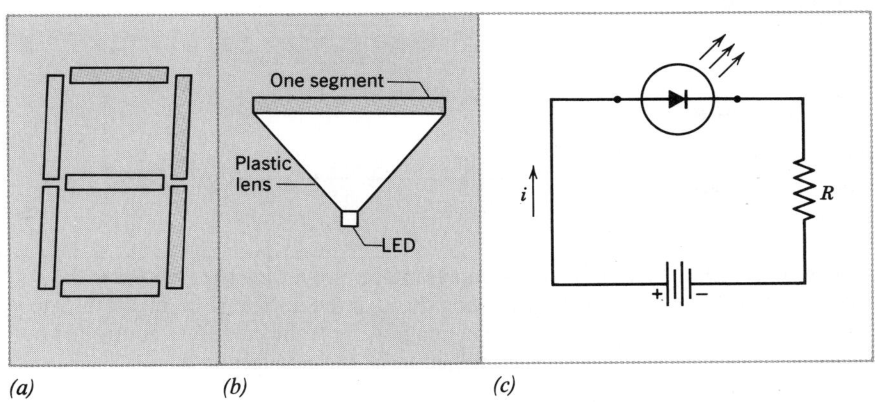

Figure 16 *(a)* The familiar seven-segment number display, activated to show the number "7." *(b)* One segment of such a display. *(c)* An LED connected to a source of emf.

Figure 17 A laser diode developed at the AT&T Bell Laboratories. The cube at the right is a grain of table salt.

intrinsic semiconducting material.* These are precisely the conditions that result when majority carriers — be they electrons or holes — are injected across the central plane of a *p-n* junction by the action of an external potential difference. That is why a simple intrinsic semiconductor will not serve as an LED. You need a *p-n* junction! To provide lots of majority carriers — and thus lots of photons — it should be heavily doped and strongly forward biased.

Apart from their use in visual displays, LEDs operating in the infrared are much used in optical communication systems based on optical fibers. The infrared region is chosen because the absorption per unit length of such fibers has a well-defined minimum at two different wavelengths in this region.† In a development of the LED, the ends of a suitable *p-n* junction crystal are polished so that a slice of the crystal across the junction plane serves as a laser. Such a device is called a *laser diode;* Fig. 17 suggests its tiny scale.‡

Sample Problem 6 An LED is constructed from a *p-n* junction based on a certain Ga-As-P semiconducting material, whose energy gap is 1.9 eV. What is the wavelength of its emitted light?

From Eq. 9 we have

$$\lambda = \frac{hc}{E_g} = \frac{(6.63 \times 10^{-34}\ \mathrm{J \cdot s})(3.00 \times 10^{8}\ \mathrm{m/s})}{(1.9\ \mathrm{eV})(1.60 \times 10^{-19}\ \mathrm{J/eV})}$$
$$= 6.54 \times 10^{-7}\ \mathrm{m} = 654\ \mathrm{nm}. \qquad \text{(Answer)}$$

Light of this wavelength is red.

46-12 The Transistor (Optional)

The devices we have discussed so far have been diodes, that is, two-terminal devices. Here we introduce a three-terminal device, a *transitor.* As Fig. 18 suggests, the function of a transistor is to control a current flowing through the device from terminal D (the *drain*) to terminal S (the *source*) by varying the potential of terminal G (the *gate*).

For many applications, particularly those involving computers, we only need to be able to turn the drain-source current ON (gate open) or OFF (gate closed.) One of these conditions corresponds to a "0" and the other to a "1" in the binary arthrimetic on which the computer logic is based. We wish further that the gate terminal draw essentially no current from the circuit to which it is attached, thus interfering with its operation as little as possible. In more formal language, we say that we wish the transistor to have a high input impedance.

The central question proves to be: "How can we control the current in a conductor without making direct electric contact with it?" The perhaps surprising answer is that, by using the variable gate potential that is at our

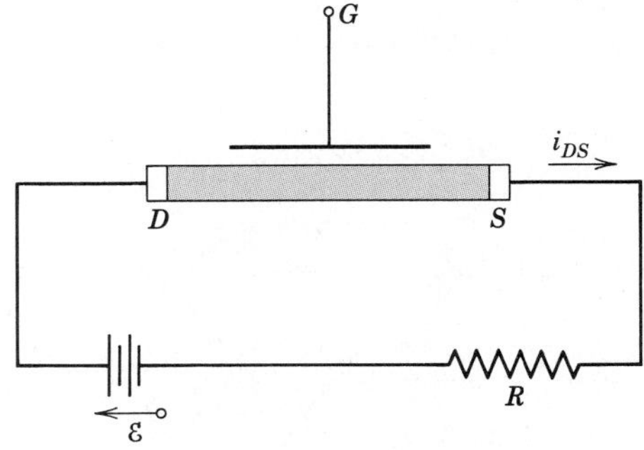

Figure 18 A representation of a transistor, showing the current i_{DS} moving through the device from the drain D to the source S. The magnitude of the current is controlled by the potential applied to G, the gate terminal.

* If the surplus of electrons and holes is great enough, there may be a population inversion so that conditions for laser action are set up.

† See Essay 17, *Optical Fibers and Communication.*

‡ See Essay 19, *Applications of Lasers.*

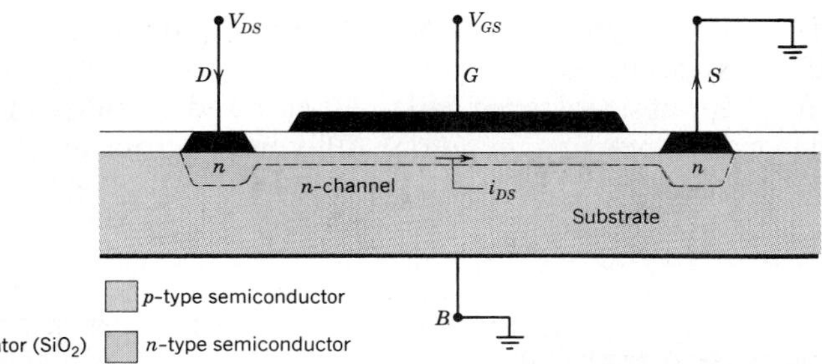

Figure 19 Showing the construction of a MOSFET. The source S and the base B are grounded and a potential V_{DS} is applied to the drain terminal D. The magnitude of the current is controlled by the gate potential V_{GS}.

disposal, we can change the effective cross-sectional area of the conductor, going even so far as to reduce it to zero.

Figure 15 gives a clue. Here we see that, by changing the bias potential of the *p-n* junction, we can control the width of the depletion zone. Simply by changing a potential we can effectively transform a conductor (*n*-type or *p*-type material) into a nonconductor (the depletion zone material). We can use the same trick in the transistor, but with a different geometrical arrangement.

Of the several types of transistor that are in common use, we choose to describe the MOSFET (Metal-Oxide-Semiconductor Field-Effect Transistor). Figure 19 shows its essential features.

A lightly doped *p*-type substrate has imbedded in it two "islands" of heavily doped *n*-type material, forming the drain *D* and the source *S*. These terminals are connected by a thin channel of *n*-type material, called the *n*-channel. An insulating layer of silicon dioxide (hence *O*xide in the acronym) is deposited on the substrate and penetrated by two metallic contacts (hence *M*etal) at *D* and *S*, so that electrical contact can be made with the drain and the source. A thin metallic layer—the gate *G*—is deposited opposite to the *n*-channel. Note that the gate makes no ohmic contact with the transistor proper, being separated from it by the insulating oxide layer. Thus, a MOSFET has the desired high input impedance, perhaps as high as $10^{15}\ \Omega$.

Consider first the situation with the source and the substrate grounded, the gate "floating" (that is, not connected electrically to a source of emf), and a positive potential V_{DS} applied to the drain. A drain-source current i_{DS} will be set up, as shown.

The potential difference across the boundary between the *n*-channel and the *p*-type substrate will vary from zero at the source end of the channel to V_{DS} at the drain end. The polarity is such (compare Fig. 15) that the *p-n* junction that exists at this boundary is back-biased for essentially its full length. A depletion zone will exist at this boundary, increasing in thickness from the source end of the channel to the drain end. For these conditions, the *n*-channel will not have the same cross-sectional area along its length, being invaded by the depletion zone to a greater and greater extent as one proceeds along the channel from the source toward the drain.

The thickness of the depletion zone along its length can be influenced by the potential that we choose to apply to the gate. If we make the gate negative with respect to the source, electrons will be repelled from the *n*-channel into the substrate, thus widening the depletion zone, constricting the channel and decreasing the drain-source current. Alternatively, a positive gate potential will attract electrons into the *n*-channel, narrow the de-

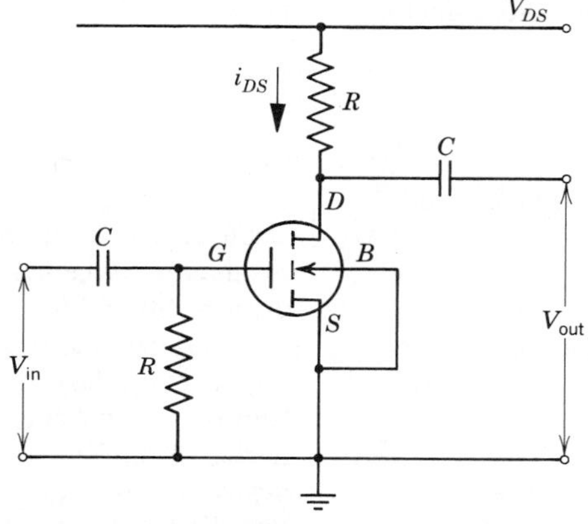

Figure 20 A MOSFET connected as an amplifier. A time-varying input signal V_{in} generates an amplified output signal V_{out}. The letters *G, D, B,* and *S* have the same meaning as in Fig. 19.

pletion zone, widen the conducting channel, and increase the drain-source current. In this way a small change in the gate potential can generate a substantial change in the drain-source current, much as a valve controls the flow of water through a pipe.

Figure 20 shows a MOSFET (note the descriptive symbol) connected into a circuit as an amplifier. The input signal is applied to the gate and the output appears as a varying potential difference across the load resistor R.

REVIEW AND SUMMARY

Insulators, Metals and Semiconductors

Quantum mechanics explains why some solids are electrical conductors, others are semiconductors, and why still others are insulators. When atoms are close to each other in a crystal lattice, their atomic energy levels become *bands* of allowed electron energies. In *insulators* the highest occupied level coincides with the top of a band; electrons are not able to accept additional kinetic energy from an applied field and so cannot conduct electricity (see Fig. 4). In *metals* the highest occupied level (the *Fermi energy*) falls somewhere in the middle of a band (Fig. 5 and Sample Problems 1 and 2). The highest occupied level in a *semiconductor* coincides, at $T = 0$, with the top of a band (the *valence band*) but the energy gap between it and the next highest band (the *conduction band*) is small enough so that charge carrying electrons can "jump" into it because of thermal agitation (Fig. 8). The higher resistivity and the decrease of resistivity with increasing temperature are both easily explained by this model.

Density of States

Assuming uniform potential, the density of energy states available to electrons in the partially-filled band of Fig. 5. is

$$n(E) = \frac{8\pi\sqrt{2}m^{3/2}}{h^3} E^{1/2} \quad \text{(density of states).} \qquad [3]$$

The Fermi energy

At $T = 0$, electrons fill all the states up to the Fermi energy with states above the Fermi energy being vacant. The Fermi energy corresponding to Eq. 3 is

$$E_F = \frac{0.121h^2}{m} n^{2/3} \qquad \text{(Fermi energy)} \qquad [6]$$

in which n is the number of conduction electrons per unit volume; see Fig. 6. At temperatures above absolute zero the distribution of occupied states is found by multiplying the density of states by the *probability function*

The Probability Function

$$p(E) = \frac{1}{e^{(E-E_F)/kT} + 1} \qquad \text{(probability function)} \qquad [7]$$

as illustrated in Fig. 7 and Sample Problem 4.

Doping—Donors and Acceptors

In practice, semiconductors are *doped* with controlled levels of selected impurities; see Table 2. Donor impurities contribute electrons to the conduction band and produce n-type semiconductors. Acceptor impurities contribute holes to the valence band and produce p-type semiconductors. Study Figs. 10 and 11 and Sample Problem 5.

A p-n Junction

A p-n junction (see Fig. 12) can serve as a diode rectifier. For forward biasing (p positive with respect to n), the potential barrier is low, the junction is thin, and the forward current is large. For back biasing, the potential barrier is high, the junction is thick, and the back current is small, usually negligibly so. The junction region itself, regardless of the applied potential difference, is called a *depletion layer.* It is virtually free of charge carriers and behaves like a somewhat leaky insulating slab. Figure 13 shows the rectifying properties of a p-n junction and Fig. 14 shows the result of introducing a p-n junction into an alternating-current circuit.

The Light-Emitting Diode (LED)

A p-n junction can, under certain circumstances, convert the energy lost by a charge carrier crossing the barrier into visible light whose wavelength is

$$\lambda = \frac{c}{\nu} = \frac{hc}{E_g}, \tag{9}$$

E_g being the energy gap width; see Sample Problem 6.

Transistors

A *transistor* is a three-terminal solid-state device in which the current flowing from the *drain* to the *source* is controlled by varying the potential of a *gate;* See Fig. 18. Figure 19 shows a MOSFET-type transistor, in which a small variation of the potential difference V_{GS} between the gate G and the source S has a major controlling effect on the current i_{DS} between the drain D and the source S.

QUESTIONS

1. Do you think that any of the properties of solids listed in Section 46–1 are related to each other? If so, which?

2. Does the Fermi energy for a given metal depend on the volume of the sample? If, for example, you compare a sample whose volume is 1 cm³ with one whose volume is twice that, the latter sample has just twice as many available conduction electrons; it might seem that you would have to go to higher energies to fill its available levels. Do you?

3. Why do the curves of Figs. 6*c* and 7*c* differ so little from each other?

4. The conduction electrons in a metallic sphere occupy states of quantized energy. Does the average energy interval between adjacent states depend on (*a*) the material of which the sphere is made, (*b*) the radius of the sphere, (*c*) the energy of the state, and (*d*) the temperature of the sphere?

5. What role does the Pauli exclusion principle play in accounting for the electrical conductivity of a metal?

6. Distinguish carefully among (*a*) the density of states $n(E)$, (*b*) the density of occupied states $n_o(E)$, and the probability function $p(E)$, all of which appear in Eq. 4.

7. In what ways do the classical model and the quantum mechanical model for the electrical conductivity of a metal differ?

8. In Chapter 21 we showed that the (molar) specific heat of an ideal monatomic gas is $\frac{3}{2}R$. If the conduction electrons in a metal behave like such a gas, we would expect them to make a contribution of about this amount to the measured specific heat capacity of a metal. However, this measured specific heat capacity can be accounted for quite well in terms of energy absorbed by the vibrations of the ion cores that form the metallic lattice. The electrons do not seem to absorb much energy as the temperature of the specimen is increased. How does Fig. 7 provide an explanation of this prequantum-days puzzle?

9. If we compare the conduction electrons of a metal with the atoms of an ideal gas we are surprised to note that so much kinetic energy is locked into the conduction electron system at the absolute zero of temperature. Would it be better to compare the conduction electrons, not with the atoms of a gas, but with the inner electrons of a heavy atom? After all, a lot of kinetic energy is also locked up in this case, and we don't seem to find that surprising. Discuss.

10. Give a physical argument to account qualitatively for the existence of allowed and forbidden energy bands in solids.

11. Is the existence of a forbidden energy gap in an insulator any harder to accept than the existence of forbidden energies for an electron in, say, the hydrogen atom?

12. On the band theory picture, what are the *essential* requirements for a solid to be (*a*) a metal, (*b*) an insulator, or (*c*) a semiconductor?

13. What can band theory tell us about solids that the classical model (see Section 28–6) cannot?

14. Distinguish between the drift speed and the Fermi speed of the conduction electrons in a metal.

15. Why is it that, in a solid, the allowed bands become wider as one proceeds from the inner to the outer atomic electrons?

16. Do pure (undoped) semiconductors obey Ohm's law?

17. At room temperature a given applied electric field will generate a drift speed for the conduction electrons of silicon that is about 40 times as great as that for the conduction electrons of copper. Why isn't silicon a better conductor of electricity than copper?

18. Consider these two statements: (*a*) At low enough temperatures silicon ceases to be a semiconductor and becomes a rather good insulator. (*b*) At high enough temperatures silicon ceases to become a semiconductor and becomes a rather good conductor. Discuss the extent to which each statement is either true or not true.

19. Which elements other than phosphorus are good candidates to use as donor impurities in silicon? Which elements other than aluminum are good candidates to use as acceptor impurities? Consult the periodic table (Appendix E).

20. Identify the following as *p*-type or *n*-type semiconductors: (*a*) Sb in Si; (*b*) In in Ge; (*c*) Al in Ge; (*d*) P in Si.

21. How do you account for the fact that the resistivity of metals increases with temperature but that of semiconductors decreases?

22. The energy gaps for the semiconductors silicon and germanium are 1.14 eV and 0.67 eV, respectively. Which substance do you expect would have the higher density of charge carriers at room temperature? At the absolute zero of temperature?

23. Discuss this sentence: "The distinction between a metal and a semiconductor is sharp and clear cut, but that between a semiconductor and an insulator is not."

24. What does a "hole" refer to in semiconductors?

25. Does the electrical conductivity of an intrinsic (undoped) semiconductor depend on the temperature? on the energy gap, E_g, between the full and empty bands?

26. Why does an *n*-type semiconductor have so many more electrons than holes? Why does a *p*-type semiconductor have so many more holes than electrons? Explain in your own words.

27. What does it mean to say that a *p-n* junction is biased in the forward direction?

28. Why does a *p-n* junction, serving as a diode rectifier, rely so centrally on doping?

29. A semiconductor contains equal numbers of donor and acceptor impurities. Do they cancel each other in their electrical effects? If so, what is the mechanism? If not, why not?

30. Germanium and silicon are similar semiconducting materials whose principal distinction is that the gap width E_g (see Fig. 8) is 0.67 eV for the former and 1.14 eV for the latter. If you wished to construct a *p-n* junction in which the back current is to be kept as small as possible, which material would you choose and why?

31. Consider two possible techniques for fabricating a *p-n* junction. (1) Prepare separately an *n*-type and a *p*-type sample and join them together, making sure that their abutting surfaces are plane and highly polished. (2) Prepare a single *n*-type sample and diffuse an excess acceptor impurity into it from one face, at high temperature. Which method is preferable and why?

32. In a *p-n* junction we have seen that electrons and holes may diffuse, in opposite directions, through the junction region. What is the eventual fate of each such particle as it diffuses into the material on the opposite side of the junction?

33. Does the diode rectifier whose characteristics are shown in Fig. 13 obey Ohm's law? What is your criterion for deciding?

34. We have seen that a simple intrinsic (undoped) semiconductor cannot be used as a light-emitting diode. Why not? Would a heavily doped *n*-type or *p*-type semiconductor work?

35. Explain in your own words how the MOSFET device of Fig. 19 works.

36. The acronym MOSFET stands for Metal-Oxide Semiconductor Field-Effect Transistor. What is the significance of each of these terms as applied to the device shown in Fig. 19?

EXERCISES AND PROBLEMS

Section 46-5 Metals—Qualitative

1E. At what pressure, in atmospheres, would an ideal gas have a density of molecules equal to the density of the conduction electrons in copper ($= 8.43 \times 10^{28}\ \mathrm{m^{-3}}$)? Assume $T = 300$ K.

2E. Gold is a monovalent metal with an atomic mass of 197 g/mol and a density of 19.3 g/cm³ (see Appendix D). Calculate the density of charge carriers.

3P. Calculate the number of particles per cubic meter for (*a*) the molecules of oxygen gas at 0°C and 1.0 atm pressure and (*b*) the conduction electrons in copper. (*c*) What is the ratio of these numbers? (*d*) What is the average distance between particles in each case? Assume that this distance is the edge of a cube whose volume is equal to the volume per particle. (See Sample Problem 3 of Chapter 28.)

4P. The density and atomic mass of sodium are 971 kg/m³ and 23 g/mol, respectively; the radius of the ion Na^+ is 98 pm. (*a*) What fraction of the volume of metallic sodium is available to its conduction electrons? (*b*) Carry out the same calculation for copper. Its density, atomic mass, and ionic radius are, respectively, 8960 kg/m³, 63.5 g/mol, and 135 pm. (*c*) For which of these two metals do you think the conduction electrons behave more like a free electron gas?

Section 46-6 Metals—Quantitative

5E. Use Eq. 6 to verify that the Fermi energy of copper is 7.0 eV. (Note, from Sample Problem 1, that the density of charge carriers in copper is $8.43 \times 10^{28}\ \mathrm{m^{-3}}$.)

6E. (*a*) Show that Eq. 3 can be written as

$$n(E) = CE^{1/2}$$

where $C = 6.78 \times 10^{27}\ \mathrm{m^{-3} \cdot eV^{-3/2}}$. (*b*) Use this relation to verify a calculation of Sample Problem 3, namely that for $E = 5.0$ eV, $n(E) = 1.52 \times 10^{28}\ \mathrm{m^{-3} \cdot eV^{-1}}$.

7E. Calculate the density $n(E)$ of conduction electron states in a metal for $E = 8.0$ eV and show that your result is consistent with the curve of Fig. 6*a*.

8E. What is the probability that a state 0.062 eV above the Fermi energy is occupied (*a*) at $T = 0$ K, and (*b*) $T = 320$ K?

9E. The Fermi energy of copper is 7.0 eV. For copper at 1000 K, (*a*) find the energy at which the occupancy probability is 0.90. (*b*) For this energy, evaluate the density of states and (*c*) the density of occupied states.

10E. Show that Eq. 6 can be written as

$$E_F = An^{2/3},$$

where the constant A has the value $3.65 \times 10^{-19}\ \mathrm{m^2 \cdot eV}$.

11E. The density of gold is 19.3 g/cm³. Each atom contributes one conduction electron. Calculate the Fermi energy of gold.

12E. Figure 7*c* shows the density of occupied states $n_o(E)$ of the conduction electrons in a metal at 1000 K. Calculate $n_o(E)$ for copper for the energies $E = 4.00$, 6.75, 7.00, 7.25 and 9.00 eV. The Fermi energy of copper is 7.00 eV.

13E. It can be shown that the conduction electrons in a metal behave like an ideal gas of the ordinary kind if the temperature is high enough. In particular, the temperature must be such that $kT \gg E_F$, the Fermi energy. What temperatures are required for copper ($E_F = 7.0$ eV) for this to be true? Study Fig. 7*c* in this connection and note that we have $kT \ll E_F$ for the conditions of that figure. This is just the reverse of the requirement cited above. Note also that copper boils at 2595°C (see Appendix D).

14E. The Fermi energy of silver is 5.5 eV. (*a*) At $T = 0$°C, what are the probabilities that states at the following energies are occupied: 4.4 eV, 5.4 eV, 5.5 eV, 5.6 eV, 6.4 eV? (*b*) At what temperature will the probability that a state at 5.6 eV is occupied be 0.16?

15E. The Fermi energy of aluminum is 11.6 eV; its density is 2.70 g/cm³ and its atomic mass is 27.0 g/mol (see Appendix D). From these data, determine the number of free electrons per atom.

16P. Show that the occupancy probabilities of two states whose energies are equally spaced above and below the Fermi energy add up to one.

17P. Show that the probability p_h that a *hole* exists at a state of energy E is given by

$$p_h = \frac{1}{e^{-(E-E_F)/kT} + 1}.$$

(*Hint:* The existence of a hole means that the state is unoccupied; convince yourself that this implies that $p_h = 1 - p$.)

18P. Zinc is a bivalent metal. Calculate (*a*) the number of conduction electrons per cubic meter, (*b*) the Fermi energy E_F, (*c*) the Fermi speed v_F, and (*d*) the de Broglie wavelength corresponding to this speed. See Appendix D for needed data on zinc.

19P. Silver is a monovalent metal. Calculate (*a*) the number of conduction electrons per cubic meter, (*b*) the Fermi energy E_F, (*c*) the Fermi speed v_F, and (*d*) the de Broglie wavelength corresponding to this speed. Extract needed data from Appendix D.

20P. White dwarf stars represent a late stage in the evolution of stars like the sun. They become dense enough and hot enough that we can analyze their structure as a solid in which all Z electrons per atom are free. For a white dwarf with a mass equal to that of the sun and a radius equal to that of the earth, calculate the Fermi energy of the electrons. Assume the atomic structure to be represented by iron atoms, and $T = 0$ K.

21P. A neutron star can be analyzed by techniques similar to those used for ordinary metals. In this case the neutrons (rather than electrons) obey the probability function, Eq. 7. Consider a neutron star of 2.0 solar masses with a radius of 10 km. Calculate the Fermi energy of the neutrons.

22P. Show that the density of states function given by Eq. 3 can be written in the form

$$n(E) = 1.5nE_F^{-3/2}E^{1/2}.$$

Explain how it can be that $n(E)$ is independent of material when the Fermi energy E_F (= 7.0 eV for copper, 9.4 eV for zinc, etc.) appears explicitly in this expression.

23P. Estimate the number N of conduction electrons in a metal that have energies greater than the Fermi energy as follows. Strictly, N is given by

$$N = \int_{E_F}^{E_T} n(E)p(E)dE,$$

where E_T is the energy at the top of the band. By studying Fig. 7*c*, convince yourself that, to a good degree of approximation, this expression can be written as

$$N = \int_{E_F}^{E_F+4kT} n(E_F)(\tfrac{1}{4})dE.$$

By substituting the density of states function, evaluated at the Fermi energy, show that this yields for the fraction f of conduction electrons excited to energies greater than the Fermi energy,

$$f = \frac{N}{n} = \frac{3kT/2}{E_F}.$$

Why not evaluate the first integral above directly without resorting to an approximation?

24P. Use the result of the preceding problem to calculate the fraction of excited electrons in copper at temperatures of (*a*) absolute zero, (*b*) 300 K, and (*c*) 1000 K.

25P. At what temperature will the fraction of excited electrons in lithium equal 0.013? The Fermi energy of lithium is 4.7 eV. See Problem 23.

26P. Silver melts at 961°C. At the melting point, what fraction of the conduction electrons are in states with energies greater than the Fermi energy of 5.5 eV? See Problem 23.

27P. Show that, at the absolute zero of temperature, the average energy $\overline{E}$ of the conduction electrons in a metal is equal to $\frac{3}{5}E_F$, where E_F is the Fermi energy. [*Hint:* Note that, by definition of average, $\overline{E} = \frac{1}{n}\int E n_o(E)dE$.]

28P. Use the result of the preceding problem to calculate the total translational kinetic energy of the conduction electrons in 1.0 cm^3 of copper at absolute zero.

29P. (*a*) Using the result of Problem 27, estimate how much energy would be released by the conduction electrons in a penny (assumed all copper; mass = 3.1 g) if we could suddenly turn off the Pauli exclusion principle. (*b*) For how long would this amount of energy light a 100-W lamp? Note that there is no known way to turn off the Pauli principle!

Section 46-8 Doping

30E. The Fermi–Dirac distribution function can be applied to semiconductors as well as to metals. In semiconductors, E is the energy above the top of the valence band. The Fermi level for an intrinsic semiconductor is nearly midway between the top of the valence band and the bottom of the conduction band. For germanium these bands are separated by a gap of 0.67 eV. Calculate the probability that (*a*) a state at the bottom of the conduction band is occupied and (*b*) a state at the top of the valence band is unoccupied at 300 K.

31E. Pure silicon at room temperature has an electron density in the conduction band of approximately $1 \times 10^{16}\ m^{-3}$ and an equal density of holes in the valence band. Suppose that one of every 10^7 silicon atoms is replaced by a phosphorus atom. (*a*) Which type will this doped semiconductor be, *n* or *p*? (*b*) What charge carrier density will the phosphorus add? (See Appendix D for needed data on silicon.) (*c*) What is the ratio of the charge carrier density in the doped silicon to that for the pure silicon?

32E. What mass of phosphorus would be needed to dope a 1.0-g sample of silicon to the extent described in Sample Problem 5?

33P. Doping changes the Fermi energy of a semiconductor. Consider silicon, with a gap of 1.11 eV between the valence and conduction bands. At 300 K the Fermi level of the pure material is nearly at the midpoint of the gap. Suppose that it is doped with donor atoms, each of which has a state 0.15 eV below the bottom of the conduction band, and suppose further that doping raises the Fermi level to 0.11 eV below the bottom of that band. (*a*) For both the pure and doped silicon, calculate the probability that a state at the bottom of the conduction band is occupied. (*b*) Also calculate the probability that a donor state in the doped material is occupied. See Fig. 21.

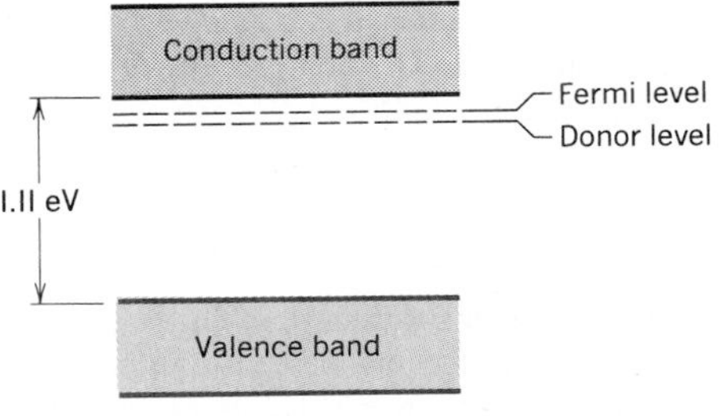

Figure 21 Problem 33.

34P. A silicon sample is doped with atoms having a donor state 0.11 eV below the bottom of the conduction band. (*a*) If each of these states is occupied with probability 5.00×10^{-5} at temperature 300 K, where is the Fermi level relative to the top of the valence band? (*b*) What then is the probability that a state at the bottom of the conduction band is occupied? The energy gap in silicon is 1.11 eV.

35P. In a simplified model of an intrinsic semiconductor (no doping), the actual distribution in energy of states is replaced by one in which there are N_v states in the valence band, all of these states having the same energy E_v, and N_c states in the conduction band, all of these states having the same energy E_c. The number of electrons in the conduction band equals the number of holes in the valence band. (*a*) Show that this last condition implies that

$$\frac{N_c}{e^{(E_c-E_F)/kT}+1} = \frac{N_v}{e^{-(E_v-E_F)/kT}+1}.$$

[*Hint:* See Problem 17.] (*b*) If the Fermi level is in the gap between the two bands and is far from both bands compared to kT, then the exponentials dominate in the denominators. Under these conditions, show that

$$E_F = \tfrac{1}{2}(E_c + E_v) + \tfrac{1}{2}kT\ln(N_v/N_c),$$

and therefore that, if $N_v \simeq N_c$, the Fermi level is close to the center of the gap.

Section 46-9 The *p-n* Junction

36E. When a photon enters the depletion region of a *p-n* junction, electron–hole pairs can be created as electrons absorb part of the photon's energy and are excited from the valence band to the conduction band. These junctions are thus often used as detectors for photons, especially for x rays and nuclear gamma rays. When a 662-keV gamma-ray photon is totally absorbed by a semiconductor with an energy gap of 1.1 eV, on the average how many electron–hole pairs are created?

37P. For an ideal *p-n*-junction diode, with a sharp boundary between the two semiconducting materials, the current i is related to the potential difference V across the diode by

$$i = i_0(e^{eV/kT} - 1),$$

where i_0, which depends on the materials but not on the current or potential difference, is called the *reverse saturation current*. V is positive if the junction is forward biased and negative if it is back biased. (*a*) Verify that this expression predicts the behavior expected of a diode by sketching i as a function of V over the range $-0.12\ V < V < +0.12\ V$. Take $T = 300$ K and $i_0 = 5.0$ nA. (*b*) For the same temperature, calculate the ratio of the current for a 0.50-V forward bias to the current for a 0.50-V back bias.

Section 46-11 The Light Emitting Diode (LED)

38E. (*a*) Calculate the maximum wavelength that will produce photoconduction in diamond, which has a band gap of 7.0 eV. (*b*) In what part of the electromagnetic spectrum does this wavelength lie?

39E. In a particular crystal, the highest occupied band of states is full. The crystal is transparent to light of wavelengths longer than 295 nm but opaque at shorter wavelengths. Calculate the width, in electron volts, of the gap between the highest occupied band and the next (empty) band.

40E. The KCl crystal has a band gap of 7.6 eV above the topmost occupied band, which is full. Is this crystal opaque or transparent to light of wavelength 140 nm?

41P. Fill in the seven-segment display shown in Fig. 16*a* to show how all 10 numbers may be generated. (*b*) If the numbers are displayed randomly, in what fraction of the displays will each of the seven segments be used?

ESSAY 20
THE LIQUID CRYSTAL STATE OF MATERIALS OR HOW DOES THE LIQUID CRYSTAL DISPLAY (LCD) IN MY WRISTWATCH WORK?

PATRICIA E. CLADIS
AT&T BELL LABORATORIES

Molecules that Align

Molecules and atoms in crystalline phases are organized into a three-dimensional structure. In general, each atom or molecule does not move far from its position in the structure. Liquid phases have no structural order so molecules or atoms move freely. Molecules are made up of many atoms and they can exhibit new states of matter, called *liquid crystals,* frequently abbreviated as LC, that are not as ordered as a solid crystal but, also, not as disordered as the usual liquid state, referred to as the *isotropic liquid* state.

Molecular shape plays an important role in determining liquid crystal phases. The shape of a molecule can be spherical, rodlike or cigar-shaped, disclike, bowl-shaped or some more complex shape. Liquid crystal phases may form, for example, when a collection of rod-shaped molecules in the liquid state spontaneously align. In the liquid crystal state called *nematic* used in displays referred to as LCDs, such as the lap-top PC in Fig. 1 and, more commonly, wristwatches, a degree of alignment is maintained despite rapid, random thermal motion of individual molecules characteristic of the liquid state. The property of certain molecules to align is called *long-range orientational order.* It is the property that characterizes the liquid crystal state.

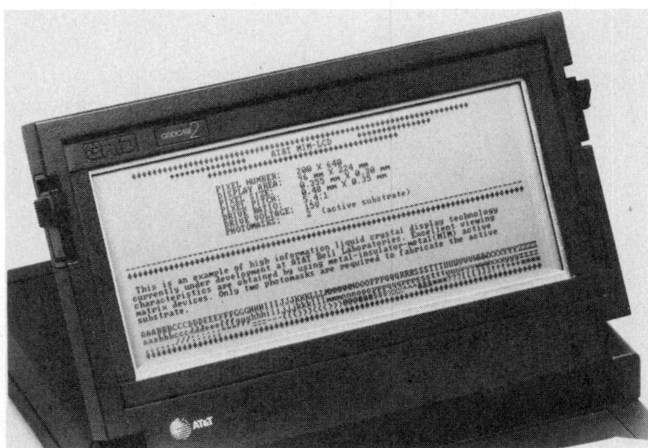

Figure 1 A lap top PC fitted with a liquid crystal screen.

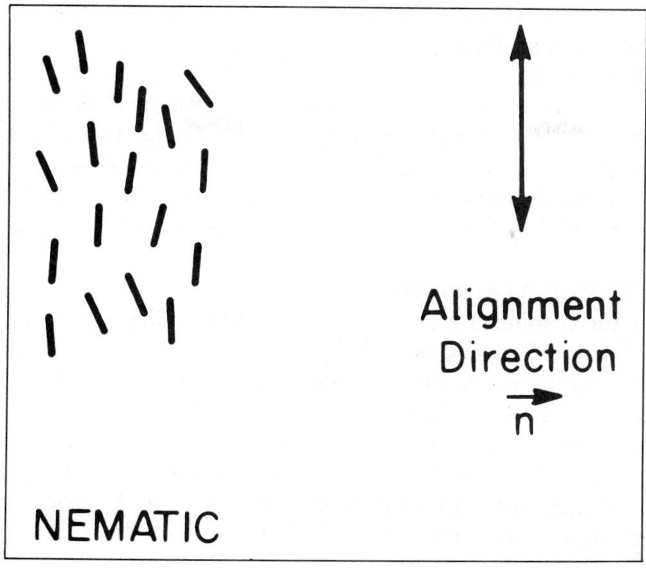

Figure 2 Structure of the nematic liquid crystal phase: rodlike molecules align. The alignment is not perfect because molecules in the liquid state are in rapid random, thermal motion.

Typically, molecules that form nematic liquid crystal states are 20 A long and about 5 Å in diameter. The nematic state is a result of many molecules aligning cooperatively in the same direction. In discussing liquid crystal phases, it is useful to think in terms of many molecules, not one. Instead of referring to a single molecule we talk about the collection of molecules in the nematic state and refer to the direction of alignment as the *director,* or, in short hand form, **n**, where **n** is a unit vector oriented in the direction chosen by many molecules for alignment. It is illustrated by lines, such as the ones shown in Fig. 2. The direction of alignment, **n**, defines the *optic axis* of the nematic state.

Two of the important physical principles that determine how liquid crystal displays work are

1. The orientation of **n** can be determined by small forces such as weak electric fields or surface forces
2. Polarized light behaves differently when traveling parallel to **n** than when traveling perpendicular to **n**.

In this essay, we describe how these two properties are used to make low-powered visual displays such as the flat screen used in the lap top PC shown in Fig. 1 and in wristwatches and calculators.

Determining the Orientation of n

Surfaces forces provide a useful way to orient **n**. Buffing or wiping glass microscope slides on a piece of white filter paper, many times in a single direction, will orient material in the nematic state in contact with the buffed surface, with **n** parallel in the buffing direction. Conventional wisdom is that this is not magic but rather that some oil or grease from the fingers is transferred to the glass surfaces. The buffing process produces microscopic grooves in the oil that provide an easy direction on the surface along which **n** aligns. Sandwiching nematic material between two of these glass plates with the buffed surfaces facing the liquid produces a single crystal with **n** nearly parallel to the surface of the glass and parallel to the direction of buffing throughout the whole sample.

Another way to orient **n** is with an *electric field,* **E**. In some materials, **n** aligns parallel to **E** and in others, it aligns perpendicular to **E**. To make an LCD, materials that align with **n** parallel to **E** are used. These are designated as *positive materials.* An ac or a dc field can be used to orient **n** since no distinction is made between the state with **n** parallel to **E** or antiparallel to **E**. Typically LCDs use ac fields.

The Twisted Nematic

This is the configuration most usually used in displays. Material in the nematic phase is sandwiched between buffed glass plates that have transparent electrodes evaporated onto the buffed sides. The two plates are oriented with their buffing directions perpendicular to each other. The director **n** twists 90° smoothly on going through the liquid crystal from one plate to the other. Typically, the distance between the two electrodes is 6 μm. The direction of polarization of incoming polarized light is rotated 90° as it passes through the liquid by the gentle twist of **n**. A sample in the twisted state looks bright when viewed between crossed polarizers. The upper part of Fig. 3 shows this.

Positive material is required for this display. An applied electric field causes the director to align parallel to the field (shown in the lower part of Fig. 3) destroying the twisted structure. Incoming polarized light is now extinguished by the second polarizer, called the *analyzer,* and the sample appears dark. This is the state used to show the characters of a display. When the electric field is turned off, the orientation of **n** is again determined by surface forces and **n** relaxes back to the initially twisted configuration.

Liquid crystal displays require little power because ambient light is used. A polarizer and analyzer are located on either side of the liquid crystal material and a mirror placed after the analyzer reflects light that has passed the analyzer back through the display (Fig. 3). When no light passes the analyzer, no light is reflected back.

An important feature of the twisted nematic structure for use as an LCD is that it works in white light. Typical switching times for these displays are 10–100 ms. For some applications, this is too slow and a major part of current LCD research is to find ways to reduce this time.

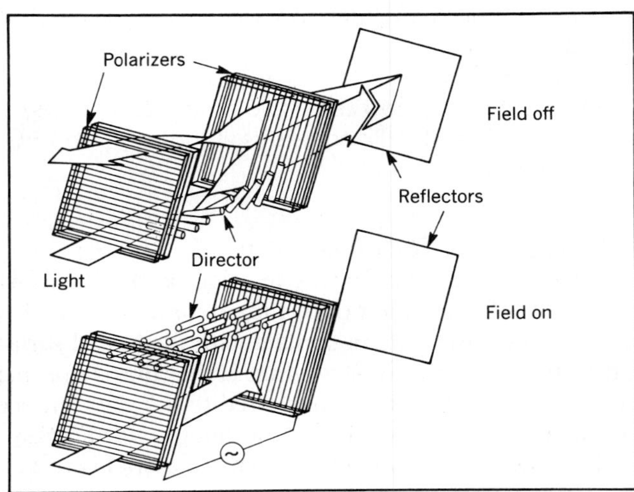

Figure 3 The most common liquid crystal display, used in watches, clocks and the PC screen of Fig. 1, uses the twisted nematic configuration. In the off-state, shown in the upper part of the figure, the director twists nearly 90° between electrodes. The twisted structure is destroyed when **n** aligns parallel to **E**. In LCDs, the twisted nematic is sandwiched between crossed polarizers. A mirror reflects light back through the twisted structure to form the white background of the display. To display a character, an applied electric field destroys the twisted structure and no light reaches the mirror to be reflected back.

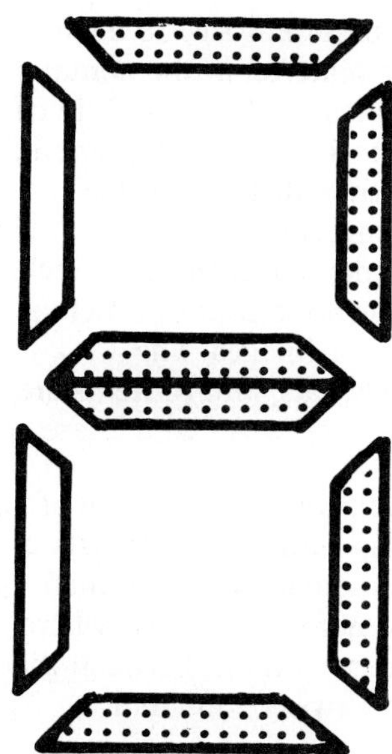

Figure 4 To observe numbers, electrodes in the pattern of a segmented "8" are used. All the numbers from one to 9 can be displayed by applying an electric field to a specific set of segments. The number "3" is shown in this figure. A dark number is observed against a white background when the polarizers on either side of the display are oriented perpendicular to each other. If the polarizers are parallel, a white character is seen against a dark background.

Devices that only display numbers, such as clocks and watches, have electrodes evaporated as a segmented figure eight shown in Fig. 4. With this clever pattern, numbers from 0 to 9 can be displayed by applying a field to the appropriate set of electrodes. More complex displays for computer screens use a matrix of intersecting row and column electrodes to form pictures as patterns of tiny dots. Recently it has been found that a 270° twisted nematic, a *supertwisted nematic,* can be read from almost any angle and has four times the contrast of the 90° twisted nematic in these complex displays.

The Effect of Temperature

Materials that exhibit liquid crystal phases are useful for studying the general problem of how order in matter is created and destroyed. When rodlike molecules align, they are more dense than when randomly oriented. For example, matches arranged in a box are more closely packed than when thrown at random onto a table. At some lower range of temperatures, rod-shaped molecules prefer to pack in the more dense, aligned state. As temperature is increased, the system becomes less dense and more energetic. The chance that each rod chooses the same alignment as its neighbors is reduced. At a special temperature, called the *transition temperature,* there is a sudden change to another liquid state—to another *phase*—without long-range orientational order. A *phase transition* takes place from the nematic liquid crystal phase to the isotropic liquid phase.

This transition is easily observed when a nematic liquid crystal* is heated. In samples a few millimeters thick, the nematic state is translucent, like frosted glass, because of fluctuations in the alignment arising from the thermal motion of the molecules. These fluctuations cause small changes in the index of refraction that scatter light. When heated into the isotropic liquid state, a thick sample becomes transparent as the molecules become uniformly disordered.

The screen material in Fig. 1 is in the nematic state from −25°C to 60°C, usually a large enough range for this application. Although nematic phases are known to exist from −50°C to about +400°C., the temperature range is much smaller in any one compound. For pure materials, composed of a single compound, the temperature range of the nematic phase is typically 1°C to 20 or 30°C. To obtain the wide temperature range for applications, several different kinds of compounds are mixed together. When the temperature is too high, the display material transforms to the isotropic liquid state and loses orientational order. Consequently, the display loses its contrast. If the temperature is too low, so the material transforms to a more ordered state, its orientation cannot be so easily changed.

Other Liquid Crystal Phases

Smectic A Many compounds exhibit more than one liquid crystal phase before the transition to the solid state. Another liquid crystal phase is called *smectic A* because its structure is the same as soap. Soap bubbles (Fig. 5*a*) are a familiar example of smectic A films. In the smectic A phase, molecules are confined to layers, like the planes shown in Fig. 5*b* with the axis of their alignment perpendicular to the planes. Whereas molecular motion is free within one layer, motion from one layer to another is hindered because the force binding molecules into layers has to be overcome.

* Liquid crystals may be purchased from practically any chemical company. One large supplier is EM Chemicals, 5 Skyline Drive, Hawthorne, NY 10532.

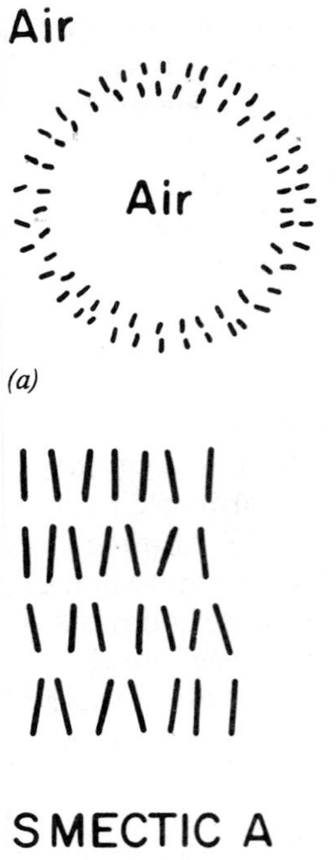

Air

Air

(a)

SMECTIC A

(b)

Figure 5 (*a*) Soap bubbles: molecules arranged on equidistant spherical shells but free to move within one shell. For simplicity, only a slice going through a diameter of the soap bubble is shown here. The name smectic is derived from the ancient Greek word for "soaplike." (*b*) The generic smectic: A structure of rodlike molecules. Molecules are in layers with the alignment direction parallel to the layer normal.

Because molecules in the more ordered smectic A phase are not as free to move as they are in the nematic phase, this phase is more *viscous.* A material in the nematic phase pours faster than in the smectic A phase. The nematic phase is the most fluid of the liquid crystal states so weaker forces are needed to influence its direction of alignment.

Chiral Liquid Crystals A *chiral* object, such as an ordinary threaded bolt or the human hand, looks different from its image in a mirror. For example, the mirror image of a left hand is a right hand. Indeed, the word *chiral* is derived from the Greek work for hand. In a chiral molecule, one or more carbon atoms bond to four different groups making chiral molecules distinguishable from their mirror images. The molecular building blocks of living systems such as amino acids, fats, and proteins are chiral.

1. ***Cholesteric Liquid Crystals.*** Many liquid crystal compounds are chiral. Some of them, in the liquid crystal phase called *cholesteric,* spontaneously form a phase with a twisted structure similar to the twisted structure forced on the nematic by the buffed glass plates in the twisted nematic display (Fig. 3). Unlike the nematic display, where the director twists only 90° over a distance determined by the separation between the glass plates, in the cholesteric phase, the director spontaneously twists 360° in a distance, called the pitch, that is characteristic of the material.

The two ways of twisting are identified as left-handed and right-handed. In the display shown in Fig. 3, the director is twisting in a left-handed manner. To see this, extend the thumb of the left hand in the direction of twist, that is, perpendicular to the glass plates and pointing toward the second plate. An important point to appreciate is that it doesn't matter which plate is chosen as the first plate. Just be sure the tip of the thumb is directed toward the second plate. Next, align the fingers of the left hand parallel to the director adjacent to the plate chosen to be the first plate. On going through the liquid from the plate called "one" to the plate called "two," fingers of the left hand naturally curl to maintain alignment with the director. Fingers of the right hand curl in the opposite sense and would follow a right-handed twist.

Figure 6 shows half a pitch of the left-handed cholesteric structure. Cylinders, rather than simple lines shown in Fig. 2 to illustrate the nematic structure, are used to depict $\mathbf{n}$ to reveal the three-dimensional property of this structure. Because of the three-dimensional liquid nature of this phase, $\mathbf{n}$ is not perfectly arranged in columns or rows. A left hand is also shown in Fig. 6 to illustrate how the handedness is determined. In the same way that the mirror image of a right hand is a left hand, the mirror image of a right-handed twisted structure is a left-handed twisted structure. The right-handed twisted structure may therefore be deduced by reflecting the left-handed one in a mirror.

Cholesteric liquid crystals have an interesting way of interacting with light. They are *optically active* meaning they rotate the plane of polarized light as the polarization follows the twisted arrangement of $\mathbf{n}$ through the material, turning 360° per pitch as the twisted nematic structure does in displays. This can be many complete rotations per millimeter. Just as it did not matter which plate is chosen to be plate "one" in the example above, the polarization of light rotates in the same sense when it travels through material in the cholesteric phase, regardless of the direction of light propagation. The rotation sense of polarization corresponds to the same hand in Fig. 3 for light going into the display as it does for light coming out of the display. This is a fundamental feature of optical activity.

Unpolarized light is thought of as light in which all directions of polarization are present at any given time. In *circularly polarized light,* the direction of polarization

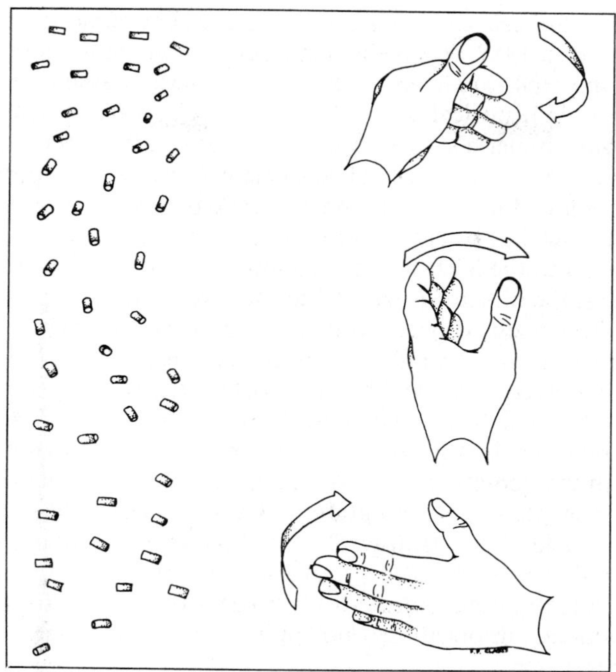

Figure 6 The structure of cholesteric liquid crystals. To capture the three-dimensional property of its helical structure, cylinders rather than lines are used to show the director orientation in the structure. The director **n** is perpendicular to the twist direction and rotates uniformly 360° in a fixed distance called the pitch. In the figure, half a pitch is shown. There are no layers in the cholesteric phase. The sense of the rotation of **n** is the same as the direction the fingers of one hand curl, shown by the arrow, when the thumb is extended in the direction **n** twists.

rotates at a particularly frequency determined by the wavelength of the light. Circularly polarized light can be right-handed or left-handed defined by pointing the thumb in the direction light is propagating. The fingers of only one hand curl naturally in the sense the polarization rotates. Looking at a hand with the tip of the thumb directed toward the observer, the fingers of the left hand curl in a clockwise sense and the fingers of the right hand, in a counterclockwise sense. Then, for left circularly polarized light the polarization rotates clockwise, and in right circularly polarized light it rotates counterclockwise. Unpolarized white light can be considered to be composed of both left and right circular polarizations rotating at all the frequencies corresponding to all the wavelengths constituting white light.

Material in the cholesteric liquid crystal phase with a pitch comparable to the wavelength of light transmits all the incident light *except* circularly polarized light with the *opposite* handedness of its twisted structure and of a particular wavelength determined by the pitch of the material and the angle of incidence of the light. For light traveling exactly in the direction of twist, the wavelength equal to the pitch is rejected and scattered back. Light traveling obliquely to the twist direction "sees" a shorter pitch so a shorter wavelength is scattered back. Viewed in ordinary diffuse daylight, the scattering of different wavelengths in different directions produces a striking display of vivid colors "recalling the appearance of a peacock's feather."

In some materials, the pitch depends sensitively on temperature. These materials are used as thermometers to translate small temperature differences into different colors. A temperature map of complicated surfaces—for example, parts of the human body—is obtained by spraying a thin layer of cholesteric liquid crystal onto the surface in question. In the medical field, temperature-sensitive cholesterics are used for the early detection of tumors, cancerous growths, and even leprosy, since these malignant growths are at a different temperature than healthy tissues.

2. ***Smectic C* and a Bit of History.*** The cholesteric liquid crystal state was first observed in 1888. It was the first liquid crystal phase observed in a chemically uniform material. There are many smectic phases named after the letters of the Roman alphabet in historical order of discovery. The smectic structure of soaps (Fig. 5) and biological materials, such as the myelinic sheath of nerves, was described in the middle of the last century. The smectic A phase in chemically uniform materials was first described in 1904, and new smectic phases have been discovered as recently as 1985. Smectic C was discovered in the 1960s. It is a layered structure with **n** aligned obliquely to the layer normal. In smectic A (Fig. 5), the phase exhibited by soap bubbles, the director is parallel to the layer normal.

The smectic C phase formed by chiral molecules is called smectic C*. Its structure is shown in Fig. 7. Similar to smectic A, **n** is arranged in layers and moves freely within a layer. Smectic C* has a twisted structure similar to the cholesteric phase except that **n** is tilted relative to the direction of twist. The significance of chirality, tilt, and layers is that the director interacts differently with an electric field than it does in the nematic phase. Nematics can be oriented with either an ac or a dc field. In these new phases, the orientation of **n** in an electric field depends on the *sign* of the electric field. A practical result is that the constrast of a display made from these materials can be changed by reversing the sign of an applied electric field. This feature is expected to make future displays faster than state-of-the-art nematic displays. The fundamental implications and applications of this idea are an active field of current LCD research.

The usefulness of liquid crystals is not restricted to displays. Another important application is made by living systems. Parts of the brain and cell membranes are composed of complicated organic molecules organized into liquid crystalline subunits. Their correct functioning depends on the body supporting the liquid crystalline state by maintaining the right composition of different elongated molecules and more spherical molecules like water. If there is too much water, for example, the liquid crystalline state transforms to the ordinary liquid state and the unit no longer functions. If there is not enough water, a transition to the solid state occurs and again, biological functions can no longer be performed. Biological membranes are composed of chiral molecules. Some of them may have a structure similar to one layer of smectic C*. Their response to electric fields transmitted by nerve impulses would then

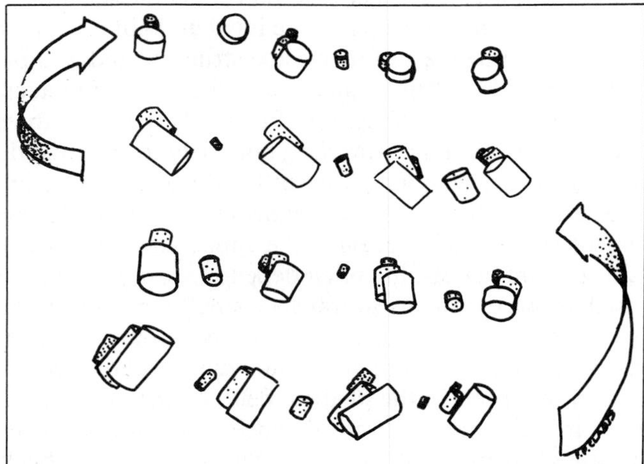

Figure 7 The structure of smectic C* is also helical. Molecules are arranged in layers with **n** at an oblique angle to the layer normal. The structure is shown for a half pitch.

also be similar and may be important for stimulating and controlling biological functions. Liquid crystals, therefore, provide a long-term link between the physical and biological sciences.

REFERENCES FOR FURTHER READING *Physics Today,* May 1982. This is a special issue devoted to liquid crystals. P. G. de Gennes, *The Physics of Liquid Crystals,* Clarendon Press, Oxford, 1975. S. Chandrasekhar, *Liquid Crystals,* Cambridge University Press, Cambridge, 1977. G. W. Gray and J. W. Goodby, *Smectic Liquid Crystals: Textures and Structures,* Blackie, Glasgow, 1984. Glenn H. Brown and Jerome J. Wolken, *Liquid Crystals and Biological Structures,* Academic Press, New York, 1979. A. R. Kmetz, "Progress in Liquid Crystal Displays" in *Journal of Imaging Technology, 11,* 236, 1985. P. E. Cladis, "Liquid Crystals—Useful New States of Materials," submitted to *The Physics Teacher.* International journals: *Liquid Crystals* (published by Taylor and Francis, London, England) and *Molecular Crystals Liquid Crystals* (published by Pergamon Press). "History of Liquid Crystals" was written by H. Kelker and published in *Molecular Crystals, Liquid Crystals 21,* 2, 1973.

CHAPTER 47

NUCLEAR PHYSICS*

Lord Ernest Rutherford and a younger colleague in the Cavendish Laboratory of Cambridge University. Rutherford discovered the atomic nucleus—that tiny concentration of matter at the center of the atom that contains all of the positive charge of the atom and most of its mass.

"If circumstances lead me, I will find where truth is hid, though it were hid indeed within the center."

—Hamlet

47-1 Discovering the Nucleus

In the first years of this century not much was known about the structure of atoms beyond the fact that they contained electrons. This particle had only been discovered (by J. J. Thomson) in 1897 and its mass was unknown in those early days. Thus it was not possible even to say just how many electrons a given atom contained. Atoms were electrically neutral so they must also contain some positive charge, but nobody knew what form this compensating positive charge took.

In 1911 Ernest Rutherford was led to propose that the positive charge of the atom was densely concentrated at the center of the atom and that, furthermore, it was responsible for most of the mass of the atom. He had discovered the atomic nucleus!

Rutherford's proposal was no mere conjecture but was based firmly on the results of an experiment suggested by him and carried out by his collaborators, Hans Geiger (of Geiger counter fame) and Ernest Marsden, a 20-year-old student who had not yet earned his Bachelor's degree.

* For a readable general reference for this and the following chapter, see Kenneth S. Krane, *Introductory Nuclear Physics* (John Wiley & Sons, New York, 1987).

Rutherford's idea was to fire energetic α particles at a thin target foil and measure the extent to which they were deflected as they passed through the foil. α particles, which are about 7300 times more massive than electrons, carry a charge of $+2e$ and are emitted spontaneously (with energies of a few MeV) by many radioactive materials. We now know that these useful projectiles are the nuclei of the atoms of ordinary helium. Figure 1 shows the experimental arrangement of Geiger and Marsden. The experiment consists in counting the number of α particles deflected through various scattering angles ϕ.

Figure 2 shows their results. Note especially that the vertical scale is logarithmic. We see that most of the α particles are scattered through rather small angles but—and this was the big surprise—a very small fraction of them are scattered through very large angles, approaching 180°. In Rutherford's words: "It was quite the most incredible event that ever happened to me in my life. It was almost as incredible as if you had fired a 15-inch shell at a piece of tissue paper and it came back and hit you."

Why was Rutherford so surprised? At the time of these experiments, most physicists believed in the so-called plum pudding model of the atom, which had been advanced by J. J. Thomson. In this view the positive charge of the atom was thought to be spread out through the entire volume of the atom. The electrons (the "plums") were thought to vibrate about fixed centers within this sphere of charge (the "pudding").

The maximum deflecting force that would act on an α particle as it passes through such a large positive sphere of charge would be far too small to deflect the α particle by even as much as one degree. (The expected deflection has been compared to that you would observe if you fired a bullet through a sack of snowballs.) The electrons in the atom would also have very little effect on the massive, energetic α particle. They would, in fact, be themselves strongly deflected, much as a swarm of gnats would be brushed aside by a stone thrown through them.

Rutherford saw that, to deflect the α particle backwards, there must be a large force and it could be provided if the positive charge, instead of being spread throughout the atom, was concentrated tightly at its center. On this model the incoming α particle can get very close to the center of the positive charge without penetrating it, resulting in a large deflecting force. (In our analogy, the snowballs now have pebbles at their centers.)

Figure 3 shows possible paths taken by typical α particles as they pass through the atoms of the target foil. As we see, most are only slightly deflected, but a few (those whose extended incoming paths pass, by chance, very close to a nucleus) are deflected through large angles. From an analysis of the data, Rutherford concluded that the radius of the nucleus must be smaller than the radius of an atom by a factor of about 10^4.

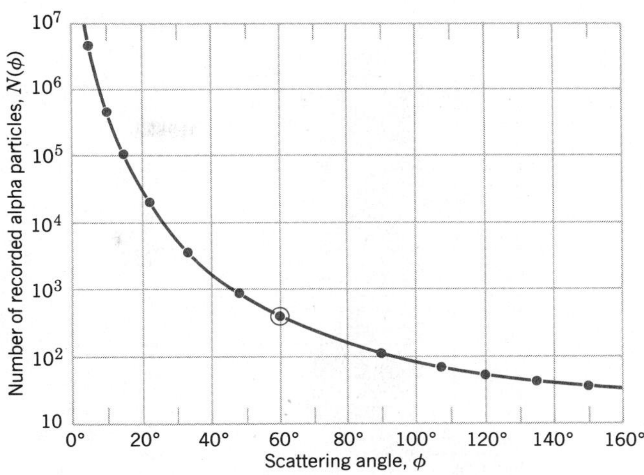

Figure 2 The dots are the α particle scattering data for a gold foil, obtained by Geiger and Marsden using the apparatus of Fig. 1. The solid curve is the theoretical prediction, based on the assumption that the atom has a small, massive, positively charged nucleus. Note that the vertical scale is logarithmic, covering six orders of magnitude. The data have been adjusted to fit the theoretical curve at the experimental point that is enclosed in a circle.

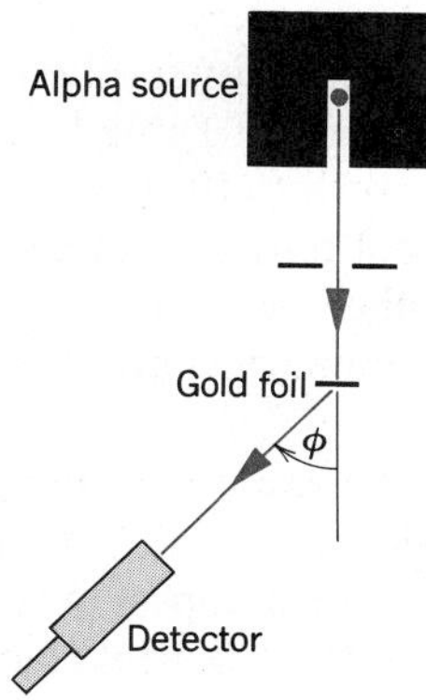

Figure 1 An arrangement (top view) used in Rutherford's laboratory in 1911–1913 to study the scattering of α particles by thin metal foils. The detector can be rotated to various values of the scattering angle ϕ. With this simple "table top" apparatus, the nucleus was discovered.

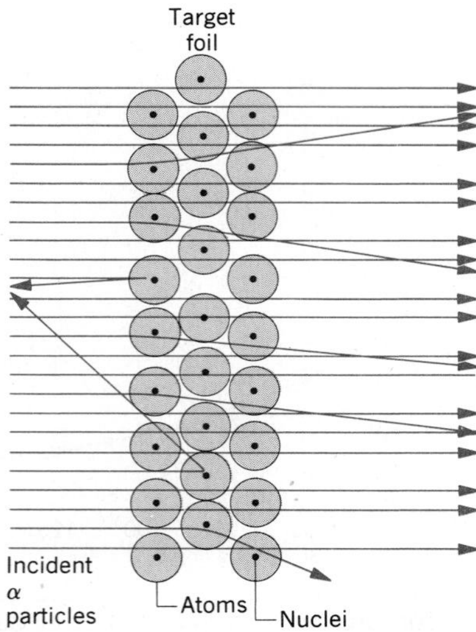

Figure 3 The angle through which an α particle is scattered depends on how close its extended incident path lies to an atomic nucleus. Large deflections result only from very close encounters.

The atom is mostly empty space! It is not often that the piercing insight of a gifted scientist, buttressed by a few simple calculations, leads to results of such importance.*

Sample Problem 1 A 5.30-MeV α particle happens, by chance, to be headed directly toward the nucleus of an atom of gold ($Z = 79$). How close does the α particle get to the center of the gold nucleus before it comes momentarily to rest and reverses its course? Neglect the recoil of the relatively massive nucleus.

Initially the total mechanical energy of the two interacting particles is just equal to K_α (= 5.30 MeV), the initial kinetic energy of the α particle. At the moment the α particle comes to rest, the total energy is the electrostatic potential energy of the system of two particles. Because energy must be conserved, these two quantities must be equal, or

$$K_\alpha = \frac{1}{4\pi\epsilon_0}\frac{Q_\alpha Q_{\mathrm{Au}}}{d},$$

in which Q_α (= $2e$) is the charge of the α particle, Q_{Au} (= $79e$) is the charge of the gold nucleus, and d is the distance between the centers of the two particles.

Substituting for the charges and solving for d yields

$$d = \frac{(2e)(79e)}{4\pi\epsilon_0 K_\alpha}$$

$$= \frac{(2 \times 79)(1.60 \times 10^{-19}\ \mathrm{C})^2}{(4\pi)(8.85 \times 10^{-12}\ \mathrm{F/m})(5.30\ \mathrm{MeV})(1.60 \times 10^{-13}\ \mathrm{J/MeV})}$$

$$= 4.29 \times 10^{-14}\ \mathrm{m} \approx 43\ \mathrm{fm}. \qquad \text{(Answer)}$$

This is a small distance by atomic standards but not by nuclear standards. As we shall see in the following section, it is considerably larger than the sum of the radii of the gold nucleus and the α particle. Thus, the α particle reverses its course without ever "touching" the gold nucleus.

47-2 Some Nuclear Properties

Table 1 shows some properties of a few selected nuclear species, or *nuclides,* as they are called. We discuss its several entries under separate headings.

Some Nuclear Terminology. Nuclei are made up of protons and neutrons. We represent the number of protons in the nucleus—called the *atomic number* (or the *proton number*) of the nucleus—by the symbol Z and the number of neutrons—the *neutron number*—by the symbol N. The total number of neutrons and protons in a nucleus is called its *mass number A,* so that

$$A = Z + N. \qquad (1)$$

Neutrons and protons, when considered collectively, are called *nucleons.*

We represent nuclides by symbols such as those displayed in the first column of Table 1. Consider $^{197}\mathrm{Au}$, for example. The superscript (197) is the mass number A. The chemical symbol tells us that this element is gold, whose atomic number (see Appendix E) is 79. From Eq. 1 we see that the neutron number of this nuclide is 197 − 79 or 118.

Nuclides with the same atomic number but different neutron numbers are called *isotopes.* As it happens, the element gold has 30 isotopes, ranging from $^{175}\mathrm{Au}$ to $^{204}\mathrm{Au}$. Only one of them ($^{197}\mathrm{Au}$) is stable, the remaining 29 being radioactive. Such *radionuclides* decay, by the spontaneous emission of a particle, with average lives (in this case) ranging from a few seconds to a few months.

Organizing the Nuclides. The neutral atoms of all isotopes for a given Z have the same number of external

* For a mathematical analysis of this scattering experiment, see Robert Resnick and David Halliday, *Basic Concepts in Relativity and Early Quantum Theory* (John Wiley & Sons, New York, 1985), 2nd ed., Supplementary Topic I.

Table 1 Some Properties of Selected Nuclides[a]

Nuclide	*Z*	*N*	*A*	*Stability*[b]	*Mass*[c] u	*Radius* fm	*Binding Energy* MeV/nucleon
^{1}H	1	0	1	99.985%	1.007825	—	—
^{7}Li	3	4	7	92.5%	7.016003	2.1	5.60
^{31}P	15	16	31	100%	30.973762	3.36	8.48
^{81}Br	35	46	81	49.3%	80.916289	4.63	8.69
^{120}Sn	50	70	120	32.4%	119.902199	5.28	8.51
^{157}Gd	64	93	157	15.7%	156.923956	5.77	8.21
^{197}Au	79	118	197	100%	196.966543	6.23	7.91
^{227}Ac	89	138	227	21.8 y	227.027750	6.53	7.65
^{239}Pu	94	145	239	24,100 y	239.052158	6.64	7.56

[a] The nuclides are chosen to lie along the stability band of Fig. 4.
[b] For stable nuclides, the *isotopic abundance* is given; this is the fraction of atoms of this type found in a typical sample of the element. For radioactive nuclides, the half-life is given.
[c] Following standard practice, the reported mass is that of the neutral atom, not that of the bare nucleus.

electrons, the same chemical properties, and fit into the same box in the chemist's periodic table of the elements. The nuclear properties of the various isotopes, however, are very different. Thus, the periodic table is of limited use to the nuclear physicist, the nuclear chemist, or the nuclear engineer.

We organize the nuclides on a *nuclidic chart* (see Fig. 4) in which a nuclide is represented by plotting its proton number against its neutron number. The stable nuclides in this figure are represented by the darker shading, the radionuclides by the lighter shading. We see that the radionuclides tend to lie on either side of—and at the upper end of—a well-defined band of stable nuclides. Note, too, that light stable nuclides tend to lie close to the $N = Z$ line, which means that they have the same numbers of neutrons and protons. Heavier nuclides, however, tend to have many more neutrons than protons. We have seen above, for example, that ^{197}Au has 118 neutrons and only 79 protons, a *neutron excess* of 39 neutrons.

Nuclidic charts are available as wall charts, in which a small box on the chart is filled with data about the nuclide it represents.* Figure 5 shows a section of such a chart, centered on ^{197}Au. Relative abundances are shown for stable nuclides, half-lives for radionuclides. The sloping line represents a typical line of constant mass number, $A = 198$ in this case.

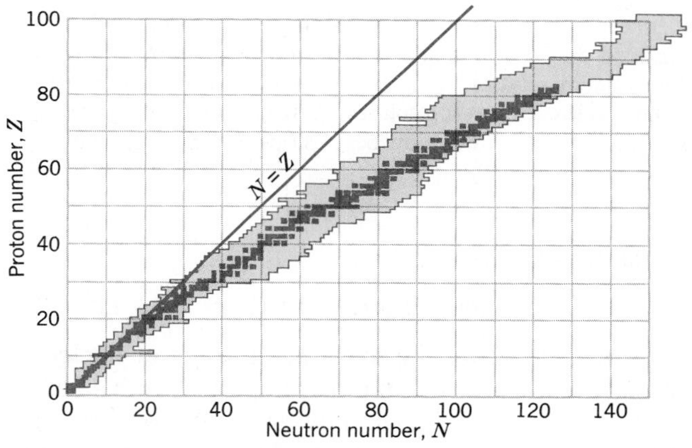

Figure 4 A plot of the known nuclides. The darker shading identifies the band of stable nuclides, the lighter shading the radionuclides. Note that light stable nuclides have essentially equal numbers of neutrons and protons but that heavier nuclides have an increasingly larger excess of neutrons. The figure shows that there are no stable nuclides with $Z > 83$ (bismuth).

Nuclear Radii. A convenient unit for measuring distances on the scale of nuclei is the *femtometer* ($= 10^{-15}$ m), the prefix *femto* coming from the Danish word for 15. This unit is often called the *fermi,* and shares

* Such charts are available, in wall chart or book form, from General Electric Company, Nuclear Energy Operations, 175 Curtner Avenue M/C 684, San Jose, CA 95125. A comprehensive listing of all properties of the nuclides is *Table of Isotopes,* C. Michael Lederer and Virginia S. Shirley (eds.), (John Wiley & Sons, New York, 1978), 7th ed.

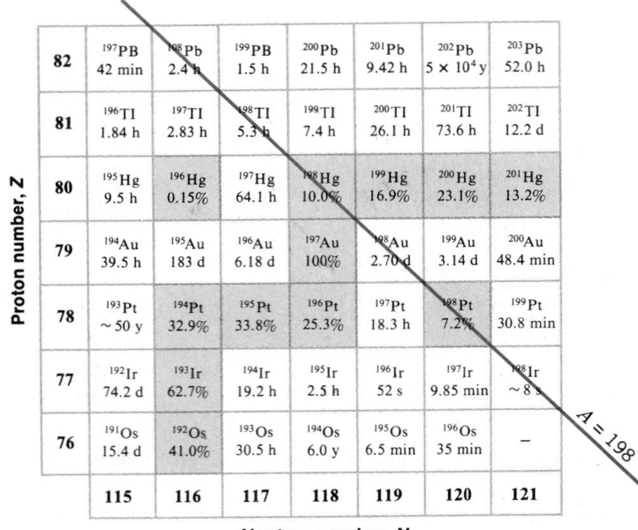

Figure 5 An enlarged section of the nuclidic chart of Fig. 4, centered on ^{197}Au. Shaded squares represent stable nuclides, their relative isotopic abundances being shown. Unshaded squares represent radionuclides, their half-lives being shown. Lines of constant mass number A slope as shown by the example for $A = 198$.

the same abbreviation. Thus

$$1 \text{ femtometer} = 1 \text{ fermi} = 1 \text{ fm} = 10^{-15} \text{ m}. \quad (2)$$

We can learn about the size and structure of nuclei—among other ways—by bombarding them with a beam of high-energy electrons and observing the way the nuclei deflect the incident electrons. The energy of the electrons must be high enough (~200 MeV) so that their de Broglie wavelength will be small enough for them to act as structure-sensitive nuclear probes.

Such experiments show that the nucleus (assumed to be spherical) has a characteristic mean radius R, given by

$$\boxed{R = R_0 A^{1/3},} \quad (3)$$

in which A is the mass number and $R_0 \approx 1.2$ fm. We see that the volume of a nucleus, which is proportional to R^3, is directly proportional to the mass number A, being independent of the separate values of Z or N.

Nuclear Masses.* Atomic masses can be measured with great precision using modern mass-spectrometer and nuclear-reaction techniques. We recall that such masses are reported in atomic mass units (abbr. u), chosen so that the atomic mass (not the nuclear mass) of ^{12}C is exactly 12 u. The relation of this unit to the SI mass unit is

$$1 \text{ u} = 1.661 \times 10^{-27} \text{ kg}. \quad (4)$$

The mass number of a nuclide is so named because the number represents the mass of the nuclide, expressed in atomic mass units, and rounded off to the nearest integer. Thus, the atomic mass of ^{197}Au is 196.966573 u, which rounds to 197.

In nuclear reactions, Einstein's $E = \Delta m\, c^2$ relation is an indispensable work-a-day tool. The energy equivalence of 1 atomic mass unit can easily be shown to be 932 MeV. Thus c^2 can be written as 932 MeV/u and we can easily find the energy equivalence (in MeV) of any given mass or mass difference (in u), or conversely.

Nuclear Binding Energies.† The total energy required to tear apart a nucleus into its constituent protons and neutrons can be calculated from Einstein's $E = \Delta m\, c^2$ relation and is called the *nuclear binding energy.* If we divide the binding energy of a nucleus by its mass number we get the *binding energy per nucleon.* Figure 6 shows a plot of this quantity as a function of mass number. The fact that this *binding energy curve* "droops" at both high and low mass numbers has practical consequences of the greatest importance.‡

The drooping of the binding energy curve at high mass numbers tells us that nucleons are more tightly bound when they are assembled into two middle-mass nuclides rather than into a single high-mass nuclide. In other words, energy can be released by the *nuclear fission* of a single massive nucleus into two smaller fragments.

The drooping of the binding energy curve at low mass numbers, on the other hand, tells us that energy will be released if two nuclides of small mass number combine to form a single middle-mass nuclide. This process, the reverse of fission, is called *nuclear fusion.* It occurs inside our sun and other stars and in thermonuclear explosions. Controlled nuclear fusion as a practical energy source is the subject of much current attention.

Nuclear Energy Levels. Nuclei, like atoms, are governed by the laws of quantum physics and exist in discrete states of well-defined energy. Figure 7 shows these

* This subject is also treated in Section 1–6, which you may wish to reread at this time.

† This subject is also treated in Section 8–9, which you may wish to reread at this time.

‡ *The Curve of Binding Energy* has even been adopted as the title of a book (by John McPhee) about the possibilities of nuclear terrorism.

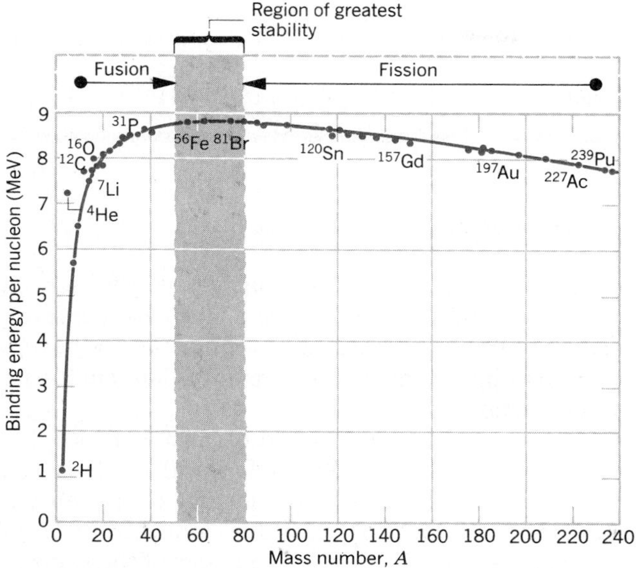

Figure 6 The curve of binding energy. The nuclides of Table 1 are shown, along with a few other nuclides of interest. Note the region of greatest stability. Fission can occur for heavier nuclides and fusion for lighter nuclides. Note that the α particle (^{4}He) lies above the binding energy curve of its neighbors and is thus particularly stable.

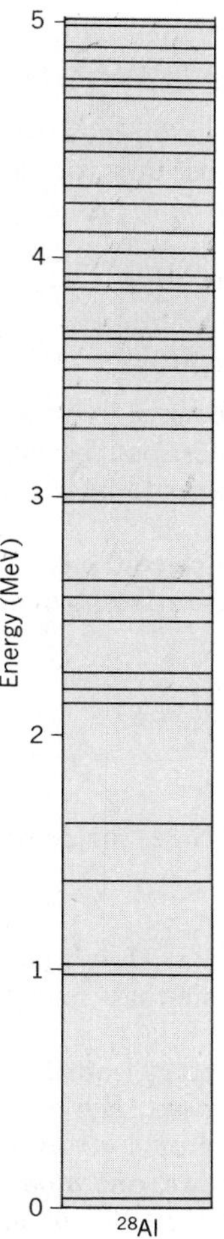

Figure 7 Energy levels for the nuclide ^{28}Al, deduced from nuclear reaction experiments.

energy levels for a typical light nuclide, ^{28}Al. Note that the energy scale is in millions of electron volts, rather than in electron volts as for atoms. When a nucleus makes a transition from one level to a level of lower energy, the emitted photon is typically in the gamma-ray region of the electromagnetic spectrum.

Nuclear Spin and Magnetism. Many nuclides have an intrinsic *nuclear angular momentum* and also an associated intrinsic *nuclear magnetic moment.* Although nuclear angular momenta are roughly the same magnitude as the angular momenta of the atomic electrons, nuclear magnetic moments are much smaller than typical atomic magnetic moments, by a factor of about 1000.

The Nuclear Force. The force that controls the motions of the atomic electrons is the familiar electromagnetic force. To bind the nucleus together, however, there must be a strong attractive nuclear force of a totally different kind, strong enough to overcome the repulsive force of the (positively charged) nuclear protons and to bind both protons and neutrons into the tiny nuclear volume. The nuclear force must also be of short range because its influence does not extend very far beyond the nuclear surface.

The present view is that the nuclear force is not a fundamental force of nature but is a residual effect of the *strong force* that binds quarks together to form neutrons and protons. In much the same way, certain electrically neutral molecules are held together to form solids by a residual effect of the electromagnetic force that acts within the individual molecules.

Sample Problem 2 We can think of all nuclides as made up of a neutron–proton mixture that we can call *nuclear matter.* What is its density?

We know that this density will be high because virtually all of the mass of the atom is found in its tiny central nucleus. The volume of a nucleus (assumed spherical) of mass number

A and radius R is

$$V = (4/3)\pi R^3 = (4/3)\pi(R_0 A^{1/3})^3 = (4/3)\pi R_0^3 A.$$

Such a nucleus contains A nucleons so that its nucleon density, expressed in nucleons per unit volume, is

$$\rho_n = \frac{A}{V} = \frac{A}{(4/3)\pi R_0^3 A}$$
$$= \frac{3}{(4\pi)(1.2\ \text{fm})^3} = 0.138\ \text{nucleons/fm}^3.$$

The fact that we can speak of nuclear matter of constant density for all nuclides comes about because A cancels in the above equation. Nucleons seems to be packed into the nucleus like marbles in a sack.

The mass of a nucleon (neutron *or* proton) is about 1.67×10^{-27} kg. The mass density of nuclear matter in SI units is then

$$\rho = \left(0.138\ \frac{\text{nucleons}}{\text{fm}^3}\right)\left(1.67 \times 10^{-27}\ \frac{\text{kg}}{\text{nucleon}}\right)\left(10^{15}\ \frac{\text{fm}}{\text{m}}\right)^3$$
$$\approx 2 \times 10^{17}\ \text{kg/m}^3. \qquad \text{(Answer)}$$

This is about 2×10^{14} times the density of water!

Sample Problem 3 (a) How much energy is required to separate a typical middle-mass nucleus such as ^{120}Sn into its constituent nucleons?

We can find this energy from $E = \Delta m\, c^2$. Following standard practice, we carry out such calculations in terms of the masses of the neutral atoms involved, not those of the bare nuclei. As Table 1 shows, one *atom* of ^{120}Sn (nucleus + 50 electrons) has a mass of 119.902199 u. This atom can be separated into 50 hydrogen atoms (50 protons + 50 electrons) and 70 neutrons. Each hydrogen atom has a mass of 1.007825 u and each neutron a mass of 1.008665 u. The combined mass of the constituent particles is

$$m = 50 \times 1.007825\ \text{u} + 70 \times 1.008665\ \text{u}$$
$$= 120.99780\ \text{u}.$$

This exceeds the atomic mass of ^{120}Sn by

$$\Delta m = 120.99780\ \text{u} - 119.902199\ \text{u}$$
$$= 1.095601\ \text{u} \approx 1.096\ \text{u}.$$

Note that the masses of the 50 electrons cancel out, so that this same mass difference applies to separating a bare ^{120}Sn nucleus into 50 (bare) protons and 70 neutrons. In energy terms this mass difference becomes

$$E = \Delta m\, c^2 = (1.096\ \text{u})(932\ \text{MeV/u})$$
$$= 1021\ \text{MeV}. \qquad \text{(Answer)}$$

(b) What is the binding energy per nucleon for this nuclide?

We have

$$E_n = \frac{E}{A} = \frac{1021\ \text{MeV}}{120} = 8.51\ \text{MeV/nucleon}, \qquad \text{(Answer)}$$

in agreement with the value shown in Table 1.

47-3 Radioactive Decay

As Fig. 4 shows, most of the nuclides that have been identified are radioactive. That is, a given nucleus spontaneously emits a particle, transforming itself in the process into a different nuclide, occupying a different square on the nuclidic chart.

Radioactive decay provided the first evidence that the laws that govern the subatomic world are statistical. Consider, for example, a one-milligram sample of uranium metal. It contains 2.5×10^{18} atoms of the very long-lived radionuclide ^{238}U. The nuclei of these atoms have existed without decaying since they were created—before the formation of our solar system—in the explosion of a supernova. During any given second about 12 of the nuclei in our sample will decay by emitting an α particle, transforming themselves into nuclei of ^{234}Th. However:

> *There is absolutely no way to predict whether any given nucleus in the sample will be among the small number that decay during the next second. All have an equal chance, namely $12/(2.5 \times 10^{18})$ or one chance in 2×10^{17} per second.*

If a sample contains N radioactive nuclei, we can express the statistical nature of the decay process by saying that the decay rate ($= -dN/dt$) is proportional to N, or

$$-\frac{dN}{dt} = \lambda N, \qquad (5)$$

in which λ, the *disintegration constant,* has a characteristic value for every radionuclide. Equation 5 integrates readily to

$$\boxed{N = N_0 e^{-\lambda t}} \quad \text{(radioactive decay)}, \qquad (6)$$

in which N_0 is the number of radioactive nuclei in the sample at $t = 0$ and N is the number remaining at any subsequent time t. Note that light bulbs (for one example) follow no such exponential decay law. If we life-test 1000 bulbs, we expect that they will all "decay" (that is, burn out) at more or less the same time. The decay of radionuclides follows quite a different law.

We are often more interested in the decay rate R $(= -dN/dt)$ than in N itself. Differentiating Eq. 6, we find

$$R = -\frac{dN}{dt} = \lambda N_0 e^{-\lambda t}$$

or

$$\boxed{R = R_0 e^{-\lambda t}} \quad \text{(radioactive decay)}, \qquad (7)$$

an alternative form of the law of radioactive decay. Here R_0 $(= \lambda N_0)$ is the decay rate at $t = 0$ and R is the rate at any subsequent time t.

A quantity of special interest is the *half-life* τ, defined as the time after which both N and R are reduced to one-half of their initial values. Putting $R = \frac{1}{2}R_0$ in Eq. 7, we have

$$\tfrac{1}{2}R_0 = R_0 e^{-\lambda\tau}.$$

Solving for τ yields

$$\boxed{\tau = \frac{\ln 2}{\lambda}}, \qquad (8)$$

a relationship between the half-life and the disintegration constant.

Sample Problem 4 The table shows some measurements of the decay rate of a sample of ^{128}I, a radionuclide often used medically as a tracer to measure the rate at which iodine is absorbed by the thyroid gland.

Time min	*R* counts/s	*Time* min	*R* counts/s
4	392.2	132	10.9
36	161.4	164	4.56
68	65.5	196	1.86
100	26.8	218	1.00

Find the disintegration constant and the half-life for this radionuclide.

If we take the natural logarithm of each side of Eq. 7, we find

$$\ln R = \ln R_0 - \lambda t.$$

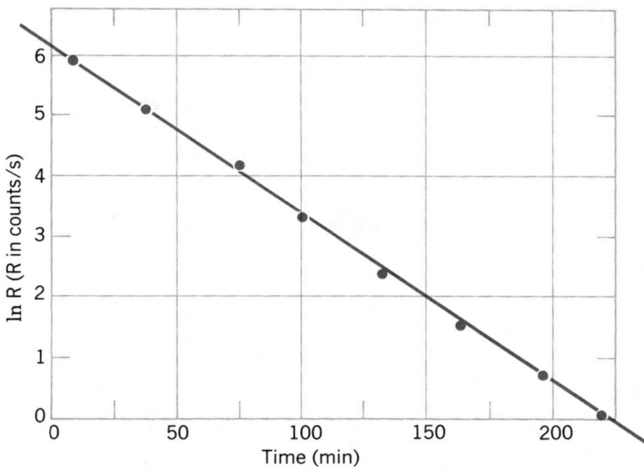

Figure 8 Sample Problem 4. A logarithmic plot of the decay of a sample of ^{128}I, based on the data in the table. The fact that it is a straight line shows that we are dealing with a simple exponential decay. The half-life of this radionuclide (= 25 min) can be found from the slope of this curve.

Thus, if we plot $\ln R$ against t, we should obtain a straight line whose slope is $-\lambda$. From Fig. 8, we find

$$-\lambda = -\frac{6.06 - 0}{220 \text{ min} - 0}$$

or

$$\lambda = 0.0275 \text{ min}^{-1}. \qquad \text{(Answer)}$$

We find the half-life readily from Eq. 8, or

$$\tau = \frac{\ln 2}{\lambda} = \frac{\ln 2}{0.0275 \text{ min}^{-1}} \approx 25 \text{ min}. \qquad \text{(Answer)}$$

The decay rate of a given sample of ^{128}I will drop to half of its initial value in 25 min, no matter what the initial value was.

Sample Problem 5 A 2.71-g sample of KCl from the chemistry stockroom is found to be radioactive and to decay at a constant rate of 4490 disintegrations/s. The decays are traced to the element potassium and in particular to the isotope ^{40}K, which constitutes 1.17% of normal potassium. Calculate the half-life of this nuclide.

We can find the half-life from Eq. 8. Since the decay rate is constant, the half-life must be very long and we cannot calculate λ by the method of Sample Problem 4. We must find it by determining both N and dN/dt in Eq. 5.

The molecular mass of KCl is 74.6 g/mol, so that the number of potassium atoms in the sample is

$$N_K = \frac{(6.02 \times 10^{23} \text{ mol}^{-1})(2.71 \text{ g})}{74.6 \text{ g/mol}} = 2.19 \times 10^{22}.$$

Of these, the number of ^{40}K atoms is

$$N_{40} = (2.19 \times 10^{22})(0.0117) = 2.56 \times 10^{20}.$$

From Eq. 5, we have

$$\lambda = \frac{-dN/dt}{N} = \frac{R_{40}}{N_{40}} = \frac{4490\ s^{-1}}{2.56 \times 10^{20}}$$
$$= 1.75 \times 10^{-17}\ s^{-1}. \qquad \text{(Answer)}$$

The half-life follows from Eq. 8, or

$$\tau = \frac{\ln 2}{\lambda} = \frac{(\ln 2)(1\ y/3.16 \times 10^7\ s)}{1.75 \times 10^{-17}\ s^{-1}}$$
$$= 1.25 \times 10^9\ y. \qquad \text{(Answer)}$$

This is of the order of magnitude of the age of the universe! No wonder we cannot measure the half-life of this radionuclide by waiting around for its decay rate to decrease. Interestingly, the potassium in our own bodies has its normal share of this radioisotope; we are all radioactive.

47-4 Alpha Decay

The radionuclide ^{238}U decays by emitting an α particle, according to the scheme

$$^{238}U \rightarrow {}^{234}Th + {}^{4}He \qquad Q = 4.25\ MeV, \qquad (9)$$

the half-life of the decay being 4.47×10^9 y. Q is the disintegration energy of the process, that is, the amount of energy released during the decay. We may well ask:

> *"If energy is released in every such decay event, why did the ^{238}U nuclei not decay shortly after they were created? Why did they wait so long?"*

To answer this question, we must study the detailed mechanism of alpha decay.

We choose a model in which the α particle is imagined to exist preformed inside the nucleus before it escapes. Figure 9 shows the approximate potential energy function $U(r)$ for the α particle and the residual ^{234}Th nucleus as a function of their separation r. It is a combination of a potential well associated with the (attractive) strong nuclear force that acts in the nuclear interior and a Coulomb potential associated with the (repulsive) electrostatic force that acts between the two particles after the decay has occurred.

The horizontal line marked $Q = 4.25$ MeV shows the disintegration energy for the process. This line is shown hatched in the region in which it lies under the potential energy curve. If the α particle found itself in that region, its potential energy U would exceed its total energy E, which would mean, classically, that its kinetic energy K $(= E - U)$ would be negative, an impossible situation.

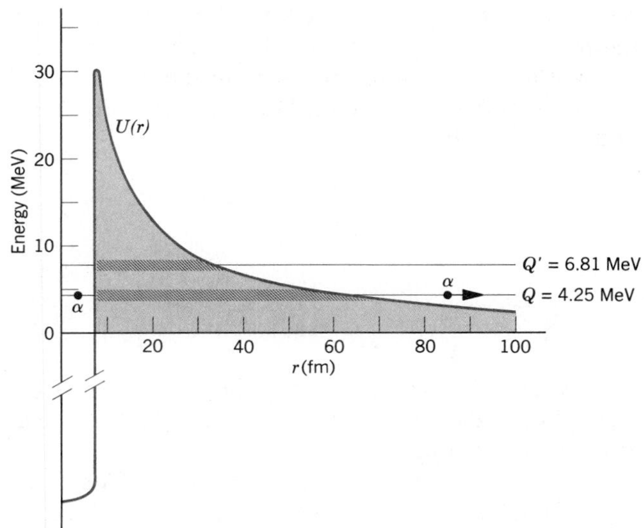

Figure 9 A potential energy function describing the emission of an α particle by ^{238}U. The horizontal line marked $Q = 4.25$ MeV shows the disintegration energy for the process. The hatched portion of this line represents a region that is classically forbidden to the α particle. The α particle is represented by a dot, both inside the barrier and outside of it, after it has tunneled through. The horizontal line marked $Q' = 6.81$ MeV shows the disintegration energy for the alpha decay of ^{228}U.

We now see why the α particle is not immediately emitted from the ^{238}U nucleus! That nucleus is surrounded by an impressive potential barrier, occupying —if you think of it in three dimensions—the volume lying between two spherical shells. This argument is so convincing that we now change our question and ask:

> *"How can the ^{238}U nucleus* ever *emit an α particle? The particle seems permanently trapped inside the nucleus by the barrier."*

The answer is that, as we learned in Section 44-7, in wave mechanics there is always a chance that a particle can tunnel through a barrier that is classically insurmountable.

For the long-lived decay of ^{238}U the barrier is actu-

Table 2 Two Alpha Emitters Compared

Radionuclide	*Q*	*Half-life*
^{238}U	4.25 MeV	4.5×10^9 y
^{228}U	6.81 MeV	9.1 min

ally not very "leaky." The α particle, presumed to be rattling back and forth within the nucleus, must present itself at the inner surface of the barrier about 10^{38} times before it succeeds in tunneling through. This is about 10^{21} times per second for about 4×10^9 years (the age of the earth)! We, of course, are waiting on the outside, able to count only those α particles that *do* manage to escape.

We can test this explanation of alpha decay by examining other alpha emitters. For an extreme contrast, consider the alpha decay of another uranium isotope, ^{228}U, which has a disintegration energy Q' of 6.81 MeV, higher than that of ^{238}U, which is 4.25 MeV. We show Q' also as a horizontal line in Fig. 9. We recall from Section 44-7 that the transmission coefficient of a barrier is very sensitive to small changes in the total energy of the particle seeking to penetrate it. Thus, we expect alpha decay to occur more readily for this nuclide than for ^{238}U. Indeed it does. As Table 2 shows, its half-life is only 9.1 min! An increase in Q by a factor of only 1.6 produces a decrease in half-life (that is, in the effectiveness of barrier tunneling) by a factor of $\sim 3 \times 10^{14}$. This is sensitivity indeed.

Sample Problem 6 Given the atomic masses:

^{238}U	238.05079 u	^4He	4.00260 u
^{234}Th	234.04363 u	^1H	1.00783 u
^{237}Pa	237.05121 u		

(a) Calculate the energy released during the alpha decay of ^{238}U. The decay process is:

$$^{238}U \rightarrow {}^{234}Th + {}^4He.$$

Note, incidentally, how nuclear charge is conserved in this equation, the atomic numbers of thorium (90) and helium (2) adding up to yield the atomic number of uranium (92). The numbers of nucleons are also conserved ($238 = 234 + 4$).

The atomic mass of the decay products in the foregoing process ($= 234.04363$ u $+ 4.00260$ u) is less than the atomic mass of ^{238}U by $\Delta m = 0.00456$ u, whose energy equivalent is

$$Q = \Delta m\, c^2 = (0.00456 \text{ u})(932 \text{ MeV/u}) = 4.25 \text{ MeV}. \qquad \text{(Answer)}$$

This disintegration energy appears as kinetic energy of the α particle and the recoiling ^{234}Th atom.

Note again that, following standard practice, we used the masses of the neutral atoms rather than those of the bare nuclei; in calculating the mass difference Δm, the masses of the extranuclear electrons cancel.

(b) Show that ^{238}U cannot decay spontaneously by emitting a proton.

If this happened, the decay process would be

$$^{238}U \rightarrow {}^{237}Pa + {}^1H.$$

(Verify that both nuclear charge and the numbers of nucleons are conserved in this process.) In this case, the mass of the decay products ($= 237.05121$ u $+ 1.00783$ u) would *exceed* the mass of ^{238}U by $\Delta m = 0.00825$ u, the energy equivalence being -7.69 MeV. The minus sign tells us that ^{238}U is stable against spontaneous proton emission.

47-5 Beta Decay

A nucleus that decays spontaneously by emitting an electron (either positive or negative) is said to undergo *beta decay*. This, like alpha decay, is a spontaneous process, with a definite disintegration energy and half-life. Again like alpha decay, beta decay is a statistical process, governed by Eqs. 6 and 7. Here are two examples:

$$^{32}P \rightarrow {}^{32}S + e^- + \nu \qquad (\tau = 14.3 \text{ d}) \qquad (10)$$

and

$$^{64}Cu \rightarrow {}^{64}Ni + e^+ + \nu \qquad (\tau = 12.7 \text{ h}). \qquad (11)$$

The symbol ν represents a *neutrino*, a massless, neutral particle that is emitted from the nucleus along with the electron during the decay process. Neutrinos interact only very weakly with matter and—for that reason—are so extremely difficult to detect that, for many years, their presence went unnoticed.*

Both charge and nucleon number are conserved in the above two processes. In the decay of Eq. 10, for example, we can write

$$(+15e) = (+16e) + (-e) + (0) \quad \text{(charge conservation)}$$

* Beta decay also includes *electron capture*, in which a nucleus decays by absorbing one of its orbital electrons, emitting a neutrino in the process. We do not consider that process here. Also, the neutral particle emitted in the decay process of Eq. 10 is actually an *antineutrino*, a distinction that we do not wish to stress in this introductory treatment. Finally, whether the mass of the neutrino is truly zero or not is a subject under current investigation.

and

$$(32) = (32) + (0) + (0) \quad \text{(nucleon conservation).}$$

Note that neither the electron or the neutrino is a nucleon and the neutrino carries no charge.

It may seem surprising that nuclei can emit electrons (and neutrinos) since we have said that nuclei are made up of neutrons and protons only. However, we saw earlier that atoms emit photons, and we certainly do not say that atoms "contain" photons. We say that the photons are created during the emission process.

So it is with the electrons and the neutrinos emitted from nuclei during beta decay. They are both created during the emission process, a neutron transforming itself into a proton within the nucleus (or conversely) according to

$$n \rightarrow p + e^- + \nu \tag{12}$$

or

$$p \rightarrow n + e^+ + \nu. \tag{13}$$

These are, in fact, the basic beta decay processes and provide evidence that—as we have pointed out earlier—neutrons and protons are not truly fundamental particles. Note (see Eqs. 10 and 11) that the mass number A of a nuclide undergoing beta decay does not change; one of its constituent nucleons simply changes its character (see Eqs. 12 and 13), the total number of nucleons remaining fixed.

In both alpha decay and beta decay, the same amount of energy is released in every individual decay process. In the alpha decay of a particular radionuclide, every emitted α particle has the same sharply defined kinetic energy. In beta decay, however, the disintegration energy Q is shared—in varying proportions—between the electron and the neutrino. Sometimes the electron gets nearly all of the energy, sometimes the neutrino does. In every case, however, the sum of the electron's energy and the neutrino's energy adds up to a constant value Q.

Thus, in beta decay the energy of the emitted *electrons* is distributed in a continuous spectrum, from zero up to a certain maximum K_{max}, as Fig. 10 shows for the beta decay of ^{64}Cu (see Eq. 11). This maximum electron energy K_{max} must equal the disintegration energy Q because the neutrino carries away no energy under these circumstances. That is

$$Q = K_{max}. \tag{14}$$

The neutrino hypothesis—put forward by Pauli in 1930—not only permitted an understanding of the nature of the continuous electron spectrum but also solved another early beta decay puzzle, involving "missing" angular momentum.

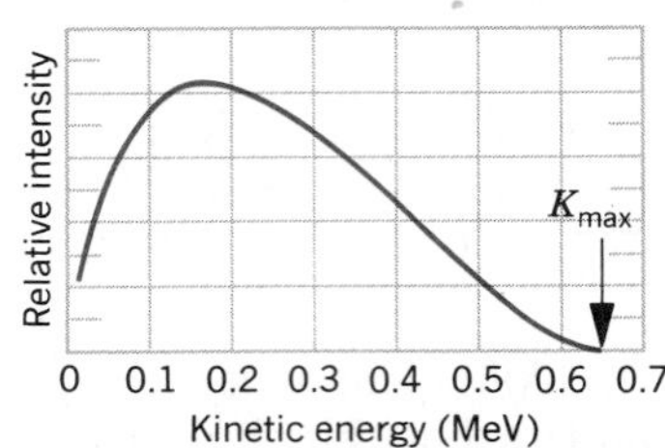

Figure 10 The distribution in kinetic energy of the positrons emitted in the beta decay of ^{64}Cu. The maximum kinetic energy of the distribution (K_{max}) is 0.653 MeV. In all decay events, this energy is shared between the positron and the neutrino, in varying proportions.

The neutrino is a truly elusive particle, the mean free path of an energetic neutrino in water being no less than several thousand light years! At the same time, neutrinos left over from the Big Bang that marked the creation of the universe are the most abundant of the particles of physics. Billions upon billions of them pass through our bodies every second, leaving no trace.

In spite of their elusive character, neutrinos have been detected in the laboratory. The first, detected in 1953 by F. Reines and C. L. Cowan, were generated in a high-power nuclear reactor. In spite of the difficulties of detection, experimental neutrino physics is now a well-developed branch of experimental physics, with avid practitioners at several major laboratories throughout the world.

The sun emits neutrinos copiously from the nuclear furnace at its core and, at night, these messengers from the center of the sun come up at us from below, the earth being totally transparent to them.* In February 1987, light from an exploding star in the Large Magellanic Cloud (a nearby galaxy) reached the earth after traveling for 170,000 years. Enormous numbers of neutrinos were generated in this explosion and about 10 of them were picked up by a sensitive neutrino detector in Japan; see Fig. 11. This is the first detection of neutrinos not only from beyond the solar system but, indeed, from beyond our own galaxy.

Radioactivity and the Nuclidic Chart. Our study of alpha and beta decay permits us to look at the nuclidic

* For a poet's view of the neutrino, see John Updike's poem "Cosmic Gall."

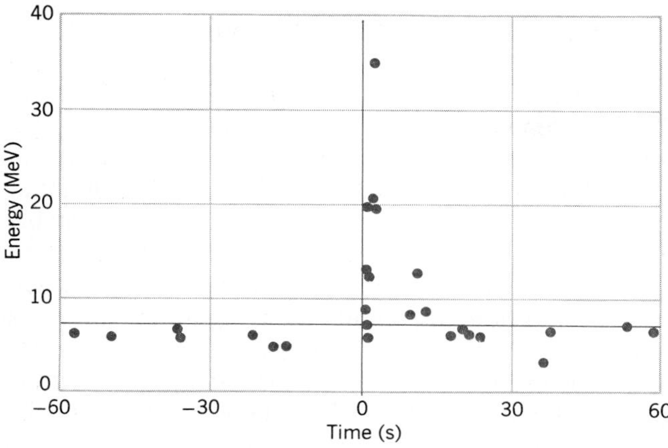

Figure 11 A burst of neutrinos from the supernova SN 1987A. (For neutrinos, ten is a "burst"!) They were detected by an elaborate detector, housed deep in a mine in Japan. The supernova was visible only in the southern hemisphere so that the neutrinos had to penetrate the earth (a trifling barrier for them!) to reach the detector.

chart of Fig. 4 in a new way. Let us construct a three-dimensional surface by plotting the mass excess (see the caption to Fig. 12) of each nuclide in a direction at right angles to the N-Z plane of that figure. The surface so formed gives a graphic representation of nuclear stability. As Fig. 12 shows (for the light nuclides) it describes a "valley of the nuclides," the stability band of Fig. 4 running along its bottom. Nuclides on the headwall of the valley (beyond the region displayed in Fig. 12) decay into it largely by chains of alpha decay and by spontaneous fission. Nuclides on the proton-rich side of the valley decay into it by emitting positive electrons and those on the neutron-rich side do so by emitting negative electrons.

Sample Problem 7 Calculate the disintegration energy Q for the beta decay of ^{32}P, as described by Eq. 10. The needed atomic masses are 31.97391 u for ^{32}P and 31.97207 u for ^{32}S.

Because of the presence of the emitted electron, we must be especially careful to distinguish between nuclear and atomic masses. Let the boldface symbols $\mathbf{m_P}$ and $\mathbf{m_S}$ represent the nuclear masses of ^{32}P and ^{32}S and let m_P and m_S represent their

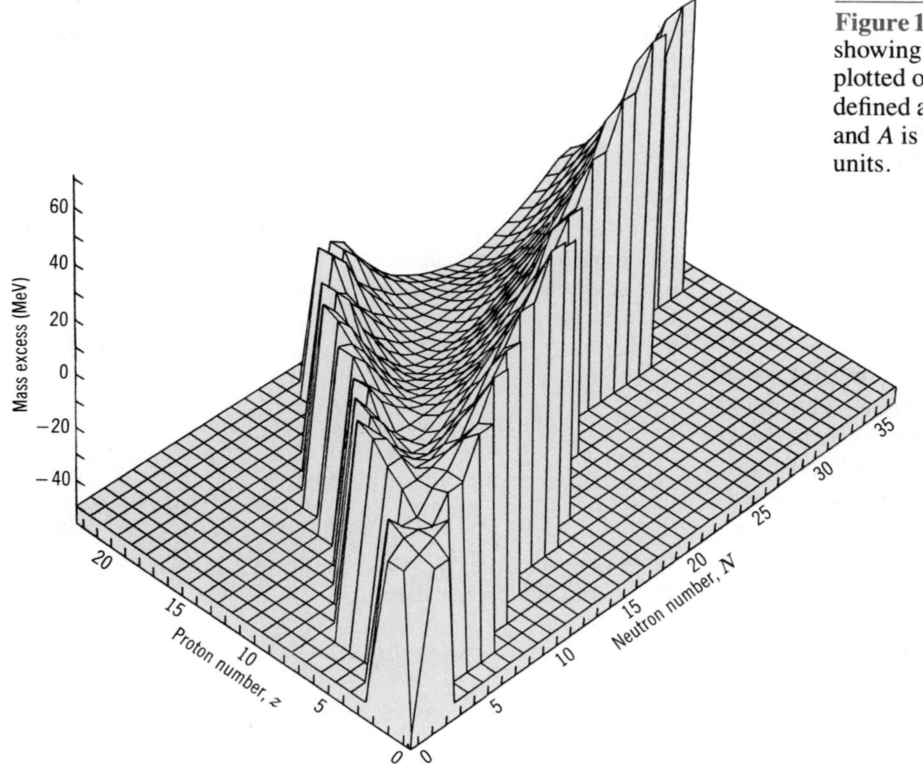

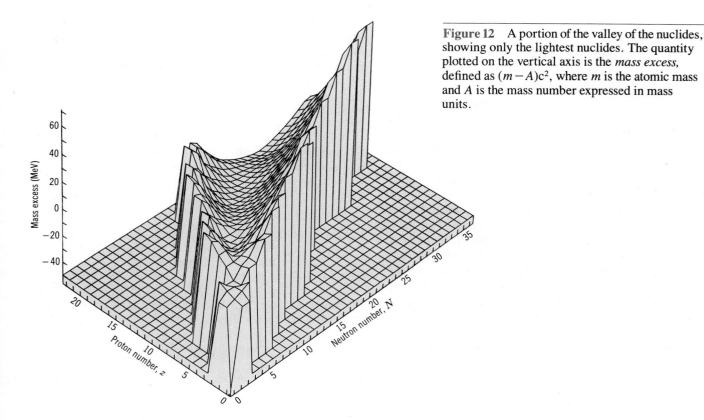

Figure 12 A portion of the valley of the nuclides, showing only the lightest nuclides. The quantity plotted on the vertical axis is the *mass excess,* defined as $(m-A)c^2$, where m is the atomic mass and A is the mass number expressed in mass units.

atomic masses. We take the disintegration energy Q to be $\Delta m\, c^2$, where (see Eq. 10)

$$\Delta m = \mathbf{m}_\mathrm{P} - (\mathbf{m}_\mathrm{S} + m_e),$$

m_e being the mass of the electron. If we add and subtract $15m_e$ to the right side of this equation, we have

$$\Delta m = (\mathbf{m}_\mathrm{P} + 15m_e) - (\mathbf{m}_\mathrm{S} + 16\, m_e).$$

The quantities in parentheses are the atomic masses, so that

$$\Delta m = m_\mathrm{P} - m_\mathrm{S}.$$

We thus see that if we subtract the atomic masses in this way, the mass of the emitted electron is automatically taken into account.*

The disintegration energy for the ^{32}P decay is then

$$\begin{aligned} Q &= \Delta m\, c^2 \\ &= (31.97391\ \mathrm{u} - 31.97207\ \mathrm{u})(932\ \mathrm{MeV/u}) \\ &= 1.71\ \mathrm{MeV}. \qquad \text{(Answer)} \end{aligned}$$

Experimentally, this calculated quantity proves to be equal (as Eq. 14 requires) to K_{max}, the maximum energy of the emitted electrons. Although 1.71 MeV is released every time a ^{32}P nucleus decays, in essentially every case the electron carries away less energy than this. The neutrino gets essentially all of the rest, carrying it undetected out of the laboratory.

47-6 Radioactive Dating

If you know the half-life of a given radionuclide, you can in principle use the decay of that radionuclide as a clock to measure a time interval. The decay of very long-lived nuclides, for example, can be used to measure the age of rocks, that is, the time that has elapsed since they were formed. Measurements for rocks from the Earth, the Moon, and for meteorites all yield a consistent age of about 4.5×10^9 y.

The radionuclide ^{40}K, for example, decays to a stable isotope ^{40}Ar of the noble gas argon. The half-life of this decay, as we saw in Sample Problem 5, is 1.25×10^9 y. By measuring the ratio of ^{40}K to ^{40}Ar found in the mineral in question, one can then calculate the age of that mineral; see Problem 65. Other long-lived decays, such as that of ^{235}U to Pb^{207} (involving a number of intermediate stages) can be used to verify these measurements.

For measuring shorter time intervals, in the range of historical interest, radiocarbon dating has proved invaluable. The radionuclide ^{14}C ($\tau = 5730$ y) is produced at a constant rate in the upper atmosphere by the bombardment of atmospheric nitrogen by cosmic rays. This radiocarbon mixes with the carbon that is normally present in the atmosphere (as CO_2) so that there is about one atom of ^{14}C for every 10^{13} atoms of ordinary stable ^{12}C. The atmospheric carbon exchanges with the carbon in every living thing on earth, including trees, tomatoes, rabbits, and humans, so that all living things contain a small fixed fraction of this nuclide.

This exchange persists as long as the organism is alive. After the organism dies, the exchange with the atmosphere stops and the amount of radiocarbon trapped in the organism, since it is no longer being replenished, dwindles away with a half-life of 5730 y. By measuring the amount of radiocarbon per gram of organic matter, it is possible to measure the time that has elapsed since the organism died. Charcoal from ancient camp fires, the Dead Sea scrolls, and many ancient artifacts have been dated in this way. (As we write this, the Shroud of Turin is being dated, by this method, in three independent laboratories.)

Sample Problem 8 Analysis of potassium and argon atoms in a Moon rock sample by a mass spectrometer shows that the ratio of the number of (stable) ^{40}Ar atoms present to the number of (radioactive) ^{40}K atoms is 10.3. Assume that all of the argon atoms were produced by the decay of the potassium atoms, the half-life for the decay being known to be 1.25×10^9 y. How old is the rock?

If N_{K_0} potassium atoms were present at the time the rock was formed by solidification from a molten form, the number of potassium atoms remaining at the time of analysis is (see Eq. 6)

$$N_\mathrm{K} = N_{\mathrm{K}_0} e^{-\lambda t}, \tag{15}$$

in which t is the age of the rock. For every potassium atom that decays, an argon atom is produced. Thus, the number of argon atoms present at the time of the analysis is

$$N_\mathrm{A} = N_{\mathrm{K}_0} - N_\mathrm{K}. \tag{16}$$

We cannot measure N_{K_0}. If we eliminate it from Eqs. 15 and 16, we find, after some algebra

$$\lambda t = \ln\left(1 + \frac{N_\mathrm{A}}{N_\mathrm{K}}\right) \tag{17}$$

in which N_A/N_K is the measured ratio. Solving for t and replacing λ by $(\ln 2)/\tau$ yields

* This is not the case for positron emission; see Problem 59.

$$t = \frac{\tau \ln (1 + N_A/N_K)}{\ln 2}$$
$$= \frac{[1.25 \times 10^9 \text{ y}][\ln (1 + 10.3)]}{\ln 2}$$
$$= 4.37 \times 10^9 \text{ y}. \qquad \text{(Answer)}$$

Lower measurements may be found for other Moon or Earth rock samples, but no substantially higher ones. This result may thus be taken as a good approximation to the age of the solar system.

47-7 Measuring Radiation Dosage*

The effects of ionizing radiations such as gamma rays, electrons, and α particles on living tissue (particularly our own!) has become a matter of public interest. Such radiations arise in nature from the cosmic rays and also from radioactive elements in the earth's crust. Radiations associated with human activity also contribute, including diagnostic and therapeutic x rays and radiations from radionuclides used in medicine and in industry. The disposal of radioactive waste and the evaluation of the probabilities of nuclear accidents continue to be dealt with at the level of national policy.

It is not our task here to explore the various sources of ionizing radiations but simply to describe the units in which the properties and effects of these radiations are expressed. There are four such units, and they are often used loosely or incorrectly in popular reporting.

The curie (abbr. Ci). This is a measure of the *activity* of a radioactive source. It is defined as

$$1 \text{ curie} = 1 \text{ Ci}$$
$$= 3.7 \times 10^{10} \quad \text{disintegrations per second.}$$

This definition says nothing about the nature of the decays. An example of the proper use of the curie is:

> *"The activity of spent reactor fuel rod #5658 on January 15, 1987, was 9.5 × 10⁴ Ci."*

The half-lives of the nuclides that make up the fuel rod and the types of radiations that they emit have no bearing.

The roentgen (abbr. R). This is a measure of *exposure*, that is, of the ability of a beam of x rays or gamma rays to deliver energy to a material through which they pass. Specifically, one roentgen is defined as that exposure that would deliver 8.78 mJ to 1 kg of dry air at standard conditions. We might say, for example:

> *"This dental x-ray beam provides an exposure of 300 mR/s."*

This says nothing about whether energy is actually delivered or whether or not there is a patient in the chair.

The rad. This is an acronym for Radiation Absorbed Dose and is a measure, as its name suggests, of the dose actually absorbed by a specific object, in terms of the energy transferred to it. An object, which might be a person (whole body) or a specific part of the body (the hands, say) is said to have received an *absorbed dose* of one rad when 10 mJ/kg have been delivered to it by ionizing radiations. A typical statement to show the usage is:

> *"A whole-body short term gamma ray dose of 300 rad will cause death in 50% of the population exposed to it."*

By way of comfort we note that the present absorbed dose of radiation from sources of both natural and human origin is about 0.2 rad (= 200 mrad) per year.

The rem. This is an acronym for Roentgen Equivalent in Man and is a measure of *dose equivalent*. It takes account of the fact that, although different types of radiation (gamma rays and neutrons, say) may deliver the same energy per unit mass to the body, they do not have the same biological effect. The dose equivalent (in rems) is found by multiplying the absorbed dose (in rads) by a *Relative Biological Effectiveness* (RBE) factor, which may be found tabulated in various reference sources. For x rays and electrons, RBE $\approx$ 1. For slow neutrons, RBE $\approx$ 5, and so on. Personnel-monitoring devices such as film badges are designed to register the dose equivalent in rems. An example of correct usage of the rem is

> *"The recommendation of the National Council on Radiation Protection is that no individual who is (nonoccupationally) exposed to radiations should receive a dose equivalent greater than 500 mrem (= 0.5 rem) in any one year."*

This includes radiations of all kinds, the appropriate RBE being used. Figure 13 should help to clarify the four radiation units.

* See "Radiation Exposure in our Daily Lives," Stewart C. Bushong, *The Physics Teacher,* March 1977.

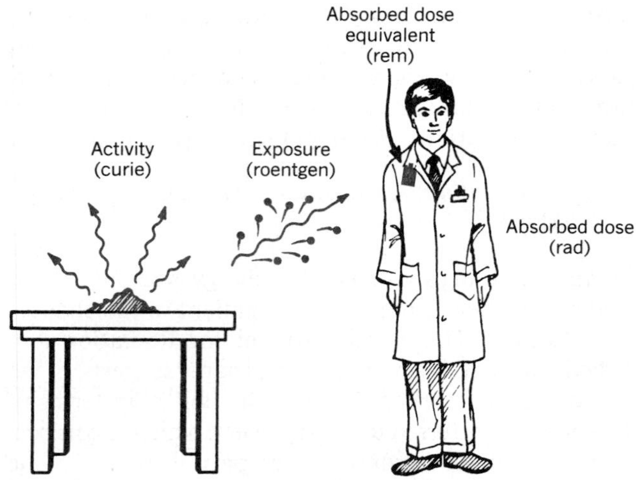

Figure 13 This sketch should help to clarify the distinction between the curie, the roentgen, the rad, and the rem.

Sample Problem 9 We have seen that a gamma-ray dose of 300 rad is lethal to half of those exposed. If the equivalent energy were absorbed as heat, what rise in body temperature would result?

An absorbed dose of 300 rad corresponds to an absorbed energy per unit mass of

$$(300 \text{ rad})\left(\frac{10 \times 10^{-3} \text{ J/kg}}{1 \text{ rad}}\right) = 3 \text{ J/kg}.$$

Assume that c, the specific heat capacity of the human body, is the same as that of water, or 4180 J/kg·K. The temperature rise follows from

$$\Delta T = \frac{Q/m}{c} = \frac{3 \text{ J/kg}}{4180 \text{ J/(kg}\cdot\text{K)}}$$
$$= 7.2 \times 10^{-4} \text{ K} \approx 700\ \mu\text{K}. \qquad \text{(Answer)}$$

We see that the damage done by ionizing radiation has nothing to do with thermal heating. The harmful effects arise because the radiation succeeds in breaking molecular bonds and thus interfering with the normal functioning of the tissues in which it is absorbed.

47-8 Nuclear Models (Optional)

The structure of atoms is now well understood: Quantum physics governs all; the force law is Coulomb's law; there is a massive force center (the nucleus) that simplifies the calculations. In principle, given enough computer time, we can calculate with confidence almost anything that we want to know about an atom.

Things are not in such a happy state for the nucleus. Quantum mechanics still governs its behavior, but the force law is complicated and cannot, in fact, be written down explicitly in full detail. Nor is there a natural force center to simplify the calculations; we are dealing with a many-body problem of great complexity.

In the absence of a comprehensive nuclear *theory,* we turn to the construction of nuclear *models.* A nuclear model is simply a way of looking at the nucleus that gives a physical insight into as wide a range of its properties as possible. The usefulness of a model is tested by its ability to make predictions that can be verified experimentally in the laboratory.

Two models of the nucleus have proved useful. Although based on assumptions that seem flatly to exclude each other, each accounts very well for a selected group of nuclear properties. After describing them separately, we shall see how these two models may be combined to form a single coherent picture of the atomic nucleus.

The Liquid Drop Model. In the liquid drop model, formulated by Niels Bohr, the nucleons are imagined to interact strongly with each other, like the molecules in a drop of liquid. A given nucleon collides frequently with other nucleons in the nuclear interior, its mean free path as it moves about being substantially less than the nuclear radius. This constant "jiggling around" reminds us of the thermal agitation of the molecules in a drop of liquid.

The liquid drop model permits us to correlate many facts about nuclear masses and binding energies; it is useful (as we shall see later) in explaining nuclear fission. As we shall see, it also provides a useful model for understanding a large class of nuclear reactions.

Consider, for example, a generalized reaction of the form

$$X + a \rightarrow C^* \rightarrow Y + b. \qquad (18)$$

Here C^* represents an excited state of a so-called *compound nucleus C.* We imagine that projectile a enters target nucleus X, forming the compound nucleus C and conveying to it a certain amount of excitation energy. The projectile, perhaps a neutron, is at once caught up by the random motions that characterize the nuclear interior. It quickly looses its identity — so to speak — and the excitation energy it carried in with it comes quickly to be shared by all of the other nucleons.

The foregoing quasistable state, represented by C^* in Eq. 18, may endure for as long as $\sim 10^{-16}$ s. By nuclear standards, however, this is a very long time, being one million times longer than the time required for a nucleon with a few MeV of energy to travel across a nucleus. The central feature of this compound-nucleus model is that the formation of this nucleus and its eventual decay are totally independent events. At the time of its decay, the nucleus has "forgotten" how it was formed. We show here three possible ways in which the compound nucleus ^{20}Ne* might be formed and three in which it might decay:

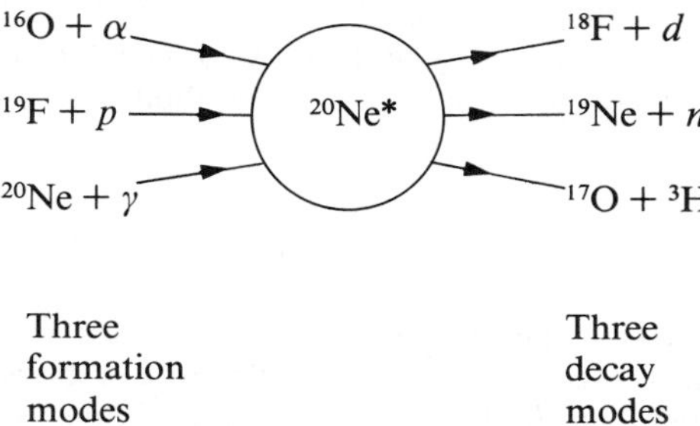

Any of the three "formation" modes can lead to any of the three "decay" modes. Analysis of any of the possible combinations leads to the same set of energy levels for ^{20}Ne*.

The Independent Particle Model. In the liquid drop model, we assumed that the nucleons moved around at random and bumped into each other frequently. The independent particle model, however, is based on just the opposite assumption, namely, that each nucleon moves in a well-defined orbit within the nucleus and hardly makes any collisions at all! The nucleus, unlike the atom, has no fixed center of charge; we assume in this model that each nucleon moves in a potential that is determined by the smeared-out motions of all the other nucleons.

A nucleon in a nucleus, like an electron in an atom, has a set of quantum numbers that defines its state of motion. Also, nucleons obey the Pauli exclusion principle, just as electrons do. That is, no two nucleons may occupy the same state at the same time. In considering nucleon states, the neutrons and the protons are treated separately, each having its own array of available quantized states.

The fact that nucleons obey the Pauli principle helps us to understand the relative stability of the nucleon states. If two nucleons within the nucleus are to collide, the energy of each of them after the collision must correspond to the energy of an *unoccupied* stationary state. If these states are filled, the collision simply cannot occur. In time, any given nucleon will find it possible to collide, but meanwhile it will have made enough revolutions in its orbit to give meaning to the notion of a stationary nucleon state with a quantized energy.

In the atomic realm, the repetitions of physical and chemical properties that we find in the periodic table are associated with the fact that the atomic electrons arrange themselves in shells that have a special stability when they are fully occupied. We can take the atomic numbers of the noble gases:

$$2, 10, 18, 36, 54, 86, \ldots$$

as *magic electron numbers* that mark the completion of such shells.

Nuclei also show such shell effects, associated with certain *magic nucleon numbers:*

$$2, 8, 20, 28, 50, 82, 126, \ldots.$$

Any nuclide whose proton number Z or neutron number N has one of these values turns out to have a special stability that may be made apparent in a variety of ways.

Examples of "magic" nuclides are ^{18}O ($Z = 8$), ^{40}Ca ($Z = 20$, $N = 20$), ^{92}Mo ($N = 50$), and ^{208}Pb ($Z = 82$, $N = 126$). Both ^{40}Ca and ^{208}Pb are said to be "doubly magic" because they contain filled shells of both protons *and* neutrons.

The magic number "2" shows up in the exceptional stability of the α particle (^{4}He), which, with $Z = N = 2$, is doubly magic. For example, the binding energy per nucleon for this nuclide stands well above that of its neighbors on the binding energy curve of Fig. 6. The α particle is so tightly bound, in fact, that it is impossible to add another particle to it; there is no stable nuclide with $A = 5$.

The central idea of a closed shell is that a single particle outside a closed shell can be relatively easily removed but that considerably more energy must be expended to remove a particle from the shell itself. The sodium atom, for example, has one (valence) electron outside a closed electron shell. It requires only ~ 5 eV to strip away the valence electron from a sodium atom; to remove a *second* electron, however (which must be plucked out of a closed shell) requires 22 eV. In a nuclear case, consider ^{121}Sb ($Z = 51$), which contains an "extra" proton outside a closed shell of 50 protons. To remove

Table 3 Atomic and Nuclear Shells Compared

Entity	*Particles Involved*	*Energy to Remove the First Particle*	*Energy to Remove the Second Particle*
Na atom	Electrons	5 eV	22 eV
^{121}Sb nucleus	Protons	5.8 MeV	11 MeV

this proton requires 5.8 MeV; to remove a *second* proton, however, requires an energy of 11 MeV. Table 3 compares these atomic and nuclear situations. We have given only a tiny part of the total experimental evidence that the nucleons in a nucleus form closed shells that exhibit stable properties.

We have seen that wave mechanics can account beautifully for the magic electron numbers, that is, for the populations of the orbitals into which the atomic electrons are grouped. It turns out that, by making certain reasonable assumptions, wave mechanics can account equally well for the magic nucleon numbers! The 1963 Nobel prize was, in fact, awarded to Maria Mayer and to Hans Jensen "for their discoveries concerning nuclear shell structure."

The Collective Model. The collective model of nuclear structure succeeds in combining the seemingly irreconcilable points of view of the liquid drop and the independent particle models.

Consider a nucleus with a small number of neutrons (or protons) outside a core of closed shells that contains a magic number of neutrons (or protons). The "extra" nucleons move in quantized orbits, in a potential established by the central core, thus preserving the central feature of the independent particle model. These extra nucleons also interact with the core, deforming it and setting up "tidal wave" motions of rotation or vibration in it. These "liquid drop" motions of the core thus preserve the central feature of that model. The collective model has been remarkably successful and perhaps represents the limits of what we can hope for in nuclear physics, given the absence of a better theory.

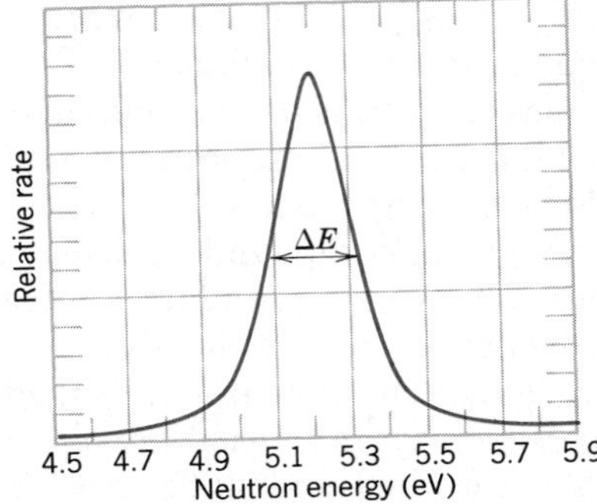

Figure 14 Sample Problem 10. A plot of the relative number of reaction events of the type described by Eq. 19, as a function of the energy of the incident neutron. The half-width ΔE of this resonance peak is about 0.20 eV.

Sample Problem 10 Consider the neutron-capture reaction

$$^{109}\text{Ag} + n \rightarrow {}^{110}\text{Ag}^* \rightarrow {}^{110}\text{Ag} + \gamma, \qquad (19)$$

in which a compound nucleus (^{110}Ag*) is formed. Figure 14 shows the relative rate at which such events take place, plotted against the energy of the incoming neutron. Use the uncertainty principle to find the mean lifetime of this compound nucleus.

We see that the relative production rate is sharply peaked at a neutron energy of about 5.2 eV. This suggests that we are dealing with a single excited level of the compound nucleus ^{110}Ag*. When the available energy just matches the energy of this level above the ^{110}Ag ground state, we have "resonance" and the reaction really "goes."

However, the resonance peak is not infinitely sharp, having an approximate half-width (see ΔE in the figure) of about 0.20 eV. We account for this by saying that the excited level is not sharply defined in energy, having an energy uncertainty ΔE of about 0.20 eV.

We can use the uncertainty principle, written in the form

$$\Delta E \cdot \Delta t \approx h \qquad (20)$$

to tell us something about any state of an atomic or nuclear system. We have seen that ΔE is a measure of the uncertainty with which the energy of the state can be defined. The quantity Δt, then, must be a measure of the time available to measure this energy. In fact, Δt is just $\bar{t}$, the mean life of the compound state before it decays to its ground state.

Numerically, we have

$$\Delta t = \bar{t} = \frac{h}{\Delta E} = \frac{4.14 \times 10^{-15}\ \text{eV} \cdot \text{s}}{0.20\ \text{eV}}$$
$$\approx 2 \times 10^{-14}\ \text{s}. \qquad \text{(Answer)}$$

This is just the order of magnitude of the lifetime that we expect for a compound nucleus.

REVIEW AND SUMMARY

The Nuclides

The nuclidic chart of Fig. 4 shows the ~2000 *nuclides* (atomic nuclei) that are known to exist. They are characterized by a *proton number Z* (the number of protons), a *neutron number N,* and a *mass number A* (the total number of *nucleons*—protons and neutrons). Thus, $A = Z + N$. Nuclides with the same atomic number but different neutron numbers are *isotopes* of each other. Nuclei have a mean radius R given by

Nuclear Radii

$$R = R_0 A^{1/3}, \qquad [3]$$

where $R_0 = 1.2$ fm (1 fm = 1 femtometer = 1 fermi = 10^{-15} m); see Sample Problems 1 and 2.

Mass–Energy Exchanges

Nuclear masses, universally reported as the masses of the corresponding neutral atoms, are useful in calculating disintegration energies, binding energies, and so on. The energy equivalent of one mass unit (u) is 932 MeV. The curve of binding energies (see Fig. 6 and Sample Problem 3) shows that middle-mass nuclides are the most stable and that energy can be released both by fission of heavy nuclei and by fusion of light nuclei.

The Nuclear Force

Nuclei are held together by an attractive force acting between the nucleons. It is thought to be a residual effect of the *strong* force acting between the quarks that make up the nucleons. Nuclei can exist in a number of discrete energy states (see Fig. 7) each with a characteristic intrinsic angular momentum and magnetic moment.

Radioactive Decay

Most known nuclides are radioactive; they spontaneously decay at a rate $R\,(= -dN/dt)$ that is proportional to the number N of radioactive atoms present, the proportionality constant being the *disintegration constant* λ. This leads to the law of exponential decay:

$$N = N_0 e^{-\lambda t} \qquad R = \lambda N = R_0 e^{-\lambda t} \quad \text{(radioactive decay)}. \qquad [6,7]$$

See Fig. 8. The half-life $\tau = (\ln 2)/\lambda$ is the time after which the decay rate R (or the number N) has dropped to half of its initial value; see Sample Problems 4 and 5.

α Decay

For some nuclides the emission of an α particle is energetically possible; see Sample Problem 6. Decay is inhibited by a potential barrier, as in Fig. 9, that cannot be penetrated according to classical mechanics but can be according to wave mechanics. The barrier penetrability, and thus the half-life for α decay, are very sensitive to the energy of the α particle, as Table 2 shows.

β Decay

In *β decay* an electron is emitted, along with a neutrino. They share the available disintegration energy between them; see Sample Problem 7. The emitted electrons have a continuous spectrum of energies up to a limit K_{max} $(= Q = \Delta m\, c^2)$; see Fig. 10.

The Valley of the Nuclides

New light is thrown on nuclear stability by constructing a "valley of the nuclides" as in Fig. 12, in which the nuclear mass excess is plotted vertically on a Z–N chart. The stability zone runs along the bottom of the valley and chains of β-decay processes descend its side walls.

Radioactive Dating

Naturally-occurring radioactive nuclides provide a means for estimating the dates of historic and prehistoric events. For example, organic materials can often be dated by measuring their ^{14}C content; rock samples can be dated using radioactive ^{40}K. See Sample Problem 8.

Radiation Dosage

Four units are used to describe exposure to ionizing radiation. The *curie* (1 Ci = 3.7×10^{10} disintegrations per second) measures the *activity* of a *source.* The *roentgen* (the amount of radiation that would deliver 8.78 mJ of energy per kg of dry air) is a unit of *exposure.* The amount of energy actually absorbed is measured in *rads,* with 1 rad corresponding to 10 mJ/kg. The estimated biological effect of the absorbed energy is quoted in *rems,* so that 1 rem of slow-neutron energy would cause the same biological effect as 1 rem of x rays even though only about $\frac{1}{5}$ as much energy was absorbed from the neutrons. See Sample Problem 9.

The Liquid Drop Model

The *liquid drop* model of nuclear structure assumes that nucleons collide constantly and that a long-lived *compound nucleus* is formed in nuclear reactions. See Eq. 18 and the formation modes–decay modes diagram on page 1091.

The Independent Particle Model

The *independent particle* model of nuclear structure assumes that each nucleon moves without collisions in a quantized orbit within the nucleus. The model predicts nucleon levels and *magic numbers* of nucleons (2, 8, 20, 28, 50, 82, and 126); these predict that nuclides with any of these numbers of neutrons or protons are particularly stable.

The Collective Model

The *collective* model, in which extra nucleons move in quantized orbits about a central core of closed shells, is highly successful in predicting many nuclear properties.

QUESTIONS

1. When a thin foil is bombarded with α particles, a few of them are scattered back toward the source. Rutherford concluded from this that the positive charge of the atom—and also most of its mass—must be concentrated in a very small "nucleus" within the atom. What was his line of reasoning?

2. In what ways do the so-called strong force and the electrostatic or Coulomb force differ?

3. Why does the *relative* importance of the Coulomb force compared to the strong nuclear force increase at large mass numbers?

4. In your body, are there more neutrons than protons? More protons than electrons? Discuss.

5. Why do nuclei tend to have more neutrons than protons at high mass numbers?

6. Why do we use atomic rather than nuclear masses in analyzing nuclear decay and reaction processes?

7. How might the equality $1\ \text{u} = 1.661 \times 10^{-27}$ kg be arrived at in the laboratory?

8. The atoms of a given element may differ in mass, have different physical characteristics, and yet not vary chemically. Why is this?

9. The deviation of isotopic masses from integer values is due to many factors. Name some. Which is most responsible?

10. How is the mass of the neutron determined?

11. The most stable nuclides have a mass number A near 60 (see Fig. 6). Why don't *all* nuclides have mass numbers near 60?

12. If we neglect the very lightest nuclides, the binding energy per nucleon in Fig. 6 is roughly constant at 7 to 8 MeV/nucleon. Do you expect the mean electronic binding energy per electron in atoms also to be roughly constant throughout the periodic table?

13. Why is the binding energy per nucleon (Fig. 6) low at low mass numbers? at high mass numbers?

14. In the binding-energy curve of Fig. 6, what is special or notable about the nuclides ^{2}H, ^{4}He, ^{56}Fe, and ^{239}Pu?

15. A particular ^{238}U nucleus was created in a massive stellar explosion, perhaps 10^{10} y ago. It suddenly decays by α emission while we are observing it. After all those years, why did it decide to decay at this particular moment?

16. Can you justify this statement: "In measuring half-lives by the method of Sample Problem 4, it is not necessary to measure the absolute decay rate R; any quantity proportional to it will suffice. However, in the method of Sample Problem 5 an absolute rate *is* needed."

17. Does the temperature affect the rate of decay of radioactive nuclides? If so, how?

18. You are running longevity tests on light bulbs. Do you expect their "decay" to be exponential? What is the essential difference between the decay of light bulbs and of radionuclides?

19. The half-life of ^{238}U is 4.47×10^{9} y, about the age of the solar system. How can such a long half-life be measured?

20. The half-life of ^{238}U is 4.47×10^{9} years. How might measurements of its activity be used to determine the age of uranium-containing rocks? How do you get around the fact that you don't know how much ^{238}U was present in the rocks to begin with? (*Hint*: What is the ultimate decay product of ^{238}U?)

21. The half-life of ^{14}C is 5730 years. This isotope is produced in the upper atmosphere at an assumed constant rate by cosmic ray bombardment. How might measurements of its activity play a role in dating ancient carbon-containing specimens, such as wooden Egyptian artifacts or the remnants of ancient camp fires?

22. Explain why, in α-decay, short half-lives correspond to large disintegration energies, and conversely.

23. A radioactive nucleus can emit a positron, e^{+}. This corresponds to a proton in the nucleus being converted to a neutron. The mass of a neutron, however, is greater than that of a proton. How then can positron emission occur?

24. In beta decay the emitted electrons form a continuous spectrum, but in alpha decay they form a discrete spectrum. What difficulties did this cause in the explanation of beta decay, and how were these difficulties finally overcome?

25. How do neutrinos differ from photons? Each has zero charge and (presumably) zero rest mass and travels at the speed of light.

26. In the development of our understanding of the atom, did we use atomic models as we now use nuclear models? Is Bohr's theory such an atomic model? Are models now used in atomic physics? What is the difference between a model and a theory?

27. What are the basic assumptions of the liquid drop and the independent particle models of nuclear structure? How do they differ? How does the collective model reconcile these differences?

28. Does the liquid drop model of the nucleus give us a picture of the following phenomena: (*a*) acceptance by the nucleus of a colliding particle; (*b*) loss of a particle by spontaneous emission; (*c*) fission; (*d*) dependence of stability on energy content?

29. What is so special ("magic") about the magic nucleon numbers?

30. Why aren't the magic nucleon numbers and the magic electron numbers the same? What accounts for each?

31. The average number of stable (or very long-lived) isotopes of the noble gases is 3.7. The average number of stable nuclides for the four magic neutron numbers, however, is 5.8, considerably greater. If the noble gases are so stable, why were not more stable isotopes of them created when the elements were formed?

EXERCISES AND PROBLEMS

Section 47-1 Discovering the Nucleus

1E. Calculate the distance of closest approach for a head-on collision between a 5.30-MeV α particle and the nucleus of a copper atom.

2E. Assume that a gold nucleus has a radius of 6.23 fm (see Table 1), and an α particle has a radius of 1.8 fm. What minimum energy must an incident α particle have to penetrate the gold nucleus?

3P. When an α particle collides elastically with a nucleus the nucleus recoils. A 5.00-MeV α particle has a head-on elastic collision with a gold nucleus, initially at rest. What is the kinetic energy (*a*) of the recoiling nucleus? (*b*) of the rebounding α particle?

Section 47-2 Some Nuclear Properties

4E. A neutron star is a stellar object whose density is about that of nuclear matter, as calculated in Sample Problem 2. Suppose that the sun were to collapse into such a star without losing any of its present mass. What would be its expected radius?

5E. The nuclide ^{14}C contains how many (*a*) protons and (*b*) neutrons?

6E. The radius of a nucleus is measured, by electron-scattering methods, to be 3.6 fm. What is the likely mass number of the nucleus?

7E. Locate the nuclides displayed in Table 1 on the nuclidic chart of Fig. 4. Verify that, as intended, they lie along the stability zone.

8E. Using the nuclidic chart of Fig. 4, write the symbols for (*a*) all stable nuclides (isotopes) with $Z = 60$, (*b*) all radioactive nuclides (isotones) with $N = 60$, and (*c*) all nuclides (isobars) with $A = 60$.

9E. The electrostatic potential energy of a uniform sphere of charge Q and radius R is

$$U = \frac{3Q^2}{20\pi\epsilon_0 R}.$$

(*a*) Find the electrostatic potential energy for the nuclide ^{239}Pu, assumed spherical; see Table 1. (*b*) Compare its electrostatic potential energy per particle with its binding energy per nucleon of 7.56 MeV. (*c*) What do you conclude?

10E. The strong neutron excess of heavy nuclei is illustrated by the fact that most heavy nuclides could never fission or break up into two stable nuclei without neutrons being left over. For example, consider the spontaneous fission of a ^{235}U nucleus. If the two daughter nuclei had atomic numbers 39 and 53, and were stable, by referring to Fig. 4, determine the daughter nuclides and the number of neutrons left over.

11E. Arrange the 25 nuclides given below in squares as a nuclidic chart similar to Fig. 5. Draw in and label (*a*) all isobaric (constant A) lines and (*b*) all lines of constant neutron excess, defined as $N-Z$. Consider nuclides $^{118-122}Te$, $^{117-121}Sb$, $^{116-120}Sn$, $^{115-119}In$, and $^{114-118}Cd$.

12E. Calculate and compare (*a*) the nuclear mass density ρ_m and (*b*) the nuclear charge density ρ_q for a fairly light nuclide such as ^{55}Mn and for a fairly heavy one such as ^{209}Bi. (*c*) Are the differences what you would expect?

13E. Verify that the binding energy per nucleon given in Table 1 for ^{239}Pu is indeed 7.56 MeV/nucleon. The needed atomic masses are 239.05216 u (^{239}Pu), 1.00783 u (^{1}H), and 1.00867 u (neutron).

14E. (*a*) Show that an approximate formula for the mass M of an atom is

$$M = Am_p,$$

where A is the mass number and m_p is the proton mass. (*b*) What percent error is committed in using this formula to calculate the masses of the atoms in Table 1? The mass of the bare proton is 1.007276 u. (*c*) Is this formula accurate enough for calculations of nuclear binding energy?

15E. *The characteristic nuclear time.* The characteristic nuclear time is a useful but loosely defined quantity, taken to be the time required for a nucleon with a few MeV of kinetic energy to travel a distance equal to the diameter of a middle-mass nuclide. What is the order of magnitude of this quantity? Consider 5-MeV neutrons traversing a nuclear diameter of ^{197}Au; see Table 1.

16E. Nuclear radii may be measured by scattering high-energy electrons from nuclei. (*a*) What is the de Broglie wavelength for 200-MeV electrons? (*b*) Are they suitable probes for this purpose?

17E. Because a nucleon is confined to a nucleus, we can take its uncertainty in position to be approximately the nuclear radius R. What does the uncertainty principle say about the kinetic energy of a nucleon in a nucleus with, say, $A = 100$? (*Hint:* Take the uncertainty in momentum Δp to be the actual momentum p.)

18E. The atomic masses of ^{1}H, ^{12}C, and ^{238}U are 1.007825 u, 12.000000 u (by definition), and 238.050785 u, respectively. (*a*) What would these masses be if the mass unit were defined so that the mass of ^{1}H was (exactly) 1.000000 u? (*b*) Use your result to suggest why this perhaps obvious choice was not made.

19P. (*a*) Convince yourself that the energy tied up in nuclear, or strong force, bonds is proportional to A, the mass number of the nucleus in question. (*b*) Convince yourself that the energy tied up in Coulomb force bonds between the protons is proportional to $Z(Z-1)$. (*c*) Show that, as we move to larger and larger nuclei (see Fig. 4), the importance of (*b*) above increases more rapidly than does that of (*a*).

20P. In the periodic table, the entry for magnesium is:

12 Mg 24.312

There are three isotopes:

^{24}Mg, atomic mass = 23.98504 u.
^{25}Mg, atomic mass = 24.98584 u.
^{26}Mg, atomic mass = 25.98259 u.

The abundance of ^{24}Mg is 78.99% by weight. Calculate the abundances of the other two isotopes.

21P. You are asked to pick apart an α particle (^{4}He) by removing, in sequence, a proton, a neutron, and a proton. Calculate (*a*) the work required for each step, (*b*) the total binding energy of the α particle, and (*c*) the binding energy per particle. Some needed atomic masses are

^{4}He 4.00260 u, ^{2}H 2.01410 u,
^{3}H 3.01605 u, ^{1}H 1.00783 u,
n 1.00867 u.

22P. Because the neutron has no charge, its mass must be found in some way other than by using a mass spectrometer. When a resting neutron and a proton meet, they combine and form a deuteron, emitting a γ ray whose energy is 2.2233 MeV. The atomic masses of the proton and the deuteron are 1.007825035 u and 2.0141019 u, respectively. Find the mass of the neutron from these data, to as many significant figures as the data warrant. (A more precise value of the mass–energy conversion factor than that used in the text is 931.502 MeV/u.)

23P. A penny has a mass of 3.0 g. Calculate the nuclear energy that would be required to separate all the neutrons and protons in this coin. Ignore the binding energy of the electrons. For simplicity assume that the penny is made entirely of ^{63}Cu atoms (mass = 62.92960 u). The atomic masses of the proton and the neutron are 1.00783 u and 1.00867 u, respectively.

24P. *Mass excess.* To simplify calculations, atomic masses are sometimes tabulated, not as the actual atomic mass m, but as $(m - A)c^2$, where A is the mass number expressed in mass units. This quantity, usually reported in MeV, is called the mass excess, symbol Δ. Using data from Sample Problem 3, find the mass excesses for (*a*) ^{1}H, (*b*) the neutron, and (*c*) ^{120}Sn.

25P. *Mass excess* (See Problem 24). Show that the total binding energy of a nuclide can be written as

$$E = Z\Delta_{\mathrm{H}} + N\Delta_n - \Delta,$$

where Δ_{H}, Δ_n, and Δ are the appropriate mass excesses. Using this method calculate the binding energy per particle for ^{197}Au. Compare your result with the value listed in Table 1. The needed mass excesses, rounded to three significant figures, are $\Delta_{\mathrm{H}} = +7.29$ MeV, $\Delta_n = +8.07$ MeV, and $\Delta_{197} = -31.2$ MeV. Δ_{H} is the mass excess of ^{1}H. Note the economy of calculation that results when mass excesses are used in place of the actual masses.

Section 47–3 Radioactive Decay

26E. The half-life of a particular radioactive isotope is 6.5 h. If there are initially 48×10^{19} atoms of this isotope, how many atoms of this isotope remain after 26 h?

27E. The half-life of a radioactive isotope is 140 d. How many days would it take for the activity of a sample of this isotope to fall to one-fourth of its initial decay rate?

28E. A radioactive nuclide has a half-life of 30 y. What fraction of an initially pure sample of this nuclide will remain undecayed after (*a*) 60 y? (*b*) 90 y?

29E. Gallium ^{67}Ga has a half-life of 78 h. Consider an initially pure 3.4-g sample of this isotope. (*a*) What is its activity? (*b*) What is its activity 48 h later?

30E. A radioactive isotope of mercury, ^{197}Hg, decays into gold, ^{197}Au, with a decay constant of 0.0108 h^{-1}. (*a*) Calculate its half-life. (*b*) What fraction of the original amount will remain after three half-lives? (*c*) after 10 days?

31E. From data presented in the first few paragraphs of Section 47–3, deduce (*a*) the disintegration constant λ and (*b*) the half-life of ^{238}U.

32E. The plutonium isotope ^{239}Pu is produced as a by-product in nuclear reactors and hence is accumulating in our environment. It is radioactive, decaying by alpha decay with a half-life of 2.41×10^4 y. But plutonium is also one of the most toxic chemicals known; as little as 2 mg is lethal to a human. (*a*) How many nuclei constitute a chemically lethal dose? (*b*) What is the decay rate of this amount? If you were handling that quantity would you fear being poisoned or suffering radiation sickness?

33E. Cancer cells are more vulnerable to x and γ radiation than are healthy cells. Though linear accelerators are now replacing it, in the past the standard source for radiation therapy has been radioactive ^{60}Co, which beta decays into an excited nuclear state of ^{60}Ni, which immediately drops into the ground state, emitting two γ-ray photons, each of approximate energy 1.2 MeV. The controlling beta-decay half-life is 5.27 y. How many radioactive ^{60}Co nuclei are present in a 6000-Ci source used in a hospital? (1 Ci = 1 curie = 3.7×10^{10} disintegrations/s.)

34P. After long effort, in 1902, Marie and Pierre Curie succeeded in separating from uranium ore the first substantial quantity of radium, one decigram of pure $RaCl_2$. The radium was the radioactive isotope ^{226}Ra, which has a decay half-life of 1600 y. (*a*) How many radium nuclei had they isolated? (*b*) What was the decay rate of their sample, in disintegrations/s? In curies? [The unit *curie* (abbr. Ci) was named in honor of the Curies, who received the 1903 Nobel prize in physics for their work on radiation phenomena. One curie equals 3.7×10^{10} disintegrations/s.]

35P. The radionuclide ^{64}Cu has a half-life of 12.7 h. How much of an initially pure 5.5-g sample of ^{64}Cu will decay during the two hour period beginning 14 h later?

36P. The radionuclide ^{32}P ($\tau = 14.28$ d) is often used as a tracer to follow the course of biochemical reactions involving phosphorus. (*a*) If the counting rate in a particular experimental setup is 3050 counts/s, after what time will it fall to 170 counts/s? (*b*) A solution containing ^{32}P is fed to the root system of an experimental tomato plant and the ^{32}P activity in a leaf is measured 3.48 days later. By what factor must this reading be multiplied to correct for the decay that has occurred since the experiment began?

37P. A source contains two phosphorus radionuclides, ^{32}P ($\tau = 14.3$ d) and ^{33}P ($\tau = 25.3$ d). Initially 10% of the decays come from ^{33}P. How long must one wait until 90% do so?

38P. A 1.00-g sample of samarium emits α particles at a rate of 120 particles/s. ^{147}Sm, whose natural abundance in bulk samarium is 15.0%, is the responsible isotope. Calculate the half-life for the decay process.

39P. Plutonium ^{239}Pu decays by α decay with a half-life of 24,100 y. How many grams of helium are produced by an initially pure 12-g sample of ^{239}Pu after 20,000 y? (Recall that an α particle is a helium nucleus.)

40P. Calculate the mass of 4.6 μCi of ^{40}K, which has a half-life of 1.28×10^{9} y. (1 Ci = 1 curie = 3.7×10^{10} disintegrations/s.)

41P. One of the dangers of radioactive fallout from a nuclear bomb is ^{90}Sr, which beta decays with a 29-year half-life. Because it has chemical properties much like calcium, the strontium, if eaten by a cow, becomes concentrated in its milk and ends up in the bones of whoever drinks the milk. The energetic decay electrons damage the bone marrow and thus impair the production of red blood cells. A 1-megaton bomb produces approximately 400 g of ^{90}Sr. If the fallout spreads uniformly over a 2000-km^2 area, what area would have radioactivity equal to the allowed bone burden for one person of 0.002 mCi? (One curie = 1 Ci = 3.7×10^{10} disintegrations/s.)

42P. After a brief neutron irradiation of silver, two activities are present: ^{108}Ag ($\tau = 2.42$ min) with an initial decay rate of 3.1×10^{5}/s, and ^{110}Ag ($\tau = 24.6$ s) with an initial decay rate of 4.1×10^{6}/s. Make a semilog plot similar to Fig. 8 showing the total combined decay rate of the two isotopes as a function of time from $t = 0$ until $t = 10$ min. In Fig. 8, the extraction of the half-life for simple decays was illustrated. Given only the plot of total decay rate, can you suggest a way to analyze it in order to find the half-lives of both isotopes?

43P. *Secular equilibrium.* A certain radionuclide is being manufactured, say, in a cyclotron, at a constant rate R. It is also decaying, with a disintegration constant λ. Let the production process continue for a time that is long compared to the half-life of the radionuclide. Convince yourself that the number of radioactive nuclei present at such times will be constant and will be given by $N = R/\lambda$. Convince yourself further that this result holds no matter how many of the radioactive nuclei were present initially. The nuclide is said to be in secular equilibrium with its source; in this state its decay rate is just equal to its production rate.

44P. *Secular equilibrium* (see Problem 43). The radionuclide ^{56}Mn has a half-life of 2.58 h and is produced in a cyclotron by bombarding a manganese target with deuterons. The target contains only the stable manganese isotope ^{55}Mn and the reaction that produces ^{56}Mn is

$$^{55}\text{Mn} + d \rightarrow {}^{56}\text{Mn} + p.$$

After being bombarded for a time $\gg 2.58$ h, the activity of the target, due to ^{56}Mn, is 2.4 curie (Ci). (*a*) At what constant rate R are ^{56}Mn nuclei being produced in the cyclotron during the bombardment? (*b*) At what rate are they decaying (also during the bombardment)? (*c*) How many ^{56}Mn nuclei are present at the end of the bombardment? (*d*) What is their total mass?

45P. *Secular equilibrium* (see Problems 43 and 44). A radium source contains 1.00 mg of ^{226}Ra, which decays with a half-life of 1600 y to produce ^{222}Rn, a noble gas. This radon isotope in turn decays by α-emission with a half-life of 3.82 d. (*a*) What is the rate of disintegration of ^{226}Ra in the source? (*b*) How long does it take for the radon to come to secular equilibrium with its radium parent? (*c*) At what rate is the radon then decaying? (*d*) How much radon is in equilibrium with its radium parent?

Section 47-4 Alpha Decay

46E. Consider a ^{238}U nucleus to be made up of an α particle (^{4}He) and a residual nucleus (^{234}Th). Plot the electrostatic potential energy $U(r)$, where r is the distance between these particles. Cover the range ≈ 10 fm $< r < \approx 100$ fm and compare your plot with that of Fig. 9.

47E. Generally speaking, heavier nuclides tend to be more unstable to alpha decay. For example, the most stable isotope of uranium, ^{238}U has an alpha decay half-life of 4.5×10^9 y. The most stable isotope of plutonium is ^{244}Pu with a 8.2×10^7 y half-life, and for curium we have ^{248}Cm and 3.4×10^5 y. When half of an original sample of ^{238}U has decayed, what fractions of the original isotopes of plutonium and curium are left?

48P. A ^{238}U nucleus emits an α particle of energy 4.196 MeV. Calculate the disintegration energy Q for this process, taking the recoil energy of the residual ^{234}Th nucleus into account.

49P. Consider that a ^{238}U nucleus emits (a) an α particle or (b) a sequence of neutron, proton, neutron, proton. Calculate the energy released in each case. (c) Convince yourself both by reasoned argument and also by direct calculation that the difference between these two numbers is just the total binding energy of the α particle. Find that binding energy. Some needed atomic masses are

^{238}U	238.05079 u	^{4}He	4.00260 u
^{237}U	237.04873 u	^{1}H	1.00783 u
^{236}Pa	236.04891 u	n	1.00867 u
^{235}Pa	235.04544 u		
^{234}Th	234.04363 u		

50P. Under certain circumstances, a nucleus can decay by emitting a particle heavier than an α-particle. Such decays are very rare and have only recently been observed. Consider the decays

$$^{223}Ra \rightarrow {}^{209}Pb + {}^{14}C$$

and

$$^{223}Ra \rightarrow {}^{219}Rn + {}^{4}He.$$

(a) Calculate the Q-values for these decays and determine that both are energetically possible. (b) The Coulomb barrier height for α particles in this decay is 30 MeV. What is the barrier height for ^{14}C decay?

Masses:

^{223}Ra	223.01850 u	^{14}C	14.00324 u
^{209}Pb	208.98107 u	^{4}He	4.00260 u
^{219}Rn	219.01008 u		

51P. Heavy radionuclides emit an α particle rather than other combinations of nucleons because the α particle is such a stable, tightly-bound structure. To confirm his, calculate the disintegration energies for these hypothetical decay processes and discuss the meaning of your findings:

$$^{235}U \rightarrow {}^{232}Th + {}^{3}He \qquad Q_3,$$
$$^{235}U \rightarrow {}^{231}Th + {}^{4}He \qquad Q_4,$$
$$^{235}U \rightarrow {}^{230}Th + {}^{5}He \qquad Q_5.$$

The needed atomic masses are

^{232}Th	232.0381 u	^{3}He	3.0160 u
^{231}Th	231.0363 u	^{4}He	4.0026 u
^{230}Th	230.0331 u	^{5}He	5.0122 u
^{235}U	235.0439 u		

Section 47-5 Beta Decay

52E. A certain stable nuclide, after absorbing a neutron, emits a negative electron and then splits spontaneously into two α particles. Identify the nuclide.

53E. ^{137}Cs is present in the fallout from above-ground detonations of nuclear bombs. Because it beta-decays with a slow 30.2-y half-life into ^{137}Ba, releasing considerable energy in the process, it is an environmental concern. The atomic masses of the Cs and Ba are 136.9073 u and 136.9058 u, respectively; calculate the total energy released in the decay.

54E. Heavy radionuclides, which may be either α or β emitters, belong to one of four decay chains, depending on whether their mass numbers A are of the form $4n$, $4n + 1$, $4n + 2$, or $4n + 3$, where n is a positive integer. (a) Justify this statement and show that if a nuclide belongs to one of these families, all its decay products will belong to the same family. (b) Classify these nuclides as to family: ^{235}U, ^{236}U, ^{238}U, ^{239}Pu, ^{240}Pu, ^{245}Cm, ^{246}Cm, ^{249}Cf, and ^{253}Fm.

55E. A free neutron decays according to Eq. 12. If the neutron—hydrogen atom mass difference is 840 μu, what is the maximum energy of the β-spectrum?

56E. An electron is emitted from a middle-mass nuclide ($A = 150$, say) with a kinetic energy of 1.0 MeV. (a) What is its de Broglie wavelength? (b) Calculate the radius of the emitting nucleus. (c) Can such an electron be confined as a standing wave in a "box" of such dimensions? (d) Can you use these numbers to disprove the argument (long since abandoned) that electrons actually exist in nuclei?

57P. *Electron capture.* Some radionuclides decay by capturing one of their own atomic electrons, a K-electron, say. An example is

$$^{49}V + e^- \rightarrow {}^{49}Ti + \nu \qquad \tau = 331 \text{ d}.$$

Show that the disintegration energy Q for this process is given by

$$Q = (m_V - m_{Ti})c^2 - E_K,$$

where m_V and m_{Ti} are the atomic masses of ^{49}V and ^{49}Ti, respectively, and E_K is the binding energy of the vanadium K-electron. (*Hint:* Put $\mathbf{m}_V$ and $\mathbf{m}_{Ti}$ as the corresponding nuclear masses and proceed as in Sample Problem 7.)

58P. *Electron capture.* Find the disintegration energy Q for the decay of ^{49}V by K-electron capture, as described in Problem 57. The needed data are $m_V = 48.94852$ u, $m_{Ti} = 48.94787$ u, and $E_K = 5.47$ keV.

59P. *Q for positron decay.* The radionuclide ^{11}C decays according to

$$^{11}\mathrm{C} \rightarrow {}^{11}\mathrm{B} + e^+ + \nu \qquad \tau = 20.3 \text{ min}.$$

The maximum energy of the positron spectrum is 0.960 MeV. (*a*) Show that the disintegration energy Q for this process is given by

$$Q = (m_C - m_B - 2m_e)c^2,$$

where m_C and m_B are the atomic mass of ^{11}C and ^{11}B, respectively and m_e is the electron (positron) mass. (*b*) Given that $m_C = 11.011434$ u, $m_B = 11.009305$ u, and $m_e = 0.0005486$ u, calculate Q and compare it with the maximum energy of the positron spectrum, given above. (*Hint:* Let $\mathbf{m}_C$ and $\mathbf{m}_B$ be the nuclear masses and proceed as in Sample Problem 7 for beta decay. Note that positron decay is an exception to the general rule that, if atomic masses are used in nuclear decay processes, the mass of the emitted electron is automatically taken care of.)

60P. Two radioactive materials that are unstable to alpha decay, ^{238}U and ^{232}Th, and one that is unstable to beta decay, ^{40}K, are sufficiently abundant in granite to contribute significantly to the heating of the earth through the decay energy produced. The α-unstable isotopes give rise to decay chains that stop at stable lead isotopes. ^{40}K has a single beta decay. Decay information follows:

Parent Nuclide	*Decay Mode*	*Half-life* (y)	*Stable Endpoint*	*Q* (MeV)	*f* (ppm)
^{238}U	α	4.47×10^9	^{206}Pb	51.7	4
^{232}Th	α	1.41×10^{10}	^{208}Pb	42.7	13
^{40}K	β	1.25×10^9	^{40}Ca	1.31	4

Q is the total energy released in the decay of one parent nucleus to the final stable endpoint and f is the abundance of the isotope in kilograms per kilogram of granite; ppm means parts per million. (*a*) Show that these materials give rise to a total heat production of 9.8×10^{-10} W for each kilogram of granite. (*b*) Assuming that there is 2.7×10^{22} kg of granite in a 20-km-thick, spherical shell around the earth, estimate the power this will produce over the whole earth. Compare this with the total solar power intercepted by the earth, 1.7×10^{17} W.

61P*. *Recoil during beta decay.* The radionuclide ^{32}P decays to ^{32}S as described by Eq. 10. In a particular decay event, a 1.71-MeV electron is emitted, the maximum possible value. What is the kinetic energy of the recoiling ^{32}S atom in this event? (*Hint:* For the electron it is necessary to use the relativistic expressions for the kinetic energy and the linear momentum. Newtonian mechanics may safely be used for the relatively slow-moving ^{32}S atom.)

Section 47-6 Radioactive Dating

62E. ^{238}U decays to ^{206}Pb with a half-life of 4.47×10^9 y. Although the series has many individual steps, the first has by far the longest half-life; therefore, one can often consider the decay to go directly to lead. That is,

$$^{238}\mathrm{U} \rightarrow {}^{206}\mathrm{Pb} + \text{various decay products.}$$

A rock is found to contain 4.2 mg of ^{238}U and 2.135 mg of ^{206}Pb. Assume that the rock contained no lead at formation, all of the lead now present arising from the decay of uranium. (*a*) How many atoms of ^{238}U and ^{206}Pb does the rock now contain? (*b*) How many atoms of ^{238}U did the rock contain at formation? (*c*) What is the age of the rock?

63E. A 5.00-g charcoal sample from an ancient fire pit has a ^{14}C activity of 63.0 disintegrations/min. Carbon from a living tree has an activity of 15.3 disintegrations/min for 1.00 g. The half-life of ^{14}C is 5730 y. How old is the charcoal sample?

64P. A particular rock is thought to be 260 million years old. If it contains 3.7 mg of ^{238}U, how much ^{206}Pb should it contain? See Exercise 62.

65P. A rock, recovered from far underground, is found to contain 0.86 mg of ^{238}U, 0.15 mg of ^{206}Pb, and 1.6 mg of ^{40}A. How much ^{40}K will it very likely contain? Needed half-lives are listed in Problem 60P.

Section 47-7 Measuring Radiation Dosage

66E. A Geiger counter records 8700 counts in one minute. Calculate the activity of the source in Ci, assuming that the counter records all decays.

67E. The nuclide ^{198}Au, half-life = 2.7 d, is used in cancer therapy. Calculate the mass of this isotope required to produce an activity of 250 Ci.

68E. An airline pilot spends an average of 20 h per week flying at 35,000 ft, at which altitude the equivalent dose due to cosmic radiation is 0.7 mrem/h. What is his annual equivalent dose from this source alone? Note that the maximum permitted yearly equivalent dose (from all sources) for the general population is 500 mrem and for radiation workers it is 5000 mrem.

69E. A 75-kg person receives a whole-body radiation dose of 24 mrad, delivered by α particles for which the RBE factor is 12. Calculate (*a*) the absorbed energy in joules, and (*b*) the equivalent dose in rem.

70P. A typical chest x-ray radiation dose is 25 mrem, delivered by x rays with an RBE factor of 0.85. Assuming that the mass of the exposed tissue is one-half the patient's mass of 88 kg, calculate the energy absorbed in joules.

71P. An 85-kg worker at a breeder reactor plant accidentally ingests 2.5 mg of plutonium ^{239}Pu dust. ^{239}Pu has a half-life of 24,100 y, decaying by alpha decay. The energy of the emitted α particles is 5.2 MeV, with an RBE factor of 13. Assume that the plutonium resides in the worker's body for 12 h, and that 95% of the emitted α particles are stopped within the body. Calculate (*a*) the number of plutonium atoms ingested, (*b*) the number that decay during the 12 h, (*c*) the energy absorbed by the body, (*d*) the resulting physical dose in rad, and (*e*) the equivalent biological dose in rem.

Section 47-8 Nuclear Models

72E. An intermediate nucleus in a particular nuclear reaction decays within 10^{-22} s of its formation. (*a*) What is the uncertainty ΔE in our knowledge of this intermediate state? (*b*) Can this state be called a compound nucleus? See Sample Problem 10.

73E. A typical kinetic energy for a nucleon in a middle-mass nucleus may be taken as 5 MeV. To what effective nuclear temperature does this correspond, using the assumptions of the liquid drop model of nuclear structure? (*Hint:* See Eq. 14 in Chapter 21.)

74E. From the following list of nuclides, identify (*a*) those with filled nucleon shells, (*b*) those with one nucleon outside a filled shell, and (*c*) those with one vacancy in an otherwise filled shell. Nuclides: ^{13}C, ^{18}O, ^{40}K, ^{49}Ti, ^{60}Ni, ^{91}Zr, ^{92}Mo, ^{121}Sb, ^{143}Nd, ^{144}Sm, ^{205}Tl, and ^{207}Pb.

75P. Consider the three formation modes shown for the compound nucleus ^{20}Ne* on p. 1091. What energy must (*a*) the α particle, (*b*) the proton, and (*c*) the x ray photon have to provide 25.0 MeV of excitation energy to the compound nucleus? Some needed atomic masses are

^{20}Ne	19.99244 u	α	4.00260 u
^{19}F	18.99840 u	p	1.00783 u.
^{16}O	15.99491 u		

76P. Consider the three decay modes shown for the compound nucleus ^{20}Ne* on p. 1091. If the compound nucleus is initially at rest and has an excitation energy of 25 MeV, what kinetic energy, measured in the laboratory, will (*a*) the deuteron, (*b*) the neutron, and (*c*) the ^{3}He nuclide have when the nucleus decays? Some needed atomic masses are

^{20}Ne	19.99244 u	d	2.01410 u
^{19}Ne	19.00188 u	n	1.00867 u
^{18}F	18.00094 u	^{3}He	3.01603 u.
^{17}O	16.99913 u		

77P. The nuclide ^{208}Pb is "doubly magic" in that both its proton number Z (= 82) and its neutron number N (= 126) represent filled nucleon shells. An additional proton would yield ^{209}Bi and an additional neutron ^{209}Pb. These "extra" nucleons should be easier to remove than a proton or a neutron from the filled shells of ^{208}Pb. (*a*) Calculate the energy required to move the "extra" proton from ^{209}Bi and compare it with the energy required to remove a proton from the filled proton shell of ^{208}Pb. (*b*) Calculate the energy required to remove the "extra" neutron from ^{209}Pb and compare it with the energy required to remove a neutron from the filled neutron shell of ^{208}Pb. Do your results agree with expectation? Use these atomic mass data:

Nuclide	*Z*	*N*	*Atomic Mass,* u
^{209}Bi	82 + 1	126	208.9804
^{208}Pb	82	126	207.9767
^{207}Tl	82 − 1	126	206.9774
^{209}Pb	82	126 + 1	208.9811
^{207}Pb	82	126 − 1	206.9759

The masses of the proton and the neutron are 1.00783 u and 1.00867 u, respectively.

78P. The nucleus ^{91}Zr ($Z = 40$, $N = 51$) has a single neutron outside a filled 50-neutron core. Because 50 is a magic number, this neutron should perhaps be especially loosely bound. (*a*) What is its binding energy? (*b*) What is the binding energy of the next neutron, which must be extracted from the filled core? (*c*) What is the binding energy per particle for the entire nucleus? Compare these three numbers and discuss. Some needed atomic masses are

^{91}Zr	90.90564 u	n	1.00867 u
^{90}Zr	89.90471 u	p	1.00783 u
^{89}Zr	88.90890 u.		

79P. Verify the data for ^{121}Sb presented in Table 3. That is, calculate (*a*) the energy needed to remove a proton from a ^{121}Sb nucleus, and (*b*) the energy needed to remove a proton from the resulting ^{120}Sn nucleus. Needed atomic masses are

^{121}Sb	120.9038 u
^{120}Sn	119.9022 u
^{119}In	118.9058 u.

ESSAY 21
NUCLEAR MEDICINE

RUSSELL K. HOBBIE
UNIVERSITY OF MINNESOTA

RICHARD L. MORIN
UNIVERSITY OF MINNESOTA

Nuclear medicine uses small amounts of radioactive material introduced into the body as a diagnostic tool. Images can be produced that show the distribution of the radioactive substance in different parts of the body. The spatial resolution of nuclear medicine images is typically several millimeters. An x-ray image usually has higher resolution—a fraction of a millimeter. An x-ray image shows structural information—the anatomy of a region. The nuclear-medicine image shows physiological information—the amount of the radioactive pharmaceutical that has lodged in an organ of interest. It is sometimes possible to monitor the physiological process by measuring the radioactivity in a region as a function of time.

There are many radioactive nuclei that might be used for a diagnostic study. However, most of them are not effective. Since the radioactivity must be measured by a detector located outside the body, nuclei that emit only alpha or beta particles would not be useful, since the alpha or beta particles do not travel very far and would come to rest before escaping from the body. Many nuclei emit photons—gamma rays—often in conjunction with the emission of an alpha or beta particle. Since an appreciable fraction of the gamma photons escape from the body, they can be detected.

The half-life of the nucleus is also important. It must live long enough so that it can be produced, administered to the patient, and measured. On the other hand (assuming that all the nuclei remain within the body), we would like the nuclei to decay fairly rapidly. Only those that decay during the measurement provide useful information. Those that decay later are "wasted." They contribute nothing to the image, but they increase the radiation dose received by the patient.

Two commonly used isotopes are ^{99m}Tc ("Technetium-99-m") and ^{131}I ("Iodine-131").

Technetium has a half life of 6 hours and emits a 140-keV gamma ray. (About 10% of the decay energy is in the form of other radiation that does not escape from the body.) As its name ("technical") suggests, Technetium does not occur naturally. It is formed from the decay of ^{99}Mo, which is a fission product (Chapter 48) produced in nuclear reactors. The ^{99}Mo, which has a half-life of 67 hours, is shipped once or twice a week to hospitals around the country. Each day the hospital extracts the ^{99m}Tc that has formed and uses it for the day's diagnostic studies. The site within the body to which the technetium will go is controlled by attaching it to a chemical with the desired pharmacological properties. For example, it can be attached to tiny microspheres of albumin to study blood flow in capillaries. It can be combined with organic phosphorous for bone studies, or it can be attached to colloidal sulfur particles for liver studies.

The isotope ^{131}I is an excellent radiopharmaceutical for studying the thyroid gland. The thyroid is located in the neck. It produces two hormones that regulate the rate of metabolism in the body. Both of these hormones contain iodine. Iodine-131 decays by a combination of γ and β emission with a half-life of 8 days. Most of the gamma rays have an energy of 364 keV. It is also produced as a fission product in a nuclear reactor.

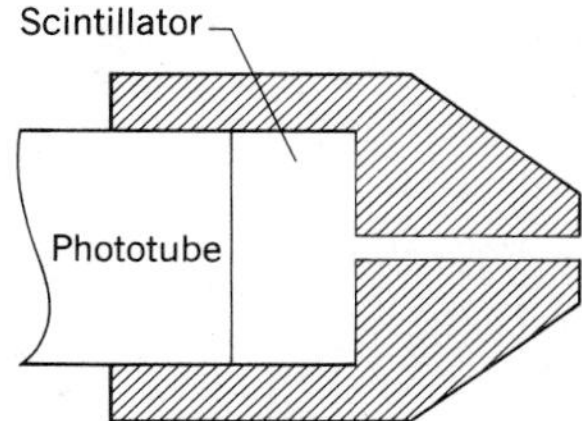

Figure 1 A scintillator is viewed by a photomultiplier tube. A cylindrical lead collimator gives directional sensitivity. Reproduced by permission from R. K. Hobbie, *Intermediate Physics for Medicine and Biology*, Wiley, 1978.

The most commonly used detector is a scintillation counter. Certain substances scintillate: they give off visible light when struck by an α or β particle or a photon. A very sensitive photomultiplier tube uses the photoelectric effect (Chapter 43) to detect this light. Figure 1 shows a photomultiplier tube viewing a piece of scintillator. The scintillator is protected by a lead collimator, so that only photons coming from a certain direction can strike the scintillator. This detector assembly could be placed to view the kidney, and the radioactivity in the kidney could be monitored after administration of a radiopharmaceutical that passes through the kidney. This kind of detector was used extensively in the early days of nuclear medicine. It is not very efficient. Because its field of view is small compared to the size of many organs, many decays go undetected, resulting in long study times and a high dose to the patient.

The *scintillation camera* operates on the same principle and is much more efficient. A large scintillator, typically 0.25 to 0.4 m in diameter, is covered on one side by a collimator with many holes in it. The other side is viewed by an array of 19 or 34 photomultiplier tubes, as shown in Fig. 2. The photomultiplier tubes are connected to electronics that can deduce the location of the flash of light due to a photon interaction within the scintillator by comparing how much of the light arrives at different photomultiplier tubes. The electronics then records the "count" in a computer memory. After the study the spatial distribution of the radioactivity is displayed. A scintillation camera is shown in Fig. 3.

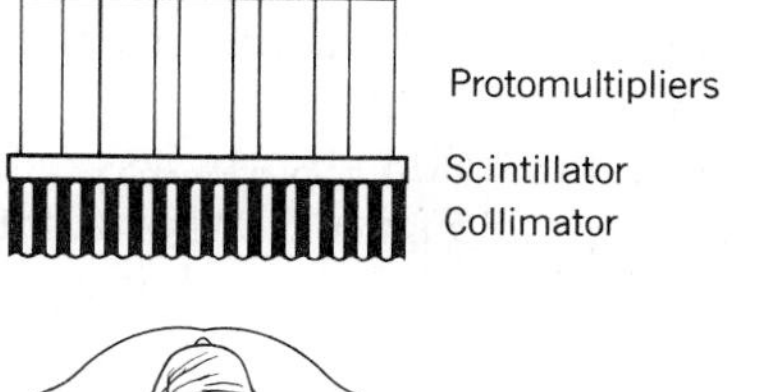

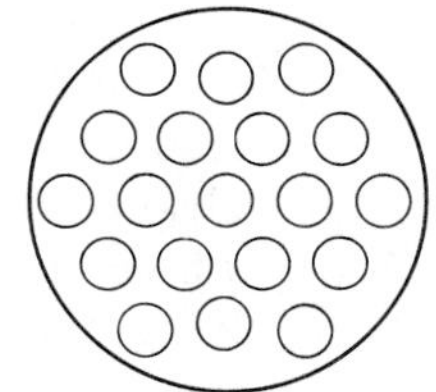

Figure 2 Schematic of a scintillation camera. A collimator allows photons from the patient to strike the scintillator directly above their source. An array of photomultiplier tubes records photon position and energy. Reproduced by permission from R. K. Hobbie, *Intermediate Physics for Medicine and Biology*, Wiley, 1978.

After a radiopharmaceutical is injected, the radioactivity usually moves to certain organs, depending on the pharmacologic properties of the radioactive substance. In a few cases, the radiopharmaceutical is absorbed more rapidly by the lesion than by

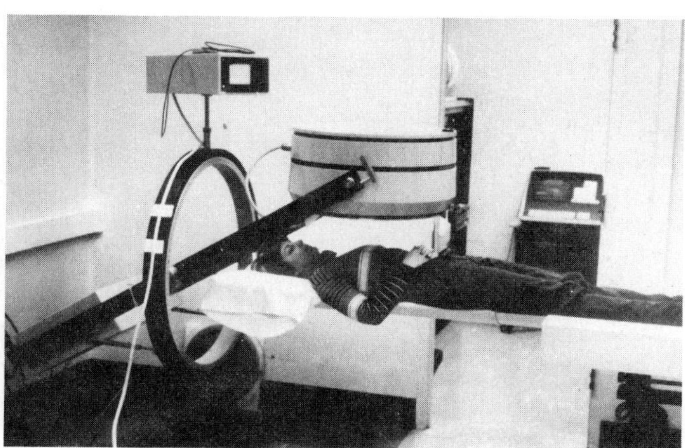

Figure 3 A scintillation camera. The collimator, scintillator and photomultiplier tubes are housed in the counterbalanced arm, which is over the patient. The patient support is made as light as possible to reduce scattering of radiation back into the detector.

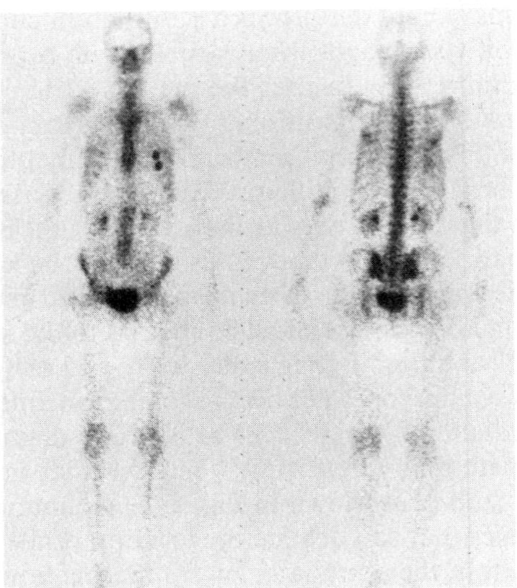

Figure 4 A whole body scan. The front view on the left shows two hot spots in the patient's ribs, corresponding to fractures. There is also a hot spot in the bladder. In the rear view, hot spots can also be seen from the patient's kidneys (at the level of the lower ribs) and the upper rear portions of the pelvis.

surrounding tissue. This leads to a "hot spot" in the image. One example is bone disease, in which there is a greater-than-normal turnover of phosphorus. A bone scan is shown in Fig. 4. Hot spots are seen in two ribs in the front view. There is also radioactivity in the patient's bladder and in the kidneys, which can be seen in the front and rear views, respectively.

In most cases, the lesion absorbs less radioactivity than surrounding tissue, leading to a "cold spot." Figure 5 shows a scan of a normal thyroid gland that has absorbed ^{131}I. This picture of the neck is at a much different magnification than the whole body scan. Figure 6 shows a cold spot in the patient's lower right thyroid (lower left in the picture), as well as a hot spot in the upper right lobe. A cold spot can result when the thyroid tumor does not produce thyroid hormone, or in a cancer that grows so rapidly that it outgrows its blood supply.

Figure 7 shows the blood distribution in normal lungs. The radioactivity in the patient's lungs is shielded from the camera by the patient's heart, which covers part of

Figure 5 A normal thyroid gland in the patient's neck. The U-shaped gland has two lobes, one on either side of the trachea.

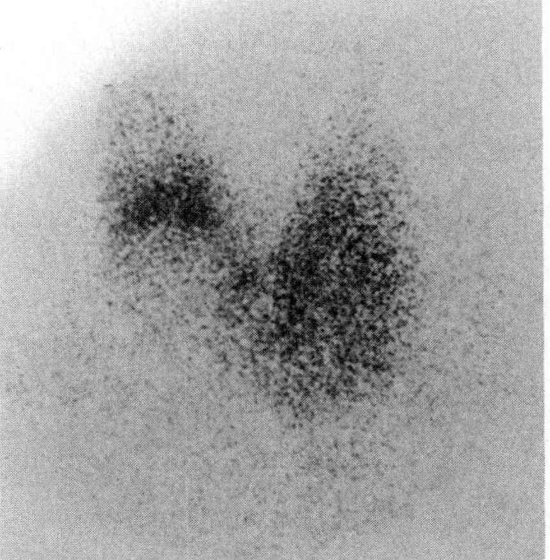

Figure 6 This thyroid gland shows a cold spot in the lower right lobe (left in the picture) and a hot spot in the upper right lobe.

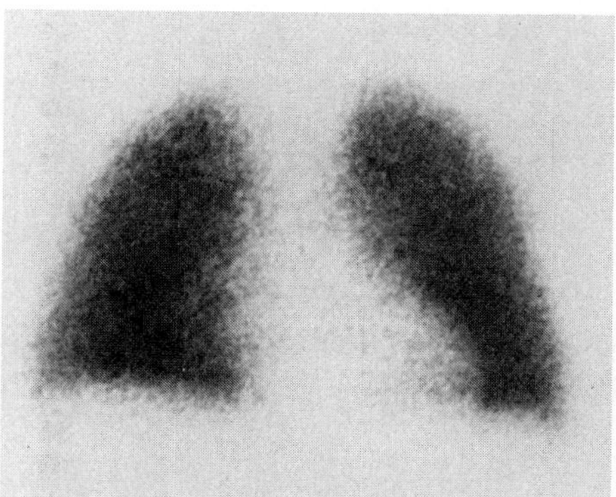

Figure 7 A normal chest. The heart shadow is seen covering the lower left lung.

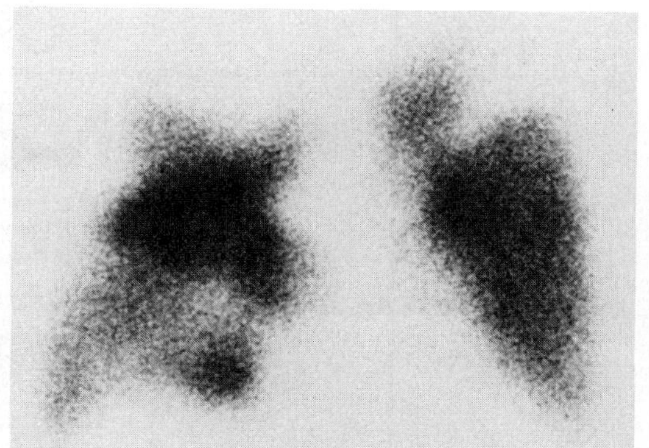

Figure 8 A chest scan of a patient with pulmonary emboli (clots). There are cold spots in the upper part of both lungs, and in the lower part of the right lung.

the lower left lung. Another example of cold spots is shown in Fig. 8, which should be compared with the normal chest of Fig. 7. The patient in Fig. 8 has several pulmonary emboli (clots). Several pieces broke off from a larger clot, probably located in a vein in the patient's leg. It passed through larger and larger veins until it entered the right side of the heart and was pumped into the lungs. In the lungs it passed through smaller and smaller arteries until it became stuck in a medium-sized artery, completely clogging it and shutting off the blood flow to a region of the lung.

The scans in Figs. 7 and 8 were made by attaching ^{99m}Tc atoms to small albumin microspheres, whose diameter is slightly larger than a capillary. These are injected into an arm vein, pass through the heart, and lodge in capillaries in the lung. (There are enough microspheres to block only a small fraction of the capillaries.) Normally the lung is uniformly radioactive. If there is a region of pulmonary embolism, the microspheres can't get to that part of the lung, and a cold spot results.

When the radiopharmaceutical is injected, it often passes through different organs or "compartments" in the body. For example, Fig. 9 shows what might happen if a radioactive substance is injected into a patient's vein. After just a few minutes the substance is uniformly mixed throughout the patient's blood. As the substance passes into the kidneys, the fraction in the blood falls and the fraction in the kidneys rises. The substance then passes from the kidneys to the bladder. The fractions of the substance in the three compartments—blood, kidneys, and bladder—are shown in Fig. 9. (Near the end of the time shown, the bladder empties.) Studies of the kidneys are best done at time t_1, bladder studies at time t_2.

All ionizing radiation may damage tissue, and x-ray or nuclear medicine procedures should not be carried out indiscriminately. However, they can provide essential information in a relatively safe and comfortable manner. The doses involved in diagnostic procedures are roughly 10^{-4} times those used for radiation therapy of cancer.

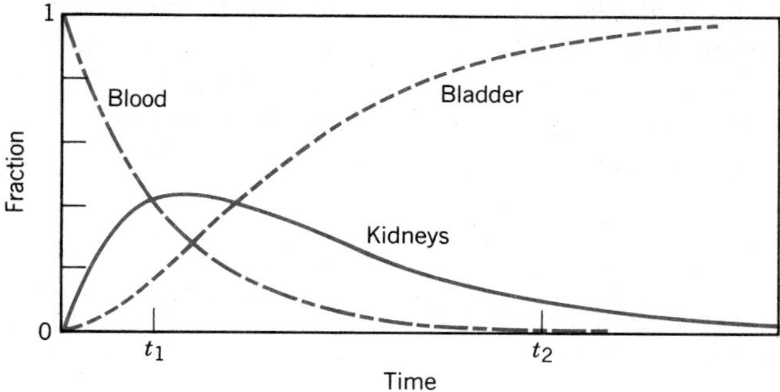

Figure 9 Plot of the distribution of a radioactive pharmaceutical in various organs after it is injected in a patient's vein. The substance passes from blood through the kidneys to the bladder.

CHAPTER 48

ENERGY FROM THE NUCLEUS*

A drawing of the first nuclear reactor, assembled on a squash court at the University of Chicago by a team headed by Enrico Fermi. It went critical on December 2, 1942. This reactor—built of lumps of uranium imbedded in blocks of graphite—served as a prototype for later power reactors whose purpose was to manufacture plutonium for the construction of nuclear bombs.

48-1 The Atom and Its Nucleus

When we get energy from coal by burning it in a furnace, we are tinkering with atoms of carbon and oxygen, rearranging their outer *electrons* into more stable combinations. When we get energy from uranium by burning it in a nuclear reactor, we are tinkering with its nucleus, rearranging its *nucleons* into more stable combinations.

Electrons are held in atoms by the electromagnetic Coulomb force and it takes a few electron volts to pull one of them out. On the other hand, nucleons are held in nuclei by the strong nuclear force and it takes a few *million* electron volts to pull one of *them* out. This factor of a few million is reflected in the fact that we can extract about that much more energy from a kilogram of uranium than we can from a kilogram of coal.

In both atomic and nuclear burning, the appearance of energy is accompanied by a decrease in rest mass, according to Einstein's $E = \Delta m\, c^2$ equation. The only difference between burning uranium and burning coal is that, in the former case, a much larger fraction of the available mass (again, by a factor of a few million) is converted to energy.

* For a more complete treatment, including many references, see Kenneth S. Krane, *Introductory Nuclear Physics* (John Wiley & Sons, New York, 1987), Chapters 13 and 14.

Table 1 Energy from 1 kg of Matter

Form of Matter	*Process*	*Time*[a]
Water	A 50-m waterfall	5 s
Coal	Burning	8 h
Enriched UO_2 (3%)	Fission in a reactor	690 y
^{235}U	Complete fission	3×10^4 y
Hot deuterium gas	Complete fusion	3×10^4 y
Matter and antimatter	Complete annihilation	3×10^7 y

[a] These numbers show how long the energy generated could power a 100-W light bulb.

You must be clear about whether your concern is for *energy* or for the rate at which the energy is delivered, that is, for *power.* In the nuclear case will you burn your kilogram of uranium slowly in a power reactor or explosively in a bomb? In the atomic case, are you thinking about exploding a stick of dynamite or digesting a jelly doughnut? (Surprisingly, the total energy release is greater in the second case than in the first!)

Table 1 shows how much energy can be extracted from one kilogram of matter by doing various things to it. Instead of reporting the energy directly, we measure it by showing how long the extracted energy could operate a 100-W light bulb. Only processes in the first three rows of the table have actually been carried out; the remaining three represent theoretical limits that may not be attainable in practice. The bottom row, the total mutual annihilation of matter and antimatter, is an ultimate goal. When you have used up all the available mass, you can do no more.

Keep in mind that the comparisons of Table 1 are on a per unit mass basis. Kilogram for kilogram you get several millions times more energy from uranium than you do from coal or from falling water. On the other hand, there is a lot of coal in the earth's crust and there is a lot of water backed up behind the Bonneville dam in the Columbia river.

48-2 Nuclear Fission: The Basic Process

In 1932 the English physicist James Chadwick discovered the neutron. A few years later Enrico Fermi and his collaborators in Rome discovered that, if various elements are bombarded by these new projectiles, new radioactive elements are produced. Fermi had predicted that the neutron, being uncharged, would be a useful nuclear projectile; unlike the proton or the α particle, it experiences no repulsive Coulomb force when it approaches a nuclear surface. *Thermal neutrons,* which are neutrons in equilibrium with matter at room temperature, have a mean kinetic energy of only about 0.04 eV but are nevertheless particularly useful projectiles.

In 1939 the German chemists Otto Hahn and Fritz Strassman, following up work initiated by Enrico Fermi and his collaborators, bombarded solutions of uranium salts with such thermal neutrons. They found by chemical analysis that after the bombardment a number of new radioactive elements were present, among them one whose chemical properties were remarkably similar to barium. Repeated tests finally convinced these able chemists that the "new" element was not new at all; it really *was* barium. How could this middle-mass element ($Z = 56$) be produced by bombarding uranium ($Z = 92$) with neutrons?

The puzzle was solved within a few weeks by Lise Meitner and her nephew Otto Frisch. They showed that a uranium nucleus, having absorbed a thermal neutron, could split, with the release of energy, into two roughly equal parts, one of which might well be barium. Frisch named the process *fission.**

Figure 1 shows the tracks left in the gas of a cloud chamber by the two energetic fission fragments that re-

* See "The Discovery of Fission," by Otto Frisch and John Wheeler, *Physics Today,* November 1967, for a fascinating account of the early days of discovery.

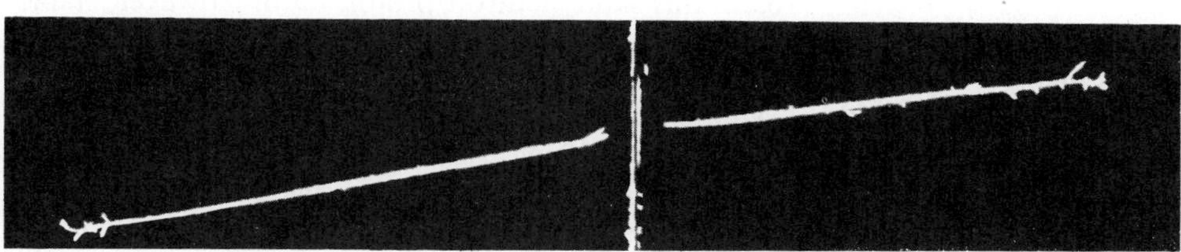

Figure 1 When a fast charged particle passes through a cloud chamber, it leaves a track of liquid droplets behind it. The two back-to-back tracks were formed by fragments produced by a fission event that took place in a thin vertical uranium foil at the center of the chamber.

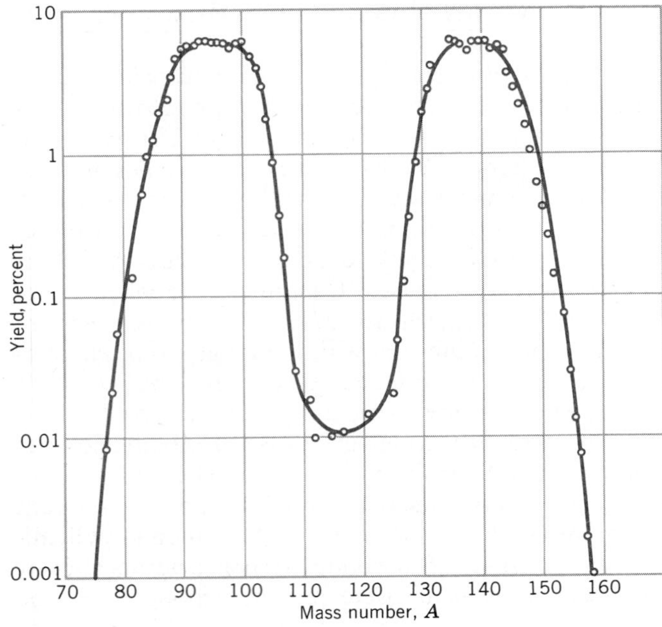

Figure 2 The nucleus ^{236}U can split apart in many ways, only one of which is described by Eq. 1. The figure shows the distribution by mass number of the fragments that are found when many fission events of this nuclide are examined. Note that the vertical scale is logarithmic.

sult from a fission event occurring near the center of the chamber. The tracks are not the same length, which suggests that the fission fragments in this case do not have the same mass and kinetic energy. They rarely do. Figure 2 shows the distribution by mass number of the fragments produced when ^{235}U is bombarded with thermal neutrons. The most probable mass numbers, occurring in about 7% of the events, are centered around $A \approx 95$ and $A \approx 137$.

A Typical Fission Event. Equation 1 shows a typical fission event, in which a ^{235}U nucleus absorbs a thermal neutron, producing a compound nucleus ^{236}U in a highly excited state. It is *this* nucleus that undergoes fission, splitting into two fragments. These fragments—between them—rapidly emit two *prompt neutrons,* leaving ^{140}Xe and ^{94}Sr as fission fragments. Thus

$$^{235}\mathrm{U} + n \rightarrow {}^{236}\mathrm{U}^* \rightarrow {}^{140}\mathrm{Xe} + {}^{94}\mathrm{Sr} + 2n. \qquad (1)$$

The fragments ^{140}Xe and ^{94}Sr are both highly unstable, undergoing beta decay (with the emission of a negative electron) until each reaches a stable end product. Thus

$$\begin{array}{cccccc} & ^{140}\mathrm{Xe} \rightarrow & ^{140}\mathrm{Cs} \rightarrow & ^{140}\mathrm{Ba} \rightarrow & ^{140}\mathrm{La} \rightarrow & ^{140}\mathrm{Ce} \\ & 14\ \mathrm{s} & 64\ \mathrm{s} & 13\ \mathrm{d} & 40\ \mathrm{h} & \mathrm{Stable} \\ Z & 54 & 55 & 56 & 57 & 58 \end{array} \qquad (2)$$

and

$$\begin{array}{cccc} & ^{94}\mathrm{Sr} \rightarrow & ^{94}\mathrm{Y} \rightarrow & ^{94}\mathrm{Zr} \\ & 75\ \mathrm{s} & 19\ \mathrm{min} & \mathrm{Stable.} \\ Z & 38 & 39 & 40 \end{array} \qquad (3)$$

The half-lives of each radionuclide are indicated. As expected, the mass numbers (140 and 94) of the fragments remain unchanged during these beta decay processes; the atomic numbers (initially 54 and 38) increase by unity at each step.

Inspection of the stability line on the nuclidic chart (Fig. 4 of Chapter 47) can show us why the fission fragments are unstable. ^{236}U, which is the fissioning nucleus in the reaction of Eq. 1, has 92 protons and $236 - 92$ or 144 neutrons, a neutron/proton ratio of ~ 1.6. The primary fragments formed immediately after fission will retain this same neutron/proton ratio. However, stable nuclides in the middle-mass region have smaller neutron/proton ratios, in the range 1.3 to 1.4. The primary fragments will thus be neutron rich and will "boil off" a small number of neutrons, two in the case of the reaction of Eq. 1. The fragments that remain are still too neutron-rich to be stable. Beta decay offers a mechanism for getting rid of these excess neutrons, namely, by changing them into protons within the nucleus; see Eq. 12 of Chapter 47.

We can use the binding energy curve of Fig. 6 of Chapter 47 to estimate the energy released in fission. From this curve, we see that for heavy nuclides ($A \approx 240$) the mean binding energy per nucleon is about 7.6 MeV. For middle-mass nuclides ($A \approx 120$) it is about 8.5 MeV. The difference in the total binding energy between a single large nucleus ($A = 240$) and two fragments (assumed equal) into which it may be split is then

$$Q = 2(8.5\text{ MeV})(\tfrac{1}{2}A) - (7.6\text{ MeV})(A) \approx 200\text{ MeV}. \qquad (4)$$

The more careful calculation of Sample Problem 1 agrees remarkably well with this rough estimate.

Sample Problem 1 Calculate the disintegration energy Q for the fission event of Eq. 1, taking into account the decay of the fission fragments as displayed in Eqs. 2 and 3.

We can calculate the disintegration energy from Einstein's $E = \Delta m\, c^2$ relation. Some atomic masses that we will need are

^{235}U	235.0439 u	^{140}Ce	139.9054 u
n	1.00867 u	^{94}Zr	93.9063 u

If we combine Eq. 1 with Eqs. 2 and 3, we see that the overall transformation is

$$^{235}\text{U} \rightarrow {}^{140}\text{Ce} + {}^{94}\text{Zr} + n. \qquad (5)$$

The single neutron comes about because the initiating neutron on the left of Eq. 1 cancels one of the two neutrons on the right of that equation. The mass difference for the reaction of Eq. 5 is

$$\Delta m = (235.0439\text{ u}) - (139.9054\text{ u} + 93.9063\text{ u} + 1.00867\text{ u}) = 0.224\text{ u},$$

and the corresponding disintegration energy is

$$Q = \Delta m\, c^2 = (0.224\text{ u})(932\text{ MeV/u}) = 209\text{ MeV}, \qquad \text{(Answer)}$$

in good agreement with our rough estimate of Eq. 4.

If the fission event takes place in a bulk solid, most of this disintegration energy appears eventually as an increase in the internal energy of that body, revealing itself as a rise in temperature. Five or six percent or so of the disintegration energy, however, is associated with neutrinos that are emitted during the beta decay of the primary fission fragments. This energy is carried out of the system and is lost to us forever.

48-3 A Model for Nuclear Fission

Soon after the discovery of fission, Niels Bohr and John Wheeler developed a model, based on the analogy between a nucleus and a charged liquid drop, that explained its main features. Figure 3 suggests how the fission process proceeds from this point of view. When a heavy nucleus—let us say ^{235}U—absorbs a slow neutron, as in Fig. 3*a*, that neutron falls into the potential well associated with the strong nuclear forces that act in the nuclear interior. Its potential energy is then transformed into internal excitation energy, as Fig. 3*b* suggests.

The amount of excitation energy that a slow neutron carries into the nucleus is equal to the work required to pull a neutron out of the nucleus, that is, to the binding energy E_n of the neutron. In much the same way, the amount of excitation energy delivered to a well when a stone is dropped into it is equal to the work required to pull the stone back up out of the well, that is, to the "binding energy" E_s of the stone.

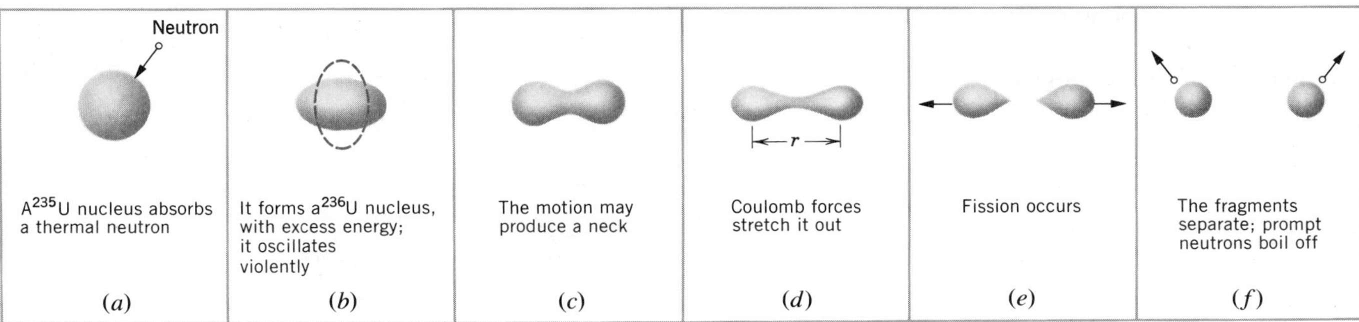
Neutron
r
A^{235}U nucleus absorbs a thermal neutron
(a)
It forms a^{236}U nucleus, with excess energy; it oscillates violently
(b)
The motion may produce a neck
(c)
Coulomb forces stretch it out
(d)
Fission occurs
(e)
The fragments separate; prompt neutrons boil off
(f)

Figure 3 The stages of a typical fission process, according to the liquid-drop fission model of Bohr and Wheeler.

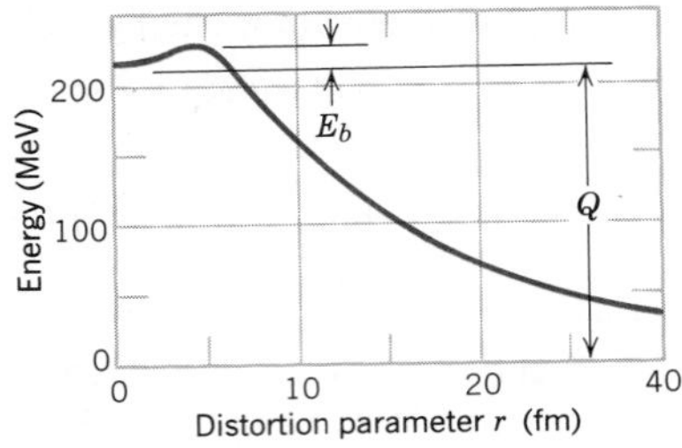

Figure 4 The potential energy at various stages in the fission process, as predicted from the liquid-drop fission model of Bohr and Wheeler. The Q of the reaction ($\approx$200 MeV) and the fission barrier height E_b are both indicated.

Figure 3*c* shows that the nucleus, behaving like an energetically-oscillating charged liquid drop, will sooner or later develop a short "neck" and will begin to separate into two charged "globs." If conditions are right, the electrostatic repulsion between these two globs will force them apart, breaking the neck. The two fragments, each still carrying some residual excitation energy, then fly apart. Fission has occurred.

Thus far this model gives a good qualitative picture of the fission process. It remains to see, however, whether it can answer a hard question: "Why are some heavy nuclides (^{235}U and ^{239}Pu, say) readily fissionable by thermal neutrons but other, equally heavy, nuclides (^{238}U and ^{243}Am, say) are not?"

Bohr and Wheeler were able to answer this question. Figure 4 shows the potential energy curve for the fission process that they derived from their model. The horizontal axis displays the *distortion parameter r*, which is a rough measure of the extent to which the oscillating nucleus departs from a spherical shape. Figure 3*d* suggests how this parameter is defined before fission occurs. When the fragments are far apart, this parameter is simply the distance between their centers.

The energy interval between the initial state and the final state of the fissioning nucleus—that is, the disintegration energy Q—is displayed in Fig. 4. The central feature of that figure, however, is that the potential energy curve passes through a maximum at a certain value of r. There is a *potential barrier* of height E_b that must be surmounted (or tunneled!) before fission can occur. This reminds us of alpha decay (see Fig. 9 of Chapter 47), which is also a process that is inhibited by a potential barrier.

We see, then, that fission will occur only if the absorbed neutron provides an excitation energy E_n great enough to overcome the barrier. E_n need not be quite as great as the barrier height E_b because of the possibility of wave-mechanical tunneling.

Table 2 shows a test of fissionability by thermal neutrons applied to four heavy nuclides, chosen from dozens of possible candidates. For each nuclide both the barrier height E_b and the excitation energy E_n are given. The former was calculated from the theory of Bohr and Wheeler; the latter was computed from the known masses, using the $E = \Delta m\, c^2$ relation.

For ^{235}U and ^{239}Pu we see that $E_n > E_b$. This means that fission by absorbing a thermal neutron is predicted to occur for these nuclides. For the other two nuclides (^{238}U and ^{243}Am), we have $E_n < E_b$, so that there is not enough energy for a thermal neutron to surmount the barrier or to tunnel through it effectively. The excited nucleus (Fig. 3*b*) prefers to get rid of its excitation energy by emitting a gamma ray instead of by breaking into two large fragments. Table 2 confirms these predictions.

^{238}U and ^{243}Am *can* be made to fission, however, if they absorb a substantially energetic (rather than a thermal) neutron. For ^{238}U, for example, the absorbed neutron must have at least 1.3 MeV of energy for this *fast fission* process to occur with meaningful probability.

48-4 The Nuclear Reactor

To make large-scale use of the energy released in fission, we must arrange to have one fission event trigger an-

Table 2 Test of the Fissionability of Four Nuclides

Target Nuclide	*Nuclide Being Fissioned*	E_n MeV	E_b MeV	$E_n - E_b$ MeV	*Fission by Thermal Neutrons?*
^{235}U	^{236}U	6.5	5.2	+1.3	Yes
^{238}U	^{239}U	4.8	5.7	−0.9	No
^{239}Pu	^{240}Pu	6.4	4.8	+1.6	Yes
^{243}Am	^{244}Am	5.5	5.8	−0.3	No

other, so that the process spreads throughout the nuclear fuel like flame through a burning log. The fact that more neutrons are produced in fission than are consumed raises just this possibility of a *chain reaction.* Such a reaction can either be rapid (as in a nuclear bomb) or controlled (as in a nuclear reactor).

Suppose that we wish to design a reactor based on the fission of ^{235}U by thermal neutrons. Natural uranium contains 0.7% of this isotope, the remaining 99.3% being ^{238}U, which is not fissionable by thermal neutrons. Let us give ourselves an edge by artificially enriching the uranium fuel so that it contains perhaps 3% ^{235}U. Three difficulties still stand in the way of a working reactor:

1. The Neutron Leakage Problem. Some of the neutrons produced by fission will leak out of the reactor and be lost to the chain reaction. Leakage is a surface effect, its magnitude being proportional to the square of a typical reactor dimension ($= 6a^2$ for a cube of edge a). Neutron production, however, occurs throughout the volume of the fuel and is thus proportional to the cube of a typical dimension ($= a^3$ for a cube). We can make the fraction of neutrons lost by leakage as small as we wish by making the reactor core large enough, thereby reducing the surface to volume ratio ($= 6/a$ for a cube).

2. The Neutron Energy Problem. The neutrons produced by fission are fast, with kinetic energies of ~ 2 MeV. However, fission is induced most effectively by thermal neutrons. The fast neutrons can be slowed down by mixing the uranium fuel with a substance—called a *moderator*—that has these properties: (*a*) It is effective in slowing down neutrons by elastic collisions and (*b*) it does not remove neutrons from the core by absorbing them in ways that do not result in fission. Most power reactors in this country use water as a moderator, the hydrogen nuclei (protons) being the effective component.

3. The Neutron Capture Problem. As the fast (~ 2 MeV) neutrons generated by fission are slowed down in the moderator to thermal energies (~ 0.04 eV), they must pass through a critical energy interval (~ 1 to ~ 100 eV) in which they are particularly susceptible to nonfission capture by ^{238}U nuclei. Such capture, which results in the emission of a gamma ray, removes the neutron from the fission chain.

To minimize such *resonance capture,* the uranium fuel and the moderator are not intimately mixed but are "clumped," occupying different regions of the reactor volume. This increases the chance that a fast neutron, produced in a uranium clump, will find itself in the moderator as it passes through the critical resonance energy range. Once the neutron has reached thermal energies, it may *still* be captured in ways that do not result in fission *(thermal capture).* However, it is much more likely that the thermal neutron will wander into a clump of fuel and produce a fission event.

Figure 5 shows the neutron balance in a typical power reactor operating with a steady power output. Let us trace the behavior of a sample of 1000 thermal neutrons in the reactor core. They produce 1330 neutrons by fission in the ^{235}U fuel and 40 more by fast fission in the ^{238}U, making a total of 370 new neutrons, all of them fast. Exactly this same number of neutrons is then lost to the chain by leakage from the core and by nonfission capture, leaving 1000 thermal neutrons to continue the chain. What has been gained in this cycle, of course, is that each of the 370 neutrons produced by fission has deposited about 200 MeV of energy in the reactor core, heating it up.

The *multiplication factor k*—an important reactor parameter—is the ratio of the number of neutrons present at the beginning of a particular generation to the number present at the beginning of the next generation. For the situation of Fig. 5, the multiplication factor is 1000/1000 or exactly unity. For $k = 1$, the operation of

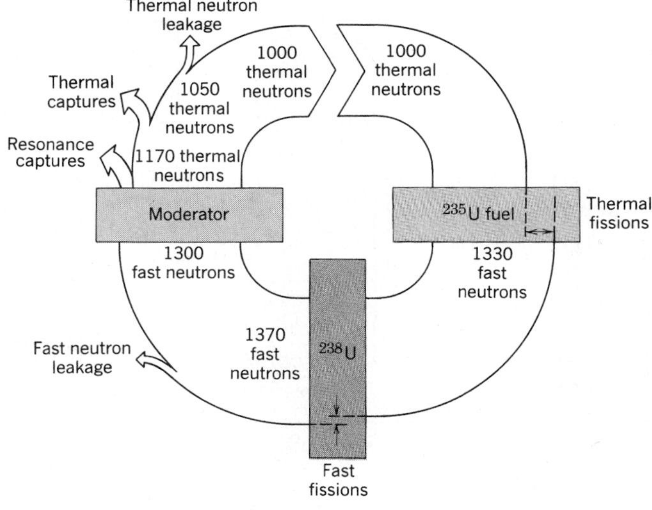

Figure 5 Neutron bookkeeping in a reactor. A generation of 1000 thermal neutrons is followed as they interact with the ^{235}U fuel, the ^{238}U matrix, and the moderator. We see that 1370 neutrons are produced by fission; 370 of these are lost, by nonfission capture or by leakage, so that exactly 1000 thermal neutrons are left to form the next generation. The figure is drawn for a reactor running at a steady power level.

the reactor is said to be exactly *critical,* which is what we wish it to be for steady power production. Reactors are designed so that they are inherently *supercritical* ($k > 1$); the multiplication factor is then adjusted to critical operation ($k = 1$) by inserting *control rods* into the reactor core. These rods, containing a material such as cadmium that absorbs neutrons readily, can then be withdrawn as needed to compensate for the tendency of reactors to go subcritical as (neutron-absorbing) fission products build up in the core during continued operation.

If you pulled out one of the control rods, how fast would the reactor power level increase? This *response time* is controlled by the fascinating circumstance that a small fraction of the neutrons generated by fission are not boiled off promptly from the newly formed fission fragments but are emitted from these fragments later, as they decay by beta emission. Of the 370 "new" neutrons analyzed in Fig. 5, for example, ~ 16 are delayed, being emitted from fragments following beta decays whose half-lives range from 0.2 to 55 s. These delayed neutrons are few in number but they serve the useful purpose of slowing down the reactor response time to match human reaction times.

Figure 6 shows the broad outlines of an electric power plant based on a *pressurized-water reactor* (PWR), a type in common use in this country. In such a reactor, water is used both as the moderator and as the heat transfer medium. In the *primary loop,* water at high temperature and pressure (possibly 600 K and 150 atm) circulates through the reactor vessel and transfers heat from the reactor core to the steam generator, which provides high-pressure steam to operate the turbine that drives the generator. To complete the *secondary loop,* low-pressure steam from the turbine is condensed to water and forced back into the steam generator by a pump. To give some idea of scale, a typical reactor vessel for a 1000-MW (electric) plant may be 40 ft high and weigh 450 tons. Water flows through the primary loop at a rate of ~ 300,000 gal/min.

An unavoidable feature of reactor operation is the accumulation of radioactive wastes, including both fission products and heavy "transuranic" nuclides such as plutonium and americium. One measure of their radioactivity is the rate at which they release energy in thermal form. Figure 7 shows the variation with time of the thermal power generated by such wastes from one year's

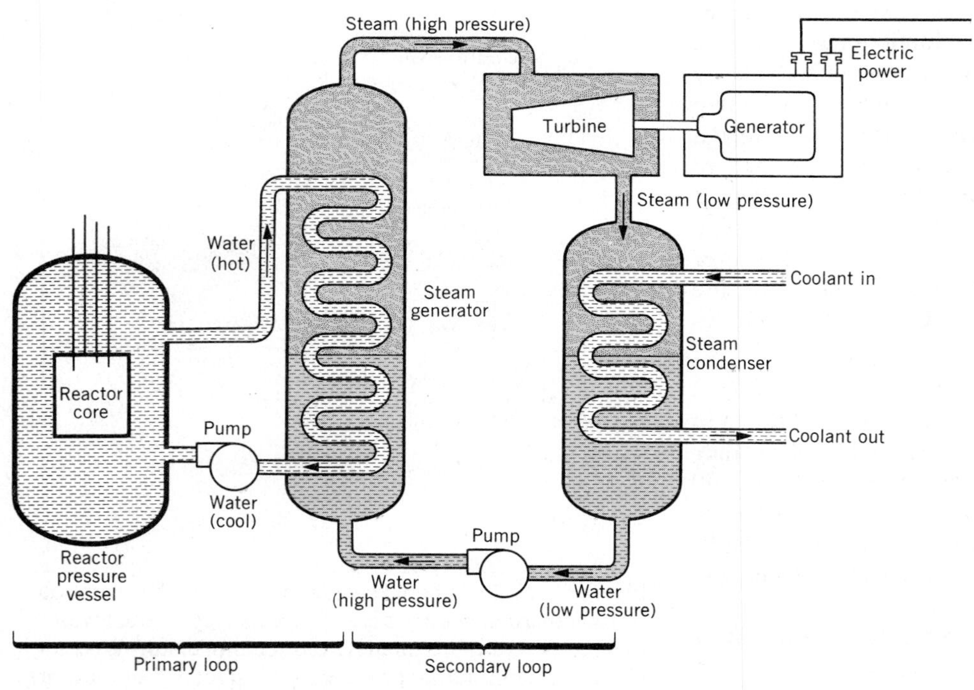

Figure 6 A simplified layout of a nuclear power plant, based on a pressurized-water reactor. Many features are omitted—among them the arrangement for cooling the reactor core in case of an emergency.

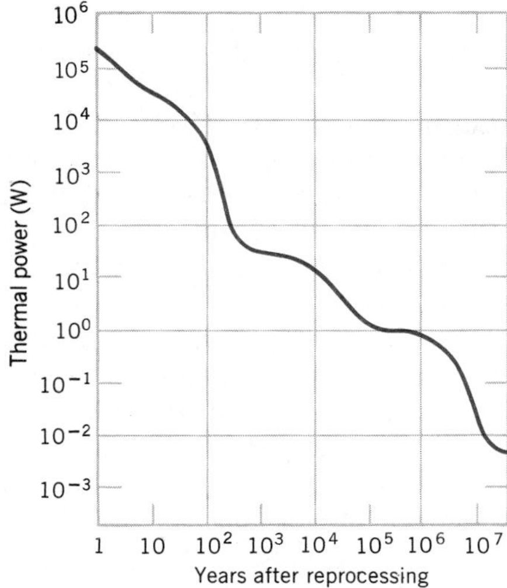

Figure 7 The thermal power released by the radioactive wastes from one year's operation of a typical large nuclear power plant is shown as a function of time. The curve is the superposition of the effects of many radionuclides, with a wide variety of half-lives. Note that both scales are logarithmic.

operation of a typical large nuclear plant. Note that both scales are logarithmic. The total activity of the waste 10 years after reprocessing is ~3×10^7 Ci.

Sample Problem 2 A large electric generating station is powered by a pressurized-water nuclear reactor. The thermal power in the reactor core is 3400 MW and 1100 MW of electricity are generated. The fuel charge is 86,000 kg of uranium, in the form of 110 tons of uranium oxide, distributed among 57,000 fuel rods. The uranium is enriched to 3.0% ^{235}U. (a) What is the plant efficiency?

$$e = \frac{\text{Useful output}}{\text{Energy input}} = \frac{1100 \text{ MW (electric)}}{3400 \text{ MW (thermal)}}$$
$$= 0.32 \text{ or } 32\%. \qquad \text{(Answer)}$$

The efficiency—as for all power plants, whether based on fossil fuel or nuclear fuel—is controlled by the second law of thermodynamics. In this plant, 3400 MW − 1100 MW or 2300 MW of power must be discharged as thermal energy to the environment.

(b) At what rate R do fission events occur in the reactor core?

If P(= 3400 MW) is the thermal power in the core and Q (= 200 MeV) is the average energy released per fission event, then, in steady-state operation,

$$R = \frac{P}{Q} = \left(\frac{3.4 \times 10^9 \text{ W}}{200 \text{ MeV/fission}}\right)\left(\frac{1 \text{ MeV}}{1.60 \times 10^{-13} \text{ J}}\right)\left(\frac{1 \text{ J/s}}{1 \text{ W}}\right)$$
$$= 1.06 \times 10^{20} \text{ fissions/s}$$
$$\approx 1.1 \times 10^{20} \text{ fissions/s}. \qquad \text{(Answer)}$$

(c) At what rate is the ^{235}U fuel disappearing? Assume conditions at start-up.

^{235}U disappears by fission at the rate calculated in (b) above. It is also consumed by (nonfission) neutron capture at a rate about one-fourth as large. The total ^{235}U consumption rate is then $(1.25)(1.06 \times 10^{20}\text{ s}^{-1})$ or $1.33 \times 10^{20}\text{ s}^{-1}$. We recast this as a mass rate as follows:

$$\frac{dM}{dt} = (1.33 \times 10^{20}\text{ s}^{-1})\left(\frac{0.235 \text{ kg/mol}}{6.02 \times 10^{23} \text{ atoms/mol}}\right)$$
$$= 5.19 \times 10^{-5} \text{ kg/s} \approx 4.5 \text{ kg/d}. \qquad \text{(Answer)}$$

(d) At this rate of fuel consumption, how long would the fuel supply last?

From the data given, we can calculate that, at start-up, about (0.03)(86,000 kg) or ~2600 kg of ^{235}U were present. Thus, a somewhat simplistic answer would be

$$T = \frac{2600 \text{ kg}}{4.5 \text{ kg/d}} = 578 \text{ d}. \qquad \text{(Answer)}$$

In practice, the fuel rods are replaced (often in batches) before their ^{235}U content is entirely consumed.

(e) At what rate is rest mass being converted to energy in the reactor core?

From Einstein's $E = \Delta m\, c^2$ relation, we can write

$$\frac{dM}{dt} = \frac{dE/dt}{c^2} = \frac{3.4 \times 10^9 \text{ W}}{(3.00 \times 10^8 \text{ m/s})^2}$$
$$= 3.8 \times 10^{-8} \text{ kg/s} \quad \text{or} \quad 3.3 \text{ g/d}. \qquad \text{(Answer)}$$

We see that the mass conversion rate is about the mass of one penny every day! This rest mass rate (conversion to energy) is quite a different quantity than the fuel consumption rate (loss of ^{235}U) calculated in (c) above.

48-5 A Natural Nuclear Reactor (Optional)*

On December 2, 1942, when the reactor assembled by Enrico Fermi and his associates first went critical, they had every right to expect that they had put into operation the first fission reactor that had ever existed on this

* See "A Natural Fission Reactor," by George A. Cowan, *Scientific American,* July 1976 for the complete story.

planet. About 30 years later it was discovered that, if they did in fact think that, they were wrong.

Some two billion years ago, in a uranium deposit now being mined in West Africa, a natural fission reactor went into operation and ran for perhaps several hundred thousand years before shutting itself off. In analyzing this claim, let us consider only two points:

1. Was There Enough Fuel? The fuel for a uranium-based fission reactor must be the easily fissionable isotope ^{235}U, which constitutes only 0.72% of natural uranium. This isotopic ratio has been measured for terrestrial samples, in moon rocks and in meteorites; the values are the same. The clue to the discovery in West Africa was that the uranium from this deposit was deficient in this isotope, some samples having a ^{235}U abundance as low as 0.44%. Investigation led to the speculation that this deficit in ^{235}U could be accounted for if, at some time in the past, this isotope was partially consumed by the operation of a natural fission reactor.

The serious problem remains that, with an isotopic abundance of only 0.72%, a reactor can be assembled (as Fermi and his team learned) only with the greatest of difficulty. There seems no chance at all that it could have happened naturally.

However, things were different in the distant past. Both ^{235}U and ^{238}U are radioactive, with half lives of 7.04×10^8 y and 44.7×10^8 y, respectively. Thus, the half-life of the readily fissionable ^{235}U is about 6.4 times shorter than that of ^{238}U. Because ^{235}U decays faster, there must have been more of it, relative to ^{238}U, in the past. Two billion years ago, in fact, this abundance was not 0.72%, as it is now, but 3.8%. This abundance happens to be just about the abundance to which natural uranium is artificially enriched to serve as fuel in modern power reactors.

With this much readily-fissionable fuel available in the distant past, the presence of a natural reactor (providing certain other conditions are met) is much less surprising. The fuel was there. Two billion years ago, incidentally, the highest order of life forms that had evolved were the blue-green algae.

2. What Is the Evidence? The mere depletion of ^{235}U in an ore deposit is not enough evidence on which to hang a claim for the existence of a natural fission reactor. One looks for more convincing proof.

If there was a reactor, there must also be fission products. Of the 30 or so elements whose stable isotopes are produced in this way, some must still remain. Study of their isotopic abundances could provide the convincing evidence we need.

Of the several elements investigated, the case of neodymium is spectacularly convincing. Figure 8*a* shows the isotopic abundances of the seven stable neo-

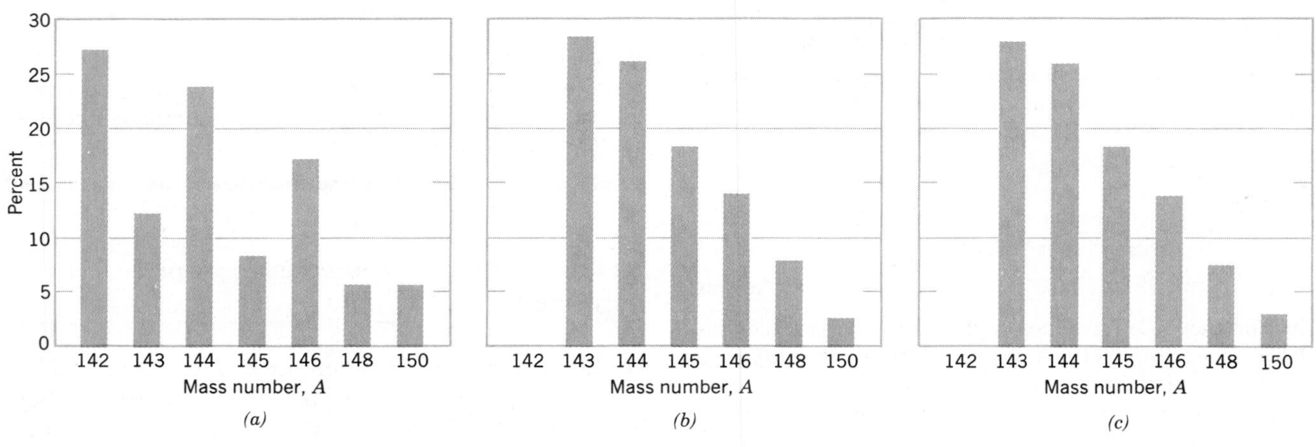

Figure 8 The distribution by mass number of the isotopes of neodymium as they occur in (*a*) natural terrestrial deposits of the ores of this element and (*b*) the spent fuel of a power reactor. (*c*) The distribution (after several corrections) found for neodymium from the uranium mine in Gabon, West Africa. Note that (*b*) and (*c*) are virtually identical and are quite different from (*a*).

dymium isotopes as they are normally found in nature. Figure 8*b* shows these abundances as they appear among the ultimate stable fission products of the fission of ^{235}U. The clear differences are not surprising, considering their totally different origins. The isotopes shown in Fig. 8*a* were formed in supernovae explosions that occurred before the formation of our solar system. The isotopes of Fig. 8*b* were cooked up in a reactor by totally different processes. Note particularly that ^{142}Nd, the dominant isotope in the natural element, is totally absent from the fission products.

The big question is, "What do the neodymium isotopes found in the uranium ore body in West Africa look like?" We must expect that, if a natural reactor operated there, isotopes from *both* sources (that is, natural isotopes as well as fission-produced isotopes) might be present. Figure 8*c* shows the results after this and other corrections have been made to the raw data. Comparison of Figs. 8*b* and 8*c* leaves little doubt that there was indeed a natural fission reactor at work.

The story of this discovery is totally fascinating at the level of the best detective thriller. More important, it provides a first-class example of the nature of the scientific evidence needed to back up what may seem to be an improbable claim. Those who hold that the earth is only a few thousand years old, for example, have here a model of the standard of evidence that is required of them if they are to convince the scientific community.

Sample Problem 3 The isotopic ratio of ^{235}U to ^{238}U in natural uranium deposits today is 0.0072. What was this ratio 2.0×10^9 y ago? The half-lives of the two isotopes are 7.04×10^8 y and 44.7×10^8 y, respectively.

Consider two samples that, at a time t in the past, contained $N_5(0)$ and $N_8(0)$ atoms of ^{235}U and ^{238}U, respectively. The numbers of atoms remaining at the present time are

$$N_5(t) = N_5(0)e^{-\lambda_5 t} \quad \text{and} \quad N_8(t) = N_8(0)e^{-\lambda_8 t},$$

respectively, in which λ_5 and λ_8 are the corresponding disintegration constants. Dividing gives

$$\frac{N_5(t)}{N_8(t)} = \frac{N_5(0)}{N_8(0)} e^{-(\lambda_5-\lambda_8)t}.$$

Expressed in terms of isotopic ratios, this becomes

$$R(0) = R(t)e^{(\lambda_5-\lambda_8)t}.$$

The disintegration constants are related to the half-lives by Eq. 8 of Chapter 47, or

$$\lambda_5 = \frac{\ln 2}{\tau_5} = \frac{\ln 2}{7.04 \times 10^8 \text{ y}} = 9.85 \times 10^{-10} \text{ y}^{-1}$$

and

$$\lambda_8 = \frac{\ln 2}{\tau_8} = \frac{\ln 2}{44.7 \times 10^8 \text{ y}} = 1.55 \times 10^{-10} \text{ y}^{-1}.$$

The exponent in the expression for $R(0)$ above is then

$$(\lambda_5 - \lambda_8)t = [(9.85 - 1.55) \times 10^{-10} \text{ y}^{-1}][2 \times 10^9 \text{ y}] = 1.66.$$

The isotopic ratio is then

$$\begin{aligned} R(0) &= R(t)e^{(\lambda_5-\lambda_8)t} \\ &= (0.0072)(e^{1.66}) \\ &= 0.0379 \text{ or } \sim 3.8\%. \qquad \text{(Answer)} \end{aligned}$$

We see that, two billion years ago, the ratio of ^{235}U to ^{238}U in natural uranium deposits was much higher than it is today. Show that when the earth was formed (4.5 billion years ago) this ratio was 30%.

48-6 Thermonuclear Fusion: The Basic Process

The binding energy curve of Fig. 6 of Chapter 47 shows that energy can be released if two light nuclei combine to form a single larger nucleus, a process called nuclear *fusion.* The process is hindered by the Coulomb repulsion that acts to prevent the two particles from getting close enough to each other to come within range of their attractive nuclear forces and "fusing." The height of the Coulomb barrier depends on the charges and the radii of the two interacting nuclei. We show in Sample Problem 4 that, for two deuterons ($Z = 1$), the barrier height is ~200 keV. For more-highly-charged particles, of course, the barrier is correspondingly higher.

To generate useful amounts of power, nuclear fusion must occur in bulk matter. The best hope for bringing this about is to raise the temperature of the material so that the particles have enough energy—due to their thermal motions alone—to penetrate the barrier. We call this process *thermonuclear fusion.*

In thermonuclear studies, temperatures are reported in terms of kinetic energy through the relation*

* It can be shown that, in a gas or a plasma in equilibrium at temperature T, kT is the kinetic energy of a particle moving with the *most probable speed;* see Section 21-7.

$$K = kT, \tag{6}$$

in which k is the Boltzmann constant. Thus, rather than saying, "The temperature at the center of the sun is 1.5×10^7 K," it is more common to say, "The temperature at the center of the sun is 1.3 keV."

Room temperature corresponds to $K \approx 0.03$ eV; a particle with only this amount of energy could not hope to overcome a barrier as high as, say 200 keV. Even at the center of the sun, where $kT = 1.3$ keV, the outlook for thermonuclear fusion does not seem promising at first glance. Yet we know that thermonuclear fusion not only occurs in the core of the sun but is the dominant feature of that body and of all other stars.

The puzzle is solved when we realize that (1) the energy calculated from Eq. 6 refers to the particle with the *most probable* speed; there is a long Maxwellian tail of particles with much higher speeds and, correspondingly, much higher energies. (2) The barrier heights that we have calculated represent the *peaks* of the barriers. Barrier tunneling can occur at energies considerably lower than these peaks, as we saw in the case of α decay (see Section 47–4).

Figure 9 sums things up. The curve marked $n(K)$ in this figure is a Maxwell distribution curve for the protons in the sun's core, drawn to correspond to the sun's central temperature. This curve differs from the Maxwell distribution curve of Fig. 8 of Chapter 21 in that it is drawn in terms of energy and not of speed. Specifically, $n(K)\,dK$ gives the probability per unit volume that a proton will have a kinetic energy lying between K and $K + dK$. The value of kT in the core of the sun is marked on the figure; note that many particles have energies greater than this.

The curve marked $p(K)$ in Fig. 9 is the relative probability of barrier penetration for two colliding protons. Study of the two curves in Fig. 9 suggests that there will be a particular proton energy at which proton–proton fusion events occur at a maximum rate. If the energy of the interacting protons is much higher than this value, the barrier is transparent enough but there are too few protons in the Maxwellian tail to sustain the reaction. If the energy is much lower than this value, there are plenty of protons but the barrier is now too formidable.

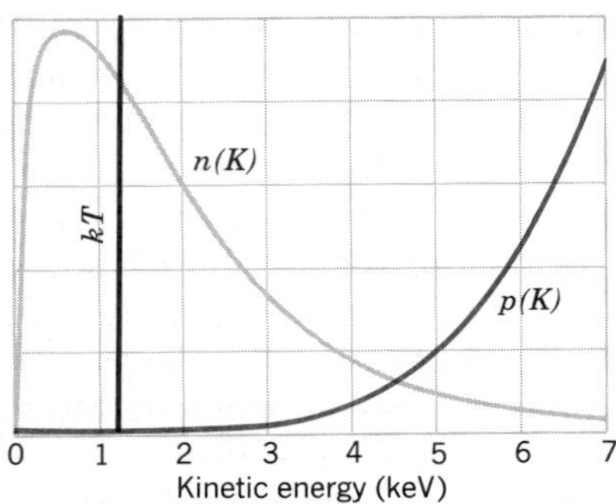

Figure 9 The curve marked $n(K)$ gives the relative distribution in energy for protons at the center of the sun. The curve marked $p(K)$ gives the relative probability of barrier penetration for proton–proton collisions at the sun's central temperature. The vertical line marks the value of kT at this temperature. Note that both curves are drawn to (separate) arbitrary vertical scales.

Sample Problem 4 The deuteron (^{2}H) has a charge $+e$ and may be taken as a sphere of effective radius $R = 2.1$ fm. Two such particles are fired at each other with the same kinetic energy K. What must K be if the particles are brought to rest by their mutual Coulomb repulsion when they are just "touching" each other? We take this value of K as a measure of the height of the Coulomb barrier.

Because the two deuterons are momentarily at rest when they touch, their initial kinetic energy has all been transformed into electrostatic potential energy. Their centers are separated by a distance $2R$ and we have

$$2K = \frac{1}{4\pi\epsilon_0}\frac{q_1 q_2}{r} = \frac{1}{4\pi\epsilon_0}\frac{e^2}{2R},$$

which yields

$$\begin{aligned} K &= \frac{e^2}{16\pi\epsilon_0 R} \\ &= \frac{(1.60 \times 10^{-19}\ \text{C})^2}{(16\pi)(8.85 \times 10^{-12}\ \text{F/m})(2.1 \times 10^{-15}\ \text{m})} \\ &= 2.74 \times 10^{-14}\ \text{J} = 171\ \text{keV} \approx 200\ \text{keV}. \quad \text{(Answer)} \end{aligned}$$

48–7 Thermonuclear Fusion in the Sun and Other Stars

The sun radiates at the rate of 3.9×10^{26} W and has been doing so for several billion years. Where does all this energy come from? Chemical burning is ruled out; if the sun had been made of coal and oxygen—in the right proportions for combustion—it would have lasted for only ~1000 y. Another possibility is that the sun is slowly shrinking, under the action of its own gravita-

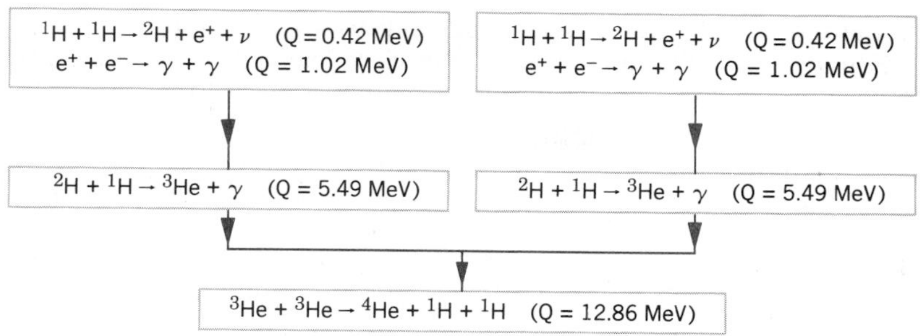

Figure 10 The proton–proton mechanism that accounts for energy production in the sun. In this process, protons fuse to form an α particle, with a net energy release of 26.7 MeV for each event.

tional forces. By transferring gravitational potential energy to thermal energy the temperature of the sun will be maintained so that it may continue to radiate. Calculation, however, shows that this mechanism also fails, producing a solar lifetime that is too short by a factor of at least 500. That leaves only thermonuclear fusion. The sun, as we shall see, does not burn coal but hydrogen, and in a nuclear furnace, not an atomic or chemical one.

The fusion reaction in the sun is a multistep process, in which hydrogen is burned into helium, hydrogen being the "fuel" and helium the "ashes." Figure 10 shows the *proton–proton* (p–p) cycle by which this is accomplished.

The p–p cycle starts with the thermal collision of two protons ($^1H + {}^1H$) to form a deuteron (^2H), with the simultaneous creation of a positron (e^+) and a neutrino (ν). The positron very quickly encounters a free electron (e^-) in the sun and both particles annihilate, their rest-mass energy appearing as two gamma-ray photons (γ).

In Fig. 10, we follow the course of a pair of such events, shown in the top row of the figure. Such events are extremely rare. In fact, only once in $\sim 10^{26}$ proton–proton collisions is a deuteron formed; in the vast majority of cases, the two protons simply rebound elastically from each other. It is the slowness of this "bottleneck" or "safety valve" process that regulates the rate of energy production and keeps the sun from exploding. Interestingly, in spite of this slowness, there are so very many protons in the huge and dense volume of the sun's core that deuterium is produced there in this way at the rate of $\sim 10^{12}$ kg/s!

Once a deuteron (^2H) has been produced it quickly collides with another proton and forms a ^3He nucleus, as the second row of Fig. 10 shows. Two such ^3He nuclei may eventually (within $\sim 10^5$ y; there is plenty of time) find each other, forming an α particle (^4He) and two protons, as the third row in the figure shows.

Taking an overall view, we see from Fig. 10 that the p–p cycle amounts to the combination of four protons and two electrons to form an α particle, two neutrinos, and four gamma rays. Thus

$$4\,^1H + 2e^- \rightarrow {}^4He + 2\nu + 4\gamma. \tag{7}$$

Now, in a formal way, let us add two electrons to each side of Eq. 7, yielding

$$(4\,^1H + 4e^-) \rightarrow ({}^4He + 2e^-) + 2\nu + 4\gamma. \tag{8}$$

The quantities in the first two parentheses then represent *atoms* (not bare nuclei) of hydrogen and of helium.

The energy release in the reaction of Eq. 8 is

$$\begin{aligned} Q &= \Delta m\, c^2 \\ &= [(4)(1.007825\text{ u}) - 4.002603\text{ u}][932\text{ MeV/u}] \\ &= 26.7\text{ MeV}, \end{aligned}$$

in which 1.007825 u is the mass of a hydrogen atom and 4.002603 u that of a helium atom; neutrinos and gamma ray photons have no rest mass and thus do not enter into the calculation of the disintegration energy.

This same value of Q follows (as it must) by adding up the Qs for the separate steps of the proton–proton cycle in Fig. 10. Thus

$$\begin{aligned} Q &= (2)(0.42\text{ MeV}) + (2)(1.02\text{ MeV}) \\ &\quad + (2)(5.49\text{ MeV}) + 12.86\text{ MeV} \\ &= 26.7\text{ MeV}. \end{aligned}$$

About 0.5 MeV of this energy is carried out of the sun by the two neutrinos in Eq. 8; the rest (= 26.2 MeV) is deposited in the core of the sun as thermal energy.

The burning of hydrogen in the sun's core is alchemy on a grand scale in the sense that one element is turned into another. The medieval alchemists, however, were more interested in changing lead into gold than in changing hydrogen into helium. In a sense, they were on the right track, except that their furnaces were not hot enough. Instead of being at ~ 600 K, they should have been at least as high as $\sim 10^8$ K!

Hydrogen burning has been going on in the sun for

about 5×10^9 y, and calculations show that there is enough hydrogen left to keep the sun going for about the same length of time into the future. The sun's core, which by that time will be largely helium, will begin to cool and the sun will start to collapse under its own gravity. This will raise the core temperature and cause the outer envelope to expand, turning the sun into what the astronomers call a *red giant.**

If the core temperature heats up to about 10^8 K, energy can be produced once more by burning helium to make carbon. As a star evolves and becomes still hotter, other elements can be formed by other fusion reactions. However, elements more massive than $A \approx 56$ (^{56}Fe, ^{56}Co, ^{56}Ni) cannot be manufactured by further fusion processes. $A = 56$ marks the peak of the binding energy curve of Fig. 6 of Chapter 47, and fusion between nuclides beyond this point involves the consumption—not the production—of energy.

Elements beyond $A = 56$ are thought to be formed by neutron capture during cataclysmic stellar explosions that we call *supernovas.* In such an event the outer shell of the star is blown outward into space where it mixes with—and becomes part of—the tenuous medium that fills the space between the stars. It is from this medium, continually enriched by debris from stellar explosions, that new stars form, by condensation under the influence of gravity.

The fact that the earth abounds in elements heavier than hydrogen and helium suggests that our solar system has condensed out of interstellar material that contained the remnants of such explosions. Thus, all the elements around us—including those in our own bodies—were manufactured in the interiors of stars that no longer exist. As one scientist put it, "In truth, we are the children of the Universe."

Sample Problem 5 At what rate is hydrogen being consumed in the core of the sun by the $p-p$ cycle of Fig. 10?

We have seen that 26.2 MeV appears as thermal energy in the sun for every four protons that are consumed, a rate of 6.6 MeV/proton. We can express this energy transfer rate as

$$\frac{dE}{dm} = (6.6 \text{ MeV/proton})\left(\frac{1 \text{ proton}}{1.67 \times 10^{-27} \text{ kg}}\right)\left(\frac{1.60 \times 10^{-13} \text{ J}}{1 \text{ MeV}}\right)$$
$$= 6.3 \times 10^{14} \text{ J/kg}.$$

This tells us that the sun radiates away 6.3×10^{14} J for every kilogram of protons consumed. The hydrogen consumption rate is then the sun's power output ($= 3.9 \times 10^{26}$ W) divided by the above quantity, or

$$R = \frac{3.9 \times 10^{26} \text{ W}}{6.3 \times 10^{14} \text{ J/k}} = 6.2 \times 10^{11} \text{ kg/s.} \quad \text{(Answer)}$$

This seems like a large mass loss per second but—to keep things in perspective—we point out that the sun's mass is 2×10^{30} kg.

48-8 Controlled Thermonuclear Fusion

The first thermonuclear reaction to take place on earth occurred at Eniwetok Atoll on October 31, 1952, when the United States exploded a fusion device, generating an energy release equivalent to 10 million tons of TNT. The high temperatures and densities needed to initiate the reaction were provided by using a fission bomb as a trigger.

A sustained and controllable fusion power source—a fusion reactor—is more difficult to achieve. The goal, however, is being pursued vigorously in many countries around the world because many look to the fusion reactor as the power source of the future, at least as far as the generation of electricity is concerned.

The $p-p$ scheme displayed in Fig. 10 is not suitable for an earth-bound fusion reactor because it is hopelessly slow. The reaction succeeds in the sun only because of the enormous density of protons in the center of the sun. The most attractive reactions for terrestrial use appear to be the deuteron-deuteron ($d-d$) and the deuteron-triton ($d-t$) reactions:*

$$^2\text{H} + {}^2\text{H} \rightarrow {}^3\text{He} + n \qquad (d-d) \qquad Q = +3.27 \text{ MeV}, \tag{9}$$

$$^2\text{H} + {}^2\text{H} \rightarrow {}^3\text{H} + {}^1\text{H} \qquad (d-d) \qquad Q = +4.03 \text{ MeV}, \tag{10}$$

and

$$^2\text{H} + {}^3\text{H} \rightarrow {}^4\text{He} + n \qquad (d-t) \qquad Q = +17.59 \text{ MeV}. \tag{11}$$

Deuterium, whose isotopic abundance in normal hydrogen is 1 part in 6700, is available in unlimited quantities as a component of sea water. Proponents of power from

* The details of this event, which promises to be rather unpleasant, are spelled out in "When the Sun Swallows the Earth," *Sky & Telescope,* December 1987, News Notes.

* The nucleus of the hydrogen isotope ^{3}H is called the *triton.* It is a radionuclide, with a half-life of 12.3 y.

the nucleus have described our ultimate power choice—when we have burned up all our fossil fuels—as either "burning rocks" (fission of uranium extracted from ores) or "burning water" (fusion of deuterium extracted from water).

There are three requirements for a successful thermonuclear reactor:

1. A High Particle Density n. The density of interacting particles (deuterons, say) must be great enough to ensure that the $d-d$ collision rate is high enough. At the high temperatures required, the deuterium gas would be completely ionized into a neutral *plasma* consisting of deuterons and electrons.

2. A High Plasma Temperature T. The plasma must be hot. Otherwise the colliding deuterons will not be energetic enough to penetrate the Coulomb barrier that tends to keep them apart. In fusion research, temperatures are often reported by giving the value of kT (not $\frac{3}{2}kT$). A plasma ion temperature of 20 keV, corresponding to 23×10^7 K, has been achieved in the laboratory. This is more than 15 times higher than the sun's central temperature (1.3 keV or 1.5×10^7 K).

3. A Long Confinement Time τ. A major problem is containing the hot plasma long enough to ensure that its density and temperature remain sufficiently high for enough of the fuel to be fused. It is clear that no solid container can withstand the high temperatures that are necessary, so that clever confining techniques are called for; we shall describe them below.

It can be shown that, for the successful operation of a thermonuclear reactor, it is necessary to have

$$\boxed{n\tau > \sim 10^{20}\ \mathrm{s \cdot m^{-3}},} \tag{12}$$

a condition known as *Lawson's criterion,* the quantity $n\tau$ being known as the *Lawson number.* Equation 12 tells us that we have a choice between confining a lot of particles for a short time or confining fewer particles for a longer time. Beyond meeting this criterion, it is also necessary that the plasma temperature be high enough.

48-9 The Tokamak

Tokamak, a Russian-language acronym for "toroidal magnetic chamber," refers to a type of thermonuclear fusion device first developed in the USSR. Large tokamaks have been built and operated in several countries and several major new machines are in the design stage.

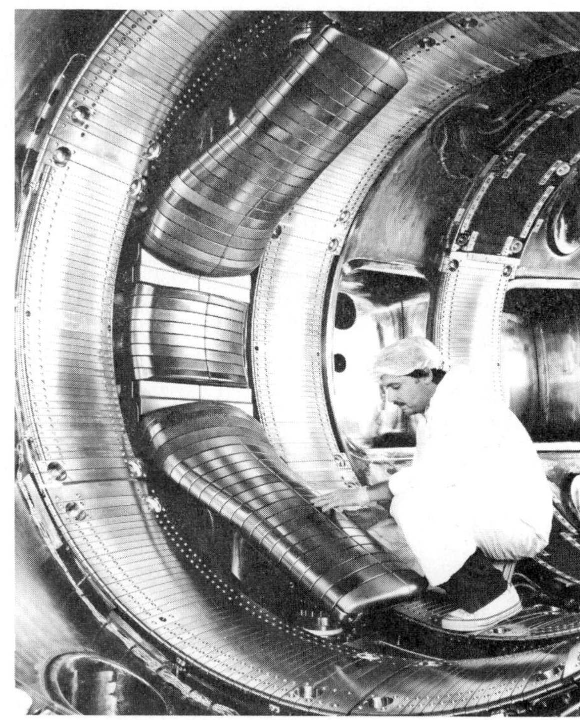

Figure 11 A technician making adjustments inside the torus of the Toroidal Fusion Test Reactor at Princeton.

In a tokamak the charged particles that make up the hot plasma are confined by a magnetic field configuration in the geometry of a doughnut or torus. Figure 11 shows a technician making adjustments inside the torus of the Toroidal Fusion Test Reactor at the Princeton Plasma Physics Laboratory.

As Fig. 12*a* suggests, the confining magnetic field in a tokamak is a sheath of helical lines of force—only one of which is shown in the figure—that spiral around the plasma "doughnut." The magnetic forces acting on the moving charges of the plasma keep the hot plasma from touching the walls of the vacuum chamber. Figures 12*b* and 12*c* show how the helical confining field is made up by combining a toroidal field (Fig. 12*b*) and a so-called poloidal field (Fig. 12*c*). The currents required to produce these fields are also shown. The current that generates the poloidal field is induced in the plasma itself, serving also to heat the plasma.

Figure 13 shows a plot of Lawson number ($= n\tau$) versus plasma temperature for various tokamaks and other magnetic confinement devices, in various coun-

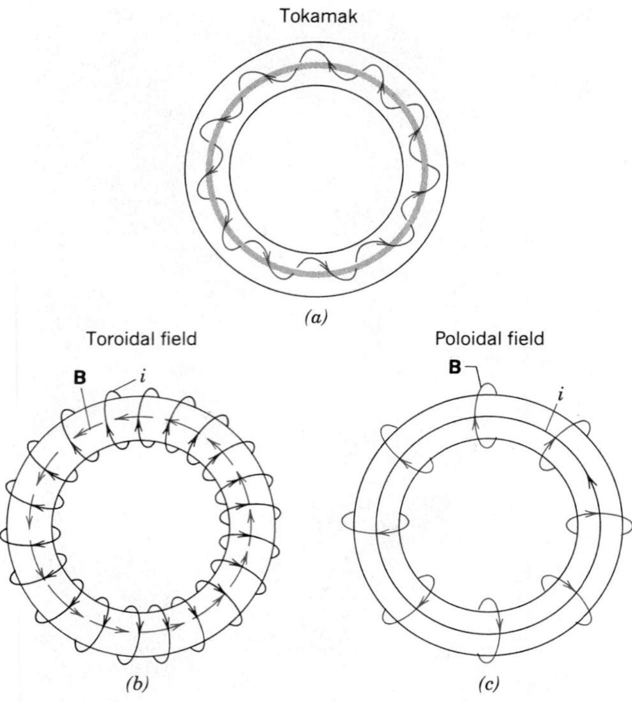

Figure 12 (a) The shaded ring suggests the confined plasma in a tokamak. The wavy line suggests the nature of the confining magnetic field. (b) The toroidal component of this magnetic field is established by currents that loop around the torus, as shown. (c) The poloidal component of this field is established by a current induced in the plasma, as shown.

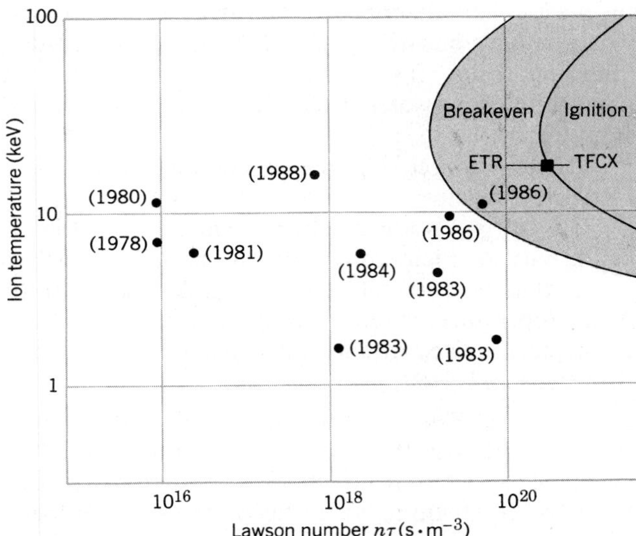

Figure 13 A plot of the Lawson number versus plasma temperature for a number of magnetic confinement fusion devices. The dots show the year of first successful operation, The square shows the expected performance of two machines that are under construction but not operational as of 1988.

tries. *Breakeven* corresponds to exceeding the Lawson criterion with a sufficiently hot, thermalized plasma; *ignition* corresponds to a self-sustaining thermonuclear reaction. Breakeven has been achieved (in the USSR) but no device has yet achieved ignition. In spite of the rapid progress being made at present, many formidable engineering problems remain, and a practical thermonuclear power plant does not seem possible before the early decades of the next century.

Sample Problem 6 Suppose that a tokamak achieves ignition with a plasma temperature of 10 keV and a confinement time of 980 ms. What would the particle density of its plasma have to be?

From Fig. 13 we see that the 10-keV temperature line intersects the curve marked "ignition" at a value of the Lawson number of about 4×10^{20} s · m^{-3}. (In making this last estimate, bear in mind that the scale is logarithmic.) The necessary particle density is then

$$n = \frac{4 \times 10^{20}\ \text{s} \cdot \text{m}^{-3}}{980 \times 10^{-3}\ \text{s}} \approx 4 \times 10^{20}\ \text{m}^{-3}. \quad \text{(Answer)}$$

(b) How does this number compare with the particle density of the atoms of an ideal gas at standard conditions?

The number density of atoms in an ideal gas at standard conditions is given by $n = N_A/V$, where N_A is the Avogadro constant. V (= 22,460 cm^3) is the volume occupied by one mole of an ideal gas at standard conditions. Thus

$$n = \frac{N_A}{V} = \frac{6.02 \times 10^{23}\ \text{mol}^{-1}}{22{,}460 \times 10^{-6}\ \text{m}^3} \approx 2.7 \times 10^{25}\ \text{m}^{-3}. \quad \text{(Answer)}$$

This is larger than the particle density of the plasma in (a) above by a factor of about 70,000.

48-10 Laser Fusion

A second technique for confining plasma so that thermonuclear reactions can take place is called *inertial confinement.* It involves compressing a fuel pellet by "zapping" it from all sides by laser beams (or particle beams), thus compressing it and increasing the temperature (to perhaps 10^8 K) and the particle density (by perhaps a factor of 10^3) so that thermonuclear fusion can occur. By comparison with magnetic confinement devices such as tokamaks, inertial confinement involves working with much higher particle densities for much shorter times.

Laser fusion is being investigated in many laboratories, in the United States and elsewhere. At the Lawrence Livermore Laboratory, for example, in the NOVA laser-fusion arrangement deuterium–tritium fuel pellets, each smaller than a grain of sand (see Fig. 14), are to be "zapped" by 10 synchronized high-powered laser pulses, symmetrically arranged around the pellet in a chamber shown in Fig. 15. The laser pulses are designed to deliver, in total, some 200 kJ of energy to each fuel pellet in less than a nanosecond. This is a delivered power of $\sim 2 \times 10^{14}$ W during the pulse, which is roughly 100 times the total installed electric power generating capacity of the world!

Figure 15 The target chamber of the NOVA inertial confinement fusion facility at the Lawrence Livermore Laboratory. The photo shows several of the 10 laser beam tubes. The fuel pellets shown in Fig. 14 are to be exploded at the center of this chamber by the simultaneous impact of energetic pulses from the laser beams.

Figure 14 The small spheres on the dime are deuterium–tritium fuel pellets, designed to be used in the laser fusion chamber of Fig. 15.

In an operating thermonuclear reactor of the laser-fusion type, it is visualized that fuel pellets would be exploded—like miniature hydrogen bombs—at a rate of perhaps 10–100 per second. The feasibility of laser fusion as the basis of a thermonuclear power reactor has not been demonstrated as of 1988, but research is continuing at a vigorous pace.

Sample Problem 7 Suppose that a fuel pellet in a laser-fusion device is made of a liquid deuterium–tritium mixture containing equal numbers of deuterium and tritium atoms. The density d (= 200 kg/m³) of the pellet is increased by a factor of 10^3 by the action of the laser pulses. (a) How many particles per unit volume (either deuterons or tritons) does the pellet contain in its compressed state?

We can write, for the density d^* of the compressed pellet,

$$d^* = 10^3\, d = m_d\left(\frac{n}{2}\right) + m_t\left(\frac{n}{2}\right)$$

in which n is the number of particles per unit volume (either deuterons or tritons) in the compressed pellet, m_d is the mass of a deuterium atom, and m_t is the mass of a tritium atom. These atomic masses are related to the Avogadro constant N_A and to the corresponding atomic masses (A_d and A_t) by

$$m_d = \frac{A_d}{N_A} \quad \text{and} \quad m_t = \frac{A_t}{N_A}.$$

Combining the foregoing equations and solving for n leads to

$$n = \frac{2000\, d\, N_A}{A_d + A_t}$$
$$= \frac{(2000)(200\ \text{kg/m}^3)(6.02 \times 10^{23}\ \text{mol}^{-1})}{2.0 \times 10^{-3}\ \text{kg/mol} + 3.0 \times 10^{-3}\ \text{kg/mol}}$$
$$\doteq 4.8 \times 10^{31}\ \text{m}^{-3}. \qquad \text{(Answer)}$$

(b) According to Lawson's criterion, for how long must the pellet maintain this particle density if breakeven operation is to take place?

From the foregoing we have, where L is the Lawson number,

$$\tau = \frac{L}{n} = \frac{\sim 10^{20}\ \text{s} \cdot \text{m}^{-3}}{4.8 \times 10^{31}\ \text{m}^{-3}} \approx 10^{-12}\ \text{s}. \qquad \text{(Answer)}$$

The pellet must remain compressed for at least this long if breakeven operation is to occur. (It is also necessary for the effective temperature to be suitably high.)

REVIEW AND SUMMARY

Energy from the Nucleus

Table 1 shows that nuclear processes are about a million times more effective, per unit mass, than chemical processes in transforming matter into energy.

Induced Fission

Equation 1 shows a *fission* of ^{235}U induced by thermal neutrons. Equations 2 and 3 show the beta-decay chains of the primary fragments. Sample Problem 1 shows that $Q \approx 200$ MeV for such fission events.

A Model for Fission

Fission can be understood in terms of the model of a charged liquid drop carrying a certain excitation energy: see Fig. 3. A potential barrier (see Fig. 4) must be tunneled if fission is to occur. Table 2 shows that fissionability depends on the relationship between the barrier height E_b and the excitation energy E_n.

Chain Reactions and Reactors

The free neutrons released during fission make possible a fission *chain reaction.* In a practical reactor the three problems raised starting on page 1107 must be satisfactorily solved. Figure 5 shows the neutron balance for a typical reactor. Figure 6 suggests the outlines of a complete nuclear power plant; see also Sample Problem 2.

Fusion

The release of energy by the *fusion* of two light nuclei is inhibited by their mutual Coulomb barrier; see Sample Problem 4. Fusion can occur in bulk matter only if the temperature is high enough (that is, if the particle energy is high enough) for appreciable barrier tunneling to occur. Figure 9 shows (1) the energy distribution $n(K)$ for protons at the central temperature of the sun, and (2) the proton–proton barrier penetrability factor $p(K)$. The figure deserves careful study.

The Proton-proton Cycle

The sun's energy arises mainly from the thermonuclear "burning" of hydrogen to form helium by the *proton–proton cycle* outlined in Fig. 10. The overall Q is 26.7 MeV per cycle; see Sample Problem 5.

Element Building

Elements up to $A = 56$ (the peak of the binding-energy curve of Fig. 6 in Chapter 47) can be built up by other thermonuclear processes once the hydrogen fuel supply of a star has been exhausted. Heavier elements are probably formed by successive neutron captures in supernova explosions.

Controlled Fusion

Controlled *thermonuclear fusion* for power generation has not yet been achieved, even on a laboratory scale. The d–d and the d–t reactions (Eqs. 9–11) are contemplated. A successful fusion reactor must satisfy *Lawson's criterion,*

Lawson's Criterion

$$n\tau > \sim 10^{20}\ \text{s} \cdot \text{m}^{-3}, \qquad [12]$$

and must have a plasma temperature T greater than about 10^8 K ($kT \approx 9$ keV). Confining such a hot plasma is a major problem. In the tokamak approach the plasma is generated as a torus and is confined by a helical magnetic field; see Fig. 12. Plasma heating is done by an induced toroidal current. Figure 13 shows the progress made by tokamaks toward achieving break-even performance. See Sample Problem 6.

Laser Fusion

In *laser fusion* a deuterium-tritium fuel pellet is "zapped" with extremely intense laser pulses, compressing it so that the d–t fusion reaction (Eq. 11) occurs before the interacting particles separate; this is called *inertial confinement.* See Sample Problem 7.

QUESTIONS

1. Can you say, from examining Table 1, that one source of energy, or of power, is better than another? If not, what other considerations enter?

2. To which of the processes in Table 1 does the relationship $E = \Delta m\, c^2$ apply?

3. Of the two fission fragment tracks shown in Fig. 1, which fragment has the larger (*a*) momentum? (*b*) kinetic energy? (*c*) speed? (*d*) mass?

4. In the generalized equation for the fission of ^{235}U by thermal neutrons, $^{235}\text{U} + n \rightarrow \text{X} + \text{Y} + bn$, do you expect the Q of the reaction to depend on the identity of X and Y?

5. Is the fission fragment curve of Fig. 2 necessarily symmetrical about its central minimum? Explain your answer.

6. In the chain decays of the primary fission fragments, see Eqs. 2 and 3, why do no e^+ decays occur?

7. The half-life of ^{235}U is 7.0×10^8 y. Discuss the assertion that if it had turned out to be shorter by a factor of ten or so, there would not be any atomic bombs today.

8. ^{238}U is not fissionable by thermal neutrons. What minimum neutron energy do you think would be necessary to induce fission in this nuclide?

9. The half-life for the decay of ^{235}U by α emission is 7×10^8 y; by spontaneous fission, acting alone, it would be 3×10^{17} y. Both are barrier tunneling processes, as Figure 9 in Chapter 47 and Figure 4 in Chapter 48 reveal. Why this enormous difference in barrier tunneling probability?

10. Compare a nuclear reactor with a coal fire. In what sense does a chain reaction occur in each? What is the energy-releasing mechanism in each case?

11. Not all neutrons produced in a reactor are destined to initiate a fission event. What happens to those which do not?

12. Explain just what is meant by the statement that in a reactor core neutron leakage is a surface effect and neutron production is a volume effect.

13. Explain the purpose of the moderator in a nuclear reactor. Is it possible to design a reactor that does not need a moderator? If so, what are some of the advantages and disadvantages of such a reactor?

14. Describe how to operate the control rods of a nuclear reactor (*a*) during initial start-up, (*b*) to reduce the power level, and (*c*) on a long-term basis, as fuel is consumed.

15. A reactor is operating at full power with its multiplication factor k adjusted to unity. If the reactor is now adjusted to operate stably at half power, what value must k now assume?

16. Separation of the two isotopes ^{238}U and ^{235}U from natural uranium requires a physical method, such as diffusion, rather than a chemical method. Explain why.

17. A piece of pure ^{235}U (or ^{239}Pu) will spontaneously explode if it is larger than a certain "critical size." A smaller piece will not explode. Explain.

18. The earth's core is thought to be made of iron because, during the formation of the earth, heavy elements such as iron would have sunk toward the earth's center and lighter elements, such as silicon, would have floated upward to form the earth's crust. However, iron is far from the heaviest element. Why isn't the earth's core made of uranium?

19. The sun's energy is assumed to be generated by nuclear reactions such as the proton–proton cycle. What alternative ways of generating solar energy were proposed in the past, and why were they rejected?

20. Elements up to mass number 56 are created by thermonuclear fusion in the cores of stars. Why are heavier elements not also created by this process?

21. Do you think that the thermonuclear fusion reaction controlled by the two curves plotted in Figure 9 necessarily has its maximum effectiveness for the energy at which the two curves cross each other? Explain your answer.

22. Why does it take so long ($\sim 10^6$ y!) for γ-ray photons generated by nuclear reactions in the sun's central core to diffuse to the surface? What kinds of interactions do they have with the protons, α particles, and electrons that make up the core?

23. The primordial matter of the early universe is thought to have been largely hydrogen. Where did all the silicon in the earth come from? All the gold?

24. Do conditions at the core of the sun satisfy Lawson's criterion for a sustained thermonuclear fusion reaction? Explain.

25. To achieve ignition in a tokamak, why do you need a high plasma temperature? A high density of plasma particles? A long confinement time?

26. Which would generate more radioactive waste products, a fission reactor or a fusion reactor?

27. Does Lawson's criterion hold both for tokamaks and for laser fusion devices?

EXERCISES AND PROBLEMS

Section 48-2 Nuclear Fission: The Basic Process

1E. (*a*) How many atoms are contained in 1.0 kg of pure ^{235}U? (*b*) How much energy, in joules, is produced by the complete fissioning of 1.0 kg of ^{235}U? Assume $Q = 200$ MeV. (*c*) For how long would this energy light a 100-W lamp?

2E. The fission properties of the plutonium isotope ^{239}Pu are very similar to those of ^{235}U. The average energy released per fission is 180 MeV. How much energy, in MeV, is liberated if all the atoms in 1.0 kg of pure ^{239}Pu undergo fission?

3E. At what rate must ^{235}U nuclei undergo fission by neutrons to generate 1.0 W? Assume that $Q = 200$ MeV.

4E. Fill in the following table, which refers to the generalized fission reaction

$$^{235}\text{U} + n \rightarrow \text{X} + \text{Y} + bn.$$

X	Y	b
^{140}Xe	—	1
^{139}I	—	2
—	^{100}Zr	2
^{141}Cs	^{92}Rb	—

5E. Verify that, as stated in Section 48-2, neutrons in equilibrium with matter at room temperature, 300 K, have an average kinetic energy of about 0.04 eV.

6E. Calculate the disintegration energy Q for the fission of ^{52}Cr into two equal fragments. The needed masses are ^{52}Cr, 51.94051 u; and ^{26}Mg, 25.98259 u. Discuss your result.

7E. Calculate the disintegration energy Q for the fission of ^{98}Mo into two equal parts. The needed masses are ^{98}Mo, 97.90541 u; and ^{49}Sc, 48.95002 u. If Q turns out to be positive, discuss why this process does not occur spontaneously.

8E. Calculate the energy released in the fission reaction

$$^{235}\text{U} + n \rightarrow {}^{141}\text{Cs} + {}^{93}\text{Rb} + 2n.$$

Needed atomic masses are

^{235}U	235.04392 u	^{93}Rb	92.92157 u
^{141}Cs	140.91963 u	n	1.00867 u.

9E. ^{235}U decays by α emission with a half-life of 7.0×10^8 y. It also decays (rarely) by spontaneous fission, and if the α decay did not occur, its half-life due to this process alone would be 3.0×10^{17} y. (*a*) At what rate do spontaneous fission decays occur in 1.0 g of ^{235}U? (*b*) How many α-decay events are there for every spontaneous fission event?

10P. Verify that, as reported in Table 1, the fission of the ^{235}U in 1.0 kg of UO_2 (enriched so that ^{235}U is 3.0% of the total uranium) could keep a 100-W lamp burning for 690 y.

11P. Consider the fission of ^{238}U by fast neutrons. In one fission event no neutrons were emitted and the final stable end products, after the beta-decay of the primary fission fragments, were ^{140}Ce and ^{99}Ru. (*a*) How many beta-decay events were there in the two beta-decay chains, considered together. (*b*) Calculate Q. The relevant atomic masses are

^{238}U	238.05079 u	^{140}Ce	139.90543 u
n	1.00867 u	^{99}Ru	98.90594 u.

12P. In a particular fission event of ^{235}U by slow neutrons, it happens that no neutron is emitted and that one of the primary fission fragments is ^{83}Ge. (*a*) What is the other fragment? (*b*) How is the disintegration energy $Q = 170$ MeV split between the two fragments? (*c*) Calculate the initial speed of each fragment.

13P. Assume that just after the fission of $^{236}U^*$ according to Eq. 1, the resulting ^{140}Xe and ^{94}Sr nuclei are just touching at their surfaces. (*a*) Assuming the nuclei to be spherical, calculate the coulomb potential energy (in MeV) of repulsion between the two fragments. (*Hint:* Use Equation 3 in Chapter 47 to calculate the radii of the fragments.) (*b*) Compare this energy with the energy released in a typical fission process. In what form will this energy ultimately appear in the laboratory?

14P. A $^{236}U^*$ nucleus undergoes fission and breaks up into two middle-mass fragments, ^{140}Xe and ^{96}Sr. (*a*) By what percentage does the surface area of the ^{236}U nucleus change during this process? (*b*) By what percentage does its volume change? (*c*) By what percentage does its electrostatic potential energy change? The potential energy of a uniformly charged sphere of radius r and charge Q is given by

$$U = \frac{3}{5}\left(\frac{Q^2}{4\pi\epsilon_0 r}\right).$$

Section 48-4 The Nuclear Reactor

15E. A 200-MW fission reactor consumes half its fuel in 3 years. How much ^{235}U did it contain initially? Assume that all the energy generated arises from the fission of ^{235}U and that this nuclide is consumed only by the fission process. See Sample Problem 2.

16E. Repeat Exercise 15 taking into account nonfission neutron capture by the ^{235}U. See Sample Problem 2.

17E. ^{238}Np has an activation energy for fission of 4.2 MeV. To remove a neutron from this nuclide requires an energy expenditure of 5.0 MeV. Is ^{237}Np fissionable by thermal neutrons?

18P. *Radionuclide power sources.* The thermal energy generated when radiations from radionuclides are absorbed in matter can be used as the basis for a small power source for use in satellites, remote weather stations, and so on. Such radionuclides are manufactured in abundance in nuclear power reactors and may be separated chemically from the spent fuel. One suitable radionuclide is ^{238}Pu ($\tau = 87.7$ y) which is an α-emitter with $Q = 5.50$ MeV. At what rate is thermal energy generated in 1.0 kg of this material?

19P. *Radionuclide power sources* (see Problem 18). Among the many fission products that may be extracted chemically from the spent fuel of a nuclear power reactor is ^{90}Sr ($\tau = 29$ y). It is produced in typical large reactors at the rate of about 18 kg/y. By its radioactivity it generates thermal energy at the rate of 0.93 W/g. (*a*) Calculate the effective disintegration energy Q_{eff} associated with the decay of a ^{90}Sr nucleus. (Q_{eff} includes contributions from the decay of the ^{90}Sr daughter products in its decay chain but not from neutrinos, which escape totally from the sample.) (*b*) It is desired to construct a power source generating 150 W (electric) to use in operating electronic equipment in an underwater acoustic beacon. If the source is based on the thermal energy generated by ^{90}Sr and if the efficiency of the thermal-electric conversion process is 5.0%, how much ^{90}Sr is needed?

20P. *The Breeder Reactor.* Many fear that helping additional nations develop nuclear power reactor technology will increase the likelihood of nuclear war because reactors can be used not only to produce energy but, as a by-product through neutron capture with inexpensive ^{238}U, to make ^{239}Pu, which is a 'fuel' for nuclear bombs. What simple series of reactions involving neutron capture and beta-decay would yield this plutonium isotope?

21P. In an atomic bomb (A-bomb), energy release is due to the uncontrolled fission of plutonium ^{239}Pu (or ^{235}U). The magnitude of the released energy is specified in terms of the mass of TNT required to produce the same energy release (bomb "rating"). One megaton (10^6 tons) of TNT produces 2.6×10^{28} MeV of energy. (*a*) Calculate the rating, in tons of TNT, of an atomic bomb containing 95 kg of ^{239}Pu, of which 2.5 kg actually undergoes fission. See Exercise 2. (*b*) Why is the other 92.5 kg of ^{239}Pu needed if it does not fission?

22P. A 66-kiloton A-bomb (see Problem 21) is fueled with pure ^{235}U, 4.0% of which actually undergoes fission. (*a*) How much uranium is in the bomb? (*b*) How many primary fission fragments are produced? (*c*) How many neutrons generated in the fissions are released to the environment? (On the average, each fission produces 2.5 neutrons.)

23P. The neutron generation time t_{gen} in a reactor is the average time needed for a fast neutron emitted in one fission to be slowed down to thermal energies by the moderator and to initiate another fission. Suppose that the power output of a reactor at time $t = 0$ is P_0. Show that the power output a time t later is $P(t)$ where

$$P(t) = P_0 k^{t/t_{\text{gen}}},$$

where k is the multiplication factor. Note that for constant power output $k = 1$.

24P. The neutron generation time (see Problem 23) of a particular power reactor is 1.3 ms. It is generating energy at the rate of 1200 MW. To perform certain maintenance checks, the power level must be temporarily reduced to 350 MW. It is desired that the transition to the reduced power level take 2.6 s. To what (constant) value should the multiplication factor be set to effect the transition in the desired time?

25P. The neutron generation time t_{gen} (see Problem 23) in a particular reactor is 1.0 ms. If the reactor is operating at a power level of 500 MW, about how many free neutrons are present in the reactor at any moment?

26P. A reactor operates at 400 MW with a neutron generation time of 30 ms. If its power increases for 5.0 min with a multiplication factor of 1.0003, find the power output at the end of the 5.0 min. See Problem 23.

27P. *Energy loss in a moderator.* (*a*) A neutron with initial kinetic energy K makes a head-on elastic collision with a resting atom of mass m. Show that the fractional energy loss of the neutron is given by

$$\frac{\Delta K}{K} = \frac{4m_n m}{(m + m_n)^2},$$

in which m_n is the neutron mass. (*b*) Find $\Delta K/K$ if the resting atom is hydrogen, deuterium, carbon, or lead. (*c*) If $K = 1.00$ MeV initially, how many such collisions would it take to reduce the neutron energy to thermal values (0.025 eV) if the material is deuterium, a commonly used moderator? (*Note:* In actual moderators, most collisions are not "head-on.")

Section 48-5 A Natural Nuclear Reactor

28E. How long ago was the ratio ^{235}U/^{238}U in natural uranium deposits equal to 0.15? See Sample Problem 3.

29E. The natural fission reactor discussed in Section 48-5 is estimated to have generated 15 gigawatt-years of energy during its lifetime. (*a*) If the reactor lasted for 200,000 y, at what average power level did it operate? (*b*) How much ^{235}U did it consume during its lifetime?

30P. In addition to ^{238}U, uranium mined today contains 0.72% of fissionable ^{235}U, too little to make reactor fuel for slow neutrons. For this reason, the natural uranium must be enriched or concentrated in ^{235}U. Both ^{235}U ($\tau = 7.0 \times 10^8$ y) and ^{238}U ($\tau = 4.5 \times 10^9$ y) are radioactive. How far back in time would natural uranium have been a practical reactor fuel, with a ^{235}U/^{238}U ratio of 3%?

31P. Some uranium samples from the natural reactor site described in Section 48-5 were found to be slightly *enriched* in ^{235}U, rather than depleted. Account for this in terms of neutron absorption by the abundant isotope ^{238}U and the subsequent beta and alpha decay of its products.

Section 48-6 Thermonuclear Fusion: The Basic Process

32E. Calculate the height of the Coulomb barrier for the head-on collision of two protons. The effective radius of a proton may be taken to be 0.80 fm. See Sample Problem 4.

33E. From information given in the text, collect and write down the approximate heights of the Coulomb barriers for (*a*) the alpha decay of ^{238}U, (*b*) the fission of ^{235}U by thermal neutrons, and (*c*) the head-on collision of two deuterons.

34E. Verify that the fusion of 1.0 kg of deuterium by the reaction

$$^2H + ^2H \rightarrow ^3He + n \qquad Q = +3.27 \text{ MeV},$$

could keep a 100-W lamp burning for 2.5×10^4 y.

35E. Methods other than heating the material have been suggested for overcoming the Coulomb barrier for fusion. For example, one might consider using particle accelerators. If you were to use two of them to accelerate two beams of deuterons directly toward each other so as to collide "head-on," (*a*) what voltage would each require to overcome the Coulomb barrier? (*b*) Would this voltage be difficult to achieve? (*c*) Why do you suppose this method is not presently used?

36P. The equation of the curve $n(K)$ in Fig. 9 is

$$n(K) = 1.13n \frac{K^{1/2}}{(kT)^{3/2}} e^{-K/kT},$$

where n is the total density of particles. At the center of the sun the temperature is 1.5×10^7 K and the mean proton energy $\overline{K}$ is 1.94 keV. Find the ratio of the density of protons at 5.00 keV to that at the mean proton energy.

37P. Calculate the Coulomb barrier height for two ^7Li nuclei, fired at each other with the same initial kinetic energy K. See Sample Problem 4. (*Hint:* Use Equation 3 in Chapter 47 to calculate the radii of the nuclei.)

38P. Expressions for the Maxwell speed and energy distributions for the molecules in a gas are given in Chapter 21. (*a*) Show that the *most probable energy* is given by

$$K_p = \tfrac{1}{2}kT.$$

Verify this result from the energy distribution curve of Fig. 9 for which $T = 1.5 \times 10^7$ K. (*b*) Show that the *most probable speed* is given by

$$v_p = \sqrt{\frac{2kT}{m}}.$$

Find its value for protons at $T = 1.5 \times 10^7$ K. (*c*) Show that *the energy corresponding to the most probable speed* (which is not the same as the most probable energy) is

$$K_{v,p} = kT.$$

Locate this quantity on the energy-distribution curve of Fig. 9.

Section 48-7 Thermonuclear Fusion in the Sun and Other Stars

39E. We have seen that Q for the overall proton–proton cycle is 26.7 MeV. How can you relate this number to the Q-values for the three reactions that make up this cycle, as displayed in Fig. 10?

40E. Show that the energy released when three alpha particles fuse to form ^{12}C is 7.27 MeV. The atomic mass of ^4He is 4.0026 u, and of ^{12}C is 12.0000 u.

41E. At the center of the sun the density is 1.5×10^5 kg/m^3 and the composition is essentially 35% hydrogen by mass and 65% helium. (*a*) What is the density of protons at the sun's center? (*b*) How much larger is this than the density of particles for an ideal gas at standard conditions of temperature and pressure?

42P. Verify the values of Q_1, Q_2, and Q_3 reported for the three reactions in Fig. 10. The needed atomic masses are

$$\begin{array}{llll} ^1H & 1.007825 \text{ u}, & ^3He & 3.016029 \text{ u}, \\ ^2H & 2.014102 \text{ u}, & ^4He & 4.002603 \text{ u}, \\ & e^{\pm} \ 0.0005486 \text{ u}. & & \end{array}$$

(*Hint:* Distinguish carefully between atomic and nuclear masses, and take the positrons properly into account.)

43P. Calculate and compare the energy released by (*a*) the fusion of 1.0 kg of hydrogen deep within the sun and (*b*) the fission of 1.0 kg of ^{235}U in a fission reactor.

44P. The sun has a mass of 2.0×10^{30} kg and radiates energy at the rate of 3.9×10^{26} W. (*a*) At what rate does the sun transfer its mass into energy? (*b*) What fraction of its original mass has the sun lost in this way since it began to burn hydrogen, about 4.5×10^9 y ago?

45P. (*a*) Calculate the rate at which the sun is generating neutrinos. Assume that energy production is entirely by the proton-proton cycle. (*b*) At what rate do solar neutrinos impinge on the earth?

46P. Coal burns according to

$$C + O_2 \rightarrow CO_2.$$

The heat of combustion is 3.3×10^7 J/kg of atomic carbon consumed. (*a*) Express this in terms of energy per carbon atom. (*b*) Express it in terms of energy per kilogram of the initial reactants, carbon and oxygen. (*c*) Suppose that the sun (mass $= 2.0 \times 10^{30}$ kg) were made of carbon and oxygen in combustible proportions and that it continued to radiate energy at its present rate of 3.9×10^{26} W. How long would it last?

47P. In certain stars the *carbon cycle* is more likely than the proton-proton cycle to be effective in generating energy. This cycle is

$$\begin{array}{rl} ^{12}C + ^1H \rightarrow ^{13}N + \gamma, & Q_1 = 1.95 \text{ MeV}, \\ ^{13}N \rightarrow ^{13}C + e^+ + \nu, & Q_2 = 1.19, \\ ^{13}C + ^1H \rightarrow ^{14}N + \gamma, & Q_3 = 7.55, \\ ^{14}N + ^1H \rightarrow ^{15}O + \gamma, & Q_4 = 7.30, \\ ^{15}O \rightarrow ^{15}N + e^+ + \nu, & Q_5 = 1.73, \\ ^{15}N + ^1H \rightarrow ^{12}C + ^4He, & Q_6 = 4.97. \end{array}$$

(*a*) Show that this cycle of reactions is exactly equivalent in its overall effects to the proton–proton cycle of Fig. 10. (*b*) Verify that both cycles, as expected, have the same Q.

48P. Let us assume that the core of the sun has one-eighth its mass and is compressed within a sphere whose radius is one-fourth of the solar radius. We assume further that the composition of the core is 35% hydrogen by mass and that essentially all of the sun's energy is generated there. If the sun continues to burn hydrogen at the rate calculated in Sample Problem 5, how long will it be before the hydrogen is entirely consumed? The sun's mass is 2.0×10^{30} kg.

49P. The effective Q for the proton–proton cycle of Fig. 10 is 26.2 MeV. (*a*) Express this as energy per kilogram of hydrogen consumed. (*b*) The luminosity of the sun is 3.9×10^{26} W. If its energy derives from the proton-proton cycle, at what rate is it losing hydrogen? (*c*) At what rate is it losing mass? Account for the difference in the results for (*b*) and (*c*). (*d*) The sun's mass is 2.0×10^{30} kg. If it loses mass at the constant rate calculated in (*c*), how long will it take before it loses 0.10% of its mass?

50P. After converting all of its hydrogen to helium, a particular star is 100% helium in composition. It now proceeds to convert the helium to carbon via the triple–alpha process

$$^4\text{He} + {}^4\text{He} + {}^4\text{He} \rightarrow {}^{12}\text{C} + 7.27 \text{ MeV};$$

see Exercise 40. The mass of the star is 4.6×10^{32} kg, and it generates energy at the rate of 5.3×10^{30} W. How long will it take to convert all the helium to carbon?

51P. Figure 16 shows an idealized schematic of a hydrogen bomb (H-bomb). The fusion fuel is deuterium ^{2}H. The high temperature and particle density needed for fusion are provided by an A-bomb "trigger," so arranged to yield an imploding, compressive shock wave upon the deuterium. The operative fusion reaction is

$$5\,{}^2\text{H} \rightarrow {}^3\text{He} + {}^4\text{He} + {}^1\text{H} + 2n.$$

(*a*) Calculate Q for the fusion reaction. For needed atomic masses see Problem 42. (*b*) Calculate the "rating" (see Problem 21) of the fusion part of the bomb if it contains 500 kg of deuterium, 30% of which undergoes fusion.

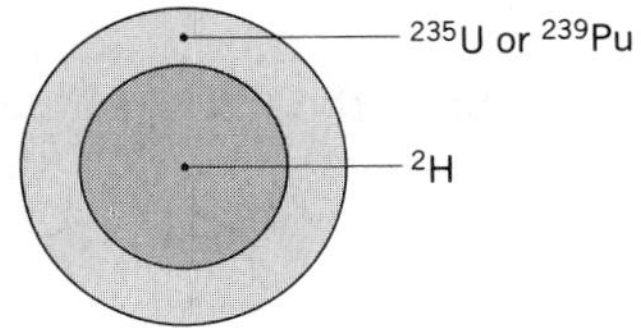

Figure 16 Problem 51.

Section 48-8 Controlled Thermonuclear Fusion

52E. Verify the Q values reported in Eqs. 9, 10, and 11. The needed masses are

^{1}H	1.007825 u	^{3}He	3.016029 u
^{2}H	2.014102 u	^{4}He	4.002603 u
^{3}H	3.016049 u	n	1.008665 u

53P. In the deuteron-triton fusion reaction of Eq. 11, how is the reaction energy Q shared between the α particle and the neutron? Neglect the relatively small kinetic energies of the two combining particles.

54P. Ordinary water consists of roughly 0.015% by mass of "heavy water," in which one of the two hydrogens is replaced with deuterium, ^{2}H. How much average fusion power could be obtained if we "burned" all of the ^{2}H in one liter of water in one day through the reaction $^2\text{H} + {}^2\text{H} \rightarrow {}^3\text{He} + n$?

Section 48-9 The Tokamak

55E. By the summer of 1985, the TFTR tokamak at the Princeton Plasma Physics Laboratory could be run consistently with plasma number densities of 3×10^{13} cm^{-3}, confinement times of 400 ms, and ion temperatures of 3 keV. Under special experimental conditions these same parameters are 6×10^{12} cm^{-3}, 100 ms, and 10 keV. Plot the two points representing these data on Fig. 13. Realizing that this represents just one machine while the figure represents many, what is your impression of progress being made in this type of fusion research?

Section 48-10 Laser Fusion

56E. Assume that a plasma temperature of 1×10^8 K is reached in a laser-fusion device. (*a*) What is the most probable speed of a deuteron at this temperature? (*b*) How far would such a deuteron move in the confinement time calculated in Sample Problem 7?

57P. The uncompressed radius of the fuel pellet of Sample Problem 7 is 20 μm. Suppose that the compressed fuel pellet "burns" with an efficiency of 10%. That is, only 10% of the deuterons and 10% of the tritons participate in the fusion reaction of Eq. 11. (*a*) How much energy is released in each such microexplosion of a pellet? (*b*) To how much TNT is each such pellet equivalent? The heat of combustion of TNT is 4.6 MJ/kg. (*c*) If a fusion reactor is constructed on the basis of 100 microexplosions per second, what power would be generated? (Note that part of this power must be used to operate the lasers.)

CHAPTER 49

QUARKS, LEPTONS, AND THE BIG BANG

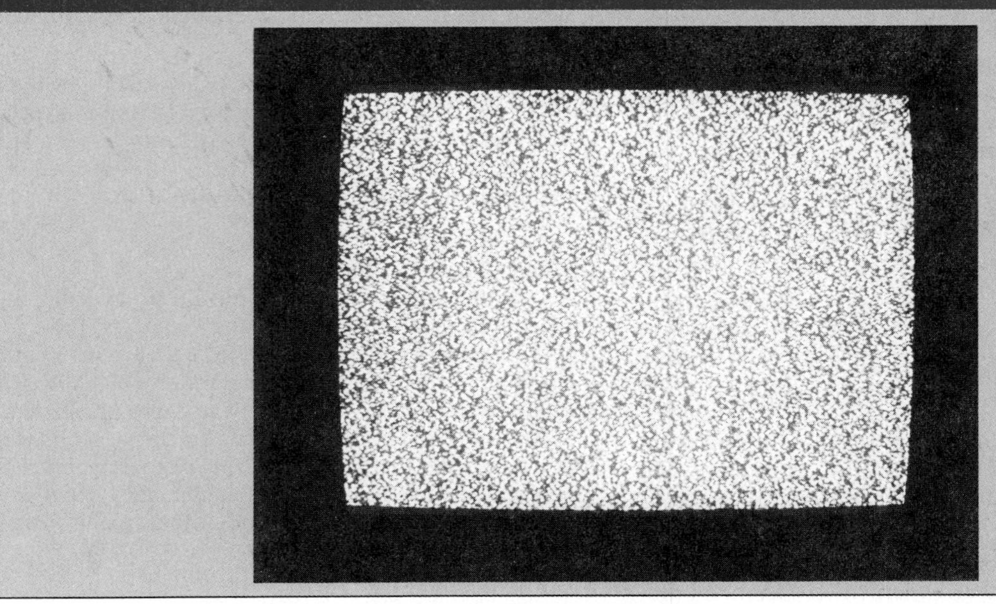

You can watch the birth pangs of the universe on any empty TV channel. A few percent of the flickering "snowflakes" that you see on the screen were triggered by photons of the microwave background radiation that, generated in the early universe, have been traveling toward your TV antenna for perhaps 15 billion years.

49-1 Life at the Cutting Edge

Physicists often refer to the theory of relativity and to the quantum theory as "modern physics," to distinguish them from the theories of Newtonian mechanics and Maxwellian electromagnetism, which are lumped together as part of "classical physics." As the years go by, the word "modern" seems less and less appropriate for theories whose foundations were laid down in the opening years of this century. Nevertheless, the label hangs on.

We present in this closing chapter two lines of investigation that are at once truly "modern" but at the same time have the most ancient of roots. They center around the questions:

What is the universe made of?

and

How did the universe come to be the way it is?

Progress in answering these questions has been rapid in the last few decades and there are those who think (perhaps unwisely) that definitive answers lie not far beyond our present horizons.

These two questions are not independent. As physicists bang particles together at higher and higher energies, using larger and larger accelerators, they come to realize that no conceivable earth-bound accelerator can generate particles with energies in the region toward which their theories are tending. There is only one source of particles with these energies and that is the universe itself within the first few minutes of its existence. The "quark soup" that constituted the universe at these early times on the cosmic clock is the ultimate testing ground for the theories of particle physics!

This chapter is different from the others in that we plan to bring you close to the frontier of current research and to identify some of the physicists whose efforts have brought us to this point. You will encounter a host of new terms and a veritable flood of particles with names that you should not try to remember. If you are temporarily bewildered, you are sharing the bewilderment of the physicists who lived through these developments and who at times saw nothing but increasing complexity with little hope of understanding. If you stick with us, however, you will come to share the excitement physicists felt as marvelous new accelerators poured out new results, as the theorists put forth ideas each more daring than the last, and as clarity finally sprang from obscurity. Hang in there!

S. Harris
"PARTICLES, PARTICLES, PARTICLES."

Figure 1 A cartoon by Sidney Harris, suggesting that dismay at the large numbers of particles found in nature is not restricted to physicists.

49-2 Particles, Particles, Particles*

In the 1930s, there were many who thought that the problem of the ultimate structure of matter was well on the way to being solved. The atom could be understood in terms of only three particles, the electron, the proton, and the neutron, "that trim little trio," in Martin Gardner's words. Quantum theory accounted well for the structure of the atom and for radioactive alpha-decay. The neutrino had been postulated and, although not yet observed, had been incorporated by Fermi into a successful theory of beta decay. There was hope that quantum theory, applied to protons and neutrons, would soon account for the structure of the nucleus. What else was there?

The euphoria did not last. The end of that same decade saw the beginning of a period of discovery of new particles that continues to this day. The new particles have names and symbols such as the *muon* (μ), the *pion* (π), the *kaon* (K), the *sigma* (Σ), and so on. Figure 1 accurately reflects the tenor of the times. All of the new particles are unstable, their half-lives ranging from $\sim 10^{-6}$ s to $\sim 10^{-23}$ s. This last value is so small that the very existence of such particles can be established only by indirect methods.

The new particles were first found in reactions triggered by the high-energy protons that stream in from space (the *cosmic rays*) and make nuclear collisions in the upper atmosphere. Increasingly, however, the new particles were produced in head-on collisions between protons or electrons accelerated to high energies in accelerators at places like Fermilab (near Chicago), CERN (near Geneva), and SLAC (at Stanford). The particle detectors grew in sophistication until (see Fig. 2) they rivaled in size and complexity the accelerators themselves of only a decade or so ago.

* You may wish to reread Section 2-8. See References 1-4 in the reference listing at the end of this chapter for readable general treatments of particle physics.

(a)

(b)

Figure 2 (*a*) The BEBC (Big European Bubble Chamber) at CERN, the European particle physics laboratory near Geneva. (*b*) A neutrino detector at CERN. Note the human figures in each case.

Today there are several hundred known particles. Naming them has strained the resources of the Greek alphabet and most are known only by an assigned number in a periodically-issued compilation. Fermi is said to have remarked that if he had known that there were so many particles whose properties he was expected to memorize, he would have taken up botany!

To make sense of this array of particles, we look for one or more simple criteria that will allow us to put the particles into one or the other of two categories. We can make such a preliminary "rough cut" in at least three ways:

1. Spin. All particles have an intrinsic angular momentum given by

$$\boxed{L = s\,\hbar} \qquad (1)$$

in which s, the *spin quantum number,* can have either half-integral ($\frac{1}{2}$, $\frac{3}{2}$, . . .) or integral (0, 1, . . .) values.

Particles with half-integral spins are called *fermions,* after Enrico Fermi, who (simultaneously with Dirac) developed the statistical rules that govern their behavior. Electrons, protons, and neutrons, all of which have $s = \frac{1}{2}$, are fermions.

Particles with integral spins are called *bosons,* after the Indian physicist Satendra Nath Bose, who (simultaneously with Einstein) developed the governing statistical rules for *these* particles. Photons, which have $s = 1$, are bosons; we shall soon meet other important particles in this class.

This may seem a trivial way to classify particles but it is very important for this reason:

> *Fermions obey the Pauli exclusion principle, which asserts that only a single particle can be assigned to a given quantum state. Bosons do* not *obey this principle. Any number of bosons can be assigned to a given state. Since particles prefer to be in states of lowest energy, bosons tend to cluster together in the lowest possible states.*

We have seen how important the Pauli principle is in assigning electrons (fermions) to quantum states of the atom.

2. Forces. We can also classify particles in terms of the forces that act on them. In Section 6–5 (which you may wish to reread) we outlined the four known fundamental forces. The *gravitational force* acts on *all* particles but its effects at the level of subatomic particles are so weak that we need not (yet!) consider them. The *electromagnetic force* acts on all *charged* particles; its effects are well-known and we can take them into account when we need to; we will largely ignore this force in the rest of this chapter.

We are left with the *strong force,* which is the force that binds the nucleus together, and the *weak force,* which is involved in beta decay and similar processes. The weak force acts on all particles, the strong force only on some particles.

Table 1 The Forces that Act on Each Category of Particle[a]

	Leptons	Hadrons: Mesons	Hadrons: Baryons
Fermions	WEAK		STRONG WEAK
Bosons		STRONG WEAK	

[a] No particles exist in categories corresponding to the shaded boxes. Thus, all leptons and all baryons are fermions and all mesons are bosons.

We can make a rough cut of the particles on the basis of whether or not the strong force acts on them.

Particles on which the *strong force* acts are called *hadrons.* Particles on which the strong force does *not* act, leaving the weak force as the dominant force, are called *leptons.* Protons, neutrons, and pions are hadrons; electrons and neutrinos are leptons. We shall soon meet other members in each class.

We can make a further rough cut among the hadrons because some of them (we call them *mesons*) are bosons; the pion is an example. Other hadrons (we call them *baryons*) are fermions; the proton is an example. Table 1 summarizes these two criteria for classifying particles.

3. Particles and Antiparticles. In 1928, Dirac predicted that the electron should have a positively charged counterpart. This particle, the *positron,* was discovered in the cosmic radiation in 1932 by Carl Anderson. It gradually became clear that *every* particle has a corresponding *antiparticle* with the same mass and spin but (if it is a charged particle) with a charge of the opposite sign.* We often represent an antiparticle by putting a bar over the symbol for the particle. Thus, p is the symbol for the proton and $\bar{p}$ for the antiproton.

When a particle meets its antiparticle, they can annihilate each other. That is, the particles can disappear, their combined rest energies becoming available to appear in other forms. For an electron annihilating with its antiparticle, this energy appears as two gamma-ray photons. Thus

$$e^- + e^+ \rightarrow \gamma + \gamma \quad (Q = 1.02\ \text{MeV}). \tag{2}$$

If the two particles are at rest when they annihilate, the photons share the disintegration energy equally between them and—to conserve momentum—they fly off in opposite directions.

One of the current puzzles in particle physics is the fact that the world we live in is a world of *particles,* not of antiparticles. This dominance of matter over antimatter certainly extends throughout our own galaxy. One can speculate that there are distant antimatter galaxies in which the atoms have nuclei with a negative charge, surrounded by clouds of positrons. One can contemplate the disaster that would occur if an antiphysicist from such a galaxy met a physicist from our galaxy in deep space and shook hands! The present view, however, is that the dominance of matter over antimatter extends throughout the universe and that there are no antiphysicists.

49-3 An Interlude

Before pressing on with the task of classifying the particles, let us step aside for a moment and capture some of the spirit of the particle enterprise by analyzing a typical particle event, that shown in the bubble-chamber photograph of Fig. 3*a*.

The tracks in this figure are streams of bubbles formed in the wake of energetic charged particles as they move through a chamber filled with liquid hydrogen. We can identify the particle that leaves a particular track—among other ways—by measuring the relative spacing between the bubbles. A magnetic field permeates the chamber, deflecting the tracks of positively charged particles counterclockwise and those of negatively charged particles clockwise. By measuring the radius of curvature of a track, we can calculate the momentum of the particle that made it. Table 2 shows some of properties of the particles that participate in the event of Fig. 3*a*.

Our tools for analysis are the laws of conservation of energy, of linear momentum, of angular momentum, and of charge, along with other conservation laws that we have not yet met. Figure 3*a* is one member only of a stereo pair so that, in practice, these analyses can be carried out in three dimensions.

* Particles and antiparticles also differ in the signs of other quantum numbers, which we have not yet identified.

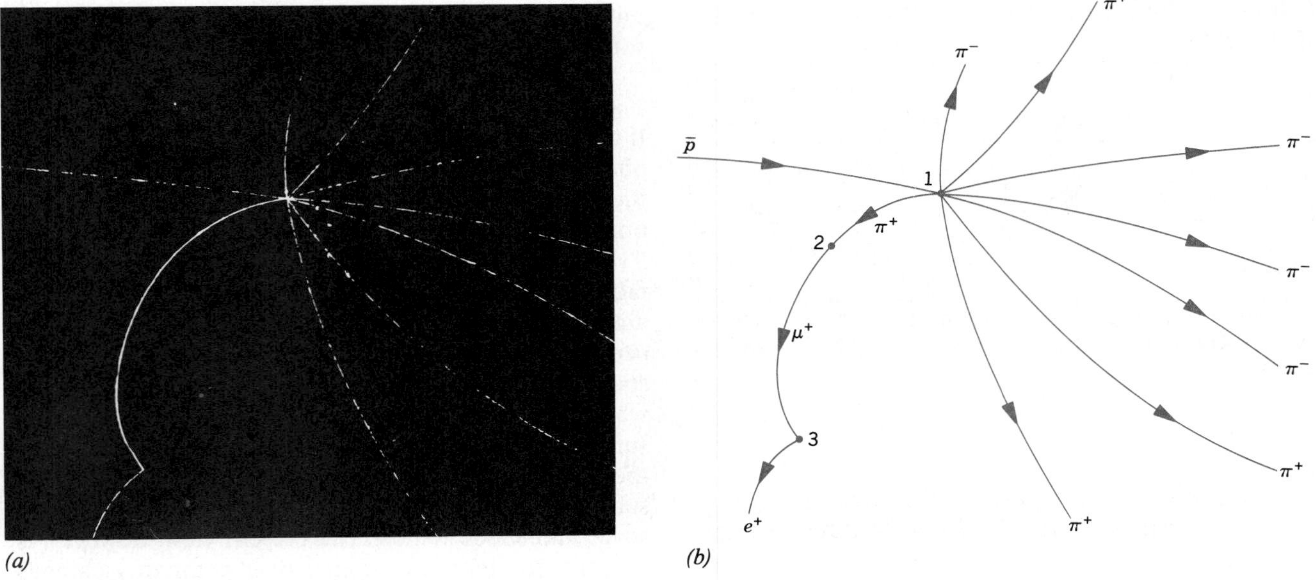

Figure 3 *(a)* A bubble-chamber photo of a series of events initiated by an antiproton that enters the chamber from the left. *(b)* The tracks are redrawn and labeled for clarity. The dots at points 1, 2, and 3 indicate the sites of a sequence of specific subevents that are described in the text. The tracks are curved because a magnetic field is present that acts on the moving charged particles.

The event of Fig. 3*a* is triggered by an energetic antiproton ($\bar{p}$) that, generated in an accelerator at the Lawrence Berkeley Laboratory, enters the chamber from the left. There are three separate subevents, occurring at points 1, 2, and 3 in Fig. 3*b*; we discuss each in turn.

1. Proton–Antiproton Annihilation. At point 1 in Fig. 3*b*, the initiating antiproton bangs into a proton of the chamber fluid and they annihilate each other. We can tell that the annihilation process occurred while the incoming antiproton was in flight because most of the particles generated in the encounter move in the forward direction, that is, toward the right in Fig. 3. From conservation of linear momentum, then, the incoming antiproton must also have had a forward momentum before the collision.

The available energy is twice the rest energy of the proton ($= 2 \times 938.3$ MeV or 1876.6 MeV) to which must be added the kinetic energy of the incoming antiproton at the moment of collision. This is enough energy to create a number of lighter particles and to endow them with kinetic energy. In this case, the annihilation pro-

Table 2 The Particles that Appear in the Event of Figure 3

Particle	*Symbol*	*Charge*	*Rest Energy* (MeV)	*Spin*	*Identity*	*Mean life*[a] (s)	*Antiparticle*
Neutrino	ν	0	0	$\frac{1}{2}$	lepton	stable	$\bar{\nu}$
Electron	e^-	-1	0.511	$\frac{1}{2}$	lepton	stable	e^+
Muon	μ^-	-1	105.7	$\frac{1}{2}$	lepton	2.2×10^{-6}	μ^+
Pion	π^+	$+1$	139.6	0	meson	2.6×10^{-8}	π^-
Proton	p	$+1$	938.3	$\frac{1}{2}$	baryon	stable	$\bar{p}$

[a] The *mean life* ($= 1/\lambda$) differs from the *half-life* [$= (\ln 2)/\lambda$]; see Section 47–3.

duces eight pions, four of them carrying a positive charge and four a negative charge.* The process is

$$p + \bar{p} \rightarrow 4\pi^+ + 4\pi^-. \quad (3)$$

The reaction of Eq. 3 is a *strong* interaction, all of the particles involved being hadrons.

Note that charge is conserved. We can write the charge of a particle as Qe in which Q is a *charge quantum number.* These numbers for the interaction of Eq. 3 are

$$(+1) + (-1) = 4 \times (+1) + 4 \times (-1),$$

which tells us that the net charge is zero before the interaction and remains zero afterwards.

As for energy conservation, we recall that the energy available from the annihilation process is a minimum of 1876.6 MeV. The rest energy of a pion is 139.6 MeV so that the rest energies of the eight pions amounts to 8×139.6 MeV or 1116.8 MeV. This leaves a substantial amount of energy (at least 1876.6 MeV − 1116.8 MeV or about 760 MeV) to distribute among the eight pions as kinetic energy.

2. Pion Decay. Pions are unstable particles, charged pions decaying with a mean life of 2.6×10^{-8} s. At point 2 in Fig. 3*b* one of the positive pions comes to rest in the chamber and decays spontaneously into a (anti)muon and a neutrino:

$$\pi^+ \rightarrow \mu^+ + \nu, \quad (4)$$

the latter particle (being uncharged) leaving no track. Both the muon and the neutrino are leptons, that is, they are particles on which the strong force does not act. The muon rest energy is 105.7 MeV, so that an energy of 139.6 MeV − 105.7 MeV or 33.9 MeV is available to share between the muon and the neutrino as kinetic energy.

The spin of the pion is zero and that of the muon and the neutrino are each one-half; therefore angular momentum can be easily conserved if the spins of the muon and the neutrino are aligned antiparallel to each other.

3. Muon Decay. Muons are also unstable, decaying with a mean life of 2.2×10^{-6} s. At point 3 in Fig. 3*b*, the muon produced in the reaction of Eq. 4 comes to rest in the chamber and decays spontaneously according to

$$\mu^+ \rightarrow e^+ + \nu + \bar{\nu}. \quad (5)$$

* For simplicity, we assume that no gamma ray photons or neutral particles (which would leave no track) are produced.

The track of the positron is clearly visible; again, we cannot see the neutrinos because they leave no tracks. The rest energy of the muon is 105.7 MeV and that of the electron is only 0.511 MeV, leaving 105.2 MeV to be shared as kinetic energy among the three particles produced in Eq. 5.

You may wonder: Why *two* neutrinos in Eq. 5? Why not just one, as in the decay of the pion in Eq. 4? One answer is that the spins of the muon, the electron, and the neutrino are each one-half; therefore, with only one neutrino, angular momentum would not be conserved in muon decay.

Sample Problem 1 In 1964 some experiments at the Brookhaven National Laboratory employed a focused beam of kaons (K^-). The kaons, whose kinetic energy was 5000 MeV, were generated in the Brookhaven Synchrotron and traveled a distance of 140 m through a highly evacuated beam tube to a bubble chamber, where the experiments took place.

The rest energy mc^2 of a kaon is 494 MeV and its half-life τ_0 against decay is 8.6×10^{-9} s. By what factor had the intensity of the kaon beam decayed while the particles were traveling from the synchrotron to the bubble chamber?

The kinetic energy of a kaon is related to its rest energy mc^2 by (see Eq. 37 of Chapter 42)

$$K = mc^2 (\gamma - 1)$$

so that the Lorentz factor γ is

$$\gamma = \frac{K}{mc^2} + 1 = \frac{5000 \text{ MeV}}{494 \text{ MeV}} + 1 = 11.1.$$

The half-life of these kaons in the reference frame of the laboratory is related to their half-life at rest by the time dilation factor (Eq. 8 of Chapter 42) or

$$\tau = \gamma\tau_0 = (11.1)(8.6 \times 10^{-9} \text{ s}) = 9.55 \times 10^{-8} \text{ s}.$$

These energetic kaons travel at essentially the speed of light. At this speed, a kaon beam could cover a distance of

$$L = c\tau = (3.00 \times 10^8 \text{ m/s})(9.55 \times 10^{-8} \text{ s}) = 28.7 \text{ m},$$

in a time τ, after which its intensity would have fallen to half its initial value. Over the full beam length the beam intensity will drop to

$$\left(\frac{1}{2}\right)^{(140/28.7)} = 0.034 \text{ or } 3.4\% \quad \text{(Answer)}$$

of its initial value, due to particle decay alone.

Such a beam loss—though unwelcome—is acceptable. Note, however, the *if it had not been for the time dilation effect,* the beam would have weakened by a factor of

$$\left(\frac{1}{2}\right)^{(140/28.7)(11.1)} \approx 5 \times 10^{-17}!!$$

Thus the time dilation effect resulted in a beam increase by a factor of nearly a million billion!

Sample Problem 2 A resting pion decays as described by Eq. 4, or

$$\pi^+ \rightarrow \mu^+ + \nu.$$

What is the kinetic energy of the muon? Of the neutrino?

From Table 2 the rest energies of the pion and the muon are 139.6 MeV and 105.7 MeV, respectively. The difference between these quantities must appear as kinetic energy of the muon and the neutrino, or

$$139.6 \text{ MeV} - 105.7 \text{ MeV} = 33.9 \text{ MeV} = K_\mu + K_\nu. \quad (6)$$

To conserve momentum, we must have

$$p_\mu = p_\nu$$

in which p_μ is the magnitude of the momentum of the muon and p_ν that of the neutrino. For convenience, we cast this in the form

$$(p_\mu c)^2 = (p_\nu c)^2 \quad (7)$$

Equation 41 of Chapter 42,

$$(pc)^2 = K^2 + 2Kmc^2, \quad (8)$$

gives the relativistic relation between the kinetic energy K of a particle and its momentum p. If we apply this relation to Eq. 7 we find

$$K_\mu^2 + 2K_\mu m_\mu c^2 = K_\nu^2; \quad (9)$$

note that $mc^2 = 0$ for the neutrino. Combining this result with Eq. 6 and solving for K_μ, we find

$$\begin{aligned} K_\mu &= \frac{(33.9 \text{ MeV})^2}{(2)(33.9 \text{ MeV} + m_\mu c^2)} \\ &= \frac{(33.9 \text{ MeV})^2}{(2)(33.9 \text{ MeV} + 105.7 \text{ MeV})} \\ &= 4.1 \text{ MeV}. \quad \text{(Answer)} \end{aligned}$$

The kinetic energy of the neutrino is then, from Eq. 6

$$\begin{aligned} K_\nu &= 33.9 \text{ MeV} - K_\mu = 33.9 \text{ MeV} - 4.1 \text{ MeV} \\ &= 29.8 \text{ MeV}. \quad \text{(Answer)} \end{aligned}$$

We see that, although the magnitude of the momentum of the two recoiling particles is the same, the neutrino gets the larger share (88%) of the kinetic energy.

Sample Problem 3 Protons in a bubble chamber are bombarded by energetic negative pions and the following reaction occurs

$$\pi^- + p \rightarrow K^- + \Sigma^+.$$

The rest energies of the particles involved are:

π^-	139.6 MeV	K^-	493.7 MeV
p	938.3 MeV	Σ^+	1189.4 MeV

What is the disintegration energy of the reaction?

The disintegration energy is given by

$$\begin{aligned} Q &= (m_\pi c^2 + m_p c^2) - (m_K c^2 + m_\Sigma c^2) \\ &= (139.6 \text{ MeV} + 938.3 \text{ MeV}) \\ &\quad - (493.7 \text{ MeV} + 1189.4 \text{ MeV}) \\ &= -605 \text{ MeV}. \quad \text{(Answer)} \end{aligned}$$

The reaction is *endothermic.* That is, if the proton is at rest, the incoming pion (π^-) must have a kinetic energy larger than a certain threshold value to make the reaction go. The threshold energy is larger than 605 MeV because linear momentum must be conserved, which means that the kaon (K^-) and the sigma (Σ^+) must not only be created but also be endowed with some kinetic energy. It can be shown by a relativistic calculation whose details are beyond our scope that the threshold energy for the incident pion is 907 MeV.

49-4 The Leptons

Now let us press on with our classification program for the particles. We turn first to the leptons, that is, to those particles on which the strong force does *not* act.

So far, we have encountered the familiar electron and the neutrino that accompanies it in beta decay. The muon, whose decay is described in Eq. 5, is another member of this family. Physicists gradually learned that the neutrino that appears in Eq. 4, associated with the production of a muon, is *not the same particle* as the neutrino produced in beta decay, associated with the appearance of an electron; see Section 47-5. We call the former the *muon neutrino* (symbol ν_μ) and the latter the *electron neutrino* (symbol ν_e) when it is necessary to distinguish between them. In Eq. 5, for example, one of the two neutrinos is a muon neutrino, the other being an electron neutrino.

These two neutrinos are known to be different particles because, if a beam of muon neutrinos (produced from pion decay as in Eq. 4) is allowed to strike a solid target, *only muons*—and never electrons—are produced. On the other hand, if electron neutrinos (pro-

Table 3 The Leptons[a]

Particle	*Symbol*	*Rest Energy* (MeV)	*Charge*	*Antiparticle*
Electron	e^-	0.511	-1	e^+
Electron neutrino[b]	ν_e	0	0	$\bar{\nu}_e$
Muon	μ^-	105.7	-1	μ^+
Muon neutrino[b]	ν_μ	0	0	$\bar{\nu}_\mu$
Tauon	τ^-	1784	-1	τ^+
Tauon neutrino[b]	ν_τ	0	0	$\bar{\nu}_\tau$

[a] All leptons have spin $\frac{1}{2}$ and are thus fermions.
[b] If the neutrino masses are not zero, they are at least very small. This is an open question as of 1988.

duced by the beta decay of fission products in a nuclear reactor) are allowed to fall on a solid target, *only electrons*—and never muons—are produced.

In addition to the electron and the muon, a third lepton, the *tauon,* was discovered in 1975.* It has its own associated neutrino, different still from the other two. Table 3 lists the known leptons. There are reasons for dividing the leptons into three "generations," each consisting of a particle (electron, muon, or tauon) and its associated neutrino. Leptons have no discernible internal structure, no measurable dimensions, and are believed to be truly fundamental particles.

49-5 A New Conservation Law

We are now ready to consider those particles, the baryons and the mesons, whose interactions are governed by the *strong* force. We start by adding another conservation law to the list of conservation laws that are more familiar to us, such as conservation of charge, energy, linear momentum, and angular momentum. It is the *conservation of baryon number.*

The decay process

$$p \rightarrow e^+ + \gamma \quad (Q = 937.8 \text{ MeV}) \tag{10}$$

never happens. We should be glad that it does not because otherwise all protons in the universe would gradually change into positrons, with disastrous consequences for the environment. Yet the decay process of Eq. 10 violates none of the conservation laws that we have so far put forward.

* See Reference 5.

We account for the apparent stability* of the proton—and for the absence of many other processes that might otherwise occur—by introducing a new quantum number, the *baryon number B,* and a new conservation law, the conservation of baryon number.

To every baryon we assign $B = +1$. To every antibaryon we assign $B = -1$. To mesons and leptons we assign $B = 0$.

We see that the process of Eq. 10 violates the law of conservation of baryon number:

$$(+1) \neq (0) + (0).$$

Baryon number conservation will prove useful in accounting for the many particle decays and reactions that—though not otherwise forbidden—simply do not occur.

Sample Problem 4 Analyze the proposed decay of the proton according to this scheme

$$p \rightarrow \pi^0 + \pi^+ \quad \text{(doesn't happen!)}$$

by testing it against the various conservation laws. (Both pions are mesons, with spin and baryon number both equal to zero. The rest energy of the π^0 meson is 135.0 MeV.)

We see at once that charge is conserved and that linear momentum can also be readily conserved. All that is necessary

* The proton may yet prove to be unstable, as some current theories predict. Its predicted halflife, however, is ~ 10^{30} y, many orders of magnitude greater than the age of the universe. Attempts to detect proton decay have so far proved unsuccessful.

is that the two pions move in opposite directions from the site of the resting proton, with momenta of equal magnitude.

The disintegration energy is found by subtracting the rest energies of the particles. Thus

$$\begin{aligned} Q &= (m_p c^2) - (m_0 c^2 + m_+ c^2) \\ &= (938.3 \text{ MeV}) - (135.0 \text{ MeV} + 139.6 \text{ MeV}) \\ &= 663.7 \text{ MeV}. \end{aligned}$$

The fact that Q is positive shows that the process cannot be ruled out on energy conservation grounds; the energy is there.

We have said that both pions have zero spin. The proton, however, has a spin of one-half. Thus angular momentum is *not* conserved and—for that reason alone—the process cannot occur.

Beyond that, baryon number is not conserved. For the proton, we have $B = +1$ and for the two pions we have $B = 0$. The process is thus doubly forbidden, violating two of the five conservation laws that we have examined.

Sample Problem 5 A particle identified as Ξ^- decays as follows:

$$\Xi^- \rightarrow \Lambda^\circ + \pi^-.$$

Both of these decay products are unstable. The following reactions occur in cascade until, ultimately, only stable products remain:

$$\begin{aligned} \Lambda^\circ &\rightarrow \eta + \pi^\circ \\ \eta &\rightarrow p + e^- + \nu \\ \pi^\circ &\rightarrow \gamma + \gamma \\ \pi^- &\rightarrow \mu^- + \nu \\ \mu^- &\rightarrow e^- + \nu + \nu \end{aligned}$$

(a) Write down the overall decay scheme for the Ξ^- particle.

Study of the above decay schemes shows that the overall decay scheme is

$$\Xi^- \rightarrow p + 4\nu + 2e^- + 2\gamma. \qquad \text{(Answer)}$$

All of the products on the right side are stable. Note that charge is conserved, the net charge quantum number being -1 on each side.

(b) Is the Ξ^- particle a meson or a baryon?

The proton in the above equation is a baryon (baryon number $= +1$). All of the other particles on the right side of the above equation have $B = 0$. Thus, from conservation of baryon number, the baryon number of the Ξ^- must be $+1$. Thus the particle is a *baryon.* If it were a meson, its baryon number would have been zero.

(c) What can you say about the spin of the Ξ^- particle?

All particles on the right side of the above equation except the gamma photons have a spin of $\frac{1}{2}$; the photon has a spin of 1. These can only add up to a half-integral spin for the Ξ^- particle. This is additional evidence that this particle is a baryon. If it were a meson, its spin would have been integral. (Actually, the spin of the Ξ^- particle is $\frac{1}{2}$; this particle is listed with other spin-$\frac{1}{2}$ baryons in Table 4.)

49-6 Another New Conservation Law!

Particles have more intrinsic properties than mass, charge, spin, and baryon number that we have listed so far. The first of these new properties emerged when it was observed that certain new particles, such as the kaon (K) and the sigma (Σ), always seemed to be produced in pairs. It seemed impossible to produce only one of them at a time. Thus, if a beam of energetic pions interacts with the protons in a bubble chamber, the reaction

$$\pi^+ + p \rightarrow K^+ + \Sigma^+ \tag{11}$$

often occurs. The reaction

$$\pi^+ + p \rightarrow \pi^+ + \Sigma^+, \tag{12}$$

which violates no conservation law known at the time, never occurs.

It was eventually proposed (by Murray Gell-Mann and independently by K. Nishijima in Japan) that certain particles possess a new property, called *strangeness,* with its own quantum number S and its own conservation law. The name arises from the fact that, before the identities of these new particles were pinned down, they became known in particle physics jargon (for reasons that do not concern us here) as "strange particles" and the label stuck.

The proton, neutron, and pion have $S = 0$, that is, they are not "strange." It was proposed, however, that the K^+ particle has a strangeness denoted by $S = +1$ and that Σ^+ has $S = -1$. Thus, strangeness is conserved in Eq. 11:

$$(0) + (0) = (+1) + (-1) \quad \text{(values of } S\text{)}$$

but is *not* conserved in Eq. 12:

$$(0) + (0) \neq (0) + (-1) \quad \text{(values of } S\text{).}$$

We say that the reaction of Eq. 12 does not occur because it violates the law of *conservation of strangeness.**

It may seem heavy-handed to invent a new property of particles just to account for a little puzzle like that posed by Eqs. 11 and 12. However, strangeness and its

* Conservation of strangeness holds for strong interactions only.

quantum number soon revealed themselves in many other areas in particle physics and it is now fully accepted as a legitimate particle attribute, on a par with charge and spin. To those who know and love particles, "strangeness" is no longer "strange."

Do not be misled by the whimsical character of the name. "Strangeness" is no more mysterious a property of particles than is "charge." Both are properties that particles may (or may not) have; each is described by an appropriate quantum number. Each obeys a conservation law. Still other properties that particles may possess appeared, with even more whimsical names, such as *charm* and *bottomness*. Before looking further into these matters, let us see how the new property of strangeness "earns its keep" by leading us to uncover important regularities in the properties of the particles.

49–7 The Eightfold Way

There are eight baryons — the neutron and the proton among them — that have a spin of one-half. Table 4 shows some of their properties. Figure 4*a* shows the fascinating pattern that emerges if we plot the strangeness of these baryons against their charge, using a sloping coordinate system. Six of the eight form a hexagon with the two remaining baryons at its center.

Let us turn now from the baryons to the mesons. There are nine of these (listed in Table 5) that have a spin of zero. If we plot them on a strangeness–charge diagram, as in Fig. 4*b*, the same fascinating pattern emerges! These and related plots, called the *Eightfold Way* patterns,* were proposed independently in 1961 by Murray Gell-Mann at the California Institute of Technology and by the Yuval Ne'eman at Imperial College, London (Fig. 5). The two patterns of Fig. 4 are only representative of a larger number of symmetrical patterns in which various groups of baryons and mesons can be displayed.

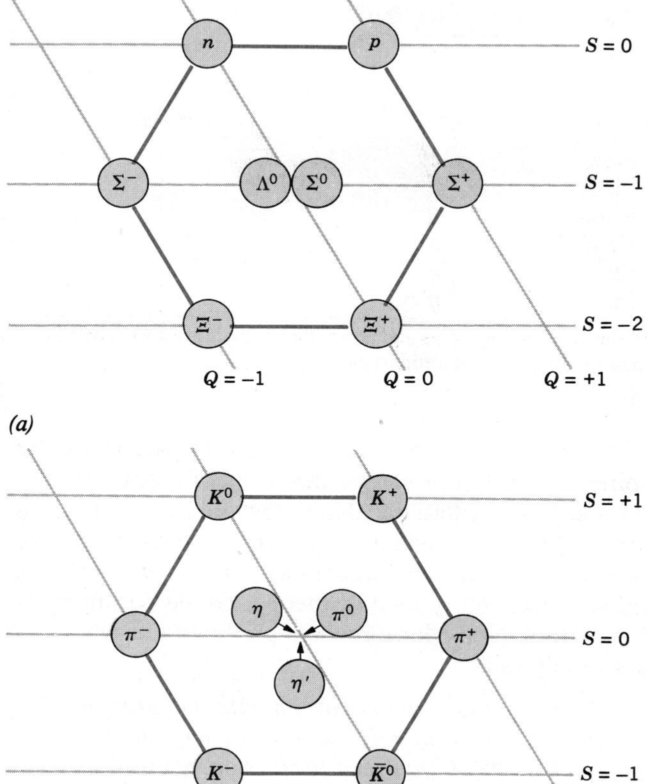

Figure 4 (*a*) The Eightfold Way pattern for the eight spin-$\frac{1}{2}$ baryons listed in Table 4. The particles are represented as points on a strangeness–charge plot, using a sloping axis for the charge quantum number. (*b*) A similar pattern for the nine spin-zero mesons listed in Table 5. Note in this case that particles lie opposite to their antiparticles; the three mesons in the center form their own antiparticles.

Table 4 The Eight Spin-One-Half Baryons[a]

Particle	*Symbol*	*Rest Energy* (MeV)	*Quantum Numbers* *Charge*	*Strangeness*
Proton	p	938.3	+1	0
Neutron	n	939.6	0	0
Lambda	Λ^0	1115.6	0	−1
Sigma	Σ^+	1189.4	+1	−1
Sigma	Σ^0	1192.5	0	−1
Sigma	Σ^-	1197.3	−1	−1
Xi	Ξ^0	1314.9	0	−2
Xi	Ξ^-	1321.3	−1	−2

[a] There are eight baryons with a spin of one-half; all are listed here. See also Figs. 4*a* and 7*a*.

* A borrowing from Eastern mysticism. The "Eight" refers to the eight quantum numbers (only a few of which we have defined here) that are involved in the symmetry-based theory that predicts the existence of the patterns.

Table 5 The Nine Spin-Zero mesons[a]

Particle	Symbol	Rest Energy (MeV)	Quantum Numbers: Charge	Quantum Numbers: Strangeness	Antiparticle
Pion	π^0	135.0	0	0	π^0
Pion	π^+	139.6	+1	0	π^-
Kaon	K^+	493.7	+1	+1	K^-
Kaon	K^0	497.7	0	+1	$\bar{K}^0$
Eta	η	548.8	0	0	η
Eta′	η'	957.6	0	0	η'

[a] Counting both particles and antiparticles, there are a total of nine spin-zero mesons. Note that the π^0, the η and the η' mesons are their own antiparticles.

The symmetry of the Eightfold Way pattern for the spin-$\frac{3}{2}$ baryons (not shown here) calls for *ten* particles arranged in a pattern like that of the tenpins in a bowling alley. However, when the pattern was first proposed, only *nine* such particles were known; the "head pin" was missing. In 1962, guided by theory and the symmetry of the pattern, Gell-Mann made a prediction in which he essentially said:

> *There exists a spin-$\frac{3}{2}$ baryon with a charge of -1, a strangeness of -3, and a rest energy of about 1680 MeV. If you look for this* omega minus *particle (as I propose to call it), I think you will find it.*

Figure 5 Murray Gell-Mann (left) and Yuval Ne'eman discovered the Eightfold Way patterns simultaneously and independently. Ne'eman, incidently, was serving as an Israeli Military Attache in London while working on this problem. Both men have made many substantial contributions to particle physics.

A team of physicists headed by Nicholas Samios of the Brookhaven National Laboratory took up the challenge and promptly found the "missing" particle, confirming all of its predicted properties.* There is nothing like the prompt experimental confirmation of a prediction to build confidence in a theory!

The Eightfold Way patterns bear the same relationship to particle physics that the periodic table does to chemistry. In each case, there is a pattern of organization in which vacancies (missing particles or missing elements) stick out like sore thumbs, guiding experimenters in their searches. In the case of the periodic table, its very existence strongly suggests that the atoms of the elements are not fundamental particles but have a common underlying structure. In the same way, the Eightfold Way patterns strongly suggest that the mesons and the baryons must have an underlying structure, in terms of which their properties can be understood. That structure is the *quark model,* which we now describe.

49-8 The Quark Model†

In 1964 Murray Gell-Mann and George Zweig independently pointed out that the Eightfold Way patterns can be understood in a simple way if the mesons and the baryons are built up out of subunits that Gell-Mann called *quarks.*‡ We deal first with three of them, called the *up quark* (symbol u), the *down quark* (symbol d), and

* See Reference 6.

† See Reference 7.

‡ The name comes from Joyce's *Finnegans Wake* ("Three quarks for Muster Mark!"). Joyce is said to have coined this word after listening to the squawks of seagulls.

the *strange quark* (symbol s), and we assign to them the properties displayed in the first three rows of Table 6. (The names of the quarks, along with those assigned to three other quarks that we shall meet later, have no meanings other than as convenient labels. Collectively, they are called the *quark flavors.* We could, for all the difference it would make, have called them vanilla, chocolate, and pistachio instead of up, down, and strange.)

The fractional charges of the quarks may jar you a little. However, withhold judgment until you see how neatly these fractional charges account for the observed integral charges of the mesons and the baryons. Quarks have not (yet) been convincingly observed in the laboratory as free particles, and theorists have put forward plausible reasons why this should be the case. In any event, the quark model is so useful that the failure to see free quarks is not regarded as a hindrance to their acceptance.

We have seen how we can put atoms together by combining electrons and nuclei. Now let us see how we can put mesons and baryons together by combining quarks. We state in advance that success will be complete. That is, for particles formed from the up, down, and the strange quark:

> *There is no known particle whose properties cannot be understood in terms of an appropriate combination of quarks. On the other hand, there is no possible quark combination to which there does not correspond an observed particle.*

Let us look first at the baryons.

Quarks and Baryons. Baryons are combinations of three quarks; see Fig. 6*a*. We see at once that we can add up three quarks ($B = +\frac{1}{3}$) to yield $B = +1$ (for a baryon). The spins work out also. With three spins of $\frac{1}{2}$ to work with, we can arrange them so that two spins are parallel and one antiparallel. This leads to $s = \frac{1}{2}$, which is the spin of all the baryons displayed in Table 4 and Fig. 4*a*. If all three spins are parallel, we have $s = \frac{3}{2}$; there are 10 baryons with a spin of this amount.*

Charges also work out, as we can see from three examples. The proton has a quark composition of *uud* so that its charge is

$$Q(uud) = (+\tfrac{2}{3}) + (+\tfrac{2}{3}) + (-\tfrac{1}{3}) = +1.$$

The neutron has a quark composition of *udd* and its charge is

$$Q(udd) = (+\tfrac{2}{3}) + (-\tfrac{1}{3}) + (-\tfrac{1}{3}) = 0.$$

The Σ^- particle has a quark composition of *dds* and its charge is

$$Q(dds) = (-\tfrac{1}{3}) + (-\tfrac{1}{3}) + (-\tfrac{1}{3}) = -1.$$

As Fig. 6*a* shows, the charge and the strangeness quantum numbers of all of the baryons displayed in Fig. 4*a* and Table 4 work out in exact agreement with the known values.

Quarks and Mesons. Mesons are quark–antiquark pairs; see Fig. 6*b*. This is consistent with the fact that the

* In the ground state of the baryon, we assume that its constituent quarks have no *orbital* angular momentum.

Table 6 The Quarks[a]

			Quantum Numbers			
Particle[b]	*Symbol*	*Mass*[c]	*Charge*	*Strangeness*	*Baryon number*	*Antiparticle*
Up	u	10	$+\frac{2}{3}$	0	$+\frac{1}{3}$	$\bar{u}$
Down	d	20	$-\frac{1}{3}$	0	$+\frac{1}{3}$	$\bar{d}$
Strange	s	200	$-\frac{1}{3}$	-1	$+\frac{1}{3}$	$\bar{s}$
Charm	c	3,000	$+\frac{2}{3}$	0	$+\frac{1}{3}$	$\bar{c}$
Bottom	b	9,000	$-\frac{1}{3}$	0	$+\frac{1}{3}$	$\bar{b}$
Top[d]	t	60,000	$+\frac{2}{3}$	0	$+\frac{1}{3}$	$\bar{t}$

[a] Quarks (as far as we know!) are truly fundamental particles. All quarks have spin $s = \frac{1}{2}$ and thus are fermions.
[b] These names are known as the quark "flavors."
[c] Masses are reported in terms of the electron mass, which is taken as unity.
[d] The top quark has not yet been observed as of 1988.

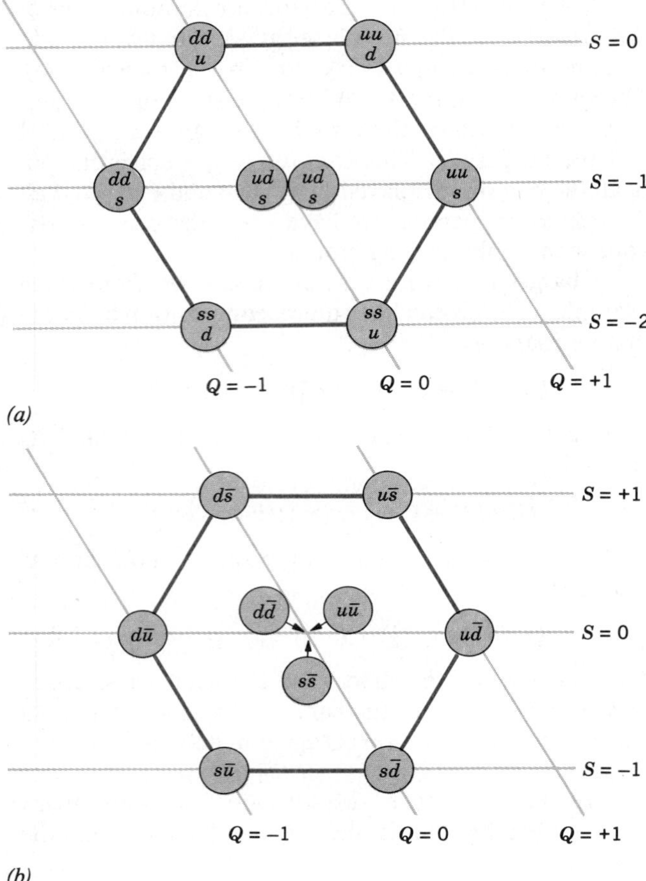

Figure 6 (*a*) The quark compositions of the baryons plotted in Fig. 4*a*. (Although the two central baryons share the same quark structure, the sigma is an excited state of the lambda, decaying into it by emission of a gamma-ray photon.) (*b*) The quark compositions of the mesons plotted in Fig. 4*b*.

spins of all the mesons displayed in Fig. 4*b* and Table 5 are zero. Both quarks and antiquarks have $s = \frac{1}{2}$ and, if these two quark spins are aligned antiparallel, we have spin zero for the meson.

The quark–antiquark model is also consistent with the fact that mesons are not baryons, that is, mesons have a baryon number $B = 0$. The baryon number for a quark is $+\frac{1}{3}$ and for an antiquark is $-\frac{1}{3}$, the combination adding to zero.

Consider the meson π^+, which is made up of an up quark (u) and an antidown quark ($\bar{d}$). We see from Table 6 that the charge of the up quark is $+\frac{2}{3}$ and of the antidown quark is $+\frac{1}{3}$ (that is, it is opposite in sign to the charge of the down quark). This adds nicely to a charge of $+1$ for the π^+ meson. Thus:

$$Q(u\bar{d}) = (+\tfrac{2}{3}) + (+\tfrac{1}{3}) = +1.$$

Figure 6*b* shows how the nine mesons of Table 5 and Fig. 4*b* are built up out of quarks. Convince yourself that all possible quark–antiquark combinations are used and that all known mesons with $s = 0$ are accounted for. Everything fits perfectly.

A New Look at Beta Decay. Let us see how beta decay appears from the quark point of view. In Eq. 10 of Chapter 47, we presented a typical example of this process:

$$^{32}\text{P} \rightarrow {}^{32}\text{S} + e^- + \nu.$$

After the neutron was discovered and Fermi had worked out his theory of beta decay, we then came to view the fundamental beta decay process as the changing of a neutron into a proton inside the nucleus. Thus

$$n \rightarrow p + e^- + \nu.$$

Today we look deeper and see that a neutron (udd) can change into a proton (uud) by changing a down quark into an up quark. We now view the fundamental beta decay process as

$$d \rightarrow u + e^- + \nu.$$

Thus, as we come to know more and more about the fundamental nature of matter, we can look at familiar processes at deeper and deeper levels. We see, too, that the quark model not only helps us to understand the structure of particles but also throws light on their interactions.

Still More Quarks. There are other particles and other Eightfold Way patterns that we have not described. To account for them, it turns out that we need to postulate three more quarks, the *charmed quark* (c), the *top quark* (t), and the *bottom quark* (b).

We note in Table 6 that these three new quarks are exceptionally heavy, the lightest of them (charm) being almost twice as heavy as a proton. To generate particles that contain such quarks, we must go to higher and higher energies, which is the reason that these three new quarks were not discovered earlier and do not enter as fully as their lighter siblings into the particle enterprise.

The first observed particle that contains a charmed quark was the J/Ψ meson, whose quark structure is ($c\bar{c}$). It was discovered simultaneously and independently in 1974 by groups headed by Samuel Ting at the Brookha-

Figure 7 Burton Richter (left) and Samuel Ting. In 1974 each led a team of physicists that independently and simultaneously discovered the J/Ψ meson, the first evidence for the existence of the charmed quark. The chart that Ting is holding displays the evidence for this meson. Richter, of Stanford University, and Ting, of MIT and the Brookhaven National Laboratory, shared the 1976 Nobel prize for this effort.

ven National Laboratory and by Burton Richter at Stanford University (see Fig. 7). These two workers shared the 1976 Nobel prize for this effort.

Compare Table 6 (the quark family) and Table 3 (the lepton family) carefully, noting the neat symmetry of these two "six packs" of particles, each dividing naturally into three corresponding two-particle "generations." In terms of what we know today, these two families, the quarks and the leptons, seem to be truly fundamental particles.

Sample Problem 6 The Ξ^- particle has a spin of $\frac{1}{2}$ and quantum numbers $Q = -1$ and $S = -2$. It is known to be a three-quark combination involving only up, down, and strange quarks. What must this combination be?

Because its strangeness is -2, this particle must contain two strange quarks, each of which (see Table 6) has $S = -1$. The third quark must then be either an up quark or a down quark. The two strange quarks have a combined charge of $(-\frac{1}{3}) + (-\frac{1}{3})$ or $-\frac{2}{3}$. We require a charge of -1 for the Ξ^- particle so that we must add a third quark whose charge is $-\frac{1}{3}$; that is the down quark. Thus the quark composition of Ξ^- is (*dss*).

49-9 Forces and Messenger Particles (Optional)

We turn now from cataloging the particles to considering the forces that act between them.

The Electromagnetic Force. Two electrons exert electromagnetic forces on each other according to Coulomb's law. At a deeper level, this interaction is described by a highly successful theory called *quantum electrodynamics* (QED). From this point of view we say that each electron learns of the presence of the other by exchanging photons, the photon being described as the quantum of the electromagnetic field.

We do not see these photons because they are emitted by one electron and absorbed by the other a very short time later. Because of their transitory existence, we call them *virtual photons.* Because of their role in communicating between the two interacting charges, we sometimes call these photons *messenger particles.*

If a resting electron emits a photon and remains itself unchanged, energy is not conserved. The conservation of energy law is saved, however, by the uncertainty principle, written in the form

$$\Delta E \cdot \Delta t \simeq h. \tag{13}$$

We interpret this relation to mean that you can "borrow" an amount of energy ΔE *provided* that you "return" it within an interval Δt given by $h/\Delta E$. The virtual photons do just that. It is something like cashing a check for more money than you have in your account but then quickly making a deposit to set things right before the bank finds out!

The Weak Force. A field theory of the weak force was developed by analogy with the field theory of the electromagnetic force. The messenger particles that transmit the force between leptons, however, are not (massless) photons but massive particles, identified by the symbols W and Z. The theory was so successful that it revealed the electromagnetic force and the weak force as different aspects of a single *electroweak force.* This accomplishment is a logical extension of the work of Maxwell, who revealed the electric and the magnetic force as different aspects of a single *electromagnetic* force.

Figure 8 From left to right, Sheldon Glashow, Abdus Salam, and Steven Weinberg. They are about to receive the Nobel prize for 1979 for their development of the electroweak theory that predicted the existence of the W and Z messenger particles. Salam is wearing the formal attire of his native Pakistan.

The electroweak theory was specific in predicting the properties of the messenger particles. Their charges and rest energies, for example, were predicted to be

Particle	*Charge*	*Rest Energy*
W	$\pm e$	82 ± 2 GeV
Z	0	92 ± 2 GeV

Recall that the proton rest energy is only 0.938 GeV; these are massive particles! The Nobel prize for 1979 was awarded to Sheldon Glashow, Steven Weinberg, and Abdus Salam for their development of the electroweak theory; see Fig. 8.

The theory was confirmed in 1983 by Carlo Rubbia and his group at CERN. Both messenger particles were observed, their rest energies agreeing exactly with the predicted values. The Nobel prize for 1984 went to Rubbia and to Simon van der Meer (Fig. 9) for this brilliant experimental work.*

Figure 9 Carlo Rubbia (left) and Simon van der Meer toasting each other at CERN after learning that they had won the 1984 Nobel prize for the experimental discovery of the W and Z particles. These experiments verified the predictions of the electroweak theory of Glashow, Salam, and Weinberg; see Fig. 8.

Some notion of the complexity of particle physics in this day and age can be found by looking at an earlier Nobel prize particle physics experiment—the discovery of the neutron in 1932. This vitally important discovery was a "table-top" experiment, employing particles emitted by naturally-occurring radioactive materials as projectiles; it was reported under the title "Possible Exis-

* See Reference 8.

tence of a Neutron," the single author being James Chadwick.

The discovery of the W and Z messenger particles in 1983, by contrast, was carried out at a large particle accelerator, about 7 km in circumference and operating in the several-hundred GeV range. The principal detector alone weighed 2000 tons. The experiment employed more than 130 physicists from 12 institutions in 8 countries.

The Strong Force. A theory of the strong force, that is, the force that acts between quarks, has also been developed. The messenger particles in this case are called *gluons* and, like the photon, they are predicted to be massless. The theory assumes that each "flavor" of quark comes in three varieties that, for convenience, have been labeled *red, yellow,* and *blue.* Thus, there are three up quarks, one of each color, and so on. The antiquarks also come in three colors, which we call *antired, antiyellow,* and *antiblue.* You must not think that quarks are actually colored, like tiny jelly beans. The names are labels of convenience but (for once!) they do have a certain formal justification, as we shall see.

The force acting between quarks is called a *color force* and the underlying theory, by analogy with quantum electrodynamics (QED) is called *quantum chromodynamics* (QCD). An important prediction of the theory is that quarks can only be assembled in *color-neutral* combinations.

There are two ways to bring this about. In the theory of actual colors, red + yellow + blue yields white, which is color-neutral; thus we can assemble three quarks to form a baryon. Antired + antiyellow + antiblue is also white, so that we can assemble three antiquarks to form an antibaryon. Finally, red + antired, or yellow + anityellow, or blue + antiblue also yield white. Thus we can assemble quark-antiquark combinations to form a meson. The color-neutral rule does not permit any other combinations of quarks and none are observed.

Unification—Einstein's Dream. The attempt to unify the fundamental forces of nature—which occupied Einstein's attention for much of his later life—is very much a current problem. Table 2 of Chapter 6 summarizes the current status. We have seen that the weak force has been successfully combined with electromagnetism so that they may be jointly viewed as aspects of a single *electroweak force.* Theories that attempt to add the strong force to this combination—called *Grand Unification Theories* (GUTs)—are being pursued actively, with considerable success. Theories that seek to complete the job by adding gravity—sometimes called *Theories of Everything* (TOE)—are at an encouraging but speculative stage at this time.

49-10 A Pause for Reflection

Let us put what we have learned in perspective. If all we are interested in is the structure of the world around us, we can get along nicely with the electron, the neutrino, the neutron, and the proton. As someone has said, we can operate "Spaceship Earth" quite well with just these particles. We can see a few of the more exotic particles by looking for them in the cosmic rays but, to see most of them, we must build massive accelerators and look for them carefully at great effort and expense. In short, most of what we now understand so well we had first to discover.

The reason for this dominance of a few particles is that—measured in energy terms—we live in a world of very low temperatures. Even at the center of the sun, the value of kT is only about 1 keV. To produce the new particles, we must be able to accelerate protons or electrons to energies in the GeV and TeV range and higher. Once upon a time, however, the temperature *was* high enough to provide just such energies, and far beyond. We are speaking of the early universe. Let us turn our attention in that direction.

The most distant object that we can "see" from Earth as of 1988 is a quasar* whose distance from us is 13×10^9 light years. As we look out in space we are also looking back in time. Thus we see the quasar as it was 13×10^9 y ago, when the photons that reach us now left it. Since the Big Bang that represents the creation of the universe occurred about 15×10^9 y ago, we are looking back to its earliest days.

We cannot begin to understand the universe until we have absorbed a central hypothesis, called the *Cosmological Principle:*

> *On a large enough scale the universe looks pretty much the same no matter where the observer is located. The universe has no unique center and therefore no unique boundary.*

An observer on our distant quasar would, we believe, see the universe much as we see it from our home galaxy, the

* *Quasars* (quasistellar objects) are extremely luminous objects whose nature is not yet fully understood.

Milky Way. Now we are ready to look at some other interesting properties of the universe.

49-11 The Universe Is Expanding

As we have seen, it is possible to measure the relative speeds at which galaxies are approaching us or receding from us by measuring the Doppler shift of the light that they emit. If we look only at distant galaxies, beyond our immediate galactic neighbors, we find an astonishing fact. They are all moving away from us!

In 1929, Edwin P. Hubble (Fig. 10) established a connection between the speed of recession of a galaxy and its distance from us, namely, that they are proportional. Thus

$$\boxed{v = Hr} \quad \text{Hubble's law,} \tag{14}$$

in which H, the *Hubble parameter,* has the value

$$H \approx 17 \times 10^{-3}\ \text{m/(s} \cdot \text{ly)}. \tag{15}$$

The value of the Hubble parameter is somewhat uncertain because of the difficulty of measuring the distances of remote galaxies. These are established by an involved chain of measurements and assumptions, its starting point (once the dimensions of our solar system have been established) being the measurement of the distances of our closest neighboring stars by parallax methods.

Figure 10 Edwin Hubble (1889–1953) shown at the controls of the 100-in. telescope at Mount Wilson, where he carried out much of the work that led him to propose the expanding universe concept.

We interpret Hubble's law to mean that the universe is expanding, much as the raisins in what is to be a loaf of raisin bread grow farther apart as the dough rises. Consistent with the Cosmological Principle, observers on all other galaxies would find that distant galaxies were rushing away from them also, in accord with Hubble's law. From the point of view of our analogy, all raisins are alike.

Hubble's law fits in well with the Big Bang hypothesis. What we are seeing are the outward-flying fragments of that primordial explosion.

Sample Problem 7 As judged by Doppler shift measurements of the light that it emits, the most distant object that we can see from earth (the quasar referred to in the foregoing) is receding from us at 2.2×10^8 m/s. (Note that this is 73% of the speed of light!) How far away is that object?

From Hubble's law (Eq. 14)

$$r = \frac{v}{H} = \frac{2.2 \times 10^8\ \text{m/s}}{17 \times 10^{-3}\ \text{m/(s} \cdot \text{ly)}}$$
$$= 13 \times 10^9\ \text{ly}. \quad \text{(Answer)}$$

This result is only approximate because the quasar has not always been receding from us at the same speed.

Sample Problem 8 Assume that the quasar in Sample Problem 7 has been moving at its calculated speed with respect to us ever since the Big Bang. What minimum limit does this impose on how long ago the Big Bang occurred? That is, what is the minimum age of the universe?

We can find the time from

$$t = \frac{r}{v} = \frac{1}{H}$$
$$= \frac{1}{17 \times 10^{-3}\ \text{m/(s} \cdot \text{ly)}} = 58.8\ \text{s} \cdot \text{ly/m}$$
$$= (58.8\ \text{s} \cdot \text{ly/m})(9.46 \times 10^{15}\ \text{m/ly})(1\ \text{y}/3.16 \times 10^7\ \text{s})$$
$$\approx 18 \times 10^9\ \text{y}. \quad \text{(Answer)}$$

This is in reasonable agreement with the age of the universe as computed by other methods.

49-12 The Microwave Background Radiation*

In 1965, Arno Penzias and Robert Wilson (Fig. 11) of what is now AT&T Bell Laboratories, were testing a sensitive microwave receiver used for communications research. They discovered a faint background "hiss" that remained unchanged in intensity no matter in what direction their antenna was pointed. It soon became clear that Penzias and Wilson were observing a *microwave background radiation,* generated in the early universe and filling all space uniformly. This background radiation, whose maximum intensity occurs at a wavelength of 1.1 mm, has the same distribution in wavelength as does cavity radiation at a temperature of 3.0 K, the "cavity" in this case being the entire universe. Penzias and Wilson were awarded the 1978 Nobel prize for their discovery.

This radiation finds its origin some 500,000 years after the Big Bang, when (as we shall see) the universe suddenly became transparent to electromagnetic waves. The radiation at that time corresponded to cavity radiation at a temperature of perhaps 10^5 K. As the universe expanded, however, this radiation cooled down to its present value of 2.7 K, much as the temperature of a gas expanding under adiabatic conditions will fall.

* See Reference 17.

49-13 The Mystery of the Dark Matter*

Vera Rubin of the Carnegie Institution of Washington (see Fig. 12) and her associates have been measuring the rotation rates of distant galaxies for a number of years. They do so by measuring the Doppler shifts of bright clusters of stars located within the galaxy at various distances from the galactic center; see Fig. 13. Her conclusion is surprising: The orbital speed of objects at the visible edge of the rotating galaxy is about the same as for objects close to the galactic center.

That is *not* what we find in the solar system. The orbital speed of Pluto (the planet most distant from the sun) is only about one-tenth that of Mercury (the planet closest to the sun).

* See Reference 14.

Figure 11 Arno Penzias (right) and Robert Wilson, standing in front of the large horn antenna with which they first detected the microwave background radiation. Curiously, when they made their discovery, Robert Dicke of nearby Princeton University was in the process of building an antenna to look for just this radiation. Penzias and Wilson shared the 1978 Nobel prize for their work.

Figure 12 Vera Rubin and her associates using the 4-M telescope at Kitt Peak National Observatory have used Doppler shift methods to measure the rotation rates of many galaxies. Incidentally, astronomers no longer peer through telescopes; they look at images on a video monitor in a warm room!

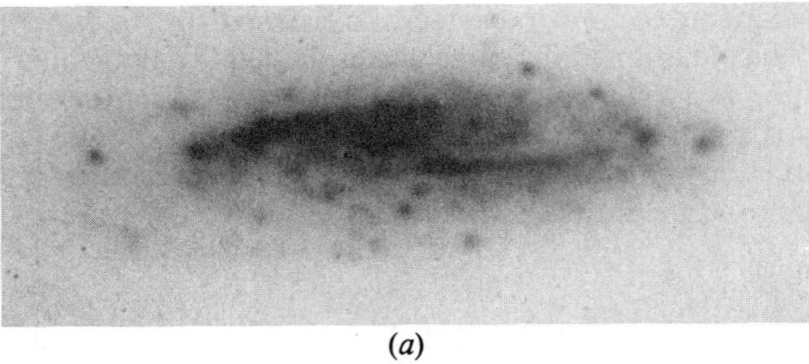

(a)

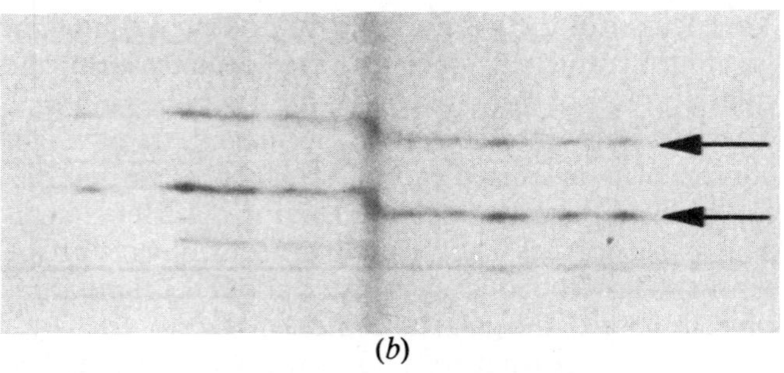

(b)

Figure 13 (*a*) The galaxy NGC7541, one of more than a hundred whose rotation curves have been measured by Vera Rubin and her collaborators. (*b*) Wavelength increases downward on this plot. The two broken lines indicated by the arrows show that, relative to the center of the galaxy, the right-hand side of the galaxy is receding from us (red shifted) and the left side moving toward us (blue shifted).

The only explanation consistent with Newtonian mechanics is that there is much more matter in a typical rotating galaxy than we can account for in terms of what we can actually see. In fact, the visible galaxy represents only about 5–10% of the total galactic mass.

What is this *dark matter* that permeates and surrounds a typical galaxy? If neutrinos have even a small rest mass, that might account for it. If not, theoretical physicists have an abundant supply of predicted new particles that might do the job, many with exotic names such as axions, winos, and wimps. At present, however, there is no experimental evidence to support any of these predictions. Philip Morrison, after describing the evidence for the existence of dark matter on a television series ("The Ring of Truth," Episode 6) puts it well:

> *If you ask me what the universe as a whole is made of, I must admit that as of now I remain in great doubt. I just do not know. But one thing I do know: we will try very hard to find out.*

49-14 The Big Bang*

In 1985, a physicist remarked at a scientific meeting:

> *It is as certain that the universe started with a Big Bang about 15 billion years ago as it is that the Earth goes around the Sun.*

This strong statement suggests the level of confidence in which this theory, first advanced by George Gamow and his colleagues, is held by those who study these matters.

You must not imagine that the Big Bang was like the explosion of some giant firecracker and that, in principle at least, you could have stood to one side and watched. There was no "one side" because the Big Bang represents the generation of spacetime itself. From the point of view of our present universe, there is no position in the space to which you can point and say, "The Big Bang happened there." It happened everywhere. Let us see what went on during various time intervals.

Big Bang Zero to 10^{-43} s. Of this tiny but important period** we know little, because the laws of physics as we know them do not hold. As one author has written: "How you do physics in a situation like this, when space and time are disconnected, is *not* described in Halliday and Resnick." †

At 10^{-43} s, the temperature of the universe was about 10^{23} K and the universe was expanding rapidly. As the expansion proceeded, the temperature dropped steadily until it reached its present value of about 3 K.

10^{-43} s to 10^{-35} s. During this period, the strong, the weak, and the electromagnetic force act as a single force, described by a Grand Unified Theory (GUT). Gravity acts separately, as it does today. ‡

10^{-35} s to 10^{-10} s. The strong force "freezes out," leaving the electroweak force and gravity, still acting as a single force.

10^{-10} s to 10^{-5} s. All four forces appear separately, just as they do today. The universe consists of a hot "soup" of quarks, leptons, and photons.

10^{-5} s to 3 min. Quarks join to form mesons and baryons. Matter and antimatter annihilate, wiping out

* See References 9–13.

** 10^{-43} s is known as the Planck time; see Sample Problem 6 of Chapter 23.

† See Reference 9.

‡ We omit a discussion of the exponential inflation of the universe that might well have occurred at this time; see Reference 15.

the antimatter and leaving the slight excess of matter† from which our present universe is formed.

3 min to 10^5 y. Protons and neutrons join to form the light nuclides, such as ^{4}He, ^{3}He, ^{2}H, and ^{7}Li, in just the abundances that we find today. The universe consists of a plasma of nuclei and electrons.

10^5 y to the present. At the beginning of this period, atoms form. The universe then becomes transparent to photons and the radiation that now reaches us as the microwave background radiation starts on its long journey. Atoms cluster to form galaxies, then stars and planets, and (in good time) us.

49-15 A Summing Up

Let us, in these closing paragraphs, step aside for a moment and consider where our rapidly accumulating store of knowledge about the universe is leading us. That it provides satisfaction to a host of curiosity-motivated physicists is beyond dispute.

However, some view it as a humbling experience in that each increase in knowledge seems to reveal more clearly our own relative insignificance in the grand scheme of things. Thus, in rough chronological order, we came to realize that:

> Our Earth is not the center of the solar system.
>
> Our Sun is but one star among many.
>
> Our Galaxy is but one of many and our Sun is an insignificant star near its outer edge.
>
> Our Earth has existed for perhaps only a third of the age of the universe and will surely disappear when our Sun burns up its fuel and becomes a Red Giant.
>
> We have lived on the Earth, as a species, for less than a million years, a blink in cosmological time.
>
> The last crushing blow: The neutrons and protons of which we are made are not the predominant form of matter in the universe. As someone has said, we are not even made of the right stuff!

However, the bright side is that it is we ourselves who discovered all these facts. Although our position in the universe may be insignificant, the laws of physics that we have discovered (uncovered?) seem to hold throughout the universe and—as far as we know—for all past and future time. At least, there is no evidence that other laws hold in other parts of the universe. Thus, until someone complains, we are entitled to stamp the laws of physics DOE for "Discovered on Earth." There remains much more to be discovered. We close our text with the forward-looking words of the philosopher:

> The universe is full of magical things patiently waiting for our wits to grow sharper.

† See Reference 17 for an account of why there is an excess.

REFERENCES

General References on Particle Physics:

1. "Particle Physics for Everybody," by Paul Davies, *Sky & Telescope,* December 1987.
2. James S. Trefil, *From Atoms to Quarks* (Charles Scribner's Sons, New York, 1980)
3. Frank Close, *The Cosmic Onion* (The American Institute of Physics, New York, 1983).
4. *Building the Universe* (Basil Blackwell and New Scientist, London, 1985), Christine Sutton, ed. (Short annotated reports from the *New Scientist.* Read about these events as they were reported at the time they occurred.)

Special Topics in Particle Physics:

5. "Heavy Leptons," by Martin L. Perl and William T. Kirk, *Scientific American,* March 1978.
6. George L. Trigg, *Landmark Experiments in Twentieth Century Physics* (Crane, Russak & Company, New York, 1975) Chapter 15.
7. "Quarks with Color and Flavor," by Sheldon Glashow, *Scientific American,* October 1975.
8. "The Discovery of the Intermediate Vector Bosons," by Anne Kernan, *American Scientist,* January-February 1986.

General References on Big Bang Physics:

9. "The Early Universe and High-Energy Physics," by David N. Schramm, *Physics Today,* April 1983.
10. James S. Trefil, *The Moment of Creation* (Macmillan Publishing Company, New York, 1983).
11. Steven Weinberg, *The First Three Minutes* (Basic Books, New York, second edition, 1988).

12. John Gribbin, *In Search of the Big Bang* (Bantam Books, 1986).

13. Joseph Silk, *The Big Bang* (W. H. Freeman and Company, 1980).

Special Topics in Big Bang Physics:

14. "Dark Matter in the Spiral Galaxies," by Vera C. Rubin, *Scientific American,* June 1983.

15. "The Inflationary Universe," by Alan H. Guth and Paul J. Steinhardt, *Scientific American,* May 1984.

16. "The Cosmic Background Radiation and the New Aether Drift," by Richard A. Muller, *Scientific American,* May 1978.

17. "A Flaw in a Universal Mirror," by Robert K. Adair, *Scientific American,* February 1988.

REVIEW AND SUMMARY

Leptons and Quarks

What is the universe made of?

Current research is consistent with the view that all of matter is made of 6 kinds of *leptons* (Table 3) and 6 kinds of *quarks* (Table 6). All of these have spin quantum numbers equal to $\frac{1}{2}$ and are thus *fermions* (particles with half-integral spin—$\frac{1}{2}$, $\frac{3}{2}$, etc.). There are also 12 *antiparticles,* one corresponding to each of the leptons and quarks.

The Interactions

Particles with electric charge interact by the electromagnetic force by exchanging "messenger-particle" *virtual photons.* The leptons interact with each other and with quarks only through the *weak force* with the massive W and Z particles as messengers. In addition, quarks interact with each other by the *color* force. The electromagnetic and weak forces have now been shown to be different manifestations of the same force, now called the *electroweak* force.

Leptons

Three of the leptons (the *electron, muon, and tauon*) have electric charge equal to $-1e$; these also have nonzero rest energy. There are uncharged *neutrinos* (also leptons), one corresponding to each of the charged leptons. The neutrinos have very small, possibly zero, rest energy. The antiparticles for the charged leptons have positive charge. See Table 3.

Quarks

The six quarks (up, down, strange, charm, bottom, and top, listed in order of increasing mass) each have baryon number $+\frac{1}{3}$ and charge equal to either $+(\frac{2}{3})e$ or $-(\frac{1}{3})e$; see Table 6. The strange quark has strangeness -1 while the others all have strangeness 0. These algebraic signs are reversed for the antiparticles.

Hadrons, Baryons, Mesons

Quarks combine into strongly-interacting particles called *hadrons. Baryons* are hadrons with half-integral spin quantum numbers ($\frac{1}{2}$ or $\frac{3}{2}$). *Mesons* are hadrons with integral spin quantum numbers (0 or 1); see Table 1. Because of their spin, baryons are fermions and mesons are bosons. There are nine zero-spin mesons, each of which can be understood in terms of a quark–antiquark combination involving up, down, and strange quarks; see Table 5 and the Eightfold Way patterns of Figs. 4*b* and 6*b*. All of these have baryon number equal to zero. These same quarks combine into the eight three-quark spin-$\frac{1}{2}$ hadrons listed in Table 4 (among which are the familiar proton and neutron) and arranged into the Eightfold Way patterns of Figs. 4*a* and 6*a*; these have baryon number equal to $+1$. Sample Problem 5 illustrates the identification procedure. The current theory, called *quantum chromodynamics,* predicts that the possible combinations of quarks are either a quark with an antiquark, three quarks, or three antiquarks. So far, this prediction is consistent with experiment. All of the hadrons, except for protons, are unstable.

Particle Interactions and Decay

These particles are studied by observing their decays and their interactions with each other. Figure 3, with Table 2 and the discussion of Section 49–3, illustrates some of the possibilities. Sample Problems 1, 2, and 3 show some sample calculations involved in analysis of these experiments. The reactions are governed by conservation laws: energy, angular momentum, electric charge, baryon number, lepton number, and strangeness. Sections 49–5 and 49–6 and Sample Problem 4 illustrate the kinds of limitation imposed by the conservation laws.

Conservation Laws

Expansion of the Universe

Current evidence strongly suggests that the universe is expanding, with the distant galaxies moving away from us at a rate given by *Hubble's law:*

$$v = Hr \quad \text{(Hubble's law),} \qquad [14]$$

Hubble's Law

in which H, the *Hubble parameter*, has the value

$$H = 17 \times 10^{-3}\ \mathrm{m/(s \cdot ly)}. \qquad [15]$$

Sample Problems 6 and 7 show, using the Hubble law and speeds calculated from Doppler shifts, that the most distant quasar is about 13×10^9 ly away and that it would have taken about 18 billion years to get that far away moving at its present speed.

History of the Universe

How did the universe come to be the way it is?
The expansion described by the Hubble law and the presence of ubiquitous background microwave radiation suggest that the universe began in a "big bang" about 15 billion years ago. A general outline of its history, as we now understand it, is given in Section 49–14.

QUESTIONS

1. What is really meant by an elementary particle? In arriving at an answer, consider such properties as lifetime, mass, size, decays into other particles, fusion to make other particles, and reactions.

2. Why do particle physicists want to accelerate particles to higher and higher energies?

3. The words *chemistry, Mendeleev, periodic table, missing elements,* and *wave mechanics* suggest a line of development in our understanding of the structure of atoms. What words suggest a corresponding line of development in our understanding of the particles of physics?

4. Name two particles that have neither rest mass nor charge. What properties do these particles have?

5. Why do neutrinos leave no tracks in detecting chambers?

6. Neutrinos have (presumably) no rest mass and travel with the speed of light. How, then, can they carry varying amounts of energy?

7. Do all particles have antiparticles? What about the photon?

8. Photons and neutrinos are alike in that they have zero charge, zero rest mass (presumably), and travel with the speed of light. What are the differences between these two particles? How would you produce them? How would you detect them?

9. Explain why we say that the π^0 meson is its own antiparticle.

10. Why can't an electron decay by distintegrating into two neutrinos?

11. Why is the electron stable? That is, why does it not decay spontaneously into other particles?

12. Why cannot a resting electron emit a single gamma-ray photon and disappear? Could a moving electron do so?

13. A neutron is massive enough to decay by the emission of a proton and two neutrinos. Why does it not do so?

14. A positron invariably finds an electron and they annihilate each other. How then can we call the positron a stable particle?

15. What is the mechanism by which two electrons exert forces on each other?

16. Do the eight pions whose tracks appear in Fig. 3*a* all have the same initial kinetic energy? If so, what is that energy? If not, which of them have the greatest energy?

17. Is the magnetic field that is present in Fig. 3 directed into the page or out of the page?

18. A particle that responds to the strong force is either a meson or a baryon. You can tell which it is by allowing the particle to decay until only stable end products remain. If there is a proton among these products, the original particle was a baryon. If there is no proton, the original particle was a meson. Explain this classification rule.

19. How many kinds of stable leptons are there? Stable mesons? Stable baryons? In each case, name them.

20. Most particle physics reactions are endothermic, rather than exothermic. Why?

21. What is the lightest strongly interacting particle? What is the heaviest particle unaffected by the strong interaction?

22. For each of the following particles, state which of the four types of interaction are influential: (*a*) electron; (*b*) neutrino; (*c*) neutron; (*d*) pion.

23. Just as x rays are used to discover internal imperfections in a metal casting caused by gas bubbles, so cosmic-ray muons have been used in an attempt to discover hidden burial chambers in Egyptian pyramids. Why were muons used?

24. Are strongly-interacting particles affected by the weak interaction?

25. Do all weak-interaction decays produce neutrinos?

26. The messengers for quantum electrodynamics are photons and they are virtual. The messengers for the weak force are the W and Z particles and they are observed. Comment on this difference.

27. What is the difference between a boson and a fermion? A hadron and a lepton?

28. Baryons and leptons are both fermions. In what ways are they different?

29. Mesons and baryons are each sensitive to the strong force. In what way are they different?

30. By comparing Tables 3 and 6, point out as many similarities between leptons and quarks as you can and also as many differences.

31. Quarks are not observed directly. What is the indirect evidence for them?

32. We can explain the "ordinary" world around us with two leptons and two quarks. Name them.

33. The neutral pion has a quark structure of $(u\bar{u})$ and decays with a mean life of only 8.3×10^{-17} s. The charged pion, on the other hand, has a quark structure of $(u\bar{d})$ and decays with a mean life of 2.6×10^{-8} s. Explain, in terms of their quark structures, why the mean life of the neutral pion should be so much shorter (by a factor of 3×10^8) than that of the charged pion. (*Hint:* Think of annihilation.)

34. Do leptons contain quarks? Do mesons? Do photons? Do baryons?

35. The ratio of the magnitude of the gravitational force between the electron and the proton in the hydrogen atom to the magnitude of the electromagnetic force of attraction between them is about 10^{-40}. If the gravitational force is so very much weaker than the electromagnetic force, how was it that the gravitational force was discovered first and is so much more apparent to us?

36. Quasars are found only at very great distances from our galaxy; there are no quasars nearby. Why is this observation not in contradiction to the Cosmological Principle, as applied to us and to an observer situated on a quasar?

37. Why can't we find the center of the expanding universe? Are we looking for it?

38. Due to the effect of gravity, the rate of expansion of the universe must have decreased in time following the Big Bang. Show that this implies that the age of the universe is less than $1/H$.

39. It is not possible, using telescopes that are sensitive in any part of the electromagnetic spectrum, to "look back" any farther than about 500,000 y from the Big Bang. Why?

40. How does one arrive at the conclusion that the visible galaxy represents about only 10% of the galactic mass?

41. Are we always looking back into time as we observe the distant universe? Does the direction in which we look make a difference?

EXERCISES AND PROBLEMS

Section 49-3 An Interlude

1E. Calculate the difference in rest mass, in kg, between the muon and pion of Sample Problem 2.

2E. A neutral pion decays into two gamma rays: $\pi^0 \rightarrow \gamma + \gamma$. Calculate the wavelengths of the gamma rays produced by the decay of a neutral pion at rest.

3E. An electron and a positron are separated by a distance r. Find the ratio of the gravitational force to the electrostatic force between them. What do you conclude from the result concerning the forces acting between particles detected in a bubble chamber or similar detector?

4E. The positively-charged pion decays by Eq. 4: $\pi^+ \rightarrow \mu^+ + \nu$. What, then, must be the decay scheme of the negatively charged pion? (*Hint:* The π^- is the antiparticle of the π^+.)

5E. How much energy is "created" if our earth is annihilated by collision with an anti-earth? (Fortunately, there is no evidence that an anti-earth actually exists.)

6P. A neutral pion has a rest energy of 135 MeV and a mean life of 8.3×10^{-17} s. If it is produced with an initial kinetic energy of 80 MeV and it decays after one mean lifetime, what is the longest possible track that this particle could leave in a bubble chamber? Take relativistic time dilation into account. (*Hint:* See Sample Problem 1.)

7P. Observations of neutrinos emitted by the supernova SN1987a in the Large Magellanic Cloud, see Fig. 14, place an upper limit on the rest energy of the electron neutrino of 20 eV. Suppose that the rest energy of the neutrino, rather than being zero, is in fact equal to 20 eV. How much slower than light is a 1.5-MeV neutrino, emitted in a β-decay, moving?

8P. Certain theories predict that the proton is unstable, with a half-life of about 10^{32} years. Assuming that this is true, calculate the number of proton decays you would expect to occur in one year in the water of an Olympic-sized swimming pool holding 114,000 gallons of water.

9P. A positive tauon (τ^+, rest energy = 1784 MeV) is moving with 2200 MeV of kinetic energy in a circular path perpendicular to a uniform 1.2-T magnetic field. (*a*) Calculate the momentum of the tauon in kg · m/s. Relativistic effects must be considered. (*b*) Find the radius of the circular path. (*Hint:* See Problem 57 in Chapter 42.)

10P. The rest energy of many short-lived particles cannot be measured directly, but must be inferred from the measured

Figure 14 Problem 7.

momenta and known rest energies of its decay products. Consider the ρ^0 meson, which decays by the reaction $\rho^0 \rightarrow \pi^+ + \pi^-$. Calculate the rest energy of the ρ^0 meson given that the oppositely directed momenta of the created pions each has magnitude 358.3 MeV/c. See Table 5 for the rest energies of the pions.

11P. (*a*) A particle m_0 at rest decays into two particles m_1 and m_2, which move off with equal but oppositely directed momenta. Show that the kinetic energy K_1 of m_1 is given by

$$K_1 = \frac{1}{2E_0}[(E_0 - E_1)^2 - E_2^2],$$

where m_0, m_1, and m_2 are rest masses and E_0, E_1, and E_2 are the corresponding rest energies. (*Hint:* Follow the arguments of Sample Problem 2 except that, in this case, neither of the created particles has zero rest mass.) (*b*) Show that the result above yields the kinetic energy of the muon as calculated in the reaction of Sample Problem 2.

Section 49-5 A New Conservation Law

12E. Verify that the hypothetical proton decay scheme given in Eq. 10 does not violate the conservation laws of (*a*) charge, (*b*) energy, (*c*) linear momentum, (*d*) angular momentum.

13E. What conservation law is violated in each of these proposed decays? Assume that the decay products have zero orbital angular momentum. (*a*) $\mu^- \rightarrow e^- + \nu$; (*b*) $\mu^- \rightarrow e^+ + \nu + \bar{\nu}$; (*c*) $\mu^+ \rightarrow \pi^+ + \nu$.

14P. The A_2^+ particle and its products decay according to the following schemes:

$$\begin{aligned} A^+ &\rightarrow \rho^0 + \pi^+, & \mu^+ &\rightarrow e^+ + \nu + \bar{\nu}, \\ \rho^0 &\rightarrow \pi^+ + \pi^-, & \pi^- &\rightarrow \mu^- + \bar{\nu}, \\ \pi^+ &\rightarrow \mu^+ + \nu, & \mu^- &\rightarrow e^- + \nu + \bar{\nu}. \end{aligned}$$

(*a*) What are the final stable decay products? (*b*) From the evidence thus supplied, is the A_2^+ particle a fermion or a boson? Is it a meson or a baryon? What is its baryon number? (*Hint:* See Sample Problem 5.)

Section 49-7 The Eightfold Way

15E. The reaction $\pi^+ + p \rightarrow p + p + \bar{n}$ proceeds by the strong interaction. By applying the conservation laws, deduce the charge, baryon number, and strangeness of the antineutron.

16E. By examining strangeness, determine which of the following decays or reactions proceed via the strong interaction. (*a*) $K^0 \rightarrow \pi^+ + \pi^-$; (*b*) $\Lambda^0 + p \rightarrow \Sigma^+ + n$; (*c*) $\Lambda^0 \rightarrow p + \pi^-$; (*d*) $K^- + p \rightarrow \Lambda^0 + \pi^0$.

17E. What conservation law is violated in each of these proposed reactions and decays? Assume that the products have zero orbital angular momentum. (*a*) $\Lambda^0 \rightarrow p + K^-$; (*b*) $\Omega^- \rightarrow \Sigma^- + \pi^0$ ($S = -3$, $Q = -1$ for Ω^-); (*c*) $K^- + p \rightarrow \Lambda^0 + \pi^+$.

18E. Calculate the disintegration energy of the reactions (*a*) $\pi^+ + p \rightarrow \Sigma^+ + K^+$; (*b*) $K^- + p \rightarrow \Lambda^0 + \pi^0$.

19E. A Σ^- particle moving with 220 MeV of kinetic energy decays according to $\Sigma^- \rightarrow \pi^- + n$. Calculate the total kinetic energy of the decay products.

20P. Use the conservation laws to identify the particle labeled x in the following reactions, which proceed by means of the strong interaction. (*a*) $p + p \rightarrow p + \Lambda^0 + x$; (*b*) $p + \bar{p} \rightarrow n + x$; (*c*) $\pi^- + p \rightarrow \Xi^0 + K^0 + x$.

21P. Show that, if instead of plotting S versus Q for the spin-$\frac{1}{2}$ baryons in Fig. 4*a* and for the spin-0 mesons in Fig. 4*b*, the quantity $Y = B + S$ is plotted against the quantity $T_z = Q - \frac{1}{2}B$, then the hexagonal patterns emerge with the use of nonsloping (perpendicular) axes. (The quantity Y is called *hypercharge* and T_z is related to a quantity called *isospin*.)

22P. Consider the decay $\Lambda^0 \rightarrow p + \pi^-$ with the Λ^0 at rest. (*a*) Calculate the disintegration energy. (*b*) Find the kinetic energy of the proton. (*c*) What is the kinetic energy of the pion? (*Hint:* See Problem 11.)

Section 49-8 The Quark Model

23E. The quark composition of the proton and the neutron are *uud* and *udd*, respectively. What are the quark compositions of (*a*) the antiproton and (*b*) the antineutron?

24E. From Tables 4 and 6, determine the identity of the baryons formed from the following combinations of quarks. Check your answers with the baryon octet shown in Fig. 4*a*. (*a*) *ddu*; (*b*) *uus;* (*c*) *ssd.*

25E. What quark combinations form (*a*) a Λ^0; (*b*) a Ξ^0?

26E. Using the up, down, and strange quarks only, construct, if possible, a baryon (*a*) with $Q=+1$ and $S=-2$. (*b*) With $Q=+2$ and $S=0$.

27E. There are 10 baryons with spin $\frac{3}{2}$. Their symbols and quantum numbers are as follows:

	Q	S		Q	S
Δ^-:	-1	0	Σ^{*0}	0	-1
Δ^0:	0	0	Σ^{*+}	$+1$	-1
Δ^+:	$+1$	0	Ξ^{*-}	-1	-2
Δ^{++}	$+2$	0	Ξ^{*0}	0	-2
Σ^{*-}	-1	-1	Ω^-	-1	-3

Make a charge–strangeness plot for these baryons, using the sloping coordinate system of Fig. 4. Compare your plot with this figure.

28P. There is no known meson with $Q=+1$ and $S=-1$ or with $Q=-1$ and $S=+1$. Explain why, in terms of the quark model.

29P. The spin-$\frac{3}{2}$ Σ^{*0} baryon (see Exercise 27) has a rest energy of 1385 MeV (with an intrinsic uncertainty ignored here); the spin-$\frac{1}{2}$ Σ^0 baryon has a rest energy of 1192.5 MeV. Suppose that each of these particles has a kinetic energy of 1000 MeV. Which, if either, is moving faster and by how much?

Section 49-11 The Universe Is Expanding

30E. If Hubble's law can be extrapolated to very large distances, at what distance would the recessional speed become equal to the speed of light?

31E. What is the observed wavelength of the 656.3-nm H_α line of hydrogen emitted by a galaxy at a distance of 2.4×10^8 ly?

32E. In the laboratory, one of the lines of sodium is emitted at a wavelength of 590.0 nm. When observing the light from a particular galaxy, however, this line is seen at a wavelength of 602.0 nm. Calculate the distance to the galaxy, assuming that Hubble's law holds.

33P. The recessional speeds of galaxies and quasars at great distances are close to the speed of light, so that the relativistic doppler shift formula (see Eq. 26 in Chapter 42) must be used. The redshift is reported as z, where $z=\Delta\lambda/\lambda_0$ is the (fractional) red shift. (*a*) Show that, in terms of z, the recessional speed parameter $\beta = v/c$ is given by

$$\beta = \frac{z^2+2z}{z^2+2z+2}.$$

(*b*) The most distant quasar detected (as of 1987) has $z=4.43$. Calculate its speed parameter. (*c*) Find the distance to the quasar, assuming that Hubble's law is valid to these distances.

34P. Will the universe continue to expand forever? To attack this question, make the (reasonable?) assumption that the recessional speed v of a galaxy a distance r from us is determined only by the matter that lies inside a sphere of radius r centered on us; see Fig. 15. If the total mass inside this sphere is M, the escape speed v_e is given by $v_e=\sqrt{2GM/r}$ (see Sample Problem 6 in Chapter 15). (*a*) Show that the average density ρ inside the sphere must be at least equal to the value given by

$$\rho = 3H^2/8\pi G$$

to prevent unlimited expansion. (*b*) Evaluate this "critical density" numerically; express your answer in terms of H-atoms/m³. Measurements of the actual density are difficult and complicated by the presence of dark matter.

Figure 15 Problem 34.

Section 49-12 The Microwave Background Radiation

35P. Due to the presence everywhere of the microwave radiation background, the minimum temperature possible of a gas in interstellar or intergalactic space is not 0 K but 2.7 K. This implies that a significant fraction of the molecules in space that possess excited states of low excitation energy may, in fact, be in those excited states. Subsequent de-excitation leads to the emission of radiation that could be detected. Consider a (hypothetical) molecule with just one excited state. (*a*) What would the excitation energy have to be in order that 25% of the molecules be in the excited state? (*Hint:* See Equation 24 of Chapter 45.) (*b*) What is the wavelength of the photon emitted in a transition to the ground state?

Section 49-13 The Mystery of Dark Matter

36E. What would the mass of the sun have to be if Pluto (the outermost planet most of the time) were to have the same orbital speed that Mercury (the innermost planet) has now? Use data from Appendix C and express your answer in terms of the sun's current mass M. (Assume circular orbits.)

37P. Suppose that the radius of the sun were increased to 5.9×10^{12} m (the average radius of the orbit of the planet Pluto, the outermost planet), that the density of this expanded sun were constant, and that the planets revolved within this tenuous object. (*a*) Calculate the earth's orbital speed in this new configuration and compare this with its present orbital speed of 29.8 km/s. Assume that the radius of the earth's orbit remains unchanged. (*b*) What would be the new period of revolution of the earth? (The sun's mass remains unchanged.)

38P. Suppose that the matter (stars, gas, dust) of a particular galaxy, total mass M, is distributed uniformly throughout a sphere of radius R. A star, mass m, is revolving about the center of the galaxy in a circular orbit of radius $r < R$. (*a*) Show that the orbital speed v of the star is given by

$$v = r\sqrt{GM/R^3},$$

and therefore that the period T of revolution is

$$T = 2\pi\sqrt{R^3/GM},$$

independent of r. Ignore any resistive forces. (*b*) What is the corresponding formula for the orbital period assuming that the mass of the galaxy is strongly concentrated toward the center of the galaxy, so that essentially all of the mass is at distances from the center less than r.

Section 49-14 The Big Bang

39E. From Planck's radiation law it is possible to derive the following relation between the temperature T of a cavity radiator and the wavelength λ_{max} at which it radiates most strongly:

$$\lambda_{max} T = 2898\ \mu\text{m} \cdot \text{K}.$$

(This is Wien's law; see Problem 48 in Chapter 43.) (*a*) The microwave background radiation peaks in intensity at a wavelength of 1.1 mm. To what temperature does this correspond? (*b*) About 10^5 years after the Big Bang, the universe became transparent to electromagnetic radiation. Its temperature then was about 10^5 K. What was the wavelength at which the background radiation was most intense at that time?

40E. The wavelength of the photons at which a radiation field of temperature T radiates most intensely is given by $\lambda_{max} = (2898\ \mu\text{m} \cdot \text{K})/T$ (see Problem 39). (*a*) Show that the energy E in MeV of such a photon can be computed from

$$E = (4.28 \times 10^{-10}\ \text{MeV/K})T.$$

(*b*) At what minimum temperature can this photon create an electron-positron pair?

APPENDIX A

THE INTERNATIONAL SYSTEM OF UNITS (SI)*

1. The SI Base Units

Quantity	*Name*	*Symbol*	*Definition*
length	meter	m	". . . the length of the path traveled by light in vacuum in 1/299,792,458 of a second." (1983)
mass	kilogram	kg	". . . this prototype [a certain platinum-iridium cylinder] shall henceforth be considered to be the unit of mass." (1889)
time	second	s	". . . the duration of 9,192,631,770 periods of the radiation corresponding to the transition between the two hyperfine levels of the ground state of the cesium-133 atom." (1967)
electric current	ampere	A	". . . that constant current which, if maintained in two straight parallel conductors of infinite length, of negligible circular cross section, and placed 1 meter apart in vacuum, would produce between these conductors a force equal to 2×10^{-7} newton per meter of length." (1946)
thermodynamic temperature	kelvin	K	". . . the fraction 1/273.16 of the thermodynamic temperature of the triple point of water." (1967)
amount of substance	mole	mol	". . . the amount of substance of a system which contains as many elementary entities as there are atoms in 0.012 kilogram of carbon 12." (1971)
luminous intensity	candela	cd	". . . the luminous intensity, in the perpendicular direction, of a surface of 1/600,000 square meter of a blackbody at the temperature of freezing platinum under a pressure of 101.325 newton per square meter." (1967)

* Adapted from " The International System of Units (SI)," National Bureau of Standards Special Publication 330, 1972 edition. The definitions above were adopted by the General Conference of Weights and Measures, an international body, on the dates shown. In this book we do not use the candela.

2. Some SI Derived Units

Quantity	*Name of Unit*	*Symbol*	
area	square meter	m^2	
volume	cubic meter	m^3	
frequency	hertz	Hz	s^{-1}
mass density (density)	kilogram per cubic meter	kg/m^3	
speed, velocity	meter per second	m/s	
angular velocity	radian per second	rad/s	
acceleration	meter per second squared	m/s^2	
angular acceleration	radian per second squared	rad/s^2	
force	newton	N	$kg \cdot m/s^2$
pressure	pascal	Pa	N/m^2
work, energy, quantity of heat	joule	J	$N \cdot m$
power	watt	W	J/s
quantity of electricity	coulomb	C	$A \cdot s$
potential difference, electromotive force	volt	V	W/A
electric field strength	volt per meter	V/m	
electric resistance	ohm	Ω	V/A
capacitance	farad	F	$A \cdot s/V$
magnetic flux	weber	Wb	$V \cdot s$
inductance	henry	H	$V \cdot s/A$
magnetic flux density	tesla	T	Wb/m^2
magnetic field strength	ampere per meter	A/m	
entropy	joule per kelvin	J/K	
specific heat capacity	joule per kilogram kelvin	$J/(kg \cdot K)$	
thermal conductivity	watt per meter kelvin	$W/(m \cdot K)$	
radiant intensity	watt per steradian	W/sr	

3. The SI Supplementary Units

Quantity	*Name of Unit*	*Symbol*
plane angle	radian	rad
solid angle	steradian	sr

APPENDIX B
SOME FUNDAMENTAL CONSTANTS OF PHYSICS

Constant	Symbol	Computational Value	Best (1986) value	
			Value[a]	Uncertainty[b]
Speed of light in a vacuum	c	3.00×10^8 m/s	2.99792458	exact
Elementary charge	e	1.60×10^{-19} C	1.60217738	0.30
Electron rest mass	m_e	9.11×10^{-31} kg	9.1093897	0.59
Permittivity constant	ϵ_0	8.85×10^{-12} F/m	8.85418781762	exact
Permeability constant	μ_0	1.26×10^{-6} H/m	1.25663706143	exact
Electron rest mass[c]	m_e	5.49×10^{-4} u	5.48579902	0.023
Neutron rest mass[c]	m_n	1.0087 u	1.008664704	0.014
Hydrogen atom rest mass[c]	$m_{^1\mathrm{H}}$	1.0078 u	1.007825035	0.011
Deuterium atom rest mass[c]	$m_{^2\mathrm{H}}$	2.0141 u	2.0141019	0.053
Helium atom rest mass[c]	$m_{^4\mathrm{He}}$	4.0026 u	4.0026032	0.067
Electron charge to mass ratio	e/m_e	1.76×10^{11} C/kg	1.75881961	0.30
Proton rest mass	m_p	1.67×10^{-27} kg	1.6726230	0.59
Ratio of proton mass to electron mass	m_p/m_e	1840	1836.152701	0.020
Neutron rest mass	m_n	1.68×10^{-27} kg	1.6749286	0.59
Muon rest mass	m_μ	1.88×10^{-28} kg	1.8835326	0.61
Planck constant	h	6.63×10^{-34} J·s	6.6260754	0.60
Electron Compton wavelength	λ_c	2.43×10^{-12} m	2.42631058	0.089
Universal gas constant	R	8.31 J/mol·K	8.314510	8.4

Constant	Symbol	Computational Value	Best (1986) value	
			Value[a]	Uncertainty[b]
Avogadro constant	N_A	6.02×10^{23} mol^{-1}	6.0221367	0.59
Boltzmann constant	k	1.38×10^{-23} J/K	1.380657	11
Molar volume of ideal gas at STP[d]	V_m	2.24×10^{-2} m^3/mol	2.241409	8.4
Faraday constant	F	9.65×10^4 C/mol	9.6485309	0.30
Stefan–Boltzmann constant	σ	5.67×10^{-8} W/m$^2 \cdot$K^4	5.67050	34
Rydberg constant	R	1.10×10^7 m^{-1}	1.0973731534	0.0012
Gravitational constant	G	6.67×10^{-11} m^3/s$^2 \cdot$kg	6.67260	100
Bohr radius	r_B	5.29×10^{-11} m	5.29177249	0.045
Electron magnetic moment	μ_e	9.28×10^{-24} J/T	9.2847700	0.34
Proton magnetic moment	μ_P	1.41×10^{-26} J/T	1.41060761	0.34
Bohr magneton	μ_B	9.27×10^{-24} J/T	9.2740154	0.34
Nuclear magneton	μ_N	5.05×10^{-27} J/T	5.0507865	0.34

[a] Same unit and power of ten as the computational value.
[b] Parts per million.
[c] Mass given in unified atomic mass units, where 1 u = $1.6605402 \times 10^{-27}$ kg.
[d] STP—standard temperature and pressure = 0°C and 1.0 bar.
* The values in this table were largely selected from a longer list in *Symbols, Units and Nomenclature in Physics* (IUPAP), prepared by E. Richard Cohen and Pierre Giacomo, 1986.

APPENDIX C

SOME ASTRONOMICAL DATA

Some Distances from the Earth

To the moon*	3.82×10^{8} m
To the sun*	1.50×10^{11} m
To the nearest star (Proxima Centauri)	4.04×10^{16} m
To the center of our galaxy	2.2×10^{20} m
To the Andromeda Galaxy	2.1×10^{22} m
To the edge of the observable universe	$\sim 10^{26}$ m

* Mean distance.

The Sun, the Earth and the Moon

Property	*Unit*	*Sun*[a]	*Earth*	*Moon*
Mass	kg	1.99×10^{30}	5.98×10^{24}	7.36×10^{22}
Mean radius	m	6.96×10^{8}	6.37×10^{6}	1.74×10^{6}
Mean density	kg/m^3	1410	5520	3340
Free fall acceleration at the surface	m/s^2	274	9.81	1.67
Escape velocity	km/s	618	11.2	2.38
Period of rotation[c]	—	37 d—poles[b] 26 d—equator	23 h 56 min	27.3 d

[a] The sun radiates energy at the rate of 3.90×10^{26} W; just outside the earth's atmosphere solar energy is received, assuming normal incidence, at the rate of 1340 W/m^2.
[b] The sun—a ball of gas—does not rotate as a rigid body.
[c] Measured with respect to the distant stars.

Some Properties of the Planets

	Mercury	*Venus*	*Earth*	*Mars*	*Jupiter*	*Saturn*	*Uranus*	*Neptune*	*Pluto*
Mean distance from sun, 10^6 km	57.9	108	150	228	778	1,430	2,870	4,500	5,900
Period of revolution, y	0.241	0.615	1.00	1.88	11.9	29.5	84.0	165	248
Period of rotation[a], *d*	58.7	−243[b]	0.997	1.03	0.409	0.426	−0.451[b]	0.658	6.39
Orbital speed, km/s	47.9	35.0	29.8	24.1	13.1	9.64	6.81	5.43	4.74
Inclination of axis to orbit	<28°	~3°	23.5°	24.0°	3.08°	26.7°	82.1°	28.8°	?
Inclination of orbit to earth's orbit	7.00°	3.39°	—	1.85°	1.30°	2.49°	0.77°	1.77°	17.2°
Eccentricity of orbit	0.206	0.0068	0.0167	0.0934	0.0485	0.0556	0.0472	0.0086	0.250
Equatorial diameter, km	4,880	12,100	12,800	6,790	143,000	120,000	51,800	49,500	3,000(?)
Mass (earth = 1)	0.0558	0.815	1.000	0.107	318	95.1	14.5	17.2	0.01(?)
Density (water = 1)	5.60	5.20	5.52	3.95	1.31	0.704	1.21	1.67	?
Surface gravity[c], m/s^2	3.78	8.60	9.78	3.72	22.9	9.05	7.77	11.0	0.3(?)
Escape velocity[c], km/s	4.3	10.3	11.2	5.0	59.5	35.6	21.2	23.6	0.9(?)
Known satellites	0	0	1	2	16 + ring	17 + rings	15 + rings	2 + rings(?)	1

[a] Measured with respect to the distant stars.
[b] The sense of rotation is opposite to that of the orbital motion
[c] Measured at the planet's equator.

APPENDIX D

PROPERTIES OF THE ELEMENTS

Element	*Symbol*	*Atomic number, Z*	*Atomic mass,* g/mol	*Density,* g/cm³ *at 20°C*	*Melting point,* °C	*Boiling point,* °C	*Specific heat,* J/(g·C°) *at 25°C*
Actinium	Ac	89	(227)	—	1323	(3473)	0.092
Aluminum	Al	13	26.9815	2.699	660	2450	0.900
Americium	Am	95	(243)	11.7	1541	—	—
Antimony	Sb	51	121.75	6.62	630.5	1380	0.205
Argon	Ar	18	39.948	1.6626×10^{-3}	−189.4	−185.8	0.523
Arsenic	As	33	74.9216	5.72	817 (28 at.)	613	0.331
Astatine	At	85	(210)	—	(302)	—	—
Barium	Ba	56	137.34	3.5	729	1640	0.205
Berkelium	Bk	97	(247)	—	—	—	—
Beryllium	Be	4	9.0122	1.848	1287	2770	1.83
Bismuth	Bi	83	208.980	9.80	271.37	1560	0.122
Boron	B	5	10.811	2.34	2030	—	1.11
Bromine	Br	35	79.909	3.12 (liquid)	−7.2	58	0.293
Cadmium	Cd	48	112.40	8.65	321.03	765	0.226
Calcium	Ca	20	40.08	1.55	838	1440	0.624
Californium	Cf	98	(251)	—	—	—	—
Carbon	C	6	12.01115	2.25	3727	4830	0.691
Cerium	Ce	58	140.12	6.768	804	3470	0.188
Cesium	Cs	55	132.905	1.9	28.40	690	0.243
Chlorine	Cl	17	35.453	3.214×10^{-3} (0°C)	−101	−34.7	0.486
Chromium	Cr	24	51.996	7.19	1857	2665	0.448
Cobalt	Co	27	58.9332	8.85	1495	2900	0.423
Copper	Cu	29	63.54	8.96	1083.40	2595	0.385
Curium	Cm	96	(247)	—	—	—	—
Dysprosium	Dy	66	162.50	8.55	1409	2330	0.172
Einsteinium	Es	99	(254)	—	—	—	—
Erbium	Er	68	167.26	9.15	1522	2630	0.167
Europium	Eu	63	151.96	5.245	817	1490	0.163
Fermium	Fm	100	(237)	—	—	—	—
Fluorine	F	9	18.9984	1.696×10^{-3} (0°C)	−219.6	−188.2	0.753
Francium	Fr	87	(223)	—	(27)	—	—
Gadolinium	Gd	64	157.25	7.86	1312	2730	0.234
Gallium	Ga	31	69.72	5.907	29.75	2237	0.377
Germanium	Ge	32	72.59	5.323	937.25	2830	0.322
Gold	Au	79	196.967	19.32	1064.43	2970	0.131

Element	*Symbol*	*Atomic number, Z*	*Atomic mass, g/mol*	*Density, g/cm³ at 20°C*	*Melting point, °C*	*Boiling point, °C*	*Specific heat, J/(g·C°) at 25°C*
Hafnium	Hf	72	178.49	13.09	2227	5400	0.144
Helium	He	2	4.0026	0.1664×10^{-3}	−269.7	−268.9	5.23
Holmium	Ho	67	164.930	8.79	1470	2330	0.165
Hydrogen	H	1	1.00797	0.08375×10^{-3}	−259.19	−252.7	14.4
Indium	In	49	114.82	7.31	156.634	2000	0.233
Iodine	I	53	126.9044	4.94	113.7	183	0.218
Iridium	Ir	77	192.2	22.5	2447	(5300)	0.130
Iron	Fe	26	55.847	7.87	1536.5	3000	0.447
Krypton	Kr	36	83.80	3.488×10^{-3}	−157.37	−152	0.247
Lanthanum	La	57	138.91	6.189	920	3470	0.195
Lawrencium	Lw	103	(257)	—	—	—	—
Lead	Pb	82	207.19	11.36	327.45	1725	0.129
Lithium	Li	3	6.939	0.534	180.55	1300	3.58
Lutetium	Lu	71	174.97	9.849	1663	1930	0.155
Magnesium	Mg	12	24.312	1.74	650	1107	1.03
Manganese	Mn	25	54.9380	7.43	1244	2150	0.481
Mendelevium	Md	101	(256)	—	—	—	—
Mercury	Hg	80	200.59	13.55	−38.87	357	0.138
Molybdenum	Mo	42	95.94	10.22	2617	5560	0.251
Neodymium	Nd	60	144.24	7.00	1016	3180	0.188
Neon	Ne	10	20.183	0.8387×10^{-3}	−248.597	−246.0	1.03
Neptunium	Np	93	(237)	19.5	637	—	1.26
Nickel	Ni	28	58.71	8.902	1453	2730	0.444
Niobium	Nb	41	92.906	8.57	2468	4927	0.264
Nitrogen	N	7	14.0067	1.1649×10^{-3}	−210	−195.8	1.03
Nobelium	No	102	(255)	—	—	—	—
Osmium	Os	76	190.2	22.57	3027	5500	0.130
Oxygen	O	8	15.9994	1.3318×10^{-3}	−218.80	−183.0	0.913
Palladium	Pd	46	106.4	12.02	1552	3980	0.243
Phosphorus	P	15	30.9738	1.83	44.25	280	0.741
Platinum	Pt	78	195.09	21.45	1769	4530	0.134
Plutonium	Pu	94	(244)	—	640	3235	0.130
Polonium	Po	84	(210)	9.24	254	—	—
Potassium	K	19	39.102	0.86	63.20	760	0.758
Praseodymium	Pr	59	140.907	6.769	931	3020	0.197
Promethium	Pm	61	(145)	—	(1027)	—	—
Protactinium	Pa	91	(231)	—	(1230)	—	—
Radium	Ra	88	(226)	5.0	700	—	—
Radon	Rn	86	(222)	9.96×10^{-3} (0°C)	(−71)	−61.8	0.092
Rhenium	Re	75	186.2	21.04	3180	5900	0.134
Rhodium	Rh	45	102.905	12.44	1963	4500	0.243
Rubidium	Rb	37	85.47	1.53	39.49	688	0.364
Ruthenium	Ru	44	101.107	12.2	2250	4900	0.239
Samarium	Sm	62	150.35	7.49	1072	1630	0.197
Scandium	Sc	21	44.956	2.99	1539	2730	0.569
Selenium	Se	34	78.96	4.79	221	685	0.318
Silicon	Si	14	28.086	2.33	1412	2680	0.712
Silver	Ag	47	107.870	10.49	960.8	2210	0.234
Sodium	Na	11	22.9898	0.9712	97.85	892	1.23
Strontium	Sr	38	87.62	2.60	768	1380	0.737

Element	*Symbol*	*Atomic number, Z*	*Atomic mass, g/mol*	*Density, g/cm³ at 20°C*	*Melting point, °C*	*Boiling point, °C*	*Specific heat, J/(g·C°) at 25°C*
Sulfur	S	16	32.064	2.07	119.0	444.6	0.707
Tantalum	Ta	73	180.948	16.6	3014	5425	0.138
Technetium	Tc	43	(99)	11.46	2200	—	0.209
Tellurium	Te	52	127.60	6.24	449.5	990	0.201
Terbium	Tb	65	158.924	8.25	1357	2530	0.180
Thallium	Tl	81	204.37	11.85	304	1457	0.130
Thorium	Th	90	(232)	11.66	1755	(3850)	0.117
Thulium	Tm	69	168.934	9.31	1545	1720	0.159
Tin	Sn	50	118.69	7.2984	231.868	2270	0.226
Titanium	Ti	22	47.90	4.507	1670	3260	0.523
Tungsten	W	74	183.85	19.3	3380	5930	0.134
Uranium	U	92	(238)	19.07	1132	3818	0.117
Vanadium	V	23	50.942	6.1	1902	3400	0.490
Xenon	Xe	54	131.30	5.495×10^{-3}	−111.79	−108	0.159
Ytterbium	Yb	70	173.04	6.959	824	1530	0.155
Yttrium	Y	39	88.905	4.472	1526	3030	0.297
Zinc	Zn	30	65.37	7.133	419.58	906	0.389
Zirconium	Zr	40	91.22	6.489	1852	3580	0.276

The values in parentheses in the column of atomic masses are the mass numbers of the longest-lived isotopes of those elements that are radioactive. Melting points and boiling points in parentheses are uncertain.

All the physical properties are given for a pressure of one atmosphere except where otherwise specified.

The data for gases are valid only when these are in their usual molecular state, such as H_2, He, O_2, Ne, etc. The specific heats of the gases are the values at constant pressure.

Source: Adapted from Wehr, Richards, Adair, *Physics of the Atom,* 4th ed., Addison-Wesley, Reading, MA, 1984.

APPENDIX E

PERIODIC TABLE OF THE ELEMENTS

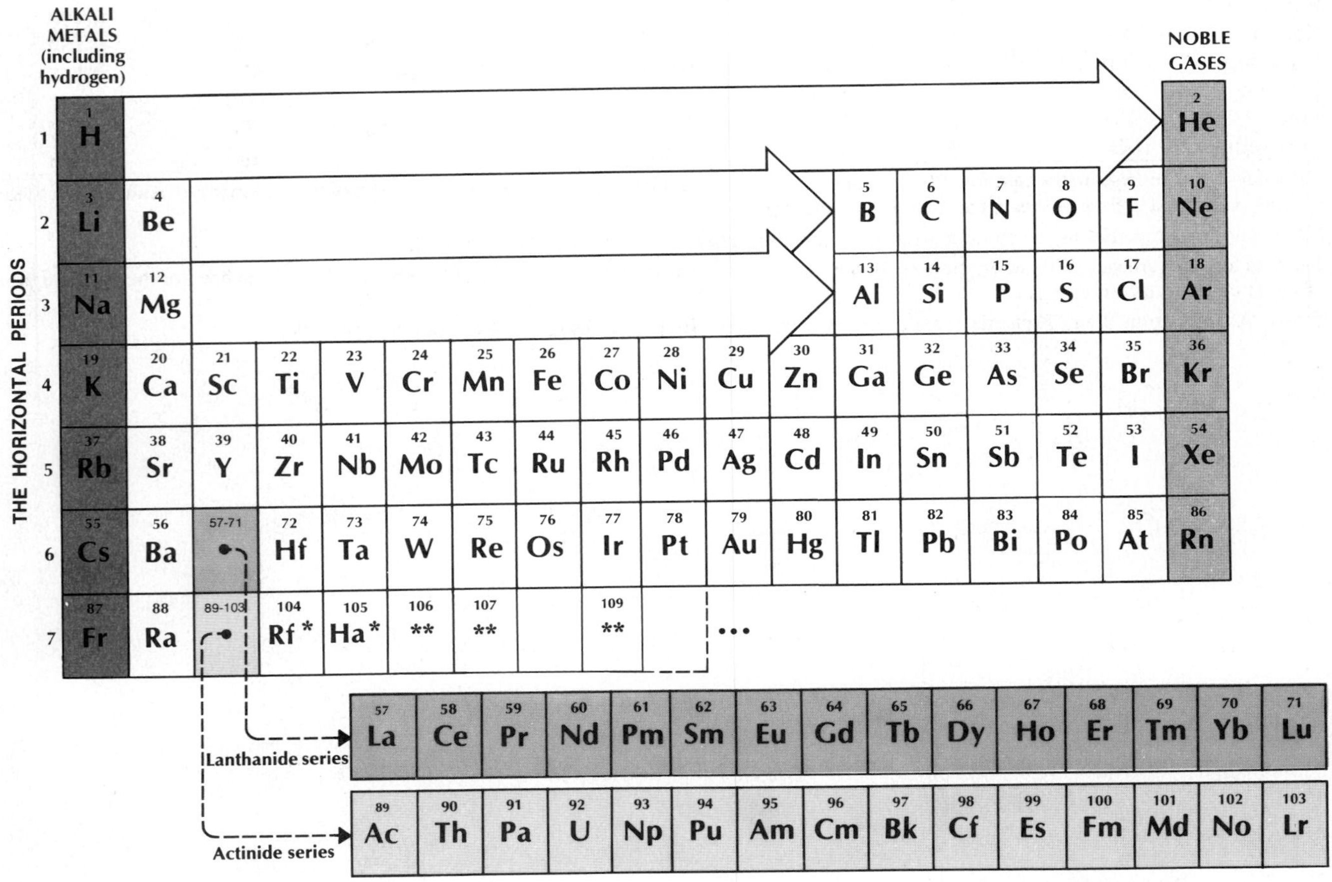

ALKALI METALS (including hydrogen)
NOBLE GASES
THE HORIZONTAL PERIODS
1 2 3 4 5 6 7
1 H
2 He
3 Li
4 Be
5 B
6 C
7 N
8 O
9 F
10 Ne
11 Na
12 Mg
13 Al
14 Si
15 P
16 S
17 Cl
18 Ar
19 K
20 Ca
21 Sc
22 Ti
23 V
24 Cr
25 Mn
26 Fe
27 Co
28 Ni
29 Cu
30 Zn
31 Ga
32 Ge
33 As
34 Se
35 Br
36 Kr
37 Rb
38 Sr
39 Y
40 Zr
41 Nb
42 Mo
43 Tc
44 Ru
45 Rh
46 Pd
47 Ag
48 Cd
49 In
50 Sn
51 Sb
52 Te
53 I
54 Xe
55 Cs
56 Ba
57-71
72 Hf
73 Ta
74 W
75 Re
76 Os
77 Ir
78 Pt
79 Au
80 Hg
81 Tl
82 Pb
83 Bi
84 Po
85 At
86 Rn
87 Fr
88 Ra
89-103
104 Rf *
105 Ha*
106 **
107 **
109 **
• • •
Lanthanide series
57 La
58 Ce
59 Pr
60 Nd
61 Pm
62 Sm
63 Eu
64 Gd
65 Tb
66 Dy
67 Ho
68 Er
69 Tm
70 Yb
71 Lu
Actinide series
89 Ac
90 Th
91 Pa
92 U
93 Np
94 Pu
95 Am
96 Cm
97 Bk
98 Cf
99 Es
100 Fm
101 Md
102 No
103 Lr

* The names of these elements (Rutherfordium and Hahnium) have not been accepted because of conflicting claims of discovery. A group in the USSR has proposed the names Kurchatovium and Neils-bohrium.

** Discovery of these three elements has been reported but names for them have not yet been proposed.

APPENDIX F
CONVERSION FACTORS

Conversion factors may be read off directly from the tables. For example, 1 degree = 2.778×10^{-3} revolutions, so 16.7° = $16.7 \times 2.778 \times 10^{-3}$ rev. The SI quantities are capitalized.

Adapted in part from G. Shortley and D. Williams, *Elements of Physics,* Prentice-Hall, Englewood Cliffs, NJ, 1971.

Plane Angle

	°	′	″	RADIAN	rev
1 degree =	1	60	3600	1.745×10^{-2}	2.778×10^{-3}
1 minute =	1.667×10^{-2}	1	60	2.909×10^{-4}	4.630×10^{-5}
1 second =	2.778×10^{-4}	1.667×10^{-2}	1	4.848×10^{-6}	7.716×10^{-7}
1 RADIAN =	57.30	3438	2.063×10^{5}	1	0.1592
1 revolution =	360	2.16×10^{4}	1.296×10^{6}	6.283	1

Solid Angle

1 sphere = 4π steradians = 12.57 steradians

Length

	cm	METER	km	in.	ft	mi
1 centimeter =	1	10^{-2}	10^{-5}	0.3937	3.281×10^{-2}	6.214×10^{-6}
1 METER =	100	1	10^{-3}	39.37	3.281	6.214×10^{-4}
1 kilometer =	10^{5}	1000	1	3.937×10^{4}	3281	0.6214
1 inch =	2.540	2.540×10^{-2}	2.540×10^{-5}	1	8.333×10^{-2}	1.578×10^{-5}
1 foot =	30.48	0.3048	3.048×10^{-4}	12	1	1.894×10^{-4}
1 mile =	1.609×10^{5}	1609	1.609	6.336×10^{4}	5280	1

1 angström = 10^{-10} m
1 nautical mile = 1852 m
= 1.151 miles = 6076 ft
1 fermi = 10^{-15} m

1 light-year = 9.460×10^{12} km
1 parsec = 3.084×10^{13} km
1 fathom = 6 ft
1 Bohr radius = 5.292×10^{-11} m

1 yard = 3 ft
1 rod = 16.5 ft
1 mil = 10^{-3} in.
1 nm = 10^{-9} m

Area

	METER^2	cm^2	ft^2	in.^{2}
1 SQUARE METER =	1	10^4	10.76	1550
1 square centimeter =	10^{-4}	1	1.076×10^{-3}	0.1550
1 square foot =	9.290×10^{-2}	929.0	1	144
1 square inch =	6.452×10^{-4}	6.452	6.944×10^{-3}	1

1 square mile = 2.788×10^7 ft^2 = 640 acres
1 barn = 10^{-28} m^2
1 acre = 43,560 ft^2
1 hectare = 10^4 m^2 = 2.471 acre

Volume

	METER^3	cm^3	L	ft^3	in.^{3}
1 CUBIC METER =	1	10^6	1000	35.31	6.102×10^4
1 cubic centimeter =	10^{-6}	1	1.000×10^{-3}	3.531×10^{-5}	6.102×10^{-2}
1 liter =	1.000×10^{-3}	1000	1	3.531×10^{-2}	61.02
1 cubic foot =	2.832×10^{-2}	2.832×10^4	28.32	1	1728
1 cubic inch =	1.639×10^{-5}	16.39	1.639×10^{-2}	5.787×10^{-4}	1

1 U.S. fluid gallon = 4 U.S. fluid quarts = 8 U.S. pints = 128 U.S. fluid ounces = 231 in.^{3}
1 British imperial gallon = 277.4 in^3 = 1.201 U.S. fluid gallons

Mass

Quantities in the colored areas are not mass units but are often used as such. When we write, for example, 1 kg "=" 2.205 lb this means that a kilogram is a *mass* that *weighs* 2.205 pounds under standard condition of gravity ($g = 9.80665$ m/s^2).

	g	KILOGRAM	slug	u	oz	lb	ton
1 gram =	1	0.001	6.852×10^{-5}	6.022×10^{23}	3.527×10^{-2}	2.205×10^{-3}	1.102×10^{-6}
1 KILOGRAM =	1000	1	6.852×10^{-2}	6.022×10^{26}	35.27	2.205	1.102×10^{-3}
1 slug =	1.459×10^4	14.59	1	8.786×10^{27}	514.8	32.17	1.609×10^{-2}
1 u =	1.661×10^{-24}	1.661×10^{-27}	1.138×10^{-28}	1	5.857×10^{-26}	3.662×10^{-27}	1.830×10^{-30}
1 ounce =	28.35	2.835×10^{-2}	1.943×10^{-3}	1.718×10^{25}	1	6.250×10^{-2}	3.125×10^{-5}
1 pound =	453.6	0.4536	3.108×10^{-2}	2.732×10^{26}	16	1	0.0005
1 ton =	9.072×10^5	907.2	62.16	5.463×10^{29}	3.2×10^4	2000	1

1 metric ton = 1000 kg

Density

Quantities in the colored areas are weight densities and, as such, are dimensionally different from mass densities. See note for mass table.

	slug/ft^3	KILOGRAM/ METER^3	g/cm^3	lb/ft^3	lb/in.^{3}
1 slug per ft^3 =	1	515.4	0.5154	32.17	1.862×10^{-2}
1 KILOGRAM per METER^3 =	1.940×10^{-3}	1	0.001	6.243×10^{-2}	3.613×10^{-5}
1 gram per cm^3 =	1.940	1000	1	62.43	3.613×10^{-2}
1 pound per ft^3 =	3.108×10^{-2}	16.02	1.602×10^{-2}	1	5.787×10^{-4}
1 pound per in.^{3} =	53.71	2.768×10^4	27.68	1728	1

Time

	y	d	h	min	SECOND
1 year =	1	365.25	8.766×10^3	5.259×10^5	3.156×10^7
1 day =	2.738×10^{-3}	1	24	1440	8.640×10^4
1 hour =	1.141×10^{-4}	4.167×10^{-2}	1	60	3600
1 minute =	1.901×10^{-6}	6.944×10^{-4}	1.667×10^{-2}	1	60
1 SECOND =	3.169×10^{-8}	1.157×10^{-5}	2.778×10^{-4}	1.667×10^{-2}	1

Speed

	ft/s	km/h	METER/ SECOND	mi/h	cm/s
1 foot per second =	1	1.097	0.3048	0.6818	30.48
1 kilometer per hour =	0.9113	1	0.2778	0.6214	27.78
1 METER per SECOND =	3.281	3.6	1	2.237	100
1 mile per hour =	1.467	1.609	0.4470	1	44.70
1 centimeter per second =	3.281×10^{-2}	3.6×10^{-2}	0.01	2.237×10^{-2}	1

1 knot = 1 nautical mi/h = 1.688 ft/s   1 mi/min = 88.00 ft/s = 60.00 mi/h

Force

Quantities in the colored areas are not force units but are often used as such. For instance, if we write 1 gram-force "=" 980.7 dynes, we mean that a gram-mass experiences a force of 980.7 dynes under standard conditions of gravity ($g = 9.80665$ m/s^2)

	dyne	NEWTON	lb	pdl	gf	kgf
1 dyne =	1	10^{-5}	2.248×10^{-6}	7.233×10^{-5}	1.020×10^{-3}	1.020×10^{-6}
1 NEWTON =	10^5	1	0.2248	7.233	102.0	0.1020
1 pound =	4.448×10^5	4.448	1	32.17	453.6	0.4536
1 poundal =	1.383×10^4	0.1383	3.108×10^{-2}	1	14.10	1.410×10^{-2}
1 gram-force =	980.7	9.807×10^{-3}	2.205×10^{-3}	7.093×10^{-2}	1	0.001
1 kilogram-force =	9.807×10^5	9.807	2.205	70.93	1000	1

Pressure

	atm	$dyne/cm^2$	inch of water	cm Hg	PASCAL	$lb/in.^2$	lb/ft^2
1 atmosphere =	1	1.013×10^{6}	406.8	76	1.013×10^{5}	14.70	2116
1 dyne per cm^2 =	9.869×10^{-7}	1	4.015×10^{-4}	7.501×10^{-5}	0.1	1.405×10^{-5}	2.089×10^{-3}
1 inch of water[a] at 4°C =	2.458×10^{-3}	2491	1	0.1868	249.1	3.613×10^{-2}	5.202
1 centimeter of mercury[a] at 0°C =	1.316×10^{-2}	1.333×10^{4}	5.353	1	1333	0.1934	27.85
1 PASCAL =	9.869×10^{-6}	10	4.015×10^{-3}	7.501×10^{-4}	1	1.450×10^{-4}	2.089×10^{-2}
1 pound per $in.^2$ =	6.805×10^{-2}	6.895×10^{4}	27.68	5.171	6.895×10^{3}	1	144
1 pound per ft^2 =	4.725×10^{-4}	478.8	0.1922	3.591×10^{-2}	47.88	6.944×10^{-3}	1

[a] Where the acceleration of gravity has the standard value $9.80665\ m/s^2$.

1 bar = $10^{6}\ dyne/cm^2$ = 0.1 MPa   1 millibar = $10^{3}\ dyne/cm^2$ = 10^{2} Pa   1 torr = 1 millimeter of mercury

Energy, Work, Heat

Quantities in the colored areas are not properly energy units but are included for convenience. They arise from the relativistic mass-energy equivalence formula $E = mc^2$ and represent the energy released if a kilogram or unified atomic mass unit (u) is completely converted to energy.

	Btu	erg	ft·lb	hp·h	JOULE	cal	kW·h	eV	MeV	kg	u
1 British thermal unit =	1	1.055×10^{10}	777.9	3.929×10^{-4}	1055	252.0	2.930×10^{-4}	6.585×10^{21}	6.585×10^{15}	1.174×10^{-14}	7.070×10^{12}
1 erg =	9.481×10^{-11}	1	7.376×10^{-8}	3.725×10^{-14}	10^{-7}	2.389×10^{-8}	2.778×10^{-14}	6.242×10^{11}	6.242×10^{5}	1.113×10^{-24}	670.2
1 foot-pound =	1.285×10^{-3}	1.356×10^{7}	1	5.051×10^{-7}	1.356	0.3238	3.766×10^{-7}	8.464×10^{18}	8.464×10^{12}	1.509×10^{-17}	9.037×10^{9}
1 horsepower-hour =	2545	2.685×10^{13}	1.980×10^{6}	1	2.685×10^{6}	6.413×10^{5}	0.7457	1.676×10^{25}	1.676×10^{19}	2.988×10^{-11}	1.799×10^{16}
1 JOULE =	9.481×10^{-4}	10^{7}	0.7376	3.725×10^{-7}	1	0.2389	2.778×10^{-7}	6.242×10^{18}	6.242×10^{12}	1.113×10^{-17}	6.702×10^{9}
1 calorie =	3.969×10^{-3}	4.186×10^{7}	3.088	1.560×10^{-6}	4.186	1	1.163×10^{-6}	2.613×10^{19}	2.613×10^{13}	4.660×10^{-17}	2.806×10^{10}
1 kilowatt-hour =	3413	3.6×10^{13}	2.655×10^{6}	1.341	3.6×10^{6}	8.600×10^{5}	1	2.247×10^{25}	2.247×10^{19}	4.007×10^{-11}	2.413×10^{16}
1 electron volt =	1.519×10^{-22}	1.602×10^{-12}	1.182×10^{-19}	5.967×10^{-26}	1.602×10^{-19}	3.827×10^{-20}	4.450×10^{-26}	1	10^{-6}	1.783×10^{-36}	1.074×10^{-9}
1 million electron volts =	1.519×10^{-16}	1.602×10^{-6}	1.182×10^{-13}	5.967×10^{-20}	1.602×10^{-13}	3.827×10^{-14}	4.450×10^{-20}	10^{6}	1	1.783×10^{-30}	1.074×10^{-3}
1 kilogram =	8.521×10^{13}	8.987×10^{23}	6.629×10^{16}	3.348×10^{10}	8.987×10^{16}	2.146×10^{16}	2.497×10^{10}	5.610×10^{35}	5.610×10^{29}	1	6.022×10^{26}
1 unified atomic mass unit =	1.415×10^{-13}	1.492×10^{-3}	1.101×10^{-10}	5.559×10^{-17}	1.492×10^{-10}	3.564×10^{-11}	4.146×10^{-17}	9.32×10^{8}	932.0	1.661×10^{-27}	1

Power

	Btu/h	ft·lb/s	hp	cal/s	kW	WATT
1 British thermal unit per hour =	1	0.2161	3.929×10^{-4}	6.998×10^{-2}	2.930×10^{-4}	0.2930
1 foot-pound per second =	4.628	1	1.818×10^{-3}	0.3239	1.356×10^{-3}	1.356
1 horsepower =	2545	550	1	178.1	0.7457	745.7
1 calorie per second =	14.29	3.088	5.615×10^{-3}	1	4.186×10^{-3}	4.186
1 kilowatt =	3413	737.6	1.341	238.9	1	1000
1 WATT =	3.413	0.7376	1.341×10^{-3}	0.2389	0.001	1

Magnetic Flux

	maxwell	WEBER
1 maxwell =	1	10^{-8}
1 WEBER =	10^{8}	1

Magnetic Field

	gauss	TESLA	milligauss
1 gauss =	1	10^{-4}	1000
1 TESLA =	10^{4}	1	10^{7}
1 milligauss =	0.001	10^{-7}	1

1 tesla = 1 weber/meter^2

APPENDIX G
MATHEMATICAL FORMULAS

Geometry

Circle of radius r: circumference $= 2\pi r$; area $= \pi r^2$.

Sphere of radius r: area $= 4\pi r^2$; volume $= \frac{4}{3}\pi r^3$.

Right circular cylinder of radius r and height h: area $= 2\pi r^2 + 2\pi rh$; volume $= \pi r^2 h$.

Triangle of base a and altitude h: area $= \frac{1}{2}ah$.

Pythagorean Theorem

$a^2 + b^2 = c^2$

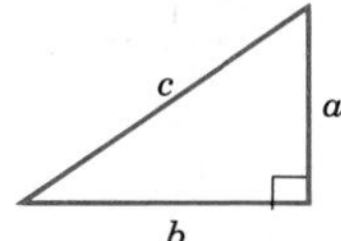

Quadratic Formula

If $ax^2 + bx + c = 0$, then $x = \dfrac{-b \pm \sqrt{b^2 - 4ac}}{2a}$.

Trigonometric Functions of Angle θ

$\sin\theta = \dfrac{y}{r}$ $\quad\cos\theta = \dfrac{x}{r}$

$\tan\theta = \dfrac{y}{x}$ $\quad\cot\theta = \dfrac{x}{y}$

$\sec\theta = \dfrac{r}{x}$ $\quad\csc\theta = \dfrac{r}{y}$

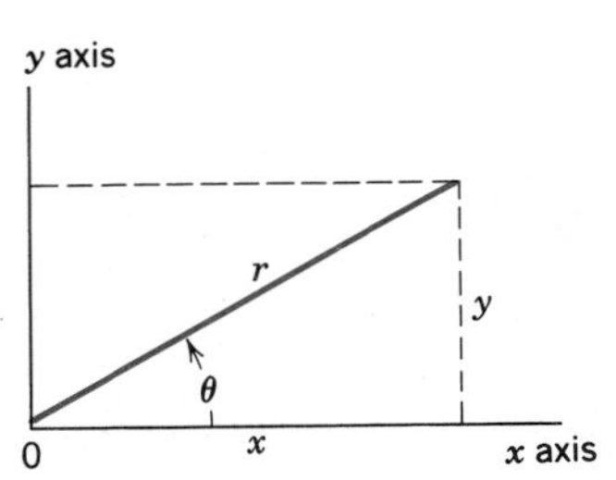

Triangles

Angles A, B, C

Opposite sides a, b, c

$A + B + C = 180°$

$$\frac{\sin A}{a} = \frac{\sin B}{b} = \frac{\sin C}{c}$$

$$c^2 = a^2 + b^2 - 2ab\cos C$$

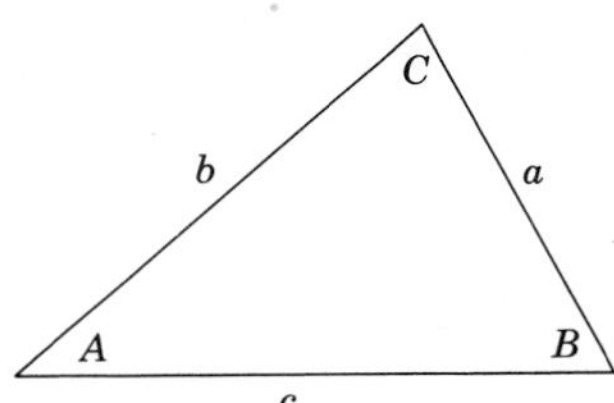

Mathematical Signs and Symbols

$=$ equals
$\approx$ equals approximately
$\neq$ is not equal to
$\equiv$ is identical to, is defined as
$>$ is greater than ($\gg$ is much greater than)
$<$ is less than ($\ll$ is much less than)
$\geq$ is greater than or equal to (or, is no less than)
$\leq$ is less than or equal to (or, is no more than)
$\pm$ plus or minus ($\sqrt{4} = \pm 2$)
$\propto$ is proportional to
Σ the sum of
$\bar{x}$ the average value of x

Trigonometric Identities

$\sin(90° - \theta) = \cos\theta$
$\cos(90° - \theta) = \sin\theta$
$\sin\theta/\cos\theta = \tan\theta$
$\sin^2\theta + \cos^2\theta = 1 \quad \sec^2\theta - \tan^2\theta = 1 \quad \csc^2\theta - \cot^2\theta = 1$
$\sin 2\theta = 2\sin\theta\cos\theta$
$\cos 2\theta = \cos^2\theta - \sin^2\theta = 2\cos^2\theta - 1 = 1 - 2\sin^2\theta$
$\sin(\alpha \pm \beta) = \sin\alpha\cos\beta \pm \cos\alpha\sin\beta$
$\cos(\alpha \pm \beta) = \cos\alpha\cos\beta \mp \sin\alpha\sin\beta$

$$\tan(\alpha \pm \beta) = \frac{\tan\alpha \pm \tan\beta}{1 \mp \tan\alpha\tan\beta}$$

$\sin\alpha \pm \sin\beta = 2\sin\frac{1}{2}(\alpha \pm \beta)\cos\frac{1}{2}(\alpha \mp \beta)$

Binomial Theorem

$$(1 \pm x)^n = 1 \pm \frac{nx}{1!} + \frac{n(n-1)}{2!}x^2 + \cdots \qquad (x^2 < 1)$$

$$(1 \pm x)^{-n} = 1 \mp \frac{nx}{1!} + \frac{n(n+1)x^2}{2!} + \cdots \qquad (x^2 < 1)$$

Exponential Expansion

$$e^x = 1 + x + \frac{x^2}{2!} + \frac{x^3}{3!} + \cdots$$

Logarithmic Expansion

$$\ln(1 + x) = x - \tfrac{1}{2}x^2 + \tfrac{1}{3}x^3 - \cdots \qquad (|x| < 1)$$

Trigonometric Expansions (θ in radians)

$$\sin\theta = \theta - \frac{\theta^3}{3!} + \frac{\theta^5}{5!} - \cdots$$

$$\cos\theta = 1 - \frac{\theta^2}{2!} + \frac{\theta^4}{4!} - \cdots$$

$$\tan\theta = \theta + \frac{\theta^3}{3} + \frac{2\theta^5}{15} + \cdots$$

Products of Vectors

Let **i**, **j**, **k** be unit vectors in the x, y, z directions. Then

$$\mathbf{i}\cdot\mathbf{i} = \mathbf{j}\cdot\mathbf{j} = \mathbf{k}\cdot\mathbf{k} = 1, \qquad \mathbf{i}\cdot\mathbf{j} = \mathbf{j}\cdot\mathbf{k} = \mathbf{k}\cdot\mathbf{i} = 0,$$

$$\mathbf{i}\times\mathbf{i} = \mathbf{j}\times\mathbf{j} = \mathbf{k}\times\mathbf{k} = 0,$$

$$\mathbf{i}\times\mathbf{j} = \mathbf{k}, \qquad \mathbf{j}\times\mathbf{k} = \mathbf{i}, \qquad \mathbf{k}\times\mathbf{i} = \mathbf{j}.$$

Any vector **a** with components a_x, a_y, a_z along the x, y, z axes can be written

$$\mathbf{a} = a_x\mathbf{i} + a_y\mathbf{j} + a_z\mathbf{k}.$$

Let **a**, **b**, **c** be arbitrary vectors with magnitudes a, b, c. Then

$$\mathbf{a}\times(\mathbf{b}+\mathbf{c}) = (\mathbf{a}\times\mathbf{b}) + (\mathbf{a}\times\mathbf{c})$$

$$(s\mathbf{a})\times\mathbf{b} = \mathbf{a}\times(s\mathbf{b}) = s(\mathbf{a}\times\mathbf{b}) \qquad (s = \text{a scalar}).$$

Let θ be the smaller of the two angles between **a** and **b**. Then

$$\mathbf{a}\cdot\mathbf{b} = \mathbf{b}\cdot\mathbf{a} = a_xb_x + a_yb_y + a_zb_z = ab\cos\theta$$

$$\mathbf{a}\times\mathbf{b} = -\mathbf{b}\times\mathbf{a} = \begin{vmatrix} \mathbf{i} & \mathbf{j} & \mathbf{k} \\ a_x & a_y & a_z \\ b_x & b_y & b_z \end{vmatrix} = (a_yb_z - b_ya_z)\mathbf{i} + (a_zb_x - b_za_x)\mathbf{j} + (a_xb_y - b_xa_y)\mathbf{k}$$

$$|\mathbf{a}\times\mathbf{b}| = ab\sin\theta$$

$$\mathbf{a}\cdot(\mathbf{b}\times\mathbf{c}) = \mathbf{b}\cdot(\mathbf{c}\times\mathbf{a}) = \mathbf{c}\cdot(\mathbf{a}\times\mathbf{b})$$

$$\mathbf{a}\times(\mathbf{b}\times\mathbf{c}) = (\mathbf{a}\cdot\mathbf{c})\mathbf{b} - (\mathbf{a}\cdot\mathbf{b})\mathbf{c}$$

Derivatives and Integrals

In what follows, the letters u and v stand for any functions of x, and a and m are constants. To each of the indefinite integrals should be added an arbitrary constant of integration. The *Handbook of Chemistry and Physics* (CRC Press Inc.) gives a more extensive tabulation.

1. $\dfrac{dx}{dx} = 1$
2. $\dfrac{d}{dx}(au) = a\dfrac{du}{dx}$
3. $\dfrac{d}{dx}(u+v) = \dfrac{du}{dx} + \dfrac{dv}{dx}$
4. $\dfrac{d}{dx}x^m = mx^{m-1}$
5. $\dfrac{d}{dx}\ln x = \dfrac{1}{x}$
6. $\dfrac{d}{dx}(uv) = u\dfrac{dv}{dx} + v\dfrac{du}{dx}$
7. $\dfrac{d}{dx}e^x = e^x$
8. $\dfrac{d}{dx}\sin x = \cos x$
9. $\dfrac{d}{dx}\cos x = -\sin x$
10. $\dfrac{d}{dx}\tan x = \sec^2 x$
11. $\dfrac{d}{dx}\cot x = -\csc^2 x$
12. $\dfrac{d}{dx}\sec x = \tan x \sec x$
13. $\dfrac{d}{dx}\csc x = -\cot x \csc x$
14. $\dfrac{d}{dx}e^u = e^u\dfrac{du}{dx}$
15. $\dfrac{d}{dx}\sin u = \cos u\dfrac{du}{dx}$
16. $\dfrac{d}{dx}\cos u = -\sin u\dfrac{du}{dx}$

1. $\displaystyle\int dx = x$
2. $\displaystyle\int au\,dx = a\int u\,dx$
3. $\displaystyle\int (u+v)\,dx = \int u\,dx + \int v\,dx$
4. $\displaystyle\int x^m\,dx = \frac{x^{m+1}}{m+1} \quad (m \neq -1)$
5. $\displaystyle\int \frac{dx}{x} = \ln|x|$
6. $\displaystyle\int u\frac{dv}{dx}\,dx = uv - \int v\frac{du}{dx}\,dx$
7. $\displaystyle\int e^x\,dx = e^x$
8. $\displaystyle\int \sin x\,dx = -\cos x$
9. $\displaystyle\int \cos x\,dx = \sin x$
10. $\displaystyle\int \tan x\,dx = \ln|\sec x|$
11. $\displaystyle\int \sin^2 x\,dx = \tfrac{1}{2}x - \tfrac{1}{4}\sin 2x$
12. $\displaystyle\int e^{-ax}\,dx = -\frac{1}{a}e^{-ax}$
13. $\displaystyle\int xe^{-ax}\,dx = -\frac{1}{a^2}(ax+1)e^{-ax}$
14. $\displaystyle\int x^2e^{-ax}\,dx = -\frac{1}{a^3}(a^2x^2+2ax+2)e^{-ax}$
15. $\displaystyle\int_0^\infty x^ne^{-ax}\,dx = \frac{n!}{a^{n+1}}$
16. $\displaystyle\int_0^\infty x^{2n}e^{-ax^2}\,dx = \frac{1\cdot 3\cdot 5\cdots(2n-1)}{2^{n+1}a^n}\sqrt{\frac{\pi}{a}}$

APPENDIX H

WINNERS OF THE NOBEL PRIZE IN PHYSICS*

1901	Wilhelm Konrad Röntgen	1845–1923	for the discovery of x-rays
1902	Hendrik Antoon Lorentz Pieter Zeeman	1853–1928 1865–1943	for their researches into the influence of magnetism upon radiation phenomena
1903	Antoine Henri Becquerel	1852–1908	for his discovery of spontaneous radioactivity
	Pierre Curie Marie Sklowdowska-Curie	1859–1906 1867–1934	for their joint researches on the radiation phenomena discovered by Professor Henri Becquerel
1904	Lord Rayleigh (John William Strutt)	1842–1919	for his investigations of the densities of the most important gases and for his discovery of argon
1905	Philipp Eduard Anton von Lenard	1862–1947	for his work on cathode rays
1906	Joseph John Thomson	1856–1940	for his theoretical and experimental investigations on the conduction of electricity by gases
1907	Albert Abraham Michelson	1852–1931	for his optical precision instruments and metrological investigations carried out with their aid
1908	Gabriel Lippmann	1845–1921	for his method of reproducing colors photographically based on the phenomena of interference
1909	Guglielmo Marconi Carl Ferdinand Braun	1874–1937 1850–1918	for their contributions to the development of wireless telegraphy
1910	Johannes Diderik van der Waals	1837–1932	for his work on the equation of state for gases and liquids
1911	Wilhelm Wien	1864–1928	for his discoveries regarding the laws governing the radiation of heat
1912	Nils Gustaf Dalén	1869–1937	for his invention of automatic regulators for use in conjunction with gas accumulators for illuminating lighthouses and buoys
1913	Heike Kamerlingh Onnes	1853–1926	for his investigations of the properties of matter at low temperatures which led, *inter alia,* to the production of liquid helium
1914	Max von Laue	1879–1960	for his discovery of the diffraction of Röntgen rays by crystals
1915	William Henry Bragg William Lawrence Bragg	1862–1942 1890–1971	for their services in the analysis of crystal structure by means of x-rays
1917	Charles Glover Barkla	1877–1944	for his discovery of the characteristic x-rays of the elements
1918	Max Planck	1858–1947	for his discovery of energy quanta
1919	Johannes Stark	1874–1957	for his discovery of the Doppler effect in canal rays and the splitting of spectral lines in electric fields
1920	Charles-Édouard Guillaume	1861–1938	for the service he has rendered to precision measurements in Physics by his discovery of anomalies in nickel steel alloys
1921	Albert Einstein	1879–1955	for his services to Theoretical Physics, and especially for his

* See *Nobel Lectures, Physics,* 1901–1970, Elsevier Publishing Company for biographies of the awardees and for lectures given by them on receiving the prize.

			discovery of the law of the photoelectric effect
1922	Niels Bohr	1885–1962	for the investigation of the structure of atoms, and of the radiation emanating from them
1923	Robert Andrews Millikan	1868–1953	for his work on the elementary charge of electricity and on the photoelectric effect
1924	Karl Manne Georg Siegbahn	1888–1979	for his discoveries and research in the field of x-ray spectroscopy
1925	James Franck	1882–1964	for their discovery of the laws governing the impact of an electron
	Gustav Hertz	1887–1975	upon an atom
1926	Jean Baptiste Perrin	1870–1942	for his work on the discontinuous structure of matter, and especially for his discovery of sedimentation equilibrium
1927	Arthur Holly Compton	1892–1962	for his discovery of the effect named after him
	Charles Thomson Rees Wilson	1869–1959	for his method of making the paths of electrically charged particles visible by condensation of vapor
1928	Owen Willans Richardson	1879–1959	for his work on the thermionic phenomenon and especially for the discovery of the law named after him
1929	Prince Louis-Victor de Broglie	1892–1987	for his discovery of the wave nature of electrons
1930	Sir Chandrasekhara Venkata Raman	1888–1970	for his work on the scattering of light and for the discovery of the effect named after him
1932	Werner Heisenberg	1901–1976	for the creation of quantum mechanics, the application of which has, among other things, led to the discovery of the allotropic forms of hydrogen
1933	Erwin Schrödinger	1887–1961	for the discovery of new productive forms
	Paul Adrien Maurice Dirac	1902–1984	of atomic theory
1935	James Chadwick	1891–1974	for his discovery of the neutron
1936	Victor Franz Hess	1883–1964	for the discovery of cosmic radiation
	Carl David Anderson	1905–1984	for his discovery of the positron
1937	Clinton Joseph Davisson	1881–1958	for their experimental discovery of the diffraction of electrons by
	George Paget Thomson	1892–1975	crystals
1938	Enrico Fermi	1901–1954	for his demonstrations of the existence of new radioactive elements produced by neutron irradiation, and for his related discovery of nuclear reactions brought about by slow neutrons
1939	Ernest Orlando Lawrence	1901–1958	for the invention and development of the cyclotron and for results obtained with it, especially for artificial radioactive elements
1943	Otto Stern	1888–1969	for his contribution to the development of the molecular ray method and his discovery of the magnetic moment of the proton
1944	Isidor Isaac Rabi	1898–	for his resonance method for recording the magnetic properties of atomic nuclei
1945	Wolfgang Pauli	1900–1958	for the discovery of the Exclusion Principle (Pauli Principle)
1946	Percy Williams Bridgman	1882–1961	for the invention of an apparatus to produce extremely high pressures, and for the discoveries he made therewith in the field of high-pressure physics
1947	Sir Edward Victor Appleton	1892–1965	for his investigations of the physics of the upper atmosphere, especially for the discovery of the so-called Appleton layer
1948	Patrick Maynard Stuart Blackett	1897–1974	for his development of the Wilson cloud chamber method, and his discoveries therewith in nuclear physics and cosmic radiation
1949	Hideki Yukawa	1907–1981	for his prediction of the existence of mesons on the basis of theoretical work on nuclear forces
1950	Cecil Frank Powell	1903–1969	for his development of the photographic method of studying nuclear processes and his discoveries regarding mesons made with this method
1951	Sir John Douglas Cockcroft	1897–1967	for their pioneer work on the transmutation of atomic nuclei by
	Ernest Thomas Sinton Walton	1903–	artificially accelerated atomic particles
1952	Felix Bloch	1905–1983	for their development of new methods for nuclear magnetic
	Edward Mills Purcell	1912–	precision methods and discoveries in connection therewith

Year	Name	Dates	Citation
1953	Frits Zernike	1888–1966	for his demonstration of the phase-contrast method, especially for his invention of the phase-contrast microscope
1954	Max Born	1882–1970	for his fundamental research in quantum mechanics, especially for his statistical interpretation of the wave function
	Walther Bothe	1891–1957	for the coincidence method and his discoveries made therewith
1955	Willis Eugene Lamb	1913–	for his discoveries concerning the fine structure of the hydrogen spectrum
	Polykarp Kusch	1911–	for his precision determination of the magnetic moment of the electron
1956	William Shockley	1910–	for their researches on semiconductors and their discovery of the transistor effect
	John Bardeen	1908–	
	Walter Houser Brattain	1902–1987	
1957	Chen Ning Yang	1922–	for their penetrating investigation of the parity laws which has led to important discoveries regarding the elementary particles
	Tsung Dao Lee	1926–	
1958	Pavel Aleksejevič Čerenkov	1904–	for the discovery and the interpretation of the Cerenkov effect
	Il' ja Michajlovič Frank	1908–	
	Igor' Evgen' evič Tamm	1895–1971	
1959	Emilio Gino Segrè	1905–	for their discovery of the antiproton
	Owen Chamberlain	1920–	
1960	Donald Arthur Glaser	1926–	for the invention of the bubble chamber
1961	Robert Hofstadter	1915–	for his pioneering studies of electron scattering in atomic nuclei and for his thereby achieved discoveries concerning the structure of the nucleons
	Rudolf Ludwig Mössbauer	1929–	for his researches concerning the resonance absorption of γ-rays and his discovery in this connection of the effect which bears his name
1962	Lev Davidovič Landau	1908–1968	for his pioneering theories of condensed matter, especially liquid helium
1963	Eugene P. Wigner	1902–	for his contributions to the theory of the atomic nucleus and the elementary particles, particularly through the discovery and application of fundamental symmetry principles
	Maria Goeppert Mayer	1906–1972	for their discoveries concerning nuclear shell structure
	J. Hans D. Jensen	1907–1973	
1964	Charles H. Townes	1915–	for fundamental work in the field of quantum electronics which has led to the construction of oscillators and amplifiers based on the maser-laser principle
	Nikolai G. Basov	1922–	
	Alexander M. Prochorov	1916–	
1965	Sin-itiro Tomonaga	1906–1979	for their fundamental work in quantum electrodynamics, with deep-ploughing consequences for the physics of elementary particles
	Julian Schwinger	1918–	
	Richard P. Feynman	1918–1988	
1966	Alfred Kastler	1902–1984	for the discovery and development of optical methods for studying Hertzian resonance in atoms
1967	Hans Albrecht Bethe	1906–	for his contributions to the theory of nuclear reactions, especially his discoveries concerning the energy production in stars
1968	Luis W. Alvarez	1911–	for his decisive contribution to elementary particle physics, in particular the discovery of a large number of resonance states, made possible through his development of the technique of using hydrogen bubble chamber and data analysis
1969	Murray Gell-Mann	1929–	for his contributions and discoveries concerning the classification of elementary particles and their interactions
1970	Hannes Alvén	1908–	for fundamental work and discoveries in magneto-hydrodynamics with fruitful applications in different parts of plasma physics
	Louis Néel	1904–	for fundamental work and discoveries concerning antiferromagnetism and ferrimagnetism which have led to important applications in solid state physics
1971	Dennis Gabor	1900–1979	for his discovery of the principles of holography

1972	John Bardeen	1908–	for their development of a theory of superconductivity
	Leon N. Cooper	1930–	
	J. Robert Schrieffer	1931–	
1973	Leo Esaki	1925–	for his discovery of tunneling in semiconductors
	Ivar Giaever	1929–	for his discovery of tunneling in superconductors
	Brian D. Josephson	1940–	for his theoretical prediction of the properties of a super-current through a tunnel barrier
1974	Antony Hewish	1924–	for the discovery of pulsars
	Sir Martin Ryle	1918–1984	for his pioneering work in radioastronomy
1975	Aage Bohr	1922–	for the discovery of the connection between collective motion and particle motion and the development of the theory of the structure of the atomic nucleus based on this connection
	Ben Mottelson	1926–	
	James Rainwater	1917–	
1976	Burton Richter	1931–	for their (independent) discovery of an important fundamental particle.
	Samuel Chao Chung Ting	1936–	
1977	Philip Warren Anderson	1923–	for their fundamental theoretical investigations of the electronic structure of magnetic and disordered systems
	Nevill Francis Mott	1905–	
	John Hasbrouck Van Vleck	1899–1980	
1978	Peter L. Kapitza	1894–1984	for his basic inventions and discoveries in low-temperature physics
	Arno A. Penzias	1926–	for their discovery of cosmic microwave background radiation
	Robert Woodrow Wilson	1936–	
1979	Sheldon Lee Glashow	1932–	for their unified model of the action of the weak and electromagnetic forces and for their prediction of the existence of neutral currents
	Abdus Salam	1926–	
	Steven Weinberg	1933–	
1980	James W. Cronin	1931–	for the discovery of violations of fundamental symmetry principles in the decay of neutral K mesons
	Val L. Fitch	1923–	
1981	Nicolaas Bloembergen	1920–	for their contribution to the development of laser spectroscopy
	Arthur Leonard Schawlow	1921–	
	Kai M. Siegbahn	1918–	for his contribution of high-resolution electron spectroscopy
1982	Kenneth Geddes Wilson	1936–	for his method of analyzing the critical phenomena inherent in the changes of matter under the influence of pressure and temperature
1983	Subrehmanyan Chandrasekhar	1910–	for his theoretical studies of the structure and evolution of stars
	William A. Fowler	1911–	for his studies of the formation of the chemical elements in the universe
1984	Carlo Rubbia	1934–	for their decisive contributions to the large project, which led to the discovery of the field particles W and Z, communicators of the weak interaction
	Simon van der Meer	1925–	
1985	Klaus von Klitzing	1943–	for his discovery of the quantized Hall resistance
1986	Ernst Ruska	1906–	for his invention of the electron microscope;
	Gerd Binnig	1947–	for their invention of the scanning-tunneling electron microscope
	Heinrich Rohrer	1933–	
1987	Karl Alex Müller	1927–	for their discovery of a new class of superconductors.
	J. George Bednorz	1950–	

ANSWERS TO ODD-NUMBERED EXERCISES AND PROBLEMS

CHAPTER 23

1. (*a*) 8.99×10^9 N. (*b*) 8990 N. **3**. 1.39 m. **5**. (*a*) 4.9×10^{-7} kg. (*b*) 7.1×10^{-11} C. **7**. $\frac{3}{8}$F. **9**. (*a*) $q_1 = 9q_2$. (*b*) $q_1 = -25q_2$. **11**. 1.2×10^{-5} C and 3.8×10^{-5} C. **13**. 14 cm from the positive charge, 24 cm from the negative charge. **15**. (*a*) A charge $-4q/9$ must be located on the line segment joining the two positive charges, a distance $L/3$ from the $+q$ charge. **17**. (*a*) $Q = -2\sqrt{2}q$. (*b*) No. **19**. (*b*) $\pm 2.4 \times 10^{-8}$ C. **21**. (*a*) $\frac{L}{2}\left(1 + \frac{1}{4\pi\epsilon_0}\frac{qQ}{Wh^2}\right)$. (*b*) $\sqrt{\frac{3}{4\pi\epsilon_0}\frac{qQ}{W}}$. **23**. 3.78 N. **25**. 1.93 MC. **27**. (*a*) 8.99×10^{-19} N. (*b*) 625. **29**. 11.9 cm. **31**. 1.34 days. **33**. 1.3×10^7 C. **35**. 1.52×10^8 N. **37**. (*a*) Positron. (*b*) Electron. **39**. (*a*) 510 N. (*b*) 7.7×10^{28} m/s^2. **41**. (*a*) $(Gh/2\pi c^3)^{1/2}$. (*b*) 1.61×10^{-35} m.

CHAPTER 24

1. 3.51×10^{15} m/s^2. **3**. (*a*) 4.8×10^{-13} N. (*b*) 4.8×10^{-13} N. **5**. (*a*) 1.5×10^3 N/C. (*b*) 2.4×10^{-16} N, up. (*c*) 1.6×10^{-26} N. (*d*) 1.5×10^{10}. **7**. (*a*) 6.4×10^{-18} N. (*b*) ≈ 20 N/C. **9**. To the right in the figure. **15**. $+1.00$ μC. **17**. (*a*) The larger charge produces a field of 1.3×10^5 N/C at the site of the smaller; the smaller produces a field of 5.3×10^4 N/C at the site of the larger. (*b*) 1.1×10^{-2} N. **19**. 2.86×10^{21} V/m; radially outward. **21**. (*a*) $0.172a$ to the right of the $+2q$ charge. **23**. 50 cm from q_1 and 100 cm from q_2. **25**. (*a*) 4.8×10^{-8} N/C. (*b*) 7.7×10^{-27} N. **27**. 1.02×10^5 N/C, upward. **29**. 6.63×10^{-15} N. **31**. $\frac{1}{4\pi\epsilon_0}\frac{p}{r^3}$; antiparallel to p. **41**. (*a*) 0.104 μC. (*b*) 1.31×10^{17}. (*c*) 4.96×10^{-6}. **43**. (*a*) 2.46×10^{17} m/s^2. (*b*) 0.122 ns. (*c*) 1.83 mm. **45**. (*a*) 7.12 cm. (*b*) 28.5 ns. (*c*) 11.2%. **47**. $5e$. **49**. (*a*) 0.245 N, 11.3° clockwise from the $+x$ axis. (*b*) $x = 108$ m; $y = -21.6$ m. **51**. (*a*) $-\mathbf{j}(2.1 \times 10^{15}$ m/s$^2)$. (*b*) $\mathbf{i}(1.5 \times 10^5$ m/s$)$ $-\mathbf{j}(2.8 \times 10^8m/s)$. **53**. (*a*) $2\pi\sqrt{\frac{l}{|g-qE/m|}}$. (*b*) $2\pi\sqrt{\frac{l}{g+qE/m}}$. **55**. (*a*) 9.30×10^{-15} C·m. (*b*) 2.05×10^{-11} J. **57**. $2pE\cos\theta_0$.

CHAPTER 25

1. (*a*) 693 kg/s. (*b*) 693 kg/s. (*c*) 347 kg/s. (*d*) 347 kg/s. (*e*) 575 kg/s. **3**. (*a*) Zero. (*b*) -3.92 N·m^2/C. (*c*) Zero. (*d*) Zero for each field. **7**. 2.03×10^5 N·m^2/C. **9**. $q/6\epsilon_0$. **11**. 3.54 μC. **13**. Through each of the three faces meeting at q: zero; through each of the three other faces: $q/24\epsilon_0$. **15**. (*a*) 36.6 μC. (*b*) 4.14×10^6 N·m^2/C. **17**. (*a*) $-Q$. (*b*) $-Q$. (*c*) $-(Q+q)$.

(*a*) Yes. **21**. (*a*) 0.317 μC. (*b*) 0.143 μC. **23**. (*a*) $E = \lambda/2\pi\epsilon_0 r$. (*b*) Zero. **25**. 3.61 nC. **27**. (*a*) 2.3×10^6 N/C, radially out. (*b*) 4.5×10^5 N/C, radially in. **29**. (*b*) $\rho R^2/2\epsilon_0 r$. **31**. (*a*) 5.30×10^7 N/C. (*b*) 59.9 N/C. **33**. 5.0 nC/m^2. **35**. (*a*) Zero. (*b*) $E = \sigma/\epsilon_0$, to the left. (*c*) Zero. **37**. 4.94×10^{-22} C/m^2. **39**. -7.51 nC. **41**. (*a*) Zero. (*b*) 2.9×10^4 N/C. (*c*) 200 N/C. **43**. (*a*) Zero. (*b*) $q_a/4\pi\epsilon_0 r^2$. (*c*) $(q_a + q_b)/4\pi\epsilon_0 r^2$. **45**. (*a*) $E = q/4\pi\epsilon_0 r^2$, radially outward. (*b*) Same as (*a*). (*c*) No. (*d*) Yes, charges are induced on the surfaces. (*e*) Yes. (*f*) No. (*g*) No. **47**. (*a*) $E = (q/4\pi\epsilon_0 a^3)r$. (*b*) $E = q/4\pi\epsilon_0 r^2$. (*c*) Zero. (*d*) Zero. (*e*) Inner, $-q$; outer, zero. **49**. (*a*) 4.0×10^6 N/C. (*b*) Zero.

CHAPTER 26

1. 1.2 GeV. **3**. (*a*) 3.00×10^{10} J. (*b*) 7.75 km/s. (*c*) 9.0×10^4 kg (99 tons). **5**. (*a*) 2.46 V. (*b*) 2.46 V. (*c*) Zero. **7**. 2.90 kV. **9**. 8.8 mm. **11**. (*a*) $-qr^2/(8\pi\epsilon_0 R^3)$. (*b*) $q/(8\pi\epsilon_0 R)$. **15**. (*a*) 4.5 m. (*b*) No. **17**. -1.1 nC. **21**. No. **23**. 637 mV. **25**. (*a*) 0.54 mm. (*b*) 790 V. **27**. (*a*) 38.0 s. (*b*) 280 days. **29**. (*a*) $Q/4\pi\epsilon_0 r$. (*b*) $\frac{\rho}{3\epsilon_0}\left(\frac{3}{2}r_2^2 - \frac{1}{2}r^2 - \frac{r_1^3}{r}\right)$; $\rho = Q/\frac{4\pi}{3}(r_2^3 - r_1^3)$. (*c*) $\frac{\rho}{2\epsilon_0}(r_2^2 - r_1^2)$, with ρ as in (*b*). **33**. 667 N/C. **35**. $\frac{1}{2\pi\epsilon_0}\frac{p \cos\theta}{r^3}$. **37**. 39.0 V/m, toward $x = 0$. **39**. (*a*) $\frac{\lambda}{4\pi\epsilon_0}\ln\left(\frac{L+y}{y}\right)$. (*b*) $\frac{\lambda}{4\pi\epsilon_0}\frac{L}{y(L+y)}$. (*c*) Zero. **41**. (*a*) $qd/2\pi\epsilon_0 a(a+d)$. **43**. (*a*) -6.8 MV. (*b*) -1.9 J. **45**. (*a*) 0.484 eV. (*b*) Zero. **47**. -1.24×10^{-6} J. **49**. (*a*) -7.8×10^5 V; $+0.6 \times 10^5$ V. (*b*) 2.5 J. (*c*) Work is converted into potential energy. **51**. (*a*) 27 V. (*b*) -27 eV. (*c*) 13.6 eV. (*d*) 13.6 eV. **53**. 2.48 km/s. **55**. 5.85×10^7 m/s. **57**. $qQ/4\pi\epsilon_0 K$. **59**. 0.32 km/s. **61**. 1.6×10^{-9} m. **65**. (*a*) $V_1 = V_2$. (*b*) $q_1 = q/3$, $q_2 = 2q/3$. **67**. (*a*) -0.12 V. (*b*) 1.8×10^{-8} N/C, radially inward. **69**. (*a*) 12,000 N/C. (*b*) 1800 V. (*c*) 5.77 cm. **71**. (*a*) 0.11 mC; 1.1 μC. **73**. (*a*) 3.2×10^{-13} J or 2.0 MeV. (*b*) 1.6×10^{-13} J or 1.0 MeV. (*c*) The proton. **75**. (*a*) $r > 9.0$ cm. (*b*) 2.7 kW. (*c*) 20 μC/m^2.

CHAPTER 27

1. 7.5 pC. **3**. 3.0 mC. **5**. (*a*) 144 pF. (*b*) 17.3 nC. **7**. 0.551 pF. **9**. 4.2×10^{-7} C. **15**. $5.05\pi\epsilon_0 R$. **19**. 3.16 μF. **21**. 315 mC. **23**. (*a*) 10.0 μF. (*b*) $q_4 = 0.800$ mC; $q_6 = 1.20$ mC. (*c*) 200 V. **25**. (*a*) $q_2 = q_8 = 0.48$ mC; $V_2 = 240$ V; $V_8 = 60$ V. (*b*) $q_2 = 0.19$ mC; $q_8 = 0.77$ mC; $V_2 = V_8 = 96$ V. (*c*) $q_2 = q_8 = 0$; $V_2 = V_8 = 0$. **27**. (*a*) $+7.9 \times 10^{-4}$ C. (*b*) + 79 V. **29**. (*a*) Five in series. (*b*) Three arrays as in (*a*) in parallel. There are other possibilities. **31**. 43 pF. **33**. $q_1 = \frac{C_1C_2 + C_1C_3}{C_1C_2 + C_1C_3 + C_2C_3}C_1V_0$; $q_2 = q_3 = \frac{C_2C_3}{C_1C_2 + C_1C_3 + C_2C_3}C_1V_0$. **35**. First case: 50 V; second case: zero. **37**. (*a*) 3.05 MJ. (*b*) 0.847 kW·h. **39**. (*a*) 0.204 μJ. (*b*) No. **41**. 0.27 J. **43**. 4.88%. **45**. 27.8 cents. **47**. (*a*) 2.0 J. **49**. (*a*) $2V$. (*b*) $U_i = \epsilon_0 AV^2/2d$; $U_f = 2U_i$. (*c*) $\epsilon_0 AV^2/2d$. **51**. (*a*) $e^2/8\pi\epsilon_0 R$. (*b*) 1.41 fm. **53**. Pyrex. **55**. (*a*) 6.20 cm. (*b*) 280 pF. **57**. 0.63 m^2. **59**. (*a*) 2.85 m^3. (*b*) 10.1×10^3. **61**. (*a*) $\epsilon_0 A/(d-b)$. (*b*) $d/(d-b)$. (*c*) $-q^2b/2\epsilon_0 A$; sucked in. **65**. $\frac{\epsilon_0 A}{4d}\left(\kappa_1 + \frac{2\kappa_2\kappa_3}{\kappa_2+\kappa_3}\right)$. **67**. (*a*) 13.4 pF. (*b*) 1.15 nC. (*c*) 1.13×10^4 N/C. (*d*) 4.33×10^3 N/C. **69**. (*a*) 89 pF. (*b*) 120 pF. (*c*) 11 nC; 11 nC. (*d*) 10 KV/m. (*e*) 2.1 kV/m. (*f*) 88 V. (*g*) 0.17 μJ.

CHAPTER 28

1. (*a*) 1200 C. (*b*) 7.5×10^{21}. **3**. 5.6 ms. **5**. (*a*) 6.4 A/m^2, north. **7**. 0.380 mm. **9**. 0.67 A, toward the negative terminal. **11**. (*a*) 0.654 μA/m^2. (*b*) 83.4 MA. **13**. 13.4 min. **15**. (*a*) $J_0A/3$. (*b*) $2J_0A/3$. **17**. 1.96×10^{-8} Ω·m. **19**. 2.40 Ω. **21**. 2.0×10^6 (Ω·m)$^{-1}$. **23**. 57.2°C. **25**. (*a*) 0.384 mV. (*b*) Negative. (*c*) 3 min 58 s. **27**. 54 Ω. **29**. 2.87 mm. **31**. (*a*) 2.39, iron being larger. (*b*) No. **33**. New length = $1.369L$; new area = $0.730A$. **35**. (*a*) 6.00 mA. (*b*) 1.59×10^{-8} V. (*c*) 21.2 pΩ. **37**. 8.21×10^{-4} Ω·m. **39**. (*a*) 5.32×10^5 A/m^2 for copper; 3.27×10^5 A/m^2 for aluminum. (*b*) 1.01 kg/m for copper; 0.495 kg/m for aluminum. **41**. 0.40 Ω. **43**. (*a*) $R = \rho L/\pi ab$. **45**. 14.0 kC. **47**. 11.1 Ω. **49**. (*a*) 1.0 kW. (*b*) 25 cents. **51**. (*a*) 28.8 Ω. (*b*) 2.60×10^{19} s^{-1}. **53**. (*a*) 1.74 A. (*b*) 2.15 MA/m^2. (*c*) 36.3 mV/m. (*d*) 2.09 W. **55**. (*a*) 5.85 m. (*b*) 10.4 m. **57**. (*a*) \$4.46 for a 31-day month. (*b*) 144 Ω. (*c*) 0.833 A. **59**. (*a*) 9.4×10^{13} s^{-1}. (*b*) 240 W. **61**. 710 cal/g. **63**. (*a*) 8.6%. (*b*) Smaller **65**. (*a*) 28 min. (*b*) 1.6 h.

CHAPTER 29

1. (*a*) 1.92×10^{-18} J (= 12 eV). (*b*) 6.53 W. **3**. (*a*) \$320. (*b*) 9.6¢. **5**. Counterclockwise; #1; *B*. **9**. (*a*) 14 V. (*b*) 100 W. (*c*) 600 W. (*d*) 10 V; 100 W. **11**. (*a*) 50 V. (*b*) 48 V. (*c*) *B* is the negative terminal. **13**. 2.5 V. **15**. (*a*) 990 Ω. (*b*) 9.4×10^{-4} W. **17**. 8.0 Ω. **19**. The cable. **21**. (*a*) 1000 Ω. (*b*) 300 mV. **23**. (*a*) 1.32×10^7 A/m^2 in each. (*b*) $V_A = 8.90$ V; $V_B = 51.1$ V. (*c*) *A*: Copper; *B*: Iron. **25**. Silicon: 85.0 Ω; iron: 915 Ω. **27**. 4.00 Ω. **29**. $i_1 = 50$ mA; $i_2 = 60$ mA; $V_{ab} = 9.0$ V. **31**. (*a*) 6.67 Ω. (*b*) 6.67 Ω. (*c*) Zero. **33**. (*a*) R_2. (*b*) R_1. **35**. 3*d*. **37**. 7.50 V. **39**. 38 Ω or 260 Ω. **41**. (*a*) $R = r/2$. (*b*) $P_{max} = \mathcal{E}^2/2r$. **43**. (*a*) $2\mathcal{E}/(2r + R)$, series; $2\mathcal{E}/(r + 2R)$, parallel. (*b*) Series if $r < R$; parallel if $R < r$. **45**. (*a*) Left branch, 0.67 A, down; center branch, 0.33 A, up; right branch, 0.33 A, up. (*b*) 3.3 V. **47**. $\mathcal{E}/7R$.

49. (*a*) 120 Ω. (*b*) $i_1 = 50$ mA; $i_2 = i_3 = 20$ mA; $i_4 = 10$ mA. **51**. (*a*) 19.5 Ω. (*b*) 82.3 W. (*c*) 57.6 W. **53**. (*a*) 2.50 Ω. (*b*) 3.125 Ω. **55**. $(50\,\text{kW})\left(\frac{x}{2000+10x-x^2}\right)^2$, x in cm. **57**. (*a*) 13.5 kΩ. (*b*) 1500 Ω. (*c*) 167 Ω. (*d*) 1480 Ω. **59**. 0.9% **65**. (*a*) 2.52 s. (*b*) 21.6 μC. (*c*) 3.40 s. **67**. 2.18 ms; no. **69**. (*a*) 2.17 s. (*b*) 39.6 mV. **71**. (*a*) 10^{-3} C. (*b*) 10^{-3} A. (*c*) $V_C = 10^3e^{-t}$, $V_R = -10^3e^{-t}$, volts. **73**. 0.72 MΩ. **77**. Decreases by 13.3 μC.

CHAPTER 30

1. M/QT; ML^2/QT. **3**. (*a*) 9.56×10^{-14} N; zero. (*b*) 27.8°. **5**. (*a*) $6.2 \times 10^{-14}\mathbf{k}$, N. (*b*) $-6.2 \times 10^{-14}\mathbf{k}$, N. **7**. (*a*) East. (*b*) 6.28×10^{14} m/s². (*c*) 2.98 mm. **9**. 2 **11**. (*a*) 3.75 km/s. **13**. (*b*) 680 kV/m. **17**. (*b*) 2.84×10^{-3}. **21**. 1.6×10^{-8} T. **23**. (*a*) 1.11×10^7 m/s. (*b*) 0.316 mm. **25**. (*a*) 2.60×10^6 m/s. (*b*) 0.217 μs. (*c*) 0.140 MeV. (*d*) 70 kV. **29**. (*a*) $K_p = K_d = \frac{1}{2}K_\alpha$ (*b*) $R_d = R_\alpha = 14$ cm. **33**. (*a*) 495 mT. (*b*) 22.7 mA. (*c*) 8.17 MJ. **35**. (*a*) 0.357 ns. (*b*) 0.17 mm. (*c*) 1.5 mm. **37**. (*a*) 2.9998×10^8 m/s. **41**. Neutron moves tangent to original path; proton moves in a circular orbit of radius 25 cm. **43**. (*b*). **45**. 20.1 N. **47**. $(-2.5\mathbf{j} + 0.75\mathbf{k}) \times 10^{-3}$ N. **51**. (*a*) 3.3×10^9 A. (*b*) 1.0×10^{17} W. (*c*) Totally unrealistic. **53**. (*a*) 0; 1.38 mN; 1.38 mN. **55**. (*a*) 20 min. (*b*) 5.94×10^{-2} N·m. **59**. $2\pi aiB\sin\theta$, normal to the plane of the loop (up). **61**. 2.45 A. **63**. 2.08 GA. **67**. (*a*)$(14\text{ N·m})(-\frac{2}{3}\mathbf{i} - \frac{1}{2}\mathbf{j} + \frac{5}{9}\mathbf{k})$. (*b*) 6.0×10^{-4} J.

CHAPTER 31

1. 7.69 mT. **3**. Yes. **5**. (*a*) 0.24**i**, nT. (*b*) Zero. (*c*) 43.3**k**, pT. (*d*) −0.144**k**, nT. **7**. (*a*) 15.6 A. (*b*) West to east. **9**. (*a*) 3.2×10^{-16} N, parallel to the current. (*b*) 3.2×10^{-16} N, radially outward if **v** is parallel to the current. (*c*) Zero. **13**. (*a*) 1.03 mT, out of figure. (*b*) 0.80 mT, out of figure. **15**. $\frac{\mu_0 i\theta}{4\pi}\left(\frac{1}{b} - \frac{1}{a}\right)$, out of page. **25**. 200 μT, into the page. **27**. (*a*) It is impossible to have other than $B = 0$ midway between them. (*b*) 30.0 A. **29**. At all points between the wires, on a line parallel to them, at a distance $d/4$ from the wire carrying current i. **35**. $0.338\mu_0 i^2/a$, towards the center of the square. **37**. (*b*) To the right. **39**. (*b*) 3.27 km/s. **41**. $+5\mu_0 i_0$. **43**. 4.52×10^{-6} T·m. **47**. (*a*) $\mu_0 ir/2\pi c^2$. (*b*) $\mu_0 i/2\pi r$. (*c*) $\frac{\mu_0 i}{2\pi(a^2-b^2)}\left(\frac{a^2-r^2}{r}\right)$. (*c*) Zero. **49**. $3i_0/8$, into the page. **53**. 5.71 mT. **55**. 108 m. **61**. 2.72 kA. **63**. 0.471 A·m². **65**. $8\mu_0 Ni/5\sqrt{5}R$. **67**. (*b*) ia^2. **71**. (*a*) 78.5 μT. (*b*) 1.08×10^{-6} N·m. **73**. (*a*) $(\mu_0 i/2R)(1 + 1/\pi)$, out of page. (*b*) $(\mu_0 i/2\pi R)\sqrt{1+\pi^2}$, 18° out of page.

CHAPTER 32

1. 57.2 μWb. **3**. 1.52 mV. **5**. (*a*) 31 mV. (*b*) Right to left. **7**. $A^2B^2/R\Delta t$. **9**. (*b*) 58.0 mA. **11**. 1.2 mV. **13**. 1.15 μWb. **15**. 51.4 mV; clockwise when viewed along the direction of **B**. **17**. (*b*) No. **19**. (*a*) 21.74 V. (*b*) Counterclockwise. **21**. (*a*) 13 μWb/m. (*b*) 17%. **23**. 104 mV. **25**. (*a*) 48.1 mV. (*b*) 2.67 mA. **27**. $BiLt/m$, away from G. **29**. (*a*) 85.2 T·m². (*b*) 56.8 V. (*c*) Linearly. **31**. Design it so that $Nab = \dfrac{5}{2\pi}$ m². **33**. 268 W. **35**. 1.55×10^{-5} C. **37**. (*a*) 0.598 μV. (*b*) Counterclockwise. **39**. (*a*) $\frac{\mu_0 ia}{2\pi}\ln\left(\frac{2r+b}{2r-b}\right)$. (*b*) $2\mu_0 iabv/\pi R(4r^2 - b^2)$. **43**. (*a*) 71.5 μV/m. (*b*) 143 μV/m. **45**. 0.151 V/m. **49**. (*a*) 33.8 V/m. (*b*) 5.94×10^{12} m/s².

CHAPTER 33

1. 10 μWb. **3**. (*a*) 800. (*b*) 2.53×10^{-4} H. **7**. (*a*) $\mu_0 i/W$. (*b*) $\pi\mu_0 R^2/W$. **9**. (*a*) Decreasing. (*b*) 0.68 mH. **11**. (*a*) 0.101 H. (*b*) 1.31 V. **13**. (*a*) 1.34 kA. (*b*) 256 A. (*c*) 1.92 kA. **15**. 6.91. **17**. 1.54 s. **19**. (*a*) 8.45 ns. (*b*) 7.73 mA. **21**. $42 + 20t$, V. **23**. 12.0 A/s. **25**. (*a*) $i_1 = i_2 = 3.33$ A. (*b*) $i_1 = 4.55$ A; $i_2 = 2.73$ A. (*c*) $i_1 = 0$; $i_2 = 1.82$ A. (*d*) $i_1 = i_2 = 0$. **27**. (*a*) 1.50 s. **29**. (*a*) 13.9 H. (*b*) 120 mA. **31**. $1.23\,\tau_L$. **33**. (*a*) 240 W. (*b*) 150 W. (*c*) 390 W. **35**. (*a*) 97.9 H. (*b*) 0.196 mJ. **37**. (*a*) $\mu_0 i^2N^2/8\pi^2r^2$. (*b*) 0.306 mJ. (*c*) 0.306 mJ. **41**. 1.5×10^8 V/m. **43**. $(\mu_0 l/2\pi)\ln(b/a)$. **45**. (*a*) 1.0 J/m³. (*b*) 4.8×10^{-15} J/m³. **47**. (*a*) 1.67 mH. (*b*) 6.01 mWb. **53**. (*a*) $\frac{\mu_0 Nl}{2\pi}\ln\left(1 + \frac{b}{a}\right)$. (*b*) 13.2 μH.

CHAPTER 34

5. (*b*) In the direction of **r × v**, **r and v being the position and velocity of any point on the ring.** **7**. +3 Wb. **9**. $(\mu_0 iL/\pi)\ln 3$. **11**. 12.7 MWb, outward. **15**. 1660 km. **17**. 61.1 μT; 84.2°. **19**. 20.8 mJ/T. **21**. Yes. **23**. (*a*) 3.7 K. (*b*) 1.3 K. **29**. (*a*) 3.00 μT. (*b*) 5.63×10^{-10} eV. **31**. (*a*) 8.9 A·m². (*b*) 13 N·m.

CHAPTER 35

1. 9.14 nF. **3**. 45.2 mA. **5**. (*a*) 6.00 μs. (*b*) 167 kHz. (*c*) 3.00 μs. **7**. (*a*) 1.25 kg. (*b*) 3.72×10^4 N/m. (*c*) 1.75×10^{-4} m. (*d*) 3.02×10^{-2} m/s. **9**. 1.59 μF. **15**. (*a*) 5770 rad/s. (*b*) 1.09 ms. **17**. (*a*) 275 Hz. (*b*) 364 mA. **19**. (*a*) 0.689 μH. (*b*) 17.9 pJ. (*c*) 0.110 μC. **21**. $Q/2$. (*b*) $0.866I$. **23**. (*a*) 6.0:1. (*b*) 36 pF; 0.22 mH. **25**. (*a*) 0.18 mC. (*b*) $T/8$. (*c*) 66.7 W. **27**. ω_0. **29**. Let $T_2 = 0.596$ s be the period of the inductor and 900 μF capacitor and $T_1 = 0.199$ s the period of inductor and 100 μF capacitor. Close S_2, wait $T_2/4$; quickly close S_1, then open S_2; wait $T_1/4$ and then open S_1. **33**. 5.85 μC; 5.52 μC; 1.93 μC.

CHAPTER 36

1. 377 rad/s. **3**. (*a*) 0.955 A. (*b*) 0.119 A. **5**. (*a*) 4.60 kHz. (*b*) 26.6 pF. (*c*) $X_L = 2.6$ kΩ; $X_C = 0.65$ kΩ. **7**. (*a*) 0.65 kHz. (*b*) 24 Ω. **9**. (*a*) 39.1 mA. (*b*) Zero. (*c*) 33.8 mA. (*d*) Supplying energy. **11**. (*a*) 6.73 ms. (*b*) 2.24 ms. (*c*) Capacitor. (*d*) 59.0 μF. **13**. (*a*) $X_C = 0$; $X_L = 87$ Ω; $Z = 182$ Ω; $I = 198$ mA; $\phi = 28.5°$. **15**. (*a*) $X_C = 37.9$ Ω; $X_L = 87$ Ω; $Z = 167$ Ω; $I = 216$ mA; $\phi = 17.1°$. **19**. 89.0 Ω. **21**. (*a*) 224 rad/s. (*b*) 6.00 A. (*c*) 228 rad/s; 219 rad/s. (*d*) 0.039. **23**. (*a*) 45°. (*b*) 70.7 Ω. **29**. 141 V. **31**. Zero; zero; 3.14 W; 182 W. **33**. 177 Ω. **35**. 7.61 A. **41**. (*a*) 117 F. (*b*) Zero. (*c*) 90 W; zero. (*d*) 0°; 90°. (*e*) 1; 0. **43**. (*a*) 2.59 A. (*b*) 38.8 V; 159 V, 224 V, 64.2 V; 75 V. (*c*) 100 W for R; zero for L and C. **45**. (*a*) 2.4 V. (*b*) 3.2 mA; 0.16 A. **47**. (*a*) 1.9 V; 5.8 W. (*b*) 19 V; 0.58 kW. (*c*) 0.19 kV; 58 kW.

CHAPTER 37

3. At $r = 27.5$ mm and $r = 110$ mm. **7**. Change the potential difference between the plates at the rate of 1.0 MV/s. **11**. (*a*) 0.63 μT. (*b*) 2.3×10^{12} V/m·s. **13**. (*a*) 2.0 A. (*b*) 2.3×10^{11} V/m·s. (*c*) 0.50 A. (*d*) 0.63 μT·m. **15**. (*a*) 7.60 μA. (*b*) 859 kV·m/s. (*c*) 3.39 mm. (*d*) 5.16 pT.

CHAPTER 38

1. (*a*) 4.71×10^{-3} Hz. (*b*) 3 min 32 s. **3**. (*a*) 4.5×10^{24} Hz. (*b*) 1.0×10^4 km, or 1.6 earth radii. **7**. (*a*) It would steadily increase. (*b*) The summed discrepancies between the apparent times of eclipse and those observed from x; the radius of the earth's orbit. **9**. 5.0×10^{-21} H. **11**. 1.07 pT. **17**. 4.8×10^{-29} W/m^2. **19**. 4.51×10^{-10}. **21**. 89.2 cm. **23**. 1.2 MW/m^2. **25**. 817 m. **27**. (*a*) 1.03 kV/m; 3.43 μT. **29**. (*a*) $\pm EBa^2/\mu_0$ for faces parallel to the xy plane; zero through each of the other four faces. (*b*) Zero. **31**. (*a*) 82.6 W/m^2. (*b*) 1.66 MW. **33**. (*a*) 1.53×10^{-17} W/m^2. (*b*) 107 nV/m. (*c*) 0.252 fT. **37**. 10 MPa. **39**. (*a*) 6.0×10^8 N. (*b*) $F_{grav} = 3.6 \times 10^{22}$ N. **41**. (*a*) 100 MHz. (*b*) 1.0 μT along the z axis. (*c*) 2.1 m^{-1}; 6.3×10^8 rad/s. (*d*) 120 W/m^2. (*e*) 8.0×10^{-7} N; 4.0×10^{-7} N/m^2. **43**. 491 nm. **47**. 1.92 mm/s. **49**. (*b*) 600 nm. **51**. (*a*) 1.94 V/m. (*b*) 1.67×10^{-11} N/m^2. **53**. 1/4. **55**. 20° or 70°. **57**. 19.0 W/m^2. **59**. (*b*) 5 sheets.

CHAPTER 39

1. (*a*) 38.0°. (*b*) 52.9°. **3**. 1.56. **5**. 1.9×10^8 m/s. **7**. 1.26. **9**. 1.07 m. **11**. 22.0°. **15**. 401 cm beneath the mirror surface. **21**. 33.7°. **23**. (*a*) 49°. (*b*) 28°. **25**. 182 cm. **27**. (*a*) Yes. (*b*) No. (*c*) 43°. **29**. (*a*) 35.6°. (*b*) 53.1°. **31**. (*b*) 23.2°. **35**. (*a*) 53°. (*b*) Yes. **37**. 55.50° to 55.77°. **39**. 40 cm. **41**. (*a*) $2v$. (*b*) v. **43**. 1.50 m. **47**. Three. **49**. New illumination is 10/9 of the old. **51**. (*a*) 7. (*b*) 5. (*c*) 2. **53**. 10.5 cm. **55**. For alternate vertical columns. (*a*) +, +40, −20, +2, no, yes. (*c*) Concave, +40, +60, −2, yes, no. (*e*) Convex, −20, +20, +0.5, no, yes. (*g*) −20, −, −, +5, +0.8, no, yes. **57**. (*b*) 0.556 cm/s. (*c*) 11.25 m/s. (*d*) 5.14 cm/s. **59**. For alternate vertical columns: (*a*) −18, no. (*c*) +71, yes. (*e*) +30, no. (*g*) −26, no. **61**. $i = -12.0$ cm. **63**. f_1^2/f_2^2. **65**. 1.85 mm. **67**. (*a*) Converging. (*b*) Diverging. (*c*) Converging. (*d*) Diverging. **69**. 5.14 cm. **72**. Alternate vertical columns (an X means that the quantity cannot be found from the data given): (*a*) +, X, X, +20, X, −1, yes, no. (*c*) Converging, +, X, X, −10, X, no, yes. (*e*) Converging, +30, −15, +1.5, no, yes. (*g*) Diverging, −120, −9.2, +0.92, no, yes. (*i*) Converging, +3.3, X, X, +5, X, −, no. **73**. Upright, virtual, 30 cm to the left of the second lens. **75**. (*a*) The final image coincides in location with the object. It is real, inverted and $m = -1.0$. **77**. (*a*) Coincides in location with the original object and is enlarged 5.0 times. (*c*) Virtual and inverted. **83**. (*a*) 13.0 cm. (*b*) 1.23 cm. (*c*) −3.25. (*d*) 3.13. (*e*) −10.2. **89**. (*b*) 8.4 mm. (*c*) 2.5 cm.

CHAPTER 40

1. (*a*) 5.1×10^{14} Hz. (*b*) 388 nm. (*c*) 1.98×10^8 m/s. **7**. $(2m + 1)\pi$. **9**. 2.25 mm. **11**. 648 nm. **13**. 1.6 mm. **15**. 0.072 mm. **17**. 600 nm. **19**. 6.6 μm. **21**. (*a*) $2h \sin\theta = m\lambda$ (minimum); $2h \sin\theta = (m + \frac{1}{2})\lambda$ (maximum). **23**. 9.0

μm. **25**. 8.0 μm. **29**. $y = 27\sin(\omega t + 8.5°)$. **33**. (*a*) 1.17; 3.00; 7.50 m. (*b*) No. **37**. $\lambda/5$. **39**. 492 nm. **41**. Bright **43**. 70.0 nm. **45**. 673 nm. **47**. (*a*) 169 nm. (*c*) Blue-violet will be sharply reduced. **49**. 840 nm. **51**. (*a*) $\lambda/2n_2$. (*b*)$\lambda/2n_2$. (*c*) $\lambda/4n_2$. (*d*)$\lambda/4n_2$. (*e*) $\lambda/2n_2$. **53**. 141. **55**. (*a*) 1800 nm. (*b*) 8. **57**. 1.89 μm. **59**. 1.00025. **61**. (*a*) 34. (*b*) 46. **65**. 588 nm. **67**. 1.003.

CHAPTER 41

1. 691 nm. **3**. 60.4 μm. **5**. (*a*) 2.19×10^{-4} rad. (*b*) 2.51 mm. **7**. (*a*) 70.0 cm. (*b*) 1.03 mm. **9**. 41.2 m from perpendicular to speaker. **11**. 160°. **15**. (*d*) 53°; 10°; 5.1°. **19**. (*a*) 1.32×10^{-4} rad. (*b*) 21.1 m. **21**. 30 m. **23**. (*a*) 1.1×10^4 km. (*b*) 11 km. **25**. 52.6 m. **27**. 4.5 m. **29**. 4.73 cm. **31**. (*a*) 0.347°. (*b*) 0.973°. **33**. (*b*) 0.07 mm. (*c*) Three times the lunar diameter; if water droplets of various sizes are present, a number of rings of different colors are present, giving a whitish appearance. (*d*) These halos are a diffraction effect, a rainbow being formed by refraction. **35**. 5. **39**. $\lambda D/d$. **41**. (*a*) 5.05 μm. (*b*) 20.2 μm. **43**. (*a*) 3330 nm. (*b*) ±10.2°; 20.7°; 32.0°; 45.0°; 62.2°. **45**. All wavelengths shorter than 635 nm. **47**. 13,600. **49**. (*a*) 6.0 μm. (*b*) 1.5 μm. (*c*) $m = 0, 1, 2, 3, 5, 6, 7, 9$. **51**. 1100. **61**. (*a*) 55.6 pm. (*b*) None. **63**. (*a*) 23,100. (*b*) 28.7°. **67**. (*a*) 2400 nm. (*b*) 800 nm. (*c*) $m = 0, 1, 2$. **71**. 2.87°. **73**. 26.2 pm; 39.4 pm. **75**. 39.8 pm. **77**. Yes; $m = 3$ for $\lambda = 0.124$ nm; $m = 4$ for $= \lambda$ 0.097 nm. **79**. (*a*) $a_0/\sqrt{2}$; $a_0/\sqrt{5}$; $a_0/\sqrt{10}$; $a_0/\sqrt{13}$; $a_0/\sqrt{17}$.

CHAPTER 42

1. (*a*) 3×10^{-18}. (*b*) 2×10^{-12}. (*c*) 8.2×10^{-8}. (*d*) 6.4×10^{-6}. (*e*) 1.1×10^{-6}. (*f*) 3.7×10^{-5}. (*g*) 9.9×10^{-5}. (*h*) 0.10. **3**. 0.750c. **5**. 0.991c. **7**. 54.7 m. **9**. 1.32 m. **11**. 1.53 cm. **13**. (*a*) 87.4 m. (*b*) 394 ns. **15**. (*a*) 2.21×10^{-12}. (*b*) 5.25 d. **17**. $x' = 138$ km; $t' = -374\,\mu$s. **19**. $t_1' = 0$; $t_2' = -250\,\mu$s. **23**. (*a*) S' must move towards S, along their common axis, at a speed of 0.480c. (*b*) The 'red' flash (suitably Doppler shifted). (*c*) 4.39 μs. **25**. 0.806c. **27**. 0.946c. **29**. 0.588c, recession. **31**. (*a*) 34,000 mi/h. (*b*) 6.4×10^{-10}. **33**. 22.9 MHz. **35**. +2.97 nm. **39**. (*a*) 0.134c. (*b*) 4.65 keV. (*c*) 1.94%. **41**. (*a*) 0.9988; 20.6. (*b*) 0.145; 1.01. (*c*) 0.073; 1.0027. **43**. (*a*) 5.71 GeV; 6.65 GeV; 6.58 GeV/c. (*b*) 3.11 MeV; 3.62 MeV; 3.59 MeV/c. **45**. 18 smu/y. **47**. (*a*) 0.943c. (*b*) 0.866c. **49**. (*a*) 0.707c. (*b*) 1.414m. (*c*) 0.414mc^2. **51**. (*a*) The photon. (*b*) The proton. (*c*) The proton. (*d*) The photon. **53**. (*a*) 207m_e; the particle is a muon. **55**. (*a*) 0.948c. (*b*) 649m_e. (*c*) 226 MeV. (*d*) 316 MeV/c. **57**. (*a*) 4.85 mm. (*b*) 16.0 mm. (*c*) 0.335 ns; no. **59**. 660 km. **61**. (*a*) 534. (*b*) 0.99999825. (*c*) 2.23 T.

CHAPTER 43

3. 2.1 μm; infrared. **5**. (*a*) 35.4 keV. (*b*) 8.57×10^{18} Hz. (*c*) 35.4 keV/c $= 1.89 \times 10^{-23}$ kg·m/s. **7**. (*a*) 1.24×10^{20} Hz. (*b*) 2.43 pm. (*c*) 2.73×10^{-22} kg·m/s = 0.511 MeV/c. **9**. (*a*) The infrared bulb. (*b*) 6.0×10^{20}. **11**.. 4.66×10^{26}. **13**. (*a*) 2.96×10^{20} s^{-1}. (*b*) 48,600 km. (*c*) 280 m. (*d*) 5.89×10^{18} m^{-2}·s^{-1}; 1.96×10^{10} m^{-3}. **15**. 233 nm. **17**. 10.1 eV. **19**. 676 km/s. **21**. (*a*) 1.3 V. (*b*) 680 km/s. **23**. (*a*) 382 nm. (*b*) 1.82 eV. **27**. (*a*) 3.1 keV. (*b*) 14.4 keV. **29**. (*a*) 2.73 pm. (*b*) 6.05 pm. **31**. (*a*) +4.8 pm. (*b*) −41 keV. (*c*) 41 keV. **33**. (*a*) 8.1×10^{-9}%. (*b*) 4.9×10^{-4} %. (*c*) 9.6 %. (*d*) 68%. **35**. 2.65 fm. **43**. 9.99 μm. **45**. 91 K. **47**. (*a*) 0.97 mm; microwave. (*b*) 9.9 μm; infrared. (*c*) 1.6 μm; infrared. (*d*) 0.26 nm; x ray. (*e*) 2.9×10^{-41} m; hard gamma ray. **53**. $4\Delta T/T$; 0.0130. **55**. 2.57 eV. **57**. −80.7 keV. **59**. (*a*) 121.5 nm. (*b*) 91.2 nm. **61**. 661 nm; 486 nm; 437 nm. **63**. (*a*) 12.7 eV. (*b*) 12.7 eV ($4 \to 1$); 2.55 eV ($4 \to 2$); 0.66 eV ($4 \to 3$); 12.1 eV ($3 \to 1$); 1.89 eV ($3 \to 2$); 10.2 eV ($2 \to 1$). **65**. (*a*) 30.5 nm; 291 nm; 1050 nm. (*b*) 8.25×10^{14} Hz; 3.65×10^{14} Hz; 2.06×10^{14} Hz. **69**. 4.1 m/s. **75**. (*a*) 3×10^{74}. (*b*) No. **77**. (*b*) n^2. (*c*) n. (*d*) $1/n$. (*e*) $1/n^3$. (*f*) $1/n$. (*g*) $1/n^4$. (*h*) $1/n^4$. (*i*) $1/n^2$. (*j*) $1/n^2$. (*k*) $1/n^2$.

CHAPTER 44

1. (*a*) 1.7×10^{-35} m. **3**. 7.75 pm. **5**. (*a*) 3.3×10^{-24} kg·m/s for each. (*b*) 38 eV for the electron; 6.2 keV for the photon. **7**. (*a*) 38.8 meV. (*b*) 146. pm. **9**. (*a*) 73 pm; 3.4 nm. (*b*) Yes. **11**. (*a*) 1.24 keV; 1.50 eV. (*b*) 1.24 GeV; 1.24 GeV. **13**. 0.025 fm. **15**. A neutron. **17**. 9.70 kV (relativistic calculation); 9.79 kV (classical calculation). **19**. 11.5°, 23.6°, 36.9°, 53.1°. **21**. (*a*) 20.5 meV. (*b*) 37.7 eV. **23**. (*a*) 1900 MeV. (*b*) No. **25**. 90.5 eV. **27**. 18.1, 36.2, 54.3, 66.3, 72.4 μeV. **29**. (*a*) 0.196. (*b*) 0.608. (*c*) 0.196. **31**. 0.323. **33**. 0.439. **37**. Proton: 9.2×10^{-6}; deuteron: 7.6×10^{-8}. **39**. 10^{104} y. 6.63×10^{-23} kg·m/s. **43**. 1.2 m. 45. 0.414 μeV; $E_2 = -3.4$ eV. **45**. 0.414 μeV; $E_2 = -3.4$ eV. **47**. (*a*) 124 keV. (*b*) 40.5 keV.

CHAPTER 45

3.. 3.64×10^{-34} J·s. **5**. 24.1°. **7**. $n > 3$; $m_\ell = +3, +2, +1, 0, -1, -2, -3$; $m_s = +1/2, -1/2$. **9**. 50. $L_z = 3\hbar, 2\hbar, \hbar, 0, -\hbar, -2\hbar, -3\hbar$; $\mu_z = 3\mu_B, 2\mu_B, \mu_B, 0, -\mu_B, -2\mu_B, -3\mu_B$; $\theta = 30°, 55°, 73°, 90°, 107°, 125°, 150°$; $L = \sqrt{12}\hbar$; $\mu = \sqrt{12}\mu_B$. **17**. 5.54 nm^{-1}. **19**. 0.0054. **21**. (*b*) 16.4 $\text{nm}^{-3/2}$. **25**. 0.981 nm^{-1}; 3.61 nm^{-1}. **27**. 0.0019. **29**. 54.7°; 125°. **31**. (*a*) 58 μeV. (*b*) 14 GHz. (*c*) 2.1 cm; short radio wave region. **33**. 5.35 cm. **35**. 19.4 mT. **37**. All statements are true. **39**. (*a*) (2, 0, 0, $\pm 1/2$). (*b*) $n = 2$, $\ell = 1$; $m_\ell = 1, 0, -1$; $m_s = \pm 1/2$. **45**. 12.4 kV. **47**. 49.6 pm; 99.2 pm. **49**. (*a*) 24.8 pm. (*b*) and (*c*) remain unchanged. **51**. 6.39 keV; 10.2 eV. **53**. 9/16. **55**. (*a*) 69.5 kV. (*b*) 17.8 pm. (*c*) 21.3 pm. **57**. 282 pm. **59**. (*b*) 24%; 15%; 11%; 7.9%; 6.5%; 4.7%; 3.5%; 2.5%; 2.0%; 1.5%. **61**. 10,000 K. **63**. 4.40×10^{17}. **65**. $2.02 \times 10^{16}\ \text{s}^{-1}$. **67**. 4.8 km. **69**. 1.8 pm. **71**. (*a*) 7.33 μm. (*b*) 707 kW/m^2. (*c*) 24.9 GW/m^2. **73**. (*a*) 53.1 GPa. (*b*) 1.20×10^8 K.

CHAPTER 46

1. 3560 atm. **3**. (*a*) 2.7; 25; m^{-3}. (*b*) $8.43 \times 10^{28}\ \text{m}^{-3}$. (*c*) 3100. (*d*) 3.3 nm for oxygen, 0.228 nm for the electrons. **7**. $1.92 \times 10^{28}\ \text{m}^{-3}\cdot\text{eV}^{-1}$. **9**. (*a*) 6.81 eV. (*b*) $1.77 \times 10^{28}\ \text{m}^{-3}\cdot\text{eV}^{-1}$. (*c*) $1.59 \times 10^{28}\ \text{m}^{-3}\cdot\text{eV}^{-1}$. **11**. 5.53 eV. **13**. $T \gg 10^5$ K. **15**. 3. **19**. (*a*) $5.86 \times 10^{28}\ \text{m}^{-3}$. (*b*) 5.52 eV. (*c*) 1390 km/s. (*d*) 0.522 nm. **21**. 137 MeV. **25**. 200°C. **29**. (*a*) 19.8 kJ. (*b*) 3 min 18 s. **31**. (*a*) *n*-type. (*b*) $5 \times 10^{21}\ \text{m}^{-3}$. (*c*) 2.5×10^5. **33**. (*a*) Pure: 4.78×10^{-10}; doped: 0.0141. (*b*) 0.824. **37**. (*b*) 2.49×10^8. **39**. 4.20 eV.

CHAPTER 47

1. 15.8 fm. **3**. (*a*) 0.39 MeV. (*b*) 4.61 MeV. **5**. (*a*) Six. (*b*) Eight. **9**. (*a*) 1150 MeV. (*b*) 4.8 MeV/nucleon; 12 MeV/proton. **15**. 4×10^{-22} s. **17**. $K \approx 30$ MeV. **21**. (*a*) 19.8 MeV, 6.26 MeV, 2.22 MeV. (*b*) 28.3 MeV. (*c*) 7.07 MeV. **23**. 1.58×10^{25} MeV. **25**. 7.92 MeV. **27**. 280 d. **29**. (*a*) $7.55 \times 10^{16}\ \text{s}^{-1}$. (*b*) $4.93 \times 10^{16}\ \text{s}^{-1}$. **31**. (*a*) $4.80 \times 10^{-18}\ \text{s}^{-1}$. (*b*) 4.57×10^9 y. **33**. 5.3×10^{22}. **35**. 265 mg. **37**. 209 d. **39**. 87.8 mg. **41**. 730 cm^2. **45**. (*a*) $3.66 \times 10^7\ \text{s}^{-1}$. (*b*) $t \gg 3.82$ d. (*c*) $3.66 \times 10^7\ \text{s}^{-1}$. (*d*) 6.42 ng. **47**. Pu: 5.5×10^{-9}; Cm: zero. **49**. (*a*) 4.25 MeV. (*b*) -24.1 MeV. (*c*) 28.3 MeV. **51**. $Q_3 = -9.50$ MeV; $Q_4 = 4.66$ MeV; $Q_5 = -1.30$ MeV. **53**. 1.21 MeV. **55**. 0.782 MeV. **59**. 0.961 MeV. **61**. 78.4 eV. **63**. 1600 y. **65**. 1.72 mg. **67**. 1.02 mg. **69**. (*a*) 18.0 mJ. (*b*) 0.288 rem. **71**. (*a*) 6.30×10^{18}. (*b*) 2.48×10^{11}. (*c*) 0.196 J. (*d*) 0.231 rad. (*e*) 3.00 rem. **73**. 3.87×10^{10} K. **75**. (*a*) 25.35 MeV. (*b*) 12.80 MeV. (*c*) 25.00 MeV. **77**. (*a*) 3.85 MeV, 7.95 MeV. (*b*) 3.98 MeV, 7.33 MeV.

CHAPTER 48

1. (*a*) 2.56×10^{24}. (*b*) 8.19×10^{13} J. (*c*) 2.59×10^4 y. **3**. $3.1 \times 10^{10}\ \text{s}^{-1}$. **7**. +5.00 MeV. **9**. (*a*) 16 fissions/day. (*b*) 4.3×10^8. **11**. (*a*) 10. (*b*) 231 MeV. **13**. (*a*) 252 MeV. (*b*) Typical fission energy = 200 MeV. **15**. 463 kg. **17**. Yes. **19**. (*a*) 1.15 MeV. (*b*) 3.2 kg. **21**. (*a*) 43.7 kton. **25**. 1.6×10^{17}. **27**. (*b*) 1.0, 0.89, 0.28, 0.019. (*c*) 8. **29**. (*a*) 75 kW. (*b*) 5770 kg. **33**. (*a*) 30 MeV. (*b*) Zero. (*c*) 170 keV. **35**. (*a*) 170 kV. **37**. 1.41 MeV. **41**. (*a*) 3.1×10^{31} photons/m^3. (*b*) 1.2×10^6 times. **43**. (*a*) 4.0×10^{27} MeV. (*b*) 5.1×10^{26} MeV. **45**. (*a*) $1.83 \times 10^{38}\ \text{s}^{-1}$. (*b*) $8.25 \times 10^{28}\ \text{s}^{-1}$. **49**. (*a*) 6.3×10^{14} J/kg. (*b*) 6.2×10^{11} kg/s. (*c*) 4.3×10^9 kg/s. (*d*) 15×10^9 y. **51**. (*a*) 24.9 MeV. (*b*) 8.65 Mton. **53**. $K_\alpha = 3.5$ MeV; $K_n = 14.1$ MeV. **57**. (*a*) 35 MJ. (*b*) 17 lb. (*c*) 3500 MW.

CHAPTER 49

1. 6.03×10^{-29} kg. **3**. 2.4×10^{-43}. **5**. 1.08×10^{42} J. **7**. 2.67 cm/s. **9**. (*a*) 1.90×10^{-18} kg·m/s. (*b*) 9.90 m. **13**. (*a*) Angular momentum. (*b*) Charge. (*c*) Energy. **15**. $Q = 0$; $B = -1$; $S = 0$. **17**. (*a*) Energy. (*b*) Angular momentum. (*c*) Charge. **19**. 338 MeV. **23**. (*a*) $\bar{u}\bar{u}\bar{d}$. (*b*) $\bar{u}\bar{d}\bar{d}$. **25**. (*a*) *sud*. (*b*) *uss*. **29**. Σ^0; 7530 km/s. **31**. 665.2 nm. **33**. (*b*) 0.934. (*c*) 1.65×10^{10} ly. **35**. (*a*) 256 μeV. (*b*) 4.84 nm. **37**. (*a*) 121 m/s. (*b*) 246 y. **39**. (*a*) 2.63 K. (*b*) 29.0 nm.

PHOTO CREDITS

Chapter 23 Opener: The Seattle Times. Fig. 2: Courtesy Xerox Corp. Fig. 3: AT&T Bell Laboratories. Fig. 7: From *Introduction to the Dectection of Nuclear Particles in a Bubble Chamber,* Ealing Press, 1969. Courtesy Lawrence Berkeley Radiation Laboratories, University of California at Berkeley. **Chapter 24** Opener: Courtesy Research-Cottrell. Figs. 13 and 14: *Scientific American,* copyright © 1979 by IBM Corp. Reprinted by permission. **Chapter 25** Opener: The Art Museum, Princeton University. The John B. Putnam, Jr., Memorial Collection. **Chapter 26** Opener: Copyright © Michael Philip Manheim. Fig. 17: Courtesy E. Philip Krider, Institute of Atmospheric Physics, University of Arizona. Fig. 18: Courtesy Ford Motor Co., Technical Photographic Services. Fig. 20: Courtesy Purdue University. Fig. 27: Courtesy NASA. **Chapter 27** Opener: Roger Ressmeyer/Wheeler Pictures. Fig. 1: Courtesy Sprague Electric Co. Fig. 10: Courtesy of the Director of The Royal Institution. **Chapter 28** Opener: Courtesy IBM Corp. Fig. 7: Courtesy Allen-Bradley Co. Fig. 15: Courtesy AT&T. **Chapter 29** Opener: Courtesy Union Carbide. Fig. 13: Courtesy Simpson Electric Co. **Essay 11** Page E11-1: Courtesy A. A. Bartlett, University of Colorado. **Chapter 30** Opener: Courtesy Dornier-System, GmBH, Friedrichschafen. Fig. 1: D. C. Heath and Co., with Education Development Center. Fig. 2: Hugh Rogers/Monkmeyer. Fig. 3: Courtesy Varian Associates. Fig. 5: Courtesy Professor J. le P. Webb, University of Sussex, Brighton, England. Fig. 6: Courtesy Lawrence Berkeley Radiation Laboratory, University of California. Fig. 11: Courtesy Professor J. le P. Webb, University of Sussex, Brighton, England. Fig. 14: Courtesy Dr. L. A. Frank, University of Iowa. Fig. 15: Courtesy A. J. Allen. Figs. 17 and 18: Courtesy Fermi National Accelerator Laboratory. Fig. 19: Courtesy NASA. **Chapter 31** Opener: D. C. Heath and Co. with Education Development Center. **Chapter 32** Opener: Courtesy Alice Halliday. **Essay 12** Page E12-1: Courtesy Brian Holton. Fig. 1: Drawing by nephew of H. Kamerlingh Onnes, from *Superconductivity* by D. Shoenberg, Ph. D, Cambridge University Press, 1952. Fig. 3: From *Physics Today,* March 1986, p. 37. Courtesy New York University Medical Center and Biomagnetic Technologies, Inc. Photo by Hank Morgan. Fig. 5: From *SCC,* 1987, p. 37. Courtesy Japanese National Railways, and Railways Graphics Co., Ltd. Fig. 7: From *Physics Bulletin,* Vol. 38, No. 6, June 1987. Courtesy Birmingham Superconductivity Consortium. Fig. 8: From *Physics Today,* August 1971, p. 32. **Chapter 33** Opener: Courtesy the Director of The Royal Institution. **Essay 13** Page E13-1: Courtesy Dr. Gerard O'Neill. Figs. 1 and 2: *2081: A Hopeful View of the Human Future* by Gerard K. O'Neill , pp. 126 and 127. Simon & Schuster, 1981. **Chapter 34** Opener: Courtesy Dr. James U. Lemke, Recording Physics, Inc., San Diego. Fig. 6: Courtesy Colchester and Essex Museum. Fig. 14: Courtesy R. W. DeBlois. Fig. 16: Courtesy General Electric Medical Systems, Inc. **Essay 14** E14-1: Courtesy Charles Bean. Fig. 1: Courtesy R. Blakemore and N. Blakemore. **Chapter 35** Opener: Courtesy Federal Aviation Administration. **Chapter 36** Opener: Courtesy Con Edison. **Chapter 37** Opener: Anne Manning. Fig. 3: "Ajax Defying the Lightning." From the cover of Electrical Plant, December 1888. By permission, British Post Office. **Chapter 38** Opener: Courtesy American Institute of Physics, Niels Bohr Library. Photo by Roy L. Bishop. Fig. 20: Courtesy NASA. **Essay 15** Page E15-1 and Figs. 1, 2, 4–6, 8, and 10: Courtesy Raymond C. Turner. **Chapter 39** Opener: Piergiorgio Scharandis/Black Star. Fig. 1: PSSC *Physics,* 2nd Ed., copyright © 1965, D. C. Heath & Co. with Education Development Center, Newton, Mass. Fig. 5: Courtesy Olympus Corp. Fig. 6: Courtesy AT&T Bell Labs. Fig. 22: Courtesy Minolta Corp. **Essay 16** Page E16-1: Courtesy Dr. Jearl Walker. **Chapter 40** Opener: Courtesy Raytheon Co. Photo by Eli Brookner. Fig. 6: From *Atlas of Optical Phenomena* by Cagnet, et al., Springer-Verlag, Prentice-Hall, 1962. Fig. 7: Education Development Center, Newton, Mass. **Essay 17** Page E17-1: Courtesy Suzanne R. Nagel. **Chapter 41** Opener: Courtesy European Southern Observatory. Fig. 1: Courtesy *Atlas of Optical Phenomena* by Cagnet, et al., Springer-Verlag. Fig. 2: From Sears, Zemansky and Young, *University Physics,* 5th Ed., copyright © 1976, Addison-Wesley, Reading, Mass. Fig. 3: *Atlas of Optical Phenomena* by Cagnet, et al., Springer-Verlag, 1962. Fig. 7: Courtesy Viking Yacht Co. Inc. Fig. 13: *Atlas of Optical Phenomena* by Cagnet, et al., Springer-Verlag, 1962. Fig. 15: Courtesy Professor L. M. Beidleu, Florida State University, Tallahassee. Figs. 17 and 19: *Atlas of Optical Phenomena* by Cagnet, et al., Springer-Verlag, 1962. **Essay 18** Page E18-1: Courtesy Richard Smith. Fig. 4: Courtesy Edward Wesly. Fig. 6: Tong H. Jeong. **Chapter 42** Opener: Courtesy Alice Halliday. Fig. 1: With permission of The Hebrew University of Jerusalem, Israel. Fig. 4: From *Spacetime Physics* by Edwin Taylor and John Archibald Wheeler. Copyright © 1963, 1966, W. H. Freeman and Company. Reprinted with permission. Fig. 13: Courtesy Los Alamos National Laboratory. Fig. 15: The Washington Post. **Chapter 43** Opener: Courtesy Alice Halliday. Fig. 12: Courtesy American Institute of Physics, Niels Bohr Library, Margarethe Bohr Collection. **Chapter 44** Opener: Courtesy University of California, Lawrence Berkeley Laboratory. Fig. 5: Courtesy of the Riber Division of Instruments SA, Inc. Fig. 15: Courtesy Alice Halliday. **Chapter 45** Opener: Courtesy Warren Nagourney, University of Washington. Fig. 1: Courtesy Professor N. Uyeda, Kyoto University. Figs. 14 and 20: Courtesy American Institute of Physics, Niels Bohr Library, Margarethe Bohr Collection. Fig. 16: Courtesy General Electric Medical Systems Inc. Fig. 22: Courtesy Coherent Medical Group. Fig. 23: AP/Wide World Photos. Fig. 24: Movie Star News. **Chapter 46** Opener: Courtesy Alice Halliday. Figs. 9 and 17: Courtesy AT&T. **Essay 20** Opener: Courtesy AT &T Bell Laboratories. Fig. 1: Courtesy R. Filas, AT&T Bell Laboratories. Fig. 3: Courtesy A. R. Kmetz, AT&T Bell Laboratories. Figs. 6 and 7: Courtesy F. P. Cladis. **Chapter 47** Opener: Photo by C. E. Wynn-Williams. **Essay 21** Figs. 3–8: Courtesy Russell K. Hobbie. **Chapter 48** Opener: Courtesy U.S. Atomic Energy Commission. Fig. 11: Courtesy Princeton University Plasma Physics Laboratory. Fig. 14: Courtesy Los Alamos National Laboratory. Fig. 15: Courtesy Lawrence Livermore National Laboratory. **Chapter 49** Opener: Courtesy Alice Halliday. Fig. 1: Copyright Sidney Harris. Figure 2*a*, 2*b*, and 9: Courtesy CERN. Fig. 3*a*: Courtesy University of California, Lawrence Berkeley Laboratory. Figs. 5 and 10: Courtesy American Institute of Physics, Niels Bohr Library, Margarethe Bohr Collection. Fig. 7*a*: Courtesy Stanford University, Stanford Linear Accelerator Center. Fig. 7*b*: Courtesy Brookhaven National Laboratory. Fig. 8: Tobbe Gustavsson/Reportagebild. Fig. 11: Courtesy AT&T. Fig. 12: Courtesy Mark Godfrey. Fig. 13: Courtesy Dr. Vera Rubin. Ex. 7P: Original plate courtesy of Ian Shelton, Las Campanas Observatory, Chile; print by Steve Padilla, Mount Wilson and Las Campanas Observatories.

INDEX

N

S

T